Separation Process Engineering
Includes Mass Transfer Analysis

THIRD EDITION

Separation Process Engineering
Includes Mass Transfer Analysis

Third Edition

PHILLIP C. WANKAT

PRENTICE
HALL

Upper Saddle River, NJ • Boston • Indianapolis • San Francisco
New York • Toronto • Montreal • London • Munich • Paris • Madrid
Capetown • Sydney • Tokyo • Singapore • Mexico City

Many of the designations used by manufacturers and sellers to distinguish their products are claimed as trademarks. Where those designations appear in this book, and the publisher was aware of a trademark claim, the designations have been printed with initial capital letters or in all capitals.

The author and publisher have taken care in the preparation of this book, but make no expressed or implied warranty of any kind and assume no responsibility for errors or omissions. No liability is assumed for incidental or consequential damages in connection with or arising out of the use of the information or programs contained herein.

The publisher offers excellent discounts on this book when ordered in quantity for bulk purchases or special sales, which may include electronic versions and/or custom covers and content particular to your business, training goals, marketing focus, and branding interests. For more information, please contact:

United States Corporate and Government Sales
(800) 382-3419
corpsales@pearsontechgroup.com

For sales outside the United States, please contact:

International Sales
international@pearsoned.com

Visit us on the Web: informit.com/ph

Library of Congress Cataloging-in-Publication Data

Wankat, Phillip C., 1944–
  Separation process engineering : includes mass transfer analysis / Phillip
C. Wankat.—3rd ed.
        p.   cm.
  Includes index.
  ISBN 0-13-138227-6 (hardcover : alk. paper)
  1. Separation (Technology) I. Title.
  TP156.S45W36 2011
  660'.2842—dc23

                                                                    2011019427

ISBN 13: 978-0-13-138227-5
ISBN 10:   0-13-138227-6

Text printed in the United States on recycled paper at Courier in Westford, Massachusetts.
Second printing, September 2013

To Dot, Chuck, and Jennie

Contents

Preface

In the twenty-first century, separations remain as important, if not more important, than in the previous century. The development of new industries such as biotechnology and nanotechnology, the increased importance of removing traces of compounds, and the probable need to recover and sequester carbon dioxide have brought new separations to the fore. Chemical engineers must understand and design new separation processes such as membrane separations, adsorption, and chromatography in addition to the standard equilibrium-staged separations including distillation, absorption, and extraction. Since membrane separations, adsorption, chromatography, and ion exchange were included, I changed the title of the second edition from *Equilibrium Staged Separations* to *Separation Process Engineering* to reflect this broader coverage. The new title has been retained for the third edition with the addition of a subtitle, *Includes Mass Transfer Analysis,* which reflects the addition of Chapter 15.

The second edition was unavoidably longer than the first, and the third edition is longer than the second. The first major addition to the third edition is the extensive Chapter 15, which includes mass transfer and diffusion. Both the Fickian and Maxwell-Stefan approaches to diffusion are covered in detail with examples and homework assignments. The old Chapter 15, which applied mass transfer techniques to equilibrium-staged separations, is now Chapter 16 with the removal of Section 15.1, which is now incorporated in the new Chapter 15.

The second major change is a much more extensive analysis of liquid-liquid extraction. Chapters 13 and 14 in the second edition both covered extraction, washing, and leaching. In the third edition, the material is reorganized so that Chapter 13 covers only extraction and Chapter 14 covers washing and leaching. In addition to the McCabe-Thiele, triangle, and computer-simulation analyses of extraction, Chapter 13 now includes a section on the detailed design of mixer-settlers and a shorter section on the design of Karr columns. Mass transfer analysis of liquid-liquid extraction systems has been added to Chapter 16.

All of the chapters have many new homework questions and problems. More than 300 new questions and problems are included. Since all of the problems were created and solved as I continued to teach this material at Purdue University, a Solutions Manual is available to professors who adopt this textbook for their course. A number of spreadsheet problems have been added, and the answers are provided in the Solutions Manual.

Since process simulators are used extensively in commercial practice, I have continued to include process simulation examples and homework problems throughout the text. I now teach the required three-credit, junior-level separations course at Purdue as two lectures and a two-hour computer lab every week. The computer lab includes a lab test to assess the ability of the students to use the simulator. Although I use Aspen Plus as the simulator, any process simulator can be used. Chapters 2, 6, 8, 10, 12, 13, and 16 include appendices that present instructions for operation of Aspen Plus. The appendices to Chapters 2, 4, 5, 15, and 17 have Excel spreadsheets, some of which use Visual Basic programs. I chose to use spreadsheets instead of a higher-level mathematical program because spreadsheets are universally available. The appendix to Chapter 18 includes brief instructions for operation of the commercial Aspen Chromatography simulator—more detailed instruction sheets are available from the author: wankat@purdue.edu.

The material in the third edition has been extensively tested in the required junior-level course on separations at Purdue University. Although I teach the material at the junior level, Chapters 1 to 14 could be taught to sophomores, and all of the material is suitable for seniors. The book is too long to cover in one semester, but almost complete coverage is probably feasible in two quarters. If mass transfer is included, this text could easily be used for a two-semester sequence. Many schools, including Purdue, allocate a single three-credit semester course for separations. Because there is too much material, topics must be selected in this case. Several course outlines are included in the Solutions Manual. Instructors may register at www.pearsonhighered.com for access to this book's Solutions Manual and PowerPoint slides of figures in the book.

Acknowledgments

Many people were very helpful in the writing of the first edition. Dr. Marjan Bace and Prof. Joe Calo got me started writing. A. P. V. Inc., Glitsch Inc., and The Norton Co. kindly provided photographs. Chris Roesel and Barb Naugle-Hildebrand did the original artwork. The secretarial assistance of Carolyn Blue, Debra Bowman, Jan Gray, and Becky Weston was essential for completion of the first edition. My teaching assistants Magdiel Agosto, Chris Buehler, Margret Shay, Sung-Sup Suh, and Narasimhan Sundaram were very helpful in finding errors. Professors Ron Andres, James Caruthers, Karl T. Chuang, Alden Emery, and David P. Kessler, and Mr. Charles Gillard were very helpful in reviewing portions of the text. I also owe a debt to the professors who taught me this material: Lowell Koppel, who started my interest in separations as an undergraduate; William R. Schowalter, who broadened my horizons beyond equilibrium staged separations in graduate school; and C. Judson King, who kept my interests alive while I was a professor and administrator through his articles, book, and personal example.

The assistance of Lee Meadows, Jenni Layne, and Karen Heide in preparing the second edition is gratefully acknowledged. I thank the reviewers, John Heydweiller, Stewart Slater, and Joe Shaeiwitz, for their very helpful reviews. The encouragement and occasional prodding from my editor, Bernard Goodwin, was very helpful in getting me to prepare the long-overdue second edition and kept me from procrastinating on the third edition.

The assistance of Karen Heide in preparing the third edition is gratefully acknowledged. The hospitality of the Department of Chemical and Process Engineering at the University of Canterbury, Christchurch, New Zealand, which gave me time to complete Chapter 15 and an opportunity to teach from the ion exchange and extraction sections, is greatly appreciated. I thank the reviewers, Ken Morison, Leonard Pease, David Rockstraw, and Joe Shaeiwitz, for their very helpful reviews. I thank my editor, Bernard Goodwin, for his continued support and encouragement.

Finally, I could not have finished any of the editions without the love and support of my wife, Dot, and my children, Chuck and Jennie.

About the Author

Phillip C. Wankat is the Clifton L. Lovell Distinguished Professor of Chemical Engineering and the Director of Undergraduate Degree Programs in the Department of Engineering Education at Purdue University. He has been involved in research and teaching of separations for more than forty years and is very interested in improving teaching and learning in engineering. He is the author of two books on how to teach and three books on separation processes. He has received a number of national research awards as well as local and national teaching awards.

Nomenclature

Chapters 1 through 18

a	interfacial area per volume, ft^2/ft^3 or m^2/m^3
a_j^I	interfacial area for heat transfer on stage j, m^2
$a_{flow}, a_{heat}, a_{mass}$	eddy diffusion parameters, Eqs. (15-48)
a_p	surface area/volume, m^2/m^3
$a_{p1}, a_{p2}, a_{p3}, a_{T1}, a_{T2}, a_{T6}$	constants in Eq. (2-30) and Table 2-3
A	area, m^2
A,B,C	constants in Antoine Eq. (2-34)
A,B,C,D,E	constants in Eq. (2-60)
A,B,C,D	constants in matrix form of mass balances, Eqs. (6-13) and (12-58)
A_E, B_E, C_E, D_E	constants in matrix form of energy balances, Eq. (6-34)
A_{active}	active area of tray, ft^2 or m^2
A_c	cross-sectional area of column, ft^2 or m^2
A_d	downcomer area, ft^2 or m^2
A_{du}	flow area under downcomer apron, Eq. (10-28), ft^2
A_f	area for flow, m^2
A_{hole}	area of holes in column, ft^2
A_I	interfacial area between two phases, ft^2 or m^2
A_{mixer}	cross-sectional area of mixer, m^2
A_{net}	net area, Eq. (10-13), ft^2 or m^2
A_{total}	total area in horizontal drum, ft^2 or m^2
A_{vap}	Area for Vapor flow in horizontal drum, ft^2 or m^2
b	empirical constant, Eq. (13-63b)
b	equilibrium constant for linear equilibrium, $y = mx + b$
$b_{flow}, b_{heat}, b_{mass}$	eddy diffusion parameters, Eqs. (15-48)
B	bottoms flow rate, kmol/h or lbmol/h
C	number of components
C_{BM}	bare module cost, Chapter 11
C_C	concentration of solute in continuous phase, $kmol/m^3$ continuous phase
C_C^*	concentration of solute in continuous phase in equilibrium with C_D, $kmol/m^3$
C_D	concentration of solute in dispersed phase, $kmol/m^3$ dispersed phase
C_{fL}	vapor load coefficient, Eq. (15-38)
C_A, C_B, C_m	molar concentrations, of A, B, and mixture, mol/m^3
C_o	orifice coefficient, Eq. (10-25)

C_p	heat capacity, Btu/lb°F or Btu/lbmol°F or cal/g°C or cal/mol°C, etc.
C_p	base purchase cost, Chapter 11
$C_{p,size}$	packing size factor, Table 10-5
C_{pW}	water heat capacity
C_s	capacity factor at flood, Eq. (10-48)
C_{sb}	capacity factor, Eq. (10-8)
d	dampening factor, Eq. (2-57)
D	diffusivity, Fickan m²/s or ft²/h
D	distillate flow rate, kmol/h or kg/h
D, Dia	diameter of column, ft or m
D'_{col}	column diameter, see Table 16-1, ft
$d_{hydraulic}$	hydraulic diameter of drop, Eq. (13-62), m
d_i	impeller diameter, m
d_p, d_d	drop diameter, m
d_p^o	characteristic drop diameter, Eq. (16-97b), m
d_{tube}	tube diameter, m
$d_{settler}$, D_s	diameter of horizontal settler, m
D_{large}, D_{pilot}	diameters of Karr columns for scale-up, Eq. (13-66), m
$D°$	infinite dilution Fickian diffusivity, m²/s
$Đ$	Maxwell-Stefan diffusivity, m²/s
D_{eddy}	eddy diffusivity, Eq. (16-111a, b), m/s
D_{total}	total amount of distillate (Chapter 9), moles or kg
e	absolute entrainment, mol/h
e	plate fractional free area in Karr column
erf	error function, Eq. (18-70)
E	extract flow rate (Chapters 13 and 14), kg/h
$\hat{E}$	mass extract, kg
E_j^V	energy transfer rate on stage j from bulk liquid to bulk vapor, J/s
E_k	value of energy function for trial k, Eq. (2-51)
E_{ML}, E_{MV}	Murphree liquid and vapor efficiencies, Eqs. (4-58) and (4-59)
E_0	activation energy, Kcal/mol
E_o	Overall efficiency, Eq. (4-56)
E_{pt}	point efficiency, Eq. (10-5) or (15-76a)
$\hat{E}_t$	holdup extract phase in tank plus settler, kg
f	friction factor
f_{AB}	friction coefficient between molecules A and B
$f = V/F$	fraction vaporized
f	fractional approach to flooding
f	frequency of reciprocation of Karr column, strokes/s
$f(x)$	equilibrium function, Chapter 9
$f_k(V/F)$	Rachford-Rice function for trial K, Eq. (2-43)
F	packing factor, Tables 10-3 and 10-4
F	degrees of freedom, Eq. (2-4)
F	charge to still pot (Chapter 9), moles or kg
F	mass of feed in batch extraction, kg
F	feed flow rate, kmol/h or lbmol/h or kg/h etc.
F_D	diluent flow rate (Chapter 13), kg/h

F_{lv}, FP
$$\frac{W_L}{W_V}\sqrt{\frac{\rho_L}{\rho_V}} = \frac{L'}{G'}\sqrt{\frac{\rho_L}{\rho_V}}$$
, flow parameter

F_m	material factor for cost, Table 11-2
F_p	pressure factor for cost, Eqs. (11-5) and (11-6)
F_q	quantity factor for cost, Eq. (11-7)
F_s, F_{solv}	flow rate solvent (Chapter 13), kg/h
F_{solid}	solids flow rate in leaching, kg insoluble solid/h
F_{weir}	weir modification factor, Eq. (10-26) and Figure 10-22
gap	gap from downcomer apron to tray, Eq. (10-28), ft
g	acceleration due to gravity, 32.2 ft/s², 9.81 m/s²
g_c	conversion factor in English units, 32.2 ft·lbm/(lbf·s²)
G	flow rate carrier gas, kmol/h or kg/h
G'	gas flux, lb/s ft²
h	pressure drop in head of clear liquid, inches liquid
h	height of liquid on stage (Chapter 16), ft
h	height, m or ft
h	height of liquid in mixer, m
h	liquid enthalpy, kcal/kg, Btu/lbmol, etc.
h	step size in Euler's method = Δt, Eq. (8-29)
$\tilde{h}$	pure component enthalpy
h_f	enthalpy of liquid leaving feed stage
h_F	feed enthalpy (liquid, vapor or two-phase)
$h_{heat\ transfer}$	heat transfer coefficient
h_L	clear liquid height on stage, m or cm
h_o	hole diameter, inches
h_p	packing height, ft or m
h_{total}	height of flash drum, ft or m
h_w	height of weir, m or cm
H	Henry's law constant, Eqs. (8-9), (8-10), and (12-1)
H	molar holdup of liquid on tray, Eqs. (8-27) and (8-28)
H	stage height in Karr column, m
H	vapor enthalpy, kcal/kg, Btu/lbmol, etc.
$H_{i,j}^V$	partial molar enthalpy of component i in vapor on stage j, J/kmol
$\overline{H}_t$	height of tank, m
$H_{t,OD}$	overall height of a transfer unit for mass transfer driving force in concentration units, Eq. (16-83a analog), m
H_G	height of gas phase transfer unit, ft or m
H_L	height of liquid phase transfer unit, ft or m
H_{OG}	height of overall gas phase transfer unit, ft or m
H_{OL}	height of overall liquid phase transfer unit, ft or m
HETP	height equivalent to a theoretical plate, ft or m
HTU	height of a transfer unit, ft or m
j_D, j_H	j-functions, Eqs. (15-50)
J_A	flux with respect to molar average velocity of fluid
k_1, k_2	empirical constants, Eq. (13-63b)
k_B	Botzmann's constant, J/k

$k_{conduction}$	thermal conductivity, J/(ms K)
$\overline{k}_x, \overline{k}_y$	individual mass transfer coefficients in liquid and vapor phases, see Table 15-4
k_c	mass transfer coefficient with concentration driving force, m/s, Eq. (15-25b)
k'_y	mass transfer coefficient in concentrated solutions, Eq. (15-32f)
k_x, k_y	individual mass transfer coefficient in molar units
$k_{x,c}, k_{xD}$	individual mass transfer coefficients in continuous and dispersed phases, kg/(s·m^3) or kmol/(s·m^3)
k_{LD}, k_{LC}	individual mass transfer coefficients in continuous and dispersed phases with driving force in concentration units, m/s
k_L, k_V	individual liquid and vapor mass transfer coefficients in distillation, Eq. (16-108), m/s
k	mass transfer coefficient in Maxwell-Stefan analysis, Đ/Δz, m/s
K_d	y/x, distribution coefficient for dilute extraction
K, K_i	y_i/x_i, equilibrium vapor-liquid ratio
K_{drum}	parameter to calculate u_{perm} for flash drums, Eq. (2-64)
K_x, K_y	overall mass transfer coefficient in liquid or vapor, lbmol/ft^2 h, or kmol/h·m^2
K_{LD}	overall mass transfer coefficient in extraction based on dispersed phase in concentration units, Eq. (16-80b analog), m/s
K_{O-ED}	overall mass transfer coefficient in extraction based on dispersed phase, Eq. (16-80a), kg/(s·m^3) or kmol/(s·m^3)
l_w	weir length, ft
L	length, m
L	liquid flow rate, kmol/h or lbmol/h
$\overline{L}$	mass liquid flow rate, lb/h (Chapter 15)
L'	liquid flux, lb/(s)(ft^2)
L_g	liquid flow rate in gal/min, Chapter 10
m	linear equilibrium constant, y = mx + b
m	local slope of equilibrium curve, Eq. (15-30b)
M	ratio HETP$_{practical}$/HETP$_{packing}$ Eq. (10-46)
m_{CD}	slope of equilibrium curve of continuous versus dispersed phase mass or mole fractions, Eq. (16-80c)
$m_{CD,conc\_units}$	slope of equilibrium curve of continuous versus dispersed phase in concentration units, Eq. (16-80c analog)
M	flow rate of mixed stream (Chapter 13), kg/h
M	multiplier times (L/D)$_{min}$ (Chapter 7)
MW	molecular weight
$\overline{MW}$	average molecular weight
n	moles
n	number of drops
n_1, n_2	empirical constants, Eq. (13-65)
n_G	number of gas phase transfer units
n_L	number of liquid phase transfer units
n_{O-ED}, n_{O-EC}	number of overall extraction transfer units in dispersed and continuous phases, Eq. (16-81)

n_{OG}	number of overall gas phase transfer units
n_{OL}	number of overall liquid phase transfer units
n_{org}	moles organic in vapor in steam distillation
n_w	moles water in vapor in steam distillation
N	impeller revolutions per second
N	number of stages
N_A	flux of A, lbmol/(h)(ft^2) or kmol/(h)(m^2)
N_f, N_{feed}	feed stage
$N_{f,j}^L$	transfer to liquid from vapor on stage j, mol component i/s
$N_{f,j}^V$	transfer to vapor from liquid on stage j, mol component i/s
N_{min}	number of stages at total reflux
$N_{feed,min}$	estimated feed stage location at total reflux
N_{Po}	power number, Eq. (13-52)
N_{tOD}	number of overall extraction transfer units for mass transfer driving force in concentration units, Eq. (16-81a analog)
Nu	Nusselt number, Eq. (15-33g)
NTU	number of transfer units
O	total overflow rate in washing, kg/h
p	pitch of sieve plate holes, m
p, p_{tot}	pressure, atm, kPa, psi, bar etc.
$\bar{p}, p_B$	partial pressure
P	Number of phases
P	power, W
Pe	dimensionless Peclet number in terms of molecular diffusivity, Eq. (15-33c)
Pe	dimensionless Peclet number in terms of eddy diffusivity, Eq. (16-111a)
Per_f	flow perimeter, Figure 13-33B, m
Pr	dimensionless Prandt number, Eq. (15-33f)
q	$L_F/F = (L - L)/F$, feed quality
q	volumetric flow rate/plate width, m^2/s
Q	amount of energy transferred, Btu/h, kcal/h, etc.
Q_c	condenser heat load
Q_c, Q_C	volumetric flow rate continuous phase, m^3/s
Q_d, Q_D	volumetric flow rate dispersed phase, m^3/s
Q_{flash}	heat loss from flash drum
Q_L	volumetric flow rate of liquid, m^3/s
Q_R	reboiler heat load
Q_z	heat flux in z direction, J/s
r	radius of column, ft or m
R	gas constant, 1.9859 cal/(mol·K) or 8.314 m^{3}Pa/(mol·K)
R	raffinate flow rate (Chapter. 13), kg/h
R_A	solute radius, m
$\hat{R}$	mass raffinate, kg
$\hat{R}_t$	Holdup raffinate phase in tank plus settler, kg
Re	dimensionless Reynolds number, Eq. (15-33b)
$Re_{settler}$	Reynold's number for settler, Eq. (13-60a)
S	solvent flow rate kmol/h or lbmol/h
S	tray spacing, inches, Eq. (10-47)

S	moles second solvent in constant-level batch distillation
$\hat{S}$	mass of solvent, kg
S	solvent flow rate, kg/h
Sc_L	Schmidt number for liquid $= \mu/(\rho D)$
Sc_v	Schmidt number for vapor $= \mu/(\rho D)$
Sh_c, Sh_x, Sh_y	dimensionless Sherwood numbers, Eq. (15-33a)
St_c, St_x, St_y	dimensionless Stanton numbers, Eq. (15-33d)
t	time, s, min, or h
t_{batch}	period for batch distillation, Eq. (9-28)
t_{down}	down time in batch distillation
$(t_{f,95} - t_0)$	residence time in extractor for 95% extraction, Eq. (16-105), s
t_L, t_V	average residence time per pass for liquid and vapor, s
$\bar{t}_{L,residence}$	liquid residence time, Eq. (16-111c), s
$t_{residence,dispersed}$	residence time of dispersed phase in settler, s
$t_{operating}$	operating time in batch distillation
t_{res}	residence time in downcomer, Eq. (10-30), s, or on plate, Eq. (16-35e)
t_{tray}	tray thickness, inches
T	temperature, °C, °F, K, or °R
T_j^{IL}, T_j^{IV}	liquid and vapor temperatures on stage j at the interface, K
T_{ref}	reference temperature
u	vapor velocity, cm/s or ft/s
u_{flood}	flooding velocity, Eq. (10-8)
u_{op}	operating velocity, Eq. (10-11)
u_{perm}	permissible vapor velocity, Eq. (2-64)
$u_{t,hindered}$	hindered settling velocity, Eq. (13-58)
$u_t, u_{t,Stokes}$	Stokes' law terminal velocity, Eq. (13-57), m/s
U	underflow liquid rate, (Chapter 14), kg/h
U_a	superficial vapor velocity in active area of tray, m/s
v	superficial vapor velocity, ft/s
$v_{characteristic}$	characteristic velocity of Karr column, Eq. (13-68), m/s
$v_{c,flood}, v_{d,flood}$	continuous and dispersed phase flooding velocities, m/s
v_o	vapor velocity through holes, Eq. (10-29), ft/s
$v_{o,bal}$	velocity where valve is balanced, Eq. (10-36)
V_A, V_B	component transfer velocities, Eqs, (15-15e, f)
v_{ref}	reference or basis velocity, Eqs. (15-15c, d)
v_y	vertical velocity
V	vapor flow rate, kmol/h or lbmol/h
V_i	molal volume Eq. (13-1)
V_A	molar volume solute at normal boiling point, m³/kmol
$V_{liq,tank}$	volume of liquid in tank, m³
V_{max}	maximum vapor flow rate
V_{mixer}	volume of liquid in mixing tank, m³
$V_{settler}$	volume settler, m³
V_{tank}	volume tank, m³
V_{surge}	surge volume in flash drum, Eq. (2-68), ft³
VP	vapor pressure, same units as p
w	plate width, m
W_L	liquid flow rate, kg/h or lb/h

W_L	liquid mass flux, lb/s ft² or lb/h ft², (Chapter 16)
W_V	vapor flow rate, kg/h or lb/h
x	weight or mole fraction in liquid
x	$[L/D - (L/D)_{min}]/(L/D + 1)$ in Eqs. (7-42)
x^*	equilibrium mole fraction in liquid
$x_{A,ref}, x_{B,ref}$	fractions to calculate velocity of center of total flux, Eq. (15-17)
$x_{i,k}, x_{i,k+1}$	trials for integration, Eq. (8-29)
x_I	interfacial mole fraction in liquid
x^*_{out}	liquid mole fraction in equilibrium with inlet gas, Eq. (16-35b)
X	weight or mole ratio in liquid
y	weight or mole fraction in vapor
y_{vol}	volume fraction in vapor
y^*	equilibrium mole fraction in vapor
y^*_{out}	vapor mole fraction in equilibrium with inlet liquid in countercurrent system, Eq. (16-35a) or in equilibrium with outlet liquid in cocurrent contactor, Eq. (16-71)
y_{lm}	log mean difference, Eq. (15-32d)
y_I	interfacial mole fraction in vapor
$\bar{y}$	mass fraction in vapor
Y	weight or mole ratio in vapor
z	weight or mole fraction in feed
z	axial distance in bed (Chapters 15 and 16)
z_l	distance from downcomer exit to weir, m

Greek

α_{AB}	K_A/K_B, relative volatility
$\alpha_{thermal}$	thermal diffusity, m²/s
β	A_{hole}/A_{active}
γ	activity coefficient
δ	thickness of mass transfer film or thickness of falling film, m
δ_p	characteristic dimension of packing, inch, Eq. (10-38)
δ_i	solubility parameter, Eq. (13-1)
Δ	change in variable or difference operator
ΔE_v	latent energy of vaporization, Eq. (13-1)
ΔH	steady state height of dispersion band in settler, m
$\Delta\rho$	$|\rho_C - \rho_D|$
ε	limit for convergence
$\varepsilon_A, \varepsilon_B, \varepsilon_{AB}$	Lennard-Jones interaction energies, Table 15-2 and Eq. (15-22c)
η	fraction of column available for vapor flow
η	parameter, Eq. (15-42b)
θ	angle of downcomer, Figure 10-20B
λ	latent heat of vaporization, kcal/kg, Btu/lb, Btu/lbmol, etc.
μ	viscosity, cp or Pa·s = kg/(m s)
μ_w	viscosity of water, cp
ρ_L	liquid density, g/cm³ or lb/ft³ or kg/m³
ρ_V	vapor density
σ, γ	surface tension, dynes/cm or interfacial tension

ς	dimensionless distance, Eq. (15-14a)
χ	term defined in Eq. (13-49)
φ_c, φ_d	volumetric fraction of continuous and dispersed phases
$\varphi_{d,feed}$	volumetric fraction of dispersed phase in feed
ϕ	liquid phase packing parameter, Eq. (16-38)
ϕ_B	solvent interaction parameter, Eq. (15-23b)
ϕ_{dc}	relative froth density in downcomer, Eq. (10-29)
φe	effective relative froth density, Eq. (16-109d)
ψ	ρ_{water}/ρ_L, Chapter 10
ψ	$e/(e + L)$, fractional entrainment, Chapter 10
ψ	packing parameter for gas phase, Eq. (16-37)
Ω_D	collision integral, Table 15-2
$\mu_C \ \mu_D$	viscosity of continuous and dispersed phases, Pa·s
$\mu_H \ \mu_L$	viscosity of heavy and light phases, Pa·s
μ_m	mixture viscosity, Eq. (13-55), Pa·s
ρ_C, ρ_D	densities of continuous and dispersed phases, g/m³
ρ_m	mixture density, Eq. (13-53), g/m³
ρ_m	molar density, mol/m³
ω	revolutions per second

Chapter 17

a, a_j	term in quadratic equations for well-mixed membrane systems, Eqs. (17-10b), (17-74a), and $\overline{(17\text{-}74a)}$
$\hat{a}$	constant in expression to calculate osmotic pressure, kPa/mole fraction, Eq. (17-15a)
a'	constant in expression to calculate osmotic pressure, kPa/weight fraction, Eq. (17-15b)
a_i	activities, Eq. (17-51)
A	membrane area available for mass transfer, cm² or m²
b, b_j	term in quadratic equations for well-mixed membrane systems, Eqs. (17-10c), (17-74b), and $\overline{(17\text{-}74b)}$
c, c_j	term in quadratic equations for well-mixed membrane systems, Eqs. (17-10d), (17-74c) and $\overline{(17\text{-}74c)}$
c	concentration, g solute/L solution
c_{out}	outlet concentration of solute, g/L
c_p	permeate concentration of solute, g/L
c_w	concentration of solute at wall, g/L
c'	water concentration in permeate in Figure 17-17
$C_{PL,p}$	liquid heat capacity of permeate, kJ/(kg °C)
$C_{PV,p}$	vapor heat capacity of permeate, kJ/(kg °C)
d_t	diameter of tube, cm
d_{tank}	tank diameter, cm
D	diffusivity in solution, cm²/s
D_m	diffusivity in the membrane, cm²/s
F_p	volumetric flow rate of permeate, cm³/s
F_{out}	volumetric flow rate of exiting retentate, cm³/s
F_{solv}	volumetric flow rate of solvent in RO, cm³/s

$\hat{F}$	molar flow rate, mol/s, mol/min, etc.
F'	mass flow rate, g/s, g/min, kg/min, etc.
h	½ distance between parallel plates, cm
h_{in}	enthalpy of inlet liquid stream in pervaporation, kJ/kg
h_{out}	enthalpy of outlet liquid retentate stream in pervaporation, kJ/kg
H_A	solubility parameter, cc(STP)/[cm^3 (cm Hg)]
H_p	enthalpy of vapor permeate stream in pervaporation, kJ/kg
k	mass transfer coefficient, typically cm/s, Eq. (17-33)
K'_{solv}	permeability of the solvent through membrane, L/(atm m^2 day) or similar units
j	counter for stage location in staged models in Figure 17-19
J	volumetric flux, cm^3/(s cm^2) or m^3/(m^2 day), Eq. (17-1b)
J'	mass flux, g/(s cm^2) or g/(m^2 day), Eq. (17-1c)
$\hat{J}$	mole flux, mol/(s cm^2) or kmol/(day m^2), Eq. (17-1d)
K'_A	solute permeability, g/(m s wt frac)
$K_{m,i}$	rate transfer term for multicomponent gas permeation, dimensionless, Eq. (17-11d)
L	tube length, cm
M	concentration polarization modulus in wt fraction units, dimensionless, Eq. (17-17)
M_c	concentration polarization modulus in concentration units, dimensionless, Eq. (17-48)
MW	molecular weight, g/mol or kg/kmol
N	number of well-mixed stages in models in Figure 17-19
p	pressure, Pa, kPa, atm, mm Hg, etc.
p_A	partial pressure of species A, Pa, atm, mm Hg, etc.
p_p	total pressure on the permeate (low pressure) side, Pa, kPa, atm, mm Hg, etc.
p_r	total pressure on the retentate (high pressure) side, Pa, kPa, atm, mm Hg, etc.
P_A	permeability of species A in the membrane, cc(STP) cm/[cm^2 s cm Hg]
R	rejection coefficient in wt frac units, dimensionless, Eq. (17-24a)
R^o	inherent rejection coefficient (M = 1), dimensionless
R_c	rejection coefficient in conc. units, dimensionless, Eq. (17-48)
R	tube radius, cm
Re	Reynolds number, dimensionless, Eq. (17-35b)
Sc	Schmidt number, dimensionless, Eq. (17-35c)
Sh	Sherwood number, dimensionless, Eq. (17-35a)
t_{ms}	thickness of membrane skin doing separation, μm, mm, cm, or m
T	temperature, °C
T_{ref}	reference temperature, °C
u_b	bulk velocity in tube, cm/s
$v_{solvent}$	partial molar volume of the solvent, cm^3/gmole
x	wt frac of retentate in pervaporation. In binary system refers to more permeable species.
x_g	wt frac at which solute gels in UF
x_p	wt frac solute in liquid permeate in RO and UF
x_r	wt frac solute in retentate in RO and UF
y	wt frac of permeate in pervaporation. In binary system refers to more permeable species.

y_p	mole fraction solute in gas permeate for gas permeation
y_r	mole fraction solute in gas retentate for gas permeation
$y_{r,w}$	mole fraction solute in gas retentate at membrane wall
$y_{t,A}$	mole fraction solute A in gas that transfers through the membrane

Greek letters

α	selectivity, dimensionless, Gas Permeation: Eq. (17-4b), RO: Eq. (17-20), pervaporation: Eq. (17-53a)
Δx	difference in wt frac of solute across the membrane
$\Delta\pi$	difference in the osmotic pressure across the membrane, Pa, atm, mm Hg, etc.
π	osmotic pressure, Pa, kPa, atm, mm Hg, etc.
θ	cut $= \hat{F}_p/\hat{F}_{in}$, with flows in molar units, dimensionless
θ'	cut $= F'_p/F'_{in}$ in flows in mass units, dimensionless
μ	viscosity, centipoise or g/(cm s)
$\nu = \mu/\rho$	kinematic viscosity, cm^2/s
ρ_{solv}	mass solvent density, kg/m^3
$\hat{\rho}_{solv}$	molar solvent density, kmol/m^3
λ_p	mass latent heat of vaporization of the permeate in pervaporation determined at the reference temperature, kJ/kg
ω	stirrer speed in radians/s

Chapter 18

a	constant in Langmuir isotherm, same units as q/c, Eq. (18-6c)
a	argument for error function, dimensionless, Eq. (18-70), Table 18-7
a_p	surface area of the particles per volume, m^{-1}
A_c	cross sectional area of column, m^2
A_w	wall surface area per volume of column for heat transfer, m^{-1}
b	constant in Langmuir isotherm, (concentration)$^{-1}$, Eq. (18-6c)
c_A	concentration of species A, kg/m^3, kmol/m^3, g/L, etc.
c_i	concentration of species i, kg/m^3, kmol/m^3, g/L, etc., or
c_i	concentration of ion i in solution, typically equivalents/m^3
c_i^*	concentration of species i that would be in equilibrium with $\overline{q}_i$, same units as c_i
$\overline{c}_i$	average concentration of solute in pore, same units as c_i
c_{pore}	fluid concentration at surface of adsorbent pores, same units as c_i
$c_{i,surface}$	fluid concentration at surface of particles, $\varepsilon_p = 0$, same units as c_i
c_{Ri}	concentration of ion i on the resin, typically equivalents/m^3
c_{RT}	total concentration of ions on the resin, typically equivalents/m^3
c_T	total concentration of ions in solution, typically equivalents/m^3
C_i	constant relating solute velocity to interstitial velocity, dimensionless, Eq. (18-15e)
$C_{P,f}$	heat capacity of the fluid, cal/(g °C), cal/(mol °C), J/(g K), etc.
$C_{P,p}$	heat capacity of particle including pore fluid, same units $C_{P,f}$
$C_{P,s}$	heat capacity of the solid, same units as $C_{P,f}$
$C_{P,w}$	heat capacity of the wall, same units as $C_{P,f}$
d_p	particle diameter, cm or m
D	desorbent rate in SMB, same units as F

D/F	desorbent to feed ratio in SMB, dimensionless
D_{col}	column diameter, m or cm
D	diffusivity including both molecular and Knudsen diffusivities, m^2/s or cm^2/s
$D_{effective}$	effective diffusivity, m^2/s or cm^2/s, Eq. (18-4)
D_K	Knudsen diffusivity, m^2/s or cm^2/s, Eq. (18-51)
$D_{molecular}$	molecular diffusivity in free solution, m^2/s or cm^2/s
D_s	surface diffusivity, m^2/s or cm^2/s, Eq. (18-53)
erf	error function, Eq. (18-70) and Table 18-7
E_D	axial dispersion coefficient due to both eddy and molecular effects, m^2/s or cm^2/s
E_{DT}	thermal axial dispersion coefficient, m^2/s or cm^2/s
E_{eff}	effective axial dispersion coefficient, same units E_D, Eq. (18-68)
F	volumetric feed rate, e.g., m^3/h, cm^3/min, liter/h
h_p	particle heat transfer coefficient, $J/(K\ s\ m^2)$ or similar units
h_w	wall heat transfer coefficient, $J/(K\ s\ m^2)$ or similar units
HETP	height of equilibrium plate, cm/plate, Eq. (18-78b)
k_f	film mass transfer coefficient, m/s or cm/s
$k_{m,c}$	lumped parameter mass transfer coefficient with concentration driving force, m/s or cm/s, Eqs. (18-56a) and (18-57a)
$k_{m,q}$	lumped parameter mass transfer coefficient with amount adsorbed driving force, m/s or cm/s, Eqs. (18-56b) and (18-57b)
K_{AB}	mass action equilibrium constant for monovalent-monovalent ion exchange, dimensionless, Eq. (18-40a)
$K_{A,c}$	adsorption equilibrium constant in terms of concentration, units are $(concentration)^{-1}$
$K'_{i,c}$	linearized adsorption equilibrium constant in terms of concentration, units are units of q/c, Eq. (18-6b)
K_{Ao}	pre-exponential factor in Arrhenius Eq, (18-7a), same units as K_A
$K_{A,p}$	adsorption equilibrium constant in terms of partial pressure, units are $(pressure)^{-1}$
$K'_{A,p}$	linearized adsorption equilibrium constant in terms of partial pressure, units are units of q_A/p_A, Eq. (18-5b)
K_d	size exclusion parameter, dimensionless
K_{DB}	mass action equilibrium constant for divalent-monovalent ion exchange, same units as c_T/c_{RT}, Eq. (18-41)
K_{DE}	Donnan exclusion factor, dimensionless, following Eq. (18-44)
L	length of packing in column, m or cm
L_{MTZ}	length of mass transfer zone, Figure 18-23, m or cm
M	molecular weight of solute, g/mol or kg/kmol
M_i	multipliers in Eqs. (18-29), dimensionless
N	equivalent number of plates in chromatography, Eq. (18-78)
N_{Pe}	Peclet number, dimensionless, Eq. (18-62)
p_A	partial pressure of species A, mm Hg, kPa, or other pressure units
p_h	high pressure, mm Hg, kPa, or other pressure units
p_L	low pressure, mm Hg, kPa, or other pressure units
Pe_L	Peclet number based on length, dimensionless, Eq. (18-78a)
q_A	amount of species A adsorbed, kg/kg adsorbent, mol/kg adsorbent, or kg/L
$q_{A,max}$	maximum amount of species A that can adsorb, kg/kg adsorbent, mol/kg adsorbent, or kg/L

q_F	amount adsorbed in equilibrium with feed concentration, same units as q_A
$\overline{q}_i$	average amount of species i adsorbed, kg/kg adsorbent, mol/kg adsorbent, or kg/L
q_i^*	amount adsorbed that would be in equilibrium with fluid of concentration c_i, same units as q_A
Q	volumetric flow rate, m^3/s, L/min, etc.
r_p	pore radius, m or cm
R	resolution, dimensionless, Eq. (18-82)
R	gas constant (e.g., R = 8.314 $\dfrac{m^3 Pa}{mol\ K}$)
Re	Reynolds number, dimensionless, Eq. (18-60)
Sc	Schmidt number, dimensionless, Eq. (18-60)
Sh	Sherwood number, dimensionless, Eq. (18-60)
t	time, s, min, or h
t_{br}	breakthrough time, s, min, or h
t_{center}	time center of pattern exits column, s, min, or h, Eq. (18-85b)
$t_{elution}$	elution time, s, min, or h
t_F, t_{feed}	feed time, s, min, or h
t_{MTZ}	time of mass transfer zone, Figure 18-23, s, min, or h
t_R	retention time, s, min, or h
t_{sw}	switching time in SMB, s, min, or h
T	temperature, °C or K
T_{amb}	ambient temperature, °C or K
T_s	solid temperature, °C or K
$u_{ion,i}$	velocity of ion i, m/s or cm/s
u_s	average solute velocity, m/s or cm/s
$\overline{u}_s$	average of solute velocities for A and B, cm/s, Eq. (18-83)
$u_{s,ion,i}$	diffuse wave velocity of ion i, m/s or cm/s
u_{sh}	shock wave velocity, m/s or cm/s
$u_{sh,ion,i}$	shock wave velocity of ion i, m/s or cm/s
u_{th}	thermal wave velocity, m/s or cm/s
u_{total_ion}	velocity of total ion wave, m/s or cm/s
$v_{A,product}$	interstitial velocity of A Product if it was in the column, m/s or cm/s = (A Product)/($\varepsilon_e\ A_c$)
$v_{B,product}$	interstitial velocity of B Product if it was in the column, m/s or cm/s = (B Product)/($\varepsilon_e\ A_c$)
v_D	interstitial velocity of desorbent if it was in the column, m/s or cm/s = D/($\varepsilon_e\ A_c$)
v_{Feed}	interstitial velocity of feed if it was in the column, m/s or cm/s = F/($\varepsilon_e\ A_c$)
v_{inter}	interstitial velocity, m/s or cm/s, Eq. (18-2b)
v_{super}	superficial velocity, m/s or cm/s, Eq. (18-2a)
$V_{available}$	volume available to molecule, m^3, Eq. (18-1c)
V_{column}	column volume, m^3
V_{feed}	volume feed gas, m^3
V_{fluid}	volume available to fluid, m^3, Eq. (18-1a)
V_{purge}	volume purge gas, m^3
w_A, w_B	width of chromatographic peak, s, min or hours
W	weight of the column per length, kg/m

x	deviation from the location of the peak maximum, dimensionless Eq. (18-79)
x_l	deviation from peak maximum in length units, Eq. (18-80b)
x_t	deviation from peak maximum in time units, Eq. (18-80a)
x	weight or mole fraction solute in liquid, kg solute/kg liquid or kmol solute/kmol liquid, dimensionless
x_i	$= c_i/c_T$ equivalent fraction of ion in solution, dimensionless
$X_{breakthrough}$ (z,t)	general solution for column breakthrough for linear isotherms, same units as c, Eq. (18-72)
y	weight or mole fraction solute in gas, kg solute/kg gas, or kmol solute/kmol gas, dimensionless
y_i	$= c_{Ri}/c_{RT}$ equivalent fraction of ion on resin, dimensionless
z	axial distance in column, m or cm. (Measured from closed end for PSA pressure change calculations)

Greek letters

β_{strong}	ratio velocities of strong and weak solutes, Eq. (18-27), dimensionless
Δc	change in solute concentration, same units as c
ΔH_{ads}	heat of adsorption, J/kg, cal/gmole, etc.
Δp_A	change in partial pressure, kPa, atm, etc.
Δq	change in amount adsorbed, kmol/kg adsorbent, kg/kg adsorbent, kmol/m^3, or kg/m^3
Δt	change in time, s, min, or h
ΔT_f	change in fluid temperature, °C or K
Δz	increment of column length, m
γ	volumetric purge to feed ratio in PSA, dimensionless, Eq. (18-26)
ε_e	external porosity, dimensionless
ε_p	internal or pore porosity, dimensionless
ε_T	total porosity, dimensionless, Eq. (18-1b)
ρ_b	bulk density of adsorbent, kg/m^3, Eq. (18-3b)
ρ_f	fluid density, kg/m^3
$\bar{\rho}_f$	molar density of fluid, kmol/m^3
ρ_p	particle density, kg/m^3, Eq. (18-3a)
ρ_s	structural density of solid, kg/m^3
σ	standard deviation of Gaussian chromatographic peak, Eq. (18-79)
σ_l	standard deviation in length units, m or cm, Eq. (18-80b)
σ_t	standard deviation in time units, min or s, Eq. (18-80a)
τ	tortuosity, dimensionless, Eq. (18-4)
ζ	Greek letter zeta used as dummy variable in Eq. (18-70)

CHAPTER 1

Introduction to Separation Process Engineering

1.1 IMPORTANCE OF SEPARATIONS

Why does chemical engineering require the study of separation techniques? Because separations are crucial in chemical engineering. A typical chemical plant is a chemical reactor surrounded by separators, as diagramed in the schematic flow sheet of Figure 1-1. Raw materials are prepurified in separation devices and fed to the chemical reactor; unreacted feed is separated from the reaction products and recycled back to the reactor. Products must be further separated and purified before they can be sold. This type of arrangement is very common. Examples for a variety of traditional processes are illustrated by Biegler et al. (1997), Chenier (2002), Couper et al. (2005), Matar and Hatch (2001), Shreve and Austin (1984), Speight (2002), and Turton et al. (2003), whereas recent processes often are shown in *Chemical Engineering* magazine. Chemical plants commonly have from 40% to 70% of both capital and operating costs in separations (Humphrey and Keller, 1997).

Since separations are ubiquitous in chemical plants and petroleum refineries, chemical engineers must be familiar with a variety of separation methods. We will first focus on some of the most common chemical engineering separation methods: flash distillation, continuous column distillation, batch distillation, absorption, stripping, and extraction. These separations all contact two phases and can be designed and analyzed as equilibrium stage processes. Several other separation methods that can also be considered equilibrium stage processes will be briefly discussed. Chapters 17 and 18 explore two important separations—membrane separators and adsorption processes—that do not operate as equilibrium stage systems.

The *equilibrium stage* concept is applicable when the process can be constructed as a series of discrete stages in which the two phases are contacted and then separated. The two separated phases are assumed to be in equilibrium with each other. For example, in distillation, a vapor and a liquid are commonly contacted on a metal plate with holes in it. Because of the intimate contact between the two phases, solute can transfer from one phase to another. Above the plate the vapor disengages from the liquid. Both liquid and vapor can be sent to additional stages for further separation. Assuming that the stages are equilibrium stages, the engineer can calculate concentrations and temperatures without detailed knowledge of

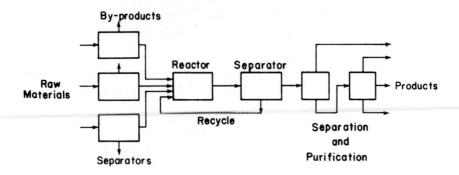

FIGURE 1-1. *Typical chemical plant layout*

flow patterns and heat and mass transfer rates. Although this example shows the applicability of the equilibrium stage method for equipment built with a series of discrete stages, we will see that the staged design method can also be used for packed columns where there are no discrete stages. This method is a major simplification in the design and analysis of chemical engineering separations that is used in Chapters 2 to 14.

A second useful concept is that of a *unit operation*. The idea here is that although the specific design may vary depending on what chemicals are being separated, the basic design principles for a given separation method are always the same. For example, the basic principles of distillation are always the same whether we are separating ethanol from water, separating several hydrocarbons, or separating liquid metals. Consequently, distillation is often called a unit operation, as are absorption, extraction, etc.

A more general idea is that design methods for related unit operations are similar. Since distillation and absorption are both liquid-vapor contacting systems, the design is much the same for both. This similarity is useful because it allows us to apply a very few design tools to a variety of separation methods. We will use *stage-by-stage* methods where calculation is completed for one stage and then the results are used for calculation of the next stage to develop basic understanding. Matrix solution of the mass and energy balances will be used for detailed computer simulations.

1.2 CONCEPT OF EQUILIBRIUM

The separation processes we are studying in Chapters 1 to 14 are based on the equilibrium stage concept, which states that streams leaving a stage are in equilibrium. What do we mean by equilibrium?

Consider a vapor and a liquid that are in contact with each other as shown in Figure 1-2. Liquid molecules are continually vaporizing, while vapor molecules are continually condensing. If two chemical species are present, they will, in general, condense and vaporize at different rates. When not at equilibrium, the liquid and the vapor can be at different pressures and temperatures and be present in different mole fractions. At equilibrium the temperatures, pressures, and fractions of the two phases cease to change. Although molecules continue to evaporate and condense, the rate at which each species condenses is equal to the rate at which

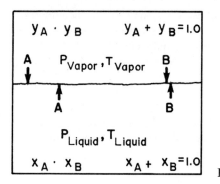

FIGURE 1-2.   *Vapor-liquid contacting system*

it evaporates. Although on a molecular scale nothing has stopped, on the macroscopic scale, where we usually observe processes, there are no further changes in temperature, pressure, or composition.

Equilibrium conditions can be conveniently subdivided into thermal, mechanical, and chemical potential equilibrium. In thermal equilibrium, heat transfer stops and the temperatures of the two phases are equal.

$$T_{liquid} = T_{vapor} \qquad \text{(at equilibrium)} \qquad \textbf{(1-1)}$$

In mechanical equilibrium, the forces between vapor and liquid balance. In the staged separation processes we will study, this usually implies that the pressures are equal. Thus for the cases in this book,

$$p_{liquid} = p_{vapor} \qquad \text{(at equilibrium)} \qquad \textbf{(1-2)}$$

If the interface between liquid and vapor is curved, equal forces do not imply equal pressures. In this case the Laplace equation can be derived (e.g., see Levich, 1962).

In phase equilibrium, the rate at which each species is vaporizing is just equal to the rate at which it is condensing. Thus there is no change in composition (mole fraction in Figure 1-2). However, in general, the compositions of liquid and vapor are *not* equal. If the compositions were equal, no separation could be achieved in any equilibrium process. If temperature and pressure are constant, equal rates of vaporization and condensation require a minimum in the free energy of the system. The resulting condition for phase equilibrium is

$$(\text{chemical potential } i)_{liquid} = (\text{chemical potential } i)_{vapor} \qquad \textbf{(1-3)}$$

The development of Eq. (1-3), including the necessary definitions and concepts, is the subject of a large portion of many books on thermodynamics (e.g., Balzhiser et al., 1972; Denbigh, 1981; Elliott and Lira, 1999; Sandler, 2006; Smith et al., 2005; Walas, 1985) but is beyond the scope of this book. However, Eq. (1-3) does require that there be some relationship between liquid and vapor compositions. In real systems this relationship may be very complex and experimental data may be required. We will assume that the equilibrium data or appropriate correlations are known (see Chapter 2), and we will confine our discussion to the *use* of the equilibrium data in the design of separation equipment.

1.3 MASS TRANSFER

In the vapor-liquid contacting system shown in Figure 1-2 the vapor and liquid will not be initially at equilibrium. By transferring mass from one phase to the other we can approach equilibrium. The basic mass transfer equation in words is

$$\text{Mass transfer rate} = (\text{area}) \times (\text{mass transfer coefficient}) \times (\text{driving force}) \qquad \textbf{(1-4)}$$

In this equation the mass transfer rate will typically have units such as kmol/h or lbmol/h. The area is the area across which mass transfer occurs in m^2 or ft^2. The driving force is the concentration difference that drives the mass transfer. This driving force can be represented as a difference in mole fractions, a difference in partial pressures, a difference in concentrations in kmol/L, and so forth. The value and units of the mass transfer coefficient depend upon which driving forces are selected. The details are discussed in Chapter 15.

For equilibrium staged separations we would ideally calculate the mass transfer rate based on the transfer within each phase (vapor and liquid in Figure 1-2) using a driving force that is the concentration difference between the bulk fluid and the concentration at the interface. Since this is difficult, we often make a number of simplifying assumptions (see Section 15.4 for details) and use a driving force that is the difference between the actual concentration and the concentration we would have if equilibrium were achieved. For example, for the system shown in Figure 1-2 with concentrations measured in mole fractions, we could use the following rate expressions.

$$\text{Rate / volume} = K_y a(y_A^* - y_A) \qquad \textbf{(1-5a)}$$

$$\text{Rate / volume} = K_x a(x_A - x_A^*) \qquad \textbf{(1-5b)}$$

In these equations K_y and K_x are overall gas and liquid mass transfer coefficients, y_A^* is the mole fraction in the gas in equilibrium with the actual bulk liquid of mole fraction x_A, x_A^* is the mole fraction in the liquid in equilibrium with the actual bulk gas of mole fraction y_A, and the term "a" is the interfacial area per unit volume (m^2/m^3 or ft^2/ft^3).

By definition, at equilibrium we have $y_A^* = y_A$ and $x_A^* = x_A$. Note that as $y_A \rightarrow y_A^*$ and $x_A \rightarrow x_A^*$ the driving forces in Eqs. (1-5) approach zero and mass transfer rates decrease. In order to be reasonably close to equilibrium, the simplified model represented by Eqs. (1-5) shows that we need high values of K_y and K_x and/or "a." Generally speaking, the mass transfer coefficients will be higher if diffusivities are higher (details are in Chapter 15), which occurs with fluids of low viscosity. Since increases in temperature decrease viscosity, increasing temperature is favorable as long as it does not significantly decrease the differences in equilibrium concentrations and the materials are thermally stable. Mass transfer rates will also be increased if there is more interfacial area/volume between the gas and liquid (higher "a"). This can be achieved by having significant interfacial turbulence or by using a packing material with a large surface area (see Chapter 10).

Although some knowledge of what affects mass transfer is useful, we don't need to know the details as long as we are willing to assume we have equilibrium stages. Thus, we will delay discussing the details until we need them (Chapters 15 through 18).

1.4 PROBLEM-SOLVING METHODS

To help develop your problem-solving abilities, an explicit strategy, which is a modification of the strategy developed at McMaster University (Woods et al., 1975), is used throughout this book. The seven stages of this strategy are:

0. I want to, and I can
1. Define the problem
2. Explore or think about it
3. Plan
4. Do it
5. Check
6. Generalize

Step 0 is a motivation and confidence step. It is a reminder that you got this far in chemical engineering because you can solve problems. The more different problems you solve, the better a problem solver you will become. Remind yourself that you *want* to learn how to solve chemical engineering problems, and you *can* do it.

In step 1 you want to *define* the problem. Make sure that you clearly understand all the words. Draw the system and label its parts. List all the known variables and constraints. Describe what you are asked to do. If you cannot define the problem clearly, you will probably be unable to solve it.

In step 2 you *explore* and *think about* the problem. What are you *really* being asked to do? What basic principles should be applied? Can you find a simple limiting solution that gives you bounds to the actual solution? Is the problem over- or underspecified? Let your mind play with the problem and chew on it, and then go back to step 1 to make sure that you are still looking at the problem in the same way. If not, revise the problem statement and continue. Experienced problem solvers always include an *explore* step even if they don't explicitly state it.

In step 3 the problem solver *plans* how to subdivide the problem and decides what parts to attack first. The appropriate theory and principles must be selected and mathematical methods chosen. The problem solver assembles required resources such as data, paper, and calculator. While doing this, new subproblems may arise; you may find there are not enough data to solve the problem. Recycle through the problem-solving sequence to solve these subproblems.

Step 4, *do it,* is often the first step that inexperienced problem solvers try. In this step the mathematical manipulations are done, the numbers are plugged in, and an answer is generated. If your plan was incomplete, you may be unable to carry out this step. In that case, return to step 2 or step 3, the *explore* or *plan* steps, and recycle through the process.

In step 5, *check* your answer. Is it the right order of magnitude? For instance, commercial distillation columns are neither 12 centimeters nor 12 kilometers high. Does the answer seem reasonable? Have you avoided blunders such as plugging in the wrong number or incorrectly punching the calculator? Is there an alternative solution method that can serve as an independent check on the answer? If you find errors or inconsistencies, recycle to the appropriate step and solve the problem again.

The last step, *generalize,* is important but is usually neglected. In this step you try to learn as much as possible from the problem. What have you learned about the physical situation? Did including a particular phenomenon have an important effect, or could you have ignored it? Generalizing allows you to learn and become a better problem solver.

At first these steps will not "feel" right. You will want to get on with it and start calculating instead of carefully defining the problem and working your way through the procedure. Stick with a systematic approach. It works much better on difficult problems than a "start calculating, maybe something will work" method. The more you use this or any other strategy, the more familiar and less artificial it will become.

In this book, example problems are solved using this strategy. To avoid repeating myself, I will not list step 0, but it is always there. The other six steps will usually be explicitly listed and developed. On the simpler examples some of the steps may be very short, but they are always present.

I strongly encourage you to use this strategy and write down each step as you do homework problems. In the long run this method will improve your problem-solving ability.

A problem-solving strategy is useful, but what do you do when you get stuck? In this case *heuristics* or rules of thumb are useful. A heuristic is a method that is often helpful but is not guaranteed to help. A large number of problem-solving heuristics have been developed. I have listed ten (Wankat and Oreovicz, 1993) that are often helpful to students.

Problem-Solving Heuristics:

1. Try solving simplified, limiting cases.
2. Relate the problem to one you know how to solve. This heuristic encapsulates one of the major reasons for doing homework.
3. Generalize the problem.
4. Try putting in specific numbers. Heuristics 3 and 4 are the opposite of each other. Sometimes it is easier to see a solution path without all the details, and sometimes the details help.
5. Solve for ratios. Often problems can be solved for ratios, but there is not enough information to solve for individual values.
6. Do the solvable parts of the problem. This approach may provide information that allows you to solve previously unsolvable parts.
7. Look for information that you haven't used.
8. Try to guess and check. If you have a strong hunch, this may lead to an answer, but you *must* check your guess.
9. Take a break. Don't quit, but do something else for a while. Coming back to the problem may help you see a solution path.
10. Ask someone for a *little* help. Then complete the problem on your own.

Ten heuristics is probably too many to use on a regular basis. Select four or five that fit you, and make them a regular part of your problem-solving method. If you want to read more about problem solving and heuristics, I recommend *How to Model It: Problem Solving for the Computer Age* (Starfield et al., 1994) and *Strategies for Creative Problem Solving* (Fogler and LeBlanc, 1995).

1.5 PREREQUISITE MATERIAL

No engineering book exists in a vacuum, and some preparatory material is always required. The first prerequisite, which is often overlooked, is that you must be able to read well. If you don't read well, get help immediately.

A second set of prerequisites involves certain mathematical abilities. You need to be comfortable with algebra and the manipulation of equations, as these skills are used throughout the text. Another required mathematical skill is graphical analysis, since many of the design methods are graphical methods. You need to be competent and to feel comfortable plotting curves and straight lines and solving simultaneous algebraic equations graphically. Familiarity with exponential and logarithmic manipulations is required for Chapter 7. The only chapters requiring calculus are Section 8.5.2, and Chapters 9 and 15 through 18.

The third area of prerequisites concerns mass balances, energy balances, and phase equilibria. Although the basics of mass and energy balances can be learned in a very short time, facility with their use requires practice. Thus, this book will normally be preceded by a course on mass and energy balances. A knowledge of the basic ideas of phase equilibrium, including the concept of equilibrium, Gibbs' phase rule, distribution coefficients, familiarity with graphical representations of equilibrium data, and a working knowledge of vapor-liquid equilibrium (VLE) correlations will be helpful.

Units are a fourth critically important area. The United States' NASA program crashed a space craft into Mars because of failure to convert between the metric and English systems of units. Because conversion of units will remain necessary throughout your career, I have used data in the units in which they were originally presented. Thus, you must do conversions throughout the book. Although problem solutions and Appendix C show conversion factors, it is assumed that you are very familiar and proficient with unit conversions. This includes conversion from weight to mole fractions, and vice versa.

A fifth area of prerequisites is problem-solving skills. Because the chemical engineer must be a good problem solver, it is important to develop skills in this area. The ability to solve problems is a prerequisite for all chemical engineering courses.

In general, later chapters depend on the earlier chapters, as shown schematically in Figure 1-3. Chapters 11, 14, 16, and 17 are not required for the understanding of later chapters and can be skipped if time is short. Figure 1-3 should be useful in planning the order in which to cover topics and for adapting this book for special purposes.

1.6 OTHER RESOURCES ON SEPARATION PROCESS ENGINEERING

Since students have different learning styles, you need to customize the way you use this book to adapt to your learning style. Of course, you will have to take charge of your learning and do this for yourself. If you are interested in exploring your learning style, a good place to start is the Index of Learning Styles, which was developed by Richard M. Felder and Linda K. Silverman. This index is available free on the Internet at www4.ncsu.edu/unity/lockers/users/felder/public/ILSpage.html. Alternatively, you may search on the term "Felder" using a search engine such as Google.

Since students (and professors) have different learning styles, no single approach to teaching or writing a book can be best for all students. Thus, there will undoubtedly be parts of this book that do not make sense to you. Many students use other students, then the teaching

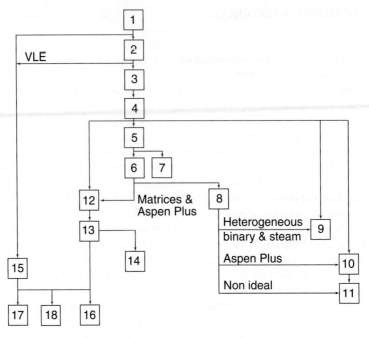

FIGURE 1-3.   *Chapter interdependency*

assistant, and finally the professor as resources. Fortunately, a number of good textbooks and Web pages exist that can be helpful because their presentations differ from those in this textbook. Table 1-1 presents a short annotated bibliography of some of the available handbook and textbook resources. A large number of useful Web sites are available but are not listed because URLs change rapidly. They can be accessed by searching on the term "separation processes" using any popular search engine.

TABLE 1-1.   *Annotated bibliography of resources on separation process engineering*

Belter, P. A., E. L. Cussler, and W.-S. Hu, *Bioseparations. Downstream Processing for Biotechnology,* Wiley-Interscience, New York, 1988. Separations textbook with emphasis on bioseparations.

Cussler, E. L., *Diffusion: Mass Transfer in Fluid Systems*, 3rd ed., Cambridge University Press, Cambridge, UK, 2009. Textbook on basics of diffusion and mass transfer with applications to a variety of separation processes in addition to other applications.

Doherty, M. F., and M. F. Malone, *Conceptual Design of Distillation Systems,* McGraw-Hill, New York, 2001. Advanced distillation textbook that uses residue curve maps to analyze complex distillation processes.

Geankoplis, C. J., *Transport Processes and Separation Process Principles,* 4th ed., Prentice Hall PTR, Upper Saddle River, NJ, 2003. Unit operations textbook that has expanded coverage of separation processes and transport phenomena.

TABLE 1-1.   *Annotated bibliography of resources on separation process engineering*
            (continued)

Harrison, R. G., P. Todd, S. R. Rudge, and D. P. Petrides, *Bioseparations Science and Engineering,* Oxford University Press, New York, 2003. Separations textbook with emphasis on bioseparations.

Hines, A. L., and R. M. Maddox, *Mass Transfer: Fundamentals and Applications,* Prentice-Hall PTR, Upper Saddle River, NJ, 1985. Textbook on basics of diffusion and mass transfer with applications to separation processes.

Humphrey, J. L,. and G. E. Keller II, *Separation Process Technology,* McGraw-Hill, New York, 1997. Industrially oriented book that includes performance, selection and scaleup information.

King, C. J., *Separation Processes,* 2nd ed., McGraw-Hill, New York, 1980. Textbook that seeks to integrate knowledge of separation processes and has extensive case studies.

McCabe, W. L., J. C. Smith, and P. Harriott, *Unit Operations of Chemical Engineering*, 7th ed., McGraw-Hill, New York, 2004. Unit operations textbook that includes extensive coverage of separations and transport phenomena.

Noble, R. D., and P. A. Terry, *Principles of Chemical Separations with Environmental Applications,* Cambridge University Press, Cambridge, UK, 2004. Basic separation principles with environmental examples and problems in a non-calculus based format.

Perry, R. H., and D. W. Green (Eds.), *Perry's Chemical Engineers' Handbook,* 8th ed., McGraw-Hill, New York, 2008. General handbook that has extensive coverage on separations, but coverage often assumes reader has some prior knowledge of technique.

Rousseau, R.W. (Ed.), *Handbook of Separation Process Technology*, Wiley-Interscience, New York, 1987. Handbook containing detailed information on a number of different separation methods.

Schweitzer, P. A. (Ed.), *Handbook of Separation Techniques for Chemical Engineers,* 3rd ed., McGraw-Hill, New York, 1997. Handbook containing detailed information on many separations. Coverage often assumes reader has some prior knowledge of technique.

Seader, J. D., E. J. Henley, and D. J. Roper, *Separation Process Principles,* 3rd ed., Wiley, New York, 2011. Textbook covering an introduction to mass transfer and a large variety of separation processes.

Seidel, A. (Ed.), *Kirk-Othmer Encyclopedia of Chemical Technology*, 5th Ed., Wiley-Interscience, New York, 2004. Extensive encyclopedia with many entries by authorities on separation processes.

Treybal, R. E., *Mass-Transfer Operations,* 3rd ed., McGraw-Hill, New York, 1980. Textbook on basics of diffusion and mass transfer with detailed applications to separation processes.

Wankat, P. C., *Mass Transfer Limited Separations*, Springer, Berlin, 1990. Advanced textbook on crystallization, adsorption, chromatography, ion exchange, and membrane separations.

http://www.engineeringtoolbox.com/ and http://www.cheric.org/research/kdb/ (click on box Korean Physical Properties Data Bank) are excellent sources for data needed for separation problems.

1.7 SUMMARY—OBJECTIVES

We have explored some of the reasons for studying separations and some of the methods we will use. At this point you should be able to satisfy the following objectives:

1. Explain how separations are used in a typical chemical plant
2. Define the concepts of equilibrium stages and unit operations
3. Explain what is meant by phase equilibrium
4. Explain the basic concepts of mass transfer
5. List the steps in the structured problem-solving approach and start to use this approach
6. Have some familiarity with the prerequisites

Note: In later chapters you may want to turn to the Summary—Objectives section first to help you see where you are going. Then when you've finished the chapter, the Summary—Objectives section can help you decide if you got there.

REFERENCES

Balzhiser, R. E., M. R. Samuels, and J. D. Eliassen, *Chemical Engineering Thermodynamics: The Study of Energy, Entropy, and Equilibrium,* Prentice Hall, Upper Saddle River, NJ, 1972.

Biegler, L. T., I. E. Grossman, and A. W. Westerburg, *Systematic Methods of Chemical Process Design*, Prentice Hall, Upper Saddle River, NJ, 1997.

Chenier, P. J., *Survey of Industrial Chemistry,* Springer-Verlag, Berlin, 2002.

Couper, J. R., W. R. Penney, J. R. Fair, and S. M. Walas, *Chemical Process Equipment: Selection and Design*, 2nd ed., Elsevier, Amsterdam, 2005.

Denbigh, K., *The Principles of Chemical Equilibrium*, 4th ed., Cambridge University Press, Cambridge, UK, 1981.

Elliott, J. R., and C. T. Lira, *Introductory Chemical Engineering Thermodynamics,* Prentice Hall, Upper Saddle River, NJ, 1999.

Fogler, H. S., and S. E. LeBlanc, *Strategies for Creative Problem Solving*, Prentice Hall, Englewood Cliffs, NJ, 1995.

Humphrey, J. L., and G. E. Keller II, *Separation Process Technology*, McGraw-Hill, New York, 1997.

Levich, V. G., *Physiochemical Hydrodynamics*, Prentice Hall, Upper Saddle River, NJ, 1962.

Matar, S., and L. F. Hatch, *Chemistry of Petrochemical Processes*, 2nd ed., Gulf Publishing Co., Houston, TX, 2001.

Sandler, S. I., *Chemical and Engineering Thermodynamics*, 4th ed. Wiley, New York, 2006.

Shreve, R. N., and G. T. Hatch, *Chemical Process Industries,* 5th ed., McGraw-Hill, New York, 1984.

Smith, J. M., H. C. Van Ness, and M. M. Abbott, *Introduction to Chemical Engineering Thermodynamics*, 7th ed., McGraw-Hill, New York, 2005.

Speight, J. G., *Chemical and Process Design Handbook,* McGraw-Hill, New York, 2002.

Starfield, A. M., K. A. Smith, and A. L. Bleloch, *How to Model It: Problem Solving for the Computer Age,* 2nd ed., McGraw-Hill, New York, 1994.

Turton, R., R. C. Bailie, W. B. Whiting, and J. A. Shaeiwitz, *Analysis, Synthesis and Design of Chemical Processes,* 2nd ed., Prentice Hall, Upper Saddle River, NJ, 2003.

Walas, S. M., *Phase Equilibria in Chemical Engineering,* Butterworth, Boston, 1985.

Wankat, P. C., and F. S. Oreovicz, *Teaching Engineering,* McGraw-Hill, New York, 1993. Available free at https://engineering.purdue.edu/ChE/News_and_Events/teaching_engineering/index.html.

Woods, D. R., J. D. Wright, T. W. Hoffman, R. K. Swartman, and I. D. Doig, "Teaching Problem Solving Skills," *Engineering Education, 66* (3), 238 (Dec. 1975).

HOMEWORK

A. *Discussion Problems*

A1. Return to your successful solution of a fairly difficult problem in one of your previous technical courses (preferably chemical engineering). Look at this solution but from the point of view of the *process* used to solve the problem instead of the technical details. Did you follow a structured method? Most people don't at first. Did you eventually do most of the steps listed? Usually, the *define, explore, plan,* and *do it* steps are done sometime during the solution. Rearrange your solution so that these steps are in order. Did you check your solution? If not, do that now. Finally, try *generalizing* your solution.

A2. Without returning to the book, answer the following:
   a. Define a unit operation. Give a few examples.
   b. What is the equilibrium stage concept?
   c. What are the steps in the systematic problem solving approach? Explain each step in your own words.

A3. The equilibrium stage concept
   a. is a hypothetical construct.
   b. assumes that phases leaving the stage are in equilibrium.
   c. is useful even when phases are not in equilibrium.
   d. all of the above.

A4. If you have studied heat transfer, relate Eq. (1-4) to the similar basic definition of heat transfer by conduction and convection.

A5. Do you satisfy the prerequisites? If not, how can you remedy this situation?

A6. Develop a key relations chart (one page or less) for this chapter. A key relations chart is a summary of everything you need to solve problems or answer questions from the chapter. In general, it will include equations, sketches, and key words. Organize it in your own way. The purpose of developing a key relations chart is to force your brain to actively organize the material. This will greatly aid you in remembering the material.

B. *Generation of Alternatives*

B1. List as many products and how they are purified or separated as you can. Go to a large supermarket and look at some of the household products. How many of these could you separate? At the end of this course you will know how to purify most of the liquid products.

B2. Some separation methods are common in homes in the United States. Most of these are concerned with water treatment. List the separations that you are familiar with and briefly describe how you think they work.

B3. The body uses several membrane separation methods. List as many of these as you can and describe how you think they work.

B4. Separation operations are very common in chemistry laboratories. List the separations that you employed in various chemistry labs.

C. *Derivations*

C1. Write the mass and energy balances (in general form) for the separator shown in Figure 1-1. If you have difficulty with this, review a book on mass and energy balances.

D. *Problems*

D1. One of the prerequisites for study of separations is the ability to convert from weight to mole fractions and vice versa. As a refresher in this conversion, solve the following problem: We have a flow rate of 1500 kmol/h of a feed that is 40 mol% ethanol and 60 mol% water. What is the weight fraction of ethanol, and what is the total flow rate in pounds per hour?

E. *Complex Problems*

There are no complex problems for this chapter.

F. *Problems Using Other Resources*

F1. Look through several recent issues of *Chemical Engineering* magazine or similar technical magazines and find an article that contains a process flow chart. Read the article and write a short (less than one page) critique. Explicitly comment on whether the flow sheet for the process fits (at least approximately) the general flow sheet shown in Figure 1-1.

F2. Arrange a tour of the unit operations laboratory in your institution to observe the different types of separation equipment. Note that although this equipment is often much larger than the separation equipment that you used in chemistry laboratory, it is much smaller than industrial-scale equipment.

G. *Simulator Problems*

There are no simulator problems for this chapter.

H. *Computer Spreadsheet Problems*

There are no computer spreadsheet problems for this chapter.

CHAPTER 2

Flash Distillation

2.1 BASIC METHOD OF FLASH DISTILLATION

One of the simplest separation processes commonly employed is flash distillation. In this process, part of a liquid feed stream vaporizes in a flash chamber or part of a vapor feed condenses, and the vapor and liquid in equilibrium with each other are separated. The more volatile component will be more concentrated in the vapor. Usually a large degree of separation is not achieved; however, in some cases, such as the desalination of seawater, complete separation results.

The equipment needed for flash distillation is shown in Figure 2-1 for a liquid feed. The fluid is pressurized and heated and is then passed through a throttling valve or nozzle into the flash drum. Because of the large drop in pressure, part of the fluid vaporizes. The vapor is taken off overhead, while the liquid drains to the bottom of the drum, where it is withdrawn. A demister or entrainment eliminator is often employed to prevent liquid droplets from being entrained in the vapor. The system is called "flash" distillation because the vaporization is extremely rapid after the feed enters the drum. Because of the intimate contact between liquid and vapor, the system in the flash chamber is very close to an equilibrium stage. Figure 2-1 shows a vertical flash drum, but horizontal drums are also common. Partial condensation is similar to Figure 2-1, except the vapor is cooled before entering the drum.

The designer of a flash system needs to know the pressure and temperature of the flash drum, the size of the drum, and the liquid and vapor compositions and flow rates. He or she also wishes to know the pressure, temperature, and flow rate of the feed entering the drum. In addition, he or she needs to know how much the original feed has to be pressurized and heated. The pressures must be chosen so that at the feed pressure, p_F, the feed is below its boiling point and remains liquid, while at the pressure of the flash drum, p_{drum}, the feed is above its boiling point and some of it vaporizes. Because the energy for vaporization comes from the hot feed, $T_F > T_{drum}$, if the feed is already hot and/or the pressure of the flash drum is quite low, the pump and heater shown in Figure 2-1 may not be needed.

The designer has six degrees of freedom to work with for a binary separation. Usually, the original feed specifications take up four of these degrees of freedom:

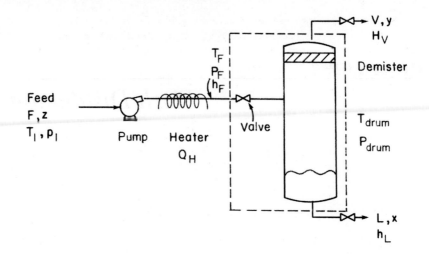

FIGURE 2-1. *Flash distillation system*

Feed flow rate, F
Feed composition, z (mole fraction of the more volatile component)
Temperature, T_1
Pressure, p_1

Of the remaining, the designer will usually select first:

Drum pressure, p_{drum}

The drum pressure must be below the critical pressure for the mixture so that a liquid phase can exist. An approximate value of the critical pressure can be calculated from

$$p_{C,mixture,approx} = \sum_{i=1}^{n} x_i p_{C,i}$$

(2-1)

where x_i are the liquid mole fractions and $p_{C,i}$ and $p_{C,mixture,approx}$ are the critical pressures of the pure components and of the mixture (Biegler et al., 1997). In addition, as the pressure increases, the pressure and temperature of the feed and the condensation temperature of the vapor increase. We prefer a feed temperature that can be readily obtained with the available steam ($T_F < T_{steam} + 5°C$), and if the vapor product will be condensed, we prefer a condensation temperature that is at least 5°C above the available cooling water temperature.

A number of other variables are available to fulfill the last degree of freedom.

As is true in the design of many separation techniques, the choice of specified design variables controls the choice of the design method. For the flash chamber, we can use either a sequential solution method or a simultaneous solution method. In the sequential procedure, we solve the mass balances and equilibrium relationships first and then solve the energy balances and enthalpy equations. In the simultaneous solution method, all equations must be solved at the same time. In both cases, we solve for flow rates, compositions, and temperatures before we size the flash drum.

We will assume that the flash drum shown in Figure 2-1 acts as an equilibrium stage. Then vapor and liquid are in equilibrium. For a binary system the mole fraction of the more volatile component in the vapor y and its mole fraction in the liquid x and T_{drum} can be determined from the equilibrium expressions:

$$y = y_v(x, p_{drum})$$ **(2-2a)**

$$T_{drum} = T(x, p_{drum})$$ **(2-2b)**

To use Eq. (2-2) in the design of binary flash distillation systems, we must take a short tangent and first discuss binary vapor-liquid equilibrium (VLE).

2.2   FORM AND SOURCES OF EQUILIBRIUM DATA

Equilibrium data are required to understand and design the separations in Chapters 1 to 16 and 18. In principle, we can always experimentally determine the VLE data we require. For a simple experiment, we could take a chamber similar to Figure 1-2, fill it with the chemicals of interest, and at different pressures and temperatures, allow the liquid and vapor sufficient time to come to equilibrium and then take samples of liquid and vapor and analyze them. If we are very careful, we can obtain reliable equilibrium data. In practice, the measurement is fairly difficult and a variety of special equilibrium stills have been developed. Marsh (1978) and Van Ness and Abbott (1982, Section 6-7) briefly review methods of determining equilibrium. With a static equilibrium cell, concentration measurements are not required for binary systems. Concentrations can be calculated from pressure and temperature data, but the calculation is complex.

If we obtained equilibrium measurements for a binary mixture of ethanol and water at 1 atm, we would generate data similar to those shown in Table 2-1. The mole fractions in each phase must sum to 1.0. Thus for this binary system,

$$x_E + x_W = 1.0, \qquad y_E + y_W = 1.0$$ **(2-3)**

TABLE 2-1.   *Vapor-liquid equilibrium data for ethanol and water at 1 atm y and x in mole fractions*

x_{Etoh}	x_w	y_{Etoh}	y_w	$T, °C$
0	1.0	0	1.0	100
0.019	0.981	0.170	0.830	95.5
0.0721	0.9279	0.3891	0.6109	89.0
0.0966	0.9034	0.4375	0.5625	86.7
0.1238	0.8762	0.4704	0.5296	85.3
0.1661	0.8339	0.5089	0.4911	84.1
0.2337	0.7663	0.5445	0.4555	82.7
0.2608	0.7392	0.5580	0.4420	82.3

(continues)

TABLE 2-1. *Vapor-liquid equilibrium data for ethanol and water at 1 atm y and x in mole fractions (continued)*

x_{Etoh}	x_w	y_{Etoh}	y_w	$T, °C$
0.3273	0.6727	0.5826	0.4174	81.5
0.3965	0.6035	0.6122	0.3878	80.7
0.5198	0.4802	0.6599	0.3401	79.7
0.5732	0.4268	0.6841	0.3159	79.3
0.6763	0.3237	0.7385	0.2615	78.74
0.7472	0.2528	0.7815	0.2185	78.41
0.8943	0.1057	0.8943	0.1057	78.15
1.00	0	1.00	0	78.30

R. H. Perry, C. H. Chilton, and S.O. Kirkpatrick (Eds.), *Chemical Engineers Handbook,* 4th ed., New York, McGraw-Hill, p. 13–5, 1963.

where x is mole fraction in the liquid and y is mole fraction in the vapor. Very often only the composition of the most volatile component (ethanol in this case) will be given. The mole fraction of the less volatile component can be found from Eqs. (2-3). Equilibrium depends on pressure. (Data in Table 2-1 are specified for a pressure of 1 atm.) Table 2-1 is only one source of equilibrium data for the ethanol-water system, and over a dozen studies have explored this system (Wankat, 1988), and data are contained in the more general sources listed in Table 2-2. The data in different references do not agree perfectly, and care must be taken in choosing good data. We will refer back to this (and other) data quite often. If you have difficulty finding it, either look in the index under ethanol data or water data, or look in Appendix D under ethanol-water VLE.

TABLE 2-2. *Sources of vapor-liquid equilibrium data*

Chu, J. C., R. J. Getty, L. F. Brennecke, and R. Paul, *Distillation Equilibrium Data*, Reinhold, New York, 1950.

Engineering Data Book, Natural Gasoline Supply Men's Association, 421 Kennedy Bldg., Tulsa, Oklahoma, 1953.

Hala, E., I. Wichterle, J. Polak, and T. Boublik, *Vapor-Liquid Equilibrium Data at Normal Pressures*, Pergamon, New York, 1968.

Hala, E., J. Pick, V. Fried, and O. Vilim, *Vapor-Liquid Equilibrium*, 3rd ed., 2nd Engl. ed., Pergamon, New York, 1967.

Horsely, L. H., *Azeotropic Data*, ACS Advances in Chemistry, No. 6, American Chemical Society, Washington, DC, 1952.

Horsely, L. H. *Azeotropic Data (II),* ACS Advances in Chemistry, No. 35, American Chemical Society, Washington, DC, 1952.

Gess, M. A., R. P. Danner, and M. Nagvekar, *Thermodynamic Analysis of Vapor-Liquid Equilibria: Recommended Models and a Standard Data Base*, DIPPR, AIChE, New York, 1991.

TABLE 2-2. *Sources of vapor-liquid equilibrium data (continued)*

Gmehling, J., J. Menke, J. Krafczyk, and K. Fischer, Azeotropic Data, VCH Weinheim, Germany, 1994.

Gmehling, J., U. Onken, W. Arlt, P. Grenzheuser, U. Weidlich, B. Kolbe, J. R. Rarey-Nies, DECHEMA Chemistry Data Series, Vol. I, *Vapor-Liquid Equilibrium Data Collection*, DECHEMA, Frankfurt (Main), Germany, 1977–1984.

Maxwell, J. B., *Data Book on Hydrocarbons*, Van Nostrand, Princeton, NJ, 1950.

Perry, R. H., and D. Green, (Eds.), *Perry's Chemical Engineer's Handbook*, 7th ed., McGraw-Hill, New York, 1997.

Prausnitz, J. M., T. F. Anderson, E. A. Grens, C. A. Eckert, R. Hsieh, and J. P. O'Connell, *Computer Calculations for Multicomponent Vapor-Liquid and Liquid-Liquid Equilibria*, Prentice-Hall, Upper Saddle River, NJ, 1980.

Stephan, K., and H. Hildwein, DECHEMA Chemistry Data Series, Vol. IV, *Recommended Data of Selected Compounds and Binary Mixtures*, DECHEMA, Frankfurt (Main), Germany, 1987.

Timmermans, J., *The Physico-Chemical Constants of Binary Systems in Concentrated Solutions*, 5 vols., Interscience, New York, 1959–1960.

Van Winkle, M., *Distillation*, McGraw-Hill, New York, 1967.

Wichterle, I., J. Linek, and E. Hala, *Vapor-Liquid Equilibrium Data Bibliography*, Elsevier, Amsterdam, 1973.

www.cheric.org/research/kdb/ (click on box Korean Physical Properties Data Bank).

We see in Table 2-1 that if pressure and temperature are set, then there is only one possible vapor composition for ethanol, y_{Etoh}, and one possible liquid composition, x_{Etoh}. Thus we cannot arbitrarily set as many variables as we might wish. For example, at 1 atm we cannot arbitrarily decide that we want vapor and liquid to be in equilibrium at 95 °C and $x_{Etoh} = 0.1$.

The number of variables that we can arbitrarily specify, known as the degrees of freedom, is determined by subtracting the number of thermodynamic equilibrium equations from the number of variables. For nonreacting systems the resulting *Gibbs' phase rule* is

$$F = C - P + 2 \qquad\qquad (2\text{-}4)$$

where F = degrees of freedom, C = number of components, and P = number of phases. For the binary system in Table 2-1, C = 2 (ethanol and water) and P = 2 (vapor and liquid). Thus,

$$F = 2 - 2 + 2 = 2$$

When pressure and temperature are set, all the degrees of freedom are used, and *at equilibrium* all compositions are determined from the experiment. Alternatively, we could set pressure and x_{Etoh} or x_w and determine temperature and the other mole fractions.

The amount of material and its flow rate are not controlled by the Gibbs' phase rule. The phase rule refers to *intensive variables* such as pressure, temperature, or mole fraction, which do not depend on the total amount of material present. The *extensive variables,* such as number of moles, flow rate, and volume, do depend on the amount of material and are not included in the degrees of freedom. Thus a mixture in equilibrium must follow Table 2-1 whether there are 0.1, 1.0, 10, 100, or 1,000 moles present.

Binary systems with only two degrees of freedom can be conveniently represented in tabular or graphical form by setting one variable (usually pressure) constant. VLE data have been determined for many binary systems. Sources for these data are listed in Table 2-2; you should become familiar with several of these sources. Note that the data are not of equal qual ity. Methods for testing the thermodynamic consistency of equilibrium data are discussed in detail by Barnicki (2002), Walas (1985), and Van Ness and Abbott (1982, pp. 56–64, 301–348). Errors in the equilibrium data can have a profound effect on the design of the separation method (e.g., see Carlson, 1996, or Nelson et al., 1983).

2.3 GRAPHICAL REPRESENTATION OF BINARY VLE

Binary VLE data can be represented graphically in several ways. The most convenient forms are temperature-composition, y-x, and enthalpy-composition diagrams. These figures all represent the same data and can be converted from one form to another.

Table 2-1 gives the equilibrium data for ethanol and water at 1 atmosphere. With pressure set, there is only one degree of freedom remaining. Thus we can select any of the intensive variables as the independent variable and plot any other intensive variable as the dependent variable. The simplest such graph is the y vs. x graph shown in Figure 2-2. Typically, we plot the mole fraction of the more volatile component (the component that has y > x; ethanol in this case). This diagram is also called a McCabe-Thiele diagram when it is used for calculations. Pressure is constant, but the temperature is different at each point on the equilibrium curve. Points on the equilibrium curve represent two phases in equilibrium. Any point not on the equilibrium curve represents a system that may have both liquid and vapor, but they are not in equilibrium. As we will discover later, y-x diagrams are extremely convenient for calculation.

The data in Table 2-1 can also be plotted on a temperature-composition diagram as shown in Figure 2-3. The result is actually two graphs: One is liquid temperature vs. x_{Etoh}, and

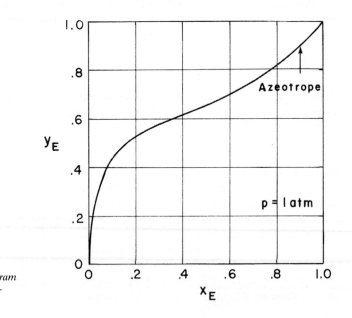

FIGURE 2-2.   *y vs. x diagram for ethanol-water*

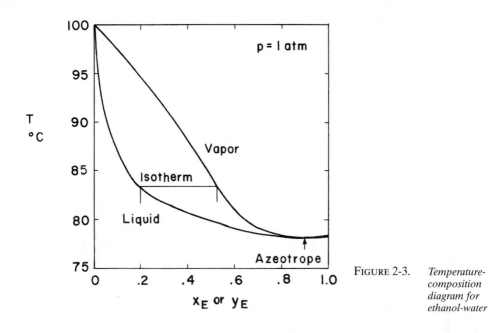

FIGURE 2-3. *Temperature-composition diagram for ethanol-water*

the other is vapor temperature vs. y_{Etoh}. These curves are called *saturated liquid* and *saturated vapor* lines, because they represent all possible liquid and vapor systems that can be in equilibrium at a pressure of 1 atm. Any point below the saturated liquid curve represents a sub-cooled liquid (liquid below its boiling point) whereas any point above the saturated vapor curve would be a superheated vapor. Points between the two saturation curves represent streams consisting of both liquid and vapor. If allowed to separate, these streams will give a liquid and vapor in equilibrium. Liquid and vapor in equilibrium must be at the same temperature; therefore, these streams will be connected by a horizontal isotherm as shown in Figure 2-3 for $x_{Etoh} = 0.2$.

Even more information can be shown on an enthalpy-composition or Ponchon-Savarit diagram, as illustrated for ethanol and water in Figure 2-4. Note that the units in Figure 2-4 differ from those in Figure 2-3. Again, there are really two plots: one for liquid and one for vapor. The isotherms shown in Figure 2-4 show the change in enthalpy at constant temperature as weight fraction varies. Because liquid and vapor in equilibrium must be at the same temperature, these points are connected by an isotherm. Points between the saturated vapor and liquid curves represent two-phase systems. An isotherm through any point can be generated using the auxiliary line with the construction shown in Figure 2-5. To find an isotherm, go vertically from the saturated liquid curve to the auxiliary line. Then go horizontally to the saturated vapor line. The line connecting the points on the saturated vapor and saturated liquid curves is the isotherm. If an isotherm is desired through a point in the two-phase region, a simple trial-and-error procedure is required.

Isotherms on the enthalpy-composition diagram can also be generated from the y-x and temperature-composition diagrams. Since these diagrams represent the same data, the vapor composition in equilibrium with a given liquid composition can be found from either the y-x or temperature-composition graph, and the value transferred to the enthalpy-composition diagram. This procedure can also be done graphically as shown in Figure 2-6 if the units are

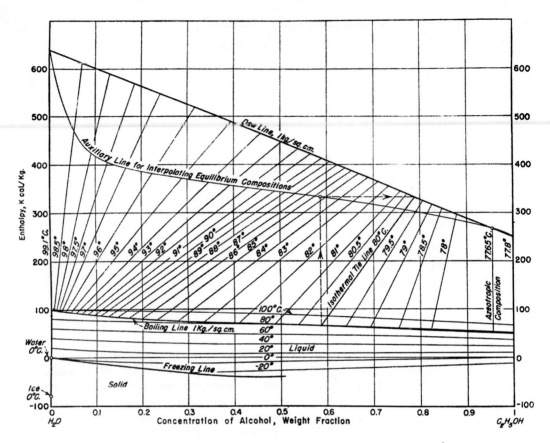

FIGURE 2-4. *Enthalpy-composition diagram for ethanol-water at a pressure of 1 kg/cm² (Bosnjakovic, Technische Thermodynamik, T. Steinkopff, Leipzig, 1935)*

the same in all figures. In Figure 2-6a we can start at point A and draw a vertical line to point A′ (constant x value). At constant temperature, we can find the equilibrium vapor composition (point B′). Following the vertical line (constant y), we proceed to point B. The isotherm connects points A and B. A similar procedure is used in Figure 2-6b, except now the y-x line

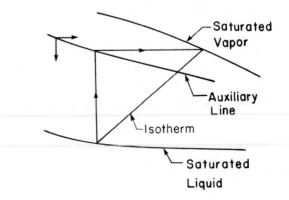

FIGURE 2-5. *Use of auxiliary line*

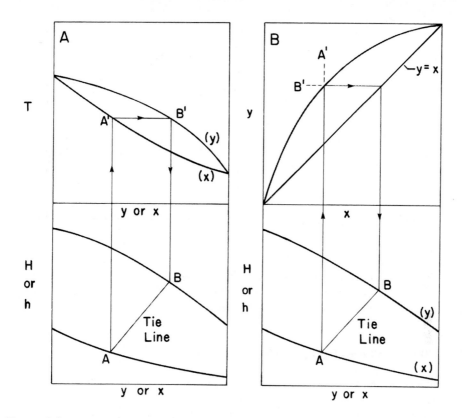

FIGURE 2-6.   *Drawing isotherms on the enthalpy-composition diagram (A) from the temperature-composition diagram; (B) from the y-x diagram*

must be used on the McCabe-Thiele graph. This is necessary because points A and B in equilibrium appear as a single point, A′/B′, on the y-x graph. The y = x line allows us to convert the ordinate value (y) on the y-x diagram to an abscissa value (also y) on the enthalpy-composition diagram. Thus the procedure is to start at point A and go up to point A′/B′ on the y-x graph. Then go horizontally to the y = x line and finally drop vertically to point B on the vapor curve. The isotherm now connects points A and B.

The data presented in Table 2-1 and illustrated in Figures 2-2, 2-3, and 2-4 show a minimum-boiling *azeotrope,* i.e., the liquid and vapor are of exactly the same composition at a mole fraction ethanol of 0.8943. This can be found from Figure 2-2 by drawing the y = x line and finding the intersection with the equilibrium curve. In Figure 2-3 the saturated liquid and vapor curves touch, while in Figure 2-4 the isotherm is vertical at the azeotrope. Note that the azeotrope composition is numerically different in Figure 2-4, but actually it is essentially the same, since Figure 2-4 is in weight fractions, whereas the other figures are in mole fractions. Below the azeotrope composition, ethanol is the more volatile component; above it, ethanol is the less volatile component. The system is called a minimum-boiling azeotrope because the azeotrope boils at 78.15°C, which is less than that of either pure ethanol or pure water. The azeotrope location is a function of pressure. Below 70 mm Hg no azeotrope exists for ethanol-water. Maximum-boiling azeotropes, although rare, also occur (see Figure 2-7).

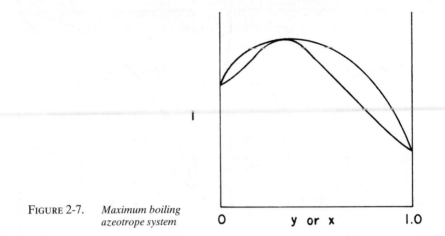

FIGURE 2-7.   *Maximum boiling*
azeotrope system

0   y or x   1.0

Only the temperature-composition diagram will look significantly different. Another type of azeotrope occurs when there are two partially miscible liquid phases. Equilibrium for partially miscible systems is considered in Chapter 8.

2.4   BINARY FLASH DISTILLATION

We will now use the binary equilibrium data to develop graphical and analytical procedures to solve the combined equilibrium, mass balance and energy balance equations. Mass and energy balances are written for the balance envelope shown as a dashed line in Figure 2-1. For a binary system there are two independent mass balances. The standard procedure is to use the overall mass balance,

$$F = V + L \qquad (2\text{-}5)$$

and the component balance for the more volatile component,

$$Fz = Vy + Lx \qquad (2\text{-}6)$$

The energy balance is

$$Fh_F + Q_{flash} = VH_v + Lh_L \qquad (2\text{-}7)$$

where h_F, H_v, and h_L are the enthalpies of the feed, vapor, and liquid streams. Usually $Q_{flash} = 0$, since the flash drum is insulated and the flash is considered to be adiabatic.

To use the energy balance equations, we need to know the enthalpies. Their general form is

$$h_F = h_F(T_F, z), \ H_v = H_v(T_{drum}, y) \, , \ h_L = h_L(T_{drum}, x) \qquad (2\text{-}8)$$

For binary systems it is often convenient to represent the enthalpy functions graphically on an enthalpy-composition diagram such as Figure 2-4. For ideal mixtures the enthalpies can be calculated from heat capacities and latent heats. Then,

$$h_L(T, x) = x_A C_{PL,A}(T - T_{ref}) + x_B C_{PL,B}(T - T_{ref}) \tag{2-9a}$$

$$h_F(T_F, z) = z_A C_{pL,A}(T_F - T_{ref}) + z_B C_{PL,B}(T_F - T_{ref}) \tag{2-9b}$$

$$H_v(T, y) = y_A[\lambda_A + C_{PV,A}(T - T_{ref})] + y_B[\lambda_B + C_{PV,B}(T - T_{ref})] \tag{2-10}$$

where x_A and y_A are mole fractions of component A in liquid and vapor, respectively. C_p is the molar heat capacity, T_{ref} is the chosen reference temperature, and λ is the latent heat of vaporization at T_{ref}. For binary systems, $x_B = 1 - x_A$, and $y_B = 1 - y_A$.

2.4.1   Sequential Solution Procedure

In the sequential solution procedure, we first solve the mass balance and equilibrium relationships, and then we solve the energy balance and enthalpy equations. In other words, the two sets of equations are uncoupled. The sequential solution procedure is applicable when the last degree of freedom is used to specify a variable that relates to the conditions in the flash drum. Possible choices are:

Vapor mole fraction, y
Liquid mole fraction, x
Fraction feed vaporized, f = V/F
Fraction feed remaining liquid, q = L/F
Temperature of flash drum, T_{drum}

If one of the equilibrium conditions, y, x, or T_{drum}, is specified, then the other two can be found from Eqs. (2-2a) and (2-2b) or from the graphical representation of equilibrium data. For example, if y is specified, x is obtained from Eq. (2-2a) and T_{drum} from Eq. (2-2b). In the mass balances, Eqs. (2-5) and (2-6), the only unknowns are L and V, and the two equations can be solved simultaneously.

If either the fraction vaporized or fraction remaining liquid is specified, Eqs. (2-2a), (2-2b), and (2-6) must be solved simultaneously. The most convenient way to do this is to combine the mass balances. Solving Eq. (2-6) for y, we obtain

$$y = -\frac{L}{V} x + \frac{F}{V} z \tag{2-11}$$

Equation (2-11) is the *operating equation,* which for a single-stage system relates the compositions of the two streams leaving the stage. Equation (2-11) can be rewritten in terms of either the fraction vaporized, f = V/F, or the fraction remaining liquid, q = L/F.

From the overall mass balance, Eq. (2-5),

$$L/V = \frac{F - V}{V} = \frac{1 - V/F}{V/F} = \frac{1 - f}{f} \tag{2-12}$$

Then the operating equation becomes

$$y = -\frac{1-f}{f} x + \frac{z}{f}$$

(2-13)

The alternative in terms of L/F is

$$\frac{L}{V} = \frac{L}{F-L} = \frac{L/F}{1-L/F} = \frac{q}{1-q}$$

(2-14)

and the operating equation becomes

$$y = -\frac{q}{1-q} x + \left(\frac{1}{1-q}\right) z$$

(2-15)

Although they have different forms Eqs. (2-11), (2-13), and (2-15) are equivalent means of obtaining y, x, or z. We will use whichever operating equation is most convenient.

Now the equilibrium and the operating equation (Eq. 2-11, 2-13, or 2-15) must be solved simultaneously. The exact way to do this depends on the form of the equilibrium data. For binary systems a graphical solution is very convenient. Equations (2-11), (2-13), and (2-15) represent a single straight line, called the *operating line,* on a graph of y vs. x. This straight line will have

$$\text{slope} = -\frac{L}{V} = -\frac{1-f}{f} = -\frac{q}{1-q}$$

(2-16)

and

$$y \text{ intercept } (x = 0) = \frac{F}{V} z = \frac{1}{f} z = \frac{1}{1-q} z$$

(2-17)

The equilibrium data at pressure p_{drum} can also be plotted on the y-x diagram. The intersection of the equilibrium curve and the operating line is the simultaneous solution of the mass balances and equilibrium. This plot of y vs. x showing both equilibrium and operating lines is called a *McCabe-Thiele diagram* and is shown in Figure 2-8 for an ethanol-water separation. The equilibrium data are from Table 2-1 and the equilibrium curve is identical to Figure 2-2. The solution point gives the vapor and liquid compositions leaving the flash drum. Figure 2-8 shows three different operating lines as V/F varies from 0 (line a) to ⅔ (line c) to 1.0 (line b) (see Example 2-1). T_{drum} can be found from Eq. (2-4), from Table 2-1, or from a temperature-composition diagram.

Two other points often used on the McCabe-Thiele diagram are the x intercept (y = 0) of the operating line and its intersection with the y = x line. Either of these points can also be located algebraically and then used to plot the operating line.

The intersection of the operating line and the y = x line is often used because it is simple to plot. This point can be determined by simultaneously solving Eq. (2-11) and the equation y = x. Substituting y = x into Eq. (2-11), we have

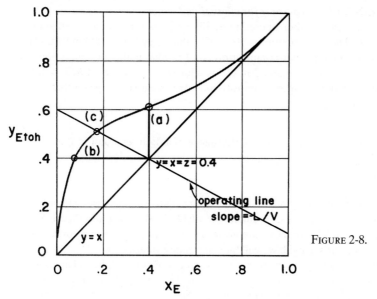

FIGURE 2-8. *McCabe-Thiele diagram for binary flash distillation; illustrated for Example 2-1*

$$y = \frac{L}{V} \, y + \frac{F}{V} \, z$$

or

$$y \left(1 + \frac{L}{V} \right) = \frac{F}{V} \, z$$

or

$$y \left(\frac{V + L}{V} \right) = \frac{F}{V} \, z$$

since $V + L = F$, the result is $y = z$ and therefore

$$x = y = z \qquad\qquad (2\text{-}18)$$

The intersection is at the feed composition.

It is important to realize that the $y = x$ line has no fundamental significance. It is often used in graphical solution methods because it simplifies the calculation. However, do *not* use it blindly.

Obviously, the graphical technique can be used if y, x, or T_{drum} is specified. The order in which you find points on the diagram will depend on what information you have to begin with.

EXAMPLE 2-1. Flash separator for ethanol and water

A flash distillation chamber operating at 101.3 kPa is separating an ethanol-water mixture. The feed mixture is 40 mol% ethanol and F = 100 kmol/h. (a) What is the maximum vapor composition and (b) what is the minimum liquid composition that can be obtained if V/F is allowed to vary? (c) If V/F = 2/3, what are the liquid and vapor compositions? (d) Repeat step c, given that F is specified as 1000 kmol/h.

Solution

A. Define. We wish to analyze the performance of a flash separator at 1 atm.
   a. Find y_{max}.
   b. Find x_{min}.
   c. and d. Find y and x for V/F = 2/3.

B. Explore. Note that p_{drum} = 101.3 kPa = 1 atm. Thus we must use data at this pressure. These data are conveniently available in Table 2-1 and Figure 2-2. Since p_{drum} and V/F for part c are given, a sequential solution procedure will be used. For parts a and b we will look at limiting values of V/F.

C. Plan. We will use the y-x diagram as illustrated in Figure 2-2. For all cases we will do a mass balance to derive an operating line [we could use Eqs. (2-11), (2-13), or (2-15), but I wish to illustrate deriving an operating line]. Note that $0 \leq V/F \leq 1.0$. Thus our maximum and minimum values for V/F must lie within this range.

D. Do It. Sketch is shown.

F = 100
Z_H = 0.4
→ p = 101.3 kPa → V, y
$\frac{V}{F}$ = 0, 2/3, or 1.0
→ L, x

Mass Balances:
$$F = V + L$$
$$Fz = Vy + Lx$$

Solve for y:

$$y = -\frac{L}{V} x + \frac{F}{V} z$$

From the overall balance, L = F – V. Thus
   when V/F = 0.0, V = 0, L = F, and L/V = F/0 = ∞
   when V/F = 2/3, V = (2/3)F, L = (1/3)F, and L/V = (1/3)F/[(2/3)F] = 1/2
   when V/F = 1.0, V = F, L = 0, and L/V = 0/F = 0
Thus the slopes (–L/V) are –∞, –1/2, and –0.
   If we solve for the y = x interception, we find it at y = x = z = 0.4 for all cases. Thus we can plot three operating lines through y = x = z = 0.4, with slopes of –∞, –1/2 and –0. These operating lines were shown in Figure 2-8.
   a. Highest y is for V/F = 0:   y = 0.61 [x = 0.4]

b. Lowest x is for V/F = 1.0:   x = 0.075 [y = 0.4]
c. When V/F is 2/3, y = 0.52 and x = 0.17
d. When F = 1,000 with V/F = 2/3, the answer is exactly the same as in part c. The feed rate will affect the drum diameter and the energy needed in the preheater.
E. Check. We can check the solutions with the mass balance, Fz = Vy + Lx.
    a. (100)(0.4) = 0(0.61) + (100)(0.4) checks
    b. (100)(0.4) = 100(0.4) + (0)(0.075) checks
    c. 100(0.4) = (66.6)(0.52) + (33.3)(0.17)
    Note V = (2/3)F and L = (1/3)F
    This is 40 = 39.9, which checks within the accuracy of the graph
    d. Check is similar to c : 400 = 399
    We can also check by fitting the equilibrium data to a polynomial equation and then simultaneously solve equilibrium and operating equations by minimizing the residual. These spread sheet calculations agree with the graphical solution.
F. Generalization. The method for obtaining bounds for the answer (setting the V/F equation to its extreme values of 0.0 and 1.0) can be used in a variety of other situations. In general, the feed rate will not affect the compositions obtained in the design of stage separators. Feed rate does affect heat requirement and equipment diameters.

Once the conditions within the flash drum have been calculated, we proceed to the energy balance. With y, x, and T_{drum} known, the enthalpies H_v and h_L are easily calculated from Eqs. (2-8) or (2-9) and (2-10). Then the only unknown in Eq. (2-7) is the feed enthalpy h_F. Once h_F is known, the inlet feed temperature T_F can be obtained from Eq. (2-8) or (2-9b).

The amount of heat required in the heater, Q_h, can be determined from an energy balance around the heater.

$$Q_h + Fh_1(T_1, z) = Fh_F(T_F, z) \qquad \text{(2-19)}$$

Since enthalpy h_1 can be calculated from T_1 and z, the only unknown is Q_h, which controls the size of the heater.

The feed pressure, p_F, required is semi-arbitrary. Any pressure high enough to prevent boiling at temperature T_F can be used.

One additional useful result is the calculation of V/F when all mole fractions (z, y, x) are known. Solving Eqs. (2-5) and (2-6), we obtain

$$V/F = (z - x)/(y - x) \qquad \text{(2-20)}$$

Except for sizing the flash drum, which is covered later, this completes the sequential procedure. Note that the advantages of this procedure are that mass and energy balances are uncoupled and can be solved independently. Thus trial and error is not required.

If we have a convenient equation for the equilibrium data, then we can obtain the simultaneous solution of the operating equation (2-9, 2-11, or 2-13) and the equilibrium equation analytically. For example, ideal systems often have a constant relative volatility α_{AB} where α_{AB} is defined as,

$$\alpha_{AB} = K_A / K_B = (y_A/x_A)/(y_B/x_B) \qquad \textbf{(2-21)}$$

For binary systems,

$$y_B = 1 - y_A, x_B = 1 - x_A$$

and the relative volatility is

$$\alpha_{AB} = y_A(1 - x_A)/[(1 - y_A)x_A] \qquad \text{(binary)} \qquad \textbf{(2-22a)}$$

Solving Eq. (2-22) for y_A, we obtain

$$y_A = \alpha_{AB} x_A/[1 + (\alpha_{AB} - 1)x_A] \qquad \text{(binary)} \qquad \textbf{(2-22b)}$$

If Raoult's law is valid, then we can determine relative volatility as

$$\alpha_{AB} = (VP)_A/(VP)_B \qquad \textbf{(2-23)}$$

The relative volatility α may also be fit to experimental data.

If we solve Eqs. (2-21) and (2-11) simultaneously, we obtain

$$\frac{(1-f)}{f}(\alpha - 1)x^2 + \left[\alpha + \frac{1-f}{f} - (\alpha - 1)z/f\right]x - z/f = 0 \qquad \textbf{(2-24)}$$

which is easily solved with the quadratic equation. This can be done conveniently with a spread sheet.

2.4.2 Simultaneous Solution and Enthalpy-Composition Diagram

If the temperature of the feed to the drum, T_F, is the specified variable, the mass and energy balances and the equilibrium equations must be solved simultaneously. You can see from the energy balance, Eq. (2-7) why this is true. The feed enthalpy, h_F, can be calculated, but the vapor and liquid enthalpies, H_v and h_L, depend upon T_{drum}, y, and x, which are unknown. Thus a sequential solution is not possible.

We could write Eqs. (2-3) to (2-8) and solve seven equations simultaneously for the seven unknowns y, x, L, V, H_v, h_L, and T_{drum}. This is feasible but rather difficult, particularly since Eqs. (2-3) and (2-4) and often Eqs. (2-8) are nonlinear, so we resort to a trial-and-error procedure. This method is: Guess the value of one of the variables, calculate the other variables, and then check the guessed value of the trial variable. For a binary system, we can select any one of several trial variables, such as y, x, T_{drum}, V/F, or L/F. For example, if we select the temperature of the drum, T_{drum}, as the trial variable, the calculation procedure is:

1. Calculate $h_F(T_F, z)$ [e.g., use Eq. (2-9b)].
2. Guess the value of T_{drum}.
3. Calculate x and y from the equilibrium equations (2-3) and (2-4) or graphically (use temperature-composition diagram).
4. Find L and V by solving the mass balance equations (2-5) and (2-6), or find L/V from Figure 2-8 and use the overall mass balance, Eq. (2-5).

5. Calculate $h_L(T_{drum}, x)$ and $H_v(T_{drum}, y)$ from Eqs. (2-8) or (2-9a) and (2-10) or from the enthalpy-composition diagram.
6. Check: Is the energy balance equation (2-7) satisfied? If it is satisfied, we are finished. Otherwise, return to step 2.

The procedures are similar for other trial variables.

For binary flash distillation, the simultaneous procedure can be conveniently carried out on an enthalpy-composition diagram. First calculate the feed enthalpy, h_F, from Eq. (2-8) or Eq. (2-9b); then plot the feed point as shown on Figure 2-9 (see Problem 2-A1). In the flash drum the feed separates into liquid and vapor in equilibrium. Thus the isotherm through the feed point, which must be the T_{drum} isotherm, gives the correct values for x and y. The flow rates, L and V, can be determined from the mass balances, Eqs. (2-5) and (2-6), or from a graphical mass balance.

Determining the isotherm through the feed point requires a minor trial-and-error procedure. Pick a y (or x), draw the isotherm, and check whether it goes through the feed point. If not, repeat with a new y (or x).

A graphical solution to the mass balances and equilibrium can be developed for Figure 2-9. Substitute the overall balance Eq. (2-5) into the more volatile component mass balance Eq. (2-6),

$$Lz + Vz = Lx + Vy$$

Rearranging and solving for L/V

$$\frac{L}{V} = \frac{y - z}{z - x} \tag{2-25}$$

Using basic geometry, $(y - z)$ is proportional to the distance $\overline{FV}$ on the diagonal line and $(z - x)$ is proportional to the distance FL. Then,

$$\frac{L}{V} = \frac{\overline{FV}}{\overline{FL}} \tag{2-26}$$

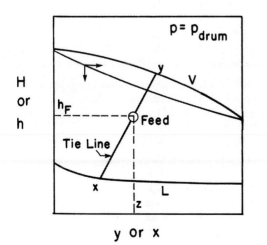

FIGURE 2-9. *Binary flash calculation in enthalpy-composition diagram*

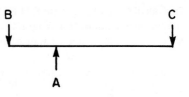

FIGURE 2-10.   *Illustration of lever-arm rule*

Equation (2-26) is called the *lever-arm rule* because the same result is obtained when a moment-arm balance is done on a seesaw. Thus if we set moment arms of the seesaw in Figure 2-10 equal, we obtain

$$(\text{wt B}) \, (\overline{BA}) = (\text{wt C}) \, (\overline{AC})$$

or

$$\text{wt B}/\text{wt C} = \overline{AC}/\overline{BA}$$

which gives the same result as Eq. (2-26). The seesaw is a convenient way to remember the form of the lever-arm rule.

The lever-arm rule can also be applied on ternary diagrams for extraction, where it has several other uses (see Chapter 13).

2.5   MULTICOMPONENT VLE

If there are more than two components, an analytical procedure is needed. The basic equipment configuration is the same as Figure 2-1.

The equations used are equilibrium, mass and energy balances, and stoichiometric relations. The mass and energy balances are very similar to those used in the binary case, but the equilibrium equations are usually written in terms of K values. The equilibrium form is

$$y_i = K_i \, x_i \qquad\qquad\qquad \textbf{(2-27)}$$

where in general

$$K_i = K_i \, (T_{drum}, \, p_{drum}, \, \text{all } x_i) \qquad\qquad \textbf{(2-28)}$$

Equations (2-27) and (2-28) are written once for each component. In general, the K values depend on temperature, pressure, and composition. These nonideal K values are discussed in detail by Smith (1963) and Walas (1985), in thermodynamics textbooks, and in the references in Table 2-2.

Fortunately, for many systems the K values are approximately independent of composition. Thus,

$$K = K(T, p) \qquad (\text{approximate}) \qquad\qquad \textbf{(2-29)}$$

For light hydrocarbons, the approximate K values can be determined from the nomographs prepared by DePriester. These are shown in Figures 2-11 and 2-12, which cover different temperature ranges. If temperature and/or pressure of the equilibrium mixture are unknown, a trial-and-error procedure is required. DePriester charts in other temperature and pressure units are given by Green and Perry (2008), Perry and Green (1997), and Smith and Van Ness (1975). The DePriester charts have been fit to the following equation (McWilliams, 1973):

$$\ln K = a_{T1}/T^2 + a_{T2}/T + a_{T6} + a_{p1} \ln p + a_{p2}/p^2 + a_{p3}/p \qquad \textbf{(2-30)}$$

Note that T is in °R and p is in psia in Eq. (2-30). The constants a_{T1}, a_{T2}, a_{T6}, a_{p1}, a_{p2}, and a_{p3} are given in Table 2-3. The last line gives the mean errors in the K values compared to the values from the DePriester charts. This equation is valid from −70°C (365.7°R) to 200°C (851.7°R) and for pressures from 101.3 kPa (14.69 psia) to 6000 kPa (870.1 psia). If K and p are known, then Eq. (2-30) can be solved for T. The obvious advantage of an equation

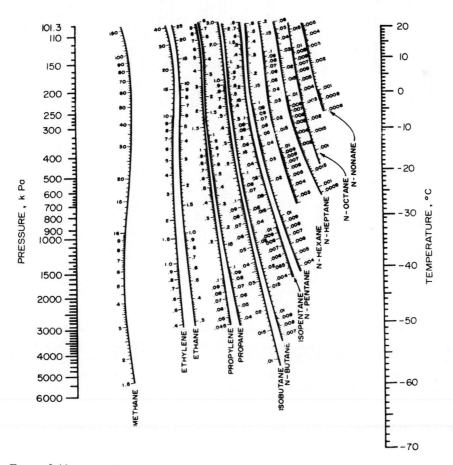

FIGURE 2-11.   *Modified DePriester chart (in S.I. units) at low temperatures (D. B. Dadyburjor,* Chem. Eng. Prog., *85, April 1978; copyright 1978, AIChE; reproduced by permission of the American Institute of Chemical Engineers)*

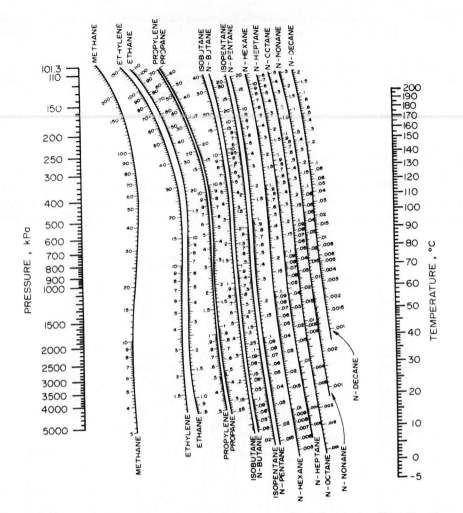

FIGURE 2-12. *Modified DePriester chart at high temperatures (D. B. Dadyburjor,* Chem. Eng. Prog., *85, April 1978; copyright 1978, AIChE; reproduced by permission of the American Institute of Chemical Engineers)*

compared to the charts is that it can be programmed into a computer or calculator. Equation (2-30) can be simplified for all components except n-octane and n-decane (see Eq. (5-35a)).

The K values are used along with the stoichiometric equations which state the mole fractions in liquid and vapor phases must sum to 1.0.

$$\sum_{i=1}^{C} y_i = 1.0, \ \sum_{i=1}^{C} x_i = 1.0 \tag{2-31}$$

where C is the number of components. Bubble-point and dew-point calculations are discussed in detail in Section 5.4.

TABLE 2-3.   *Constants for fit to K values using Eq. (2-30)*

Compound	a_{T1}	a_{T2}	a_{T6}	a_{p1}	a_{p2}	a_{p3}	Mean Error
Methane	−292,860	0	8.2445	−.8951	59.8465	0	1.66
Ethylene	−600,076.875	0	7.90595	−.84677	42.94594	0	2.65
Ethane	−687,248.25	0	7.90694	−.88600	49.02654	0	1.95
Propylene	−923,484.6875	0	7.71725	−.87871	47.67624	0	1.90
Propane	−970,688.5625	0	7.15059	−.76984	0	6.90224	2.35
Isobutane	−1,166,846	0	7.72668	−.92213	0	0	2.52
n-Butane	−1,280,557	0	7.94986	−.96455	0	0	3.61
Isopentane	−1,481,583	0	7.58071	−.93159	0	0	4.56
n-Pentane	−1,524,891	0	7.33129	−.89143	0	0	4.30
n-Hexane	−1,778,901	0	6.96783	−.84634	0	0	4.90
n-Heptane	−2,013,803	0	6.52914	−.79543	0	0	6.34
n-Octane	0	−7646.81641	12.48457	−.73152	0	0	7.58
n-Nonane	−2,551,040	0	5.69313	−.67818	0	0	9.40
n-Decane	0	−9760.45703	13.80354	−.71470	0	0	5.69

Note: T is in °R, and p is in psia

Source: McWilliams (1973)

If only one component is present, then y = 1.0 and x = 1.0. This implies that K_i = y/x = 1.0. This gives a simple way of determining the boiling temperature of a pure compound at any pressure. For example, if we wish to find the boiling point of isobutane at p = 150 kPa, we set our straightedge on p = 150 and at 1.0 on the isobutane scale on Figure 2-11. Then read T = −1.5°C as the boiling point. Alternatively, Eq. (2-30) with values from Table 2-3 can be solved for T. This gives T = 488.68°R or −1.6°C.

For ideal systems Raoult's law holds. Raoult's law states that the partial pressure of a component is equal to its vapor pressure multiplied by its mole fraction in the liquid. Thus,

$$p_A = x_A (VP)_A \qquad \text{(2-32a)}$$

where vapor pressure (VP) depends on temperature. By Dalton's law of partial pressures,

$$y_A = p_A/p \qquad \text{(2-32b)}$$

Combining these equations,

$$y_A = (VP)_A x_A/p \qquad \text{(2-32c)}$$

Comparing Eqs. (2-32c) and (2-27), the Raoult's law K value is

$$K_A = (VP)_A/p \qquad \text{(2-33)}$$

This is handy, since extensive tables of vapor pressures are available (e.g., Boublik et al., 1984; Dean, 1985; Green and Perry, 2008). Vapor pressure is often correlated in terms of the Antoine equation

$$\log_{10}(VP) = A - B/(T + C) \tag{2-34}$$

where A, B, and C are constants for each pure compound. These constants are tabulated in various data sources (Boublik et al., 1984; Yaws et al., 2005). The equations based on Raoult's law should be used with great care, since deviations from Raoult's law are extremely common.

Nonidealities in the liquid phase can be taken into account with a liquid-phase activity coefficient, γ_i. Then Eq. (2-33) becomes

$$K_A = \gamma_A(VP_A)/p_{total} \tag{2-35}$$

The activity coefficient depends on temperature, pressure, and concentration. Excellent correlation procedures for activity coefficients such as the Margules, Van Laar, Wilson, NRTL, and UNIQUAC methods have been developed (Poling et al., 2001; Prausnitz et al., 1999; Sandler, 2006; Tester and Modell, 1997; Van Ness and Abbott, 1982; Walas, 1985). The coefficients for these equations for a wide variety of mixtures have been tabulated along with the experimental data (see Table 2-2). When the binary data are not available, one can use infinite dilution coefficients (Table 2-2; Carlson, 1996; Lazzaroni et al., 2005; Schad, 1998) or the UNIFAC group contribution method (Fredenslund et al., 1977; Prausnitz et al., 1980) to predict the missing data. Many distillation simulators use Eqs. (2-34), (2-35), and an appropriate activity coefficient equation. Although a detailed description of these methods is beyond the scope of this book, a guide to choosing VLE correlations for use in computer simulations is presented in Table 2-4. For final designs, you must be confident in the VLE correlation used. Check the predictions with experimental data such as VLE data, flash distillation results, or distillation column results.

2.6 MULTICOMPONENT FLASH DISTILLATION

Equations (2-27) and (2-28) are solved along with the stoichiometric equations (2-31), the overall mass balance Eq. (2-5), the component mass balances,

$$Fz_i = Lx_i + Vy_i \tag{2-36}$$

and the energy balance, Eq. (2-7). Equation (2-36) is very similar to the binary mass balance, Eq. (2-6).

Usually the feed flow rate, F, and the feed mole fractions z_i for C – 1 of the components will be specified. If p_{drum} and T_{drum} or one liquid or vapor composition are also specified, then a sequential procedure can be used. That is, the mass balances, stoichiometric equations, and equilibrium equations are solved simultaneously, and then the energy balances are solved.

Now consider for a minute what this means. Suppose we have 10 components (C = 10). Then we must find 10 K's, 10 x's, one L, and one V, or 32 variables. To do this we must

TABLE 2-4.   *Approximate guides for selection of K-value methods.*

Chemical Systems

Low MW alcohol and hydrocarbons	Wilson
Higher MW alcohol and hydrocarbons	NRTL
Hydrogen bonding systems	Margules
Liquid-liquid equilibrium	NRTL/UNIQUAC
Water as a second liquid phase	NRTL
Components in a homologous family	UNIQUAC
Low pressure systems with associating vapor phase	Hayden-O'Connell

Light Hydrocarbon and Oil Systems

Natural gas systems w/sweet and sour gas	SRK/PR
Cryogenic systems	SRK/PR
Refinery mixtures with p<5000 psia	SRK/PR
Hydrotreaters and reformers	Grayson-Stread
Simple paraffinic systems	SRK/PR
Heavy components w/ NBP>1000°F	BK10
Aromatics (near critical region) + H_2	PR/SRK

Based on Polarity and Ideality

nonpolar–nonpolar	ideal & nonideal	any activity coefficient model
nonpolar–weakly polar	ideal	any activity coefficient model
nonpolar–weakly polar	nonideal	UNIQUAC
nonpolar–strongly polar	ideal	UNIQUAC
nonpolar–strongly polar	nonideal	Wilson
weakly polar–weakly polar	ideal	NRTL
weakly polar–weakly polar	nonideal	UNIQUAC
weakly polar–strongly polar	ideal	NRTL
weakly polar–strongly polar	nonideal	UNIQUAC
strongly polar–strongly polar	ideal	UNIQUAC
strongly polar–strongly polar	nonideal	NO RECOMMENDATION
aqueous–strongly polar		UNIQUAC

Chemical systems and light hydrocarbon and oil systems suggestions courtesy of Dr. William Walters. Based on polarity and ideality suggestions from Gess et al. (1991) (see Table 2-2).

Key: NRTL = nonrandom two liquid model; SRK = Soave-Redlich-Kwong model; PR = Peng-Robinson; BK10 = Braun K10 for petroleum.

solve 32 equations [10 Eq. (2-27), 10 Eq. (2-31), and 10 independent mass balances, Eq. (2-36)] simultaneously. And this is the simpler sequential solution for a relatively simple problem.

How does one solve 32 simultaneous equations? In general, the K value relations could be nonlinear functions of composition. However, we will restrict ourselves to ideal solutions where Eq. (2-29) is valid and

$$K_i = K_i(T_{drum}, p_{drum})$$

Since T_{drum} and p_{drum} are known, the 10 K_i can be determined easily [say, from the DePriester charts or Eq. (2-30)]. Now there are only 22 linear equations to solve simultaneously. This can be done, but trial-and-error procedures are simpler.

To simplify the solution procedure, we first use equilibrium, $y_i = K_i x_i$, to remove y_i from Eq. (2-36):

$$Fz_i = Lx_i + VK_i x_i \qquad i = 1, C$$

Solving for x_i, we have

$$x_i = \frac{Fz_i}{L + VK_i} \qquad i = 1, C$$

If we solve Eq. (2-5) for L, L = F − V, and substitute this into the last equation we have

$$x_i = \frac{Fz_i}{F - V + K_i V} \qquad i = 1, C \qquad \textbf{(2-37)}$$

Now if the unknown V is determined, all of the x_i can be determined. It is usual to divide the numerator and denominator of Eq. (2-37) by the feed rate F and work in terms of the variable V/F. Then upon rearrangement we have

$$x_i = \frac{z_i}{1 + (K_i - 1)\dfrac{V}{F}} \qquad i = 1, C \qquad \textbf{(2-38)}$$

The reason for using V/F, the fraction vaporized, is that it is bounded between 0 and 1.0 for all possible problems. Since $y_i = K_i x_i$, we obtain

$$y_i = \frac{K_i z_i}{1 + (K_i - 1)\dfrac{V}{F}} \qquad i = 1, C \qquad \textbf{(2-39)}$$

Once V/F is determined, x_i and y_i are easily found from Eqs. (2-38) and (2-39).

How can we derive an equation that allows us to calculate V/F?

To answer this, first consider what equations have not been used. These are the two stoichiometric equations, $\Sigma x_i = 1.0$ and $\Sigma y_i = 1.0$. If we substitute Eqs. (2-38) and (2-37) into these equations, we obtain

$$\sum_{i=1}^{C} \frac{z_i}{1 + (K_i - 1)\dfrac{V}{F}} = 1.0 \tag{2-40}$$

and

$$\sum_{i=1}^{C} \frac{K_i z_i}{1 + (K_i - 1)\dfrac{V}{F}} = 1.0 \tag{2-41}$$

Either of these equations can be used to solve for V/F. If we clear fractions, these are Cth-order polynomials. Thus, if C is greater than 3, a trial-and-error procedure or root-finding technique must be used to find V/F. Although Eqs. (2-40) and (2-41) are both valid, they do not have good convergence properties. That is, if the wrong V/F is chosen, the V/F that is chosen next may not be better.

Fortunately, an equation that does have good convergence properties is easy to derive. To do this, subtract Eq. (2-40) from (2-41).

$$\sum \frac{K_i z_i}{1 + (K_i - 1)\dfrac{V}{F}} - \sum \frac{z_i}{1 + (K_i - 1)\dfrac{V}{F}} = 0$$

Subtracting the sums term by term, we have

$$f\left(\frac{V}{F}\right) = \sum_{i=1}^{C} \frac{(K_i - 1) z_i}{1 + (K_i - 1)\dfrac{V}{F}} = 0 \tag{2-42}$$

Equation (2-42), which is known as the Rachford-Rice equation, has excellent convergence properties. It can also be modified for three-phase (liquid-liquid-vapor) flash systems (Chien, 1994).

Since the feed compositions, z_i, are specified and K_i can be calculated when T_{drum} and p_{drum} are given, the only variable in Eq. (2-42) is the fraction vaporized, V/F. This equation can be solved by many different convergence procedures or root finding methods. The Newtonian convergence procedure will converge quickly. Since f(V/F) in Eq. (2-42) is a function of V/F that should have a zero value, the equation for the Newtonian convergence procedure is

$$f_{k+1} - f_k = \frac{df_k}{d(V/F)} \, \Delta(V/F) \tag{2-43}$$

where f_k is the value of the function for trial k and $df_k/d(V/F)$ is the value of the derivative of the function for trial k. We desire to have f_{k+1} equal zero, so we set $f_{k+1} = 0$ and solve for $\Delta(V/F)$:

$$\Delta(V/F) = (V/F)_{k+1} - (V/F)_k = \frac{-f_k}{\left[\dfrac{df_k}{d(V/F)}\right]} \tag{2-44}$$

This equation gives us the best next guess for the fraction vaporized. To use it, however, we need equations for both the function and the derivative. For f_k, use the Rachford-Rice equation, (2-42). Then the derivative is

$$\frac{df_k}{d(V/F)} = - \sum_{i=1}^{C} \frac{(K_1 - 1)^2 z_i}{[1 + (K_i - 1) V/F]^2} \qquad \textbf{(2-45)}$$

Substituting Eqs. (2-42) and (2-45) into (2-44) and solving for $(V/F)_{k+1}$, we obtain

$$\left(\frac{V}{F}\right)_{k+1} = \left(\frac{V}{F}\right)_k - f_k \Bigg/ \left[\frac{df_k}{d(V/F)}\right] \qquad \textbf{(2-46)}$$

Equation (2-46) gives a good estimate for the next trial. Once $(V/F)_{k+1}$ is calculated the value of the Rachford-Rice function can be determined. If it is close enough to zero, the calculation is finished; otherwise repeat the Newtonian convergence for the next trial.

Newtonian convergence procedures do not always converge. One advantage of using the Rachford-Rice equation with the Newtonian convergence procedure is that there is always rapid convergence. This is illustrated in Example 2-2.

Once V/F has been found, x_i and y_i are calculated from Eqs. (2-38) and (2-39). L and V are determined from the overall mass balance, Eq. (2-5). The enthalpies h_L and H_v can now be calculated. For ideal solutions the enthalpies can be determined from the sum of the pure component enthalpies multiplied by the corresponding mole fractions:

$$H_v = \sum_{i=1}^{C} y_i \, \tilde{H}_{vi} \, (T_{drum}, p_{drum}) \qquad \textbf{(2-47a)}$$

$$h_L = \sum_{i=1}^{C} x_i \, \tilde{h}_{Li} \, (T_{drum}, p_{drum}) \qquad \textbf{(2-47b)}$$

where $\tilde{H}_{vi}$ and $\tilde{h}_{Li}$ are enthalpies of the pure components. If the solutions are not ideal, heats of mixing are required. Then the energy balance, Eq. (2-7), is solved for h_F, and T_F is determined.

If V/F and p_{drum} are specified, then T_{drum} must be determined. This can be done by picking a value for T_{drum}, calculating K_i, and checking with the Rachford-Rice equation, (2-42). A plot of f(V/F) vs. T_{drum} will help us select the temperature value for the next trial. Alternatively, an approximate convergence procedure similar to that employed for bubble- and dew-point calculations can be used (see Section 5-4). The new K_{ref} can be determined from

$$K_{ref}(T_{new}) = \frac{K_{ref}(T_{old})}{1 + (d)f(T_{old})} \qquad \textbf{(2-48)}$$

where the damping factor $d \le 1.0$. In some cases this may overcorrect unless the initial guess is close to the correct answer. The calculation when V/F = 0 gives us the bubble-point temperature (liquid starts to boil) and when V/F = 1.0 gives the dew-point temperature (vapor starts to condense).

EXAMPLE 2-2. Multicomponent flash distillation

A flash chamber operating at 50°C and 200 kPa is separating 1000 kmol/h of a feed that is 30 mol% propane, 10 mol% n-butane, 15 mol% n-pentane and 45 mol% n-hexane. Find the product compositions and flow rates.

Solution

A. Define. We want to calculate y_i, x_i, V, and L for the equilibrium flash chamber shown in the diagram.

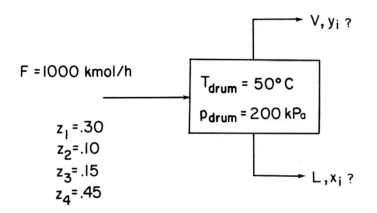

B. Explore. Since T_{drum} and p_{drum} are given, a sequential solution can be used. We can use the Rachford-Rice equation to solve for V/F and then find x_i, y_i, L, and V.

C. Plan. Calculate K_i from DePriester charts or from Eq. (2-30). Use Newtonian convergence with the Rachford-Rice equation, Eq. (2-46), to converge on the correct V/F value. Once the correct V/F has been found, calculate x_i from Eq. (2-38) and y_i from Eq. (2-39). Calculate V from V/F and L from overall mass balance, Eq. (2-5).

D. Do it. From the DePriester chart (Fig. 2-11), at 50°C and 200 kPa we find

$$K_1 = 7.0 \qquad C_3$$
$$K_2 = 2.4 \qquad n\text{-}C_4$$
$$K_3 = 0.80 \qquad n\text{-}C_5$$
$$K_4 = 0.30 \qquad n\text{-}C_6$$

Calculate f (V/F) from the Rachford-Rice equation:

$$f\left(\frac{V}{F}\right) = \sum_{i=1}^{4} \frac{(K_i - 1)z_i}{1 + (K_i - 1)V/F}$$

Pick V/F = 0.1 as first guess (this illustrates convergence for a poor first guess).

$$f(0.1) = \frac{(7.0 - 1)(0.3)}{1 + (7.0 - 1)(0.1)} + \frac{(2.4 - 1)(0.1)}{1 + (2.4 - 1)(0.1)} + \frac{(0.8 - 1)(0.15)}{1 + (0.8 - 1)(0.1)} + \frac{(0.3 - 1)(0.45)}{1 + (0.3 - 1)(0.1)}$$

$$= 1.125 + 0.1228 + (-0.0306) + (-0.3387) = 0.8785$$

Since f(0.1) is positive, a higher value for V/F is required. Note that only one term in the denominator of each term changes. Thus we can set up the equation so that only V/F will change. Then f(V/F) equals

$$\frac{1.8}{1 + 6\left(\dfrac{V}{F}\right)} + \frac{0.14}{1 + 1.4\left(\dfrac{V}{F}\right)} + \frac{-0.03}{1 - 0.2\left(\dfrac{V}{F}\right)} + \frac{-0.315}{1 - 0.7\left(\dfrac{V}{F}\right)}$$

Now all subsequent calculations will be easier.

The derivative of the R-R equation can be calculated for this first guess

$$(\frac{df}{d(\frac{V}{F})})_1 = -\left\{\left\langle \frac{(K_1 - 1)^2\, z_1}{[1 + (K_1 - 1)\,\frac{V}{F}]^2} + \frac{(K_2 - 1)^2\, z_2}{[1 + (K_2 - 1)\,\frac{V}{F}]^2} + \frac{(K_3 - 1)^2\, z_3}{[1 + (K_3 - 1)\,\frac{V}{F}]^2} + \frac{(K_4 - 1)^2\, z_4}{[1 + (K_4 - 1)\,\frac{V}{F}]^2} \right\rangle\right\}$$

$$= -\left\{ \frac{10.8}{[1 + (6.0)\,\frac{V}{F}]^2} + \frac{0.196}{[1 + 1.4\,\frac{V}{F}]^2} + \frac{0.006}{[1 - 0.2\,\frac{V}{F}]^2} + \frac{0.2205}{[1 - 0.7\,\frac{V}{F}]^2} \right\}$$

With V/F = 0.1 this is $(\dfrac{df}{d(V/F)})_1 = -4.631$

From Eq. (2-46) the next guess for V/F is $(V/F)_2 = 0.1 + 0.8785/4.631 = 0.29$. Calculating the value of the Rachford-Rice equation, we have f(0.29) = 0.329. This is still positive and V/F is still too low.

Second Trial: $(\dfrac{df}{d(V/F)})|_{0.29} = -1.891$

which gives $(V/F)_3 = 0.29 + 0.329/1.891 = 0.46$
and the Rachford-Rice equation is f(0.46) = 0.066. This is closer, but V/F is still too low. Continue convergence.

Third Trial: $(\dfrac{df}{d(V/F)})|_{0.46} = -1.32$

which gives $(V/F)_4 = 0.46 + 0.066/1.32 = 0.51$

We calculate that f(0.51) = 0.00173, which is close to zero and is within the accuracy of the DePriester charts. Thus V/F = 0.51.

Now we calculate x_i from Eq. (2-38) and y_i from $K_i x_i$. For example,

$$x_1 = \frac{z_1}{1 + (K_1 - 1)\dfrac{V}{F}} = \frac{0.30}{1 + (7.0 - 1)(0.51)} = 0.0739$$

Chapter 2   Flash Distillation

$$y_1 = K_1 x_1 = (7.0)(0.0739) = 0.5172$$

By similar calculations, $x_2 = 0.0583$, $y_2 = 0.1400$
$x_3 = 0.1670$, $y_3 = 0.1336$
$x_4 = 0.6998$, $y_4 = 0.2099$

since F = 1000 and V/F = 0.51, V = 0.51F = 510 kmol/h, and L = F − V = 1000 − 510 = 490 kmol/h.

E. Check. We can check $\Sigma\, y_i$ and $\Sigma\, x_i$.

$$\sum_{i=1}^{4} x_i = 0.999, \qquad \sum_{i=1}^{4} y_i = 1.0007$$

These are close enough. They aren't perfect, because V/F wasn't exact. Essentially the same answer is obtained if Eq. (2-30) is used for the K values. Note: Equation (2-30) may seem more accurate since one can produce a lot of digits; however, since it is a fit to the DePriester chart it can't be more accurate.

F. Generalize. Since the Rachford-Rice equation is almost linear, the Newtonian convergence routine gives rapid convergence. Note that the convergence was monotonic and did not oscillate. Faster convergence would be achieved with a better first guess of V/F. This type of trial-and-error problem is easy to program on a spreadsheet (see Appendix B in this chapter).

If the specified variables are F, z_i, p_{drum}, and either x or y for one component, we can follow a sequential convergence procedure using Eq. (2-38) or (2-39) to relate to the specified composition (the reference component) to either K_{ref} or V/F. We can do this in either of two ways. The first is to guess T_{drum} and use Eq. (2-38) or (2-39) to solve for V/F. The Rachford-Rice equation is then the check equation on T_{drum}. If the Rachford-Rice equation is not satisfied, we select a new temperature—using Eq. (2-49)—and repeat the procedure. In the second approach, we guess V/F and calculate K_{ref} from Eq. (2-38) or (2-39). We then determine the drum temperature from this K_{ref}. The Rachford-Rice equation is again the check. If it is not satisfied, we select a new V/F and continue the process.

If there are nonvolatile compounds present, the K_i values for these compounds are zero. The presence of these compounds will cause no difficulties for Eqs. (2-38) to (2-49). However, if there are noncondensable compounds present, the K_i for these compounds will be very large, particularly if the solubilities are small. It is tempting to set these K_i values to infinity, but then Eq. (2-42) becomes undefined. This difficulty is easily handled by rearranging Eq. (2-42) (Hatfield, 2008). If we divide numerator and denominator of the noncondensable term of Eq. (2-42) by K_{NC}, this term becomes

$$\frac{\left(1 - \dfrac{1}{K_{NC}}\right) z_{NC}}{\dfrac{1}{K_{NC}} + \left(1 - \dfrac{1}{K_{NC}}\right)\dfrac{V}{F}} = \frac{z_{NC}}{V/F}$$

(2-49)

Substitution of this term into Eq. (2-42) results in a well-behaved equation in the presence of noncondensable compounds. Equations (2-38) and (2-39) become $x_{NC} = 0$ and $y_{NC} = Fz_{NC}/V$.

2.7 SIMULTANEOUS MULTICOMPONENT CONVERGENCE

If the feed rate F, the feed composition consisting of $(C - 1)$ z_i values, the flash drum pressure p_{drum}, and the feed temperature T_F are specified, the hot liquid will vaporize when its pressure is dropped. This "flashing" cools the liquid to provide energy to vaporize some of the liquid. The result T_{drum} is unknown; thus, we must use a simultaneous solution procedure. First, we choose a feed pressure such that the feed will be liquid. Then we can calculate the feed enthalpy in the same way as Eqs. (2-47) and (2-48):

$$h_F = \sum_{i=1}^{C} z_i \, \tilde{h}_{Fi} \, (T_F, p_F)$$

(2-50)

Although the mass and energy balances, equilibrium relations, and stoichiometric relations could all be solved simultaneously, it is again easier to use a trial-and-error procedure. This problem is now a double trial and error.

The first question to ask in setting up a trial-and-error procedure is: What are the possible trial variables and which ones shall we use? Here we first pick T_{drum}, since it is required to calculate all K_i, $\tilde{H}_{vi}$ and $\tilde{h}_{Li}$ and since it is difficult to solve for. The second trial variable is V/F, because then we can use the Rachford-Rice approach with Newtonian convergence.

The second question to ask is: Should we converge on both variables simultaneously (that is change both T_{drum} and V/F at the same time), or should we converge sequentially? Both techniques will work, but sequential convergence tends to be more stable. If we use sequential convergence, then a third question is: Which variable should we converge on first, V/F or T_{drum}? To answer this question we need to consider the chemical system we are separating. If the mixture is wide-boiling, that is, if the dew point and bubble point are far apart (say more than 80 to 100°C), then a small change in T_{drum} cannot have much effect on V/F. In this case we wish to converge on V/F first. Then when T_{drum} is changed, we will be close to the correct answer for V/F. For a significant separation in a flash system, the volatilities must be very different, so this is the typical situation for flash distillation. The narrow-boiling procedure is shown in Figure 6-1 for distillation.

The procedure for wide-boiling feeds is shown in Figure 2-13. Note that the energy balance is used last. This is standard procedure since accurate values of x_i and y_i are available to calculate enthalpies for the energy balance.

The fourth question is: How should we do the individual convergence steps? For the Rachford-Rice equation, linear interpolation or Newtonian convergence will be satisfactory. Several methods can be used to estimate the next flash drum temperature. One of the fastest and easiest to use is a Newtonian convergence procedure. To do this we rearrange the energy balance (Eq. 2-7) into the functional form,

$$E_k(T_{drum}) = VH_v + Lh_L - Fh_F - Q_{flash} = 0$$

(2-51)

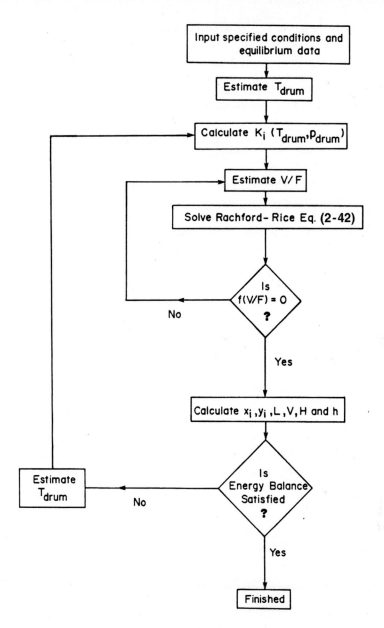

FIGURE 2-13. *Flowsheet for wide-boiling feed*

The subscript k again refers to the trial number. When E_k is zero, the problem has been solved. The Newtonian procedure estimates $E_{k+1}(T_{drum})$ from the derivative,

$$E_{k+1} - E_k = \frac{dE_k}{dT_{drum}} (\Delta T_{drum})$$

(2-52)

where ΔT_{drum} is the change in T_{drum} from trial to trial,

$$\Delta T_{drum} = T_{k+1} - T_k \qquad (2\text{-}53)$$

and dE_k/dT_{drum} is the variation of E_k as temperature changes. Since the last two terms in Eq. (2-51) do not depend on T_{drum}, this derivative can be calculated as

$$\frac{dE_k}{dT_{drum\,k}} = V\frac{dH_v}{dT_{drum}} + L\frac{dH_L}{dT_{drum}} = VC_{Pv} + Lc_{PL} \qquad (2\text{-}54)$$

where we have used the definition of the heat capacity. In deriving Eq. (2-54) we set both dV/dT and dL/dT equal to zero since a sequential convergence routine is being used and we do not want to vary V and L in this loop. We want the energy balance to be satisfied after the next trial. Thus we set $E_{k+1} = 0$. Now Eq. (2-52) can be solved for ΔT_{drum}:

$$\Delta T_{drum} = \frac{-E_k(T_{drum\,k})}{\dfrac{dE_k}{dT_{drum\,k}}} \qquad (2\text{-}55)$$

Substituting the expression for ΔT_{drum} into this equation and solving for $T_{drum\,k+1}$, we obtain the best guess for temperature for the next trial,

$$T_{drum\,k+1} = T_{drum\,k} - \frac{E_k(T_{drum\,k})}{\dfrac{dE_k}{dT_{drum\,k}}} \qquad (2\text{-}56a)$$

In this equation E_k is the calculated numerical value of the energy balance function from Eq. (2-51) and dE_k/dT_{drum} is the numerical value of the derivative calculated from Eq. (2-54).

The procedure has converged when

$$|\Delta T_{drum}| < \varepsilon \qquad (2\text{-}56b)$$

For computer calculations, $\varepsilon = 0.01°C$ is a reasonable choice. For hand calculations, a less stringent limit such as $\varepsilon = 0.2°C$ would be used. This procedure is illustrated in Example 2-3.

It is possible that this convergence scheme will predict values of ΔT_{drum} that are too large. When this occurs, the drum temperature may oscillate with a growing amplitude and not converge. To discourage this behavior, ΔT_{drum} can be damped.

$$\Delta T_{drum} = (d)(\Delta T_{drum,\,calc\ Eq.\ (2\text{-}55)}) \qquad (2\text{-}57)$$

where the damping factor d is about 0.5. Note that when d = 1.0, this is just the Newtonian approach.

The drum temperature should always lie between the bubble- and dew-point temperature of the feed. In addition, the temperature should converge toward some central value. If either of these criteria is violated, then the convergence scheme should be damped or an alternative convergence scheme should be used.

EXAMPLE 2-3. Simultaneous solution for flash distillation

We have a liquid feed that is 20 mol% methane, 45 mol% n-pentane, and 35 mol% n-hexane. Feed rate is 1500 kmol/h, feed temperature is 45°C, and pressure is 100.0 psia. The flash drum operates at 30 psia and is adiabatic. Find: T_{drum}, V/F, x_i, y_i, L, V.

Solution

A. Define. The process is sketched in the diagram.

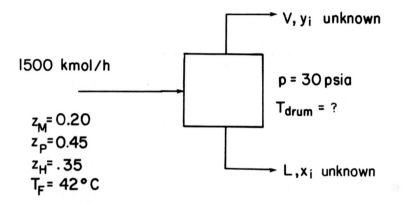

B. Explore. Since T_F is given, this will be a double trial and error. K values from the DePriester charts or from Eq. (2-30) can be used. For energy balances, enthalpies can be calculated from heat capacities and latent heats. The required data are listed in Table 2-5.

C. Plan. Since this is a double trial and error, all calculations will be done on the computer and summarized here. Newtonian convergence will be used for both the Rachford-Rice equation and the energy balance estimate of new

TABLE 2-5. *Data for methane, n-pentane, and n-hexane*

Component	λ, kcal/mol,	Normal boiling pt., °C	C_{PL}, cal/(mol °C)
1. Methane	1.955	−161.48	11.0 (est.)
2. n-Pentane	6.160	36.08	39.66
3. n-Hexane	6.896	68.75	45.58

Vapor heat capacities in cal/(mol °C); T in °C:

$C_{PV,1} = 8.20 + 0.01307\ T + 8.75 \times 10^{-7}T^2 - 2.63 \times 10^{-9}T^3$

$C_{PV,2} = 27.45 + 0.08148\ T - 4.538 \times 10^{-5}T^2 + 10.1 \times 10^{-9}T^3$

$C_{PV,3} = 32.85 + 0.09763\ T - 5.716 \times 10^{-5}T^2 + 13.78 \times 10^{-9}T^3$

Source: Himmelblau (1974)

drum temperature. $\varepsilon = 0.02$ is used for energy convergence (Eq. 2-56b). The Rachford-Rice equation is considered converged when

$$| (V/F)_{k+1} - (V/F)_k | < \varepsilon_R \tag{2-58}$$

$\varepsilon_R = 0.005$ is used here.

D. **Do It.** The first guess is made by arbitrarily assuming that $T_{drum} = 15°C$ and $V/F = 0.25$. Since convergence of the program is rapid, more effort on an accurate first guess is probably not justified. Using Eq. (2-46) as illustrated in Example 2-2, the following V/F values are obtained [K values from Eq. (2-30)]:

$V/F = \quad$ 0.25, 0.2485, 0.2470, 0.2457, 0.2445, 0.2434, 0.2424, 0.2414, 0.2405, 0.2397, 0.2390, 0.2383, 0.2377, 0.2371, 0.2366, 0.2361

Note that convergence is monotonic. With V/F known, x_i and y_i are found from Eqs. (2-38) and (2-39).

Compositions are: $x_m = 0.0124$, $x_p = 0.5459$, $x_H = 0.4470$, and $y_m = 0.8072$, $y_p = 0.1398$, $y_H = 0.0362$.

Flow rates L and V are found from the mass balance and V/F value. After determining enthalpies, Eq. (2-56a) is used to determine $T_{drum,2} = 27.9°C$. Obviously, this is still far from convergence.

The convergence procedure is continued, as summarized in Table 2-6. Note that the drum temperature oscillates, and because of this the converged V/F oscillates. Also, the number of trials to converge on V/F decreases as the calculation proceeds. The final compositions and flow rates are:

$x_m = 0.0108$, $x_p = 0.5381$, $x_H = 0.4513$
$y_m = 0.7531$, $y_P = 0.1925$, $y_H = 0.0539$
$V = 382.3$ kmol/h and $L = 1117.7$ kmol/h

E. **Check.** The results are checked throughout the trial-and-error procedure. Naturally, they depend upon the validity of data used for the enthalpies and Ks. At least the results appear to be self-consistent (that is, $\Sigma x_i = 1.0$, $\Sigma y_i = 1.0$) and are of the right order of magnitude. This problem was also solved

TABLE 2-6. *Iterations for Example 2-3*

Iteration No.	Initial T_{drum}	Trials to find V/F	V/F	Calc T_{drum}
1	15.00	16	0.2361	27.903
2	27.903	13	0.2677	21.385
3	21.385	14	0.2496	25.149
4	25.149	7	0.2567	23.277
5	23.277	3	0.2551	24.128
6	24.128	2	0.2551	23.786
7	23.786	2	0.2550	23.930
8	23.930	2	0.2550	23.875
9	23.875	2	0.2549	23.900
10	23.900	2	0.2549	23.892

using Aspen Plus with the Peng-Robinson equation for VLE (see Chapter 2 Appendix A). The results are $x_m = 0.0079$, $x_p = 0.5374$, $x_H = 0.4547$, $L = 1107.8$, and $y_m = 0.7424$, $y_p = 0.2032$, $y_H = 0.0543$, $V = 392.2$, and $T_{drum} = 27.99°C$. With the exception of the drum temperature these results, which use different data, are close.

F. Generalization. The use of the computer greatly reduces calculation time on this double trial-and-error problem. Use of a process simulator that includes VLE and enthalpy correlations will be fastest. However, any software package must be validated (Shacham et al., 2008).

2.8 THREE-PHASE FLASH CALCULATIONS

Many systems, particularly mixtures of nonpolar organics and polar compounds such as water, will form two liquid phases and one vapor phase. A binary example, n-butanol and water, is discussed later (see Figure 8-2 and Problem 8.D3). In this section we consider calculations for multicomponent liquid-liquid-vapor systems. For example, if a vapor mixture of gasoline and water is partially condensed, the result will be an aqueous layer with a high mole fraction of water, an organic phase containing very little water, and a vapor phase. The different components of gasoline will distribute between the three phases differently.

With three phases, the component mass balance for a flash distillation system is

$$Fz_i = L_{liquid\_1}\, x_{i,liquid\_1} + L_{liquid\_2}\, x_{i,liquid\_2} + Vy_i \tag{2-59}$$

which is equivalent to Eq. (2-36). There are i independent component mass balances, but the overall balance is not independent, since it is obtained by summing all of the component balances.

When there are three phases, we can write three equilibrium distribution relationships for each component i,

$$K_{i,vapor\text{-}liquid\_1} = y_i/x_{i,liquid\_1}, \; K_{i,vapor-liquid\_2} = y_i/x_{i,liquid\_2}, \; K_{i,liquid\_1-liquid\_2} = x_{i,liquid\_1}/x_{i,liquid\_2} \tag{2-60a,b,c}$$

Solving Eq. (2-60a) for y_i, and substituting this into Eq. (2-60b), rearranging and comparing to Eq. (2-60c) we obtain

$$x_{i,liquid\_1}/x_{i,liquid\_2} = K_{i,vapor-liquid\_2}/\, K_{i,vapor-liquid\_1} = K_{i,liquid\_1-liquid\_2} \tag{2-61}$$

Thus, only two of the three K values for each component are independent. Of course, all the K values are, in general, functions of temperature, pressure, and composition.

We can now follow exactly the same steps as were used to derive the Rachford-Rice equation [Eqs. (2-37) to (2-42)] to derive two equations for the three-phase flash.

$$0 = \sum_{i=1}^{C} \frac{\left(K_{i,vapor-liquid\_2}-1\right)z_i}{\left[1+\left(K_{i,liquid\_1-liquid2}-1\right)\dfrac{L_{liquid\_1}}{F}+\left(K_{i,vapor-liquid\_2}-1\right)\dfrac{V}{F}\right]} \tag{2-62}$$

$$0 = \sum_{i=1}^{C} \frac{\left(K_{i,\text{liquid}\_1-\text{liquid}\_2} - 1\right) z_i}{\left[1 + \left(K_{i,\text{liquid}\_1-\text{liquid}2} - 1\right) \dfrac{L_{\text{liquid}\_1}}{F} + \left(K_{i,\text{vapor}-\text{liquid}\_2} - 1\right) \dfrac{V}{F}\right]}$$

(2-63)

If temperature and pressure are specified and correlations for the equilibrium parameters are available, these two forms of the Rachford-Rice equation can be solved simultaneously for L_{liquid_1}/F and V/F. Then y_i, x_{i,liquid_1}, and x_{i,liquid_2} can be calculated from the three-phase equations equivalent to Eqs. (2-37) and (2-38) (see Problem 2.C9). If only liquid 1 and vapor are present, then an equation equivalent to Eq. (2-62) but written for vapor and liquid 1 must be used (Chien, 1994). Process simulators can do these calculations (e.g., problem 2.G5), but the equilibrium correlations, particularly the liquid-liquid equilibrium correlations, need to be checked against data.

2.9   SIZE CALCULATION

Once the vapor and liquid compositions and flow rates have been determined, the flash drum can be sized. This is an empirical procedure. We will discuss the specific procedure first for vertical flash drums (Figure 2-1) and then adjust the procedure for horizontal flash drums.

Step 1. Calculate the permissible vapor velocity, u_{perm},

$$u_{\text{perm}} = K_{\text{drum}} \sqrt{\frac{\rho_L - \rho_v}{\rho_v}}$$

(2-64)

u_{perm} is the maximum permissible vapor velocity in feet per second at the maximum cross-sectional area. ρ_L and ρ_v are the liquid and vapor densities. K_{drum} is in ft/s.

K_{drum} is an empirical constant that depends on the type of drum. For vertical drums the value has been correlated graphically by Watkins (1967) for 85% of flood with no demister. Approximately 5% liquid will be entrained with the vapor. Use of the same design with a demister will reduce entrainment to less than 1%. The demister traps small liquid droplets on fine wires and prevents them from exiting. The droplets then coalesce into larger droplets, which fall off the wire and through the rising vapor into the liquid pool at the bottom of the flash chamber. Blackwell (1984) fit Watkins' correlation to the equation

$$K_{\text{drum}} = (\text{Const.}) \exp[A + B \ln F_{lv} + C(\ln F_{lv})^2 + D(\ln F_{lv})^3 + E(\ln F_{lv})^4] \qquad \text{(2-65)}$$

where $F_{lv} = \dfrac{W_L}{W_v} \sqrt{\dfrac{\rho_v}{\rho_L}}$ and const = 1.0 ft/s,

with W_L and W_v being the liquid and vapor flow rates in weight units per hour (e.g., lb/h). The constants are (Blackwell, 1984):

A = –1.877478097   C = –0.1870744085   E = –0.0010148518
B = –0.8145804597   D = –0.0145228667

The resulting value for K_{drum} typically ranges from 0.1 to 0.35.

Step 2. Using the known vapor rate, V, convert u_{perm} into a horizontal area. The vapor flow rate, V, in lbmol/h is

$$V\left(\frac{lbmol}{h}\right) = \frac{u_{perm}\left(\frac{ft}{s}\right)\left(\frac{3600\ s}{h}\right) A_c\ (ft^2)\ \rho_v\left(\frac{lbm}{ft^3}\right)}{MW_{vapor}\left(\frac{lbm}{lbmol}\right)}$$

Solving for the cross-sectional area,

$$A_c = \frac{V(MW_v)}{u_{perm}\ (3600)\rho_v} \tag{2-66}$$

For a vertical drum, diameter D is

$$D = \sqrt{\frac{4A_c}{\pi}} \tag{2-67}$$

Usually, the diameter is increased to the next largest 6-in. increment.

Step 3. Set the length/diameter ratio either by rule of thumb or by the required liquid surge volume. For vertical flash drums, the rule of thumb is that h_{total}/D ranges from 3.0 to 5.0. The appropriate value of h_{total}/D within this range can be found by minimizing the total vessel weight (which minimizes cost).

Flash drums are often used as liquid surge tanks in addition to separating liquid and vapor. The design procedure for this case is discussed by Watkins (1967) for petrochemical applications. The height of the drum above the centerline of the feed nozzle, h_v, should be 36 in. plus one-half the diameter of the feed line (see Figure 2-14). The minimum of this distance is 48 in.

The height of the center of the feed line above the maximum level of the liquid pool, h_f, should be 12 in. plus one-half the diameter of the feed line. The minimum distance for this free space is 18 in.

The depth of the liquid pool, h_L, can be determined from the desired surge volume, V_{surge}.

$$h_L = \frac{V_{surge}}{\pi D^2/4} \tag{2-68}$$

The geometry can now be checked, since

$$\frac{h_{total}}{D} = \frac{h_v + h_f + h_L}{D}$$

should be between 3 and 5. These procedures are illustrated in Example 2-4. If $h_{total}/D < 3$, a larger liquid surge volume should be allowed. If $h_{total}/D > 5$, a horizontal flash drum should be used.

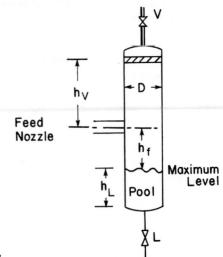

FIGURE 2-14.   *Measurements for vertical flash drum*

Horizontal flash drums (Figure 2-15) are used when large liquid surge volumes are required because additional surge volume is formed by increasing the length (Evans, 1980). The cross sectional area required for vapor flow is given by Eq. (2-66) except the area is relabeled as A_{vap} because it is not the final cross sectional area of the horizontal drum. If the time required to respond to an upset (the surge time) and the average residence time of fluid (the holding time) in the drum are known, the full liquid volume V_{liq} is,

$$V_{liq} = (\text{Volumetric liquid flow rate}) \times (\text{holding time} + \text{surge time}) \qquad \textbf{(2-69a)}$$

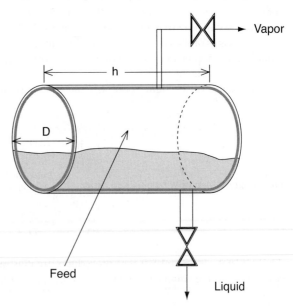

FIGURE 2-15.   *Horizontal flash drum*

If we arbitrarily choose h/D = constant C and solve for diameter D, we obtain

$$D = \sqrt{\frac{V \, MW_v}{u_{perm} \, 3600 \, \rho_v C}}$$

(2-69b)

If the ideal gas law is valid, the molar density $\hat{\rho}_V$ and the mass density ρ_V are,

$$\hat{\rho}_V = \frac{n}{V} = \frac{p}{RT} \quad \text{and} \quad \rho_V = \hat{\rho}_V(MW_V)$$

(2-69c)

Equation (2-69b) becomes

$$D = \sqrt{\frac{VRT}{3600 \, Cpu_{perm}}} \quad \text{for ideal gas}$$

(2-69d)

The value of u is found from Eq. (2-64) with (Blackwell, 1984)

$$K_{horizontal} = 1.25 \, K_{vertical}$$

(2-69e)

where $K_{vertical}$ is determined from Eq. (2-65).

The typical range for h/D is from 3 to 5. Horizontal drums are particularly useful when large liquid surge capacities are needed. More detailed design procedures and methods for horizontal drums are presented by Evans (1980), Blackwell (1984), and Watkins (1967). Note that in industries other than petrochemicals that sizing may vary.

EXAMPLE 2-4. Calculation of drum size

A vertical flash drum is to flash a liquid feed of 1500 lbmol/h that is 40 mol% n-hexane and 60 mol% n-octane at 101.3 kPa (1 atm). We wish to produce a vapor that is 60 mol% n-hexane. Solution of the flash equations with equilibrium data gives $x_H = 0.19$, $T_{drum} = 378K$, and $V/F = 0.51$. What size flash drum is required?

Solution

A. Define. We wish to find diameter and length of flash drum.

B. Explore. We want to use the empirical method developed in Eqs. (2-64) to (2-68). For this we need to estimate the following physical properties: ρ_L, ρ_v, MW_v. To do this we need to know something about the behavior of the gas and of the liquid.

C. Plan. Assume ideal gas and ideal mixtures for liquid. Calculate average ρ_L by assuming additive volumes. Calculate ρ_v from the ideal gas law. Then calculate u_{perm} from Eq. (2-64) and diameter from Eq. (2-68).

D. Do It.

1. Liquid Density

The average liquid molecular weight is

$$\overline{MW}_L = x_H MW_H + x_O MW_O$$

where subscript H is n-hexane and O is n-octane. Calculate or look up the molecular weights. MW_H = 86.17 and MW_O = 114.22. Then $\overline{MW}_L$ = (0.19)(86.17) + (0.81)(114.22) = 108.89. The specific volume is the sum of mole fractions multiplied by the pure component specific volumes (ideal mixture):

$$\overline{V}_L = x_H \overline{V}_H + x_o \overline{V}_o = x_H \frac{MW_H}{\rho_H} + \frac{x_o MW_o}{\rho_o}$$

From the *Handbook of Chemistry and Physics*, ρ_H = 0.659 g/mL and ρ_O = 0.703 g/mL at 20°C. Thus,

$$\overline{V}_L = (0.19)\frac{86.17}{0.659} + (0.81)\frac{114.22}{0.703} = 156.45 \text{ mL/mol}$$

Then

$$\rho_L = \frac{\overline{MW}_L}{\overline{V}_L} = \frac{108.89}{156.45} = 0.6960 \text{ g/mL}$$

2. Vapor Density
Density in moles per liter for ideal gas is $\hat{\rho}_v = n/V = p/RT$, which in g/L is $\rho_v = p \, \overline{MW}_v / RT$.
The average molecular weight of the vapor is

$$\overline{MW}_v = y_H MW_H + y_O MW_O$$

where y_H = 0.60 and y_O = 0.40, and thus $\overline{MW}_v$ = 97.39 lb/lbmol. This gives

$$\rho_v = \frac{(1.0 \text{atm})(97.39 \text{g/mol})}{(82.0575 \frac{\text{mL atm}}{\text{mol K}})(378\text{K})} = 3.14 \times 10^{-3} \text{ g/mL}$$

3. K_{drum} Calculation.
   Calculation of flow parameter F_{lv}:
   V = (V/F)(F) = (0.51)(1500) = 765 lbmol/h
   $W_v = (V)(\overline{MW}_v) = (765)(97.39) = 74,503$ lb/h
   L = F − V = 735 lbmol/h
   $W_L = (L)(\overline{MW}_L) = (735)(108.89) = 80,034$ lb/h

$$F_{lv} = \frac{W_L}{W_v} \sqrt{\frac{\rho_v}{\rho_L}} = \frac{80034}{74503} \sqrt{\frac{3.14 \times 10^{-3}}{0.6960}} = 0.0722$$

K_{drum} from Eq. (2-65) gives K_{drum} = 0.4433, which seems a bit high but agrees with Watkins's (1967) chart.

4.

$$u_{perm} = K_{drum} \sqrt{\frac{\rho_L - \rho_v}{\rho_v}}$$

$$= 0.4433 \sqrt{\frac{0.6960 - 0.00314}{0.00314}} = 6.5849 \text{ft}/\text{s}$$

5.

$$A_c = \frac{V(\overline{MW_v})}{u_{perm}(3600)\rho_v}$$

$$= \frac{(765)(97.39)(454 \text{ g}/\text{lb})}{(6.5849)(3600)(0.00314 \text{ g}/\text{mL})(28316.85 \text{ mL}/\text{ft}^3)}$$

$$= 16.047 \text{ ft}^2$$

$$D = \sqrt{\frac{4A_c}{\pi}} = 4.01 \text{ ft}$$

Use a 4.0 ft diameter drum or 4.5 ft to be safe.

6. If use $h_{total}/D = 4$, $h_{total} = 4(4.5 \text{ ft}) = 18.0$ ft.

E. Check. This drum size is reasonable. Minimums for h_v and h_f are easily met. Note that units do work out in all calculations; however, one must be careful with units, particularly calculating A_c and D.

F. Generalization. If the ideal gas law is not valid, a compressibility factor could be inserted in the equation for ρ_v. Note that most of the work involved calculation of the physical properties. This is often true in designing equipment. In practice we pick a standard size drum (4.0 or 4.5 ft diameter) instead of custom building the drum.

2.10   USING EXISTING FLASH DRUMS

Individual pieces of equipment will often outlive the entire plant. This used equipment is then available either in the plant's salvage section or from used equipment dealers. As long as used equipment is clean and structurally sound (it pays to have an expert check it), it can be used instead of designing and building new equipment. Used equipment and off-the-shelf new equipment will often be cheaper and will have faster delivery than custom-designed new equipment; however, it may have been designed for a different separation. The challenge in using existing equipment is to adapt it with minimum cost to the new separation problem.

The existing flash drum has its dimensions h_{total} and D specified. Solving Eqs. (2-66) and (2-67) for a vertical drum for V, we have

$$V_{max} = \frac{\pi(D)^2 u_{perm}(3600)\rho_v}{4MW_v} \tag{2-70}$$

This vapor velocity is the maximum for this existing drum, since it will give a linear vapor velocity equal to u_{perm}.

The maximum vapor capacity of the drum limits the product of (V/F) multiplied by F, since we must have

$$(V/F) \, F < V_{Max} \tag{2-71}$$

If Eq. (2-71) is satisfied, then use of the drum is straightforward. If Eq. (2-71) is violated, something has to give. Some of the possible adjustments are:

a. Add chevrons or a demister to increase V_{Max} or to reduce entrainment (Woinsky, 1994).
b. Reduce feed rate to the drum.
c. Reduce V/F. Less vapor product with more of the more volatile components will be produced.
d. Use existing drums in parallel. This reduces feed rate to each drum.
e. Use existing drums in series (see Problems 2.D2 and 2.D5).
f. Try increasing the pressure (note that this changes everything—see Problem 2.C1).
g. Buy a different flash drum or build a new one.
h. Use some combination of these alternatives.
i. The engineer can use ingenuity to solve the problem in the cheapest and quickest way.

2.11   SUMMARY—OBJECTIVES

This chapter discussed VLE and the calculation procedures for binary and multicomponent flash distillation. At this point you should be able to satisfy the following objectives:

1. Explain and sketch the basic flash distillation process
2. Find desired VLE data in the literature or on the Web
3. Plot and use y-x, temperature-composition, enthalpy-composition diagrams; explain the relationship between these three types of diagrams
4. Derive and plot the operating equation for a binary flash distillation on a y-x diagram; solve both sequential and simultaneous binary flash distillation problems
5. Define and use K values, Raoult's law, and relative volatility
6. Derive the Rachford-Rice equation for multicomponent flash distillation, and use it with Newtonian convergence to determine V/F
7. Solve sequential multicomponent flash distillation problems
8. Determine the length and diameter of a flash drum
9. Use existing flash drums for a new separation problem

REFERENCES

Barnicki, S. D., "How Good Are Your Data?" *Chem. Engr. Progress, 98* (6), 58 (June 2002).

Biegler, L. T., I. E. Grossman, and A. W. Westerberg, *Systematic Methods of Chemical Process Design*, Prentice Hall, Upper Saddle River, NJ, 1997.

Binous, H., "Equilibrium-Staged Separations Using MATLAB and Mathmatica," *Chem. Engr. Educ., 42* (2), 69 (Spring 2008).

Blackwell, W. W., *Chemical Process Design on a Programmable Calculator*, McGraw-Hill, New York, 1984, chapter 3.

Boublik, T., V. Fried, and E. Hala, *Vapour Pressures of Pure Substances,* Elsevier, Amsterdam, 1984.

Carlson, E. C., "Don't Gamble with Physical Properties for Simulators," *Chem. Engr. Progress, 92* (10), 35 (Oct. 1996).

Chien, H. H.-Y., "Formulations for Three-Phase Flash Calculations," *AIChE Journal, 40* (6), 957 (1994).

Dadyburjor, D. B., "SI Units for Distribution Coefficients," *Chem. Engr. Progress, 74* (4), 85 (April 1978).

Dean, J. A.(Ed.), *Lange's Handbook of Chemistry,* 13th ed., McGraw-Hill, New York 1985.

Evans, F. L., Jr., *Equipment Design Handbook for Refineries and Chemical Plants,* Vol. 2, 2nd ed., Gulf Publishing Co., Houston, TX, 1980.

Fredenslund, A., J. Gmehling, and P. Rasmussen, *Vapor-Liquid Equilibria Using UNIFAC: A Group-Contribution Method,* Elsevier, Amsterdam, 1977.

Green, D. W., and R. H. Perry (Eds.), *Perry's Chemical Engineers' Handbook*, 8th ed., McGraw-Hill, New York, 2008.

Hatfield, A., "Analyzing Equilibrium When Noncondensables Are Present," *Chem. Engr. Progress, 42* (April 2008).

Himmelblau, D. M., *Basic Principles and Calculations in Chemical Engineering,* 3rd ed, Prentice Hall, Upper Saddle River, NJ, 1974.

King, C. J., *Separation Processes,* 2nd ed., McGraw-Hill, New York, 1981.

Lazzaroni, M. J., D. Bush, C. A. Eckert, T. C. Frank, S. Gupta, and J. D. Olsen, "Revision of MOSCED Parameters and Extension to Solid Solubility Calculations," *Ind. Eng. Chem. Research, 44,* 4075 (2005).

Maxwell, J. B., *Data Book on Hydrocarbons,* Van Nostrand, Princeton, NJ, 1950.

McWilliams, M. L., "An Equation to Relate K-factors to Pressure and Temperature," *Chem. Engineering, 80* (25), 138 (Oct. 29, 1973).

Marsh, K. N., "The Measurement of Thermodynamic Excess Functions of Binary Liquid Mixtures," *Chemical Thermodynamics,* Vol. 2, Chemical Society, London, 1978, pp. 1-45.

Nelson, A. R., J. H. Olson, and S. I. Sandler, "Sensitivity of Distillation Process Design and Operation to VLE Data," *Ind. Eng. Chem. Process Des. Develop., 22,* 547 (1983).

O'Connell, J. P., R. Gani, P. M. Mathias, G. Maurer, J. D. Olson, and P. A. Crafts, "Thermodynamic Property Modeling for Chemical Process and Product Engineering: Some Perspectives," *Ind. Engr. Chem. Research, 48,* 4619 (2009).

Perry, R. H., C. H. Chilton and S. D. Kirkpatrick (Eds.), *Chemical Engineer's Handbook,* 4th ed., McGraw-Hill, New York, 1981.

Perry, R. H. and D. Green (Eds.), *Perry's Chemical Engineer's Handbook,* 7th ed., McGraw-Hill, New York, 1997.

Poling, B. E., J. M. Prausnitz, and J. P. O'Connell, *The Properties of Gases and Liquids,* 5th ed., McGraw-Hill, New York, 2001.

Prausnitz, J. M., T. F. Anderson, E. A. Grens, C. A. Eckert, R. Hsieh, and J. P. O'Connell, *Computer Calculations for Multicomponent Vapor-Liquid and Liquid-Liquid Equilibria,* Prentice Hall, Upper Saddle River, NJ, 1980.

Prausnitz, J. M., R. N. Lichtenthaler, and E. G. de Azevedo, *Molecular Thermodynamics of Fluid-Phase Equilibria,* 3rd ed., Prentice Hall, Upper Saddle River, NJ, 1999.

Sandler, S. I., *Chemical and Engineering Thermodynamics,* 4th ed., Wiley, New York, 2006.

Schad, R. C., "Make the Most of Process Simulation," *Chem. Engr. Progress, 94* (1), 21 (Jan. 1998).

Seider, W. D., J. D. Seader, and D. R. Lewin, *Process Design Principles,* 3rd ed., Wiley, New York, 2009.

Shacham, M., N. Brauner, W. R. Ashurst, and M. B. Cutlip, "Can I Trust This Software Package: An Exercise in Validation of Computational Results," *Chem. Engr. Educ., 42* (1), 53 (Winter 2008).

Smith, B. D., *Design of Equilibrium Stage Processes,* McGraw-Hill, New York, 1963.

Smith, J. M., H. C. Van Ness, and M. M. Abbott, *Introduction to Chemical Engineering Thermodynamics,* 7th ed., McGraw-Hill, New York, 2005.

Smith, J. M., and H. C. Van Ness, *Introduction to Chemical Engineering Thermodynamics,* 3rd ed., McGraw Hill, New York, 1975.

Tester, J. W., and M. Modell, *Thermodynamics and Its Applications,* 3rd ed., Prentice Hall, Upper Saddle River, NJ 1997.

Van Ness, H. C., and M. M. Abbott, *Classical Thermodynamics of Non-Electrolyte Solutions. With Applications to Phase Equilibria,* McGraw-Hill, New York, 1982.

Walas, S. M., *Phase Equilibria in Chemical Engineering,* Butterworth, Boston, 1985.

Wankat, P. C., *Equilibrium-Staged Separations,* Prentice Hall, Upper Saddle River, NJ, 1988.

Watkins, R. N., "Sizing Separators and Accumulators," *Hydrocarbon Processing, 46* (1), 253 (Nov. 1967).

Woinsky, S. G., "Help Cut Pollution with Vapor/Liquid and Liquid/Liquid Separators," *Chem. Engr. Progress, 90* (10), 55 (Oct. 1994).

Yaws, C. L., P. K. Narasimhan, and C. Gabbula, *Yaws Handbook of Antoine Coefficients for Vapor Pressure* (electronic edition), Knovel, 2005.

HOMEWORK

A. *Discussion Problems*

A1. In Figure 2-9 the feed plots as a two-phase mixture, whereas it is a liquid before introduction to the flash chamber. Explain why. Why can't the feed location be plotted directly from known values of T_F and z? In other words, why does h_F have to be calculated separately from an equation such as Eq. (2-9b)?

A2. Can weight units be used in the flash calculations instead of molar units?

A3. Explain why a sequential solution procedure cannot be used when T_{feed} is specified for a flash drum.

A4. In the flash distillation of salt water, the salt is totally nonvolatile (this is the equilibrium statement). Show a McCabe-Thiele diagram for a feed water containing 3.5 wt % salt. Be sure to plot weight fraction of more volatile component.

A5. Develop your own key relations chart for this chapter. That is, on *one* page summarize everything you would want to know to solve problems in flash distillation. Include sketches, equations, and key words.

A6. In a flash drum separating a multicomponent mixture, raising the pressure will:

   a. increase the drum diameter and increase the relative volatilities.

   b. increase the drum diameter and cause no change to the relative volatilities.

   c. increase the drum diameter and decrease the relative volatilities.

   d. not change the drum diameter but increase the relative volatilities.

   e. not change the drum diameter and not change the relative volatilities.

   f. not change the drum diameter but decrease the relative volatilities.

   g. decrease the drum diameter and increase the relative volatilities.

   h. decrease the drum diameter and not change to the relative volatilities.

   i. decrease the drum diameter and decrease the relative volatilities.

A7. **a.** What would Figure 2-2 look like if we plotted y_2 vs. x_2 (i.e., plot less volatile component mole fractions)?

 b. What would Figure 2-3 look like if we plotted T vs. x_2 or y_2 (less volatile component)?

 c. What would Figure 2-4 look like if we plotted H or h vs. y_2 or x_2 (less volatile component)?

A8. For a typical straight-chain hydrocarbon, does:

 a. K increase, decrease, or stay the same when temperature is increased?

 b. K increase, decrease, or stay the same when pressure is increased?

 c. K increase, decrease, or stay the same when mole fraction in the liquid phase is increased?

 d. K increase, decrease, or stay the same when the molecular weight of the hydrocarbon is increased within a homologous series?

 Note: It will help to visualize the DePreister chart in answering this question.

A9. In the vapor-liquid equilibrium data for methanol-water, if the methanol vapor mole fraction is 0.60, what is the methanol liquid mole fraction?

A10. Is there an azeotrope in the methanol-water system at a pressure of 1.0 atmospheres?

A11. The equilibrium K value is usually defined as

 a. K = y/x, where y and x are weight fractions of the component in the vapor and liquid phases, respectively.

 b. K = x/y, where x and y are weight fractions of the component in the liquid and vapor phases, respectively.

 c. K = y/x, where y and x are mole fractions of the component in the vapor and liquid phases, respectively.

 d. K = x/y, where x and y are mole fractions of the component in the liquid and vapor phases, respectively.

A12. In a sequential solution procedure for flash distillation,

 a. the mass balances, equilibrium relationships, and energy balances are solved simultaneously.

 b. the mass balances and equilibrium relationships are solved first, and then the energy balance is solved.

 c. the energy balance is solved first, and then the mass balances and equilibrium relationship are solved.

A13. Calculations are simpler for multicomponent flash distillation if the feed flow rate and mole fractions of the feed are specified plus

 a. the drum pressure and feed temperature.

 b. the drum temperature and feed temperature.

 c. the drum temperature and the drum pressure.

 d. the feed temperature and feed pressure.

 e. all of the above; they are all equally difficult.

A14. The Rachford-Rice equation,

 a. has excellent convergence properties for flash distillation.

 b. was derived from the mass balances, equilibrium relationships, and energy balances.

 c. is only useful for binary flash distillation.

 d. all of the above.

 e. none of the above.

A15. Use the DePriester chart:

 a. What is the K value of propane at 240 kPa and 25°C?

 b. What is the normal boiling point of n-pentane?

A16. Flash distillation is usually operated adiabatically. Where does the energy to vaporize part of the feed come from?

B. *Generation of Alternatives*
B1. Think of all the ways a binary flash distillation problem can be specified. For example, we have usually specified F, z, T_{drum}, p_{drum}. What other combinations of variables can be used? (I have over 20.) Then consider how you would solve the resulting problems.
B2. An existing flash drum is available. The vertical drum has a demister and is 4 ft in diameter and 12 ft tall. The feed is 30 mol% methanol and 70 mol% water. A vapor product that is 58 mol% methanol is desired. We have a feed rate of 25,000 lbmol/h. Operation is at 1 atm pressure. Since this feed rate is too high for the existing drum, what can be done to produce a vapor of the desired composition? Design the new equipment for your new scheme. You should devise at least three alternatives. Data are given in Problem 2.D1.
B3. In principle, measuring VLE data is straightforward. In practice, actual measurement may be very difficult. Think of how you might do this. How would you take samples without perturbing the system? How would you analyze for the concentrations? What could go wrong? Look in your thermodynamics textbook for ideas.

C. *Derivations*
C1. Determine the effect of pressure on the temperature, separation and diameter of a flash drum.
C2. Solve the Rachford-Rice equation for V/F for a binary system.
C3. Assume that vapor pressure can be calculated from the Antoine equation and that Raoult's law can be used to calculate K values. For a binary flash system, solve for the drum pressure if drum temperature and V/F are given.
C4. Derive Eq. (2-24) and show that it is correct.
C5. Choosing to use V/F to develop the Rachford-Rice equation is conventional but arbitrary. We could also use L/F, the fraction remaining liquid, as the trial variable. Develop the Rachford-Rice equation as f(L/F).
C6. In flash distillation a liquid mixture is partially vaporized. We could also take a vapor mixture and partially condense it. Draw a schematic diagram of partial condensation equipment. Derive the equations for this process. Are they different from flash distillation? If so, how?
C7. Plot Eq. (2-40) vs. V/F for Example 2-2 to illustrate that convergence is not as linear as the Rachford-Rice equation.
C8. Show how to use a temperature composition diagram to solve a binary flash distillation problem when the drum temperature, feed mole fraction, drum pressure, and feed rate are specified. Show how to determine x, y, L, and V.
C9. For a vapor-liquid-liquid flash distillation, derive Eqs. (2-62) and (2-63) and the equations that allow calculation of all the mole fractions once V/F and L_{liquid_1}/F are known.

D. *Problems*
   Answers to problems with an asterisk are at the back of the book.
D1.* We are separating a mixture of methanol and water in a flash drum at 1 atm pressure. Equilibrium data are listed in Table 2-7.
   a. Feed is 60 mol% methanol, and 40% of the feed is vaporized. What are the vapor and liquid mole fractions and flow rates? Feed rate is 100 kmol/h.

TABLE 2-7. *Vapor-liquid equilibrium data for methanol water (p = 1 atm) (mol%)*

Methanol Liquid	Methanol Vapor	Temp., °C
0	0	100
2.0	13.4	96.4
4.0	23.0	93.5
6.0	30.4	91.2
8.0	36.5	89.3
10.0	41.8	87.7
15.0	51.7	84.4
20.0	57.9	81.7
30.0	66.5	78.0
40.0	72.9	75.3
50.0	77.9	73.1
60.0	82.5	71.2
70.0	87.0	69.3
80.0	91.5	67.6
90.0	95.8	66.0
95.0	97.9	65.0
100.0	100.0	64.5

Source: Perry et al. (1963), p. 13–5.

b. Repeat part a for a feed rate of 1500 kmol/h.

c. If the feed is 30 mol% methanol and we desire a liquid product that is 20 mol% methanol, what V/F must be used? For a feed rate of 1000 lbmol/h, find product flow rates and compositions.

d. We are operating the flash drum so that the liquid mole fraction is 45 mol% methanol. L = 1500 kmol/h, and V/F = 0.2. What must the flow rate and composition of the feed be?

e. Find the dimensions of a vertical flash drum for Problem 2.D1c.
Data: $\rho_w = 1.00$ g/cm³, $\rho_{m,L} = 0.7914$ g/cm³, $MW_w = 18.01$, $MW_m = 32.04$. Assume vapors are ideal gas.

f. If z = 0.4, p = 1 atm, and $T_{drum} = 77$°C, find V/F, x_m, and y_m.

g. If F = 50 mol/h, z = 0.8, p = 1 atm, and $y_m = 0.892$ mole fraction methanol, find V, L, and x_m.

D2.* Two flash distillation chambers are hooked together as shown in the diagram. Both are at 1 atm pressure. The feed to the first drum is a binary mixture of methanol and water that is 55 mol% methanol. Feed flow rate is 10,000 kmol/h. The second flash drum operates with $(V/F)_2 = 0.7$ and the liquid product composition is 25 mol% methanol. Equilibrium data are given in Table 2-7.

a. What is the fraction vaporized in the first flash drum?

b. What are y_1, y_2, x_1, T_1, and T_2?

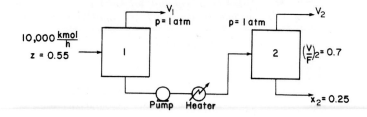

D3. We are separating a mixture of methanol and water in a flash drum at 1.0 atm pressure. Use the equilibrium data listed in Table 2-7. Feed rate is 10.0 kmol/h.

    a. The feed is 40 mol% methanol and 60% of the feed is vaporized. Find mole fractions and flow rates of vapor and liquid products. Estimate the temperatures by linear interpolation for both vapor and liquid products.

    b. The feed is 40 mol% methanol, the product temperatures are both 78.0°C. Find the mole fractions of liquid and vapor products and V/F.

    c. We desire a vapor product that is 80 mol% methanol and want to operate at V/F = 0.3. What must the feed composition be?

D4. We have a mixture that is 20 mol% propane, 35 mol% n-butane, and 45 mol% n-hexane. If a flash drum operates at 400 kPa, what is the highest temperature at which the flash drum can operate and still have vapor and liquid present? Use the DePriester chart for equilibrium.

D5. We have a feed that is a binary mixture of methanol and water (55.0 mol% methanol) that is sent to a system of two flash drums hooked together. The vapor from the first drum is cooled, which partially condenses the vapor, and then is fed to the second flash drum. Both drums operate at a pressure of 1.0 atm and are adiabatic. The feed rate to the first drum is 1000 kmol/h. We desire a liquid product from the first drum that is 30.0 mol% methanol ($x_1 = 0.30$). The second drum operates at a fraction vaporized of $(V/F)_2 = 0.25$. The equilibrium data are in Table 2-7.

    a. Sketch the process labeling the different streams.

    b. Find the following for the first drum: vapor mole fraction y_1, fraction vaporized $(V/F)_1$, and vapor flow rate V_1.

    c. Find the following for the second drum: vapor mole fraction y_2, liquid mole fraction x_2, and vapor flow rate V_2.

D6. A feed that is 45.0 mol% n-butane, 35.0 mol% n-pentane, and 20.0 mol% n-hexane is fed at a rate of 1.0 kmol/min to a flash drum operating at 50°C and 200 kPa. Find V/F, liquid mole fractions and vapor mole fractions. Use the DePriester charts.

D7. A flash drum is separating methane from propane at 0°C and 2500 kPa. The feed is 30.0 mol% methane and 70.0 mol% propane. Find V/F, y and x. Use the results of Problem 2.C2. Use the DePriester chart for K values.

D8.* You want to flash a mixture with a drum pressure of 2 atm and a drum temperature of 25°C. The feed is 2000 kmol/h. The feed is 5 mol% methane, 10 mol% propane, and the rest n-hexane. Find the fraction vaporized, vapor mole fractions, liquid mole fractions, and vapor and liquid flow rates. Use DePriester charts.

D9.* We wish to flash distill an ethanol-water mixture that is 30 wt % ethanol and has a feed flow of 1000 kg/h. Feed is at 200°C. The flash drum operates at a pressure of 1 kg/cm². Find: T_{drum}, wt frac of liquid and vapor products, and liquid and vapor flow rates.

Data: $C_{PL,EtOH} = 37.96$ at 100°C, kcal/(kmol°C)
$C_{PL,W} = 18.0$, kcal/(kmol°C)
$C_{PV,EtOH} = 14.66 + 3.758 \times 10^{-2}T - 2.091 \times 10^{-5}T^2 + 4.74 \times 10^{-9}T^3$
$C_{PV,W} = 7.88 + 0.32 \times 10^{-2}T - 0.04833 \times 10^{-5}T^2$
Both C_{PV} values are in kcal/(kmol°C), with T in °C.
$\rho_{EtOH} = 0.789$ g/mL, $\rho_W = 1.0$ g/mL, $MW_{EtOH} = 46.07$, $MW_W = 18.016$, $\lambda_{EtOH} = 9.22$
kcal/mol at 351.7 K, and $\lambda_W = 9.7171$ kcal/mol at 373.16 K.
Enthalpy composition diagram at p = 1 kg/cm² is in Figure 2-4. Note: Be careful with units.

D10. We have a mixture that is 35 mol% n-butane with unknown amounts of propane and n-hexane. We are able to operate a flash drum at 400 kPa and 70°C with $x_{C6} = 0.7$. Find the mole fraction of n-hexane in the feed, z_{C6}, and the value of V/F.

D11. A vapor stream which is 40.0 mol% ethanol and 60.0 mol% water is partially condensed and sent to a flash drum operating at 1.0 atm.
   a. What is the highest vapor mole fraction which can be produced?
   b. If 60.0 % of the feed is liquefied, what are the outlet mole fractions of liquid and vapor?

D12. Find the dimensions (h_{total} and D) for a horizontal flash drum for Problem 2.D1c. Use $h_{total}/D = 4$.

D13. The phenol-cresol system has an approximately constant relative volatility at $\alpha_{p-c} = 1.76$. Phenol is more volatile. At equilibrium, if cresol mole fraction in the liquid is 0.3, what is the mole fraction of cresol in the vapor?

D14. We are feeding 100 kmol/h of a mixture that is 30 mol% n-butane and 70 mol% n-hexane to a flash drum. We operate with V/F = 0.4 and $T_{drum} = 100$°C. Use Raoult's law to estimate K values from vapor pressures. Use Antoine's equation to calculate vapor pressure,

$$\log_{10}(VP) = A - \frac{B}{(T+C)}$$

where VP is in mm Hg and T is in °C.
n-butane: A = 6.809, B = 935.86, C = 238.73
n-hexane: A = 6.876, B = 1171.17, C = 224.41

Find p_{drum}, x_i and y_i

D15.* We have a flash drum separating 50 kmol/h of a mixture of ethane, isobutane, and n-butane. The ratio of isobutane to n-butane is held constant at 0.8 (that is, $z_{iC4}/z_{nC4} = 0.8$). The mole fractions of all three components in the feed can change. The flash drum operates at a pressure of 100 kPa and a temperature of 20°C. If the drum is operating at V/F = 0.4, what must the mole fractions of all three components in the feed be? Use Figure 2-11 or 2-12 or Eq. (2-30).

D16. A feed that is 50 mol% methane, 10 mol% n-butane, 15 mol% n-pentane, and 25 mol% n-hexane is flash distilled. F = 150 kmol/h. Drum pressure = 250 kPa, drum temperature = 10°C. Use the DePriester charts. Find V/F, x_i, y_i, V, L.

D17. We are separating a mixture of acetone (MVC) from ethanol by flash distillation at p = 1 atm. Equilibrium data are listed in Problem 4.D7. Solve graphically.

    a. 1000 kmol/day of a feed that is 70 mol% acetone is flash distilled. If 40% of the feed is vaporized, find the flow rates and mole fractions of the vapor and liquid products.

    b. Repeat part a for a feed rate of 5000 kmol/day.

    c. If feed is 30 mol% acetone, what is the lowest possible liquid mole fraction and the highest possible vapor mole fraction?

    d. If we want to obtain a liquid product that is 40 mol% acetone while flashing 60% of the feed, what must the mole fraction of the feed be?

D18.* We wish to flash distill a feed that is 10 mol% propane, 30 mol% n-butane, and 60 mol% n-hexane. Feed rate is 10 kmol/h, and drum pressure is 200 kPa. We desire a liquid that is 85 mol% n-hexane. Use DePriester charts. Find T_{drum} and V/F. Continue until your answer is within 0.5°C of the correct answer. Note: This is a single trial and error, *not* a simultaneous mass and energy balance convergence problem.

D19.* A flash drum operating at 300 kPa is separating a mixture that is fed in as 40 mol% isobutane, 25% n-pentane, and 35% n-hexane. We wish a 90% recovery of n-hexane in the liquid (that is, 90% of the n-hexane in the feed exits in the liquid product). F = 1000 kmol/h. Find T_{drum}, x_j, y_j, V/F.

D20. A flash drum operating at a pressure of 2.0 bar and a temperature of 0°C is separating a mixture of ethane (E), n-butane (B), and n-pentane (P). The fraction vaporized is V/F = 0.25. The feed mole fraction of ethane is $z_E = 0.20$. Find the mole fraction of n-butane, z_B, in the feed. Use the DePriester chart.

D21. We wish to flash distill a feed that is 55 mol% ethane and 45 mol% n-pentane. The drum operates $p_{drum} = 700 \text{ kPa}$ and $T_{drum} = 30°C$. Feed flow rate is 100,000 kg/h.

    a. Find V/F, V, L, liquid mole fraction, vapor mole fraction.

    b. Find the dimensions in metric units required for a vertical flash drum. Assume the vapor is an ideal gas to calculate vapor densities. Use DePriester chart for VLE. Be careful of units. Arbitrarily pick $h_{total}/D = 4$.

$$MW_{ethane} = 30.07, MW_{pentane} = 72.15$$

$$\text{Liquid densitites: } \rho_E = 0.54 \text{ g/ml (estimated)}, \rho_P = 0.63 \text{ g/ml}$$

D22. A flash drum is separating a feed that is 50 wt % n-propanol and 50 wt % isopropanol with F=100 kmol/h. $T_{drum} = 90°C$ and $p_{drum} = 101.3 \text{ kPa}$. Use the Rachford-Rice equation to find V/F. Note: This does not have to be trial and error. Then find y and x. Determine K values from Raoult's law using Antoine equation for vapor pressure. Watch your units.

The Antoine constants are (Dean, 1985):

    n-propanol: A = 7.84767, B = 1499.2, C = 204.64

    isopropanol: A = 8.11778, B = 1580.9, C = 219.61

    where the form of the Antoine equation is

$$\log_{10}(Vp) = A - \frac{B}{T + C} \quad \text{VP in mm Hg and T in °C.}$$

D23. We wish to flash distill 1000 kmol/h of a feed that is 40 mol% methane, 5 mol% ethylene, 35% mole ethane, and 20 mol% n-hexane. Drum pressure is 2500 kPa and drum temperature is 0°C. Use the DePriester charts. Find V/F, x_i, y_i, V, F.

D24. We plan to separate a mixture of propane and n-hexane at 300 kPa.

    a. Using the data in the DePriester charts, plot y propane versus x propane for this mixture at this pressure.

    b. If the feed is 30 mol% propane, and 40 mol% of the feed is vaporized, what are the liquid and vapor mole fractions, and what is the drum temperature? Solve graphically.

    c. What is the drum temperature in part b?

    d. If y – 0.8 and the feed is 0.6 (both mole fraction propane), what is the value of V/F?

    e. Use the Rachford-Rice equation to check the answers obtained in parts b and c.

D25. We wish to flash distill a mixture of methane and n-butane in a flash drum operating at 50°C. The feed is 20 mol% methane and 80 mol% n-butane. Feed rate is 100 kmol/h. Feed is at a pressure and temperature such that in the drum V/F = 0.40. Use the DesPriester charts.

    a. Find the drum pressure.

    b. Find the methane mole fraction in the liquid and the vapor.

D26. We are feeding 100 kmol/h of a 45 mol% propane, 55 mol% n-pentane feed to a flash distillation system. We measure the outlet vapor and liquid mole fractions leaving the flash drum, which is an equilibrium stage, and obtain, $y_{propane}$ = 0.8, $x_{propane}$ = 0.2162. Find: (a) L and V. (b) $x_{pentane}$ and $y_{pentane}$. (c) T_{drum} and p_{drum}. Use the DePriester charts. Note: This is not trial and error.

D27. For the Antoine equation in the form

$$\log_{10}(VP) = A - \frac{B}{T+C}$$

with VP in mm Hg and T in °C, the constants for n-pentane are: A = 6.853, B = 1064.8, C = 233.01. n-hexane constants are: A = 6.876, B = 1171.17, C = 224.41 (Dean, 1985)

    a. Predict the vapor pressure at 0°C for pure n-pentane

    b. Predict the boiling point of pure n-pentane at 3 atmosphere pressure.

    c. Predict the boiling pressure if pure n-pentane is boiling at 0°C.

    d. At a pressure of 500 mm Hg and temperature of 30°C, predict the K values for n-pentane and n-hexane using Raoult's law.

    e. For a binary mixture with two phases at equilibrium, F = C – P + 2 = 2 – 2 + 2 = 2 degrees of freedom. If T = 30°C and p = 500 mm Hg, determine the mole fractions in the liquid and vapor phases of an equilibrium mixture of n-pentane and n-hexane.

    f. One mole of a mixture that is 75 mol% n-pentane and 25 mol% n-hexane is placed in a closed chamber. The pressure is adjusted to 500 mm Hg and the temperature to 30°C. The vapor and liquid mole fraction were found in part e. How many moles of liquid and moles of vapor are there at equilibrium?

    g. If 1 mol/min of a mixture that is 75 mol% n-pentane and 25 mol% n-hexane is fed continuously to an equilibrium flash chamber operating at 30°C and 500 mm Hg, what are the flow rates of the liquid and vapor products?

D28. Repeat Example 2-4, but with F = 3000 lbmol/h, and use a horizontal flash drum with holding time = 55 min and surge time = 85 min. Calculate D, h, and h/D.

D29. Design a horizontal flash drum to separate 15,000 kg/h of a feed with the following mass fractions: methane 0.21, propane 0.39, n-butane 0.24, i-butane 0.11, and n-pentane 0.05. The feed is at 0°C and a pressure just high enough that it is all liquid. The drum pressure is 4.0 atm and the drum is adiabatic (heat duty = 0).

  a. Find the minimum feed pressure (to within a whole number of atmospheres) that keeps the feed a liquid. Report this pressure and the VLE correlation used, and use this as the feed pressure for the flash calculation.

  b. Find L, V (in kg/h and in kmol/h), y_i, x_i (in mole fractions), and T_{drum} (in °C).

  c. If a horizontal drum is used find D_{min}. If the holding time is 9 minutes and the ratio of the drum length to diameter is h/D = 6, find the surge time that can be accommodated.

Note: The easiest way to solve this problem is to use Aspen Plus for part a (trial and error) and once you have solved part a, obtain the solution for part b. Although Aspen Plus does not do the drum sizing, it does calculate the physical properties needed for drum sizing. Obtain these and do the drum sizing with a hand calculation.

D30. Data for the equilibrium of water and n-butanol at 1.0 atmosphere is given in Table 8-2. Plot y_w versus x_w. A feed of 100 kmol/h that is 20 mol% water is fed to flash chamber A in the figure that follows. The vapor from flash chamber A is 40 mol% water. This vapor is then partially condensed and fed to flash chamber B. In flash chamber B, 40% of the vapor from chamber A is condensed.

  a. What are the flow rates L_A and V_A and what is the value of x_A?

  b. What are the flow rates L_B and V_B and what are the values of x_B and y_B?

  c. Suppose we changed the operation of chamber B so that enough vapor was condensed to give a liquid mole fraction of x_B = 0.20 mole fraction water. If x_B = 0.20, what are y_B, L_B, and V_B? We could now easily recycle this liquid to the feed (also 20% water) and produce only $V_{B,new}$ and $L_{A,new}$ as products. Calculate these new flow rates when L_B with mole fraction 0.2 is recycled.

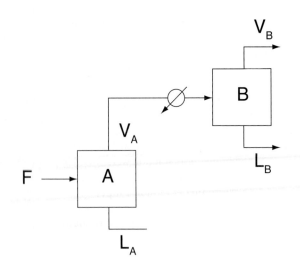

E. *More Complex Problems.*

E1. A vertical flash drum will be used to separate 1000 kmol/h of a feed that is 10 mol% isopropanol and 90 mol% water. The feed is at 9 bar and 75°C. $V/F_{drum} = 0.5$ and $p_{drum} = 1.0$ bar.

   a. Find L, V, x_i, y_i, T_{drum}, the heat duty in kW, and the drum diameter.

   b. Unfortunately, the drum available is 1.0 m in diameter, which is too small. Your boss suggests raising the drum pressure to use the existing drum with V/F = 0.5. The maximum pressure for the drum is 9.0 bar. Will the current drum work at a pressure < 9.0 bar?

   c. If yes, what is the lowest drum pressure that works (that is, for what pressure does $D_{drum} = 1.0$ m)? Find L, V, x_i, y_i, T_{drum}, and the heat duty in kW.

E2. A flash drum is to flash 10,000 lbmol/h of a feed that is 65 mol% n-hexane and 35 mol% n-octane at 1 atm pressure. V/F = 0.4.

   a. Find T_{drum}, liquid mole fraction, vapor mole fraction.

   b. Find the size required for a vertical flash drum.

   Note, y, x, T equilibrium data can be determined from the DePriester charts. Other required physical property data are given in Example 2-4.

F. *Problems Requiring Other Resources*

F1. Benzene-toluene equilibrium is often approximated as $\alpha_{BT} = 2.5$. Generate the y-x diagram for this relative volatility. Compare your results with data in the literature (see references in Table 2-2). Also, generate the equilibrium data using Raoult's law, and compare your results to these.

F2.* Ethylene glycol and water are flash distilled in a cascade of three stills connected as shown in the figure. All stills operate at 228 mm Hg. Feed is 40 mol% water. One-third of the feed is vaporized in the first still, two-thirds of the feed to the second still is vaporized, and one-half the feed to the third still is vaporized. What are the compositions of streams L_3 and V_3? Note that water is the more volatile component.

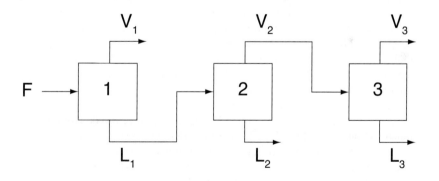

F3.* (Long Problem!) We wish to flash distill a mixture that is 0.517 mole fraction propane, 0.091 mole fraction n-pentane, and 0.392 mole fraction n-octane. The feed rate is 100 kmol/h. The feed to the flash drum is at 95°C. The flash drum will operate at 250 kPa. Find the drum temperature, the value of V/F, and the mole fractions of each component in the liquid and vapor products. Converge $| \Delta T_{drum} |$ to < 0.20°C.

F4. We wish to do a flash distillation of 10 kmol/h of a feed that is 25 mol% water and and 75 mol% n-propanol. The flash chamber is at 1.0 atm and is adiabatic. We vaporize

40% of the feed in the flash chamber. Find the flow rates of the liquid and the vapor and the mole fraction of water in the liquid and vapor products.

G. Computer Problems

G1. We are operating a flash drum at a pressure of 1000 kPa. The flash drum is adiabatic. The feed to the flash drum is 100 kmol/h of a multicomponent mixture. This feed is 5.3 mol% ethylene, 22.6 mol% ethane, 3.7 mol% propylene, 36.2 mol% propane, and 32.2% n-butane. The feed is at 10,000 kPa and 100°C (this is at the point after the pump and heater—immediately before the valve to the drum where the feed is flashed).
Find the values of L, V, x_i, and y_i.
Use a process simulator and report the VLE package you use and briefly (one sentence) explain why you made this choice.

G2. We are operating a flash drum at a pressure of 1000 kPa. The flash drum is adiabatic. The feed to the flash drum is 100 kmol/h of a multicomponent mixture. This feed is 8.3 mol% ethylene, 23.6 mol% ethane, 4.7 mol% propylene, 34.2 mol% propane, and 29.2% n-butane. The feed is at 10,000 kPa and 100°C (this is at the point after the pump and heater—immediately before the valve to the drum where the feed is flashed). Use a process simulator and report the VLE package you use and briefly (one sentence) explain why you made this choice.
Find the values of L, V, x_i, and y_i.

G3. Use a process simulator to solve Problem 2.D3. Compare answers and comment on differences. Report the VLE package you use and briefly (one sentence) explain why you made this choice.

G4. Use a process simulator to solve Problem 2.D16. Do "what if?" simulations to see what happens to V/F and product compositions as temperature and/or pressure vary.

G5. One kmol/s of a feed containing 0.20 mole fraction furfural, 0.75 mole fraction water, and 0.05 mole fraction ethanol at 105°C and 3.0 bar is fed to a three-phase flash drum. The drum is at 1 bar and operates with V/F = 0.4. The key component in the second liquid (bottom of drum) is water. Find the outlet temperature, heat duty in kW, the ratio of the first liquid/total liquid, and the compositions of the three phases.

G6. Use a process simulator to solve the following flash distillation problem. Feed is 2 mol% methane, 30 mol% n-butane, 47 mol% n-pentane, and 21 mol% n-hexane and is a liquid. The flash drum is at 1.0 atm, it is adiabatic (heat duty = 0), and there is no heat exchanger. Initially, for the feed, set vapor fraction = 0 (a saturated liquid) and set the temperature. You want to obtain V/F in the drum of 0.4000 (rounded off to the fourth decimal). Find the feed temperature that gives this value of V/F. (As you do runs, note that feed pressure for a saturated liquid increases as feed temperature is increased. Why?) Once you find the correct feed temperature, remove the feed specification that vapor fraction = 0 and specify a feed pressure that is 1.0 atm above the pressure reported by Aspen Plus when vapor fraction of feed = 0. Liquids that are saturated liquids (this is what Aspen means by vapor fraction = 0) cannot be pumped easily. By raising the pressure, we make pumping easy. Rerun simulation one last time to check that you have met all requirements. Expect to do several Aspen Plus runs to solve this problem. Report the feed temperature and pressure, drum temperature, heat duty in drum, x, and y values.

H. *Computer Spreadsheet Problems*

H1. Use a spreadsheet to solve Problem 2.D.16, but with iso-butane replacing n-butane. See Appendix B for Chapter 2 before starting on this problem.

H2. Show that the spreadsheet in Figures 2.B-3 and 2.B-4. has convergence difficulties if Goal Seek is used to make cell B19 ($\Sigma x_i = 1$) equal 1 by changing cell B9 (V/F).

H3. A feed containing 30 mol% isobutane, 25% n-pentane, and 45% n-hexane is flashed at a drum pressure of 50 psia and drum temperature of 650°R. Find V/F, liquid mole fractions, and vapor mole fractions. Use Eq. (2-30) and parameters in Table 2-3. Use a spreadsheet to solve this problem. (Turn in two copies of the spreadsheet—one with numbers in the cells and one with equations in the cells.)

CHAPTER 2 APPENDIX A.   COMPUTER SIMULATION OF FLASH DISTILLATION

Multicomponent flash distillation is a good place to start learning how to use a process simulator. The problems can easily become so complicated that you don't want to do them by hand, but are not so complicated that the working of the simulator is a mystery. In addition, the simulator is unlikely to have convergence problems. Although the directions in this appendix are specific to Aspen Plus, the procedures and problems are adaptable to any process simulator. The directions were written for Aspen Plus V 7.2, 2010 but will probably apply with little change to newer versions when they are released. Additional details on operation of process simulators are available in the book by Seider et al. (2009) and in the manual and help for your process simulator.

As you use the simulator take notes on what you do and what works. If someone shows you how to do something, insist on doing it yourself—and then make a note of how to do it. Without notes, you may find it difficult to repeat some of the steps even if they were done just 15 minutes earlier. If difficulties persist, see the Appendix A, "Aspen Plus Separations Trouble Shooting Guide," at the end of the book.

Lab 1.

The purposes of this lab are to become familiar with your simulator and to explore flash distillation. This includes drawing and specifying flowsheets and choosing the appropriate physical properties packages.

1. Start-up of Aspen Plus

First, log into your computer. Once you are logged in, use the specific steps for your computer to log into the simulator and ask for a blank simulation. This should give you an Aspen Plus (or other simulator) blank screen.

2. Drawing a Flowsheet in Aspen Plus

Go to the bottom menu and left-click Separators (flash drums). (If you can't see a bottom menu, go to VIEW and click on Model Library.) After clicking on Separators, left-click Flash2 (That is Flash drums with two outlets). Drag your cursor to the center of the blank space and left-click. This gives the basic module for a flash drum. You can deselect the

Flash2 option (to avoid getting extra copies by accident) by clicking on the arrow in the lower-left corner. This is the Cancel Insert Mode button—it is a good idea to click it after completing each step of setting up the flowsheet.

Try left-clicking on the module (in the working space) to select it and right-clicking the mouse to see the menu of possibilities. Rename the block by left-clicking Rename Block, "typing in a name such as "FLASH" and clicking OK.

The basic flash drum needs to have a feed line and two outlets. Left-click on the icon labeled Material Streams in the bottom menu to get possible ports (after you move the cursor into the white drawing screen). Move cursor to one of the arrows until it lights up. Left-click (take your finger off the button) and move cursor away from flash drum. Then left-click again to obtain a labeled material stream. Additional streams can be obtained the same way. Put the feed, vapor product, and liquid product streams on your flowsheet. After clicking the Cancel Insert Mode button, highlight the stream names by clicking the left mouse button, and use the right button to obtain a menu. Rename the streams as desired. An Aspen Plus screen shot is shown in Figure 2-A1.

Note that Aspen Plus will happily create mass streams and modules that you do not want (and can hide one behind another), and thus some deletion may be necessary. If a stream is created in error, highlight it with the left button. Then click the right button to obtain a

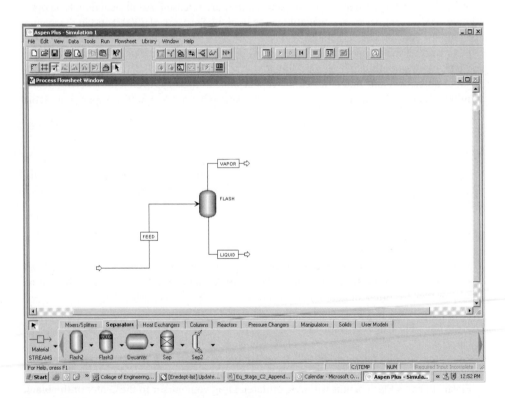

FIGURE 2-A1.   *Screenshot of Aspen Plus for flash distillation*

menu that allows you to delete the stream. Click OK with the left button. Play with the functions until you determine how everything works. If things appear that you don't want, click the Cancel Insert Mode arrow and then delete.

Once you are happy with your flowsheet, click the Next button (blue N with an arrow in the menu bar). This will tell you if the flowsheet connectivity is OK. If it is, click OK. If not OK, either complete connectivity or delete blocks and streams that are not being used.

3. Input Data

You will now obtain a Setup Specifications Data Browser. Set the units you want (e.g., MET). Run should be set at Flowsheet input mode at steady state, and stream class at CONVEN. Use mole units, Ambient pressure = 1 bar, Valid phases = Vapor – liquid, and Free water = No. When happy with this page, click on the Next button. The next page lets you select the components. First, pick a binary mixture of interest (use ethanol and water). Component ID is whatever you want to call it. Type should be conventional. Then click on the component name box or hit ENTER. If you used ethanol and water as the component IDs, Aspen will complete that row. If you used a different ID, such as E or W, give the appropriate component name (ethanol or water). Then do the next component. Aspen Plus will recognize these two components. If Aspen Plus does not fill in the formula, click on the Find button and proceed. When done with components, click on Next. Note that Aspen Plus can be picky about the names or the way you write formulas. Note: If you use W as component ID, Aspen will think this is tungsten. Type water as component name instead.

Click Next. You will probably now see the Properties Specifications-Data Browser screen. You must select the appropriate physical properties package to predict the equilibrium for *your* chemical system. *There is no choice that is always best.* The choice is made through a menu item on the right side labeled "Property Method." You may need to click twice to get the complete menu. This choice is very important (Carlson, 1996; O'Connell et al., 2009; Schad, 1998). *If you pick the wrong model, your results are garbage.* A brief selection guide is given in Table 2-4.

We will try different models and compare them to *data*. [Note that data also needs to be checked for consistency (Barnicki, 2002; O'Connel et al., 2009; Van Ness and Abbott, 1982).] First, try the IDEAL model. Before allowing you to check the VLE data, Aspen (in flowsheet mode) requires you to complete the input information. Thus, we need to continue. Left-click on the *Next* button after you have selected a VLE model (IDEAL). If you get a data bank with binary parameters, left-click Next again. When you get a box that says, "Required Properties Input Complete," click either "go to next required input step" or "modify required property specifications." Assuming you are happy with what you have, left-click on OK to go to next required input step.

You should now get a data browser for your input stream. Fill this out. Try a pressure of 1.0 bar, vapor fraction of 0.4 (click on arrow next to Temperature and select vapor fraction), total flow of 100.0 kmol/h, ethanol mole fraction of 0.9, and water mole fraction of 0.1 (use menu under Composition to select mole fraction). Then left-click on the Next button.

Note, always use mole fraction or mass fractions for the units for the composition of the stream (use the menu). Other choices will often lead to inadvertent errors.

Fill out the conditions for the flash drum block (same pressure and vapor fraction as in the feed) and click on the Next button. At this point you will probably get a screen that says input is complete and asks if you want to run the simulation now. Don't. Click Cancel.

4. Analysis of VLE Data

Aspen will now let us look at the VLE so that we can determine if it makes sense. To do this, go to the menu bar to tools. Choose, in order, Analysis-Property-Binary and left-click. Click OK on box that says you will disable the Interactive Load. On the menu under Analysis Type, pick Txy plot, make sure Valid Phases menu is Vapor-Liquid, change pressure unit to bar, and then click on GO. Cancel or minimize this plot to reveal the data table below it. Compare the vapor and liquid mole fractions to the ethanol water data in Table 2-1. To generate a y-x plot, click on Plot Wizard and follow the instructions. Compare the y-x plot to Figure 2-2. Note that ideal is very far off for systems such as ethanol and water that have azeotropes. *Since Aspen Plus is quite willing to let you be stupid in picking the wrong properties package, it is your responsibility to check that the equilibrium data make sense.* Later, we will find that some VLE packages closely match the actual data.

5. Doing a Flash Run with Ideal Model

Cancel the screens from the analysis. Click the Next button, and when the dialogue box asks if you want to do a run, click OK, and watch Aspen Plus as it calculates (this takes very little time). When it says "Simulation calculations completed" or "Generating results" you can either click on the blue check box on the menu or go to Run on the menu bar and click Check Results or the Next button.

The most important item in the Results Summary is the line that hopefully says, "Calculations were completed normally." If it says anything else, you may have a problem. To scroll through the results, use the << or >> buttons near the top of the screen, bracketing the word "results." If you want to see the input, use the menu to scroll to input. Another useful way to look at results is to go to the tool bar and click on View, then scroll to the bottom and click on Report, then in the window click on the box next to the block you want to see the report for and then click on OK. This report can be printed using the file column in Notepad. By scrolling in the window labeled "Display report for," you can obtain other reports—Streams is very useful.

Note that Aspen Plus gives a huge amount of results. Spend some time exploring these. Write down the values for vapor and liquid mole flow rates and drum temperature. Also look at the phase equilibrium and record the x and y values or print the xy graph. Of course, all these numbers are *wrong*, since we used the wrong VLE model.

6. Rerun with Better VLE Model

Now, cancel screens until you get back to the flowsheet (simulators run faster if there are not a large number of open screens). Go to Data in the menu and click on properties. In the Global screen, change the base method for VLE to NRTL-2. Click on Next. Continue clicking OK and redo the run. Check the results. Write these results down or print the plots. Compare the vapor and liquid products with this equilibrium data to the previous run. Go to Analysis and look at the T-x-y and the x-y plots. Compare to the VLE data in the textbook (the most accurate comparison is with Table 2-1).

7. Try Different Inputs

Once you are happy with the previous runs, change the input conditions (using the same flowsheet) to look at different feed compositions and different fraction vaporized. There are at least two ways to input new data. (1) Left-click on the block for the flash. Go to Data in the menu and left-click on input. Put in the desired numbers using the << or >> buttons (not Next button) to move to different screens. (2) Go to Data in the menu and click on Streams. Click on the << or >> button until you get the Stream (Material) Input-Data Browser. Change the data as desired. Try feeds that are 10, 30, 50, and 70 mol% ethanol and vary vapor fraction.

8. Adiabatic Flash

The most common way to operate a flash system is to feed a hot liquid at elevated pressure through a valve into an insulated (adiabatic) flash chamber that operates at lower pressure. Try a feed of 100 kmol/h that is 30 mol% ethanol and 70 mol%. The flash chamber operates at 1.0 atm and is adiabatic (set Heat duty = 0).

a. The feed is at 110°C and a pressure of 50 psia. Repeat for 100 psia.
b. The feed is at 130°C and a pressure of 100 psia.
c. The feed is at 150°C and a pressure of 100 psia.
d. The feed is at 151.5°C and a pressure of 100 psia.
e. The feed is at 151.5°C and a pressure of 200 psia.

For all of these cases, look at the feed stream and the two product streams (e.g., View→ Report→Streams→All). Is the feed stream entirely liquid? When the feed is not entirely liquid, what happens to the liquid and vapor product flow rates? What is the effect of the feed pressure? Why are the two runs for part a essentially identical, but runs d and e give very different results?

9. Switch to a Ternary Problem

Remove any leftover dialog boxes or screens. Go to Data in the menu and click on components. Use Edit to delete the two rows. You may need to click on the row to obtain the Delete Row command in edit. Then add propane, n-butane, and n-pentane as the three components. Using the menu, go to Data Properties and change the choice of VLE model. Peng-Robinson is a good choice for hydrocarbons. Use the Next button and input the mole fractions (propane 0.2, n-butane 0.3 and n-pentane 0.5). Keep fraction vaporized at 0.4 and pressure of 1.0 bar for now. Click on the Next button and do the run when ready. Record your results (component flow rates, T, y and x) or print out the report. If you have time, try different fraction vaporized, different feed compositions, and different temperatures.

10. What Does It All Mean?

Reflect on the meaning of your results for both the binary and the ternary flash systems.

a. Binary: How are the compositions of the vapor and liquid streams from the flash system related? What is the role of the fraction vaporized? How can you do the calculation by hand?

b. Ternary: How are the compositions of the vapor and liquid streams from the flash system related? What is the role of the fraction vaporized? How can you do the calculation by hand?

Note that the calculation methods used for hand calculations will be different for the binary and ternary systems since the equilibrium data are available in different forms (graphically for the binary and DePriester chart for the ternary).

11. **Finish.** Exit Aspen Plus and log out. There is *no* formal lab report for this lab.

Lab 2. Find Equilibrium Data before Lab

For part 1a, use Flash2 with the Peng-Robinson VLE correlation. For part 1b, use Flash3 and find a VLE correlation that fits the data. For part 1c, use NRTL and Flash3. Use F = 1.0 kmol/h for all cases. Compare the predictions of Analysis with equilibrium data from the literature.

Part 1. Simulations

a. The feed is 45 mol% n-butane and 55 mol% n-hexane at 1.0 atm. The feed is a saturated liquid (vapor fraction = 0.0). The drum pressure is 0.8 atm. Do for V/F in the drum = 0.2, 0.4, 0.6, and 0.8 (four runs). Report feed temperature, Q, drum T, y, x, V, L.

b. The feed is 55 mol% benzene and 45 mol% water at 5.0 atm and is a saturated liquid. Drum temperature is 120°C. Do for V/F in the drum = 0.2, 0.4, 0.6, and 0.8 (four runs). Note: Delete the block for Flash2, redraw with Flash3, and then reconnect streams and add another liquid product. Be sure to list valid phases as Vapor-liquid-liquid in both SETUP and ANALYSIS. Use the y-x diagram to explain your results. Report VLE correlation used, feed temperature, drum pressure, Q, and compositions of all three phases for each V/F.

c. Feed is initially 60 mol% furfural (a cyclic alcohol) and 40 mol% water at 1.0 atm and 50°C. The drum operates at 105°C and V/F = 0.4. Use NRTL, Flash3, and list valid phases as Vapor-liquid-liquid. With 40 mol% water, the two liquid phases should be identical, which means there is only one liquid phase. Try 60, 70, 90, and 99 mol% water in the feed. Generate a y-x diagram with ANALYSIS, and use it to explain your results. Report the drum pressure, Q, and the compositions of all three phases for each feed composition.

Part 2. Design.

Before starting, delete Flash 3 and the second liquid product, add Flash2, and connect streams. List valid phases as Vapor-liquid in SETUP. Feed is 100 kmol/h that is 3 mol% methane, 25 mol% n-butane, 44 mol% n-pentane, and 28 mol% n-hexane and is a liquid. The flash drum is at 1.0 atm, it is adiabatic (heat duty = 0), and there is no heat exchanger. Choose an appropriate property method. Initially, for the feed, set vapor fraction = 0 (a saturated liquid) and set the temperature. You want to obtain V/F in the drum of 0.4000 (rounded off to fourth decimal). Find the feed temperature that gives this value of V/F. (As you do runs, note that feed pressure for a saturated liquid increases as feed temperature is increased. Why?) Once you find the correct feed temperature, remove the feed specification that vapor fraction = 0 and specify a feed pressure that is 1.0 atm. above the pressure reported by Aspen Plus when vapor fraction of feed = 0. Liquids that are saturated liquids (this is what Aspen means by vapor fraction = 0) cannot be pumped easily. By raising the pressure, we make pumping easy. Rerun simulation one last time to check that you have met all requirements. Expect to do several Aspen Plus runs to solve this problem. Report the property method, feed temperature and pressure, drum temperature, heat duty in drum, vapor and liquid flow rates, and x and y values.

ASSIGNMENT TO HAND IN

Each group should write a one- to two-page memo addressed to the professor or teaching assistant from the entire group of members. You may attach a few appropriate graphs and tables on a third page (do *not* attach the entire Aspen results printout). Anything beyond three pages will *not* be looked at or graded (this includes cover pages). The memo needs to have words in addition to numbers. Give a short introduction. Present the numbers in a way that is clearly identified. Mention the graphs or figures you have attached as backup information. (If a group member is absent for the lab and does not help in preparation of the memo, leave his/her name off the memo. Attach a very short note explaining why this member's name is not included. For example, "Sue Smith did not attend lab and never responded to our attempts to include her in writing the memo.") Prepare this memo on a word processor. Graphs should be done on the computer (e.g., Aspen or Excel). Proofread everything *and* do a spell check also!

CHAPTER 2 APPENDIX B.   SPREADSHEETS FOR FLASH DISTILLATION

2.B.1. Regression of Binary VLE with Excel

McCabe-Thiele calculations are easiest to do on spreadsheets if the y versus x VLE data are expressed in an equation. The form $y = f(x)$ is most convenient for flash distillation and for distillation columns (see Chapter 4) if stepping off stages from the bottom of the column up. The form $x = g(y)$ is most convenient for distillation columns if stepping off stages from the top down. Built-in functions in Excel will determine polynomials that fit the data, although the fit will usually not be perfect. This will be illustrated for fitting the ethanol-water equilibrium data in Table 2-1 in the form $y_E = f(x_E)$. (Note: An additional data point $x_E = 0.5079$, $y_E = 0.6564$ was added to the numbers in the table.) Enter the data in the spreadsheet with x_E

values in one column and the corresponding y_E values in the adjacent column. Highlight all the the x-y data. In Excel 2007 in the menu bar, click on the Insert tab. Select a Scatter chart and then select the icon showing "scatter with data points connected by smooth lines." This creates a figure of the data plotted as y versus x. If desired, you can add labels, grid lines, and other touches by clicking on Layout in the menu and then following the instructions. If you are not familiar with these techniques, try it step-by-step in Excel.

To find a polynomial that fits the data, first highlight the x-y data. Then go to the Excel tool bar and click on Layout→Analysis→Trendline→More Trendline Options. Choose polynomial as type and in the menu select the desired order of the polynomial. (You can try different orders to find which has the best fit.) Make sure the boxes "Display Equation on Chart" and "Set intercept = 0" are checked. Then click Close.

For the ethanol-water data at 1.0 atm, the best fit was a 6th-order polynomial. This result is:

$$yeq = -48.087x^6 + 161.76x^5 - 212.75x^4 + 138.81x^3 - 46.669x^2 + 7.9335x \qquad \textbf{(2.B-1)}$$

By repeating the steps with the x and y columns reversed, the function xeq = g(y) can easily be generated.

$$xeq = 71.76y^6 - 211.92y^5 + 227.47y^4 - 107.39y^3 + 22.653y^2 - 1.5768y \qquad \textbf{(2.B-2)}$$

An alternative to fitting the data is to input the table of data into Excel and then use the "Lookup" function built into Excel to linearly interpolate between data points.

2.B.2. Binary Flash Distillation with Excel

Once an equation form of the equilibrium data is available, it is relatively easy to develop a spreadsheet to solve binary flash problems. We need to input the known values, which we will assume are the mole fraction of the more volatile component in the feed, z, and the fraction vaporized, V/F. Then input the constants for the VLE data in the form $y_{MVC} = f(x_{MVC})$. For this example we will separate ethanol and water at 1.0 atm with an ethanol feed that is 30 mol% ethanol and V/F = 0.4. The VLE equation for ethanol-water at 1.0 atm is given by Eq. (2.B-1). Next, input a guessed value for $x_{ethanol}$. Now calculate yeq using this value of xguess from Eq. (2.B-1) and calculate yop from Eq. (2-13). Since yop = yeq at the intersection of the equilibrium and operating curves, eq0 = yeq – yop = 0. Thus, calculate eq0 and use Goal Seek to make it equal zero by changing the value of xguess. [Note: Goal Seek is hidden in Excel 2007. In the spreadsheet go to Data Tab→What if Analysis→Goal Seek.] The resulting spreadsheet with cell formulas is listed in Figure 2-B1.

The numerical results are presented in Figure 2-B2. Goal Seek was used to set cell B12 to zero by changing cell B8.

Since eq0 ~ 0, the answer to this flash problem is x = 0.1555 and y = 0.5167.

	A	B	C	D	E	F	G
1	Table 2.B-1	Flash Distillation Binary	z	0.3			
2			f = V/F	0.4			
3		Ethanol-water VLE					
4	6th	5th	4th	3rd	2nd	1st	0th
5	-48.087	161.76	-212.75	138.81	-46.669	7.9335	0
6							
7	Use	Goal Seek to make cell	B12 = 0.0	change	cell B8		
8	xguess	0.155497293317958					
9	yeq	=A5*B8^6+B5*B8^5+C5*B8^4+D5*B8^3+E5*B8^2+F5*B8					
10	yop	=-((1-D2)/D2)*B8+D1/D2					
11	eq0	=B9-B10					
12	check	=1000*B11					
13							
14							

Sheet1 **Sheet2** Sheet3

FIGURE 2-B1. *Spreadsheet with Equations for Binary Flash*

	A	B	C	D	E	F	G	H	I	J	K
1	Table 2.B	Flash Distillation Binary	z	0.3							
2			f = V/F	0.4							
3		Ethanol-water VLE									
4	6th	5th	4th	3rd	2nd	1st	0th				
5	-48.087	161.76	-212.75	138.81	-46.669	7.9335	0				
6											
7	Use	Goal Seek to make cell	B12 = 0.0	change	cell B8						
8	xguess	0.155497293									
9	yeq	0.516754276									
10	yop	0.51675406									
11	eq0	2.15601E-07									
12	check	0.000215601									
13											
14											

Sheet1 **Sheet2** Sheet3

FIGURE 2-B2. *Spreadsheet with Numbers for Binary Flash*

2.B.3. Multicomponent Flash Distillation with Excel

If VLE data are available in equation form, spreadsheet calculations can also be used for multicomponent flash distillation. These calculations are illustrated for a chemical mixture that follows Eq. (2-16) for Problem 2.D16. First, the spreadsheet is shown in Figure 2-B3 with the equations in each cell. Cells B3 to B6, C3 to C6, D3 to D6, E3 to E6, F3 to F6, and G3 to G6 are the constants for Eq. (2-16) from Table 2-3. Conditions for the operation are input in cells B7, D7, and F7, and the feed mole fractions are in cells B8, C8, F8, and G9. Eq. (2-16) is programmed for each component in cells B10, B11, B12, and B13. Then the liquid mole fractions are determined from Eq. (2-38) in cells B15 to B18. These four numbers are summed in cell B19. The Rachford-Rice terms from Eq. (2-42) for each component are calculated in cells B20 to B23, and the sum is in B24.

	A	B	C	D	E	F	G	H
1	Table 2.B-3.	MC flash, App 2B.			Eq 2-16			
2	K const.	aT1	aT2	aT6	ap1	ap2	ap3	
3	M	-292860	0	8.2445	-0.8951	59.8465	0	
4	nB	-1280557	0	7.94986	-0.96455	0	0	
5	nPentane	-1524891	0	7.33129	-0.89143	0	0	
6	nHex	-1778901	0	6.96783	-0.84634	0	0	
7	T deg R	521	p psia	36.258	F	150		
8	zM	0.5	z nB	0.1	z np	0.15	znhex	
9	V/F guess	0.616449235845455					0.25	
10	KM	=EXP(B3/(B7*B7)+D3+E3*LN(D7)+F3/(D7*D7))						
11	KnB	=EXP(B4/(B7*B7)+D4+E4*LN(D7)+F4/(D7*D7))						
12	KnPen	=EXP(B5/(B7*B7)+D5+E5*LN(D7)+F5/(D7*D7))						
13	KnHex	=EXP(B6/(B7*B7)+D6+E6*LN(D7)+F6/(D7*D7))						
14		Goal seek cell B25 = 0.0 change B9						
15	xM	=B8/(1+(B10-1)*B9)						
16	xnb	=D8/(1+(B11-1)*B9)						
17	xnPen	=F8/(1+(B12-1)*B9)						
18	xnHex	=G9/(1+(B13-1)*B9)						
19	Sum	=SUM(B15:B18)						
20	RR M	=(B8*(B10-1))/(1+(B10-1)*B9)						
21	RR nB	=(D8*(B11-1))/(1+(B11-1)*B9)						
22	RRnP	=(F8*(B12-1))/(1+(B12-1)*B9)						
23	RRnHex	=(G9*(B13-1))/(1+(B13-1)*B9)						
24	sum RR	=SUM(B20:B23)						
25	1000 sumRR	=1000*B24						
26								
27								
28								
29								
30								

Sheet1 / Sheet2 / Sheet3

FIGURE 2-B3.　*Spreadsheet for Multicomponent Flash for Problem 2.D16 with Equations in Each Cell.*

Goal Seek was used to find the value of V/F that makes cell B25 equal to 0.0 by changing the value in B9. Cell B25 multiplies cell B24 by 1000 to make the result obtained from Goal Seek more accurate. Goal Seek converged for any guess of V/F from 0 to 1.0. The results are given in Figure 2-B4 for the conditions given in Problem 2.D16: T = 10°C and p = 250 kPa. Since the constants in Eq. (2-3) are for temperature units in °R and pressure units of psia, the temperature is input in °R (cell B7) and pressure is input in psia (cell D7).

We could also try not writing the Rachford-Rice terms and use Goal Seek to set the sum of x_i in cell B19 = 1.0. In this problem, Goal Seek works for $\Sigma x_i = 1.0$ with $V/F_{guess} > \sim.5$ but does not work for low values of V/F_{guess} (See Problem 2.H2.). This difficulty reinforces the need to check results from any software package, even one as common and robust as a spreadsheet (Shacham et al., 2008).

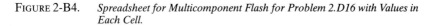

	A	B	C	D	E	F	G
1	Table 2.B-4.	MC flash, App 2B.			Eq 2-16		
2	K const.	aT1	aT2	aT6	ap1	ap2	ap3
3	M	-292860	0	8.2445	-0.8951	59.8465	0
4	nB	-1280557	0	7.94986	-0.96455	0	0
5	nPentane	-1524891	0	7.33129	-0.89143	0	0
6	nHex	-1778901	0	6.96783	-0.84634	0	0
7	T deg R	521	p psia	36.258	F	150	
8	zM	0.5	z nB	0.1	z np	0.15	znhex
9	V/F guess	0.616449236					0.25
10	KM	54.44084615					
11	KnB	0.793630962					
12	KnPen	0.225983266					
13	KnHex	0.072466207					
14		Goal seek cell B25 = 0.0 change B9					
15	xM	0.014730331					
16	xnb	0.114575891					
17	xnPen	0.286884789					
18	xnHex	0.583808648					
19	Sum	0.999999658					
20	RR M	0.787201347					
21	RR nB	-0.023644916					
22	RRnP	-0.222053627					
23	RRnHex	-0.541502249					
24	sum RR	5.54968E-07					
25	1000 sumRR	0.000554968					
26							
27							
28							
29							
30							

Sheet1 Sheet2 Sheet3

FIGURE 2-B4. *Spreadsheet for Multicomponent Flash for Problem 2.D16 with Values in Each Cell.*

Note that there are many other possible approaches to solve this problem with a spreadsheet, and other software tools such as MATLAB or Mathematica could be used (Binous, 2008)

CHAPTER 3

Introduction to
Column Distillation

Distillation is by far the most common separation technique in the chemical process industry, accounting for 90% to 95% of the separations. The approximately 40,000 distillation columns in use account for approximately 40% of the energy use by the United States' chemical process industry—equivalent to a staggering 1.2 million barrels of crude oil a day (Humphrey and Keller, 1997).

This chapter introduces how continuous distillation columns work and serves as the lead to a series of nine chapters on distillation. The basic calculation procedures for binary distillation are developed in Chapter 4. Multicomponent distillation is introduced in Chapter 5, detailed computer calculation procedures for these systems are developed in Chapter 6, and simplified shortcut methods are covered in Chapter 7. More complex distillation operations such as extractive and azeotropic distillation are the subject of Chapter 8. Chapter 9 switches to batch distillation, which is commonly used for smaller systems. Detailed design procedures for both staged and packed columns are discussed in Chapter 10. Finally, Chapter 11 looks at the economics of distillation and methods to save energy (and money) in distillation systems.

3.1 DEVELOPING A DISTILLATION CASCADE

In Chapter 2, we learned how to do the calculations for flash distillation. Flash distillation is a very simple unit operation, but in most cases it produces a limited amount of separation. In Problems 2.D2, 2.D5, and 2.F2 we saw that more separation could be obtained by adding on (or *cascading*) more flash separators. The cascading procedure can be extended into a process that produces one pure vapor and one pure liquid product. First, we could send the vapor streams to additional flash chambers at increasing pressures and the liquid streams to flash chambers with decreasing pressures, as shown in Figure 3-1. Stream V_1 will have a high concentration of the more volatile component, and stream L_5 will have a low concentration of the more volatile component. Each flash chamber in Figure 3-1 can be analyzed by the methods developed previously.

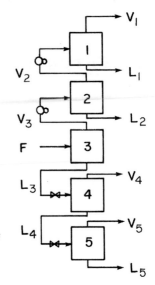

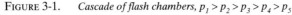

FIGURE 3-1.   *Cascade of flash chambers, $p_1 > p_2 > p_3 > p_4 > p_5$*

One difficulty with the cascade shown in Figure 3-1 is that the intermediate product streams, L_1, L_2, V_4, and V_5, are of intermediate concentration and need further separation. Of course, each of these streams could be fed to another flash cascade, but then the intermediate products from those cascades would have to be sent to additional cascades, and so forth. A much cleverer solution is to use the intermediate product streams as additional feeds within the same cascade.

Consider stream L_2, which was generated by flashing part of the feed stream and then condensing part of the resulting vapor. Since the material in L_2 has been vaporized once and condensed once, it probably has a concentration close to that of the original feed stream. (To check this, you can do the appropriate flash calculation on a McCabe-Thiele diagram.) Thus, it is appropriate to use L_2 as an additional feed stream to stage 3. However, since $p_2 > p_3$, its pressure must first be decreased.

Stream L_1 is the liquid obtained by partially condensing V_2, the vapor obtained from flashing vapor stream V_3. After one vaporization and then a condensation, stream L_1 will have a concentration close to that of stream V_3. Thus it is appropriate to use stream L_1 as an additional feed to stage 2 after pressure reduction.

A similar argument can be applied to the intermediate vapor products below the feed, V_4 and V_5. V_4 was obtained by partially condensing the feed stream and then partially vaporizing the resulting liquid. Since its concentration is approximately the same as the feed, stream V_4 can be used as an additional feed to stage 3 after compression to a higher pressure. By the same reasoning, stream V_5 can be fed to stage 4.

Figure 3-2 shows the resulting *countercurrent cascade*, so called because vapor and liquid streams go in opposite directions. The advantages of this cascade over the one shown in Figure 3-1 are that there are no intermediate products, and the two end products can both be pure *and* obtained in high yield. Thus V_1 can be almost 100% of the more volatile components and contain almost all of the more volatile component of the feed stream.

Although a significant advance, this variable pressure (or isothermal distillation) system is seldom used commercially. Operation at different pressures requires a larger number of

Chapter 3   Introduction to Column Distillation

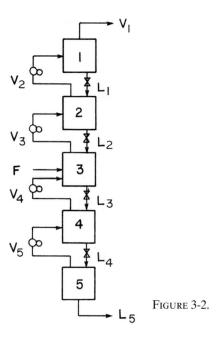

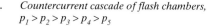

FIGURE 3-2.   *Countercurrent cascade of flash chambers,*
$p_1 > p_2 > p_3 > p_4 > p_5$

compressors, which are expensive. It is much cheaper to operate at constant pressure and force the temperature to vary. Thus, in stage 1 of Figure 3-2 a relatively low temperature would be employed, since the concentration of the more volatile component, which boils at a lower temperature, is high. For stage 5, where the less volatile component is concentrated, the temperature would be high. To achieve this temperature variation, we can use heat exchangers (reboilers) to partially vaporize the liquid streams. This is illustrated in Figure 3-3, where partial condensers and partial reboilers are used.

The cascade shown in Figure 3-3 has a decreasing vapor flow rate as we go from the feed stage to the top stage and a decreasing liquid flow rate as we go from the feed stage to the bottom stage. Operation and design will be easier if part of the top vapor stream V_1 is condensed and returned to stage 1 and if part of the bottom liquid stream L_5 is vaporized and returned to stage 5, as illustrated in Figure 3-4. This allows us to control the internal liquid and vapor flow rates at any desired level. Stream D is the distillate product, while B is the bottom product. Stream L_0 is called the *reflux* while V_6 is the *boilup*.

The use of reflux and boilup allows for a further simplification. We can now apply all of the heat required for the distillation to the bottom reboiler, and we can do all of the required cooling in the top condenser. The required partial condensation of intermediate vapor streams and partial vaporization of liquid streams can be done with the same heat exchangers as shown in Figure 3-5. Here stream V_2 is partially condensed by stream L_1 while L_1 is simultaneously partially vaporized. Since L_1 has a higher concentration of more volatile component, it will boil at a lower temperature and heat transfer is in the appropriate direction. Since the heat of vaporization per mole is usually approximately constant, condensation of 1 mole of vapor will vaporize approximately 1 mole of liquid. Thus liquid and vapor flow rates tend to remain constant. Heat exchangers can be used for all other pairs of *passing streams:* L_2 and V_3, L_3 and V_4, and L_4 and V_5.

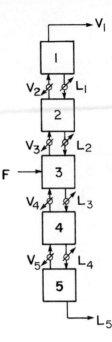

FIGURE 3-3.   *Countercurrent cascade of flash chambers with intermediate reboilers and condensers, p = constant; $T_1 < T_2 < T_3 < T_4 < T_5$*

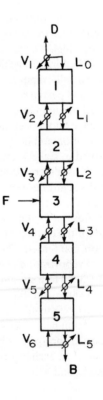

FIGURE 3-4.   *Countercurrent cascade of flash chambers with reflux and boilup, p = constant; $T_1 < T_2 < T_3 < T_4 < T_5$*

Chapter 3   Introduction to Column Distillation

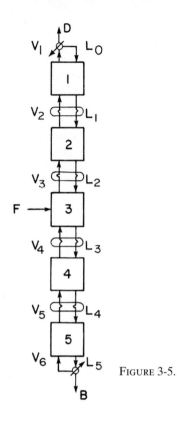

FIGURE 3-5. *Countercurrent cascade with intermediate heat exchangers*

Note that reflux and boilup are *not* the same as recycle. Recycle returns a stream to the feed to the process. Reflux (or boilup) first changes the phase of a stream and then returns the stream to the *same* stage the vapor (or liquid) was withdrawn from. This return at the same location helps increase the concentration of the component that is concentrated at that stage.

The cascade shown in Figure 3-5 can be further simplified by building the entire system in a column instead of as a series of individual stages. The intermediate heat exchange can be done very efficiently with the liquid and vapor in direct contact on each stage. The result is a much simpler and cheaper device. A schematic of such a distillation column is shown in Figure 3-6.

The cascade shown in Figure 3-6 is the usual form in which distillation is done. Because of the repeated vaporizations and condensations as we go up the column, the top product (distillate) can be highly concentrated in the more volatile component. The section of the column above the feed stage is known as the *enriching* or *rectifying* section. The bottom product (bottoms) is highly concentrated in the less volatile component, since the more volatile component has been stripped out by the rising vapors. This section is called the *stripping* section.

The distillation separation works because every time we vaporize material the more volatile component tends to concentrate in the vapor, and the less volatile component in the liquid. As the relative volatility α, Eq. (2-21), of the system decreases, distillation becomes more difficult. If $\alpha = 1.0$, the liquid and vapor will have the same composition, and no separation will occur. Liquid and vapor also have the same composition when an azeotrope occurs. In this case one can approach the azeotrope concentration at the top or bottom of the column but cannot get past it except with a heterogeneous azeotrope (see Chapter 8). The third limit to distillation is the presence of either chemical reactions between components or decomposition reactions. This problem can often be controlled by operating at lower temperatures and using vacuum or steam distillation (see Chapter 8).

While we are still thinking of flash distillation chambers, a simple but useful result can be developed. In a flash chamber a component will tend to exit in the vapor if $y_i V > x_i L$. Rearranging this, if $K_i V/L > 1$ a component tends to exit in the vapor. In a distillation column this means that components with $K_i V/L > 1$ tend to exit in the distillate, and components with $K_i V/L < 1$ tend to exit in the bottoms. This is only a qualitative guide, since the separation on each stage is far from perfect, and K_i, V, and L all vary in the column; however, it is useful to remember.

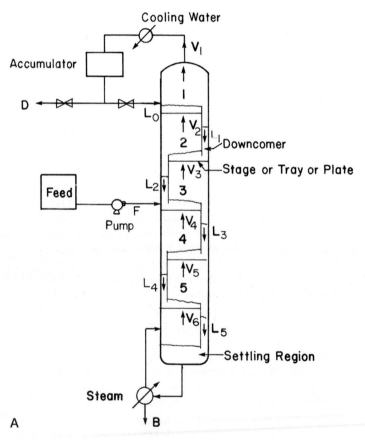

FIGURE 3-6. *Distillation column: (A) schematic of five-stage column*
($T_1 < T_2 < T_3 < T_4 < T_5 < T_6$), (B) photograph of distillation
columns courtesy of APV Equipment, Inc. Tonawanda, NY.

Chapter 3   Introduction to Column Distillation

DISTILLATION SYSTEM
Three column distillation system with a solvent recovery unit at the extreme right.

B

FIGURE 3-6.   *(Continued)*

3.2. DISTILLATION EQUIPMENT

It will be helpful for you to have a basic understanding of distillation equipment before studying the design methods. A detailed description of equipment is included in Chapter 10. Figure 3-6A is a schematic of a distillation column, and Figure 3-6B is a photograph of several columns.

The column is usually metal and has a circular cross section. It contains trays (plates or stages) where liquid-vapor contact occurs. The simplest type of tray is a sieve tray, which is a sheet of metal with holes punched into it for vapor to pass through. This is illustrated in Figure 3-7. The liquid flows down from the tray above in a downcomer and then across the sieve tray where it is intimately mixed with the vapor. The vapor flowing up through the holes prevents the liquid from dripping downward, and the metal weir acts as a dam to keep a sufficient level of liquid on the plate. The liquid that flows over the weir is a frothy mixture containing a lot of vapor. This vapor disengages in the downcomer so that clear liquid flows into the stage below. The space above the tray allows for disengagement of liquid from vapor and needs to be high enough to prevent excessive *entrainment* (carryover of liquid from one stage to the next). Distances between trays vary from 2 to 48 inches and tend to be greater the larger the diameter of the column.

To say that there is liquid on the tray is an oversimplification. In practice, any one of four distinct flow regimes can be observed on trays, depending on the gas flow rate. In the *bubble* regime the liquid is close to being a stagnant pool with distinct bubbles rising through it. This regime occurs at low gas flow rates. The poor mixing causes poor liquid and vapor contact, which results in low stage efficiency. Because of the low gas flow rate and low efficiency, the bubble regime is undesirable in commercial applications.

At higher gas flow rates the stage will often be in a *foam* regime. In this regime, the liquid phase is continuous and has fairly distinct bubbles rapidly rising through it. There is a distinct foam similar to the head resting atop the liquid in a mug of beer. Because of the large surface area in a foam, the area for vapor-liquid mass transfer is large, and stage efficiency may be quite high. However, if the foam is too stable it can fill the entire region between stages. When this occurs, entrainment becomes excessive, stage efficiency drops, and the column may *flood* (fill up with liquid and become inoperative). This may require the use of a chemical antifoam agent. The foam regime is usually at vapor flow rates that are too low for most industrial applications.

At even higher vapor flow rates the *froth* regime occurs. In this regime the liquid is continuous and has large, pulsating voids of vapor rapidly passing through it. The surface of the liquid is boiling violently, and there is considerable splashing. The liquid phase is thoroughly mixed, but the vapor phase is not. In most distillation systems where the liquid-phase mass transfer controls, this regime has good efficiency. Because of the good efficiency and reasonable vapor capacity, this is usually the flow regime used in commercial operation.

At even higher gas flow rates the vapor-liquid contact on the stage changes markedly. In the *spray* regime the vapor is continuous and the liquid occurs as a discontinuous spray of droplets. The vapor is very well mixed, but the liquid droplets usually are not. Because of this poor liquid mixing, the mass transfer rate is usually low and stage efficiencies are low. The significance of this is that relatively small increases in vapor velocity can cause the column to go from the froth to the spray regime and cause a significant decrease in stage efficiency (for example, from 65% to 40%).

A variety of other configurations and modifications of the basic design shown in Figures 3-6 and 3-7 are possible. Valve trays (see Chapter 10) are popular. Downcomers can be chords

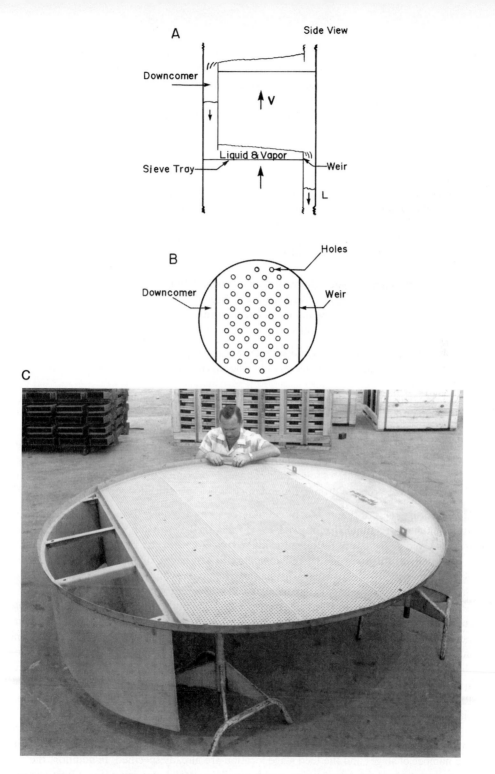

FIGURE 3-7. *Sieve trays: (A) schematic side view, (B) schematic top view, (C) photograph courtesy of Glitsch, Inc.*

3.2. Distillation Equipment

of a circle as shown or circular pipes. Both partial and total condensers and a variety of reboilers are used. The column may have multiple feeds, sidestream withdrawals, intermediate reboilers or condensers, and so forth. The column also usually has a host of temperature, pressure, flow rate, and level measurement and control devices. Despite this variety, the operating principles are the same as for the simple distillation column shown in Figure 3-6.

3.3   SPECIFICATIONS

In the design or operation of a distillation column, a number of variables must be specified. For both design and simulation problems we usually specify column pressure (which sets the equilibrium data); feed composition, flow rate and feed temperature or feed enthalpy or feed quality; and temperature or enthalpy of the reflux liquid. The usual reflux condition set is a saturated liquid reflux. These variables are listed in Table 3-1. The other variables set depend upon the type of problem.

Selecting an appropriate column pressure for distillation is an important decision that is usually done early in the design. As discussed in Section 2.1, in order to have a liquid phase, the condenser pressure must be below the critical pressure of the distillate mixture. In addition, if possible we would like to meet the following heuristics (Biegler et al., 1997):

1. Because vacuum columns are more expensive, column pressure should be greater than or equal to 1 bar. There is little increase in column costs for pressures between 1 and 7 bar (Keller, 1987).
2. The condenser pressure should be set so that cooling water can be used in the condenser. Assuming that ΔT in the heat exchanger is ~5°C, the minimum temperature of the condenser will typically range from ~30° to 50°C depending on the location.
3. The reboiler pressure should be set so that available steam or other hot utility can be used for heating. Thus, the boiling temperature of the bottoms should be ~5°C below the steam temperature. Note that reboiler and condenser pressures are not independent but are related by the pressure drop in the column (~0.1 psi per stage for column with p > 1 bar).

Heuristics are rules of thumb that often disagree with each other. If they disagree, the most important heuristic should be followed. For distillation the most important pressure heuristic is often avoiding the use of refrigeration in the condenser because refrigeration is expensive.

In *design* problems, the desired separation is set, and a column is designed that will achieve this separation. For binary distillation we would usually specify the mole fraction of the more volatile component in the distillate and bottoms products. In addition, the external reflux ratio, L_0/D in Figure 4-6, is usually specified. Finally, we usually specify that the *optimum feed location* be used; that is, the feed location that will result in the fewest total number of stages. The designer's job is to calculate distillate and bottoms flow rates, the heating and cooling requirements in the reboiler and condenser, the number of stages required and the optimum feed stage location, and finally the required column diameter. Alternative specifications such as the splits (fraction of a component recovered in the distillate or bottoms) or distillate or bottoms flow rates are common. Four sets of possibilities are summarized in Table 3-2.

In *simulation* problems, the column has already been built and we wish to predict how much separation can be achieved for a given feed. Since the column has already been built, the number of stages and the feed stage location are already specified. In addition, the column diameter and the reboiler size, which usually control a maximum vapor flow rate, are

TABLE 3-1.   *Usual specified variables for binary distillation*

1. Column pressure
2. Feed flow rate
3. Feed composition
4. Feed temperature or enthalpy or quality
5. Reflux temperature or enthalpy (usually saturated liquid)

TABLE 3-2.   *Specifications and calculated variables for binary distillation for design problems*

Specified Variables

A. 1. Mole fraction more volatile component in distillate, x_D
    2. Mole fraction more volatile component in bottoms, x_B
    3. External reflux ratio, L_0/D
    4. Use optimum feed plate
B. 1,2. Fractional recoveries of components in distillate and bottoms, $(FR_A)_{dist}$, $(Fr_B)_{bot}$
    3. External reflux ratio, L_0/D
    4. Use optimum feed plate
C. 1. D or B
    2. x_D or x_B
    3. External reflux ratio, L_0/D
    4. Use optimum feed plate
D. 1,2. x_D and x_B
    3. Boilup ratio, V/B
    4. Use optimum feed plate

Designer Calculates

A. Distillate and bottoms flow rates, D and B
    Heating and cooling loads, Q_R and Q_c
    Number of stages, N
    Optimum feed plate
    Column diameter
B. x_B, x_D, D, B
    Q_R, Q_c
    N
    N_{feed}
    Column diameter
C. B or D
    x_B or x_D
    Q_R, Q_c
    N and N_{feed}
    Column diameter
D. D and B, Q_R and Q_c
    N, N_{feed}
    Column diameter

TABLE 3-3. *Specifications and calculated variables for binary distillation for simulation problems*

Specified Variables	Designer Calculates
A. 1,2. N, N_{feed} 3,4. x_D and x_B Column diameter	A. L_0/D B, D, Q_c, Q_R Check $V < V_{max}$
B. 1,2. N, N_{feed} 3,4. L_0/D, x_D (or x_B) Column diameter	B. x_B (or x_D) B, D, Q_c, Q_R Check $V < V_{max}$
C. 1,2. N, N_{feed} 3. x_D (or x_B) 4. Column diameter (set V = fraction × V_{max})	C. L_0/D, x_B (or x_D) B, D, Q_c, Q_R D. B, D, Q_c, x_B (or x_D), L_0/D Check $V < V_{max}$
D. 1,2. N, N_{feed} 3. Q_R 4. x_D (or x_B) Column diameter	

set. There are a variety of ways to specify the remainder of the problem (see Table 3-3). The desired composition of more volatile component in the distillate and bottoms could be specified, and the engineer would then have to determine the external reflux ratio, L_0/D, that will produce this separation and check that the maximum vapor flow rate will not be exceeded. An alternative is to specify L_0/D and either distillate or bottoms composition, in which case the engineer determines the unknown composition and checks the vapor flow rate. Another alternative is to specify the heat load in the reboiler and the distillate or bottoms composition. The engineer would then determine the reflux ratio and unknown product composition and check the vapor flow rate. The thread that runs through all these alternatives is that since the column has been built, some method of specifying the separation must be used.

The engineer always specifies variables that can be controlled. Several sets of possible specifications and calculated variables are outlined in Tables 3-1 to 3-3. Study these tables to determine the difference between design-type and simulation-type problems. Note that other combinations of specifications are possible.

In Table 3-1 we find five specified variables common to both types of problems. For design problems (Table 3-2), four additional variables must be set. Note that whereas column diameter is a specified variable in simulation problem C, it serves as a constraint in simulation problems A, B, and D (Table 3-3). Column diameter will allow us to calculate V_{max} and then we can check that $V < V_{max}$. However, we have not specified a variable for simulation. In problem C, where we specify V = fraction × V_{max}, the column diameter serves as a variable for simulation.

Chapter 4 starts with the simple design problem and progresses to simulation and other more complicated problems.

3.4 EXTERNAL COLUMN BALANCES

Once the problem has been specified, the engineer must calculate the unknown variables. Often it is not necessary to solve the entire problem, since only limited answers are required. The first step is to do mass and energy balances around the entire column. For binary design problems, these balances can usually be solved without doing stage-by-stage calculations. Figure 3-8 shows the schematic of a distillation column. The specified variables for a typical design problem are circled. We will assume that the column is well insulated and can be considered adiabatic. All of the heat transfer takes place in the condenser and reboiler. Column pressure is assumed to be constant.

From the balances around the entire column we wish to calculate distillate and bottoms flow rates, D and B, and the heat loads in the condenser and reboiler, Q_c and Q_R. We can start with mass balances around the entire column using the balance envelope shown by the dashed outline in the figure. The overall mass balance is

$$F = B + D \tag{3-1}$$

and the more volatile component mass balance is

$$Fz = Bx_B + Dx_D \tag{3-2}$$

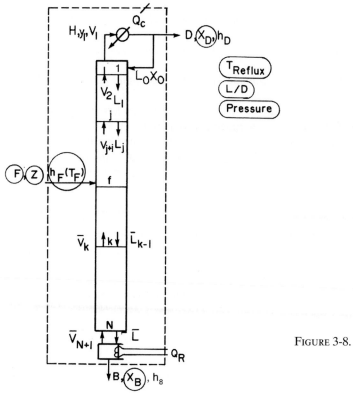

FIGURE 3-8. *Binary distillation column: Circled variables are typically specified in design problems*

For the design problem shown in Figure 3-8, Eqs. (3-1) and (3-2) can be solved immediately, since the only unknowns are B and D. Solving Eq. (3-1) for B, substituting this into Eq. (3-2), and solving for D, we obtain

$$D = (\frac{z - x_B}{x_D - x_B})F \tag{3-3}$$

and

$$B = F - D = (\frac{x_D - z}{x_D - x_B})F \tag{3-4}$$

Don't memorize equations like these; they can be derived as needed.

For the energy balance we will use the convention that all heat loads will be treated as inputs. If energy is removed, then the numerical value of the heat load will be negative. The steady-state energy balance around the entire column is

$$Fh_F + Q_c + Q_R = Dh_D + Bh_B \tag{3-5}$$

where we have assumed that kinetic and potential energy and work terms are negligible. The column is assumed to be well insulated and adiabatic. Q_R will be positive and Q_c negative. The enthalpies in Eq. (3-5) can all be determined from an enthalpy-composition diagram (e.g., Figure 2-4) or from the heat capacities and latent heats of vaporization. In general,

$$h_F = h_F(z, T_F, p), h_D = h_D(x_D, T_{reflux}, p), h_B = h_B(x_B, \text{saturated liquid}, p) \textbf{(3-6a, b, c)}$$

These three enthalpies can all be determined. We would find h_B on Figure 2-4 on the saturated liquid (boiling-line) at $x = x_B$.

Since F was specified and D and B were just calculated, we are left with two unknowns, Q_R and Q_c, in Eq. (3-5). Obviously another equation is required.

For the total condenser shown in Figure 3-8 we can determine Q_c. The total condenser changes the phase of the entering vapor stream but does not affect the composition. The splitter after the condenser changes only flow rates. Thus composition is unchanged and

$$y_1 = x_D = x_0 \tag{3-7}$$

The condenser mass balance is

$$V_1 = L_0 + D \tag{3-8}$$

Since the external reflux ratio, L_0/D, is specified, we can substitute its value into Eq. (3-8).

$$V_1 = (\frac{L_0}{D})D + D = (1 + \frac{L_0}{D})D \tag{3-9}$$

Then, since the terms on the right-hand side of Eq. (3-9) are known, we can calculate V_1. The condenser energy balance is

$$V_1 H_1 + Q_c = D h_D + L_0 h_0 \tag{3-10}$$

Since stream V_1 is a vapor leaving an equilibrium stage in the distillation column, it is a saturated vapor. Thus,

$$H_1 = H_1(y_1, \text{saturated vapor, p}) \tag{3-11}$$

and the enthalpy can be determined (e.g., on the saturated vapor (dew-line) of Figure 2-4 at $y = y_1$). Since the reflux and distillate streams are at the same composition, temperature, and pressure, $h_0 = h_D$. Thus,

$$V_1 H_1 + Q_c = (D + L_0) h_D = V_1 h_D \tag{3-12}$$

Solving for Q_c we have

$$Q_c = V_1 (h_D - H_1) \tag{3-13}$$

or, substituting in Eq. (3-9) and then Eq. (3-3),

$$Q_c = (1 + \frac{L_0}{D}) D (h_D - H_1) = (1 + \frac{L_0}{D})(\frac{z - x_B}{x_D - x_B}) F (h_D - H_1) \tag{3-14}$$

Note that $Q_c < 0$ because the liquid enthalpy, h_D, is less than the vapor enthalpy, H_1. This agrees with our convention. If the reflux is a saturated liquid, $H_1 - h_D = \lambda$, the latent heat of vaporization per mole. With Q_c known we can solve the column energy balance, Eq. (3-5), for Q_R.

$$Q_R = D h_D + B h_B - F h_F - Q_c \tag{3-15a}$$

or

$$Q_R = D h_D + B h_B - F h_F + (1 + \frac{L_0}{D}) D (H_1 - h_D) \tag{3-15b}$$

or

$$Q_R = (\frac{z - x_B}{x_D - x_B}) F h_D + (\frac{x_D - z}{x_D - x_B}) F h_B - F h_F + (1 + \frac{L_0}{D})(\frac{z - x_B}{x_D - x_B}) F (H_1 - h_D) \tag{3-16}$$

Q_R will be a positive number. Use of these equations is illustrated in Example 3-1.

EXAMPLE 3-1. External balances for binary distillation

A steady-state, countercurrent, staged distillation column is to be used to separate ethanol from water. The feed is a 30 wt % ethanol, 70 wt % water mixture at

40°C. Flow rate of feed is 10,000 kg/h. The column operates at a pressure of 1 kg/cm². The reflux is returned as a saturated liquid. A reflux ratio of $L/D = 3.0$ is being used. We desire a bottoms composition of $x_B = 0.05$ (weight fraction ethanol) and a distillate composition of $x_D = 0.80$ (weight fraction ethanol). The system has a total condenser and a partial reboiler. Find D, B, Q_c, and Q_R.

Solution

A. Define. The column and known information are sketched in the following figure.

Find D, B, Q_c, Q_R.

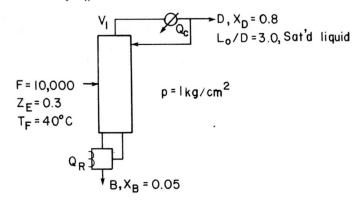

B. Explore. Since there are only two unknowns in the mass balances, B and D, we can solve for these variables immediately. Either solve Eqs. (3-1) and (3-2) simultaneously, or use Eqs. (3-3) and (3-4). For the energy balances, enthalpies must be determined. These can be read from the enthalpy-composition diagram (Figure 2-4). Then Q_c can be determined from the balance around the condenser and Q_R from the overall energy balance.

C. Plan. Use Eqs. (3-3) and (3-4) to find D and B, Eq. (3-14) to determine Q_c, and Eq. (3-15a) to determine Q_R.

D. Do It. From Eq. (3-3),

$$D = F\left(\frac{z - x_B}{x_D - x_B}\right) = 10,000\left[\frac{0.3 - 0.05}{0.8 - 0.05}\right] = 3333 \text{ kg/h}$$

From Eq. (3-4), $B = F - D = 10,000 - 3333 = 6667$ kg/h
From Figure 2-4 the enthalpies are
$h_D(x_D = 0.8, \text{ saturated liquid}) = 60 \text{ kcal/kg}$
$h_B(x_B = 0.05, \text{ saturated liquid}) = 90 \text{ kcal/kg}$
$h_f(z = 0.3, 40 \degree \text{ C}) = 30 \text{ kcal/kg}$
$H_1(y_1 = x_D = 0.8, \text{ saturated vapor}) = 330 \text{ kcal/kg}$

From Eq. (3-14),

$$Q_c = \left(1 + \frac{L_0}{D}\right)D(h_D - H_1) = (1 + 3)(3333)(60 - 330) = -3,559,640 \text{ kcal/h}$$

From Eq. (3-15a), $Q_R = Dh_D + Bh_B - Fh_F - Q_C$

$Q_R = (3333)(60) + (6667)(90) - (10,000)(30) - (-3,599,640) = 4,099,650$ kcal/h

E. Check. The overall balances, Eqs. (3-1) and (3-5), are satisfied. If we set up this problem on a spreadsheet without explicitly solving for D, B, Q_c, and Q_R we obtain identical answers.

F. Generalize. In this case we could solve the mass and energy balances sequentially. This is not always the case. Sometimes the equations must be solved simultaneously (see Problem 3.D3). Also, the mass balances and energy balances derived in the text were for the specific case shown in Figure 3-8. When the column configuration is changed, the mass and energy balances change (see Problem 3.D2, 3.D3, and 3.D5). For binary distillation we can usually determine the external flows and energy requirements from the external balances. Exceptions will be discussed in Chapter 4.

3.5   SUMMARY—OBJECTIVES

In this chapter we introduced the idea of distillation columns and saw how to do external balances. At this point you should be able to satisfy the following objectives:

1. Explain how a countercurrent distillation column physically works
2. Sketch and label the parts of a distillation system; explain the operation of each part and the flow regime on the trays
3. Explain the difference between design and simulation problems; list the specifications for typical problems
4. Write and solve external mass and energy balances for binary distillation systems

REFERENCES

Biegler, L. T., I. E. Grossman, and A. W. Westerberg, *Systematic Methods of Chemical Process Design*, Prentice Hall, Upper Saddle River, NJ, 1997.

Felder, R. M. and R. W. Rousseau, *Elementary Principles of Chemical Processes,* 3rd Updated ed., Wiley, New York, 2004.

Humphrey, J. L., and G. E. Keller II, *Separation Process Technology,* McGraw-Hill, New York, 1997.

Keller, G. E., II, "Separations: New Directions for an Old Field," *AIChE Monograph Series, 83* (17), 1987.

HOMEWORK

A.   *Discussion Problems*
A1.   Explain how a distillation column works.
A2.   Without looking at the text, define the following:
    a. Isothermal distillation
    b. The four flow regimes in a staged distillation column
    c. Reflux and reflux ratio
    d. Boilup and boilup ratio

     e. Rectifying (enriching) and stripping sections

     f. Simulation and design problems

     Check the text for definitions you did not know

A3. Explain the reasons a constant pressure distillation column is preferable to:

     a. An isothermal distillation system.

     b. A cascade of flash separators at constant temperature.

     c. A cascade of flash separators at constant pressure.

A4. In a countercurrent distillation column at constant pressure, where is the temperature highest? Where is it lowest?

A5. Develop your own key relations chart for this chapter. In one page or less draw sketches, write equations, and include all key words you would want for solving problems.

A6. What type of specifications will lead to simultaneous solution of the mass and energy balances?

A7. Specifications for a distillation column cannot include all three flow rates F, D, and B. Why not?

A8. What are the purposes of reflux? How does it differ from recycle?

A9. Without looking at the text, name the streams or column parts labeled A to H in the following figure.

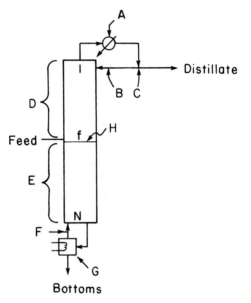

A10. Explain in what ways reflux and boilup have similar functions.

A11. For a binary separation, is $K_{MVC}V/L$ usually less than, equal to, or greater than 1.0? For a binary separation, is $K_{LVC}V/L$ usually less than, equal to, or greater than 1.0?

A12. Explain how to do mass balances if the percentage recovery of the more volatile component is specified in the distillate and the percentage recovery of the less volatile component is specified in the bottoms.

  B. *Generation of Alternatives*

B1. There are ways in which columns can be specified other than those listed in Tables 3-1 to 3-3.

a. Develop alternative specifications for design problems.

b. Develop alternative specifications for simulation problems.

C. *Derivations*

C1. For the column shown in Problem 3.D2, derive equations for D, B, Q_c, and L/D.

C2. For the column shown in Problem 3.D3, derive equations for D, B, $\overline{V}$, and Q_R.

C3. Show that Eqs. (3-3) and (3-4) are valid for a column with two feeds (e.g., shown in Figure 4-18) as long as we define $F = F_1 + F_2$ and $z = (F_1 z_1 + F_2 z_2)/F$.

C4. A partial condenser takes vapor leaving the top of a distillation column and condenses a portion of it. The vapor portion of mole fraction y_D is removed as the distillate product. The liquid portion of mole fraction x_0 is returned to the column as reflux. The liquid and vapor leaving the partial condenser can be assumed to be in equilibrium. Derive the mass and energy balances for a partial condenser.

D. *Problems*

Answers to problems with an asterisk are at the back of the book.

D1. A distillation column with two feeds is separating ethanol from water. The first feed is 60 wt % ethanol, has a total flow rate of 1000 kg/h, and is a mix of liquid and vapor at 81°C. The second feed is 10 wt % ethanol, has a total flow rate of 500 kg/h, and is liquid at 20°C. We desire a bottoms product that is 0.01 wt % ethanol and a distillate product that is 85 wt % ethanol. The column operates at 1 kg/cm² and is adiabatic. The column has a partial reboiler, which acts as an equilibrium contact, and a total condenser. The distillate and reflux are saturated liquids. Find B and D in kg/h, and find Q_c and Q_R in kcal/h. Use data in Figure 2-4. Do both parts a and b.

a. External reflux ratio, $L_0/D = 3.0$

b. Boilup ratio, $\overline{V}/B = 2.5$.

D2.* A distillation column separating ethanol from water is shown. Pressure is 1 kg/cm². Instead of having a reboiler, steam (pure water vapor) is injected directly into the bottom of the column to provide heat. The injected steam is a saturated vapor. The feed is 30 wt % ethanol and is at 20 °C. Feed flow rate is 100 kg/min. Reflux is a saturated liquid. We desire a distillate concentration of 60 wt % ethanol and a bottoms product that is 5 wt % ethanol. The steam is input at 100 kg/min. What is the external reflux ratio, L/D?

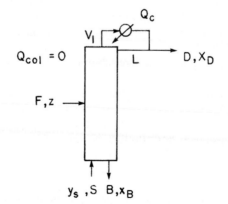

D3.* A distillation column separating ethanol from water is shown. Pressure is 1 kg/cm². Instead of having a condenser, a stream of pure liquid ethanol is added directly to the column to serve as the reflux. This stream is a saturated liquid. The feed is 40 wt % ethanol and is at –20 °C. Feed flow rate is 2000 kg/h. We desire a distillate concentration of 80 wt % ethanol and a bottoms composition of 5 wt % ethanol. A total reboiler is used, and the boilup is a saturated vapor. The cooling stream is input at C – 1000 kg/h. Find the external boilup rate, $\overline{V}$. Note: Set up the equations, solve in equation form for $\overline{V}$ including explicit equations for all required terms, read off all required enthalpies from the enthalpy composition diagram (Figure 2-4), and then calculate a numerical answer.

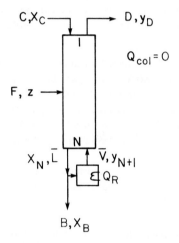

D4. A partial condenser takes vapor leaving the top of a distillation column and condenses a portion of it. The vapor portion of mole fraction y_D is removed as the distillate product. The liquid portion of mole fraction x_0 is returned to the column as reflux. The liquid and vapor leaving the partial condenser can be assumed to be in equilibrium.

A distillation column with a partial condenser and a partial reboiler is separating 100 kmol/h of a mixture that is 30 mol% methanol and 70 mol% water and is a saturated liquid. Column pressure is 1.0 atm. We desire a 99% recovery of the methanol in the vapor distillate and a 98% recovery of water in the bottoms. Equilibrium data are in Table 2-7 (in Problem 2.D1), and other data are in Problem 3.E1.

   a. Find D, B, $y_{D,M}$, and $x_{B,M}$.
   b. If L/D = 2.0, find $x_{0,M}$ and L_0 where subscript 0 refers to the reflux stream.
   c. If L/D = 2.0, find Q_c and Q_R.

D5.* A distillation column is separating ethanol from water at a pressure of 1 kg/cm². A two-phase feed of 20 wt % ethanol at 93°C is input at 100 kg/min. The column has a total condenser and a partial reboiler. The distillate composition is 90 wt % ethanol. Distillate and reflux are at 20°C. Bottoms composition is 1 wt % ethanol. Reflux ratio is $L_0/D = 3$. A liquid side stream is withdrawn above the feed stage. Side stream is 70 wt % ethanol, and side stream flow rate is 10 kg/min. Find D, B, Q_c, and Q_R. Data are in Figure 2-4.

D6.* A distillation column receives a feed that is 40 mol% n-pentane and 60 mol% n-hexane. Feed flow rate is 2500 lbmol/h, and feed temperature is 30°C. The column is at

1 atm. A distillate that is 99.9 mol% n-pentane is desired. A total condenser is used. Reflux is a saturated liquid. The external reflux ratio is $L_0/D = 3$. Bottoms from the partial reboiler is 99.8 mol% n-hexane. Find D, B, Q_R, Q_c. Note: Watch your units on temperature.

Data: λ_{c5} = 11,369 Btu/lbmol
λ_{c6} = 13,572 Btu/lbmol, both λ at boiling points.
$C_{PL,C5}$ = 39.7 (assume constant)
$C_{PL,C6}$ = 51.7 (assume constant)
$C_{PV,C5}$ = 27.45 + 0.08148 T − 4.538 × 10^{-5} T^2 + 10.1 × 10^{-9} T^3
$C_{PV,C6}$ = 32.85 + 0.09763 T − 5.716 × 10^{-5} T^2 + 13.78 × 10^{-9} T^3
where T is in °C and C_{PV} and C_{PL} are

$$\frac{cal}{mol\ °C} \quad or \quad \frac{Btu}{1bmol\ °F}$$

Source for λ and C_p data is Felder and Rousseau (2004).

D7. A continuous, steady-state distillation column is fed a mixture that is 70 mol% n-pentane and 30 mol% n-hexane. Feed rate is 1000 kmol/h. Feed is at 35°C. Column is at 101.3 kPa. The vapor distillate product is 99.9 mol% n-pentane and the bottoms product is 99.9 mol% n-hexane. The system has a partial condenser (thus the distillate product is a saturated vapor) and operates at an external reflux ratio of $L/D = 2.8$. The reboiler is a partial reboiler. Find D, B, Q_c & Q_R. Data are given in Problem 3.D6. Use DePriester chart to determine boiling points.

D8. A distillation column with a partial condenser and a partial reboiler is separating methanol and water. Column pressure is 1.0 atm. We desire 120 kmol/h of a bottoms product that is 0.0001 mole fraction methanol. Boilup ratio $\overline{V}/B$ is 1.5. Equilibrium data are in Table 2-7 (in Problem 2.D1), and other data are in Problem 3.E1. Assume that all streams entering and leaving the partial reboiler contain very little methanol. Find Q_R.

D9. A distillation column is separating 500 kmol/h of a mixture that is 76 mol% methanol and 24 mol% water. The bottoms product contains 0.00002 mole fraction methanol, and the distillate is 0.9999 mol% methanol. The boilup ratio $\overline{V}/B = 1.5$. Estimate the reboiler heat duty Q_R. Data are available in Problem 3.E1.

D10. A distillation column operating at 2.0 atm. is separating a feed that is 55.0 mol% n-pentane and 45.0 mol% n-hexane. The feed is at 65°C, and its flow rate is 1000 kmol/h. The distillate is 99.93 mol% n-pentane, and we want a 99.50% recovery of n-pentane. The system uses a total condenser, and reflux is a saturated liquid with an external reflux ratio of $L/D = 2.8$. There is a partial reboiler. Data are available in Problem 3.D6 and in the DePriester charts. Find D, B, x_B, Q_c, Q_R.

E. *More Complex Problems*

E1. A mixture of methanol and water is being separated in a distillation column with open steam (see figure in Problem 3.D2). The feed rate is 100 kmol/h. Feed is 60.0 mol% methanol and is at 40 °C. The column is at 1.0 atm. The steam is pure water vapor ($y_M = 0$) and is a saturated vapor. The distillate product is 99.0 mol% methanol and leaves as a saturated liquid. The bottoms is 2.0 mol% methanol and, since it leaves an equilibrium stage, must be a saturated liquid. The column is adiabatic. The column has a total condenser. External reflux ratio is $L/D = 2.3$.

Equilibrium data are in Table 2-7 in Problem 2.D1. Data for water and methanol is available in Felder and Rousseau (2004) (C_p, λ and steam tables) and in Perry's. Find D, B, Q_c, and S. Be careful with units and in selecting basis for energy balance. Data:

$\lambda_{methanol} = \Delta H_{vap} = 8.43$ kcal/mol $= 35.27$ kJ/mol (at boiling point)

$\lambda_{water} = \Delta H_{vap} = 9.72$ kcal/mol $= 40.656$ kJ/mol

$C_{p,w,liquid} = 1.0$ cal/(g °C) $= 75.4$ J/mol °C

$C_{PL,Meoh} = 75.86 + 0.1683T$ J/(gmole °C)

$C_{p,w,vapor} = 33.46 + 0.006880\,T + 0.7604 \times 10^{-5}\,T^2 - 3.593 \times 10^{-9}\,T^3$

$C_{p,meoh,vapor} = 42.93 + 0.08301\,T - 1.87 \times 10^{-5}\,T^2 - 8.03 \times 10^{-9}\,T^3$

For Vapor T is in °C, C_p is in J/mol °C

VLE data: Table 2-7. Density and MW data Problem 2.D1

E2. A mixture of methanol and water is being separated in a distillation column with open steam (see figure in Problem 3.D2). The feed rate is 500 kmol/h. Feed is 60.0 mol% methanol and is a saturated liquid. The column is at 1.0 atm. The steam is pure water vapor ($y_M = 0$) and is a saturated vapor. The distillate product is 99.8 mol% methanol and leaves as a saturated liquid. The bottoms is 0.13 mol% methanol and since it leaves an equilibrium stage must be a saturated liquid. The column is adiabatic. The column has a total condenser. External reflux ratio is $L/D = 3$. Data for water and methanol are available in Problem 3.E1.

Find D, B, Q_c, and S. Be careful with units and in selecting basis for energy balance.

F. *Problems Requiring Other Resources*

F1.* A mixture of oxygen and nitrogen is to be distilled at low temperature. The feed rate is 25,000 kmol/h and is 21 mol% oxygen and 79 mol% nitrogen. An ordinary column (as shown in Figure 3-8) will be used. Column pressure is 1 atm. The feed is a superheated vapor at 100 K. We desire a bottoms composition of 99.6 mol% oxygen and a distillate that is 99.7 mol% nitrogen. Reflux ratio is $L_0/D = 4$, and reflux is returned as a saturated liquid. Find D, B, Q_R, and Q_c.

F2.* A mixture of water and ammonia is to be distilled in an ordinary distillation system (Figure 3-8) at a pressure of 6 kg/cm². The feed is 30 wt % ammonia and is at 20 °C. We desire a distillate product that is 98 wt % ammonia and a 95% recovery of the ammonia in the distillate. The external reflux ratio is $L_0/D = 2.0$. Reflux is returned at −20 °C. Find D, B, x_B, Q_R, and Q_c per mole of feed.

G. *Computer Problems*

G1. Solve for Q_c and Q_R in Problem 3.D1 with a process simulator.
  a. Part a.
  b. Part b.
  Note: With Aspen Plus, use RADFRAC (see Appendix to Chapter 6, Lab 3) with an arbitrary (but large) number of stages and feed location = N/2. Do calculation for D by hand and input correct values for D and L/D (or *V/B*).

G2. [This problem should be done after studying the Appendix A of Chapter 4.] Solve Problem 3.D6 using AspenPlus to find Q_c and Q_R. To do this, do a hand calculation to find the value of D. Then arbitrarily set N = 40 and $N_{feed} = 20$ in RADFRAC and do the simulation.

Binary Column Distillation: Internal Stage-by-Stage Balances

4.1 INTERNAL BALANCES

In Chapter 3 we introduced column distillation and developed the external balance equations. In this chapter we start looking inside the column. For binary systems the number of stages required for the separation can conveniently be obtained by use of stage-by-stage balances. We start at the top of the column and write the balances and equilibrium relationship for the first stage, and then once we have determined the unknown variables for the first stage we write balances for the second stage. Utilizing the variables just calculated we can again calculate the unknowns. We can now proceed down the column in this stage-by-stage fashion until we reach the bottom. We could also start at the bottom and proceed upwards. This procedure assumes that each stage is an equilibrium stage, but this assumption may not be true. Ways to handle nonequilibrium stages are discussed in Section 4.11.

In the enriching section of the column it is convenient to use a balance envelope that goes around the desired stage and around the condenser. This is shown in Figure 4-1. For the first stage the balance envelope is shown in Figure 4-1A. The overall mass balance is then

$$V_2 = L_1 + D \qquad \text{(4-1, stage 1)}$$

The more volatile component mass balance is

$$V_2 y_2 = L_1 x_1 + D x_D \qquad \text{(4-2, stage 1)}$$

For a well-insulated, adiabatic column ($Q_{column} = 0$), the energy balance is

$$V_2 H_2 + Q_c = L_1 h_1 + D h_D \qquad \text{(4-3, stage 1)}$$

Assuming that each stage is an equilibrium stage, we know that the liquid and vapor leaving the stage are in equilibrium. For a binary system, the Gibbs phase rule becomes

$$\text{Degrees of freedom} = C - P + 2 = 2 - 2 + 2 = 2$$

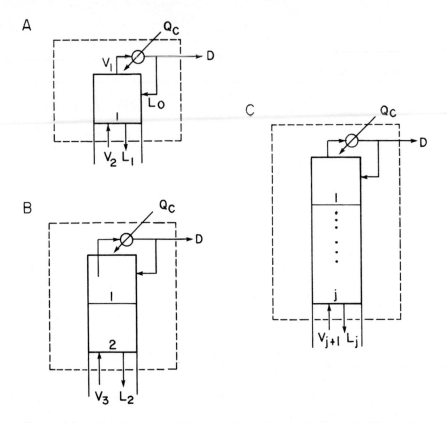

FIGURE 4-1.   *Enriching section balance envelopes; (A) stage 1, (B) stage 2, (C) stage j*

Since pressure has been set, there is one remaining degree of freedom. Thus for the equilibrium stage the variables are all functions of a single variable. For the saturated liquid we can write

$$h_1 = h_1(x_1) \qquad \text{(4-4a, stage 1)}$$

and for the saturated vapor,

$$H_2 = H_2(y_2) \qquad \text{(4-4b, stage 1)}$$

The liquid and vapor mole fractions leaving a stage are also related:

$$x_1 = x_1(y_1) \qquad \text{(4-4c, stage 1)}$$

Equations (4-4) for stage 1 represent the equilibrium relationship. Their exact form depends on the chemical system being separated. Equations (4-1, stage 1) to (4-4c, stage 1) are six equations with six unknowns: L_1, V_2, x_1, y_2, H_2, and h_1.

Since we have six equations and six unknowns, we can solve for the six unknowns. The exact methods for doing this are the subject of the remainder of this chapter. For now we will

Chapter 4   Binary Column Distillation: Internal Stage-by-Stage Balances

just note that we can solve for the unknowns and then proceed to the second stage. For the second stage we will use the balance envelope shown in Figure 4-1B. The mass balances are now

$$V_3 = L_2 + D \qquad \text{(4-1, stage 2)}$$

$$V_3 y_3 = L_2 x_2 + D x_D \qquad \text{(4-2, stage 2)}$$

while the energy balance is

$$Q_c + V_3 H_3 = L_2 h_2 + D h_D \qquad \text{(4-3, stage 2)}$$

The equilibrium relationships are

$$h_2 = h_2(x_2), \qquad H_3 = H_3(y_3), \qquad x_2 = x_2(y_2) \qquad \text{(4-4, stage 2)}$$

Again we have six equations with six unknowns. The unknowns are now L_2, V_3, x_2, y_3, H_3, and h_2.

We can now proceed to the third stage and utilize the same procedures. After that, we can go to the fourth stage and then the fifth stage and so forth. For a general stage j (j can be from 1 to f − 1, where f is the feed stage) in the enriching section, the balance envelope is shown in Figure 4-1C. For this stage the mass and energy balances are

$$V_{j+1} = L_j + D \qquad \text{(4-1, stage j)}$$

$$V_{j+1} y_{j+1} = L_j x_j + D x_D \qquad \text{(4-2, stage j)}$$

and

$$Q_c + V_{j+1} H_{j+1} = L_j h_j + D h_D \qquad \text{(4-3, stage j)}$$

while the equilibrium relationships are

$$h_j = h_j(x_j), \qquad H_{j+1} = H_{j+1}(y_{j+1}), \qquad x_j = x_j(y_j) \qquad \text{(4-4, stage j)}$$

When we reach stage j, the values of y_j, Q_c, D, and h_D will be known, and the unknown variables will be L_j, V_{j+1}, x_j, y_{j+1}, H_{j+1}, and h_j. At the feed stage, the mass and energy balances will change because of the addition of the feed stream.

Before continuing, we will stop to note the symmetry of the mass and energy balances and the equilibrium relationships as we go from stage to stage. A look at Eqs. (4-1) for stages 1, 2, and j will show that these equations all have the same structure and differ only in subscripts. Equations (4-1, stage 1) or (4-1, stage 2) can be obtained from the general Eq. (4-1, stage j) by replacing j with 1 or 2, respectively. The same observations can be made for the other Eqs. (4-2, 4-3, 4-4a, 4-4b, and 4-4c). The unknown variables as we go from stage to stage are also similar and differ in subscript only.

In addition to this symmetry from stage to stage, there is symmetry between equations for the same stage. Thus Eqs. (4-1, stage j), (4-2, stage j), and (4-3, stage j) are all steady-state balances that state

Input = output

In all three equations the output (of overall mass, solute, or energy) is associated with streams L_j and D. The input is associated with stream V_{j+1} and (for energy) with the cooling load, Q_c.

Below the feed stage the balance equations must change, but the equilibrium relationships in Eqs. (4-4a, b, c) will be unchanged. The balance envelopes in the stripping section are shown in Figure 4-2 for a column with a partial reboiler. The bars over flow rates signify that they are in the stripping section. It is traditional and simplest to write the stripping section balances around the bottom of the column using the balance envelope shown in Figure 4-2. Then these balances around stage $f + 1$ (immediately below the feed plate) are

$$\overline{V}_{f+1} = \overline{L}_f - B \qquad\qquad \textbf{(4-5, stage f + 1)}$$

$$\overline{V}_{f+1} y_{f+1} = \overline{L}_f x_f - B x_B \qquad\qquad \textbf{(4-6, stage f + 1)}$$

$$\overline{V}_{f+1} H_{f+1} = \overline{L}_f h_F - B h_B + Q_R \qquad\qquad \textbf{(4-7, stage f + 1)}$$

The equilibrium relationships are Eqs. (4-4) written for stage $f + 1$.

$$h_f = h_f(x_f), \qquad H_{f+1} = H_{f+1}(y_{f+1}), \qquad x_f = x_f(y_f) \qquad \textbf{(4-4, stage f + 1)}$$

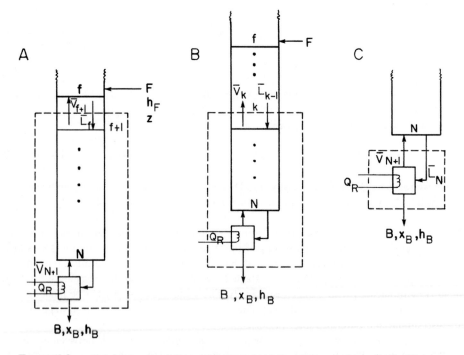

FIGURE 4-2. *Stripping section balance envelopes; (A) below feed stage (stage f + 1), (B) stage k, (C) partial reboiler*

Chapter 4   Binary Column Distillation: Internal Stage-by-Stage Balances

These six equations have six unknowns: L_f, $\overline{V}_{f+1}$, x_f, y_{f+1}, H_{f+1}, and h_f. x_B is specified in the problem statement; B and Q_R were calculated from the column balances; and y_f (required for the last equation) was obtained from the solution of Eqs. (4-1, stage j) to (4-4c, stage j) with j = f – 1. At the feed stage we change from one set of balance envelopes to another.

Note that the same equations will be obtained if we write the balances above stage f + 1 and around the top of the distillation column (use a different balance envelope). This is easily illustrated with the overall mass balance, which is now

$$\overline{V}_{f+1} + F = D + \overline{L}_f$$

Rearranging, we have

$$\overline{V}_{f+1} = \overline{L}_f - (F - D)$$

However, since the external column mass balance says F – D = B, the last equation becomes

$$\overline{V}_{f+1} = \overline{L}_f - B$$

which is Eq. (4-5, stage f + 1). Similar results are obtained for the other balance equations. Thus the balance envelope we use is arbitrary.

Once the six Eqs. (4-4a) to (4-7) for stage f + 1 have been solved, we can proceed down the column to the next stage, f + 2. For a balance envelope around general stage k as shown in Figure 2-2B, the equations are

$$\overline{V}_k = \overline{L}_{k-1} - B \qquad\qquad \textbf{(4-5, stage k)}$$

$$\overline{V}_k y_k = \overline{L}_{k-1} x_{k-1} - Bx_B \qquad\qquad \textbf{(4-6, stage k)}$$

$$\overline{V}_k H_k = \overline{L}_{k-1} h_{k-1} - Bh_B + Q_R \qquad\qquad \textbf{(4-7, stage k)}$$

the equilibrium expression will correspond to Eqs. (4-4, stage f + 1) with k – 1 replacing f as a subscript. Thus,

$$h_{k-1} = h_{k-1}(x_{k-1}), \qquad H_k = H_k(y_k), \qquad x_{k-1} = x_{k-1}(y_{k-1}) \qquad \textbf{(4-4, stage k)}$$

A partial reboiler as shown in Figure 4-2C acts as an equilibrium contact. If we consider the reboiler as stage N + 1, the balances for the envelope shown in Figure 4-2C can be obtained by setting k = N + 1 and k – 1 = N in Eqs. (4-5, stage k), (4-6, stage k) and (4-7, stage k).

If $x_{N+1} = x_B$, the N + 1 equilibrium contacts gives us exactly the specified separation, and the problem is finished. If $x_{N+1} < x_B$ while $x_N > x_B$, the N + 1 equilibrium contacts gives slightly more separation than is required.

Just as the balance equations in the enriching section are symmetric from stage to stage, they are also symmetric in the stripping section.

4.2   BINARY STAGE-BY-STAGE SOLUTION METHODS

The challenge for any stage-by-stage solution method is to solve the three balance equations and the three equilibrium relationships simultaneously in an efficient manner. This problem was first solved by Sorel (1893), and graphical solutions of Sorel's method were developed independently by Ponchon (1921) and Savarit (1922). These methods all solve the complete mass and energy balance and equilibrium relationships stage by stage. Starting at the top of the column as shown in Figure 4-1A, we can find the liquid composition, x_1, in equilibrium with the leaving vapor composition, y_1, from Eq. (4-4c, stage 1). The liquid enthalpy, h_1, is easily found from Eqs. (4-4a, stage 1). The remaining four Eqs. (4-1) to (4-3) and (4-4b) for stage 1 are coupled and must be solved simultaneously. The Ponchon-Savarit method does this graphically. The Sorel method uses a trial-and-error procedure on each stage.

The trial-and-error calculations on every stage of the Sorel method are obviously slow and laborious. Lewis (1922) noted that in many cases the molar vapor and liquid flow rates in each section (a region between input and output ports) were constant. Thus in Figures 4-1 and 4-2,

$$L_1 = L_2 = \cdots = L_j = \cdots = L_{f-1} = L$$

$$V_1 = V_2 = \cdots = V_{j+1} = \cdots = V_f = V \tag{4-8}$$

and

$$\overline{L}_f = \overline{L}_{f+1} = \cdots = \overline{L}_{k-1} = \cdots = \overline{L}_N = \overline{L}$$

$$\overline{V}_{f+1} = \cdots = \overline{V}_k = \cdots = \overline{V}_{N+1} = \overline{V} \tag{4-9}$$

For each additional column section there will be another set of equations for constant flow rates. Note that in general $L \neq \overline{L}$ and $V \neq \overline{V}$. Equations (4-8) and (4-9) will be valid if every time a mole of vapor is condensed a mole of liquid is vaporized. This will occur if:

1. The column is adiabatic.
2. The specific heat changes are small compared to latent heat changes.

$$|H_{j+1} - H_j| \ll \lambda \text{ and } |h_{j+1} - h_j| \ll \lambda \tag{4-10}$$

3. The heat of vaporization per mole, λ, is constant; that is, λ does not depend on concentration. Condition 3 is the most important criterion. Lewis called this set of conditions *constant molal overflow* (CMO). An alternative to conditions 2 and 3 is
4. The saturated liquid and vapor lines on an enthalpy-composition diagram (in molar units) are parallel.

For some systems, such as hydrocarbons, the latent heat of vaporization per kilogram is approximately constant. Then the mass flow rates are constant, and constant *mass* overflow should be used.

The Lewis method assumes before the calculation is done that CMO is valid. Thus Eqs. (4-8) and (4-9) are valid. With this assumption, the energy balance, Eqs. (4-3) and (4-7), will be automatically satisfied. Then only Eqs. (4-1), (4-2), and (4-4c), or (4-5), (4-6), and (4-4c) need be solved. Eqs. (4-1, stage j) and (4-2, stage j) can be combined. Thus,

$$V_{j+1}y_{j+1} = L_jx_j + (V_{j+1} - L_j)x_D \tag{4-11}$$

Solving for y_{j+1}, we have

$$y_{j+1} = \frac{L_j}{V_{j+1}} x_j + (1 - \frac{L_j}{V_{j+1}})x_D \tag{4-12a}$$

Since L and V are constant, this equation becomes

$$y_{j+1} = \frac{L}{V} x_j + (1 - \frac{L}{V})x_D \tag{4-12b}$$

Eq. (4-12b) is the *operating equation* in the enriching section. It relates the concentrations of two passing streams in the column and thus represents the mass balances in the enriching section. Eq. (4-12b) is solved sequentially with the equilibrium expression for x_j, which is Eq. (4-4c, stage j).

To start we first use the column balances to calculate D and B. Then $L_0 = (L_0/D)D$ and $V_1 = L_0 + D$. For a saturated liquid reflux, $L_0 = L_1 = L_2 = L$ and $V_1 = V_2 = V$. At the top of the column we know that $y_1 = x_D$. The vapor leaving the top stage is in equilibrium with the liquid leaving this stage (see Figure 4-1A). Thus x_1 can be calculated from Eq. (4-4c, stage j) with j = 1. Then y_2 is found from Eq. (4-12b) with j = 1. We then proceed to the second stage, set j = 2, and obtain x_2 from Eq. (4-4c, stage j) and y_3 from Eq. (4-12b). We continue this procedure down to the feed stage.

In the stripping section, Eqs. (4-5, stage k) and (4-6, stage k) are combined to give

$$y_k = \frac{\overline{L}_{k-1}}{\overline{V}_k} x_{k-1} - (\frac{\overline{L}_{k-1}}{\overline{V}_k} - 1)x_B \tag{4-13}$$

With CMO, $\overline{L}$ and $\overline{V}$ are constant, and the resulting stripping section operating equation is

$$y_k = (\overline{L}/\overline{V})x_{k-1} - (\overline{L}/\overline{V} - 1)x_B \tag{4-14}$$

Once we know $\overline{L}/\overline{V}$ we can obviously alternate between the operating Eq. (4-14) and the equilibrium Eq. (4-4c, stage k).

The phase and temperature of the feed obviously affect the vapor and liquid flow rates in the column. For instance, if the feed is liquid, the liquid flow rate below the feed stage must be greater than liquid flow above the feed stage, $\overline{L} > L$. If the feed is a vapor, $V > \overline{V}$. These effects can be quantified by writing mass and energy balances around the feed stage. The feed stage is shown schematically in Figure 4-3. The overall mass balance and the energy balance for the balance envelope shown in Figure 4-3 are

$$F + \overline{V} + L = \overline{L} + V \tag{4-15}$$

and

$$Fh_F + \overline{V} H_{f+1} + Lh_{f-1} = \overline{L}h_f + VH_f \tag{4-16}$$

(Despite the use of "h_F" as the symbol for the feed enthalpy, the feed can be a liquid or vapor or a two-phase mixture.) If we assume CMO neither the vapor enthalpies nor the liquid

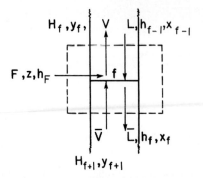

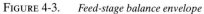

FIGURE 4-3.　*Feed-stage balance envelope*

enthalpies vary much from stage to stage. Thus $H_{f+1} \sim H_f$ and $h_{f-1} \sim h_f$. Then Eq. (4-16) can be written as

$$Fh_F + (\overline{V} - V)H \sim (\overline{L} - L)h$$

The mass balance Eq. (4-15) can be conveniently solved for $\overline{V} - V$,

$$\overline{V} - V = \overline{L} - L - F$$

which can be substituted into the energy balance to give us

$$Fh_F + (\overline{L} - L)H - FH \sim (\overline{L} - L)h$$

Combining terms, this is

$$(\overline{L} - L)(H - h) \sim F(H - h_F)$$

or

$$q \equiv \frac{\overline{L} - L}{F} \sim \frac{H - h_F}{H - h} \tag{4-17}$$

In words, the "quality" q is

$$q \equiv \frac{\text{liquid flow rate below feed stage} - \text{liquid flow rate above feed stage}}{\text{feed rate}}$$

$$q \sim \frac{\text{vapor enthalpy on feed plate} - \text{feed enthalpy}}{\text{vapor enthalpy on feed plate} - \text{liquid enthalpy on feed plate}} \tag{4-18}$$

This result is analogous to the use of q in flash distillation. Since the liquid and vapor enthalpies can be estimated, we can calculate q from Eq. (4-17). Then

$$\overline{L} = L + qF \tag{4-19}$$

The quality q is the fraction of feed that is liquid. For example, if the feed is a saturated liquid, $h_F = h$, $q = 1$, and $\overline{L} = L + F$. Once $\overline{L}$ has been determined, $\overline{V}$ is calculated from either Eq. (4-15) or Eq. (4-5, stage f + 1) or from

$$\overline{V} = V - (1 - q)F \qquad (4\text{-}20)$$

Which can be derived from Eqs. (4-15) and (4-19).

EXAMPLE 4-1. Stage-by-stage calculations by the Lewis method

A steady-state countercurrent, staged distillation column is to be used to separate ethanol from water. The feed is a 30 wt % ethanol, 70 wt % water mixture that is a saturated liquid at 1 atm pressure. Flow rate of feed is 10,000 kg/h. The column operates at a pressure of 1 atm. The reflux is returned as a saturated liquid. A reflux ratio of $L/D = 3.0$ is being used. We desire a bottoms composition of $x_B = 0.05$ (weight fraction ethanol) and a distillate composition of $x_D = 0.80$ (weight fraction ethanol). The system has a total condenser and a partial reboiler. The column is well insulated.

Use the Lewis method to find the number of equilibrium contacts required if the feed is input on the second stage from the top.

Solution

A. Define. The column and known information are shown in the following figure. Find the number of equilibrium contacts required.

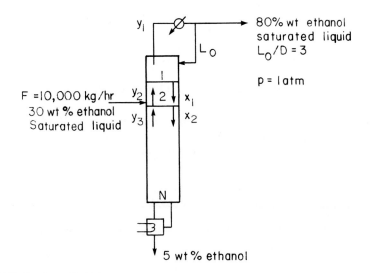

B. Explore. Except for some slight changes in the feed temperature and column pressure, this problem is very similar to Example 3-1. The solution for B and D obtained in that example is still correct. B = 6667 kg/h, D = 3333 kg/h. Equilibrium data are available in weight fractions in Figure 2-4 and in mole fraction units in Figure 2-2 and Table 2-1. To use the Lewis method we must have CMO. We can check this by comparing the latent heat per mole of pure ethanol and pure water. (This checks the third and most important criterion for CMO. Since the column is well insulated, the first criterion, adiabatic, will be satisfied.) The latent heats are (Himmelblau, 1974):

$$\lambda_E = 9.22 \text{ kcal/mol}, \qquad \lambda_W = 9.7171 \text{ kcal/mol}$$

The difference of roughly 5% is reasonable particularly since we always use the ratio of L/V or $\overline{L}/\overline{V}$. (Using the ratio causes some of the change in L and V to divide out.) Thus we will assume CMO. Now we must convert flows and compositions to molar units.

C. Plan. First, convert to molar units. Carry out preliminary calculations to determine L/V and $\overline{L}/\overline{V}$. Then start at the top, alternating between equilibrium (Figure 2-2 or Eq. (2.B-2)) and the top operating Eq. (4-12b). Since stage 2 is the feed stage, calculate y_3 from the bottom operating Eq. (4-14).

D. Do It. *Preliminary Calculations:* To convert to molar units:

$$MW_W = 18, \qquad MW_E = 46, \qquad z_E = \frac{0.3/46}{0.3/46 + 0.7/18} = 0.144$$

Average molecular weight of feed is
$\overline{MW}_F = (0.144)(46) + (0.856)(18) = 22.03$
Feed rate = (10,000 kg/h)/(22.03 kg/kmol) = 453.9 kmol/h

$$x_{D,E} = \frac{0.8/46}{0.8/46 + 0.2/18} = 0.61, \qquad x_{BE} = 0.02$$

For distillate, the average molecular weight is

$$\overline{MW}_{dist} = (0.61)(46) + (0.39)(18) = 35.08$$

which is also the average for the reflux liquid and vapor stream V since they are all the same composition.
   Then D = (3333 kg/h)/35.08 = 95.2 kmol/h and

$$L = \left(\frac{L}{D}\right)D = (3)(95.23) = 285.7 \text{ kmol/h}$$

while   V = L + D = 380.9. Thus,

$$\frac{L}{V} = \frac{285.7}{380.9} = 0.75$$

Because of CMO, L/V is constant in the rectifying section.
   Since the feed is a saturated liquid,

$$\overline{L} = L + F = 285.7 + 453.9 = 739.6 \text{ kmol/h}$$

where we have converted F to kmol/h. Since a saturated liquid feed does not affect the vapor, $\overline{V} = V = 380.9$. Thus,

$$\frac{\overline{L}}{\overline{V}} = \frac{739.6}{380.9} = 1.94$$

An internal check on consistency is $L/V < 1$ and $\overline{L}/\overline{V} > 1$.

Stage-by-Stage Calculations: At the top of the column, $y_1 = x_D = 0.61$. Liquid stream L_1 of concentration x_1 is in equilibrium with the vapor stream y_1. From Figure 2-2, $x_1 = 0.4$. (Note that $y_1 > x_1$ since ethanol is the more volatile component.) Vapor stream y_2 is a passing stream relative to x_1 and can be determined from the operating Eq. (4-12).

$$y_2 = \frac{L}{V} x_1 + (1 - \frac{L}{V}) x_D = (0.75)(0.4) + (0.25)(0.61) = 0.453$$

Stream x_2 is in equilibrium with y_2. From Figure 2-2 we obtain $x_2 = 0.11$.
    Since stage 2 is the feed stage, use bottom operating Eq. (4-14) for y_3.

$$y_3 = \frac{\overline{L}}{\overline{V}} x_2 + (1 - \frac{\overline{L}}{\overline{V}}) x_B = (1.942)(0.11) + (-0.942)(0.02) = 0.195$$

Stream x_3 is in equilibrium with y_3. From Figure 2-2, this is $x_3 = 0.02$. Since $x_3 = x_B$ (in mole fraction), we are finished.
    The third equilibrium contact would be the partial reboiler. Thus the column has two equilibrium stages plus the partial reboiler.
E. Check. This is a small number of stages. However, not much separation is required, the external reflux ratio is large, and the separation of ethanol from water is easy in this concentration range. Thus the answer is reasonable. We can check the calculation of L/V with mass balances.
        Since $V_1 = L_0 + D$,

$$\frac{L_0}{V_1} = \frac{L_0}{D + L_0} = \frac{L_0/D}{\dfrac{D}{D} + \dfrac{L_0}{D}} = \frac{L_0/D}{1 + \dfrac{L_0}{D}} = \frac{3}{4}$$

Since L_0, V_1, and D, are the same composition, L_0/D and L_0/V_1 have the same values in mass and molar units. We can check the equilibrium calculation with Eq. (2.B-2). For example, for $y_1 = 0.61$ we obtain $x_1 = 0.385$.
F. Generalizations. We should always check that CMO is valid. Then convert all flows and compositions into molar units. The procedure for stepping off stages is easily programmed on a spreadsheet (Burns and Sung, 1996). We could also have started at the bottom and worked our way up the column stage by stage. Going up the column we calculate y values from equilibrium and x values from the operating equations.
        Note that $L/V < 1$ and $\overline{L}/\overline{V} > 1$. This makes sense, since we must have a net flow of material upwards in the rectifying section (to obtain a distillate

product) and a net flow downwards in the stripping section. We must also have a net upward flow of ethanol in the rectifying section ($Lx_j < Vy_{j+1}$) and in the stripping section ($\overline{L}x_j < \overline{V}y_{j+1}$). These conditions are satisfied by all pairs of passing streams.

The Lewis method is obviously much faster and more convenient than the Sorel method. It is also easier to program on a computer or in a spreadsheet. In addition, it is easier to understand the physical reasons why separation occurs instead of becoming lost in the algebraic details. However, remember that the Lewis method is based on the assumption of CMO. If CMO is not valid, the answers will be incorrect.

If the calculation procedure in the Lewis method is confusing to you, continue on to the next section. The graphical McCabe-Thiele procedure explained there is easier for many students to understand. After completing the McCabe-Thiele procedure, return to this section and study the Lewis method again.

4.3   INTRODUCTION TO THE McCABE-THIELE METHOD

McCabe and Thiele (1925) developed a graphical solution method based on Lewis' method and the observation that the operating Eqs. (4-12b) and (4-14) plot as straight lines (the *operating lines*) on a y-x diagram. On this graph the equilibrium relationship can be solved from the y-x equilibrium curve and the mass balances from the operating lines.

To illustrate, consider a typical design problem for a binary distillation column such as the one illustrated in Figure 3-8. We will assume that equilibrium data are available at the operating pressure of the column. These data are plotted as shown in Figure 4-4. At the top of the column is a total condenser. As noted in Chapter 3 in Eq. (3-7), this means that $y_1 = x_D = x_0$. The vapor leaving the first stage is in equilibrium with the liquid leaving the first stage. This liquid composition, x_1, can be determined from the equilibrium curve at $y = y_1$. This is illustrated in Figure 4-4.

Liquid stream L_1 of composition x_1 passes vapor stream V_2 of composition y_2 inside the column (Figures 3-8 and 4-1A). When the mass balances are written around stage 1 and the

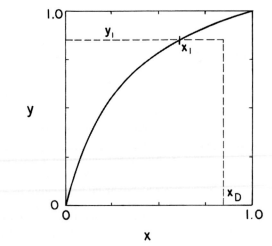

FIGURE 4-4.   *Equilibrium for top stage on McCabe-Thiele diagram*

top of the column (see balance envelope in Figure 4-1A), the result after assuming CMO and doing some algebraic manipulations is Eq. (4-12) with j = 1. This equation can be plotted as a straight line on the y-x diagram. Suppressing the subscripts j+1 and j, we write Eq. (4-12b) as

$$y = \frac{L}{V} x + (1 - \frac{L}{V})x_D \tag{4-21}$$

which is understood to apply to passing streams. Eq. (4-21) plots as a straight line (the *top operating line*) with a slope of L/V and a y intercept (x = 0) of $(1 - L/V)x_D$. Once Eq. (4-12) has been plotted, y_2 is easily found from the y value at $x = x_1$. This is illustrated in Figure 4-5. Note that the top operating line goes through the point (y_1, x_D) since these coordinates satisfy Eq. (4-21).

With y_2 known we can proceed down the column. Since x_2 and y_2 are in equilibrium, we easily obtain x_2 from the equilibrium curve. Then we obtain y_3 from the operating line (mass balances), since x_2 and y_3 are the compositions of passing streams. This procedure of *stepping off stages* is shown in Figure 4-6. It can be continued as long as we are in the rectifying section. Note that this produces a staircase on the y-x, or McCabe-Thiele, diagram. Instead of memorizing this procedure, you should follow the points on the diagram and compare them to the schematics of a distillation column (Figures 3-8 and 4-1). Note that the horizontal and vertical lines have no physical meaning. The points on the equilibrium curve (squares) represent liquid and vapor streams leaving an equilibrium stage. The points on the operating line (circles) represent the liquid and vapor streams passing each other in the column.

In the stripping section the top operating line is no longer valid, since different mass balances and, hence, a different operating equation are required. The stripping section operating equation was given in Eq. (4-14). When the subscripts k and k − 1 are suppressed, this equation becomes

$$y = \frac{\overline{L}}{\overline{V}} x - (\frac{\overline{L}}{\overline{V}} - 1)x_B \tag{4-22}$$

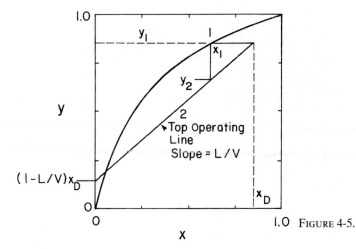

FIGURE 4-5.   *Stage 1 calculation on McCabe-Thiele diagram*

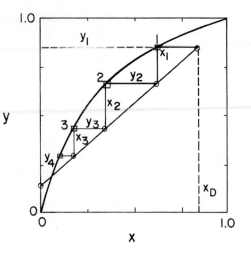

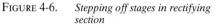

FIGURE 4-6.   *Stepping off stages in rectifying section*

Eq. (4-22) plots as a straight line with slope $\bar{L}/\bar{V}$ and y intercept $-(\bar{L}/\bar{V}-1)x_B$, as shown in Figure 4-7. This *bottom operating line* applies to passing streams in the stripping section. Starting with the liquid leaving the partial reboiler, of mole fraction $x_B = x_{N+1}$, we know that the vapor leaving the partial reboiler is in equilibrium with x_B. Thus we can find y_{N+1} from the equilibrium curve. x_N is easily found from the bottom operating line, since liquid of composition x_N is a passing stream to vapor of composition y_{N+1} (compare Figures 4-2 and 4-7). We can continue alternating between the equilibrium curve and the bottom operating line as long as we are in the stripping section.

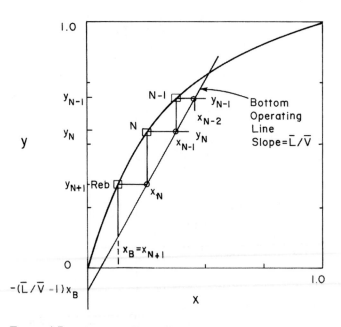

FIGURE 4-7.   *Stepping off stages in stripping section*

Chapter 4   Binary Column Distillation: Internal Stage-by-Stage Balances

If we are stepping off stages down the column, at the feed stage f we switch from the top operating line to the bottom operating line (refer to Figure 4-3, a schematic of the feed stage). Above the feed stage, we calculate x_{f-1} from equilibrium and y_f from the top operating line. Since liquid and vapor leaving the feed stage are assumed to be in equilibrium, we can determine x_f from the equilibrium curve at $y = y_f$ and then find y_{f+1} from the bottom operating line. This procedure is illustrated in Figure 4-8A, where stage 3 is the feed stage. The separation shown in Figure 4-8A would require 5 equilibrium stages plus an equilibrium partial reboiler, or 6 equilibrium contacts, when stage 3 is used as the feed stage. In this problem, stage 3 is the *optimum feed stage*. That is, a separation will require the fewest total number of stages when feed stage 3 is used. Note in Figure 4-8B and 4-8C that if stage 2 or stage 5 is used, more total stages are required. For binary distillation the optimum feed plate is easy to determine; it will always be the stage where the step in the staircase includes the point of intersection of the two operating lines (compare Figure 4-8A to Figures 4-8B and 4-8C). A mathematical analysis of the optimum feed plate location suitable for computer calculation with the Lewis method is developed later.

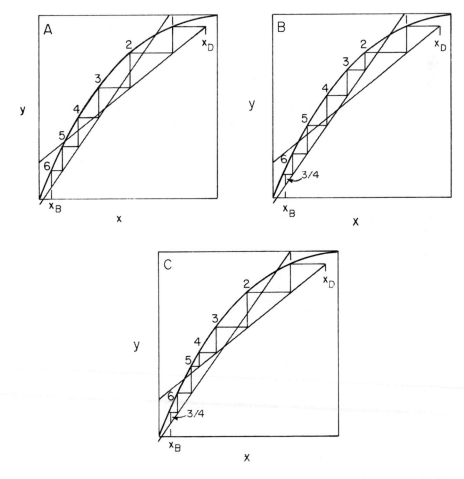

FIGURE 4-8.   *McCabe-Thiele diagram for entire column; (A) optimum feed stage (stage 3); (B) feed stage too high (stage 2); (C) feed stage too low (stage 5)*

When stepping off stages from the top down, the fractional number of stages can be calculated as (see Figure 4-8B and 4-8C),

$$\text{Fraction} = \frac{\text{distance from operating line to } x_B}{\text{distance from operating line to equilibrium curve}} \qquad \textbf{(4-23)}$$

where the distances are measured horizontally on the diagram. The fraction has no physical meaning because we build either five or six stages; however, the fraction is somewhat useful when the stage efficiency is < 1.

Now that we have seen how to do the stage-by-stage calculations on a McCabe-Thiele diagram, let us consider how to start with the design problem given in Figure 3-8 and Tables 3-1 and 3-2. The known variables are F, z, q, x_D, x_B, L_0/D, p, saturated liquid reflux, and we use the optimum feed location. Since the reflux is a saturated liquid, there will be no change in the liquid or vapor flow rates on stage 1 and $L_0 = L_1$ and $V_1 = V_2$. This allows us to calculate the internal reflux ratio, L/V, from the external reflux ratio, L_0/D, which is specified.

$$\frac{L}{V} = \frac{L}{L+D} = \frac{L/D}{L/D+1} \qquad \textbf{(4-24)}$$

With L/V and x_D known, the top operating line is fully specified and can be plotted.

Since the boilup ratio, $\overline{V}/B$, was not specified, we cannot directly calculate $\overline{L}/\overline{V}$, which is the slope of the bottom operating line. Instead, we need to utilize the condition of the feed to determine flow rates in the stripping section. The same procedure used with the Lewis method can be used here. The feed quality, q, is calculated from Eq. (4-17), which is repeated below:

$$q \equiv \frac{\overline{L} - L}{F} \sim \frac{H - h_F}{H - h} \qquad \textbf{(4-17)}$$

Then $\overline{L}$ is given by Eq. (4-19), $\overline{L} = L + qF$, and $\overline{V} = \overline{L} - B$. We can calculate L as (L/D)D, where D and B are found from mass balances around the entire column. Alternatively, for a simple column Eqs. (3-3) and (3-4) can be substituted into the equations for $\overline{L}$ and $\overline{V}$. When this is done, we obtain

$$\frac{\overline{L}}{\overline{V}} = \frac{\dfrac{L_0}{D}(z - x_B) + q(x_D - x_B)}{\dfrac{L_0}{D}(z - x_B) + q(x_D - x_B) - (x_D - z)} \qquad \textbf{(4-25)}$$

With $\overline{L}/\overline{V}$ and x_B known, the bottom operating equation is fully specified, and the bottom operating line can be plotted. Eq. (4-25) is convenient for computer calculations but is specific for the simple column shown in Figure 3-8. For graphical calculations the alternative procedure shown in the next section is usually employed.

4.4 FEED LINE

In any section of the column between feeds and/or product streams the mass balances are represented by the operating line. In general, the operating line can be derived by drawing a mass balance envelope through an arbitrary stage in the section and around the top or bottom of the column. When material is added or withdrawn from the column the mass balances will change and the operating lines will have different slopes and intercepts. In the previous section the effect of a feed on the operating lines was determined from the feed quality and mass balances around the entire column or from Eq. (4-25). Here we will develop a graphical method for determining the effect of a feed on the operating lines.

Consider the simple single-feed column with a total condenser and a partial reboiler shown in Figure 3-8. The mass balance in the rectifying section for the more volatile component is

$$yV = Lx + Dx_D \tag{4-26}$$

while the balance in the stripping section is

$$y\overline{V} = \overline{L}x - Bx_B \tag{4-27}$$

where we have assumed that CMO is valid. At the feed plate we switch from one mass balance to the other. We wish to find the point at which the top operating line—representing Eq. (4-26)—intersects the bottom operating line—representing Eq. (4-27).

The intersection of these two lines means that

$$y_{top\ op} = y_{bot\ op}, \qquad x_{top\ op} = x_{bot\ op} \tag{4-28}$$

Equations (4-28) are valid only at the point of intersection. Since the y's and x's are equal at the point of intersection, we can subtract Eq. (4-26) from Eq. (4-27) and obtain

$$y(\overline{V} - V) = (\overline{L} - L)x - (Dx_D + Bx_B) \tag{4-29}$$

From the overall mass balance around the entire column, Eq. (3-2), we know that the last term is $-Fz_F$. Then, solving Eq. (4-29) for y,

$$y = -\left(\frac{\overline{L} - L}{V - \overline{V}}\right)x + \frac{Fz_F}{V - \overline{V}} \tag{4-30}$$

Eq. (4-30) is one form of the feed equation. Since L, $\overline{L}$, V, $\overline{V}$, F, and z_F are constant, it represents a straight line (the feed line) on a McCabe-Thiele diagram. Every possible intersection point of the two operating lines *must* occur on the feed line.

For the special case of a feed that flashes in the column to form a vapor and a liquid phase, we can relate Eq. (4-30) to flash distillation. In this case we have the situation shown in Figure 4-9. Part of the feed, V_F, vaporizes, while the remainder is liquid, L_F. Looking at the terms in Eq. (4-30), we note that $\overline{L} - L$ is the change in liquid flow rates at the feed stage. In this case,

$$\overline{L} - L = L_F \tag{4-31}$$

The change in vapor flow rates is

$$V - \overline{V} = V_F \tag{4-32}$$

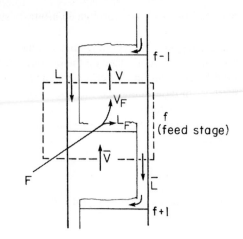

FIGURE 4-9.   *Two-phase feed*

Equation (4-30) then becomes

$$y = -\frac{L_F}{V_F} x + \frac{F}{V_F} z_F \qquad (4\text{-}33)$$

which is essentially the same as Eq. (2-11), the operating equation for flash distillation. Thus the feed line represents the flashing of the feed into the column. Equation (4-33) can also be written in terms of the fraction vaporized, $f = V_F/F$, as [see Eqs. (2-12) and (2-13)]

$$y = -\frac{1-f}{f} x + \frac{1}{f} z_F \qquad (4\text{-}34)$$

In terms of the fraction remaining liquid, $q = L_F/F$ [see Eqs. (2-14) and (2-15)], Eq. (4-33) is

$$y = \frac{q}{q-1} x + \frac{1}{1-q} z_F \qquad (4\text{-}35)$$

Equations (4-33) to (4-35) were all derived for the special case where the feed is a two-phase mixture, but they can be used for any type of feed. For example, if we want to derive Eq. (4-35) for the general case, we can start with Eq. (4-30). An overall mass balance around the feed stage (balance envelope shown in Figure 4-9) is

$$F + \overline{V} + L = V + \overline{L}$$

which can be rearranged to

$$V - \overline{V} = F - (\overline{L} - L)$$

Chapter 4   Binary Column Distillation: Internal Stage-by-Stage Balances

Substituting this result into Eq. (4-30) gives

$$y = -\frac{\overline{L} - L}{F - (\overline{L} - L)} x + \frac{Fz_F}{F - (\overline{L} - L)}$$

and dividing numerator and denominator of each term by the feed rate F, we get

$$y = -\frac{(\overline{L} - L)/F}{F/F - (\overline{L} - L)/F} x + \frac{(F/F)z_F}{F/F - (\overline{L} - L)/F}$$

which becomes Eq. (4-35), since q is defined to be $(\overline{L} - L)/F$. Equation (4-34) can be derived in a similar fashion.

Previously, we solved the mass and energy balances and found that

$$q \equiv \frac{\overline{L} - L}{F} \sim \frac{H - h_F}{H - h} \tag{4-17}$$

From Eq. (4-17) we can determine the value of q and hence the slope, $q/(q-1)$, of the feed line. For example, if the feed enters as a saturated liquid (that is, at the liquid boiling temperature at the column pressure), then $h_F = h$ and the numerator of Eq. (4-17) equals the denominator. Thus $q = 1.0$ and the slope of the feed line, $q/(q-1) = \infty$. The feed line is vertical.

The various types of feeds and the slopes of the feed line are illustrated in Table 4-1 and Figure 4-10. Note that all the feed lines intersect at one point, which is at $y = x$. If we set $y = x$ in Eq. (4-35), we obtain

$$y = x = z_F \tag{4-36}$$

as the point of intersection (try this derivation yourself). The feed line is easy to plot from the points $y = x = z_F$ or y intercept $(x = 0) = z_F/(1-q)$ or x intercept $(y = 0) = z_F/q$, and the slope, which is $q/(q-1)$. (This entire process of plotting the feed line should remind you of graphical binary flash distillation.)

The feed line was derived from the intersection of the top and bottom operating lines. It thus represents *all possible locations at which the two operating lines can intersect* for a given feed (z_F, q). Thus if we change the reflux ratio we change the points of intersection, but they all lie on the feed line. This is illustrated in Figure 4-11A. If the reflux ratio is fixed (the top

TABLE 4-1   *Feed conditions*

Type Feed	T*	h_F	q	f	Slope
Subcooled liquid	$T_F < T_{BP}$	$h_F < h$	$q > 1$	$f < 0$	> 1.0
Saturated liquid	$T_F = T_{BP}$	h	1	0	∞
Two-phase mixture	$T_{DP} > T_F > T_{BP}$	$H > h_F > h$	$1 > q > 0$	$0 < f < 1$	Negative
Saturated vapor	$T_F = T_{DP}$	H	0	1	0
Superheated vapor	$T_F > T_{DP}$	$h_F > H$	$q < 0$	$f > 1$	$1 > $ slope > 0

* T_{BP} = bubble point of feed; T_{DP} = dew point of feed.

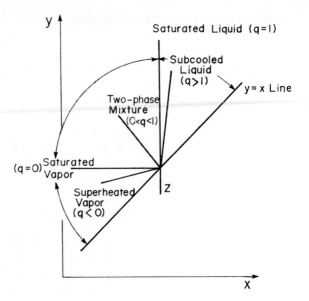

FIGURE 4-10.   *Feed lines*

operating line is fixed) but q varies, the intersection point varies as shown in Figure 4-11B. The slope of the bottom operating line, $\overline{L}/\overline{V}$, depends upon L_0/D, x_D, x_B, and q as was shown in Eq. (4-25).

In Figure 4-8 we illustrated how to determine the optimum feed stage graphically. For computer applications an explicit test is easier to use. If the point of intersection of the two operating lines (y_I, x_I), is determined, then the optimum feed plate, f, is the one for which

$$y_{f-1} < y_I < y_f \tag{4-37a}$$

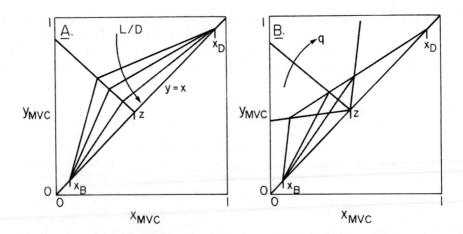

FIGURE 4-11.   *Operating line intersection; (A) changing reflux ratio with constant q;*
(B) changing q with fixed reflux ratio. Boilup ratio varies.

Chapter 4   Binary Column Distillation: Internal Stage-by-Stage Balances

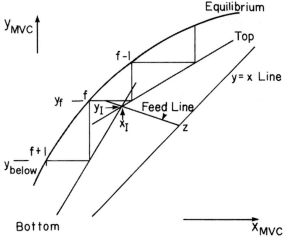

FIGURE 4-12. *Optimum feed plate calculation*

and

$$x_f < x_I < x_{f-1} \tag{4-37b}$$

This is illustrated in Figure 4-12. The intersection point can be determined by straightforward but tedious algebraic manipulation as

$$x_I = \frac{-(q-1)(1-L/V)x_D - z_F}{(q-1)(L/V) - q} \tag{4-38a}$$

$$y_I = \frac{L}{V}x_I + \left(1 - \frac{L}{V}\right)x_D \tag{4-38b}$$

for the simple column shown in Figure 3-8.

The feed equations were developed for this simple column; however, Eqs. (4-30), and (4-33) through (4-35) are valid for any column configuration if we generalize the definitions of f and q. In general,

$$q \equiv \frac{L_{below\ feed} - L_{above\ feed}}{feed\ flow\ rate} \tag{4-39a}$$

$$f \equiv \frac{V_{above\ feed} - V_{below\ feed}}{feed\ flow\ rate} \tag{4-39b}$$

EXAMPLE 4-2. Feed line calculations

Calculate the feed line slope for the following cases.

a. A two-phase feed where 80% of the feed is vaporized under column conditions.

Solution. The slope is $q/(q-1)$, where $q = (L_{\text{below feed}} - L_{\text{above feed}})/F$ (other expressions could also be used). With a two-phase feed we have the situation shown.

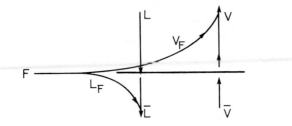

$\bar{L} = L + L_F$. Since 80% of the feed is vapor, 20% is liquid and $L_F = 0.2F$. Then

$$q = \frac{\bar{L} - L}{F} = \frac{(L + 0.2F) - L}{F} = \frac{0.2F}{F} = 0.2$$

$$\text{Slope} = \frac{q}{q-1} = \frac{0.2}{0.2 - 1} = -\frac{1}{4}$$

This agrees with Figure 4-10.

b. A superheated vapor feed where 1 mole of liquid will vaporize on the feed stage for each 9 moles of feed input.

Solution. Now the situation is shown in the following figure.

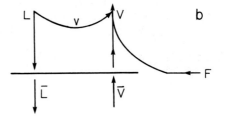

When the feed enters, some liquid must be boiled to cool the feed. Thus,

$$\bar{L} = L - \text{amount vaporized, } v \qquad\qquad \textbf{(4-40)}$$

and the amount vaporized is $v = (1/9)F$. Thus,

$$q = \frac{\bar{L} - L}{F} = \frac{L - \frac{1}{9}F - L}{F} = -\frac{\frac{1}{9}F}{F} = -\frac{1}{9}$$

$$\text{Slope} = \frac{q}{q-1} = \frac{-\frac{1}{9}}{-\frac{1}{9} - 1} = \frac{-\frac{1}{9}}{-\frac{10}{9}} = \frac{1}{10}$$

which agrees with Figure 4-10.

c. A liquid feed subcooled by 35°F. Average liquid heat capacity is 30 Btu/lbmol°F and $\lambda = 15,000$ Btu/lbmol.

Solution. Here some vapor must be condensed by the entering feed. Thus the situation can be depicted as shown.

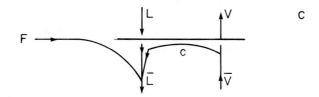

and $\overline{L} = L + F + c$, where c is the amount condensed.

Since the column is insulated, the source of energy to heat the feed to its boiling point is the condensing vapor.

$$F\,C_p\,(\Delta T) = c\lambda \qquad \text{(4-41a)}$$

where $\Delta T = T_{BP} - T_F = 35°$
or

$$c = \frac{C_p\,(\Delta T)}{\lambda}\,F \qquad \text{(4-41b)}$$

$$= \frac{(30)(35)}{15,000}\,F = 0.07F$$

$$q = \frac{\overline{L} - L}{F} = \frac{L + F + 0.07F - L}{F} = 1.07$$

$$\text{Slope} = \frac{q}{q-1} = \frac{1.07}{1.07 - 1} = 15.29$$

This agrees with Figure 4-10. Despite the large amount of subcooling, the feed line is fairly close to vertical, and the results will be similar to a saturated liquid feed. If T_F is given instead of ΔT, we need to estimate T_{BP}. This can be done with a temperature composition graph (Figure 2-3), an enthalpy-composition graph (Figure 2-4), or a bubble point calculation (Section 5.4).

d. A mixture of ethanol and water that is 40 mol% ethanol. Feed is at 40°C. Pressure is 1.0 kg/cm².

Solution. We can now use Eq. (4-17):

$$q \sim \frac{H - h_F}{H - h}$$

The enthalpy data are available in Figure 2-4. To use that figure we must convert to weight fraction. 0.4 mole fraction is 0.63 wt frac. Then from Figure 2-4 we have

$$h_F(0.63, 40°C) = 20 \text{ kcal/kg}$$

The vapor (represented by H) and liquid (represented by h) will be in equilibrium at the feed stage, but the concentrations of the feed stage are not known. Comparing the feed stage locations in Figures 4-8A, 4-8B, and 4-8C, we see that liquid and vapor concentrations on the feed stage can be very different and are usually not equal to the feed concentration z or to the concentrations of the intersection point of the operating line, y_I and x_I (Figure 4-12). However, since CMO is valid, H and h in molal units will be constant. We can calculate all enthalpies at a weight fraction of 0.63, convert the enthalpies to enthalpies per kilogram mole, and estimate q. From Figure 2-4 H (0.63, satd vapor) = 395, h (0.63, satd liquid) = 65 kcal/kg, and

$$q \sim \frac{H(MW) - h_F(MW)}{H(MW) - h(MW)}$$

Since all the molecular weights are at the same concentration, they divide out.

$$q \sim \frac{H - h_F}{H - h} = \frac{395 - 20}{395 - 65} = 1.136$$

$$\text{Slope} = \frac{q}{q - 1} \sim \frac{1.136}{0.136} = 8.35$$

This agrees with Figure 4-10. Despite considerable subcooling, this feed line is also steep. Note that feed rate was not needed to calculate q or the slope for any of these calculations.

4.5 COMPLETE McCABE-THIELE METHOD

We are now ready to put all the pieces together and solve a design distillation problem by the McCabe-Thiele method. We will do this in the following example.

EXAMPLE 4-3. McCabe-Thiele method
A distillation column with a total condenser and a partial reboiler is separating an ethanol-water mixture. The feed is 20 mol% ethanol, feed rate is 1000 kmol/h, and feed temperature is 80°F. A distillate composition of 80 mol% ethanol and a bottoms composition of 2 mol% ethanol are desired. The external reflux ratio is 5/3. The reflux is returned as a saturated liquid and CMO can be assumed. Find the optimum feed plate location and the total number of equilibrium stages required. Pressure is 1 atm.

Solution

A. Define. The column is sketched in the figure

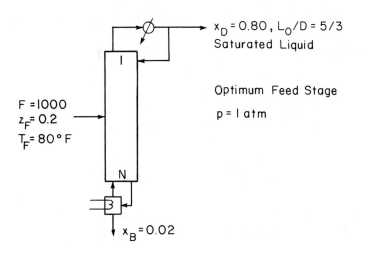

$x_D = 0.80$, $L_0/D = 5/3$
Saturated Liquid

F = 1000
$z_F = 0.2$
$T_F = 80°F$

Optimum Feed Stage

$p = 1$ atm

$x_B = 0.02$

Find the optimum feed plate location and the total number of equilibrium stages.

B. Explore. Equilibrium data at 1 atm are given in Figure 2-2. An enthalpy-composition diagram at 1 atm will be helpful to estimate q. These are available in other sources (e.g., Brown et al., 1950, or Foust et al., 1980, p. 36), or a good estimate of q can be made from Figure 2-4 despite the pressure difference. In Example 4-1 we showed that CMO is valid. Thus we can apply the McCabe-Thiele method.

C. Plan. Determine q from Eq. (4-17) and the enthalpy-composition diagram at 1 atm. Plot the feed line. Calculate L/V. Plot the top operating line; then plot the bottom operating line and step off stages.

D. Do It. *Feed Line:* To find q, first convert feed concentration, 20 mol%, to wt % ethanol = 39 wt %. Two calculations in different units with different data are shown.

Exact Calculation
Data at p = 1 atm
from Brown et al. (1950)

Approx Calc.
(p = 1 kg/cm²)
using Figure 2-4

$h_F = 25$ Btu/lb (80 °F)
H = 880 Btu/lb (sat'd vapor)
h = 125 Btu/lb (sat'd liquid)

$h_F = 15$ kcal/kg (30 °C)
H = 485 kcal/kg (sat'd vapor)
h = 70 kcal/kg (sat'd liquid)

$$q = \frac{880 - 25}{880 - 125} = 1.13$$

$$q \sim \frac{485 - 15}{485 - 70} = 1.13$$

Thus small differences caused by pressure differences in the diagrams do not change the value of q. Note that molecular weight terms divide out as in Example 4-2d. Then

$$\text{Slope of feed line} = \frac{q}{q-1} \sim 8.7$$

Feed line intersects y=x line at feed concentration z=0.2. Feed line is plotted in Figure 4-13.

Top Operating Line:

$$y = \frac{L}{V} x + \left(1 - \frac{L}{V}\right)x_D$$

$$\text{Slope} = \frac{L}{V} = \frac{L/D}{1+L/D} = \frac{5/3}{1+5/3} = 5/8$$

$$y \text{ intercept} = \left(1 - \frac{L}{V}\right)x_D = (3/8)(0.8) = 0.3$$

Alternative solution: Intersection of top operating line and y = x (solve top operating line and y = x simultaneously) is at y = x = x_D. The top operating line is plotted in Figure 4-13.

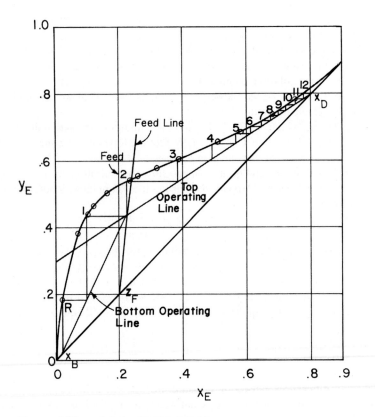

FIGURE 4-13.   *Solution for Example 4-3*

Chapter 4 Binary Column Distillation: Internal Stage-by-Stage Balances

Bottom Operating Line:

$$y = \frac{\overline{L}}{\overline{V}} x - (\frac{\overline{L}}{\overline{V}} - 1)x_B$$

We know that the bottom operating line intersects the top operating line at the feed line; this is one point. We could calculate $\overline{L}/\overline{V}$ from mass balances or from Eq. (4-25), but it easier to find another point. The intersection of the bottom operating line and the y = x line is at y = x = x_B (see Problem 4.C9). This gives a second point.

The feed line, top operating line, and bottom operating line are shown in Figure 4-13. We stepped off stages from the bottom up (this is an arbitrary choice). The optimum feed stage is the second above the partial reboiler. 12 equilibrium stages plus a partial reboiler are required.

E. Check. We have a built-in check on the top operating line, since a slope and two points are calculated. The bottom operating line can be checked by calculating $\overline{L}/\overline{V}$ from mass balances and comparing it to the slope. The numbers are reasonable, since L/V < 1, $\overline{L}/\overline{V}$ > 1, and q > 1 as expected. The most likely cause of error in Figure 4-13 (and the hardest to check) is the equilibrium data.

F. Generalization. If constructed carefully, the McCabe-Thiele diagram is quite accurate. Note that there is no need to plot parts of the equilibrium diagram that are greater than x_D or less than x_B. Specified parts of the diagram can be expanded to increase the accuracy.

We did not have to use external balances in this example, while in Example 4-1 we did. This is because we used the feed line as an aid in finding the bottom operating line. The y = x intersection points are useful, but when the column configuration is changed their location may change.

4.6 PROFILES FOR BINARY DISTILLATION

Figure 4-13 essentially shows the complete solution of Example 4-3; however, it is useful to plot compositions, temperatures, and flow rates leaving each stage (these are known as *profiles*). From Figure 4-13 we can easily find the ethanol mole fractions in the liquid and vapor leaving each stage. Then $x_W = 1 - x_E$ and $y_W = 1 - y_E$. The temperature of each stage can be found from equilibrium data (Figure 2-3) because the stages are equilibrium stages. Since we assumed CMO, the flow rates of liquid and vapor will be constant in the enriching and stripping sections, and we can determine the changes in the flow rates at the feed stage from the calculated value of q.

The profiles are shown in Figure 4-14. As expected, the water concentration in both liquid and vapor streams decreases monotonically as we go up the column, while the ethanol concentration increases. Since the stages are discrete, the profiles are not smooth curves. Compare Figures 4-13 and 4-14. Note where the operating line and equilibrium curve are close together. When these two lines almost touch, we have a *pinch point*. Then the composition and temperature profiles will become almost horizontal and there will be very little

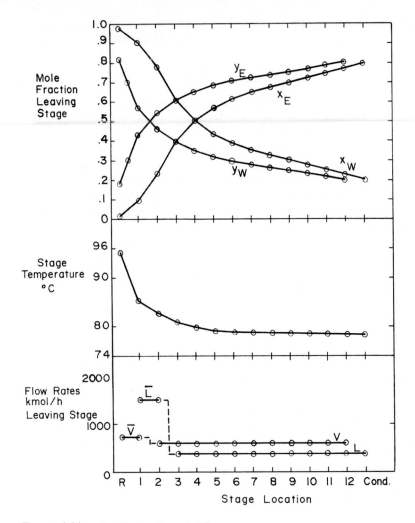

FIGURE 4-14.   *Profiles for Example 4-3*

change in composition from stage to stage. The location of a pinch point within the column depends on the system and the operating conditions.

In this ethanol-water column the temperature decreases rapidly for the first few contacts above the reboiler but is almost constant for the last eight stages. This occurs mainly because of the shape of the temperature-composition diagram for ethanol-water (see Figure 2-3).

Since we assumed CMO, the flow profiles are flat in each section of the column. As expected, $\overline{L} > \overline{V}$ and $V > L$ (a convenient check to use). Since stage 2 is the feed stage, L_2 is in the stripping section while V_2 is in the enriching section (draw a sketch of the feed stage if this isn't clear). Different quality feeds will have different changes at the feed stage. Liquid and vapor flow rates can increase, decrease, or remain unchanged in passing from the stripping to the enriching section.

Figure 4-13 illustrates the main advantage of McCabe-Thiele diagrams. They allow us to visualize the separation. Before the common use of digital computers, large (sometimes covering a wall) McCabe-Thiele diagrams were used to design distillation columns. McCabe-Thiele diagrams cannot compete with the speed and accuracy of process simulators (see this chapter's appendix) or for binary separations with spreadsheets; however, McCabe-Thiele diagrams still provide superior visualization of the separation (Kister, 1995). Ideally, McCabe-Thiele diagrams will be used in conjunction with process simulator results for both analysis and troubleshooting.

4.7   OPEN STEAM HEATING

We now have all the tools required to solve any binary distillation problem with the graphical McCabe-Thiele procedure. As a specific example, consider the separation of methanol from water in a staged distillation column.

EXAMPLE 4-4. McCabe-Thiele analysis of open steam heating

The feed is 60 mol% methanol and 40 mol% water and is input as a two-phase mixture that flashes so that $V_F/F = 0.3$. Feed flow rate is 350 kmol/h. The column is well insulated and has a total condenser. The reflux is returned to the column as a saturated liquid. An external reflux ratio of $L_0/D = 3.0$ is used. We desire a distillate concentration of 95 mol% methanol and a bottoms concentration of 8 mol% methanol. Instead of using a reboiler, saturated steam at 1 atm is sparged directly into the bottom of the column to provide boilup. (This is called direct or open steam.) Column pressure is 1 atm. Calculate the number of equilibrium stages and the optimum feed plate location.

Solution

A. Define. It helps to draw a schematic diagram of the apparatus, particularly since a new type of distillation is involved. This is shown in Figure 4-15. We wish to find the optimum feed plate location, N_F, and the total number of equilibrium stages, N, required for this separation. We could also calculate Q_c, D, B, and the steam rate S, but these were not asked for. We assume that the column is adiabatic since it is well insulated.

B. Explore. The first thing we need is equilibrium data. Fortunately, these are readily available (see Table 2-7 in Problem 2.D1).

Second, we would like to assume CMO so that we can use the McCabe-Thiele analysis procedure. An easy way to check this assumption is to compare the latent heats of vaporization per mole (Himmelblau, 1974).

$$\Delta H_{vap} \text{ methanol (at bp)} = 8.43 \text{ kcal/mol}$$

$$\Delta H_{vap} \text{ water (at bp)} = 9.72 \text{ kcal/mol}$$

These values are not equal, and in fact, water's latent heat is 15.3% higher than methanol's. Thus, CMO is not strictly valid; however, we will solve this problem assuming CMO and will check our results with a process simulator.

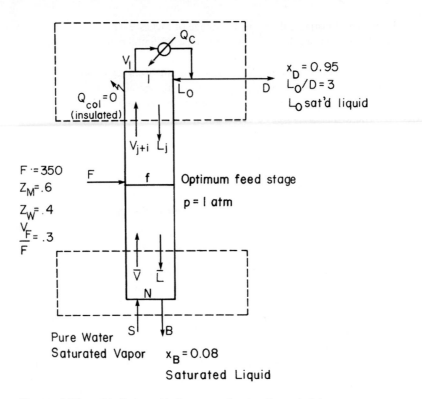

FIGURE 4-15.   *Distillation with direct steam heating, Example 4-4*

A look at Figure 4-15 shows that the configuration at the bottom of the column is different than when a reboiler is present. Thus we should expect that the bottom operating equations will be different from those derived previously.

C. Plan. We will use a McCabe-Thiele analysis. Plot the equilibrium data on a y-x graph.

Top Operating Line: Mass balances in the rectifying section (see Fig. 4-15) are

$$V_{j+1} = L_j + D$$

$$y_{j+1}V_{j+1} = L_j x_j + Dx_D$$

Assume CMO and solve for y_{j+1}.

$$y_{j+1} = (L/V)x_j + (1 - L/V)x_D$$

Slope = L/V,   y intercept (x = 0) = $(1 - L/V) x_D$

Intersection y = x= x_D

Since the reflux is returned as a saturated liquid,

$$L/V = \frac{L_0}{L_0 + D} = \frac{L_0/D}{L_0/D + 1}$$

Enough information is available to plot the top operating line.
Feed Line:

$$y = \frac{q}{(q-1)} \, x + \frac{z}{(1-q)}$$

$$\text{Slope} = \frac{q}{q-1}, \qquad y \text{ intercept } (x = 0) = \frac{z}{q-1}$$

Intersection: $y = x = z$

$$q = \frac{\overline{L} - L}{F} = \frac{\overline{V} - V + F}{F} = \frac{F}{F} - \frac{V_F}{F} = 1 - \frac{V_F}{F}$$

Once we substitute in values, we can plot the feed line.
Bottom Operating Line: The mass balances are

$$\overline{V} + B = \overline{L} + S$$
$$\overline{V}y + Bx_B = \overline{L}x + Sy_S \text{ (for methanol)}$$

Solve for y:

$$y = (\overline{L}/\overline{V})x + (S/\overline{V}) \, y_s - (B/\overline{V}) \, x_B$$

Simplifications: Since the steam is pure water vapor, $y_s = 0.0$ (contains no methanol). Since steam is saturated, $S = V$ and $B = L$ (constant molal overflow). Then

$$y = (\overline{L}/\overline{V})x - (\overline{L}/\overline{V}) \, x_B \qquad \textbf{(4-42)}$$

Note this *is* different from the operating equation for the bottom section when a reboiler is present. Slope $= \overline{L}/\overline{V}$ (unknown), y intercept $= -(\overline{L}/\overline{V})x_B$ (unknown),

$$y = x = \frac{-(\overline{L}/\overline{V})x_B}{(1 - \overline{L}/\overline{V})} \text{ (unknown)}$$

One known point is the intercept of the top operating line with the feed line. We still need a second point, and we can find it at the x intercept. When y is set to zero, $x = x_B$ (this is left as Problem 4.C1).
D. Do It. Equilibrium data are plotted on Figure 4-16.
Top Operating Line:

$$L/V = \frac{L/D}{1 + L/D} = 3/4 = \text{slope}$$

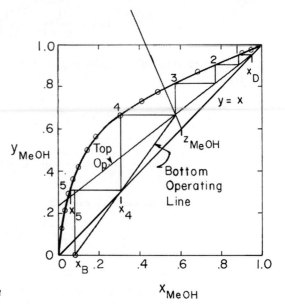

FIGURE 4-16.   *Solution for Example 4-4*

$$y = x = x_D = 0.95$$

$$y \text{ intercept} = (1 - L/V)x_D = 0.2375$$

We can plot this straight line as shown in Figure 4-16.

Feed Line: Slope = $q/(q - 1) = 0.7/(0.7 - 1) = -7/3$.

Intersects at $y = x = z = 0.6$. Plotted in Figure 4-16.

Bottom Operating Line: We can plot this line between two points, the intercept of top operating line and feed line, and

$$x \text{ intercept } (y = 0) = x_B = 0.08$$

This is also shown in Figure 4-16.

Step off stages, starting at the top. x_1 is in equilibrium with y_1 at x_D. Drawing a horizontal line to the equilibrium curve gives value x_1. y_2 and x_1 are related by the operating line. At constant y_2 (horizontal line), go to the equilibrium curve to find x_2. Continue this stage-by-stage procedure.

Optimum feed stage is determined as in Figure 4-8A. Optimum feed in Figure 4-16 is on stage 3 or 4 (since by accident x_3 is at intersection point of feed and operating lines). Since the feed is a two-phase feed, we would introduce it above stage 4 in this case.

Number of stages: Five is more than enough. We can calculate a fractional number of stages.

$$\text{Frac} = \frac{\text{distance from operating line to product}}{\text{distance from operating line to equilibrium curve}} \qquad \textbf{(4-43)}$$

In Figure 4-16,

$$\text{Frac} = \frac{x_B - x_4}{x_5 - x_4} = \frac{0.08 - 0.305}{0.06 - 0.305} = 0.9$$

We need $4 + 0.9 = 4.9$ equilibrium contacts.

E. Check. There are a series of internal consistency checks that can be made. Equilibrium should be a smooth curve. This will pick up misplotted points. $L/V < 1$ (otherwise no distillate product), and $\overline{L/V} > 1$ (otherwise no bottoms product). The feed line's slope is in the correct direction for a two-phase feed. A final check on the assumption of CMO would be advisable since the latent heats vary by 15%.

This problem was also run on the Aspen Plus process simulator (see Problem 4.G1 and chapter appendix). Aspen Plus does not assume CMO and with an appropriate vapor-liquid equilibrium (VLE) correlation (the nonrandom two-liquid model was used) should be more accurate than the McCabe-Thiele diagram, which assumes CMO. With 5 equilibrium stages and feed on stage 4 (the optimum location), $x_D = 0.9335$ and $x_B = 0.08365$, which doesn't meet the specifications. With 6 equilibrium stages and feed on stage 5 (the optimum), $x_D = 0.9646$ and $x_B = 0.0768$, which is slightly better than the specifications. The differences in the McCabe-Thiele and process simulation results are due to the error involved in assuming CMO and, to a lesser extent, differences in equilibrium. Note that the McCabe-Thiele diagram is useful since it visually shows the effect of using open steam heating.

F. Generalize. Note that the $y = x$ line is not always useful. Don't memorize locations of points. Learn to derive what is needed. The total condenser does not change compositions and is not counted as an equilibrium stage. The total condenser appears in Figure 4-16 as the single point $y = x = x_D$. Think about why this is true. In general, all inputs to the column can change flow rates and hence slopes inside the column. The purpose of the feed line is to help determine this effect. The reflux stream and open steam are also inputs to the column. If they are not saturated streams the flow rates are calculated differently; this is discussed later.

Note that the open steam can be treated as a feed with $q = (L_{below} - L_{above})/S = (B - \overline{L})/S = 0$. Thus, $B = \overline{L}$. The slope of this feed line is $q/(q - 1) = 0$ and it intersects the $y = x$ line at $y = x = z = 0$, which means the feed line for saturated steam is the x-axis.

Ludwig (1997) states that one tray is used to replace the reboiler and 1/3 to one and possibly more trays to offset the water dilution; however, since reboilers are typically much more expensive than trays (see Chapter 11) this practice is economical. Open steam heating can be used even if water is not one of the original components.

Note to Students: When you read the description of developing the bottom operating equation and plotting the bottom operating line in Example 4-4, it probably appears a lot easier than doing the development by yourself will prove to be. You need to practice deriving operating equations. Then practice simplifying the operating equation, realizing that "pure steam" means $y_s = 0$ and that "saturated steam" means $S = V$ and thus $B = L$ are nontrivial steps. How did we know that the bottom operating line could be plotted using the point $x = 0$

(the y intercept)? We did not know this in advance, but tried it and it worked. You will become better at solving these problems as you work additional problems. To aid in this, there are a large number of problems at the end of this chapter.

4.8 GENERAL McCABE-THIELE ANALYSIS PROCEDURE

The open steam example illustrated one specific case. It is useful to generalize this analysis procedure. A *section* of the column is the segment of stages between two input or exit streams. Thus in Figure 4-15 there are two sections: top and bottom. Figure 4-17 illustrates a column with four sections. Each section's operating equation can be derived independently. Thus, the secret (if that's what it is) is to treat each section as an independent subproblem connected to the other subproblems by the feed lines (which are also independent).

An algorithm for any problem is the following:

1. Draw a figure of the column and label all known variables (e.g., as in Figure 4-15). Check to see if CMO is valid.
2. For each section:
    a. Draw a mass balance envelope. We desire this envelope to cut the unknown liquid and vapor streams in the section and known streams (feeds, specified products or specified side-streams). The fewer streams involved, the simpler the mass balances will be. This step is important, since it controls how easy the following steps will be.
    b. Write the overall and most volatile component mass balances.
    c. Derive the operating equation.
    d. Simplify.
    e. Calculate all known slopes, intercepts, and intersections.

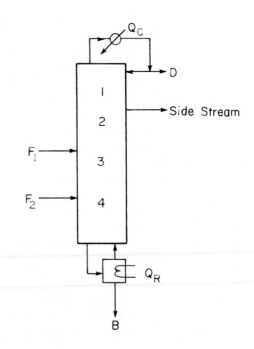

FIGURE 4-17.   *Distillation column with four sections*

3. Develop feed line equations. Calculate q values, slopes, and y = x intersections.
4. For operating and feed lines:
   a. Plot as many of the operating lines and feed lines as you can.
   b. If all operating lines cannot be plotted, step off stages if the stage location of any feed or side stream is specified.
   c. If needed, do external mass and energy balances (see Example 4-5). Use the values of D and B in step 2.
5. When all operating lines have been plotted, step off stages, determine optimum feed plate locations and the total number of stages. If desired, calculate a fractional number of stages.

Not all of these general steps were illustrated in the previous examples, but they will be illustrated in the examples that follow.

This problem-solving algorithm should be used as a guide, not as a computer code to be followed exactly. The wide variety of possible configurations for distillation columns will allow a lot of practice in using the McCabe-Thiele method for solving problems.

EXAMPLE 4-5. Distillation with two feeds
We wish to separate ethanol from water in a distillation column with a total condenser and a partial reboiler. We have 200 kmol/h of feed 1, which is 30 mol% ethanol and is saturated vapor. We also have 300 kmol/h of feed 2, which is 40

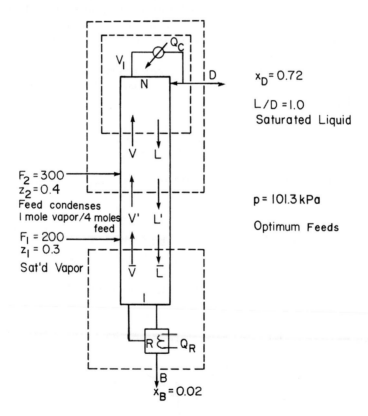

FIGURE 4-18.   *Two-feed distillation column for Example 4-5*

mol% ethanol. Feed 2 is a subcooled liquid. One mole of vapor must condense inside the column to heat up 4 moles of feed 2 to its boiling point. We desire a bottoms product that is 2 mol% ethanol and a distillate product that is 72 mol% ethanol. External reflux ratio is $L_0/D = 1.0$. The reflux is a saturated liquid. Column pressure is 101.3 kPa, and the column is well insulated. The feeds are to be input at their optimum feed locations. Find the optimum feed locations (reported as stages above the reboiler) and the total number of equilibrium stages required.

Solution

A. Define. Again a sketch will be helpful; see Figure 4-18. Since the two feed streams are already partially separated, it makes sense to input them separately to maintain the separation that already exists. We have made an inherent assumption in Figure 4-18. That is, feed 2 of higher mole fraction ethanol should enter the column higher up than feed 1. This assumption will be checked when the optimum feed plate locations are calculated, but it will affect the way we do the preliminary calculations. Since the feed plate locations were asked for as stages above the reboiler, the stages have been numbered from the bottom up.

B. Explore. Obviously, equilibrium data are required, and they are available from Figure 2-2. We already checked (Example 4-1) that CMO is a reasonable assumption. A look at Figure 4-18 shows that the top section is the same as top sections used previously. The bottom section is also familiar. Thus the new part of this problem is the middle section. There will be two feed lines and three operating lines.

C. Plan. We will look at the two feed lines, top operating line, bottom operating line, and middle operating line. The simple numerical calculations will also be done here.

$$\text{Feeds: } y = \frac{q_i}{q_i - 1} x - \frac{z_i}{q_i - 1}$$

Feed 1: Saturated vapor, $q_1 = 0$, slope = 0, y intercept = $z_1/(q_1 - 1) = 0.3$, intersection at $y = x = z_1 = 0.3$.
Feed 2:

$$q_2 = \frac{L_{below} - L_{above}}{F_2}$$

The feed stage for feed 2 looks schematically as shown in the figure:

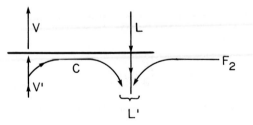

Then, $L_{below\ feed} = L' = L + F_2 + c$
Amount condensed = $c = (1/4)\ F_2$,

$$q_2 = \frac{(L + F_2 + c) - L}{F_2} = \frac{F_2 + (1/4)F_2}{F_2} = 5/4$$

$$\text{Slope} = \frac{q_2}{q_2 - 1} = \frac{5/4}{5/4 - 1} = 5$$

Intersection: $y = x = z_2 = 0.4$
Top Operating Line: We can derive the top operating equation:

$$y = \frac{L}{V}\ x + (1 - \frac{L}{V})x_D$$

This is the usual top operating line. With saturated liquid reflux the slope is,

$$\frac{L}{V} = \frac{L}{L + D} = \frac{L/D}{L/D + 1} = 1/2$$

Intersection: $y = x = x_D = 0.72$
Bottom Operating Line: We can derive the bottom operating equation:

$$y = \frac{\overline{L}}{\overline{V}}\ x - (\frac{\overline{L}}{\overline{V}} - 1)x_B$$

This is the usual bottom operating line. Then, slope $= \overline{L}/\overline{V}$ is unknown.
    $y = x$ intersection is at $y = x = x_B = 0.02$. One other point is the intersection of the F_1 feed line and the middle operating line, if we can find the middle operating line. *Middle Operating Line:* To derive an operating equation for the middle section, we can write a mass balance around the top or the bottom of the column. The resulting equations will look different, but they are equivalent. We arbitrarily use the mass balance envelope around the top of the column as shown in Figure 4-18. These mass balances are

$$F_2 + V' = L' + D$$
$$F_2 z_2 + V'\ y = L'\ x + Dx_D\ \text{(MVC)}$$

Solve this equation for y to develop the middle operating equation,

$$y = \frac{L'}{V'}\ x + \frac{Dx_D - F_2 z_2}{V'} \qquad\qquad \textbf{(4-44)}$$

    One known point is the intersection of the F_2 feed line and the top operating line. A second point is needed. We can try y intercept =

$$\frac{Dx_D - F_2 z_2}{V'}$$

but this is unknown.

The intersection of the middle operating line with the y = x line is found by setting y = x in Eq. (4-44) and solving,

$$x = y = \frac{Dx_D - F_2 z_2}{V' - L'}$$

From the overall balance equation,

$$V' - L' = D - F_2$$

Thus,

$$x = y = \frac{Dx_D - F_2 x_2}{D - F_2} \qquad \text{(4-45)}$$

This point is not known, but it can be calculated once D is known.

Slope = L'/V' is unknown, but it can be calculated from the feed-stage calculation. From the definition of q_2,

$$L' = F_2 q_2 + L \qquad \text{(4-46)}$$

where $L = (L/D)D$. Once D is determined, L and then L' can be calculated. Then the mass balance gives

$$V' = L' + D - F_2$$

and the slope L'/V' can be calculated.

The conclusion from these calculations is, we have to calculate D. To do this we need external mass balances:

$$F_1 + F_2 = D + B \qquad \text{(4-47a)}$$

$$F_1 z_1 + F_2 z_2 = Dx_D + Bx_B \text{ (MVC)} \qquad \text{(4-47b)}$$

These two equations have two unknowns, D and B, and we can solve for them. Rearranging the first, we have $B = F_1 + F_2 - D$. Then, we substitute this into the more volatile component (MVC) mass balance,

$$F_1 z_1 + F_2 z_2 = Dx_D + (F_1 + F_2)x_B - Dx_B$$

and we solve for D:

$$D = \frac{F_1 z_1 + F_2 z_2 - (F_1 + F_2)x_B}{x_D - x_B} \qquad \text{(4-48)}$$

D. Do It. The two feed lines and the top operating line can immediately be plotted on a y-x diagram. This is shown in Figure 4-19. Before plotting the middle operating line we must find D from Eq. (4-48).

$$D = \frac{(200)(0.3) + (300)(0.4) - (500)(0.02)}{0.72 - 0.02}$$

$$= 242.9 \text{ kmol/h}$$

Now the middle operating line y=x intercept can be determined from Eq. (4-45)

$$y = x = \frac{Dx_D - F_2x_2}{D - F_2}$$

$$= \frac{(242.9)(0.72) - (300)(0.4)}{242.9 - 300} = -0.96$$

This intersection point could be used, but it is off of the graph in Figure 4-19. Instead of using a larger sheet of paper, we will calculate the slope L'/V'.

$$L' = F_2q_2 + L = F_2q_2 + (\frac{L}{D})(D)$$

$L' = 300(5/4) + (1.0)(242.9) = 617.9 \text{ kmol/h}$
$V' = L' + D - F_2 = 617.9 + 242.9 - 300 = 560.8$
$L'/V' = 1.10$

    Now the middle operating line is plotted in Figure 4-19 from the intersection of feed line 2 and the top operating line, with a slope of 1.10. The bottom operating

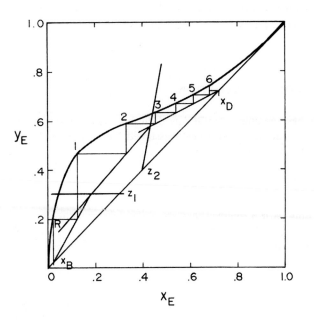

line then goes from $y = x = x_B$ to the intersection of the middle operating line and feed line 1 (see Figure 4-19).

Since the feed locations were desired as stages above the reboiler, we step off stages from the bottom up, starting with the partial reboiler as the first equilibrium contact. The optimum feed stage for feed 1 is the first stage, while the optimum feed stage for feed 2 is the second stage. Six stages + partial reboiler are more than sufficient.

$$\text{Fraction} = \frac{x_D - \text{operating eq. value}}{\text{equilibrium value} - \text{operating eq. value}} \qquad \textbf{(4-49)}$$

$$\text{Frac} = \frac{x_D - y_5}{y_6 - y_5} = 1/2$$

We need 5½ stages + partial reboiler.

E. Check. The internal consistency checks all make sense. Note that L'/V' can be greater than or less than 1.0. Since the latent heats of vaporization per mole are close, CMO is probably a good assumption. Our initial assumption that feed 2 enters the column higher up than feed 1 is shown to be valid by the McCabe-Thiele diagram. We could also calculate $\overline{L}$ and $\overline{V}$ and check that the slope of the bottom operating line is correct.

F. Generalize. The method of inserting the overall mass balance to simplify the intersection of the $y = x$ line and middle operating line to derive Eq. (4-45) can be used in other cases. The method for calculating L'/V' can also be generalized to other situations. That is, we can calculate D (or B), find flow rate in section above (or below), and use feed conditions to find flow rates in the desired section. Since we stepped off stages from the bottom up, the fractional stage is calculated from the difference in y values (that is, vertical distances) in Eq. (4-49). Industrial problems use lower reflux ratios and have more stages. A relatively large reflux ratio is used in this example to keep the graph simple.

4.9   OTHER DISTILLATION COLUMN SITUATIONS

A variety of modifications of the basic columns are often used. In this section we will briefly consider the unique aspects of several of these. CMO will be assumed. Detailed examples will not be given but will be left to serve as homework problems.

4.9.1   Partial Condensers

A partial condenser condenses only part of the overhead stream and returns this as reflux. This distillate product is removed as vapor as shown in Figure 4-20. If a vapor distillate is desired, then a partial condenser will be very convenient. The partial condenser acts as one equilibrium contact.

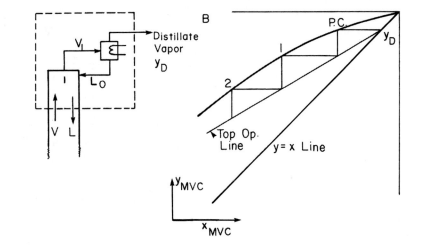

FIGURE 4-20. *Partial condenser; (A) balance envelope, (B) top operating line*

If a mass balance is done on the more volatile component using the mass balance envelope shown in Figure 4-20, we obtain

$$Vy = Lx + Dy_D$$

Removing D and solving for y, we obtain the operating equation

$$y = \frac{L}{V}x + (1 - \frac{L}{V})y_D$$

(4-50)

This is essentially the same as the equation for a top operating line with a total condenser except that y_D has replaced x_D. The top operating line will intersect the y = x line at y = x = y_D. The top operating line is shown in Figure 4-20. The major difference between this case and that for a total condenser is that the partial condenser serves as the first equilibrium contact.

4.9.2 Total Reboilers

A total reboiler vaporizes the entire stream sent to it; thus, the vapor composition is the same as the liquid composition. This is illustrated in Figure 4-21. The mass balance and the bottom operating equation with a total reboiler are exactly the same as with a partial reboiler. The only difference is that a partial reboiler is an equilibrium contact and is labeled as such on the McCabe-Thiele diagram. The total reboiler is not an equilibrium contact and appears on the McCabe-Thiele diagram as the single point y = x = x_B.

Some types of partial reboilers may act as more or less than one equilibrium contact. In these cases, exact details of the reboiler construction are required.

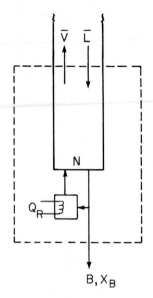

FIGURE 4-21. *Total reboiler*

B, X$_B$

4.9.3   Side Streams or Withdrawal Lines

If a product of intermediate composition is required, a vapor or liquid side stream may be withdrawn. This is commonly done in petroleum refineries and is illustrated in Figure 4-22A for a liquid side stream. Three additional variables such as flow rate, S, type of side draw (liquid or vapor), and location or composition x_S or y_S, must be specified. The operating equation for the middle section can be derived from mass balances around the top or bottom of the column. For the situation shown in Figure 4-22A, the middle operating equation is

$$y = \frac{L'}{V'} x + \frac{Dx_D + Sx_s}{V'}$$

(4-51)

The y = x intercept is

$$x = y = \frac{Dx_D + Sx_s}{D + S}$$

(4-52)

This point can be plotted if S, x_S, D, and x_D are known. Derivation of Eqs. (4-51) and (4-52) is left as Problem 4.C3.

A second point can be found where the side stream is withdrawn. A saturated liquid withdrawal is equivalent to a negative feed of concentration x_S. Thus there must be a vertical feed line at x = x_S. The top and middle operating lines must intersect at this feed line.

Side-stream calculations have one difference that sets them apart from feed calculations. The stage must hit exactly at the point of intersection of the two operating lines. This is illustrated in Figure 4-22B. Since the liquid side stream is withdrawn from tray 2, we must have $x_S = x_2$. If the stage location is given, x_S can be found by stepping off the required number of stages.

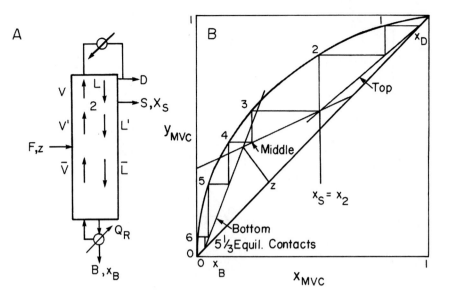

FIGURE 4-22. *Liquid side stream (A) column, (B) McCabe-Thiele diagram*

For a liquid withdrawal, a balance on the liquid gives

$$L = L' + S \tag{4-53}$$

while vapor flow rates are unchanged, $V = V'$. Thus slope, L'/V', of the middle operating line can be determined if L and V are known. L and V can be determined from L/D and D, where D can be found from external balances once x_S is known.

For a vapor side stream, the feed line is horizontal at $y = y_S$. A balance on vapor flow rates gives

$$V' = V + S_V \tag{4-54}$$

while liquid flow rates are unchanged. Again L'/V' can be calculated if L and V are known.

If a specified value of x_S (or y_S) is desired, the problem is trial and error. The top operating line is adjusted (change L/D) until a stage ends exactly at x_S or y_S.

Calculations for side streams below the feed can be developed using similar principles (see Problem 4.C4).

4.9.4 Intermediate Reboilers and Intermediate Condensers

Another modification that is used occasionally is to have an intermediate reboiler or an intermediate condenser. The intermediate reboiler removes a liquid side stream from the column, vaporizes it, and reinjects the vapor into the column. An intermediate condenser removes a vapor side stream, condenses it, and reinjects it into the column. Figure 4-23A illustrates an intermediate reboiler.

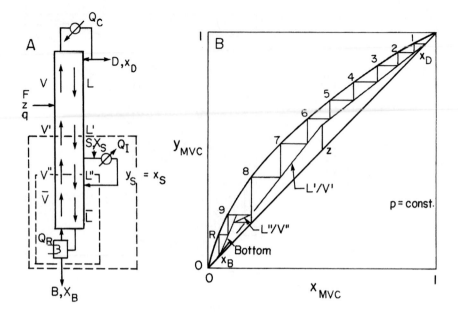

FIGURE 4-23. *Intermediate reboiler; (A) balance envelopes, (B) McCabe-Thiele diagram*

An energy balance around the column will show that Q_R without an intermediate reboiler is equal to $Q_R + Q_I$ with the intermediate reboiler (F, z, q, x_D, x_B, p, L/D constant). Thus the amount of energy required is unchanged; what changes is the temperature at which it is required. Since $x_S > x_B$, the temperature of the intermediate reboiler is lower than that of the reboiler, and a cheaper heat source can be used. (Check this out with equilibrium data.)

Since the column shown in Figure 4-23A has four sections, there will be four operating lines. This is illustrated in the McCabe-Thiele diagram of Figure 4-23B. One would specify that the liquid be withdrawn at flow rate S at either a specified concentration x_S or a given stage location. The saturated vapor is at concentration $y_S = x_S$. Thus there is a horizontal feed line at y_S. If the optimum location for inputting the vapor is immediately below the stage where the liquid is withdrawn, the L″/V″ line will be present, but no stages will be stepped off on it, as shown in Figure 4-23B. (The optimum location for vapor feed may be several stages below the liquid withdrawal point.) Development of the two middle operating lines is left as Problem 4.C5. Use the mass balance envelopes shown in Figure 4-23A.

Intermediate condensers are useful since the coolant can be at a higher temperature. See Problem 4.C6. Intermediate reboilers and condensers are fairly common because they allow for more optimum use of energy resources, and they can help balance column diameters (see Section 10.4). During normal operation, they should cause no problems; however, startup may be difficult (Sloley, 1996).

4.9.5 Stripping and Enriching Columns

Up to this point we have considered complete distillation columns with at least two sections. Columns with only a stripping section or only an enriching section are also commonly used. These are illustrated in Figures 4-24A and B. When only a stripping section is used, the feed must be a subcooled or saturated liquid. No reflux is used. A very pure bottoms product can be obtained but the vapor distillate will not be pure. In the enriching or rectifying column, on the other hand, the feed is a superheated vapor or a saturated vapor, and the distillate can be very pure but the bottoms will not be very pure. Striping columns and enriching columns are used when a pure distillate or a pure bottoms, respectively, is not needed.

Analysis of stripping and enriching columns is similar. We will analyze the stripping column here and leave the analysis of the enriching column as a homework assignment (Problem 4.C8). The stripping column shown in Figure 4-24A can be thought of as a complete distillation column with zero liquid flow rate in the enriching section. Then the top operating line is $y = y_D$. The bottom operating line can be derived as

$$y = (\overline{L}/\overline{V})x - (\overline{L}/\overline{V} - 1)x_B$$

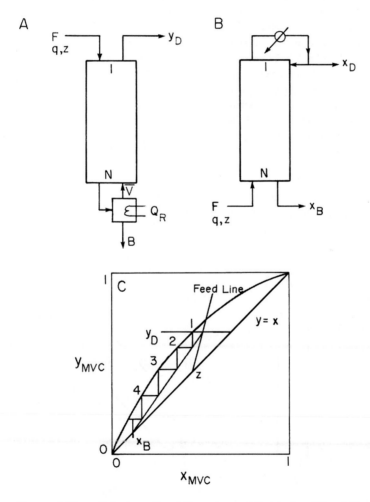

FIGURE 4-24.   *Stripping and enriching columns; (A) stripping, (B) enriching, (C) McCabe-Thiele diagram for stripping column*

which is the usual equation for a bottom operating equation with a partial reboiler. Top and bottom operating lines intersect at the feed line. If the specified variables are F, q, z, p, x_B, and y_D, the feed line can be plotted and then the bottom operating line can be obtained from its intersection at $y = x = x_B$ and its intersection with the feed line at y_D. (Proof is left as Problem 4.C7.) If the boilup rate, $\overline{V}/B$, is specified, then y_D will not be specified and can be solved for. The McCabe-Thiele diagram for a stripping column is shown in Figure 4-24C.

4.10 LIMITING OPERATING CONDITIONS

It is always useful to look at limiting conditions. For distillation, two limiting conditions are total reflux and minimum reflux. In total reflux, all of the overhead vapor is returned to the column as reflux, and all of the underflow liquid is returned as boilup. Thus distillate and bottoms flow rates are zero. At steady state the feed rate must also be zero. Total reflux is used for starting up columns, for keeping a column operating when another part of the plant is shut down, and for testing column efficiency.

The analysis of total reflux is simple. Since all of the vapor is refluxed, L = V and L/V = 1.0. Also, $\overline{L} = \overline{V}$ and $\overline{L}/\overline{V} = 1.0$. Thus both operating lines become the y = x line. This is illustrated in Figure 4-25B. Total reflux represents the maximum separation that can be obtained with a given number of stages but zero throughput. Total reflux also gives the minimum number of stages required for a given separation. Although simple, total reflux can cause safety problems. Leakage near the top of the column can cause concentration of high boilers with a corresponding increase in temperature. This can result in polymerization, fires, or explosions (Kister, 1990). Thus, the temperature at the top of the column should be monitored, and an alarm should sound if this temperature becomes too high.

Minimum reflux, $(L/D)_{min}$, is defined as the external reflux ratio at which the desired separation could just be obtained with an infinite number of stages. This is obviously not a real condition, but it is a useful hypothetical construct. To have an infinite number of stages,

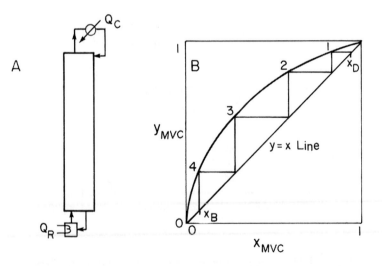

FIGURE 4-25.   *Total reflux; (A) column, (B) McCabe-Thiele diagram*

the operating and equilibrium lines must touch. In general, this can happen either at the feed or at a point tangent to the equilibrium curve. These two points are illustrated in Figures 4-26A and 4-26B. The point where the operating line touches the equilibrium curve is called the *pinch point*. At the pinch point the concentrations of liquid and vapor do not change from stage to stage. This is illustrated in Figure 4-26C for a pinch at the feed stage. If the reflux ratio is increased slightly, then the desired separation can be achieved with a finite number of stages.

For binary systems the minimum reflux ratio is easily determined. The top operating line is drawn to a pinch point as in Figure 4-26A or 4-26B. Then $(L/V)_{min}$ is equal to the slope of this top operating line (which cannot be used for an actual column, since an infinite number of stages are needed), and

$$(\frac{L}{D})_{min} = (\frac{L}{V-L})_{min} = \frac{(L/V)_{min}}{1 - (L/V)_{min}} \qquad \textbf{(4-55)}$$

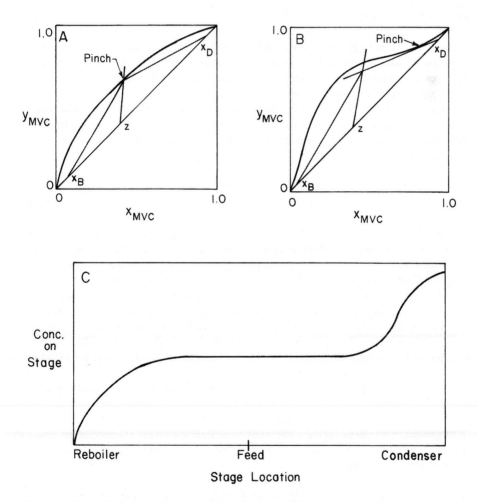

FIGURE 4-26. *Minimum reflux; (A) pinch at feed stage, (B) tangent pinch, (C) concentration profile for $L/D \sim (L/D)_{min}$*

Note that the minimum reflux ratio depends on x_D, z, and q and can depend on x_B. The calculation of minimum reflux may be more complex when there are two feeds or a sidestream. This is explored in the homework problems.

The minimum reflux ratio is commonly used in specifying operating conditions. For example, we may specify the reflux ratio as $L/D = 1.2(L/D)_{min}$. Minimum reflux would use the minimum amount of reflux liquid and hence the minimum amount of heat in the reboiler, but the maximum (infinite) number of stages and a maximum (infinite) diameter for a given separation. Obviously, the best operating conditions lies somewhere between minimum and total reflux. As a rule of thumb, the optimum external reflux ratio is between 1.05 and 1.25 times $(L/D)_{min}$. (See Chapter 11 for more details.)

A maximum $\overline{L/V}$ and hence a minimum boilup ratio $\overline{V}/B$ can also be defined. The pinch points will look the same as in Figure 4-26A or 4-26B. Problem 4.C12 looks at this situation further.

4.11   EFFICIENCIES

Up until now we have always assumed that the stages are equilibrium stages. Stages that are very close to equilibrium can be constructed, but they are only used for special purposes, such as determining equilibrium concentrations. To compare the performance of an actual stage to an equilibrium stage, we use a measure of efficiency.

Many different measures of efficiency have been defined. Two that are in common use are the overall efficiency and the Murphree efficiency. The overall efficiency, E_o, is defined as the number of equilibrium stages required for the separation divided by the actual number of stages required:

$$E_o = \frac{N_{equil}}{N_{actual}}$$

(4-56)

Partial condensers and partial reboilers are not included in either the actual or equilibrium number of stages, since they will not have the same efficiency as the stages in the column.

The overall efficiency lumps together everything that happens in the columns. What variables would we expect to affect column efficiency? The hydrodynamic flow properties such as viscosity and gas flow rate would affect the flow regime, which affects efficiency. The mass transfer rate, which is affected by the diffusivity, will in turn affect efficiency. Overall efficiency is usually smaller as the separation becomes easier (α_{AB} increases). The column size can also have an effect. Correlations for determining the overall efficiency will be discussed in Chapter 10. For now, we will consider that the overall efficiency is determined from operating experience with similar distillation columns.

The overall efficiency has the advantage of being easy to use but the disadvantage that it is difficult to calculate from first principles. Stage efficiencies are defined for each stage and may vary from stage to stage. The stage efficiencies are easier to estimate from first principles or to correlate with operating data. The most commonly used stage efficiencies for binary distillation are the Murphree vapor and liquid efficiencies (Murphree, 1925). The Murphree vapor efficiency is defined as

$$E_{MV} \equiv \frac{\text{actual change in vapor concentration}}{\text{change in vapor concentration for equilibrium stage}}$$

(4-57)

Murphree postulated that the vapor between trays is well mixed, that the liquid in the down-comers is well mixed, and that the liquid on the tray is well mixed and is of the same composition as the liquid in the downcomer leaving the tray. For the nomenclature illustrated in Figure 4-27A, the Murphree vapor efficiency is

$$E_{MV} = \frac{y_j - y_{j+1}}{y_j^* - y_{j+1}}$$

(4-58)

where y_j^* is vapor mole fraction in equilibrium with actual liquid mole fraction x_j.

Once the Murphree vapor efficiency is known for every stage, it can easily be used on a McCabe-Thiele diagram. (In fact, Murphree adjusted his paper to use the newly developed McCabe-Thiele diagram.) The denominator in Eq. (4-58) represents the vertical distance from the operating line to the equilibrium line. The numerator is the vertical distance from the operating line to the actual outlet concentration. Thus the Murphree vapor efficiency is the fractional amount of the total vertical distance to move from the operating line. If we step off stages from the bottom up, we get the result shown in Figure 4-27B. Note that the partial reboiler is treated separately since it will have a different efficiency than the remainder of the column.

The Murphree efficiency can be used as a ratio of distances as shown in Figure 4-27B. If the Murphree efficiencies are accurate, the locations labeled by the stage numbers represent the actual vapor and liquid compositions leaving a stage. These points can be connected to form a pseudo-equilibrium curve, but this curve depends on the operating lines used and thus has to be redrawn for each new set of operating lines. Figure 4-27B allows us to calculate the real optimum feed plate location and the real total number of stages.

A Murphree liquid efficiency can be defined as

$$E_{ML} = \frac{x_j - x_{j-1}}{x_j^* - x_{j-1}}$$

(4-59)

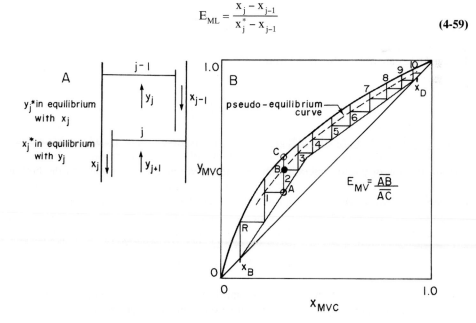

FIGURE 4-27.   *Murphree efficiency; (A) stage nomenclature, (B) McCabe-Thiele diagram for E_{MV}*

which is the actual change in mole fraction divided by the change for an equilibrium stage. The Murphree liquid efficiency is similar to the Murphree vapor efficiency except that it uses horizontal distances. Note that $E_{ML} \neq E_{MV}$.

For binary mixtures the Murphree efficiencies are the same whether they are written in terms of the more volatile or least volatile component. For multicomponent mixtures they can be different for different components and can even be negative.

4.12   SIMULATION PROBLEMS

In a simulation problem the column has already been built, and we want to know how much separation can be obtained. As noted in Tables 3-1 and 3-3, the real number of stages, the real feed location, the column diameter, and the types and sizes of reboiler and condenser are known. The engineer does the detailed stage-by-stage calculation and the detailed diameter calculation, and finally he or she checks that the operation is feasible.

To be specific, consider the situation where the known variables are F, z, q, x_D, T_{reflux} (saturated liquid), p, $Q_{col} = 0$, N_{actual}, $N_{feed\,actual}$, diameter, L_0/D, the overall efficiency E_o, and CMO. This column is illustrated in Figure 4-28A. The engineer wishes to determine the bottoms composition, x_B.

We start by deriving the top and bottom operating equations. These are the familiar forms,

$$y = \frac{L}{V} x + (1 - \frac{L}{V})x_D$$

and

$$y = \frac{\overline{L}}{\overline{V}} x - (\frac{\overline{L}}{\overline{V}} - 1)x_B$$

The feed line equation is also unchanged

$$y = \frac{q}{(q - 1)} x + \frac{z_F}{(1 - q)}$$

Since the reflux is a saturated liquid and the external reflux ratio, L_0/D, is known, we calculate L/V and plot the top operating line (Figure 4-28B). The feed line can also be plotted. The intersection of the top operating line and the feed line gives one point on the bottom operating line. Unfortunately, the bottom operating line cannot be plotted because neither $\overline{L}/\overline{V}$ nor x_B is known (why won't external balances give one of these variables?).

To proceed we must use the three items of information that have not been used yet. These are N_{actual}, $N_{feed\,actual}$, and E_o. We can estimate the equilibrium number of stages as

$$N = E_o N_{actual}, \qquad N_F = \text{Integer } [N_{feed\,actual}\, E_o] \qquad \textbf{(4-60)}$$

The feed location in equilibrium stages, N_F, must be estimated as an integer, but the total number of equilibrium stages, N, could be a fractional number. Now we can step off equilib-

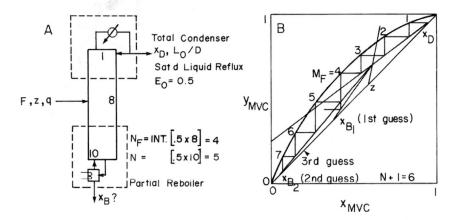

FIGURE 4-28.   *Simulation problems; (A) existing column, (B) McCabe-Thiele diagram*

rium stages on the top operating line until we reach the feed stage N_F. At this point we need to switch to the bottom operating line, which is not known. To use the final bit of information, the value N, we must guess x_B, plot the bottom operating line, and check to see if the separation is achieved with N + 1 (the +1 includes the partial reboiler) equilibrium contacts. Thus, the simulation problem is one of trial and error when a stage-by-stage computation procedure is used. This procedure is illustrated in Figure 4-28B. Note that the actual feed stage may not be (and probably is not) optimum.

Once x_B has been determined the external balances can be completed, and we can determine B, D, Q_c, and Q_R. Now L, V, $\overline{L}$, and $\overline{V}$ can be calculated and we can proceed to an exact check that V and $\overline{V}$ are less than V_{max}. This is done by calculating a permissible vapor velocity. This calculation is similar to calculating u_{perm} for a flash drum and is shown in Chapter 10. The condenser and reboiler sizes can also be checked. If the flow rates are too large or the condenser and reboiler are too small, the existing column will not satisfactorily do the desired separation. Either the feed rate can be decreased or L/D can be decreased. This latter change obviously requires that the entire solution be repeated.

When other variables are specified, the stage-by-stage calculation is still trial and error. The basic procedure remains the same. That is, calculate and plot everything you can first, guess the needed variable, and then check whether the separation can be obtained with the existing number of stages. Murphree stage efficiencies are easily employed in these calculations.

Simulation problems are probably easier to do using the matrix approach developed in Chapter 6, which is easily adapted to computer use.

4.13   NEW USES FOR OLD COLUMNS

Closely related to simulation is the use of existing or used distillation systems for new separations. The new use may be debottlenecking—that is, increasing capacity for the same separation. With increasing turnover of products, the problem of using equipment for new separations is becoming much more common.

Why would we want to use an existing column for a problem it wasn't designed for? First, it is usually cheaper to modify a column that has already been paid for than to buy a new one. Second, it is usually quicker to do minor modifications than to wait for construction of a new column. Finally, for many engineers solving the often knotty problems involved in adapting a column to a new separation is an interesting challenge.

The first thing to do when new chemicals are to be separated is clean the entire system and inspect it thoroughly. Is the system in good shape? If not, will minor maintenance and parts replacement put the equipment in working order? If there are major structural problems such as major corrosion, it will probably be cheaper and less of a long-term headache to buy new equipment.

Do simulation calculations to determine how close the column will come to meeting the new separation specifications. Rarely will the column provide a perfect answer to the new problem. Difficulties can be classified as problems with the separation required and problems with capacity.

What can be done if the existing column cannot produce the desired product purities? The following steps can be explored (they are listed roughly in the order of increasing cost).

1. Find out whether the product specifications can be relaxed. A purity of 99.5% is much easier to obtain than 99.99%.
2. See if a higher reflux ratio will do the separation. Remember to check if column vapor capacity and the reboiler and condenser are large enough. If they are, changes in L/D affect only operating costs.
3. Change the feed temperature. This may make a nonoptimum feed stage optimum.
4. Will a new feed stage at the optimum location (the existing feed stage is probably nonoptimum) allow you to meet product specifications?
5. Consider replacing the existing trays (or packing) with more efficient or more closely spaced trays (or new packing). This is relatively expensive but is cheaper than buying a completely new system.
6. Check to see if two existing columns can be hooked together in series to achieve the desired separation. Feed can be introduced at the feed tray of either column or in between the two columns. Since vapor loading requirements are different in different sections of the column (see Chapter 10), the columns do not have to be the same diameter.

What if the column produces product much purer than specifications? This problem is pleasant. Usually the reflux ratio can be decreased, which will decrease operating expenses.

Problems with vapor capacity are discussed in more detail in Chapter 10. Briefly, if the column diameter is not large enough, the engineer can consider:

1. Operating at a reduced L/D, which reduces V. This may make it difficult to meet the product specifications.
2. Operating at a higher pressure, which increases the vapor density. Note that the column must have been designed for these higher pressures and the chemicals being separated must be thermally stable.
3. Using two columns in parallel.
4. Replacing the downcomers with larger downcomers (see Chapter 10).
5. Replacing the trays or packing with trays or packing with a higher capacity. Major increases in capacity are unlikely.

If the column diameter is too large, vapor velocities will be low. The trays will operate at tray efficiencies lower than designed, and in severe cases they may not operate at all since liquid may dump through the holes. Possible solutions include:

1. Decrease column pressure to decrease vapor density. This increases the linear vapor velocity.
2. If the column has sieve trays, cover some of the holes. This increases the vapor velocity in the open holes reducing weeping.
3. Increase L/D to increase V.
4. Recycle some distillate and bottoms product to effectively increase F.

Using existing columns for new uses often requires a creative solution. Such problems can be both challenging and fun; they are also often assigned to engineers just out of school.

4.14 SUBCOOLED REFLUX AND SUPERHEATED BOILUP

What happens if the reflux liquid is subcooled or the boilup vapor is superheated? We have already looked at two similar cases where we have a subcooled liquid or a superheated vapor *feed*. In those cases we found that a subcooled liquid would condense some vapor in the column, while a superheated vapor would vaporize some liquid. Since reflux and boilup are inputs to the column, we should expect exactly the same behavior if these streams are subcooled or superheated.

Subcooled reflux often occurs if the condenser is at ground level. Then a pump is required to return the reflux to the top of the column. A saturated liquid will cause cavitation and destroy the pump; thus, the liquid *must* be subcooled if it is to be pumped. To analyze the effect of subcooled reflux, consider the top of the column shown in Figure 4-29. The cold liquid stream, L_0, must be heated up to its boiling point. This energy must come from condensing vapor on the top stage, stream c in Figure 4-29. Thus, the flow rates on the first stage are different from those in the rest of the rectifying section. CMO is valid in the remainder of the column. The internal reflux ratio in the rectifying column is $L_1/V_2 = L/V$, and the top operating line is

$$y = \frac{L}{V}x + (1 - \frac{L}{V})x_D$$

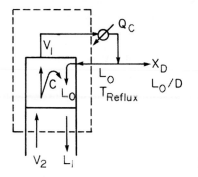

FIGURE 4-29. *Balance envelope for subcooled reflux*

Now, L/V cannot be directly calculated from the external reflux ratio L_0/D, since L and V change on the top stage.

Balances on vapor and liquid streams give

$$V_2 = V_1 + c \tag{4-61}$$

$$L_1 = L_0 + c \tag{4-62}$$

An energy balance using the balance envelope in Figure 4-29 is

$$V_1 H_1 + L_1 h_1 = L_0 h_{reflux} + V_2 H_2 \tag{4-63}$$

With CMO, $H_1 \sim H_2$. Then Eq. (4-63) becomes

$$(V_2 - V_1)H \sim L_1 h_1 - L_0 h_{reflux}$$

Using Eqs. (4-61) and (4-62), this becomes

$$cH \sim (L_0 + c)h_1 - L_0 h_{reflux}$$

Solving for amount condensed, c, we obtain

$$c \sim \left(\frac{h_{liq} - h_{reflux}}{II_{vap} - h_{liq}} \right) L_0 = f_c L_0 \tag{4-64a}$$

which can also be written as

$$c \sim \frac{\overline{C}_{PL} (T_{BP} - T_{reflux})}{\lambda} L_0 = f_c L_0 \tag{4-64b}$$

or

$$c \sim \left(\frac{\text{energy to heat reflux to boiling}}{\text{latent heat of vaporization}} \right) L_0 = f_c L_0 \tag{4-64c}$$

where f_c is the fraction condensed per mole of reflux. We can now calculate the internal reflux ratio, $L/V = L_1/V_2$. To do this, we start with the ratio we desire and use Eqs. (4-61), (4-62), (4-64), and (4-65).

$$\frac{L}{V} = \frac{L_1}{V_2} = \frac{L_0 + c}{V_1 + c} \sim \frac{L_0 + f_c L_0}{V_1 + f_c L_0} \sim \frac{(1 + f_c)L_0/V_1}{1 + f_c L_0/V_1} \tag{4-65}$$

The ratio L_0/V_1 is easily found from L_0/D as

$$\frac{L_0}{V_1} = \frac{L_0/D}{1 + L_0/D}$$

Using this expression in Eq. (4-65) we obtain

$$\frac{L}{V} = \frac{L_1}{V_2} \sim \frac{(1 + f_c)L_0/D}{1 + (1 + f_c)L_0/D} \tag{4-66}$$

Note that when $f_c = 0$, Eqs. (4-65) and (4-66) both say $L_1/V_2 = L_0/V_1$. As the fraction condensed increases (reflux is subcooled more), the internal reflux ratio, L_1/V_2, becomes larger. Thus the net result of subcooled reflux is equivalent to increasing the reflux ratio. Numerical calculations (such as Problem 4.D5) show that a large amount of subcooling is required to have a significant effect on L/V. With highly subcooled reflux, an extra tray should be added for heating the reflux (Kister, 1990).

A superheated direct steam input or a superheated boilup from a total reboiler will cause vaporization of liquid inside the column. This is equivalent to a net increase in the boilup ratio, $\overline{V}/B$, and makes the slope of the stripping section operating line approach 1.0. Since superheated vapor inputs can be analyzed in the same fashion as the subcooled liquid reflux, it will be left as homework assignments 4.C14 and 4.C15.

4.15 COMPARISONS BETWEEN ANALYTICAL AND GRAPHICAL METHODS

Both the Lewis and McCabe-Thiele methods are based on the CMO assumption. When the CMO assumption is valid, the energy balances are automatically satisfied, which greatly simplifies the stage-by-stage calculations. The reboiler and condenser duties, Q_R and Q_c, are determined from the balances around the entire column. If CMO is not valid, Q_R and Q_c will be unaffected if the same external reflux ratio can be used, but this may not be possible. The exact calculation, which can be done using the methods developed in Chapter 6, may show that a higher or lower L_0/D is required if CMO is invalid.

When a programmable calculator or a computer is to be used, the analytical method is more convenient, particularly if a spreadsheet is used (Burns and Sung, 1996). The computer calculations are obviously more convenient if a very large number of stages are required, if a trial-and-error solution is required, or if many cases are to be run to explore the effect of many variables (e.g., for economic analysis). However, since accuracy is limited by the equilibrium data, the computer calculations assuming CMO are *not* more accurate than the graphical method.

If calculations are to be done by hand, the graphical method is faster than alternating between the analytical forms of the equilibrium relationship and the operating equations. In the McCabe-Thiele method, solution of the equilibrium relationship is simply a matter of drawing a line to the equilibrium curve. In the analytical solution we must first either fit the equilibrium data to an analytical expression or develop an interpolation routine. Then we must solve this equation, which may be nonlinear, each time we do an equilibrium calculation. Since the operating equation is linear, it is easy to solve analytically. With a sharp pencil, large graph paper, and care, the McCabe-Thiele technique can easily be as accurate as the equilibrium data (two significant figures).

When doing the stage-by-stage calculations from the top down, we solve for x values from the equilibrium relationship and for y values from the operating equation. If we step off stages from the bottom up, then we calculate x values from the operating equation and y values from equilibrium. Check this statement out on a McCabe-Thiele diagram. When going from the bottom up, we want to solve all the operating equations for x. As noted previously, the optimum feed stage can be determined from the test in Eq. (4-37), and the point of intersection (Eq. 4-38) is more convenient to use than the feed line. For more complex situations, the point of intersection of a feed line and an operating equation can be found by simultaneously solving the equations. The biggest problem in using the Lewis method on the computer is obtaining a good fit for the equilibrium data. The data can be fit to curves, or an interpolation routine can be used to interpolate between data points. Burns and Sung (1996) report that linear interpolation between data points is accurate if enough data points are entered for the equilibrium data.

One of the most convenient ways to discuss computer and calculator methods is by reference to a flowchart. The flowchart is fairly general while computer code is very specific to the language used. Consider the specific design problem we solved in Example 4-3. Assume that we decide to step off stages from the top down. Now the computer or calculator program can proceed as shown in Figure 4-30. Other flowcharts are possible. If we step off stages from the bottom up, we will calculate y_i from equilibrium and x_i from the operating equation. Note that a McCabe-Thiele diagram is very useful for following the logic of flowcharts. Try this with Figure 4-30.

For a more complex column the flowchart in Figure 4-30 would be modified by:

1. Including equations for intermediate operating lines.
2. Including additional tests for optimum feeds [change Eqs. (4-37a,b)].
3. Including side streams

The principles again follow McCabe-Thiele calculations step by step.

The development of digital computers has made the graphical McCabe-Thiele technique obsolete for detailed design calculations. Engineers used to cover an entire wall with graph paper to do a McCabe-Thiele diagram when a very large number of stages were required. The method is still useful for one or two calculations, but its major uses are as a teaching tool and as a visualization tool. The graphical procedure presents a very clear visual picture of the calculation that is easier to understand than the interactions of the equations. The graphs are also extremely useful as a tool to help determine what the effect of changing variables will be, as a diagnostic when the computer program appears to be malfunctioning, and as a diagnostic when the column appears to be malfunctioning (Kister, 1995). Because of its visual impact, we have used the McCabe-Thiele diagram extensively to explore a variety of distillation systems.

4.16 SUMMARY—OBJECTIVES

In this long chapter we developed the stage-by-stage balances for distillation columns and showed how to solve these equations when CMO is valid. You should now be able to satisfy the following objectives:

1. Write the mass and energy balances and equilibrium expressions for any stage in a column
2. Explain what CMO is, and determine if CMO is valid in a given situation
3. Derive the operating equations for CMO systems
4. Calculate the feed quality and determine its effect on flow rates. Plot the feed line on a y-x diagram

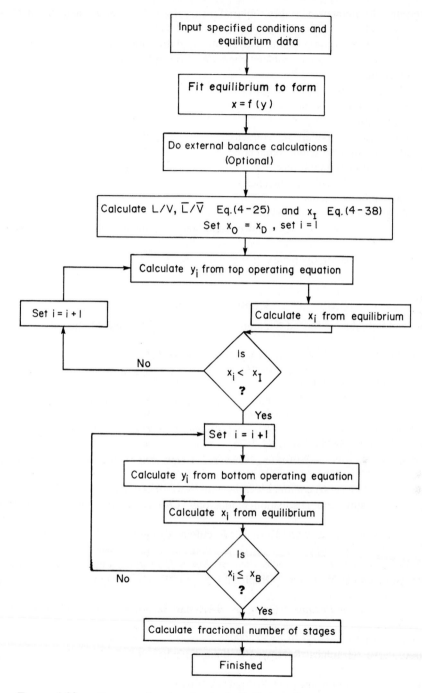

FIGURE 4-30.   *Computer flowchart for simple distillation column*

5. Determine the number of stages required, using the Lewis method and the McCabe-Thiele method
6. Develop and explain composition, temperature, and flow profiles
7. Solve any binary distillation problem where CMO is valid. This includes:
    a. Open steam
    b. Multiple feeds
    c. Partial condensers and total reboiler
    d. Side streams
    e. Intermediate reboilers and condensers
    f. Stripping and enriching columns
    g. Total and minimum reflux
    h. Overall and Murphree efficiencies
    i. Simulation problems
    j. Any combination of the above
8. Include the effects of subcooled reflux or superheated boilup in your McCabe-Thiele and Lewis analyses
9. Develop flowcharts for any binary distillation problem

REFERENCES

Binous, H., "Equilibrium-Staged Separations Using MATLAB and Mathematica," *Chem. Engr. Educ., 42* (2), 69 (Spring 2008).

Brown, G. G. et al., *Unit Operations*, Wiley, New York, 1950.

Burns, M. A., and J. C. Sung, "Design of Separation Units Using Spreadsheets," *Chem. Engr. Educ., 30* (1), 62 (Winter 1996).

Foust, A. S., L. A. Wenzel, C. W. Clump, L. Maus, and L. B. Andersen, *Principles of Unit Operations,* 2nd ed., Wiley, New York, 1980.

Himmelblau, D. M., *Basic Principles and Calculations in Chemical Engineering,* 3rd ed., Prentice Hall, Upper Saddle River, NJ, 1974.

Horwitz, B. A., "Hardware, Software, Nowhere," *Chem. Engr. Progress, 94* (9), 69 (Sept. 1998).

Kister, H. Z., *Distillation Operation,* McGraw-Hill, New York, 1990.

Kister, H. Z., "Troubleshoot Distillation Simulations," *Chem. Engr. Progress, 91* (6), 63 (June 1995).

Lewis, W. K., "The Efficiency and Design of Rectifying Columns for Binary Mixtures," *Ind. Engr. Chem., 14,* 492 (1922).

Ludwig, E. E., *Applied Process Design,* 3rd ed., Vol. 2, Gulf Publishing Co., Houston, 1997.

McCabe, W. L., and E. W. Thiele, "Graphical Design of Fractionating Columns," *Ind. Engr. Chem., 17,* 602 (1925).

McFedries, P., *Absolute Beginner's Guide to VBA*, Safari Tech Books Online (Pearson), Indianapolis, 2004.

Microsoft, *Step by Step Microsoft Excel 2000*, Microsoft Press, Redmond, WA, 1999.

Murphree, E. V., "Graphical Rectifying Column Calculations," *Ind. Engr. Chem., 17,* 960 (1925).

Perry, R. H., C. H. Chilton, and S. D. Kirkpatrick (Eds.), *Chemical Engineer's Handbook*, 4th ed., McGraw-Hill, New York, 1963.

Ponchon, M., *Tech. Moderne, 13,* 20 and 53 (1921).

Savarit, R., *Arts et Metiers, 65,* 142, 178, 241, 266, and 307 (1922).

Shacham, M., N. Brauner, W. R. Ashurst, and M. B. Cutlip, "Can I Trust This Software Package: An Exercise in Validation of Computational Results," *Chem. Engr. Educ., 42* (1), 53 (Winter 2008).

Sloley, A. W., "Avoid Problems During Distillation Column Startups," *Chem. Engr. Progress, 92* (7), 30 (July 1996).

Sorel, E., *La Retification de l'Alcohol,* Gauthier-Villars, Paris, 1893.

HOMEWORK

A. *Discussion Problems*

A1. In the figure shown, what streams are represented by point A? By point B? How would you determine the temperature of stage 2? How about the temperature in the reboiler? If feed composition is as shown, how can the liquid composition on the optimum feed stage be so much less than z?

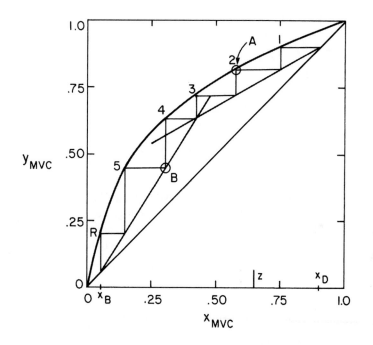

A2. For this McCabe-Thiele diagram answer the following questions.

    a. (1) The actual feed tray is?

        (2) The mole fraction MVC in the feed is?

        (3) The vapor composition on the feed tray is?

        (4) The liquid composition on the feed tray is?

    b. Is the feed a superheated vapor feed, saturated vapor feed, two-phase feed, saturated liquid feed, or subcooled liquid feed?

    c. Is the temperature at stage 7 higher, lower, or the same as at stage 1?

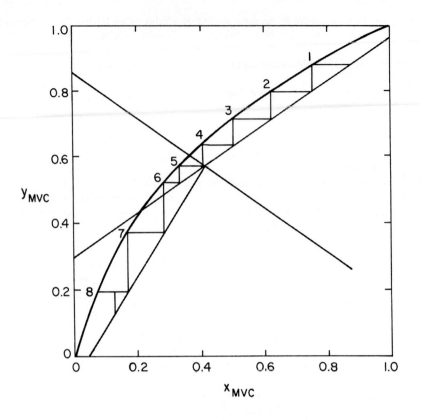

A3. Suppose that constant mass overflow is valid instead of CMO. Explain how to carry out the Lewis and McCabe-Thiele procedures in this case.

A4. Drawing the McCabe-Thiele graph as y_{MVC} vs. x_{MVC} is traditional but not necessary. Repeat Example 4-3, but plot y_w vs. x_w. Note the differences in the diagram. Do you expect to get the same answer?

A5. For distillation at CMO, show the flow profiles schematically (plot L_j and V_j vs. stage location) for
   a. Subcooled liquid feed
   b. Two-phase feed
   c. Superheated vapor feed

A6. A distillation column is operating under a vacuum. The column has 18 stages with the feed at stage 9, a partial reboiler, and a partial condenser. The pressure drop per stage is 1.0 mm Hg. The pressure of the condenser is controlled at 100 mm Hg. The pressure of the reboiler is: a. higher than 100 mm Hg, b. 100 mm Hg, c. lower than 100 mm Hg.

A7. What is the effect of column pressure on distillation? To explore this consider pressure's effect on the reboiler and condenser temperatures, the volumetric flow rates inside the column, and the relative volatility (which can be estimated for hydrocarbons from the DePriester charts).

A8. What happens if we try to step off stages from the top down and E_{MV} is given? Determine how to do this calculation.

A9. When would it be safe to ignore subcooling of the reflux liquid and treat the reflux as a saturated liquid? Do a few numerical calculations for either methanol and water or ethanol and water to illustrate.

A10. Eqs. (4-53) and (4-54) are mass balances on particular phases. When will these equations be valid?

A11. When might you use an intermediate condenser on a column? What are the possible advantages?

A12. When would just a stripping column be used?

A13. We have a binary distillation column. The feed flow rate, F, the mole fraction MVC in the feed, z, and the initial feed temperature are constant. The column pressure, p, and the distillate and bottoms mole fractions, x_D and x_B, are constant. The reflux is a saturated liquid, and the reflux ratio is $L/D = 1.20 \times (L/D)_{min}$. A heat exchanger uses waste heat, Q_w, to heat the feed.

As Q_w increases, does,

A)	q	a) increase	b) decrease	c) stay constant?
B)	$(L/D)_{min}$	a) increase	b) decrease	c) stay constant?
C)	$(L/V)_{actual}$	a) increase	b) decrease	c) stay constant?
D)	$(\bar{L}/\bar{V})_{actual}$	a) increase	b) decrease	c) stay constant?
E)	$(\bar{V}/B)_{actual}$	a) increase	b) decrease	c) stay constant?
F)	$|Q_c|$	a) increase	b) decrease	c) stay constant?
G)	Q_R	a) increase	b) decrease	c) stay constant?

Think your way through this problem using the tools developed in this course (definition of q and feed line, mass and energy balances, and McCabe-Thiele diagrams) as reasoning tools.

Hint: The parts of this problem are listed in a logical order of solution.

A14. Explain with a McCabe-Thiele diagram how changing feed temperature (or equivalently, q) may help an existing column achieve the desired product specifications.

A15. Develop a key relations chart for binary McCabe-Thiele distillation. That is, on one sheet of paper summarize everything you need to know about binary distillation. You will probably want to include information about operating lines, feed lines, efficiencies, subcooled reflux, and so forth.

B. *Generation of Alternatives*

B1. Invent your own problem that is distinctly different from those discussed in this chapter. Show how to solve this problem.

B2. Several ways of adapting existing columns to new uses were listed. Generate *new* methods that might allow existing systems to meet product specifications that could not be met without modification. Note that you can postulate a complex existing column such as one with an intermediate reboiler.

C. *Derivations*

C1. For Example 4-4 (open steam), show that the x intercept (y = 0) is at x = x_B.

C2. Derive the bottom operating line for a column with a total reboiler. Show that this is the same result as is obtained with a partial reboiler.

C3. Derive Eqs. (4-51) and (4-52).

C4. For a side stream below the feed:
   a. Draw a sketch corresponding to Figure 4-22A.
   b. Derive the operating equation and y = x intercept.
   c. Sketch the McCabe-Thiele diagram.

C5. Derive the operating equations for the two middle operating sections when an intermediate reboiler is used (see Figures 4-23A and 4-23B). Show that the operating line with slope of L'/V' goes through the point y = x = x_B.

C6. Show that the total amount of cooling needed is the same for a column with one total condenser (Q_c) as for a column with a total condenser and an intermediate total condenser ($Q_c + Q_I$). F, z, q, x_D, x_B, and Q_R are constant for the two cases. Sketch a system with an intermediate condenser. Derive the operating equations for the two middle operating lines, and sketch the McCabe-Thiele diagram.

C7. For the stripping column shown in Figures 4-24A and 4-24C, show formally that the intersection of the bottom operating line and the feed line is at y_D. In other words, solve for the intersection of these two lines.

C8. Develop the McCabe-Thiele procedure for the enriching column shown in Figure 4-24B.

C9. For Example 4-3 prove that:
   a. The top operating line and the y=x line intersect at y=x=x_D.
   b. The bottom operating line and the y=x line intersect at y=x=x_B.

C10. For a continuous distillation column with a saturated liquid feed, a total condenser that produces a saturated liquid reflux, and a partial reboiler, show that $Q_R/D = (1 + L/D)\lambda$ if CMO is valid.

C11. The boilup ratio $\overline{V}/B$ may be specified. Derive an expression for $\overline{L}/\overline{V}$ as a function of $\overline{V}/B$ for a partial reboiler.

C12. Show how to determine $(\overline{V}/B)_{min}$. Derive an equation for calculation of $(\overline{V}/B)_{min}$ from $(\overline{L}/\overline{V})_{max}$.

C13. Sketch the McCabe-Thiele diagram if the Murphree *liquid* efficiency is constant and $E_{ML} = 0.75$.

C14. Derive the equations to calculate $\overline{L}/\overline{V}$ when a superheated boilup is used.

C15. Derive the equations to calculate $\overline{L}/\overline{V}$ when direct superheated steam is used.

C16. Part a. In a binary distillation column with two feeds, show that the intersection of the top and bottom operating lines occurs at the feed line for fictitious feed F_T where $F_T = F_1 + F_2$, $z_T F_T = z_1 F_1 + z_2 F_2$, and $h_{FT} F_T = h_{F1} F_1 + h_{F2} F_2$.
Suggestion: Draw the McCabe-Thiele diagram for the actual column with three operating lines using both actual feed lines. On the same diagram draw the two operating lines for a column with the single mixed feed F_T (they are unchanged from top and bottom operating lines of two-feed column) and then determine the feed line for this mixed column.
b. Assuming that CMO is valid, show that $q_T \approx (F_1 q_1 + F_2 q_2)/F_T$

C17. Derive the operating equation for section 2 of Figure 4-17. Show that the equations are identical whether the mass balance envelope is drawn around the top of the column or the bottom of the column.

D. *Problems*

*Answers to problems with an asterisk are at the back of the book.

D1. A continuous, steady-state distillation column with a total condenser and a partial reboiler is separating methanol from water at one atmosphere (see Table 2-7 for data). The feed rate is 100 kmol/h. The feed is 55 mol% methanol and 45 mol% water. We desire a distillate product that is 90 mol% methanol and a bottoms product that is 5 mol% methanol. Assume CMO.

   a. If the external reflux ratio $L/D = 1.25$ plot the top operating line.

   b. If the boilup ratio $\overline{V}/B = 2.0$ plot the bottom operating line.

   c. Step off stages starting at the bottom with the partial reboiler. Use the optimum feed stage. Report the optimum feed stage and the total number of stages.

   d. Plot the feed line. Calculate its slope. Calculate q. What type of feed is this?

D2. We are separating ethanol and water. All percentages are mol%. Find the q values and plot the feed lines.

   a. Feed is 60% ethanol and flashes in the column with $V/F = 0.37$.

   b. Feed is 40% ethanol and is a two-phase mixture with liquid and vapor in equilibrium at a temperature of 84.1°C.

   c. Feed is 40% ethanol and is a liquid at 20°C.

   d. Feed is 40% ethanol and is a vapor at 120°C.

   e. Feed is 50% ethanol and is a subcooled liquid. One mole of vapor must be condensed to heat 12 moles of feed to their boiling point.

   f. Feed is 20% ethanol and 70% is vaporized in a flash distillation system. The products of the flash system are fed to distillation column. Calculate the two q values and plot the two feed lines.

D3.* **a.** A feed mixture of ethanol and water is 40 mol% ethanol. The feed is at 200°C and is at a high enough pressure that it is a liquid. It is input into the column through a valve, where it flashes. Column pressure is 1 kg/cm². Find the slope of the feed line.

   b. We are separating ethanol and water in a distillation column at a pressure of 1 kg/cm². Feed is 50 wt % ethanol, and feed rate is 1 kg/h. The feed is initially a liquid at 250°C and then flashes when the pressure is dropped as it enters the column. Find q. Data are in Figure 2-4. You may assume that CMO is valid.

D4.* **a.** We have a superheated vapor feed of 60 mol% more volatile component at 350°C. Feed flow rate is 1000 kmol/h. On the feed plate the temperature is 50°C. For this mixture the average heat capacities are

   $C_{PL} = 50$ cal/(mol- °C), $C_{PV} = 25$ cal/(mol- °C)

   while the latent heat of vaporization is $\lambda = 5000$ cal/mol. Plot the feed line for this feed.

   b. If a feed to a column is a two-phase feed that is 40 mol% vapor, find the value of q and the slope of the feed line.

   c. If the feed to a column is a superheated vapor and 1 mole of liquid is vaporized on the feed plate to cool 5 moles of feed to a saturated vapor, what is the value of q? What is the slope of the feed line?

D5.* A distillation column is operating with a subcooled reflux. The vapor streams have an enthalpy of $H_1 = H_2 = 17{,}500$ Btu/lbmol, while the saturated liquid $h_1 = 3100$ Btu/lbmol. Enthalpy of the reflux stream is $h_0 = 1500$ Btu/lbmol. The external reflux ratio is set at $L_0/D = 1.1$. Calculate the internal reflux ratio inside the column, L_1/V_2.

D6. We are separating a mixture of acetone and ethanol in a distillation column operating at one atmosphere pressure. The column has a total condenser and a partial reboiler. The distillate is 90 mol% acetone, and the bottoms is 10 mol% acetone. The reflux is returned as a saturated liquid. Use a boilup ratio of $\overline{V}/B = 2.0$. Two feeds are fed to the column. The first feed has a flow rate of 75 kmol/h, it is a saturated liquid, and it is 60 mol% acetone. The second feed has a flow rate of 100 kmol/h, it is a two-phase mixture that is 60% vapor, and it is 40 mol% acetone. Use the optimum feed location for each feed. Assume CMO. Equilibrium data are in Problem 4.D7.

  a. Find the distillate and bottoms flow rates, D and B.

  b. Plot the two feed lines and the three operating lines.

  c. Step off stages and find the optimum feed location for each feed stream and the total number of equilibrium stages required.

D7.* a. A distillation column with a total condenser is separating acetone from ethanol. A distillate concentration of $x_D = 0.90$ mole fraction acetone is desired. Since CMO is valid, $L/V = $ constant. If L/V is equal to 0.8, find the composition of the liquid leaving the fifth stage below the total condenser.

  b. A distillation column separating acetone and ethanol has a partial reboiler that acts as an equilibrium contact. If the bottoms composition is $x_B = 0.13$ mole fraction acetone and the boilup ratio $\overline{V}/B = 1.0$, find the vapor composition leaving the second stage above the partial reboiler.

  c. The distillation column in parts a and b is separating acetone from ethanol and has $x_D = 0.9$, $x_B = 0.13$, $L/V = 0.8$, and $\overline{V}/B = 1.0$. If the feed composition is $z = 0.3$ (all concentrations are mole fraction of more volatile component), find the optimum feed plate location, total number of stages, and required q value of the feed. Equilibrium data for acetone and ethanol at 1 atm (Perry et al., 1963, pp. 13–4) are

x_A	.10	.15	.20	.25	.30	.35	.40	.50	.60	.70	.80	.90
y_A	.262	.348	.417	.478	.524	.566	.605	.674	.739	.802	.865	.929

D8.* For Problem 4.D7c for separation of acetone from ethanol, determine:

  a. How many stages are required at total reflux?

  b. What is $(L/V)_{min}$? What is $(L/D)_{min}$?

  c. The L/D used is how much larger than $(L/D)_{min}$?

  d. If $E_{MV} = 0.75$, how many real stages are required for $L/V = 0.8$?

D9. A distillation column with two feeds is separating ethanol (E) and water at a pressure of 1.0 atm. The column has a total condenser with saturated liquid reflux and a partial reboiler. Feed 1 is a saturated liquid and is 42 mol% ethanol. Feed 2 flow rate is $F_2 = 100$ kmol/h. Feed 2 is 18 mol% ethanol and is a two-phase mixture that is 30% vapor. The external reflux ratio is $L/D = 1/2$, and the distillate flow rate is $D = 80$ kmol/h. We desire a distillate mole fraction of $x_D = 0.66$ mole fraction ethanol and a bottoms that is $x_B = 0.04$ mole fraction ethanol. You can assume that CMO is valid. Equilibrium data are in Table 2-1.

  a. Find the flow rates F_1 and B.

  b. Find the liquid and vapor flow rates in the middle section, L' and V'.

  c. Determine and plot the operating lines. Be neat.

  d. Find both optimum feed locations (above partial reboiler) and the total number of equilibrium stages needed. Step off stages from the bottom up. Be neat.

D10.* A distillation column is separating phenol from p-cresol at 1 atm pressure. The distillate composition desired is 0.96 mole fraction phenol. An external reflux ratio of

$L/D = 4$ is used, and the reflux is returned to the column as a saturated liquid. The equilibrium data can be represented by a constant relative volatility, $\alpha_{phenol-cresol} = 1.76$ (Perry et al., 1963, pp. 13–3). CMO can be assumed.

a. What is the vapor composition leaving the third equilibrium stage below the total condenser? Solve this by an analytical stage-by-stage calculation alternating between the operating equation and the equilibrium equation.

b. What is the liquid composition leaving the sixth equilibrium stage below the total condenser? Solve this problem graphically using a McCabe-Thiele diagram plotted for $\alpha_{p-c} = 1.76$.

D11. A mixture of methanol and water is being separated in a distillation column with open steam. The feed is 100.0 kmol/h. Feed is 60.0 mol% methanol and is at 40°C. The column is at 1.0 atm. The steam is pure steam ($y_M = 0$) and is a saturated vapor. The distillate product is 99.0 mol% methanol and leaves as a saturated liquid. The bottoms is 2.0 mol% methanol and since it leaves an equilibrium stage must be a saturated liquid. The column is adiabatic. The column has a total condenser. External reflux ratio is $L/D = 2.3$. Assume CMO is valid. Equilibrium data are in Table 2-7. Data for water and methanol are available in Problem 3.E1.

a. Estimate q.

b. Find optimum feed stage and total number of equilibrium stages (step off stages from top down). Use a McCabe-Thiele diagram.

c. Find $(L/D)_{min}$. Use a McCabe-Thiele diagram.

D12. Solve Problem 3.E2 for the optimum feed location and the total number of stages. Assume CMO and use a McCabe-Thiele diagram. Expand the portions of the diagram near the distillate and bottoms to be accurate.

D13. A distillation column is separating a 30% methanol–70% water feed. The feed rate is 237 kmol/h and is a saturated liquid. The column has a partial reboiler and a partial condenser. We desire a distillate mole fraction of $y_{D,M} = 0.95$ and a bottoms mole fraction of $x_{B,M} = 0.025$. Assume CMO is valid. Data are in Table 2-7 and Problem 3.E1.

a. Find N_{min}.

b. Find $(L/V)_{min}$ and $(L/D)_{min}$.

c. If $L/D = 2.0 (L/D)_{min}$, find the optimum feed plate location and the total number of equilibrium stages required.

d. Determine the boilup ratio used.

D14. A distillation column with open steam heating is separating a feed that is 80.0 mol% methanol and 20.0 mol% water in a steady state operation. The column has 10 stages, a total condenser, and the feed is on stage 5. Operation is at 1.0 atm. The steam is pure water and is a saturated vapor. CMO can be assumed to be valid. At 2:16 a.m. 25 days ago the feed and distillate flows were shut off ($D = F = 0$), but the steam rate was unchanged and the total condenser is still condensing the vapor to a saturated liquid. The column has now reached a new steady state operation.

a. What is the current methanol mole fraction in the bottoms?

b. At the new steady state estimate the methanol mole fraction in the liquid leaving the total condenser.

D15. A partial condenser takes vapor leaving the top of a distillation column and condenses a portion of it. The vapor portion of mole fraction y_D is removed as the distillate product. The liquid portion of mole fraction x_0 is returned to the column as reflux. The liquid and vapor leaving partial condensers and partial reboilers can be assumed to be in equilibrium. A distillation column with a partial condenser and a partial reboiler is separating 300 kmol/h of a mixture that is 30 mol% ethanol and 70 mol% water and is a saturated liquid. We desire a 98% recovery of the ethanol in the vapor distillate and an 81% recovery of water in the bottoms. If $L_0/D = 2.0$ find the optimum feed location and the total number of equilibrium stages.

D16.* Estimate q for Problem 3.D6. Estimate that the feed stage is at same composition as the feed.

D17. A mixture of acetone and ethanol (acetone is more volatile) is fed to an enriching column that has a liquid side stream withdrawn. The feed flow rate is 100.0 mol/min. Feed is 60.0 mol% acetone and is a saturated vapor. The liquid side product is withdrawn from the second stage below the total condenser at a flow rate of S = 15.0 mol/min. The reflux is returned as a saturated liquid. The distillate should be 90.0 mol% acetone. The external reflux ratio is L/D = 7/2. Column pressure is 1.0 atm. Column is adiabatic and CMO is valid. Equilibrium data are in Problem 4.D7. Note: Trial and error is **not** required.

Find the mole fraction of acetone in the sidestream x_S, the mole fraction of acetone in the bottoms x_B, and the number of equilibrium stages required.

D18. We have a stripping column with two feeds separating acetone and ethanol at 1 atm. Feed F_1 is a saturated liquid and is fed into the top of column (no condenser). Flow rate of F_1 is 100 kgmol/h, and this feed is 60 mol% acetone. Feed F_2 is 40 mol% acetone, it is a two-phase feed that is 80% vapor, and flow rate is 80 kmol/h. We desire a bottoms mole fraction that is 0.04 mole fraction acetone. The column has a partial reboiler. Equilibrium data are in Problem 4.D7.

a. Calculate $(\overline{V}/B)_{min}$.

b. If $(\overline{V}/B) = 1.5$, find D and y_D, the optimum feed stage for feed F_2, and the total number of stages. Please step off stages from the bottom upwards.

D19.* A distillation column is separating acetone and ethanol. The column effectively has six equilibrium stages plus a partial reboiler. Feed is a two-phase feed that is 40% liquid and 75 mol% acetone. Feed rate is 1000 kmol/h, and the feed stage is fourth from the top. The column is now operating at a steady state with the bottoms flow valve shut off. However, a distillate product is drawn off, and the vapor is boiled up in the reboiler. $L_0/D = 2$. Reflux is a saturated liquid. CMO can be assumed. p = 1 atm. Equilibrium data are in Problem 4.D7. Find the distillate composition. If one drop of liquid in the reboiler is withdrawn and analyzed, predict x_B.

D20.* A distillation column with a total condenser and a partial reboiler is separating ethanol and water at 1 kg/cm² pressure. Feed is 0.32 mole fraction ethanol and is at 30° C. Feed flow rate is 100 kmol/h. The distillate product is a saturated liquid, and the distillate is 80 mol% ethanol. The condenser removes 2,065,113 kcal/h. The bottoms is 0.04 mole fraction ethanol. Assume that CMO is valid. Find the number of stages and the optimum feed stage location.

Data: $C_{PL\ EtOH} = 24.65$ cal/(mol- °C) at 0 °C

The enthalpy-composition diagram is given in Figure 2-4, and the y-x diagram is in Figure 2-2. Note: Watch your units.

D21. An enricher column has two feeds. Feed F_1 (input at the bottom) is a saturated vapor. Flow rate F_1 = 100.0 kmol/h. This feed is 20.0 mol% methanol and 80.0 mol% water. Feed F_2 (input part way up the column) is a two-phase mixture that is 90 % liquid. Flow rate F_2 = 80.0 kmol/h. Feed F_2 is 45.0 moles % methanol and 55.0 mol% water. We desire a distillate that is 95.0 mol% methanol. Reflux is returned as a saturated liquid. Pressure is one atmosphere. $L/D = 1.375$. Assume CMO. Data are available in Table 2-7.

Find: D, B, x_B, optimum feed location and number of equilibrium stages required.

D22.* When water is the more volatile component we do not need a condenser but can use direct cooling with boiling water. This was shown in Problem 3.D3. We set $y_D = 0.92$, $x_B = 0.04$, $z = 0.4$ (all mole fractions water), feed is a saturated vapor, feed rate is 1000 kmol/h, p = 1 atm, CMO is valid, the entering cooling water (C) is a saturated liquid and is pure water, and $C/D = 3/4$. Derive and plot the top operating line. Note that external balances (that is, balances around the entire column) are not required.

D23.* When water is the more volatile component we do not need a condenser but can use direct cooling. This was illustrated in Problem 3.D3. We set $y_D = 0.999$, $x_B = 0.04$, $z = 0.4$ (all mole fractions water), feed is a saturated liquid, feed rate is 1000 kmol/h, p = 1 atm, CMO is valid, the entering cooling water (C) is pure water and $C/D = 3/4$. The entering cooling water is at 100 °F while its boiling temperature is 212 °F.

$$C_{PL,W} = 18.0 \ \frac{Btu}{lb\ mol\ °F}, \qquad MW_w = 18, \qquad \lambda_w = 17{,}465.4 \ \frac{Btu}{lb\ mol}$$

Find the slope of the top operating line, L/V.

Note: Equilibrium data are not needed.

D24. We have a distillation column with a partial condenser and a total reboiler separating a feed of 200.0 kmol/h. The feed is 40.0 mol% acetone and 60.0 mol% ethanol. The feed is a two-phase mixture that is 80% liquid. We desire a distillate vapor that is 85.0 mol% acetone and a bottoms that is 5.0 mol% acetone. Column pressure is one atmosphere. Reflux is returned as a saturated liquid and $L/D = 3.25$. Assume CMO. Equilibrium data are in Problem 4.D7.
  a. Find the optimum feed location and the total number of equilibrium stages.
  b. What is the minimum external reflux ratio?
  c. What is the minimum number of stages (total reflux)?

D25. We are separating methanol and water. Calculate the internal reflux ratio inside the column, L_1/V_2, for the following cases. The column is at 101.3 kPa. Data are available in Table 2-7 and in Problem 3.E1.
  a. Distillate product is 99.9 mol% methanol. External reflux ratio is $L_0/D = 1.2$. Reflux is cooled to 40.0°C. (i.e., it is subcooled).
  b. Repeat part a except for a saturated liquid reflux. Compare with Part a.

D26. A distillation column with a partial condenser and a total reboiler is separating acetone and ethanol. There are two feeds. One feed is 50.0 mol% acetone, flows at 100.0 mol/min, and is a superheated vapor where approximately 1 mole of liquid will vaporize on the feed stage for each 20 moles of feed. The other feed is a saturated liquid, flows at 150.0 mol/min and is 35.0 mol% acetone. We desire a distillate product that is $y_D = 0.85$ mole fraction acetone and a bottoms product that is $x_B = 0.10$ mole fraction acetone. The column has a partial condenser and a total reboiler. Boilup is returned as a saturated vapor. Column operates at a pressure of 1.0 atm. Assume CMO and use a McCabe-Thiele diagram. VLE data are given in Problem 4.D7.
  a. Find $(\overline{V}/B)_{min}$. (Plot both feed lines to decide which one to use.)
  b. Operate at $\overline{V}/B = 3 \times (\overline{V}/B)_{min}$. The partial condenser is an equilibrium contact. If the stages each have a Murphree liquid efficiency of 0.75, find optimum feed loca-

tions for each feed and total number of real stages. (It is easiest to start at top and step down.)

D27.* A distillation column is separating methanol from water at 1 atm pressure. The column has a total condenser and a partial reboiler. In addition, a saturated vapor stream of pure steam is input on the second stage above the partial reboiler (see figure).

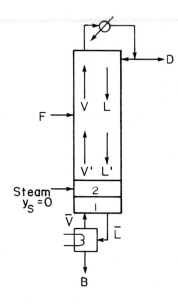

The feed flow rate is 2000 kmol/day. Feed is 48 mol% methanol and 52 mol% water and is a subcooled liquid. For every 4 moles of feed, 1 mole of vapor must condense inside the column. Distillate composition is 92 mol% methanol. Reflux is a saturated liquid, and $L_0/D = 1.0$. Bottoms composition is 8 mol% methanol. Boilup ratio is $\overline{V}/B = 0.5$. Equilibrium data are given in Table 2-7. Assume that CMO is valid. Find the optimum feed plate location and the total number of equilibrium stages required.

D28. A stripping column is separating 10,000.0 kmol/day of a saturated liquid feed. The column has a partial reboiler. The feed is 60.0 mol% acetone and 40.0 mol% ethanol. We desire a bottoms that is 2.0 mol% acetone. Operate at a boilup rate $\overline{V}/B = 1.5(\overline{V}/B)_{min}$. Find y_D, N, D, and B. Equilibrium data are in Problem 4.D7.

D29. A distillation column will use the optimum feed stage. A liquid side stream is withdrawn on the third stage below the total condenser at a rate of 15.0 kmol/h. The feed is a two phase mixture that is 20% vapor. Feed to the column is 100.0 kmol/h. The feed is 60.0 mol% acetone and 40.0 mol% ethanol. We desire a distillate composition that is 90.0 mol% ethanol. We operate with an external reflux ratio of L/D = 3. The bottoms product is 10.0 mol% acetone. A partial reboiler is used. Find the mole fraction ethanol in the side stream x_s, the optimum feed location, and the total number of equilibrium contacts needed. Equilibrium data are available in Problem 4.D7.

D30.* A distillation column is separating methanol from water. The column has a total condenser that subcools the reflux so that 1 mole of vapor is condensed in the column for each 3 moles of reflux. $L_0/D = 3$. A liquid side stream is withdrawn from the second stage below the condenser. This side stream is vaporized to a saturated vapor and then mixed with the feed and input on stage 4. The side withdrawal rate is S = 500 kmol/h. The feed is a saturated vapor that is 48 mol% methanol. Feed rate is F = 1000 kmol/h. A total reboiler is used, which produces a saturated vapor boilup. We desire a distillate 92 mol% methanol and a bottoms 4 mol% methanol. Assume CMO. Equilibrium data are given in Table 2-7. Find:

a. The total number of equilibrium stages required.

b. The value of $\overline{V}/B$.

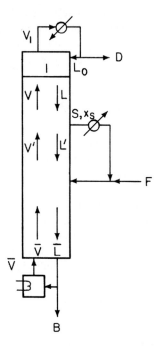

D31.* A distillation column with a total condenser and a partial reboiler is separating ethanol from water. Feed is a saturated liquid that is 25 mol% ethanol. Feed flow rate is 150 mol/h. Reflux is a saturated liquid and CMO is valid. The column has three equilibrium stages (i.e., four equilibrium contacts), and the feed stage is second from the condenser. We desire a bottoms composition that is 5 mol% ethanol and a distillate composition that is 63 mol% ethanol. Find the required external reflux ratio. Data are in Table 2-1 and Figure 2-2.

D32.* A distillation column is separating acetone from ethanol. Feed is a saturated liquid that is 40 mol% acetone. Feed rate is 50 kmol/h. Operation is at 1 atm and CMO can be assumed. The column has a total condenser and a partial reboiler. There are eight equilibrium stages in the column, and the feed is on the third stage above the reboiler. Three months ago the distillate flow was shut off (D = 0), but the column kept running. The boilup ratio was set at the value of $\overline{V}/B = 1.0$. Equilibrium data are given in Problem 4.D7. What is x_B?

D33.* A distillation column is separating 1000 mol/h of a 32 mol% ethanol, 68 mol% water mixture. The feed enters as a subcooled liquid that will condense 1 mole of vapor on the feed plate for every 4 moles of feed. The column has a partial condenser and uses open steam heating. We desire a distillate product $y_D = 0.75$ and a bottoms product

$x_B = 0.10$. CMO is valid. The steam used is pure saturated water vapor. Data are in Table 2-1 and Figure 2-2.

a. Find the minimum external reflux ratio.

b. Use $L/D = 2.0(L/D)_{min}$, and find the number of real stages and the real optimum feed location if the Murphree vapor efficiency is 2/3 for all stages.

c. Find the steam flow rate used.

D34.* A distillation column with two equilibrium stages and a partial reboiler (three equilibrium contacts) is separating methanol and water. The column has a total condenser. Feed, a 45 mol% methanol mixture, enters the column on the second stage below the condenser. Feed rate is 150 mol/h. The feed is a subcooled liquid. To heat 2 moles of feed to the saturated liquid temperature, 1 mole of vapor must condense at the feed stage. A distillate concentration of 80 mol% methanol is desired. Reflux is a saturated liquid, and CMO can be assumed. An external reflux ratio of $L/D = 2.0$ is used. Find the resulting bottoms concentration x_B. Data are in Table 2-7.

E. *More Complex Problems*

E1. A system known as a pump-around is shown below. Saturated liquid is withdrawn from stage 2 above the partial reboiler, and the liquid is returned to stage 3 (assume it is still a saturated liquid). Pump-around rate is $P = 40.0$ kmol/h. The column is separating methanol and water at 101.3 kPa. The feed flow rate is 100.0 kmol/h. The feed is 60.0 mol% methanol and 40.0 mol% water. The feed is saturated liquid. We desire a bottoms product that is 2.5 mol% methanol. The distillate product should be 95.0 mol% methanol. The column has a total condenser and the reflux is a saturated liquid. Assume CMO. Use $(L/D) = 2.0 \times (L/D)_{min}$. Data are given in Table 2-7. Find x_p, the optimum feed stage and total number of equilibrium stages required.

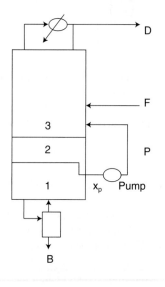

E2.* A distillation column is separating methanol and water. The column has open (direct) steam heating and a total of five stages. A liquid side stream is withdrawn from the second plate above the bottom of the column. The feed is 30 mol% methanol and is a subcooled liquid. One mole of vapor is condensed to heat 2 moles of feed to the saturated liquid temperature on the feed plate. Feed rate is 1000 mol/h. A bottoms concentration of 1.5 mol% is desired. The steam used is pure saturated water vapor and the steam flow rate is adjusted so that Steam flow rate/Bottoms flow rate = 0.833. The side stream is removed as a saturated liquid. The side-stream flow rate is adjusted so that Side-stream flow rate/Bottoms flow rate = 0.4. A total condenser is used. Reflux is a saturated liquid, and CMO can be assumed. Find the side-stream concentration and the distillate concentration. Data are given in Table 2-7.

E3.* A distillation column with at total condenser and a total reboiler is separating ethanol from water. Reflux is returned as a saturated liquid, and boilup is returned as a saturated vapor. CMO can be assumed. Assume that the stages are equilibrium stages. Column pressure is 1 atm. A saturated liquid feed that is 32 mol% ethanol is fed to the column at 1000 kmol/h. The feed is to be input on the optimum feed stage. We desire a distillate composition of 80 mol% ethanol and a bottoms composition that is 2 mol% ethanol. A liquid side stream is removed on the eighth stage from the top of the column at a flow rate of S = 457.3 kmol/h. This liquid is sent to an intermediate reboiler and vaporized to a saturated vapor, which is returned to the column at its optimum feed location. The external reflux ratio is $L_0/D = 1.86$. Find the optimum feed locations of the feed and of the vapor from the intermediate reboiler. Find the total number of equilibrium stages required. Be very neat! Data are in Table 2-1.

E4. A distillation column is separating methanol and water at a pressure of 1 atm. Feed rate of the mixture is 100 kmol/h. The feed is 25 mol% methanol and 75 mol% water. The feed is a subcooled liquid. To heat the feed to the boiling point, one mole of vapor condenses on the feed stage for every 10 moles of feed. We desire a distillate mole fraction that is 90 mol% methanol and a bottoms mole fraction that is 2.5 mol% methanol. We operate with a boilup ratio of 1.0. The column is operated with two reflux streams. The reflux to the first stage operates with a flow rate of $L_0 = 21.4$ kmol/h. Sufficient reflux (L_R) is returned to stage 2 for the column to operate. Both reflux streams are saturated liquids. The column has a total condenser and a partial reboiler. Assume CMO. Equilibrium data are in Table 2-7.

  a. Find D and B (distillate and bottoms flow rates).
  b. Determine the reflux rate to stage 2 (L_R) in kmol/h.
  c. Find the optimum feed plate location and the total number of stages.

E5. A distillation column is separating a feed that is 30 mol% acetone and 70 mol% ethanol. The column has a partial condenser. Operation is at p = 1 atm. The feed flow rate is 1000 kmol/day, the feed is a saturated liquid, and feed is input at the optimum location. We desire a distillate that is 90 mol% acetone and a bottoms that is 10 mol% acetone. Use a boilup ratio of $\overline{V}/B = 1.25$. On the fourth stage above the partial reboiler, a vapor sidestream with flow rate S = 200 kmol/day is withdrawn and then condensed to a saturated liquid, which is returned to the column as feed at the optimum location. Assume CMO. Find the mole fraction of vapor side stream y_s, optimum feed location, optimum location to input saturated liquid from intermediate condenser, and total number of equilibrium stages. Equilibrium data are in Problem 4.D7.

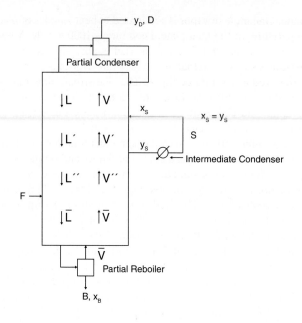

F. *Problems Requiring Other Resources*

F1. A distillation column separating ethanol from water uses open steam heating. The bottoms composition is 0.00001 mole fraction ethanol. The inlet steam is pure water vapor and is superheated to 700°F. The pressure is 1.0 atm. The ratio of bottoms to steam flow rate is B/S = 2.0. Find the slope of the bottom operating line.

F2. The paper by McCabe and Thiele (1925) is a classic paper in chemical engineering. Read it.

a. Write a one-page critique of the paper.

b. McCabe and Thiele (1925) show a method for finding the feed lines and middle operating lines for a column with two feeds that is not illustrated here. Generalize this approach when $q \neq 1.0$.

F3.* If we wish to separate the following systems by distillation, is CMO valid?

a. Methanol and water

b. Isopropanol and water

c. Acetic acid and water

d. n-Butane from n-pentane

e. Benzene from toluene

G. *Computer Problems*

G1. a.* Solve Example 4-4 with a process simulator.

b. Repeat this problem but with L/D = 1.5.

G2. Use a process simulator and find the optimum feed stage and total number of equilibrium stages for Problem 3.D6. Report the VLE correlation used. Record the values of $Q_{condenser}$ and $Q_{reboiler}$ (in Btu/h).

G3.* Write a computer, spreadsheet, or calculator program to find the number of equilibrium stages and the optimum feed plate location for a binary distillation with a constant relative volatility. System will have CMO, saturated liquid reflux, total condenser, and a partial reboiler. The given variables will be F, z_F, q x_B, x_D, α, and L_0/D. Test your program by solving the following problems:

a. Separation of phenol from p-cresol. F = 100, z_F = 0.6, q = 0.4, x_B = 0.04, x_D = 0.98, α = 1.76, and L/D = 4.00.

b. Separation of benzene from toluene. F = 200, z_F = 0.4, q = 1.3, x_B = 0.0005, x_D = 0.98, α = 2.5, and L/D = 2.00.

c. Since the relative volatility of benzene and toluene can vary from 2.61 for pure benzene to 2.315 for pure toluene, repeat part b for α = 2.315, 2.4, and 2.61.

Write your program so that it will calculate the fractional number of stages required. Check your program by doing a hand calculation.

H. *Spreadsheet Problems*

H1. [VBA required] Using the spreadsheet in Appendix B of Chapter 4 or your own spreadsheet, solve the following problem. A methanol water mixture is being distilled in a distillation column with a total condenser and a partial reboiler. The pressure is 1.0 atm, and the reflux is returned as a saturated liquid. The feed rate is 250 kmol/h and is a saturated vapor. The feed is 40 mol% methanol. We desire a bottoms product that is 1.1 mol% methanol and a distillate product that is 99.3 mol% methanol. L/D = 4.5. Find the optimum feed stage, the total number of stages, D and B. Assume CMO is valid. Equilibrium data are available in Table 2-7. Use Excel to fit this data with a 6th-order polynomial. After solving the problem, do "What if?" simulations to see what happens if the products are made purer and if L/D is decreased.

H2. [VBA required] Using the spreadsheet in Appendix B of Chapter 4 or your own spreadsheet, find the minimum L/D by trial and error for a saturated vapor feed of ethanol and water with z = 0.1, x_D = 0.7, and x_B = 0.0001.

H3. [VBA required] Write a spreadsheet for binary distillation with CMO that automatically calculates $(L/D)_{min}$, determines L/D = (Multiplier) $(L/D)_{min}$, then calculates the number of stages when the feed stage is specified. Use this program to find $(L/D)_{min}$, the optimum feed stage, and the total number of stages for distillation of a 0.17 mole fraction ethanol and 0.83 mole fraction water feed with q = 0.5, x_D = 0.7, x_B = 0.0001, and Multiplier = 1.05.

CHAPTER 4 APPENDIX A.   COMPUTER SIMULATIONS FOR BINARY DISTILLATION

Although binary distillation problems can be done conveniently on a McCabe-Thiele diagram, Chapter 6 will show that multicomponent distillation problems are easiest to solve as matrix solutions for simulation problems (the number of stages and feed locations are known). Commercial simulators typically solve all problems this way. Lab 3 in this appendix provides an opportunity to use a process simulator for binary distillation. Although the instructions discuss Aspen Plus, other simulators will be similar.

Prerequisite: This appendix assumes you are familiar with Appendix 2A [in Chapter 2], which included Labs 1 and 2, and that you are able to do basic steps with your simulator. If you need to, use the instructions in Lab 1 as a refresher on how to use Aspen Plus. If problems persist while trying to run the simulations, see Appendix A, "Aspen Plus Separations Trouble Shooting Guide," at the end of the book.

Lab 3. The goals of this lab are: 1) to become familiar with Aspen Plus simulations using RADFRAC for binary distillation systems, and 2) to explore the effect of changing operating variables on the results of the binary distillation. There is no assignment to hand in. However, understanding this material should help you understand the textbook and will help you do later labs.

Start a blank simulation with general metric units. At the bottom of the flowsheet page click on the tab labeled "columns." Use the column simulator RADFRAC. Draw a flowsheet of a simple column with one feed, a liquid distillate product (red arrow), and a liquid bottoms product. (This is the most common configuration for distillation columns.) Click on the arrow to the right of the RADFRAC button to see alternative sketches of the column. *Do not* use a packed bed system. The screen should look similar to Figure 4-A1.

Once the drawing is finished, click on the Next button. Aspen Plus will then take you to the Setup page. In the Global Settings on the right-hand side are a number of menus. Select Flowsheet for the Run Type, Steady-State for the Input Type, Conven for the Stream Class, Mole for Flow Basis, 1.0 atm for Ambient Pressure, Vapor-Liquid for Valid Phases, and No for Free Water. Then click on Next. Input the components, properties, and the input stream in the same way as in Labs 1 and 2. After the input stream, the Next button will take you to Setup for the distillation Block.

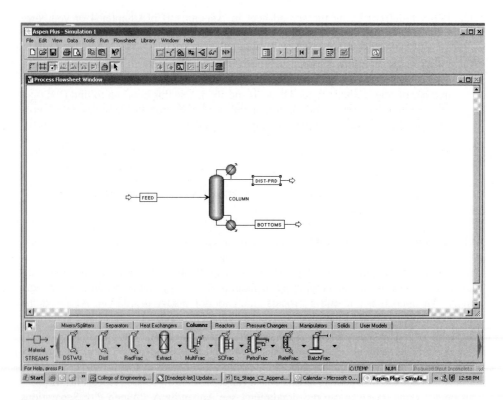

FIGURE 4-A1.   *Aspen Plus screenshot for simple distillation*

On the Configuration tab in Setup are a number of menus. Select Equilibrium for Calculation Type. For the number of stages, you need to follow Aspen Plus' method of numbering stages (illustrated in Figure 6-2). Aspen Plus calls the condenser stage 1 and the reboiler stage N. Thus, if you specify N = 40, there are 38 equilibrium stages, a condenser, and a reboiler. Use 21 stages for now. In the next menu, if you drew a column with a liquid distillate, select Total for Condenser. In the Reboiler menu select Kettle. Select Vapor-Liquid for Valid Phases, and select Standard for Convergence. In the Operating Specifications there are two menus with a variety of options. For now select Distillate Rate (use 60 kmol/h) and reflux ratio (use 2.0). Be sure to select reflux ratio, not reflux rate.

In the Streams tab you need to select a feed location. Aspen Plus counts down from the top with the condenser as stage 1. Input 9, and in the Convention menu select Vapor. In the Pressure tab it is only necessary to put a pressure for the top stage. Input 1.0 atm. Ignore the Condenser and Reboiler tabs.

In the runs that follow, if RADFRAC is completed with errors, try running a similar separation that converges, and then run the desired simulation again. Aspen Plus uses the previous converged run to initialize the new run. A column that will converge one time may not converge when run with different initial conditions. If the run is close to converged, the values may still be quite reasonable. You can try changing the conditions and run again. Runs with critical errors should not be used (don't look at the results since they have no meaning). *If Aspen Plus has a critical error, such as the column drying up, reinitialize before you do another run.* The systems for this lab were chosen so that they should converge.

When you do runs, record L/D, feed rate, feed condition (fraction vaporized or temperature), the feed stage location and the number of stages (you set these values), and the results: distillate and bottoms compositions, the temperatures of distillate and bottoms, the maximum and minimum flow rates of liquid and vapor, and the heat loads in condenser and reboiler. *Do **not** print out the entire Aspen Plus report.*

> Binary Distillation. Separate 1,2-dichloroethane from 1,1,2-trichloroethane. Feed is 100 kmol/h of 60% dichloroethane. Operation is at 1.0 atm. Column has a total condenser and a kettle type reboiler. Peng-Robinson is a reasonable VLE package. Be sure to select the correct components from the AspenPlus menu. Note that 1,1,1-trichloroethane should not be used since it has very different properties.
>
> **a.** Plot the x-y diagram from Aspen Plus and compare to an approximate plot with a constant a relative volatility of 2.24 (an over-simplification, but close). Note: Input the data for Step 1b first.
> **b.** The feed is 100% vapor at 1.0 atm. (Set the vapor fraction in the feed to 1.0.) We want a distillate product that is 92 mol% dichloroethane and a bottoms that is 8 mol% dichloroethane. Since Aspen Plus is a simulator program, it will not allow you to specify these concentrations directly. Thus, you need to try different columns (change number of stages and feed location) to find some that work. Use L/D = 2.0. Set Distillate flow rate that will satisfy the mass balance. (Find D and B from mass balances. Do this calculation *accurately* or you may never get the solution you want.) Pick a reasonable number of stages and a reasonable feed stage (Try N = 21 and feed at 9 [vapor] to start). Simulate the column and check the distillate and bottoms mole fractions. This is easiest

to do with the *liquid* composition profiles for stage 1 and N. They should be purer than you want. Also, check the K values and calculate the relative volatility at the top, feed stage, and bottom of the column to see how much it varies.

c. Continue part 1b: Find the optimum feed stage by trying different feed stages to see which one gives the greatest separation (highest dichloroethane mole fraction in distillate and lowest in bottoms). Then decrease the total number of stages (using optimum feed stage each time) until you have the lowest total number of stages that still satisfies the specifications for distillate and bottoms. Note that since the ratio (optimum feed stage)/(total number) is approximately constant you probably need to check only three or four feed stages for each new value of the total number of stages. If your group works together and have different members check different runs, this should not take too long.

d. Triple the pressure, and repeat steps a and b. Determine the effect of higher pressure on the relative volatility.

e. Return to p = 1 atm and your answer for part c. Now increase the temperature of the feed instead of specifying a vapor fraction of 1. What happens?

f. Return to the simulation in part c. Determine the boilup ratio ($\overline{V}_{reboiler}/B$). Now, run the simulation with this boilup ratio instead of specifying Distillate rate. Then start decreasing the boilup ratio in the simulation. What happens?

g. Return to the simulation in part e. Instead of setting fraction vaporized in the feed, set the feed temperature. (Specify L/D = 2 and boilup ratio, N and feed location.) Raise the feed temperature and see what happens.

h. Return to part b, but aim for a distillate product that is 99% dichloro and a bottoms that is 2% dichloro. Do the mass balances (accurately) and determine settings for Distillate flow rate. Try to find the N and feed location that just achieves this separation with an L/D = 2. Note that this separation should not be possible since the external reflux ratio (L/D) is too low. See what happens as you increase L/D (try 4 and 8).

i. Try varing different settings for Operating Specifications in the Radfrac Setup-Data-Browser. For example, pick a condenser duty instead of distillate flow rate that will give the same specification.

j. Add 1,1,1 trichloroethane to your component list. Run analysis for 1,2-dichloroethane and 1,1,1 trichloroethane. Calculate $\alpha_{Di-1,1,1tri}$.

In the appendix to Chapter 6 we will use process simulators for multicomponent distillation calculations.

One caveat applies to these and all other simulations. The program can only solve the problem that you give it. If you make a mistake in the input (e.g., by not including a minor component that appears in the plant) the simulator cannot predict what will actually happen in the plant. A simulation that is correct for the problem given it is not helpful if that problem does not match plant conditions. Chemical engineering judgment must always be applied when using the process simulator (Horwitz, 1998).

CHAPTER 4 APPENDIX B. SPREADSHEETS FOR BINARY DISTILLATION

The Lewis method for binary distillation is easy to program on a spreadsheet using Visual Basic for Applications (VBA). If you are not familiar with VBA, read Appendix 4.B. Part 1. If you are familiar with VBA, you can skip Part 1 and proceed to Part 2.

Appendix 4.B. Part 1. Introduction to Spreadsheets and VBA

Most engineering students are familiar with the use of spreadsheets, but they may not be familiar with the power of spreadsheets when they are coupled with VBA. VBA is an extremely useful programming language for controlling spreadsheets, Word, and other Microsoft programs. This short introduction focuses on use of VBA for spreadsheet calculations. The particular example is binary distillation, but the VBA programming method is applicable to many of the separation methods in this book. Readers interested in more information on VBA are referred to Microsoft (1999) or McFedries (2004).

Once you have learned to program with any language (MATLAB, Mathematica, FORTRAN, C, C++, or whatever) you can learn additional languages on your own. The easiest way to learn a new programming language is usually to study a few basic concepts and have available lists that delineate different programming steps, study and run a simple program that illustrates the basic features, and then create your own program by mimicking other programs. Facility comes with practice. A simple program is presented here, and other examples are in the appendices to Chapters 5 and 17.

VBA extends the usefulness of spreadsheets by allowing the programmer to include loops and logic decisions. Input and output is done through the spreadsheet, and the VBA editor is accessed through Excel. Most math symbols and functions are the same as in Excel. The basic math operations are +, –, * (multiply), /, and ^ (exponentiation). Math functions are written as Function (variable). For example, the cosine of x is written as Cos (x). Commonly used math functions are:

Abs (x)	Absolute value of x
Cos (x)	Cosine of x
Sin (x)	Sine of x
Sqr (x)	Square root of x
Log (x)	Natural log of x (ln x)
Exp (x)	Exponential of x (x)

VBA always determines the value on the right-hand side of equations first and then assigns this value to the left-hand side. Thus, the = sign can be read as "is assigned to the value on the right hand side." Because of this convention, equations such as $x = x + 2$, or $x = 2x + y$ are perfectly valid as written and should not be simplified. Their purpose is to change the value of x.

If there is any possibility that the equation can be misread, use () to make the equation clear. For example, the equations $y = a/(b+x)$ and $y = (a/b) + x$ are not the same. For every left parenthesis, (, there needs to be a corresponding right parenthesis,), to close the equation. The VBA editor will tell you if this rule is violated.

VBA does loops in two different ways. "For....Next" loops do a set number of iterations of the loop (typically from $i = 1$ to a higher number [N in loop below] or a variable that is equal to a higher number). A simple loop to write values on the spreadsheet is,

```
For i = 1 To N
    Cells(7 + i, 7).Value = x
        x = x + deltax
Next i
```

We can also write "For....Next" loops with a logic expression (e.g., "If eqn0 > 0, Then Exit For") to exit the loop.

The other type of loops, Do loops, are discussed after a few introductory comments on the distillation program and in more detail by McFedries (2004). To essentially do a McCabe-Thiele diagram with a spreadsheet program, we need to have an equation for the binary VLE (see Chapter 2, Appendix 2.B.1. Regression of Binary VLE with Excel). We also need operating equations such as Eqs. (4-21) and (4-22). In addition, we need to be able to decide when to switch from one operating line to the other. This can be done based on either the specified feed stage or a test for the optimum feed stage based on passing the intersection points for x_I given by Eq. (4-38a) or for y_I given by Eq. (4-38b). If stepping off stages up the column, a convenient test is that the stage where y_{MVC} first becomes greater than y_I is the optimum feed stage. Finally, we need a test for when the calculation if finished. Stepping off stages up the column, the number of stages is sufficient when $y > x_D$ (or $y > y_D$ with a partial condenser) and stepping off stages down the column the number of stages is sufficient when $x < x_B$.

The use of a Do loop is illustrated for the case with a specified feed stage, and we are stepping off stages from the bottom up (partial reboiler is stage 1). After initializing i = 1 and $x = x_B$, we can set up a loop to step off stages in the stripping section,

```
Do
    y = a6 * x ^ 6 + a5 * x ^ 5 + a4 * x ^ 4 + a3 * x ^ 3 + a2 *
x ^ 2 + a1 * x + a0
    i = i + 1
    x = (y / LbaroverVbar) + (LbaroverVbar - 1) * xb /
LbaroverVbar
Loop While i < feedstage
```

This loop starts with the word "Do" and then calculates y_1 from the equilibrium expression (fit to a 6th-order polynomial) at the known value of x_1, then calculates x_2 from the operating equation (Eq. (4-22) solved for x) with the known value of y_1. This process is repeated to calculate y_2 and x_3, and so forth as long as i < the specified feed stage. When i = the feed stage, we skip the loop and calculate the steps in the enriching section using another loop. Note that Do loops always end with the word "Loop" either by itself (see next example) or with a logic statement, as shown in first Do loop example.

```
Do While y < xd
    y = a6 * x ^ 6 + a5 * x ^ 5 + a4 * x ^ 4 + a3 * x ^ 3 + a2
* x ^ 2 + a1 * x + a0
    i = i + 1
    x = (y / LoverV) - (1 - LoverV) * xd / LoverV
Loop
```

Note that when we enter this loop, we have just calculated the value of x on the feed stage. Thus, the first value calculated is the y value leaving the feed stage.

These loops illustrate two different types of Do loops. In the first example, the logic test (while i < feedstage) is done at the end of the loop, and in the second example, the logic test (the "While" statement) is done at the beginning of the loop. We could have written the stripping and enriching section equations with either type of loop. The difference is that if the logic test is at the end of the loop, the loop will run at least the first time (e.g., if the reboiler is the feed stage).

Appendix 4.B. Part 2. Binary Distillation Example Spreadsheet

In this section we look at a program for binary flash distillation to help make these statements concrete.

First, since VBA is a macro, you must enable macros.

To create a robust program, there are two conditions that we need to check for. Stepping off stages up the column, the specified feed stage may be too low. In this case, the x value will eventually become negative (try this on a McCabe-Thiele diagram). We can test for this by adding the following lines after the calculation of x in the enriching section loop.

```
If x < 0 Then
Cells(i + 7, 4).Value = "Feed stage too low"
Exit Do
End If
```

The person running the spreadsheet will then realize that a higher feed stage location needs to be specified.

The second condition we want to test for is a reflux ratio that is too low. If we assume that the pinch is at the feed stage, a reflux ratio that is too low will cause the intersection of the two operating lines to be above the equilibrium curve. At the intersection point, the values of xint and yint can be calculated from Eq. (4-38). The vapor equilibrium value yeq can be determined from the polynomial fit to the VLE data at x = xint. Then the following test is inserted into both stripping and enriching section loops.

```
If yeq < yint Then
    Cells(i + 7, 3).Value = "Reflux rate too low"
    Exit Do
    End If
```

Every time you add any tests with new variables, you need to remember to include the variables in the appropriate dimension "Dim" statements (see Figure 4-B1).

The spreadsheet with numerical results is shown in Figure 4-B1. The problem solved is to separate 1000 mol/h of a saturated vapor feed that is 0.10 mole fraction ethanol and 0.90 mole fraction water. The column has a total condenser and a partial reboiler. We desire a distillate that is 0.70 mole fraction ethanol and a bottoms that is 0.0001 mole fraction ethanol. Operate with an $L/D = 7.1$ and use stage 20 (reboiler is stage 1) as the feed. CMO is assumed. For VLE, use the constants for ethanol-water VLE for y = f(x) (6th-order polynomial) obtained in Chapter 2, Appendix B, Eq. (2.B-1). The results are given in row 8 and following rows. The terms yint and xint are the intersection of the two operating lines (since q = 0, the

	A	B	C	D	E	F	G	H	I	J	K	L
1	xd	0.7	xb	0.0001	F	1000	z	0.1				
2	L/D	7.1	q	0			feed stg	20				
3	partial reboiler		total condenser		Ethanol-water		Appendix 4B. Tables 4B1 & 4B2					
4	VLE 6th	5th	4th	3rd	2nd	1st	constant					
5	-47.949	161.42	-212.43	138.68	-46.65	7.9322	0					
6	yeqatxint	0.112201	yint		0.1	xint	0.015493					
7	stage	x	y									
8	1	0.0001	0.000793									
9	2	0.000207	0.001638									
10	3	0.000337	0.002668									
11	4	0.000496	0.00392									
12	5	0.000689	0.00544									
13	6	0.000923	0.00728									
14	7	0.001206	0.009501									
15	8	0.001549	0.012173									
16	9	0.00196	0.01537									
17	10	0.002453	0.019178									
18	11	0.00304	0.023684									
19	12	0.003734	0.028975									
20	13	0.004549	0.035133									
21	14	0.005498	0.042224									
22	15	0.006591	0.050291									
23	16	0.007834	0.059342									
24	17	0.009228	0.069335									
25	18	0.010768	0.080175									
26	19	0.012438	0.091709									
27	20	0.014215	0.103723									
28	21	0.019741	0.139444									
29	22	0.060492	0.33713									
30	23	0.286021	0.558829									
31	24	0.538946	0.678204									
32	25	0.675134	0.736928									
33												
34												

Sheet1   **Sheet2**   Sheet3

FIGURE 4-B1.   *Spreadsheet results for binary distillation.*

intersection is at yint = z = 0.1), and the term yeqatxint is the value of y in equilibrium with xint. These numbers are calculated by the VBA program.

Note: Since this spreadsheet uses a VBA program, changing a value of the input (e.g., z) does not change the spreadsheet. You must also run the VBA program by doing the following steps from the spreadsheet tool bar in Excel 2007: View→Macros→View Macros and then click on Run (if you want to see the VBA program, click on Edit).

Note that the result is quite sensitive to the equilibrium expression used.

The VBA program for this spreadsheet is shown in Table 4-B1. The first statement in the program, "Option Explicit," asks VBA to check that every variable is declared in the dimension statements, "Dim." This is useful in debugging the program. The program actually starts with the "Sub McCabeThiele()" statement. In this statement, the programmer chooses the title (McCabeThiele) and carefully avoids including any spaces. Note that lines beginning with an ' are comments and are ignored by the program. The statement "Sheets("Sheet2").Select" tells VBA where to find the spreadsheet, and the "Range("A8", "G108").Clear" statement removes old results. The "Dim ...As Integer" and "Dim...As Single" statements declare all variables used in the program. Data are input with statements such as "xd = Cells(1, 2).Value" that tells VBA that xd has the value in cell (1,2) of the spreadsheet. For this example, that value is 0.7. Note that a large part of the VBA program is reading data from the spreadsheet

Table 4-B1. VBA program for binary distillation stepping off stages from bottom

```
Option Explicit
Sub McCabeThiele()
'   Steps off stages from the bottom up. Assumes that the feed stage is specified.
'   By varying feedstage in the spreadsheet can find an optimum feed stage.
    Sheets("Sheet2").Select
    Range("A8", "G108").Clear
Dim i, feedstage As Integer
Dim D, B, xd, xb, F, z, q, LoverD, LoverV, x, y, xint, yint, yeq As Single
Dim a6, a5, a4, a3, a2, a1, a0, L, V, LbaroverVbar As Single
'   Input values from spread sheet
    xd = Cells(1, 2).Value
    xb = Cells(1, 4).Value
    F = Cells(1, 6).Value
    z = Cells(1, 8).Value
    LoverD = Cells(2, 2).Value
    q = Cells(2, 4).Value
    feedstage = Cells(2, 8).Value
'   Fit VLE data to 6th order polynomial to find y. a6 is coefficient of x to the 6th.
    a6 = Cells(5, 1).Value
    a5 = Cells(5, 2).Value
    a4 = Cells(5, 3).Value
    a3 = Cells(5, 4).Value
    a2 = Cells(5, 5).Value
    a1 = Cells(5, 6).Value
    a0 = Cells(5, 7).Value
'   Calculate flow rates and ratios.
    D = ((z - xb) / (xd - xb)) * F
    L = LoverD * D
    V = L + D
    LoverV = LoverD / (1 + LoverD)
    LbaroverVbar = (LoverV + (q * F / V)) / (1 - ((1 - q) * F / V))
'   Calculate intersection point of operating lines. If yeq < yint reflux ratio too low.
    xint = ((-(q - 1) * (1 - LoverV) * xd) - z) / (((q - 1) * LoverV) - q)
    x = xint
    yint = LoverV * xint + (1 - LoverV) * xd
'   Equilibrium y at value of x intersection. When yint=yeq, have minimum L/D.
    yeq = a6 * x ^ 6 + a5 * x ^ 5 + a4 * x ^ 4 + a3 * x ^ 3 + a2 * x ^ 2 + a1 * x + a0
'Record intersection and equilibrium values on spreadsheet.
    Cells(6, 2).Value = yeq
    Cells(6, 4).Value = yint
    Cells(6, 6).Value = xint
'   Step off stages from bottom up.  Stage 1 is partial reboiler.  Initialize
    x = xb
    i = 1
'   Loop in stripping section stepping off stages with equilibrium and operating eqs.
Do
If yeq < yint Then
    Cells(i + 7, 3).Value = "Reflux rate too low"
    Exit Do
    End If
    y = a6 * x ^ 6 + a5 * x ^ 5 + a4 * x ^ 4 + a3 * x ^ 3 + a2 * x ^ 2 + a1 * x + a0
'   Record values of stage number, x and y values on spreadsheet.
    Cells(i + 7, 1).Value = i
    Cells(i + 7, 2).Value = x
    Cells(i + 7, 3).Value = y
    i = i + 1
    x = (y / LbaroverVbar) + (LbaroverVbar - 1) * xb / LbaroverVbar
```

```
Loop While i < feedstage
'    Calculations in enriching section.
Do While y < xd
If yeq < yint Then
    Cells(i + 7, 5).Value = "Reflux rate too low"
Exit Do
End If
    y = a6 * x ^ 6 + a5 * x ^ 5 + a4 * x ^ 4 + a3 * x ^ 3 + a2 * x ^ 2 + a1 * x + a0
'    Record values of stage number, x, and y on spreadsheet
    Cells(i + 7, 1).Value = i
    Cells(i + 7, 2).Value = x
    Cells(i + 7, 3).Value = y
    i = i + 1
    x = (y / LoverV) - (1 - LoverV) * xd / LoverV
  If x < 0 Then
    Cells(i + 7, 4).Value = "Feed stage too low"
  Exit Do
  End If
Loop
End Sub
```

and then recording results on the spreadsheet. The last line of the VBA program must be the "End Sub" statement.

By varying the feed stage location, the optimum location (stage 20) can easily be found. The minimum external reflux ratio can be determined quickly by trial and error (see Problem 4.H2). The program can also be coded to find either of these values automatically (see Problems 4.H1 and 4.H3).

Obviously this problem can be solved graphically. The problem can also be solved with a hand calculation on a spreadsheet without using VBA by writing the equilibrium and operating equations in cell form and then repeating them as needed. The advantages of a programmed spreadsheet or other software tools (e.g., Binous, 2008) are a huge number of trials can be done by changing just one number (e.g., z or L/D), different systems can easily be studied by using different values for the VLE coefficients, and one does not have to manually step off 25 stages. For a single use graphical or hand-calculation with a spreadsheet may take less time. However, for multiple uses the programmed spreadsheet is obviously preferable. In all cases if the CMO assumption is not valid, the results will be incorrect. Note that this program, and all other software packages, must be checked to make sure it is producing correct results (Shacham et al., 2008).

CHAPTER 5

Introduction to Multicomponent Distillation

Binary distillation problems can be solved in a straightforward manner using a stage-by-stage calculation that can be done either on a computer or graphically using a McCabe-Thiele diagram. When additional components are added, the resulting multicomponent problem becomes significantly more difficult, and the solution may not be straightforward. In this chapter we will first consider why multicomponent distillation is more complex than binary distillation, and then we will look at the profile shapes typical of multicomponent distillation. In Chapter 6, matrix calculation methods will be applied to multicomponent distillation, and approximate methods will be developed in Chapter 7.

5.1   CALCULATIONAL DIFFICULTIES

Consider the conventional schematic diagram of a plate distillation column with a total condenser and a partial reboiler shown in Figure 5-1. Assume constant molal overflow (CMO), constant pressure, and no heat leak. With the constant pressure and zero heat leak assumptions, a degree-of-freedom analysis around the column yields $C + 6$ degrees of freedom, where C is the number of components. For binary distillation this is 8 degrees of freedom. In a design problem we would usually specify these variables as follows (see Tables 3-1 and 3-2): F, z, feed quality q, distillate composition x_D, distillate temperature (saturated liquid), bottoms composition x_B, external reflux ratio L_0/D, and the optimum feed stage. With these variables chosen, the operating lines are defined, and we can step off stages from either end of the column using the McCabe-Thiele method.

Now, if we add a third component we increase the degrees of freedom to 9. Nine variables that would most likely be specified for design of a ternary distillation column are listed in Table 5-1. Comparing this table with Tables 3-1 and 3-2, we see that the extra degree of freedom is used to completely specify the feed composition. If there are four components, there will be 10 degrees of freedom. The additional degree of freedom must again be used to completely specify the feed composition.

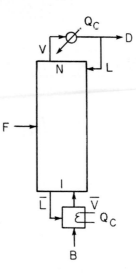

FIGURE 5-1.   *Distillation column*

Note that in multicomponent distillation neither the distillate nor the bottoms composition is completely specified because there are not enough variables to allow complete specification. This inability to completely specify the distillate and bottoms compositions has major effects on the calculation procedure. The components that do have their distillate and bottoms fractional recoveries specified (such as component 1 in the distillate and component 2 in the bottoms in Table 5-1) are called *key* components. The most volatile of the keys is called the *light key* (LK), and the least volatile the *heavy key* (HK). The other components are *non-keys* (NK). If a non-key is more volatile (lighter) than the light key, it is a *light non-key* (LNK); if it is less volatile (heavier) than the heavy key, it is a *heavy non-key* (HNK).

The external balance equations for the column shown in Figure 5-1 are easily developed. These are the overall balance equation,

TABLE 5-1.   *Specified design variables for ternary distillation*

Number of variables	Variable
1	Feed rate, F
2	Feed composition, z_1, z_2 ($z_3 = 1 - z_1 - z_2$)
1	Feed quality, q (or h_F or T_F)
1	Distillate, $x_{1, \text{dist}}$ (or $x_{3, \text{dist}}$ or D or one fractional recovery)
1	Bottoms, $x_{2, \text{bot}}$ (or $x_{3, \text{bot}}$ or one fractional recovery)
1	L_0/D or V/B or Q_R
1	Saturated liquid reflux or T_{reflux}
1	Optimum feed plate location
9	

Column pressure and $Q_{\text{col}} = 0$ are already specified.

$$F = B + D \qquad \text{(5-1)}$$

the component balance equations,

$$Fz_i = Bx_{i,bot} + Dx_{i,dist} \qquad i = 1, 2, \cdots C \qquad \text{(5-2)}$$

and the overall energy balance,

$$Fh_F + Q_c + Q_R = Bh_B + Dh_D \qquad \text{(5-3)}$$

Since we are using mole fractions, the mole fractions must sum to 1.

$$\sum_{i=1}^{C} x_{i,dist} = 1.0 \qquad \text{(5-4a)}$$

$$\sum_{i=1}^{C} x_{i,bot} = 1.0 \qquad \text{(5-4b)}$$

For a ternary system, Eqs. (5-2) can be written three times, but these equations must add to give Eq. (5-1). Thus, only two of Eqs. (5-2) plus Eq. (5-1) are independent.

For a single feed column, we can solve any one of Eqs. (5-2) simultaneously with Eq. (5-1) to obtain results analogous to the binary Eqs. (3-3) and (3-4).

$$D = \left(\frac{z_i - x_{i,bot}}{x_{i,dist} - x_{i,bot}} \right) F \qquad \text{(5-5)}$$

$$B = \left(\frac{x_{i,dist} - z_i}{x_{i,dist} - x_{i,bot}} \right) F \qquad \text{(5-6)}$$

If the feed, bottoms, and distillate compositions are specified for any component we can solve for D and B. In addition, these equations show that the ratio of concentration differences for all components must be identical (Doherty and Malone, 2001). These equations can be helpful, but do not solve the complete problem even when $x_{i,dist}$ and $x_{i,bot}$ are specified for one component since the other mole fractions in the distillate and bottoms are unknown.

Now, how do we completely solve the external mass balances? The unknowns are B, D, $x_{2,dist}$, $x_{3,dist}$, $x_{1,bot}$, and $x_{3,bot}$. There are six unknowns and five independent equations. Can we find an additional equation? Unfortunately, the additional equations (energy balances and equilibrium expressions) always add additional variables (see Problem 5-A1), so we cannot start out by solving the external mass and energy balances. This is the first major difference between binary and multicomponent distillation.

Can we do the internal stage-by-stage calculations first and then solve the external balances? To begin the stage-by-stage calculation procedure in a distillation column, we need to know all the compositions at one end of the column. For ternary systems with the variables specified as in Table 5-1, these compositions are unknown. To begin the analysis we would have to assume one of them. Thus, internal calculations for multicomponent distillation problems are trial and error. This is a second major difference between binary and multicomponent problems.

Fortunately, in many cases it is easy to make an excellent first guess that will allow one to do the external balances. If a sharp separation of the keys is required, then almost all of the HNKs will appear only in the bottoms, and almost all of the LNKs will appear only in the distillate. The obvious assumption is that all LNKs appear only in the distillate and all HNKs appear only in the bottom. Thus,

$$x_{LNK,bot} = 0 \tag{5-7a}$$

$$x_{HNK,dist} = 0 \tag{5-7b}$$

These assumptions allow us to complete the external mass balances. The procedure is illustrated in Example 5-1.

EXAMPLE 5-1. External mass balances using fractional recoveries

We wish to distill 2000 kmol/h of a saturated liquid feed. The feed is 0.056 mole fraction propane, 0.321 n-butane, 0.482 n-pentane, and the remainder n-hexane. The column operates at 101.3 kPa. The column has a total condenser and a partial reboiler. Reflux ratio is $L_0/D = 3.5$, and reflux is a saturated liquid. The optimum feed stage is to be used. A fractional recovery of 99.4% n-butane is desired in the distillate and 99.7% of the n-pentane in the bottoms. Estimate distillate and bottoms compositions and flow rates.

Solution

A. Define. A sketch of the column is shown.

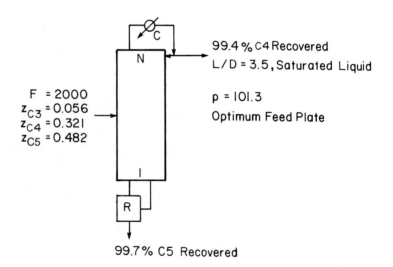

Find $x_{i,dist}$, $x_{i,bot}$, D, and B.

B. Explore. This appears to be a straightforward application of external mass balances, *except* there are two variables too many. Thus, we will have to assume the recoveries or concentrations of two of the components. A look at the DePriester charts (Figures 2-11 and 2-12) shows that the order of volatilities is propane > n-butane > n-pentane > n-hexane. Thus, n-butane is the

LK, and n-pentane is the HK. This automatically makes propane the LNK and n-hexane the HNK. Since the recoveries of the keys are quite high, it is reasonable to assume that all of the LNK collects in the distillate and all of the HNK collects in the bottoms. We will estimate distillate and bottoms based on these assumptions.

C. Plan. Our assumptions of the NK splits can be written either as

$$Dx_{C3,dist} = Fz_{C3} \quad \text{and} \quad Bx_{C6,bot} = Fz_{C6} \quad \text{(5-8a,b)}$$

or

$$Bx_{C3,bot} = 0 \quad \text{and} \quad Dx_{C6,dist} = 0 \quad \text{(5-9a,b)}$$

The fractional recovery of n-butane in the distillate can be used to write

$$Dx_{C4,dist} = (\text{frac. recovery } C_4 \text{ in distillate})(Fz_{C4}) \quad \text{(5-10)}$$

Note that this also implies

$$Bx_{C4,bot} = (1 - \text{frac. rec. } C_4 \text{ in dist.})(Fz_{C4}) \quad \text{(5-11)}$$

For n-pentane the equations are

$$Bx_{C5,bot} = (\text{frac. rec. } C_5 \text{ in bot.}) \, Fz_{C5} \quad \text{(5-12)}$$

$$Dx_{C5,dist} = (1 - \text{frac. rec. } C_5 \text{ in bot.}) \, Fz_{C5} \quad \text{(5-13)}$$

Equations (5-8) to (5-13) represent eight equations with ten unknowns (four compositions in both distillate and bottoms plus D and B). Equations (5-4) give two additional equations, which we will write as

$$\sum_{i=1}^{C} (Dx_{i,dist}) = D \quad \text{(5-14a)}$$

$$\sum_{i=1}^{C} (Bx_{i,bot}) = B \quad \text{(5-14b)}$$

These ten equations can easily be solved, since distillate and bottoms calculations can be done separately.

D. Do it. Start with the distillate.

$Dx_{C3,dist} = Fz_{C3} = (2000)(0.056) = 112$
$Dx_{C6,dist} = 0$
$Dx_{C4,dist} = (0.9940)(2000)(0.321) = 638.5$
$Dx_{C5,dist} = (0.003)(2000)(0.482) = 2.89$

Then

$$D = \sum_{i=1}^{4} Dx_{i,dist} = 753.4$$

Now the individual distillate mole fractions are

$$x_{i,dist} = \frac{(Dx_{i,dist})}{D} \qquad (5\text{-}15)$$

Thus,

$$x_{C3,dist} = \frac{112}{753.4} = 0.1487, \text{ and}$$

$$x_{C4,d} = \frac{638.15}{753.4} = 0.8474, \ x_{C5,d} = \frac{2.89}{753.4} = 0.0038, \ x_{C6,d} = 0$$

Check:

$$\sum_{i=1}^{4} x_{i,dist} = 0.9999, \text{ which is OK.}$$

Bottoms can be found from Eqs. (5-8b), (5-9a), (5-11), (5-12), and (5-14b). The results are $x_{C3,bot} = 0$, $x_{C4,bot} = 0.0031$, $x_{C5,bot} = 0.7708$, $x_{C6,bot} = 0.2260$, and B = 1246.6. Remember that these are *estimates* based on our assumptions for the splits of the NK.

E. Check. Two checks are appropriate. The results based on our assumptions can be checked by seeing whether the results satisfy the external mass balance Eqs. (5-1) and (5-2). These equations are satisfied. The second check is to check the assumptions, which requires internal stage-by-stage analysis and is much more difficult. In this case the assumptions are quite good.

F. Generalize. This type of procedure can be applied to many multicomponent distillation problems. It is more common to specify fractional recoveries rather than concentrations because it is more convenient. Note that it is important to not make specifications that violate material balances and distillation fundamentals (e.g., 99.4% recovery of C4 in the distillate and 90% mole fraction of C5 in the bottoms).

Surprisingly, the ability to do reasonably accurate external mass balances on the basis of a first guess does not guarantee that internal stage-by-stage calculations will be accurate. The problem given in Example 5-1 would be very difficult for stage-by-stage calculations. Let us explore why.

At the feed stage all components must be present at finite concentrations. If we wish to step off stages from the bottom up, we cannot use $x_{C3,bot} = 0$ because we would not get a nonzero concentration of propane at the feed stage. Thus, $x_{C3,bot}$ must be a small but nonzero value. Unfortunately, we don't know if the correct value should be 10^{-5}, 10^{-6}, 10^{-7}, or 10^{-20}. Thus, the percentage error in $x_{C3,bot}$ will be large, and it will be difficult to obtain convergence of the trial-and-error problem. If we try to step off stages from the top down, $x_{C3,dist}$ is known accurately, but $x_{C6,dist}$ is not. Thus, when both heavy and LNKs are present, stage-by-stage calculation methods are difficult. Hengstebeck (1961) developed the use of pseudo-components for approximate multicomponent calculations on a McCabe-Thiele diagram. The LK and LNKs are lumped together and the HK and HNKs are lumped together. As we will see in

Section 5.3, these results can only be approximate. Other design procedures should be used for accurate results.

If there are *only* LNKs or *only* HNKs, then an accurate first guess of compositions can be made. Suppose in Example 5-1 that we specified 99.4% recovery of propane in the distillate and 99.7% recovery of n-butane in the bottoms. This makes propane the LK, n-butane the HK, and n-pentane and n-hexane the HNKs. The assumption that all the HNKs appear in the bottoms is an excellent first guess. Then we can calculate the distillate and bottoms compositions from the external mass balances. The composition calculated in the bottoms is quite accurate. Thus, in this case we can step off stages from the bottom upward and be quite confident that the results are accurate. If only LNKs are present, the stage-by-stage calculation should proceed from the top downward.

5.2. STAGE-BY-STAGE CALCULATIONS FOR CONSTANT MOLAL OVERFLOW AND CONSTANT RELATIVE VOLATILITY

Although computer simulator programs use matrix methods (see Chapter 6), historically stage-by-stage methods were first used for multicomponent distillation. These methods work well if there are no HNK or if there are no LNK, they have an obvious direct relationship to the McCabe-Thiele approach, and they are easy to implement on a spreadsheet or in MATLAB. The stage-by-stage method also has the advantage for design applications that it is a design method. In this section we look at the simplest case—constant relative volatility. In Section 5.4 the method is expanded to include bubble-point or dew-point calculations on every stage for systems that do not have constant relative volatilities.

Consider a typical design problem for ternary distillation with an LNK, an LK, and an HK. The feed flow rate, composition, and temperature are specified, as are L_0/D, saturated liquid reflux, pressure, use of the optimum feed stage, and recoveries of the light and heavy keys in distillate and bottoms, respectively. We wish to predict the number of stages required and the separation obtained.

To start the calculation, we need to assume the fractional recovery (frac rec.) for the LNK (A) in the distillate. Then

$Dx_{A,dist} = Fz_A$ (frac rec. A in dist)

$Bx_{A,bot} = Fz_A$ (1 − frac rec. A in dist)

If, for example, (frac rec. A in dist) = 1.0, then $x_{A,bot} = 0$ and $Dx_{A,dist} = Fz_A$. Once the fractional recovery is assumed, we can find L and V in the rectifying section. Since constant molal overflow is valid,

$$L = \left(\frac{L_0}{D}\right)D, \quad V = L + D$$

(5-16)

At the feed stage, q can be estimated from enthalpies as

$$q \sim \frac{H - h_{feed}}{H - h}$$

(5-17)

or $q = L_F/F$ can be found from a flash calculation on the feed stream. Then $\bar{L}$ and $\bar{V}$ are determined from balances at the feed stage,

$$\bar{L} = L + qF, \quad \bar{V} = V - (1-q)F \qquad (5\text{-}18)$$

This completes the preliminary calculations for this assumption of how the LNK splits in the column.

If there is an LNK and no HNK, the assumed compositions are very accurate at the top of the column but not at the bottom. (At the bottom, the mole fraction cannot be exactly zero. If we assume it is 10^{-9} when it is actually 10^{-11}, then our relative error is very large.) Thus, with only an LNK present, we want to step off stages from the top down. The general procedure for stepping off stages down the column when CMO is valid is:

1. Set $j = 1$. For total condenser, $y_{i,1} = x_{i,dist}$, where i is component A, B, or C, and the second subscript is the stage location.
2. Use equilibrium to calculate $x_{i,j}$ values from known $y_{i,j}$ values for stage j.
3. Use mass balances (operating equations) to calculate $y_{i,j+1}$ values from known $x_{i,j}$ values.
4. Repeat steps 2 and 3 until the feed stage is reached. Then change to the stripping section operating equations and continue.
5. The calculation is finished when $x_{HK,N+1} \geq x_{HK,bot}$ and $x_{LK,N+1} \leq x_{LK,bot}$

If the relative volatilities are constant, the equilibrium calculations become simple. We arbitrarily choose component B as the reference. Then by definition,

$$\alpha_{AB} = \frac{K_A}{K_B} = \frac{y_A/x_A}{y_B/x_B}, \quad \alpha_{BB} = \frac{K_B}{K_B} = 1.0, \quad \alpha_{CB} = \frac{K_C}{K_B} = \frac{y_C/x_C}{y_B/x_B} \qquad (5\text{-}19)$$

As we step down the column, the y values leaving a stage will be known and the x values can be calculated from equilibrium. Thus for component i on stage j,

$$x_{i,j} = \frac{y_{i,j}}{\left(y_{i,j}/x_{i,j}\right)/\left(y_{B,j}/x_{B,j}\right)} \frac{1}{\left(y_{B,j}/x_{B,j}\right)} = \frac{y_{i,j}}{\alpha_{iB,j}} \frac{1}{K_{B,j}} \qquad (5\text{-}20)$$

Note that the first equals sign for equation (5-20) reduces to $x_{i,j} = x_{i,j}$.

In general, both $\alpha_{iB,j}$ and $K_{B,j}$ depend upon temperature and thus vary from stage to stage. When the relative volatilities are constant, only $K_{B,j}$ varies. Fortunately, because the liquid mole fractions must sum to 1.0, we can obtain an equation for $K_{B,j}$ and then remove it from Eq. (5-20). For this ternary problem,

$$x_{A,j} + x_{B,j} + x_{C,j} = \frac{y_{A,j}}{\alpha_{AB}K_{B,j}} + \frac{y_{B,j}}{\alpha_{BB}K_{B,j}} + \frac{y_{C,j}}{\alpha_{CB}K_{B,j}} = 1.0$$

Solving for $K_{B,j}$,

$$K_{B,j} = \frac{y_{A,j}}{\alpha_{AB}} + \frac{y_{B,j}}{\alpha_{BB}} + \frac{y_{C,j}}{\alpha_{CB}} = \sum_{i=1}^{3} \left(\frac{y_{i,j}}{\alpha_{iB}} \right)$$

(5-21a)

or, in general,

$$K_{ref,j} = \sum_{i=1}^{C} \left(\frac{y_{i,j}}{\alpha_{i-ref}} \right)$$

(5-21b)

where C is the number of components. If desired, the stage temperature can be determined from the calculated $K_{ref,j}$ value. This is not necessary, since Eq. (5-21a) or (5-21b) can be substituted into Eq. (5-20) to obtain an equilibrium expression for component A

$$x_{A,j} = \frac{y_{A,j}/\alpha_{AB}}{\sum_{i=1}^{3} \left(\dfrac{y_{i,j}}{\alpha_{iB}} \right)}$$

(5-22a)

or, in general,

$$x_{i,j} = \frac{y_{i,j}/\alpha_{i-ref}}{\sum_{i=1}^{C} \left(y_{i,j}/\alpha_{i-ref} \right)}$$

(5-22b)

Equations (5-22) can be used to calculate the liquid mole fractions at equilibrium. This form of the equilibrium is known as a dew-point calculation because we start with a vapor (the $y_{i,j}$ values are known) and find the concentration of a liquid in equilibrium with this vapor. Dew-point calculations are needed when stepping off stages down the column. Note that the choice of the reference component is arbitrary. If Eqs. (5-22) are expanded, K_{ref} in the relative volatility terms divides out; thus, any component can be used as the reference as long as we are consistent.

The operating equations are essentially the same as for binary systems. These are

$$y_{i,j+1} = \frac{L}{V} x_{i,j} + \left(1 - \frac{L}{V} \right) x_{i,dist}$$

(5-23)

in the enriching section and

$$y_{i,k+1} = \frac{\bar{L}}{\bar{V}} x_{i,k} - \left(\frac{\bar{L}}{\bar{V}} - 1 \right) x_{i,bot}$$

(5-24)

in the stripping section.

Stepping off stages is straightforward. In the enriching section with a total condenser $y_{i,1} = x_{i,dist}$. Start stepping off stages with equilibrium Eq. (5-22) to calculate $x_{i,1}$ in equilibrium with vapor of composition $y_{i,1}$. Then we determine $y_{i,2}$ for each component from operating Eq. (5-23). Equilibrium Eq. (5-22) is used to find $x_{i,2}$ from values $y_{i,2}$, and so on. Stepping off stages down the column we calculate $y_{i,f}$ from Eq. (5-23), $x_{i,f}$ from Eq. (5-22) with $y_{i,j} = y_{i,f}$, and $y_{i,f+1}$ from Eq. (5-24) with $x_{i,j} = x_{i,f}$, where f is the specified feed stage. In the stripping section we alternate between Eqs. (5-22) and (5-24). We stop the calculation when

$$x_{HK,N+1} \geq x_{HK,bot} \quad \text{and} \quad x_{LK,N+1} \leq x_{LK,bot} \qquad \text{(5-25)}$$

This leaves us with two unanswered questions: How do we determine the optimum feed plate, and how do we correct our initial assumption of the LNK split?

The optimum feed plate is defined as the feed plate that gives the fewest total number of stages. To be absolutely sure you have the optimum feed plate location, use this definition. That is, pick a feed plate location and calculate N. Then repeat until you find the minimum total number of stages. Note that often several stages must be stepped off before the feed can be input. The first legal feed stage may be the optimum. This procedure sounds laborious, but, as we will see, it is very easy to implement on a spreadsheet (Appendix A of Chapter 5).

If you try to switch stages too early, the stage-by-stage calculation will eventually give negative mole fractions. With a spreadsheet, you can guard against this mistake by checking that all mole fractions ($x_{i,j}$ and $y_{i,j}$) are between zero and 1 for every stage.

How do we check and correct our initial guess for the splits of the NK components? One way to do this is to use the calculated value of the LNK mole fraction to estimate the fractional recovery of the LNK,

$$\text{Calc. frac. rec. LNK in dist} = 1 - \frac{\left(x_{LNK,N+1} B \right)}{F z_{LNK}} \qquad \text{(5-26)}$$

If ε is the acceptable error, then if

$$\left| \text{Calc. frac. rec. LNK dist} - \text{previous frac rec. LNK dist} \right| > \varepsilon \qquad \text{(5-27)}$$

a new trial is required. For the next trial we can use a damped direct substitution and set

$$\begin{aligned} \text{frac rec. LNK in dist} &= \text{frac rec. LNK previous} + \\ (df)&(\text{frac rec. LNK calc} - \text{frac rec. LNK previous}) \end{aligned} \qquad \text{(5-28)}$$

where df is the damping factor ≤ 1. This procedure is shown in the spreadsheet in Appendix A of Chapter 5. If df = 1, Eq. (5-28) becomes direct substitution, which may result in oscillations.

Stage-by-stage calculations for systems with constant relative volatilities are relatively easy, and the resulting profiles illustrate most of the behaviors observed with multicomponent systems. Fortunately, extending the calculation to nonconstant relative volatility systems is not difficult and is discussed in Section 5.4. We return to these calculations in Chapter 8 for total reflux systems (see Example 8-3).

This entire discussion was for calculations down the column. If only HNKs are present, then the calculation should proceed up the column. Now renumber stages so that the partial reboiler = 1, bottom stage in column = 2, and so forth. Now we calculate y values from equilibrium and x values from the operating equations.

Stepping off stages up the column, the equilibrium calculation on every stage is a bubble-point calculation. In this case, the K value for the reference component is

$$K_{ref,j} = 1 / \sum_{i=1}^{C} \left(x_{i,j} \alpha_{i-ref} \right)$$

(5-29)

and the vapor compositions are determined from the equilibrium equation,

$$y_{i,j} = \frac{x_{i,j} \alpha_{i-ref}}{\sum_{i=1}^{C} \left(x_{i,j} \alpha_{i-ref} \right)}$$

(5-30)

Derivation of Eqs. (5-29) and (5-30) is left as an exercise. The liquid mole fractions are found from the operating equations, which are inverted to find $x_{i,j}$:

$$x_{i,k+1} = \frac{\bar{V}}{\bar{L}} y_{i,k} + \left(1 - \frac{\bar{V}}{\bar{L}} \right) x_{i,bot}$$

(5-31)

$$x_{i,j+1} = \frac{V}{L} y_j - \left(\frac{V}{L} - 1 \right) x_{i,dist}$$

(5-32)

Convergence requires checking the assumed and calculated values of the fractional recovery of the HNK in the bottoms.

For problems where both light and heavy non-keys are present, Lewis and Matheson (1932) and Thiele and Geddes (1933) calculated from both ends of the column and matched compositions at the feed stage (see Smith, 1963, Chapter 20, for details). Unfortunately, closure can be very difficult. When there are both light and heavy non-keys, and when there is a sandwich component, other calculation methods such as the matrix method discussed in Chapter 6 are preferable.

5.3   PROFILES FOR MULTICOMPONENT DISTILLATION

What do the flow, temperature, and composition profiles look like? Our intuition would tell us that these profiles will be similar to the ones for binary distillation. As we will see, this is true for the total flow rates and temperature, but not for the composition profiles.

If CMO is valid, the total vapor and liquid flow rates will be constant in each section of the column. The total flow rates can change at each feed stage or sidestream withdrawal stage. This behavior is illustrated for a computer simulation for a saturated liquid feed in Figure 5-2 and is the same behavior we would expect for a binary system. For nonconstant

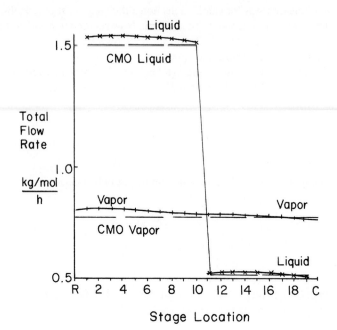

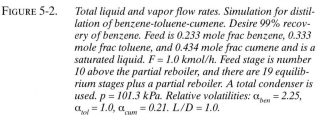

FIGURE 5-2. *Total liquid and vapor flow rates. Simulation for distillation of benzene-toluene-cumene. Desire 99% recovery of benzene. Feed is 0.233 mole frac benzene, 0.333 mole frac toluene, and 0.434 mole frac cumene and is a saturated liquid. F = 1.0 kmol/h. Feed stage is number 10 above the partial reboiler, and there are 19 equilibrium stages plus a partial reboiler. A total condenser is used. p = 101.3 kPa. Relative volatilities: α_{ben} = 2.25, α_{tol} = 1.0, α_{cum} = 0.21. L/D = 1.0.*

molal overflow, the total flow rates will vary from section to section. This is also shown in Figure 5-2. Although both liquid and vapor flow rates may vary significantly, the ratio L/V will be much more constant.

The temperature profile decreases monotonically from the reboiler to the condenser. This is illustrated in Figure 5-3 for the same computer simulation. This is again similar to the behavior of binary systems. Note that plateaus start to form where there is little temperature change between stages. When there are a large number of stages, these plateaus can be quite pronounced. They represent pinch points in the column.

The compositions in the column are much more complex. To study these, we will first look at two computer simulations for the distillation of benzene, toluene, and cumene in a column with 20 equilibrium contacts. The total flow and temperature profiles for this simulation are given in Figures 5-2 and 5-3, respectively. With a specified 99% recovery of benzene in the distillate, the liquid mole fractions are shown in Figure 5-4.

At first Figure 5-4 is a bit confusing, but it will make sense after we go through it step-by-step. Since benzene recovery in the distillate was specified as 99%, benzene is the LK. Typically, the next less volatile component, toluene, will be the HK. Thus, cumene is the HNK, and there is no LNK. Following the benzene curve, we see that benzene mole fraction is very low in the

Chapter 5 Introduction to Multicomponent Distillation

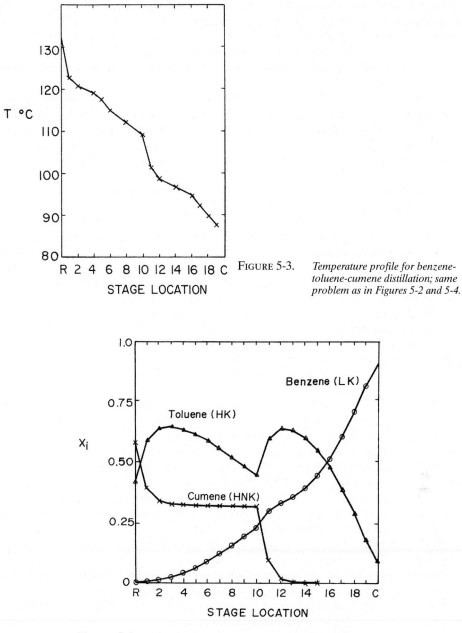

FIGURE 5-3. *Temperature profile for benzene-toluene-cumene distillation; same problem as in Figures 5-2 and 5-4.*

FIGURE 5-4. *Liquid-phase composition profiles for distillation for benzene-toluene-cumene; same conditions as Figures 5-2 and 5-3 for nonconstant molal overflow. Benzene is the LK, and toluene is the HK. Stage 10 is the feed stage.*

reboiler and increases monotonically to a high value in the total condenser. This is essentially the same behavior as that of the more volatile component in binary distillation (for example, see Figure 4-14). In this problem benzene is always most volatile, so its behavior is simple.

Since cumene is the HNK, we would typically assume that all of the cumene leaves the column in the bottoms. Figure 5-4 shows that this is essentially true (cumene distillate mole fraction was calculated as 2.45×10^{-8}). Starting at the reboiler, the mole fraction of cumene rapidly decreases and then levels off to a plateau value until the feed stage. Above the feed stage the cumene mole fraction decreases rapidly. This behavior is fairly easy to understand. Cumene's mole fraction decreases above the reboiler because it is the least volatile component. Since there is a large amount of cumene in the feed, there must be a finite concentration at the feed stage. Thus, after the initial decrease there is a plateau to the feed stage. Note that the concentration of cumene on the feed stage is *not* the same as in the feed. Above the feed stage, cumene concentration decreases rapidly because cumene is the least volatile component.

The concentration profile for the HK toluene is most complex in this example. The behavior of the HK can be explained by noting which binary pairs of components are distilling in each part of the column. In the reboiler and stages 1 and 2 there is very little benzene (LK) and the distillation is between the HK and the HNK. In these stages the toluene (HK) concentration increases as we go up the column, because toluene is the more volatile of the two components distilling. In stages 3 to 10, the cumene (HNK) concentration plateaus. Thus, the distillation is between the LK and the HK. Now toluene is the less volatile component, and its concentration decreases as we go up the column. This causes the primary maximum in HK concentration, which peaks at stage 3. Above the feed stage, in stages 11, 12, and 13, the HNK concentration plummets. The major distillation is again between the HK and the HNK. Since the HK is temporarily the more volatile of these components, its concentration increases as we go up the column and peaks at stage 12. After stage 12, there is very little HNK present, and the major distillation is between benzene and toluene. The toluene concentration then decreases as we continue up to the condenser. The secondary maximum above the feed stage is often much smaller than shown in Figure 5-4. The large amounts of cumene in this example cause a larger than normal secondary maximum.

In this example the HNK (cumene) causes the two maxima in the HK (toluene) concentration profile. Since there was no LNK, the LK (benzene) has no maxima. It is informative to redo the example of Figures 5-2 to 5-4 with everything the same except for specifying 99% recovery of toluene in the distillate. Now toluene is the LK, cumene is the HK and benzene an LNK. The result achieved here is shown in Figure 5-5. This figure can also be explained qualitatively in terms of the distillation of binary pairs (see Problem 5.A12). Note that with no HNKs, the HK concentration does not have any maxima.

What happens for a four-component distillation if there are LKs and HKs and LNKs and HNKs present? Since there is an LNK, we would expect the LK curve to show maxima; and since there is an HNK, we would expect maxima in the HK concentration profile. This is the case shown in Figure 5-6 for the distillation of a benzene-toluene-xylene-cumene mixture. Note that in this figure the secondary maxima near the feed stage are drastically repressed, but the primary maxima are readily evident.

It is interesting to compare the purities of the distillate and bottoms products in Figures 5-4 to 5-6. In Figure 5-4 (no LNK) the benzene (LK) can be pure in the distillate, but the bottoms (HK and HNK present) is clearly not pure. In Figure 5-5 (no HNK) the cumene (HK) can be pure in the bottoms, but the distillate (LK and LNK present) is not pure. In Figure 5-6 (both LNK and HNK present) neither the distillate nor the bottoms can be pure. Simple

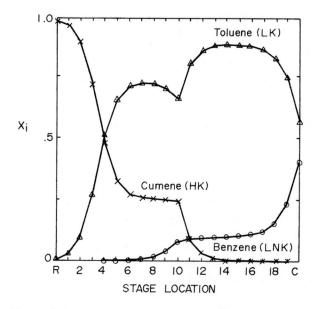

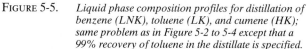

FIGURE 5-5.   *Liquid phase composition profiles for distillation of benzene (LNK), toluene (LK), and cumene (HK); same problem as in Figure 5-2 to 5-4 except that a 99% recovery of toluene in the distillate is specified.*

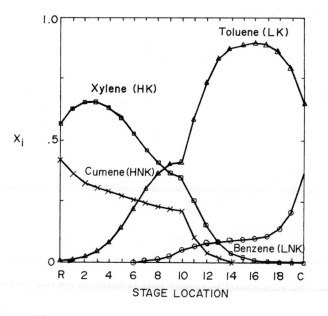

FIGURE 5-6.   *Liquid composition profiles for distillation of benzene (LNK), toluene (LK), xylene (HK), and cumene (HNK). Feed is 0.125 benzene, 0.225 toluene, 0.375 xylene, and 0.275 cumene. Recovery of toluene in distillate 99%. Relative volatilities:* $\alpha_{ben} = 2.25$, $\alpha_{tol} = 1.0$, $\alpha_{xy} = 0.33$, $\alpha_{cum} = 0.21$.

distillation columns separate the feed into two fractions, and a single column can produce either pure most volatile component as distillate by making it the LK, or it can produce pure least volatile component as bottoms by making it the HK. If we want to completely separate a multicomponent mixture, we need to couple several columns together (see Lab 6 in the Appendix to Chapter 6 and Section 11.5).

There is one other case to consider. Suppose we have the four components benzene, toluene, cumene, and xylene, and we choose cumene as the HK and toluene as the LK. This makes benzene the LNK, but what is xylene? In this case xylene is an *intermediate* or *sandwich component*, which is an NK component with a volatility between the two key components. Sandwich components will tend to concentrate in the middle of the column since they are less volatile than the LK in the rectifying section and more volatile than the HK in the stripping section. Prediction of their final distribution requires a complete simulation.

If the top temperature is too cold and the bottom temperature is too hot to allow sandwich components to exit at the rate they enter the column, they become trapped in the center of the column and accumulate there (Kister, 2004). This accumulation can be quite large for trace components in the feed and can cause column flooding and development of a second liquid phase. The problem can be identified from the simulation if the engineer knows all the trace components that occur in the feed, accurate vapor-liquid equilibrium (VLE) correlations are available, and the simulator allows two liquid phases and one vapor phase. Unfortunately, the VLE may be very nonideal and trace components may not accumulate where we think they will. For example, when ethanol and water are distilled, there often are traces of heavier alcohols present. Alcohols with four or more carbons (butanol and heavier) are only partially miscible in water. They are easily stripped from a water phase (relative volatility >> 1), but when there is little water present they are less volatile than ethanol. Thus, they collect somewhere in the middle of the column where they may form a second liquid phase in which the heavy alcohols have low volatility. The usual solution to this problem is to install a side withdrawal line, separate the intermediate component from the other components, and return the other components to the column. These heterogeneous systems are discussed in more detail in Chapter 8.

The differences in the composition profiles for multicomponent and binary distillation for relatively ideal VLE with no azeotropes can be summarized as follows:

1. In multicomponent distillation the key component concentrations can have maxima.
2. The NK usually do *not* distribute. That is, HNKs usually appear only in the bottoms, and LNKs only in the distillate.
3. The NK often go through a plateau region of nearly constant composition.
4. All components must be present at the feed stage, but at that stage the primary distillation changes. Thus, discontinuities occur at the feed stage.

Understanding the differences between binary and multicomponent distillation will be helpful when you are doing calculations for multicomponent distillation.

5.4. BUBBLE-POINT AND DEW-POINT EQUILIBRIUM CALCULATIONS

Although convenient mathematically, few systems have constant or almost constant relative volatilities. In this case we need to do a complete equilibrium calculation on each stage. If the distillation has HNKs, HK, and LK, we should step off stages from the bottom up and do a bubble-point calculation on each stage. The bubble-point temperature is the temperature of stage j at which the liquid mixture on the stage begins to boil. The pressure, p_j, and the mole fractions of the liquid, $x_{i,j}$, will be known. We wish to find the temperature, T_j, at which $\Sigma \, y_{i,j} = 1.0$, where the $y_{i,j}$ are calculated as $y_{i,j} = K_{i,j}(T_j)x_{i,j}$. Stepping off stages down the column (HK, LK, and LNKs present), we use dew-point calculations with known $y_{i,j}$ to determine T_j and $x_{i,j}$.

Consider the distillation column shown in Figure 5-7, where all the non-keys are HNKs. Note that the column is now numbered from the bottom up, since that is the direction in which we will step off stages. With no LNK, a good first guess of concentrations can be made at the bottom of the column, and we can start the stage-by-stage calculations by calculating the reboiler temperature and the values of $y_{i,R}$ from a bubble-point calculation. For the bubble-point calculation, we want to find the temperature that satisfies the stoichiometric equation.

$$\sum_{i=1}^{C} y_{i,j} = \sum_{i=1}^{C} K_{i,j}(T_i)x_{i,j} = 1.0$$

(5-33)

If a simple expression for K_i as a function of temperature is available, we may be able to solve the resulting equation for temperature. Otherwise, a root-finding technique or a trial-and-error procedure will be required. When graphs or charts are used, a trial-and-error procedure is always needed. When we are finished with the calculation, the $y_{i,j}$ calculated are the mole fractions of the vapor leaving stage j.

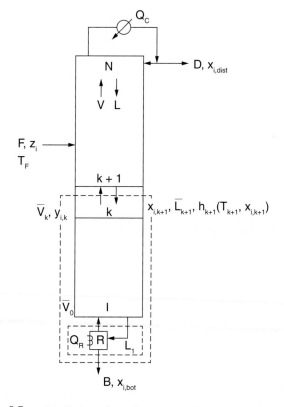

FIGURE 5-7   *Distillation column for stepping off stages from the bottom up*

How can we make a good initial guess for the temperature T_j? Note that $\sum\limits_{i=1}^{C} x_{i,j} = 1.0$

and $\sum\limits_{i=1}^{C}\left(K_{i,j}x_{i,j}\right) = 1.0$. If all $K_{i,j} > 1.0$, then $\sum\limits_{i=1}^{C}\left(K_{i,j}x_{i,j}\right) > 1.0$. If all $K_{i,j} < 1.0$, then

$\sum\limits_{i=1}^{C}\left(K_{i,j}x_{i,j}\right) < 1.0$.

Therefore, we should choose a temperature T_j so that some $K_{i,j}$ are greater than 1.0 and some are less than 1.0. This procedure is similar to the procedure we used for a first guess of the drum temperature in flash distillation.

How do we pick a new temperature when the previous trial was not accurate? If we look at the DePriester charts or Eq. (2-30), we see that the K_i are complex functions of temperature. However, the function is quite similar for all K's. Thus, the change in K for one component (the reference component) and the change in the value of $\sum\limits_{i=1}^{C}\left(K_{i,j}x_{i,j}\right)$ will be quite similar when temperature changes. By setting the ratio of the new and old values of the reference component equal to the ratio of the desired new value of the summation (1.0) and the calculated value of the summation, we can estimate the appropriate K value for the reference component as

$$K_{ref,j}(T_{new,j}) = \frac{K_{ref,j}(T_{old,j})}{\sum\limits_{i=1}^{C}(K_{i,j}x_{i,j})_{calc}}$$

(5-34)

The temperature for the next trial is determined from the new value, $K_{ref,j}(T_{newj})$. This procedure, which is again similar to the flash distillation procedure, should converge quite rapidly. As an alternative, a Newtonian convergence scheme can be used.

Example 5–2. Bubble-point calculation

At 1.0 atm, what is the temperature and vapor mole fraction in the reboiler (stage 1) if the bottoms is 15 mol% isopentane, 30 mol% n-pentane, and 55 mol% n-hexane?

Solution

A. Define. We want T_1 for which $\sum\limits_{i=1}^{C} y_{i,1} = \sum\limits_{i=1}^{C}\left(K_{i,1}x_{i,1}\right) = 1.0$.

B. Explore. To illustrate a trial-and-error procedure, we use the DePriester charts, Figures 2-11 and 2-12. Equation (2-30) could be used instead of the DePriester charts, and we use it for the Check. First, convert atm to kPa.

$$p = 1.0 \text{ atm}(101.3 \text{ kPa}/1.0 \text{atm}) = 101.3 \text{ kPa}$$

C. Plan. Use the DePriester chart to pick a temperature T_1 for which $K_{ic5} > K_{nc5} > 1.0 > K_{nc6}$. Calculate the summation in Eq. (5-33), use Eq. (5-34) to calculate K_{ref}, and find T_1 from the DePriester chart. We pick nC_6 as the reference component because it is the most abundant component. Then repeat the calculation with this new temperature.

D. **Do It.** First guess: Using Figure 2-12 at 50°C: $K_{ic5} = 2.02$, $K_{nc5} = 1.55$, $K_{nc6} = 0.56$. Thus, 50°C (and many other temperatures) satisfies our first guess criteria. Calculate stoichiometric sum:

$$\sum_{i=1}^{3} y_i = \sum_{i=1}^{3} (K_i x_i) = (2.02)(0.15) + (1.55)(0.30) + (0.56)(0.55) = 1.076$$

Since the sum is too high, the temperature of 50°C is too high. From Eq. (5-34), calculate

$$K_{cNew} = \frac{K_{c6\ old}}{\sum_{i=1}^{3}(K_i x_i)} = \frac{0.56}{1.076} = 0.52$$

From the DePriester chart, the corresponding temperature is $T_1 = 47.5°C$. At this temperature, $K_{ic5} = 1.92$, $K_{nc5} = 1.50$. Note that all the K's are lower, so the summation will be lower. Now the summation is

$$\sum_{i=1}^{3} y_i = \sum_{i=1}^{3} (K_i x_i) = (1.92)(0.15) + (1.50)(0.30) + (0.52)(0.55) = 1.024$$

The next $K_{c6} = 0.52/1.024 = 0.508$, which corresponds to $T = 47°C$. At this temperature, $K_{ic5} = 1.89$, $K_{nc5} = 1.44$. The summation is

$$\sum_{i=1}^{3} (K_i x_i) = (1.89)(0.15) + (1.44)(0.30) + (0.508)(0.55) = 0.995$$

This is about as close as we can get with the DePriester chart. Thus, $T_1 = 47°C$.

The $y_{i,1}$ values are equal to $K_{i,1} x_{i,1}$. Thus, $y_{ic5,1} = (1.89)(0.15) = 0.284$, $y_{nc5,1} = (1.44)(0.30) = 0.432$, $y_{nc6,1} = (0.508)(0.55) = 0.279$. Because of the accuracy of the DePriester charts, the y_i values should be rounded off to two significant figures when they are reported.

E. **Check.** An alternative solution can be obtained using Eq. (2-30).

First guess: $T = 50°C = 122°F = 122 + 459.58 = 581.58°R$, and $p = 14.7$ psia. For iC5, nC5, nC6, Eq. (2-30) simplifies to

$$K = exp\left[a_{T1}(\frac{1}{T^2}) + a_{T6} + a_{p1} \ln p \right]$$

(5-35a)

which gives $K_{iC6} = 2.0065$, $K_{nC5} = 1.5325$, $K_{nC6} = 0.5676$. Then

$$\sum_{i=1}^{3} y_i = \sum_{i=1}^{3} \left(K_i x_i \right) = \left(2.0065 \right)\left(0.15 \right) + \left(1.5325 \right)\left(0.3 \right) + \left(0.5676 \right)\left(0.55 \right) = 1.0729.$$

This is too high. To find the next temperature, use Eq. (5-34).

$$K_{C6 \, new} = \frac{K_{C6 \, old}}{\sum_{i=1}^{3} \left(K_i x_i \right)} = \frac{0.5676}{1.0729} = 0.5290$$

Solving Eq. (5-35a) for T,

$$T = \left(\frac{a_{T1}}{\ln \, K - a_{T6} - a_{p1} \ln \, p - a_{p2}/p^2 - a_{p3}/p} \right)^{1/2} \tag{5-35b}$$

and we obtain $T_1 = 577.73°R$. Using this for the new guess, we can continue. The final result is $T_1 = 576.9°R = 47.4°C$. This result is within the error of Eq. (2-30) when compared to the 47.0°C found from the DePriester charts. Equation (5-35b) is valid for all the hydrocarbons covered by the DePriester charts except n-octane and n-decane.

F. Generalize. If K values depend on composition, then an extra loop in the trial-and-error procedure will be required. When K values are in equation form such as Eq. (2-30), bubble-point calculations are easy to solve with a spreadsheet. With a process simulator, one of the vapor-liquid equilibrium (VLE) correlations (see Table 2-4) will be used to find the bubble-point temperature and the y_i values.

If only LNKs are present, we should step off stages going down the column. Then the liquid mole fractions are determined from dew-point calculations and the vapor mole fractions are found from the operating equations. Otherwise, the procedure is very similar to going down the column with constant relative volatilities.

For dew-point calculations, all $y_{i,j}$ are known. We need to find the temperature T_j at which

$$\sum_{i=1}^{C} x_{i,j} = \sum_{i=1}^{C} \left[y_{i,j} / K_{i,j} \left(T_j \right) \right] = 1.0 \tag{5-36}$$

We determine the temperature for the next guess by calculating the reference component K value.

$$K_{ref,j} \left(T_{New,j} \right) = K_{ref,j} \left(T_{old,j} \right) \left\{ \sum_{i=1}^{C} \left[y_{i,j} / K_{i,j} \left(T_{old,j} \right) \right] \right\} \tag{5-37}$$

Remember that convergence of the stage-by-stage distillation calculation is easy only when the NKs are all heavy or are all light. Both the dew-point and bubble-point methods can be solved with a spreadsheet (see the Appendix for Chapter 5).

5.5 SUMMARY—OBJECTIVES

At the end of this chapter you should be able to satisfy the following objectives:

1. Explain why multicomponent distillation is trial and error
2. Make appropriate assumptions and solve the external mass balances
3. Explain the flow, temperature, and composition profiles for multicomponent distillation
4. Do stage-by-stage calculations for distillation with no HNKs or no LNKs

REFERENCES

Doherty, M. F., and M. F. Malone, *Conceptual Design of Distillation Systems*, McGraw-Hill, New York, 2001.

Dean, J. A. (Ed.), *Lange's Handbook of Chemistry,* 13th ed., McGraw-Hill, New York, 1985.

Hengstebeck, R. J., *Distillation: Principles and Design,* Reinhold, New York, 1961.

Kister, H. Z., "Component Trapping in Distillation Towers: Causes, Symptoms and Cures," *Chem. Engr. Progress, 100,* (8), 22 (August 2004).

HOMEWORK

A. *Discussion Problems*

A1. Explain why the external mass balances cannot be solved for a ternary distillation system without an additional assumption. Why aren't the equations for the following useful?
   a. External energy balance
   b. Energy balance around the condenser
   c. Equilibrium expression in the reboiler

A2. If constant relativity is valid for a ternary separation and we choose component A or C as the reference component instead of component B, what are the consequences?

A3. Define the following:
   a. Heavy key
   b. Heavy non-key
   c. Sandwich component (see Problem 5.A6)
   d. Optimum feed stage
   e. Minimum reflux ratio

A4. We are distilling a mixture with two HNKs, an HK, and an LK. Sketch the expected concentration profiles (x_i vs. stage location).

A5. How would you introduce Murphree vapor efficiencies into the stage-by-stage calculation process?

A6. A distillation column is separating methane, ethane, propane, and butane. We pick methane and propane as the keys. This means that ethane is a *sandwich component.*
   a. Show the approximate composition profiles for each of the four components. Label each curve.
   b. Explain in detail the reasoning used to obtain the profile for ethane.

A7. We are separating a mixture that is 10 mol% methanol, 20 mol% ethanol, 30 mol% n-propanol, and 40 mol% n-butanol in a distillation column. Methanol is most volatile and n-butanol is least volatile. The feed is a saturated liquid. We desire to recover 98% of the ethanol in the distillate and 97% of the n-propanol in the bottoms product. The column has a total condenser and a partial reboiler. Feed rate is 100 kmol/h. Pressure is one atmosphere. L/D = 3.

1. The column is hottest at:
   a. The condenser.
   b. The feed plate.
   c. The reboiler.

2. In the rectifying (enriching) section:
   a. Liquid flow rate > vapor flow rate
   b. Liquid flow rate = vapor flow rate
   c. Liquid Flow rate < vapor flow rate

3. Comparing the stripping section to the rectifying section:
   a. Liquid flow rate in stripping section > liquid flow rate in rectifying section.
   b. Liquid flow rate in stripping section = liquid flow rate in rectifying section.
   c. Liquid flow rate in stripping section < liquid flow rate in rectifying section.

4. The HK is:
   a. Methanol.
   b. Ethanol.
   c. n-propanol.
   d. n-butanol.

5. If you were going to do external mass balances around the column to find B and D, the best assumption to make is:
   a. All of the methanol and n-propanol are in the distillate.
   b. All of the methanol is in the distillate and all of the n-butanol is in the bottoms.
   c. All of the ethanol is in the distillate and all of the n-propanol is in the bottoms.
   d. All of the n-propanol and n-butanol are in the bottoms.

A8. Show for the problem illustrated in Figure 5-2 that L/V is more constant than either L or V when CMO is not valid. Explain why this is so.

A9. It has often been suggested that when there is the corresponding NK component present, key components should be withdrawn as sidestreams at the location where their concentration maximum occurs. If there is an LNK, can a pure LK be withdrawn? Why or why not? Can a pure LNK be obtained at the top of the column? Why or why not?

A10. Develop a key relations chart for this chapter. You will probably want to include some sketches.

A11. In Figure 5-4, a 99% recovery of benzene does not give a high benzene purity. Why not? What would you change to also achieve a high benzene purity in the distillate?

A12. Explain Figure 5-5 in terms of the distillation of binary pairs.

A13. In Figure 5-4, the HNK and HK concentrations cross near the bottom of the columns, and in Figure 5-5 the LK and LNK concentrations do not cross near the top of the column. Explain when the concentrations of HK and HNK and LK and LNK pairs will and will not cross.

A14. Figure 5-6 shows the distillation of a four-component mixture. What would you expect the profiles to look like if xylene were the LK and cumene the HK?

A15. a to e. Determine whether the following multicomponent distillation problems can or cannot be solved with a stage-by-stage calculation and if a stage-by-stage calculation can be used indicate the direction one should step off stages.

1. Have two LNKs, an LK, and an HK
    a. step off stages top down.
    b. step off stages bottom up.
    c. step off stages in either direction.
    d. stage-by stage calculation probably will not work because of extreme convergence problems.

2. Have an LNK, an LK, an HK, and an HNK
    a. step off stages top down.
    b. step off stages bottom up.
    c. step off stages in either direction.
    d. stage-by stage calculation probably will not work because of extreme convergence problems.

3. Have an LK, an HK, and two HNKs
    a. step off stages top down.
    b. step off stages bottom up.
    c. step off stages in either direction.
    d. stage-by stage calculation probably will not work because of extreme convergence problems.

4. Have an LK, a sandwich component, and an HK
    a. step off stages top down.
    b. step off stages bottom up.
    c. step off stages in either direction.
    d. stage-by stage calculation probably will not work because of extreme convergence problems.

5. Have an LK, a sandwich component, an HK, and an HNK
    a. step off stages top down.
    b. step off stages bottom up.
    c. step off stages in either direction.
    d. stage-by stage calculation probably will not work because of extreme convergence problems.

C. *Derivations*
C1. Derive Eq. (5-29).

D. *Problems*
Answers to problems with an asterisk are at the back of the book.
D1. A distillation column with a total condenser and a partial reboiler is fed a mixture of alcohols. The feed is 19 wt % methanol, 31 wt % ethanol, 27 wt % n-propanol, and 23 wt % n-butanol. (Methanol is most volatile and n-butanol is least volatile.) The feed rate is 12,000 kg/h. We desire a 97.8% recovery of ethanol in the distillate and a 99.4% recovery of n-propanol in the bottoms.
    a. Estimate D and B and distillate and bottoms weight fractions.
    b. List your assumptions, if any.

D2. What is the dew point of a vapor that is 30 mol% n-butane, 50 mol% n-pentane, and 20 mol% n-hexane at p = 760 mm Hg? Use Raoult's law to predict K values. Find vapor pressures from Antoine's equation,

$$\log_{10}(VP) = A - \frac{B}{T+C}$$

where VP is in mm Hg and T is in °C. The Antoine constants (Dean, 1985) are:

n – C$_4$:	A = 6.809	B = 935.86	C = 238.73
n – C$_5$:	A = 6.853	B = 1064.8	C = 233.01
n – C$_6$:	A = 6.876	B = 1171.17	C = 224.41

D3.* We have a feed mixture of 22 mol% methanol, 47 mol% ethanol, 18 mol% n-propanol, and 13 mol% n-butanol. Feed is a saturated liquid, and F = 10,000 kmol/day. We desire a 99.8% recovery of methanol in the distillate and a methanol mole fraction in the distillate of 0.99.

a. Find D and B.

b. Find compositions of distillate and bottoms.

D4.* We are separating a mixture that is 40 mol% isopentane, 30 mol% n-hexane, and 30 mol% n-heptane. We desire a 98% recovery of n-hexane in the bottoms and a 99% recovery of isopentane in the distillate. F = 1000 kmol/h. Feed is a two-phase mixture that is 40% vapor. L/D = 2.5.

a. Find D and B. List any required assumptions.

b. Find compositions of distillate and bottoms.

c. Calculate L, V, $\overline{L}$, and $\overline{V}$, assuming CMO.

d. Show schematically the expected composition profiles for isopentane, n-hexane, and n-heptane. Label curves. Be neat!

D5. A distillation column with a total condenser and a partial reboiler is separating two feeds. The first feed is a saturated liquid and its rate is 100.0 kmol/h. This feed is 55.0 mol% methanol, 21.0 mol% ethanol, 23.0 mol% propanol and 1.0 mol% butanol. The second feed is a saturated liquid with a flow rate of 150.0 kmol/h. This feed is 1.0 mol% methanol, 3.0 mol% ethanol, 26.0 mol% propanol and 70.0 mol% butanol. We want to recover 99.3% of the propanol in the distillate and 99.5% of the butanol in the bottoms. Find the distillate and bottoms flow rates and the mole fractions of the distillate and bottoms products.

D6. We have a mixture of benzene, toluene, and cumene distilling in a column with a partial reboiler and a total condenser. Constant molal overflow can be assumed. The bottoms product is sampled, and the following compositions are measured: $x_B = 0.1$, $x_T = 0.3$, $x_C = 0.6$. The boilup ratio is $\overline{V}/B = 1.0$. The relative volatilities are $\alpha_{BT} = 2.5$, $\alpha_{TT} = 1.0$, $\alpha_{CT} = 0.21$. What is the composition of the vapor leaving the stage above the partial reboiler? (This is several stages below the feed stage.)

D7. We are separating hydrocarbons in a column that has two feeds. The column operates at 75 psig. It has a total condenser and a partial reboiler. The first feed is 30 wt % ethane, 0.6 wt % propylene, 45 wt % propane, 15.4 wt % n-butane, and 9 wt % n-pentane. This feed is a saturated liquid at a flow rate of 1000 kg/h. The second feed is 2.0 wt % ethane, 0.1 wt % propylene, 24.9 wt % propane, 40.0 wt % n-butane, 18.0 wt % n-pentane, and 15.0 wt % n-hexane. The flow rate of this feed is 1500.0 kg/h. It is a saturated liquid. We desire 99.1% recovery of the propane in the distillate and 98% recovery of the n-butane in the bottoms. Make appropriate assumptions and calculate the distillate and bottoms flow rates. Also calculate the weight fractions of each component in the distillate and bottoms streams.

D8. A distillation column with a partial reboiler and a total condenser is being used to separate a mixture of benzene, toluene, and cumene. The feed is 40 mol% benzene, 30 mol% toluene, and 30 mol% cumene and is a saturated liquid. Feed rate is 1000 kmol/h. Reflux is returned as a saturated liquid, and $L/D = 2.0$. We desire 95% recovery of cumene in the bottoms and 95% recovery of toluene in the distillate. Pressure is 1 atm. Assume constant relative volatilities $\alpha_{BT} = 2.25$, $\alpha_{TT} = 1.0$, $\alpha_{CT} = 0.21$. Find the optimum feed stages and the total number of equilibrium contacts required.

D9. A distillation column is separating a feed that is 20 mol% methanol, 50 mol% n-propanol, and 30 mol% n-butanol. Feed is 750 kmol/h. We want a 92 mol% recovery of methanol in the distillate and 95% recovery of n-propanol in the bottoms. Use $L/D = 10$. Feed is a saturated liquid. The column has a total condenser and a partial reboiler. As a guess, assume that n-butanol does not distribute (it all exits in the bottoms). Do not do iterations to improve this guess. If we choose n-propanol as the reference, the relative volatilities are methanol = 3.58, n-propanol = 1.0, and n-butanol = 0.412. Assume relative volatilities are constant. Do not use Aspen Plus. This calculation is simple enough that it can be done by hand or with a spreadsheet (with or without Visual Basic for Applications [VBA]) or with MATLAB.
   a. Find D, B, $x_{i,dist}$, and $x_{i,bot}$.
   b. Step off stages using the second stage above the partial reboiler (the third equilibrium contact from bottom) for feed location and find the total number of stages and the calculated distillate mole fractions.

D10. What is the bubble-point temperature at 760 mm Hg pressure of a mixture that is 40 mol% n-pentane and 60 mol% n-hexane? Use Raoult's law and the Antoine coefficients given in Problem 5.D2.

D11. We have a liquid mixture that is 10 mol% ethane, 35 mol% n-pentane, and 55 mol% n-heptane at 40°C and a high pressure. As the pressure is slowly dropped, at what pressure will the mixture first start to boil? Use the DePriester chart.

D12. Suppose n-hexane, n-heptane, and n-octane are available so that the desired mole fraction of the mixture can be changed to any desired value. If the system pressure is 300 kPa,
   a. What is the highest possible bubble-point temperature?
   b. What is the lowest possible bubble-point temperature?

D13*. Find the bubble-point temperature and vapor mole fractions at the bubble-point for a mixture at 1.0 atm that is 20.0 mol% n-butane, 50.0 mol% n-pentane, and 30.0 mol% n-hexane. Use the DePriester chart.

E. *More Complex Problems*
E1. We wish to distill a mixture of ethane, propane, n-butane. The column has a partial reboiler and a partial condenser and operates at 400 kPa. The feed flow rate is 200 kmol/h. The feed is a saturated liquid and is 22 mol% ethane, 47 mol% propane, and 31 mol% n-butane. We wish to recover 97% of the ethane in the distillate and 99% of the propane in the bottoms. The reflux is a saturated liquid, and the external reflux ratio $L_0/D = 3.0$. Find the optimum feed stage and the total number of equilibrium contacts required. Assume constant molal overflow, and use the DePriester charts or Eq. (2-30) for K values.

H. *Computer Spreadsheet Problems*

H1. Do part b of this problem with a spreadsheet or with MATLAB. The use of VBA is recommended if a spreadsheet is used. We have 200 kmol/h of a saturated liquid feed that is 35 mol% methanol, 40 mol% i-propanol, and 25 mol% n-propanol at 1.0 atm. We want 96.1% recovery of methanol in the distillate and 99.6% recovery of i-propanol in the bottoms. Assume CMO and constant relative volatilities. The column has a total condenser and a partial reboiler. L/D = 6. As a first guess, assume that n-propanol does not distribute (it all exits in the bottoms). Relative volatilities are: methanol = 3.58, i-propanol = 1.86, and n-propanol = 1.0.

    a. Find D, B, $x_{i,dist}$, and $x_{i,bot}$.

    b. Step off stages and find the optimum feed location and the total number of stages.

H2. [VBA required] Either write your own program or use the program in Appendix A of Chapter 5 to solve the following problem. A feed of 100 mol/h of a saturated liquid that is 25 mol% A = benzene, 35 mol% B = toluene, and 40 mol% C = cumene is fed on the optimum feed plate to a distillation column that has a total condenser and a partial reboiler. Fractional recoveries of B (toluene) in the distillate of 0.9 and of C in the bottoms of 0.97 are desired. The relative volatilities are $\alpha_{AB} = 2.25$, $\alpha_{BB} = 1.0$, and $\alpha_{CB} = 0.21$. Use an external reflux ratio of L/D = 0.3. Find the optimum feed stage, the total number of stages, the fractional recovery of A (benzene) in the distillate, D and B. After solving the problem, try "What if?" simulations to explore the effects of changing the feed concentrations, the fractional recoveries, L/D, and the relative volatility α_{CB}.

H3. [VBA required] Distillation programs such as the ternary stage-by-stage program in Appendix A of Chapter 5 can be used to find the minimum external reflux ratio by setting an arbitrarily high feed location and an arbitrarily high maximum number of stages and then varying L/D until the lowest L/D at which the separation can be achieved is found. Arbitrarily set the feed stage to 100 (in the spreadsheet) and set the maximum number of stages to 200 (in the VBA program). For the separation of benzene (A), toluene (B), and cumene (C) with a 0.99 fractional recovery of B in the distillate and a 0.999 fractional recovery of C in the bottoms, find $(L/D)_{min}$. The feed and alpha values are the same as in Problem 5.H2. Check your answer by trying a feed stage of 125 with a maximum number of stages of 250.

H4. [VBA required] Although it is usually not possible to make an accurate guess of the fractional recovery of a sandwich component, a stage-by-stage calculation often still converges, since the fractional recovery of the sandwich component is finite for both distillate and bottoms. For a ternary system, it often does not matter at which end of the column one starts the calculation. In the process of doing the stage-by-stage calculation, one also obtains reasonably accurate values for D and B.

    a. The easiest way to run a sandwich component is probably to run a standard program (e.g., Appendix A of Chapter 5) *as if* A = LNK, B = LK, and C = HK. Since the fractional recovery of A in the distillate and of C in the bottoms are specified but the program wants B and C to be specified, guess the fractional recovery of B in the distillate. Run the program and check if the calculated value of the fractional recovery of A is equal to the specified value. If not try, another value for fractional recovery of B.

    b. Solve the following problem. A feed of 100 mol/h of a saturated liquid that is 25 mol% A, 35 mol% B, and 40 mol% C is fed on the optimum feed plate to a distillation column that has a total condenser and a partial reboiler. Fractional recoveries of A in the distillate of 0.94 and of C in the bottoms of 0.995 are desired. The relative volatilities are $\alpha_{AB} = 1.4$, $\alpha_{BB} = 1.0$, and $\alpha_{CB} = 0.7$. Use an external reflux ratio of L/D = 5.0. Find the optimum feed stage, the total number of stages, the fractional recovery of B in the distillate, distillate and bottoms flow rates.

   c. Plot the profiles for both liquid and vapor mole fractions of the sandwich component.

   d. Compositions of trace components will build up in the column if they are sandwich components. Repeat the sandwich component analysis, but with $z_A = 0.38$, $z_B = 0.02$, and $z_C = 0.6$, L/D = 3.3, 0.99 recovery of A in the distillate, and 0.995 recovery of C in the bottoms. Calculate the fractional recovery of B, the distillate and bottoms flow rates, the mole fractions on every stage, the optimum feed stage, and the number of stages.

   Note: This problem is challenging because the program is sensitive and fractional recoveries change as the feed stage is changed.

H5. [VBA required] Including a bubble-point or dew-point calculation on every stage is not substantially more work once the program for constant relative volatility has been developed.

   a. Develop a program for a system with an LK, HK, and HNK (step off stages from the bottom up) doing a bubble-point calculation on every stage.

   b. Use this program to find the optimum feed stage, the total number of stages, distillate and bottoms flow rates and compositions, and composition and temperature profiles for the following problem. We wish to separate 100 kmol/h of a saturated liquid feed that is 30 mol% n-butane, 30 mol% n-pentane, and 40 mol% n-hexane. Column pressure is 14.7 psia. The fractional recovery of n-butane in the distillate is 0.995, and the fractional recovery of n-pentane in the bottoms is 0.997. L/D = 8.0. Equation (2-30) and Table 2-3 can be used to calculate K values.

CHAPTER 5 APPENDIX. SPREADSHEET CALCULATIONS FOR TERNARY DISTILLATION WITH CONSTANT RELATIVE VOLATILITY

The spreadsheet results and the VBA program for ternary distillation calculations with constant relative volatility (Section 5.2.) are shown in Figure 5-A1 and Table 5-A1. If you are not familiar with VBA look at Appendix B of Chapter 4. The problem solved is to determine the number of stages and the optimum feed stage for the distillation of 100 mol/h of a saturated liquid feed that is 30 mol% A, 20 mol% B, and 50 mol% C. L/D = 1, and the desired fractional recoveries are B in distillate = 0.99 and C in bottoms = 0.97. Component A = benzene, component B = toluene, and component C = cumene. The constant relative volatilities with respect to toluene as the reference are $\alpha_{AB} = 2.25$, $\alpha_{BB} = 1.0$, and $\alpha_{CB} = 0.21$. By trial and error, the optimum feed stage was determined to be the second stage from the top (the total condenser is not counted as a stage).

   The VBA program that goes with this spreadsheet and calculates the values of D, B, L/V, Lbar/Vbar and everything below these values is given in Table 5-A1. If you want to understand how to use spreadsheets with VBA, you need to work Problems 5.H1 to 5.H4. Note that the VBA program in Table 5-A1 includes a number of bells and whistles (e.g., the tests for feed stage too high or reflux ratio too low or convergence of mass balance based on guess of LNK recovery). These refinements would not normally be included in a program

	A	B	C	D	E	F	G	H	I	J
1	Tables 5A1. Ternary Distillation with Constant relative volatility.					Table 5A2 is VBA program.				
2	System has A = LNK, B = LK and C = HK					Step off stages from top down.				
3	alpha A-B	2 25	alpha B-B	1	Alpha C-B	0.21	foodctage	2		
4	z A	0.3	z B	0.2	z C	0.5	epsilon (values for	1E-07		
5	F	100	q	1	L/D	1	N loop(convergence	100		
6	frac rec B in dist		0.99	frac rec C in bot		0.97	df(LNK frac recovery)	0.9		
7	Guess: frac rec A in dist		1							
8	D	51.29918	B	48.70082	L/V		0.5	Lbar/Vbar	1.47467	
9	xAdist	0.584789	xBdist	0.385971	xCdist		0.02924	Mass balance values		
10	xAbot	1.68E-05	xBbot	0.004107	xCbot		0.995877	Mass balance values		
11	Stage by stage calculations									
12	i	xA	yA	xB	yB	xC	yC	Note y1 =		
13	1	0.331041	0.584789	0.49161	0.385971	0.177348	0.029240232	xdist		
14	2	0.17944	0.457915	0.386877	0.438791	0.433684	0.103294362			
15	3	0.079431	0.264607	0.384018	0.568568	0.536551	0.166824922			
16	4	0.024403	0.117126	0.26456	0.564353	0.711037	0.318520984			
17	5	0.005082	0.035979	0.123383	0.38819	0.871535	0.575830939			
18	6	0.000821	0.007487	0.044418	0.18	0.954761	0.812513236			
19	7	0.000118	0.001203	0.014068	0.063552	0.985814	0.935244853			
20	8	1.58E-05	0.000167	0.004007	0.018796	0.995977	0.981037684			
21										
22										
23	Calc frac recovery A in distillate				0.999973	j		4		
24	Mass balance: fraction stage, A, B, C estimated at bottom, % error B									
25	0.990111	1.68E-05		0.004107		0.995877		0.000100888		

FIGURE 5-A1.   *Spreadsheet results for ternary distillation with constant relative volatilities. Step off stages from top of column*

developed to solve a single problem but are useful if a number of problems will be solved. The use of a fractional number of stages to estimate bottoms mole fractions to determine the fractional recovery of the LNK is included to prevent excessive oscillations in the answer when there are an LK, a sandwich component, and an HK (see Problem 5.H4.).

Table 5-A1.   VBA program for ternary distillation with constant relative volatility starting at top of column

```
Option Explicit
Sub Ternary_top_down()
' Ternary distillation with constant alpha. Frac recoveries of LK and HK given.
' There is a LNK present and its frac rec in distillate is guessed.
 Sheets("Sheet1").Select
 Range("A13", "G120").Clear
' Declare variables
 Dim i, j, feedstage, N As Integer
 Dim alphaAB, alphaBB, alphaCB, F, fracBdist, fracCbot, q, LoverD, LoverV As Double
 Dim LbaroverVbar, D, B, L, V, Lbar, Vbar, Eqsum, fracAdist As Double
 Dim xA, xB, xC, yA, yB, yC, zA, zB, zC, xAbot, xBbot, xCbot As Double
 Dim DxA, DxB, DxC, BxA, BxB, BxC, xAdist, xBdist, xCdist As Double
```

Chapter 5   Introduction to Multicomponent Distillation

```
 Dim fracAdistcalc, difference, epsilon, df, watch As Double
 Dim xAold, xBold, xCold, frac, xAcalc, xBcalc, xCcalc As Double
'Input data from spreadsheet
 alphaAB = Cells(3, 2).Value
 alphaBB = Cells(3, 4).Value
 alphaCB = Cells(3, 6).Value
 feedstage = Cells(3, 8).Value
 F = Cells(5, 2).Value
 q = Cells(5, 4).Value
 LoverD = Cells(5, 6).Value
 zA = Cells(4, 2).Value
 zB = Cells(4, 4).Value
 zC = Cells(4, 6).Value
 fracBdist = Cells(6, 3).Value
 fracCbot = Cells(6, 6).Value
 fracAdist = Cells(7, 4).Value
 epsilon = Cells(4, 8).Value
 N = Cells(5, 8).Value
 df = Cells(6, 8).Value
' The For loop (most of remainder of program) is to obtain convergence of guess of
' fractional recovery of A in distillate.
For j = 1 To N
' Calculate compositions and flow rates based on latest calculated value of frac
' recovery of A in distillate.
 DxA = F * zA * fracAdist
 DxB = F * zB * fracBdist
 DxC = F * zC * (1 - fracCbot)
 BxA = F * zA * (1 - fracAdist)
 BxB = F * zB * (1 - fracBdist)
 BxC = F * zC * fracCbot
 D = DxA + DxB + DxC
 B = BxA + BxB + BxC
 xAdist = DxA / D
 xBdist = DxB / D
 xCdist = DxC / D
 xAbot = BxA / B
 xBbot = BxB / B
 xCbot = BxC / B
 L = LoverD * D
 V = L + D
 LoverV = L / V
 Lbar = L + q * F
 Vbar = Lbar - B
 LbaroverVbar = Lbar / Vbar
' Record values of flowrates and mole fractions on spreadsheet.
 Cells(8, 2) = D
 Cells(8, 4) = B
 Cells(8, 6) = LoverV
 Cells(8, 8) = LbaroverVbar
 Cells(9, 2) = xAdist
 Cells(9, 4) = xBdist
 Cells(9, 6) = xCdist
 Cells(10, 2) = xAbot
 Cells(10, 4) = xBbot
 Cells(10, 6) = xCbot
' initialize (top stage =1) and start loops
 i = 1
 yA = xAdist
 yB = xBdist
 yC = xCdist
```

```
' Calculations in enriching section: first equilibrium.
Do While i < feedstage
 Eqsum = (yA / alphaAB) + (yB / alphaBB) + (yC / alphaCB)
 xA = (yA / alphaAB) / Eqsum
 xB = (yB / alphaBB) / Eqsum
 xC = (yC / alphaCB) / Eqsum
' Record values on spreadsheet
 Cells(i + 12, 1).Value = i
 Cells(i + 12, 2).Value = xA
 Cells(i + 12, 3).Value = yA
 Cells(i + 12, 4).Value = xB
 Cells(i + 12, 5).Value = yB
 Cells(i + 12, 6).Value = xC
 Cells(i + 12, 7).Value = yC
' Top operating line
 i = i + 1
 yA = LoverV * xA + (1 - LoverV) * xAdist
 yB = LoverV * xB + (1 - LoverV) * xBdist
 yC = LoverV * xC + (1 - LoverV) * xCdist
Loop
' Calculations in stripping section
Do
' Save values of x from previous stage.
 xAold = xA
 xBold = xB
 xCold = xC
' Equilibrium calculation for x values for current stage number i.
 Eqsum = (yA / alphaAB) + (yB / alphaBB) + (yC / alphaCB)
 xA = (yA / alphaAB) / Eqsum
 xB = (yB / alphaBB) / Eqsum
 xC = (yC / alphaCB) / Eqsum
' Record values on spreadsheet.
 Cells(i + 12, 1).Value = i
 Cells(i + 12, 2).Value = xA
 Cells(i + 12, 3).Value = yA
 Cells(i + 12, 4).Value = xB
 Cells(i + 12, 5).Value = yB
 Cells(i + 12, 6).Value = xC
 Cells(i + 12, 7).Value = yC
' Test for feed stage too high
If xA < 0 Or xB < 0 Or xC < 0 Then
 Cells(i + 13, 3) = "Feed stage too high"
 Exit For
End If
 i = i + 1
' Test for too many stages, which may mean reflux rate is too low.
If i > 100 Then
 Cells(i + 12, 2).Value = "Too many stages"
 Exit For
End If
 yA = (LbaroverVbar * xA) - (LbaroverVbar - 1) * xAbot
 yB = (LbaroverVbar * xB) - (LbaroverVbar - 1) * xBbot
 yC = (LbaroverVbar * xC) - (LbaroverVbar - 1) * xCbot
' Test for calculations being done.
Loop While xC < xCbot
' Fractional recovery of A based on stage-by-stage calculation with stage
' calculation using fractional stage, which is an approximate calculation. With very
' few stages this can be inaccurate. In this case adjust L/D so that frac is close
' either 0 or 1.0.
 frac = (xCbot - xCold) / (xC - xCold)
 xCcalc = xCbot
```

```
' Assume that frac calculate for C is also valid for A and B.
 xAcalc = (xA - xAold) * frac + xAold
 xBcalc = (xB - xBold) * frac + xBold
' Calculate new frac recovery of A in distillate from estimated bottoms calculation.
' Calculation is damped to prevent excessive oscillation.
 fracAdistcalc = 1 - (xAcalc * B) / (F * zA)
 difference = fracAdist - fracAdistcalc
 fracAdist = fracAdist + df * (fracAdistcalc - fracAdist)
' Test if have convergence of fractional recovery of A.
If Abs(difference) < epsilon Then Exit For
Next j
 Cells(i + 14, 1).Value = "Calc frac recovery A in distillate"
 Cells(i + 14, 5).Value = fracAdistcalc
 Cells(i + 14, 6).Value = "j"
 Cells(i + 14, 7).Value = j
' Calculate and record on spreadsheet mass balance results and error in mass balance.
 Cells(i + 15, 1).Value = "Mass bal: frac, A, B, C at bot, % error B"
 Cells(i + 16, 1).Value = frac
 Cells(i + 16, 2).Value = xAcalc
 Cells(i + 16, 4).Value = xBcalc
 Cells(i + 16, 6).Value = xCcalc
 watch = (Abs((xBbot - xBcalc) / xBbot)) * 100
 Cells(i + 16, 7).Value = watch
 If watch > 0.05 Then Cells(i + 16, 1).Value = "Mass balance mismatch"
End Sub
```

CHAPTER 6

Exact Calculation Procedures for Multicomponent Distillation

Since multicomponent calculations are trial and error, it is convenient to do them on a computer. Because stage-by-stage calculations are restricted to problems where a good first guess of compositions can be made at some point in the column, matrix methods for multicomponent distillation will be used. These methods are not restricted to cases where a good guess of compositions can be made.

6.1 INTRODUCTION TO MATRIX SOLUTION FOR MULTICOMPONENT DISTILLATION

Since distillation is a very important separation technique, considerable effort has been spent in devising better calculation procedures. Details of these procedures are available in a variety of textbooks (Seader and Henley, 2006; Holland, 1981; King, 1980; Smith, 1963; and Wankat, 1988).

The general behavior of multicomponent distillation columns (see Chapter 5) and the basic mass and energy balances and equilibrium relationships do not change when different calculation procedures are used. (The physical operation is unchanged; thus, the basic laws and the results are invariant.) What different calculation procedures do is rearrange the equations to enhance convergence, particularly when it is difficult to make a good first guess. The most common approach is to group and solve the equations by *type*, not stage by stage. That is, all mass balances for component i are grouped and solved simultaneously, all energy balances are grouped and solved simultaneously, and so forth. Most of the equations can conveniently be written in matrix form. Computer routines for solution of these equations are easily written. The advantage of this approach is that even very difficult problems can be made to converge.

A convenient set of variables to specify are F, z_i, T_F, N, N_F, p, T_{reflux}, L/D, and D. Multiple feeds can be specified. This is then a simulation problem with distillate flow rate specified. Because the matrices require that N and N_F be known, for design problems, a good first guess

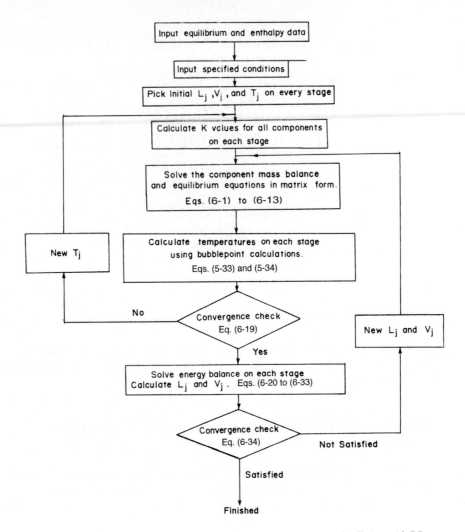

FIGURE 6-1. *Flowchart for matrix calculation for multicomponent distillation with BP method.*

of N and N_F must be made (see Chapter 7), and then a series of simulation problems are solved to find the best design.

In Section 2.7 we looked at solution methods for multicomponent flash distillation. The questions asked in that section are again pertinent for multicomponent distillation. First, what trial variables should we use? As noted, because N and N_F are required to set up the matrices, in design problems we choose these and solve a number of simulation problems to find the best design. We select the temperature on every stage T_j because temperature is needed to calculate K values and enthalpies. We also estimate the overall liquid L_j and vapor V_j flow rates on every stage because these flow rates are needed to solve the component mass balances.

Next, we need to decide if we should converge on all the trial variables in a sequential or simultaneous fashion. Since commercial simulators often allow the user to select either of these approaches, we consider both sequential and simultaneous approaches. The simultaneous approach is discussed in Section 6.6.

If we decide to use a sequential approach, we must decide which trial variable to converge on first: temperatures or flow rates. The answer depends on the type of problem we wish to solve. Distillation problems, which tend to be narrow boiling, usually converge best if temperature is converged on first. This method is illustrated in this chapter. Wide-boiling feeds such as flash distillation (Section 2.7) and absorption and stripping (Chapter 12) tend to converge best if the sum-rates method that converges on flow rates first is used.

The narrow-boiling or *bubble-point* procedure used for distillation is shown in Figure 6-1. This procedure uses the equilibrium (bubble-point) calculations to determine new temperatures. The energy balance is used to calculate new flow rates. Temperatures are calculated and converged on first, and then new flow rates are determined. This procedure makes sense, since an accurate first guess of liquid and vapor flow rates can be made by assuming constant molal overflow (CMO). Thus, temperatures are calculated using reasonable flow rate values. The energy balances are used last because they require values for $x_{i,j}$, $y_{i,j}$, and T_j.

Figure 6-1 is constructed for an ideal system where the K_i depend only on temperature and pressure. If the K_i depend on compositions, then compositions must be guessed and corrected before doing the temperature calculation. We will discuss only systems where $K_i = K_i(T, p)$.

6.2   COMPONENT MASS BALANCES IN MATRIX FORM

Amundson and Pontinen (1958) realized that the component mass balance equations for multicomponent distillation could be put into matrix form with one matrix for each component. To conveniently put the mass balances in matrix form, renumber the column as shown in Figure 6-2. Stage 1 is the total condenser, stage 2 is the top stage in the column, stage N-1 is the bottom stage, and the partial reboiler is listed as N. For a general stage j within the column (Figure 6-3), the mass balance for any component i is

$$V_j y_{i,j} + L_j x_{i,j} - V_{j+1} y_{i,j+1} - L_{j-1} x_{i,j-1} = F_j z_{i,j} \qquad (6\text{-}1)$$

The unknown vapor compositions, $y_{i,j}$ and $y_{i,j+1}$, can be replaced using the equilibrium expressions

$$y_{i,j} = K_j x_{i,j} \qquad \text{and} \qquad y_{i,j+1} = K_{j+1} x_{i,j+1} \qquad (6\text{-}2)$$

where the K values depend on T and p. If we also replace $x_{i,j}$ and $x_{i,j-1}$ with

$$x_{i,j} = \ell_{i,j}/L_j,\ x_{i,j-1} = \ell_{i,j-1}/L_{j-1} \qquad (6\text{-}3a,b)$$

where $\ell_{i,j}$ and $\ell_{i,j-1}$ are the liquid component flow rates, we obtain

$$(-1)\ell_{i,j-1} + \left(1 + \frac{V_j K_j}{L_j}\right)\ell_{i,j} + \left(-\frac{V_{j+1} K_{j+1}}{L_{j+1}}\right)\ell_{i,j+1} = F_j z_{i,j} \qquad (6\text{-}4)$$

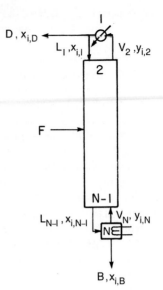

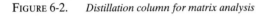

FIGURE 6-2. *Distillation column for matrix analysis*

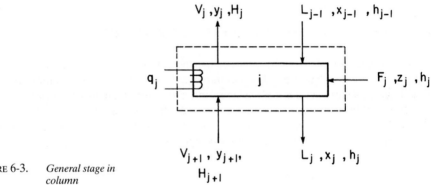

FIGURE 6-3. *General stage in column*

For each component, this equation can be written in the general form

$$A_j \ell_{j-1} + B_j \ell_j + C_j \ell_{j+1} = D_j \tag{6-5}$$

The constants A_j, B_j, C_j, and D_j are easily determined by comparing Eqs. (6-4) and (6-5).

$$A_j = -1, B_j = 1 + \frac{V_j K_j}{L_j}, C_j = -\frac{V_{j+1} K_{j+1}}{L_{j+1}}, D_j = F_j z_j \tag{6-6}$$

Equations (6-4) and (6-5) are valid for all stages in the column, $2 \leq j \leq N - 1$, and are repeated for each of the C components. If a stage has no feed, then $F_j = D_j = 0$.

For a total condenser, the mass balance is

$$L_1x_1 + Dx_D - V_2y_2 = F_1z_1 \qquad\qquad (6\text{-}7)$$

Since $x_1 = x_D$, $y_2 = K_2x_2$ and $x_1 = \ell_1/L_1$, this equation becomes

$$C_1\ell_2 + B_1\ell_1 = D_1 \qquad\qquad (6\text{-}8)$$

where

$$C_1 = -(K_2V_2/L_2),\ B_1 = 1 + D/L_1,\ D_1 = F_1z_1 \qquad\qquad (6\text{-}9)$$

Note that only B_1 does not follow the general formulas of Eq. (6-6). This occurs because the total condenser is *not* an equilibrium contact.

For the reboiler, the mass balance is

$$-L_{N-1}x_{N-1} + V_Ny_N + Bx_{bot} = F_Nz_N \qquad\qquad (6\text{-}10)$$

Substituting $y_N = K_Nx_N$ and $\ell_N = L_Nx_N = Bx_{bot}$, we get

$$A_N\ell_{N-1} + B_N\ell_N = D_N \qquad\qquad (6\text{-}11)$$

where

$$A_N = -1,\ B_N = 1 + \frac{V_NK_N}{L_N} = 1 + \frac{V_NK_N}{B},\ D_N = F_Nz_N \qquad\qquad (6\text{-}12)$$

In matrix notation, the component mass balance and equilibrium relationship for each component is

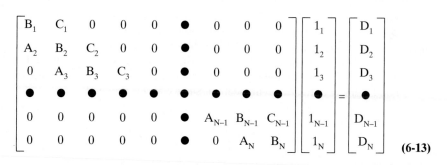

$$(6\text{-}13)$$

This set of simultaneous linear algebraic equations can be solved by inverting the ABC matrix. This can be done using any standard matrix inversion routine. The particular matrix form shown in Eq. (6-13) is a tridiagonal matrix, which is particularly easy to invert using the Thomas algorithm (see Table 6-1) (Lapidus, 1962; King, 1980). Inversion of the ABC matrix allows direct determination of the component liquid flow rate, l_j, leaving each contact. You must construct the ABC matrix and invert it for each of the components.

TABLE 6-1.   *Thomas algorithm for inverting tridiagonal matrices*

Consider the solution of a matrix in the form of Eq. (6-13) where all A_j, B_j, C_j, and D_j are known.

1. Calculate three intermediate variables for each row of the matrix starting with j = 1. For $1 \leq j \leq N$,

   $(V1)_j = B_j - A_j(V3)_{j-1}$
   $(V2)_j = [D_j - A_j(V2)_{j-1}]/(V1)_j$
   $(V3)_j = C_j/(V1)_j$
   since $A_1 \equiv 0$, $(V1)_1 = B_1$, and $(V2)_1 = D_1/(V1)_1$.

2. Initialize $(V3)_0 = 0$ and $(V2)_0 = 0$ so you can use the general formulas.

3. Calculate all unknowns U_j [l_j in Eq. (6-13), V_j in Eq. (6-28), or ΔT_j in Eq. (12-58)]. Start with j = N and calculate

   $U_N = (V2)_N$
   Then going from j = N – 1 to j = 1, calculate U_{N-1}, U_{N-2}, ... U_1 from
   $U_j = (V2)_j - (V3)_j U_{j+1}$, $1 \leq j \leq N-1$

The A, B, and C terms in Eq. (6-13) must be calculated, but they depend on liquid and vapor flow rates and temperature (in the K values) on each stage, which we don't know. To start, guess L_j, V_j, and T_j for *every* stage j! For ideal systems the K values can be calculated for each component on every stage. Then the A, B, and C terms can be calculated for each component on every stage. Inversion of the matrices for each component gives the $l_{i,j}$. The liquid-component flow rates are correct *for the assumed* L_j, V_j, and T_j.

6.3   INITIAL GUESSES FOR FLOW RATES AND TEMPERATURES

A reasonable first guess for L_j and V_j is to assume CMO. CMO was *not* assumed in Eqs. (6-1) to (6-13). With the CMO assumption, we can use overall mass balances to calculate all L_j and V_j.

To start the calculation we need to assume the split for non-key (NK) components. The obvious first assumption is that all the light non-key (LNK) exits in the distillate so that $x_{LNK,bot} = 0$ and $Dx_{LNK,dist} = Fz_{LNK}$. And all heavy non-keys (HNK) exit in the bottoms, $x_{HNK,dist} = 0$ and $Bx_{HNK,bot} = Fz_{HNK}$. Now we can do external mass balances to find all distillate and bottoms compositions and flow rates. This was illustrated in Chapter 5. Once this is done, we can find L and V in the rectifying section. Since CMO is assumed,

$$L = \left(\frac{L_0}{D}\right)D, \quad V = L + D \tag{6-14}$$

At the feed stage, q can be estimated from enthalpies as

$$q \sim \frac{H - h_{feed}}{H - h} \tag{6-15}$$

or $q = L_F/F$ can be found from a flash calculation on the feed stream. Then $\overline{L}$ and $\overline{V}$ are determined from balances at the feed stage,

$$L = L + qF, \quad \overline{V} = V - (1 - q)F \tag{6-16}$$

This completes the preliminary calculations for flow rates.

We can estimate the temperature from bubble-point calculations (Section 5.4). Often it is sufficient to do a bubble-point calculation for the feed and then use this temperature on every stage. A better first guess can be obtained by estimating the distillate and bottoms compositions (usually NKs do not distribute) and doing bubble-point calculations for both. Then assume that temperature varies linearly from stage to stage.

6.4   TEMPERATURE CONVERGENCE

After the first guess and the solution of the matrix equations, the temperature must be corrected. This is done with bubble-point calculations on each stage.

Now the component flow rates are used to determine liquid mole fractions.

$$x_{i,j} = \frac{(\ell_{i,j})}{\sum\limits_{i=1}^{C} (\ell_{i,j})}$$

(6-17)

This procedure normalizes the mole fractions on each stage so that they sum to 1.0. Once the mole fractions have been determined, the new temperatures on each stage are calculated with bubble-point calculations (Section 5.4), which are illustrated in Example 6-1. To prevent excessive oscillation of temperatures the change in temperature can be damped,

$$T_{j,new} = T_{j,old} + df(T_{j,BP\_calculation} - T_{j,old})$$

(6-18)

where df is a damping factor. When df = 1.0, this procedure becomes direct substitution.

The new temperatures are used to calculate new K values (see Figure 6-1) and then new A, B, and C coefficients for Eq. (6-13). The component mass balance matrices are inverted for all components, and new $l_{i,j}$ are determined. This procedure is continued until the temperature loop has converged, which you know has occurred when

$$\left| T_{j,new} - T_{j,old} \right| < \varepsilon_T \qquad \text{for all stages j}$$

(6-19)

where ε_T is the tolerance set for the temperature loop—typically 10^{-2} to 10^{-3}.

EXAMPLE 6-1. Matrix and bubble-point calculations

A distillation column with a partial reboiler and a total condenser is separating nC_4, nC_5, and nC_8. The column has two equilibrium stages (a total of three equilibrium contacts), and feed is a saturated liquid fed into the bottom stage of the column. The column operates at 2 atm. Feed rate is 1000 kmol/h. $z_{C4} = 0.20$, $z_{C5} = 0.35$, $z_{C8} = 0.45$ (mole fractions). The reflux is a saturated liquid, and L/D = 1.5. The distillate rate is D = 550 kmol/h. Assume CMO. Use the DePriester chart or Eq. (2-30) for K values. For the first guess, assume that the temperatures on all stages and in the reboiler are equal to the feed bubble-point temperature. Use a matrix to solve the mass balances. Use bubble-point calculations for *one* iteration toward a solution for stage compositions *and* to predict new temperatures that could be used for a second iteration. Report the compositions on each stage and in the reboiler, and the temperature of each stage and the reboiler.

Solution

This is a long and involved problem. The solution will be shown without all the intermediate calculations.

Start with a bubble-point calculation on the feed, $\sum z_i K_i (T_{bp}) = 1.0$. This converges to $T = 60°C$ at $p = 202.6$ kPa. The K values are $K_{C4} = 3.00$, $K_{C5} = 1.05$, $K_{C8} = 0.072$.

$$\sum (z_i K_i) = 1.0002$$

Next, calculate flow rates
$$L = (L/D)(D) = (1.5)(550) = 825 \text{ kmol/h} = L_1, L_2$$
$$V = L + D = 1375 \text{ kmol/h} = V_2, V_3, V_4$$
$$\overline{L} = L + F = 1825 \text{ kmol/h} = L_3$$
$$B = 450 \text{ kmol/h} = L_4$$

Calculate matrix variables A, B, C, D for each component (see Table 6-2 for n-C_4 solution).

$$j = 1 \text{(Total Condenser)}: C_1 = \frac{-K_2 V_2}{L_2}, B_1 = 1 + \frac{D}{L_1}, D_1 = 0$$

$$j = 2 : C_2 = \frac{-V_3 K_3}{L_3}, B_2 = 1 + \frac{V_2 K_2}{L_2}, A_2 = -1, D_2 = 0$$

$$j = 3 \text{(Feed Stage)} : C_3 = \frac{-K_4 V_4}{L_4}, B_3 = 1 + \frac{V_3 K_3}{L_3}, A_3 = -1, D_3 = Fz$$

$$\text{Reboiler} (j = 4): B_4 = 1 + \frac{V_4 K_4}{L_4}, A_4 = -1, D_4 = 0$$

Invert the tridiagonal matrix either with a spreadsheet or the Thomas algorithm (shown here). The Thomas parameters for each component (see Table 6-2 for n-C_4 solution) are:

$$(V1)_1 = B_1, (V1)_2 = B_2 - A_2(V3)_1, \text{ etc.}$$

$$(V2)_1 = D_1/(V1)_1, (V1)_2 = \frac{D_2 - A_2(V2)_1}{(V1)_2}, \text{ etc.}$$

$$(V3)_1 = C_1/(V1)_1, (V3)_2 = C_2/(V1)_2, \text{ etc.}$$

Calculate component flow rate on each stage for each component (see Table 6-2 for n-C_4 solution)

$$l_4 = (V2)_4, l_3 = (V2)_3 - (V3)_3 l_4$$

$$l_2 = (V2)_2 - (V3)l_3, l_1 = (V2)_1 - (V3)_1 l_2$$

The results are:

	nC$_4$	nC$_5$	nC$_8$
ℓ_1	280.217	303.037	1.993
ℓ_2	93.593	289.141	27.741
ℓ_3	124.492	623.042	549.114
ℓ_4	12.241	148.284	450.273

Calculate normalized values for component flow rates on each stage

$$x_{i,j} = \ell_{i,j} \bigg/ \left(\sum_{i=1}^{C} \ell_{i,j} \right)$$

See Tables 6-2 and 6-3.

TABLE 6-2. *Matrix calculations for Example 6-1 for n-C$_4$*

Matrix:

$$\begin{bmatrix} 1.6 & -5.00 & 0 & 0 \\ -1 & 6.00 & -2.26 & 0 \\ 0 & -1 & 3.26 & -9.17 \\ 0 & 0 & -1 & 10.17 \end{bmatrix} \begin{bmatrix} \ell_{1,C4} \\ \ell_{2,C4} \\ \ell_{3,C4} \\ \ell_{4,C4} \end{bmatrix} = \begin{bmatrix} 0 \\ 0 \\ 200 \\ 0 \end{bmatrix}$$

Intermediate values for the Thomas algorithm are:

$$(V1)_1 = B_1 = 1.67, \qquad (V2)_1 = 0, \qquad (V3)_1 = -5.0/1.67 = -2.994$$

$$(V1)_2 = 6.0 - (-1)(-2.994) = 3.006, \qquad (V2)_2 = \frac{0 - (-1)(0)}{3.006} = 0.0$$

$$(V3)_2 = -2.26/3.006 = -0.7518$$

$$(V1)_3 = 2.5082, \qquad (V2)_3 = 79.7384, \qquad (V3)_3 = -3.6560$$

$$(V1)_4 = 6.5140 \text{ and } (V2_4) = 12.241$$

Then flow rates n-C$_4$ are:

$$\ell_4 = (V1)_4 = 12.241$$
$$\ell_3 = (V2)_3 - (V3)_3 \ell_4 = 79.7384 - (-3.6560)(12.241) = 124.492$$
$$\ell_2 = 93.593, \ \ell_1 = 280.217$$

Calculate composition on each stage.

$$x_{i,j} = \frac{l_{ii,j}}{\sum_1 l_{i,j}} \text{ and } y_{i,j} = K_{i,j} x_{i,j}$$

TABLE 6-3.   *Summary of calculations after one iteration of Example 6-3*

	Reboiler (j = 4)	Feed Stage (j = 3)	Stage (j = 2)	Condenser (q = 1)
xC_4	0.020	0.096	0.228	0.497
xC_5	0.243	0.481	0.704	0.508
xC_8	0.737	0.423	0.068	0.0034
yC_4	0.134	0.344	0.496	0.747
yC_5	0.659	0.620	0.501	0.243
yC_8	0.207	0.036	0.002	6.8 E –5
$T°C$	103.4°	67.2°	44.8°	32.0°C

Calculate temperature on each stage with a bubble-point calculation.

Reboiler (j = 4). Converges to T = 103.4°C. Then the vapor mole fractions are

$$y_{C4} = (0.02)(6.7) = 0.134, \quad y_{C5} = (0.243)(2.71) = 0.659,$$

$$y_{C8} = (0.737)(0.281) = 0.207$$

$$\text{and} \sum y_{i,4} = 1.00$$

Other calculations are summarized in Table 6-3.

At this point, we would use these temperatures to determine new K values and then repeat the matrix calculation and bubble-point calculations. To speed convergence, process simulators use more advanced methods for determining the next set of temperatures (see Section 6.6). Obviously, with this amount of effort we would prefer to use a process simulator to solve the problem (see Problem 6.G1). The process simulator results for temperature are generally higher than the temperatures calculated in this example after one iteration except for Stage 1, which has a calculated temperature that is too high. Also, since this system does not follow CMO, there is considerable variation in the flow rates.

6.5   ENERGY BALANCES IN MATRIX FORM

After convergence of the temperature loop, the liquid and vapor flow rates, L_j and V_j, can be corrected using energy balances (see Figure 6-1). For the general stage shown in Figure 6-3, the energy balance is

$$L_j h_j + V_j H_j = V_{j+1} H_{j+1} + L_{j-1} h_{j-1} + F_j h_{F_j} + Q_j \qquad \textbf{(6-20)}$$

This equation is for $2 \le j \le N - 1$. The liquid flow rates can be substituted in from a mass balance around the top of the column:

$$L_j = V_{j+1} - D + \sum_{k=1}^{j} F_k \tag{6-21a}$$

$$L_{j-1} = V_j - D + \sum_{k=1}^{j-1} F_k \tag{6-21b}$$

where $2 \le j \le N - 1$.

Substituting Eqs. (6-21a) and (6-21b) into Eq. (6-20) and rearranging, we obtain

$$(h_j - H_{j+1})V_{j+1} + (H_j - h_{j-1})V_j$$

$$= F_j h_{F_j} + Q_j - D(h_{j-1} - h_j) - \left(\sum_{k=1}^{j} F_k\right)h_j + \left(\sum_{k=1}^{j-1} F_k\right)h_{j-1} \tag{6-22}$$

For a total condenser since $V_2 = L_1 + D$ and $h_1 = h_D$, the energy balance is

$$Q_1 + V_2 H_2 + F_1 h_{F_1} = V_2 h_1 \tag{6-23}$$

which upon rearrangement is

$$(h_1 - H_2)V_2 = F h_{F_1} + Q_1 \tag{6-24}$$

For the partial reboiler ($j=N$), the energy balance is

$$F_N h_{F_N} + Q_N + L_{N-1}h_{N-1} = B h_B + V_N H_N \tag{6-25}$$

and the overall flow rate can be calculated from,

$$L_{N-1} = B + V_N - F_N \tag{6-26}$$

Substituting Eq. (6-26) into (6-25) and rearranging, we have

$$(H_N - h_{N-1})V_N = F_N h_{F_N} + Q_N + B(h_{N-1} - h_N) - F_N h_{N-1} \tag{6-27}$$

If Eqs. (6-22), (6-24), and (6-27) are put in matrix form, we have

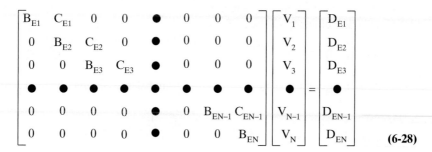

$$\tag{6-28}$$

This matrix again has a tridiagonal form (with $A_j = 0$) and can easily be inverted to obtain the vapor flow rates. The B_E and C_E coefficients in Eq. (6-28) are easily obtained by comparing Eqs. (6-22), (6-24), and (6-27) to Eq. (6-28). For $j = 1$ these values are

$$B_{E1} = 0, \ C_{E1} = h_1 - H_2, \ D_{E1} = Fh_{F_1} + Q_1 \qquad \text{(6-29a)}$$

For $2 \le j \le N - 1$

$$B_{Ej} = H_j - h_{j-1} \qquad C_{Ej} = h_j - H_{i,j+1}$$

$$D_{Ej} = F_j h_{F_j} + Q_j + D(h_{j-1} - h_j) - \left(\sum_{k=1}^{j} F_k\right) h_j + \left(\sum_{k=1}^{j-1} F_k\right) h_{j-1} \qquad \text{(6-29b)}$$

and for $j = N$

$$B_{EN} = H_N - H_{N-1}, \ D_{EN} = F_N h_{F_N} + Q_N + B(h_{N-1} - h_N) - F_N h_{N-1} \qquad \text{(6-29c)}$$

The coefficients in Eqs. (6-29) require knowledge of the enthalpies leaving each stage and the Q values. The enthalpy values can be calculated since all x's, y's, and temperatures are known from the component mass balances and the converged temperature loop. For ideal mixtures, the enthalpies are

$$h_j = \sum_{i=1}^{C} x_{i,j} \tilde{h}_i(T_j) \qquad \text{(6-30a)}$$

and

$$H_j = \sum_{i=1}^{C} y_{i,j} \tilde{H}_i(T_j) \qquad \text{(6-30b)}$$

where $\tilde{h}_i(T_j)$ and $\tilde{H}_i(T_j)$ are the pure component enthalpies.

$$\tilde{h}_i(T_j) = \tilde{h}_i(T_{ref}) + C_{p,\ell iq}(T_j - T_{ref}) \qquad \text{(6-30c)}$$

$$\tilde{H}_i(T_j) = \tilde{h}_i(T_{ref}) + \lambda_i(T_{ref}) + C_{p,vap}(T_j - T_{ref}) \qquad \text{(6-30d)}$$

They can be determined from data (for example, see Maxwell, 1950, or Smith, 1963) or from heat capacities and latent heats of vaporization.

Usually the column is adiabatic. Thus,

$$Q_j = 0 \qquad \text{for} \qquad 2 \le j \le N - 1; \qquad Q_N = Q_R, \qquad Q_1 = Q_C \qquad \text{(6-31)}$$

The condenser requirement can be determined from balances around the total condenser:

$$Q_1 = V_2(h_D - H_2) = D\left(1 + \frac{L}{D}\right)(h_D - H_2) \qquad \text{(6-32)}$$

Since D and L/D are specified and h_D and H_2 can be calculated from Eqs. (6-30), Q_1 is easily calculated. For an adiabatic column, the reboiler heat load can be calculated from an overall energy balance.

$$Q_N = Dh_D + Bh_B - \sum_{k=1}^{N} (F_k h_{F_k}) - Q_1 \qquad \text{(6-33)}$$

Inversion of Eq. (6-28) gives new guesses for all the vapor flow rates. The liquid flow rates can then be determined from mass balances such as Eqs. (6-21) and (6-26). These new liquid and vapor flow rates are compared to the values used for the previous convergence of the mass balances and temperature loop. The check on convergence is, if

$$\left| \frac{L_{j,calc} - L_{j,old}}{L_{j,calc}} \right| < \varepsilon \qquad \text{and} \qquad \left| \frac{V_{j,calc} - V_{j,old}}{V_{j,calc}} \right| < \varepsilon \qquad \text{(6-34)}$$

for all stages, then the calculation has converged. For computer calculations, an ε of 10^{-4} or 10^{-5} is appropriate.

If the problem has not converged, the new values for L_j and V_j must be used in the mass balance and temperature loop (see Figure 6-1). Direct substitution is the easiest approach. That is, use the L_j and V_j values just calculated for the next trial.

When Eqs. (6-34) are satisfied, the calculation is finished. This is true because the mass balances, equilibrium relationships, and energy balance have all been satisfied. The solution gives the liquid and vapor mole fractions and flow rates and the temperature on each stage and in the products.

6.6 INTRODUCTION TO NAPHTALI-SANDHOLM SIMULTANEOUS CONVERGENCE METHOD

One of the more robust methods for solution of multicomponent distillation and absorption problems was developed in a classic paper by Naphtali and Sandholm (1971). This method is available in many commercial simulators. Naphtali and Sandholm developed a linearized Newtonian method to solve all the equations for multicomponent distillation simultaneously.

The Newtonian procedure outlined in Eqs. (2-51) to (2-57) for the energy balance for flash distillation can be considered a simplified version of the method used by Naphtali and Sandholm. I recommend that students reread that material before proceeding.

To develop a simultaneous Newtonian procedure for multicomponent distillation, Naphtali and Sandholm (1971) first wrote the $N(2C + 1)$ equations and variables consisting of component mass balances [essentially Eq. (6-13) but without substitution of the equilibrium K values for y_i], the energy balances [essentially Eq. (6-28)], and equilibrium relationships including Murphree vapor efficiencies in functional matrix form. Their procedure is illustrated for the energy balance. Essentially, the functional matrix form for Eq. (6-28) is

$$[\text{EB function}] = [\text{Energy Bal matrix}] \times [\text{Vapor flow rate matrix}] - [\text{Energy input } (D_E) \text{ matrix}] = [0] \qquad \text{(6-35)}$$

Unfortunately, when one first starts the calculation with the initial guesses for all temperatures, vapor, and liquid flow rates, the energy balance, component mass balance, and equilibrium functions will not equal zero. The Newtonian method is used to develop new values for the variables to calculate enthalpies and K values. To use the Newtonian method, Naphtali and Sandholm developed derivatives for the changes in all variables. Note that the energy balance (and component mass balances and equilibrium relationships) on plate j depend only on the variables on plates j-1, j and j+1. For the energy balance (EB) function on plate j, the derivatives with respect to the variables on plate j-1 are

$$d(EB_j)/dv_{j-1,i} = 0$$

$$d(EB_j)/dT_{j-1} = L_{j-1} \, dh_{j-1}/dT_{j-1} \tag{6-36}$$

$$d(EB_j)/dl_{j-1} = L_{j-1} \, dh_{j-1}/dl_{j-1}$$

where $v_{j-1,i} = V_{j-1} y_{j-1,i}$. The derivatives for EB on plate j with respect to the variables on plate j are

$$d(EB_j)/dv_{j,i} = V_j \, dH_j /dv_{j,i}$$

$$d(EB_j)/dT_j = V_j \, dH_j/dT_j + L_j \, dh_j/dT_j \tag{6-37}$$

$$d(EB_j)/dl_j = L_j \, dh_j/dl_j$$

And the derivatives for the EB function on plate j with respect to the variables on plate j+1 are

$$d(EB_j)/dv_{j+1,i} = -V_{j+1} \, dH_{j+1} /dv_{j+1,i}$$

$$d(EB_j)/dT_{j+1} = -V_{j+1} \, dH_{j+1}/dT_{j+1} \tag{6-38}$$

$$d(EB_j)/dl_{j+1} = 0$$

These equations extend Eq. (2-54), developed for multicomponent flash distillation, to multicomponent column distillation.

The next value for every variable is the old value plus the calculated correction. For example, for temperature on stage j, the value for the next trial is

$$T_{j,new} = T_{j,old} + \Delta T_j \tag{6-39}$$

where the correction ΔT_j is determined from the Newtonian approximation

$$\Delta T_j = [F_{old}]/[(dF/dT)_{old}] \tag{6-40}$$

where $[F_{old}]$ is the combined matrix of all functions (energy balance, component mass balances, and equilibrium) and $[(dF/dT)_{old}]$ is the combined matrix of all the derivatives, since temperature affects all of these functions. Equation (6-40) is an extension of Eq. (2-55) to a multistage distillation with multiple variables. This procedure is followed for every variable $(T_j, v_j, l_j$ on every stage j) to obtain new values for each variable. These new values are used in the next trial to calculate the values of the energy balance, component mass balance, and equilibrium functions. These functions should all have a value of zero, and their differences from zero are the discrepancies in the answer. The convergence check is that the sum of squares of the discrepancies is less than some tolerance such as 10^{-7}.

6.7   DISCUSSION

All of the methods for solving multicomponent distillation and absorption problems have weaknesses. The bubble-point method works well for narrow-boiling feeds, but is inherently unstable for wide boiling feeds. The sum-rates method (Chapter 12) works well for wide-boiling feeds such as most absorbers but is unstable for narrow-boiling feeds. The Naphtali-Sandholm approach often works well for both narrow- and wide-boiling feeds, but Newtonian methods are notorious for diverging for nonlinear problems if the first guess is not close to the final answer. Since distillation problems can be extremely nonlinear, the first guess can be very important for all of these methods. If a good first guess is difficult to find, try finding a simpler set of conditions where convergence does occur (e.g., fewer stages or larger L/D) and then approach the desired condition by slowly changing the values of the variables. Since commercial simulators use the previous run as an initial guess for the current run, this method should make the initial guess closer to the answer, which means convergence is more likely.

A number of methods to make the convergence schemes more stable have been developed and are employed in commercial process simulators. As a result, commercial simulators are quite robust, particularly if an appropriate method (bubble-point, sum-rates or Naphtali-Sandholm) is chosen and a good first guess has been used, although they occasionally still have difficulty converging. When there is a convergence difficulty, first check that the basic solution approach chosen appears to be appropriate. Then try increasing the number of iterations allowed. If this does not work, try reducing the tolerance on convergence. Finally, most simulators have a number of options to make convergence more stable, such as damping the predicted change in variables, and these options can be tried. With a combination of these approaches, almost all equilibrium-staged multicomponent distillation problems can be solved. This is one of the great advances in chemical engineering in the last 100 years. Rate-based models for distillation are considered in Chapters 15 and 16.

The matrix approach is easily adapted to partial condensers and to columns with side streams (see Problems 6.C2 and 6.C1). The approach will converge for normal distillation problems. Extension to more complex problems such as azeotropic and extractive distillation or very wide boiling feeds is beyond the scope of this book; however, these problems will be solved with a process simulator.

In some ways, the most difficult part of writing a multicomponent distillation program has not been discussed. This is the development of a physical properties package that will accurately predict equilibrium and enthalpy relationships (Barnicki, 2002; Carlson, 1996; Sadeq et al., 1997; Schad, 1998). Sadaq et al. (1997) compared three process simulators and found that relatively small differences in the parameters and in the VLE correlation can cause major errors in the results. Fortunately, a considerable amount of research has been done (see Table 2-2 and Fredenslund et al., 1977; and Walas, 1985) to develop accurate physical property correlations. Very detailed physical properties packages can be purchased commercially and are included in the commercial process simulators.

Most companies using distillation have available computer programs using one of the advanced calculation procedures. Several software and design companies sell these programs. The typical engineer will use these routines and not go to the large amount of effort required to write his or her own routine. However, an understanding of the expected profiles and the basic mass and energy balances in the column can be very useful in interpreting the computer output and in determining when that output is garbage. Thus, it is important to understand the principles of distillation calculations even though the details of the computer program may not be understood.

6.8 SUMMARY—OBJECTIVES

In this chapter we have developed methods for multicomponent distillation. The objectives for this chapter are as follows:

1. Use a matrix approach to solve the multicomponent mass balances
2. Use bubble-point calculations to determine new temperatures on each stage
3. Use a matrix approach to solve the energy balances for new flow rates.
4. Explain the difference between the bubble-point and Naphtali-Sandholm methods

REFERENCES

Amundson, N. R., and A. J. Pontinen, "Multicomponent Distillation Calculations on a Large Digital Computer," *Ind. Engr. Chem, 50*, 730 (1958).

Barnicki, S. D., "How Good Are Your Data?" *Chem. Engr. Progress, 98* (6), 58 (June 2002).

Carlson, E. C., "Don't Gamble with Physical Properties for Simulations," *Chem. Engr. Progress, 92* (10), 35 (Oct. 1996).

Fredenslund, A., J. Gmehling, and P. Rasmussen, *Vapor-Liquid Equilibria Using UNIFAC: A Group-Contribution Method,* Elsevier, Amsterdam, 1977.

Holland, D. D., *Fundamentals of Multicomponent Distillation,* McGraw-Hill, New York, 1981.

King, C. J., *Separation Processes,* 2nd ed., McGraw-Hill, New York, 1981.

Lapidus, L. *Digital Computation for Chemical Engineers,* McGraw-Hill, New York, 1962.

Maxwell, J. B., *Data Book on Hydrocarbons,* Van Nostrand, Princeton, NJ, 1950.

Naphtali, L. M., and D. P. Sandholm, "Multicomponent Separation Calculations by Linearization," *AIChE Journal, 17* (1), 148 (1971).

Poling, B. E., J. M. Prausnitz, and J. P. O'Connell, *The Properties of Gases and Liquids,* 5th ed., McGraw-Hill, New York, 2001.

Sadeq, J., H. A. Duarte, and R. W. Serth, "Anomalous Results from Process Simulators," *Chem. Engr. Education, 31* (1), 46 (Winter 1997).

Schad, R. C., "Make the Most of Process Simulation," *Chem. Engr. Progress, 94* (1), 21 (Jan. 1998).

Seider, W. D., J. D. Seader, and D. R. Lewin, *Process Design Principles,* Wiley, New York, 1999.

Smith, B. D., *Design of Equilibrium Stage Processes,* McGraw-Hill, New York, 1963.

Walas, S. M., *Phase Equilibria in Chemical Engineering,* Butterworth, Boston, 1985.

Wankat, P. C., *Equilibrium-Staged Separations*, Prentice Hall, Upper Saddle River, NJ, 1988.

HOMEWORK

A. *Discussion Problems*

A1. In the matrix approach, we assumed $K = K(T, p)$. How would the flowchart in Figure 6-1 change if $K = K(T, p, x_i)$?

A2. The method described in this chapter is a simulation method because the number of stages and the feed and withdrawal locations must all be specified. How do you determine the optimum feed stage?

A3. Develop your key relations chart for this chapter.

A4. In a multicomponent simulation program for distillation the loops are nested. The outermost loop is mole fractions, next is flow rates and the innermost loop is temperature.

1. Mole fractions are the outermost loop because,
   a. Many distillation problems can be done without this loop.
   b. Changing mole fractions often do not have a major effect on K values.
   c. Mole fractions have a major impact on K values only for systems with complex equilibrium behavior.
   d. All of the above.

2. The temperature loop is done before the flow rate loop because,
   a. Temperatures cannot be constant in distillation.
   b. Flow rates are often very close to constant in each section of the column.
   c. A very good guess of flow rates can be made.
   d. All of the above.

3. A good initial guess of the flow rates is to
   a. Assume CMO.
   b. Assume liquid and vapor flow rates are constant throughout the column.
   c. Use a bubble-point calculation at the top and bottom of the column.

4. For a good initial guess of temperatures,
   a. Assume CMO.
   b. Use any arbitrary temperature.
   c. Use a bubble-point calculation at the top and bottom of the column.

5. The mass balances are solved by developing a matrix that is then inverted. The matrix allows one to have feed at,
   a. Only one location in the column.
   b. Two locations in the column.
   c. Any stage within the column, but not the condenser and not the reboiler.
   d. Any stage in the column, plus the reboiler and the condenser.

A5. The method described in this chapter is a simulation method because the number of stages and the feed and withdrawal locations must all be specified. How do you determine if you have enough stages for a design problem?

C. *Derivations*

C1. Suppose there is a liquid side stream of composition x_j and flow rate S_j removed from stage j in Figure 6-3.
   a. Derive the mass balance Eqs. (6-4) to (6-6) for this modified column.
   b. Develop new energy balance equations. Derive new coefficients for Eq. (6-29b).

C2. Derive the mass balance expression for the matrix approach if there is a partial condenser instead of a total condenser. Replace Eqs. (6-7) to (6-9).

C3. Show that the sum from i = 1 to C of the $x_{i,j}$ in Eq. (6-17) = 1.0.

D. *Problems*

Answers to problems with an asterisk are at the back of the book.

D1. For the first trial of Example 6-1 determine the component matrix for n-pentane and then use the Thomas algorithm to find the n-pentane liquid flow rates leaving each stage. Compare your n-pentane flow rates with the values reported in Example 6-1.

D2. We are separating a mixture of ethane, propane, n-butane and n-pentane in a distillation column operating at 5.0 atm. The column has a total condenser and a partial reboiler. The feed flow rate is 1000 kmol/h. The feed is a saturated liquid. Feed is 8.0 mol% ethane, 33.0 mol% propane, 49.0 mol% n-butane and 10.0 mol% n-pentane. The column has 4 equilibrium stages plus the partial reboiler, which is an equilibrium contact. Feed is on 2nd stage below total condenser. Reflux ratio $L_0/D = 2.5$. Distillate flow rate $D = 410$ kmoles/hr. Develop the mass balance and equilibrium matrix (Eq. 6-13) with numerical values for each element (A_j, B_j, C_j, and D_j). Do this for your first guess: T_j = bubble-point temperature of feed (use same temperature for all stages); K values are from DePriester chart; L and V are CMO values. *Do the matrix for propane only.*

D3. Do the matrix for n-butane for Problem 6.D2.

D4. A distillation column is separating 100 kmol/h of a saturated liquid feed that is 30 mol% methanol, 25 mol% ethanol, 35 mol% n-propanol, and 10 mol% n-butanol at a pressure of 1.0 atm. The column has a total condenser and a partial reboiler. We want a 98.6% recovery of i-propanol in the distillate and 99.2% recovery of n-propanol in the bottoms (but realize that this first trial will not provide this amount of separation). Operation is with $L/D = 5$, $D = 60$, $N = 4$, and $N_{feed} = 3$ (#1 = total condenser and #4 = partial reboiler), set up the mass balance matrix Eq. (6-13) for the first trial for n-butanol, and then solve. This is a hand calculation.

  a. Use CMO to estimate liquid and vapor flow rates in the column for the first trial. Report these flow rates.

  b. For a first guess of K values, assume the K values in the column are constant and equal to those found in a bubble-point calculation for the feed. The K_{np} values are $(y/x)_{np}$ where y_{np} and x_{np} are from the bubble-point calculation with constant alpha, Eq. (5-30). The other $K_i = \alpha_i K_{np}$.

  c. Calculate all the A, B, C, and D values (but for n-butanol only) and write the complete matrix.

  d. Solve the n-butanol matrix using the Thomas algorithm and find the n-butanol flow rates $l_{j,n\text{-butanol}}$ leaving each stage.

  e. The T implicitly used in the calculation to find the K values is the bubble-point temperature of the feed. To determine T, first calculate the K value for n-propanol as $(y/x)_{n\text{-propanol}}$ where y and x are found from the constant relative volatility solution for the bubble-point. Then use Raoult's law to find the T that gives this K value. Report this T.

  Do not do additional trials.

System properties: If we choose n-propanol as the reference, the relative volatilities are methanol = 3.58, ethanol = 2.17, n-propanol = 1.0, and n-butanol = 0.412. These relative volatilities can be assumed to be constant. The K value for n-propanol can be estimated from Raoult's law. The vapor pressure data for n-propanol from Perry's is:

T°C =	−15	5.0	14.7	25.3	36.4	43.5	52.8	66.8	82.0	97.8
vp mm Hg =	1.0	5.0	10.0	20.0	40.0	60.0	100.0	200.0	400.0	760.0

F. *Problems Requiring Other Resources*

F1.* A distillation column with two stages plus a partial reboiler and a partial condenser is separating benzene, toluene, and xylene. Feed rate is 100 kmol/h, and feed is a saturated vapor introduced on the bottom stage of the column. Feed compositions (mole fractions) are $z_B = 0.35$, $z_T = 0.40$, $z_X = 0.25$. Reflux is a saturated liquid and $p = 16$ psia. A distillate flow rate of $D = 30$ kmol/h is desired. Assume $K_i = VP_i/p$. Do *not* assume constant relative volatility, but do assume CMO. Use the matrix approach to solve mass balances and the bubble-point method for temperature convergence. For the first guess for temperature, assume that all stages are at the dew-point temperature of the feed. Do only one iteration. See Problem 6.C2 for handling the partial condenser.

G. *Computer Simulation Problems.*

G1. Use a process simulator to completely solve Example 6-1. Do not assume CMO. Compare temperature and mole fractions on each stage to the values obtained in Example 6-1 after one trial.

G2. You have an ordinary distillation column separating ethane, propane, and n-butane. The feed rate is 100 kmol/h and is a saturated liquid at 10.0 atm. The mole fractions in the feed are ethane = 0.2, propane = 0.35, and n-butane = 0.45. The column has 28 equilibrium stages and the feed is input on stage 8 in the column (counting from the top down). The feed is input above the stage. The column pressure is 10.0 atm. and can be considered to be constant. The column has a total condenser and a kettle type reboiler. The external reflux ratio $L/D = 2.0$ and the distillate flow rate is set at $D = 55.0$ kmol/h. Use a process simulator to simulate this system and answer the following questions:

1. What VLE package did you use?
Explain why you chose this package.

2. Report the following values:
Temperature of condenser = _____ °C
Temperature of reboiler   = _____ °C
Distillate product mole fractions _____
Bottoms product mole fractions _____

3. Was the specified feed stage the optimum feed stage? Yes   No
If no, the feed stage should be: a. closer to the condenser, or b. closer to the reboiler.
Note: Just do the minimum number of simulations to answer these questions. Do not optimize.

4. Which tray gives the largest column diameter (in meters) with sieve trays when one uses the originally specified feed stage? Tray #_____ Diameter = _____
[In Aspen Plus: Use one pass, tray spacing (0.6096 m), minimum downcomer area (0.10), foaming factor (1), and over design factor (1). Set the fractional approach to flooding at 0.7. Use the Fair design method for flooding.]

5. Which components in the original problem are the key components?

6. Change one specification in the operating conditions (keep original number of stages, feed location, feed flow, feed composition, feed pressure, feed temperature/fraction vaporized constant) to make *ethane* the light key and *propane* the heavy key.
What operating parameter did you change, and what is its new value? _____
Temperature of condenser = _____ °C
Temperature of reboiler   = _____ °C
Distillate product mole fractions _____
Bottoms product mole fractions _____

G3. Use a process simulator to simulate the separation of a mixture that is 0.25 mole fraction methanol, 0.30 mole fraction ethanol and 0.45 mole fraction n-propanol in a series of two distillation columns. The feed rate is 100 kmol/h and the feed is at 50°C and 1.0 atm. The feed goes to the first column, which has 22 equilibrium stages, a total condenser, and a kettle reboiler; on (above) stage 11 below the condenser. The distillate from this column, containing mainly methanol and ethanol is fed to the second column. Set the D value to 55.0. Vary L/D in the first column to achieve a 0.990 or better split fraction of ethanol in the distillate and a 0.990 or better split fraction of n-propanol in the bottoms. Vary L/D by increments of 0.1 (e.g., L/D = 0.8, 0.9, 1.0, and so forth) until you find the lowest L/D that gives the desired separation. (You can start with any L/D value you wish.) Both columns are at 1.0 atm pressure. Use the Wilson VLE package.

For column 1 report the following:
a. Final value of L/D _____
b. Split fractions of ethanol (distillate)_____and n-propanol (bottoms) _____
c. Mole fractions in bottoms _____
d. Mole fractions in distillate _____

The second column receives the distillate from column 1 as a saturated liquid feed at one atm. pressure. The second column is at 1.0 atm pressure and operates with D = 25.0 and L/D = 4.0. It has a total condenser and a kettle reboiler. There are 34 equilibrium stages in the column. Find the optimum feed stage location. For column 2 report the following:
a. Optimum feed location *in the column* _____
b. Mole fractions in bottoms _____
c. Mole fractions in distillate _____

G4. You have an ordinary, single feed distillation column separating benzene, toluene, and biphenyl ($C_{12}H_{10}$, also called diphenyl). There are 15 trays in column and feed location in column is tray 9 (input above stage), reflux ratio is 1.1, pressure is 1.5 atm (operate column at constant pressure), total condenser with saturated liquid reflux, kettle type reboiler, feed flow rate is 200 kmol/h, feed mole fractions: benzene = 0.2, toluene = 0.65, and biphenyl = 0.15; feed pressure is 1.5 atm, feed temperature is 100°C, D = 170 kmol/h, adiabatic column, and use the Peng-Robinson VLE package. Open a new blank file, simulate this system with RADFRAC, and answer the following questions:

1. Report the following values:
Temperature of condenser = _____ K, temperature of reboiler = _____ K
$Q_{condenser}$ = _____ cal/sec, $Q_{reboiler}$ = _____ cal/sec
Distillate product mole fractions: _____
Bottoms product mole fractions: _____

2. Was the specified feed stage the optimum feed stage? Yes No
If no, the feed stage should be: a. closer to the condenser, b. closer to the reboiler. (Note: Do minimum number of simulations to answer these questions. Do not optimize.)

3. Which tray gives the largest column diameter with sieve trays when one uses the originally specified feed stage? Aspen Tray #_____
Column diameter = _____ meters
[Use the default values for number of passes (1), tray spacing (0.6096 m), minimum downcomer area (0.10), foaming factor (1), and over design factor (1). Set the fractional approach to flooding at 0.65. Use the Fair design method for flooding.]

4. Which components in the original problem are the key components (label light and heavy keys)? _____

5. Change one specification in the operating conditions (keep N, feed location, feed flow, feed composition, feed pressure, feed temperature or fraction vaporized constant) to make benzene the light key and toluene the heavy key. *Also increase the reflux ratio to 4.0.*
What operating parameter did you change (not including the reflux ratio), and what is its new value? _____
Temperature of condenser = _____ K, temperature of reboiler = _____ K
Distillate product mole fractions: _____
Bottoms product mole fractions: _____

G5. You have an ordinary, single feed distillation column separating benzene, toluene, cumene, p-xylene (the last one is para xylene, C_8H_{10}). There are 23 trays in column and feed location in column is tray 12 (input above stage), reflux ratio is 4.0, pressure is 3.0 atm (operate column at constant pressure), total condenser with saturated liquid reflux, kettle type reboiler, feed flow rate is 400 kmol/h, feed mole fractions: benzene = 0.2, toluene = 0.45, p-xylene = 0.35, feed pressure is 3.0 atm, feed temperature is 100°C, D = 80 kmol/h, adiabatic column, and use the Peng-Robinson VLE package. Open a new blank file, simulate this system with RADFRAC, and answer the following questions:

1. Report the following values:
Temperature of condenser = _____ K, temperature of reboiler = _____ K
$Q_{condenser}$ = _____ cal/sec, $Q_{reboiler}$ = _____ cal/sec
Distillate product mole fractions: _____
Bottoms product mole fractions: _____

2. Was the specified feed stage the optimum feed stage? Yes No
If no, the feed stage should be: a. closer to the condenser, b. closer to the reboiler. (Note: Do minimum number of simulations to answer these questions. Do not optimize.)

3. Which tray gives the largest column diameter with sieve trays when one uses the originally specified feed stage? Aspen Tray #_____ Column diameter = _____ meters
[Use the default values for number of passes (1), tray spacing (0.6096 m), minimum downcomer area (0.10), foaming factor (1), and over design factor (1). Set the fractional approach to flooding at 0.7. Use the Fair design method for flooding.]

4. Which components in the original problem are the key components (label light and heavy keys)? _____

5. Change one specification in the operating conditions (keep N, feed location, feed flow, feed composition, feed pressure, feed temperature or fraction vaporized constant) to make toluene the light key and p-xylene the heavy key.
What operating parameter did you change, and what is its new value? _____
Temperature of condenser = _____ K, temperature of reboiler = _____ K
Distillate product mole fractions: _____
Bottoms product mole fractions: _____

G6. A distillation column with a partial condenser and a partial reboiler is separating 1500 kmol/h of a feed that is 10 mol% ethane, 30 mol% n-butane, and 60 mol% n-pentane. The feed is a saturated liquid at 8.1 atm. The column operates at a constant pressure of 8.0 atm. The distillate is withdrawn as a saturated vapor. In Aspen notation N = 40 and the feed is "on stage" 25. In an attempt to obtain close to three pure products, a sidestream is withdrawn between the partial condenser and the feed stage. We want a 99.9 % recovery of n-pentane in the bottoms (split fraction in bottoms stream > 0.999). Note: Setting Standard Convergence on the RADFRAC Block setup may have convergence problems. If this occurs, change convergence to Petroleum/Wide-boiling. Report the VLE correlation used.

a. Set distillate flow rate $D = 150$ kmol/h and side withdrawal flow rate $S = 450$ kmol/h (these are appropriate values if the separation is perfect—although the separation is not and cannot be perfect). Put the side withdrawal on stage 12 and make it a liquid. Increase the reflux ratio until the mole fraction of ethane in the distillate is > 0.98 (remember that distillate is a vapor, so look at y values. You do not have to be exactly at 0.980. The reflux ratio will be high). Report the reflux ratio.

b. To improve the ethane mole fraction in the distillate to > 0.99, first try increasing the reflux ratio. Find the value necessary and report this reflux ratio.

c. The reflux ratio required for part b is rather high. Reduce the reflux ratio to 20. Now set $D = 150 - \Delta$ and $S = 450 + \Delta$. Find the Δ value that gives ethane mole fraction in the distillate to > 0.99. Report D, S, and mole fractions in the three products. The reason reducing D and increasing S works is that there must be some ethane in the sidestream, since the liquid in the sidestream is in equilibrium with the upward-flowing vapor that carries ethane to the distillate. Thus, not all of the sidestream is n-butane, but our original values for D and S assumed it was pure n-butane. Since there is ethane in the sidestream, the value of $D = 150$ is too large, which forces some n-butane up into the distillate. Report D, S, and the mole fractions in the three products.

d. Keeping everything the same as in part c, change the sidestream to a vapor and run again. Report the mole fractions of the three products (remember to use y values for the sidestream). Why is the separation significantly worse than with a liquid sidestream?

e. We could also put the sidestream below the feed stage (in this case the feed stage would be closer to the condenser). However, this configuration will not work as well for this separation. Why not?

G7. You have an ordinary, single feed distillation column separating methanol, ethanol, n-propanol, and n-butanol. There are 24 trays in column and feed location in column is tray 14 (input above stage), boilup ratio is 4.0, pressure is 3.0 atm (operate column at constant pressure), total condenser with saturated liquid reflux, kettle type reboiler, feed flow rate is 200 kmol/h, feed mole fractions: methanol = 0.30; ethanol = 0.20, n-propanol = 0.25, and n-butanol = 0.25, feed pressure is 3.0 atm, feed temperature is 50.0°C, D = 100 kmol/h, adiabatic column, and use the NRTL VLE package. Open a new blank file, simulate this system with RADFRAC, and answer the following questions:

1. Report the following values:
Temperature of condenser = _____ K, temperature of reboiler = _____ K
$Q_{condenser}$ = _____cal/sec, $Q_{reboiler}$ = _____cal/sec
Distillate product mole fractions: _____
Bottoms product mole fractions: _____

2. Was the specified feed stage the optimum feed stage? Yes No
If no, the feed stage should be: a. closer to the condenser, b. closer to the reboiler.

(Note: Do minimum number of simulations to answer these questions. Do not optimize.)

3. Which tray gives the largest column diameter with sieve trays when one uses the originally specified feed stage? Aspen Tray #_____ Column diameter = _____ meters

   [Use the default values for number of passes (1), tray spacing (0.6096 m), minimum downcomer area (0.10), foaming factor (1), and overdesign factor (1). Set the fractional approach to flooding at 0.7. Use the Fair design method for flooding.]

4. Which components in the original problem are the key components (label light and heavy keys)? _____

5. Change one specification in the operating conditions (keep N, feed location, feed flow, feed composition, feed pressure, feed temperature or fraction vaporized constant) to make methanol the light key and ethanol the heavy key.

   What operating parameter did you change, and what is its new value? _____

   Temperature of condenser = _____ K, temperature of reboiler = _____ K

   Distillate product mole fractions: _____

   Bottoms product mole fractions: _____

CHAPTER 6 APPENDIX.   COMPUTER SIMULATIONS FOR MULTICOMPONENT COLUMN DISTILLATION

Lab 4. The methods to start the process simulator are discussed in Labs 1 and 2 (appendix to Chapter 2) and specific directions for distillation are discussed in Lab 3 (appendix to Chapter 4).

I. Multicomponent distillation. Aspen Plus uses the same calculation methods for binary and multicomponent distillation, but there are some differences. Draw a distillation column with a partial condenser using RADFRAC. To do this, take a vapor distillate product instead of a liquid distillate product. Then in the Block for the column, choose the following from the menu: Calculation Type: Equilibrium, Number of Stages: 40 (this is initial setting only), Condenser: Partial-Vapor, Reboiler: Kettle, Valid Phases: Vapor-Liquid, Convergence: Standard. If you get an error message in part 1a, chances are you have a liquid distillate in the drawing and are trying to take a vapor distillate product (that is what Condenser: Partial-Vapor requires). Redraw your column with a vapor distillate stream. The following system converges and allows you to explore the effect of many variables. The feed consists of: n-butane, 20.0 kmol/h, n-pentane, 30.0 kmol/h, and n-hexane, 50.0 kmol/h. Pick an appropriate VLE package to use and check the equilibrium predictions (e.g., K values should be reasonably close to predictions made with the DePriester chart—but note that the correlation should be more accurate than the DePriester charts). In part 1a use a saturated vapor feed at 1 bar. Initially try a column pressure of 1 bar. A column with 40 stages with the feed on number 20 and L/D = 3 are more than sufficient for this separation.

   a. Adjust the column settings to make the butane and pentane exit from the top of the column and the hexane from the bottom. This can be done by setting D (or equivalently B) so that an external mass balance will be satisfied if a perfect split is achieved (e.g., butane recovery in distillate of 100%, pentane recovery in distillate of 100% and hexane recovery in bottoms of 100%). Note the compositions and temperatures of the distillate and bottoms. Check the heat loads in the condenser and reboiler.

b. Change the settings so that the butane exits the top and the pentane and hexane the bottoms. (Change D or B to do this.) Under these conditions you should get an error message that the column dried up. Change the feed to a saturated liquid feed (V/F = 0 in feed), reinitialize (in tool bar click on Run and scroll down to Reinitialize. Click on OK to messages) and run again.

c. Try an intermediate cut where butane exits from the top, hexane from the bottom, and pentane distributes between the top and bottoms. (Change D or B.) Do with saturated liquid feed. Compare Q_C and Q_R in runs b (V/F = 0) and c. Explain.

d. Continue item b with V/F = 0. Find the total number of stages and the optimum feed location if we want butane mole fraction in the bottoms to be just less than 1.0 E –3 and pentane mole fraction in the distillate to be just less than 1.0 E –3. This calculation is trial and error with a simulation program (commercial simulators have a "design" option that will do this trial-and-error process for you, but it should not be used until you have a firm grasp of distillation). In the "Results Summary" Browser accurate values for the distillate and bottoms compositions can be found in the section titled Compositions. Use the menu bar to switch to "Liquid." Use the other menu bar for mole units. Since Aspen Plus calls the partial condenser #1, the vapor composition leaving stage 1 is the distillate, and the last liquid mole fraction is the bottoms. Is it more difficult to meet the C_4 or the C_5 requirement?

e. Look at your condenser and reboiler temperatures (continuing item d). The condenser temperature is low enough that refrigeration is needed. This is expensive. To prevent this, raise the column pressure until the condenser temperature is high enough that cooling water can be used for condensation. (The appropriate value depends on the plant location. Use a cooling water temperature appropriate for your location and add 5°C for ΔT in the heat exchanger.) Changing the pressure changes the VLE. Check to see if you still have the desired separation. If not, find the new values for optimum feed and total number of stages to obtain the desired mole fractions. Note: Be sure to also raise your feed pressure so that it is equal to the column pressure. Why are more stages required at the higher P?

f. Try different values for L/D or boilup ratio. See how this affects the separation. Try using different operating specifications in RADFRAC.

g. Now, try a different feed temperature. For example, try a feed at 30°C. Look at how reboiler and condenser heat loads change and compare to run 1e. Try changing the feed composition. (Remember to change D or B to satisfy mass balances.)

h. Try a different feed flow rate (but same concentrations and fraction vapor) at conditions that you optimized previously. The number of stages, optimum feed location and separation achieved should not change. The heat requirements will be different, as will outlet flow rates. Compare Q_R/(Feed rate) for the two runs. What does this say about the design of distillation for different flow rates?

i. Change the column configuration and have a liquid distillate product or two feeds. This will require redrawing your flowsheet. Compare results to 1e, but remember distillate is a liquid.

j. On the menu bar, click on Tools, then Analysis, then Property and then on Residue. Then click on Go. This gives a residue curve for your chemical system. A residue curve shows the path that a batch distillation will follow for any starting condition. The absence of nodes and azeotropes for this nearly ideal chemical system shows that the designer can obtain either the lightest or the heaviest species as pure products for any feed concentration. This is not true when there are azeotropes. (This topic is covered in detail in Chapter 8.)

II. (Optional) Try a more complicated chemical system such as methanol-ethanol-water at 1 atm (use NRTL-2 for VLE). Look at the residue curve map. Expand the size of the map and look for the ethanol-water azeotrope (where one of the curves intersects the side of the triangle running from ethanol to water). Try a distillation with this system.

Feel free to further explore Aspen Plus on your own.

Lab 5. In this lab we continue to use RADFRAC to explore distillation in more detail and to learn more about the capabilities of Aspen Plus. For parts I, II, III, and IV use the following feed: 100 kmol/h at 20°C. Mole fractions of components are: ethane 0.091, ethylene 0.034, propane, 0.322, propylene 0.064, n-butane 0.413, n-pentane 0.057, n-hexane 0.019. Use Peng-Robinson for VLE. Pressure of feed should be 0.1 atm above that of column. The column has a total condenser and a partial reboiler. There is a single feed, a liquid distillate and a liquid bottoms. Draw the flowsheet and input all data. Then save the file. For part I record the mole fractions of butane in distillate and propane in bottoms. Record the temperatures, and the K values for butane and propane in reboiler and condenser.

I. Pressure effects—temperatures.
   a. Suppose we want to split between the C3 and the C4 components. Set D = 51.1 kmoles/hr (Why is this value selected?), p = 1 atm., N = 30 (includes reboiler and condenser), Nfeed = 15 (on stage), L/D = 2. Report distillate and bottoms purities and condenser and reboiler temperatures.
   b. Repeat run Ia but at p = 10 atm. Same report as part a.
   c. Since cooling water is much cheaper than refrigeration, we want to operate with the condenser at a temperature that is high enough that cooling water can be used. Assume the minimum temperature is 30°C (25°C cooling water plus 5°C approach in heat exchanger). Have we satisfied this for either run? If not, what pressure is necessary? Try 20 atmospheres and go down.

Reduction in pressure is used if the reboiler temperature would be too high or if there is excessive thermal degradation.

II. Pressure effects—Changing split. Suppose we want to split between the C2 and the C3 compounds (ethane and ethylene are in distillate and everything else in bottoms). Change the value for distillate flow rate in Aspen Plus to achieve this. Operate at pressure of 15 atmospheres. Use the same settings as previously. Run Aspen Plus. Compare the temperature in the condenser to the temperature in run I.c. (also at 15 atm). What can you conclude? Report distillate and bottoms mole fractions and temperatures.

III. Pressure effects—column diameter. Aspen Plus will size the column diameter and set-up the tray dimensions. Click on Data (toolbar) and then Blocks. In the Data Browser open up the block that is your distillation column (click on the box with the + in it). Then click on the file "Tray Sizing." Click on New (or Edit if you have previously set this up) and OK for section 1. Section 1 can be the entire system—stages 2 to 29 (remember that Aspen Plus calls the condenser 1 and the reboiler 30). You want 1 pass (the default) and for tray type use the menu to specify "sieve." For tray spacing 2 feet (or the default) is OK. Use the default value for hole area/tray area. Then click on the tab for Design. Fractional approach to flooding should be in range 0.75 to 0.85—use 0.75. For minimum downcomer area, the default value of 0.1 is good for larger columns. Use 0.15 here. The default values for foaming and overdesign (1) are both fine. Use the menu to select the "Fair" method for flooding design (this is the procedure used in Chapter 10).

Now run the design (for Ia, D = 51.1) at pressures of 0.25, 1.0, 4.0 and 16.0 atm with constant feed pressure 16.1 atm. If convergence is a problem in the Column Block Configuration tab, change Convergence from Standard to Petroleum/Wide-boiling. Look at the column diameter and other new information now included in the results. What is the effect of increasing column pressure on column diameter? This result is explained in Chapter 10. Note: Do Part IV at the same time as Part III. Report the largest column diameter for each pressure.

Save the file. Feel free to explore the other possible alternatives in the Data Browser when the distillation block is open. If you have time, look at Convergence and Tray Rating. Downcomer backup (under profiles in tray rating) is useful. We will look at Efficiency in section VI below.

IV. Pressure effects—VLE changes and changes in separation.

Using the K-value results in Aspen Plus, calculate and report the relative volatility of propane-butane (Kpropane/Kbutane) in both reboiler and condenser at pressures of 0.25, 1.0, 4.0 and 16.0 atmospheres for the runs in Ia. At the condenser, what is the trend of relative volatility of propane-butane as the pressure is raised? Is the general trend similar in the reboiler? Do the trend results agree with the predictions from the DePriester charts?

V. Pressure effects—Azeotropes. Switch the system to isopropanol and water. Use NRTL or NRTL-2 as the VLE package. We want to look at the analysis at different pressures. To do this, you need to set up a column (dimensions and so forth are arbitrary). Run the simulation so that Aspen Plus will let you use Analysis. Look at the T-y,x and y,x diagrams at p = 1.0 atm, p = 10.0 atm, and p = 0.1 atm. Notice how the concentration of the azeotrope shifts. (In the Binary Analysis Results Table the azeotrope occurs when K_{i-p} = 1.000. Record the azeotrope mole fractions). This shift may be large enough to develop a process to separate azeotropic mixtures (see Chapter 8).

VI. Tray Efficiencies. Switch to the ethanol-water system using NRTL as the VLE package. Make the feed 60 mole% water and 40 mol% ethanol, 100 kmol/h, saturated liquid at 1 atm. Column has a total condenser and a partial reboiler. Set N = 15, Nfeed = 10, on stage. D = 44, L/D = 2. Column pressure is constant at 1.0 atm.

a. Run the system. Report the liquid compositions on stages 1 and 15 (distillate and bottoms). This run is equivalent to efficiencies of 1.0.

b. Go to the Data Browser and click on the block for your distillation column. Click on Efficiency. Click on Murphree efficiency and Specify stage efficiencies. Then click on tab for vapor-liquid. For stage 1, specify an efficiency of 1 (although for a total condenser this does not matter). Click on Enter. Then for stage 2 specify a Murphree vapor efficiency of 0.6. Click on Enter. Continue for stages 3 to 14, specifing a Murphree vapor efficiency of 0.6. Click on Enter. For stage 15 (partial reboiler) specify efficiency of 1.0. Click on the Next button and run the simulation. Compare liquid compositions on stages 1 and 15 with the run in part VI.a. Explain the effect of the lower efficiency (the effect will be larger if there is no pinch point).

Note: To save time, the runs in Lab 5 were not at the optimum feed plate. Operating at the optimum feed plate should not change any of the general trends. *Practice finding the optimum feed plate for at least one of these problems.* A general procedure to do this is: For your initial value of N, find the optimum feed location by trial and error. Then, reduce or increase the total number of contacts N to just reach the desired specifications. While you do this, choose the feed stage by noting that the ratio N_{feed}/N is approximately constant. Once you have an N that just gives the desired recoveries, redo the optimization of the feed location (your value should be reasonably close). This practice will be very helpful for Lab 6, which requires a lab report.

Lab 6. Use RADFRAC with an appropriate VLE package for this assignment. Use the direct sequence in Figure 11-9A. Do the overall mass balances for both columns to determine both distillate values before lab.

We are separating 1000 kmol/h of a feed containing propane, n-butane, and n-pentane. The feed pressure is 4.0 atm. This feed is 22.4 mol% propane, 44.7 mol% n-butane, and the remainder n-pentane. In the *overall* process we plan to recover 99.6% of the propane in the propane product, 99% of the n-butane in the n-butane product, and 99.7% of the n-pentane in the n-pentane product. In column 1 recover 99.5% of the n-butane in the bottoms product. For purposes of your initial mass balances, assume: 1) There is no n-pentane in the propane product stream, and 2) There is no propane in the n-pentane product stream. Check these guesses after you have run the simulations. Both columns operate at 4.0 atm. Operate each column at 1.15 L/D minimum. Use the optimum feed stage for each column. Use total condensers and reflux should be returned as a saturated liquid. Both columns have partial reboilers. Feeds are saturated liquids.

Use general metric units. You can choose to either simulate the two connected columns or simulate them one at a time. One advantage of doing one at a time is you produce a lot less results to wade through. Since the bottoms product from the first column keeps changing, the second column does not have the correct feed until you select the final design for the first column. Once you have the first column finished, rerunning it every time as you optimize the second column gives no additional information. But, you may prefer to work at simultaneously optimizing both columns. Thus, the procedure is up to you; however, the instructions are written assuming you will do one column at a time.

Simulate the first column. Find the minimum L/D, and the actual L/D. Then find the optimum feed stage and the number of stages that just gives the desired separation. You may choose to do this first with DSTWU (a short cut program similar to the Fenske-Underwood-Gilliland approach). Set N to some arbitrary number (e.g., 40) to find the *estimate* of min L/D. Multiply this minimum by 1.15 and use this L/D as the input to find *estimates* for N and the feed location. These numbers are then used as *first guesses* for RADFRAC. With RADFRAC, first find the actual $(L/D)_{min}$ (set N = 100 or some other large number, feed stage at $\frac{1}{2}$ this value, and vary L/D until obtain desired separation). Then do RADFRAC at 1.15 times $(L/D)_{min}$. Even though the DSTWU estimate for $(L/D)_{min}$ is not accurate, the estimates for N and feed location at L/D = 1.15 $(L/D)_{min}$ may be accurate. Alternatively, you can decide to bypass the use of DSTWU and either do the Fenske-Underwood-Gilliland approach by hand or start RADFRAC totally with guesses. The best specifications for RADFRAC are to specify the values of L/D and D. Note: You must finish $(L/D)_{min}$ calculation with RADFRAC.

Find the optimum feed location for column 1 by trial and error (see the Note at the end of Lab 5).

For column 2, use the bottoms from the first column as the feed to the second column. Repeat the procedure using DSTWU, a hand calculation with the Fenske-Underwood-Gilliland approach, a McCabe-Thiele diagram, or guesses to find estimated values of: $(L/D)_{min}$, L/D, optimum feed stage, and total number of stages. Then do exact RAD-FRAC calculations to find accurate values for $(L/D)_{min}$, L/D, optimum feed stage and N that just give desired recoveries.

Once the optimum columns have been designed, do one more RADFRAC run to determine the column diameters at 80% of flood with a tray spacing of 2.5 feet (see section III in Lab 5). Use Fair's method to calculate flooding.

If you haven't already done so, connect the columns and do a run with everything connected. This run checks to make sure you did not inadvertently change conditions when going from the bottoms of the first column to the feed of the second.

Report the number of stages and the optimum feed stages in each column, the reflux ratios, the temperatures at the top and bottom of each column (in K), the heat duties in the reboilers and condensers (kJ/s), the compositions (in mole fractions) and flow rates of the three products and of the interconnecting stream (kmol/h), the column diameters, and any other information you consider to be relevant. If you use DSTWU or the Fenske-Underwood-Gilliland approach or a McCabe-Thiele diagram for the second column, compare these results with RADFRAC.

CHAPTER 7

Approximate Shortcut Methods for Multicomponent Distillation

The previous chapters served as an introduction to multicomponent distillation. Matrix methods are efficient, but they still require a fair amount of time even on a fast computer. In addition, they are simulation methods and require a known number of stages and a specified feed plate location. Fairly rapid approximate methods are required for preliminary economic estimates, for recycle calculations where the distillation is only a small portion of the entire system, for calculations for control systems, and as a first estimate for more detailed simulation calculations.

In this chapter we first develop the Fenske equation, which allows calculation of multicomponent separation at total reflux. Then we switch to the Underwood equations, which allow us to calculate the minimum reflux ratio. To predict the approximate number of equilibrium stages we then use the empirical Gilliland correlation that relates the actual number of stages to the number of stages at total reflux, the minimum reflux ratio, and the actual reflux ratio. The feed location can also be approximated from the empirical correlation.

7.1 TOTAL REFLUX: FENSKE EQUATION

Fenske (1932) derived a rigorous solution for binary and multicomponent distillation at total reflux. The derivation assumes that the stages are equilibrium stages.

Consider the multicomponent distillation column operating at total reflux shown in Figure 7-1, which has a total condenser and a partial reboiler. For an equilibrium partial reboiler for any two components A and B,

$$\left[\frac{y_A}{y_B}\right]_R = \alpha_R \left[\frac{x_A}{x_B}\right]_R \tag{7-1}$$

Equation (7-1) is just the definition of the relative volatility applied to the reboiler. Material balances for these components around the reboiler are

$$V_R y_{A,R} = L_N x_{A,N} - B x_{A,R}$$ (7-2a)

and

$$V_R y_{B,R} = L_N x_{B,N} - B x_{B,R}$$ (7-2b)

However, at total reflux, $B = 0$ and $L_N = V_R$. Thus the mass balances become

$$y_{A,R} = x_{A,N}, \qquad y_{B,R} = x_{B,N} \qquad \text{(at total reflux)}$$ (7-3)

For a binary system this naturally means that the operating line is the $y = x$ line. Combining Eqs. (7-1) and (7-3),

$$\left[\frac{x_A}{x_B} \right]_N = \alpha_R \left[\frac{x_A}{x_B} \right]_R$$ (7-4)

If we now move up the column to stage N, the equilibrium equation is

$$\left[\frac{y_A}{y_B} \right]_N = \alpha_N \left[\frac{x_A}{x_B} \right]_N$$

The mass balances around stage N simplify to

$$y_{A,N} = x_{A,N-1} \qquad \text{and} \qquad y_{B,N} = x_{B,N-1}$$

Combining these equations, we have

$$\left[\frac{x_A}{x_B} \right]_{N-1} = \alpha_N \left[\frac{x_A}{x_B} \right]_N$$ (7-5)

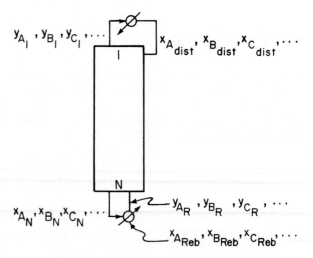

FIGURE 7-1.   *Total reflux column*

Then Eqs. (7-4) and (7-5) can be combined to give

$$\left[\frac{x_A}{x_B}\right]_{N-1} = \alpha_N \alpha_R \left[\frac{x_A}{x_B}\right]_R \tag{7-6}$$

which relates the ratio of liquid mole fractions leaving stage N−1 to the ratio in the reboiler. Repeating this procedure for stage N−1, we obtain

$$\left[\frac{x_A}{x_B}\right]_{N-2} = \alpha_{N-1} \alpha_N \alpha_R \left[\frac{x_A}{x_B}\right]_R \tag{7-7}$$

We can alternate between the operating and equilibrium equations until we reach the top stage. The result is

$$\left[\frac{x_A}{x_B}\right]_{dist} = \alpha_1 \alpha_2 \alpha_3 \cdots \alpha_{N-1} \alpha_N \alpha_R \left[\frac{x_A}{x_B}\right]_R \tag{7-8}$$

If we define α_{AB} as the geometric average relative volatility,

$$\alpha_{AB} = [\alpha_1 \alpha_2 \alpha_3 \cdots \alpha_{N-1} \alpha_N \alpha_R]^{1/N_{min}} \tag{7-9}$$

Eq. (7-8) becomes

$$\left[\frac{x_A}{x_B}\right]_{dist} = \alpha_{AB}^{N_{min}} \left[\frac{x_A}{x_B}\right]_R \tag{7-10}$$

Solving Eq. (7-10) for N_{min}, we obtain

$$N_{min} = \frac{\ln[\,(\frac{x_A}{x_B})_{dist} / (\frac{x_A}{x_B})_R\,]}{\ln \alpha_{AB}} \tag{7-11}$$

which is one form of the Fenske equation. N_{min} is the number of equilibrium contacts including the partial reboiler required at total reflux. If the relative volatility is constant, Eq. (7-11) is exact.

An alternative form of the Fenske equation that is very convenient for multicomponent calculations is easily derived. Equation (7-11) can also be written as

$$N_{min} = \frac{\ln[\frac{(Dx_A/Dx_B)_{dist}}{(Bx_A/Bx_B)_R}]}{\ln \alpha_{AB}} \tag{7-12}$$

$(Dx_A)_{dist}$ is equal to the fractional recovery of A in the distillate multiplied by the amount of A in the feed.

$$(Dx_A)_{dist} = (FR_A)_{dist} Fz_A \tag{7-13}$$

where $(FR_A)_{dist}$ is the fractional recovery of A in the distillate. From the definition of fractional recovery,

$$(Bx_A)_R = [1 - (FR_A)_{dist}] Fz_A \qquad (7\text{-}14)$$

Substituting Eqs. (7-13) and (7-14) and the corresponding equations for component B into Eq. (7-12) gives

$$N_{min} = \frac{\ln\left\{\dfrac{(FR_A)_{dist}(FR_B)_{bot}}{[1 - (FR_A)_{dist}][1 - (FR_B)_{bot}]}\right\}}{\ln\alpha_{AB}} \qquad (7\text{-}15)$$

Note that in this form of the Fenske equation, $(FR_A)_{dist}$ is the fractional recovery of A in the distillate, while $(FR_B)_{bot}$ is the fractional recovery of B in the bottoms. Equation (7-15) is in a convenient form for multicomponent systems.

The derivation up to this point has been for any number of components. If we now restrict ourselves to a binary system where $x_B = 1 - x_A$, Eq. (7-11) becomes

$$N_{min} = \frac{\ln\left(\dfrac{[x/(1-x)]_{dist}}{[x/(1-x)]_{bot}}\right)}{\ln\alpha_{AB}} \qquad (7\text{-}16)$$

where $x = x_A$ is the mole fraction of the more volatile component. The use of the Fenske equation for binary systems is quite straightforward. With distillate and bottoms mole fractions of the more volatile component specified, N_{min} is easily calculated if α_{AB} is known. If the relative volatility is not constant, α_{AB} can be estimated from a geometric average as shown in Eq. (7-9). This can be estimated for a first trial as

$$\alpha_{avg} = (\alpha_1\alpha_R)^{1/2}$$

where α_R is determined from the bottoms composition and α_1 from the distillate composition.

For multicomponent systems calculation with the Fenske equation is straightforward if fractional recoveries of the two keys, A and B, are specified. Equation (7-15) can now be used directly to find N_{min}. The relative volatility can be approximated by a geometric average. Once N_{min} is known, the fractional recoveries of the non-keys (NK) can be found by writing Eq. (7-15) for an NK component, C, and either key component. Then solve for $(FR_C)_{dist}$ or $(FR_C)_{bot}$. When this is done, Eq. (7-15) becomes

$$(FR_C)_{dist} = \frac{\alpha_{CB}^{N_{min}}}{\dfrac{(FR_B)_{bot}}{1 - (FR_B)_{bot}} + \alpha_{CB}^{N_{min}}} \qquad (7\text{-}17)$$

If two mole fractions are specified, say $x_{LK,bot}$ and $x_{HK,dist}$, the multicomponent calculation is more difficult. We can't use the Fenske equation directly, but several alternatives are possible. If we can assume that all NKs are nondistributing, we have

Chapter 7   Approximate Shortcut Methods for Multicomponent Distillation

$$Dx_{LNK,dist} = Fz_{LNK}, \qquad x_{LNK,bot} = 0 \qquad\qquad \textbf{(7-18a)}$$

$$Bx_{HNK,bot} = Fz_{HNK}, \qquad x_{HNK,dist} = 0 \qquad\qquad \textbf{(7-18b)}$$

As shown in Chapter 5, Eqs. (7-18) can be solved along with the light key (LK) and heavy key (HK) mass balances and the equations

$$\sum_{i=1}^{c} (Dx_{i,dist}) = D \text{ and } \sum_{i=1}^{c} (Bx_{i,bot}) = B \qquad\qquad \textbf{(7-19)}$$

Once all distillate and bottoms compositions or values for $Dx_{i,dist}$ and $Bx_{i,bot}$ have been found, Eqs. (7-11) or (7-12) can be used to find N_{min}. Use the key components for this calculation. The assumption of nondistribution of the NKs can be checked with Eq. (7-10) or (7-17). If the original assumption is invalid, the calculated value of N_{min} obtained for key compositions can be used to calculate the LNK and HNK compositions in distillate and bottoms. Then Eq. (7-11) or (7-12) is used again.

If NKs do distribute, a reasonable first guess for the distribution is required. This guess can be obtained by assuming that the distribution of NKs is the same at total reflux as it is at minimum reflux. The distribution at minimum reflux can be obtained from the Underwood equation and is covered later.

Accurate use of the Fenske equation obviously requires an accurate value for the relative volatility. Smith (1963) covers in detail a method of calculating α by estimating temperatures and calculating the geometric average relative volatility. Winn (1958) developed a modification of the Fenske equation that allows the relative volatility to vary. Wankat and Hubert (1979) modified both the Fenske and Winn equations for nonequilibrium stages by including a vaporization efficiency.

EXAMPLE 7-1. Fenske equation

A distillation column with a partial reboiler and a total condenser is being used to separate a mixture of benzene, toluene, and cumene. The feed is 40 mol% benzene, 30 mol% toluene, and 30 mol% cumene and is input as a saturated vapor. We desire 95% recovery of the toluene in the distillate and 95% recovery of the cumene in the bottoms. The reflux is returned as a saturated liquid, and constant molal overflow (CMO) can be assumed. Pressure is 1 atm.

Equilibrium can be represented as constant relative volatilities. Choosing toluene as the reference component, $\alpha_{benz-tol} = 2.25$ and $\alpha_{cumene-tol} = 0.21$. Find the number of equilibrium stages required at total reflux and the recovery fraction of benzene in the distillate.

Solution

A. Define. The problem is sketched below. For A = toluene (LK), B = cumene (HK), C = benzene (LNK), we have $\alpha_{CA} = 2.25$, $\alpha_{AA} = 1.0$, $\alpha_{BA} = 0.21$, $z_A = 0.3$, $z_B = 0.3$, $z_C = 0.4$, $FR_{A,dist} = 0.95$, and $FR_{B,bot} = 0.95$.

a. Find N at total reflux.

b. Find $FR_{C,dist}$ at total reflux.

B. Explore. Since operation is at total reflux and relative volatilities are constant, we can use the Fenske equation.

C. Plan. Calculate N_{min} from Eq. (7-15), and then calculate $FR_{C,dist}$ from Eq. (7-17).

D. Do It. Equation (7-15) gives

$$N_{min} = \frac{\ln\left[\left(\dfrac{FR_{A,dist}}{1-FR_{A,dist}}\right)/\left(\dfrac{1-FR_{B,bot}}{FR_{B,bot}}\right)\right]}{\ln\alpha_{AB}} = \frac{\ln\left[\dfrac{0.95}{0.05}/\dfrac{0.05}{0.95}\right]}{\ln(1/0.21)} = 3.77$$

Note that $\alpha_{AB} = \alpha_{tol-cumene} = 1/\alpha_{BA} = 1/\alpha_{cumene-tol}$. Equation (7-17) gives

$$FR_{C,dist} = \frac{(\alpha_{CB})^{N_{min}}}{\dfrac{FR_{B,bot}}{1-FR_{B,bot}} + (\alpha_{CB})^{N_{min}}} = \frac{(2.25/0.21)^{3.77}}{\dfrac{0.95}{0.05} + \left(\dfrac{2.25}{0.21}\right)^{3.77}} = 0.998$$

which is the desired benzene recovery in the distillate. Note that

$$\alpha_{CB} = \frac{K_C}{K_B} = \frac{K_C/K_A}{K_B/K_A} = \frac{\alpha_{CA}}{\alpha_{BA}} = \frac{\alpha_{benz-tol}}{\alpha_{cumene-tol}}$$

E. Check. The results can be checked by calculating $FR_{C,dist}$ using component A instead of B. The same answer is obtained.

F. Generalize. We could continue this problem by calculating $Dx_{i,dist}$ and $Bx_{i,bot}$ for each component from Eqs. (7-13) and (7-14). Then distillate and bottoms flow rates can be found from Eqs. (7-19), and the distillate and bottoms compositions can be calculated.

7.2   MINIMUM REFLUX: UNDERWOOD EQUATIONS

For binary systems, the pinch point usually occurs at the feed plate. When this occurs, an analytical solution for the limiting flows can be derived (King, 1980) that is also valid for multicomponent systems as long as the pinch point occurs at the feed stage. Unfortunately, for multicomponent systems there will be separate pinch points in both the stripping and enriching sections if there are nondistributing components. In this case an alternative analysis procedure developed by Underwood (1948) is used to find the minimum reflux ratio.

The development of the Underwood equations is quite complex and is presented in detail by Underwood (1948), Smith (1963), and King (1980). Since for most practicing engineers the details of the development are not as important as the use of the Underwood equations, we will follow the approximate derivation of Thompson (1980). Thus we will outline the important points but wave our hands about the mathematical details of the derivation.

If there are nondistributing HNKs present, a "pinch point" of constant composition will occur at minimum reflux in the enriching section above where the HNKs are fractionated out. With nondistributing LNKs present, a pinch point will occur in the stripping section. For the enriching section in Figure 7-2, the mass balance for component i is

$$V_{min}\, y_{i,j+1} = L_{min} x_{i,j} + D x_{i,dist} \tag{7-20}$$

At the pinch point, where compositions are constant,

$$x_{i,j-1} = x_{i,j} = x_{i,j+1}, \qquad \text{and} \qquad y_{i,j-1} = y_{i,j} = y_{i,j+1} \tag{7-21}$$

The equilibrium expression can be written in terms of K values as

$$y_{i,j+1} = K_i\, x_{i,j+1} \tag{7-22}$$

Combining Eqs. (7-20) to (7-22) we obtain a simplified balance valid in the region of constant compositions.

$$V_{min} y_{i,j+1} = \frac{L_{min}}{K_i} y_{i,j+1} + D x_{i,dist} \tag{7-23}$$

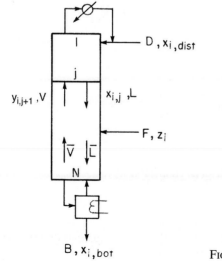

$$B, x_{i,bot}$$

FIGURE 7-2.   *Distillation column*

Defining the relative volatility $\alpha_i = K_i/K_{ref}$ and combining terms in Eq. (7-23),

$$V_{min}y_{i,j+1}\left(1 - \frac{L_{min}}{V_{min}\alpha_{i-ref}K_{ref}}\right) = Dx_{i,dist} \tag{7-24}$$

Solving for the component vapor flow rate, $V_{Min}\,y_{i,j+1}$, and rearranging

$$V_{min}y_{i,j+1} = \frac{\alpha_{i-ref}\,Dx_{ti,dis}}{\alpha_{i-ref} - \dfrac{L_{min}}{V_{min}K_{ref}}} \tag{7-25}$$

Equation (7-25) can be summed over all components to give the total vapor flow rate in the enriching section at minimum reflux.

$$V_{min} = \sum_{i=1}^{c}\left(V_{min}y_{i,j+1}\right) = \sum_{i=1}^{c}\left(\frac{\alpha_{i-ref}Dx_{ti,dis}}{\alpha_{i-ref} - \dfrac{L_{min}}{V_{min}K_{ref}}}\right) \tag{7-26}$$

In the stripping section a similar analysis can be used to derive,

$$-\overline{V}_{min} = \sum_{i=1}^{c}\left(\frac{\overline{\alpha}_{i-ref}Bx_{i,bot}}{\overline{\alpha}_{i-ref} - \dfrac{\overline{L}_{min}}{\overline{V}_{min}\overline{K}_{ref}}}\right) \tag{7-27}$$

Since the conditions in the stripping section are different than in the rectifying section, in general $\alpha_i \neq \overline{\alpha}_i$ and $K_{ref} \neq \overline{K}_{ref}$.

Underwood (1948) described generalized forms of Eqs. (7-26) and (7-27) which are equivalent to defining

$$\phi = \frac{L_{min}}{V_{min}K_{ref}} \quad \text{and} \quad \overline{\phi} = \frac{\overline{L}_{min}}{\overline{V}_{min}\overline{K}_{ref}} \tag{7-28}$$

Equations (7-26) and (7-27) then become polynomials in ϕ and $\overline{\phi}$ and have C roots. The equations are now

$$V_{min} = \sum_{i=1}^{c}\frac{\alpha_{i-ref}(Dx_{i,dist})}{\alpha_{i-ref} - \phi} \tag{7-29}$$

and

$$-\overline{V}_{min} = \sum_{i=1}^{c}\frac{\overline{\alpha}_{i-ref}(Bx_{i,bot})}{\overline{\alpha}_{i-ref} - \overline{\phi}} \tag{7-30}$$

If we assume CMO and constant relative volatilities $\alpha_{i-ref} = \overline{\alpha}_{i-ref}$, Underwood showed there are common values of ϕ and $\overline{\phi}$ which satisfy both equations. Equations (7-29) and (7-30) can now be added. Thus, at minimum reflux

Chapter 7 Approximate Shortcut Methods for Multicomponent Distillation

$$V_{min} - \overline{V}_{min} = \sum_{i=1}^{c} \left[\frac{\alpha_{i-ref}(Dx_{i,dist})}{\alpha_{i-ref} - \phi} + \frac{\alpha_{i-ref}(Bx_{i,bot})}{\alpha_{i-ref} - \phi} \right] \tag{7-31}$$

where α is now an average volatility.

Eq. (7-31) is easily simplified with the overall column mass balance

$$Fz_i = Dx_{i,dist} + Bx_{i,bot} \tag{7-32}$$

to

$$\Delta V_{feed} = V_{min} - \overline{V}_{min} = \sum_{i=1}^{c} \frac{\alpha_{i-ref} Fz_i}{\alpha_{i-ref} - \phi} \tag{7-33}$$

ΔV_{feed} is the change in vapor flow rate at the feed stage. If q is known

$$\Delta V_{feed} = F(1 - q) \tag{7-34}$$

If the feed temperature is specified a flash calculation on the feed can be used to determine ΔV_{feed}.

Equation (7-33) is known as the first Underwood equation. It can be used to calculate appropriate values of ϕ. Equation (7-29) is known as the second Underwood equation and is used to calculate V_{min}. Once V_{min} is known, L_{min} is calculated from the mass balance

$$L_{min} = V_{min} - D \tag{7-35}$$

The exact method for using the Underwood equation depends on what can be assumed. Three cases will be considered.

Case A. Assume all NKs do *not* distribute. In this case the amounts of NKs in the distillate are:

$$Dx_{HNK,dist} = 0 \qquad \text{and} \qquad Dx_{LNK,dist} = Fz_{LNK}$$

while the amounts of the keys are:

$$Dx_{LK,dist} = (FR_{LK})_{dist} Fz_{LK} \tag{7-36}$$

$$Dx_{HK,dist} = [1-(FR_{HK})_{bot}] Fz_{HK} \tag{7-37}$$

Equation (7-33) can now be solved for the one value of ϕ between the relative volatilities of the two keys, $\alpha_{HK-ref} < \phi < \alpha_{LK-ref}$. This value of ϕ can be substituted into Eq. (7-29) to immediately calculate V_{min}. Then

$$D = \sum_{i=1}^{c} (Dx_{i,dist}) \tag{7-38}$$

And L_{min} is found from mass balance Eq. (7-35).

This assumption of nondistributing NKs will probably not be valid for sloppy separations or when a sandwich component is present. In addition, with a sandwich component there are two ϕ values between α_{HK-ref} and α_{LK-ref}. Thus use Case C (discussed later) for sandwich components. The method of Shiras et al. (1950) can be used to check for distribution of NKs.

Case B. Assume that the distributions of NKs determined from the Fenske equation at total reflux are also valid at minimum reflux. In this case the $Dx_{NK,dist}$ values are obtained from the

Fenske equation as described earlier. Again solve Eq. (7-33) for the ϕ value between the relative volatilities of the two keys. This ϕ, the Fenske values of $Dx_{NK,dist}$, and the $Dx_{LK,dist}$ and $Dx_{HK,dist}$ values obtained from Eqs. (7-36) and (7-37) are used in Eq. (7-29) to find V_{min}. Then Eqs. (7-38) and (7-35) are used to calculate D and L_{min}. This procedure is illustrated in Example 7-2.

Case C. Exact solution without further assumptions. Equation (7-33) is a polynomial with C roots. Solve this equation for all values of ϕ lying between the relative volatilities of all components,

$$\alpha_{LNK,1-ref} < \phi_1 < \alpha_{LNK,2-ref} < \phi_2 < \alpha_{LK-ref} < \phi_3 < \alpha_{HK-ref} < \phi_4 < \alpha_{HNK,1-ref}$$

This gives C-1 valid roots. Now write Eq. (7-29) C-1 times; once for each value of ϕ. We now have C-1 equations and C-1 unknowns (V_{min} and $Dx_{i,dist}$ for all LNK and HNK). Solve these simultaneous equations and then obtain D from Eq. (7-38) and L_{min} from Eq. (7-35). A sandwich component problem that must use this approach is given in Problem 7.D15.

In general, Eq. (7-33) will be of order C in ϕ where C is the number of components. Saturated liquid and saturated vapor feeds are special cases and, after simplification, are of order C-1. If the resulting equation is quadratic, the quadratic formula can be used to find the roots. Otherwise, a root-finding method should be employed. If only one root, $\alpha_{LK-ref} > \phi > \alpha_{HK-ref}$, is desired, a good first guess is to assume $\phi = (\alpha_{LK-ref} + \alpha_{HK-ref})/2$.

The results of the Underwood equations will only be accurate if the basic assumption of constant relative volatility and CMO are valid. For small variations in α a geometric average calculated as

$$\alpha_{i-ref} = (\alpha_{bot}\,\alpha_{dist})^{1/2} \text{ or } \alpha_{i-ref} = (\alpha_{bot}\,\alpha_{feed}\,\alpha_{dist})^{1/3} \tag{7-39}$$

can be used as an approximation. Application of the Underwood equations to systems with multiple feeds was studied by Barnes et al. (1972).

EXAMPLE 7-2. Underwood equations
For the distillation problem given in Example 7-1 find the minimum reflux ratio. Use a basis of 100 kmol/h of feed.

Solution
A. Define. The problem was sketched in Example 7-1. We now wish to find $(L/D)_{min}$.
B. Explore. Since the relative volatilities are approximately constant, the Underwood equations can easily be used to estimate the minimum reflux ratio.
C. Plan. This problem fits into Case A or Case B. We can calculate $Dx_{i,dist}$ values as described in Cases A or B, Eqs. (7-36) and (7-37), and solve Eq. (7-33) for ϕ where ϕ lies between the relative volatilities of the two keys $0.21 < \phi < 1.00$. Then V_{min} can be found from Eq. (7-29), D from Eq. (7-38) and L_{min} from Eq. (7-35).
D. Do It. Follow Case B analysis. Since the feed is a saturated vapor, q = 0 and $\Delta V_{feed} = F(1 - q) = F = 100$ and Eq. (7-33) becomes

$$100 = \frac{(2.25)(40)}{2.25 - \phi} + \frac{(1.0)(30)}{1.0 - \phi} + \frac{(0.21)(30)}{0.21 - \phi}$$

Solving for ϕ between 0.21 and 1.00, we obtain $\phi = 0.5454$. Equation (7-29) is

$$V_{min} = \sum_{i=1}^{c} \left(\frac{\alpha_{i-ref}(Dx_{i,dist})}{\alpha_{i-ref} - \phi} \right)$$

where

$$Dx_{i,dist} = F z_i (FR)_{i,dist}$$

For benzene this is

$$Dx_{ben,dist} = 100(0.4)(0.998) = 39.92$$

where the fractional recovery of benzene is the value calculated in Example 7-1 at total reflux. The other distillate values are

$$Dx_{tol,dist} = 100(0.3)(0.95) = 28.5 \text{ and } Dx_{cum,dist} = 100(0.3)(0.05) = 1.5$$

Summing the three distillate flows, $D = 69.92$. Equation (7-29) becomes

$$V_{min} = \frac{(2.25)(39.92)}{2.25 - 0.5454} + \frac{(1.0)(28.5)}{1.0 - 0.5454} + \frac{(0.21)(1.5)}{0.21 - 0.5454} = 114.4$$

From a mass balance, $L_{min} = V_{min} - D = 44.48$, and $(L/D)_{min} = 0.636$.

E. Check. The Case A calculation gives essentially the same result.

F. Generalize. The addition of more components does not make the calculation more difficult as long as the fractional recoveries can be accurately estimated. The value of ϕ must be accurately determined since it can have a major effect on the calculation. Since the separation is easy, $(L/D)_{min}$ is quite small in this case. $(L/D)_{min}$ will not be as dependent on the exact values of ϕ as it is when $(L/D)_{min}$ is large.

7.3   GILLILAND CORRELATION FOR NUMBER OF STAGES AT FINITE REFLUX RATIO

A general shortcut method for determining the number of stages required for a multicomponent distillation at finite reflux ratios would be extremely useful. Unfortunately, such a method has not been developed. However, Gilliland (1940) noted that he could empirically relate the number of stages N at finite reflux ratio L/D to the minimum number of stages N_{min} and the minimum reflux ratio $(L/D)_{min}$. Gilliland did a series of accurate stage-by-stage calculations and found that he could correlate the function $(N-N_{min})/(N + 1)$ with the function $[L/D - (L/D)_{min}]/(L/D + 1)$. This correlation as modified by Liddle (1968) is shown in Figure 7-3. The data points are the results of Gilliland's stage-by-stage calculations and show the scatter inherent in this correlation.

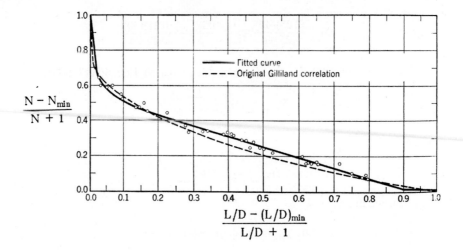

FIGURE 7-3. *Gilliland correlation as modified by Liddle (1968); reprinted with permission from* Chemical Engineering, *75(23), 137 (1968), copyright 1968, McGraw-Hill.*

To use the Gilliland correlation we proceed as follows:

1. Calculate N_{min} from the Fenske equation.
2. Calculate $(L/D)_{min}$ from the Underwood equations or analytically for a binary system.
3. Choose actual (L/D). This is usually done as some multiplier (1.05 to 1.5) times $(L/D)_{min}$.
4. Calculate the abscissa.
5. Determine the ordinate value.
6. Calculate the actual number of stages, N.

The Gilliland correlation should only be used for rough estimates. The calculated number of stages can be off by ± 30% although they are usually within ± 7%. Since L/D is usually a multiple of $(L/D)_{min}$, $L/D = M (L/D)_{min}$, the abscissa can be written as

$$abscissa = \frac{M - 1}{[1/(L/D)_{min}] + M}$$

The abscissa is not very sensitive to the $(L/D)_{min}$ value, but does depend on the multiplier M.

The optimum feed plate location can also be estimated. First, use the Fenske equation to estimate where the feed stage would be at total reflux. This can be done by determining the number of stages required to go from the feed concentrations to the distillate concentrations *for the keys.*

$$N_{F,min} = \frac{\ln\left[\left(\frac{x_{LK}}{x_{HK}}\right)_{dist} \middle/ \left(\frac{z_{LK}}{z_{HK}}\right)\right]}{\ln \alpha_{LK-HK}} \tag{7-40a}$$

Now assume that the relative feed location is constant as we change the reflux ratio from total reflux to a finite value. Thus

Chapter 7 *Approximate Shortcut Methods for Multicomponent Distillation*

$$\frac{N_{F,min}}{N_{min}} = \frac{N_F}{N}$$

(7-40b)

The actual feed stage can now be estimated from Eq. (7-40b).

An alternate procedure that is probably a more accurate estimate of the feed stage location is Kirkbride's method (Humphrey and Keller, 1997). The ratio of the number of trays above the feed, $N_f - 1$, to the number below the feed stage, $N - N_f$, can be estimated as,

$$\log\left(\frac{N_f - 1}{N - N_f}\right) = 0.260 \log\left\{\frac{B}{D}\left(\frac{z_{HK}}{z_{LK}}\right)\left[\frac{x_{LK,bot}}{x_{HK,dist}}\right]^2\right\}$$

(7-41)

Since neither procedure is likely to be very accurate, they should only be used as first guesses of the feed location for simulations.

The Gilliland correlation can also be fit to equations. Liddle (1968) fit the Gilliland correlation to three equations. Let $x = [L/D - (L/D)_{min}]/(L/D + 1)$. Then

$$\frac{N - N_{min}}{N + 1} = 1.0 - 18.5715x \qquad \text{for } 0 \le x \le 0.01$$

(7-42a)

while for $0.01 < x < 0.90$

$$\frac{N - N_{min}}{N + 1} = 0.545827 - 0.591422x + \frac{0.002743}{x}$$

(7-42b)

and for $0.90 \le x \le 1.0$

$$\frac{N - N_{min}}{N + 1} = 0.16595 - 0.16595x$$

(7-42c)

For most situations Eq. (7-42b) is appropriate. The fit to the data is shown in Figure 7-3. Naturally, the equations are useful for computer calculations. Erbar and Maddox (1961) (see King, 1980, or Hines and Maddox, 1985) developed a somewhat more accurate correlation that uses more than one curve.

As a rough rule of thumb we can estimate $N = 2.5 N_{min}$. This estimate then requires *only* a calculation of N_{min} and will be useful for very preliminary estimates.

EXAMPLE 7-3. Gilliland correlation

Estimate the total number of equilibrium stages and the optimum feed plate location required for the distillation problem presented in Examples 7-1 and 7-2 if the actual reflux ratio is set at $L/D = 2$.

Solution

A. Define. The problem was sketched in Examples 7-1 and 7-2. F = 100, L/D = 2, and we wish to estimate N and N_F.
B. Explore. An estimate can be obtained from the Gilliland correlation, while a more exact calculation could be done with a process simulator. We will use the Gilliland correlation.

C. Plan. Calculate the abscissa

$$= \frac{L/D - (L/D)_{min}}{L/D + 1},$$

determine the ordinate

$$= \frac{N - N_{min}}{N + 1}$$

from the Gilliland correlation, and then find N. $(L/D)_{min}$ was found in Example 7-2, and N_{min} in Example 7-1. The feed plate location is estimated from Eqs. (7-41) and (7-40).

D. Do It.

$$\text{abscissa} = \frac{L/D - (L/D)_{min}}{L/D + 1} = \frac{2 - 0.636}{2+1} = 0.455$$

The corresponding ordinate $(N - N_{min})/(N + 1) = 0.27$ using Liddle's curve. Since $N_{min} = 3.77$, $N = 5.53$. From Eq. (7-40a), $N_{F,min}$ is calculated as

$$N_{F,min} = \frac{\ln\left[\left(\frac{x_{LK}}{x_{HK}}\right)_{dist} \middle/ \left(\frac{z_{LK}}{z_{HK}}\right)\right]}{\ln \alpha_{LK-HK}} = \frac{\ln\left[\left(\frac{0.408}{0.021}\right) \middle/ \left(\frac{0.3}{0.3}\right)\right]}{\ln(1/0.21)} = 1.90$$

Where $x_{LK,dist}$ was found from Example 7-2 as

$$x_{LK,dist} = x_{tol,dist} = \frac{Dx_{tol,dist}}{D} = \frac{28.5}{69.92} = 0.408$$

and

$$x_{HK,dist} = x_{cum,dist} = 0.021$$

Then, from Eq. (7-40b),

$$N_F = N \frac{N_{F,min}}{N_{min}} = 5.53(\frac{1.90}{3.77}) = 2.79 \text{ or stage 3.}$$

E. Check. A check of the Gilliland correlation can be obtained from Eq. (7-42b). With x = 0.455 this is

$$\frac{N - N_{min}}{N + 1} = 0.545827 - (0.591422)(0.455) + \frac{0.002743}{0.455} = 0.283$$

or $(1 - 0.283) N = N_{min} + 0.283$, which gives N = 5.65. The 2% difference between these two results gives an idea of the accuracy of Eq. (7-42) in fitting the curve.

A check on the value of N_f can be obtained with Kirkbride's Eq. (7-41). To use this equation we need to know the terms on the RHS. From Example 7-2, F = 100 and D = 69.92. Thus, B = 100 – 69.92 = 30.08. The HK = cumene and the LK = toluene. The feed mole fractions of both are 0.30. From example 7-2:

$$Dx_{cum,dist} = 1.5. \text{ Then } x_{HK,dist} = (Dx_{cum,dist})/D = 1.5/69.92 = 0.02145.$$

$$Dx_{tol,dist} = 28.5. \text{ Then } Bx_{tol,bot} = Fz - Dx_{tol,dist} = 30.0 - 28.5 = 1.5, \text{ and}$$

$$x_{LK,bot} = Bx_{tol,bot}/B = 1.5/30.08 = 0.04987.$$

Then, Eq. (7-41) becomes,

$$\log_{10}\left(\frac{N_f - 1}{N - N_f}\right) = 0.260 \log_{10}\left\{\left(\frac{30.08}{69.92}\right)\left(\frac{0.30}{0.30}\right)\left(\frac{0.04987}{0.02145}\right)^2\right\} = 0.07549$$

and $(N_f - 1)/(N - N_f) = 1.1898$

which gives $N_f = 3.46$ if we use N = 5.53 or $N_f = 3.7$ if we use N = 6. Thus, the best estimate is to use either the 3rd or 4th stage for the feed. This agrees rather well with the previous estimate.

A complete check would require solution with a process simulator.

F. Generalize. The Gilliland correlation is a rapid method for estimating the number of equilibrium stages in a distillation column. It should not be used for final designs because of its inherent inaccuracy.

7.4   SUMMARY – OBJECTIVES

In this chapter we developed approximate shortcut methods for binary and multicomponent distillation. You should be able to satisfy the following objectives:

1. Derive the Fenske equation and use it to determine the number of stages required at total reflux and the splits of NK components
2. Use the Underwood equations to determine the minimum reflux ratio for multicomponent distillation
3. Use the Gilliland correlation to estimate the actual number of stages in a column and the optimum feed stage location

REFERENCES

Barnes, F. J., D. N. Hansen, and C. J. King, "Calculation of Minimum Reflux for Distillation Columns with Multiple Feeds," *Ind. Eng. Chem. Process Des. Develop., 11*, 136 (1972).

Erbar, J. H., and R. N. Maddox, *Petrol. Refin., 40*(5), 183 (1961).

Fenske, M. R., "Fractionation of Straight-Run Pennsylvania Gasoline," *Ind. Eng. Chem., 24*, 482 (1932).

Gilliland, E. R., "Multicomponent Rectification," *Ind. Eng. Chem., 32*, 1220 (1940).

Hines A. L., and R. N. Maddox, *Mass Transfer. Fundamentals and Applications*, Prentice Hall, Englewood Cliffs, New Jersey, 1985.

Humphrey, J. L., and G. E. Keller II, *Separation Process Technology*, McGraw-Hill, New York, 1997.

King, C. J., *Separation Processes*, 2nd ed., McGraw-Hill, New York, 1980.

Liddle, C. J., "Improved Shortcut Method for Distillation Calculations," *Chem. Eng., 75*(23), 137 (Oct. 21, 1968).

Shiras, R. N., D. N. Hansen and C. H. Gibson, "Calculation of Minimum Reflux in Distillation Columns," *Ind. Eng. Chem., 42,* 871 (1950).

Smith, B. D., *Design of Equilibrium Stage Processes*, McGraw-Hill, New York, 1963.

Thompson, R. E., "Shortcut Design Method-Minimum Reflux," *AIChE Modular Instructions*, Series B, Vol. 2, 5 (1981).

Underwood, A. J. V., "Fractional Distillation of Multicomponent Mixtures," *Chem. Eng. Prog., 44,* 603 (1948).

Wankat, P. C., and J. Hubert, "Use of the Vaporization Efficiency in Closed Form Solutions for Separation Columns," *Ind. Eng. Chem. Process Des. Develop., 18,* 394 (1979).

Winn, F. W., *Pet. Refiner, 37,* 216 (1958).

HOMEWORK

A. *Discussion Problems*

A1. The Fenske equation:
   a. Is valid only for binary systems.
   b. Was derived for minimum reflux.
   c. Requires constant molal overflow (CMO).
   d. Requires constant K values.
   e. All of the above.
   f. None of the above.

A2. If you want to use an average relative volatility, how do you calculate it for the Underwood equation?

A3. Develop your key relations chart for this chapter.

A4. In multicomponent distillation the Fenske equation can be used to:
   a. Estimate the fractional recoveries of the nonkeys at total reflux.
   b. Calculate the number of equilibrium contacts at minimum reflux.
   c. Estimate the average K value of the light key at total reflux.
   d. All of the above.
   e. None of the above.

C. *Derivations*

C1. Derive Eq. (7-17). Derive an equation for $(FR_C)_{bot}$ in terms of $(FR_A)_{dist}$.

C2. Derive Eq. (7-34).

C3. If the pinch point occurs at the feed point, mass balances can be used to find the minimum flows. Derive these equations.

C4. The choice of developing the Underwood equations in terms of V_{min} instead of solving for L_{min} is arbitrary. Rederive the Underwood equations solving for L_{min} and $\overline{L}_{min}$. Develop the equations analogous to Eqs. (7-29) and (7-33).

C5. For binary systems, Eq. (7-33) simplifies to a linear equation for both saturated liquid and saturated vapor feeds. Prove this.

D. *Problems*
   Answers to problems with an asterisk are at the back of the book.

D1.* We have 10 kmol/h of a saturated liquid feed that is 40 mol% benzene and 60 mol% toluene. We desire a distillate composition that is 0.992 mole fraction benzene and a bottoms that is 0.986 mole fraction toluene (note units). CMO is valid. Assume constant relative volatility with $\alpha_{BT} = 2.4$. Reflux is returned as a saturated liquid. The column has a partial reboiler and a total condenser.

   a. Use the Fenske equation to determine N_{min}.

   b. Use the Underwood equations to find $(L/D)_{min}$.

   c. For $L/D = 1.1(L/D)_{min}$, use the previous results and the Gilliland correlation to estimate the total number of stages and the optimum feed stage location.

D2. We are separating a mixture of ethane, propane, n-butane, and n-pentane in a distillation column operating at 5.0 atm. The column has a total condenser and a partial reboiler. The feed flow rate is 1000 kmol/h. The feed is a saturated liquid. Feed is 8 mol% ethane, 33 mol% propane, 49 mol% n-butane, and 10 mol% n-pentane.

A 99.7% recovery of propane is desired in the distillate. A 99.8% recovery of n-butane is desired in the bottoms.

   a. Find N_{min} from the Fenske equation (hand calculation).

   b. Find $(L/D)_{min}$ from the Underwood equation (hand calculation).

   c. Use $L/D = 1.15 (L/D)_{min}$ and estimate N and N_{Feed} from Gilliland correlation (hand calculation).

Use DePriester charts. Assume relative volatility is constant at value calculated at the bubble-point temperature of the feed. For bubble-point calculation, choose propane as the reference component. Choose n-butane (the heavy key) as the reference component for other calculations.

D3.* We have designed a special column that acts as exactly three equilibrium stages. Operating at total reflux, we measure vapor composition leaving the top stage and the liquid composition leaving the bottom stage. The column is separating phenol from o-cresol. We measure a phenol liquid mole fraction leaving the bottom stage of 0.36 and a phenol vapor mole fraction leaving the top stage of 0.545. What is the relative volatility of phenol with respect to o-cresol?

D4. We desire to separate 1,2 dichloroethane from 1,1,2 trichloroethane at one atmosphere. We desire 99.15 mol% dichloroethane in the distillate and 1.773% dichloroethane in the bottoms. The feed is a saturated liquid and is 60.0 mol% 1,2 dichloroethane. Assume the relative volatility is approximately constant, $\alpha = 2.4$.

   a. Find the minimum number of stages using the Fenske equation.

   b. Calculate L/D_{min}.

   c. Estimate the actual number of stages for $L/D = 2.2286$ using the Gilliland correlation.

   d. A detailed simulation gave 99.15 mol% dichloroethane in the distillate and 1.773% dichloroethane in the bottoms for $L/D = 2.2286$, N =25 equilibrium contacts, optimum feed location is 16 equilibrium contacts from the top of the column. Compare this N with part c and calculate the % error in the Gilliland prediction.

D5.* A column with 29 equilibrium stages and a partial reboiler is being operated at total reflux to separate a mixture of ethylene dibromide and propylene dibromide. Ethylene dibromide is more volatile, and the relative volatility is constant at a value of 1.30. We are measuring a distillate concentration that is 98.4 mol% ethylene dibromide. The column has a total condenser and saturated liquid reflux, and CMO can be assumed. Use the Fenske equation to predict the bottoms composition.

D6.* We are separating 1,000 mol/h of a 40% benzene, 60% toluene feed in a distillation column with a total condenser and a partial reboiler. Feed is a saturated liquid. CMO is valid. A distillate that is 99.3% benzene and a bottoms that is 1% benzene are desired. Use the Fenske equation to find the number of stages required at total reflux, a McCabe-Thiele diagram to find $(L/D)_{min}$, and the Gilliland correlation to estimate the number of stages required if $L/D = 1.15(L/D)_{min}$. Estimate that the relative volatility is constant at $\alpha_{BT} = 2.4$. Check your results with a McCabe-Thiele diagram.

D7. We are separating a mixture of ethane, propane, n-butane and n-pentane in a distillation column operating at 5.0 atm. The column has a total condenser and a partial reboiler. The feed flow rate is 1000.0 kmol/h. The feed is a saturated liquid. Feed is 8.0 mol% ethane, 33.0 mol% propane, 49.0 mol% n-butane and 10.0 mol% n-pentane. A 98.0% recovery of propane is desired in the distillate. A 99.2% recovery of n-butane is desired in the bottoms. Use the optimum feed stage. Use DePriester chart. Assume relative volatility is constant at value calculated at the bubble point temperature of the feed.

a. Find N_{min} from the Fenske equation.

b. Find $(L/D)_{min}$ from the Underwood equation.

c. Use $L/D = 1.2 (L/D)_{min}$ and estimate N and N_{Feed} from Gilliland correlation.

D8.* We wish to separate a mixture of 40 mol% benzene and 60 mol% ethylene dichloride in a distillation column with a partial reboiler and a total condenser. The feed rate is 750 mol/h, and feed is a saturated vapor. We desire a distillate product of 99.2 mol% benzene and a bottoms product that is 0.5 mol% benzene. Reflux is a saturated liquid, and CMO can be used. Equilibrium data can be approximated with an average relative volatility of 1.11 (benzene is more volatile).

a. Find the minimum external reflux ratio.

b. Use the Fenske equation to find the number of stages required at total reflux.

c. Estimate the total number of stages required for this separation, using the Gilliland correlation for $L/D = 1.2(L/D)_{min}$.

D9. We are separating a mixture of ethanol and water in a distillation column with a total condenser and a partial reboiler. Column is at 1.0 atm. pressure. The feed is 30.0 mol% ethanol. The feed is a two-phase mixture that is 80 % liquid. Feed rate is 100.0 kmol/h. We desire a bottoms concentration of 2.0 mol% ethanol, and a distillate that is 80.0 mol% ethanol.

Find **a.** $(L/D)_{min}$.

b. N_{min}.

c. If $(L/D) = 1.05 (L/D)_{min}$ estimate N and the feed location.

Do this problem by using McCabe-Thiele diagram for parts a and b and the Gilliland correlation for part c. Equilibrium data are given in Table 2-1.

D10. A distillation column is separating ethane, propane, and n-butane at 5 atm. Operation is at total reflux. We want a 98.9% recovery of ethane in the distillate and a 99.8% recovery of n-butane in the bottoms. Propane is a sandwich component (e.g., in between light and heavy keys). F = 100 kmol/h and is 30 mol% ethane, 33 mol% propane and 37 mol% n-butane. Feed is a saturated liquid. Assume relative volatilities are constant, $\alpha_{EB} = 13.14$, and $\alpha_{PB} = 3.91$.

a. At total reflux find N_{min}.

b. At total reflux find the fractional recovery of propane in the distillate.

c. At total reflux find distillate flow rate D.

D11. A distillation column with a total condenser and a partial reboiler is separating a mixture of propane (P), n-butane (B), and n-hexane (H). The feed (a saturated vapor) is 20 mol% propane, 35 mol% n-butane, and 45 mol% n-hexane. Feed rate is F = 100 kmol/h. We desire a 99% recovery of the n-butane in the distillate and a 98% recovery of n-hexane in the bottoms. CMO can be assumed to be valid. If we choose butane as the reference, the average relative volatilities are $\alpha_{PB} = 2.04$, $\alpha_{BB} = 1.0$, $\alpha_{HB} = 0.20$. Calculate the distillate flow rate D assuming all propane is in the distillate, and find the minimum external reflux ratio, $(L/D)_{min}$.

D12.* a. A distillation column with a partial reboiler and a total condenser is being used to separate a mixture of benzene, toluene, and cumene. The feed is 40 mol% benzene, 30 mol% toluene and 30 mol% cumene. The feed is input as a saturated vapor. We desire 99% recovery of the toluene in the bottoms and 98% recovery of the benzene in the distillate. The reflux is returned as a saturated liquid, and CMO can be assumed. Equilibrium can be represented as constant relative volatilities. Choosing toluene as the reference component, $\alpha_{benzene-toluene} = 2.25$ and $\alpha_{cumene-toluene} = 0.210$. Use the Fenske equation to find the number of equilibrium stages required at total reflux and the recovery fraction of cumene in the bottoms.

b. For the distillation problem given in part a, find the minimum reflux ratio by use of the Underwood equations. Use a basis of 100 moles of feed/h. Clearly state your assumptions.

c. For $L/D = 1.25(L/D)_{min}$, find the total number of equilibrium stages required for the distillation problem presented in parts a and b. Use the Gilliland correlation. Estimate the optimum feed plate location.

D13.* We have a column separating benzene, toluene, and cumene. The column has a total condenser and a total reboiler and has 9 equilibrium stages. The feed is 25 mol% benzene, 30 mol% toluene, and 45 mol% cumene. Feed rate is 100 mol/h and feed is a saturated liquid. The equilibrium data can be represented as constant relative volatilities: $\alpha_{BT} = 2.5$, $\alpha_{TT} = 1.0$, and $\alpha_{CT} = 0.21$. We desire 99% recovery of toluene in the distillate and 98% recovery of cumene in the bottoms. Determine the external reflux ratio required to achieve this separation. If $\alpha_{BT} = 2.25$ instead of 2.5, how much will L/D change?

D14. At total reflux a separation requires $N_{min} = 10$ equilibrium contacts. At a finite external reflux ratio of $L/D = 2.0$, the separation requires N = 18 equilibrium contacts. (N and N_{min} include the partial reboiler and stages in the column but do not include the total condenser.) Find $(L/D)_{min}$.

D15.* A distillation column is separating benzene ($\alpha = 2.25$), toluene ($\alpha = 1.00$), and cumene ($\alpha = 0.21$). The column is operating at 101.3 kPa. The column is to have a total condenser and a partial reboiler, and the optimum feed stage is to be used. Reflux is returned as a saturated liquid, and $L_0/D = 1.2$. Feed rate is 1000 kmol/h. Feed is 39.7 mol% benzene, 16.7 mol% toluene, and 43.6 mol% cumene and is a saturated liquid. We desire to recover 99.92% of the benzene in the distillate and 99.99% of the cumene in the bottoms. For a first guess to this design problem, use the Fenske-Underwood-Gilliland approach to estimate the optimum feed stage and the total number of equilibrium stages. Note: The Underwood equations must be treated as a case C problem.

D16.* We are separating a mixture of ethanol and n-propanol. Ethanol is more volatile and the relative volatility is approximately constant at 2.10. The feed flow rate is 1000 kmol/h. Feed is 60 mol% ethanol and is a saturated vapor. We desire $x_D = 0.99$ mole fraction ethanol and $x_B = 0.008$ mole fraction ethanol. Reflux is a saturated liquid.

There are 30 stages in the column. Use the Fenske-Underwood-Gilliland approach to determine

a. Number of stages at total reflux

b. $(L/D)_{min}$

c. $(L/D)_{actual}$

D17. A distillation column is separating toluene and xylene, $\alpha = 3.03$. Feed is a saturated liquid and reflux is returned as a saturated liquid. $p = 1.0$ atm. $F = 100.0$ kmol/h. Distillate mole fraction is $x_D = 0.996$ and bottoms $x_B = 0.008$. Use the Underwood equation to find $(L/D)_{min}$ and V_{min} at feed mole fractions of $z = 0.1, 0.3, 0.5, 0.7,$ and 0.9. Check your result at $z = 0.5$ with a McCabe-Thiele diagram. What are the trends for $|Q_{c,min}|$ and $Q_{R,min}$ as toluene feed concentration increases?

D18. A depropanizer has the following feed and constant relative volatilities:

Methane (M): $z_M = 0.229$, $\alpha_{M-P} = 9.92$
Propane (P): $z_P = 0.368$, $\alpha_{P-P} = 1.0$
n-butane (B): $z_B = 0.322$, $\alpha_{B-P} = 0.49$
n-hexane (H): $z_H = 0.081$, $\alpha_{H-P} = 0.10$

Reflux is a saturated liquid. The feed is a saturated liquid fed in at 1.0 kmol/(unit time). Assume CMO.

a. * $L/D = 1.5$, $FR_{P,dist} = 0.9854$, $FR_{B,bot} = 0.8791$. Estimate N.

b. $N = 20$, $FR_{P,dist} = 0.9854$, $FR_{B,bot} = 0.8791$. Estimate L/D.

c. Find the split of normal hexane at total reflux using N_{min}.

d. $L/D = 1.5$, $FR_{P,dist} = 0.999$, $FR_{B,bot} = 0.8791$. Estimate N.

Note: Once you have done part a, you don't have to resolve the entire problem for the other parts.

D19. Revisit Problem 7.D4. Using the process simulator we found $N_{min} = 11$ for this problem. Using this value plus the simulation data in 7.D4. part d, estimate $(L/D)_{min}$ using the Gilliland correlation.

D20. A distillation column is separating a mixture of benzene, toluene, xylene and cumene. The feed to the column is 5.0 mol% benzene, 15.0 mol% toluene, 35.0 mol% xylene and 45.0 mol% cumene. Feed rate is 100.0 kmol/h and is a saturated liquid. We wish to produce a distillate that is 0.57895 mole fraction xylene, 0.07018 mole fraction cumene, and the remainder is toluene and benzene. The bottoms should contain no benzene or toluene. If we select toluene as the reference component the relative volatilities are approximately constant in the column at the following values: benzene = 2.25, toluene = 1.0, xylene – 0.330, and cumene = 0.210.

a. Find distillate and bottoms flow rates.

b. Find the number of equilibrium contacts at total reflux.

D21. A distillation column is separating 100 kmol/h of a saturated vapor feed that is 30 mol% ethanol, 25 mol% i-propanol, 35 mol% n-propanol, and 10 mol% n-butanol at a pressure of 1.0 atm. We want a 98.6% recovery of i-propanol in the distillate and 99.2% recovery of n-propanol in the bottoms. The column has a total condenser and a partial reboiler. For parts b, c, and d, use the Fenske-Underwood-Gilliland method. If we choose n-propanol as the reference, the relative volatilities are ethanol = 2.17, i-propanol = 1.86, n-propanol = 1.0, and n-butanol = 0.412. These relative volatilities can be assumed to be constant.

a. Find D, B, $x_{i,dist}$, and $x_{i,bot}$.

b. Find N_{min} and $N_{F,min}$.

c. Find $(L/D)_{min}$. A spreadsheet is highly recommended to find ϕ.

d. If $L/D = 1.10 \ (L/D)_{min}$, find N and the feed stage.

F. *Problems Requiring Other Resources*

F1. What variables does the Gilliland correlation not include? How might some of these be included? Check the Erbar-Maddox (1961) method (or see King, 1980, or Hines and Maddox, 1985) to see one approach that has been used.

F2. A distillation column with a total condenser and a partial reboiler operates at 1.0 atm.

a. Estimate the number of stages at total reflux to separate nitrogen and oxygen to produce a nitrogen mole fraction in the bottoms of 0.001 and a nitrogen distillate mole fraction of 0.998.

b. If the feed is 79.0 mol% nitrogen and 21.0 mol% oxygen and is a saturated vapor, estimate $(L/D)_{min}$.

c. Estimate the number of stages and the feed location if $L/D = 1.1 \ (L/D)_{min}$. The column has a total condenser and a partial reboiler.

G. *Computer Simulation Problems*

G1. Repeat Problem 7.D7 on AspenPlus using RADFRAC.

a. Find N at total reflux.

b. Find $(L/D)_{min}$ accurately with a few hundred stages.

CHAPTER 8

Introduction to Complex Distillation Methods

We have looked at binary and multicomponent mixtures in both simple and fairly complex columns. However, the chemicals separated have usually had fairly simple equilibrium behavior. In this chapter you will be introduced to a variety of more complex distillation systems used for the separation of less ideal mixtures.

Simple distillation columns are not able to completely separate mixtures when azeotropes occur, and the columns are very expensive when the relative volatility is close to 1. Distillation columns can be coupled with other separation methods to break the azeotrope. This is discussed in the first section. Extractive distillation, azeotropic distillation, and two-pressure distillation are methods for modifying the equilibrium to separate these complex mixtures. These three methods are described in Sections 8.2 to 8.7 of this chapter. In Section 8.8 we discuss the use of a distillation column as a chemical reactor, to simultaneously react and separate a mixture.

8.1 BREAKING AZEOTROPES WITH OTHER SEPARATORS

Azeotropic systems normally limit the separation that can be achieved. For an azeotropic system such as ethanol and water (shown in Figures 2-2 and 4-13), it isn't possible to get past the azeotropic concentration of 0.8943 mole frac ethanol with ordinary distillation. Some other separation method is required to break the azeotrope. The other method could employ adsorption (Chapter 18), membranes (Chapter 17), extraction (Chapter 13), and so forth. It could also involve adding a third component to the distillation to give the azeotropic and extractive distillation systems discussed later in this chapter.

Two ways of using an additional separation method to break the azeotrope are shown in Figure 8-1. The simplest, but least likely to be used, is the completely uncoupled system shown in Figure 8-1A. The distillate, which is near the azeotropic concentration, is sent to another separation device, which produces both the desired products. If the other separator can completely separate the products, why use distillation at all? If the separation is not complete, what would be done with the waste stream?

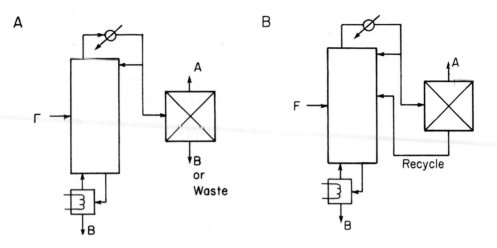

FIGURE 8-1. *Breaking azeotropes; (A) separator uncoupled with distillation, (B) recycle from separator to distillation*

A more likely configuration is that of Figure 8-1B. The incompletely separated stream is recycled to the distillation column, which now operates as a two-feed column, so the design procedures used for two-feed columns (Example 4-5) can be used. The arrangement shown in Figure 8-1B is commonly used industrially. The separator may actually be several separators.

8.2 BINARY HETEROGENEOUS AZEOTROPIC DISTILLATION PROCESSES

The presence of an azeotrope can be used to separate an azeotropic system. This is most convenient if the azeotrope is heterogeneous; that is, the vapor from the azeotrope will condense to form two liquid phases that are immiscible. Azeotropic distillation is often performed by adding a solvent or entrainer that forms an azeotrope with one or both of the components. Before discussing these more complex azeotropic distillation systems in Section 8.7, let us consider the simpler binary systems that form a heterogeneous azeotrope.

8.2.1 Binary Heterogeneous Azeotropes

Although not common, there are systems such as n-butanol and water, which form a heterogeneous azeotrope (Figure 8-2). Figures 8-2A and 8-2B plot the same data, but for different components. When the vapor of the azeotrope with mole frac y_{az} is condensed, a water-rich liquid phase α and an organic-rich liquid phase β separate from each other. You should feel comfortable converting from Figure 8-2A to 8-2B, and vice versa.

How should we distill a feed that forms a heterogeneous azeotrope? Suppose we have a saturated liquid feed that is 40 mol% water and 60 mol% n-butanol. The distillation system shown in Figure 8-3A (Luyben, 1973) consists of a column and a liquid-liquid settler. The column will produce a pure butanol product as the bottoms. The distillate product is the aqueous layer from the settler and automatically has $x_{w,dist} = x_{w,\alpha}$. We can develop the top operating line using the mass balance envelope shown in Figure 8-3A.

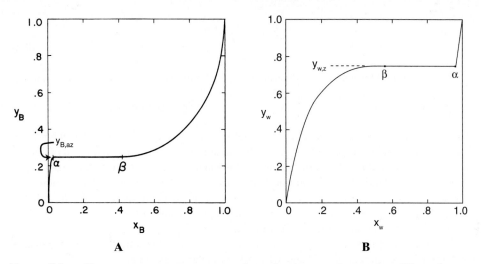

FIGURE 8-2. *Heterogeneous azeotrope system, n-butanol and water at 1 atmosphere (Chu et al., 1950). (A) Plotted as butanol mole fractons. (B) Plotted as water mole fractions.*

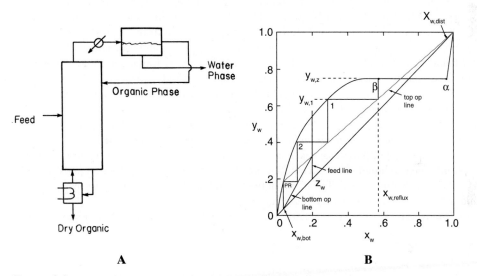

FIGURE 8-3. *Distillation column plus settler for distillation of system with heterogenous azeotrope.*

$$y_w = \frac{L}{V}x_w + \left(1 - \frac{L}{V}\right)x_{w,dist}$$

(8-1)

This looks like a normal top operating line, but there will be differences in the way it is plotted and the way it is used. The bottom operating line is the normal bottom operating line.

$$y_w = \frac{\bar{L}}{\bar{V}} x_w - \left(\frac{\bar{L}}{\bar{V}} - 1\right) x_{w,bot}$$

(8-2)

We can plot a McCabe-Thiele diagram by using Figure 8-2B. The top operating line has a slope $= L/V$, a y-intersect $= \left(1 - \dfrac{L}{V}\right) x_{w,dist}$, and a y = x intersection of y = x = $x_{w,dist}$. This is shown in Figure 8-3B. Note that $x_{w,dist} = x_{w,\alpha}$ but the top operating line goes through y = x = $x_{w,dist}$ *not* the equilibrium curve at $x_{w,\alpha}$. The top operating line cuts through the equilibrium curve, but the distillation does not operate in this range. Instead the separation from $x_{w,\alpha}$ to $x_{w,\beta}$ is done in the liquid-liquid separator. The reflux in Figure 8-3A and 8-3B has a mole fraction of $x_{w,reflux} = x_{w,\beta}$, *not* the distillate composition as is normally the case. In the column, $x_{w,reflux}$ and $y_{w,1}$ are passing steams and are on the operating line in Figure 8-3B. Stepping off the remaining stages follows the normal procedure.

The vapor of mole fraction $y_{w,1}$ is condensed and sent to the settler. The mass balances for the settler are straightforward and are illustrated later in Example 8-1.

What can we do if we want a purer water product?

For this type of heterogeneous azeotrope the two-column systems shown in Figure 8-4A can provide a complete separation. Column 2 is a stripping column that receives liquid of composition x_α from the liquid-liquid settler. It operates where n-butanol is more volatile (low n-butanol mole fractions on Figure 8-2A) and the bottoms from column 2 is almost pure water ($x_{B,bot} \sim 0$). The overhead vapor from column 2, $y_{1,col2}$ is condensed and goes to the liquid-liquid settler where it separates into two liquid phases. Liquid of composition x_α is refluxed to column 2, while liquid of composition x_β is refluxed to column 1.

The first column again operates on the left-hand side of Figure 8-3B, where water is the more volatile component. Thus, the bottoms from this column is almost pure B ($x_{w,bot1} \sim 0$). The overhead vapor, which is richer in water, is condensed and sent to the liquid-liquid separator. The McCabe–Thiele diagram for this column is shown in Figure 8-4B.

The liquid-liquid separator takes the two condensed liquids, $x_\alpha < x < x_\beta$, and separates them into the two liquid phases in equilibrium at mole fracs x_α and x_β. These liquids are used as reflux to columns 1 and 2. The liquid-liquid separator allows one to get past the azeotrope and is therefore a necessary part of the equipment.

The overall external mass balance for the two-column system shown in Figure 8-4A is

$$F = B_1 + B_2$$

(8-3a)

while the external mass balance on water is

$$Fz_w = B_1 x_{w,bot1} + B_2 x_{w,bot2}$$

(8-3b)

Solving these equations simultaneously for the unknown bottoms flow rates, we obtain

$$B_1 = \frac{F(z_w - x_{w,bot2})}{(x_{w,bot1} - x_{w,bot2})}$$

(8-4)

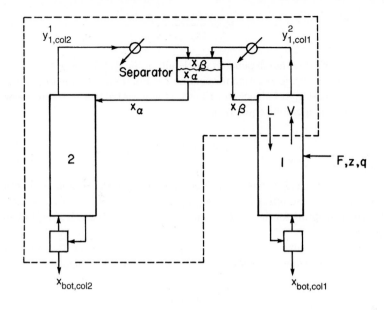

A

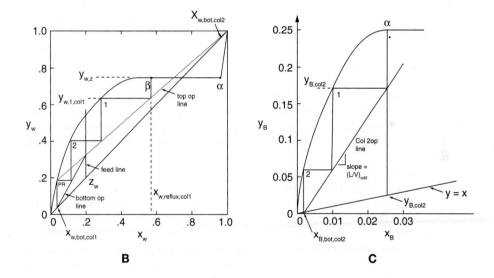

B   **C**

FIGURE 8-4.   *Binary heterogeneous azeotrope; (A) two-column distillation sys-*
tem, (B) McCabe-Thiele diagram for column 1. (C) McCabe-Thiele
diagram for column 2 with expanded coordinates.

$$B_2 = \frac{F(z_w - x_{w,bot1})}{(x_{w,bot2} - x_{w,bot1})} \tag{8-5}$$

Note that this result does not depend on the details of the distillation system.

The bottom operating equation for column 1 is Eq. (8-2). The top operating line is a bit different. The easiest mass balance to write uses the mass balance envelope shown in Figure 8-4A. Then the top operating equation is

$$y_w = \left(\frac{L}{V}\right)_{col1} x_w + \left(1 - \frac{L}{V}\right)_{col1} x_{w,bot2} \tag{8-6}$$

This is somewhat unusual, because it includes a bottoms concentration leaving the stripping column (column 2). The reflux for this top operating line is liquid of composition $x_{w,\beta}$. The McCabe-Thiele diagram for this system is shown in Figure 8-4B and is almost identical to Figure 8-3B.

Analysis of stripping column 2 is straightforward. It is easiest to use Figure 8-2A and develop the operating equation for n-butanol. This bottom operating equation is

$$y_B = \left(\frac{\overline{L}}{\overline{V}}\right)_2 x_B - \left[\left(\frac{\overline{L}}{\overline{V}}\right)_2 - 1\right] x_{B,bot2} \tag{8-7}$$

The feed to this column is the saturated liquid reflux of composition $x_{B,\alpha}$. This is a vertical feed line. Then the overhead vapor $y_{B,1,col2}$ is found on the operating line at $x_{B,\alpha}$ (Figure 8-4C).

The two dashed lines in Figures 8-4B and C show the route of the overhead vapor streams as they are condensed to saturated liquids (made into x values), and then sent to the liquid-liquid separator. The lever-arm rule [see Eq. (2-26) and Figure 2-10] can be applied to the liquid-liquid separator (see Example 8-1).

Note that each column operates in a region where either water (column 1) or n-butanol (column 2) is more volatile. McCabe-Thiele diagrams are easiest to use if plotted for the more volatile component. Thus, we plotted water mole fractions for column 1 and n-butanol mole fractions for column 2. It is also possible to do the calculations on a single diagram (*Separation Process Engineering*, 2nd ed., p. 228), which results in one of the calculations looking upside down.

The operation of distillation columns separating heterogeneous azeotropes can be quite erratic (Kovach and Seider, 1987). Small shifts in the aqueous reflux rate can cause a number of trays to shift from operation in the homogeneous region to a heterogeneous region. This erratic switching will cause large variations in the product purity. These columns need very careful control of the reflux stream flow rate to operate properly.

Several modifications of the basic arrangement shown in Figure 8-4A can be used. If the feed composition is less than $x_{butanol,\alpha}$, then column 2 should be a complete column and column 1 would be just a stripping column. The liquids may be subcooled so that the liquid-liquid separator operates below the boiling temperature. This can be advantageous, since the partial miscibility of the system depends on temperature. When the liquids are subcooled, the separator calculation must be done at the temperature of the settler. Then the reflux concentrations can be plotted on the McCabe-Thiele diagram. Design of separators (decanters) is discussed in Section 13.14. More details for heterogeneous azeotropes are explored by Doherty et al. (2008), Doherty and Malone (2001), Hoffman (1964), and Shinskey (1984).

8.2.2 Drying Organic Compounds That Are Partially Miscible with Water

For "immiscible" systems (really partially miscible systems) of organics and water, a single phase is formed only when the water concentration is low or very high. For example, a small amount of water can dissolve in gasoline. If more water is present, two phases will form. In the case of gasoline, the water phase is detrimental to the engine, and in cold climates it can freeze in gas lines, immobilizing the car. Since the solubility of water in gasoline decreases as the temperature is reduced, it is important to have dry gasoline.

Fortunately, small amounts of water can easily be removed by distillation or adsorption (Chapter 18). During distillation the water acts as a very volatile component, so a mixture of water and organics is taken as the distillate. After condensation, two liquid phases form, and the organic phase can be refluxed. The system is a type of heterogeneous azeotropic system similar to those discussed in the previous section. Drying differs from the previous systems since a pure water phase is usually not desired, the relative solubilities are often quite low, and the water phase is usually sent to waste treatment. Thus, the system will look like Figure 8-3a. With very high relative volatilities, one equilibrium stage may be sufficient and a flash system plus a separator can be used.

Because detailed equilibrium data are often unavailable, simplified equilibrium theories are useful for "immiscible" liquids. There is always a range of concentrations where the species are miscible even though the concentrations may be quite small. It is reasonable to assume that the relative volatility is constant over the small range of compositions where the liquids are miscible.

$$\alpha_{w-org\ in\ org} = \frac{\left(y_w / x_{w\ in\ org}\right)}{\left(y_{org} / x_{org\ in\ org}\right)} \tag{8-8}$$

If data for the heterogeneous azeotrope (y_w and $x_{w\ in\ org}$) are available, then $y_{org} = 1 - y_w$, $x_{org\ in\ org} = 1 - x_{w\ in\ org}$, and $\alpha_{w-org\ in\ org}$ is easily estimated from the data. This will be more accurate than assuming Raoult's law or assuming a linear relationship.

If the data are not available, we can assume in the water phase that the water follows Raoult's law and the organic components follow Henry's law (Robinson and Gilliland, 1950). Thus,

$$p_w = (VP_w)x_{w\ in\ w}, \quad p_{org} = H_{org}\ x_{org\ in\ w} \text{ (Water phase)} \tag{8-9}$$

where H_{org} is the Henry's law constant for the organic component in the aqueous phase, VP_w is the vapor pressure of water, and $x_{w\ in\ w}$ and $x_{org\ in\ w}$ are the mole fracs of water and organic in the water phase, respectively. In the organic phase, it is reasonable to use Raoult's law for the organic compounds and Henry's law for the water.

$$p_w = H_w x_{w\ in\ org}, \quad p_{org} = (VP_{org})x_{org\ in\ org} \text{ (Organic phase)} \tag{8-10}$$

where H_w is the Henry's law constant for water in the organic phase. At equilibrium, the partial pressure of water in the two phases must be equal. Thus, equating p_w in Eqs. (8-8) and (8-10) and solving for H_w we obtain

$$H_w = \frac{(VP_w)x_{w \text{ in } w}}{x_{w \text{ in org}}} \tag{8-11a}$$

Similar manipulations for the organic phase give

$$H_{org} = \frac{(VP_{org})x_{org \text{ in org}}}{x_{org \text{ in } w}} \tag{8-11b}$$

Using Eqs. (8-11a) and (8-11b), we can calculate the Henry's law constants from the known solubilities (which give the mole fracs) and the vapor pressures.

Eqs. (8-8) to (8-11) are valid for both drying organic compounds and steam distillation (Section 8.3). The ease of removing small amounts of water from an organic compound that is immiscible with water can be seen by estimating the relative volatility of water in the organic phase.

$$\alpha_{w-org \text{ in org}} \equiv \frac{y_w/x_{w \text{ in org}}}{y_{org}/x_{org \text{ in org}}} = \frac{(p_w/p_{tot})/x_{w \text{ in org}}}{(p_{org}/p_{tot})/x_{org \text{ in org}}} = \frac{p_w/x_{w \text{ in org}}}{p_{org}/x_{org \text{ in org}}} \tag{8-12}$$

In the organic phase Eqs. (8-10) and (8-11b) can be substituted into Eq. (8-12) to give

$$\alpha_{w-org \text{ in org}} = \frac{H_w}{VP_{org}} = \frac{(VP_w)x_{w \text{ in } w}}{(x_{w \text{ in org}})(VP_{org})} \tag{8-13}$$

This calculation is illustrated in Example 8-1.

Organics can be dried either by continuous distillation or by batch distillation. In both cases the vapor will condense into two phases. The water phase can be withdrawn and the organic phase refluxed to the distillation system. For continuous systems, the McCabe-Thiele design procedure can be used. The McCabe-Thiele equilibrium can be plotted from Eq. (8-13), and the analysis is the same as in the previous section. This is illustrated in Example 8-1.

EXAMPLE 8-1. Drying benzene by distillation

A benzene stream contains 0.01 mole frac water. Flow rate is 1000 kmol/h, and feed is a saturated liquid. Column has saturated liquid reflux of the organic phase from the liquid-liquid separator (see Figure 8-3A) and uses $L/D = 2 (L/D)_{min}$. We want the outlet benzene to have $x_{w \text{ in benz,bot}} = 0.001$. Design the column and the liquid-liquid settler.

Solution

A. Define. The column is the same as Figure 8-3A. Find the total number of stages and optimum feed stage. For the settler determine the compositions and flow rates of the water and organic phases.

B. Explore. Need equilibrium data. From Robinson and Gilliland (1950), $x_{benz \text{ in } w} = 0.00039$, $x_{w \text{ in benz}} = 0.015$. The solubility data give the compositions of the streams leaving the settler. Use Eq. (8-13) for equilibrium. At the boiling point of benzene (80.1°C), $VP_{benz} = 760$ mm Hg and $VP_w = 356.6$ mm Hg (Perry and Green, 1997). Operation will be at a different temperature, but the ratio of vapor pressures will be approximately constant.

C. Plan. Calculate equilibrium from Eq. (8-13):

$$\alpha_{w-benz} = \frac{(VP_w)x_{w\,in\,w}}{x_{w\,in\,benz}(VP_{benz})} = \frac{(356.6)(1-0.00039)}{(0.015)(760)} = 31.3$$

This is a good approximation of VLE for $x_{w\,in\,benz} < 0.015$. After that, we have a heterogeneous azeotrope. Plot the curve represented by this value of α_{w-benz} on a McCabe-Thiele diagram. (Two diagrams will be used for accuracy.) Solve with the McCabe-Thiele method as a heterogeneous azeotrope problem. Mass balances will be used to find flow rates leaving the settler.

D. Do it. Plot equilibrium: $$y_w = \frac{\alpha x_w}{1 + (\alpha - 1)x_w} = \frac{31.3x_w}{1 + 30.3x_w}$$

where y_w and x_w are mole fracs of water in the benzene phase. This is valid for $x_w \leq 0.015$. At the solubility limit $x_w = 0.015$, we can determine the y_w value for the azeotrope,

$$y_{w,az} = \frac{(31.3)(0.015)}{1+30.3(0.015)} = 0.32$$

See Figure 8-5. Since Figure 8-5A is obviously not accurate for stepping off stages, we use Figure 8-5B. Calculate vapor mole frac in equilibrium with feed, and then the required reflux ratio.

$$y(z) = \frac{(31.3)(0.01)}{1 + 30.3(0.01)} = 0.24$$

$$(L/V)_{min} = \frac{x_D - y(z)}{x_D - z} = \frac{0.9996 - 0.24}{0.9996 - 0.01} = 0.7676$$

$$(L/D)_{min} = \left(\frac{L/V}{1 - L/V}\right)_{min} = 3.303$$

$$(L/D)_{act} = 2(3.303) = 6.606$$

$$(L/V)_{act} = \frac{L/D}{1 + L/D} = 0.868$$

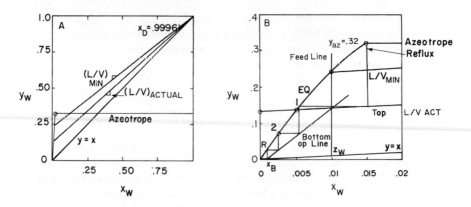

FIGURE 8-5. *Solution for Example 8-1; (A) McCabe-Thiele diagram for entire range, (B) McCabe-Thiele diagram for low concentrations*

Top Operating Line:

$$y = \frac{L}{V} x + \left(1 - \frac{L}{V}\right) x_D = 0.868x + (0.132)(0.99961)$$

where x_D is water mole frac in distillate $= 1 - x_{benz\ in\ w} = 1 - 0.00039 = 0.99961$

$$y \text{ intercept } (x = 0) = 0.132$$
$$y = x = x_D = 0.99961$$

Plot top operating line (Figure 8-5A).
Feed Line: Saturated liquid.

$$\text{Slope} = \frac{q}{q - 1} = \infty \text{ at } y = x = z_w = 0.01$$

Bottom Operating Line: $y = \frac{\overline{L}}{\overline{V}} x - \left(\frac{\overline{L}}{\overline{V}} - 1\right) x_B$

goes through $y = x = x_B = 0.001$ and intersection of top operating line and feed line.

Reflux is the benzene phase from the liquid-liquid separator (see Figure 8-3A); thus, $x_{reflux} = x_{w\ in\ benz} = 0.015$. Use this to start stepping off stages on Figure 8-5B. Optimum feed stage is top stage of column. We need 2 stages plus a partial reboiler.

Settler mass balance requires distillate flow rate. From Eq. (3-3),

$$D = \frac{z - x_B}{x_D - x_B} F = \frac{0.01 - 0.001}{0.99961 - 0.001} (1000) = 9.01 \text{ kmol/h (water phase)}$$

$$B = F - D = 1000 - 9.01 = 990.99$$

$$L = (L/D)D = 6.606 \, D = 59.54 \text{ (Organic phase)}$$

$$V_1 = L + D = 68.55$$

E. Check. All of the internal consistency checks work. The value of α_{w-benz} agrees with the calculation of Robinson and Gilliland (1950). The best check on α_{w-benz} would be comparison with data.

A check on the settler flow rates can be obtained from a water mass balance. The vapor leaving Stage 1, $y_{1,w}$, is a passing stream to the reflux,

$$y_{1,w} = (0.868)(0.015) + (0.132)(0.99961) = 0.145$$

This stream is condensed and separated into the water layer (distillate) and the organic layer (reflux).

Is $\qquad V_2 y_{1,w} = D x_{D,w} + L x_{reflux,w}?$

$$(68.55)(0.145) = (9.01)(0.99961) + (59.54)(0.015)?$$

9.94 = 9.90, which is within the accuracy of the graph.

F. Generalize. Since the solubility of organics in water is often very low, this type of heterogeneous azeotrope system requires only one distillation column.

Even though water has a higher boiling point than benzene, the relative volatility of water dissolved in benzene is extremely high. This occurs because water dissolved in an organic cannot hydrogen bond as it does in an aqueous phase, and thus, it acts as a very small molecule that is quite volatile. The practical consequence of this is that small amounts of water can easily be removed from organics if the liquids are partially immiscible. There are alternative methods for drying organics such as adsorption that may be cheaper than distillation in many cases.

8.3 STEAM DISTILLATION

In steam distillation, water (as steam) is intentionally added to the distilling organic mixture to reduce the required temperature and to keep suspended any solids that may be present. Steam distillation may be operated with one or two liquid phases in the column. In both cases the overhead vapor will condense into two phases. Thus, the system can be considered a type of azeotropic distillation where the added solvent is water and the *separation is between volatiles and nonvolatiles*. This is a pseudo-binary distillation with water and the volatile organic forming a heterogeneous azeotrope. Steam distillation is commonly used for purification of essential oils in the perfume industry, for distillation of organics obtained from coal, for hydrocarbon distillations, and for removing solvents from solids in waste disposal (Ellerbe, 1997; Ludwig, 1997; Woodland, 1978).

For steam distillation with a liquid water phase present, both the water and organic layers exert their own vapor pressures. At 1 atm pressure the temperature must be less than 100 °C even though the organic material by itself might boil at several hundred degrees. Thus, one advantage of steam distillation is lower operating temperatures. With two liquid phases present and in equilibrium, their compositions will be fixed by their mutual solubilities. Since each phase exerts its own vapor pressure, the vapor composition will be constant regardless of the average liquid concentration. A heterogeneous azeotrope is formed. As the amount of water or organic is increased, the phase concentrations do not change; only the amount of each liquid phase will change. Since an azeotrope has been reached, no additional separation is obtained by adding more stages. Thus, only a reboiler is required. This type of steam distillation is often done as a batch operation (see Chapter 9).

Equilibrium calculations are similar to those for drying organics except that now two liquid phases are present. Since each phase exerts its own partial pressure, the total pressure is the sum of the partial pressures. With one volatile organic,

$$p_{org} + p_w = p_{tot} \qquad \text{(8-14)}$$

Substituting in Eqs. (8-9) and (8-10), we obtain

$$(VP_{org})x_{volatile\ in\ org} + (VP_w)x_{w\ in\ w} = p_{tot} \qquad \text{(8-15)}$$

The compositions of the liquid phases are set by equilibrium. If total pressure is fixed, then Eq. (8-15) enables us to calculate the temperature. Once the temperature is known the vapor composition is easily calculated as

$$y_w = \frac{p_w}{p_{tot}} = \frac{(VP_w)x_{w\ in\ w}}{p_{tot}}, \quad y_{volatile,org} = \frac{p_{org}}{p_{tot}} = \frac{(VP_{org})x_{org\ in\ org}}{p_{tot}} \qquad \text{(8-16)}$$

The number of moles of water carried over in the vapor can be estimated, since the ratio of moles of water to moles organic is equal to the ratio of vapor mole fracs.

$$\frac{n_{org}}{n_w} = \frac{y_{volatile,org}}{y_w} \qquad \text{(8-17)}$$

Substituting in Eq. (8-16), we obtain

$$\frac{n_{org}}{n_w} = \frac{p_{org}}{p_w} = \frac{p_{org}}{p_{tot} - p_{org}} = \frac{(VP)_{org}x_{volatile,org}}{p_{tot} - (VP)_{org}x_{volatile,org}} \qquad \text{(8-18)}$$

If several organics are present, y_{org} and p_{org} are the sums of the respective values for all the organics. The total moles of steam required is n_w plus the amount condensed to heat and vaporize the organic (see Example 8-2).

EXAMPLE 8-2. Steam distillation

A cutting oil that has approximately the properties of n-decane ($C_{10}H_{22}$) is to be recovered from nonvolatile oils and solids in a steady-state single-stage steam distillation. Operation will be with liquid water present. The feed is 50 mol% n-decane. A bottoms that is 15 mol% n-decane in the organic phase is desired. Feed rate is 10 kmol/h. Feed enters at the temperature of the boiler. Pressure is atmospheric pressure, which in your plant is approximately 745 mm Hg. Find:

  a. The temperature of the still
  b. The moles of water carried over in the vapor
  c. The moles of water in the bottoms

Solution

A. Define. The still is sketched in the figure. Note that there is no reflux.

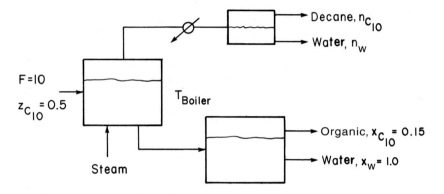

B. Explore. Equilibrium is given by Eq. (8-16). Assuming that the organic and water phases are completely immiscible, we have in the bottoms $x_{C10 \text{ in org}} = 0.15$, $x_{w \text{ in } w} = 1.0$ and in the two distillate layers $x_{C10,org,dist} = 1.0$ and $x_{w,water,dist} = 1.0$. Vapor pressure data as a function of temperature are available in Perry and Green (1997). Then Eq. (8-15) can be solved by trial and error to find T_{boiler}. Equation (8-18) and a mass balance can be used to determine the moles of water and decane vaporized. The moles of water condensed to vaporize the decane can be determined from an energy balance. Latent heat data are available in Perry and Green (1997).

C. Plan. On a water-free basis the mass balances around the boiler are

$$F = n_{C10,vapor} + B \tag{8-19a}$$

$$z_{C10} F = n_{C10,vapor} + x_{bot} B \tag{8-19b}$$

where B is the bottoms flow rate of the organic phase. Since $F = 10$, $x_{bot} = 0.15$, and $z_{C10} = 0.5$, we can solve for $n_{C10,vapor}$ and B. Eq. (8-18) gives n_w once T_{boiler} is known. Since the feed, bottoms, and vapor are all at T_{boiler}, the energy balance simplifies to

$$n_{C10}\lambda_{C10} = (\text{moles of water condensed in still})\lambda_w$$

D. Do it. a. Perry and Green (1997) give the following n-decane vapor pressure data (vapor pressure in mm Hg and T in °C):

VP	5	10	20	40	60	100	200	400	760
T	42.3	55.7	69.8	85.5	95.5	108.6	128.4	150.6	174.1

A very complete table of water vapor pressures is given in that source (see Problem 8.D10). As a first guess, try 95.5°C, where $(VP)_w = 645.7$ mm.Hg. Then Eq. (8-15),

$$(VP)_{C10}\, x_{C10\ in\ org} + (VP)_w\, x_{w\ in\ w} = P_{tot}$$

becomes

$$(60)(0.15) + (645.7)(1.0) = 654.7 < 745 = P_{tot}$$

where we have assumed completely immiscible phases so that $x_w = 1.0$. This temperature is too low. Approximate solution of Eq. (8-15) eventually gives $T_{boiler} = 99°C$ and $(VP)_{C10} = 66$ mm Hg.

b. Solving the mass balances, Eqs. (8-19), the kmol/h of vapor are

$$n_{C10,vapor} = \frac{F(z_{C10} - x_{bot})}{1 - x_{bot}} \qquad \textbf{(8-20)}$$

$$= \frac{10(0.5 - 0.15)}{1 - 0.15} = 4.12$$

which is 586.2 kg/h. Eq. (8-18) becomes

$$n_w = \frac{n_{c10,vapor}}{(VP)_{C10} x_{bot}} [P_{tot} - (VP)_{c10} x_{bot}]$$

or

$$n_w = \frac{4.12}{(68)(0.15)} [745 - (68)(0.15)] = 296.8\ kmol/h$$

which is 5347.1 kg/h.

c. The moles of water to vaporize the decane is

$$\text{Moles water condensed} = \frac{n_{C10\lambda} \lambda_{C10}}{\lambda_w}$$

$$= \frac{(4.12\,kmol/h)\left(319\dfrac{kJ}{kg}\right)\left(142.28\dfrac{kg}{kmol}\right)}{\left(2260\dfrac{kJ}{kg}\right)\left(18.016\dfrac{kg}{kmol}\right)} = 4.59\ kmol/h$$

where $\lambda = H_{vap} - h_{liq}$ and the saturated vapor and liquid enthalpies at 99 °C are interpolated from Tables 2-249 and 2-352 in Perry and Green (1997) for decane and water, respectively.

E. Check. A check for complete immiscibility is advisable since all the calculations are based on this assumption.

F. Generalize. Obviously, the decane is boiled over at a temperature well below its boiling point, but a large amount of water is required. Most of this water is carried over in the vapor. On a weight basis, the kilograms of total water required per kilogram decane vaporized is 9.26. Less water will be used if the boiler is at a higher temperature and there is no liquid water in the still. Less water is also used for higher values of $x_{org,bot}$ (see Problem 8.D10).

Additional separation can be obtained by operating without a liquid water phase in the column. Reducing the number of phases increases the degrees of freedom by one. Operation must be at a temperature higher than that predicted by Eq. (8-15), or a liquid water layer will form in the column. Thus, the column must be heated with a conventional reboiler and/or the sensible heat available in superheated steam. The latent heat available in the steam cannot be used, because it would produce a layer of liquid water. Operation without liquid water in the column reduces the energy requirements but makes the system more complex.

8.4 TWO-PRESSURE DISTILLATION PROCESSES

Pressure affects vapor-liquid equilibrium (VLE), and in systems that form azeotropes it will affect the composition of the azeotrope. For example, Table 2-1 shows that the ethanol-water system has an azeotrope at 0.8943 mole frac ethanol at 1 atm pressure. If the pressure is reduced, the azeotropic concentration increases (Seader, 1984). At pressures below 70 mm Hg, the azeotrope disappears entirely, and the distillation can be done in a simple column. Unfortunately, use of this disappearance of the azeotrope for the separation of ethanol and water is not economical because the column requires a large number of stages and has a large diameter (Black, 1980). However, the principle of finding a pressure where the azeotrope disappears may be useful in other distillations. The effect of pressure on the azeotropic composition and temperature can be estimated using the VLE correlations in process simulators (Waslkiewski, 2005).

Even though the azeotrope may not disappear, in general, pressure affects the azeotropic composition. If the shift in composition is large enough, a two-column process using two different pressures can be used to completely separate the binary mixture. Doherty et al. (2008) recommend a minimum mole fraction change of at least 5% (e.g., from 55% to 60%) with a 10% change being preferable. A schematic of the flowchart for this two-pressure distillation process is shown in Figure 8-6 (Doherty and Malone, 2001; Frank, 1997; Drew, 1997; Shinskey, 1984; Van Winkle, 1967). Column 1 usually operates at atmospheric pressure, while column 2 is usually at a higher pressure but can be at a lower pressure.

To understand the operation of this process, consider the separation of methyl ethyl ketone (MEK) and water (Drew, 1997). At 1 atm the azeotrope contains 35% water, while at 100 psia the azeotrope is 50% water. If a feed containing more than 35% water is fed to the first column, the bottoms will be pure water. The distillate from this atmospheric column will be the 35% azeotrope. When this azeotrope is sent to the high-pressure column, an azeotrope containing 50% water comes off as the distillate; this distillate is recycled to column 1. Since the feed to column 2 (the 35% azeotrope) contains less water than this distillate,

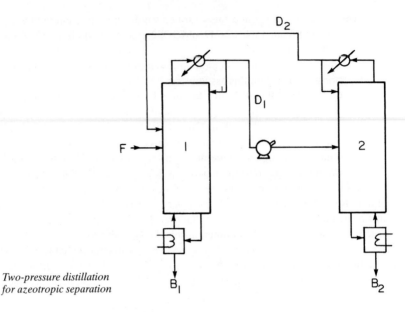

FIGURE 8-6. *Two-pressure distillation for azeotropic separation*

the bottoms from column 2 is pure MEK. Note that the water is less volatile in column 1 and the MEK is less volatile in column 2.

Mass balances for the system shown in Figure 8-6 are of interest. The external mass balances are identical to Eqs. (8-3a) and (8-3b). Thus, the bottoms flow rates are given by Eqs. (8-4) and (8-5). Although the processes shown in Figures 8-4A and 8-6 are very different, they look the same to the external mass balances. Differences in the processes become evident when balances are written for individual columns. For instance, for column 2 in Figure 8-6 the mass balances are

$$D_1 = D_2 + B_2 \qquad (8\text{-}21)$$

and

$$D_1 x_{dist1} = D_2 x_{dist2} + B_2 x_{bot2} \qquad (8\text{-}22)$$

Solving these equations simultaneously and then inserting the results in Eq. (8-4b), we obtain

$$D_2 = \frac{B_2(x_{bot2} - x_{dist1})}{x_{dist1} - x_{dist2}} = F\left(\frac{z - x_{bot1}}{x_{bot2} - x_{bot1}}\right)\left(\frac{x_{bot2} - x_{dist1}}{x_{dist1} - x_{dist2}}\right) \qquad (8\text{-}23)$$

This is of interest since D_2 is the recycle flow rate. As the two azeotrope concentrations at the two different pressures approach each other, $x_{dist1} - x_{dist2}$ will become small. According to Eq. (8-23), the recycle flow rate D_2 becomes large. This increases both operating and capital costs and makes this process too expensive if the shift in the azeotrope concentration is small.

The two-pressure system is also used for the separation of acetonitrile-water, tetrahydrofuran-water, methanol-MEK, and methanol-acetone (Frank, 1997). In the latter application the second column is at 200 torr. Realize that these applications are rare. For most azeotropic

Chapter 8 Introduction to Complex Distillation Methods

systems the shift in the azeotrope with pressure is small, and use of the system shown in Figure 8-6 will involve a very large recycle stream. This causes the first column to be rather large, and costs become excessive.

Before the relatively recent development of detailed and accurate VLE correlations, most VLE data were only available at one atmosphere; thus, many azeotropic systems have probably not been explored as candidates for two-pressure distillation. Fortunately, it is fairly easy to simulate two-pressure distillation with a process simulator (see Lab 7, part A in the appendix to Chapter 8). Methods for estimating VLE and rapidly screening possible systems are available (Frank, 1997). Because two-pressure distillation does not require a mass separating agent, it is a prefered method when it works. If two-pressure distillation were routinely considered as an option for breaking azeotropes, we would undoubtedly discover additional systems where this method is economical.

8.5   COMPLEX TERNARY DISTILLATION SYSTEMS

In Chapters 5, 6, and 7 we studied multicomponent distillation for systems with relatively ideal VLE that do not exhibit azeotropic behavior. In Chapter 4 when we studied both relatively ideal and azeotropic binary systems we found that there were significant differences between these systems. If no azeotrope forms, one can obtain essentially pure distillate and pure bottom products. If there is an azeotrope we found that one can at best obtain one pure product and the azeotrope. In Section 8.2 we found that if the binary azeotrope was heterogeneous one could usually use a liquid-liquid separator to get past the azeotrope and obtain two pure products with two columns. Ternary systems with nonideal VLE can have one or more azeotropes that may be homogeneous or heterogeneous. Since the behavior of ideal ternary distillation is more complex than that of ideal binary distillation, we expect that the behavior of nonideal ternary distillation is probably more complex than nonideal binary distillation.

Although McCabe-Thiele diagrams can be used for ternary systems, they have not been nearly as successful as the binary applications. Hengstebeck (1961) developed a pseudo-binary approach that is useful for systems with close to ideal VLE. It has also been applied to extractive distillation by assuming the solvent concentration is constant. Chambers (1951) developed a method that could be applied with fewer assumptions to systems with azeotropes and illustrated it with the ternary system methanol-ethanol-water. His approach consisted of drawing two McCabe-Thiele diagrams (e.g., one for methanol and one for ethanol). Equilibrium consists of several curves with methanol mole frac as a parameter on the ethanol diagram and ethanol mole frac as a parameter on the methanol diagram. Each equilibrium step required simultaneous solution of the two diagrams (see Wankat, 1981). The operating lines plot on these diagrams in the normal fashion. Although visually instructive, Chambers's method is awkward and has not been widely used. The conclusion is we need new visualization tools to study ternary distillation.

8.5.1   Distillation Curves

You may have noticed in Chapter 2 that enthalpy-composition and temperature-composition diagrams contain more information than the McCabe-Thiele y-x diagram. We started using the diagram with less information because it was easier to show the patterns and visualize the separation. For ternary distillation we will repeat this pattern and go to a ternary composition diagram that shows the paths taken by liquid mole fracs throughout a distillation operation.

We gain in visualization power since a variety of possible paths are easy to illustrate, but we lose power since the stages are no longer shown. For complex systems the gains are much more important than the losses.

A *distillation curve* is a plot of the mole fracs on every tray for distillation. Distillation curves can be generated at total reflux or finite reflux. If you have run a multicomponent distillation simulation or solved a ternary distillation problem with a hand calculation, you have obtained the information ($x_{i,j}$ for j = 1,...N) necesary to plot a distillation curve. Thus, what is different is the presentation. The two different formats used for these diagrams are shown in Figures 8-7 and 8-8. To generate these plots at total reflux consider a distillation column numbered from the bottom up (Figure 5-1). If we start at the reboiler, we first do a bubble-point calculation,

$$\sum_{i=1}^{c} y_{i,j} = \sum_{i=1}^{c} K_i\,(T,p,x_{i,j})x_{i,j} = 1.0 \tag{8-24}$$

Calculation of the bubble-point at every stage can be laborious; fortunately, the calculation is easily done with a process simulator. Next, at total reflux the operating equation is,

$$x_{i,j+1} = y_{i,j} \tag{8-25a}$$

Alternation of these two equations results in values for x_A, x_B, and x_C on every stage. These values (see Example 8-3) are then plotted in Figures 8-7 and 8-8. The starting mole fracs are

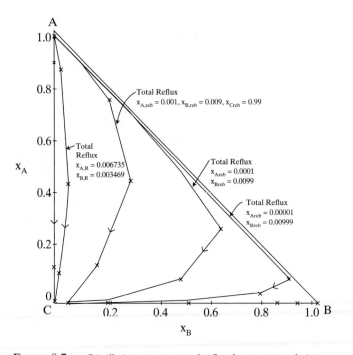

FIGURE 8-7.   *Distillation curves at total reflux for constant relative volatility system; A = benzene, B = toluene, C = cumene; $\alpha_{AB} = 2.4$, $\alpha_{BB} = 1.0$, $\alpha_{CB} = 0.21$; stages are shown as ×.*

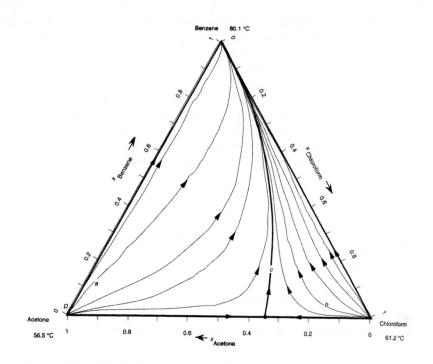

FIGURE 8-8.   *Distillation curves at total reflux for acetone, chloroform, benzene mixtures at 1.0 atm (Biegler et al., 1997), reprinted with permission of Prentice-Hall PTR, copyright 1997, Prentice-Hall.*

chosen so that the distillation curves fill the entire space of the diagrams. If the relative volatility is constant, then the vapor mole fracs can be easily calculated from Eq. (5-30). Substituting Eq. (5-30) into Eq. (8-25a) and solving for $x_{i,j+1}$, we obtain the recursion relationship for the distillation curve for constant relative volatility systems.

$$x_{i,j+1} = \frac{\alpha_{i\text{-ref}} x_{i,j}}{\sum_{i=1}^{c} (\alpha_{i\text{-ref}} x_{i,j})}$$

(8-25b)

Figure 8-7 shows the characteristic pattern of distillation curves for ideal or close to ideal VLE with no azeotropes. All of the systems considered in Chapters 5, 6, and 7 follow this pattern. The y-axis ($x_B = 0$) represents the binary A-C separation. This starts at the reboiler ($x_A = 0.01$ is an arbitrary value) and requires only the reboiler plus 4 stages to reach a distillate value of $x_A = 0.994$. The x axis ($x_A = 0$) represents the binary B-C separation, which was started at the arbitrary value $x_B = 0.01$ in the reboiler. The maximum in B concentration should be familiar from the profiles shown in Chapter 5. Distillation curves at finite reflux ratios are similar but not identical to those at total reflux. Note that the entire space of the diagram can be reached by starting with concentrations near 100% C (the heavy boiler). Distillation curves are usually plotted as smooth curves—they were plotted as discrete points in this diagram to emphasize the location of the stages. The arrows are traditionally shown in the direction of increasing temperature. Your understanding of the procedure for plotting these curves will be aided significantly by studying Example 8-3 and by doing Problem 8.H1.

Figure 8-8 (Biegler et al., 1997) shows the total reflux distillation curves for acetone, chloroform, benzene distillation on an equilateral triangle diagram. This system has a maximum boiling azeotrope between acetone ($x_{acetone} \sim 0.34$) and chloroform. To reach compositions on the right side of the diagram we need to start with locations that are to the right of *distillation boundary* curve c. For compositions on the left side of the diagram we have to start with locations that are to the left of the distillation boundary. For a given feed concentration, only part of the space in Figure 8-8 can be reached for distillate and bottoms products. If we want to produce pure acetone in a single column, the feed needs to be to the left of the distillation boundary. To produce pure chloroform, the feed needs to be to the right of the distillation boundary. Although the distillation boundary will move slightly at finite reflux ratios, this basic principle still holds. Figure 8-8 is unusual because it shows a relatively rare maximum boiling azeotrope. All of the other systems we will consider are the much more common minimum boiling azeotropes.

We can also do mass balances on the triangular diagrams. First, consider the separation of binaries, which occur along the y-axis, the x-axis, and the hypotenuse of the right triangle. In Figure 8-9 the binary separation of the heavy (highest boiling) component H from the intermediate component I occurs along the y-axis (line $B_1F_1D_1$).

In Chapter 3 we developed Eq. (3-3) for binary separations. This equation applies to each component in the binary separations. For example, for separation of intermediate and heavy components,

$$\frac{D}{F} = \frac{x_{I,feed} - x_{I,bot}}{x_{I,dist} - x_{I,bot}} = \frac{x_{H,feed} - x_{H,bot}}{x_{H,dist} - x_{H,bot}} \tag{8-26}$$

Equation (8-26) remains valid if a third (or fourth or more) components are present in the distillation, and we can write a similar equation for every component. Thus, Eq. (8-26) applies for the general ternary separation shown as line BFD. This equation also proves that the points representing bottoms, distillate and feed all lie on a straight line and that the lever-arm rule applies. [This is very similar to the graphical solution developed for the enthalpy-composition diagram of Figure 2-9 and the lever-arm rule derived in Eq. (2-26). The rules for mass balances on triangular diagrams will be developed in detail in Section 13-8.] As a first approximation, the points representing the bottoms and distillate products from a distillation column with a

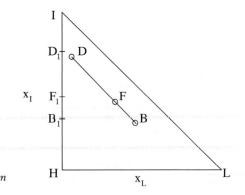

FIGURE 8-9.   *Mass balance on triangular diagram*

Chapter 8   Introduction to Complex Distillation Methods

single feed will lie on the distillation and residue curves. The path traced from distillate to bottoms must follow the same direction as the arrows (increasing temperature). Thus, these curves will show us if the separation indicated by the line BFD is feasible. If there is a distillation boundary as in Figure 8-8, not all separations will be feasible.

8.5.2 Residue Curves

We could do all of our calculations with distillation curves at total and finite reflux ratios; however, these curves depend to some extent on the distillation system. It is convenient to use a thermodynamically based curve that does not depend on the number of stages. A *residue curve* is generated by putting a mixture in a still pot and boiling it without reflux until the pot runs dry and only the residue remains. This is a simple batch distillation that will be discussed in more detail in Chapter 9. The plot of the changing mole fracs on a triangular diagram is the residue curve. It will be similar but not identical to the distillation curves shown in Figures 8-7 and 8-8. The differences between distillation curves and residue curves are explored by Widagdo and Seider (1996).

A simple equilibrium still is shown in Figure 8-10. As the distillation continues the molar holdup of liquid H decreases. The unsteady state overall mass balance is,

$$\frac{dH}{dt} = -V$$

(8-27a)

where V is the molar rate (not necessarily constant) at which vapor is removed. The component mass balances will have a similar form.

$$\frac{d(Hx_i)}{dt} = -Vy_i$$

(8-27b)

Expanding the derivative, substituting in Eq. (8-27a) and rearranging, we obtain

$$\frac{dx_i}{dt} = \frac{V}{H}(x_i - y_i)$$

(8-28)

This derivation closely follows the derivations in Biegler et al. (1997) and Doherty and Malone (2001). An alternate derivation is given in sections 9.1 and 9.2.

Integration of Eq. (8-28) gives us the values of x_i vs. time and allows us to plot the residue curve. This integration can be done with any suitable numerical integration technique; however, the vapor mole frac y in equilibrium with x must be determined at each time step. Although Doherty and Malone (2001) recommend the use of either Gear's method or a fourth order Runge-Kutta integration, they note that Eq. (8-28) is well behaved and can be integrated with Euler's method. This result is particularly simple,

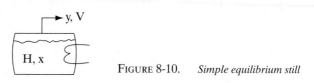

FIGURE 8-10.   *Simple equilibrium still*

$$x_{i,k+1} = x_{i,k} + h(x_{i,k} - y_{i,k})$$ **(8-29)**

where k refers to the step number and h is the step size. A step size of h = 0.01 or smaller is recommended (Doherty and Malone, 2001). A case where bubble-point calculations are required for each step is discussed in HW Problem 8.H3. If relative volatilities are constant, we can determine y from Eq. (5-30) and the recursion relationship simplifies to

$$x_{i,k+1} = x_{i,k} + h\left(x_{i,k} - \frac{\alpha_{i\text{-ref}}x_{i,k}}{\sum\limits_{i=1}^{c} \alpha_{i\text{-ref}}x_{i,k}}\right)$$ **(8-30)**

Despite the fact that process simulators will do these calculations for us, doing the integration for a simple case will greatly increase your understanding. Thus, studying Example 8-3 and doing Problem 8.H2 are highly recommended.

Siirola and Barnicki (1997) and Doherty et al. (2008) show simplified residue curve plots for all 125 possible systems. The most common residue curve is the plot for ideal distillation, which is similar to Figure 8-7. Next most common will be systems with a single minimum boiling azeotrope occurring between one of the sets of binary pairs. The three possibilities are shown in Figure 8-11 (Doherty and Malone, 2001). Figure 8-11c is of interest since it is one of the two residue curves that occur in extractive distillation (the other is Figure 8-7 with small relative volatilities). As mentioned earlier, a residue curve plot for a maximum boiling azeotrope as shown in Figure 8-8 will be rare. One can also have multiple binary and ternary azeotropes (Doherty et al., 2008; Siirola and Barnecki, 1997). Heterogeneous ternary azeotropes can also occur and are important in azeotropic distillation (Section 8.7). Figure 8-12 (Doherty and Malone, 2001) is an example of the residue curves that occur in azeotropic distillation with added solvent. The systems shown in Figures 8-11c and 8-12 are often formed on purpose by adding a solvent to a binary azeotropic system.

This completes the introduction to residue curves. Residue curves will be used in the explanation of extractive distillation (Section 8.6) and azeotropic distillation (Section 8.7). In Section 11-6 we will use residue curves to help synthesize distillation sequences for complex systems. Doherty and Malone (2001) develop the properties and applications of residue curves in much more detail than can be done in this introduction.

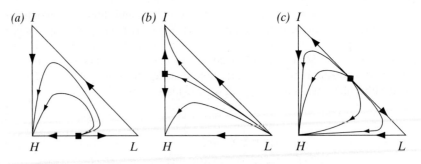

FIGURE 8-11. *Schematics of residue curve maps when there is one binary minimum-boiling azeotrope (Doherty and Malone, 2001); reprinted with permission of McGraw-Hill, copyright 2001, McGraw-Hill.*

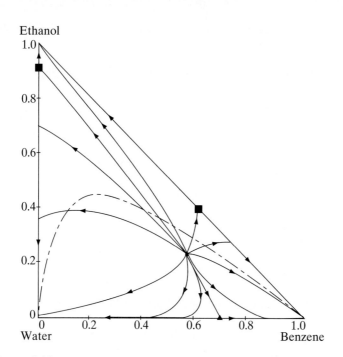

Ethanol

Water   Benzene

FIGURE 8-12.   *Calculated residue curve map for ethanol-water-benzene (Doherty and Malone, 2001). The black squares are binary homogeneous azeotropes, the black triangle is a heterogeneous binary azeotrope, the black dot in the center of the diagram is a heterogeneous ternary azeotrope, and the dot-dash line represents the solubility envelope for the two liquid layers. Reprinted with permission of McGraw-Hill, copyright 2001, McGraw-Hill.*

Example 8-3. Development of distillation and residue curves for constant relative volatility

We plan to study the distillation of the ideal system A = benzene, B = toluene, and C = cumene by generating total reflux distillation curves and residue curves. The equilibrium data for this system can be approximated by the following constant relative volatilities: $\alpha_{AB} = 2.4$, $\alpha_{BB} = 1.0$, and $\alpha_{CB} = 0.21$.

a. Generate the total reflux distillation curve starting with the following reboiler mole fractions: $x_{A,reb} = 0.001$, $x_{B,reb} = 0.009$, and $x_{C,reb} = 0.990$.

b. Generate the residue curve with the following initial mole fractions in the still pot: $x_A = 0.9905$, $x_B = 0.0085$, and $x_C = 0.001$.

Solution Part a. This calculation can be done with a process simulator or with the recursion relationship, Eq. (8-25b). Since the simulator hides the calculation procedure, we use Eq. (8-25b). Starting with the reboiler (j = 0) mole fractions, $x_{A,reb} = 0.001$, $x_{B,reb} = 0.009$, and $x_{C,reb} = 0.990$, we calculate the mole fractions on the stage above the reboiler (j = 1) from Eq. (8-25b). The denominator in Eq. (8-25b) for j + 1 = 1 is,

$\Sigma(\alpha_{i\text{-}B} \; x_{i,0}) = \alpha_{AB} \; x_{A,0} + \alpha_{BB} \; x_{B,0} + \alpha_{CB} \; x_{C,0} = 2.4(0.001) + 1.0 \; (.009) + 0.21(.990) = 0.21915$

Then the individual mole fractions are calculated from Eq. (8-25b):

$x_{A,1} = \alpha_{AB} \; x_{A,0} / \Sigma(\alpha_{i\text{-}B} \; x_{i,0}) = 2.4 \; (0.001)/0.21915 = 0.0103$

$x_{B,1} = \alpha_{BB} \; x_{B,0} / \Sigma(\alpha_{i\text{-}B} \; x_{i,0}) = 1.0 \; (0.009)/0.21915 = 0.0411$

$x_{C,1} = \alpha_{CB} \; x_{C,0} / \Sigma(\alpha_{i\text{-}B} \; x_{i,0}) = 0.21 \; (0.99)/0.21915 = 0.9487$

Increasing j by 1, we can calculate the denominator for j+1 = 2,

$\Sigma(\alpha_{i\text{-}B} \; x_{i,1}) = \alpha_{AB} \; x_{A,1} + \alpha_{BB} \; x_{B,1} + \alpha_{CB} \; x_{C,1} = 2.4(0.0103) + 1.0 \; (.0411) + 0.21(.9487) = 0.263388$

And the individual mole fractions for stage 2 are,

$x_{A,2} = \alpha_{AB} \; x_{A,1} / \Sigma(\alpha_{i\text{-}B} \; x_{i,1}) = 2.4 \; (0.010267)/0.263388 = 0.0877$

$x_{B,2} = \alpha_{BB} \; x_{B,1} / \Sigma(\alpha_{i\text{-}B} \; x_{i,1}) = 1.0 \; (0.041068)/0.263388 = 0.1559$

$x_{C,2} = \alpha_{CB} \; x_{C,1} / \Sigma(\alpha_{i\text{-}B} \; x_{i,1}) = 0.21 \; (0.948655)/0.263388 = 0.7564$

Continuing the calculation we generate the following values for the remaining stages:

Stage 3:	$x_A = 0.3854$	$x_B = 0.3045$	$x_C = 0.3102$
Stage 4:	$x_A = 0.7011$	$x_B = 0.2462$	$x_C = 0.0527$
Stage 5:	$x_A = 0.8598$	$x_B = 0.1342$	$x_C = 0.0060$
Stage 6:	$x_A = 0.9346$	$x_B = 0.0649$	$x_C = 0.0006$
Stage 7:	$x_A = 0.9700$	$x_B = 0.02991$	$x_C = 0.59 \text{ E-4}$
Stage 8:	$x_A = 0.9865$	$x_B = 0.0135$	$x_C = 0.56 \text{ E-5}$

These values were generated from a simple spreadsheet and are plotted in Figure 8-7. Obviously, the calculation can be continued for as many stages as desired. Problem 8.H1 asks that you generate additional values for Figure 8-7.

Solution Part b. Although process simulators will determine residue curves, it is instructive to do a calculation ourselves. This calculation uses Eq. (8-30), starting with still pot mole fractions $x_{A,1} = 0.9905$, $x_{B,1} = 0.0085$, and $x_{C,1} = 0.001$ (where index 1 represents k = 1, which is step 1 in the integration). We will show one hand calculation and then the VBA program to do the calculation for h = 0.01 and N (maximum number of iterations) = 1000. For k = 1,

$\Sigma(\alpha_{i\text{-}B} \; x_{i,1}) = \alpha_{AB} \; x_{A,1} + \alpha_{BB} \; x_{B,1} + \alpha_{CB} \; x_{C,1} = 2.4(0.9905) + 1.0 \; (.0085) + 0.21(.001) = 2.38591$

And Eq. (8-30) is

$$x_{i,2} = x_{i,1} + h[x_{i,1} - \alpha_{i\text{-}B} \, x_{i,1} / \Sigma(\alpha_{i\text{-}B} \, x_{i,1})],$$ which for each component is,

$$x_{A,2} = x_{A,1} + h[x_{A,1} - \alpha_{A\text{-}B} \, x_{A,1} / \Sigma(\alpha_{i\text{-}B} \, x_{i,1})] = 0.9905 + 0.01[0.9905 - 2.4(.9905)/2.38591] = 0.99044$$

$$x_{B,2} = x_{B,1} + h[x_{B,1} - \alpha_{B\text{-}B} \, x_{B,1} / \Sigma(\alpha_{i\text{-}B} \, x_{i,1})] = 0.0085 + 0.01[0.0085 - 1.0(.0085)/2.38591] = 0.008547$$

$$x_{C,2} = x_{C,1} + h[x_{C,1} - \alpha_{C\text{-}B} \, x_{C,1} / \Sigma(\alpha_{i\text{-}B} \, x_{i,1})] = 0.001 + 0.01[0.001 - 0.21(.001)/2.38591] = 0.001009$$

Setting k = 2 in Eq. (8-30), we can continue. This process is obviously laborious to do by hand but is easily programmed in VBA. A program that will do this calculation is,

```
Option Explicit
Sub Residue_Curve()
' Calculation of constant alpha residue curve. Declare variables.
Dim i, N, j As Integer
Dim alphaA, alphaB, alphaC, sumalx, h, xA, xB, xC As Double
Sheets("Sheet1").Select
Range("A4", "D1005").Clear
' Read values
alphaA = Cells(1, 2).Value
alphaB = Cells(1, 4).Value
alphaC = Cells(1, 6).Value
h = Cells(2, 4).Value
N = Cells(2, 6).Value
' Initial values in still pot
xA = Cells(3, 2).Value
xB = Cells(3, 4).Value
xC = Cells(3, 6).Value
' Do loop for calculation of Eq. (8-30) and print values
For i = 1 To N
j = i + 1
sumalx = xA * alphaA + xB * alphaB + xC * alphaC
xA = xA * (1 + h * (1 - alphaA / sumalx))
xB = xB * (1 + h * (1 - alphaB / sumalx))
xC = xC * (1 + h * (1 - alphaC / sumalx))
Cells(4 + i + 1, 1).Value = j
Cells(4 + i + 1, 2).Value = xA
Cells(4 + i + 1, 3).Value = xB
Cells(4 + i + 1, 4).Value = xC
Next i
End Sub
```

The results obtained are:

k	x_A	x_B	x_C
1	0.9905	0.0085	0.0010
100	0.9829	0.0147	0.0024
200	0.9686	0.0253	0.0060
300	0.9416	0.0436	0.0148
400	0.8898	0.0739	0.0363
500	0.7895	0.1220	0.0885
600	0.5992	0.1880	0.2127
700	0.2780	0.2332	0.4887
800	0.0103	0.0962	0.8935
900	0.14E-5	0.0036	0.9964
1000	0.5 E-11	0.75E-4	0.9999

These values can be plotted to form a residue curve that will look similar to Figure 8-7 except the curves should be drawn as smooth curves. Problem 8.H2 asks students to generate additional residue curves. Note: If too large a value of h is used, the recursion formula can give negative mole fractions or mole fractions greater than one. Obviously, these values should not be used.

8.6 EXTRACTIVE DISTILLATION

Extractive distillation is used for the separation of azeotropes and close-boiling mixtures. In extractive distillation, a solvent is added to the distillation column. This solvent is selected so that one of the components, B, is selectively attracted to it. Since the solvent is usually chosen to have a significantly higher boiling point than the components being separated, the attracted component, B, has its volatility reduced. Thus, the other component, A, becomes relatively more volatile and is easy to remove in the distillate. A separate column is required to separate the solvent and component B. The residue curves, after the solvent is added, are shown in Figure 8-7 (separation of close boiling components) and 8-11C (separation of azeotropes).

A typical flowsheet for separation of a binary mixture is shown in Figure 8-13. If an azeotrope is being separated, the feed should be close to the azeotrope concentration; thus, a binary distillation column (not shown) usually precedes the extractive distillation system. In column 1 the solvent is added several stages above the feed stage and a few stages below the top of the column. In the top section, the relatively nonvolatile solvent is removed and pure A is produced as the distillate product. In the middle section, large quantities of solvent are present and components A and B are separated from each other. It is common to use 1, 5, 10, 20, or even 30 times as much solvent as feed; thus, the solvent concentration in the middle section is often quite high. Note that the A-B separation must be complete in the middle section, because any B that gets into the top section will not be separated (there is very little solvent present) and will exit in the distillate. The bottom section strips the A from the mixture so that only solvent and B exit from the bottom of the column.

The residue curve diagram for the extractive distillation of ethanol and water using ethylene glycol as the solvent is shown in Figure 8-14. The curves start at the binary azeotrope

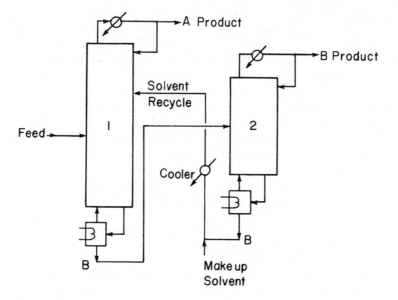

FIGURE 8-13.   *Extractive distillation flowsheet*

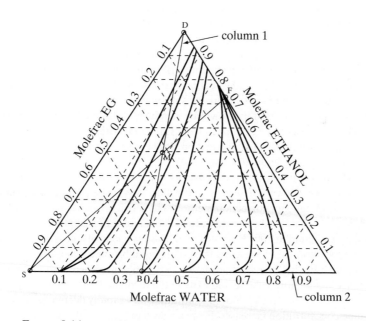

FIGURE 8-14.   *Residue curves for water-ethanol-ethylene glycol for extractive distillation to break ethanol-water azeotrope. Curves generated by Aspen Plus 2004 using NRTL for VLE. Mass balance lines FMS and BMD were added.*

(89% ethanol and 11% water), which makes this residue curve plot similar to Figure 8-11c. Since all of the residue curves have the same general shape, there is no distillation boundary in this system. (I suggest that you compare the shape of the residue curves in Figure 8-14 to those in Figure 8-8, which has a distillation boundary.) The feed to column 1 is 100 kmol/h of a mixture that is 72 mol% ethanol and 27 mol% water. Essentially pure solvent is added at a rate of 52 kmol/h. (Lab 8 in the Appendix to Chapter 8.) The solvent and feed can be combined as a mixed feed M (found from the lever-arm rule), which is then separated in column 1 into streams D and B as shown in the figure. Detailed simulations are needed to determine if the solvent flow rate and reflux ratio are large enough. Increasing the solvent flow rate will move point M towards the solvent vertex. Binary stream B is easily separated in column 2. Although this system can be used to break the ethanol-water azeotrope, azeotropic distillation (see Section 8.7) is more economical. The residue curves from D to B both stop at S. Extractive distillation works because the solvent is added to the column several stages above the feed stage. Complete analysis of extractive distillation is beyond this introductory treatment. Doherty and Malone (2001) provide detailed information on the use of residue curve maps to design extractive distillation systems.

The mixture of solvent and B is sent to column 2 where they are separated along line SBW. If the solvent is selected correctly, the second column can be quite short, since component B is significantly more volatile than the solvent. The recovered solvent can be cooled and stored for reuse in the extractive distillation column. Note that the solvent must be cooled before entering column 1, since its boiling point is significantly higher than the operating temperature of column 1.

Column 2 is a simple distillation that can be designed by the methods discussed in Chapter 4. Column 1 is considerably more complex, but the bubble-point matrix method discussed in Chapter 6 can often be adapted. Since the system is nonideal and K values depend on the solvent concentration, a concentration loop is required in the flowchart shown in Figure 6-1. Fortunately, a good first guess of solvent concentrations can be made. Solvent concentration will be almost constant in the middle section and also in the bottom section except for the reboiler. In the top section of the column, the solvent concentration will very rapidly decrease to zero. These solvent concentrations will be relatively unaffected by the temperatures and flow rates. The K values can be calculated from Eq. (2-35) with the activity coefficients determined from the appropriate VLE correlation. Process simulators are the easiest way to do these calculations (see appendix to Chapter 8).

Concentration profiles for the extractive distillation of n-heptane and toluene using phenol as the solvent are shown in Figure 8-15 (Seader, 1984). The profiles were rigorously calculated using a simultaneous correction method, and activity coefficients were calculated with the Wilson equation. The feed was a mixture of 200 lbmol/h of n-heptane and 200 lbmol/h of toluene input as a liquid at 200 °F on stage 13. The recycled solvent is input on stage 6 at a total rate of 1200 lbmol/h. There are a total of 21 equilibrium contacts including the partial reboiler. The high-boiling phenol is attractive to the toluene, since both are aromatics. The heptane is then made more volatile and exits in the distillate (component A in Figure 8-13). Note from Figure 8-15 that the phenol concentration very rapidly decreases above stage 6 and the n-heptane concentration increases. From stages 6 to 12, phenol concentration is approximately constant and the toluene is separated from the heptane. From stages 13 to 20, phenol concentration is again constant but at a lower concentration. This change in the solvent (phenol) concentration occurs because the feed is input as a liquid. A constant solvent concentration can be obtained by vaporizing the feed or by adding some recycled solvent to it. Heptane

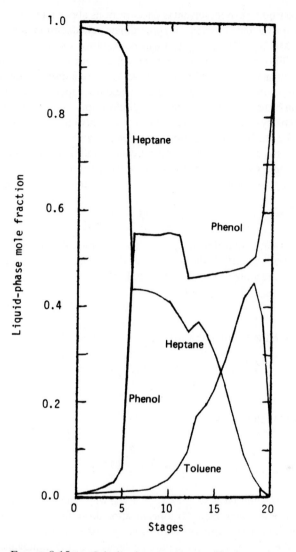

FIGURE 8-15. *Calculated composition profiles for extractive distillation of toluene and n-heptane; from Seader (1984), reprinted with permission from* Perry's Chemical Engineer's Handbook, *6th ed, copyright 1984, McGraw-Hill.*

is stripped from the mixture in stages 13 to 20. In the reboiler, the solvent is nonvolatile compared to toluene. Thus the boilup is much more concentrated in toluene than in phenol. The result is the large increase in phenol concentration seen for stage 21 in Figure 8-15. Concentration profiles for other extractive distillation systems are shown by Robinson and Gilliland (1950), Siirola and Barnicki (1997), Doherty and Malone (2001), and Doherty et al. (2008).

The almost constant phenol (solvent) concentrations above and below the feed stage in Figure 8-15 allow for the development of approximate calculation methods (Knickle, 1981). A pseudo-binary McCabe-Thiele diagram can be used for the separation of heptane and toluene. The "equilibrium" curve represents the y-x data for heptane and toluene with constant phenol concentration. The equilibrium curve will have a discontinuity at the feed stage. For the stages above the solvent feed stage, a standard heptane-phenol binary McCabe Thiele diagram can be used. Knickle (1981) also discusses modifying the Fenske-Underwood-Gilliland approach by modifying the relative volatility to represent heptane-toluene equilibrium with the phenol concentration of the feed stage.

The reason for the large increase in solvent concentration in the reboiler is easily seen if we look at an extreme case where none of the solvent vaporizes in the reboiler. Then the boilup is essentially pure component B. The liquid flow rate in the column can be split up as

$$\overline{L} = S + B_{liq} + \overline{V} \tag{8-31}$$

where S is the constant solvent flow rate; B_{liq} is the flow rate of component B, which stays in the liquid in the reboiler; and $\overline{V}$ is the flow rate of the vapor, which is mainly B. The bottoms flow rate consists of the streams that remain liquid,

$$Bot = S + B_{liq} \tag{8-32}$$

The mole frac of B in the liquid in the column can be estimated as

$$x_{B,col} = \frac{B_{liq} + \overline{V}}{S + B_{liq} + \overline{\overline{V}}} \tag{8-33}$$

and the mole frac B in the bottoms is

$$x_{B,bot} = \frac{B_{liq}}{S + B_{liq}} \tag{8-34}$$

For the usual flow rates this gives $x_{B,col} >> x_{B,bot}$. Even when the solvent is fairly volatile, as in Figure 8-15, the toluene (component B) concentration drops in the reboiler.

The behavior of even the simplest azeotropic distillations systems where an entrainer is added to recover the two components as pure products (e.g., the homogeneous systems illustrated in Figure 8-11, which include extractive distillation) is often different than the behavior of more ideal ternary distillation systems (Doherty and Malone, 2001; Laroche et al., 1992). For example, feasible operation may require recovering the middle component as distillate from the first column, not the light component as would normally be expected. We expect that adding trays will reduce the reflux ratio needed, but with homogeneous azeotropic systems including extractive distillation, adding trays may increase the required reflux ratio. Similarly, increasing the reflux ratio past an optimum value may cause the product purities to decrease, and poor separation is observed at total reflux.

Selection of the solvent is extremely important. The process is similar to that of selecting a solvent for liquid-liquid extraction, which is discussed in Chapter 13. By definition, the solvent should not form an azeotrope with any of the components (Figure 8-11C). If the solvent does form an azeotrope, the process becomes azeotropic distillation, which is discussed in the next section. Usually, a solvent is selected that is more similar to the heavy key. Then the volatility of the heavy key will be reduced. Exceptions to this rule exist; for example, in the n-butane—1-butene system, furfural decreases the volatility of the 1-butene, which is more volatile (Shinskey, 1984). Lists of extractants (Doherty et al., 2008; Van Winkle, 1967) for extractive distillation are helpful in finding a general structure that will effectively increase the volatility of the keys. Salts including ionic liquids can be used as the "solvent," particularly when there are small amounts of water to remove (Fu, 1996; Furter, 1993). Residue curve analysis is also very helpful to ensure that the solvent does not form additional azeotropes (Biegler et al., 1997; Doherty and Malone, 2001; Julka et al., 2009).

Solvent selection can be aided by considering the polarities of the compounds to be separated. A short list of classes of compounds arranged in order of increasing polarity is given in Table 8-1. If two compounds of different polarity are to be separated, a solvent can be selected to attract either the least polar or the most polar of the two. For example, suppose we wish to separate acetone (a ketone boiling at 56.5°C) from methanol (an alcohol boiling at 64.7°C). This system forms an azeotrope.We could add a hydrocarbon to attract the acetone, but if enough hydrocarbon were added, the methanol would become more volatile. A simpler alternative is to add water, which attracts the methanol and makes acetone more volatile. The methanol and water are then separated in column 2. In this example, we could also add a higher molecular weight alcohol such as butanol to attract the methanol.

When two hydrocarbons are to be separated, the larger the difference in the number of double bonds the better a polar solvent will work to change the volatility. For example, furfural will decrease the volatility of butenes compared to butanes. Furfural (a cyclic alcohol) is used instead of water because the hydrocarbons are miscible with furfural. A more detailed analysis of solvent selection shows that hydrogen bonding is more important than polarity (Berg, 1969; Doherty et al., 2008; Smith, 1963).

Once a general structure has been found, homologs of increasing molecular weight can be checked to find which has a high enough boiling point to be easily recovered in column 2. However, too high a boiling point is undesirable, because the solvent recovery column would have to operate at too high a temperature. The solvent should be completely miscible with both components over the entire composition range of the distillation.

It is desirable to use a solvent that is nontoxic, nonflammable, noncorrosive, and nonreactive. In addition, it should be readily available and inexpensive since solvent makeup and inventory costs can be relatively high. Environmental effects and life-cycle costs of various solvents need to be included in the decision (Allen and Shonnard, 2002). As usual, the

TABLE 8-1.   *Increasing polarities of classes of compounds*

| Hydrocarbons |
| Ethers |
| Aldehydes |
| Ketones |
| Esters |
| Alcohols |
| Glycols |
| Water |

designer must make tradeoffs in selecting a solvent. One common compromise is to use a solvent that is used elsewhere in the plant or is a by-product of a reaction even if it may not be the optimum solvent otherwise.

For isomer separations, extractive distillation usually fails, since the solvent has the same effect on both isomers. For example, Berg (1969) reported that the best entrainer for separating m- and p-xylene increased the relative volatility from 1.02 to 1.029. An alternative to normal extractive distillation is to use a solvent that preferentially and reversibly reacts with one of the isomers (Doherty and Malone, 2001). The process scheme will be similar to Figure 8-14, with the light isomer being product A and the heavy isomer product B. The forward reaction occurs in the first column, and the reaction product is fed to the second column. The reverse reaction occurs in column 2, and the reactive solvent is recycled to column 1. This procedure is quite similar to the combined reaction-distillation discussed in Section 8.8.

8.7   AZEOTROPIC DISTILLATION WITH ADDED SOLVENT

When a homogeneous azeotrope is formed or the mixture is very close boiling, the procedures shown in Section 8.2 cannot be used. However, the engineer can add a solvent (or entrainer) that forms a binary or ternary azeotrope and use this to separate the mixture. The trick is to pick a solvent that forms an azeotrope that is either heterogeneous (then the procedures of Section 8.2 are useful) or easy to separate by other means such as extraction with a water wash. Since there are now three components, it is possible to have one or more binary azeotropes or a ternary azeotrope. The flowsheet depends upon the equilibrium behavior of the system, which can be investigated with distillation curves and residue curves (Section 8.5). A few typical examples will be illustrated here.

Figure 8-16 shows a simplified flowsheet (extensive heat exchange is not shown) for the separation of butadiene from butylenes using liquid ammonia as the entrainer (Poffenberger et al., 1946). Note the use of the intermediate reboiler in the azeotropic distillation column to minimize polymerization. At 40°C the azeotrope is homogeneous. The ammonia can be recovered by cooling, since at temperatures below 20°C two liquid phases are formed. The colder the operation of the settler the purer the two liquid phases. At the −40°C used in commercial plants during World War II, the ammonia phase contained about 7 wt % butylene. This ammonia is recycled to the azeotropic column either as reflux or on stage 30. The top phase is fed to the stripping column and contains about 5 wt % ammonia. The azeotrope produced in the stripping column is recycled to the separator. This example illustrates the following general points: (1) The azeotrope formed is often cooled to obtain two phases and/or to optimize the operation for the liquid-liquid settler. (2) Streams obtained from a settler are seldom pure and have to be further purified. This is illustrated by the stripping column in Figure 8-16. (3) Product (butylene) can often be recovered from solvent (NH_3) in a stripping column instead of a complete distillation column because the azeotrope is recycled.

Another system with a single binary azeotrope is shown in Figure 8-17 (Smith, 1963). In the azeotropic column, component A and the entrainer form a minimum boiling azeotrope, which is recovered as the distillate. The other component, B, is recovered as a pure bottoms product. In this case the azeotrope formed is homogeneous, and a water wash (extraction using water) is used to recover the solvent from the desired component with which it forms an azeotrope. Pure A is the product from the water wash column. A simple distillation column is required to recover the solvent from the water. Chemical systems using flow diagrams similar

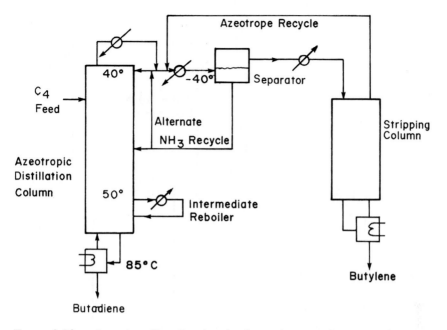

FIGURE 8-16. *Separation of butadiene from butylenes using ammonia as an entrainer (Poffenberger et al., 1946)*

to this include the separation of cyclohexane (A) and benzene (B), using acetone as the solvent, and the removal of impurities from benzene with methanol as the solvent.

A third example that is quite common is the separation of the ethanol water azeotrope using a hydrocarbon as the entrainer. Benzene used to be the most common entrainer (the residue curve is shown in Figure 8-12), but because of its toxicity it has been replaced by diethyl ether, n-pentane, or n-hexane. A heterogeneous ternary azeotrope is removed as the distillate product from the azeotropic distillation column. A typical flowsheet for this system is shown in Figure 8-18 (Black, 1980; Doherty and Malone, 2001; Doherty et al., 2008; Robinson and Gilliland, 1950; Seader, 1984; Shinskey, 1984; Smith, 1963; Widagdo and Seider, 1996). The feed to the azeotropic distillation column is the distillate product from a binary ethanol-water column and is close to the azeotropic composition. The composition of the ternary azeotrope will vary slightly depending upon the entrainer chosen. For example, when n-hexane is the entrainer the azeotrope contains 85 wt % hexane, 12 wt % ethanol, and 3 wt % water (Shinskey, 1984). The water/ethanol ratio in the ternary azeotrope must be greater than the water/ethanol ratio in the feed so that all the water can be removed with the azeotrope and excess ethanol can be removed as a pure bottoms product. The upper layer in the separator is 96.6 wt % hexane, 2.9 wt % ethanol and 0.5 wt % water, while the bottom layer is 6.2 wt % hexane, 73.7 wt % ethanol and 20.1 wt % water. The upper layer from the separator is refluxed to the azeotropic distillation column, while the bottom layer is sent to a stripping column to remove water.

Calculations for any of the azeotropic distillation systems are considerably more complex than for simple distillation or even for extractive distillation. The complexity arises from the obviously very nonideal equilibrium behavior and from the possible formation of three

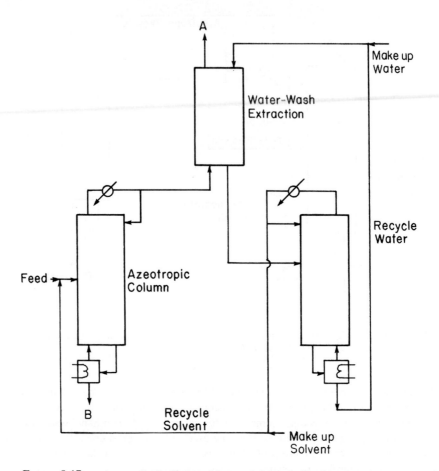

FIGURE 8-17.   *Azeotropic distillation with one minimum boiling binary azeotrope;*
use of water wash for solvent recovery from Smith (1963)

phases (two liquids and a vapor) inside the column. The residue curve shown in Figure 8-12 clearly demonstrates the complexity of these systems. Calculation procedures for azeotropic distillation are reviewed by Doherty and Malone (2001), Prokopakis and Seider (1983), and Widago Seider (1996).

Results of simulations have been presented by Black (1980), Hoffman (1964), Prokopakis and Seider (1983), Robinson and Gilliland (1950), Seader (1984), Smith (1963), and Widago and Seider (1996). Seader's (1984) results for the dehydration of ethanol using n-pentane as the solvent are plotted in Figure 8-19. The system used is similar to the flowsheet shown in Figure 8-18. The feed to the column contained 0.8094 mole frac ethanol. The column operated at a pressure of 331.5 kPa to allow condensation of the distillate with cooling water and had 18 stages plus a partial reboiler and a total condenser. The third stage below the condenser was the feed stage. Note that the profiles are different from those shown in Chapter 5. The pentane appears superficially to be a light key except that none of it appears in the bottoms. Instead, a small amount of the water exits in the bottoms with the ethanol.

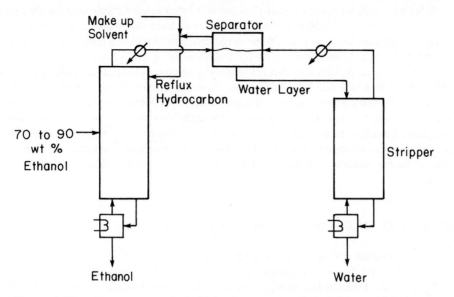

FIGURE 8-18.   *Ternary azeotropic distillation for separation of ethanol-water with hydro-carbon entrainer*

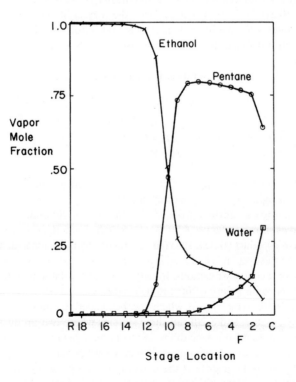

FIGURE 8-19.   *Composition profiles for azeotropic distillation column separating water and ethanol with n-pentane entrainer (Seader, 1984).*

Selecting a solvent for azeotropic distillation is often more difficult than for extractive distillation. There are usually fewer solvents that will form azeotropes that boil at a low enough temperature to be easy to remove in the distillate or boil at a high enough temperature to be easy to remove in the bottoms. Distillation curve and residue curve analyses are useful for screening prospective solvents and for developing new processes (Biegler et al., 1997; Doherty and Malone, 2001; Julka et al., 2009; Widago and Seider, 1996). In addition, the binary or ternary azeotrope formed must be easy to separate. In practice, this requirement is met by heterogeneous azeotropes and by azeotropes that are easy to separate with a water wash. The chosen entrainer must also satisfy the usual requirements of being nontoxic, noncorrosive, chemically stable, readily available, inexpensive, and green. Because of the difficulty in finding suitable solvents, azeotropic distillation systems with unique solvents are patentable.

8.8 DISTILLATION WITH CHEMICAL REACTION

Distillation columns are occasionally used as chemical reactors. The advantage of this approach is that distillation and reaction can take place simultaneously in the same vessel, and the products can be removed to drive the reversible reaction to completion. The most common industrial application is for the formation of esters from a carboxylic acid and an alcohol. For example, the manufacture of methyl acetate by reactive distillation was a major success that conventional processes could not compete with (Biegler et al., 1997). Reactive distillation was first patented by Backhaus in 1921 and has been the subject of many patents since then (see Doherty and Malone, 2001; Doherty et al., 2008; and Siirola and Barnicki, 1997, for references). Reaction in a distillation column may also be undesirable when one of the desired products decomposes.

Distillation with reaction is useful for reversible reactions. Examples would be reactions such as

$$A = C$$
$$A = C + D$$
$$A + B = C + D$$

The purposes of the distillation are to separate the product(s) from the reactant(s) to drive the reactions to the right, and to recover purified product(s).

Depending on the equilibrium properties of the system, different distillation configurations can be used as shown in Figure 8-20. Figure 8-20A shows the case where the reactant is less volatile than the product (Belck, 1955). If several products are formed, no attempt is made to separate them in this system. The bleed is used to prevent the buildup of nonvolatile impurities or products of secondary reactions. If the feed is more volatile than the desired product, the arrangement shown in Figure 8-20B can be used (Belck, 1955). This column is essentially at total reflux except for a small bleed, which may be needed to remove volatiles or gases.

Figures 8-20C, 8-20D, and 8-20E all show systems where two products are formed and the products are separated from each other and from the reactants in the distillation column. In Figure 8-20C the reactant(s) are of intermediate volatility between the two products. Then the reactants will stay in the middle of the column until they are consumed, while the products are continuously removed, driving the reaction to the right. If the reactants are not of

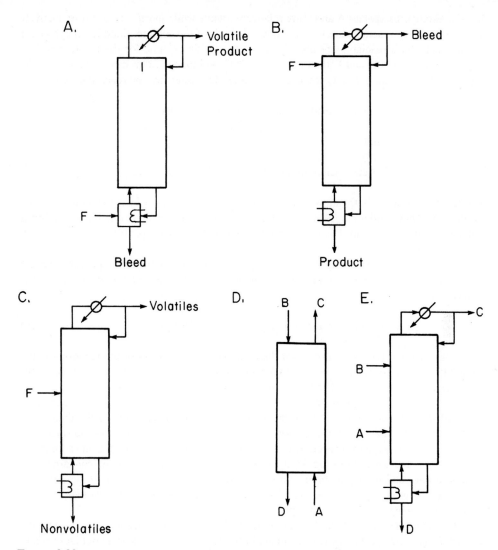

FIGURE 8-20. *Schemes for distillation plus reaction: (A) Volatile product, reaction is A = C; (B) non-volatile product, reaction is A = D; (C) Two products, reactions are A = C + D or A + B = C + D; (D, E) Reaction A + B = C + D with B and D nonvolatile.*

intermediate volatility, some of the reactants will appear in each product stream (Suzuki et al., 1971). The alternative schemes shown in Figures 8-20D and 8-20E (Siirola and Barnicki, 1997; Suzuki et al., 1971) will often be advantageous for the reaction

$$A + B = C + D$$

In these two figures, species A and C are relatively volatile while species B and D are relatively nonvolatile. Since reactants are fed in at opposite ends of the column, there is a much larger region where both reactants are present. Thus, the residence time for the reaction will be larger in Figures 8-20D and E than in Figure 8-20C, and higher yields can be expected. The systems shown in Figures 8-20C and D have been used for esterification reactions such as

$$\text{Acetic acid + ethanol} \rightarrow \text{ethyl acetate + water}$$

(Suzuki et al., 1971) and

$$\text{Acetic acid + methanol} \rightarrow \text{methyl acetate + water}$$

Siirola and Barnicki (1997) show a four-component residue curve map for methyl acetate production. They used a modification of Figure 8-20E where a nonvolatile liquid catalyst is fed between the acetic acid (B) and methanol (A) feeds. They show profiles for the four components and the catalyst.

When a reaction occurs in the column, the mass and energy balance equations must be modified to include the reaction terms. The general mass balance equation for stage j (Eq. 6-1) can be modified to

$$V_j y_{i,j} + L_j x_{i,j} - V_{j-1} y_{i,j-1} - L_{j+1} x_{i,j+1} = F_j z_{i,j} + r_{i,j} \qquad (8\text{-}35)$$

where the reaction term $r_{i,j}$ is positive if the component is a product of the reaction. To use Eq. (8-35), the appropriate rate equation for the reaction must be used for $r_{i,j}$. In general, the reaction rate will depend on both the temperature and the liquid compositions.

When the mass balances are in matrix form, the reaction term can conveniently be included with the feed in the D term in Eqs. (6-6) and (6-13). This retains the tridiagonal form of the mass balance, but the D term depends upon liquid concentration and stage temperature. The convergence procedures to solve the resulting set of equations must be modified, because the procedures outlined in Chapter 6 may not be able to converge. The equations have become highly nonlinear because of the reaction rate term. Modern process simulators are usually able to converge for reactive distillation problems, although it may be necessary to change the convergence properties.

A sample of composition and temperature profiles for the esterification of acetic acid and ethanol is shown in Figure 8-21 (Suzuki et al., 1971) for the distillation system of Figure 8-20C. The distillation column is numbered with 13 stages including the total condenser (No. 1) and the partial reboiler (No. 13). Reaction can occur on every stage of the column and in both the condenser and the reboiler. Feed is introduced to stage 6 as a saturated liquid. The feed is mainly acetic acid and ethanol with a small amount of water. A reflux ratio of 10 is used. The top product contains most of the ethyl acetate produced in the reaction plus ethanol and a small amount of water. All of the nonreacted acetic acid appears in the bottoms along with most of the water and a significant fraction of the ethanol. Reaction is obviously not complete.

A somewhat different type of distillation with reaction is "catalytic distillation" (Doherty et al., 2008; Parkinson, 2005). In this process bales of catalyst are stacked in the column. The bales serve both as the catalyst and as the column packing (see Chapter 10). This process was used commercially for production of methyl tert-butyl ether (MTBE) from the liquid-phase reaction of isobutylene and methanol. The heat generated by the exothermic reaction is used

Chapter 8 Introduction to Complex Distillation Methods

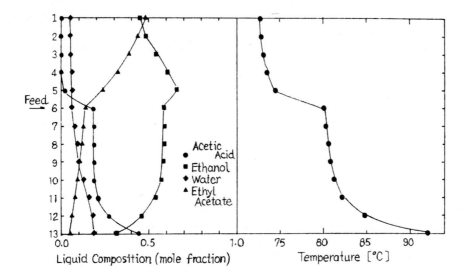

FIGURE 8-21. *Composition and temperature profiles for the reaction acetic acid + ethanol = ethyl acetate + water from Suzuki* et al. *(1971), copyright 1971. Reprinted with permission from* Journal of Chemical Engineering of Japan.

to supply much of the heat required for the distillation. Since MTBE use as a gasoline additive has been outlawed because of pollution problems from leaky storage tanks, these units are shut down. Other applications of catalytic distillation include desulfurization of gasoline, separation of 2-butene from a mixed C_4 stream, esterification of fatty acids and etherification.

Although many reaction systems do not have the right reaction equilibrium or VLE characteristics for distillation with reaction, for those that do this technique is a very valuable industrial tool.

8.9 SUMMARY—OBJECTIVES

In this chapter we have looked at azeotropic and extractive distillation systems plus distillation with simultaneous chemical reaction. At the end of this chapter you should be able to satisfy the following objectives:

1. Analyze binary distillation systems using other separation schemes to break the azeotrope
2. Solve binary heterogeneous azeotrope problems, including the drying of organic solvents, using McCabe-Thiele diagrams
3. Explain and analyze steam distillation
4. Use McCabe-Thiele diagrams or process simulators to solve problems where two pressures are used to separate azeotropes
5. Determine the possible products for a ternary distillation using residue curves
6. Explain the purpose of extractive distillation, select a suitable solvent, explain the expected concentration profiles, and do the calculations with a process simulator

7. Use a residue curve diagram to determine the expected products for an azeotropic distillation with an added solvent
8. Explain qualitatively the purpose of doing a reaction in a distillation column, and discuss the advantages and disadvantages of the different column configurations

REFERENCES

Allen, D. T., and D. R. Shonnard, *Green Engineering: Environmentally Conscious Design of Chemical Processes*, Prentice Hall, Upper Saddle River, NJ, 2002.

Belck, L. H., "Continuous Reactions in Distillation Equipment," *AIChE J.*, *1*, 467 (1955).

Berg, L., "Selecting the Agent for Distillation Processes," *Chem. Engr. Progr., 65* (9), 52 (Sept. 1969).

Biegler, L. T., I. E. Grossmann, and A. W. Westerberg, *Systematic Methods of Chemical Process Design,* Prentice Hall, Upper Saddle River, NJ, 1997.

Black, C., "Distillation Modeling of Ethanol Recovery and Dehydration Processes for Ethanol and Gasohol," *Chem. Engr. Progr., 76* (9), 78 (Sept. 1980).

Chambers, J. M., "Extractive Distillation, Design and Application," *Chem. Engr. Progr., 47*, 555 (1951).

Chu, J. C., R. J. Getty, L. F. Brennecke, and R. Paul, *Distillation Equilibrium Data,* Reinhold, New York, 1950.

Doherty, M. F., Z. T. Fidkowski, M. F. Malone, and R. Taylor. "Distillation," in D. W. Green and R. H. Perry (Eds.), *Perry's Chemical Engineer's Handbook*, 8th ed., McGraw-Hill, New York, Section 13 (2008).

Doherty, M., and M. Malone, *Conceptual Design of Distillation Systems,* McGraw-Hill, New York, 2001.

Drew, J. W., "Solvent Recovery," in P. A. Schweitzer (Ed.), *Handbook of Separation Techniques for Chemical Engineers,* 3rd ed., McGraw-Hill, New York, 1997, Section 1.6.

Ellerbe, R. W., "Steam Distillation/Stripping," in P. A. Schweitzer (Ed.), *Handbook of Separation Techniques for Chemical Engineers,* 3rd ed., McGraw-Hill, New York, 1997, Section 1.4.

Frank, T. C., "Break Azeotropes with Pressure-Sensitive Distillation," *Chem. Engr. Prog., 93* (4) 52 (April 1997).

Fu, J., "Salt-Containing Model for Simulation of Salt-Containing Extractive Distillation," *AIChE Journal, 42* (12), 3363 (Dec. 1996).

Furter, W. F., "Production of Fuel-Grade Ethanol by Extractive Distillation Employing the Salt Effect," *Separ. Purific. Methods, 22,* 1 (1993).

Hengstebeck, R. J., *Distillation: Principles and Design,* Reinhold, New York, 1961.

Hoffman, E. J., *Azeotropic and Extractive Distillation,* Interscience, New York, 1964.

Julka, V., M. Chiplunkar, and L. O'Young, "Selecting Entainers for Azeotropic Distillation," *Chem. Engr. Progress, 105* (3), 47 (March 2009).

Knickle, H. N., "Extractive Distillation," *AIChE Modular Instruction,* Series B: *Stagewise and Mass Transfer Operations.* Vol. 2: *Multicomponent Distillation,* AIChE, New York, 1981, pp. 50–59.

Kovach, J. W., III, and W. D. Seider, "Heterogeneous Azeotropic Distillation: Experimental and Stimulation Results," *AIChE J., 33* (8), 1300 (Aug. 1987).

Laroche, L., N. Bekiaris, H. W. Anderson, and M. Morari, "The Curious Behavior of Azeotropic Distillations—Implications for Entrainer Selection," *AIChE J.*, 38, (9), 1309 (Sept.1992).

Ludwig, E. E., *Applied Process Design,* 3rd ed., Vol. 2, Gulf Publishing Co., Houston, 1997.

Luyben, W. L., "Azeotropic Tower Design by Graph," *Hydrocarbon Processing, 52* (1), 109 (Jan. 1973).

Nelson, P. A., "Countercurrent Equilibrium Stage Separation with Reaction," *AIChE J., 17,* 1043 (1971).

Parkinson, G., "Distillation: New Wrinkles for an Age-Old Technology," *Chem. Engr. Prog., 101* (7) 10 (July 2005).

Perry, R. H., and D. W. Green (Eds.), *Perry's Chemical Engineers' Handbook*, 7th ed., McGraw-Hill, New York, 1997.

Poffenberger, N., L. H. Horsley, H. S. Nutting, and E. C. Britton, "Separation of Butadiene by Azeotropic Distillation with Ammonia," *Trans. Amer. Inst. Chem. Eng., 42,* 815 (1946).

Prokopakis, G. J., and W. D. Seider, "Dynamic Simulation of Azeotropic Distillation Towers," *AIChE J., 29,* 1017 (1983).

Robinson, C. S., and E. R. Gilliland, *Elements of Fractional Distillation,* 4th ed., McGraw-Hill, New York, 1950, chap. 10.

Seader, J. D., "Distillation," in R. H. Perry and D. W. Green (Eds.), *Perry's Chemical Engineers' Handbook*, 6th ed., McGraw-Hill, New York, 1984, Section 13.

Shinskey, F. G., *Distillation Control, For Productivity and Energy Conservation,* 2nd ed., McGraw-Hill, New York, 1984, chaps. 9 and 10.

Siirola, J. J., and S. D. Barnicki, "Enhanced Distillation," in Perry, R. H. and D. W. Green (Eds.), *Perry's Chemical Engineers' Handbook*, 7th ed., McGraw-Hill, New York, pp. 13-54 to 13-85, 1997.

Smith, B. D., *Design of Equilibrium Stage Processes*, McGraw-Hill, New York, 1963, chap. 11.

Suzuki, I., H. Yagi, H. Komatsu, and M. Hirata, "Calculation of Multicomponent Distillation Accompanied by a Chemical Reaction," *J. Chem. Eng. Japan, 4,* 26 (1971).

Van Winkle, M., *Distillation,* McGraw-Hill, New York, 1967.

Wankat, P. C., "Graphical Multicomponent Distillation," *AIChE Modular Instruction,* Series B: *Stagewise and Mass Transfer Operations.* Vol. 2: *Multicomponent Distillation,* AIChE, New York, 1981, p. 13-21.

Wasylkiewicz, S. K., "Troubleshooting Distillation Simulation," *Chem. Engr. Prog, 102* (6), 22 (June 2006).

Widago, S., and W. D. Seider, "Azeotropic Distillation," *AIChE Journal, 42,* 96 (1996).

Woodland, L. R., "Steam Distillation," in D. J. DeRenzo (Ed.), *Unit Operations for Treatment of Hazardous Industrial Wastes,* Noyes Data Corp., Park Ridge, NJ, 1978, pp. 849–868.

HOMEWORK

A. *Discussion Problems*

A1. Given a homogeneous azeotrope to separate, if you know nothing about the system, which method is least likely to be economical?
   a. Two-pressure distillation
   b. Extractive distillation
   c. Azeotropic distillation

A2. Which method(s) are commonly used for separating systems with relative volatilities close to 1.0 in addition to separating azeotropes?
   a. Two-pressure distillation
   b. Extractive distillation
   c. Azeotropic distillation

d. All of the above

e. a and b

f. a and c

g. b and c

h. None of the above.

A3. Why is a cooler required in Figure 8-13? Can this energy be reused in the process?

A4. Explain the purpose of the liquid-liquid settler in Figures 8-3A, 8-4A, and 8-18.

A5. Explain why the external mass balances are the same for Figures 8-3A and 8-6.

A6. Why are makeup solvent additions shown in Figures 8-13, 8-17, and 8-18?

A7. Explain why extra stages are not usually helpful in steam distillation.

A8. If a liquid mixture of n-butanol and water that is 20 mol% n-butanol is vaporized, what is the vapor composition? (See Figure 8-2.) Repeat for mixtures that are 10, 30, and 40 mol% n-butanol. Explain what is happening.

A9. When doing distillation with reaction, the column should be designed both as a reactor and as a distillation column. In what ways might these columns differ from normal distillation columns?

A10. Reactions are usually not desirable in distillation columns. If there is a reaction occurring, what can be done to minimize it?

A11. Develop your key relations chart for this chapter.

A12. In both steam distillation and heterogeneous azeotropic distillation, a component is added that forms a second liquid phase. Explain differences in the ways these two systems are normally operated.

A13. We plan to use extractive distillation (Figure 8-13) to separate ethanol from water with ethylene glycol as the solvent. The A product will be ethanol and the B product water. Stages are counted with condenser = 1, and reboiler = N. Feed plate for solvent recycle in column 1 is = NS. Feed stages are NF1 in column 1 and NF2 in column 2. Pick the best solution of the listed items.

  1. If too much solvent is found in the water product,

   a. Increase L/D in column 1.

   b. Increase the value of NS.

   c. Increase the number of stages in the stripping section of column 1.

   d. Increase the number of stages in the stripping section of column 2.

   e. Increase the number of stages in the enriching section of column 2.

  2. If excessive water is found in the ethanol product when fresh solvent is added (with no solvent recycle),

   a. Increase L/D in column 1.

   b. Increase the solvent rate.

   c. Increase the value of NS.

   d. Increase the number of stages in the stripping section of column 1.

  3. If excessive water is found in the ethanol product when solvent recycle is used, but was not when fresh solvent was used,

   a. Increase L/D in column 1.

   b. Increase the value of NS.

   c. Increase the number of stages in the stripping section of column 1.

   d. Reduce the water in the bottoms of column 2.

  4. If excessive ethanol is found in the water product,

   a. Increase L/D in column 1.

   b. Increase the value of NS.

c. Increase the number of stages in the stripping section of column 1.

d. Reduce the water in the bottoms of column 2.

e. Increase the number of stages in the enriching section of column 2.

5. If too much solvent occurs in the ethanol product,

    a. Increase L/D in column 1.

    b. Increase the value of NF1.

    c. Increase the number of stages in the stripping section of column 1.

    d. Reduce the water in the bottoms of column 2.

B. *Generation of Alternatives*

B1. We wish to separate two organics that form a homogeneous azeotrope. This will be done in a two-column azeotropic system with water as the solvent. Sketch possible system arrangements to do this.

C. *Derivations*

C1. Derive Eq. (8-7) for the two-column, binary, heterogeneous azeotrope system.

C2. An equation for $\alpha_{org-w\,in\,w}$ similar to Eq. (8-13) is easy to derive; do it. Compare the predicted equilibrium in water with the butanol-water equilibrium data given in Problem 8.D2. Comment on the fit. Vapor pressure data are in Perry and Green (1984). Use the data in Table 8-2 for solubility data.

C3. For a binary heterogeneous azeotrope separation, the feed can be introduced into the liquid-liquid separator. In this case two stripping columns are used.

    a. Sketch the column arrangement.

    b. Draw the McCabe-Thiele diagram for this system.

    c. Compare this system to the system in Figure 8-3A.

C4. Sketch the McCabe-Thiele diagram for a two-pressure system similar to that of Figure 8-6.

D. *Problems*

Answers to problems with an asterisk are at the back of the book.

D1. A mixture of water and n-butanol is being processed in an enriching column coupled to a total condenser and a liquid-liquid settler. The feed is a saturated vapor that is 28 mol% water. Feed rate is 100 kmol/h. The distillate product is the water phase from the liquid-liquid settler. The butanol phase in the settler is refluxed to the enriching column as a saturated liquid. Operation is at 1.0 atm and CMO is valid. An external reflux ratio of L/D = 4.0 is used. Equilibrium data are in Table 8-2.

    a. Find the mole fraction water in the bottoms product.

    b. How many equilibrium stages are needed?

D2.* VLE data for water-n-butanol are given in Table 8-2. We wish to distill 5000 kmol/h of a mixture that is 28 mol% water and 30% vapor in a two-column azeotropic distillation system. A butanol phase that contains 0.04 mole frac water and a water phase that is 0.995 mole frac water are desired. Pressure is 101.3 kPa. Reflux is a saturated liquid. Use $L/V = 1.23(L/V)_{min}$ in the column producing almost pure butanol. Both columns have partial reboilers. $(V/B)_2 = 0.132$ in the column producing water.

    a. Find flow rates of the products.

    b. Find the optimum feed location and number of stages in the columns.

    Note: Draw two McCabe-Thiele diagrams.

D3. The VLE data for water and n-butanol are given in Table 8-2. We have flash distillation systems separating 100.0 kmol/h of two different water and n-butanol mixtures.

TABLE 8-2. *Vapor-liquid equilibrium data for water and n-butanol at 1 atm mol% water*

Liquid	Vapor	T °C	Liquid	Vapor	T °C
0	0	117	57.3	75.0	92.8
3.9	26.7	111.5	97.5	75.2	92.7
4.7	29.9	110.6	98.0	75.6	93.0
5.5	32.3	109.6	98.2	75.8	92.8
7.0	35.2	108.8	98.5	77.5	93.4
25.7	62.9	97.9	98.6	78.4	93.4
27.5	64.1	97.2	98.8	80.8	93.7
29.2	65.5	96.7	99.2	84.3	95.4
30.5	66.2	96.3	99.4	88.4	96.8
49.6	73.6	93.5	99.7	92.9	98.3
50.6	74.0	93.4	99.8	95.1	98.4
55.2	75.0	92.9	99.9	98.1	99.4
56.4	75.2	92.9	100	100	100
57.1	74.8	92.9			

(*Source*: Chu et al. (1950). mol% water. Data at 0% water are from Perry and Green (1984).)

a. The feed is 20 mol% water and the vapor product is 40 mol% water. Find L, x, V, and T_{drum}.

b. The feed is 99 mol% water and 30% of the feed is vaporized. Find L, x, V, y and T_{drum}.

D4. A mixture containing 88.0 mol% water and 12.0 mol% n-butanol is allowed to settle into two liquid phases. What are the compositions and the flow rates of these two phases? VLE data are in Table 8-2. F = 100 kmol/h.

D5. We have a saturated vapor feed that is 80.0 mol% water and 20.0 mol% butanol. Feed rate is 200.0 kmol/h. This feed is condensed and sent to a liquid-liquid separator. The water layer is taken as the water product, W, and the butanol (top) layer is sent to a stripping column which has a partial reboiler. The bottoms from this stripping column is the butanol product, which should contain 4.0 mol% water. The vapor leaving the stripping column is condensed and sent to the settler. Equilibrium data are in Table 8-2. Find:

a. Flow rates W and B

b. If $\overline{V}/B = 4.0$, find the number of equilibrium stages (step off from bottom up).

c. Determine $(\overline{V}/B)_{min}$ for this separation.

D6. We are separating water from n-butanol in a stripping column. The feed [F = 100 kmol/h, z = 0.65(mole fraction water), a saturated vapor] is mixed with the vapor leaving the top of the column before the combined stream is sent to the total condenser and then to a liquid-liquid settler. The column has a partial reboiler and CMO is valid. The top layer from the liquid-liquid settler ($x_\alpha = x_{dist} = 0.573$ mole fraction water) is sent as a saturated liquid reflux to the distillation column. The bottom layer ($x_\beta = x_{dist} = 0.975$ mole fraction water) is the distillate product. Pressure p = 1.0 atm, reflux is a saturated liquid, the boilup rate $\overline{V}/B = 4$, $x_{bot} = 0.02$ mole fraction water.

a. Find D and B in kmol/h.

b. What is the reflux flow rate, kmol/h?

c. How many stages are required? Step off stages from the top down.

d. What is the minimum value of the boilup rate?

D7. Aniline and water are partially miscible and form a heterogeneous azeotrope. At $p = 778$ mm Hg the azeotrope concentrations are: The vapor is 0.0364 mole frac aniline, the liquid aqueous phase contains 0.0148 mole frac aniline, and the liquid organic phase contains 0.628 mole frac aniline. Estimate the relative volatility of water with respect to aniline in the organic phase. Note: Be careful to use the mole fracs of the correct component.

D8.* We have a feed of 15,000 kg/h of diisopropyl ether ($C_6H_{14}O$) that contains 0.004 wt frac water. We want a diisopropyl ether product that contains 0.0004 wt frac water. Feed is a saturated liquid. Use the system shown in Figure 8-4, operating at 101.3 kPa. Use $L/D = 1.5 \ (L/D)_{min}$. Determine $(L/D)_{min}$, L/D, optimum feed stage, and total number of stages required. Assume that CMO is valid. The following data for the diisopropyl ether–water azeotrope are given (*Trans. AIChE, 36*, 593, 1940):

$y = 0.959$, Separator: Top layer $x = 0.994$; bottom layer $x = 0.012$; at 101.3 kPa and 62.2°C. All compositions are weight fractions of diisopropyl ether.

Estimate $\alpha_{w-ether \ in \ ether}$ from these data (in mole frac units). Assume that this relative volatility is constant.

D9. We are using an enriching column to dry 100.0 kmol/h of diisopropyl ether that contains 0.02 mole frac water. This feed enters as a saturated vapor. A heterogeneous azeotrope is formed. After condensation in a total condenser and separation of the two liquid layers, the water layer is withdrawn as product and the diisopropyl ether layer is returned as reflux. We operate at an external reflux ratio that is 2.0 times the minimum external reflux ratio. Operation is at 1.0 atm. Find the minimum external reflux ratio, the actual L/D, the distillate flow rate and mole frac water, the bottoms flow rate and mole frac water, and the number of equilibrium stages required (number the top stage as number 1). Use an expanded McCabe-Thiele diagram to determine the number of stages. Data for the azeotropic composition (Problem 8.D8) can be used to find the mole fracs of water in the two layers in the separator and the relative volatility of water with respect to ether at low water concentrations. The weight fractions have to be converted to mole fracs first.

D10.* A single-stage steam distillation system is recovering n-decane from a small amount of nonvolatile organics. Pressure is 760 mm Hg. If the still is operated with liquid water present and the organic layer in the still is 99 mol% n-decane, determine:

a. The still temperature

b. The moles of water vaporized per mole of n-decane vaporized.

Decane vapor pressure is in Example 8-2. Water vapor pressures are (T in °C and VP in mm Hg) (Perry and Green, 1997)

T	95.5	96.0	96.5	97.0	97.5	98.0	98.5	99.0	99.5
VP	645.67	657.62	669.75	682.07	694.57	707.27	720.15	733.24	746.52

D11. We are doing a single-stage, continuous steam distillation of 1-octanol. The unit operates at 760 mm Hg. The steam distillation is operated with liquid water present. The distillate vapor is condensed and two immiscible liquid layers form. The entering organic stream is 90.0 mol% octanol and the rest is nonvolatile compounds. Flow rate of feed is 1.0 kmol/h. We desire to recover 95% of the octanol.

Vapor pressure data for water are given in Problem 8.D10. This data can be fit to an Antoine equation form with $C = 273.16$. The vapor pressure of 1-octanol is predicted by the Antoine equation:

$$Log_{10} (VP) = A - B/(T+C)$$

With A = 6.8379   B = 1310.62   C = 136.05
Where VP is vapor pressure in mm Hg, and T is in Celsius.
a. Find the operating temperature of the still.
b. Find the octanol mole frac in the vapor leaving the still.
c. Find the moles of octanol recovered and the moles of water condensed in the distillate product.

D12. The two-pressure distillation system shown in Figure 8-6 is being used to separate compounds P and Q. $F = 100$, $z_P = .4$, $z_Q = 0.6$, saturated liquid. At the pressure of column 1, p_1, component P is more volatile from $x_P = 0$ to $x_P = x_{P,Az@P1} = 0.65$. Thus, stream B_1 is the Q product. Assuming that B_1 is pure Q and B_2 is pure P, first calculate the flow rates of B_1 and B_2. Then assume that the distillate product from column 1, $x_{P, dist,1} = x_{P,Az@P1}$, and the distillate product from column 2, $x_{P, dist,2} = x_{P,Az@P2}$, and calculate D_1 and D_2 for the following values of the azeotrope in column 2, $x_{P,Az@P2}$:
a. $x_{P,Az@P2} = 0.55$
b. $x_{P,Az@P2} = 0.64$

D13. We are separating water from n-butanol in a system with two feeds. Feed 1[$F_1 = 100$ kmol/h, $z_1 = 0.84$(mole fraction water), saturated vapor] is mixed with the vapor leaving the top of the column before it is sent to the total condenser and then to the liquid-liquid settler. Feed 2 [$F_2 = 80$ kmol/h, $z_2 = 0.20$(mole fraction water), saturated liquid] is fed within the column. The column has a partial reboiler and CMO is valid. The top layer from the liquid-liquid settler ($x_0 = 0.573$ mole fraction water) is sent as reflux to the distillation column. The bottom layer ($x_{dist} = 0.975$ mole fraction water) is the distillation product. Pressure p = 1.0 atm, reflux is a saturated liquid, $\overline{V}/B = 1.5$, $x_{bot} = 0.04$ mole fraction water. Equilibrium data are in Table 8-2.
a. Find D and B in kmol/h
b. Find $\overline{V}, \overline{L}$ in the bottom section of the column V' and L' in the section above feed 2, both in kmol/h.
c. Derive the operating equation for the distillation section above feed 2 (flow rates V' and L').
d. Plot the operating lines and step off stages starting at the reboiler and going up the column.
   Use the optimum feed stage for feed 2. Report the optimum feed stage above the partial reboiler and total number of stages needed.

D14. We plan to use a two-pressure system (similar to Figure 8-6) to separate a feed that is 15.0 mol% benzene and 85.0 mol% ethanol. The feed rate is 100.0 kmol/h, and is a saturated liquid. The two columns will be at 101.3 kPa and 1333 kPa. We desire an ethanol product that is 99.2 mol% ethanol and a benzene product that is 99.4 mol% benzene. Assume the two distillate products are at the azeotrope concentrations.
a. Draw the flowsheet for this process. Label streams to correspond to your calculations.

b. Find the flow rates of the ethanol and benzene products.

c. Find the flow rates of the two distillate streams that are fed to the other column.

Data (Seader, 1984): Azeotrope is homogeneous. At 101.3 kPa the azeotrope temperature is 67.9°C and is 0.449 mole frac ethanol. At 1333 kPa the azeotrope temperature is 159°C and is 0.75 mole frac ethanol. Ethanol is more volatile at mole fracs below the azeotrope mole frac.

D15. We wish to use continuous steam distillation to recover 1-octanol from 100 kg/h of a mixture that is 15 wt % 1-octanol and the remainder consists of nonvolatile organics and solids of unknown composition. The feed will be preheated to the same temperature as the still pot, which operates at 1.0 atm pressure. The pot is operated with liquid water in the pot. Assume the still pot is well mixed and liquid and vapor are in equilibrium. Ninety-five percent of the 1-octanol should be recovered in the distillate. Assume that water is completely immiscible with 1-octanol and with the nonvolatile organics. Because the composition of the non-volatile organics is not known, we do a simple experiment and boil the feed mixture under a vacuum with no water present. The result is at 0.05 atm pressure the mixture boils at 129.8°C.

    a. Find the mole fraction of 1-octanol in the feed and the effective average molecular weight of the nonvolatile organics and the solids.

    b. Find the kg/h and kmol/h of 1-octanol in the distillate, the kg/h of total organics in the waste, and the 1-octanol weight fraction and 1-octanol mole fraction in the waste.

    c. Find the temperature of the still pot. A spreadsheet or MATLAB is highly recommended for finding T.

    d. Find the kg/h and kmol/h of water in the distillate.

    Octanol boils at about 195°C. The formula for octanol is $CH_3(CH_2)_6CH_2OH$ and its molecular weight is 130.23. Vapor pressure formula for octanol is available in Problem 8.D11. Water vapor pressure data can be fit to the following, $\log^{10}(VP)water = 8.68105 - 2164.42/(T + 273.16)$

    Where VP is in mm Hg and T is in °C.

    Note: We will see in Problem 9.E2 that batch steam distillation requires less steam than continuous steam distillation.

D16. An extractive distillation system is separating ethanol from water using ethylene glycol as the solvent. The makeup solvent stream is pure ethylene glycol. The flowsheet is shown in Figure 8-13. The feed flow rate is 100.0 kmol/h. The feed is 0.810 mole frac ethanol and 0.190 mole frac water. We desire the ethanol product (distillate from column 1) to be 0.9970 mole frac ethanol, 0.0002 mole frac ethylene glycol, and remainder water. The water product (distillate from column 2) should be 0.9990 mole frac water, 0.00035 mole frac ethylene glycol, and remainder ethanol.

Find flow rates of makeup solvent, distillate from column 1, and distillate from column 2.

D17. An extractive distillation system is separating compound A from C with a solvent. Solvent flow rate is S, feed flow rate is F, entering solvent is pure, and entering feed is a saturated liquid with mole fractions z_A and z_C. Because of plant operating problems, the distillate flow rate has to be shut down (D = 0). The values of F, S, Q_c, and Q_R are not changed. After a new steady state is reached, predict the values of B and the mole fractions in the bottoms stream.

D18. An extractive distillation system is separating ethanol from water using ethylene glycol as the solvent. The makeup solvent stream is pure ethylene glycol. The diagram is in Figure 8-13 with ethanol as the A product and water as the B product. The feed flow rate is 100.0 kmol/h. The feed is 0.2000 mole frac ethanol and 0.8000 mole frac water. We desire the ethanol product (distillate from column 1) to be 0.9970 mole frac ethanol, 0.0002 mole frac ethylene glycol, and remainder water. The water product (distillate from column 2) should be 0.9990 mole frac water, 0.00035 mole frac ethylene glycol, and remainder ethanol.

Find flow rates of Makeup solvent, A Product, and B Product.

D19. We wish to use n-hexane as an entrainer to separate a feed that is 80.0 wt % ethanol and 20 wt % water into ethanol and water. The system shown in Figure 8-18 will be used. The feed is 10,000.0 kg/h and is a saturated liquid. The ethanol product is 99.999 wt % ethanol, 0.001 wt % hexane, and a trace of water. The water product is 99.998% water, 0.002 wt % ethanol and a trace of hexane. Do external mass balances and calculate the flow rates of make-up solvent (n-hexane), ethanol product, and water product. (Assume that trace = 0.)

Note: Watch your decimal points when using weight fractions and wt %.

D20. This problem explores the azeotropic distillation column shown in Figure 8-18 with mole fraction profiles shown in Figure 8-19. The feed to the distillation column in Figure 8-18 is a saturated liquid that is 0.8094 mole fraction ethanol with the remainder being water. Do calculations on a basis of 1000 kmol/h. The ethanol product is 0.998 mole fraction ethanol and 0.002 mole fraction water. The water product is 0.9999 mole fraction water and 0.0001 mole fraction ethanol. In practice there will be traces of pentane in both products, but ignore this (assume Makeup Solvent = 0). The vapor mole fractions of the compounds leaving the top stage of the distillation column in Figure 8-18 (correspond to the profiles in Figure 8-19) are: ethanol $y_{E,1}$ = .0555, water y_{w1} = 0.3000, pentane $y_{p,1}$ = 0.6445. Assume that the water and pentane are totally immiscible; thus, there is no water in the reflux to the distillation column and there is no pentane in the feed to the stripping column. The ethanol distributes between the water and pentane layers in the settler. In the absence of data assume the distribution coefficient

$K_{d,ethanol} = x_{ethanol\_in\_pentane} / x_{ethanol\_in\_water} = 1.0$ (the x are mole fractions).
CMO is valid in both columns, and operation of the stripping column is at 1.0 atm.

a. Calculate the flow rates of the ethanol product E and the water product W.

b. What is V_1 in the distillation column, and what is the boilup rate in this column? (Note that Vy_{w1} must equal the water in the feed minus the small amount of water lost in the ethanol product.) Also calculate the L_0 = reflux rate.

c. Since no pentane enters the stripper, at steady state all of the pentane that enters the settler must be returned to the distillation column in the reflux stream. In addition, all of the ethanol entering the settler minus the small amount of ethanol that leaves with the water product must be returned to the distillation column in the reflux stream. Calculate the steady state values of $x_{ethanol\_in\_pentane}$ and $x_{ethanol\_in\_water}$ and the mole fractions of the reflux stream. Note that $x_{ethanol\_in\_pentane}$ and $x_{ethanol\_in\_water}$ are not equal to $y_{E,1}$.

d. If the boilup ratio in the stripper is 0.5, what is the flow rate of liquid fed to the stripper, and what is the flow rate of the vapor leaving the stripper? What is the mole fraction ethanol in the vapor leaving the top stage of the stripper ($y_{E,1,stripper}$)?

D21. A distillation column is separating water from n-butanol at 1 atmosphere pressure. Equilibrium data are in Table 8-2. The distillation system is similar to Figure 8-3A and has a partial reboiler, a total condenser and a liquid-liquid settler. The bottom layer from the settler (rich in water with $x_w = 0.975$) is taken as the distillate product. The top layer ($x_w = 0.573$) is returned to the column as a saturated liquid reflux. The feed is 40.0 mol% water, is a saturated vapor and flows at 500.0 kmol/h. The bottoms is 0.04 mole frac water. Use a boilup ratio of $\overline{V}/B = 0.5$. Assume CMO is valid. Step off stages from the bottom up. Find the optimum feed stage location and the total number of equilibrium stages needed.

D22. We are separating water from n-butanol in a system with an enriching column, a total condenser, and a liquid-liquid settler. The feed has F = 100 kmol/h, z = 0.20(mole fraction water) and is a saturated vapor. CMO is valid. The top layer from the liquid-liquid settler ($x_\alpha = 0.573$ mole fraction water) is sent as a saturated liquid reflux to the enriching column. The bottom layer ($x_\beta = x_{dist} = 0.975$ mole fraction water) is the distillate product. Pressure p = 1.0 atm, reflux is a saturated liquid, $x_{bot} = 0.08$ mole fraction water.
   a. Find D and B in kmol/h.
   b. Derive and plot the operating line.
   c. How many stages are required ? Step off stages from the top down.
   d. What is the minimum value of $x_{bot,water}$ that can be obtained as N→ infinity?

D23. 100 kmol/h of a saturated vapor feed that is 25 mol% nitromethane (NM) and 75 mol% water is to be separated in a system with two distillation columns and a liquid-liquid separator. The feed is sent to column W that produces a water product that is 0.01 mole fraction NM. The boilup ratio in column W is ¼. The optimum feed stage is used. The vapor from column W is condensed and sent to the decanter. The water phase from the decanter (0.086 mole fraction NM) is refluxed to column W. The NM phase from the decanter (0.312 mole fraction water) is sent to stripping column NM. The nitromethane product from the bottom of stripping column NM is 0.02 mole fraction water. The boilup ratio in column NM = 3.0. Assume both columns operate at 1.0 atm. pressure, that CMO is valid, that both condensers are total condensers, and that both reboilers are partial reboilers. Equilibrium data are in Table 8-3. Find:
   a. The flow rates of both products.
   b. The optimum feed plate location and the number of stages in column W. Step off stages from the bottom up and calculate a fractional number of stages.
   c. The number of stages required in stripping column NM. Step off stages from the top down and calculate a fractional number of stages.
   d. The vapor flow rates entering each of the condensers.
   Note: Be careful with the component that you are using for mole fractions in mass balances and in operating lines.
   Safety note: Nitromethane (aka methyl nitrate, CH_3ONO_2) is a dangerous compound. At 1 atm, it boils at 77.04°C. If one tries to freeze the pure compound, it explodes.

D24. We feed 100 kmol/h of water that is saturated with n-butanol at 1.0 atm. This saturated liquid feed is sent to a still pot (a reboiler). We want $x_{butanol,bottoms} = 0.004$. Find $y_{butanol}$, V, and B. Equilibrium data are in Table 8-2.

D25. We wish to use continuous steam distillation to recover benzene from 100 kg/h of a mixture that is 20 wt % benzene and the remainder consists of nonvolatile organics and solids of unknown composition. The feed will be preheated to the same temperature as the still pot, which operates at 1.0 atm pressure. The pot is operated with liquid water in the pot. Ninety percent of the benzene should be recovered in the distillate. Because the composition of the nonvolatile organics is not known, we do a simple experiment and boil the feed mixture with no water present. The result is that at 1.0 atm pressure, the mixture boils at 93°C.

Assumptions and data: Assume that the nonvolatile organics are completely immiscible in water. Although not totally valid (see Example 8-1), assume that benzene and water are completely immiscible. Enthalpy data for benzene and water are available in Tables 2-237 and 2-352 of the seventh edition of Perry's Handbook (Perry and Green, 1997), respectively, and also available in Tables 2-193 and 2-305 in the eighth edition. In the seventh edition, $H = h_g$ and $h = h_f$. In the eighth edition, read the instructions at the bottom of Table 2-193 to find the values you need. Data from these sources is very slightly different. Vapor pressure data for benzene follows the Antoine equation (Lange, pp. 10–37),

$$\log_{10} (VP)_{benzene} = 6.90565 - 1211.033/(T + 220.790)$$

Water vapor pressure data can be fit to the following,

$$\log_{10} (VP)_{water} = 8.68105 - 2164.42/(T + 273.16)$$

where VP is in mm Hg and T is in °C. Note: It is always a good idea to check equations like this by using known points (e.g., benzene boils at 80.1°C when pressure is 760 mm Hg).

a. Find the mole fraction of benzene in the feed and the effective average molecular weight of the nonvolatile organics and the solids.

b. Find the kg/h and kmol/h of benzene in the distillate, the kg/h of total organics in the waste, and the benzene weight fraction and benzene mole fraction in the waste.

c. Find the temperature of the still pot. A spreadsheet or MATLAB is highly recommended for finding T.

d. Find the kg/h and kmol/h of water in the distillate.

e. Find the kg/h and kmol/h of water in the waste stream.

E. *More Complex Problems*

E1. Two feeds containing water (W) and nitromethane (NM) are to be separated in a system with two distillation columns and a liquid-liquid separator. Feed $F_1 = 100$ kmol/h of a saturated liquid feed that is 8 mol% nitromethane and 92 mol% water is fed to column W that produces a water product that is 0.01 mole fraction NM. The boilup ratio in column W is ⅓. The optimum feed stage is used. The vapor from column W is condensed and sent to a decanter. The water phase from the decanter, which is 0.086 mole fraction NM, is refluxed to column W. The NM phase from the decanter (0.312 mole fraction water) is sent to column NM as reflux. Feed $F_2 = 150$ kmol/h of a saturated liquid that is 15 mol% water and 85 mol% nitromethane is also fed to column NM. The nitromethane product from column NM is 0.02 mole fraction water. The boilup ratio in column NM = 1.0. Assume both columns operate at 1.0 atm pressure, that CMO is valid, that both condensers are total condensers, and that both reboilers are partial reboilers. Equilibrium data are in Table 8-3. Find:

a. The flow rates of both products.

b. The optimum feed plate location and the number of stages in column W. Step off stages from the top down.

c. The optimum feed stage location and number of stages required in column NM. Step off stages from the bottom up.

TABLE 8-3.   *Water-nitromethane equilibrium data at 1 atm*

Liquid	Vapor	Liquid	Vapor
7.9	34.0	81.4	50.1
16.2	45.0	88.0	50.2
23.2	49.7	91.4	50.3
31.2	50.0	92.5	51.3
39.2	50.2	93.5	52.9
45.0	50.2	95.1	55.9
55.2	50.2	95.9	57.5
63.8	50.2	97.5	65.8
73.0	50.2	97.8	68.4
77.0	50.2	98.6	79.2

Source: Chu et al. (1950). (mol% water)

d. The vapor flow rates entering each of the condensers.

e. What is the minimum boilup ratio in column NM?

Note: Be careful with the component that you are using for mole fractions in mass balances and in operating lines. Also, the top operating lines are different than what you probably expect.

Safety note: Nitromethane (aka methyl nitrate, CH_3NO_2) is a dangerous compound. If one tries to freeze the pure compound, it explodes.

E2. The VLE data for water and n-butanol at 1.0 atm are given in Table 8-2. The fresh feed is 100.0 kmol/h of a saturated liquid that is 30.0 mol% water. This fresh feed is mixed with the return line from the separator to form the total feed, F_T, which is fed to a stripping column that produces the butanol product (see figure). Assume this total feed is a saturated liquid. The butanol product is 2.0 mol% water. The water product is produced from another stripping column and is 99.5 mol% water. In the butanol column the boilup ratio is 1.90. In the water column the boilup ratio is 0.1143. Assume CMO. Both reboilers are partial reboilers. Both reflux streams are returned as saturated liquids. Operation is at 1.0 atm. Since this problem is challenging, this list coaches you through one solution approach.

a. Find the flow rates of the two products (use external mass balances).

b. Find vapor flow rate in butanol column, then liquid flow rate in butanol column, which = F_T.

c. Calculate z_T from mass balance at mixing tee for fresh feed and reflux.

d. Plot the bottom operating line and find value of vapor mole frac leaving the butanol column and the number of equilibrium stages required for this column.

e. Use an *expanded* McCabe-Thiele plot to find the vapor mole frac leaving the water column and the number of equilibrium stages required.

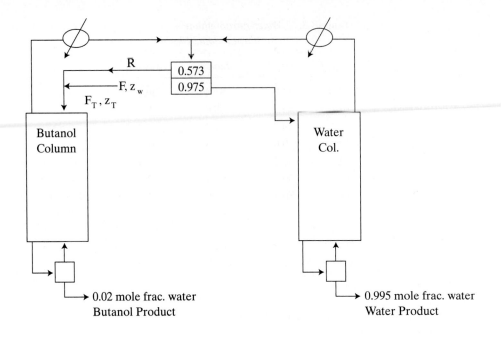

0.02 mole frac. water
Butanol Product

0.995 mole frac. water
Water Product

E3. We wish to produce pure water by boiling it with n-decane vapor (see figure). This is sort of reverse of steam distillation. Seawater is roughly 3.5 wt % salt which can be approximated as NaCl. The feed is 1000 kg of seawater/h. The feed temperature is 30°C. The water is heated with pure saturated n-decane vapor at 760 mm Hg. Most of the n-decane vapor condenses while the remainder is carried overhead with the water vapor. Pressure is 760 mm Hg. We wish to recover 60% of the water as condensate.
a. Find the approximate still temperature.
b. Find the moles of n-decane carried over in the vapor/h.
Data: VP of n-decane: Example 8-2, VP of water: Problem 8.D10, MW_{water} = 18.016, MW_{NaCl} = 58.45, MW_{C10} = 142.28, Salt is nonvolatile. Water and n-decane are immiscible. NaCl dissolves only in water.

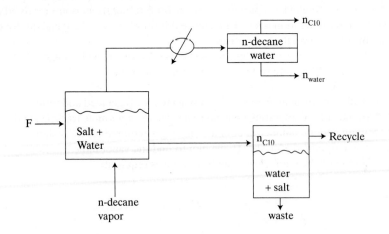

F. *Problems Requiring Other Resources*

F1. A single-stage steam distillation apparatus is to be used to recover n-nonane (C_9H_{20}) from nonvolatile organics. Operation is at 1 atm (760 mm Hg), and the still will operate with liquid water present. If the still bottoms are to contain 99 mol% nonane in the organic phase (the remainder is nonvolatiles), determine:

   a. The temperature of the distillation

   b. The moles of water vaporized per mole of nonane vaporized

   c. The moles of water condensed per mole of nonane vaporized for a liquid entering at the still temperature

   d. Repeat parts a and b if still bottoms will contain 2.0 mol% nonane

   Data are available in Perry and Green (1984).

F2. We wish to recover a gasoline component, n-nonane (C_9H_{20}), from a nonvolatile mixture of oils, grease, and solids. This will be done in a steady-state, single-stage, steam distillation system operating with liquid water present and at a total pressure of 102.633 kPa. The feed is 95.0 mol% n-nonane and we desire to recover 90% of the n-nonane in the distillate. The feed enters at the temperature of the boiler and the feed rate is 10.0 kmol/h. Find:

   a. The bottoms mole frac n-nonane in the organic layer.

   b. The still temperature.

   c. The kmoles of nonane in the distillate.

   d. The kmoles of water in the distillate.

   e. The kmoles of water in the bottoms.

   Assume water and n-nonane are completely immiscible. Vapor pressure data for water are given in Problem 8.D10. Use the DePriester chart or Raoult's law to obtain K_{c9} (T, p_{tot}). Then $p_{org} = K_{c9} x_{c9,org} p_{tot}$. Latent heat values are available in Perry's and similar sources.

G. *Computer Simulation Problems*

G1. We wish to separate a mixture of pyridine (C_5H_5N) and water, which forms a homogeneous azeotrope, by using extractive distillation using bisphenol-A ($C_{15}H_{16}O_2$) as the solvent. When finished, the system will look similar to Figure 8-13 with the water as the A product and the pyridine as the B product. The feed to column 1 is 200 kmol/h of a liquid at 80°C and 1.0 atm. The feed is 30 mol% pyridine and 70 mol% water. The makeup solvent is pure bisphenol-A at 1.0 atm and 100°C, and the recycle solvent is also cooled to 100°C with a heat exchanger operating at 1.0 atm. We want to produce a water product in D_1 99.0 mol% water or slightly higher, and a pyridine product in D_2 98.0 mol% pyrdine or slightly higher. Use NRTL for the VLE correlation.

Both columns have total condensers, kettle-type reboilers, pressure = 1.0 atm and is constant in column, and have saturated liquid reflux.

Column 1: N = 40, makeup solvent and, when connected, recycle solvent both "on stage" 3, the feed is initially "on stage 30" (all in Aspen notation), reflux ratio = 0.1. Set an appropriate value for the distillate flow rate that would produce pure water and there would be no water in the bottoms if the separation was perfect. Note: N and solvent feed stage will not be changed.

Column 2: N = 8, feed above stage 3 (all in Aspen notation), reflux ratio = 1.0. Set B_2 initially to 1000 kmol/h. For the final design stream B_2 should have less than 10^{-5} mole fraction of water and pyridine. Note: N and reflux ratio will not be changed.

The initial run will probably not give the desired purities. Find the optimum feed stage in column 1. Using this optimum feed stage and $N_1 = 40$, increase the value of solvent added by increments of 50 kmol/h until the desired purities of water and pyridine are obtained.

Estimate the desired makeup rate with an external balance. Once the system is operating at the desired makeup flow rate, if necessary to obtain the desired purities, make small adjustments in the following: D_1, recycle flow rate B_2, L/D in column 1, and Makeup flow rate.

Report the following:

1. Final makeup solvent flow rate _____ kmol/h
2. Final value solvent recycle rate (B_2) _____ kmol/h and L/D in col 1 _____
3. Final values of flow rates D_1 _____, B_1 _____, and D_2 _____ kmol/h
4. Mole fractions in stream D_1 _____
5. Mole fractions in stream D_2 _____
6. Mole fractions in stream B_1 _____
7. Mole fractions in stream B_2 (solvent recycle stream) _____
8. Heat load in cooler on solvent recycle line _____ cal/s

G2. Pressure can have a large effect on the equilibrium data for nonideal systems. Your company plans to process a mixture of acetone, methyl ethyl ketone, and methyl isobutyl ketone in an ordinary, single-feed distillation column with a total condenser and a partial reboiler. Feed flow rate is 200 kmol/h; feed mole fractions: acetone = 0.25; methyl ethyl ketone = 0.45; and methyl isobutyl ketone = 0.3. Feed can be made available at any desired pressure. Feed temperature is 75°C, D = 50 kmol/h, and adiabatic column. The purpose is to produce pure acetone.

Two of the senior engineers are arguing over the proper column pressure to use. One claims that 1.0 atm is better and the other that 4.0 atm is better. They both agree that the NRTL VLE package is satisfactory. Use Aspen Plus to determine which pressure is better, and explain why it is better. There are at least two different approaches to solving this problem, and one is considerably less work than the other.

G3. We will use a system similar to Figure 8-4a except both columns have direct reflux from the condensers in addition to the reflux from the decanter. Distill a feed that is 17 mol% water and 83 mol% furfural ($C_5H_4O_2$). Feed flow rate is 200 kmol/h and is at 42°C and 1.0 atm. Furfural-water forms a heterogeneous azeotrope. Use NRTL-RK for the equilibrium. The bottoms from column 1 is the furfural product. The total condenser produces a saturated liquid, and the saturated liquid distillate from column 1 is sent to a decanter. The water layer (liquid 1) from the decanter is the feed to column 2. Operate the decanter as an adiabatic system at 1.0 atm. Make furfural the second liquid with threshold of 0.5 and return the furfural layer as a feed (above) stage 2 (in Aspen notation) of column 1.

Distillation column 1 operates at 1.0 atm with constant pressure. Column 1 has 6 stages (Aspen notation) and has a kettle reboiler and a total condenser. The feed is above stage 3, and the furfural layer from the decanter is sent to above stage 2 (both Aspen notation). Use the total condenser to condense the distillate before it is sent to the decanter. Final values of the reflux ratios in both columns should be 0.01.

Distillation column 2 operates at 1.0 atm with constant pressure. Column 2 has 6 stages (Aspen notation) and has a kettle reboiler and a total condenser. The furfural layer from the decanter is sent to above stage 2 (Aspen notation). The boilup *rate* in column 2 should initially be set to 4.0 kmol/h. Use the total condenser to condense the distillate and then send it as a second feed to the decanter.

Once you have convergence and the final values of the reflux ratios, change the boilup rate in column 2 to obtain 0.995 furfural or slightly higher and 0.990 water or slightly higher in the bottoms to columns 1 and 2, respectively.

Report the following:

a. Final reflux ratio column 1 _____ and final reflux ratio column 2 _____.
If these values are not 0.01, you are not finished with Part B.

b. Flow rates furfural product _____ kmol/h and water product _____ kmol/h.

c. Boilup rate in column 2 (column producing water product) _____ kmol/h.

d. Mole fraction water in furfural product _____ and mole fraction furfural in water product _____.

e. Flow rate of distillate from column 1 (column producing furfural product) _____ kmol/h.

f. Column 1 condenser temperature _____ K and column 1 reboiler temp. _____ K.

g. Outlet temperature of decanter _____ K.

h. Molar ratio of water phase/total liquid in decanter _____.

G4. The system water-acetonitrile (C_2H_3N) forms an azeotrope. This system can be separated using a two-pressure distillation system. Both columns have total condensers and kettle-type reboilers. Valid phases are vapor-liquid. Use NRTL for VLE data. The feed to the system is at 1.0 atm and 60°C. The feed is 15 mol% water and 85 mol% acetonitrile, and the flow rate of the feed is 200 kmol/h. This feed and the recycle stream labeled D_2 in Figure 8-6 are input into column 1 (see Figure 8-6) "above" the same feed stage. Column 1 operates at 1.0 atm. As a start, column 1 has $N = 15$ and $N_{feed} = 8$ (in Aspen notation) for both fresh feed and D_2. This will provide more separation than necessary once a large enough value of D_2 is determined. The molar reflux ratio for column 1 is 2.0. Set the bottoms flow rate for column 1 at a value that will produce pure acetonitrile. Column 1 will eventually produce a bottoms product that is 99.5 mol% acetonitrile or higher. *There is no need to change the specifications for column 1 (assuming they are input correctly).*

Column 2 is at 200 torr, which is a vacuum. This column should eventually produce a bottoms product that is 99.5 mol% water or slightly higher. Column 2 with $N = 14$ (in Aspen notation) and the feed "on stage" 7 (in Aspen notation) with a molar reflux ratio of 2.0 is a good starting point for the calculations. Put a pump (discharge pressure of 1.0 atm and efficiencies for pump and driver of 1.0) between column 2 and column 1. Find the value of D_2 (accurate to an increment of 10.0 in D_2) that has column 1 meeting or slightly exceeding the desired purity of 99.5 mol% acetonitrile in bottoms and column 2 meeting or slightly exceeding the desired purity of 99.5 mol% water in bottoms.

After correct D_2 has been determined, determine the lowest reflux ratio in column 2 (accurate to an increment of 0.1 in L/D) (but keep $N = 14$ and the feed "on stage" 7) as necessary to produce the desired bottoms product that is 99.5 mol% water or very slightly higher. Once you have this, report values for both columns 1 and 2 based on this run.

Report the following:

Column 1:

a. Bottoms product mole fraction acetonitrile._____.

b. Distillate flow rate _____ kmol/h and bottoms flow rate _____ kmol/h.

c. Distillate mole fraction acetonitrile _____.

Column 2:
a. Distillate flow rate _____ kmol/h, and reflux ratio _____.
b. Bottoms product mole fraction water _____.
c. Distillate mole fraction acetonitrile _____.

G5. a. Aspen Plus automatically generates residue curves for ternary mixtures. Generate the residue curve at 5.0 atm for a mixture propane, n-butane, and n-pentane. Report the VLE correlation used.

b. To obtain some idea of what this means, simulate a column with a total condenser and kettle type reboiler at 5.0 atm with 50 stages, saturated liquid feed on stage 25 with F = 100 kmol/h, and mole fractions of all three components = 0.333333, and initially far from total reflux (L/D = 0.999). Set D = 66.6667 to put pentane in the bottoms. Generate a plot of liquid compositions versus stage location. To generate the plot, after the run, click on the blue box with a checkmark. Then use the button with << to go to compositions. In the tool bar you will now see a button labeled "Plot." Click on this and click on "Plot Wizard" in the menu. Click Next, then click on the plot type, "Comp" and click Next again. To plot all three components click on the >> button. To plot liquid compositions, click on the liquid button in the middle of the page. Click on Next again and then "Finish." This gives the desired plot. Print it, and label the conditions (L/D, mole fraction of feed, D and N) on the plot. Note that complete separation is obtained.

c. Simulate the same system as in part b, but with D = 33.3333, L/D = 0.999, N = 100 and N feed = 50. Print the plot for this case. For the following "essentially complete separation" means propane mole fraction in distillate > 0.990.
Is it possible to obtain essentially complete separation with L/D = 0.999, N = 100 and N feed = 50?
Is it possible to obtain essentially complete separation with N = 50 and N feed = 25 with L/D = 2.0?
Is it possible to obtain essentially complete separation with N = 10 and N feed = 5 with L/D = 2.0?
Is it possible to obtain essentially complete separation with N = 10 and N feed = 5 with L/D = 10.0?
Is it possible to obtain essentially complete separation with N = 6 and N feed = 3 with L/D = 50.0?
Print the plots for the cases with complete separation and label the conditions on the plots.

d. The residue curve obtained implies that one can obtain pure propane or pure n-pentane for any feed composition by: setting D appropriately and using a large enough L/D (> $(L/D)_{min}$), and a large enough N (> N_{min}). To convince yourself that this is true, try two different feed compositions and show that essentially pure propane or pure n-pentane can be obtained by changing conditions.

H. *Computer Spreadsheet Problems*

H1. Generation of distillation curves for systems with constant relative volatility is fairly straightforward and is an excellent learning experience. Generate distillation curves in Figure 8-7 for the benzene (A), toluene (B), cumene (C) system with $\alpha_{AB} = 2.4$, $\alpha_{BB} = 1.0$ and $\alpha_{CB} = 0.21$. Operation is at total reflux.
a. Mole fraction of A in the reboiler is 0.006735 and mole fraction of B is 0.003469.*
b. Mole fraction of A in the reboiler is 0.001 and mole fraction of B is 0.009.*

c. Mole fraction of A in the reboiler is 0.0003 and mole fraction of B is 0.0097. Remember the sum of the mole fractions of A, B, and C in the reboiler is 1.0. *Solution is shown in Figure 8-7.

H2. Generation of residue curves for systems with constant relative volatility is fairly straightforward and is an excellent learning experience. Generate residue curves for the benzene (A), toluene (B), cumene (C) system with $\alpha_{aB} = 2.4$, $\alpha_{BB} = 1.0$, and $\alpha_{CB} = 0.21$. Mole fraction of A in the stillpot is 0.990, and mole fraction of B in the stillpot is 0.001. Compare your results with the distillation curves in Figure 8-7. Remember the sum of the mole fractions of A + B + C in the stillpot is equal to 1.0.

H3. Usually, relative volatilities are not constant, and determination of the residue curve requires a bubble-point calculation at each time used to integrate Eq. (8-28) to determine the temperature T and the vapor mole fractions. The bubble-point calculation was illustrated in Example 5-3 for light hydrocarbons. The K values for these compounds can be determined from Eq. (2-16) with the constants tabulated in Table 2-3. Develop a spreadsheet that can be used to determine the residue curves for any three of the following light hydrocarbons: i-butane, n-butane, i-pentane, n-pentane, and n-hexane. Note: If Euler's method, Eq. (8-29), is used, the tolerance on the sum of the $y_{i,k}$ values must be quite small (e.g., E -9). Find the residue curve for the following problems.

a. n-butane, i-butane and n-pentane

b. i-pentane, n-pentane and n-hexane

c. any other combination of interest.

CHAPTER 8 APPENDIX.   SIMULATION OF COMPLEX DISTILLATION SYSTEMS

This appendix follows the instructions in the appendices to Chapters 2 and 6. Although the Aspen Plus simulator is referred to, other process simulators can be used. The three problems in this appendix all employ recycle streams in distillation columns. The procedures shown here to obtain convergence are all forms of *stream tearing*. Since these are not the only methods that will work, you are encouraged to experiment with other approaches. If problems persist while running the simulator, see Appendix A: Aspen Plus Separations Troubleshooting Guide, at the end of the book.

The vapor-liquid equilibrium for the very nonideal systems studied in this chapter may not be fit well with any of the correlations in Aspen Plus if the parameters embedded in Aspen Plus are used. An alternative is to use Aspen Plus to fit the parameter values to give the best fit to VLE data. This procedure is explained in Appendix B at the end of the book.

Lab 7.   *Two-pressure distillation for separating azeotropes.* A modification of the arrangement of columns shown in Figure 8-6 (or the very similar arrangement where the feed is input into the higher pressure column 2) can be used to separate azeotropes *if* the azeotrope concentration shifts significantly when pressure is changed. You should develop a system that inputs D_2 and F on separate stages. Use "Strongly non-ideal" for convergence of both columns.

We want to separate a feed that is 60 mol% water and 40 mol% methyl ethyl ketone (MEK). Feed is a saturated liquid and is input into column 1 as a saturated liquid at 1.0 atm. Fresh feed rate to column 1 is 100 kmol/h. Column 1 operates at 1.0 atm. Start with N = 10, both feeds on $N_{feed} = 5$, and L/D = 1.0 in column 1. Distillate is a saturated liquid. Use a total condenser and a kettle-type reboiler. Set convergence at 75 iterations. We want products that are > 99% purity. Do external mass balances for 100% pure products.

Column 2 is at 100 psia, and has a total condenser, saturated liquid distillate, and a kettle-type reboiler. Start with N = 15 and feed at $N_{feed} = 7$, with L/D = 2.0. Set convergence at 75 iterations.

When you draw the flowchart, put a pump (select under "pressure changers") on the distillate line going from column 1 that becomes the feed to column 2. The resulting flowchart should be similar to Figure 8-A1. Pump efficiency = 1.0, and Outlet pressure = 100 psia.

To get started, do external mass balances and calculate accurate values for the two bottoms flow rates assuming that the water and MEK products are both pure. Specify the bottoms flow rate of column 1 = 60, but do NOT specify bottoms in column 2. In column 2 start with D = 20 and increase D in steps, 30, 40, 50, 60, 70, and so forth without reinitializing. If you don't step up, Aspen Plus will have errors. Note that B in column 2 = 40 every time even though you did not specify it (Why?). Continue increasing D_2 until both purities are ≥ 99.0%. Use lowest D_2 that gives desired purities.

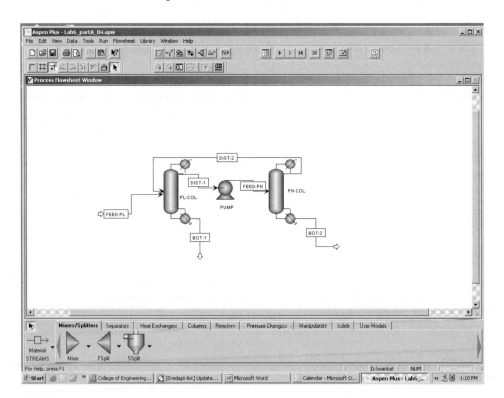

FIGURE 8-A1.   *Aspen Plus screen shot for two-pressure distillation system*

Chapter 8   Introduction to Complex Distillation Methods

Although not a perfect fit, the Wilson equation gives reasonable VLE data. Use analysis to look at the T-y,x and y-x plots at both pressures. Feel free to look for better VLE packages.

Once an appropriate value for D_2 is found, the designer can firm up the designs for the two columns. For example, in column 2 try reducing the reflux ratio to 1.75, 1.5, 1.25, and 1.0. The reflux ratio has only a small effect, although at a value of 1.0 the purity requirement for the MEK product is probably not satisfied. At this reflux ratio with $N_2 = 15$, find the optimum feed stage. Then if the purity requirements are satisfied, reduce the reflux ratio further. Once the purity requirement is not satisfied, find the optimum feed stage again (it may shift a bit). The goal is to have a low reflux ratio with $N_2 = 15$ and still satisfy the purity requirements for both products. If you have time, a similar analysis can be done for column 1. Complete optimization of the system requires an economic analysis (see Chapter 11).

After you are happy with the design, run one more time, to size the trays for both columns. Use sieve trays with the "Fair"method and default values for the other variables.

When done, look at the distillate flow rates. Why are they so large? *Note* that they make the column diameters and heat loads large for this very modest feed rate. These large recycle rates and hence large diameters and heat loads make this design economical only when the shift of the azeotrope with pressure is rather large. MEK and water is an example that is done commercially.

Lab 8. *Binary distillation of systems with heterogeneous azeotropes.* The purpose of this lab is to design a system similar to Figure 8-4A for separating n-butanol and water. Open a new blank file. Under Setup, check Met and make the valid phases vapor-liquid-liquid. List n-butanol and water as the components. Finding a suitable VLE model for heterogeneous azeotropes is a challenge. NRTL-RK was the best model of the half dozen I tried. (Feel free to fit the NRTL-RK constants using the method in Appendix B.) NRTL-RK fits the vapor composition of the azeotrope and the liquid composition of the water phase quite well (see Table 8-2, in Problem 8.D3.) but misses a bit on the composition of the butanol phase. After setting up the decanter, use Analysis to look at the T-y,x and y-x plots. Compare with the data. Remember to allow vapor-liquid-liquid as the phases in Analysis.

The problem we want to solve is to separate 100 kmol/h of a saturated liquid feed at 1 atm pressure. The feed is 78 mol% n-butanol and 22 mol% water. We want the purity of both products to be 99% or higher. A system similar to Figure 8-4A is to be used. The columns will operate at 1.0 atm pressure. Use external mass balances to determine the flow rates of the two bottoms products assuming that they are essentially pure butanol and pure water (< E -04 mole frac of the other component).

Draw the flow diagram. Use total condensers for both columns. Both distillate products should be taken off as liquids and then be connected to the feed to the decanter. The hydrocarbon layer from the decanter should be connected to the feed of the distillation column producing pure butanol. The decanter (or liquid-liquid settler) on the flow diagram is listed under separators. Pick a decanter pressure of 1.0 atm., a heat duty of 0.0, the key component for the second phase is water, and set the threshold at ~.7 (the equilibrium data show the highest water content in the organic phase is < .6). Set convergence at 75 iterations. The water layer from the decanter should be connected to the feed of the distillation column that produces pure water. Note that the Aspen Plus columns in Figure 8-A2 do not look exactly like the arrangement in Figure 8-4A. We are using the condensers on the columns to condense the liquids. We will use the specification sheets to make the system behave like Figure 8-4A.

On the specification sheets for both distillation columns the lines from the decanter should be input as feed on stage 2 (first actual stage in the column). Use kettle-type reboilers. Set the bottoms rate for the butanol column at the values calculated from the external mass balances. Initially, set Col-1 with N = 10 and feed at 5. Initially set the stripping column with 10 stages. Set the boilup rate to 1.0 kmol/hour in the stripping column. NOTE: This rate is not the boilup ratio. Set convergence in both columns to "Strongly non-ideal liquid." Set convergence at 75 iterations for both columns.

Set the reflux ratios for both columns to 0.5. We don't really want to reflux from the condensers, but want to take our reflux from the decanter; however, if you try setting the reflux ratio to 0.0 Aspen Plus will not run. Reduce both reflux ratios in a few steps down to about 0.025. These ratios are small enough that the results will be very close to using only reflux from the decanter phases. If you don't do this in steps, the condenser will dry up and the run will have errors. Pamper Aspen Plus, and make it happy! After you finish the lab, you might try reinitializing and see what happens if you set reflux ratio to 0.025. Results when the column dried up are not useable.

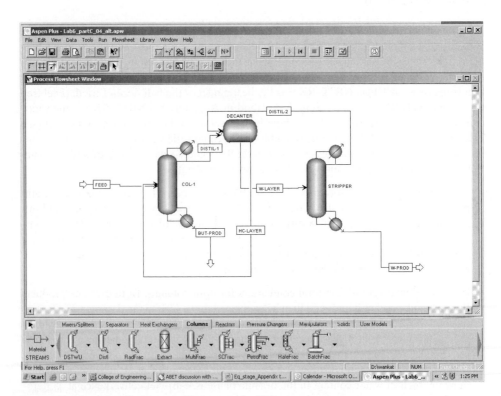

FIGURE 8-A2.   *Distillation system for separation of heterogeneous azeotrope*

After you have successfully run the simulation, note that both columns are probably not producing products as pure as required. Increase the boilup rate modestly in steps of 0.2 kmol/h until both products meet the purity specifications, but the butanol product is purer than necessary. Since there is more separation than required, this implies that column 1 can have fewer stages and/or a lower boilup ratio. However, before changing conditions, save your results and note that there is a difficulty inherent in the set of specified variables for the flowsheet in Figure 8-A2. The loop from Col-1 to the Decanter and back to Col-1 via the HC-layer has no positive control on flow rates. (The other loop from Stripper to Decanter and back does because the boilup rate in column 2 is specified). Thus, it is possible for the flow rates of Distil-1 and HC-Layer to become very large. This will happen if the stages in Col-1 are reduced drastically. The cure is to set a flow rate that will control this loop; however, since the number of variables one can specify is set, another variable has to made free. A possibility that works once AspenPlus is giving good results is to set reflux ratio (.025) and boilup rate in Col-1 (use the value from the run with 10 stages and N_{feed} = 5 with bottoms in Col-1 specified). To be sure this works, repeat previous run but with boilup rate in Col-1 replacing specification of bottoms rate. Then reduce the number of stages and the feed location. Increase the boilup rate by a maximum of 10% to obtain the desired separation. If there are convergence difficulties as you do this, try decreasing the error tolerance for each column to 1.0E-5. Also try increasing the number of maximum flowsheet iterations in the Wegstein method to 75 (below Block, Utilities, and Reactions in AspenPlus click on Convergence → Conv Options → Methods and then Wegstein tab and Convergence Parameters).

Report your final results for number of stages, optimum feed location, boilup rate in Col-1, and product purities.

When done, make the column not converge! Return to the original number of stages, original feed location and original specification of bottoms flow rate in the butanol column. Then increase the boilup rate in the water stripper to 10 kmol/h or higher and run the simulation. The messages in the control panel, in the run summary and the reports for the blocks (obtained from View) all state there is a problem. However, the reports for the streams (obtained from View) do not point out the convergence problem. Do NOT save the run that did not converge.

Lab 9.   *Extractive Distillation.* This assignment is more prescriptive and involves less exploration than other labs since convergence is often a problem with extractive distillation. The basic algorithm (Figure 6-1) assumes that the concentration loop will have little effect on the other loops. Extractive distillation systems have very nonideal VLE and this assumption is often not true. As always with Aspen Plus, it may be possible to obtain convergence by starting with a set of conditions that converges and slowly changing the variable of interest (e.g., L/D) to approach the desired value. When you have a convergence problem, reinitialize RADFRAC and then return to a condition that converged previously.

Problem. Use extractive distillation to break the ethanol-water azeotrope. Use the two-column system shown in Figure 8-13. The solvent is ethylene glycol. Both columns operate at a pressure of 1.0 atmosphere. The feed to column 1 is 100 kmol/h. This feed is a saturated liquid. It is 74 mol% ethanol and 26 mol% water. Use NRTL.

In Figure 8-13 the A product will be the ethanol product and the B product (distillate from column 2) will be water. We want the A product to be 0.9975 mole frac ethanol (this exceeds requirements for ethanol used in gasoline). Use an external reflux ratio in column 1 L/D = 1.0. (Normally these would be optimized, but to save time leave it constant.) The reflux is returned as a saturated liquid. This set of conditions should remove sufficient ethylene glycol from the distillate to produce an ethanol of suitable purity (check to make sure that this happens). The bottoms product from column 1 should have less than 0.00009 mole frac ethanol. This number is low to increase the recovery of ethanol.

The distillate from column 2 should contain less than 0.0001 mole frac ethylene glycol. The bottoms product from column 2 should contain less than 0.0001 mole frac water since any water in this stream will probably end up in the ethanol product when solvent is recycled.

Steps 1 and 2. First design the two columns in RADFRAC as columns in series with no solvent recycle. Thus, all of the solvent used must be added as the make-up solvent. For this part use a makeup solvent that is close to the expected concentration of the recycle solvent. That is, the solvent stream should be 0.9999 mole frac ethylene glycol and 0.0001 mole frac water. Use a solvent temperature of 80°C to approximately match the temperature of the solvent feed stage NS = 5 in column 1. Solvent pressure is 1.0 atmosphere. Use total condensers and kettle reboilers.

In the convergence section of the input block set the number of iterations to 75 (for both columns). (Go to Data in the menu bar and click on Blocks. On the left hand side of the screen click on the + sign next to the block for the distillation column you want to increase the number of iterations for. Then find Convergence and click on the blue check mark for convergence. This gives a table. Increase the maximum number of iterations to 75. Higher values don't help.)

Step 1. Design column 1. The solvent is treated as a second feed. On the flow diagram, use the feed port to add a solvent feed to the column. Then specify its location in the Table of input conditions. Add solvent at stage NS = 5. Use a solvent rate of S = 52.0 kmol/h. Start with N of column 1 = 50 and feed location NF1 = 23. List convergence as strongly non-ideal liquid. Specify the *distillate flow rate* that will give you the desired purity of the ethanol product and the desired ethanol mole fraction in the bottoms product (do external balances). Find the approximate optimum feed stage and the lowest total number of stages that will do the desired separation. Start by keeping NF1 fixed while you reduce N (total number of stages in column 1) to approximately obtain the desired bottoms and distillate concentrations. Then find the optimum feed stage, and try reducing N more. Low values of the feed stage (e.g., NF1 = 10) will probably not converge. Thus, start with a high number for the feed stage location and reduce NF1 slowly. If your ethanol distillate mole fraction is > 0.975 but the ethanol bottoms mole fraction > 0.00009, the cause may be a distillate flow rate that is too low. The mass balance to calculate the distillate flow rate has to be accurate. If you exceed the purity requirements, that is okay at this point, since the system is not coupled.

Step 2. Design column 2. The feed to column 2 is the bottoms from column 1. It is a saturated liquid at 1.0 atmosphere. Specify the *bottoms flow rate* that will give you the desired purity of distillate and bottoms (do a very accurate external balance). Very small adjustments in the bottoms flow rate may be necessary to meet the specifications. Find an approximate $(L/D)_{min}$ (N2 = 50 and NF2 = 25 is sufficient; start with L/D = 1.0 and move down).

AspenPlus has convergence problems near the value of $(L/D)_{min}$. Do runs with $L/D > (L/D)_{min}$ where the purity is better than required and runs where the purity is less than required, which means $L/D < (L/D)_{min}$. Then approximate $(L/D)_{min}$. Operate with $L/D = 1.15 (L/D)_{min}$. Then find the optimum value of the feed stage (start with N2 = 20 and NF2 = 10). Note that column 2 is quite simple and a low value of L/D works.

Step 3. Eventually we will connect the solvent recycle loop from column 2 to column 1. But first include a heat exchanger to cool the solvent. If you don't do this column 1 will not work after you connect the solvent recycle loop (why not?). (To put the heat exchanger in the solvent line, go to Heat Exchangers and put a HEATER—used as a cooler in this case—in your flowsheet. Then left click on the solvent stream and right click on Reconnect Destination. Then connect the solvent stream to the arrow on the heat exchanger. Hit the Next button. You will get a window for the heat exchanger. Use 80°C. Pressure is 1.0 atmosphere.) For Valid Phases use "liquid only." To check that the HEATER is hooked up properly, try a run without changing the solvent makeup flow rate.

Step 4. Connect the line leaving the heater to column 1 (use reconnect destination and connect it to the feed port for column 1). In the table of input conditions list the solvent makeup stream and the recycle line at the same stage (stage number 5).

To calculate the makeup solvent flow rate that you eventually want to use, do an external mass balance around the entire system,

Pure Solvent Makeup flow rate = Solvent out in A product + solvent out in B product.

The resulting Makeup flow rate will be extremely small since losses of solvent are small (after all, no one wants to drink ethylene glycol with their alcohol or their water). If you immediately use this value of solvent makeup as a feed to the system, Aspen Plus will not converge. Start with the value you were using previously and rapidly decrease it (say by *factors* of roughly 5 or 10). Until the solvent makeup stream is at the desired value for the external mass balance, the "extra" ethylene glycol will exit with the distillate (water product) from column 2. This occurs because Bottoms flow rate in column 2 is specified and the only place for the extra ethylene glycol to go is with the distillate from column 2. (This is why you must set the bottoms rate in column 2. If you set the distillate rate in column 2 there is no place for the "extra" ethylene glycol to go and the system will not converge.) Ignore the values of the distillate flow rate and distillate compositions from column 2 until you use the Makeup solvent rate that satisfies the overall balance. Once you are close to the correct flow for makeup, change it to pure ethylene glycol. The final appearance of the system should be similar to Figure 8-A3.

If for some reason (not recommended) you have to uncouple the recycle loop, remember to reset the makeup flow rate to the desired value.

Step 5. If necessary, make minor adjustments in stages, flows or reflux ratios to achieve desired purities.

As a minimum, record the following: the mole fracs of the two products and the two bottom streams; the heat duties of the two condensers, the two reboilers and the heat exchanger; the flow rates of all the streams in the process; L/D in each column; N and feed locations in each column; and the temperatures of the reboilers and condensers.

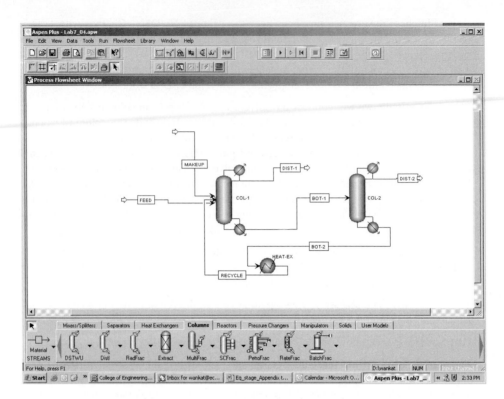

FIGURE 8-A3.   *Aspen Plus screen shot of completed extractive distillation system*

Chapter 8   Introduction to Complex Distillation Methods

CHAPTER 9

Batch Distillation

Continuous distillation is a thermodynamically efficient method of producing large amounts of material of constant composition. When small amounts of material or varying product compositions are required, batch distillation has several advantages. In batch distillation a charge of feed is loaded into the reboiler, the steam is turned on, and after a short startup period, product can be withdrawn from the top of the column. When the distillation is finished, the heat is shut off and the material left in the reboiler is removed. Then a new batch can be started. Usually the distillate is the desired product.

Batch distillation is a much older process than continuous distillation. Mesopotamian clay distillation pots have been dated to around 3500 BCE (RT, 2007), and alchemists in Alexandria used simple batch retorts in the first century A.D. (Davies, 1980). Batch distillation was developed to concentrate alcohol by Arab alchemists around 700 A.D. (Vallee, 1998). It was adopted in Western Europe, and the first known book on the subject was Hieronymus Brunschwig's *Liber de arte distillandi*, published in Latin in the early 1500s. This book remained a standard pharmaceutical and medical text for more than a century. The first distillation book written for a literate but not scholarly community was Walter Ryff's *Das New gross Distillier Buch* published in German in 1545 (Stanwood, 2005). This book included a "listing of distilling apparatus, techniques, and the plants, animals, and minerals able to be distilled for human pharmaceutical use." Advances in batch distillation have been associated with its use to distill alcohol, pharmaceuticals, coal oil, petroleum oil, and fine chemicals.

Batch distillation is versatile. A run may last from a few hours to several days. Batch distillation is the choice when the plant does not run continuously and the batch must be completed in one or two shifts (8 to 16 hours). It is often used when the same equipment distills several different products at different times. If distillation is required only occasionally, batch distillation would again be the choice.

Equipment can be arranged in a variety of configurations. In simple batch distillation (Figure 9-1), the vapor is withdrawn continuously from the reboiler. The system differs from flash distillation in that there is no continuous feed input and the liquid is drained only at the end of the batch. An alternative to simple batch distillation is constant-level batch distillation where solvent is fed continually to the still pot to keep the liquid level constant (Gentilcore, 2002).

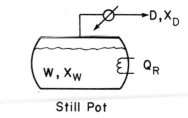

Still Pot

FIGURE 9-1. *Simple batch distillation*

In a multistage batch distillation, a staged or packed column is placed above the reboiler as in Figure 9-2. Reflux is returned to the column. In the usual operation, distillate is withdrawn continually (Barton and Roche, 1997; Diwekar, 1995; Luyben, 1971; Mujtaba, 2004; Pratt, 1962; Robinson and Gilliland, 1950) until the column is shut down and drained. In an alternative method (Treybal, 1970), no distillate is withdrawn; instead, the composition of liquid in the accumulator changes. When the distillate in the accumulator is of the desired composition in the desired amount, both the accumulator and the reboiler are drained. Luyben

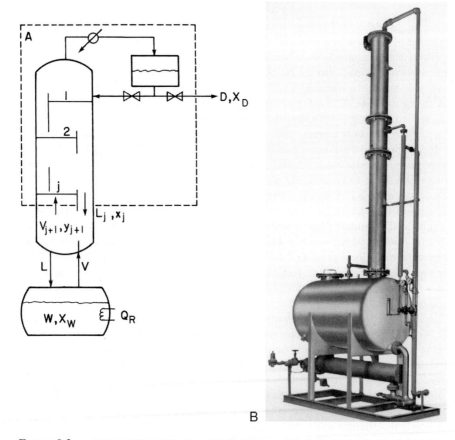

FIGURE 9-2. *Multistage batch distillation; (A) schematic, (B) photograph of packaged batch distillation/solvent recovery system of approximately 400-gallon capacity. Courtesy of APV Equipment, Inc., Tonowanda, New York*

(1971) indicated that the usual method should be superior; however, the alternative method may be simpler to operate.

Another alternative is called inverted batch distillation (Diwekar, 1995; Pratt, 1967; Robinson and Gilliland, 1950) because bottoms are withdrawn continuously while distillate is withdrawn only at the end of the distillation (see Figure 9-8 and Problem 9.C2). In this case the charge is placed in the accumulator and a reboiler with a small holdup is used. Inverted batch distillation is seldom used, but it is useful when quite pure bottoms product is required.

9.1. BINARY BATCH DISTILLATION: RAYLEIGH EQUATION

The mass balances for batch distillation are somewhat different from those for continuous distillation. In batch distillation we are more interested in the total amounts of bottoms and distillate collected than in the rates. For a binary batch distillation, mass balances around the entire system for the entire operation time are

$$F = W_{final} + D_{total} \tag{9-1}$$

$$Fx_F = x_{W,final}W_{final} + D_{total}x_{D,avg} \tag{9-2}$$

The feed into the column is F kg moles of mole fraction x_F of the more volatile component. The final moles in the reboiler at the end of the batch is W_{final} of mole fraction $x_{W,final}$. The symbol W is used since the material left in the reboiler is often a waste. D_{total} is the total kilogram moles of distillate of average concentration $x_{D,avg}$. Equations (9-1) and (9-2) are applicable to simple batch and normal multistage batch distillation. Some minor changes in variable definitions are required for inverted batch distillation.

Usually F, x_F, and the desired value of either $x_{W,final}$ or $x_{D,avg}$ are specified. An additional equation is required to solve for the three unknowns D_{total}, W_{final}, and $x_{W,final}$ (or $x_{D,avg}$). This additional equation, known as the Rayleigh equation (Rayleigh, 1902), is derived from a differential mass balance. Assume that the holdup in the column and in the accumulator is negligible. Then if a differential amount of material, –dW, of concentration x_D is removed from the system, the differential component mass balance is

$$- \text{Out} = \text{accumulation in reboiler} \tag{9-3}$$

or

$$- x_D \, dW = - d \, (Wx_W) \tag{9-4}$$

Expanding Eq. (9-4),

$$- x_D \, dW = - W \, dx_W - x_W dW \tag{9-5}$$

Then rearranging and integrating,

$$\int_{W=F}^{W_{final}} \frac{dW}{W} = \int_{x_F}^{x_{W,final}} \frac{dx_W}{x_D - x_W} \tag{9-6}$$

which is

$$\ln\left[\frac{W_{final}}{F}\right] = - \int_{x_{W,final}}^{x_F} \frac{dx_W}{x_D - x_W} \tag{9-7}$$

The minus sign comes from switching the limits of integration. Equation (9-7) is a form of the Rayleigh equation that is valid for both simple and multistage batch distillation. Of course, to use this equation we must relate x_D to x_W and do the appropriate integration. This is covered in sections 9.2 and 9.5.

Time does not appear explicitly in the derivation of Eq. (9-7), but it is implicitly present since W, x_W, and usually x_D are all time-dependent.

9.2 SIMPLE BINARY BATCH DISTILLATION

In the simple binary batch distillation system shown in Figure 9-1 the vapor product is in equilibrium with liquid in the still pot at any given time. Since we use a total condenser, $y = x_D$. Substituting this into Eq. (9-7), we have

$$\ln\left(\frac{W_{final}}{F}\right) = - \int_{x_{W,final}}^{x_F} \frac{dx}{y - x} = - \int_{x_{W,final}}^{x_F} \frac{dx}{f(x) - x} \tag{9-8}$$

where y and x are now in equilibrium and the equilibrium expression is $y = f(x,p)$. For any given equilibrium expression, Eq. (9-8) can be integrated analytically, graphically, or numerically.

The general graphical or numerical integration procedure for Eq. (9-8) is:

1. Plot or fit y-x equilibrium curve.
2. At a series of x values, find $y - x$.
3. Plot $1/(y - x)$ vs. x or fit it to an equation.
4. Graphically or numerically integrate from x_F to $x_{W,final}$. Graphical integration is shown in Figure 9-3.
5. From Eq. (9-8), find the final charge of material in the still pot:

$$W_{final} = F \exp\left(- \int_{x_{W,final}}^{x_F} \frac{dx}{y - x}\right) = Fe^{-Area} \tag{9-9a,b}$$

where the area is shown in Figure 9-3.
6. The average distillate concentration, $x_{D\ avg}$, can be found from the mass balances. Solving Eqs. (9-1) and (9-2),

$$x_{D,avg} = \frac{Fx_F - W_{final}x_{W,final}}{F - W_{final}} \tag{9-10}$$

$$D_{total} = F - W_{final} \tag{9-11}$$

The Rayleigh equation can also be integrated numerically. One convenient method for doing this is to use Simpson's rule (e.g., see Mickley et al., 1957, pp. 35-42). If the ordinate in Figure 9-3 is called f(x), then one form of Simpson's rule is

$$\int_{x_{W,final}}^{x_F} f(x)dx = \frac{x_F - x_{W,final}}{6}\left[f(x_{W,final}) + 4f\left(\frac{x_{W,final} + x_F}{2}\right) + f(x_F)\right]$$

(9-12)

where terms are shown in Figure 9-3. Simpson's rule is exact if f(x) is cubic or lower order. For smooth curves, such as in Figure 9-3, Simpson's rule will be quite accurate (see Example 9-1). For more complex shapes, Simpson's rule will be more accurate if the integration is done in two or more pieces (see Example 9-2). Simpson's rule is also the MATLAB command quad (for "quadrature") (Pratap, 2006). Other integration formulas that are more accurate can be used.

If the average distillate concentration is specified, a trial-and-error procedure is required. This involves guessing the final still pot concentration, $x_{W,final}$, and calculating the area in Figure 9-3 either graphically or using Simpson's rule. Then Eq. (9-9) gives W_{final} and Eq. (9-10) is used to check the value of $x_{D,avg}$. For a graphical solution, the trial-and-error procedure can be conveniently carried out by starting with a guess for $x_{W,final}$ that is too high. Then every time $x_{W,final}$ is decreased, the additional area is added to the area already calculated.

If the equilibrium expression is given as a constant relative volatility, α, the Rayleigh equation can be integrated analytically. In this case the equilibrium expression is Eq. (2-22), which is repeated here.

$$y = \frac{\alpha x}{1 + (\alpha - 1)x}$$

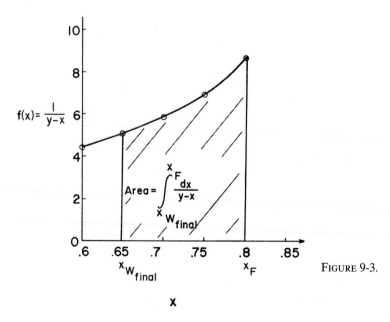

FIGURE 9-3. *Graphical integration for simple batch distillation, Example 9-1*

Substituting this equation into Eq. (9-8) and integrating, we obtain

$$\ln(\frac{W_{final}}{F}) = \frac{1}{\alpha - 1} \ln(\frac{x_{W,final}(1 - x_F)}{x_F(1 - x_{W,final})}) + \ln(\frac{1 - x_F}{1 - x_{W,final}}) \qquad \textbf{(9-13)}$$

When it is applicable, Eq. (9-13) is obviously easier to apply than graphical or numerical integration.

EXAMPLE 9-1. Simple Rayleigh distillation

We wish to use a simple batch still (one equilibrium stage) to separate methanol from water. The feed charge to the still pot is 50 moles of an 80 mol% methanol mixture. We desire an average distillate concentration of 89.2 mol% methanol. Find the amount of distillate collected, the amount of material left in the still pot, and the concentration of material in the still pot. Pressure is 1 atm. Methanol-water equilibrium data at 1 atm are given in Table 2-7 in Problem 2.D1.

Solution

A. Define. The apparatus is shown in Figure 9-1. The conditions are: p = 1 atm, F = 50, x_F = 0.80, and $x_{D,avg}$ = 0.892. We wish to find $x_{W,final}$, D_{tot}, and W_{final}.

B. Explore. Since the still pot acts as one equilibrium contact, the Rayleigh equation takes the form of Eqs. (9-8) and (9-9). To use these equations, either a plot of $1/(y - x)_{equil}$ vs. x is required for graphical integration or Simpson's rule can be used. Both will be illustrated. Since $x_{W,final}$ is unknown, a trial-and-error procedure will be required for either integration routine.

C. Plan. First plot $1/(y - x)$ vs. x from the equilibrium data. The trial-and-error procedure is as follows:
Guess $x_{W,final}$
Integrate to find

$$Area = \int_{x_{W,final}}^{x_F} \frac{dx}{y - x}$$

Calculate W_{final} from Rayleigh equation and $x_{D,calc}$ from mass balance.
Check: Is $x_{D,calc} = x_{D,avg}$? If not, continue trial and error.

D. Do it. From the equilibrium data the following table is easily generated:

x	y	y − x	$\frac{1}{y-x}$
.8	.915	.115	8.69
.75	.895	.145	6.89
.70	.871	.171	5.85
.65	.845	.195	5.13
.60	.825	.225	4.44
.50	.780	.280	3.57

These data are plotted in Figure 9-3. For the numerical solution a large graph on millimeter graph paper was constructed.
First guess: $x_{W,final} = 0.70$.

From Figure 9-3,

$$\text{Area} = \int_{x_{W,final}}^{x_F} \frac{dx}{y - x} = 0.7044$$

Then, $W_{final} = F \exp(-\text{Area}) = 50\, e^{-0.7044} = 24.72$

$$D_{calc} = F - W_{final} = 25.28$$

$$x_{D,calc} = \frac{Fx_F - W_{final}x_{W,final}}{D_{calc}} = 0.898$$

The alternative integration procedure using Simpson's rule gives

$$\text{Area} = \left(\frac{x_F - x_{W,final}}{6}\right)\left[\left.\left(\frac{1}{y-x}\right)\right|_{x_{W,final}} + 4\left.\left(\frac{1}{y-x}\right)\right|_{(x_{W,final} + x_F)/2} + \left.\left(\frac{1}{y-x}\right)\right|_{x_F}\right]$$

where $\left.\left(\dfrac{1}{y-x}\right)\right|_{x_{W,final}}$

is the value of $1/(y-x)$ calculated at $x_{W,final}$.

$$\text{Area} = \left(\frac{0.1}{6}\right)[5.85 + 4(6.89) + 8.69] = 0.70166$$

Then for Simpson's rule, $W_{final} = 24.79$, $D_{calc} = 25.21$, and $x_{D,calc} = 0.898$. Simpson's rule appears to be quite accurate. W_{final} is off by 0.3%, and $x_{D,calc}$ is the same as the more exact calculation. These values appear to be close to the desired value, but we don't yet know the sensitivity of the calculation.
Second guess: $x_{W,final} = 0.60$. Calculations similar to the first trial give Area = 1.2084, $W_{final} = 14.93$, $D_{calc} = 35.07$, $x_{D,calc} = 0.885$ from Figure 9-3, and $x_{D,calc} = 0.884$ from the Simpson's rule calculation. These are also close, but they are low. For this problem the value of x_D is insensitive to $x_{W,final}$.
Third guess: $x_{W,final} = 0.65$. Calculations give Area = 0.971, $W_{final} = 18.94$, $D_{calc} = 31.06$, $x_{D,calc} = 0.891$ from Figure 9-3 and $x_{D,calc} = 0.890$ from the Simpson's rule calculation of the area, which are both close to the specified value of 0.892.

Thus, use $x_{W,final} = 0.65$ as the answer.
E. Check. The overall mass balance should check. This gives: $W_{final} x_{W,final} + D_{calc}$ $x_{D,calc} = 39.985$ as compared to $Fx_F = 40$. Error is $(40–39.985) / 40 \times 100$, or 0.038%, which is acceptable.
F. Generalize. The integration can also be done numerically on a computer using Simpson's rule or an alternative integration method. This is an advantage, since then the entire trial-and-error procedure can be programmed. Note that large differences in $x_{W,final}$ and hence in W_{final} cause rather small differences in

$x_{D,avg}$. Thus, for this problem, exact control of the batch system may not be critical. This problem illustrates a common difficulty of simple batch distillation—a pure distillate and a pure bottoms product cannot be obtained unless the relative volatility is very large. Note that although Eq. (9-13) is strictly not applicable since methanol-water equilibrium does not have a constant relative volatility, it could be used over the limited range of this batch distillation.

It is interesting to compare the simple batch distillation result to a flash distillation of the same feed producing y = 0.892 mole fraction methanol. This flash distillation problem was solved previously as Problem 2.D1. The results were x = 0.756 (compare to $x_{W,final}$ = 0.65), V = 16.18 (compare to D_{total} = 31.06), and L = 33.82 (compare to W_{final} = 18.94). The simple batch distillation gives a greater separation with more distillate product because the bottoms product is an average of liquids in equilibrium with vapor in the entire range from y = 0.915 (in equilibrium with x = z = 0.8) to y = 0.845 (in equilibrium with x = $x_{W,final}$ = 0.65), while for the flash distillation the liquid is always in equilibrium with the final vapor y = 0.892. In general, a simple batch distillation will give more separation than flash distillation with the same operating conditions. However, because flash distillation is continuous, it is much easier to integrate into a continuous plant.

A process that is closely related to simple batch distillation is differential condensation (Treybal, 1980). In this process vapor is slowly condensed and the condensate liquid is rapidly withdrawn. A derivation similar to the derivation of the Raleigh equation for a binary system gives,

$$\ln\left[\frac{F}{D_{final}}\right] = \int_{y_F}^{y_{D,final}}\left[\frac{dy}{y-x}\right] \tag{9-14}$$

where F is the moles of vapor fed of mole fraction y_F, and D_{final} is the left over vapor distillate of mole fraction $y_{D,final}$.

9.3 CONSTANT-LEVEL BATCH DISTILLATION

One common application of batch distillation (or evaporation) is to switch solvents in a production process. For example, solvents may need to be exchanged prior to a crystallization or reaction step. Solvent switching can be done in a simple batch system by charging the still pot with feed, concentrating the solution (if necessary) by boiling off most of the original solvent (some solvent needs to remain to maintain agitation, keep the solute in solution, and keep the heat transfer area covered), adding the new solvent and doing a second batch distillation to remove the remainder of the original solvent. If desired the solution can be diluted by adding more of the desired solvent. Although this process mimics the procedure used by a bench chemist, a considerable amount of the second solvent is evaporated in the second batch distillation. An alternative is to do the second batch distillation by constant-level batch distillation in which the pure second solvent is added continuously during the second batch distillation at a rate that keeps the moles in the still-pot constant (Gentilcore, 2002). We will focus on the constant-level batch distillation step.

For a constant-level batch distillation the general mole balance is

In – Out = accumulation in still pot (9-15a)

For the total mole balance this is: In – Out = 0, since the moles in the still pot are constant. Thus, if dS moles of the second solvent are added, the overall mole balance for a constant-level batch distillation is

$$dV = dS \tag{9-15b}$$

where dV is the moles of vapor withdrawn.

If we do a component mole balance on the original solvent (solute is assumed to be non-volatile and is ignored), we obtain the following for a constant-level system,

$$0 - ydV = Wdx_w \tag{9-16a}$$

where y and x_w are the mole fraction of the first solvent in the vapor and liquid, respectively. Substituting in Eq. (9-15b), this becomes

$$-ydS = Wdx_w \tag{9-16b}$$

Note that since the amount of liquid W is constant, there is no term equivalent to the last term in Eq. (9-4). Integration of Eq. (9-16b) and minor rearrangement gives us,

$$S/W = \int_{x_{w,final}}^{x_{w,initial}} \frac{dx_w}{y} \tag{9-17}$$

Since vapor and liquid are assumed to be in equilibrium, y is related to x_w by the equilibrium relationship. Equation (9-17) can be integrated graphically or numerically in a procedure that is quite similar to that used for simple batch distillation. Problem 9.D17 will lead you through these calculations for constant-level batch distillation.

If constant relative volatility between the two solvents can be assumed, Eq. (2-22) can be substituted into Eq. (9-17) and the equation can be integrated analytically (Gentilcore, 2002),

$$S / W = \frac{1}{\alpha} \ln\left[\frac{x_{w,initial}}{x_{w,final}}\right] + \frac{\alpha - 1}{\alpha} (x_{w,initial} - x_{w,final}) \tag{9-18}$$

Gentilcore (2002) presents a constant relative volatility sample calculation that illustrates the advantage of constant-level batch distillation when it is used to exchange solvents.

9.4 BATCH STEAM DISTILLATION

In batch steam distillation, steam is sparged directly into the still pot as shown in Figure 9-4. This is normally done for systems that are immiscible with water. The reasons for adding steam directly to the still pot are that it keeps the temperature below the boiling point of water, it eliminates the need for heat transfer surface area and it helps keep slurries and sludges well mixed so that they can be pumped. The major use is in treating wastes that contain valuable volatile organics. These waste streams are often slurries or sludges that would be difficult to process in an ordinary batch still. Compounds that are often steam distilled include glycerine, lube oils, fatty acids, and halogenated hydrocarbons (Woodland, 1978). Section 8-3 is a prerequisite for this section.

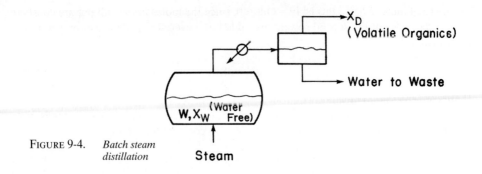

FIGURE 9-4. *Batch steam distillation*

Batch steam distillation is usually operated with liquid water present in the still. Then both the liquid water and the liquid organic phases exert their own partial pressure. Equilibrium is given by Eqs. (8-14) to (8-18) when there is one volatile organic and some nonvolatile organics present. As long as there is minimal entrainment, there is no advantage to having more than one stage. For low-molecular-weight organics, vaporization efficiencies, defined as the actual partial pressure divided by the partial pressure at equilibrium, p*,

$$E = \frac{p_{volatile}}{p_{volatile}^{*}} = \frac{p_{volatile}}{(VP)_{volatile}\, x_{volatile}} \tag{9-19}$$

are often in the range from 0.9 to 0.95 (Carey, 1950). This efficiency is close enough to equilibrium that equilibrium calculations are adequate.

The system shown in Figure 9-4 can be analyzed with mass balances on a water-free basis. The mass balances are Eqs. (9-1) and (9-2), which can be solved for W_{final} if $x_{W,final}$ is given.

$$W_{final} = F(\frac{x_D - x_F}{x_D - x_{w,final}}) \tag{9-20}$$

With a single volatile organic, $x_D = 1.0$ if entrainment is negligible. Then,

$$W_{final} = F(\frac{1 - x_F}{1 - x_{w,final}}) \tag{9-21}$$

and the flow rate of the organic distillate product is

$$D = F - W_{final} \tag{9-22}$$

The Rayleigh equation can also be used and will give the same results.

At any moment the instantaneous moles of water dn_w carried over in the vapor can be found from Eq. (8-18). This becomes

$$dn_w = dn_{org} \frac{P_{tot} - (VP)_{volatile} x_{volatile\ in\ org}}{(VP)_{volatile} x_{volatile\ in\ org}} \tag{9-23a}$$

The total moles of water carried over in the vapor can be obtained by integrating this equation:

$$n_w = \int_0^{D_{total}} \frac{P_{tot} - (VP)_{volatile} x_{volatile\ in\ org}}{(VP)_{volatile} x_{volatile\ in\ org}} \, dn_{org} \qquad \textbf{(9-23b)}$$

During the batch steam distillation, the mole fraction of the volatile organics in the still varies, and thus, the still temperature determined by Eq. (8-15) varies. Equation (9-23b) can be integrated numerically in steps. The total moles of water required is n_w plus the moles of water condensed to heat the feed and still pot, and to vaporize the volatile organics.

To integrate Eq. (9-23b) we must relate the moles of volatile organic in the distillate n_{org} to its mole fraction in the still pot, $x_{pot} = x_{volatile\ in\ org}$. From a mass balance on the volatile organic very similar to that used to derive Eq. (9-21), we obtain

$$n_{org} = F(x_F - x_{pot})/(1 - x_{pot}) \qquad \textbf{Eq. (9-24a)}$$

This equation is valid at any time. As the mole fraction of the volatile organic in the still pot, x_{pot}, decreases, the moles of volatile organic collected in the distillate, n_{org}, increases.

For step-by-step numerical integration, Eq. (9-23b) can be written as,

$$\Delta n_w = \Delta n_{org} (P_{total} - (VP)_{volatile}\ x_{pot,average})/[(VP)_{volatile}\ x_{pot,average}] \qquad \textbf{Eq. (9-24b)}$$

$$n_w = \Sigma\,(\Delta n_w) \qquad \textbf{Eq. (9-24c)}$$

The total change in mole fraction of the volatile organic as x_{pot} goes from x_F to $x_{w,final}$, can be broken up into a number of intervals Δx_{pot}, and then n_{org}, $x_{pot,average}$ and Δn_w are calculated for each interval. The amount of water carried over into the distillate with the volatile organic is then given by Eq. (9-24c). The larger the number of intervals from x_F to $x_{w,final}$, the more accurate the calculation.

A short example for the simplified case when temperature in the still pot does not vary much and $(VP)_{volatile}$ is essentially constant will help to clarify the procedure. Suppose $F = 1.0$ kmol, $x_F = 0.35$, $P_{total} = 760$ mm Hg, $(VP)_{volatile} = 50$ mm Hg, and we want to calculate Δn_w for a change in x_{pot} from 0.25 to 0.20. Then from Eq. (9-24a) at $x_{pot} = 0.25$,

$$n_{org} = F(x_F - x_{pot})/(1 - x_{pot}) = 1.0(.35 - .25)/(1 - .25) = 0.13333$$

At $x_{pot} = 0.20$, $n_{org} = 0.18750$. In the interval from $x_{pot} = 0.25$ to $x_{pot} = 0.20$, $x_{pot,average} = 0.225$ and $\Delta n_{org} = 0.18750 - 0.13333 = 0.05417$. Then from Eq. (9-24b),

$$\Delta n_w = 0.05417[760 - (50)(0.225)]/[(50)(0.225)] = 3.605 \text{ kmol water.}$$

If the still pot temperature varies significantly, the temperature at each value of $x_{pot,average}$ needs to be determined from Eq. (8-15) so that the value of $(VP)_{volatile}$ can be calculated. This is easiest to do with a spreadsheet.

During most of the batch operation, the mole fraction of the volatile organic is considerably higher than it is at the end of the batch. In continuous steam distillation the mole fraction of the volatile organic is always at its lowest value. Thus, batch steam distillation requires less steam for a given separation than continuous steam distillation.

9.5 MULTISTAGE BATCH DISTILLATION

The separation achieved in a single equilibrium stage is often not large enough to both obtain the desired distillate concentration and a low enough bottoms concentration. In this case a distillation column is placed above the reboiler as shown in Figure 9-2. The calculation procedure will be detailed here for a staged column, but packed columns can easily be designed using the procedures explained in Chapter 10.

For multistage systems x_D and x_W are no longer in equilibrium. Thus, the Rayleigh equation, Eq. (9-7), cannot be integrated until a relationship between x_D and x_W is found. This relationship can be obtained from stage-by-stage calculations. We will assume that there is negligible holdup on each plate, in the condenser, and in the accumulator. Then at any specific time we can write mass and energy balances around stage j and the top of the column as shown in Figure 9-2A. These balances simplify to

$$\text{Input} = \text{output}$$

since accumulation was assumed to be negligible everywhere except the reboiler. Thus, at any given time t,

$$V_{j+1} = L_j + D \tag{9-25a}$$

$$V_{j+1}y_{j+1} = L_jx_j + Dx_D \tag{9-25b}$$

$$Q_c + V_{j+1}H_{j+1} = L_jh_j + Dh_D \tag{9-25c}$$

In these equations V, L and D are now molal flow rates. These balances are essentially the same equations we obtained for the rectifying section of a continuous column except that Eqs. (9-25) are time-dependent. If we can assume constant molal overflow (CMO), the vapor and liquid flow rates will be the same on every stage and the energy balance is not needed. Combining Eqs. (9-25a) and (9-25b) and solving for y_{j+1}, we obtain the operating equation for CMO:

$$y_{j+1} = \frac{L}{V} x_j + (1 - \frac{L}{V})x_D \tag{9-26}$$

At any specific time Eq. (9-26) represents a straight line on a y-x diagram. The slope will be L/V, and the intercept with the y = x line will be x_D. Since either x_D or L/V will have to vary during the batch distillation, the operating line will be continuously changing.

9.5.1 Constant Reflux Ratio

The most common operating method is to use a constant reflux ratio and allow x_D to vary. This procedure corresponds to a simple batch operation where x_D also varies. The relationship between x_D and x_W can now be found from a stage-by-stage calculation using a McCabe-Thiele analysis. Operating Eq. (9-26) is plotted on a McCabe-Thiele diagram for a series of x_D values. Then we step off the specified number of equilibrium contacts on each operating line starting at x_D to find the x_W value corresponding to that x_D. This procedure is shown in Figure 9-5 and Example 9-2.

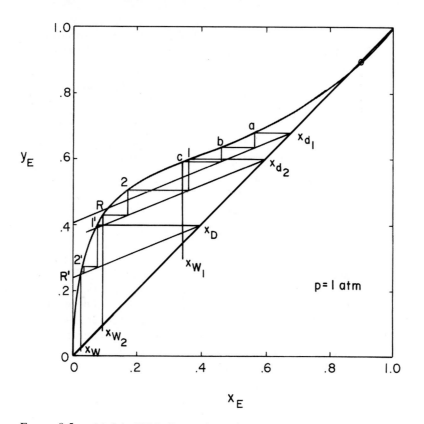

FIGURE 9-5. *McCabe-Thiele diagram for multistage batch distillation with constant L/D, Example 9-2*

The McCabe-Thiele analysis gives x_W values for a series of x_D values. We can now calculate $1/(x_D - x_W)$. The integral in Eq. (9-7) can be determined by either numerical integration such as Simpson's rule given in Eq. (9-12) or by graphical integration. Once x_W values have been found for several x_D values, the same procedure used for simple batch distillation can be used. Thus, W_{final} is found from Eq. (9-7), $x_{D\,avg}$ from Eq. (9-10), and D_{total} from Eq. (9-11). If $x_{D\,avg}$ is specified, a trial-and-error procedure will again be required.

EXAMPLE 9-2. Multistage batch distillation

We wish to batch distill 50 kmol of a 32 mol% ethanol, 68 mol% water feed. The system has a still pot plus two equilibrium stages and a total condenser. Reflux is returned as a saturated liquid, and we use $L/D = 2/3$. We desire a final still pot composition of 4.5 mol% ethanol. Find the average distillate composition, the final charge in the still pot, and the amount of distillate collected. Pressure is 1 atm.

Solution

A. Define. The system is shown in the figure

Find W_{final}, D_{total}, $x_{D,avg}$.

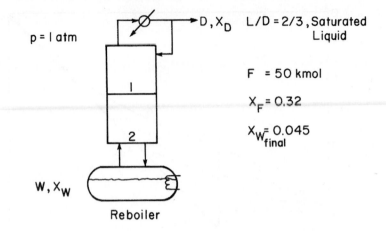

$p = 1$ atm

D, X_D $L/D = 2/3$, Saturated Liquid

$F = 50$ kmol

$X_F = 0.32$

$X_{W,final} = 0.045$

W, X_W

Reboiler

B and C. Explore and Plan. Since we can assume CMO, a McCabe-Thiele diagram (Figure 2-2) can be used. This will relate x_D to x_W at any time. Since x_F and $x_{W,final}$ are known, the Rayleigh Eq. (9-7) can be used to determine W_{final}. Then $x_{D,avg}$ and D_{total} can be determined from Eqs. (9-10) and (9-11), respectively. A trial-and-error procedure is not needed for this problem.

D. Do it. The McCabe-Thiele diagram for several arbitrary values of x_D is shown in Figure 9-5. The top operating line is

$$y = \frac{L}{V} x + \left(1 - \frac{L}{V}\right) x_D$$

where

$$\frac{L}{V} = \frac{L/D}{1 + L/D} = \frac{2/3}{5/3} = \frac{2}{5}$$

The corresponding x_W and x_D values are used to calculate $x_D - x_W$ and then $1/(x_D - x_W)$ for each x_W value. These values are plotted in Figure 9-6 (some values not shown in Figure 9-5 are shown in Figure 9-6). The area under the curve (going down to an ordinate value of zero) from $x_F = 0.32$ to $x_{W,final} = 0.045$ is 0.608 by graphical integration.

Then from Eq. (9-7):

$$W_{final} = Fe^{-Area} = (50) \exp(-0.608) = 27.21$$

From Eq. (9-11): $D_{total} = F - W_{final} = 22.79$
and from Eq. (9-10):

$$x_{D,avg} = \frac{Fx_F - W_{final} x_{W,final}}{F - W_{final}} = 0.648$$

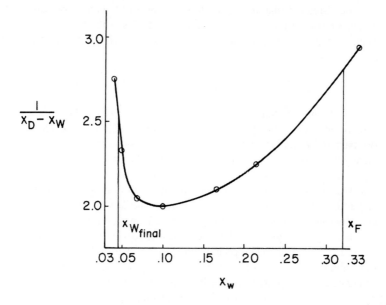

FIGURE 9-6.   *Graphical integration, Example 9-2*

The area can also be determined by Simpson's rule. However, because of the shape of the curve in Figure 9-6 it will probably be less accurate than in Example 9-1. Simpson's rule gives

$$\text{Area} = \left(\frac{x_F - x_{W,final}}{6}\right)\left[\left(\frac{1}{x_D - x_W}\right)\Big|_{x_{W,f}} + 4\left(\frac{1}{x_D - x_W}\right)\Big|_{\frac{(x_{W,f} + x_F)}{2}} + \left(\frac{1}{x_D - x_W}\right)\Big|_{x_F}\right]$$

where $(x_{W,final} + x_F)/2 = 0.1825$ and $1/(x_D - x_W) = 2.14$ at this midpoint.

$$\text{Area} = \left(\frac{0.275}{6}\right)[2.51 + 4(2.14) + 2.82] = 0.6366$$

This can be checked by breaking the area into two parts and using Simpson's rule for each part. Do one part from $x_{W,final} = 0.045$ to $x_W = 0.10$ and the other part from 0.1 to $x_F = 0.32$. Each of the two parts should be relatively easy to fit with a cubic. Then,

$$\text{Area part 1} = \left(\frac{0.10 - 0.045}{6}\right)[2.51 + 4(2.03) + 2.00] = 0.1158$$

$$\text{Area part 2} = \left(\frac{0.32 - 0.10}{6}\right)[2.00 + 4(2.23) + 2.82] = 0.5038$$

Total area = 0.6196
Note that Figure 9-6 is very useful for finding the values of $1/(x_D - x_W)$ at the intermediate points $x_W = 0.0725$ (value = 2.03) and $x_W = 0.21$ (value = 2.23).

The total area calculated is closer to the answer obtained graphically (1.9% difference compared to 4.7% difference for the first estimate).

Then, doing the same calculations as previously [Eqs. (9-7), (9-11), and (9-10)] with Area = 0.6196,

$$W_{final} = 26.91, \qquad D_{total} = 23.09, \qquad x_{D,avg} = 0.640$$

E. Check. The mass balances for an entire cycle, Eqs. (9-1) and (9-2), should be and are satisfied. Since the graphical integration and Simpson's rule (done as two parts) give similar results, this is another reassurance.

F. Generalize. Note that we did not need to find the exact value of x_D for x_F or $x_{W,final}$. We just made sure that our calculated values went beyond these values. This is true for both integration methods. Our axes in Figure 9-6 were selected to give maximum accuracy; thus, we did not graph parts of the diagram that we didn't use. The same general idea applies if fitting data—only fit the data in the region needed. For more accuracy, Figure 9-5 should be expanded. Note that the graph in Figure 9-6 is very useful for interpolation to find values for Simpson's rule. If Simpson's rule is to be used for very sharply changing curves, accuracy will be better if the curve is split into two or more parts. Comparison of the results obtained with graphical integration to those obtained with the two-part integration with Simpson's rule shows a difference in $x_{D,avg}$ of 0.008. This is within the accuracy of the equilibrium data.

Once $x_{W,final}$, D and W_{final} are determined, we can calculate the values of Q_c, Q_R and operating time (see Section 9.6).

9.5.2 Variable Reflux Ratio

The batch distillation column can also be operated with variable reflux ratio to keep x_D constant. The operating Eq. (9-26) is still valid. Now the slope will vary, but the intersection with the y=x line will be constant at x_D. The McCabe-Thiele diagram for this case is shown in Figure 9-7. This diagram relates x_W to x_D. Since x_D is kept constant, the calculation procedure is somewhat different.

With x_D and the number of stages specified, the initial value of L/V is found by trial and error to give the feed concentration x_F. The operating line slope L/V is then increased until the specified number of equilibrium contacts gives $x_W = x_{W,final}$. This gives $(L/V)_{final}$. W_{final} is found from mass balance Eqs. (9-1) and (9-2). The required maximum values of Q_c and Q_R and the operating time can be determined next (see Section 9.6). Note that the Rayleigh equation is not required when x_D is constant. However, if used, the Rayleigh equation gives exactly the same answer as the mass balances [see Problem 9.C1 which includes Eq. (9-27)].

9.6 OPERATING TIME

The operating time and batch size may be controlled by economics or other factors. For instance, it is not uncommon for the entire batch including startup and shutdown to be done in one eight-hour shift. If the same apparatus is used for several different chemicals, the batch

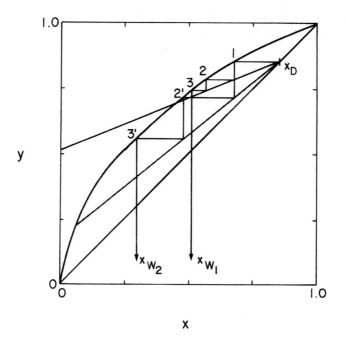

FIGURE 9-7. *McCabe-Thiele diagram for multistage batch distillation with constant x_D and variable reflux ratio*

sizes may vary. Also the time to change over from one chemical to another may be quite long, since a rigorous cleaning procedure may be required.

The total batch time, t_{batch}, is

$$t_{batch} = t_{down} + t_{op} \qquad \text{(9-28)}$$

The down time, t_{down}, includes dumping the bottoms, cleanup, loading the next batch, and heating the next batch until reflux starts to appear. This time can be estimated from experience. The operating time, t_{op}, is the actual period during which distillation occurs, so it must be equal to the total amount of distillate collected divided by the distillate flow rate.

$$t_{op} = \frac{D_{total}}{D} \qquad \text{(9-29)}$$

D_{total} is calculated from the Rayleigh equation calculation procedure, with F set either by the size of the still pot or by the charge size. For an existing apparatus the distillate flow rate, D in kmol/h, cannot be set arbitrarily. The column was designed for a given maximum vapor velocity, u_{flood}, which corresponds to a maximum molal flow rate, V_{max} (see Chapter 10). Then, from the mass balance around the condenser,

$$D_{max} = \frac{V_{max}}{1 + \dfrac{L}{D}} \qquad \text{(9-30)}$$

We usually operate at some fraction of this flow rate such as $D = 0.75 \, D_{max}$. Then Eqs. (9-29) and (9-30) can be used to estimate t_{op}. If the resulting t_{batch} is not convenient, adjustments must be made.

The energy requirements in the reboiler or still pot Q_R and the total condenser Q_c can be estimated from energy balances. For a total condenser Eq. (3-13) is valid, but V_1, h_D and H_1 may all be functions of time (if x_D varies the enthalpies will vary). If the reflux is a saturated liquid reflux, then $H_1 - h_D = \lambda$. In this case the total condenser just condenses the vapor to saturated liquid. Likewise, the still pot vaporizes a saturated liquid to a vapor. Thus,

$$Q_c = -V_1 \lambda_1 \qquad Q_R = V_{pot} \lambda_{pot} \qquad \text{(9-31a,b)}$$

If CMO is valid, then $V_1 = V_{pot}$ and $\lambda_1 = \lambda_{pot}$, and

$$Q_R = -Q_c = V\lambda \qquad \text{(9-31c)}$$

Since $V = (1 + L/D)D$, we obtain

$$Q_R/D = (1 + L/D)\lambda \qquad \text{(9-32)}$$

During operation (the charge and still pot have been heated, and vapor is flowing throughout the column), the energy balance around the entire system is

$$Q_R + Q_c - Dh_D = d(V_{pot}h_{pot})/dt \qquad \text{(9-33a)}$$

If CMO is valid, then Eq. (9-31c) holds and this result simplifies to,

$$-Dh_D = d(V_{pot}h_{pot})/dt \qquad \text{(9-33b)}$$

For an existing batch distillation apparatus we must check that the condenser and reboiler are large enough to handle the calculated values of $|Q_c|$ and Q_R. If $|Q_c|$ or Q_R are too large, then the rate of vaporization needs to be decreased. Either the operating time t_{op} will need to be increased or the charge to the still pot F will have to be decreased.

If the assumption of negligible holdup is not valid, then the holdup on each stage and in the accumulator acts like a flywheel and retards changes. A different calculational procedure is required for this case and for multicomponent systems (Barton and Roche, 1997; Diwekar, 1995; Mujtabe, 2004).

Batch distillation also has somewhat different design and process control requirements than continuous distillation. In addition, startup and troubleshooting are somewhat different. These aspects are discussed by Ellerbe (1979).

9.7   SUMMARY—OBJECTIVES

In this chapter we have explored binary batch distillation calculations. At this time you should be able to satisfy the following objectives:

1. Explain the operation of simple and multistage batch distillation systems

2. Discuss the differences between batch and continuous operation
3. Derive and use the Rayleigh equation for simple batch distillation
4. Solve problems for constant-level batch distillation
5. Solve problems in batch steam distillation
6. Use the McCabe-Thiele method to analyze multistage batch distillation for:
   a. Batch distillation with constant reflux ratio
   b. Batch distillation with constant distillate composition
   c. Inverted batch distillation
7. Determine the operating time and energy requirements for a batch distillation

REFERENCES

Barton, P., and E. C. Roche, Jr., "Batch Distillation," Section 1.3, in Schweitzer, P. A. (Ed.), *Handbook of Separation Techniques for Chemical Engineers,* 3rd ed., McGraw-Hill, New York, 1997.

Carey, J. S., "Distillation," in J. H. Perry (Ed.), *Chemical Engineer's Handbook,* 3rd ed., McGraw-Hill, New York, 1950, pp. 582–585.

Davies, J. T., "Chemical Engineering: How Did It Begin and Develop?" in W. F. Furter (Ed.), *History of Chemical Engineering,* Washington, D.C., American Chemical Society, *Adv. Chem. Ser, 190,* 15–43 (1980).

Diwekar, U. W., *Batch Distillation: Simulation, Optimal Design and Control,* Taylor and Francis, Washington, D.C., 1995.

Ellerbe, R. W., "Batch Distillation," in P. A. Schweitzer (Ed.), *Handbook of Separation Techniques for Chemical Engineers,* McGraw-Hill, New York, 1979, p. 1.147.

Gentilcore, M. J., "Reduce Solvent Usage in Batch Distillation," *Chem. Engr. Progress, 98*(1), 56 (Jan. 2002).

Luyben, W. L., "Some Practical Aspects of Optimal Batch Distillation," *Ind. Eng. Chem. Process Des. Develop., 10,* 54 (1971).

Matab, I. M., *Batch Distillation. Design and Operation,* Imperial College Press, London, 2004.

Mickley, H. S., Sherwood, T. K. and Reed, C. E., *Applied Mathematics in Chemical Engineering,* McGraw-Hill, New York, 1957, pp. 35–42.

Pratap, R., *Getting Started with MATLAB7,* Oxford University Press, Oxford, 2006.

Pratt, H. R. C., *Countercurrent Separation Processes,* Elsevier, New York, 1967.

Rayleigh, Lord (J. Strutt), "On the Distillation of Binary Mixtures," *Phil. Mag.* [vi], *4*(23), 521 (1902).

Robinson, C. S., and E. R. Gilliland, *Elements of Fractional Distillation,* 4th ed., McGraw-Hill, New York, 1950, Chaps. 6 and 16.

R T, "Distillation through the Ages," *Chem. Heritage, 25* (2), 40 (Summer 2007).

Stanwood, C., "Found in the Othmer Library," *Chemical Heritage, 23*(3), 34 (Fall 2005).

Treybal, R. E., *Mass-Transfer Operations,* 3rd ed., McGraw-Hill, New York, 1980.

Treybal, R. E., "A Simple Method for Batch Distillation," *Chem. Eng., 77*(21), 95 (Oct. 5, 1970).

Vallee, B. L., "Alcohol in the Western World," *Scientific American,* 80 (June 1998).

Woodland, L. R., "Steam Distillation," in D. J. DeRenzo (Ed.), *Unit Operations for Treatment of Hazardous Industrial Wastes,* Noyes Data Corp., Park Ridge, NJ, 1978, pp. 849–868.

HOMEWORK

A. *Discussion Problems*

A1. Why is the still pot in Figure 9-2B much larger than the column?

A2. In the derivation of the Rayleigh equation:

    a. In Eq. (9-4), why do we have $-x_D\,dW$ instead of $-x_D\,dD$?

    b. In Eq. (9-4), why is the left-hand side $-x_D\,dW$ instead of $-d(x_D W)$?

A3. Explain how the graphical integration shown in Figures 9-3 and 9-5 could be done numerically on a computer.

A4. Suppose you have two feeds containing methanol and water that you want to batch distill. One feed is 60 mol% methanol, and the other is 32 mol% methanol. How do you do the batch distillation to obtain the largest amount of distillate of given mole fraction? N is constant.

    a. Mix the two feeds together and batch distill.

    b. Start the batch with the feed with higher methanol mole fraction, and the add second feed (32 mol% methanol) when the still pot concentration equals that feed concentration.

    c. Do two separate batches—one for the more concentrated feed (60 %) and the other for the less concentrated feed (32 %).

    d. All of the above.

    e. a and b.

    f. a and c.

    g. b and c.

    h. None of the above.

    Explain your answer.

A5. Which system(s) require less energy for batch distillation than continuous distillation with the same amount of separation?

    a. Simple batch compared to one-stage continuous.

    b. Multi-stage batch compared to multi-stage continuous.

    c. Batch steam distillation compared to continuous steam distillation.

    d. All of the above.

    e. a and b.

    f. a and c.

    g. b and c.

    h. None of the above.

    Explain your answer.

A6. Batch-by-Night, Inc. has developed a new simple batch with reflux system (Figure 9-1 with some of stream D refluxed to the still pot) that they claim will outperform the normal simple batch (Figure 9-1). Suppose you want to batch distill a mixture similar to methanol and water where there is no azeotrope and two liquid phases are not formed. Both systems are loaded with the same charge F moles with the same mole fraction x_F, and distillation is done to the same value $x_{w,final}$. Will the value of $x_{d,avg}$ from the simple batch with reflux be

    a. greater than

    b. the same

    c. less than

    the $x_{d,avg}$ from the normal simple batch distillation? Explain your answer.

A7. When there are multiple stages in batch distillation, the calculation looks like the McCabe-Thiele diagram for several

    a. stripping columns.

    b. enriching columns.

    c. columns with a feed at the optimum location.

A8. Develop a key relations chart for this chapter.

B. *Generation of Alternatives*

B1. List all the different ways a binary batch or inverted batch problem can be specified. Which of these will be trial and error?

B2. What can be done if an existing batch system cannot produce the desired values of x_D and x_W even at total reflux? Generate ideas for both operating and equipment changes.

C. *Derivations*

C1. For a binary, multistage batch distillation with constant x_{dist}, prove that the mass balances over the entire batch period and the Raleigh equation give identical results.

$$W_{final} = [(x_{dist} - x_F)/(x_{dist} - x_{W,final})]F \qquad \textbf{(9-27)}$$

C2. Assume that holdup in the column and in the total reboiler is negligible in an inverted batch distillation (Figure 9-8).

a.*Derive the appropriate form of the Rayleigh equation.

b. Derive the necessary operating equations for CMO. Sketch the McCabe-Thiele diagrams.

C3. Derive Eq. (9-14).

D. *Problems*

*Answers to problems with an asterisk are at the back of the book.

D1. A simple batch distillation is done for a binary mixture of ethanol and water. The feed is 0.5 kmol of liquid that is 0.10 mole fraction ethanol. The simple batch distillation is continued until the liquid remaining in the still pot is $x_{W,final} = 0.00346$ mole fraction ethanol. Find the values of W_{final} and the distillate D_{total} and the average mole fraction of ethanol in the distillate, $x_{D,avg}$.

D2.* We wish to use a simple batch still (one equilibrium stage) to separate methanol from water. The feed charge to the still pot is 100 moles of a 75 mol% methanol mixture. We desire a final bottoms concentration of 55 mol% methanol. Find the amount of distillate

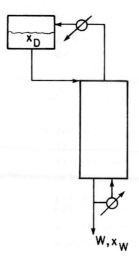

$$W, x_W$$

FIGURE 9-8. *Inverted batch distillation*

collected, the amount of material left in the still pot, and the average concentration of distillate. Pressure is 1 atm. Equilibrium data are given in Table 2-7.

D3.* We wish to use a distillation system with a still pot plus a column with one equilibrium stage to batch distill a mixture of methanol and water. A total condenser is used. The feed is 57 mol% methanol. We desire a final bottoms concentration of 15 mol% methanol. Pressure is 101.3 kPa. Reflux is a saturated liquid, and L_0/D is constant at 1.85. Find W_{final}, D_{total}, and $x_{D,avg}$. Methanol-water equilibrium data are given in Table 2-7 Calculate on the basis of 1 kmol of feed.

D4. We wish to do a simple batch distillation (1 equilibrium contact) of a mixture of acetone and ethanol. The feed charge to the still pot is 80 mol% acetone. The final concentration in the still pot will be 40 mol% acetone. The final amount of material in the still pot is 2.0 kmole. Vapor-liquid equilibrium (VLE) data are in Problem 4.D7. Find the feed amount F, the average mole fraction of the distillate, and the kmoles of distillate collected.

D5. A simple batch distillation is being done to process a feed containing water and n-butanol that is 52 mol% water. The feed to the still pot is 3.0 kmol. The final still pot concentration should be 28 mol% water. Equilibrium data are in Table 8-2.

    a. Find W_{final} (final amount in the still pot, kmole), $D_{v,tot}$ (total amount of distillate vapor collected, kmole), and $y_{D,avg}$ (average mole fraction water in the vapor distillate).

    b. After the distillate vapor is condensed in the total condenser, the liquid is sent to the settler. Two distillate liquid products are withdrawn. Find the total amounts of each distillate liquid collected, D_1 (with $x_{D1,water} = 0.573$) and D_2 (with $x_{D2,water} = 0.975$), in kmoles.

D6. We have a simple batch still separating 1,2 dichloroethane from 1,1,2 trichloroethane. Pressure is 1 atm. The relative volatility for this system is very close to constant at a value of 2.4. The use of Eq. (9-13) is recommended.

    a. The charge (F) is 1.3 kmol. This feed is 60 mol% dichloroethane. We desire a final still pot concentration of 30 mol% dichloroethane. Find the final moles in the still pot and the average mole fraction of the distillate product.

    b. Repeat part a if the feed charge is 3.5 kmol.

    c. If the feed charge is 2.0 kmol, the feed is 60 mol% dichloroethane, and we want a distillate with an average concentration of 75 mol% dicholoroethane, find the final kmoles in the still pot and the final mole fraction of dichloroethane in the still pot.

D7. We will use a batch distillation system with a still pot and one equilibrium stage (2 equilibrium contacts total) to distill a feed that is 10 mol% water and 90 mol% n-butanol (see Table 8-2 for VLE data). Pressure is one atmosphere. The charge is 4.0 kmoles. We desire a final still concentration that is 2.0 mol% water. The system has a total condenser and the reflux is returned as a saturated liquid. The reflux ratio L/D = 1/2. Find the final number of moles in the still and the average concentration of the distillate.

D8. We plan to batch distill a mixture of methanol and water in a batch distillation system at 1.0 atm pressure. The distillation system consists of a large still pot that is an equilibrium contact, a distillation column that acts as two equilibrium contacts (total 3 equilibrium contacts), and a total condenser. The feed to the system is F = 10.0 kmol and x_F = 0.4 (mole fraction methanol). We operate with a constant x_D = 0.8 as we increase L/D. The batch operation is continued until L/D = 4.0. CMO is valid. Equi-

librium data are in Table 2-7. Find:
  a. $x_{w,final}$.
  b. W_{final} and D_{total}.
  c. Initial L/D.

D9.* We wish to batch distill a mixture of 1-butanol and water. Since this system has a heterogeneous azeotrope (see Chapter 8), we will use the system shown in the figure. The bottom liquid layer with 97.5 mol% water is removed as product, and the top liquid layer, which is 57.3 mol% water, is returned to the still pot. Pressure is 1 atm. The feed is 20 kmol and is 40 mol% water. Data are given in Problem 8.D2.

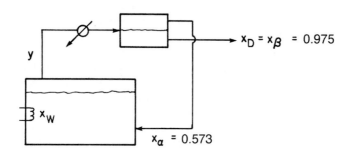

  a. If the final concentration in the still pot should be $x_{w,final} = 0.28$, what is the final amount of distillate D_{total} collected, in kmoles?
  b. If the batch distillation is continued what is the lowest value of $x_{w,final}$ that can be obtained while still producing a distillate $x_D = 0.975$?

D10. We plan to batch distill a mixture of ethanol and water. The feed contains 0.52 mole fraction ethanol and we wish to continue the distillation until $x_{w,final} = 0.20$ mole fraction ethanol. The initial charge is $F = 10.0$ kmol. The batch system has a still pot, which acts as an equilibrium contact and a column with one equilibrium stage (total of 2 equilibrium contacts). The distillate vapor is condensed in a total condenser and we use an external reflux ratio $L/D = \frac{1}{4}$. VLE is in Table 2-1 Find: W_{final}, D_{total}, and $x_{D,avg}$.

D11. We plan on doing a constant-level batch distillation to change from a solvent that is pure butanol to a solvent that is 60 mol% butanol and 40 mol% water. We do this by adding pure water to the batch while removing vapor that is concentrated in butanol. (Adding water without removing vapor would dilute the nonvolatile solute in the solvent.) If the initial charge is $W = 2.0$ kmol, how much water S must be added to achieve the desired constant-level batch distillation? Do the integration with Simpson's rule dividing the entire integration from 0.6 to 1.0 first into one step and then into two steps.

D12. Acetone-ethanol VLE data are given in Problem 4.D7. We have 3.0 kmol of a mixture of acetone and ethanol that we wish to separate in a simple batch still. The feed is 48 mol% acetone. The final still pot should be 16 mol% acetone. Find the final amount in the still pot, the amount of distillate collected, and the average distillate mole fraction.

D13. We are doing a simple batch distillation of butanol and water. The feed is 8.00 kmol and is 40.0 mol% water and 60.0 mol% butanol. The batch distillation is continued until the still pot contains 0.080 mole fraction water.

a. Find W_{final}, D_{total}, and $x_{D,avg}$.

b. After settling, the final distillate product is two liquid phases. What are the mole fractions and the amounts (kmole) of each liquid phase?

D14.* A simple steam distillation is being done in the apparatus shown in Figure 9-4. The organic feed is 90 mol% n-decane and 10% nonvolatile organics. The system is operated with liquid water in the still. Distillation is continued until the organic layer in the still is 10 mol% n-decane. $F = 10$ kmole. Pressure is 760 mm Hg.

a. At the final time, what is the temperature in the still?

b. What is W_{final}? What is D_{total}?

c. Estimate the moles of water passed overhead per mole of n-decane at the end of the distillation.

Data: Assume that water and n-decane are completely immiscible. Vapor pressure data for nC_{10} is in Example 8-2. Vapor pressure data for water are listed in Problem 8.D10.

D15. We wish to batch distill 100 kmol of a mixture of n-butanol and water. The system consists of a batch still pot plus 1 equilibrium stage. The system is at one atmosphere. The feed is 48 mol% water and 52 mol% butanol. The distillate vapor is condensed and sent to a liquid-liquid settler. The water rich product (0.975 mole fraction water) is taken as the distillate product and the butanol rich layer (0.573 mole fraction water) is refluxed to the column. We desire a final still pot mole fraction of 0.08 water. Energy is added at a constant rate to the still pot; thus, V = constant. Note that the distillate product is a constant mole fraction. The reflux ratio increases as the distillate vapor mole fraction decreases during the course of the batch distillation. Equilibrium data are given in Table 8-2.

a. Find the final amount of material in the still pot and the amount of distillate collected.

b. Find the initial and final internal reflux ratios (L/V values).

D16.* We wish to do a normal batch distillation of methanol and water. The system has a still pot that acts as an equilibrium stage and a column with two equilibrium stages (total of three equilibrium contacts). The column has a total condenser, and reflux is returned as a saturated liquid. The column is operated with a varying reflux ratio so that x_D is held constant. The initial charge is $F = 10$ kmol and is 40 mol% methanol. We desire a final still-pot concentration of 8 mol% methanol, and the distillate concentration should be 84 mol% methanol. Pressure is 1 atm and CMO is valid. Equilibrium data are given in Table 2-7.

a. What initial external reflux ratio, L_0/D, must be used?

b. What final external reflux ratio must be used?

c. How much distillate product is withdrawn, and what is the final amount, W_{final}, left in the still pot?

D17. A nonvolatile solute is dissolved in 1.0 kmol of methanol. We wish to switch the solvent to water. Because the solution is already concentrated, a first batch distillation to concentrate the solution is not required. We desire to have the solute in 1.0 kmol of solution that is 99.0 mol% water and 1.0 mol% methanol. This can be done either with a constant-level batch distillation or by diluting the mixture with water and then doing a simple batch distillation. VLE data (ignore the effect of the solute) are in Table 2-7. Do a constant-level batch distillation from $x_{M,initial} = 1.0$ (pure methanol) to $x_{M,final} = 0.01$. Find the moles of water added during the constant-level batch distillation and the moles of water evaporated with the methanol in the distillate. Results can be compared with dilution followed by simple batch distillation in Problem 9.E4.

Note: Since $x_{M,final} = 0.01$ is quite small, $1/y$ is quite large at this limit. Straightforward application of Simpson's rule from $x_{M,initial} = 1.0$ to $x_{M,final} = 0.01$ will not be accurate. The integral needs to be divided into at least two sections.

D18. A differential condensation [see Eq. (9-14)] is done for a binary mixture of ethanol and water. The feed is 0.5 kmol of vapor that is 0.10 mole fraction ethanol. The differential condensation is continued until the vapor remaining is $y_{D,final} = 0.5$ mole fraction ethanol. Find the values of D_{final} and the condensate C_{total}, and the average mole fraction of ethanol in the condensate, $x_{C,avg}$. Note that this operation can only be approximated in practice (Treybal, 1980). Compare the result to the solution to Problem 9.D1.

D19. Batch steam distillation is being done to recover octanol from nonvolatile organics. Assume that water and octanol, and water and the nonvolatile organics, are completely immiscible. The water layer is pure. The organic layer in the still pot contains 0.6 mole fraction octanol and 0.4 mole fraction nonvolatile organics. The still pot temperature is 90°C. The Antoine equations are:

Octanol: $\log_{10} (VP) = 6.8379 - 1310.62 / (T + 135.05)$

Water: $\log_{10} (VP) = 8.68105 - 2164.42 / (T + 273.16)$

In these equations T is in °C and VP is the vapor pressure in mm Hg. What is the pressure of the still pot?

D20. We will use a simple batch still (one equilibrium contact—the still pot) to separate 1.5 kmol of a mixture of n-pentane and n-octane at a pressure of 101.3 kPa. The feed is 35 mol% n-pentane. We desire a final still pot concentration of 5 mol% n-pentane. Find the final amount in the still pot, the amount of distillate collected, and the average distillate mole fraction.

D21. A mixture of ethanol and water is batch distilled in a system with a still pot, a column with 20 equilibrium stages, and a total condenser. Operation is at a constant external reflux ratio of $L/D = 1.0$. The feed charge is 2.5 kmol that is 0.06 mole fraction ethanol. We want a final still pot that is 0.02 mole fraction ethanol. Find W_{final}, D_{total}, and $x_{D,avg}$. Work smart, and this problem is not as much work as it appears at first.

D22. We are doing a simple batch distillation of n-pentane from n-hexane. The feed is 2.0 lbmol and is 40 mol% n-pentane. We collect 0.6 lbmol of distillate. The heating rate in the still pot is 5000 Btu/h. The average latent heat of the mixture is 12,470 Btu/lbmol. What is the operating time for the batch distillation?

Note: This does not include the down time for dumping, cleaning, filling and so forth.

D23. We wish to batch distill a mixture of ethanol and water. The feed is 10.0 mol% ethanol. Operation is at 1.0 atmosphere. The batch distillation system consists of a still pot plus a column with the equivalent of 9 equilibrium contacts and a total condenser. We operate at a constant external reflux ratio of 2/3. The initial charge to the still pot is 1000.0 kg. We desire a final still pot concentration of 0.004 mole fraction ethanol. Equilibrium data are in Table 2-1. Convert the amount of feed to kmole using an average molecular weight. Find W_{final} and D_{total} in kmole, and $x_{D,avg}$.

Note: Stepping off 10 stages for a number of operating lines sounds like a lot of work. However, after you try it once, you will notice that there is a short cut to determining the relationship between x_D and x_w.

D24. A mixture that is 62 mol% methanol and 38 mol% water is batch distilled in a system with a still pot and a column with 1 equilibrium stage (2 equilibrium contacts total). $F = 3.0$ kmol. The system operates with a constant distillate concentration that is

85 mol% methanol. We desire a final still pot concentration that is 45 mol% methanol. Reflux is a saturated liquid and the external reflux ratio L/D varies. Assume CMO. VLE data are in Table 2-7.

a. Find D_{total} and W_{final} (kmoles).

b. Find the final value of the external reflux ratio L/D.

D25. We have 1.5 kmol of feed 1 that is 40 mol% methanol and 60 mol% water. We also have 1.0 kmol of feed 2 that is 20 mol% methanol and 80 mol% water. We want to do a simple batch distillation of these mixtures. The following three approaches have been proposed:

a. Mix the two feeds together and do the batch distillation.

b. Do a batch distillation of feed 1 until the still pot concentration is 20% methanol (same as feed 2), add feed 2, and complete the batch distillation.

c. Do a batch distillation of feed 1 and a separate batch distillation of feed 2. Add the two distillate products and the two still pot products.

If we want $x_{W,final,total} = 0.10$, do the calculations for each method. Determine W_{final}, D_{total}, and $x_{D,average}$ and compare the three methods. Equilibrium data are in Table 2-7. Logically, parts b and c should give the same result. If they don't, there is probably a numerical error from the use of Simpson's rule. Try dividing an area into two parts for more accuracy.

E. *More Complex Problems*

E1. We are doing a single-stage, batch steam distillation of 1-octanol. The unit operates at 760 mm Hg. The batch steam distillation is operated with liquid water present. The distillate vapor is condensed and two immiscible liquid layers form. The feed is 90 mol% octanol and the rest is nonvolatile organic compounds. The feed is 1.0 kmol. We desire to recover 95% of the octanol.

Vapor pressure data for water are given in Problem 8.D10. For small ranges in temperature, these data can be fit to an Antoine equation form with C = 273.16. The vapor pressure equation for 1-octanol is in Problem 8.D11.

a. Find the operating temperature of the still at the beginning and end of the batch.

b. Find the amount of organics left in the still pot at the end of the batch.

c. Find the kmoles of octanol recovered in the distillate.

d. Find the kmoles of water condensed in the distillate product. To do this, use the average still temperature to estimate the average octanol vapor pressure which can be assumed to be constant. To numerically integrate Eq. (9-23b) relate x_{org} to n_{org} with a mass balance around the batch still.

e. Compare with your answer to Problem 8.D11. Which system produces more water in the distillate? Why?

E2. We wish to use batch steam distillation to recover 1-octanol from 100 kg of a mixture that is 15 wt % 1-octanol and the remainder consists of nonvolatile organics and solids of unknown composition. The still pot operates at 1.0 atm pressure. The pot is operated with liquid water in the pot. Assume the still pot is well mixed and liquid and vapor are in equilibrium. Ninety-five percent of the 1-octanol should be recovered in the distillate. Assume that water is completely immiscible with 1-octanol and with the nonvolatile organics. Because the composition of the nonvolatile organics is not known, we do a simple experiment and boil the feed mixture under a vacuum with no water present. The result is that at 0.05 atm pressure, the mixture boils at 129.8°C.

a. Find the mole fraction of 1-octanol in the feed and the effective average molecular weight of the non-volatile organics and the solids. (Note: This is identical to the solution of part a of Problem 8.D15.)

b. Find the kg and kmole of 1-octanol in the distillate, the kg of total organics in the waste, and the 1-octanol weight fraction and 1-octanol mole fraction in the waste.

c. Find the initial and final values of temperature and of $VP_{octanol}$ in the still pot. A spreadsheet or MATLAB is highly recommended for finding T and VP.

d. Find the kg and kmole of water in the distillate. This requires a numerical integration of Eq. (9-23b); however, the problem is simplified in this case. Because the still pot temperature does not change much, the value of $(VP)_{octanol}$ is very close to constant and the average value can be used.

e. Compare the solution with the solution of Problem 8.D15. Why does batch steam distillation require less steam than continuous steam distillation?

Octanol boils at about 195°C. The formula for octanol is $CH_3(CH_2)_6CH_2OH$, and its molecular weight is 130.23. Vapor pressure formulas for octanol and water are available in Problems 8.D11 and 8.D15, respectively.

E3. In *inverted batch distillation* the charge of feed is placed in the accumulator at the top of the column (Figure 9-8). Liquid is fed to the top of the column. At the bottom of the column bottoms are continuously withdrawn and part of the stream is sent to a total reboiler, vaporized and sent back up the column. During the course of the batch distillation the less volatile component is slowly removed from the liquid in the accumulator and the mole fraction more volatile component x_d increases. Assuming that holdup in the total reboiler, total condenser and the trays is small compared to the holdup in the accumulator, the Rayleigh equation for inverted batch distillation is,

$$\ln (D_{final}/F) = -_{xfeed}\int^{xd,final} [(dx_d)/(x_d - x_B)]$$

We feed the inverted batch system shown in the Figure 9-8 with F = 10 mole of a feed that is 50 mol% ethanol and 50% water. We desire a final distillate mole fraction of 0.63. There are 2 equilibrium stages in the column. The total reboiler, the total condenser and the accumulator are *not* equilibrium contacts. VLE are in Table 2-1. Find D_{final}, B_{total} and $x_{B,avg}$ if the boilup ratio is 1.0.

E4. A nonvolatile solute is dissolved in 1.0 kmol of methanol. We wish to switch the solvent to water. Because the solution is already concentrated, a first batch distillation to concentrate the solution is not required. We desire to have the solute in 1.0 kmol of solution that is 99.0 mol% water and 1.0 mol% methanol. This can be done either with a constant-level batch distillation or by diluting the mixture with water and then doing a simple batch distillation. VLE data (ignore the effect of the solute) are in Table 2-7. Dilute the original pure methanol (plus solute) with water and then do a simple batch distillation with the goal of having W_{final} = 1.0 and $x_{M,final}$ = 0.01. Find the moles of water added and the moles of water evaporated during the batch distillation. Compare with constant-level batch distillation in Problem 9.D17.

H. *Computer Spreadsheet Problems*

H1. Solve the following problems with a spread sheet. A mixture of benzene (A) and cumene (C) is to be distilled in a simple batch system. The initial charge of 5.0 kmol is 37 mol% benzene and 63 mol% cumene. At the system pressure of one atmosphere and choosing cumene (C) as the reference component, the relative volatility is $\alpha_{A-C} = 10.71$. Use Eq. (9-13).

a. We desire a fractional recovery of benzene in the distillate of 75%. Find $x_{A,Wfinal}$, W_{final}, D_{total}, and $x_{A,D,avg}$.

b. If we want $x_{A,Wfinal} = 0.05$, what fractional recovery of benzene in the distillate is required? Find W_{final}, D_{total}, and $x_{A,D,avg}$.

CHAPTER 10

Staged and Packed Column Design

In previous chapters we saw how to determine the number of equilibrium stages and the separation in distillation columns. In the first part of this chapter we will discuss the details of staged column design such as tray geometry, determination of column efficiency, calculation of column diameter, downcomer sizing, and tray layout. We will start with a qualitative description of column internals and then proceed to a quantitative description of efficiency prediction, determination of column diameter, sieve tray design, and valve tray design. In the second part (sections 10.6 to 10.10) we will discuss packed column design including the selection of packed materials and determination of the length and diameter of the column. New engineers are expected to be able to do these calculations. The internals of distillation columns are usually designed under the supervision of experts with many years of experience.

This chapter is not a shortcut to becoming an expert. However, upon completion of this chapter you should be able to finish a preliminary design of the column internals for both staged and packed columns, and you should be able to discuss distillation designs intelligently with the experts.

10.1 STAGED COLUMN EQUIPMENT DESCRIPTION

A very basic picture of staged column equipment was presented in Chapter 3. In this section, a much more detailed qualitative picture will be presented. Much of the material included here is from the series of articles and books by Kister (1980, 1981, 1992, 2003), the book by Ludwig (1997), and the chapter by Larson and Kister (1997). These sources should be consulted for more details.

Sieve trays, which were illustrated in Figure 3-7, are easy to manufacture and are inexpensive. The holes are punched or drilled (a more expensive process) in the metal plate. Considerable design information is available, and since the designs are not proprietary, anyone can build a sieve tray column. The efficiency is good at design conditions. However, *turndown* (the performance when operating below the designed flow rate) is relatively poor. This means that operation at significantly lower rates than the design condition will result in low efficiencies. For sieve plates, efficiency drops markedly for gas flow rates that are less than about 60% of the design value. Thus, these trays are not extremely flexible. Since sieve trays are easy to clean, they are very good in fouling applications and are the normal choice when solids are present. Sieve trays are a standard item in industry, but new columns are more likely to have valve trays (Kister, 1992).

Valve trays are designed to have better turndown properties than sieve trays, and thus, they are more flexible when the feed rate varies. There are many different proprietary valve tray designs, of which one type is illustrated in Figure 10-1. The valve tray is similar to a sieve tray in that it is has a deck with holes in it for gas flow and downcomers for liquid flow. The difference is that the holes, which are quite large, are fitted with "valves," covers that can move up and down as the pressures of the vapor and the liquid change. Each valve has feet or a cage that restrict its upward movement. Round valves with feet are most popular but wearing of the feet can be a problem (Kister, 1990). At high vapor velocities, the valve will be fully open, providing a maximum slot for gas flow (see Figure 10-1). When the gas velocity drops,

A

B

FIGURE 10-1.   (A) Valve assembly for Glitsch A-1 valve, and (B) small Gitsch A-1 ballast tray; courtesy of Glitsch, Inc., Dallas, Texas

the valve will drop. This keeps the gas velocity through the slot close to constant, which keeps efficiency close to constant and prevents weeping. An individual valve is stable only in the fully closed or fully opened position. At intermediate velocities some of the valves on the tray will be open and some will be closed. Usually, the valves alternate between the open and closed positions. The Venturi valve has the lip of the hole facing upwards to produce a Venturi opening, which will minimize the pressure drop.

At the design vapor rate, valve trays have about the same efficiency as sieve trays. However, their turndown characteristics are generally better, and the efficiency remains high as the gas rate drops. They can also be designed to have a lower pressure drop than sieve trays, although the standard valve tray will have a higher pressure drop. The disadvantages of valve trays are they are about 20% more expensive than sieve trays (Glitsch, 1985) and they are more likely to foul or plug if dirty solutions are distilled.

Bubble-cap trays are illustrated in Figure 10-2. In a bubble-cap there is a riser, or weir, around each hole in the tray. A cap with slots or holes is placed over this riser, and the vapor bubbles through these holes. This design is quite flexible and will operate satisfactorily at very high and very low liquid flow rates. However, entrainment is about three times that of a sieve tray, and there is usually a significant liquid gradient across the tray. The net result is that tray spacing must be significantly greater than for sieve trays. In columns less than about 1.2 m in diameter, a tray spacing of 0.45 m is satisfactory because workers can reach through a side entry for maintenance (Torzewski, 2009). For columns from 1.2 to 3.0 m, workers need to be able to crawl through the column, and a minimum 0.6 m spacing is required. For columns greater than 3.0 m in diameter, the tray spacing must be greater than 0.6 m because the heavy support beams required to support the trays restrict access. Efficiencies are usually the same or less than for sieve trays, and turndown characteristics are often worse. The bubble-cap has problems with coking, polymer formation, or high fouling mixtures. Bubble-cap trays are approximately four times as expensive as valve trays (Glitsch, 1985). Very few new bubble-cap columns are being built. However, new engineers are likely to see older bubble-cap

FIGURE 10-2. *Different bubble-cap designs made by Glitsch, Inc., courtesy of Glitsch, Inc., Dallas, Texas*

columns still operating. Lots of data are available for the design of bubble-cap trays. Since excellent discussions on the design of bubble-cap columns are available (Bolles, 1963; Ludwig, 1997), details will not be given here.

Perforated plates without downcomers look like sieve plates but with significantly larger holes. The plate is designed so that liquid weeps through the holes at the same time that vapor is passing through the center of the hole. The advantage of this design is that the cost and space associated with downcomers are eliminated. Its major disadvantage is that it is not robust. That is, if something goes wrong the column may not work at all instead of operating at a lower efficiency. These columns are usually designed by the company selling the system. Some design details are presented by Ludwig (1997).

10.1.1 Trays, Downcomers, and Weirs

In addition to choosing the type of tray, the designer must select the flow pattern on the trays and design the weirs and downcomers. This section will continue to be mainly qualitative.

The most common flow pattern on a tray is the cross-flow pattern shown in Figure 3-7 and repeated in Figure 10-3A. This pattern works well for average flow rates and can be designed to handle suspended solids in the feed. Cross-flow trays can be designed by the user on the basis of information in the open literature (Bolles, 1963; Fair, 1963, 1984, 1985; Kister, 1980, 1981, 1992; Ludwig, 1997), from information in company design manuals (Glitsch, 1974; Koch, 1982), or from any of the manufacturers of staged distillation columns. Design details for cross-flow trays are discussed later.

Multiple-pass trays are used in large-diameter columns with high liquid flow rates. Double-pass trays (Figure 10-3B) are common (Pilling, 2005). The liquid flow is divided into two sections (or passes) to reduce the liquid gradient on the tray and to reduce the downcomer loading. With even larger liquid loadings, four-pass trays are used; three-pass trays are not used because of liquid maldistribution (Kister et al, 2010; Pilling, 2005; Summers, 2010). This type of tray is usually designed by experts, although preliminary designs can be obtained by following the design manuals published by some of the equipment manufacturers (Glitsch, 1974; Koch, 1982).

Which flow pattern is appropriate for a given problem? As the gas and liquid rates increase, the tower diameter increases. However, the ability to handle liquid flow increases with weir length, while the gas flow capacity increases with the square of the tower diameter. Thus, eventually multiple-pass trays are required. A selection guide is given in Figure 10-4

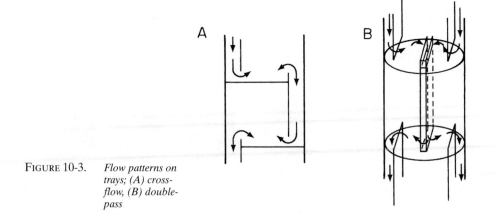

FIGURE 10-3.   *Flow patterns on trays; (A) cross-flow, (B) double-pass*

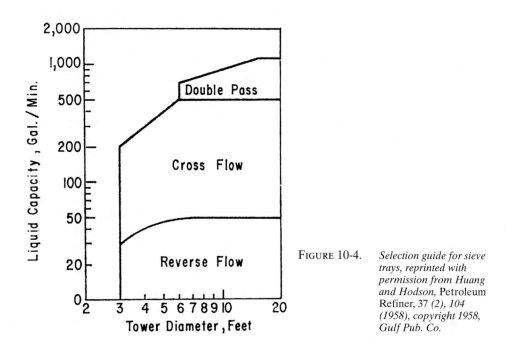

FIGURE 10-4. *Selection guide for sieve trays, reprinted with permission from Huang and Hodson,* Petroleum Refiner, 37 (2), 104 (1958), copyright 1958, Gulf Pub. Co.*

(Huang and Hodson, 1958), but it is only approximate, particularly near the lines separating different types of trays.

Downcomers and weirs are very important for the proper operation of staged columns, since they control the liquid distribution and flow. A variety of designs are used, four are shown in Figure 10-5 (Kister, 1980e, 1992). In small columns and pilot plants the circular pipe shown in

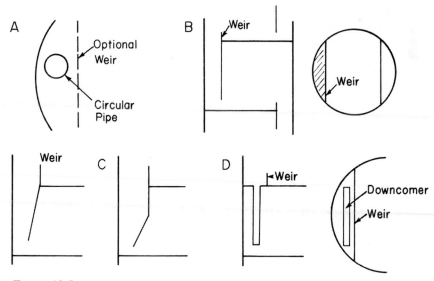

FIGURE 10-5. *Downcomer and weir designs; (A) circular pipe, (B) straight segmental, (C) sloped downcomers, (D) envelope*

Figure 10-5A is commonly used. The pipe may stick out above the tray floor to serve as the weir, or a separate weir may be used. In the Oldershaw design commonly used in pilot plants, the pipe is in the center of the sieve plate and is surrounded by holes. The most common design in commercial columns is the segmented vertical downcomer shown in Figure 10-5B. This type is inexpensive to build, easy to install, almost impossible to install incorrectly, and can be designed for a wide variety of liquid flow rates. If liquid-vapor disengagement is difficult, one of the sloped segmental designs shown in Figure 10-5C can be useful. These designs help retain the active area of the tray below. Unfortunately, they are more expensive and are easy to install backwards. For very low liquid flow rates, the envelope design shown in Figure 10-5D is occasionally used.

The simplest weir design is the straight horizontal weir from 2 to 4 inches high (Figures 3-7 and 10-5). This type is the cheapest but does not have the best turndown properties. The adjustable weir shown in Figure 10-6A is a very seductive design, since it appears to solve the problem of turndown. Unfortunately, if maladjusted, this weir can cause lots of problems such as excessive weeping or trays running dry, so it should be avoided. When flexibility in liquid rates is desired, one of the notched (or picket-fence) weirs shown in Figure 10-6B will work well (Pilling, 2005); they are not much more expensive than a straight weir. Notched weirs are particularly useful with low liquid flow rates.

Trays, weirs, and downcomers need to be mechanically supported. This is illustrated in Figure 10-7 (Zenz, 1997). The trick is to adequately support the weight of the tray plus the highest possible liquid loading it can have without excessively blocking either the vapor flow area or the active area on the tray. As the column diameter increases, tray support becomes more critical. See Ludwig (1997), or Kister (1980d, 1990, 1992) for more details.

10.1.2   Inlets and Outlets

Inlet and outlet ports must be carefully designed to prevent problems (Glitsch, 1985; Kister, 1980a, b, 1990, 1992, 2005; Ludwig, 1997). Inlets should be designed to avoid both excessive weeping and entrainment when a high-velocity stream is added. Several acceptable designs for a feed or reflux to the top tray are shown in Figure 10-8. The baffle plate or pipe elbow prevents high-velocity fluid from shooting across the tray. The designs shown in Figures 10-8D and E can be used if there is likely to be vapor in the feed. These two designs will not allow excessive entrainment. Intermediate feed introduction is somewhat similar, and several common designs are shown in Figure 10-9. Low-velocity liquid feeds can be input through the side of the column as shown in Figure 10-9A. Higher velocity feeds and feed containing vapor require baffles as shown in Figures 10-9B and C. The vapor is directed sideways or downward to prevent excessive entrainment. When there is a large quantity of vapor in the feed, the feed tray should have extra space for disengagement of liquid and vapor. For large diameter columns some type of distributor such as the one shown in Figure 10-9D is often used.

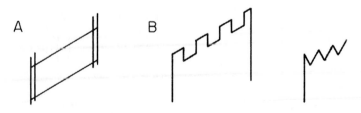

FIGURE 10-6.   *Weir designs (A) adjustable, and (B) notched*

Chapter 10   Staged and Packed Column Design

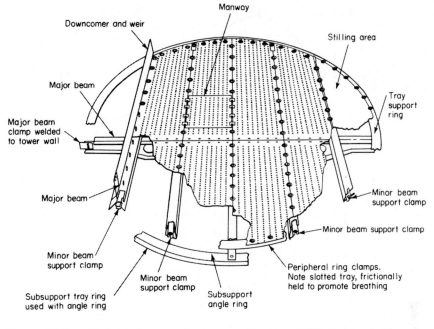

FIGURE 10-7.   *Mechanical supports for sieve trays from Zenz (1997). Reprinted with permission from Schweitzer,* Handbook of Separation Techniques for Chemical Engineers, *3rd ed., copyright 1997, McGraw-Hill, New York.*

The vapor return at the bottom of the column should be at least 12 inches above the liquid surge level. The vapor inlet should be parallel to the seal pan and parallel to the liquid surface, as shown in Figure 10-10A. The purpose of the seal pan is to keep liquid in the downcomer. The vapor inlet should *not* el-down to impinge on the liquid, as shown in Figure 10-10B. When a thermosiphon reboiler (a common type of total reboiler) is used, the split drawoff shown in Figure 10-10C is useful. Since there is usually no pump between the column

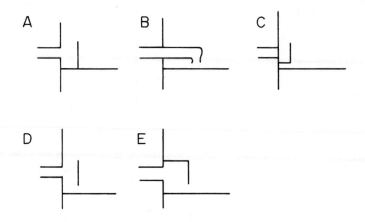

FIGURE 10-8.   *Inlets for reflux or feed to top tray*

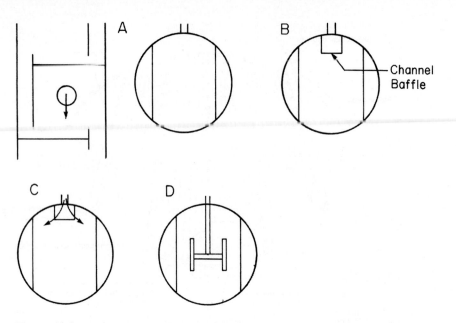

FIGURE 10-9. *Intermediate feed systems; (A) side inlet, (B) and (C) baffles, (D) distributor*

and the reboiler, the driving pressure to move the liquid from the column into the reboiler is based on the head of liquid in the column. This head must be large enough to overcome the pressure drop in the lines and in the reboiler. For kettle type reboilers the liquid will be on the shell side, and the effect of shell-side fouling needs to be included in the pressure-drop calculation. The recommended height from the overflow baffle in the kettle to the column vapor

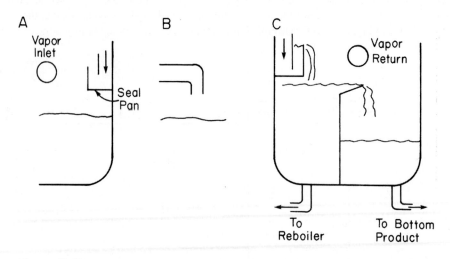

FIGURE 10-10. *Bottom vapor inlet and liquid drawoffs; (A) correct—inlet vapor parallel to seal pan, (B) incorrect—inlet vapor el-down onto liquid, (C) bottom draw-off with thermosiphon reboiler*

Chapter 10 Staged and Packed Column Design

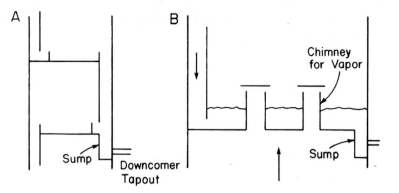

FIGURE 10-11. *Intermediate liquid draw-off; (A) downcomer sump, (B) chimney tray (downcomer to next tray not shown)*

inlet is approximately 2 m (Lieberman and Lieberman, 2008). Smaller clearances can eventually result in flooding of the tower because of high liquid levels in the bottom of the column.

Intermediate liquid drawoffs require some method for disengaging the liquid and vapor. The cheapest way to do this is with a downcomer tapout as shown in Figure 10-11A. A more expensive but surer method is to use a chimney tray (Figure 10-11B). The chimney tray provides enough liquid volume to fill lines and start pumps. There are no holes or valves in the deck of this tray; thus, it doesn't provide for mass transfer and should not be counted as an equilibrium stage. Chimney trays must often support quite a bit of liquid; therefore, mechanical design is important (Kister, 1990). An alternative to the chimney tray is to use a downcomer tapout with an external surge drum.

The design of vapor outlets is relatively easy and is less likely to cause problems than liquid outlets (Kister, 1990). The main consideration is that the line must be of large enough diameter to have a modest pressure drop. At the top of the column a demister may be used to prevent liquid entrainment. An alternative is to put a knockout drum in the line before any compressors.

10.2 TRAY EFFICIENCIES

Tray efficiencies were introduced in Chapter 4. In this section they will be discussed in more detail, and simple methods for estimating the value of the efficiency will be explored. The effect of mass transfer rates on the stage efficiency is discussed in Chapter 16.

The overall efficiency, E_O, is

$$E_O = \frac{N_{equil}}{N_{actual}}$$

(10-1)

The determination of the number of equilibrium stages required for the given separation should not include a partial reboiler or a partial condenser. The overall efficiency is extremely easy to measure and use; thus, it is the most commonly used efficiency value in the plant. However, it is difficult to relate overall efficiency to the fundamental heat and mass transfer processes occurring on the tray, so it is not generally used in fundamental studies.

The Murphree vapor and liquid efficiencies were also introduced in Chapter 4. The Murphree vapor efficiency is defined as

$$E_{MV} = \frac{y_{out} - y_{in}}{y^*_{out} - y_{in}} = \frac{\text{actual change in vapor}}{\text{change in vapor at equilibrium}} \qquad \textbf{(10-2)}$$

while the Murphree liquid efficiency is

$$E_{ML} = \frac{x_{out} - x_{in}}{x^*_{out} - x_{in}} = \frac{\text{actual change in liquid}}{\text{change in liquid at equilibrium}} \qquad \textbf{(10-3)}$$

The physical model used for both of these efficiencies is shown in Figure 10-12. The gas streams and the downcomer liquids are assumed to be perfectly mixed. Murphree also assumed that the liquid on the tray is perfectly mixed, which means $x = x_{out}$. The term y^*_{out} is the vapor mole fraction that would be in equilibrium with the actual liquid mole fraction leaving the tray, x_{out}. In the liquid efficiency, x^*_{out} is in equilibrium with the actual leaving vapor mole fraction, y_{out}. The Murphree efficiencies are popular because they are relatively easy to measure and they are very easy to use in calculations (see Figure 4-27B). Unfortunately, there are some difficulties with their definitions. In large columns the liquid on the tray is not well mixed; instead there will be a cross-flow pattern. If the flow path is long, the more volatile component will be preferentially removed as liquid flows across the tray. Thus, in Figure 10-12,

$$x_{out} < x_4 < x_3 < x_2 < x_1$$

Note that x_{out} and hence y^*_{out} are based on the lowest concentration on the tray. Thus, it is possible to have $y^*_{out} < y_{out}$, because y_{out} is an average across the tray. Then the numerator in Eq. (10-2a) will be greater than the denominator and $E_{MV} > 1$. This is often observed in large-diameter columns. Although not absolutely necessary, it is desirable to have efficiencies defined so that they range between zero and 1. A second and more serious problem with Murphree efficiencies is that the efficiencies of different components must be different for multicomponent systems. Fortunately, for binary systems the Murphree efficiencies are the same for the two components. In multicomponent systems not only are the efficiencies different, but on some trays they may be negative. This is both disconcerting and extremely difficult to predict. Despite these problems, Murphree efficiencies remain popular. The overall efficiency and the Murphree vapor efficiency are related by the following equation, which is derived in Problem 12.C4.

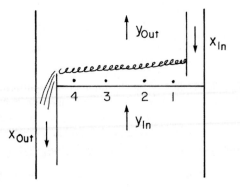

FIGURE 10-12. *Murphree efficiency model*

$$E_o = \frac{\log[1 + E_{MV}(\frac{mV}{L} - 1)]}{\log(mV/L)} \tag{10-4}$$

The point efficiency is defined in a fashion very similar to the Murphree efficiency,

$$E_{pt} = \frac{y'_{out} - y'_{in}}{y'^* - y'_{in}} \tag{10-5}$$

where the prime indicates that all the concentrations are determined at a specific point on the tray. The Murphree efficiency can be determined by integrating all of the point efficiencies (which will vary from location to location) on the tray. Typically, the point and Murphree efficiencies are not equal. The point efficiency is difficult to measure in a commercial column, but it can often be predicted from heat and mass transfer calculations. Thus, it is used for prediction and for scale-up.

In general, the efficiency of a tray depends on the vapor velocity, which is illustrated schematically in Figure 10-13. The trays are designed to give a maximum efficiency at the design condition. At higher vapor velocities, entrainment increases. When entrainment becomes excessive, the efficiency plummets. At vapor velocities less than the design rate the mass transfer is less efficient. At very low velocities the tray starts to weep and efficiency again plummets. Trays with good turndown characteristics have a wide maximum, so there is little loss in efficiency when vapor velocity decreases.

The best way to determine efficiency is to have data for the chemical system in the same type of column of the same size at the same vapor velocity. If velocity varies, then the efficiency will follow Figure 10-13. The Fractionation Research Institute (FRI) has reams of efficiency data, but until recently, most of the data were available to members only. Most large chemical and oil companies belong to FRI. The second best approach is to have efficiency data for the same chemical system but with a different type of tray. Much of the data available in the literature are for bubble-cap or sieve trays. Usually, the efficiency of valve trays is equal to or better than sieve tray efficiency, which is equal to or better than bubble-cap tray efficiency. Thus, if bubble-cap efficiencies are used for a valve tray column, the design will be conservative. The third best approach is to use efficiency data for a similar chemical system.

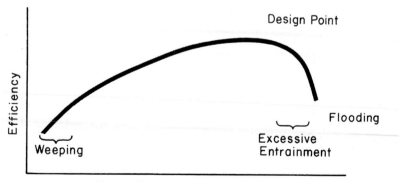

FIGURE 10-13.   *Efficiency as a function of vapor velocity*

If data are not available, a detailed calculation of the efficiency can be made on the basis of fundamental mass and heat transfer calculations. With this method, you first calculate point efficiencies from heat and mass transfer calculations and then determine Murphree and overall efficiencies from flow patterns on the tray. Unfortunately, the results are often not extremely accurate. A simple application of this method is developed in Chapter 16.

The simplest approach is to use a correlation to determine the efficiency. The most widely used is the O'Connell correlation shown in Figure 10-14 (O'Connell, 1946), which gives an estimate of the overall efficiency as a function of the relative volatility of the key components times the liquid viscosity at the feed composition. Both α and μ are determined at the average temperature and pressure of the column. Efficiency drops as viscosity increases, since mass transfer rates are lower. Efficiency drops as relative volatility increases, since the mass that must be transferred to obtain equilibrium increases. The scatter in the 38 data points is evident in the figure. O'Connell was probably studying bubble-cap columns; thus, the results are conservative for sieve and valve trays (Walas, 1988). Walas (1988) surveys a large number of tray efficiencies for a variety of tray types. O'Brien and Schultz (2004) recently reported that many petrochemical towers are designed with stage efficiencies in the 75 to 80% range. Although these columns are probably valve trays, the efficiency range agrees with O'Connell's correlation. Seider et al. (2009) recommend a first estimate of 70%. Recent FRI data are discussed by Kister (2008), who notes that for low volatility systems, errors in VLE data have an enormous effect on the measured overall efficiency. Vacuum systems, which are not included in Figure 10-14, will typically have efficiencies of 15 to 20% (O'Brien and Schultz, 2004).

For computer and calculator use it is convenient to fit the data points to an equation. When this was done using a nonlinear least squares routine, the result (Kessler and Wankat, 1987) was

$$E_O = 0.52782 - 0.27511 \log_{10}(\alpha\mu) + 0.044923[\log_{10}(\alpha\mu)]^2 \qquad \textbf{(10-6)}$$

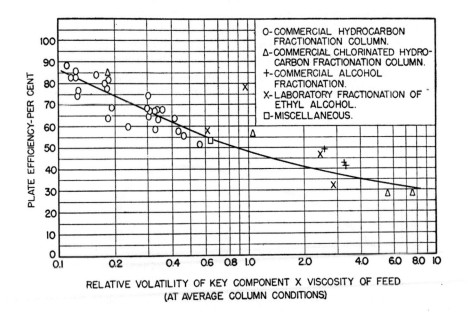

FIGURE 10-14. *O'Connell correlation for overall efficiency of distillation columns from O'Connell (1946). Reprinted from* Transactions Amer. Inst. Chem. Eng., *42, 741 (1946), copyright 1946, American Institute of Chemical Engineers.*

Chapter 10 Staged and Packed Column Design

Viscosity is in centipoise (cP) in Eq. (10-6) and in Figure 10-14. This equation is not an exact fit to O'Connell's curve, since O'Connell apparently used an eyeball fit. Ludwig (1997) discusses other efficiency correlations.

Efficiencies can be scaled up from laboratory data taken with an Oldershaw column (a laboratory-scale sieve-tray column) (Fair et al., 1983; Kister, 1990). The overall efficiency measured in the Oldershaw column is often very close to the point efficiency measured in the large commercial column. This is illustrated in Figure 10-15, where the vapor velocity has been normalized with respect to the fraction of flooding (Fair et al., 1983). The point efficiency can be converted to Murphree and overall efficiencies once a model for the flow pattern on the tray has been adopted (see section 16.6).

For very complex mixtures, the entire distillation design can be done using the Oldershaw column by changing the number of trays and the reflux rate until a combination that does the job is found. Since the commercial column will have an overall efficiency equal to or greater than that of the Oldershaw column, this combination will also work in the commercial column. This approach eliminates the need to determine vapor-liquid equilibrium (VLE) data (which may be quite costly), and it also eliminates the need for complex calculations. The Oldershaw column also allows one to observe foaming problems (Kister, 1990).

EXAMPLE 10-1. Overall efficiency estimation

A sieve-plate distillation column is separating a feed that is 50 mole % n-hexane and 50 mole % n-heptane. Feed is a saturated liquid. Plate spacing is 24 in. Average column pressure is 1 atm. Distillate composition is $x_D = 0.999$ (mole fraction n-hexane) and $x_B = 0.001$. Feed rate is 1000 lbmol/h. Internal reflux ratio $L/V = 0.8$. The column has a total reboiler and a total condenser. Estimate the overall efficiency.

Solution

To use the O'Connell correlation we need to estimate α and μ at the average temperature and pressure of the column. The column temperature can be estimated

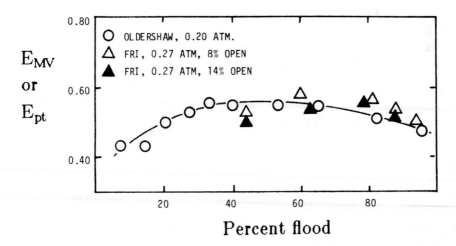

FIGURE 10-15. *Overall efficiency of 1-inch-diameter Oldershaw column compared to point efficiency of 4-foot-diameter FRI column. System is cyclohexane/n-heptane from Fair* et al. *(1983). Reprinted with permission from* Ind. Eng. Chem. Process Des. Develop., 22, 53 (1983), copyright 1983, American Chemical Society.

from equilibrium (the DePriester chart). The following values are easily generated from Figure 2-12.

x_{C6}	0	0.341	0.398	0.50	1.0
y_{C6}	0	0.545	0.609	0.70	1.0
T, °C	98.4	85	83.7	80	69

Relative volatility is $\alpha = (y/x)/[(1-y)/(1-x)]$. The average temperature can be estimated several ways:

Arithmetic average T = (98.4 + 69)/2 = 83.7, α = 2.36
Average at x = 0.5, T = 80, α = 2.33
Not much difference. Use α = 2.35 corresponding to approximately 82.5°C.

The liquid viscosity of the feed can be estimated (Reid et al., 1977, p. 462) from

$$\ln \mu_{mix} = x_1 \ln \mu_1 + x_2 \ln \mu_2 \qquad \text{(10-7a)}$$

The pure component viscosities can be estimated from

$$\log_{10}\mu = A[\frac{1}{T} - \frac{1}{B}] \qquad \text{(10-7b)}$$

where μ is in cP and T is in kelvins (Reid et al., 1977, App. A)

$$nC_6: A = 362.79, B = 207.08$$
$$nC_7: A = 436.73, B = 232.53$$

These equations give $\mu_{C6} = 0.186$, $\mu_{C7} = 0.224$, and $\mu_{mix} = 0.204$. Then $\alpha\mu_{mix} = 0.480$. From Eq. (10-6), $E_O = 0.62$, while from Figure 10-14, $E_O = 0.59$. To be conservative, the lower value would probably be used.

Note that once T_{avg}, α_{avg}, and μ_{feed} have been estimated, calculating E_O is easy.

In many respects the most difficult part of determining design conditions for staged and packed columns is determining the physical properties. For hand calculations good sources are Green and Perry (2008), Perry and Green (1997), Poling et al. (2001), Reid et al. (1977), Smith and Srivastra (1986), Stephan and Hildwein (1987), Woods (1995), and Yaws (1999). Commercial simulators have physical property packages to do this grunt work.

10.3 COLUMN DIAMETER CALCULATIONS

To design a sieve tray column we need to calculate the column diameter that prevents flooding, design the tray layout, and design the downcomers. Several procedures for designing column diameters have been published in the open literature (Fair, 1963, 1984, 1985; Kister, 1992; Ludwig, 1997; McCabe et al., 2005). In addition, each equipment manufacturer has its own procedure. We will follow Fair's procedure, since it is widely known and is an option in the Aspen Plus simulator (see Chapter 6 Appendix). This procedure first estimates the vapor velocity that will cause flooding due to excessive entrainment, then uses a rule of thumb to determine the operating velocity, and from this calculates the column diameter. Column diameter is very important in controlling costs and has to be estimated even for preliminary designs. The method is applicable to sieve, valve, and bubble-cap trays. In Sections 10.3 and

10.5 where hand calculations for diameter and tray layout are discussed, we are essentially forced to use English units because many of the graphs and equations that present required empirical data are in English units. Fortunately, commercial simulators allow one to work in a variety of units, and in Lab 10 in the Appendix to Chapter 10 SI units are used.

The flooding velocity based on net area for vapor flow is determined from

$$u_{flood} = C_{sb,f}\left(\frac{\sigma}{20}\right)^{0.2}\sqrt{\frac{\rho_L - \rho_v}{\rho_v}}, \text{ ft/s} \tag{10-8}$$

Where σ is the surface tension in dynes/cm and $C_{sb,f}$ is the capacity factor. This is similar to Eq. (2-64), which was used to size vertical flash drums. $C_{sb,f}$ is a function of the flow parameter

$$FP = F_{lv} = \frac{W_L}{W_v}\sqrt{\frac{\rho_v}{\rho_L}} = \frac{Q_L}{Q_v}\sqrt{\frac{\rho_L}{\rho_v}} \tag{10-9}$$

where W_L and W_v are the mass flow rates of liquid and vapor, Q_L and Q_v are volumetric flow rates, and densities are mass densities. The correlation for $C_{sb,f}$ is shown in Figure 10-16 (Fair and Matthews, 1958). For computer use it is convenient to fit the curves in Figure 10-16 to equations. The results of a nonlinear least squares regression analysis (Kessler and Wankat, 1987) for 6-inch tray spacing are

$$\log_{10} C_{sb,f} = -1.1977 - 0.53143 \log_{10} F_{lv} - 0.18760 (\log_{10} F_{lv})^2 \tag{10-10a}$$

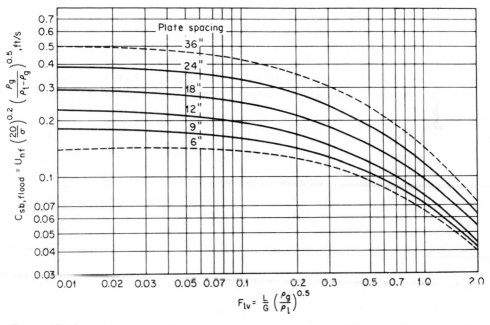

FIGURE 10-16.   *Capacity factor for flooding of sieve trays from Fair and Matthews (1958). Reprinted with permission from* Petroleum Refiner, 37 (4), 153 (1958), copyright 1958, Gulf Pub. Co.

for 9-inch tray spacing

$$\log_{10} C_{sb,f} = -1.1622 - 0.56014 \log_{10} F_{lv} - 0.18168 \ (\log_{10} F_{lv})^2 \qquad \textbf{(10-10b)}$$

for 12-inch tray spacing

$$\log_{10} C_{sb,f} = -1.0674 - 0.55780 \log_{10} F_{lv} - 0.17919 \ (\log_{10} F_{lv})^2 \qquad \textbf{(10-10c)}$$

for 18-inch tray spacing

$$\log_{10} C_{sb,f} = -1.0262 - 0.63513 \log_{10} F_{lv} - 0.20097 \ (\log_{10} F_{lv})^2 \qquad \textbf{(10-10d)}$$

for 24-inch tray spacing

$$\log_{10} C_{sb,f} = -0.94506 - 0.70234 \log_{10} F_{lv} - 0.22618 \ (\log_{10} F_{lv})^2 \qquad \textbf{(10-10e)}$$

and for 36-inch tray spacing

$$\log_{10} C_{sb,f} = -0.85984 - 0.73980 \log_{10} F_{lv} - 0.23735 \ (\log_{10} F_{lv})^2 \qquad \textbf{(10-10f)}$$

The same flooding correlation but in metric units is available in graphical form (Fair, 1985). Figure 10-16 and Eqs. (10-10) predict conservative values for the flooding velocity.

The flooding correlation assumes that β, the ratio of the area of the holes, A_{hole}, to the active area of the tray, A_{active}, is equal to or greater than 0.1. If $\beta < 0.1$, then the flooding velocity calculated from Eq. (10-8) should be multiplied by a correction factor (Fair, 1984). If $\beta = 0.08$, the correction factor is 0.9; while if $\beta = 0.06$, the correction factor is 0.8. Note that this is a linear correction and can easily be interpolated.

Tray spacing, which is required for the flooding correlation, is usually selected according to maintenance requirements because it has little effect on tray efficiency (Kister, 2008). Sieve trays are spaced 6 to 36 inches apart with 12 to 16 inches a common range for smaller (less than 5 feet) towers. Tray spacing is usually greater in large-diameter columns. A minimum of 0.4572 m, with 0.6096 m typical, is used if it is desirable to have a worker crawl through the column for inspection. With the typical 24-inch (0.6096 m) tray spacing, $C_{sb,f} \sim 0.33$ ft/s (Biegler et al., 1997).

The operating vapor velocity is determined as

$$u_{op} = (\text{fraction}) u_{flood}, \qquad \text{ft/s} \qquad \textbf{(10-11)}$$

where the fraction can range from 0.65 to 0.9. Jones and Mellbom (1982) suggest using a value of 0.75 for the fraction for all cases. Higher fractions of flooding do not greatly affect the overall system cost, but they do restrict flexibility. The operating velocity u_{op} can be related to the molar vapor flow rate,

$$u_{op} = \frac{V \ \overline{MW}_v}{\rho_V A_{net}(3600)}, \qquad \text{ft/s} \qquad \textbf{(10-12)}$$

where the 3600 converts from hours (in V) to seconds (in u_{op}). The net area for vapor flow is

$$A_{net} = \frac{\pi (\text{Dia})^2}{4} \eta, \qquad \text{f}^2 \qquad \textbf{(10-13)}$$

where η is the fraction of the column cross-sectional area that is available for vapor flow above the tray. Then $1 - \eta$ is the fraction of the column area taken up by one downcomer. Typically η lies between 0.85 and 0.95; its value can be determined exactly once the tray layout is finalized. Equations (10-12) and (10-13) can be solved for the diameter of the column.

$$\text{Dia} = \sqrt{\frac{4\,V(\overline{MW}_v)}{\pi\,\eta\,\rho_v(\text{fraction})u_{\text{flood}}(3600)}}, \quad \text{ft} \tag{10-14}$$

If the ideal gas law holds,

$$\rho_v = \frac{p\,\overline{MW}_v}{R\,T} \tag{10-15}$$

and Eq. (10-14) becomes

$$\text{Dia} = \sqrt{\frac{4VRT}{\pi\,\eta\,(3600)p(\text{fraction})u_{\text{flood}}}}, \quad \text{ft} \tag{10-16}$$

Note that these equations are dimensional, since C_{sb} is dimensional.

The terms in Eqs. (10-8) to (10-16) vary from stage to stage in the column. If the calculation is done at different locations, different diameters will be calculated. The largest diameter should be used and rounded off to the next highest ½-foot increment. (For example, a 9.18-foot column is rounded off to 9.5 feet). Ludwig (1997) and Kister (1990) suggest using a minimum column diameter of 2.5 feet; that is, if the calculated diameter is 2.0 feet, use 2.5 feet instead, since it is usually no more expensive. These small columns typically use cartridge trays that are prefabricated outside the column (Ludwig, 1997). Columns with diameters less than 2.5 feet are usually constructed as packed columns. If diameter calculations are done at the top and bottom of the column and above and below the feed, one of these locations will be very close to the maximum diameter, and the design based on the largest calculated diameter will be satisfactory. For columns operating at or above atmospheric pressure, the pressure is essentially constant in the column. If we substitute Eqs. (10-8) and (10-15) into Eq. (10-16), calculate the conditions at both the top and bottom of the column, take the ratio of these two equations and simplify, the result is:

$$\frac{\text{Dia}_{\text{top}}}{\text{Dia}_{\text{bot}}} = \left(\frac{V_{\text{top}}}{V_{\text{bot}}}\right)^{\frac{1}{2}} \left(\frac{MW_{V,\text{top}}}{MW_{V,\text{bot}}}\right)^{\frac{1}{4}} \left(\frac{\rho_{L,\text{bot}}}{\rho_{L,\text{top}}}\right)^{\frac{1}{4}} \left(\frac{T_{\text{top}}}{T_{\text{bot}}}\right)^{\frac{1}{4}} \left(\frac{\sigma_{\text{bot}}}{\sigma_{\text{top}}}\right)^{0.1} \left(\frac{C_{sb,f,\text{bot}}}{C_{sb,f,\text{top}}}\right)^{\frac{1}{2}} \tag{10-17}$$

The last two terms on the right-hand side are usually close to 1.0 although when water and organics are separated the surface tension term could be approximately 1.1 (water is bottom product) or 0.9 (water is distillate product). The ratio of temperatures is always < 1 with a range for binary separations typically from 0.9 to 0.99 although it can be lower for distillation at low temperatures. For a saturated vapor feed the vapor velocity term is > 1.0 and can range from approximately 1.1 to 2. For a saturated liquid feed this term is very close to 1.0. The molecular weight and liquid density terms can be greater than or less than one. For distillation of homologous series, the molecular weight term is < 1 while the liquid density term is probably > 1. Since one would not expect large differences, the product of these terms is probably ~ 1. For distillation of water and an organic if water is the bottom product, the molecular weight term is < 1 while the density term is > 1 and the product of the two terms is probably close to one. If water is the distillate product these two terms flip, and the product of the two terms probably remains close to one.

As a rule of thumb, if feed is a saturated liquid and a homologous series is being distilled, the ratio of diameters is probably less than one and the diameter calculated at the bottom is probably larger. This is also true with a saturated liquid feed if the distillation is water from organics, and water is the distillate product, but if water is the bottoms product (more common), either top or bottom diameter can be larger. If feed is a saturated vapor feed the ratio of diameters is probably greater than one and one should design at the top; however, when water and an organic are distilling with water as the distillate product and for cryogenic distillation the ratio of diameters could be greater than or less than one.

If there is a very large change in the vapor velocity in the column, the calculated diameters can be quite different. Occasionally, columns are built in two sections of different diameter to take advantage of this situation, but this solution is economical only for large changes in diameter. If a column with a single diameter is constructed, the efficiencies in different parts of the column may vary considerably (see Figure 10-13). This variation in efficiency has to be included in the design calculations. Another alternative is to attempt to balance the required diameters by adjusting vapor velocities. This method is discussed in Section 10.4.

The design procedure sizes the column to prevent flooding caused by excessive entrainment. Flooding can also occur in the downcomers, and this case is discussed later. Excessive entrainment can also cause a large drop in stage efficiency because liquid that has not been separated is mixed with vapor. The effect of entrainment on the Murphree vapor efficiency can be estimated from

$$E_{MV\text{entrainment}} = E_{MV}\left[\frac{1}{1 + E_{MV}\,\psi/(1-\psi)}\right] \tag{10-18a}$$

where EMV is the Murphree efficiency without entrainment and ψ is the fractional entrainment defined as

$$\psi = \frac{e}{L+e} = \frac{\text{absolute entrainment}}{\text{total liquid flow rate}} \tag{10-18b}$$

where e is the mol/h of entrained liquid. The relative entrainment ψ for sieve trays can be estimated from Figure 10-17 (Fair, 1963). Once the corrected value of E_{MV} is known, the overall efficiency can be determined from Eq. (10-7). Usually, entrainment is not a problem until fractional entrainment is above 0.1, which occurs when operation is in the range of 85 to 100% of flood (Ludwig, 1997). Thus, a 75% of flood value should have a negligible correction for entrainment. This can be checked during the design procedure (see Example 10-3).

EXAMPLE 10-2. Diameter calculation for tray column
Determine the required diameter at the top of the column for the distillation column in Example 10-1.

Solution
We can use Eq. (10-16) with 75% of flooding. Since the distillate is almost pure n-hexane, we can approximate properties as pure n-hexane at 69°C. Physical properties are from Perry and Green (1984). T = 69°C = 342 K; liquid sp grav. = 0.659 (at 20°); viscosity = 0.22 cP; MW = 86.17.

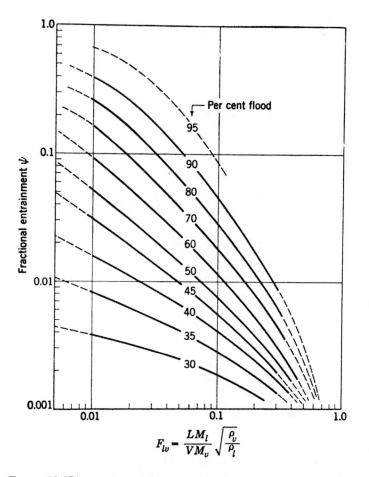

$$F_{lv} = \frac{LM_l}{VM_v} \sqrt{\frac{\rho_v}{\rho_l}}$$

FIGURE 10-17. *Entrainment correlation from Fair (1963). Reprinted with permission from Smith, B.D.,* Design of Equilibrium Stage Processes, *copyright 1963, McGraw-Hill, New York.*

$$\rho_v = \frac{p(MW)}{RT} = \frac{(1\,atm)(86.17\,\dfrac{lb}{lbmol})}{(1.314\,\dfrac{atm\,ft^3}{K\,lbmol})(342\,K)} = 0.1917\,lb/ft^3$$

$\rho_L = (0.659)(62.4) = 41.12\,lb/ft^3$ (will vary, but not a lot)

$$\frac{W_L}{W_v} = \frac{L}{V}\frac{MW_L}{MW_v} = \frac{L}{V} = 0.8$$

Surface tension $\sigma = 13.2$ dynes/cm (Reid et al., 1977, p. 610)
Flow parameter,

$$F_{lv} = \frac{W_L}{W_v}(\frac{\rho_v}{\rho_L})^{0.5} = 0.0546$$

Ordinate from Figure 10-16 for 24 inch tray spacing, $C_{sb} = 0.36$ while $C_{sb} = 0.38$ from Eq. (10-10e). Then

$$K = C_{sb}(\frac{\sigma}{20})^{0.2} = 0.36(\frac{13.2}{20})^{0.2} = 0.331$$

$K = 0.35$ if Eq. (10-10e) is used. The lower value is used for a conservative design. From Eq. (10-8),

$$u_{flood} = K\sqrt{\frac{\rho_L - \rho_v}{\rho_v}} = (0.331)(\frac{41.12 - 0.1917}{0.1917})^{0.5} = 4.836$$

We will estimate η as 0.90. The vapor flow rate $V = L + D$. From external mass balances, $D = 500$. Since $L = V (L/V)$,

$$V = \frac{D}{1 - L/V} = \frac{500}{0.2} = 2500 \text{ lbmol/h}$$

The diameter Eq. (10-16) becomes

$$Dia = \left[\frac{4(2500)(1.314)(342)}{\pi(0.90)(3600)(1)(0.75)(4.836)}\right]^{\frac{1}{2}} = 11.03 \text{ ft}$$

Since Fair's diameter calculation procedure is conservative, an 11-foot diameter column would probably be used. If $\eta = 0.95$, Dia = 10.74 feet; thus, the value of η is not extremely important. Note that this is a large-diameter column for this feed rate. The reflux rate is quite high, and thus, V is high, which leads to a larger diameter. The effect of location on the diameter calculation can be explored by doing Problem 10.D2. Since the result obtained in that problem is a 12-foot diameter, the column would be designed at the bottom. This result agrees with the rule of thumb given after Eq. (10-17) (hexane and heptane are part of the homologous series of alkanes). The value of $\eta = 0.9$ will be checked in Example 10-3. The effect of column pressure is explored in Problem 10.C1.

10.4 BALANCING CALCULATED DIAMETERS

Occasionally the calculated diameter at different points in the column will vary by a large factor. For example, for distillation of a vapor feed of ethanol and water, the calculated diameter at the top of the column was 3.7 times as large as the calculated diameter at the bottom of the column (Wankat, 2007a). For distillation of a liquid feed of acetic acid and water (water is more volatile), the calculated diameter at the bottom was 1.7 times the calculated diameter at the top (Wankat, 2007b).

Instead of building the entire column at the larger diameter, it may be possible to balance the column diameters. Equation (10-14) shows that the column diameter is directly proportional to the square root of the vapor velocity. If we can reduce V at the location with the largest calculated diameter and increase V elsewhere, we may be able to decrease column diameter significantly with very little effect on purity.

For a vapor feed, the column diameter is often largest in the enriching section. We can reduce the diameter by reducing V by condensing all or part of the feed or by condensing vapor inside the column with an intermediate condenser. Three methods for doing this are shown in Figure 10-18. In all cases, there will be a reduction in purity if L/D and N are constant. (Why?) Often, adding a few stages will restore purity to the desired value. If the calculated diameters are markedly different, then the net column volume and column cost will probably be reduced despite the use of more stages. For ethanol-water distillation with a vapor feed, a reduction in column volume of 58% was achieved with the two-enthalpy feed (Figure 10-18B) and of 42% with an intermediate condenser with constant heating and cooling loads. Cooling the entire feed (Figure 10-18A) could reduce volume significantly, but $|Q_c|$ and Q_R had to be increased (Wankat, 2007a).

For a liquid feed system with the largest diameter at the bottom of the column, we want to reduce V at the bottom of the column and increase V higher up the column. The reboiler duty can be reduced (V decreases) if we heat the entire feed (Figure 10-19A), or use two-enthalpy feed vaporizing a portion of the feed (Figure 10-19C), or use an intermediate reboiler (Figure 4-23A). If number of stages and reflux ratio are constant, these methods will all result in lower purity. Again, adding a few stages will often solve this difficulty. For acetic acid-water distillation with constant total heat loads, a reduction of column volume of 54% was obtained with an intermediate reboiler (Figure 4-23A), vapor bypass (Figure 10-19C)

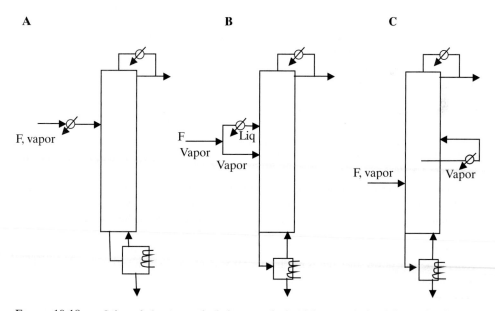

A   **B**   **C**

F, vapor

F. Vapor — Liq / Vapor

F, vapor   Vapor

FIGURE 10-18.   *Column balancing methods for vapor feed with largest calculated diameter at the top. (A) Cool entire feed. (B) Use two-enthalpy feed and cool portion of feed. (C) Intermediate condenser.*

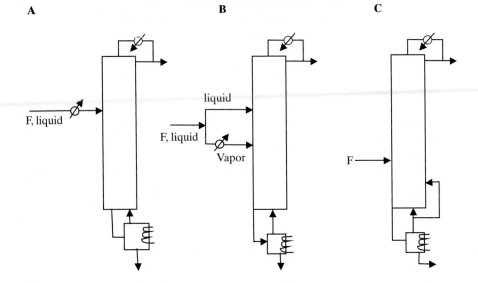

A   B   C

FIGURE 10-19. *Column balancing methods for liquid feed with largest calculated diameter at the bottom. (A) Heat entire feed. (B) Use two-enthalpy feed and heat portion of feed. (C) Vapor bypass.*

reduction was 9.6%, two-enthalpy feed (Figure 10-19B) reduction was 12%, and heating the entire feed (Figure 10-19A) reduction was 10% (Wankat, 2007b).

The addition of heat exchangers and an additional feed stage (Figures 10-18A,B and 10-19A,B) are straightforward and should cause no operating problems. Intermediate condensers and reboilers (Figures 10-18C and 4-23A) are commonly used; however, they can make start-up significantly more difficult (Sloley, 1996). The best system can be determined with computer simulations (see Problem 10.G1).

10.5 SIEVE TRAY LAYOUT AND TRAY HYDRAULICS

Tray layout is an art with its own rules. This section follows the presentations of Ludwig (1997), Bolles (1963), Fair (1963, 1984, 1985), Kister (1990, 1992, 2008), and Lockett (1986), and more details are available in those sources. The holes on a sieve plate are not scattered randomly on the plate. Instead, a detailed pattern is used to ensure even flow of vapor and liquid on the tray. The punched holes in the tray usually range in diameter from 1/8 to 1.0 inch. In the range from ½ to 1 inch, efficiency is constant and at its highest level. The 1/8-inch holes with the holes punched from the bottom up are often used in vacuum operation to reduce entrainment and minimize pressure drop. In normal operation, holes are punched from the bottom down since this is much safer for maintenance personnel. In fouling applications, holes are ½ inch or larger. For clean service, ½ inch is a reasonable first guess for hole diameter.

A common tray layout is the equilateral triangular pitch shown in Figure 10-20. The use of standard punching patterns will be cheaper than use of a non-standard pattern. The holes

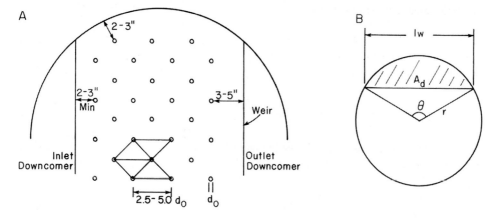

FIGURE 10-20. *Tray geometry; (A) equilateral triangular pitch, (B) downcomer area geometry*

are spaced from $2.5d_o$ to $5d_o$ apart, with $3.8d_o$ a reasonable average. The region containing holes should have a minimum 2- to 3-inches clearance from the column shell and from the inlet downcomer. A 3- to 5-inch minimum clearance is used before the downcomer weir because it is important to allow for disengagement of liquid and vapor. Since flow on the tray is very turbulent, the vapor does not go straight up from the holes. The active hole area is considered to be 2- to 3-inches from the peripheral holes; thus, the area up to the column shell is active. The fraction of the column that is taken up by holes depends upon the hole size, the pitch, the hole spacing, the clearances, and the size of the downcomers. Typically, 4% to 15% of the entire tower area is hole area. This corresponds to a value of $\beta = A_{hole}/A_{active}$ of 6% to 25%. The average value of β is between 7% and 16%, with 10% a reasonable first guess.

The value of β is selected so that the vapor velocity through the holes, v_o, lies between the weep point and the maximum velocity. The exact design point should be selected to give maximum flexibility in operation. Thus, if a reduction in feed rate is much more likely than an increase in feed rate, the design vapor velocity will be close to the maximum. The vapor velocity through the holes, v_o, in feet per second (ft/sec) can be calculated from

$$v_o = \frac{V\,\overline{MW}_v}{3600\,\rho_V\,A_{hole}}$$
(10-19)

where V is the lbmol/h of vapor, ρ_v is the vapor density in lb/ft³, and A_{hole} is the total hole area on the tray in ft². Obviously, A_{hole} can be determined from the tray layout.

$$A_{hole} = (\text{No. of holes})(\pi\frac{d_o^2}{4})$$
(10-20a)

or

$$A_{hole} = \beta\,A_{active}$$
(10-20b)

The active area can be estimated as

$$A_{active} \sim A_{total}(1 - 2(1 - \eta)) = A_{total}(2\eta - 1) \tag{10-20c}$$

Obviously,

$$A_{total} = \frac{\pi(Dia)^2}{4} \tag{10-20d}$$

The downcomer geometry is shown in Figure 10-18B. From this and geometric relationships, the downcomer area A_d can be determined from

$$A_d = \frac{1}{2}r^2(\theta - \sin\theta) \tag{10-21}$$

where θ is in radians. The downcomer area can also be calculated from

$$A_d = (1 - \eta)A_{total} \tag{10-22}$$

Combining Eqs. (10-20d), (10-21), and (10-22), we can solve for angle θ and the length of the weir. We obtain the results given in Table 10-1. Typically the ratio of l_{weir}/Dia falls in the range of 0.6 to 0.75.

If the liquid is unable to flow down the downcomer fast enough, the liquid level will increase, and if it keeps increasing until it reaches the top of the weir of the tray above, the tower will flood. This downcomer flooding must be prevented. Downcomers are designed on the basis of pressure drop and liquid residence time, and their cost is relatively small. Thus, downcomer design is done only in the final equipment sizing.

The tray and downcomer are drawn schematically in Figure 10-21, which shows the pressure heads caused by various hydrodynamic effects. The head of clear liquid in the downcomer, h_{dc}, can be determined from the sum of heads that must be overcome.

$$h_{dc} = h_{\Delta p,dry} + h_{weir} + h_{crest} + h_{grad} + h_{du} \tag{10-23}$$

The head of liquid required to overcome the pressure drop of gas on a dry tray, $h_{\Delta p,dry}$, can be measured experimentally or estimated (Ludwig, 1995) from

$$h_{\Delta p,dry} = 0.003\, v_o^2\, \rho_v\, (\frac{\rho_{water}}{\rho_L})(1 - \beta^2)/C_o^2 \tag{10-24}$$

where v_o is the vapor velocity through the holes in ft/sec from Eq. (10-19). The orifice coefficient, C_o, can be determined from the correlation of Hughmark and O'Connell (1957). This correlation can be fit by the following equation (Kessler and Wankat, 1987):

TABLE 10-1.   *Geometric relationship between η and l_{weir}/diameter*

η	0.8	0.825	0.85	0.875	0.900	0.925	0.95	0.975
l_{weir}/Dia	0.871	0.843	0.811	0.773	0.726	0.669	0.593	0.478

Chapter 10   Staged and Packed Column Design

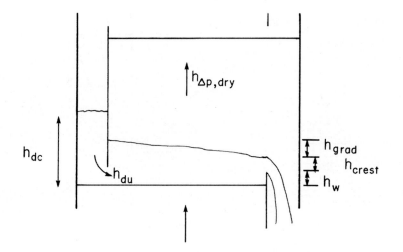

FIGURE 10-21. *Pressure heads on sieve trays*

$$C_o = 0.85032 - 0.04231 \frac{d_o}{t_{tray}} + 0.0017954(\frac{d_o}{t_{tray}})^2 \qquad \textbf{(10-25)}$$

where t_{tray} is the tray thickness. The minimum value for d_o/t_{tray} is 1.0. Equation (10-24) gives $h_{\Delta p,dry}$ in inches.

The weir height, h_{weir}, is the actual height of the weir. The minimum weir height is 0.5 inch with 2 to 4 inches more common. The weir must be high enough that the opposite downcomer remains sealed and always retains liquid. The height of the liquid crest over the weir, h_{crest}, can be calculated from the Francis weir equation.

$$h_{crest} = 0.092 \, F_{weir} \, (L_g/l_{weir})^{2/3} \qquad \textbf{(10-26)}$$

where h_{crest} is in inches. In this equation, L_g is the liquid flow rate in gal/min that is due to both L and e. The entrainment e can be determined from Figure 10-17. l_{weir} is the length of the straight weir in feet. The factor F_{weir} is a modification factor to take into account the curvature of the column wall in the downcomer (Bolles, 1946, 1963; Ludwig, 1997). This is shown in Figure 10-22 (Bolles, 1946). An equation for this figure is available (Bolles, 1963). For large columns where l_{weir} is large, F_{weir} approaches 1.0. On sieve trays, the liquid gradient h_{grad}, across the tray is often very small and is usually ignored.

There is a frictional loss due to flow in the downcomer and under the downcomer onto the tray. This term, h_{du}, can be estimated from the empirical equation (Ludwig, 1997; Bolles, 1963).

$$h_{du} = 0.56(\frac{L_g}{449 \, A_{du}})^2 \qquad \textbf{(10-27)}$$

where h_{du} is in inches and A_{du} is the flow area under the downcomer apron in ft². The downcomer apron typically has a 1-in gap above the tray.

$$A_{du} = (gap)l_{weir} \qquad \textbf{(10-28)}$$

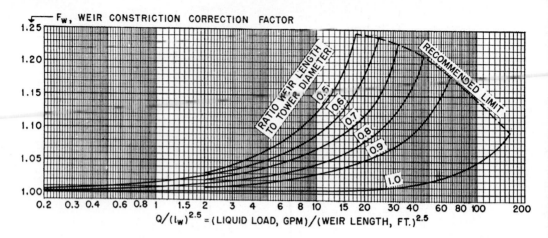

FIGURE 10-22. *Weir correction factor, F_{weir}, for segmental weirs from Bolles (1946). Reprinted with permission of Petroleum Processing*

The value of h_{dc} calculated from Eq. (10-23) is the head of clear liquid in inches. In an operating distillation column the liquid in the downcomer is aerated. The density of this aerated liquid will be less than that of clear liquid, and thus, the height of aerated liquid in the downcomer will be greater than h_{dc}. The expected height of the aerated liquid in the downcomer, $h_{dc,aerated}$, can be estimated (Fair, 1984) from the equation

$$h_{dc,aerated} = h_{dc}/\phi_{dc} \tag{10-29}$$

where ϕ_{dc} is the relative froth density. For normal operation, a value of $\phi_{dc} = 0.5$ is satisfactory, while 0.2 to 0.3 should be used in difficult cases (Fair, 1984). To avoid downcomer flooding, the tray spacing must be greater than $h_{dc,aerated}$. Thus, in normal operation the tray spacing must be greater than $2h_{dc}$.

The downcomer is designed to give a liquid residence time of three to seven seconds. Minimum residence times are listed in Table 10-2 (Kister, 1980e, 1990; Ludwig, 1997). The residence time in a straight segmental downcomer is

$$t_{res} = \frac{A_d h_{dc}(3600)\rho_L}{(L+e)(\overline{MW}_L)(12)}, \qquad \text{sec} \tag{10-30}$$

TABLE 10-2. *Minimum residence times in downcomers*

Foaming Tendency	Examples	Residence Time(s)
Low	Alcohols, low-MW hydrocarbons	3
Medium	Medium-MW hydrocarbons	4
High	Mineral oil absorbers	5
Very high	Amines, glycols	7

Source: Kister (1980e, 1990) and Ludwig (1997)

Chapter 10 Staged and Packed Column Design

where the 3600 converts hours to seconds (from $L + e$) and the 12 converts h_{dc} in inches to feet. Density is the density of clear liquid, and h_{dc} is the height of clear liquid. Equation (10-30) is used to make sure there is enough time to disengage liquid and vapor in the downcomers. Kister (1990) recommends a minimum downcomer area of 5% of the column area. With saturated liquid feeds downcomers should be designed for the stripping section where liquid flow rate is largest.

Figure 10-13 showed that the two limits to acceptable tray operation are excessive entrainment and excessive weeping. Weep and dump points are difficult to determine exactly. An approximate analysis can be used to ensure that operation is above the weep point. Liquid will not drain through the holes as long as the sum of heads due to surface tension, h_σ, and gas flow, $h_{\Delta p,dry}$, are greater than a function depending on the liquid head. This condition for avoiding excessive weeping can be determined from Fair's (1963) graphical correlation or estimated (Kessler and Wankat, 1987) as

$$h_{\Delta p,dry} + h_\sigma \geq 0.10392 + 0.25119x - 0.021675\, x^2 \tag{10-31}$$

where $x = h_{weir} + h_{crest} + h_{grad}$. Equation (10-31) is valid for β ranging from 0.06 to 0.14. The dry tray pressure drop is determined from Eqs. (10-24) and (10-25). The surface tension head h_σ can be estimated (Fair, 1963) from

$$h_\sigma = \frac{0.040\,\sigma}{\rho_L d_o} \tag{10-32}$$

where σ is in dynes/cm, ρ_L in lb/ft³, d_o in inches, and h_σ in inches of liquid. Equation (10-31) is conservative.

EXAMPLE 10-3. Tray layout and hydraulics
Determine the tray layout and pressure drops for the distillation column in Examples 10-1 and 10-2. Determine if entrainment or weeping is a problem. Determine if the downcomers will work properly. Do these calculations only at the top of the column.

Solution
This is a straightforward application of the equations in this section. We can start by determining the entrainment. In Example 10-2 we obtained $F_{lv} = 0.0546$. Then Figure 10-17 at 75% of flooding gives $\psi = 0.045$. Solving for e,

$$e = \frac{\psi L}{1 - \psi} \tag{10-33}$$

Since $L = (L/V)V = 2000$ lbmol/h,

$$e = \frac{(0.045)(2000)}{1 - 0.045} = 94.24 \text{ lbmol/h}$$

and $L + e = 2094.24$ lbmol/h. This amount of entrainment is quite reasonable.

The geometry calculations proceed as follows for a 11.0-foot diameter column:
Eq. (10-20d), $A_{total} = (\pi)(11.0)^2/4 = 95.03$ ft²
Eq. (10-22), $A_d = (1 - 0.9)(95.03) = 9.50$ ft²

Table 10-1 gives $l_{weir}/Dia = 0.726$ or $l_{weir} = 8.0$ feet.
Eq. (10-20c), $A_{active} = (95.03)(1 - 0.2) = 76$ ft^2
Eq. (10-20b), $A_{hole} = (0.1)(76) = 7.6$ ft^2
    We will use 14 gauge standard tray material ($t_{tray} = 0.078$ in) with 3/16-inch holes. Thus, $d_o/t_{tray} = 2.4$.
    From Eq. (10-19)

$$v_o = \frac{V \overline{MW}_v}{3600 \, \rho_V \, A_{hole}} = \frac{2500(86.17)}{3600(0.192)(7.6)} = 41.1 \text{ ft/s}$$

where ρ_v is from Example 10-2. This hole velocity is reasonable.
    The individual pressure drop terms can now be calculated. The orifice coefficient C_o is from Eq. (10-25),

$$C_o = 0.85032 - 0.04231 \, (2.4) + 0.0017954 \, (2.4)^2 = 0.759$$

From Eq. (10-24),

$$h_{\Delta p,dry} = (0.003)(41.1)^2 (0.192)(\frac{61.03}{41.12}) \frac{(1 - 0.01)}{(0.759)^2} = 2.472 \text{ in}$$

In this equation ρ_w at 69 °C was found from Perry and Green (1984, p 3-75). A weir height of $h_{weir} = 2$ inch will be selected.
    The correlation factor F_{weir} can be found from Figure 10-22. L_g in the abscissa is the liquid flow rate including entrainment in gallons per minute.

$$L_g = (2094.24)(86.17 \frac{lb}{lbmole})(\frac{1 \text{ ft}^3}{41.12 \text{ lb}})(7.48 \frac{gal}{ft^3})(\frac{1 \text{ h}}{60 \text{ min}}) = 547.1 \text{ gal/min}$$

$$\text{Abscissa} = \frac{L_g}{1_{weir}^{2.5}} = \frac{547.1}{(8)^{2.5}} = 3.022$$

$$\text{Parameter } 1_{weir}/Dia = 0.727. \text{ Then } F_{weir} = 1.025$$

From Eq. (10-26),

$$h_{crest} = (0.092)(1.025)(\frac{547.1}{8})^{2/3} = 1.577 \text{ in.}$$

We will assume $h_{grad} = 0$. The area under the downcomer is determined with an 1-in gap.
From Eq. (10-28), $A_{du} = (1/12)(8) = 2/3$ ft^2.
From Eq. (10-27),

$$h_{du} = 0.56 \left[\frac{547.1}{449(2/3)} \right]^2 = 1.871 \text{ in.}$$

Total head (total pressure drop across the tray) from Eq. (10-23),

$$h_{dc} = 2.472 + 2 + 1.577 + 0 + 1.871 = 7.92$$

This is inches of clear liquid of density 41.12 lbm/ft^3.

This value can be converted to a pressure drop with two different methods. First,

$$\Delta p = \rho g h / g_c = [(41.12 \text{ lbm/ft}^3)(32.2 \text{ ft/s}^2)(7.92 \text{ in})/$$
$$(32.2 \text{ ft lbm/lbf s}^2)](1 \text{ ft}^3/1728 \text{ in}^3) = 0.189 \text{ lbf/in}^2$$

which is 1.30 kPa. We can also convert the inches of clear liquid of density 41.12 lbm/ft^3 to cm Hg. Since the density of mercury is 845.3 lbm/ft^3, we have

$$(7.92 \text{ in})(41.12/845.3)(2.54 \text{ cm/in}) = 0.9786 \text{ cm Hg} = 1.30 \text{ kPa}.$$

Since the typical pressure drop for sieve or valve trays is in the range from 0.7 to 1.4 kPa per tray (Woods, 2007), this total pressure drop is reasonable.

$$\text{For the aerated system,} \quad h_{dc,aerated} = \frac{7.92}{0.5} = 15.84 \text{ inches}$$

Since this is much less than the 24-inch tray spacing, there should be no problem. The residence time is, from Eq. (10-30),

$$t_{res} = \frac{(9.51)(7.92)(3600)(41.12)}{(2094.24)(86.17)(12)} = 5.15 \text{ s}$$

This is greater than the minimum residence time of 3 s.

Weeping can be checked. From Eq. (10-32),

$$h_\sigma = \frac{(0.040)(13.2)}{(41.12)(3/16)} = 0.068$$

Then the left-hand side of Eq. (10-31) is

$$h_{\Delta p,dry} + h_\sigma = 2.472 + 0.068 = 2.54$$

The function $x = 2 + 1.577 + 0 = 3.577$, and the right-hand side of Eq. (10-31) is 0.725. The inequality is obviously satisfied. Weeping should not be a problem.

Note that the design should be checked at other locations in the column. Since Problem 10.D2 calculated a 12-foot diameter is needed in the stripping section, calculations need to be repeated for the stripping section (see Problem 10.D3). Problem 10.D3 shows that backup of liquid in the downcomers might be a problem in the bottom of the column even with a 12-foot diameter. This occurs because $L = L + F = 3000$ lbmol/h, which is significantly greater than the liquid flow in the top of the column. This problem can be handled by an increase in the gap between the downcomer and the tray.

NOTE: Aspen Plus will do these calculations—see Lab 10 in Appendix to Chapter 10.

10.6   VALVE TRAY DESIGN

Valve trays, which were illustrated in Figure 10-1, are proprietary devices, and the final design would normally be done by the supplier. However, the supplier will not do the optimization studies that the buyer would like without receiving compensation for the additional work. In addition, it is always a good idea to know as much as possible about equipment before making a major purchase. Thus, the nonproprietary valve tray design procedure of Bolles (1976) is very useful for estimating performance. Ludwig (1997) discusses design of proprietary value trays.

Bolles's (1976) design procedure uses the sieve tray design procedure as a basis and modifies it as necessary. One major difference between valve and sieve trays is in their pressure drop characteristics. The dry tray pressure drop in a valve tray is shown in Figure 10-23 (Bolles, 1976). As the gas velocity increases, Δp first increases and then levels off at a plateau level. In the first range of increasing Δp, all valves are closed. At the closed balance point, some of the valves open. Additional valves open in the plateau region until all valves are open

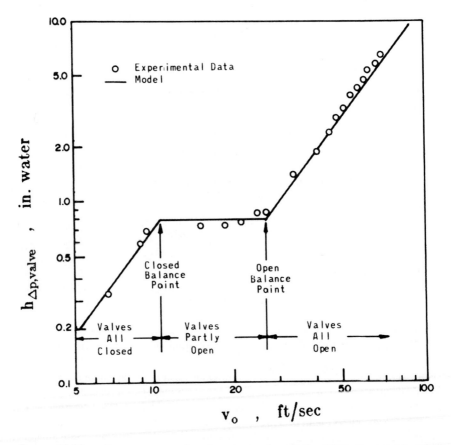

FIGURE 10-23.   *Dry tray pressure drop for valve tray from Bolles (1976). Reprinted from* Chemical Engineering Progress, *Sept. 1976, copyright 1976, American Institute of Chemical Engineers.*

at the open balance point. With all valves open, Δp increases as the gas velocity increases further. The head loss in inches of liquid for both closed and open valves can be expressed in terms of the kinetic energy.

$$h_{\Delta p,valve} = K_v \frac{\rho_v}{\rho_L} \frac{v_o^2}{2g}$$

(10-34a)

where v_o is the velocity of vapor through the holes in the deck in ft/s, g = 32.2 ft/s^2, and K_v is different for closed and open valves. For the data shown in Figure 10-23,

$$K_{v,closed} = 33, \qquad K_{v,open} = 5.5$$

(10-34b)

Note that Eq. (10-34a) has the same dependence on $v_o^2 \rho_v / \rho_L$ as $h_{\Delta p,dry}$ for sieve trays in Eq. (10-24).

The closed balance point can be determined by noting that the pressure must support the weight of the valve, W_{valve}, in pounds. Pressure is W_{valve}/A_v, where A_v is the valve area in square feet. The pressure drops in terms of feet of liquid density ρ_L is then

$$h_{\Delta p,valve} = C_v \frac{W_{valve}}{A_v \rho_L}$$

(10-35)

where the valve coefficient C_v is introduced to include turbulence losses. For the data in Figure 10-23, C_v = 1.25. Setting Eqs. (10-33) and (10-35) equal, allows solution of both the closed and open balance points.

$$v_{o,bal} = \sqrt{(C_v W_{valve} 2g)/(K_v A_v \rho_v)}$$

(10-36)

where $v_{o,bal}$ is the closed balance point velocity if $K_{v,closed}$ is used. The values of $K_{v,closed}$, $K_{v,open}$, and C_v depend upon the thickness of the deck and, to a small extent, on the type of valve (Bolles, 1976).

Much of the remainder of the preliminary design of valve trays is the same or slightly modified from the sieve tray design (Kister, 1990, 1992; Larson and Kister, 1997; Lockett, 1986). Flooding and diameter calculations are the same except that the correction factor for $A_{hole}/A_a < 0.1$ is replaced by a correction factor for $A_{slot}/A_a < 0.1$. The same values for the correction factors are used. The slot area A_{slot} is the vertical area between the tray deck and the top of the valve through which the vapor passes in a horizontal direction. In the region between the balance points the slot area is variable and can be determined from the fraction of valves that are open. The pressure drop Eq. (10-23) is the same, while Eq. (10-24) for $h_{\Delta p,dry}$ is replaced by Eq. (10-33) or (10-35). Equations (10-26) to (10-30) are unchanged. The gradient across the valve tray, h_{grad}, is probably larger than on a sieve tray but is often ignored. Valve trays usually operate with higher weirs. An h_{weir} of 3 inches is normal. There are typically 12 to 16 valves per ft^2 of active area (Kister, 1990).

The efficiency of valve tray depends upon the vapor velocity, the valve design, and the chemical system being distilled. Except at vapor flow rates near flooding, the efficiencies of valve trays are equal to or higher than sieve tray efficiencies, which are equal to or higher than bubble-cap tray efficiencies. Thus, the use of the efficiency correlations discussed earlier will result in a conservative design.

10.7 INTRODUCTION TO PACKED COLUMN DESIGN

Instead of staged columns we often use packed columns for distillation, absorption, stripping, and occasionally extraction. Packed columns are used for smaller diameter columns since it is expensive to build a staged column that will operate properly in small diameters. Packed columns are definitely more economical for columns less than 2.5 feet in diameter. In larger packed columns the liquid may tend to channel, and without careful design randomly packed towers may not operate very well; in many cases large-diameter staged columns are cheaper. Packed towers have the advantage of a smaller pressure drop and are therefore useful in vacuum fractionation.

In designing a packed tower, the choice of packing material is based on economic considerations. A wide variety of packings including random and structured are available. Once the packing has been chosen it is necessary to know the column diameter and the height of packing needed. The column diameter is sized on the basis of either the approach to flooding or the acceptable pressure drop. Packing height can be found either from an equilibrium stage analysis or from mass transfer considerations. The equilibrium stage analysis using the height equivalent to a theoretical plate (HETP) procedure will be considered here; the mass transfer design method is discussed in Chapter 16.

10.8 PACKED COLUMN INTERNALS

In a packed column used for vapor-liquid contact, the liquid flows over the surface of the packing and the vapor flows in the void space inside the packing and between pieces of packing. The purpose of the packing is to provide for intimate contact between vapor and liquid with a very large surface area for mass transfer, which according to Eq. (1-4) will increase the rate of mass transfer. At the same time, the packing should provide for easy liquid drainage and have a low pressure drop for gas flow. Since packings are often randomly dumped into the column, they also have to be designed so that one piece of packing will not cover up and mask the surface area of another piece.

Packings are available in a large variety of styles, some of which are shown in Figure 10-24. The simpler styles such as Raschig rings are usually cheaper on a volumetric basis but will often be more expensive on a performance basis, since some of the proprietary packings are much more efficient. The individual rings and saddles are dumped into the column and are distributed in a random fashion. The structured or arranged packings (e.g., Glitsch grid, Goodloe, and Koch Sulzer) are placed carefully into the column. The structured packings usually have lower pressure drops and are more efficient than dumped packings, but they are often more expensive. Packings are available in a variety of materials including plastics, metals, ceramics, and glass. One of the advantages of packed columns is they can be used in extremely corrosive service.

The packing must be properly held in the column to fully utilize its separating power. A schematic diagram of a packed distillation column is shown in Figure 10-25. In addition to the packed sections where separation occurs, sections are needed for distribution of the reflux, feed, and boilup and for disengagement between liquid and vapor. The liquid distributors are very important for proper operation of the column. Small random and structured packings need better liquid and vapor distribution (Kister, 1990, 2005). If (column diameter)/(packing diameter) > ~40 maldistribution of liquid and vapor is more probable. Maldistribution also reduces the turndown capability of the packing and causes a large increase in HETP with low liquid rates. Liquid distributors typically have a manifold with a number of drip points.

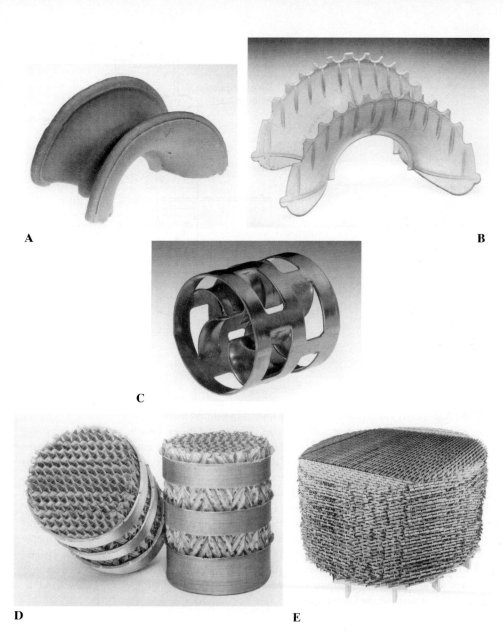

A   B

C

D   E

FIGURE 10-24.   *Types of column packing; (A) ceramic Intalox saddle, (B) plastic super Intalox saddle, (C) Pall ring, (D) GEMPAK cartridge, (E) Glitsch EF-25A grid. Figures A, B, and C courtesy of Norton Chemical Process Products, Akron, Ohio. Figures D and E courtesy of Glitsch, Inc., Dallas, Texas*

Bonilla (1993) recommends six (ten for high purity fractionation) drip points per square foot for large packings (random packings ≥ 2.5 inches or structured packings with crimps > ½ inch). For small packings (random packings ≤ 1 inch or structured packings with crimps ≥ ¼ inch) he recommends eight (12 for high purity) drip points per square foot at high liquid loads and ten

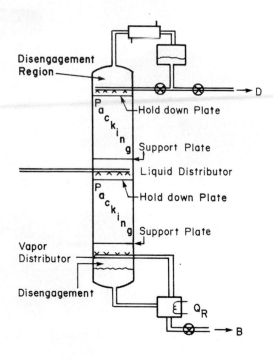

Disengagement Region

Hold down Plate

Support Plate

Liquid Distributor

Hold down Plate

Support Plate

Vapor Distributor

Disengagement

Q_R

FIGURE 10-25. *Packed distillation column*

(14 for high purity) drip points per square foot at low liquid loads. Requirements for medium size packings are between the small and large recommendations. Low liquid loads need more drip points because the liquid tends to spread less. The effects of liquid maldistribution are explored in Problem 10.D10. Scale-up of distributors is done by keeping the number of drip points per square foot constant. Redistribution systems may be required on large columns. The packing is supported by a support plate, which may be a grid or series of bars. A hold-down plate is often employed to prevent packing movement when surges in the gas rate occur. Since liquid and vapor are flowing countercurrently throughout the column, there are no downcomers. The packed tower internals must be carefully designed to obtain good operation (see Fair, 1985; Green and Perry, 2008; Kister, 1990, 2005; Ludwig, 1997; and Strigle, 1994). Additional details are given in the manufacturers' literature. The internals of a packed column for absorption or stripping would be similar to the distillation column shown in Figure 10-23 but without the center feed, reboiler, or condenser.

10.9   HEIGHT OF PACKING: HETP METHOD

Even though a packed tower has continuous instead of discontinuous contact of liquid and vapor, it can be analyzed like a staged tower. We assume that the packed portion of the column can be divided into a number of segments of equal height. Each segment acts as an equilibrium stage, and liquid and vapor leaving the segment are in equilibrium. It is important to note that this staged model is *not* an accurate picture of what is happening physically in the column, but the model can be used for design. The staged model for designing packed columns was first used by Peters (1922).

We calculate the number of stages from either a McCabe-Thiele or Lewis analysis and then calculate the height as

$$\text{Height} = \text{number of equilibrium stages} \times \text{HETP} \qquad \textbf{(10-37a)}$$

The HETP, which is measured experimentally, is the height of packing needed to obtain the change in composition obtained with one theoretical equilibrium contact. HETPs can vary from ½-inch (very low gas flow rates in self-wetting packings) to several feet (large Raschig rings). In normal industrial equipment the HETP varies between 1 and 4 feet. The smaller the HETP, the shorter the column and the more efficient the packing.

To measure the HETP, determine the top and bottom compositions at total reflux and then calculate the number of equilibrium stages.

Then

$$\text{HETP} = \frac{\text{height of packing}}{\text{number of theoretical stages}} \qquad \textbf{(10-37b)}$$

A partial reboiler is usually used but should not be included in the calculation of HETP.

The HETP determined at total reflux is then used at the actual reflux ratio. Ellis and Brooks (1971) found that there is an increase in the HETP for internal reflux ratios below 1.0, but the increase is usually quite small until L/V approaches 1/2. Thus, the usual measurement procedure can be used for most design situations.

The HETP varies with the packing type and size, chemicals being separated, and gas flow rate. Some typical HETP curves are shown in Figure 10-26. The HETP values for several types of packing are listed in the manufacturers' bulletins and have been compared by Ellis and Brooks (1971), Kister and Larson (1997), Kister et al. (1994), Ludwig (1997), Perry (1950, p. 620), and Walas (1988). Mass transfer results are compared by Furter and Newstead (1973), Green and Perry, 2008, Ludwig (1997), and Strigle (1994). Correlations to determine HETP values were developed by Murch (1953) and Whitt (1959), but these do not include modern packings (see Ludwig, 1995). An improved mass transfer model for packings and HETP data for a variety of packings are presented by Bolles and Fair (1982) and discussed in

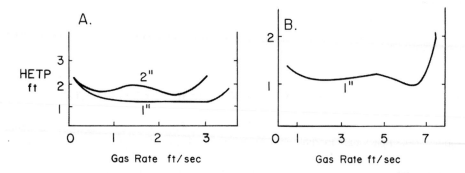

FIGURE 10-26.   *HETP vs. vapor rate for metal Pall rings; (A) Iso-Octane-Toluene, (B) acetone-water. Reprinted with permission from "Pall Rings in Mass Transfer Operations," 1968 courtesy of Norton Chemical Process Products, Akron, Ohio*

Chapter 16. The HETPs are different for different chemical systems and are higher for larger size packing. Most packings of the same size will have approximately the same HETP. Note from Figure 10-26 that the HETP for a given system and packing size is roughly constant over a wide range of gas flow rates. This constant value in the usual design range makes the design procedure convenient. As flooding is approached, the efficiency of the contact decreases, and the HETP increases. Also, at very low gas flow rates, the HETP often increases. This occurs because the packing is not completely wet. For self-wetting packings, where capillary action keeps the packing wet, the HETP usually drops at very low gas flow rates.

HETP values are most accurate if determined from data. If no data are available, generalized mass transfer correlations are used (see Chapter 16). If no information is available, Ludwig (1997) suggests using an average of 1.5 to 2.0 feet for dumped packings. If the column diameter is greater than 1 foot, an HETP greater than 1 foot should be used. Another approximate approach is to set HETP equal to column diameter (Ludwig, 1997). Eckert (1979) notes that HETP values for 1, 1½, and 2-inch Pall rings are 1, 1½, and 2 feet, respectively. Woods (2007) gives ranges of 0.4 to 0.8 m and 0.7 to 0.9 m for 1 and 2 inch Pall rings, respectively. These HETP values are almost independent of the system distilled. Approximate values of HETP for structured packings can be obtained from the following approximate equation (Geankoplis, 2003; Kister, 1992)

$$HETP = 100/a_p + 0.10 \qquad \textbf{(10-37c)}$$

where HETP is in meters and a_p, the surface area per volume, is in m^2/m^3. This result is restricted to low viscosity fluids at moderate or low pressures. Typical HETP values range from 0.3 to 0.6 m. Systems with high surface tensions will have higher HETP values. For example, for systems (amines, glycols) with $\sigma \sim 40$ dynes/cm multiply HETP from Eq. (10-37c) by 1.5 and for aqueous systems with $\sigma \sim 70$ dynes/cm multiply HETP from Eq. (10-37c) by 2 (Anon., 2005). A number of other shortcut HETP relationships are summarized by Wang et al. (2005). If liquid distribution is not excellent, a 30% to 50% safety factor is suggested.

Although packed columns operate with a continuous change in vapor and liquid concentrations, the staged model is still a useful design method. Since the HETP is often almost constant throughout the usual design range for gas flow rates, concentrations, and reflux ratios, a single HETP value can usually be used in comparing many different designs. This greatly facilitates design. In certain cases HETP can vary significantly within the column because of changes in composition; it can then be estimated for each stage from the mass transfer coefficient (see Sherwood et al., 1965, pp. 521–523 for an example). Alternatively, a mass transfer design approach can be used and is preferred (see Chapter 16).

10.10 PACKED COLUMN FLOODING AND DIAMETER CALCULATION

The column diameter is sized to operate at 65% to 90% of flooding or to have a given pressure drop per foot of packing. Flooding can be more easily measured in a packed column than in a plate column and is usually signaled by a break in the curve of pressure drop vs. gas flow rate.

The generalized flooding correlation developed by Sherwood et al. (1938) as modified by Eckert (1970, 1979) is shown in Figure 10-27. Note that the ordinate of Figure 10-27 has units and these units need to be used in the calculation. A graph of the same data with an arithmetic scale for the ordinate is available (Geankoplis, 2003; Strigle, 1994), and a graph

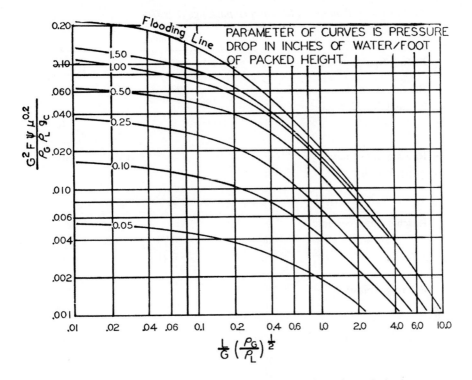

FIGURE 10-27.   *Generalized flooding and pressure drop correlation for packed columns. Reprinted with permission from Eckert, Chem. Eng. Prog., 66(3), 39 (1970), copyright 1970 AIChE. Reproduced by permission of the American Institute of Chemical Engineers.*

updated with new data is available (Kister et al, 2007). The packing factor, F, depends on the type and size of the packing. The higher the value of F, the larger the pressure drop per foot of packing. F values for several types of dumped packing are given in Table 10-3 (Eckert, 1970, 1979; Ludwig, 1997), and F values for structured packings are in Table 10-4 (Fair, 1985; Geankoplis, 2003). As the packing size increases, the F value decreases, and thus, pressure drop per foot will decrease. More extensive lists of F values are available in *Perry's Handbook* (Perry and Green, 1997) and in Strigle (1994). The effect of packing size on the packing factor can be fit reasonably well with the equation (Bennett, 2000),

$$F = C_{p,size}(\delta_p)^{-1.1} \qquad \textbf{(10-38)}$$

where the packing size factor $C_{p,size}$ is given in Table 10-5 and δ_p is the characteristic dimension of the packing in inches. For random packing δ_p = the nominal size or diameter of the packing and for structured packing δ_p = the height of the corrugation. Ceramic packings have thicker walls than plastic, and the plastic have thicker walls than metal; ceramic packings thus, have the lowest free space, highest pressure drops, and highest F values. Generally, the lower the F value the smaller the column diameter. Figure 10-27 is not a perfect fit of all the data. Better results can be obtained using pressure drop curves measured for a given packing, (e.g., see Kister et al., 2007). Specifically, predicted pressure drop for non-aqueous systems at high flow parameter values is too low (Kister and Gill, 1991).

TABLE 10-3. *Parameters for dumped packings (F is 1/ft)*

Packing Type		Nominal			
		1/4	3/8	1/2	5/8
Raschig rings	F	700	390	300	170
(metal, 1/32″ wall)	α				1.20
	β				0.28
Raschig rings (metal	F	—	—	410	290
1/16″ wall)	α				
	β				
Raschig rings (ceramic)	F	1600	1000	580	380
	α		4.70	3.10	2.35
	β		0.41	0.41	0.26
Pall rings (plastic)	F	—	—	—	97
	α				
	β				
Pall rings (metal)	F	—	—	—	70
	α				0.43
	β				0.17
Berl saddles (ceramic)	F	900	—	240	—
	α			1.2	
	β			0.21	
Intalox saddles	F	725	330	200	—
(ceramic)	α			1.04	
	β			0.37	
Intalox saddles (plastic)	F	—	—	—	—
Intalox saddles (metal)	F				
Flexirings (plastic)	F				78
Ballast ring (plastic)	F				
Cascade miniring (plastic)	F				—
Hy-Pak (plastic)	F				

(continues)

TABLE 10-3. *Parameters for dumped packings (F is 1/ft)* (continued)

Packing Size, in					
3/4	*1*	*1 1/4*	*1 1/2*	*2*	*3*
155	115	—	—	—	—
220	137	110	83	57	32
0.80	0.42	—	0.29	0.23	—
0.30	0.21	—	0.20	0.14	—
255	155	125	95	65	37
1.34	0.97	0.57	0.39	0.24	0.18
0.26	0.25	0.23	0.23	0.17	0.15
—	52	—	32	25	—
	0.22			0.10	—
	0.14		—	0.12	—
—	48	—	28	20	—
	0.15		0.08	0.06	—
	0.16		0.15	0.12	—
170	110	—	65	45	—
0.62	0.39	—	0.21	0.16	—
0.17	0.17	—	0.13	0.12	—
145	98	—	52	40	22
0.52	0.52		0.13	0.14	—
0.25	0.16		0.15	0.10	—
—	33	—	—	21	16
	41			18	
—	45	—	28	22	—
97	—	52	32	25	—
30	25	18	—	15	—
	25			12	

Source: Eckert (1970), Ludwig (1997), Coker (1991), Geankoplis (2003)

10.10 Packed Column Flooding and Diameter Calculation

TABLE 10-4. *F values (1/ft) for structured packings (Fair, 1985; Geankoplis, 2003)*

	Flexipac 2	Flexipac 4	Gempak 2A	Gempak 4A	Sulzer CY	Sulzer BX	Munters 12060	Munters 19060	Intalox 2T	Intalox 3T
F	22	6	16	32	70	21	27	15	17	13

TABLE 10-5. *Size factors for Eq. (10-38) (Bennett, 2000), copyright 2000, AIChE. Reproduced by permission of the American Institute of Chemical Engineers.*

Packing Type	$C_{p,size}$
Raschig Rings	140
Metal Pall Rings	62
Intalox Metal Tower Packing	39
Cascade Mini-Rings	42
Nutter Rings	35
Hiflow Rings	34
Structured Packing	8

The Flooding curve can be fit by the equation (Kessler and Wankat, 1988)

$$\log_{10}\left[\frac{G'^2 F \psi \mu^{0.2}}{\rho_G \rho_L g_c}\right] = -1.6678 - 1.085 \log (F_{lv}) - 0.29655 [\log (F_{lv})]^2 \tag{10-39a}$$

where μ is the liquid viscosity in cP, $\psi = \rho_{water}/\rho_L$, $g_c = 32.2$, and F_{lv} is the abscissa of Figure 10-27 and is given in Eq. (10-9), densities are mass densities and L and G or W_L and W_v are mass flow rates.

In the region below the flooding curves, the pressure drop can be correlated with an equation of the form

$$\Delta p = \alpha (10^{\beta L'})(\frac{G'^2}{\rho_G}) \tag{10-39b}$$

where Δp is the pressure drop in inches of water per foot of packing. L' and G' are fluxes in lb/s-ft². Constants α and β are also given in Table 10-3 for dumped packings (Ludwig, 1997). Alternative flooding and pressure drop correlations are given by Kister and Larson (1997), Kister et al., 2007, Ludwig (1997), and Strigle (1994).

The generalized correlation in Figure 10-27 for Eqs. (10-39) are used as follows. The designer first picks a point in the column and determines gas and liquid densities (ρ_G and ρ_L), viscosity (μ), value of ψ, and packing factor for the packing of interest. The ratio of liquid to vapor fluxes, L'/G', is equal to the internal reflux ratio, L/V, if the liquid and vapor are of the same composition, because the area terms divide out and molecular weights cancel. If liquid and vapor mole fractions are significantly different at this point, then

$$\frac{L' \text{ lb/(s − ft}^2)}{G' \text{ lb/(s − ft}^2)} = \frac{L}{V} \frac{\dfrac{\text{mol}}{s} \ (\text{MW liquid} \ \dfrac{\text{lb}}{\text{mol}})}{\dfrac{\text{mol}}{s} \ (\text{MW vapor} \ \dfrac{\text{lb}}{\text{mol}})}$$

(10-40)

In the first design method the designer chooses the pressure drop per unit length of packing. This number ranges from 0.1 to 0.4 inches of water per foot for vacuum columns, from 0.25 to 0.4 inches of water per foot for absorbers and strippers, and from 0.4 to 0.8 inches of water per foot for atmospheric and high-pressure columns (Coker, 1991). With the value of the abscissa and the parameter known, the ordinate can be determined. G' is the only unknown in the ordinate. Once G' is known, the area is

$$\text{Area} = \frac{(V \dfrac{\text{lbmol}}{s})(\text{M.W. vapor} \ \dfrac{\text{lb}}{\text{lbmol}})}{G' \dfrac{\text{lb}}{\text{sft}^2}}$$

(10-41)

In the second design method, the flooding curve in Figure 10-27 or Eq. (10-39a) is used. Then G'_{flood} is calculated from the ordinate. The actual operating vapor flux will be some percent of G'_{flood}. The usual range is 65% to 90% of flooding with 70% to 80% being most common. The area is then determined from Eq. (10-41). The flooding correlation is not perfect. To have 95% confidence a safety factor of 1.32 should be used for the calculated cross-sectional area (Bolles and Fair, 1982).

The diameter is easily calculated once the area is known. Since the liquid and vapor properties and gas and liquid flow rates all vary, the designer must calculate the diameter at several locations and use the largest value. Usually, variations in vapor flow rate dominate diameter calculations.

EXAMPLE 10-4. Packed column diameter calculation

A distillation column is separating n-hexane from n-heptane using 1-inch ceramic Intalox saddles. The allowable pressure drop in the column is 0.5 inches of water per foot. Average column pressure is 1 atm. Separation in the column is essentially complete, so the distillate is almost pure hexane and the bottoms is almost pure heptane. Feed is a 50-50 mixture and is a saturated liquid. In the top, L/V = 0.8. If F = 1000 lbmol/h and D = 500 lbmol/h, estimate the column diameter required at the top.

Solution
A. Define. Find the diameter that gives Δp = 0.5 for top of column.
B. Explore. Need physical properties, most of which are in Examples 10-1 to 10-3:
   n-Hexane: MW 86.17, bp 69°C = 342 K, sp grav = 0.659, viscosity (at 69°) = 0.22 cP
   n-Heptane: MW 100.2, bp 98.4°C = 371.4 K, sp grav = 0.684, viscosity (at 98.4°) = 0.205 cP
   The ideal gas law can be used to estimate vapor densities,

$$\rho_v = \frac{n(\text{MW})}{V} = \frac{p(\text{MW})}{RT}.$$

Water density at 69°C = 0.9783 g/ml. We can use Figure 10-27 with F from Table 10-3 or Eq. (10-39b) with α and β from Table 10-3. We will use both methods.

C. Plan. Figure 10-27 and Eq. (10-39b) can both be used to determine the required diameter.

D. Do it. The *top* is essentially pure n-hexane. Then,

$$\rho_v = \frac{p(MW)}{RT} = \frac{(1 \text{ atm})(86.17 \frac{\text{lb}}{\text{lbmol}})}{(1.314 \frac{\text{atm ft}^3}{\text{K lbmol}})(342 \text{ K})} = 0.1917 \frac{\text{lb}}{\text{ft}^3}$$

$$\frac{L'}{G'} = (\frac{L}{V})(\frac{MW \text{ liquid}}{MW \text{ vapor}}) = (0.8)(\frac{86.17}{86.17}) = 0.8$$

Abscissa for Figure 10-27 is

$$\frac{L'}{G'} (\frac{\rho_v}{\rho_L})^{1/2} = (0.8) \left[\frac{0.1917 \text{ lb/ft}^3}{(0.659 \frac{\text{g}}{\text{cm}^3})(\frac{62.4 \text{ lb/ft}^3}{\text{g/cm}^3})} \right]^{1/2} = 0.055$$

From Figure 10-27 at (Δp = 0.5), ordinate = $\dfrac{G'^2 F \varphi \mu^{0.2}}{\rho_G \rho_L g_c}$ = 0.055

(Obtaining the same value for ordinate and abscissa is an accident!) Then

$$G' = (\frac{0.055 \rho_G \rho_L g_c}{F \psi \mu^{0.2}})^{1/2}$$

From Table 10-3, F = 98. Thus

$$G' = \left[\frac{(0.055)(0.1917)(0.659)(62.4)(32.2)}{98(\frac{0.9783}{0.659})(0.22)^{0.2}} \right]^{1/2} = 0.360 \frac{\text{lb}}{\text{s ft}^2}$$

From Eq. (10-41),

$$\text{Area} = \frac{(V \frac{\text{lbmol}}{\text{s}})(MW)}{G'}$$

Calculate V from V = L + D = (L/D + 1) D, where

$$\frac{L}{D} = \frac{L/V}{1 - L/V} = \frac{0.8}{0.2} = 4$$

$$V = (5) D = 5(500 \frac{\text{lbmol}}{\text{h}})(\frac{1 \text{ h}}{3600 \text{ s}}) = 0.6944 \frac{\text{lbmol}}{\text{s}}$$

This gives

$$\text{Area} = \frac{(0.6944)(86.17)}{0.360} = 166 \text{ ft}^2$$

$$\text{Diameter} = (\frac{4\,\text{Area}}{\pi})^{1/2} = (\frac{4}{\pi}166)^{1/2} = 14.54 \text{ ft}$$

Alternative: Use Eq. (10-39b). First we must rearrange the equation. Since $L'/G' = L/V$, have $L' =(L/V)G'$. Then Eq. (10-39b) becomes

$$\Delta p = \alpha(10^{\beta(\frac{L}{V})G'})(\frac{G'^2}{\rho_G}) \tag{10-42}$$

From Table 10-3, $\alpha = 0.52$ and $\beta = 0.16$. Then the equation is

$$0.50 = (0.52)(10^{(0.16)(0.8)G'})(\frac{G'^2}{0.1917})$$

This is an equation with one unknown, G', so it can be solved for G'. Rearranging the equation,

$$G' = (\frac{0.1843}{10^{0.128G'}})^{1/2}$$

Using our previous answer, $G' = 0.360$, as the first guess and using direct substitution, we obtain $G' = 0.404$ as the answer in two trials. This equation is also easy to solve on a spreadsheet with Goal Seek.
Then,

$$\text{Area} = \frac{V(MW)}{G'} = \frac{(0.6944)(86.17)}{0.404} = 148.1 \text{ ft}^2$$

$$\text{Diameter} = (\frac{4\,\text{Area}}{\pi})^{1/2} = 13.73 \text{ ft}$$

Note that there is a 6% difference between this answer and the one we obtained graphically. Since α and β in Eqs. (10-39b) and (10-42) are specific for this packing and are not based on generalized curves, the lower value is probably more accurate and a 14-foot diameter would be used. If we wanted to be conservative (safe), a 14.5-foot diameter would be used. Additional safety factors (see Fair, 1985) might be employed if the pressure drop is critical.

E. Check. Solving the problem two different ways is a good, but incomplete, check. The check is incomplete because the same values for several variables (e.g., ρ_G, V, and MW) were used in both solutions. Errors in these variables will not be evident in the comparison of the two solutions.

F. Generalize. Either Figure 10-27 or Eq. (10-39b) can be used for pressure drop calculations in packed beds. The use of both is a good check procedure the first time you calculate a diameter (or Δp). Remember, the required diameters should also be estimated at other locations in the column. It is interesting to compare this design with a design using the same packing at 75% of flooding. The 75% of flooding design requires a diameter of 12.4 feet and has pressure drop of approximately 1.5 inches of water per foot of packing (0.37 kPa/ft). It is also interesting to compare this example with Example 10-2, which is a sieve-tray column for the same distillation problem. At 75% of flooding the sieve tray was 11.03 feet in diameter. The packed column is a larger diameter because a small packing was used. If a larger diameter packing were used, the packed column would be smaller (see Problem 10.D16). The effect of location on the calculated column diameter is explored in Problem 10.D17. If calculated diameters are very different, the column balancing method explored in Section 10.4 will help lower capital costs. If pressure drop is absolutely critical, the column area should be multiplied by a safety factor of 2.2 (Bolles and Fair, 1982).

There are three alternatives that can overcome the flooding limitations of counter-current packed columns. The rotating packed bed or Higee process (Lin et al., 2003; Ramshaw, 1983) puts the packing inside an annular-flow column that is rotated at high rpm. This device achieves high rates of mass transfer, and because of the effective increase in gravity floods at much higher velocities. The column size is considerably smaller than for normal distillation or absorption systems, but the device is more complicated. A few commercial units have been built. The second approach is to do absorption, stripping, or extraction in counter-current flow with hollow fiber membrane phase contactors (Humphrey and Keller, 1997; Reed et al., 1995). One phase flows inside the hollow fibers and the other flows counter-current to it in the shell. High flow rates are observed because flooding is unlikely, and relatively low HETP values have been reported because of the very large surface area of hollow fibers. This process has been commercialized by Celanese as Liqui-Cel. The third approach is to avoid counter-current operation altogether and operate in co-current flow where there is no flooding (see Section 12.8).

10.11   ECONOMIC TRADE-OFFS FOR PACKED COLUMNS

In the design of a packed column the designer has many trade-offs that are ultimately reflected in the operating and capital costs. After deciding that a packed column will be used instead of a staged column, the designer must choose the packing type. There is no single packing that is most economical for all separations. For most distillation systems the more efficient packings (low HETP and low F) are most expensive per volume but may be cheaper overall. The designer must then pick the material of construction. Since random commercial packings of the same size will all have an HETP in the range of 1 to 2 feet, the major difference between them is the packing factor, F. Perusal of Table 10-3 shows that there is a very large effect of packing size and a lesser but still up to fourfold effect of packing type on F value. The material of construction can also change F by a factor that can be as high as 3. From Figure 10-27,

$$G' = (\frac{(\text{ordinate}) \, \rho_G \rho_L g_c}{F \psi \mu^{0.2}})^{1/2} \qquad \textbf{(10-43)}$$

A fourfold increase in F would cause a halving of G' and a doubling of the required area [(Eq. (10-41)]. The diameter can then be calculated,

$$\text{Diameter} \propto (F)^{1/4} \qquad \textbf{(10-44)}$$

Note that F is the packing factor, not the feed rate! Thus, the major advantage of more efficient packings, structured packings, and the larger size packings is that they can be used with a smaller diameter column, which is not only less expensive but will also require less packing.

At this point, the designer can pick the packing size and determine both the HETP and packing factor. Larger size packing will have a larger HETP (require a larger height), but a smaller F factor and hence a smaller diameter. Thus, there is a trade-off between packing sizes. The larger size packings are cheaper per cubic foot but can't be used in very small diameter columns. As a rule of thumb,

$$\text{Column diameter/Packing diameter} > 8 \text{ to } 12 \qquad \textbf{(10-45)}$$

depending on whose thumb you are using. The purpose of this rule is to prevent excessive channeling in the column. Structured packings are purchased for the desired diameter column and are not restrained by Eq. (10-45). In small-diameter columns (say, less than 6 inches), structured packings allow low F factors and hence low Δp without violating Eq. (10-45).

The designer can also select the pressure drop per foot. Operating costs in absorbers and strippers will increase as $\Delta p/ft$ increases, but the diameter decreases and hence capital costs for the column decrease. Operation should be in the range of 20% to 90% of flood and is usually in the range of 65% to 90% of flooding. Since columns are often made in standard diameters, the pressure drop per foot is usually adjusted to give a standard size column.

The reflux ratio is a critical variable for packed columns as it is for staged columns. An L/D between 1.05 $(L/D)_{min}$ and 1.25 $(L/D)_{min}$ would be an appropriate value for the reflux ratio. The exact optimum point depends upon the economics of the particular case.

In many applications packed columns appear to be the wave of the future. Columns packed with structured packings have a combination of low pressure drop and high efficiency that often makes them less expensive than either randomly packed columns or staged columns. Random packings can be made in a wide variety of materials so that practically any chemicals can be processed. Many of the flow distribution problems that limited the size of packed columns appear to have been solved.

10.12   CHOICE OF COLUMN TYPE

You're a young engineer asked to design a distillation column. Do you use trays, random packing, or structured packing?

Kister et al. (1994) considered which column to use in considerable detail. They concluded that in vacuum columns, particularly high vacuum, the lower pressure drop of packing is a major advantage and packed columns will have higher capacities and lower reflux ratios than tray columns. For atmospheric and pressure columns pressure drop is usually unimportant, and one needs to compare optimally designed tray columns to optimal packed columns. An optimally designed tray column balances tray and downcomer areas so that both restrict capacity simultaneously. An optimally designed packed column has good liquid and vapor distribution and capacity restrictions are due to the inherent qualities of the packing not due to supports or distributors.

Kister et al. (1994) compared a large amount of data generated by FRI in a 4-foot diameter column and for the structured packing at p < 90 psia by the Separation Research Program (SRP) at the University of Texas-Austin in a 17-inch diameter column. Nutter rings were chosen to represent state-of the-art random packings. The only structured packing measured at high pressures was Norton's Intalox 2T, which was chosen to represent structured packing. Test data were available on both sieve and valve trays with 24-inch tray spacing. All comparisons were at total reflux. For efficiency they compared *practical* HETP values that include height consumed by distribution and redistribution equipment.

$$\text{HETP}_{\text{practical}} = M \text{ HETP}_{\text{packing}} \qquad \textbf{(10-46)}$$

where multiplier M > 1. For 2-inch random packing M = 1.1, while for structured packing (Intalox 2T) M = 1.2. For trays

$$\text{HETP}_{\text{tray}} = S/E_0 \qquad \textbf{(10-47)}$$

where S is the tray spacing in inches and E_o is the fractional overall efficiency. For capacity they compared the capacity factor at flooding,

$$C_s = U_g \sqrt{\frac{\rho_G}{\rho_L - \rho_G}} \qquad \textbf{(10-48)}$$

where U_g is the gas velocity in ft/sec and the densities are mass densities. Both efficiency and capacity results were plotted against the flow parameter FP given in Eq. (10-49) and repeated here.

$$FP = \frac{W_L}{W_v} \sqrt{\frac{\rho_v}{\rho_L}} \qquad \textbf{(10-9)}$$

Since the valve trays were superior to the sieve trays (higher $C_{s,\text{flood}}$ and lower HETP), valve trays were compared to the packings. The overall comparison of tray and packing efficiencies (as measured by the adjusted HETP) is shown in Figure 10-28. The adjusted HETP for the structured packing is lower over most of the range of FP while the HETPs for trays and ran-

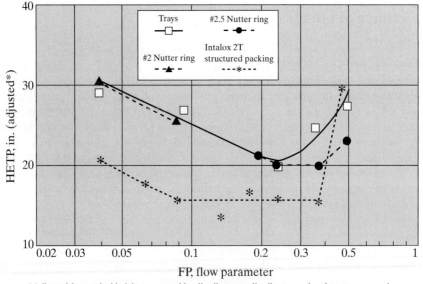

*Adjusted for vertical height consumed by distributor, redistributor, and end tray; see equations (10-46) and (10-47).

FIGURE 10-28.   *Overall comparison of efficiency of trays, random packing, and structured packing (Kister et al., 1994). Reprinted with permission from* Chemical Engineering Progress, *copyright AICHE 1994.*

dom packings are almost identical. At high FP values, which are at high pressures, the random packing had the lowest HETP and the structured packing the highest. Capacities, as measured by $C_{s,flood}$ values, are compared in Figure 10-29. At low FP (low pressures) the structured packing has a 30% to 40% advantage in capacity. At FP values around 0.2 all devices have similar capacities. At high pressures (high FP values) the trays clearly had higher capacities.

In addition to efficiency and capacity, engineers need to be vitally concerned with safety. Many fires are related to distillation and absorption columns. The most dangerous periods are during abnormal operation—upsets, shutdowns, maintenance, total reflux operation, and startups. Structured packing has been implicated in a number of fires, mainly occurring during shutdown and maintenance of units (Ender and Laird, 2003). Several different mechanisms were identified. 1) Hydrocarbons may coat the packing with a thin film that is very difficult to remove. If the packing is at an elevated temperature when it is exposed to air, the hydrocarbons may self-ignite. 2) Coke or polymers may form on the packing during normal operation. Although coking or polymerization will eventually be signaled by increased pressure drop, a large fraction of the bed will contain coke or polymer by time this is noticed. Cooling the interior of the coke or polymer is very difficult because the thermal conductivity is low. Thus, the interior of the coke or polymer may be much hotter than the vapor, which is where the temperature measuring device is located. When the column is opened to air, the coke or polymer may catch fire. 3) If sulfur is present in the feed, corrosion of carbon steel components up- or downstream of the column can form iron sulfide (FeS) that can settle on the packing. This iron sulfide is difficult to remove and will autoignite at ambient temperature when exposed to air. 4) Packing usually arrives from the factory coated with lubricating oil.

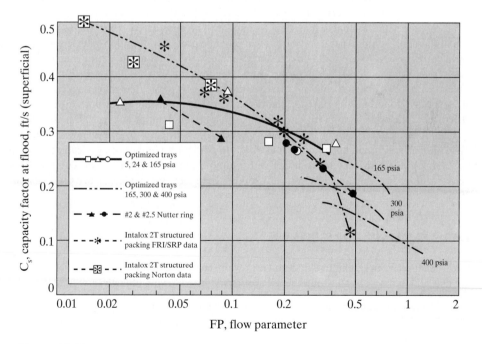

FIGURE 10-29.   Overall comparison of capacity of trays, random packing, and structured packing *(Kister et al, 1994). Reprinted with permission from* Chemical Engineering Progress, *copyright AICHE 1994.*

This oil can catch fire if hot work (e.g., welding) is conducted on the column. 5) If the hydrocarbons, coke, or FeS on the packing catches fire, the very thin metal packing can burn. This is more likely with reactive metals such as aluminum, titanium, and zirconium. The danger of metal fires is because they burn at very high temperatures (up to 1500°C), they can be very destructive.

Standard operating procedures should be designed to prevent these fires (Ender and Laird, 2003).

1. Cool the column to ambient before opening the column. Continuously monitor the temperature.
2. Wash the column extensively to remove residues and deposits.
3. If sulfur is present in the feed assume FeS formed and wash the column with permanganate or percarbonate.
4. Purge with inert gas (nitrogen, carbon dioxide, or steam), and be sure personnel are equipped with respirators before opening the column.
5. Minimize the number of open manways to reduce air entry and allow for rapid closure of the column if there is a fire. Do not force air circulation into the column. This may sound counterintuitive since air circulation will help cool the column if there is no fire; however, fires can be prevented or stopped by starving them of oxygen.

Ultimately, the decision of which column design to use is often economic. Currently, unless there are exceptional circumstances, it appears that structured packings are most economical in low-pressure operation and valve trays are most economical in high-pressure operation. At atmospheric and slightly elevated pressures structured packings have an efficiency but not a capacity advantage; however, structured packings are often considerably more expensive than valve trays (Humphrey and Keller, 1997). The slight added safety risk of structured packing with flammable materials also needs to be considered. Thus, random packings (small-diameter columns) and valve trays will often be preferred for atmospheric and slightly elevated pressures. Exceptional circumstances include the need to use exotic materials such as ceramics, which favors random packings and the need for high holdup for reactive systems, which favors trays. Packed columns are dynamically more stable than staged columns. They are less likely to fail during start-up, shutdown, or other upsets in operation. This is particularly true of absorption and stripping columns (Gunaseelan and Wankat, 2002), and should be considered if the column will be used for processing vent gases in emergencies. Note that tray columns still retain advantages in addition to cost for a number of applications such as very large column diameters, columns with multiple drawoffs, varying feed compositions, highly fouling systems, and systems with solids where sieve trays are usually used (Anon., 2005).

10.13 SUMMARY—OBJECTIVES

In this chapter both qualitative and quantitative aspects of column design are discussed. At the end of this chapter you should be able to satisfy the following objectives:

1. Describe the equipment used for staged distillation columns
2. Define different definitions of efficiency, predict the overall efficiency, and scale-up the efficiency from laboratory data
3. Determine the diameter of sieve and valve tray columns, and adjust vapor flow rates to balance column diameters

4. Determine tray pressure drop terms for sieve and valve trays and design downcomers
5. Lay out a tray that will work
6. Describe the parts of a packed column and explain the purpose of each part
7. Use the HETP method to design a packed column. Determine the HETP from data
8. Calculate the required diameter of a packed column
9. Determine an appropriate range of operating conditions for a packed column
10. Select the appropriate design (sieve tray, valve tray, random packing, or structured packing) for a separation problem

REFERENCES

Anon., "Facts at Your Fingertips," *Chemical Engineering,* 59 (April 2005).

Biegler, L. T., I. E. Grossman, and A. W. Westerberg, *Systematic Methods of Chemical Process Design*, Prentice Hall, Upper Saddle River, NJ, 1997.

Bolles, W. L., *Pet Refiner, 25,* 613 (1946).

Bolles, W. L., "Tray Hydraulics, Bubble-Cap Trays," in B. D. Smith, (Ed.), *Design of Equilibrium Stage Processes,* McGraw-Hill, New York, 1963, Chap. 14.

Bolles, W. L. "Estimating Valve Tray Performance," *Chem. Eng. Prog., 72* (9), 43 (Sept. 1976).

Bolles, W. L. and J. R. Fair, "Improved Mass-Transfer Model Enhances Packed-Column Design," *Chem. Eng., 89* (14), 109 (July 12, 1982).

Bonilla, J. A., "Don't Neglect Liquid Distributors," *Chem. Engr. Progress, 89,* (3), 47 (March 1993).

Coker, A. K., "Understand the Basics of Packed-Column Design," *Chem. Engr. Progress, 87,* (11), 93 (November 1991).

Dean, J. A., *Lange's Handbook of Chemistry*, 13th ed., McGraw-Hill, New York, 1985.

Eckert, J. S., "Selecting the Proper Distillation Column Packing," *Chem. Eng. Prog., 66* (3), 39 (1970).

Eckert, J. S., "Design of Packed Columns," in P. A. Schweitzer (Ed.), *Handbook of Separation Techniques for Chemical Engineers,* McGraw-Hill, New York, 1979, Section 1.7.

Ellis, S. R. M., and F. Brooks, "Performance of Packed Distillation Columns Under Finite Reflux Conditions," *Birmingham Univ. Chem. Eng., 22,* 113 (1971).

Ender, C., and D. Laird, "Minimize the Risk of Fire During Column Maintenance," *Chem. Engr. Progress, 99,* (9), 54 (September 2003).

Fair, J. R., "Tray Hydraulics. Perforated Trays," in B. D. Smith, *Design of Equilibrium Stage Processes,* McGraw-Hill, New York, 1963, Chap. 15.

Fair, J. R., "Gas-Liquid Contacting," in R. H. Perry and D. Green (Eds.), *Perry's Chemical Engineer's Handbook,* 6th ed., McGraw-Hill, New York, 1984, Section 18.

Fair, J. R., "Stagewise Mass Transfer Processes," in A. Bisio and R. L. Kabel (Eds.), *Scaleup of Chemical Processes,* Wiley, New York, 1985, Chap. 12.

Fair, J. R., H. R. Null, and W. L. Bolles, "Scale-up of Plate Efficiency from Laboratory Oldershaw Data," *Ind. Eng. Chem. Process Des. Develop., 22,* 53 (1983).

Furter, W. F., and W. T. Newstead, "Comparative Performance of Packings for Gas-Liquid Contacting Columns," *Can. J. Chem. Eng., 51,* 326 (1973).

Geankoplis, C. J., *Transport Processes and Separation Process Principles,* 4th ed., Prentice Hall, Upper Saddle River, NJ, 2003.

Green, D. H,. and R. H. Perry (Eds.), *Perry's Chemical Engineers' Handbook*, 8th ed., McGraw-Hill, New York, 2008

Gunaseelan, P., and P. C. Wankat, "Transient Pressure and Flow Predictions for Concentrated Packed Absorbers Using a Dynamic Non-Equilibrium Model," *Ind. Engr Chem. Research, 41,* 5775 (2002).

Huang, C.-J., and J. R. Hodson, "Perforated Trays," *Pet. Ref., 37*(2), 104 (1958).

Hughmark, G. A., and H. E. O'Connell, "Design of Perforated Plate Fractionating Towers," *Chem. Eng. Prog., 53,* 127-M (1957).

Humphrey, J. L., and G. E. Keller II, *Separation Process Technology,* McGraw-Hill, New York, 1997.

Jones, E. A., and M. E. Mellborn, "Fractionating Column Economics," *Chem. Eng. Prog., 75*(5), 52(1982).

Kessler, D. P., and P. C. Wankat, "Correlations for Column Parameters," *Chem. Eng., 71* (Sept. 26, 1988).

Kister, H. Z., "Guidelines for Designing Distillation—Column Intervals," *Chem. Eng., 87* (10), 138 (1980a).

Kister, H. Z., "Outlets and Internal Devices for Distillation Columns," *Chem. Eng., 87* (15), 79 (1980b).

Kister, H.Z., "Design and Layout for Sieve and Valve Trays," *Chem. Eng., 87*(18), 119 (1980c).

Kister, H. Z., "Mechanical Requirements for Sieve and Valve Trays," *Chem. Eng., 87* (23), 283 (1980d).

Kister, H. Z., "Downcomer Design for Distillation Tray Columns," *Chem. Eng., 87* (26), 55 (1980e).

Kister, H. Z., "Inspection Assures Trouble-Free Operation," *Chem. Eng., 88*(3), 107 (1981a).

Kister, H. Z., "How to Prepare and Test Columns Before Startup," *Chem. Eng., 88* (7), 97 (1981b).

Kister, H. Z., *Distillation Operation,* McGraw-Hill, New York, 1990.

Kister, H. Z., *Distillation Design,* McGraw-Hill, New York, 1992.

Kister, H. Z., *Distillation Troubleshooting,* McGraw-Hill, New York, 2005.

Kister, H. Z., "Effects of Design on Tray Efficiency in Commercial Towers," *Chem. Engr. Progress, 104* (6), 39 (June 2008).

Kister, H. Z., R. Dionne, W. J. Stupin, and M. R. Olsson, "Preventing Maldistribution in Multi-Pass Trays," *Chem. Engr. Progress, 106* (4), 32 (April 2010).

Kister, H. Z., and D.F. Gill, "Predict Flood Point and Pressure Drop for Modern Random Packings," *Chem. Engr. Progress, 87,* (2), 32 (February 1991).

Kister, H. Z., and K. F. Larson, "Packed Distillation Tower Design," in Schweitzer, P.A. (Ed.), *Handbook of Separation Techniques for Chemical Engineers,* 3rd ed., McGraw-Hill, New York, 1997, section 1.6.

Kister, H. Z., K. F. Larson, and T. Yanagi, "How Do Trays and Packings Stack Up?" *Chem. Engr. Progress, 90,* (2), 23 (February 1994).

Kister, H. Z., J. Scherffius, K. Afshar, and E. Abkar, "Realistically Predict Capacity and Pressure Drop for Packed Columns," *Chem. Eng., 28* (7), 28 (July 2007).

Larson, K. F., and H. Z. Kister, "Distillation Tray Design," in Schweitzer, P. A. (Ed.), *Handbook of Separation Techniques for Chemical Engineers,* 3rd ed., McGraw-Hill, New York, 1997, section 1.5.

Lieberman, N. P., and E. T. Lieberman, *A Working Guide to Process Equipment*, 3rd ed., McGraw-Hill, New York, 2008.

Lin, C.-C., W.-T. Liu, and C.-S. Tan, "Removal of Carbon Dioxide by Absorption in a Rotating Packed Bed," *Ind. Engr. Chem. Res., 42,* 2381 (2003).

Lockett, M. J., *Distillation Tray Fundamentals,* Cambridge University Press, Cambridge, UK, 1986.

Ludwig, E. E., *Applied Process Design for Chemical and Petrochemical Plants,* Vol. 2, second ed., Gulf Pub. Co., Houston, TX, 1979.

MacFarland, S. A., P. M. Sigmund, and M. Van Winkle, "Predict Distillation Efficiency," *Hydrocarbon Proc.,* 51 (7), 111 (1972).

McCabe, W. L., J. C. Smith, and P. Harriott, *Unit Operations of Chemical Engineering,* 4th ed., McGraw-Hill, New York, 1985.

Murch, D. P., "Height of Equivalent Theoretical Plate in Packed Fractionation Columns," *Ind. Eng. Chem., 45,* 2616 (1953).

O'Brien, D., and M. A. Schulz, "Ask the Experts: Distillation," *Chem. Engr. Progress, 100,* (2), 14 (Feb. 2004).

O'Connell, H. E., "Plate Efficiency of Fractionating Columns and Absorbers," *Trans. Amer. Inst. Chem. Eng., 42,* 741 (1946).

Perry, J. H. (Ed.), *Chemical Engineer's Handbook,* 3rd ed., McGraw-Hill, New York, 1950.

Perry, R. H., C. H. Chilton, and S. D. Kirkpatrick (Eds.), *Chemical Engineer's Handbook,* 4th ed., McGraw-Hill, New York, 1963.

Perry, R. H., and C. H. Chilton (eds.), *Chemical Engineer's Handbook,* 5th ed., McGraw-Hill, New York, 1973.

Perry, R. H., and D. W. Green (Eds.), *Perry's Chemical Engineers' Handbook,* 6th ed., McGraw-Hill, New York, 1984.

Perry, R. H., and D. W. Green (Eds.), *Perry's Chemical Engineers' Handbook,* 7th ed., McGraw-Hill, New York, 1997.

Pilling, M., "Ensure Proper Design and Operation of Multi-Pass Trays," *Chem. Engr. Progress, 101,* (6), 22 (June 2005).

Poling, B. E., J. M. Prausnitz, and J. P. O'Connell, *The Properties of Gases and Liquids,* 5th ed., McGraw-Hill, New York, 2001.

Peters, W. A., "The Efficiency and Capacity of Fractionating Columns," *Ind. Eng. Chem., 14, 476* (1922).

Ramshaw, C., "Higee Distillation—An Example of Process Intensification," *Chem. Engr.,* (389), 13 (1983).

Reed, B. W., et al., "Membrane Contactors," in R. D. Noble and S. A. Stern, eds., *Membrane Separations Technology—Principles and Applications,* Chap. 10, Elsevier, 1995.

Reid, R. C., J. M. Prausnitz, and T. K. Sherwood, *The Properties of Gases and Liquids,* 3rd ed., McGraw-Hill, New York, 1977.

Seider, W. D., J. D. Seader, and D. R. Lewin, *Product & Process Design Principles*, 3rd ed., Wiley, New York, 2009.

Sherwood, T. K., G. H. Shipley, and F. A. L. Holloway, "Flooding Velocities in Packed Columns," *Ind. Eng. Chem., 30,* 765 (1938).

Sherwood, T. K., R. L. Pigford, and C. R. Wilke, *Mass Transfer,* McGraw-Hill, New York, 1975.

Sloley, A. W., "Avoid Problems During Distillation Column Startups," "*Chem. Engr. Progress, 92* (7), 30 (July 1996).

Smith, B. D., *Design of Equilibrium Stage Processes,* McGraw-Hill, New York, 1963.

Smith, B. D., and R. Srivastava, *Thermodynamic Data for Pure Compounds,* Elsevier, Amsterdam, 1986.

Stephan, K. and H. Hildwein, *Recommended Data of Selected Compounds and Binary Mixtures: Tables, Diagrams and Correlations,* DECHEMA, Frankfurt am Main, Germany, 1987.

Strigle, R. F. Jr., *Packed Tower Design and Applications. Random and Structured Packings,* 2nd ed., Gulf Publishing Co., Houston, 1994.

Summers, D. R., "Designing Four-Pass Trays, *Chem. Engr. Progress, 106* (4), 26 (April 2010).

Torzewski, K. (Dept. Ed.), "Facts at Your Fingertips, Tray Column Design," *Chemical Engineering, 116* (1), 17 (January 2009).

Walas, S. M., *Chemical Process Equipment: Selection and Design,* Butterworth, Boston, 1988.

Wang, G. Q., X. G. Yuan, and K. T. Yu, "Review of Mass-Transfer Correlations for Packed Columns," *Ind. Engr. Chem. Res., 44,* 8715 (2005).

Wankat, P. C., "Balancing Diameters of Distillation Column with Vapor Feeds," *Ind. Engr. Chem. Research, 46*, 8813-8826 (2007).

Wankat, P. C., "Reducing Diameters of Distillation Column with Largest Calculated Diameter at the Bottom," *Ind. Engr. Chem. Research, 46*, 9223–9231 (2007).

Whitt, F. R., "A Correlation for Absorption Column Packing," *Brit. Chem. Eng.,* July, 1959, p. 395.

Woods, D. R., *Data for Process Design and Engineering Practice,* Prentice Hall, Englewood Cliffs, NJ, 1995.

Woods, D. R., *Rules of Thumb in Engineering Practice*, Wiley-VCH, Weinheim, 2007.

Yaws, C. L., *Chemical Properties Handbook: Physical, Thermodynamic, Environmental, Transport, Safety, and Health Related Properties for Organic and Chemicals,* McGraw-Hill, New York, 1999.

Zenz, F. A., "Design of Gas Absorption Towers," in P. A. Schweitzer (Ed.), *Handbook of Separation Techniques for Chemical Engineers,* 3rd ed., McGraw-Hill, New York, 1997, Section 3.2.

HOMEWORK

A. *Discussion Problems*

A1. What effect does increasing the spacing between trays have on
  a. Column efficiency?
  b. $C_{sb,f}$ and column diameter?
  c. Column height?

A2. Several different column areas are used in this chapter. Define and contrast: total cross-sectional area, net area, downcomer area, active area, and hole area.

A3. Relate the head of clear liquid to a pressure drop in psig.

A4. Another type of weir not shown in Figure 10-6 is the adjustable weir. In an adjustable weir the height can be changed by loosening bolts, moving the weir, and retightening the bolts. List as many reasons as you can think of why this type of weir is generally not recommended.

A5. As shown by the Leaning Tower of Pisa, towers that do not have a proper footing can develop a significant lean. Even if a distillation tower is structurally sound, leaning away from being vertical affects the operation. What is likely to go wrong for a tray tower if it leans?

A6. Valve trays are often constructed with two different weight valves. What would this do to Figure 12-21? What are the probable advantages of this design?

A7. Beer stills are distillation columns that process the raw feed from fermentation. This feed includes cells and cell debris in addition to ethanol and water. Why are sieve plates with larger-than-normal holes the preferred column internals for beer stills?

A8. Intermediate feeds should not be introduced into a downcomer. Explain why not.

A9. What are the characteristics of a good packing? Why are marbles a poor packing material?

A10. Develop your key relations chart for this chapter.

A11. If HETP varies significantly with the gas rate, how would you design a packed column?

A12. Structured packings work very well in vacuum and atmospheric pressure distillation columns, but sometimes structured packings do not work well in high-pressure columns. What is different about high-pressure columns that may make the structured packings not work as well?

A13. What effect will an increase in viscosity have on pressure drop in a packed column?

A14. Refer to Table 10-3.

    a. Which is more desirable, a high or low packing factor, F?

    b. As packing size increases, does F increase or decrease? What is the functional form of this change (linear, quadratic, cubic, etc.)?

    c. Why do ceramic packings have higher F factors than plastic or metal packings of the same type and size? When would you choose a ceramic packing?

A15. Why can't large-size packings be used in small-diameter columns? What is the reason for the rule of thumb given in Eq. (10-45)?

A16. We have designed a sieve tray distillation column for p = 3 atm. We decide to look at the design for p − 1 atm. Assume feed flow rate, mole fractions in the feed, and L/D are the same for both designs. Feed is a saturated liquid for both designs. We want same recoveries of light and heavy key for both designs. Both designs have a total condenser and a partial reboiler. Compared to the original design at 3 atm, the design at 1 atm will:

    1. Have   **a.** fewer stages

                         b. more stages

                         c. same number of stages

    2. Require   **a.** smaller diameter column

                         b. larger diameter column

                         c. same diameter column

    3. Have   **a.** lower reboiler temperature

                         b. higher reboiler temperature

                         c. same reboiler temperature

A17. If we compare 1-inch Pall rings to 3-inch Pall rings,

    a. The 1-inch rings have higher HETP and higher Δp/ft than 3-inch rings.

    b. The 1-inch rings have same HETP and higher Δp/ft than 3-inch rings.

    c. The 1-inch rings have lower HETP and higher Δp/ft than 3-inch rings.

    d. The 1-inch rings have higher HETP and same Δp/ft than 3-inch rings.

    e. The 1-inch rings have same HETP and same Δp/ft than 3-inch rings.

    f. The 1-inch rings have lower HETP and same Δp/ft than 3-inch rings.

    g. The 1-inch rings have higher HETP and lower Δp/ft than 3-inch rings.

    h. The 1-inch rings have same HETP and lower Δp/ft than 3-inch rings.

    i. The 1-inch rings have lower HETP and lower Δp/ft than 3-inch rings.

B. *Generation of Alternatives*

B1. One type of valve is shown in Figure 10-1. Brainstorm alternative ways in which valves could be designed.

B2. What other ways of contacting in packed columns can you think of?

B3. **a.** A farmer friend of yours is going to build his own distillation system to purify ethanol made by fermentation. He wants to make his own packing. Suggest 30 different things he could make or buy cheaply to use as packing (set up a brainstorming group to do this—make no judgments as you list ideas).

    b. Look at your list in part a. Which idea is the craziest? Use this idea as a trigger to come up with 20 more ideas (some of which may be reasonable).

    c. Go through your two lists from parts a and b. Which ideas are technically feasible? Which ideas are also cheap and durable? List about 10 ideas that look like the best for further exploration.

C. Derivations

C1. You need to temporarily increase the feed rate to an existing column without flooding. Since the column is now operating at about 90% of flooding, you must vary some operating parameter. The column has a 0.4572 m tray spacing, is operating at 1 atm, and has a flow parameter

$$F_{lv} = \frac{L}{G}(\frac{\rho_v}{\rho_L})^{0.5} = 0.05$$

The column is rated for a total pressure of 10 atm. L/D = constant. The relative volatility for this system does not depend on pressure. The condenser and reboiler can easily handle operation at a higher pressure. Downcomers are large enough for larger flow rates. Will increasing the column pressure increase the feed rate that can be processed? It is likely that:

$$\text{Feed rate} \propto (\text{pressure})^{\text{exponent}} \tag{10-49}$$

Determine the value of the exponent for this situation. Use the ideal gas law.

C2. Show that staged column diameter is proportional to (feed rate)$^{1/2}$ and to $(1 + L/D)^{1/2}$.

C3. If the packing factor were unknown, you could measure Δp at a series of gas flow rates. How would you determine F from these data?

C4. Using a McCabe-Thiele diagram for a binary system, show why the purity will be reduced compared to the base case if N and L/D are constant and one of the diameter balancing methods in Figure 10-18 (base case is saturated vapor feed) or Figure 10-19 (base case is saturated liquid feed) is used to reduce the column volume. Although this demonstration is for a binary system, the logic is true for multicomponent systems also.

C5. Using a McCabe-Thiele diagram for a binary system, show why increasing N may not be sufficient to keep constant purity compared to the base case if the diameter balancing method in Figure 10-18A (base case is saturated vapor feed) or 10-19A (base case is saturated liquid feed) is used to reduce the column volume. Although this demonstration is for a binary system, the logic is true for multicomponent systems also.

D. Problems

Answers to problems with an asterisk are at the back of the book.

D1.* Repeat Example 10-1 for an average column pressure of 700 kPa.

D2.* Repeat Example 10-2, except calculate the diameter at the bottom of the column. For n-heptane: MW = 100.2, bp 98.4°C, sp gravity = 0.684, viscosity (98.4°C) = 0.205 cP, σ (98.4°C) = 12.5 dynes/cm.

D3. The calculations in Example 10-3 were done for conditions at the top of the column. Physical properties will vary throughout the column, but columns are normally constructed with identical trays, downcomers, weirs, etc., on every stage (this is simpler and cheaper). For a 12-foot diameter column, calculate entrainment, pressure drops, downcomer residence time, and weeping at the bottom of the column. The results of Problem 10.D2 are required. If the column will not operate, will it work if the gap between the tray and downcomer apron is increased to 1.5 inches?

D4.* We wish to repeat the distillation in Examples 10-2 and 10-3 except that valve trays will be used. The valves have a 2-inch diameter head. For the top of the column, estimate the pressure drop vs. hole velocity curve. Assume that K_v and C_v values are the same as in Figure 10-21. Each valve weighs approximately 0.08 pound.

D5.* We are testing a new type of packing. A methanol-water mixture is distilled at total reflux and a pressure of 101.3 kPa. The packed section is 1 meter long. We measure a concentration of 96 mol% methanol in the liquid leaving the condenser and a composition of 4 mol% methanol in the reboiler liquid. What is the HETP of this packing at this gas flow rate? Equilibrium data are in Table 2-7.

D6.* We are testing a new packing for separation of benzene and toluene. The column is packed with 3.5 meters of packing and has a total condenser and a partial reboiler. Operation is at 760 mm Hg, where α varies from 2.61 for pure benzene to 2.315 for pure toluene (Perry et al., 1963, p. 13-3). At total reflux we measure a benzene mole fraction of 0.987 in the condenser and 0.008 in the reboiler liquid. Find HETP:
   a. Using $\alpha = 2.315$
   b. Using $\alpha = 2.61$
   c. Using a geometric average α.
   Use either the Fenske equation or a McCabe-Thiele diagram.

D7. You have designed a sieve tray column with 0.3048 m tray spacing to operate at a pressure of 1.0 atm. The value of the flow parameter is $F_{lv} = 0.090$ and the flooding velocity was calculated as 1.83 m/s. Unfortunately, your boss dislikes your design. She thinks that 0.3048 m tray spacing is not enough and that your reflux ratio is too low. You must redesign for a 0.6096 m tray spacing and increase L/V by 11%. Estimate the new flooding velocity.
   Assumptions: Ideal gas, σ, ρ_L, and ρ_G are unchanged.

D8. You have designed a sieve tray column with 0.4572 m tray spacing to operate at a pressure of 2.0 atm. The value of the flow parameter is $F_{lv} = 0.5$, and the flooding velocity was calculated as $u_{flood} = 1.83$ m/s. Unfortunately, the latest lab results show that the distillation temperature is too high and unacceptable thermal degradation occurs. To reduce the operating temperature, you plan to reduce the column pressure to 0.5 atm. Estimate the new flooding velocity.
   Assumptions: Ideal gas, σ and ρ_L are unchanged, $\rho_L \gg \rho_V$ (thus, $\rho_L - \rho_V = \rho_L$). Ignore the effect of temperature change in calculation of ρ_V.

D9.* We are distilling methanol and water in a sieve plate column operating at 75% of flooding velocity. The distillate composition is 0.999 mole fraction methanol. The bottoms composition is 0.01 mole fraction methanol. The column operates at 1.0 atmosphere pressure. Use L/V = 0.6. The flow rate of the feed is 1000.0 kmol/h. The feed is a saturated vapor that is 60 mol% methanol. Use an 0.4572 m tray spacing and $\eta = 0.90$. Density of pure liquid methanol is 0.79 g/ml. Data are available in Table 2-7. Assume an ideal gas. The surface tension of pure methanol can be estimated as $\sigma = 24.0 - 0.0773$ T with T in °C (Dean, 1985, p. 10–110). Calculate the diameter based on the conditions at the top of the column.

D10. The effect of liquid maldistribution in packed columns can be explored with a McCabe-Thiele diagram. Assume that a packed distillation column is separating a saturated liquid binary feed that is 40.0 mol% MVC. A distillate product, D = 100.0 kmol/h, that is 90% MVC is desired. Relative volatility = 3.0 and is constant. We operate at an $L/D = 2(L/D)_{min}$. If there is liquid maldistribution, the actual L/V represents an average for the entire column. Assume that vapor is equally distributed throughout the column, but there is more liquid on one side than the other. The slope of the operating line on the low liquid side will be L_{low}/V not $(L/V)_{avg}$. If the low liquid flow rate is small enough, the operating line on the low side will be pinched at the feed point, and the desired separation will not be obtained. What fraction of the average liquid flow rate must the low liquid flow rate be to just pinch at the feed concentration? Then generalize your result for $L/D = M(L/D)_{min}$, where $M > 1$.

D11.* We wish to distill an ethanol-water mixture to produce 2250 lb of distillate product per day. The distillate product is 80 mol% ethanol and 20 mol% water. An L/D of 2.0 is to be used. The column operates at 1 atm. A packed column will use 5/8-inch plastic Pall rings. Calculate the diameter at the top of the column.

Physical properties: $MW_E = 46$, $MW_W = 18$, assume ideal gas, $\mu_L = 0.52$ cP at 176°F, $\rho_L = 0.82$ g/ml.

a. Operation is at 75% of flooding. What diameter is required?

b. Operation is at a pressure drop of 0.25 inches of water per foot of packing. What diameter is required?

c. Repeat part a but for a feed that is 22,500 lb of distillate product per day. Note: It is *not* necessary to redo the entire calculation, since D and hence V and hence diameter are related to the feed rate.

D12.* A distillation system is a packed column with 1.524 m of packing. A saturated vapor feed is added to the column (which is only an enriching section). Feed is 23.5 mole % water with the remainder nitromethane. $F = 10$ kmol/h. An L/V of 0.8 is required. $x_D = x_\beta = 0.914$. Find HETP and water mole fraction in bottoms. Water-nitromethane data are given in Problem 8.E1.

D13. Repeat Problem 10.D9 calculating the diameter at the top, but condense the feed to a saturated liquid. Adjust the reflux ratio so that the multiplier M used to determine $L/D = M(L/D)_{min}$ is the same for the saturated vapor feed (in 10.D9) and the saturated liquid feed in this problem. Compare the diameter of this column to the diameter of 10.27 feet calculated for Problem 10.D9. Calculate the ratio $Q_{R,liquid\_feed} / Q_{R,vapor\_feed}$ for this problem.

D14. a. We are distilling methanol and water in a column packed with 1-inch ceramic Berl saddles. The bottoms composition is 0.0001 mole fraction methanol. The column operates at 1 atm pressure. The feed to the column is a saturated liquid at 1000.0 kmol/day and is 40 mol% methanol. An $L/D = 2.0$ will be used. The distillate product is 0.998 mole fraction methanol. If we will operate at a vapor flow rate that is 80% of flooding, calculate the diameter based on conditions at the bottom of the column. Data are available in Tables 2-7 and 10-3. You may assume the vapors are an ideal gas. Viscosity of water at 100°C is 0.26 cP.

b. Suppose we wanted to use 1-inch plastic Intalox saddles. What diameter is required?

Note: Part b can be done in one line once part a is finished.

D15. Repeat Problem 10.D14a except determine the diameter of a sieve plate column operating at 80% of flooding velocity. Use a 0.3048 m tray spacing and $\eta = 0.85$. The liquid surface tension of pure water is $\sigma = 58.9$ dynes/cm at 100°C.

D16.* Repeat Example 10-4, except use 3-inch Intalox saddles.

D17.* Repeat Example 10-4, except calculate the diameter at the bottom of the column.

D18. You have designed a packed column at 1.0 atm. The flow parameter F_{lv} has a value of 0.2. The calculated gas flux at flooding is 0.50 lbm/[(sec)(ft²)]. Your boss now wants to increase the column pressure to 4.0 atm. Assume that the vapor in the column follows the ideal gas law. What is the new gas flux at flooding?

D19. Repeat Problems 10.D9 and 10.D13 using the two-enthalpy feed method in which a portion of the feed is condensed to a saturated liquid and the remainder is still a saturated vapor. Select the amount of feed to condense so that the column has pinch points at minimum reflux at both the liquid and the vapor feed simultaneously. Use the same multiplier M as in Problem 10.D13. Compare the diameter of this column to the diameter of 10.27 feet calculated for Problem 10.D9 and to the diameter calculated for Problem 10.D13. Calculate the ratio $Q_{R,2-enthalpy\_feed} / Q_{R,vapor\_feed}$ for this problem.

D20. *Do after studying Chapter 12.* If the column uses sieve plates, what column diameter is required for the absorber in Problem 12.D14? Operate at 75% of flood. Use a 0.6096 m tray spacing. Assume $\eta = 0.85$. The density of liquid ammonia is approximately 0.61 gm/ml. Assume that nitrogen is an ideal gas. Note that you will have to extrapolate the graph or the equation for 24-inch tray spacing to find C_{sb}. Since surface tension data are not reported, assume that $\sigma = 20$ dynes/cm. Watch your units!

D21. Repeat Problem 10.D14, except calculate the diameter at the top of the column.

E. *More Complex Problems*

E1. You need a solution to Problem 9.D23 (batch distillation) for this problem.
The batch distillation column in Problem 9.D23 is a 6-inch diameter packed column that is packed with 5/8-inch metal Pall rings. Operate at a vapor flux that is 70% of flooding. Design for conditions at the end of the batch calculated at the bottom of the column. Estimate the viscosity as that of pure water at 100°C (0.26 cP). Find the operating time for the batch distillation with the packed column.

F. Problems Requiring Other Resources

F1. Calculate the expected overall efficiency for the column in Problem 10.D9. Viscosities are available in *Perry's Chemical Engineers Handbook* (7th Edition, p. 2–322 and 2–323). Note that O'Connell's correlation uses the liquid viscosity of the same composition as the feed at the average temperature of the column. Estimation of the viscosity of a mixture is shown in Eq. (10-7a).

F2. Estimate the overall plate efficiency for Problem 10.D15. See Problem 10.F1 for hints.

F3.* We are separating an ethanol-water mixture in a column operating at atmospheric pressure with a total condenser and a partial reboiler. Constant molal overflow (CMO) can be assumed, and the reflux is a saturated liquid. The feed rate is 100 lbmol/h of a 30 mole % ethanol mixture. The feed is a subcooled liquid, and 3 moles of feed will condense 1 mole of vapor at the feed plate. We desire an $x_D = 0.8$, $x_B = 0.01$ and use $L/D = 2.0$. Use a plate spacing of 0.4572 m. What diameter is necessary if we will operate at 75% of flooding? How many real stages are required, and how tall is the column?

The downcomers can be assumed to occupy 10% of the column cross-sectional area. Surface tension data are available in Dean (1985) and in the *Handbook of Chemistry and Physics*. The surface tension may be extrapolated as a linear function of temperature. Liquid densities are given in *Perry's*. Vapor densities can be found from the perfect gas law. The overall efficiency can be estimated from the O'Connell correlation. Note that the diameter calculated at different locations in the column will vary. The largest diameter calculated should be used. Thus, you must either calculate a diameter at several locations in the column or justify why a given location will give the largest diameter.

F4.* Repeat Problem 10.F3, except design a packed column using 1-inch metal Pall rings. Approximate HETP for ethanol-water is 0.366 m.

G1. [Note: This problem is quite extensive.] A saturated vapor feed at 1000 kmol/h of methanol (5 mol%) and water (95 mol%) is fed to a distillation column with 18 stages plus a kettle reboiler and a total condenser (N = 20 in Aspen Plus notation). Use the NRTL VLE correlation. Operate at 80% of flooding using Fair's diameter calculation method and a tray spacing = 0.4572 m. Use an external reflux ratio of 24. Pressure is 1.0 atm.

Calculate the optimum feed plate location, product purities, Q_c and Q_R, and the column diameters at different locations for the base case. Then determine which of the methods in Figure 10-18 will do the best job of balancing the diameters and reducing the column volume with constant product purities, Q_c and Q_R. Report the results for your system with the largest volume reduction.

G2. [Note: This problem is quite extensive.] Biorefineries producing ethanol by fermentation have several distillation columns to separate the ethanol from the water. The first column, the beer still, is a stripping column that takes the dilute liquid fermenter product containing up to 15% solids and produces a clean vapor product that is sent to the main distillation column. The main column produces a distillate product between about 65 mole % and the ethanol azeotrope, and a bottoms product with very little ethanol. The calculated diameter of the main distillation column is much greater at the top than elsewhere. To reduce the size and hence the cost of the main column, one can use a two-enthalpy feed system: split the vapor feed into two parts and condense one part, then feed both parts to the main column at their optimum feed locations. This method reduces the vapor velocity in the top of the column, which reduces the calculated diameter; however, a few additional stages may be required to obtain the desired purity.

Simulate the following base case and two-enthalpy feed system with Aspen Plus using NRTL VLE correlation. The feed for the base case is 1000 kmol/h of a 10 mol% ethanol saturated vapor stream. The feed and the column are at 2.5 atm. A vapor distillate product from a partial condenser with y_D = 79.01 mol% ethanol and a liquid bottoms product with 99.86 mol% water are desired. The column has 34 equilibrium stages plus a partial condenser and a kettle reboiler (N = 36 in Aspen Plus notation). Tray spacing = 18 inches = 0.4572 meters, operation is at 80% of flooding and the flooding design method developed by Jim Fair was used. For the base case (all feed is a vapor) the feed stage is 23 (in Aspen Plus notation). Q_R = 902 kW, and D = 125 kmol/h.

For the two-enthalpy feed case, 600 kmol/h of the feed is condensed to a saturated liquid and is fed to the column on stage 17. Vapor feed remains on stage 23. Other parameters are unchanged from the base case.

a. Use Aspen Plus to simulate the base case and calculate the product mole fractions and the column diameter.

b. Use Aspen Plus to simulate the two-enthalpy feed system and calculate the product mole fractions and the column diameter.

c. To explain the reduction in column diameter, assume constant molal overflow and calculate V in the top of the column for both cases. Then explain, using the equations in Chapter 10, how this will reduce the diameter.

G3. We wish to distill 0.10 kmol/s of a feed at 25°C and 15.0 atm. The feed is 0.100 mole fraction ethane, 0.350 mole fraction propane, 0.450 mole fraction n-butane, and 0.100 mole fraction n-pentane. Use the Peng-Robinson VLE correlation. Design a column with N (Aspen notation) = 35 and the feed on stage 16. The column operates at 15.0

atm, has a partial condenser, and produces a vapor distillate with D = 0.0450 kmol/s. A kettle type reboiler is used. Reflux ratio is L/D = 2.4.

a. Do tray sizing using the Fair flooding method with 70% flooding, 0.6 m distance between plates, and default values for the other tray parameters. Find the largest column diameter and mole fractions of distillate (vapor) and bottoms.

b. Do tray rating for this single-section column with one pass, with weir height of 0.0508 m, same diameter calculated in Part a, and default values for other parameters. Leave the Downcomers tab blank. Report the worst downcomer backup (amount and stage) and the weir loading m²/s.

c. Design a two-pass system and compare to the one-pass system.

G4. **a.** Repeat Problem 10.G3 except use the method shown in Figure 10-19B to partially balance the column diameters. The liquid and vapor feeds have the same mole fractions as the feed in Lab 10, and both are at 15.0 atm. The liquid feed is at 25°C, and the vapor feed is a saturated vapor. Input the liquid feed on stage 15. Input the vapor feed as a vapor on stage 18. Treat the column as a single section with one pass. Vary the vapor feed and liquid feed and record the values in the following table.

V feed kmol/s	L feed kmol/s	Q_c kW	Q_R kW	Max Dia m	Stage max dia	$y_{D,C4}$	$x_{B,C3}$
0*	.1 (N_F =16)						
0 part d	.1 (N_F = 15)						
.01	.09						
.02	.08						
.03	.07						
.04	.06						
.05	.05						
.06	.04						
Part b .03	.07 (N=41)						

* Values from Problem 10.G3.

Note that Q_R decreases, but since energy will be required to vaporize the feed the net change in heating requirement is negligible. The maximum diameter decreases, but the purity of the distillate and the bottoms decreases.

b. In order to obtain a purer product, increase N (Aspen notation) to 41, use $N_{F,liq}$ = 18 and $N_{F,vap}$ = 21. Then repeat the run with 0.03 kmol/s of vapor feed and 0.07 kmol/s of liquid feed. Compare with the all-liquid feed run. How much volume decrease is there in the column?

c. Do a tray rating run with the diameter calculated in part b and defaults for tray spacing and for the downcomer. Record the maximum downcomer backup and the tray it occurs at and the maximum weir loading and the tray it occurs at.

d. Since Problem 10.G3 has not been optimized, try feed of the liquid on stage 15 with no vapor feed. Record the maximum downcomer backup and the tray it occurs at and the maximum weir loading and the tray it occurs at.

CHAPTER 10 APPENDIX. TRAY AND DOWNCOMER DESIGN WITH COMPUTER SIMULATOR

This appendix shows how the Aspen Plus simulator can be used to do detailed tray and downcomer design.

Lab 10. Aspen Plus uses RADFRAC with the Tray Rating option to do detailed tray and downcomer design. Start by setting up the problem below with RADFRAC.

1. We wish to distil a feed of 1.0 kmol/s of a feed at 25°C and 15.0 atm. The feed is 0.100 mol fraction ethane, 0.300 mole fraction propane, 0.500 mole fraction n-butane and 0.100 mole fraction n-pentane. Use the Peng-Robinson VLE correlation. Design a column with N (Aspen notation) = 35 and the feed on stage 16. The column operates at 15.0 atm, has a partial condenser and produces a vapor distillate with D = 0.400 kmol/s. A kettle type reboiler is used. Reflux ratio is L/D = 2.5.*

2. In the Tray Sizing section divide the column into two sections. (When you click on Tray Sizing and see the Object Manager, click on New and call this section 1. Once you have completed tray sizing section 1, click on Tray Sizing again, in the Object Manager click on New and call this section 2.) Section 1 is stages 2 to 15 and section 2 is stages 16 to 34. In both sections use sieve plates with the Fair flooding calculation method with flooding at 70%. For other variables initially use the default numbers for the variables in both sections.

3. Click Next and run the simulation and look at the Report (View→Report→check block). Note that the two sections have different diameters. If the column was designed with a single section it would all have to be the larger diameter. [Note: the methods in Section 10.4 might allow design of a column with constant diameter between the two values calculated.]

4. Now click on Tray Rating. In the Object Manager Click on New and call this section 1. Input the same information for section 1 as you did for section 1 in Tray Sizing. Use the same diameter as you calculated previously for section 1. Use a weir height of 0.050 m. On the Design/Pdrop tab pick the Fair flooding calculation method and do NOT check Update section pressure drop. In the Layout tab the tray type is sieve and use the defaults for the other items. At this point do not put any numbers into the Downcomers tab (this means default values will be used). Return to Tray Sizing→Object Manager→New→call it section 2. Put in same values as for section 1, except use the diameter that was calculated for section 2.

5. Click Next and run the simulation and look at the Report (View→Report→check block). Note that the report now contains a Tray Rating section after the Tray Sizing section. Look at the Downcomer (DC) backup for both sections. Section 1 should be OK (backup < tray spacing), but section 2 is not OK. Section 2 would flood due to excessive downcomer backup and will not operate. [The flooding calculation done in tray sizing is for entrainment flooding—that calculation does not consider flooding caused by downcomer backup.]

6. We must adjust section 2 so that the downcomer does not backup excessively. This can be done by increasing the size of the downcomer and/or increasing the tray spacing. First, in Tray Sizing for section 2, in the Design tab change the Minimum Downcomer Area (fraction of total tray area) to 0.15. Rerun the simulation. Look at the Report for

　　　　　　　　　　　　　　　Chapter 10   Staged and Packed Column Design

both tray sizing and tray rating for section 2. Note that the tray sizing diameter has changed, but (since you input the value) the diameter in tray rating has not changed. Change the diameter in the tray rating for section 2, rerun the simulation, open the block report, and look at the downcomer backup in section 2. It should be better, but not good enough.

7. Now try increasing the tray spacing in section 2 to 0.700 m. Do this in both tray spacing and tray rating (although tray rating is the critical location). Run the simulation again, obtain the block report, and look at section 2 in both tray sizing and tray rating parts. Note that the diameter required in tray sizing has dropped. Keep the diameter you have from step 6 in tray sizing. The downcomer backup is better, but probably not good enough.

8. Increase the downcomer dimensions in tray rating for section 2 (use the Downcomer tab in tray rating) and in Downcomer Geometry list a side Clearance of 0.06 m. Run the simulation again, obtain the block report, and look at section 2 in both tray sizing and tray rating parts. Note that the tray sizing still lists 0.15 for Minimum Downcomer Area (fraction of total tray area). *The tray sizing and tray rating sections do not communicate with each other.* Look at the downcomer backup in section 2. It now should be OK. Report the distillate (remember it is a vapor) and bottoms mole fractions, the diameters and largest percentages of flooding and largest fractional downcomer backups in both sections 1 and 2.

9. Now we need to look at a number of other aspects based on rules of thumb for design (Torzewski, 2009).

   A. For column diameter > 3.0 m need tray spacing > 0.6 m. OK.

   B. For diameter > 3.0 m can use up to four passes. The number of tray passes depends on the weir loading. The maximum weir loading is 70 m^3 of liquid/(h m weir length) = 0.01944 m^2/s. Both of the sections have weir loadings that are too high. We must increase the number of passes.

   C. The downercomer clearance < weir height. This is violated by section 2.

10. Increase both sections 1 and 2 to two passes. Use section 2 downcomer clearances (both downcomers) of 0.05 m and increase weir height to 0.052 m. Run the program, look at the downcomer backup, and compare your results with the items in item 9. The weir loading is still too high.

11. Increase section 2 to four passes and run again. Look at the downcomer backup, and compare your results with the items in item 9. The weir loading is still a bit too high, but everything else is fine.

12. Since the downcomer backups are now quite low in section 2, reduce the tray spacing to 0.6096 m, and use a 0.05 m weir with the default value 0.0373 m for downcomer clearance. The resulting design is satisfactory except for a slightly high weir loading. Probably accept this as an initial design.

   Note that this column is not optimized, but is a workable design.

* This problem was adapted from Example 14.3.1 in R. Taylor and R. Krishna, *Multicomponent Mass Transfer,* Wiley, New York (1993).

CHAPTER 11

Economics and Energy Conservation in Distillation

There are an estimated 40,000 distillation columns in the United States, which have a combined capital value in excess of $8 billion. These columns are estimated to have at least a 30-year life, and they are used in more than 95% of all chemical processes. The total energy use of these columns is approximately 3% of total U.S. energy consumption (Humphrey and Keller, 1987). Thus, estimating and, if possible, reducing capital and operating costs of distillation are important. Because they are a major energy user, saving energy in distillation systems is particularly important in a time of high and uncertain energy costs.

11.1 DISTILLATION COSTS

Now that we have considered the design of the entire column we can explore the effect of design and operating parameters on the cost of operation. A brief review of economics will be helpful (for complete coverage, see a design or economics text such as Peters et al., 2003; Seider, et al., 2009; Sullivan et al., 2006; Turton et al., 2009; Ulrich, 1984; or Woods, 1976).

Capital costs can be determined by estimating delivered equipment costs and adding on installation, building, piping, engineering, contingency, and indirect costs. These latter costs are often estimated as a factor times the delivered equipment cost for major items of equipment.

$$\text{Total capital cost} = (\text{Lang factor})(\text{delivered equipment cost}) \qquad \textbf{(11-1)}$$

where the Lang factor ranges from approximately 3.1 to 4.8. The Lang factor for a plant processing only fluids is 4.74 (Turton et al., 2009). Thus, these "extra" costs greatly increase the capital cost.

Costs of major equipment are often estimated from a power law formula:

$$\text{Cost for size A} = (\text{cost size B})(\frac{\text{size A}}{\text{size B}})^{\text{exponent}} \qquad \textbf{(11-2)}$$

Some equipment will not follow this power law. The appropriate size term depends on the type of equipment. For example, for shell and tube heat exchangers such as condensers, the size used is the area of heat exchanger surface. The exponent has an "average" value of 0.6, but varies widely. For shell and tube heat exchangers the exponent has been reported as 0.41 (Seider et al., 2009), 0.59 (Turton et al., 2009), and 0.48 (Rudd and Watson, 1968). Thus,

$$\text{Condenser cost, size A} = (\text{cost, size B})(\frac{\text{area A}}{\text{area B}})^{0.41 \text{ to } 0.59} \tag{11-3}$$

As the area becomes larger, the cost per square meter decreases.

The exponent in Eq. (11-2) is usually less than 1. This means that as size increases, the cost per unit size decreases. This will be translated into a lower cost per kilogram of product. This "economy of scale" is the major reason that large plants have been built in the past. However, there is currently a trend toward smaller, more flexible plants that can change when the economy changes.

The current cost can be estimated by updating published sources or from current vendors' quotes. The method for updating costs is to use a cost index.

$$\text{Cost at time 2} = (\text{cost at time 1})(\frac{\text{index, time 2}}{\text{index, time 1}}) \tag{11-4}$$

The Marshall and Stevens equipment cost index or the *Chemical Engineering* magazine plant cost index are usually used. Current values are given in each issue of *Chemical Engineering* magazine. Values for several years' cost indices are tabulated in Table 11-1. The total uninstalled equipment cost will be the sum of condenser, reboiler, tower casing, and tray costs. The total capital cost is then found from Eq. (11-1). The capital cost per year equals the capital cost times the depreciation rate. Note that cost estimates can easily vary by up to 35%.

The final bare module cost C_{BM} of distillation equipment can be estimated from charts and equations. The calculation starts with the base purchase cost C_p for systems built of carbon steel and operating at ambient pressure, and then adjusts this cost with factors for additional costs. The charts that we need to estimate the base purchase cost C_p of distillation systems are shown in Figures 11-1 to 11-3 (Turton et al., 2003). Since the shell of distillation columns is a vertical vessel, Figure 11-1 can be used to determine $C_p = C_p^\circ \times (\text{Volume})$ for towers. The base purchase costs for sieve and valve trays and tower packing are shown in Figure 11-2, $C_p = C_p^\circ \times (\text{Area})$ for trays and $C_p = C_p^\circ \times (\text{Volume})$ for packing. In the heat exchanger costs (Figure 11-3), distillation systems commonly use kettle type reboilers and fixed tube sheet shell and tube exchangers, $C_p = C_p^\circ \times (\text{Area})$. The Chemical Engineering Plant Cost Index (CEPCI) listed on these figures is the CEPCI at the time the figures were prepared.

All types of equipment are affected by increased costs due to expensive materials, and the material factors F_m for these increased costs in distillation systems are given in Table 11-2 (Turton et al., 2003). Pressure effects on the costs of process vessels and heat exchangers are included through the pressure factor F_p. The pressure factor for process vessels is given by (Turton et al., 2003),

$$F_p = \frac{pD}{(10.71 - 0.00756p)} + 0.5 \tag{11-5}$$

TABLE 11-1. *Values of Cost Indices. Values are averages for entire year, which is approximately the June-July value. The base year (index value of 100) of the Chemical Engineering Plant Cost Index is 1957-59 and of the Marshall & Swift Index is 1926.*

Year	Chemical Engineering Plant Cost Index	Marshall & Swift Equipment Cost Index
1975	182	452
1980	261	675
1985	325	813
1990	358	935
1995	381	1037
2000	394	1103
2001	395	1110
2002	396	1104
2003	402	1124
2004	444	1179
2005	468	1245
2006	500	1302
2007	525	1373
2008	575	1449
2009	522	1469
2010	556 (June)	1473 (3rd quarter)

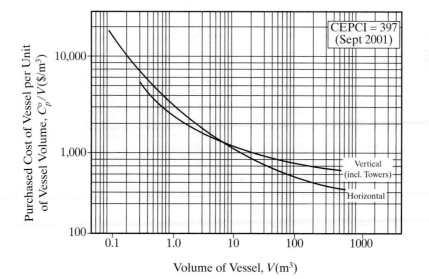

FIGURE 11-1. *Purchased costs of process vessels (Turton et al., 2003), reprinted with permission, copyright 2003 Prentice Hall*

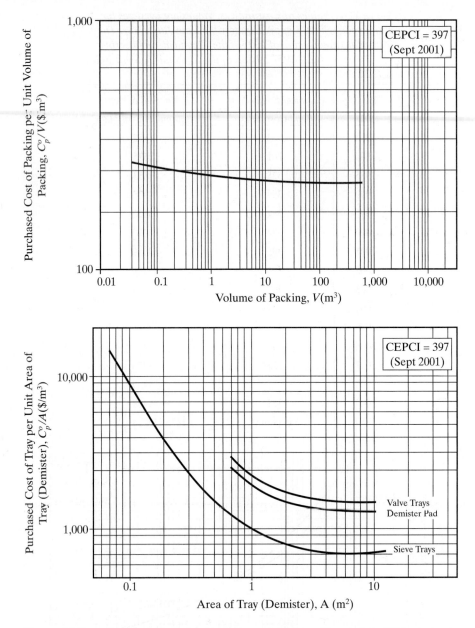

FIGURE 11-2. *Purchased costs of packing, trays, and demisters (Turton* et al.*, 2003), reprinted with permission, copyright 2003 Prentice Hall*

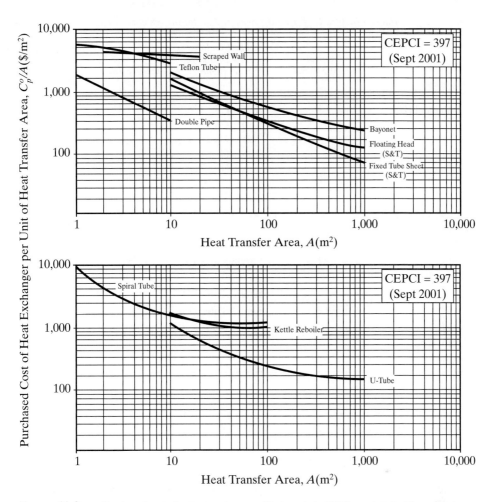

FIGURE 11-3. *Purchased costs for heat exchangers (Turton et al., 2003), reprinted with permission, copyright 2003, Prentice Hall.*

which is valid for a vessel wall thickness >0.0063 m. In this equation p = absolute pressure in bars and D = vessel diameter in meters. If F_p < 1 set F_p = 1.0. For vacuum operation with p < 0.5 bar, use F_p = 1.25. Further limitations on the use of this equation are discussed by Turton et al. (2003, p. 925). The pressure factor for condensers and kettle type reboilers can be determined from (Turton et al., 2003),

$$\log_{10} F_p = 0.03881 - 0.11272 \log_{10}(p - 1) + 0.08183[\log_{10}(p - 1)]^2 \qquad \textbf{(11-6)}$$

In this equation p = absolute pressure in bars and the range of validity of the equation is from 6 < p < 141 bar. If p ≤ 6, set F_p = 1.0.

Sieve and valve tray and packing costs do not depend directly on pressure although they depend indirectly on pressure since column diameter depends on pressure. Sieve tray costs

TABLE 11-2. *Material factors F_m for equipment in distillation systems (Turton et al., 2003)*

Material of Construction	Vertical Process Vessels F_m	Shell-&-Tube Heat Exchange & Kettle Reboilers F_m	Sieve Trays F_m	Tower Packings F_m
Carbon steel (CS)	1.0	Shell=CS Tube=CS 1.0 Tube=Cu 1.3 Tube=SS 1.8 Tube=Ni 2.7 Alloy Tube=Ti 4.7	1.0	1.0
Stainless steel (SS)	3.1	Shell&Tube 2.7	1.8	
SS clad	1.7			
304 SS				7.1
Ni alloy	7.1	Shell&Tube 3.7	5.6	
Ni alloy clad	3.6			
Cu		Shell&Tube 1.7		
Ti	9.4	Shell&Tube 11.4		
Ti Clad	4.7			
Polyethylene				1.0
Ceramic				4.1

depend on the number of trays ordered through the quantity factor F_q, which can be calculated from the following equation (Turton et al., 2003).

$$\log_{10} F_q = 0.4771 + 0.08516\log_{10} N - 0.3473(\log_{10} N)^2 \qquad \textbf{(11-7)}$$

where N is the number of trays. This equation is valid for N < 20. For N ≥ 20 set $F_q = 1.0$. Packing costs also depend on the amount of packing, which is inherently included in Figure 11-2.

The final bare module cost C_{BM} for vertical process vessels is given by (Turton et al., 2003),

$$C_{BM} = C_p(2.25 + 1.82F_m F_p) \qquad \textbf{(11-8)}$$

For condensers and kettle type reboilers the final bare module cost is,

$$C_{BM} = C_p(1.63 + 1.66F_m F_p) \qquad \textbf{(11-9)}$$

For sieve and valve trays the final bare module cost is determined from,

$$C_{BM} = C_p N F_q F_m \qquad \textbf{(11-10)}$$

where N = number of trays. For packings the final bare module cost is

$$C_{BM} = C_p F_m \qquad \textbf{(11-11)}$$

Chapter 11 Economics and Energy Conservation in Distillation

The use of these graphs and equations is illustrated in Example 11-1.

Since the cost per-volume does not vary much, Seider et al. (2009) report a single value for installed costs per unit volume of dumped packings based on the packing size, type, and materials of construction. As expected, the cost per volume decreases for larger-size packings (e.g., carbon steel Pall rings were \$49/ft^3 for 1 inch and \$32/ft^3 for 2 inch). Costs depend greatly on the material of construction. For example, for 1-inch Pall rings costs (in 2006 with CEPCI = 500) ranged from \$37/ft^3 for polypropylene to \$49 for carbon steel to \$168 for stainless steel. Note that the ratio of costs for stainless steel packing compared to carbon steel is 3.4, while Table 11-2 lists $F_m = 1.0$ for carbon steel packing and $F_m = 7.1$ for stainless steel, which is a ratio of 7.1. This type of difference is not unusual for cost data from different sources. The installed cost of structured packing varies greatly. Seider et al. (2009) suggest \$250/ft^3 as a rough average installed cost for stainless steel corrugated-sheet structured packing.

The total operating costs per year can be determined as

$$\text{Total operating cost } \$/\text{yr} = [(\text{kg steam/h})(\text{cost of steam, \$/kg})$$

$$+ (\text{kg water/kw})(\text{cost water, \$/kg}) + (\text{kw elec./h})(\text{cost elec./kw})$$

$$+ (\text{labor costs/h})(\text{labor h/h operation})] \times [\text{hours operation/yr}] \qquad \textbf{(11-12)}$$

For most continuous distillation columns, the electricity costs are modest, and the labor costs are the same regardless of the values of operating variables.

A useful rule of thumb is the annual operating cost of a distillation system is half operating costs and half capital costs (Keller, 1987).

11.2 OPERATING EFFECTS ON COSTS

We can use the methods developed in Chapters 4, 6, and 7 to calculate the number of equilibrium stages, N_{equil}, required. Then $N_{actual} = N_{equil}/E_o$. The height of a staged column is

$$H = (N_{actual})(\text{tray spacing}) + \text{disengagement heights} \qquad \textbf{(11-13)}$$

The column diameter is found using the methods in Chapter 10. Equation (10-16) shows that for higher pressures the diameter will be somewhat reduced. Conversely, for vacuum operation the diameter will be increased. Increases in tray spacing increase $C_{sb, flood}$, which also increases u_{op}, and thus, the diameter will decrease while the column height increases.

As $L/D \rightarrow \infty$ (total reflux), the number of stages approaches a minimum that minimizes the column height, but the diameter goes to infinity. As $L/D \rightarrow (L/D)_{min}$, the number of stages and the height become infinite while the diameter becomes a minimum. Both these limits will have infinite capital costs. Thus, we expect an optimum L/D to minimize capital costs. Column height is independent of feed flow rate, while diameter is proportional to $F^{1/2}$ and $(1 + L/D)^{1/2}$.

Pressure effects on distillation columns are extremely important and were discussed previously in Section 3.3. At first glance Equation (10-16) shows that the diameter of staged columns is proportional to $(1/p)^{1/2}$. However, as p increases, T also increases and u_{flood} decreases. Thus, the effect is less than $(1/p)^{1/2}$. Operating at higher pressures reduces the column diameter, although the height may increase because relative volatility typically decreases as pressure increases. Usually the effect of pressure on diameter is significantly larger than its effect on column height.

Based on Figure 11-1 we would expect the base purchased cost of the column to decrease. Typically the pressure factor $F_p = 1.0$ below approximately 6 bar and then rises modestly for pressures below about 20 bar. If the column is made of an expensive material (high F_m), cost factors for pressure increases are accelerated since the bare module factor depends on the product of F_p and F_m, Eq. (11-8). Since sieve tray purchase cost C_p decreases as diameter decreases (Figure 11-2), we would expect the purchased cost per tray to drop although a few more trays may be required. Keller (1987) reported on detailed economic analyses and found that the net result is that bare module costs of distillation columns (shell and trays) generally decrease up to a pressure of approximately 6.8 atm (100 psia). This result assumes that there is no thermal degradation at these pressures. Compounds that degrade must often be processed under vacuum (at lower pressures the boiling points and hence column temperatures are lower) even though the columns are more expensive. If the chemicals being distilled are flammable, special safety precautions must be taken to prevent air from leaking into vacuum columns (Biegler et al., 1997). The effect of pressure on purchase costs above 6.8 atm need to be determined on a case-by-case basis.

Condenser and reboiler sizes depend on Q_c and Q_R. These values can be determined from external mass and energy balances around the column. Equations (3-14) and (3-16) allow us to calculate Q_c and Q_R for columns with a single feed. Q_c is proportional to $(1 + L/D)F$ and Q_R is proportional to F and increases linearly with $(1 + L/D)$. [The subtle distinction between proportional and linearly dependent can be important.] The amount of cooling water is determined from an energy balance.

$$\text{kg cooling water / h} = |Q_c| / (C_{p,w} \Delta T_w) \tag{11-14}$$

where $\Delta T_w = T_{w,hot} - T_{w,cold}$. Seider et al. (2009) recommend assuming $T_{w,cold} = 90°F$ and $T_{w,hot} = 120°F$. A water condenser can easily cool to 100°F (O'Brien and Schultz, 2004). The cooling water cost per year is

$$\text{Cooling water cost, \$/yr} = (\text{kg water /h})\left(\frac{\text{cost, \$}}{\text{kg}}\right)\left(\frac{\text{h}}{\text{yr}}\right) \tag{11-15}$$

This cost is proportional to $(1 + L/D)F$ if cost per kilogram is constant.

In the reboiler the steam is usually condensed from a saturated or superheated vapor to a saturated liquid. Then the steam rate is

$$\text{Steam rate, kg/h} = \frac{Q_R}{H_{steam} - h_{liquid}} \tag{11-16a}$$

In many applications, $H_{steam} = H_{saturated\ vapor}$, and

$$\text{Steam rate, kg/h} = \frac{Q_R}{\lambda} \tag{11-16b}$$

where λ is the latent heat of vaporization of water at the operating pressure. The steam rate is proportional to F and increases linearly with $(1 + L/D)$. The value of λ can be determined from the steam tables. Then the steam cost per year is

$$\text{Steam cost, \$/yr} = (\text{kg steam/h})\left(\frac{\text{cost, \$}}{\text{kg steam}}\right)\left(\frac{\text{h operation}}{\text{yr}}\right) \tag{11-17}$$

At higher pressures, λ decreases and the cost/kg steam increases. Thus, higher pressures will probably have a modest increase in steam cost.

Because Q_c and Q_R are proportional to feed rate F, doubling F doubles cooling water and steam rates. Q_c is also proportional to $(1 + L/D)$. Q_R is linearly dependent on $(1 + L/D)$, but is not proportional to $(1 + L/D)$.

Although detailed design of condensers and particularly reboilers is specialized (Ludwig, 2001; McCarthy and Smith, 1995) and is beyond the scope of this book, an estimate of the heat transfer area is sufficient for preliminary cost estimates. The sizes of the heat exchangers can be estimated from the heat transfer equation

$$| Q | = UA \, \Delta T \qquad\qquad (11\text{-}18)$$

where U = overall heat transfer coefficient, A = heat transfer areas, and ΔT is the temperature difference between the fluid being heated and the fluid being cooled. Use of Eq. (11-18) is explained in detail in books on transport phenomena and heat transfer (e.g., Chengel, 2003; Geankoplis, 2003; Greenkorn and Kessler, 1972; Griskey, 2002; Kern, 1950; Ludwig, 2001). For condensers and reboilers, the condensing fluid is at constant temperature. Then

$$\Delta T = \Delta T_{avg} = T_{hot} - T_{cold,avg} \qquad\qquad (11\text{-}19)$$

where T_{hot} = condensing temperature of fluid or of steam, and $T_{cold,avg} = (1/2)(T_{cold,1} + T_{cold,2})$. For a reboiler, the cold temperature will be constant at the boiling temperature, and $\Delta T = T_{steam} - T_{bp}$. For multicomponent mixtures in the bottoms, $T_{bp, \text{partial reboiler}} < T_{bp, \text{total reboiler}}$. Thus, partial reboilers can use lower steam temperatures and are naturally preferred. The values of the heat transfer coefficient U depend upon the fluids being heated and cooled and the condition of the heat exchangers. Tabulated values and methods of calculating U are given in the references. Approximate ranges are given in Table 11-3.

If we use average values for U and for the water temperature in the condenser, we can estimate the condenser area. With the steam pressure known, the steam temperature can be found from the steam tables; then, with an average U, the area of the reboiler can be found. Condenser and reboiler costs can then be determined. Area is directly proportional to Q, and thus is directly proportional to F. Q_c is also directly proportional to $(1 + L/D)$ and Q_R depends linearly on $(1 + L/D)$.

If the column pressure is raised, the condensation temperature in the condenser will be higher. This is desirable, since ΔT in Eqs. (11-18) and (11-19) will be larger and required condenser area will be less. In addition, higher pressures will often allow the designer to cool

TABLE 11-3.   *Approximate heat transfer coefficients*

		U, Btu/h-ft²– °F (Greenkorn, & Kessler, 1972)	U, Btu/h-ft²– °F Biegler et al (1997)
Reboiler:	Steam to boiling aqueous solution	300 to 800 (average ~ 600)	400 to 1000
	Steam to boiling oil	20 to 80	15 to 90
Condenser:	Condensing aqueous mixture to water	150 to 800	—
	Condensing organic vapor to water	60 to 300	20 to 200
Air-cooled	Aqueous solution	—	10 to 50

with water instead of using refrigeration (see lab 4 in Appendix A of Chapter 6). This can result in a large decrease in cooling costs because refrigeration is expensive. With increased column pressure, the boiling point in the reboiler will be raised. Since this is the cold temperature in Eq. (11-18), the value of ΔT in Eqs. (11-18) and (11-19) is reduced and the reboiler area will be increased. An alternative solution is to use a higher pressure steam so that the steam temperature is increased and a larger reboiler won't be required. This approach does increase operating costs, though, since higher pressure steam is more expensive.

The total operating cost per year is given in Eq. (11-12). This value can be estimated as steam costs [(Eq. (11-17)] plus cooling water costs [Eq. (11-15)]. The effects of various variables are summarized in Table 11-4.

The capital cost per year is

$$\text{Capital cost per year} = (\text{depreciation rate}) (\text{Lang factor}) \qquad \textbf{(11-20)}$$

$$\times \, [\text{condenser cost} + \text{reboiler cost}$$

$$+ \text{ tower casing cost} + (N)(\text{sieve tray cost})]$$

The individual equipment costs depend on the condenser area, reboiler area, tower size, and tray diameter. Some of the variable effects on the capital costs are complex. These are outlined in Table 11-5. The net result of increasing L/D is shown in Figure 11-4; capital cost goes through a minimum.

The total cost per year is the sum of capital and operating costs and is also illustrated in Figure 11-4. Note that there is an optimum reflux ratio. As operating costs increase (increased energy costs), the optimum will shift closer to the minimum reflux ratio. As capital cost increases due to special materials or very high pressures, the total cost optimum will

TABLE 11-4. *Effect of changes in operating variables on operating costs*

Cost Item	Change in Variable	Effect on Other Variables	Effect on Cost
Cooling water	L/D increases	$|Q_c|$ proportional to (1 + L/D)	Up
	F increases	$|Q_c|$ proportional to F	Up
	Column pressure increases	ΔT_{water} may become larger since T_{bp} in condenser increases or may be unchanged	Down or no effect
		Cooling water may be used instead of refrigeration	Down significantly
	Water costs increase		Up
Steam	L/D increases	Q_R linearly depends (1 + L/D)	Up
	F increases	Q_R proportional to F	Up
	Column pressure increases	Temp. in reboiler increases; steam pressure increases	Up
	Steam costs increase		Up

TABLE 11-5.   *Effect of changes in design variables on capital costs*

Equipment Item	Change in Design Variable	Effect on Other Variables	Effect on Cost
Condenser	L/D increases	$\lvert Q_c \rvert$ and Area proportional to $(1 + L/D)$	Up
	F increases	$\lvert Q_c \rvert$ and Area proportional to F	Up
	Column pressure increases	ΔT (Eqs. 11-18 and 11-19) increases, area decreases, but more expensive construction may be needed	Down or up
	U increases	Area decreases	Down
Reboiler	Increase L/D	Q_R and Area linearly depend on $(1 + L/D)$	Up
	Increase F	Q_R and Area proportional to F	Up
	Increase feed temp.	h_F increases and Q_R drops	Down
	Column pressure increases	ΔT (Eqs. 11-18 and 11-19) drops and area increases (if steam pressure constant).	Up
		If more expensive construction	Up
	U increases	Required area drops	Down
Tower casing	Increase L/D	Col. Area proportional $(1 + L/D)$, weight increases, height drops	Down, then up
	Increase F	Col. Area proportional to F, weight increases	Up
	Column pressure up	Diameter down, thickness up, weight up	Down, then up
	Higher tray spacing	u_{op} up, diameter down, height up, weight up	Up
Sieve tray costs	Increase L/D	Number of trays down then diameter becomes excessive	Down, then up
	Increase F	Area proportional to F	Up
	Column pressure up	Diameter drops	Down
	Feed temperature increases	More trays at constant (L/D) (see Problem 11.A3)	Up
	Higher tray spacing	Entrainment down, better efficiency	Down
	Trays fancier than sieve trays with increased tray efficiency	Cost per tray up Fewer trays but may cost more per tray	Calculate for each case

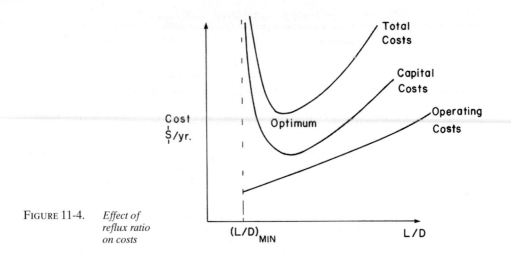

FIGURE 11-4. *Effect of reflux ratio on costs*

shift toward the capital cost optimum. The optimum L/D is usually in the range from 1.05 $(L/D)_{min}$ to 1.25 $(L/D)_{min}$.

The column pressure also has complex effects on the costs. If two pressures both above 1 atm are compared and cooling water can be used for both pressures, then total costs can be either higher or lower for the higher pressure. The effect depends on whether tower costs or tray costs dominate. If refrigeration would be required for condenser cooling at the lower pressure and cooling water can be used at the higher pressure, then the operating costs and the total costs will often be less at the higher pressure (see Tables 11-4 and 11-5).

The effects of other variables are somewhat simpler than the effect of L/D or pressure. For example, when the design feed rate increases, all costs go up; however, the capital cost per kilogram of feed drops significantly. Thus, total costs per kilogram can be significantly cheaper in large plants than in small plants. The effects of other variables are also summarized in Tables 11-4 and 11-5.

EXAMPLE 11-1. Cost estimate for distillation

Estimate the cost in June 2010 of the distillation column (shell and trays) designed in Examples 10-1 to 10-3.

Solution

The number of equilibrium stages can be calculated from a McCabe-Thiele diagram or estimated by the Fenske-Underwood-Gilliland approach. We will use the latter approach. In Example 10-1, $\alpha = 2.35$ was used. Then from the Fenske Eq. (7-16),

$$N_{min} = \frac{\left[\left(\dfrac{x}{1-x}\right)_{dist} \middle/ \left(\dfrac{x}{1-x}\right)_{bot}\right]}{\ln \alpha}$$

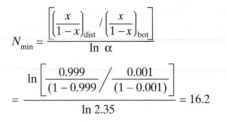

$$= \frac{\ln\left[\dfrac{0.999}{(1-0.999)} \middle/ \dfrac{0.001}{(1-0.001)}\right]}{\ln 2.35} = 16.2$$

From Example 10-1, y = 0.7 when x = 0.5, which is the feed concentration. Then for a saturated liquid feed,

$$\left(\frac{L}{V}\right)_{min} = \frac{x_D - y}{x_D - z} = \frac{0.999 - 0.7}{0.999 - 0.5} = 0.599$$

$$\left(\frac{L}{D}\right)_{min} = \frac{(L/V)_{min}}{1 - (L/V)_{min}} = 1.495$$

Since L/V = 0.8,

$$\left(\frac{L}{D}\right)_{act} = \frac{L/V}{1 - L/V} = \frac{0.8}{0.2} = 4$$

Using Eq. (7-42b) for the Gilliland correlation, we have

$$x = \frac{L/D - (L/D)_{min}}{L/D + 1} = \frac{4 - 1.495}{5} = 0.501$$

and

$$\frac{N - N_{min}}{N + 1} = 0.545827 - 0.591422(0.501) + \frac{0.002743}{(0.501)} = 0.255$$

$$N = \frac{0.255 + N_{min}}{(1 - 0.255)} = 22.09 \text{ equil. stages}$$

Subtract 1 for a partial reboiler. From Example 10-1, the overall efficiency E_o = 0.59. Then

$$N_{act} = \frac{N_{equil}}{0.59} = \frac{21.09}{0.59} = 35.7 \text{ or } 36 \text{ stages}$$

The calculated column diameter was 11 feet in Example 10-2 and 12 feet in Problem 10.D2. Use the larger value.

With 24 in = 0.6096 m tray spacing, we need (36)(0.6096) = 21.95 m. In addition, add approximately 3 m for vapor disengagement and a liquid pool at bottom. Do calculation in meters, since Figures 11-1 and 11-2 are in metric units:

Height = 24.95 m, Diameter = (12 ft) (1 m/3.2808 ft) = 3.66 m, Tray area = $\pi D^2/4$ = 10.5 m^2, and Volume of tower = $(\pi D^2/4)(L)$ =262.5 m^3.

From Figure 11-1, C_p^o = Cost/volume ~ $680/m^3, and $C_{p,tower}$ = (680)(262.5) = $178,000 in 2001. (Note: First horizontal line below 1000 is 800.)

From Figure 11-2, C_p^o = Tray cost/area ~ $720/m^2, and $C_{p,tray}$ = (720)(10.5) = $7560/tray in 2001.

We now need to include the extra cost factors and calculate the bare module costs. Since the column and trays are probably of carbon steel, F_m = 1.0. At 1 atm,

$F_p = 1.0$. Since $N > 20$, $F_q = 1.0$ for the trays. Then from Eqs. (11-8) and (11-9), respectively,

$$C_{BM,tower} = (\$178{,}000)(2.25 + 1.82\,(1.0)(1.0)) = \$726{,}000$$

$$C_{BM,trays} = (\$7560)(36)(1.0)(1.0) = \$272{,}000$$

Total bare module cost for tower and trays was \$998,000 in 2001.

This result should be compared with Problem 11.D1 for the same separation, but at a pressure of 7 bar. That column costs slightly more.

The cost in June 2010 can be estimated from Eq. (11-4) and the CEPCI indices. The base cost was in September 2001 with the CEPCI = 397. The June 2010 CEPCI index = 556.

$$\text{June 2010 cost} = (\$998{,}000)(556/397) = \$1{,}398{,}000$$

This cost does not include pumps, instrumentation and controls, reboiler, condenser, or installation.

Note: For aircraft safety and structural reasons, towers rarely exceed 200 feet (62.3 m) (Sieder et al., 2009). If a taller tower is required, it will be constructed as two columns in series.

11.3 CHANGES IN PLANT OPERATING RATES

Plants are designed for some maximum nameplate capacity but commonly produce less. The operating cost per kilogram of feed can be found by dividing Eq. (11-2) by the feed rate.

$$
\begin{aligned}
\text{Operating cost}/\text{kg} = {}& (\frac{\text{kg steam/h}}{F})\,(\text{cost steam}) \\
& + (\frac{\text{kg water/h}}{F})\,(\text{water cost}) + (\frac{\text{kW elec/h}}{F})\,(\$/\text{kW}) \\
& + (\frac{\text{worker h labor/h}}{F})\,(\text{labor cost, \$/man hour})
\end{aligned}
\tag{11-21}
$$

The kg steam/h, kg water/h, and kW elec/h are all directly proportional to F. Thus, except for labor costs, the operating cost per kilogram will be constant regardless of the feed rate. Since labor costs are often a small fraction of total costs in automated continuous chemical plants, we can treat the operating cost per kilogram as constant.

Capital cost per kilogram depends on the total amount of feed processed per year. Then, from Eq. (11-20),

$$\text{Capital cost}/\text{kg} = \frac{(\text{depreciation rate})(\text{Lang factor})(\text{sum of costs})}{(F, \text{kg/h})(\text{No. h operation/yr})} \tag{11-22}$$

Operation at half the designed feed rate doubles the capital cost per kilogram.

The total cost per kilogram is

$$\text{Total cost}/\text{kg} = \text{cap. cost}/\text{kg} + \text{op. cost}/\text{kg} + \text{admin. cost}/\text{kg} \qquad \textbf{(11-23)}$$

The effect of reduced feed rates depends on what percent of the total cost is due to capital and administrative costs.

If the cost per kilogram for the entire plant is greater than the selling price, the plant will be losing money. However, this does not mean that it should be shut down. If the selling cost is greater than the operating cost plus administrative costs, then the plant is still helping to pay off the capital costs. Since the capital charges are present even if $F = 0$, it is usually better to keep operating (Seidler et al., 2009; Sullivan et al., 2006). Of course, a new plant would not be built under these circumstances.

11.4 ENERGY CONSERVATION IN DISTILLATION

Distillation columns are often the major user of energy in a plant. Mix et al. (1978) estimated that approximately 3% of the total U.S. energy consumption is used by distillation! Thus, energy conservation in distillation systems is extremely important, regardless of the current energy price. Although the cost of energy oscillates, the long-term trend has been up and will probably continue to be up for many years. This passage, originally written in 1986, is obviously also true for the twenty-first century. Several energy-conservation schemes have already been discussed in detail. Most important among these are optimization of the reflux ratio and choice of the correct operating pressure.

What can be done to reduce energy consumption in an existing, operating plant? Since the equipment already exists, there is an incentive to make rather modest, inexpensive changes. Retrofits like this are a favorite assignment to give new engineers, since they serve to familiarize the new engineer with the plant and failure will not be critical. The first thing to do is to challenge the operating conditions (Geyer and Kline, 1976). If energy can be saved by changing the operating conditions the change may not require any capital. When the feed rate to the column changes, is the column still operating at vapor rates that are near those for optimum efficiency? If not, explore the possibility of varying the column pressure to change the vapor flow rate and thus, operate closer to the optimum. This will allow the column to have the equivalent of more equilibrium contacts and allow the operator to reduce the reflux ratio. Reducing the reflux ratio saves energy in the system. Challenge the specifications for the distillate and bottoms products. When products are very pure, rather small changes in product purities can mean significant changes in the reflux ratio. If a product of intermediate composition is required, side withdrawals require less energy than mixing top and bottom products (Allen and Shonnard, 2002).

Second, look at modifications that require capital investment. Improving the controls and instrumentation can increase the efficiency of the system (Geyer and Kline, 1976; Mix et al., 1978; Shinskey, 1984). Better control allows the operator to operate much closer to the required specifications, which means a lower reflux ratio. Payback on this investment can be as short as 6 months. Distillation column control is an extremely important topic that is beyond the scope of this textbook. Rose (1985) provides a qualitative description of distillation control systems while Shinskey (1984) is more detailed. If the column has relatively inefficient trays (e.g., bubble-caps) or packing (e.g., Raschig rings), putting in new, highly efficient trays (e.g., valve trays) or new high-efficiency packing (e.g., modern rings, saddles, or structured packing) will usually pay even though it is fairly expensive (Kenney, 1988). Certainly, any damaged trays or packing should be replaced with high-efficiency/high-capacity trays or packing (Kenney, 1988). Any column with damaged insulation should have the insulation removed and replaced. Heat exchange and integration of columns may be far from optimum in existing distillation systems. An upgrade of these facilities should be considered to determine whether it is economical.

When designing new facilities, many energy conservation approaches can be used that might not be economical in retrofits. Heat exchange between streams and integration of processes should be used extensively to minimize overall energy requirements. These methods have been known for many years (e.g., see Robinson and Gilliland, 1950, or Rudd and Watson, 1968), but they were not economical when energy costs were very low. When energy costs shot up in the 1970s and early 1980s and again after 2004, many energy conservation techniques suddenly became very economical (Doherty and Malone, 2001; Geyer and Kline, 1976; King, 1981; Mix et al., 1978; Null, 1976; O'Brien, 1976; Seider et al., 2009; Shinskey, 1984; and Siirola, 1996). These sources and many other references discussed by these authors give more details. Current concerns over global warming and existing or pending carbon dioxide cap and trading legislation in many parts of the world make energy reduction in distillation columns even more important.

The basic idea of heat exchange is to use hot streams that need to be cooled to heat cold streams that need to be heated. The optimum way to do this depends upon the configuration of the entire plant, since streams from outside the distillation system can be exchanged. The goal is to use exothermic reactions to supply all or at least as much as possible of the heat energy requirements in the plant. Pinch technology has been developed that allows analysis of optimum heat exchange for very complicated plants (Seider et al., 2009). If only the distillation system is considered, there are two main heat exchange locations, as illustrated in Figure 11-5. The cold feed is preheated by heat exchange with the hot distillate; this partially or totally condenses the distillate. The trim condenser is used for any additional cooling that's

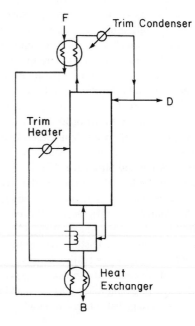

FIGURE 11-5.   *Heat exchange for an isolated distillation column*

needed and for improved control of the system. The feed is then further heated with the sensible heat from the bottoms product. The heat exchange is done in this order since the bottoms is hotter than the distillate. Further heating of the feed is done in a trim heater to help control the distillation. The system shown in Figure 11-5 may not be optimum, though, particularly if several columns are integrated. Nevertheless, the heat exchange ideas shown in Figure 11-5 are quite basic.

A technique similar to that of Figure 11-5 is to produce steam in the condenser. If there is a use for this low-pressure steam elsewhere in the plant, this can be a very economical use of the energy available in the overhead vapors. Intermediate condensers (Sections 4.9.4 and 10.4) can also be used to produce steam, and they have a hotter vapor.

Heat exchange integration of columns is an important concept for reducing energy use (Andrecovich and Westerberg, 1985; Biegler et al., 1997; Doherty and Malone, 2001; Douglas, 1988; King, 1981; Linnhoff et al., 1982; Robinson and Gilliland, 1950; Seider et al., 2009). The basic idea is to condense the overhead vapor from one column in the reboiler of a second column. This is illustrated in Figure 11-6. (In practice, heat exchanges like those in Figure 11-5 will also be used, but they have been left off Figure 11-6 to keep the figure simple.) Obviously, the condensation temperature of stream D_1 must be higher than the boiling temperature of stream B_2. When distillation is used for two rather different separations the system shown in Figure 11-6 can be used without modification. However, in many cases stream D_1 is the feed to the second column or B_2 is the feed to column 1 as shown in Figure 11-7. The system shown

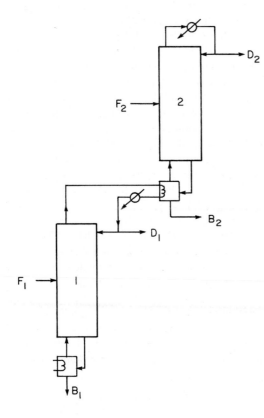

FIGURE 11-6.   *Integration of distillation columns*

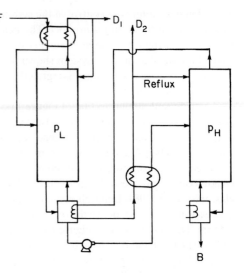

FIGURE 11-7. *Multieffect distillation system*

in Figure 11-6 will work if the first column is at a higher pressure than the second column so that stream D_1 condenses at a higher temperature than that at which stream B_2 boils.

Many variations of the basic idea shown in Figure 11-6 have been developed. If a solvent is recovered from considerably heavier impurities, some variant of the multieffect system shown in Figure 11-7 is useful (Agrawal, 2000; Andrecovich and Westerberg, 1985; Doherty and Malone, 2001; King, 1981; O'Brien, 1976; Robinson and Gilliland, 1950; Siirola, 1996; Wankat, 1993). After preheating, the solvent is first recovered as the distillate product in the first column, which operates at low pressure. The bottoms from this column is pumped to a higher pressure, preheated, and fed to the second column. Since the second column is at a higher pressure, the overheads can be used in the reboiler of the low-pressure column. Thus, the steam used in the reboiler of the higher pressure column serves to heat both columns. The steam efficiency is almost doubled. Since the separation is easy, not too many stages are required, and the two distillate products are both essentially pure solvent. Multieffect systems can also have feed to the high-pressure column in addition to or instead of feed to the low-pressure column. Liquified oxygen and nitrogen from air are produced on very large scales in Linde double columns, which are multieffect distillation columns. Multieffect distillation is closely related to multieffect evaporation (Mehra, 1986).

The condensing vapor from overhead can be used to heat the reboiler of the same column if vapor recompression or a heat pump is used (Humphrey and Keller, 1997; King, 1981; Meili, 1990; Null, 1976; Robinson and Gilliland, 1950). One arrangement for this is illustrated in Figure 11-8. The overhead vapors are compressed to a pressure at which they condense at a higher temperature than that at which the bottoms boil. Vapor recompression works best for close-boiling distillations, since modest pressure increases are required. Generally, vapor recompression is more expensive than heat integration or multieffect operation of columns. Thus, vapor recompression is used when the column is an isolated installation or is operating at extremes of high or low temperatures. O'Brien and Schultz (2004) report that UOP (formerly Universal Oil Products, Inc.) uses heat pumps for the difficult propane-propylene separation.

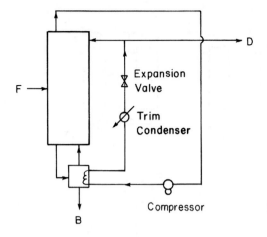

F

B

Compressor

D

FIGURE 11-8.   *Vapor recompression or heat pump system*

11.5   SYNTHESIS OF COLUMN SEQUENCES FOR ALMOST IDEAL MULTICOMPONENT DISTILLATION

A continuous distillation column is essentially a binary separator; that is, it separates a feed into two parts. For binary systems, both parts can be the desired pure products. However, for multicomponent systems, a simple single column is unable to separate all the components. For ternary systems, two columns are required to produce pure products; for four-component systems, three columns are required; and so forth. There are many ways in which these multiple columns can be coupled together for multicomponent separations. The choice of cascade can have a large effect on both capital and operating costs. In this section we will briefly look at the coupling of columns for systems that are almost ideal. More detailed presentations are available in other books (Biegler et al., 1997; Doherty and Malone, 2001; Douglas, 1988; King, 1981; Rudd et al., 1973; Woods, 1995), in reviews (Nishida et al., 1981; Siirola, 1996), and in a huge number of papers only a few of which are cited here (Agrawal, 2000; Garg et al., 1991; Kim and Wankat, 2004; Schultz et al., 2002; Shah, 2002; Tedder and Rudd, 1978; Thompson and King, 1972). Synthesis of sequences for nonideal systems are considered in the next section.

How many ways can columns be coupled for multicomponent distillation? Lots! For example, Figure 11-9 illustrates nine ways in which columns can be coupled for a ternary system that does not form azeotropes. With more components, the number of possibilities increases geometrically. Figure 11-9A shows the "normal" sequence, where the more volatile components are removed in the distillate one at a time. This is probably the most commonly used sequence, particularly in older plants. Scheme B shows an inverted sequence, where products are removed in the bottoms one at a time. With more components, a wide variety of combinations of these two schemes are possible. The scheme in Figure 11-9C is similar to the one in part A except that the reboiler has been removed and a return vapor stream from the second column supplies boilup to the first column. Capital costs will be reduced, but the columns are coupled, which will make control and startup more difficult. Scheme D is similar to B, except that a return liquid stream supplies reflux.

The scheme in Figure 11-9E uses a side enricher, while the one in part F uses a side stripper to purify the intermediate component B. The stream is withdrawn at the location where

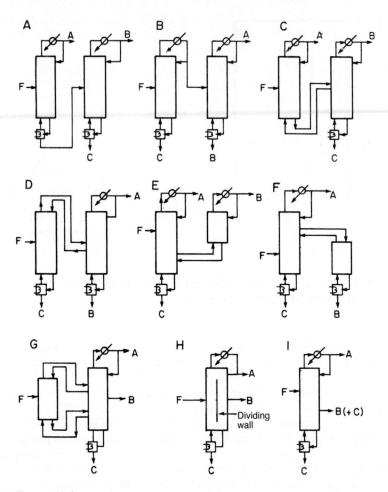

FIGURE 11-9. *Sequences for the distillation of ternary mixtures; no azeotropes.*
Component A is most volatile, and component C is least volatile

component B has a concentration maximum. These schemes are often used in petroleum refineries. Figure 11-9G illustrates a thermally coupled system (sometimes called Petyluk columns). The first column separates A from C, which is the easiest separation, and the second column then produces three pure products. The system will often have relatively low energy requirements, but it will be more difficult to start up and control. This system may also require an excessive number of stages if either the A-B or B-C separations are difficult. Sometimes this scheme can be achieved in a single "divided wall" column shown in Figure 11-9H (Parkinson, 2007; Schultz et al., 2002). This four-section column has a vertical wall between two parts of the column in the middle two sections. The feed enters on the left side of the wall and the two sections on the left side of the wall do the same separation done by the first column in Figure 11-9G. At the top of the wall a mixture of A and B would spill over the wall and go into the rectifying and the top intermediate sections. At the bottom of the wall a mixture of B and C would flow under the wall and into the stripping and bottom intermediate sections. Thus, the two intermediate sections on the right side of the wall do the separation done in the middle two sections of the second column in Figure 11-9G. Divided wall columns can be designed as if they were three simple columns; for example, with the Fenske-Underwood-Gilliland approach (Muralikrishna et al., 2002). Compared to Figure 11-9G, divided wall columns have been reported to have savings up to 30% for energy costs and 25% for capital costs (O'Brien and Schultz, 2004). Note that this arrangement appears to be somewhat sensitive to upsets. Divided-wall columns can also be used for four-component separations and extractive distillation (Parkinson, 1997).

The scheme in Figure 11-9I is quite different from the others, since a single column with a side stream is used (Tedder and Rudd, 1978). The side stream cannot be completely pure B, although it may be pure enough to meet the product specifications. This or closely related schemes are most likely to be useful when the concentration of C in the feed is quite low. Then at the point of peak B concentration there will not be much C present. Methods using side streams to connect columns (Doherty and Malone, 2001) and for four-component separations (Kim and Wankat, 2004) can also be used.

These sequences are only the start of what can be done. For example, the heat exchange and energy integration schemes discussed in the previous section can be interwoven with the separation scheme. Agrawal (2000) shows multieffect distillation cascades for ternary separations. These schemes are currently the most economical methods to produce oxygen, nitrogen, and argon from air by cryogenic distillation. Obviously, the system becomes quite complex.

Which method is the best to use depends upon the separation problem. The ease of the various separations, the required purities, and the feed concentrations are all important in determining the optimum configuration. The optimum configuration may also depend upon how "best" is defined. The engineer in charge of operating the plant will prefer the uncoupled systems, while the engineer charged with minimizing energy consumption may prefer the coupled and integrated systems. The only way to be assured of finding the best method is to model all the systems and try them. This is difficult to do, because it involves a large number of interconnected multicomponent distillation columns. Shortcut methods are often used for the calculations to save computer time and money. Unfortunately, the result may not be optimum. Many studies have ignored some of the arrangements shown in Figure 11-9; thus, they may not have come up with the optimum scheme. Conditions are always changing, and a distillation cascade may become nonoptimum because of changes in plant operating conditions such as feed rates and feed or product concentrations. Changes in economics such as energy costs or interest rates may also alter the optimality of the system. Sometimes it is best to build a nonoptimum system because it is more versatile.

An alternative approach to design is to use heuristics, which are rules of thumb used to exclude many possible systems. The heuristic approach may not result in the optimum separation scheme, but it usually produces a scheme that is close to optimum. Heuristics have been developed by doing a large number of simulations and then looking for ideas that connect the best schemes. Some of the most common heuristics listed in approximately the order of importance (Biegler et al., 1997; Doherty and Malone, 2001; Douglas, 1988; Garg et al., 1991; King, 1981; Thompson and King, 1972) include:

1. Remove dangerous, corrosive, and reactive components first.
2. Do not use distillation if $\alpha_{LK-HK} < \alpha_{min}$, where $\alpha_{min} \sim 1.05$ to 1.10.
3. Remove components requiring very high or very low temperatures or pressures first.
4. Do the easy splits (large α) first.

5. The next split should remove components in excess.
6. The next split should remove the most volatile component.
7. Do the most difficult separations as binary separations.
8. Favor 50:50 splits.
9. If possible, final product withdrawals should be as distillate products.

Two additional heuristics not listed in order of importance that can be used to force the designer to look at additional sequences are:

10. Consider side stream withdrawals for sloppy separations.
11. Consider thermally coupled and multieffect columns, particularly if energy is expensive.

There are rational reasons for each of the heuristics. Heuristic 1 will minimize safety concerns, remove unstable compounds, and reduce the need for expensive materials of construction in later columns. Heuristic 2 eliminates the need for excessively tall columns. Since very high and very low temperatures and pressures require expensive columns or operating conditions, heuristic 3 will keep costs down. Since the easiest split will require a shorter column and low reflux ratios, heuristic 4 says to do this when there are a number of components present and feed rates are large. The next heuristic suggests reducing feed rates as quickly as possible. Removing the most volatile component (heuristic 6) removes difficult to condense materials, probably allowing for reductions in column pressures. Heuristic 7 forces the lowest feed rate and hence the smallest diameter for the large column required for the most difficult separation. Heuristic 8 balances columns so that flow rates don't change drastically. Since thermal degradation products are usually relatively nonvolatile, heuristic 9 is likely to result in purer products. The purpose of heuristics 10 and 11 is to force the designer to think outside the usual box and look at schemes that are known to be effective in certain cases.

Each heuristic should be preceded with the words "All other things being equal." Unfortunately, all other things usually are not equal, and the heuristics often conflict with each other. For example, the most concentrated component may not be the most volatile. When there are conflicts between the heuristics, the cascade schemes suggested by both of the conflicting heuristics should be generated and then compared with more exact calculations.

EXAMPLE 11-2. Sequencing columns with heuristics

A feed with 25 mol% ethanol, 15 mol% isopropanol, 35 mol% n-propanol, 10 mol% isobutanol, and 15 mol% n-butanol is to be distilled. Purity of 98% for each alcohol is desired. Determine the possible optimum column configurations.

Solution

A, B, C. Define, explore, plan. With five components, there are a huge number of possibilities; thus, we will use heuristics to generate possible configurations. Equilibrium data can be approximated as constant relative volatilities (King, 1981) with n-propanol as the reference component: ethanol, $\alpha = 2.09$, isopropanol, $\alpha = 1.82$, n-propanol, $\alpha = 1.0$; isobutanol, $\alpha = 0.677$; n-butanol, $\alpha = 0.428$.

To use the heuristics it is useful to determine the relative volatilities of all adjacent pairs of compounds:

$\alpha_{E\text{-}IP} = 2.09/1.82 = 1.15$, $\alpha_{IP\text{-}nP} = 1.82/1.0 = 1.82$, $\alpha_{nP\text{-}IB} = 1.0/0.677 = 1.48$, $\alpha_{IB\text{-}nB} = 0.677/0.428 = 1.58$.

Since the easiest separation, isopropanol—n-propanol, is not that much easier than the other separations, heuristic 4 can probably be ignored.

D. Do it.

Case 1. Heuristics 6 and 9 give the direct sequence

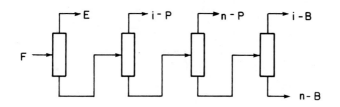

(Reboilers and condensers are not shown.) This will certainly work, but it is not very inventive.

Case 2. Heuristic 7 is often very important. Which separation is most difficult? From the list of relative volatilities of adjacent pairs of compounds ethanol-isopropanol is the hardest separation. If we also use heuristic 8 for column A, heuristic 7 for column B, and heuristics 5 and 8 for column C, we obtain the scheme shown in the figure.

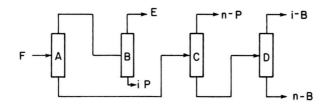

Naturally, other alternatives are possible.

Case 3. Heuristic 11 can be used to generate an entirely thermally coupled system (see Problem 11.A17). This would be difficult to operate. However, we can use heuristics 7, 8, and 11 to obtain a modification of case 2 (see figure).

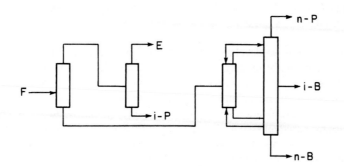

Case 4. Heuristic 11 can also be used to develop a system with one or more multieffect columns. If we use heuristic 7 to do the ethanol-isopropanol separation by itself, one option is to pressurize the liquid feed to column D in

case 2 and operate column D at a higher pressure. This multieffect arrangement is shown below. There are many other options possible for using multieffect columns for this separation.

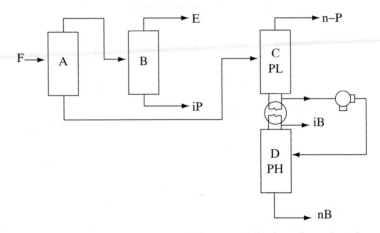

Other systems can be generated, but one of the four shown here is probably reasonably close to optimal.

E. Check. Finding the optimum configuration requires a simulation of each alternative. This can be done for cases 1 and 2 using the Fenske-Underwood-Gilliland approach. For cases 3 and 4 the thermally coupled and multieffect columns are more complex and probably should be simulated in detail.

F. Generalize. It is likely that one of these designs is close to optimum. Because of the low relative volatility between ethanol and isopropanol, heuristic 7 is important. Use of the heuristics does avoid having to look at several hundred other alternatives.

These heuristics have been developed for systems that have no azeotropes. When azeotropes are present, the methods developed in the next section should be used.

11.6 SYNTHESIS OF DISTILLATION SYSTEMS FOR NONIDEAL TERNARY SYSTEMS

In the previous section we developed heuristics for synthesis of distillation sequences for almost ideal systems; unfortunately, many of these heuristics do not apply to nonideal systems. Instead, we must use a different set of operational suggestions and the tools developed in section 8.5, distillation and residue curves. The purpose of the operational suggestions is to first develop a feasible separation scheme and then work to improve it.

Heuristics for nonideal systems have not been formalized and agreed upon to the same degree as for ideal systems. The following operational suggestions are from Biegler et al. (1997), Doherty and Malone (2001), Doherty et al. (2008), and common sense.

Operational Suggestions: Preliminary
1. Obtain reliable equilibrium data and/or correlations for the system.
2. Develop residue curves or distillation curves for the system.

3. Classify the system as A) almost ideal; B) nonideal without azeotropes; C) one homogeneous binary azeotrope without a distillation boundary; D) one homogeneous binary azeotrope with a distillation boundary; E) two or more homogeneous binary azeotropes with possibly a ternary azeotrope; F) heterogeneous azeotrope, which may include several binary and ternary azeotropes. Although solutions for cases D to F are beyond the scope of this introductory treatment, they will be discussed briefly.

Operational Suggestions: Case by Case

A. Almost ideal. If the system is reasonably close to ideal (Figure 8-7), rejoice and use the heuristics in Section 11.5.

B. Nonideal systems without azeotropes. These systems are often similar to ideal, and a variety of column sequences will probably work. However, it may be easier to do the most difficult separation with a non-key present instead of the binary separation recommended for ideal mixtures. Doherty and Malone (2001) present a detailed example for the acetaldehyde-methanol-water system.

    1. Generate the y-x curves for each binary pair. If all separations are relatively easy (reasonable relative volatilities and no inflection points causing tangent pinches for one of the pure components) use the ideal heuristics in Section 11.5.

    2. If one of the binary pairs has a small relative volatility or a tangent pinch, determine if this separation is easier in the presence of the third component. This can be done by generating distillation curves or y-x "binary" equilibrium at different constant concentrations of the third component. If the presence of the third component aids the separation, separate the difficult pair first. The concentration of the third component can be adjusted by recycling it from the column where it is purified. Note that this approach is very similar to using the third component as an extractive distillation solvent for separation of close boiling components (Section 8.6).

C. One homogeneous binary azeotrope, without a distillation boundary. The residue curve maps look like Figure 8-11a or 8-11c.

    1. If the binary azeotrope is between the light and intermediate components (Figure 8-11c), the situation is very similar to using extractive distillation to separate a binary azeotrope, and the heavy component is used instead of an added solvent. In general, the flowsheet is similar to Figure 8-13 except the heavy component product is withdrawn where the makeup solvent is added. If there is sufficient heavy component in the feed, a heavy recycle may not be required. The residue curve map will be similar to the extractive distillation residue map, Figure 8-14, but since the feed contains all three components point F will be inside the triangle.

    2. If the binary azeotrope is between the heavy and light components (Figure 8-11a), a separation can be obtained with an intermediate recycle as shown in Figures 11-10a and 11-10b. If there is sufficient intermediate component in the feed, the intermediate recycle may not be required. Separation using the flowchart in Figure 11-10a is illustrated in Example 11-3.

D. One homogeneous binary azeotrope with distillation boundary. There can either be a maximum boiling azeotrope (Figure 8-8) or a minimum boiling azeotrope (Figure 8-11b).

    1. If the distillation boundary is straight, complete separation of the ternary feed is not possible without addition of a mass-separating agent (Doherty and Malone, 2001).

    2. If the distillation boundary is curved, complete separation of the ternary feed may be possible. Distillation boundaries can often be crossed by mixing a feed with a recycle stream. An example of the synthesis of feasible flowsheets is given by

Biegler et al. (1997) for the system in Figure 8-8. Detailed solution of this case and the remaining two cases is beyond the scope of this section.

E. Two or more homogeneous binary azeotropes, with possibly a ternary azeotrope. These systems are messy, and there are invariably one or more distillation boundaries. If there is a single curved distillation boundary, it may be possible to develop a scheme to separate the mixture without addition of a mass-separating agent. If there are only two binary azeotropes and no ternary azeotropes, look for a separation method (e.g., extraction) that will remove the component that occurs in both azeotropes.

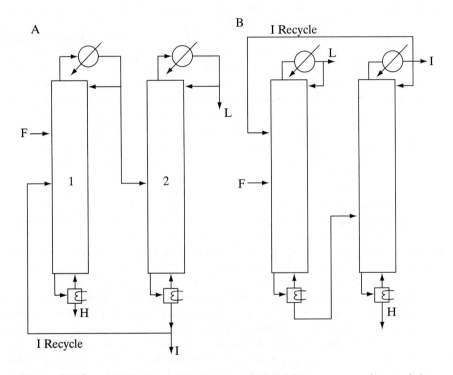

FIGURE 11-10.   *Distillation cascades for ternary feed with binary azeotrope between light L and heavy H components; (A) Indirect sequence, (B) direct sequence*

F. Heterogeneous azeotrope, which may include several binary and ternary azeotropes. An example was shown in Figure 8-12 for an azeotropic distillation scheme. The residue curve map developed by a process simulator probably will not show the envelope of the two-phase region, and this region will have to be added. Since liquid-liquid separators can cross distillation boundaries, there is a good chance that a separation can be achieved without adding an additional mass-separating agent. Distillation boundaries can be crossed by mixing, decanting, and reaction. If possible, use components already in the feed as an extractive distillation solvent or entrainer for azeotropic distillation. Also explore using components in the feed as extraction solvents. These systems are discussed in detail by Doherty and Malone (2001), and a process example is discussed by Biegler et al. (1997).

EXAMPLE 11-3. Process development for separation of complex ternary mixture

We have 100.0 kmol/h of a saturated liquid feed that is 50.0 mol% methanol, 10.0 mol% methyl butyrate and 40 mol% toluene. We want to separate this feed into three pure products (99.7⁺% purities). Develop a feasible distillation cascade for this system. Prove that your system is feasible.

Solution

A. Define. We want to develop a sequence of distillation columns including recycle that we claim can produce 99.7⁺% pure methanol, methyl butyrate, and toluene. Proof requires running a process simulator to show that the separation is achieved. Note that optimization is not required.

B. Explore. These components are in the Aspen Plus data bank and residue curves were generated with Aspen Plus using NRTL (Figure 11-11) (obviously, other process simulators could be used). Since there is one minimum boiling binary azeotrope between methanol (light component) and toluene (heavy) component without a distillation boundary, this residue curve map is similar to Figure 8-11a. We expect that the flowchart in either Figure 11-10a or 11-10b will do the separation.

C. Plan. The fresh feed point is plotted in Figure 11-11. Since the methyl butyrate concentration in the fresh feed is low, this point is close to the binary toluene-methanol line. If we don't recycle intermediate (methyl butyrate) we may have a problem with the binary azeotrope. Thus, for a feasible design, it is safer to start with recycle of intermediate.

Is Figure 11-10a or 11-10b likely to be better? Probably either one will work. Comparing the boiling points, the separation of methanol from methyl

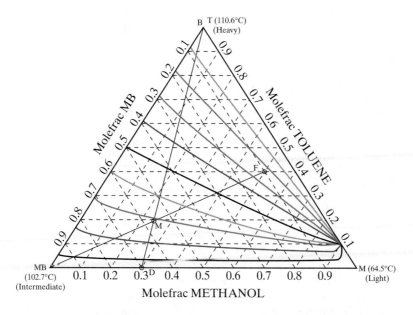

FIGURE 11-11.   Residue curves and mass balances for methanol, toluene, and methyl butyrate distillation. Key: M = methanol, MB = methyl butyrate, T = toluene

butyrate is probably simpler than separating methyl butyrate from toluene. Thus, the second column is likely to be considerably smaller if we use the flowchart in Figure 11-10a, although the first column will be larger. We will use Figure 11-10a and leave exploration of Figure 11-10b to Problems 11.D6 and 11.G1.

D. Do it. Use the flowchart in Figure 11-10a with recycle of intermediate. Arbitrarily, pick a recycle rate of 100 kmol/h. (The purpose of this example is to show feasibility. This initial assumption would be varied as the design is polished.) Since the methyl butyrate needs to be 99.7+% pure, we will assume the recycle stream is pure. Mixing the fresh feed and the recycle stream and using the lever arm rule, we find point M as shown in Figure 11-11. This combined feed can then be split into a distillate that contains essentially no toluene and a bottoms product that is 99.7+% pure toluene. This path is feasible because the straight lines going from D to MB to B (pure toluene) are always in the direction of increasing temperatures and these represent a valid residue curve. The distillate product can then be separated in the second column into a 99.7+% pure methanol distillate and a 99.7+% pure methyl butyrate bottoms. Thus, the separation appears to be possible.

Proof: The mixed feed to the column is 55% methyl butyrate, 20% toluene and 25% methanol. Flow rate is 200 kmol/h, and it was assumed to be a saturated liquid. The system in Figure 11-10a was simulated on Aspen Plus using NRTL for equilibrium. For the feasibility study pure methyl butyrate (instead of the recycle stream) and fresh feed were mixed together and input on the same stage of the first column. After some trial-and-error, the following results were obtained.

Column 1: $N = 81$ (including total condenser and partial reboiler), $N_{feed} = 41$, $D = 160$ kmol/h, $L/D = 8$, $p = 1.0$ atm.
Distillate mole fractions: Methanol = 0.3125, Methyl butyrate = 0.6869907, Toluene = 0.00050925.
Bottoms mole fractions: Methanol = 6.345 E -35, Methyl butyrate = 0.002037, Toluene = 0.997963.
Column 2: $N = 20$ (including total condenser and partial reboiler), $N_{feed} = 10$, $D = 50$ kmol/h, $L/D = 1.5$, $p = 1.0$ atm.
Distillate mole fractions: Methanol = 0.99741, Methyl butyrate = 0.002315, Toluene = 0.0002753.
Bottoms mole fractions: Methanol =0.0011751, Methyl butyrate = 0.998207, Toluene = 0.00061559.
    Thus, the separation is feasible.

E. Check. The residue curve map plotted from the process simulator agrees with the map in Doherty and Malone (2001). The predicted distillate composition for the first column determined from the mass balance calculation on Figure 11-11 is 31% methanol and 69% methyl butyrate. This agrees quite well with the results from the simulator. The literature states that Figure 11-10a should be successful for separating this type of mixture. Thus, we are quite confident that the process is feasible.

F. Generalization. The development of a feasible process is the first major step in the design. We would also need to do a preliminary design of Figure 11-10b to make sure that process is not better (see Problems 11.D6 and

11.G1). We need to optimize the better process to find appropriate values for the recycle rate, and optimum values for N, N_{feed}, and L/D for each column. We should also try the processes without recycle of MB to see if either of these processes are feasible (see Problems 11.D7a and 11.G2). In the first column the use of separate feed locations for the fresh feed and the recycle stream should be explored. If neither process is clearly better than the other, we would optimize both processes to determine which is more economical.

11.7 SUMMARY—OBJECTIVES

In this chapter we looked briefly at the economics of distillation, energy reduction, and multicomponent cascades. At the end of this chapter you should be able to satisfy the following objectives:

1. Estimate the capital and operating costs for a distillation column
2. Predict the effect of the following variables on column capital and operating costs:
   a. Feed rate
   b. Column pressure
   c. External reflux ratio
3. Estimate the effects that external factors have on capital and operating costs; external factors would include:
   a. Energy costs
   b. The general state of the economy
4. Discuss methods for reducing energy in distillation systems; develop flowsheets with appropriate heat exchange
5. Use heuristics to develop alternative cascades for the distillation of almost ideal multicomponent mixtures.
6. Use distillation or residue curves to develop a feasible separation scheme for nonideal mixtures.

REFERENCES

Agrawal, R., "Multieffect Distillation for Thermally Coupled Configurations," *AIChE J., 46,* 2211 (Nov. 2000).

Allen, D. T., and D. R. Shonnard, *Green Engineering: Environmentally Conscious Design of Chemical Processes*, Prentice Hall, Upper Saddle River, NJ, 2002.

Andrecovich, M. J., and A. W. Westerberg, "A Simple Synthesis Based on Utility Bounding for Heat-Integrated Distillation Sequences," *AIChE J., 31,* 363 (1985).

Biegler, L. T., I. E. Grossmann, and A. W. Westerberg, *Systematic Methods of Chemical Process Design,* Prentice Hall, Upper Saddle River, NJ, 1997.

Chengel, Y. A., *Heat Transfer: A Practical Approach,* McGraw-Hill, New York, 2003.

Doherty, M. F., and M. F. Malone, *Conceptual Design of Distillation Systems,* McGraw-Hill, New York, 2001.

Douglas, J. M., *The Conceptual Design of Chemical Processes,* McGraw-Hill, New York, 1988.

Garg, M. K., P. L. Douglas, and J. G. Linders, "An Expert System for Identifying Separation Processes," *Can. J. Chem. Eng., 69,* 67 (Feb. 1991).

Geankoplis, C. J., *Transport Processes and Separation Process Principles,* 4th ed., Prentice Hall, Upper Saddle River, NJ 2003.

Geyer, G. R., and P. E. Kline, "Energy Conservation Schemes for Distillation Processes," *Chem. Eng. Progress, 72,* (5), 49 (May 1976).

Greenkorn, R. A., and D. P. Kessler, *Transfer Operations,* McGraw-Hill, New York, 1972.

Griskey, R. G., *Transport Phenomena and Unit Operations: A Combined Approach,* Wiley, New York, 2002.

Humphrey, J. L., and G. E. Keller II, *Separation Process Technology,* McGraw-Hill, New York, 1997.

Keller, G. E., II, *Separations: New Directions for an Old Field,* AIChE Monograph Series, *83,* (17) 1987.

Kenney, W. F., "Strategies for Conserving Energy," *Chem. Eng. Progress, 84* (3), 43 (March 1988).

Kern, D. Q., *Process Heat Transfer,* McGraw-Hill, New York, 1950.

Kim, J. K., and P. C. Wankat, "Quaternary Distillation Systems with Less than N − 1 Columns," *Ind. Eng. Chem. Res., 43,* 3838 (2004).

King, C. J., *Separation Processes,* 2nd ed., McGraw-Hill, New York, 1981.

Linnhoff, B., D. W. Townsend, D. Boland, and G. F. Hewitt, *A User's Guide on Process Integration for the Efficient Use of Energy,* Institution of Chemical Engineers, Rugby, United Kingdom, 1982.

Ludwig, E. E., *Applied Process Design,* 3rd ed., Vol. 3, Gulf Pub Co., Boston, 2001.

McCarty, A. J., and B. R. Smith, "Reboiler System Design: The Tricks of the Trade," *Chem. Eng. Progress, 91,* (5), 34 (May 1995).

Mehra, D. K., "Selecting Evaporators," *Chem. Eng., 93* (3), 56 (Feb. 3, 1986).

Meili, A., "Heat Pumps for Distillation Columns," *Chem. Eng. Progress, 86,* (6), 60 (June 1990).

Mix, T. J., J. S. Dweck, M. Weinberg, and R. C. Armstrong, "Energy Conservation in Distillation," *Chem. Eng. Progress, 74,* (4), 49 (April 1978).

Muralikrishna, K., K. P. Madhavan, and S. S. Shah, "Development of Dividing Wall Distillation Column Design Space for a Specified Separation," *Chem. Engr. Res. & Des., 80* (#A2) 155 (March 2002).

Nishida, N., G. Stephanopoulous, and W. W. Westerberg, "A Review of Process Synthesis," *AIChE J., 27,* 321 (1981).

Null, H. R., "Heat Pumps in Distillation," *Chem. Eng. Progress, 72,* (7), 58 (July 1976).

O'Brien, D., and M. A. Schultz, "Ask the Experts: Distillation," *Chem. Eng. Progress, 100* (2), 14 (Feb. 2004).

O'Brien, N. G., "Reducing Column Steam Consumption," *Chem. Eng. Progress, 72,* (7), 65 (July 1976).

Parkinson, G., "Dividing-Wall Columns Find Greater Appeal," *Chem. Engr. Progress, 103* (5), 8 (May 2007).

Perry, R. H., C. H. Chilton, and S. D. Kirkpatrick (Eds.), *Chemical Engineer's Handbook,* 4th ed., McGraw-Hill, New York, 1963.

Peters, M. S., K. D. Timmerhaus, and R. E. West, *Plant Design and Economics for Chemical Engineers,* 5th ed., McGraw-Hill, New York, 2003.

Robinson, C. S., and E. R. Gilliland, *Elements of Fractional Distillation,* McGraw-Hill, New York, 1950.

Rose, L. M., *Distillation Design in Practice,* Elsevier, Amsterdam, 1985.

Rudd, D. F., F. J. Powers, and J. J. Siirola, *Process Synthesis,* Prentice Hall, Upper Saddle River, NJ, 1973.

Rudd, D. F., and C. C. Watson, *Strategy of Process Engineering,* Wiley, New York, 1968.

Schultz, M. A., D. G. Stewart, J. M. Harris, S.P. Rosenblum, M. S. Shakur, and D. E. O'Brien, "Reduce Costs with Dividing-Wall Columns, "*Chem. Eng. Progress, 98,* (5) 64 (May 2002).

Seider, W. D., J. D. Seader, D. R. Lewin, and S. Widago, *Design Process Principles,* 3rd ed., Wiley, New York, 2009.

Shah, P. B., "Squeeze More Out of Complex Columns," *Chem. Eng. Progress, 98,* (7) 46 (July 2002).

Shinskey, F. G., *Distillation Control, For Productivity and Energy Conservation,* 2nd ed., McGraw-Hill, New York, 1984.

Siirola, J. J., "Industrial Applications of Chemical Process Synthesis," *Adv. Chem. Engng., 23,* 1-62 (1996).

Sullivan, W. G., E. M. Wicks, and J. T. Luxhoj, *Engineering Economy,* 13th Ed., Prentice Hall, Upper Saddle River, NJ, 2006.

Tedder, D. W. and D. F. Rudd, "Parametric Studies in Industrial Distillation: Part 1. Design Comparisons," *AIChE J., 24,* 303 (1978).

Thompson, R. W., and C. J. King, "Systematic Synthesis of Separation Systems," *AIChE J., 18,* 941 (1972).

Turton, R., R. C. Bailie, W. B. Whiting, and J. A. Shaeiwitz, *Analysis, Synthesis, and Design of Chemical Processes,* 2nd ed., Prentice Hall, Upper Saddle River, NJ, 2003.

Turton, R., R. C. Baile, W. B. Whiting, and J. A. Shaeiwitz, *Analysis, Synthesis, and Design of Chemical Processes,* 3rd Ed., Prentice Hall, Upper Saddle River, NJ, 2009.

Ulrich, G. D., *A Guide to Chemical Engineering Process Design and Economics,* Wiley, New York, 1984.

Wankat, P. C., "Multieffect Distillation Processes," *Ind. Eng. Chem. Res., 32,* 894 (1993).

Woods, D. R., *Financial Decision Making in the Process Industries,* Prentice Hall, Upper Saddle River, NJ, 1976.

Woods, D. R., *Process Design and Engineering Practice,* Prentice Hall, Upper Saddle River, NJ, 1995.

HOMEWORK

A. *Discussion Problems*

A1. If valve trays cost more than sieve trays, why are they often advertised as a way of decreasing tower costs?

A2. Develop your key relations chart for this chapter.

A3. What is the effect of increasing the feed temperature if $L/D = 1.15 \, (L/D)_{min}$. Note that $(L/D)_{min}$ will change.
Include effects on Q_R and number of stages. Use a McCabe-Thiele diagram.

A4. Optimums usually occur because there are two major competing effects. For the optimum L/D these two major effects for capital cost are (select two answers):

a. The number of stages is infinite as L/D goes to infinity and has a minimum value at $(L/D)_{min}$.

b. The number of stages is infinite at $(L/D)_{min}$ and approaches a minimum value as L/D goes to infinity.

c. The column diameter becomes infinite at $(L/D)_{min}$ and approaches a minimum value as L/D approaches infinity.

d. The column diameter becomes infinite as L/D approaches infinity and has a minimum value at $(L/D)_{min}$.

A5. How does the general state of the economy affect:

a. Design of new plants

b. Operation of existing plants

A6. We are separating ethanol, i-propanol, and n-propanol in a distillation column with a sidestream. The relative volatilities can be assumed to be constant: $\alpha_{E\text{-}nP} = 2.17$, $\alpha_{iP\text{-}nP} = 1.86$, $\alpha_{nP\,nP} = 1.0$. We want to recover ethanol in the distillate, n-propanol in the bottoms, and as pure a sidestream of i-propanol as possible. The feed has equal amounts of ethanol and n-propanol. To obtain maximum sidestream purity, the best location and type of sidestream is:

    a. A vapor sidestream between the feed and the distillate.

    b. A liquid sidestream between the feed and the distillate.

    c. A vapor sidestream between the feed and the bottoms.

    d. A liquid sidestream between the feed and the bottoms.

    Explain your answer.

A7. Why is the dependence on size usually less than linear—in other words, why is the exponent in Eq. (11-2) less than 1?

A8. It is common to design columns at reflux ratios slightly above $(L/D)_{opt}$. Use a curve of total cost/yr vs. L/D to explain why an $L/D > (L/D)_{opt}$ is used. Why isn't there a large cost penalty?

A9. Discuss the concept of economies of scale. What happens to economies of scale if the feed rate is half the design value?

A10. Sketch how a divided wall column could be set up for separation of a four-component mixture.

A11. Referring to Figure 11-6 if D1 is the feed to column 2, explain what conditions are necessary for this system to work.

A12. Sometimes a list of what not to do is as valuable as a list of what to do. For separation of a close to ideal system by distillation, develop a list of heuristics of what not to do.

A13. To estimate future values of the cost indices, one is tempted to assume that the average value for the year occurred at midyear (June 30–July 1) and that the linear fit to the recent data can be extrapolated past the last data point. Based on Table 11-1, for which years would this procedure work fairly well, and for which years will it fail?

A14. The use of components in the feed as solvents for extractive or azeotropic distillation or extraction is recommended even if they are not the best solvents for stand-alone separations. Explain the reasoning behind this recommendation.

A15. Residue curves and distillation curves have similar shapes but are not identical. Even though the residue curves might be misleading in some cases, they are still useful for screening possible distillation separations. Explain why.

A16. Preheating the feed will often increase the number of stages required for the separation ($F, z, x_D, x_B, L/D$ constant). Use a McCabe-Thiele diagram to explain why this happens.

A17. Draw the entirely thermally coupled system (an extension of Figure 11-9G) for Example 11-2.

B. *Generation of Alternatives*

B1. Sketch possible column arrangements for separation of a four-component system. Do not include sidestream products. Note that there are a large number of possibilities.

B2. We wish to generate additional arrangements for quaternary mixtures (see Problem 11.B1). Sketch possible arrangements that use one or two columns that have sidestreams for one or both of the intermediate component product streams.

B3. Multieffect distillation or column integration can be done with more than two columns. Use the basic ideas in Figures 11-5 and 11-6 to sketch as many ways of thermally connecting three columns as you can.

B4. We wish to separate a feed that is 10 mol% benzene, 55 mol% toluene, 10 mol% xylene, and 25 mol% cumene. Use heuristics to generate desirable alternatives. Average relative volatilities are $\alpha_{BT} = 2.5$, $\alpha_{TT} = 1.0$, $\alpha_{XT} = 0.33$, $\alpha_{CT} = 0.21$. 98% purity of all products is required.

B5. Repeat Problem 11.B4 for an 80% purity of the xylene product.

C. *Derivations*

C1. Show that Eq. (11-1) will plot as a straight line on log-log paper, and show that the exponent can be determined from a slope.

C2. For large volumes Figure 11-2 shows that packing costs are directly proportional to the volume of packing. Show that packing costs go through a minimum as L/D increases.

D. *Problems*

Answers to problems with an asterisk are at the back of the book.

D1.* Repeat Example 11-1 except at a pressure of 700 kPa. At this pressure $E_o = 0.73$, D = 9 ft, and the relative volatility will be a function of pressure.

   a.* Find $(L/D)_{min}$.

   b.* Find N_{min}.

   c.* Estimate N_{equil}.

   d.* Estimate N_{actual}.

   e.* Find cost of shells and trays as of Sept. 2001.

   f. Update cost of shells and trays to current date.

D2.* Estimate the cost of the condenser and reboiler (both fixed tube sheet, shell and tube) for the distillation of Example 11-1. Pressure is 101.3 kPa. $C_{PL,C7} = 50.8$ Btu/(lbmol–°F), $\lambda_{C7} = 14{,}908$ Btu/lbmol. Data for hexane are given in Problem 3.D6. The saturated steam in the reboiler is at 110°C. $\lambda_{steam} = 958.7$ Btu/lb. Cooling water enters at 70°F and leaves at 110°F. $C_{Pw} = 1.0$ Btu/(lb–°F). Use heat transfer coefficients from Table 11-2. Watch your units.

D3. Determine the steam and water operating costs per hour for Problem 11.D2. Cost of steam is $20.00/1000 lb, and cost of cooling water is $3.00/1000 gal.

D4. Example 10-4 and Problem 10.D17 sized the diameter of a packed column doing the separation in Example 11-1. Suppose a 15-foot diameter column is to be used. The 1-in Intalox saddles have an HETP of 0.37 m. Estimate the packing and tower costs. Pressure is 101.3 kPa.

D5. The cost data for the double pipe heat exchanger plots as a straight line (Figure 11-3), which indicates that Eq. (11-2) is valid. Determine the exponent in Eq. (11-2) for the double pipe heat exchanger.

Note: If your exponent is negative, you have made a mistake.

D6. Repeat the residue curve analysis for Example 11-3, but using the flowsheet in Figure 11-10b. Arbitrarily use a recycle flow rate of 100 kmol/h.

D7. Repeat the residue curve analysis for Example 11-3 but with no recycle.

   a. For process in Figure 11-10a.

   b. For process in Figure 11-10b.

D8. **a.** If the feed rate in Example 11-1 is doubled, what is the total bare module cost of the column plus trays in June 2010?

   b. Compare the June 2010 total bare module cost of the column plus trays per lbmol of product for the two different feed rates.

F. *Problems Requiring Other Resources*

F1. Look up the current cost index in *Chemical Engineering* magazine. Use this to update Table 11-1 and the ordinates in Figures 11-1 to 11-3.

G. *Computer Problems*

G1. Repeat the computer simulation proof of feasibility for Example 11-3, but using the flowsheet in Figure 11-10b. The input for the simulator should be based on the solution to Problem 11.D6. If desired, you may input the recycle stream and the fresh feed to different stages in the first column.

G2. Repeat the computer simulation proof of feasibility for Example 11-3, but with no recycle. The input for the simulator should be based on the solution to Problem 11.D7.
   a. For process in Figure 11-10a.
   b. For process in Figure 11-10b.

G3. A distillation column is being designed to process a feed that is 10 mol% ethanol and 90 mol% water. The feed rate is 100 kmol/h, and the feed is a saturated liquid at a pressure of 5.0 atm. The column has a partial reboiler, a total condenser, N = 10 in Aspen notation, D = 10, and L/D = 2. Do tray sizing for a sieve try with one pass, 0.4572 m tray spacing, 85% approach to flood (use Jim Fair's flooding calculation method), and default values for other design variables.
   a. Operate the column at 1.0 atm. Find and report the optimum feed stage (based on maximum separation). For the optimum feed stage, report $Q_{reboiler}$, $Q_{condenser}$, the distillate and bottoms mole fractions ethanol, and the maximum tray diameter.
   b. Operate the column at 3.0 atm. Find and report the optimum feed stage (based on maximum separation). For the optimum feed stage, report $Q_{reboiler}$, $Q_{condenser}$, the distillate and bottoms mole fractions ethanol, and the maximum tray diameter.
   c. Operate the column at 5.0 atm. Find and report the optimum feed stage (based on maximum separation). For the optimum feed stage, report $Q_{reboiler}$, $Q_{condenser}$, the distillate and bottoms mole fractions ethanol, and the maximum tray diameter.
   d. Compare your results.
      D1. Which pressure gives the best separation? Why?
      D2. Which pressure has the lowest $Q_{reboiler}$? Why?
      D3. Which pressure has the lowest absolute value of $Q_{condenser}$? Why?
      D4. Which pressure has the smallest diameter column? Why?
   e. When operating and capital costs are considered, the optimum pressure to obtain the same purity is usually above 1.0 atm, but is usually less than 6.8 atm. Speculate on the reasons for this.
   Note: A useful rule of thumb is the annual operating cost of a distillation column is half operating costs and half capital costs (Keller, 1987).

G4. [This problem is extensive.] A plant needs to distill 1000 kmol/h of a feed that is 5.0 mol% ethanol and 95 mol% water and a temperature of 76°C. We desire a bottoms that is 0.01 mol% ethanol or slightly less and a distillate that is 75.0 mol% ethanol or slightly more. Use NRTL for equilibrium data.
   a. Design (find number of equilibrium stages, optimum feed stage, and column diameter) a distillation column at 1.0 atm to do this separation. The column has a total condenser and partial reboiler. Operate with a saturated liquid reflux and with L/D = 1.1 × (L/D)$_{min}$. Determine (L/D)$_{min}$ from simulations with 100 stages. Operate at 80% of flooding (based on Jim Fair's method) with sieve plates and 0.6096 m spac-

ing between trays. Report number of trays, optimum feed tray, column diameter, L/D, Q_R, and Q_c.

b. Design a multieffect distillation system to produce two bottoms products that are 0.01 mol% ethanol or slightly less and two distillate products that are 75.0 mol% ethanol or slightly more. Both columns receive fresh feed. The low-pressure column operates at 1.0 atm. The high-pressure column is at 3.0 atm. The condenser of the high-pressure column is also the reboiler of the low-pressure column; thus, $Q_{c,high}$ $_{pressure column} = - Q_{R,low\ pressure\ column}$. Use the same optimum feed stage and the total number of stages for the low-pressure column as determined in part a. Add one stage to the high-pressure column but use the same feed location as in the low-pressure column. Make Q_R in the high-pressure column $0.7555 \times Q_R$ determined in part a. Adjust the feed flow rate in the two columns so that

$$F(\text{low-pressure column}) + F(\text{high-pressure column}) = F_{total} = 1000 \text{ kmol/h}.$$

Operate both columns at 80% of flooding (based on Jim Fair's method) with sieve plates and 0.6096 m spacing between trays. Report F, number of trays, feed tray, column diameter, L/D, Q_R, and Q_c for both columns.

c. Compare the results of parts a and b.

CHAPTER 12

Absorption and Stripping

Up to now we have talked almost entirely about distillation. There are other unit operations that are very useful in the processing of chemicals or in pollution control. *Absorption* is the unit operation where one or more components of a gas stream are removed by being taken up (absorbed) in a nonvolatile liquid (solvent). In this case the liquid solvent must be added as a *separating* agent. Absorption is one of the methods used to remove CO_2 from natural gas and flue gasses so that the CO_2 is not added to the atmosphere where it helps cause global warming (Socolow, 2005).

Stripping is the opposite of absorption. In stripping, one or more components of a liquid stream are removed by being vaporized into an insoluble gas stream. Here the gas stream (stripping agent) must be added as a separating agent.

What was the separating agent for distillation? Energy.

Absorption can be either physical or chemical. In *physical absorption* the gas is removed because it has greater solubility in the solvent than other gases. An example is the removal of butane and pentane ($C_4 - C_5$) from a refinery gas mixture with a heavy oil. In *chemical absorption* the gas to be removed reacts with the solvent and remains in solution. An example is the removal of CO_2 or H_2S by reaction with NaOH or with monoethanolamine (MEA). The reaction can be either irreversible (as with NaOH) or reversible (as with MEA). For irreversible reactions the resulting liquid must be disposed of, whereas in reversible reactions the solvent can be regenerated (in stripper or distillation columns). Thus, reversible reactions are often preferred. Chemical absorption systems are discussed in more detail by Astarita et al. (1983), Kister et al. (2008), Kohl (1987), Kohl and Nielsen (1995), and Zarycki and Chacuk (1993).

Chemical absorption usually has a much more favorable equilibrium relationship than physical absorption (solubility of most gases is usually very low) and is therefore often preferred. However, the Murphree efficiency is often quite low (10% is not unusual), and this must be taken into account.

Both absorption and stripping can be operated as equilibrium stage operations with contact of liquid and vapor. Since distillation is also an equilibrium stage operation with contact of liquid and vapor, we would expect the equipment to be quite similar. This is indeed the

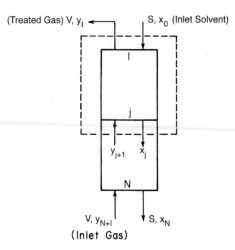

FIGURE 12-1.   *Gas absorber*

case; both absorption and stripping are operated in packed and plate towers. Plate towers can be designed by following an adaptation of the McCabe-Thiele method. Packed towers can be designed by use of HETP or preferably by mass transfer considerations (see Chapter 16).

In both absorption and stripping a separate phase is added as the separating agent. Thus, the columns are simpler than those for distillation in that reboilers and condensers are normally not used. Figure 12-1 is a schematic of a typical absorption column. In this column solute B entering with insoluble carrier gas C in the inlet gas stream is absorbed into the non-volatile solvent A.

A gas treatment plant often has both absorption and stripping columns as shown in Figure 12-2. In this operation the solvent is continually recycled. The heat exchanger heats the saturated solvent, changing the equilibrium characteristics of the system so that the solvent can be stripped. A very common type of gas treatment plant is used for the removal of CO_2 and/or H_2S from refinery gas or natural gas. In this case MEA or other amine solvents in water are used as the solvent, and steam is used as the stripping gas (for more details see Kohl

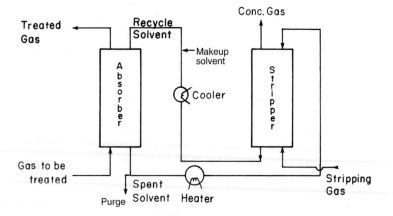

FIGURE 12-2.   *Gas treatment plant*

and Nielsen, 1997 or Ball and Veldman, 1991). Both random packings and sieve trays, but not structured, packings are commonly used for these acid gas systems (Kister, 2006).

12.1   ABSORPTION AND STRIPPING EQUILIBRIA

For absorption and stripping in three component systems we often assume that

1. Carrier gas is insoluble.
2. Solvent is nonvolatile.
3. The system is isothermal and isobaric.
   The Gibbs phase rule is

$$F = C - P + 2 = 3(A, B, \text{ and } C) - 2(\text{vapor and liquid}) + 2 = 3$$

If we set T and p constant, there is one remaining degree of freedom. The equilibrium data are usually represented either by plotting solute composition in vapor vs. solute composition in liquid or by giving a Henry's law constant. Henry's law for dilute solute B is

$$p_B = H_B x_B \tag{12-1}$$

where H_B is Henry's law constant, in atm/mole frac, H = H(p, T, composition); x_B is the mole frac B in the liquid; and p_B is the partial pressure of B in the vapor.

Henry's law is valid only at low concentrations of B. Since partial pressure is defined as

$$y_B \equiv \frac{p_B}{p_{tot}} \tag{12-2}$$

Henry's law becomes

$$y_B = \frac{H_B}{p_{tot}} x_B \tag{12-3}$$

This will plot as a straight line if H_B is a constant. Equilibrium data for absorption are given by Hwang (1981), Hwang et al. (1992a, b), Kohl (1987), Kohl and Nielsen (1997), Perry et al. (1963, pp. 14-2 to 14-12), Perry and Chilton (1973, p. 14-3), Perry and Green (1997, pp. 2-125 to 2-128), and Yaws et al. (2005). For example, the values given for CO_2, CO, and H_2S are shown in Table 12-1 (Perry et al., 1963). The large H values in Table 12-1 show that CO_2 and H_2S are very sparingly soluble in water. Since H is roughly independent of p_{tot}, this means that more gas is absorbed at higher pressure. This phenomenon is commonly taken advantage of to make carbonated beverages. When the bottle or can is opened the pressure drops and the gas desorbs, forming little bubbles. Selected Henry's law constants for chlorinated compounds in water are listed in Table 12-2 (Yaws et al., 2005) at 25°C. These values are useful for developing processes for removal of these compounds from contaminated water by stripping. Note that these compounds are much more soluble than the gases listed in Table 12-1. Obviously, Henry's law is only valid when x is less than the solubility limit.

The Henry's law constants depend upon temperature and usually follow an Arrhenius relationship. Thus,

TABLE 12-1. *Henry's law constants, H for CO₂, CO, and H₂S in water. H is in atm/mole frac.*

T° C	CO₂	CO	H₂S
0	728	35,200	268
5	876	39,600	315
10	1040	44,200	367
15	1220	48,900	423
20	1420	53,600	483
25	1640	58,000	545
30	1860	62,000	609
35	2090	65,900	676
40	2330	69,600	745
45	2570	72,900	814
50	2830	76,100	884
60	3410	82,100	1030
70	—	84,500	1190
80	—	84,500	1350
90	—	84,600	1440
100	—	84,600	1480

Source: Perry et al. (1963), pp. 14-4 and 14-6, and Green and Perry (2008), pp. 2–130 (H₂S).

$$H = H_0 \exp\left(\frac{-E}{RT}\right)$$

(12-4)

A plot of log H vs. 1/T will often give a straight line.

The effect of concentration is shown in Table 12-3, where the absorption of ammonia in water (Perry et al., 1963) is illustrated. Note that the solubilities are nonlinear and $H = p_{NH3}/X$ is not a constant. This behavior is fairly general for soluble gases.

We will convert equilibrium data to the concentration units required for calculations. If mole or mass ratios are used, equilibrium *must* be converted into ratios.

TABLE 12-2. *Henry's law constants and solubilities for chlorinated compounds in water at 25°C and 1 atm (Yaws et al., 2005)*

Compound	Carbon Tetra-chloride CCl_4	Chloro-form $CHCl_3$	Dichloro-methane CH_2Cl_2	Vinyl Chloride C_2H_2Cl	1,2,3,-trichloro propane $C_3H_5Cl_3$	1 chloro-heptane $C_7H_{15}Cl$	1 chloro-naphthalene $C_{10}H_7Cl$
$H \dfrac{atm}{mole\ frac.}$	1589.36	211.19	137.00	1243.84	23.78	2130.28	10.96
Solubility, ppm (mol)	93.68	1190.1	4175.4	778.9	213.99	1.8196	2.48

TABLE 12-3. *Absorption of ammonia in water*

Weight NH₃ per 100 weight H₂O	Partial pressure of NH₃, mm Hg			30°C	40°C	50°C	60°C
	0°C	10°C	20°C				
100	947						
90	785						
80	636	987	1450	—	—	3300	
70	500	780	1170	—	—	2760	
60	380	600	945	—	—	2130	
50	275	439	686	—	—	1520	
40	190	301	470	—	719	1065	
30	119	190	298	—	454	692	
25	89.5	144	227	—	352	534	825
20	64	103.5	166	—	260	395	596
15	42.7	70.1	114	—	179	273	405
10	25.1	41.8	69.6	—	110	167	247
7.5	17.7	29.9	50.0	—	79.7	120	179
5	11.2	19.1	31.7	—	51.0	76.5	115
4	—	16.1	24.9	—	40.1	60.8	91.1
3	—	11.3	18.2	23.5	29.6	45	67.1
2.5	—	—	15.0	19.4	24.4	(37.6)*	(55.7)
2	—	—	12.0	15.3	19.3	(30.0)	(44.5)
1.6	—	—	—	12.0	15.3	(24.1)	(35.5)
1.2	—	—	—	9.1	11.5	(18.3)	(26.7)
1.0	—	—	—	7.4	—	(15.4)	(22.2)
0.5	—	—	—	3.4			

*Extrapolated values.

Source: Perry et al. (1963). Copyright 1963. Reprinted with permission of McGraw-Hill.

12.2 MCCABE-THIELE SOLUTION FOR DILUTE ABSORPTION

The McCabe-Thiele diagram is most useful when the operating line is straight. This requires that the energy balance is automatically satisfied and liquid flow rate/vapor flow rate = constant. In order for energy balances to be automatically satisfied, we must assume that

1. The heat of absorption is negligible.
2. Operation is isothermal.

These two assumptions will guarantee satisfaction of the energy balances. When the gas and liquid streams are both fairly dilute, the assumptions will probably be satisfied.

If the solute mole fraction in the feed $y_{B,N+1}$ is very low, then transferring most or even all of the solute to the liquid will have very little effect on the overall vapor flow rate V or on the overall liquid flow rate L. Thus, we can assume that L and V are both constant, and the operating line on a McCabe-Thiele diagram will be straight. Using the mass balance envelope around the top of the absorption column shown in Figure 12-1, we can write the solute B mass balance for constant L and V.

$$y_{j+1} \, V + x_0 \, L = y_1 \, V + x_j \, L \tag{12-5}$$

We dropped the subscript B because this mass balance plus L = constant and V = constant are the only mass balances we need for this simple dilute absorber.

Solving for y_{j+1} we obtain the equation for the McCabe-Thiele operating line.

$$y_{j+1} = \left(L/V\right)x_j + \left[y_1 - \left(L/V\right)x_0\right] \tag{12-6}$$

This operating line is a straight line with a slope of L/V and a y-intercept of $[y_1 - (L/V)x_0]$. All possible passing streams with compositions (x_j, y_{j+1}) must lie on the operating line. This includes the two streams at the top of the absorber (x_0, y_1) and the two streams at the bottom of the absorber (x_N, y_{N+1}).

The procedure for solving a dilute absorption problem is:

1. Plot the y versus x equilibrium data.
2. For a design problem, typically x_0, y_{N+1}, y_1, and L/V will be known. Point (x_0, y_1) is on the operating line and the slope is L/V. Plot the operating line.
3. Start at stage 1 and step off stages by alternating between the equilibrium and operating lines.

This procedure is illustrated in Figure 12-3 for Example 12-1.

Note that the operating line is above the equilibrium line. This occurs because solute is being transferred from gas to liquid. In distillation the more volatile component was transferred from liquid to gas and the operating line was below the equilibrium curve.

EXAMPLE 12-1. McCabe-Thiele analysis for dilute absorber

1000 kmol/h of air containing 100 ppm (mol) of chloroform at 25°C and 2.0 atmosphere is to be processed. We plan to absorb the chloroform with pure water at 25°C. Operation is at L/V = 1.4 (L/V)$_{min}$. If we want an outlet air stream containing 10.0 ppm chloroform, how many equilibrium stages are required?

Solution

A. Define. The operation uses an absorber similar to Figure 12-1 with V = 1000 kmol/h, y_{N+1} = 100 ppm, y_1 = 10 ppm, x_0 = 0, and L/V = 1.4 (L/V)$_{min}$. Find the number of equilibrium stages, N.

B. Explore. Equilibrium data in the form of a Henry's law constant are given in Table 12-2. Since concentrations are low, the total flow rates L and V will be constant. The McCabe-Thiele diagram can be plotted in terms of ppm. Then (L/V)$_{min}$, L/V, and N can be determined.

C. Plan. Plot the equilibrium data. Plot known point (x_0, y_1). Find (L/V)$_{min}$ and L/V. Plot the operating line and step off stages.

D. Do It. Equilibrium is $y = Hx/p = 211.19x/2.0 = 105.6x$ with y and x in ppm. This is a straight line of slope 105.6 that goes through the origin. With reasonable care we can plot this line on a y versus x plot, as shown in Figure 12-3. (Suggested procedure: calculate x_{equil} at $y = 100$ as $x_{equil} = 100/105.6 = 0.947$ ppm and plot this point.)

The operating line goes through point $(x_0, y_1) = (0, 10 \text{ ppm})$. The minimum operating line goes through this point and the point on the equilibrium line at $y_{N+1} = 100$ ppm. At this value of y, $x_{min\_L/V} = y_{N+1}/\text{slope} = 100/105.6 = 0.947$ ppm. (Note that it is more accurate to calculate the value than to determine it from the value of equilibrium at y_{N+1} on Figure 12-3.) The slope of the minimum operating line is

$$(L/V)_{min} = (y_N - y_1)/(x_{min\_L/V} - x_0) \tag{12-7}$$

$$(L/V)_{min} = (110 - 10)/(0.947 - 0) = 95.0, \text{ and}$$

$$L/V = 1.4 \, (L/V)_{min} = (1.4)(95) = 133.$$

Equation (12-6) is valid when $j = N$. Solving for x_N with $x_0 = 0$, we obtain,

$$x_N = (y_{N+1} - y_1)/(L/V) = (100 - 10)/133 = 0.68 \text{ ppm}$$

The operating line is most easily plotted without error by drawing the straight line from point $(x_0, y_1) = (0, 10 \text{ ppm})$ to point $(x_N, y_{N+1}) = (0.68, 100 \text{ ppm})$. This is shown in Figure 12-3. Step off stages as shown in the figure. Five equilibrium stages are more than sufficient. If desired, we can estimate a fractional number of equilibrium contacts,

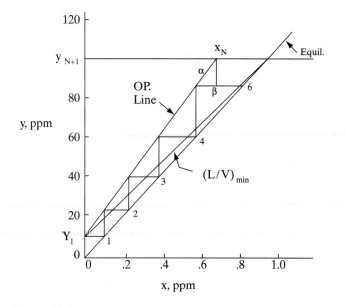

FIGURE 12-3.   *McCabe-Thiele diagram for absorption, Example 12-1*

Fraction = (distance from op line to x_N)/(distance from op line to equil line)

Fraction = (distance a to b)/(distance a to c) ≈ 0.4

     Thus, we need 4.4 equilibrium contacts.
- E. Check. We can check the result with the Kremser equation (see Section 12.4). This check gives N = 4.5, which is within the accuracy of the graph and is left as Problem 12.D19.
- F. Generalize. Note that the gas concentration is higher than the liquid concentration. Stripping chloroform from water with air would work quite well (see Section 12.3), but the physical absorption step does not work as well.

A fair amount of attention in this example went into proper plotting of the equilibrium and operating lines. With unequal axes, many students will plot these lines incorrectly if they do not calculate the values of points on the lines.

12.3 STRIPPING ANALYSIS FOR DILUTE SYSTEMS

Since stripping is very similar to absorption we expect the method to be similar. The mass balance for the column shown in Figure 12-4 is the same as for absorption and the operating line is still

$$y_{j+1} = \frac{L}{V} x_j + (y_1 - \frac{L}{V} x_0)$$

    For stripping we know x_0, x_N, y_{N+1}, and L/V. Since (x_N, y_{N+1}) is a point on the operating line, we can plot the operating line and step off stages. This is illustrated in Figure 12-5.
    Note that the operating line is below the equilibrium curve because solute is transferred from liquid to gas. This is therefore similar to the stripping section of a distillation column. A maximum L/V ratio can be defined; this corresponds to the minimum amount of stripping gas. Start from the known point (y_{N+1}, x_N), and draw a line to the intersection of x = x_0 and the equilibrium curve. Alternatively, there may be a tangent pinch point. For a stripper, $y_1 > y_{N+1}$, while the reverse is true in absorption. Thus, the top of the column is on the right side in

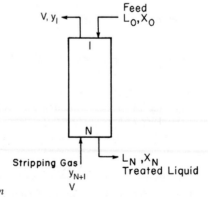

FIGURE 12-4.   *Stripping column*

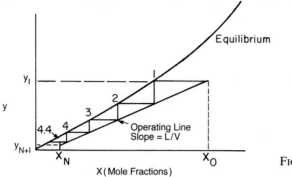

FIGURE 12-5.   *McCabe-Thiele diagram for stripping*

Figure 12-5 but on the left side in Figure 12-3. Stripping often has large temperature changes, so the calculation method used here is often appropriate only for very dilute systems.

Murphree efficiencies can be used on these diagrams.

Efficiencies for absorption and stripping are often quite low.

12.4   ANALYTICAL SOLUTION FOR DILUTE SYSTEMS: KREMSER EQUATION

When the solution is quite dilute (say less than 1% solute in both gas and liquid), the total liquid and gas flow rates will not change significantly since little solute is transferred. The column was shown in Figure 12-1 and the operating equation was given in Eq. (12-6).

To use Eq. (12-6) in a McCabe-Thiele diagram, we assume the following:

1. L/V (total flows) is constant.
2. Isothermal system.
3. Isobaric system.
4. Negligible heat of absorption.

These are reasonable assumptions for dilute absorbers and strippers.

If one additional assumption is valid, the stage-by-stage problem can be solved analytically. This additional assumption is that the

5. Equilibrium line is straight.

$$y_j = mx_j + b \qquad\qquad (12\text{-}8)$$

This assumption is reasonable for very dilute solutions and agrees with Henry's law, Eq. (12-3), if $m = H_B/p_{tot}$ and $b = 0$.

An analytical solution for absorption is easily derived for the special case shown in Figure 12-6, where the operating and equilibrium lines are parallel. Now the distance between operating and equilibrium lines, Δy, is constant. To go from outlet to inlet concentrations with N stages, we have

$$N \, \Delta y = y_{N+1} - y_1 \qquad\qquad (12\text{-}9)$$

since each stage causes the same change in vapor composition. Δy can be obtained by subtracting the equilibrium Eq. (12-8) from the operating Eq. (12-6).

$$(\Delta y)_j = y_{j+1} - y_j = (\frac{L}{V} - m)x_j + (y_1 - \frac{L}{V}x_0 - b) \tag{12-10}$$

For the special case shown in Figure 12-6 $L/V = m$ (the lines are parallel), Eq. (12-10) becomes

$$\Delta y = (\Delta y)_j = y_j - \frac{L}{V}x_0 - b = \text{constant} \tag{12-11}$$

Combining Eqs. (12-9) and (12-11), we get

$$N = \frac{y_{N+1} - y_1}{(y_1 - \frac{L}{V}x_0 - b)} \qquad \text{for } \frac{L}{mV} = 1 \tag{12-12}$$

Equation (12-12) is a special case of the Kremser equation. When this equation is applicable, absorption and stripping problems can be solved quite simple and accurately without the need for a stage-by-stage calculation.

Figure 12-6 and the resulting Eq. (12-12) were for a special case. The more general case is shown in Figure 12-7. Now Δy_j varies from stage to stage. The Δy_j values can be determined from Eq. (12-10). Equation (12-10) is easier to use if we replace x_j with the equilibrium Eq. (12-8),

$$x_j = \frac{y_j - b}{m} \tag{12-13}$$

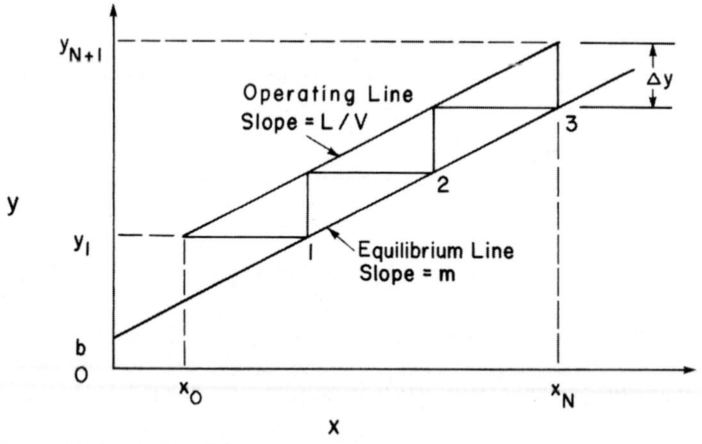

FIGURE 12-6.   *McCabe-Thiele diagram for dilute absorber with parallel equilibrium and operating lines*

Then

$$(\Delta y)_j = (\frac{L}{mV} - 1)y_j + (y_1 - \frac{L}{mV}b - \frac{L}{V}x_0)$$

(12-14a)

$$(\Delta y)_{j+1} = (\frac{L}{mV} - 1)y_{j+1} + (y_1 - \frac{L}{mV}b - \frac{L}{V}x_0)$$

(12-14b)

Subtracting Eq. (12-14a) from (12-14b), and solving for $(\Delta y)_{j+1}$

$$(\Delta y)_{j+1} = \frac{L}{mV}(\Delta y)_j$$

(12-15)

Equation (12-15) relates the change in vapor composition from stage to stage to (L/mV), which is known as the *absorption factor*. If either the operating or equilibrium line is curved, this simple relationship no longer holds and a simple analytical solution does not exist.

The difference between inlet and outlet gas concentrations must be the sum of the Δy_j values shown in Figure 12-7. Thus,

$$\Delta y_1 + \Delta y_2 + \cdots + \Delta y_N = y_{N+1} - y_1$$

(12-16a)

Applying Eq. (12-15)

$$\Delta y_1(1 + \frac{L}{mV} + (\frac{L}{mV})^2 + \cdots + (\frac{L}{mV})^{N-1}) = y_{N+1} - y_1$$

(12-16b)

The summation in Eq. (12-16b) can be calculated. The general formula is

$$\sum_{i=0}^{k} aA^i = \frac{a(1 - A^{k+1})}{(1 - A)} \qquad \text{for} \quad |A| < 1$$

(12-17)

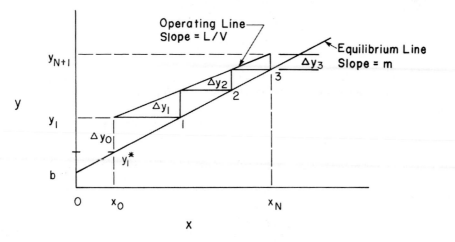

FIGURE 12-7. *McCabe-Thiele diagram for dilute absorber. (L/V) < m*

Then Eq. (12-16b) is

$$\frac{y_{N+1} - y_1}{\Delta y_1} = \frac{1 - (\frac{L}{mV})^N}{1 - (\frac{L}{mV})}$$

(12-18)

If $L/mV > 1$, then divide both sides of Eq. (12-16b) by $(L/mV)^{N-1}$ and do the summation in terms of mV/L. The resulting equation will still be Eq. (12-18). From Eq. (12-15), $\Delta y_1 = \Delta y_0 L/mV$ where $\Delta y_0 = y_1 - y_1^*$ is shown in Figure 12-7. The vapor composition y_1^* is the value that would be in equilibrium with the inlet liquid, x_0. Thus,

$$y_1^* = mx_0 + b$$

(12-19)

Removal of Δy_1 from Eq. (12-18) gives

$$\frac{y_{N+1} - y_1}{y_1 - y_1^*} = \frac{\frac{L}{mV} - (\frac{L}{mV})^{N+1}}{1 - \frac{L}{mV}}$$

(12-20)

Equation (12-20) is one form of the Kremser equation (Kremser, 1930; Souders and Brown, 1932). A large variety of alternative forms can be developed by algebraic manipulation. For instance, if we add 1 to both sides of Eq. (12-20) and rearrange, we have

$$\frac{y_{N+1} - y_1^*}{y_1 - y_1^*} = \frac{1 - (\frac{L}{mV})^{N+1}}{1 - \frac{L}{mV}}$$

(12-21)

which can be solved for N. After manipulation, this result is

$$N = \frac{\ln[(1 - \frac{mV}{L})(\frac{y_{N+1} - y_1^*}{y_1 - y_1^*}) + \frac{mV}{L}]}{\ln(\frac{L}{mV})}$$

(12-22)

where $L/(mV) \neq 1$. Equations (12-21) and (12-22) are also known as forms of the Kremser equation. Alternative derivations of the Kremser equation are given by Brian (1972) and King (1980).

A variety of forms of the Kremser equation for $L/(mV) \neq 1$ can be developed. Several alternative forms in terms of the gas-phase composition are

$$\frac{y_{N+1} - y_1}{y_{N+1} - y_1^*} = \frac{(L/(mV)) - (L/(mV))^{N+1}}{1 - (L/(mV))^{N+1}}$$

(12-23)

$$\frac{y_{N+1} - y_{N+1}^*}{y_1 - y_1^*} = (\frac{L}{mV})^N$$

(12-24)

Chapter 12   Absorption and Stripping

$$N = \frac{\ln[(y_{N+1} - y_{N+1}^*)/(y_1 - y_1^*)]}{\ln(L/(mV))} \qquad (12\text{-}25)$$

$$N = \frac{\ln[(y_{N+1} - y_{N+1}^*)/(y_1 - y_1^*)]}{\ln[(y_{N+1} - y_1)/(y_{N+1}^* - y_1^*)]} \qquad (12\text{-}26)$$

where

$$y_{N+1}^* = mx_N + b \quad \text{and} \quad y_1^* = mx_0 + b \qquad (12\text{-}27)$$

Alternative forms in terms of the liquid phase composition are

$$N = \frac{\ln\left[(1 - \dfrac{L}{mV})(\dfrac{x_0 - x_N^*}{x_N - x_N^*}) + \dfrac{L}{mV}\right]}{\ln(mV/L)} \qquad (12\text{-}28)$$

$$N = \frac{\ln[(x_N - x_N^*)/(x_0 - x_0^*)]}{\ln(L/(mV))} \qquad (12\text{-}29)$$

$$N = \frac{\ln[(x_N - x_N^*)/(x_0 - x_0^*)]}{\ln[(x_0^* - x_N^*)/(x_0 - x_N)]} \qquad (12\text{-}30)$$

$$\frac{x_N - x_N^*}{x_0 - x_N^*} = \frac{1 - (mV/L)}{1 - (mV/L)^{N+1}} \qquad (12\text{-}31)$$

$$\frac{x_N - x_N^*}{x_0 - x_0^*} = \left(\frac{L}{mV}\right)^N \qquad (12\text{-}32)$$

where

$$x_N^* = \frac{y_{N+1} - b}{m} \quad \text{and} \quad x_0^* = \frac{y_1 - b}{m} \qquad (12\text{-}33)$$

A form including a constant Murphree vapor efficiency is (King, 1980)

$$N_{actual} = -\frac{\ln\{[1 - mV/L][(y_{N+1} - y_1^*)/(y_1 - y_1^*)] + mV/L\}}{\ln[1 + E_{MV}(mV/L - 1)]} \qquad (12\text{-}34)$$

Forms for systems with three phases where two phases flow co-currently and countercurrent to the third phase were developed by Wankat (1980). Forms of the Kremser equation for columns with multiple sections are developed by Brian (1972, Chap. 3) and by King (1980, pp. 371-376). Forms for reboiled absorbers are given by Hwang et al. (1992a).

When the assumptions required for the derivation are valid, the Kremser equation has several advantages over the stage-by-stage calculation procedure. If the number of stages is large, the Kremser equation is much more convenient to use, and it is easy to program on a computer or calculator. When the number of stages is specified, the McCabe-Thiele stage-by-stage procedure is trial-and-error, but the use of the Kremser equation is not. Because calculations can be done faster, the effects of varying y_1, x_0, L/V, m etc. are easy to determine. The major disadvantage of the Kremser equation is that it is accurate only for dilute solutions where L/V is constant, equilibrium is linear, and the system is isothermal. The appropriate form of the Kremser equation depends on the context of the problem.

The optimum value of mV/L for absorption is approximately 0.7 and for stripping is approximately 1.4 (Woods, 2007). For an absorber removing most of the solute $y_1 \approx 0$ and the operating line with the minimum L/V is essentially collinear with the equilibrium curve. Thus, $(L/V)_{min} = m$ and

$$(L/V)_{\text{optimum for absorption}} \approx m/0.7 \approx (L/V)_{min}/0.7 = 1.4 \, (L/V)_{min} \qquad \textbf{(12-35a)}$$

For stripping with most of the solute removed, $x_N \approx 0$ and the operating line with minimum V and a maximum slope is collinear with the equilibrium curve. Then

$$(L/V)_{\text{optimum for stripping}} \approx m/1.4 \approx (L/V)_{max}/1.4 = 0.7 \, (L/V)_{max} \qquad \textbf{(12-35b)}$$

EXAMPLE 12-2. Stripping analysis with Kremser equation

A plate tower providing six equilibrium stages is employed for stripping ammonia from a wastewater stream by means of countercurrent air at atmospheric pressure and 80°F. Calculate the concentration of ammonia in the exit water if the inlet liquid concentration is 0.1 mol% ammonia in water, the inlet air is free of ammonia, and 30 standard cubic feet (scf) of air are fed to the tower per pound of wastewater.

Solution

A. Define. The column is sketched in the figure.

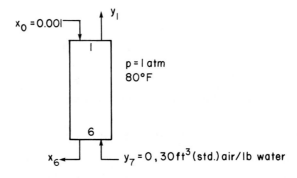

We wish to find the exit water concentration, x_6.

B. Explore. Since the concentrations are quite low we can use the Kremser equation. Equilibrium data are available in several sources. From King (1971, p. 273) we find $y_{NH3} = 1.414 \, x_{NH3}$ at 80°F.

C. Plan. We have to convert flow to molar units. Since we want a concentration of liquid, forms (12-31) or (12-32) of the Kremser equation will be convenient. We will use Eq. (12-31).

D. Do it. We can calculate ratio V/L,

$$\frac{V}{L} = \frac{30 \text{ scf air}}{1 \text{ lb water}} \times \frac{1 \text{ lbmol air}}{379 \text{ scf air}} \times \frac{18 \text{ lb water}}{1 \text{ lbmol water}}$$

$$= 1.43 \text{ mol air/mol water}$$

Note that the individual flow rates are not needed.
The Kremser equation [form (12-31)] is

$$\frac{x_N - x_N^*}{x_0 - x_N^*} = \frac{1 - \dfrac{mV}{L}}{1 - (\dfrac{mV}{L})^{N+1}}$$

Where $x_N = x_6$ is unknown, $x_0 = 0.001$, $m = 1.414$, $b = 0$, $x_N^* = y_7/m = 0$, $V/L = 1.43$, $N = 6$
Rearranging,

$$x_N = \frac{1 - mV/L}{1 - (mV/L)^{N+1}} x_0 = \frac{1 - (1.414)(1.43)}{1 - [(1.414)(1.43)]^7} (0.001) = 7.45 \times 10^{-6} \text{ mole fraction}$$

Most of the ammonia is stripped out by the air.

E. Check. We can check with a different form of the Kremser equation or by solving the results graphically; both give the same result. We should also check that the major assumptions of the Kremser equation (constant flow rates, linear equilibrium, and isothermal) are satisfied. In this dilute system they are.

F. Generalize. This problem is trial-and-error when it is solved graphically, but a graphical *check* is not trial-and-error. Also, the Kremser equation is very easy to set up on a computer or calculator. Thus, *when it is applicable,* the Kremser equation is very convenient.

12.5 EFFICIENCIES

Usually the best way to determine efficiencies is to measure them on commercial-scale equipment. In the absence of such data a rough prediction of the overall efficiency E_0 can be obtained from O'Connell's correlation shown in Figure 12-8 (O'Connell, 1946). Although originally done for bubble-cap systems, the results can be used for a first estimate for sieve and valve trays. The data in Figure 12-6 are fit by (Kessler and Wankat, 1988)

$$E_0 = 0.37237 + 0.19339 \log (Hp/\mu) + 0.024816 (\log Hp/\mu)^2 \qquad \textbf{(12-36)}$$

which is valid for $(Hp/\mu) > 0.000316$. The Henry's law constant H is in lb mole/(atm · ft³), pressure p in atm, and liquid viscosity μ in centipoise (cP). More detailed estimates can be made using a mass transfer analysis (see Chapter 16). Correlations for very dilute strippers are given by Hwang (1981). As a first estimate, Seider et al. (2009) recommend overall efficiencies of 30% for absorbers and 50% for strippers.

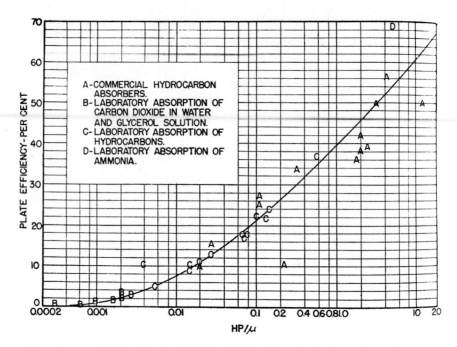

FIGURE 12-8. *O'Connell's correlation for overall efficiency of bubble-cap absorbers, reprinted from O'Connell, Trans. AIChE, 42, 741 (1946), copyright 1946, AIChE*

12.6 MCCABE-THIELE ANALYSIS FOR MORE CONCENTRATED SYSTEMS

If absorption or stripping can be assumed 1) to be isothermal and 2) to have negligible heat of absorption, then the energy balances will be satisfied. In this case the McCabe-Thiele analysis procedure can be adapted to more concentrated systems where the total flow rates L and V are not constant. We will have the desired straight operating line if we define

$$S/G = \frac{\text{mol nonvolatile solvent/h}}{\text{mol insoluble carrier gas/h}}$$

and if we also assume that:

3. Solvent is nonvolatile
4. Carrier gas is insoluble

Assumptions 3 and 4 are often closely satisfied. The results of these last two assumptions are that the mass balance for solvent becomes

$$S_N = S_j = S_0 = S = \text{constant} \tag{12-37a}$$

while the mass balance for the carrier gas is

$$G_{N+1} = G_j = G_1 = G = \text{constant} \tag{12-37b}$$

Note that we cannot use overall flow rates of gas and liquid in concentrated mixtures because a significant amount of solute may be absorbed which would change gas and liquid flow rates and give a curved operating line. Since we want to use S = moles nonvolatile solvent/hr and G = moles insoluble carrier gas (C)/hr, we must define our compositions in such a way that we can write a mass balance for solute B. How do we do this?

After some manipulation we find that the correct way to define our compositions is as *mole ratios*. Define

$$Y = \frac{\text{mol B in gas}}{\text{mol pure carrier gas C}} \quad \text{and} \quad X = \frac{\text{mol B in liquid}}{\text{mol pure solvent S}} \quad \textbf{(12-38a)}$$

The mole ratios Y and X are related to our usual mole fracs by

$$Y = \frac{y}{1-y} \quad \text{and} \quad X = \frac{x}{1-x} \quad \textbf{(12-38b)}$$

Note that both Y and X can be greater than 1.0. With the mole ratio units, we have

$$Y_j G = \left[\frac{\text{mol B in gas stream } j}{\text{mol carrier}}\right]\left[\frac{\text{mol carrier gas}}{h}\right] = \frac{\text{mol B in gas stream } j}{h}$$

and

$$X_j S = \left[\frac{\text{mol B in liquid stream } j}{\text{mol solvent}}\right]\left[\frac{\text{mol solvent}}{h}\right] = \frac{\text{mol B in solvent stream } j}{h}$$

Thus, we can easily write the steady-state mass balance, input = output, in these units. The mass balance around the top of the column using the mass balance envelope shown in Figure 12-1 is

$$Y_{j+1} G + X_0 S = X_j S + Y_1 G \quad \textbf{(12-39)}$$

or

$$\text{Mol B in}/h = \text{moles B out}/h$$

Solving for Y_{j+1} we obtain

$$Y_{j+1} = \frac{S}{G} X_j + \left[Y_1 - \frac{S}{G} X_0\right] \quad \textbf{(12-40)}$$

This is a straight line with slope S/G and intercept $(Y_1 - (S/G) X_0)$. It is our *operating line* for absorption. Thus, if we plot ratios Y vs. X we have a McCabe-Thiele type of graph as shown in Figure 12-9. Note that this graph looks similar to Figure 12-3 except different variables are used.

The steps in this procedure are very similar to those used for dilute systems:

1. Plot Y vs. X equilibrium data (convert from fraction to ratios).

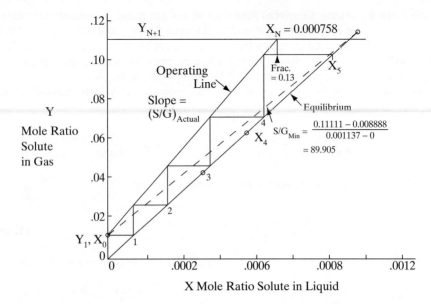

FIGURE 12-9. *McCabe-Thiele diagram for absorption, Example 12-3*

2. Values of X_0, Y_{N+1}, Y_1 and S/G are known. Point (X_0, Y_1) is on operating line, since it represents passing streams.
3. Slope is S/G. Plot operating line.
4. Start at stage 1 and step off stages by alternating between the equilibrium curve and the operating line.

Equilibrium data must be converted to ratio units, Y vs. X. These values can be greater than 1.0, since $Y = y/(1 - y)$ and $X = x/(1 - x)$. The Y = X line has no significance in absorption. As usual the stages are counted at the equilibrium curve. A minimum S/G ratio can be defined as shown in Figure 12-9. If the system is not isothermal, the operating line will not be affected, but the equilibrium line will be. Then the McCabe-Thiele method must be modified to include changing equilibrium curves.

EXAMPLE 12-3. Graphical analysis for more concentrated absorber

A gas stream is 90 mol% N_2 and 10 mol% CO_2. We wish to absorb the CO_2 into water. The inlet water is pure and is at 5°C. Because of cooling coils, operation can be assumed to be isothermal. Operation is at 10 atm. If the liquid flow rate is 1.5 times the minimum liquid flow rate, how many equilibrium stages are required to absorb 92% of the CO_2? Choose a basis of 1 mol/h of entering gas.

Solution

A. Define. See the sketch. We need to find the minimum liquid flow rate, the value of the outlet gas concentration, and the number of equilibrium stages required.

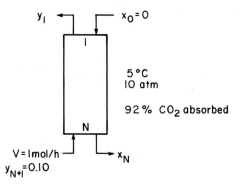

B. Explore. First we need equilibrium data. These are available in Table 12-1. Since concentrations are fairly high, the problem should be solved in mole ratios. Thus, we need to convert all compositions including equilibrium data to mole ratios.

C. Plan. Derive the equilibrium equation from Henry's law. Convert compositions from mole fracs to mole ratios using Eq. (12-38). Calculate Y_1 by a percent recovery analysis. Plot mole ratio equilibrium data on a Y-X diagram, and determine $(S/G)_{min}$ and hence S_{min}. Calculate actual S/G, plot operating line, and step off stages.

The problem appears to be straightforward.

D. Do it. *Equilibrium:*

$$y = \frac{H}{P_{tot}} x = \frac{876}{10} x = 87.6x$$

Change the equilibrium data to mole ratios with a table as shown below. (The equation can also be converted, but it is easier to avoid a mistake with a table.)

x	X = x/(1–x)	y = 87.6 x	Y = y/(1–y)
0	0	0	0
0.0001	0.0001	0.00876	0.00884
0.0004	0.0004	0.0350	0.0363
0.0006	0.0006	0.0526	0.0555
0.0008	0.0008	0.0701	0.0754
0.0010	0.0010	0.0876	0.0960
0.0012	0.0012	0.10512	0.1175

Note that x = X in this concentration range, but y ≠ Y. The inlet gas mole ratio is

$$Y_{N+1} = \frac{y_{N+1}}{1 - y_{N+1}} = \frac{0.1}{0.9} = 0.1111 \frac{\text{mol CO}_2}{\text{mol N}_2}$$

$$G = (1 \text{ mol total gas/h})(1 - y_{N+1}) = 0.9 \frac{\text{mol N}_2}{\text{h}}$$

Percent Recovery Analysis: 8% of CO_2 exits.

(0.1 mole in)(0.08 exits) = 0.008 moles CO_2 out

Thus,

$$Y_1 = \frac{mol\,CO_2\,out}{mol\,N_2\,out} = \frac{0.008\,CO_2}{0.9\,mol\,N_2} = 0.008888$$

Operating Line:

$$Y_{j+1} = \frac{S}{G}\,X_j + (Y_1 - \frac{S}{G}\,X_0)$$

Goes through point $(Y_1, X_0) = (0.008888, 0)$.

$(S/G)_{min}$ is found as the slope of the operating line from point (Y_1, X_0) to the intersection with the equilibrium curve at Y_{N+1}. This is shown on Figure 12-9.

The fraction was calculated as

$$Frac = \frac{X_{out} - X_4}{X_5 - X_4} = \frac{0.000758 - 0.00071}{0.00108 - 0.00071} = 0.13$$

E. Check. The overall mass balances are satisfied by the outlet concentrations. The significant figures carried in this example are excessive compared with the equilibrium data. Thus, they should be rounded off when reported (e.g., N = 4.1). The concentrations used were quite high for Henry's law. Thus, it would be wise to check the equilibrium data.

F. Generalize. Note that the gas concentration is considerably greater than the liquid concentration. This situation is common for physical absorption (solubility is low). Chemical absorption is used to obtain more favorable equilibrium. The liquid flow rate required for physical absorption is often excessive. Thus, in practice, this type of operation uses chemical absorption.

If we had assumed that total gas and liquid flow rates were constant (dilute solutions), the result would be in error. An estimate of this error can be obtained by estimating $(L/V)_{min}$ with the incorrect mole fraction units. The minimum operating line in these units goes from $(y_1, x_0) = (0.00881, 0)$ to $(y_{N+1}, x_{equil,N+1})$. $y_{N+1} = 0.1$ and $x_{equil,N+1} = y_{N+1}/87.6 = 0.1/87.6 = 0.0011415$. Then

$$(L/V)_{min,dilute} = \frac{y_{N+1} - y_1}{x_{equil,N+1} - x_0} = \frac{0.1 - 0.00881}{0.0011415 - 1} = 79.9$$

This is in error by more than 10%.

12.7   COLUMN DIAMETER

For absorption and stripping, the column diameter is designed the same way as for a staged or packed distillation column (Chapter 10). However, note that the gas flow rate, G, must now be converted to the total gas flow rate, V. The carrier gas flow rate, G, is

$$G, \text{mol carrier gas/h} = (1 - y_j)V_j \qquad \textbf{(12-41)}$$

Since

$$y_j = \frac{Y_j}{1 + Y_j}$$

this is

$$G = \left(\frac{1}{1 + Y_j}\right)V_j \qquad \textbf{(12-42a)}$$

or

$$(1 + Y_j)G = V_j \qquad \textbf{(12-42b)}$$

The total liquid flow rate, L_j, can be determined from an overall mass balance. Using the balance envelope shown in Figure 12-1, we obtain

$$L_j - V_{j+1} = L_0 - V_1 = \text{constant for any } j \qquad \textbf{(12-43)}$$

Note that the difference between total flow rates of passing streams is constant. This is illustrated in Figure 12-10.

Both total flows V_j and L_j will be largest where Y_j and X_j are largest. This is at the bottom of the column for absorption, and therefore you design the diameter at the bottom of the column. In strippers, flow rates are highest at the top of the column, so you design the diameter for the top of the column. Specific design details for absorbers and strippers are discussed by Kister et al. (2008) and Zenz (1997).

An order of magnitude estimate of the column diameter can be made quite easily (Reynolds et al., 2002). The superficial gas velocity (velocity in an empty column) is typically in the range v = 0.9144 to 1.8288 m/s (3 to 6 ft/sec). The volumetric flow rate of the gas is $V/\hat{\rho}$ ($\hat{\rho}$ is the molar gas density), the required cross sectional area is $A_c = V/(\hat{\rho}\, v)$ and the column diameter is $D = \sqrt{4A_c/\pi}$. Substituting an average superficial gas velocity of 1.37 m/s (4.5 ft/s) with other appropriate values gives an estimate of the column diameter.

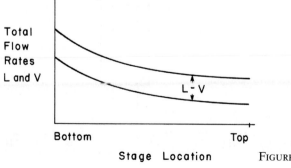

FIGURE 12-10. *Total flow rates in absorber*

12.8 DILUTE MULTISOLUTE ABSORBERS AND STRIPPERS

Up to this point we have been restricted to cases where there is a single solute to recover. Both the stage-by-stage McCabe-Thiele procedures and the Kremser equation can be used for multisolute absorption and stripping if certain assumptions are valid. The single-solute analysis by both procedures required systems that 1) are isothermal, 2) are isobaric, 3) have a negligible heat of absorption, and 4) have constant flow rates. These assumptions are again required.

To see what other assumptions are required consider the Gibbs phase rule for a system with three solutes plus a solvent and a carrier gas. The phase rule is

$$F = C - P + 2 = 5 - 2 + 2 = 5$$

Five degrees of freedom is a large number. In order to represent equilibrium as a single curve or in a linear form like Eq. (12-16), four of these degrees of freedom must be specified. Constant temperature and pressure utilize two degrees of freedom. The other two degrees of freedom can be specified by assuming that 5) solutes are independent of each other; in other words, that equilibrium for any solute does not depend on the amounts of other solutes present. This assumption requires dilute solutions. In addition, the analysis must be done in terms of mole or mass *fractions* and total flow rates for which dilute solutions are also required. An analysis using ratio units will not work because the ratio calculation

$$Y_i = \frac{y_1}{1 - y_1 - y_2 - y_3}$$

involves other solute concentrations that will be unknown.

The practical effect of the fifth assumption is that we can solve the multisolute problem once for each solute, treating each problem as a single-component problem. This is true for both the stage-by-stage solution method and the Kremser equation. Thus, for the absorber shown in Figure 12-11 we solve three single-solute problems.

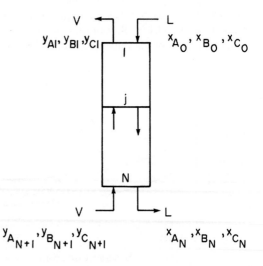

FIGURE 12-11. *Dilute multisolute absorber* $y_{A_{N+1}}, y_{B_{N+1}}, y_{C_{N+1}}$ $x_{A_N}, x_{B_N}, x_{C_N}$

Each additional solute increases the degrees of freedom for the absorber by two. These two degrees of freedom are required to specify the inlet gas and inlet liquid compositions. For the usual design problem, as shown in Figure 12-11, the inlet gas and inlet liquid compositions and flow rates will be specified. With temperature and pressure also specified, one degree of freedom is left. This is usually used to specify one of the outlet solute concentrations such as $y_{B,1}$. The design problem for solute B is now fully specified. To solve for the number of stages, we can plot the equilibrium data, which are of the form $y_B = f_B(x_B)$ on a McCabe-Thiele diagram. When assumptions 1 to 5 are satisfied, the equilibrium expression will usually be linear. The operating equation

$$y_{B,j+1} = \frac{L}{V}x_{B,j} + (y_{B,1} - \frac{L}{V}x_{B,0})$$

(12-44a)

is the same as Eq. (12-6) and can also be plotted on the McCabe-Thiele diagram. Then the number of stages is stepped off as usual. This is shown in Figure 12-12.

Once the number of stages has been found from the solute B calculation, the concentrations of solutes A and C can be determined by solving two fully specified simulation problems. That is, the number of stages is known and the outlet compositions have to be calculated. Simulation problems require a trial-and-error procedure when a stage-by-stage calculation is used. One way to do this calculation for component A is:

1. Plot the A equilibrium curve, $y_A = f_A(x_A)$.
2. Guess $y_{A,1}$ for solute A.
3. Plot the A operating line,

$$y_{A,j+1} = \frac{L}{V}x_{A,j} + (y_{A,1} - \frac{L}{V}x_{A,0})$$

(12-44b)

Slope = L/V, which is same as for solute B. Point $(y_{A,1}, x_{A,0})$ is on the operating line. This is shown in Figure 12-12.
4. Step off stages up to $y_{A,N+1}$ (see Figure 12-12).
5. Check: Are the number of stages the same as calculated? If yes, you have the answer. If no, return to step 2.

The procedure for solute C is the same.

The three diagrams shown in Figure 12-12 are often plotted on the same y-x graph. This saves paper but tends to be confusing.

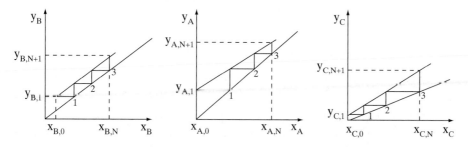

FIGURE 12-12. *McCabe-Thiele solution for dilute three-solute absorber where solute B is specified, and solutes A and C are trial and error*

If the equilibrium function for each solute is linear as in Eq. (12-8), the Kremser equation can be used. First the design problem for solute B is solved by using Eq. (12-22), (12-25), or (12-26) to find N. Then separately solve the two simulation problems (for solutes A and C) using equations such as (12-20), (12-21), or (12-24). Remember to use m_A and m_C when you use the Kremser equation. Note that when the Kremser equation can be used, the simulation problems are *not* trial-and-error.

These solution methods are restricted to very dilute solutions. In more concentrated solutions, flow rates are not constant, solutes may not have independent equilibria, and temperature effects become important. When this is true, more complicated computer solution methods involving simultaneous mass and energy balances plus equilibrium are required. These methods are discussed in the next section and the computer simulation is explored in the chapter appendix.

12.9 MATRIX SOLUTION FOR CONCENTRATED ABSORBERS AND STRIPPERS

For more concentrated solutions, absorbers and strippers are usually not isothermal, total flow rates are not constant, and solutes may not be independent. The matrix methods discussed for multicomponent distillation (reread section 6.2) can be adapted for absorption and stripping.

Absorbers, like flash distillation, are equivalent to very wide boiling feeds. Thus, in contrast with distillation, a wide-boiling feed (sum rates) flowchart such as Figure 2-13 should be used. The flow rate loop is now solved first, since flow rates are never constant in absorbers. The energy balance, which requires the most information, is used to calculate new temperatures, since this is done last. Figure 12-13 shows the sum-rates flow diagram for absorbers and strippers when $K_i = K_i(T, p)$. If $K_i = K_i (T, p, x_1, x_2, \cdots x_c)$ a concentration correction loop is added. The initial steps are very similar to those for distillation, and usually the same physical properties package is used.

The mass balance and equilibrium equations are very similar to those for distillation and the column is again numbered from the top down as shown in Figure 12-14. To fit into the matrix form, streams V_{N+1} and L_0 are relabeled as feeds to stages N and 1, respectively. The stages from 2 to N have the same general shape as for distillation (Figure 6-3). Thus, the mass balances and the manipulations [Eqs. (6-1) to (6-6) and (6-10) to (6-12)] are the same for distillation and absorption.

For stage 1 the mass balance becomes

$$B_1 \ell_1 + C_1 \ell_2 = D_1 \tag{12-45}$$

where

$$B_1 = 1 + \frac{V_1 K_1}{L_1}, \qquad C_1 = \frac{-V_2 K_2}{L_2}, \qquad D_1 = F_1 z_1 = L_0 x_0 \tag{12-46}$$

and the component flow rates are $l_1 = L_1 x_1$ and $l_2 = L_2 x_2$. These equations are repeated for each component. These equations differ from Eqs. (6-7) to (6-9) since the absorber equations are for an equilibrium stage, not a total condenser.

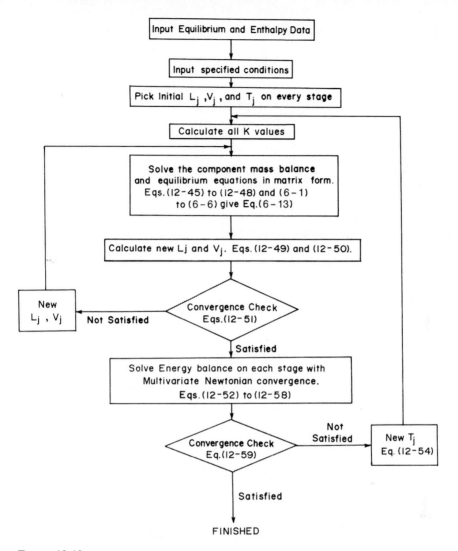

FIGURE 12-13.   *Sum rates convergence procedure for absorption and stripping*

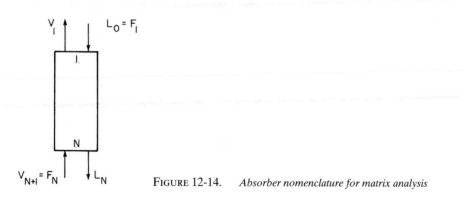

FIGURE 12-14.   *Absorber nomenclature for matrix analysis*

For stage N the mass balance is

$$A_n \ell_{N-1} + B_N \ell_N = D_N \tag{12-47}$$

where

$$A_N = -1, \qquad B_N = 1 + \frac{V_N K_N}{L_N}, \; D_N = F_N z_N = V_{N+1} y_{N+1} \tag{12-48}$$

These equations are essentially identical to Eqs. (6-10) to (6-12).

Combining Eqs. (12-45), (6-5), and (12-47) results in a tridiagonal matrix, Eq. (6-13), with all terms defined in Eqs. (12-46), (6-6), and (12-48). There is one matrix for each component. These tridiagonal matrices can each be inverted with the Thomas algorithm (Table 6-1). The results are liquid component flow rates, $l_{i,j}$, that are valid for the assumed L_j, V_j, and T_j.

The next step is to use the summation equations to find new total flow rates L_j and V_j. The new liquid flow rate is conveniently determined as

$$L_{j,new} = \sum_{i=1}^{C} \ell_{i,j} \tag{12-49}$$

The vapor flow rates are determined by summing the component vapor flow rates,

$$V_{j,new} = \sum_{i=1}^{C} \left[\left(\frac{K_{i,j} V_j}{L_j} \right)_{old} \ell_{i,j} \right] \tag{12-50}$$

Convergence can be checked with

$$\left| \frac{L_{j,old} - L_{j,new}}{L_{j,old}} \right| < \varepsilon \quad \text{and} \quad \left| \frac{V_{j,old} - V_{j,new}}{V_{j,old}} \right| < \varepsilon \tag{12-51}$$

for all stages. For computer calculations, an ε of 10^{-4} or 10^{-5} can be used. If convergence has not been reached, new liquid and vapor flow rates are determined, and we return to the component mass balances (see Figure 12-13). Direct substitution ($L_j = L_{j,new}$, $V_j = V_{j,new}$) is usually adequate.

Once the flow rate loop has converged, the energy balances are used to solve for the temperatures on each stage. This can be done in several different ways (King, 1980; Smith, 1963). We will discuss only a multivariate Newtonian convergence procedure. For the general stage j, the energy balance was given as Eq. (6-20). This can be rewritten as

$$E_j(T_{j-1}, T_j, T_{j+1}) = $$
$$L_j h_j + V_j H_j - L_{j-1} h_{j-1} - V_{j+1} H_{j+1} - F_j h_{Fj} - q_j = 0 \tag{12-52}$$

The multivariate Newtonian approach is an extension of the single variable Newtonian convergence procedure. The change in the energy balance is

$$(E_j)_{k+1} - (E_j)_k = \frac{\partial E_j}{\partial T_{j-1}}(\Delta T_{j-1}) + \frac{\partial E_j}{\partial T_j}(\Delta T_j) + \frac{\partial E_j}{\partial T_{j+1}}(\Delta T_{j+1}) \tag{12-53}$$

where k is the trial number. The ΔT values are defined as

$$\Delta T_{j-1} = (T_{j-1})_{k+1} - (T_{j-1})_k, \quad \Delta T_j = (T_j)_{k+1} - (T_j)_k, \quad \Delta T_{j+1} = (T_{j+1})_{k+1} - (T_{j+1})_k \tag{12-54}$$

The partial derivatives can be determined from Eq. (12-52) by determining which terms in the equation are direct functions of the stage temperatures T_{j-1}, T_j, and T_{j+1}. The partial derivatives are

$$\frac{\partial E_j}{\partial T_{j-1}} = -L_{j-1}\frac{\partial h_{j-1}}{\partial T_{j-1}} = -L_{j-1}c_{PL,j-1} = A_{E,j} \tag{12-55a}$$

$$\frac{\partial E_j}{\partial T_j} = L_j\frac{\partial h_j}{\partial T_j} + V_j\frac{\partial H_j}{\partial T_j} = L_jc_{PL,j} + V_jC_{PV,j} = B_{E,j} \tag{12-55b}$$

$$\frac{\partial E_j}{\partial T_{j+1}} = -V_{j+1}\frac{\partial H_{j+1}}{\partial T_{j+1}} = -V_{j+1}C_{PV,j+1} = C_{E,j} \tag{12-55c}$$

where we have identified the terms as A, B, and C terms for a matrix. The total stream heat capacities can be determined from individual component heat capacities. For ideal mixtures this is

$$C_{PL,j} = \left(\sum_{i=1}^{c} C_{PL,i} x_i\right)_j, \quad C_{PL,j} = \left(\sum_{i=1}^{c} C_{PV,i} y_i\right)_j \tag{12-56}$$

For the next trial we hope to have $(E_j)_{k+1} = 0$. If we define

$$-(E_j)_k = D_j \tag{12-57}$$

where $(E_j)_k$ is the numerical value of the energy balance for trial k on stage j, then the equations for ΔT_j can be written as

$$\begin{bmatrix} B_{E,1} & C_{E,1} & \cdot & & & & \\ A_{E,2} & B_{E,2} & C_{E,2} & \cdot & & & \\ 0 & A_{E,3} & B_{E,3} & C_{E,3} & \cdot & & \\ \cdots\cdots & & & & & & \\ & & & & A_{E,N-1} & B_{E,N-1} & C_{E,N-1} \\ 0 & & & & 0 & A_{E,N} & B_{E,N} \end{bmatrix} \begin{bmatrix} \Delta T_1 \\ \Delta T_2 \\ \Delta T_3 \\ \cdot \\ \Delta T_{N-1} \\ \Delta T_N \end{bmatrix} = \begin{bmatrix} D_1 \\ D_2 \\ D_3 \\ \cdot \\ D_{N-1} \\ D_N \end{bmatrix} \tag{12-58}$$

Equation (12-58) can be inverted using any computer inversion program or the Thomas algorithm shown in Table 6-1. The result will be all the ΔT_j values.

Convergence can be checked from the ΔT_j. If

$$|\Delta T_j| < \varepsilon_T \qquad \textbf{(12-59)}$$

for all stages, then convergence has been achieved. The problem is finished! If Eq. (12-59) is not satisfied, determine new temperatures from Eq. (12-54) and return to calculate new K values and redo the component mass balances (see Figure 12-13). A reasonable range for ε_T for computer solution is 10^{-2} to 10^{-3}. Computer solution methods are explored further in the appendix to this chapter.

12.10   IRREVERSIBLE ABSORPTION AND CO-CURRENT CASCADES

Absorption with an irreversible chemical reaction is often used in small facilities for removing obnoxious chemicals. For example, NaOH is used to remove both CO_2 and H_2S; it reacts with the acid gas in solution and forms a nonvolatile salt. This is convenient in small facilities, because the absorber is usually small and simple and no regeneration facilities are required. However, the cost of the reactant (NaOH) can make operation expensive. In addition, the salt formed has to be disposed of responsibly. In large-scale systems it is usually cheaper and more sustainable to use a solvent that can be regenerated.

Consider a simple absorber where the gas to be treated contains carrier gas C and solute B. The solvent contains a nonvolatile solvent S and a reagent R that will react irreversibly with B according to the irreversible reaction

$$R + B \rightarrow RB \qquad \textbf{(12-60)}$$

The resulting product RB is nonvolatile. At equilibrium, x_B (in the free form) = 0 since the reaction is irreversible and any B that is in solution will form product RB. Thus, $y_B = 0$ at equilibrium. As long as there is any reagent R present, the equilibrium expression is $y_B = 0$. From the stoichiometry of the reaction shown in Eq. (12-60), there will be reagent available as long as $Lx_{R,0} > Vy_{B,N+1}$.

For a dilute countercurrent absorber, the mass balance was given by Eq. (12-5), where x is the total mole frac of B in the liquid (as free B and as bound RB). The operating and equilibrium diagrams can be plotted on a McCabe-Thiele diagram as shown in Figure 12-15. One equilibrium stage will give $y_1 = 0$, which is more than sufficient. Unfortunately, the stage efficiency is often very low because of low mass transfer rates of the solute into the liquid. If the Murphree vapor efficiency

$$E_{MV} = \frac{y_{in} - y_{out}}{y_{in} - y_{out}^*} \qquad \textbf{(12-61)}$$

is used in Figure 12-15, the number of real stages can be stepped off. Murphree vapor efficiencies less than 30% are common.

Since only one equilibrium stage is required, alternatives to countercurrent cascades may be preferable. A co-current cascade is shown in Figure 12-16A. Packed columns would normally be used for the co-current cascade. The advantage of the co-current cascade is that

Chapter 12   Absorption and Stripping

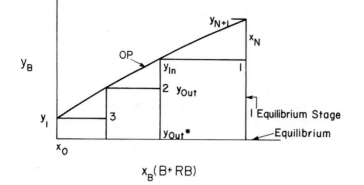

FIGURE 12-15.   *McCabe-Thiele diagram for countercurrent irreversible absorption*

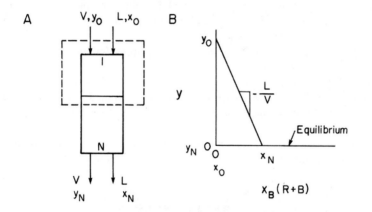

FIGURE 12-16.   *Co-current irreversible absorption; (A) apparatus, (B) McCabe-Thiele diagram*

it cannot flood, so smaller diameter columns with higher vapor velocities can be used. The higher vapor velocities give higher mass transfer rates (see Chapter 15), and less packing will be required (Gianetto et al., 1973). The mass balance for the co-current system using the mass balance envelope shown in Figure 12-16A is

$$y_0 V + L x_0 = yV + Lx \tag{12-62}$$

Solving for y, we obtain the operating equation

$$y = -\frac{L}{V}x + \frac{y_0 V - L x_0}{V} \tag{12-63}$$

This is a straight line with a slope of $-L/V$, and is plotted in Figure 12-16B. At equilibrium, $y_N = 0$, and x_N can be found from the operating equation as shown in Figure 12-16B. When equilibrium is not attained, the system can be designed with the mass transfer analysis discussed in Chapter 16. Co-current absorbers are used commercially for irreversible absorption.

For reversible chemical absorption, the higher flow rates and lack of flooding available in co-current absorbers is still desirable, but one equilibrium contact is rarely sufficient. Connecting a co-current absorber and a countercurrent absorber in series or parallel can provide more flexibility for operation (Isom and Rogers, 1994). This combination is explored in Problem 12.B3.

12.11   SUMMARY—OBJECTIVES

In this chapter we studied absorption and stripping. At the end of this chapter you should be able to achieve the following objectives:

1. Explain what absorption and stripping do and describe a complete gas treatment plant
2. Use the McCabe-Thiele method to analyze absorption and stripping systems for both concentrated and dilute systems
3. Design the column diameter for an absorber or stripper for staged columns and packed columns
4. Derive the Kremser equation for dilute systems
5. Use the Kremser equation for dilute absorption and stripping problems
6. Solve problems for dilute multicomponent absorbers and strippers both graphically and analytically
7. Use the matrix solution method for nonisothermal multicomponent absorption or stripping
8. Discuss how irreversible absorption differs from reversible systems, and design irreversible absorbers

REFERENCES

Astarita, G., D. W. Savage, and A. Bisio, *Gas Treating with Chemical Solvents*, Wiley, New York, 1983.

Ball, T., and R. Veldman, "Improve Gas Treating," *Chem. Engr. Progress, 87* (1) 67 (Jan. 1991).

Brian, P. L. T., *Staged Cascades in Chemical Processes*, Prentice Hall, Upper Saddle River, NJ, 1972.

Gianetto, A., V. Specchia, and G. Baldi, "Absorption in Packed Towers with Concurrent Downward High-Velocity Flows—II: Mass Transfer," *AIChE J., 19* (5), 916 (Sept. 1973).

Hwang, S. T., "Tray and Packing Efficiencies at Extremely Low Concentrations," in N. N. Li (Ed.), *Recent Developments in Separation Science*, Vol. 6, CRC Press, Boca Raton, FL, 1981, pp. 137–148.

Hwang, Y. L., G. E. Keller II, and J. D. Olson, "Steam Stripping for Removal of Organic Pollutants from Water. 1. Stripping Effectiveness and Stripping Design," p. 1753; "Part 2, Vapor-Liquid Equilibrium Data," p. 1759, *Ind. Eng. Chem. Research, 31*, (1992 a, b).

Isom, C., and J. Rogers, "Sour Gas Treatment Gets More Flexible," *Chem. Engr.*, 147 (July 1994).

Kessler, D. P., and P. C. Wankat, "Correlations for Column Parameters," *Chem Engr.*, 72 (Sept. 26, 1988).

King, C. J., *Separation Processes*, McGraw-Hill, New York, 1971.

King, C. J., *Separation Processes,* 2nd ed., McGraw-Hill, New York, 1981.

Kister, H. Z., "Ask the Experts. Acid-Gas Absorption," *Chem. Engr. Progress, 102* (6), 16 (June 2006).

Kister, H. Z., P. M. Mathias, D. E. Steinmeyer, W. R. Penney, B. B. Crocker, and J. R. Fair, "Equipment for Distillation, Gas Absorption, Phase Dispersion, and Phase Separation," in D. W. Green and R. H. Perry (Eds.), *Perry's Chemical Engineers' Handbook*, 8th ed., McGraw-Hill, New York, 2008, Section 14.

Kohl, A. L., "Absorption and Stripping," in R. W. Rousseau (Ed.), *Handbook of Separation Process Technology*, Wiley, New York, 1987, Chap. 6.

Kohl, A. L., and R. B. Nielsen, *Gas Purification,* 5th ed., Gulf Publishing Co., Houston, 1997.

Kremser, A., *Nat. Petrol. News,* 43 (May 30, 1930).

O'Connell, H. E., "Plate Efficiency of Fractionating Columns and Absorbers," *Trans. AIChE, 42,* 741 (1946).

Perry, R. H., C. H. Chilton, and S. D. Kirkpatrick (Eds.), *Chemical Engineer's Handbook,* 4th ed., McGraw-Hill, New York, 1963.

Perry, R. H., and C. H. Chilton (Eds.), *Chemical Engineer's Handbook,* 5th ed., McGraw-Hill, New York, 1973.

Perry, R. H., and D. W. Green (Eds.), *Perry's Chemical Engineers' Handbook*, 7th ed., McGraw-Hill, New York, 1997.

Reynolds, J., J. Jeris, and L. Theodore, *Handbook of Chemical and Environmental Engineering Calculations,* Wiley, New York, 2002.

Seider, W. D., J. D. Seader, D. R. Lewin, and S. Widago, *Design Process Principles*, 3rd ed., Wiley, New York, 2009.

Smith, B. D., *Design of Equilibrium Stage Processes,* McGraw-Hill, New York, 1963.

Socolow, R. H., "Can We Bury Global Warming?" *Scientific American,* 49–55 (July 2005).

Souders, M., and G. G. Brown, "Fundamental Design of High Pressure Equipment Involving Paraffin Hydrocarbons. IV. Fundamental Design of Absorbing and Stripping Columns for Complex Vapors," *Ind. Eng. Chem., 24,* 519 (1932).

Woods, D. R., *Rules of Thumb in Engineering Practice*, Wiley-VCH, Weinheim, 2007.

Yaws, C. L., P. K. Narasimhan, H. H. Lou, and R. W. Pike, "Solubility & Henry's Law Constants for Chlorinated Compounds in Water," *Chem. Engr.,* 50 (Feb. 2005).

Yaws, C. L., and M. Rahate, "Solubility of Water in Benzene Derivatives," *Chemical Engineering, 117* (3), 44–47 (March 2010).

Zarzycki, R. and A. Chacuk, *Absorption: Fundamentals & Applications,* Pergamon Press, Oxford, 1993.

Zenz, F. A., "Design of Gas Absorption Towers," in P. A. Schweitzer (Ed.), *Handbook of Separation Techniques for Chemical Engineers,* 3rd ed., McGraw-Hill, New York, 1997, Section 3.2.

HOMEWORK

A. *Discussion Problems*

A1. How can the direction of mass transfer be reversed as it is in a complete gas plant? What controls whether a column is a stripper or an absorber?

A2. Why is the Murphree efficiency often lower in chemical absorption than in physical absorption? (What additional resistances are present?)

A3. After reviewing Chapter 10, outline the method of determining the column diameter for an absorber or stripper in a packed column.

A4. As the system becomes dilute, $L/G \rightarrow L/V$, $Y \rightarrow y$, and $X \rightarrow x$. At what concentration levels could you safely work in terms of fractions and total flows instead of ratios and flows of solvent and carrier gas? What variable will this depend on? Explore numerically.

A5. Which methods will give accurate results for concentrated, nonisothermal absorber (circle all correct answers)?
  a. McCabe-Thiele diagram
  b. Kremser equation
  c. Aspen Plus
  d. None of the above

A6. A stripper is unable to obtain the specified purity of the outlet liquid. The outlet liquid concentration of the impurity can be decreased (which might allow one to reach or exceed the specified purity) by doing which of the following changes? Circle all correct answers. There are five correct answers—one in each pair (A,B,C,D,E). Note: Think of this as doing each change by itself and determine if it decreases the concentration of impurity in the outlet liquid.
  A. **a.** Increase inlet gas flow rate.
      b. Decrease inlet gas flow rate.
  B. **c.** Increase inlet liquid flow rate
      d. Decrease inlet liquid flow rate.
  C. **e.** Increase number of stages.
      f. Decrease number of stages.
  D. **g.** Increase column pressure.
      h. Decrease column pressure.
  E. **i.** Increase temperature of inlet gas and inlet liquid.
      j. Decrease temperature of inlet gas and inlet liquid.

A7. Explain how the single assumption that "solutes are independent of each other" can specify more than one degree of freedom.

A8. Develop your key relations chart for this chapter.

B. *Generation of Alternatives*

B1. The Kremser equation can be used for more than just determining the number of stages. List as many types of problems (where a different variable is solved for) as you can. What variables would be specified? How would you solve the equation?

B2. Many other configurations of absorbers and strippers can be devised. For example, there could be two feeds. Generate as many as possible.

B3. You want to use both co-current and countercurrent absorbers in a process. Sketch as many ways of doing this as you can think of. What are the advantages and disadvantages of each method?

C. *Derivations*

C1. Derive Eq. (12-22) starting with Eq. (12-21).

C2. Derive an equation that is equivalent to Eq. (12-12) for $L/(mV) = 1$, but in terms of liquid mole fractions.

C3. Derive an operating equation similar to Eq. (12-40), but draw your balance envelope around the bottom of the column. Show that the result is equivalent to Eq. (12-40).

C4. Derive Eq. (10-4), which relates the overall efficiency to the Murphree vapor efficiency, for dilute systems by determining N_{equil} and N_{actual} in Eq. (10-1) from appropriate forms of the Kremser equation.

C5. Occasionally it is useful to apply the Kremser equation to systems with a constant relative volatility. Where on the y vs. x diagram for distillation can you do this? Derive the appropriate values for m and b in the two regions where the Kremser equation can be applied.

C6. For dilute systems show that $(L/V)_{min}$ for absorbers and $(L/V)_{max}$ for strippers calculated from the Kremser equation agrees with a graphical calculation.

D. *Problems*

Answers to problems with an asterisk are at the back of the book.

D1. A 5-stage countercurrent absorber is used to absorb acetone from air into water at 3 atm pressure and 20°C. The total inlet gas flow rate is 100 kmol/h. The inlet gas is 0.004 mole fraction acetone. The inlet liquid contains 0.0001 mole fraction acetone. Outlet gas is 0.0002 mole fraction acetone. Assume total liquid and gas flow rates are constant. At 20°C the Henry's law constant for acetone in water is H = 1.186 atm/(mole fraction). Find the liquid flow rate required and the mole fraction acetone in the outlet liquid.

D2. We are absorbing hydrogen sulfide at 15°C into water. The entering water is pure. The feed gas contains 0.0012 mole frac hydrogen sulfide and we want to remove 97% of this in the water. The total gas flow rate is 10 kmol/h. The total liquid flow rate is 2000 kmol/h. Total pressure is 2.5 atm. You can assume that total liquid and gas flow rates are constant. Equilibrium data are in Table 12-1.

    a. Calculate the outlet gas and liquid mole fracs of hydrogen sulfide.

    b. Calculate the number of equilibrium stages required using a McCabe-Thiele diagram.

    c. If $L/V = M \times (L/V)_{min}$, find the multiplier M (M > 1).

    d. Why is this operation not practical? What would an engineer do to develop a practical process?

D3. A stripper with one equilibrium stage is stripping 1-chloro-naphthalene from water into air. The liquid feed is $x_{in} = 2.0 \times 10^{-6}$ mole fraction 1-chloro-naphthalene and the total liquid flow rate is L = 100 kmol/h. The inlet air is pure $y_{in} = 0$ and total gas flow rate is V = 10.0 kmol/h. The column operates at T = 25°C, and the Henry's law constant for 1-chloro-naphthalene is given in Table 12-2. If an outlet liquid mole fraction of $x_{out} = 0.4 \times 10^{-6}$ is desired, what is the pressure (in atm) of the stripper, and what is the value of y_{out}?

Assume H does not depend on pressure and that both L and V are constant. Note: there are multiple approaches to solve this problem.

D4. We have a steam stripper operating isothermally at 100°C. The entering liquid stream contains 0.0002 mole frac nitrobenzene in water at 100°C. Flow rate of entering liquid is 1 kmol/min. The entering steam is pure water at 100°C. We desire an outlet liquid mole frac of 0.00001 nitrobenzene. L/V = 12.0. You can assume that total liquid and gas flow rates are constant. At 100°C equilibrium in terms of nitrobenzene mole frac is y = 28 x. Find the outlet mole fraction of the nitrobenzene in the vapor stream and the number of stages.

D5.* A packed column 0.0762 m in diameter with 3.048 m of Intalox saddle packing is being run in the laboratory. P is being stripped from nC_9 using methane gas. The methane can be assumed to be insoluble and the nC_9 is nonvolatile. Operation is isothermal. The laboratory test results are:

$$X_{in} = 0.40 \; \frac{\text{lb P}}{\text{lb n} - C_9}, \qquad Y_{out} = 0.50 \; \frac{\text{lbP}}{\text{lb methane}}$$

$$X_{out} = 0.06 \frac{\text{lb P}}{\text{lb n} - C_9}, \qquad Y_{in} = 0.02 \frac{\text{lbP}}{\text{lb methane}}$$

Equilibrium data can be approximated as $Y = 1.5X$. Find the HETP for the packing.

D6.* We wish to design a stripping column to remove carbon dioxide from water. This is done by heating the water and passing it countercurrent to a nitrogen stream in a staged stripper. Operation is isothermal and isobaric at 60°C and 1 atm pressure. The water contains 9.2×10^{-6} mole frac CO_2 and flows at 100,000 lb/h. Nitrogen (N_2) enters the column as pure nitrogen and flows at 2500 ft³/h. Nitrogen is at 1 atm and 60°C. We desire an outlet water concentration that is 2×10^{-7} mole frac CO_2. Ignore nitrogen solubility in water and ignore the volatility of the water. Equilibrium data are in Table 12-1. Use a Murphree vapor efficiency of 40%. Find outlet vapor composition and number of real stages needed.

D7. We wish to absorb ammonia from an air stream using water at 0°C and a total pressure of 1.30 atm. The entering water stream is pure water. The entering vapor is 17.2 wt % ammonia. We desire to recover 98% of the ammonia in the water outlet stream. The total gas flow rate is 1050 kg/h. We want to use a solvent rate that is 1.5 times the minimum solvent rate. Assume that temperature is constant at 0°C, water is nonvolatile, and air does not dissolve in water. Equilibrium data are available in Table 12-3. Find L_{min}, L, and N.

D8. HCl is being absorbed from two air streams into water in a countercurrent staged laboratory absorber at 10°C and a pressure of 2.0 atm. Feed rate of gas feed 1 is 1.0 kmol/h of total gas, and this gas is 20 mol% HCl. Feed rate of gas feed 2 is 0.5 kmol/h of total gas, and this gas is 5 mol% HCl. The entering water is pure. The outlet gas should be 0.002 mole fraction HCl. Equilibrium data are given below. Use the optimum feed locations for both gas feeds.

Note: Use ratio units. If careful with units, the liquid units can be in mass and the gas units in moles, which is effectively the form of the equilibrium data. Derive the operating equation and external mass balances to determine where to include the molecular weight of HCl (36.46). Because HCl has a very large heat of absorption in water, the column will have to be well-cooled to maintain the temperature at 10°C. Commercial units are not isothermal.

a. Find the minimum water flow rate required.

b. If L = 0.4 kg water/h, find the outlet liquid concentration, the number of stages required (including a fractional number of stages) and the optimum feed location for gas feeds 1 and 2.

Equilibrium data from Perry and Green (1997, p. 2-127):

Grams HCl/g water	HCl partial pressure, mm Hg
0	0
0.0870	0.000583
0.1905	0.016
0.316	0.43
0.47	11.8
0.563	56.4
0.667	233
0.786	840

D9. Problem 12.D2 satisfies the criteria for using the Kremser equation. Repeat Problem 12.D2b, but use the Kremser equation. Compare your answer with the McCabe-Thiele solution.

D10.* A stripping tower with four equilibrium stages is being used to remove ammonia from wastewater using air as the stripping agent. Operation is at 80°F and 1 atm. The inlet air is pure air, and the inlet water contains 0.02 mole frac ammonia. The column operates at $L/V = 0.65$. Equilibrium data in mole fracs are given as $y = 1.414x$. Find the outlet concentrations.

D11.* An absorption column for laboratory use has been carefully constructed so that it has exactly 4 equilibrium stages and is being used to measure equilibrium data. Water is used as the solvent to absorb ammonia from air. The system operates isothermally at 80°F and 1 atm. The inlet water is pure distilled water. The ratio of $L/V = 1.2$, inlet gas concentration is 0.01 mole frac ammonia, and the measured outlet gas concentration is 0.0027 mole frac ammonia. Assuming that equilibrium is of the form $y = mx$, calculate the value of m for ammonia. Check your result.

D12.* Read Section 13.4 on cross flow in Chapter 13 before proceeding. We wish to strip CO_2 from a liquid solvent using air as the carrier gas. Since the air and CO_2 mixtures will be vented and since cross flow has a lower pressure drop, we will use a cross-flow system. The inlet liquid is 20.4 wt % CO_2, and the total inlet liquid flow rate is 1000 kg/h. We desire an outlet liquid composition that is 2.5 wt % (0.025 wt frac) CO_2. The gas flow to each stage is 25,190 kg air/h. In the last stage a special purified air is used that has no CO_2. Find the number of equilibrium stages required. In weight fraction units, equilibrium is $y = 0.04x$. Use unequal axes for your McCabe-Thiele diagram. Note: With current environmental concerns, venting the CO_2 is not a good idea even if it is legal.

D13. A water cleanup is stripping vinyl chloride from contaminated ground water at 25°C and 850 mm Hg using a countercurrent, staged stripper. The feed is 5.0 ppm (molar) vinyl chloride. An outlet water that contains 0.1 ppm (molar) vinyl chloride is required. The inlet air used for stripping is pure. For a base liquid flow rate of $L = 1.0$ kmol/h, determine the following:

a. The minimum gas flow rate G_{min} in kmol/h.

b. If $G = 2 G_{min}$, use a McCabe-Thiele diagram to determine the number of equilibrium stages needed (including a fractional number). Note: Scales on your y and x axes should be different. Calculate two points to plot the straight lines.

c. If $G = 2 G_{min}$, use one of the forms of the Kremser equation to determine the number of equilibrium stages needed (including a fractional number).

d. For parts b and c, what is the concentration of vinyl chloride in the outlet gas? What would you propose doing with this gas (it cannot be vented to the atmosphere)?

e. Are the assumptions required for the solution methods satisfied?

Data are in Table 12-2.

D14. In an ammonia plant we wish to absorb traces of argon and methane from a nitrogen stream using liquid ammonia. Operation is at 253.2 K and 175 atm pressure. The feed flow rate of gas is 100 kmol/h. The gas contains 0.00024 mole frac argon and 0.00129 mole frac methane. We desire to remove 95% of the methane. The entering liquid ammonia is pure. Operate with $L/V = 1.4(L/V)_{min}$. Assume the total gas and liquid rates are constant. The equilibrium data at 253.2 K are:

Methane: partial pressure methane atm = 3600(methane mole frac in liquid)
Argon: partial pressure argon atm = 7700(argon mole frac in liquid)

Reference: Alesandrini et al., *Ind. Eng. Chem. Process Design and Develop.*, *11*, 253 (1972)

a. Find outlet methane mole frac in the gas.

b. Find $(L/V)_{min}$ and actual L/V.

c. Find outlet methane mole frac in the liquid.

d. Find the number of equilibrium stages required.

e. Find the outlet argon mole fracs in the liquid and the gas, and the % recovery of argon in the liquid.

D15.* We want to remove traces of propane and n-butane from a hydrogen stream by absorbing them into a heavy oil. The feed is 150 kmol/h of a gas that is 0.0017 mole frac propane and 0.0006 mole frac butane. We desire to recover 98.8% of the butane. Assume that H_2 is insoluble. The heavy oil enters as a pure stream (which is approximately C_{10} and can be assumed to be nonvolatile). Liquid flow rate is 300 kmol/h. Operation is at 700 kPa and 20°C. K values can be obtained from the DePriester charts. Find the equilibrium number of stages required and the compositions of gas and liquid streams leaving the absorber.

D16. You have a water feed at 25°C and 1 atm. The water has been in contact with air and can be assumed to be in equilibrium with the normal CO_2 content of air (about 0.035 volume %) (Note: This dissolved CO_2 will decrease the pH of the water.) We wish to strip out the CO_2. A countercurrent stripping column will be used operating at 25°C and 50 mm Hg pressure. The stripping gas used will be nitrogen gas saturated with pure water at 25°C. Assume the nitrogen is insoluble in water. Assume ideal gas (vol % = mol%). Data: See Table 12-1. We want to remove 95% of the initial CO_2 in the water.

a. Calculate the inlet and outlet mole fracs of CO_2 in water

b. Determine $(L/G)_{max}$. If $L = 1$ kgmol/h, calculate G_{min} (corresponds to $(L/G)_{max}$).

c. If we operate at $G = 1.5 \times G_{min}$ find the CO_2 mole frac in the outlet nitrogen and the number of equilibrium stages needed.

D17.* We wish to absorb ammonia from air into water. Equilibrium data are given as $y_{NH3} = 1.414 x_{NH3}$ in mole fracs. The countercurrent column has three equilibrium stages. The entering air stream has a total flow rate of 10 kmol/h and is 0.0083 mole frac NH_3. The inlet water stream contains 0.0002 mole frac ammonia. We desire an outlet gas stream with 0.0005 mole frac ammonia. Find the required liquid flow rate, L.

D18. Dilute amounts of ammonia are to be absorbed from two air streams into water. The absorber operates at 30°C and 2 atm pressure, and the equilibrium expression is $y = 0.596x$ where y is mole frac of ammonia in the gas and x is mole frac of ammonia in the liquid. The solvent is pure water and flows at 100 kmol/h. The main gas stream to be treated has a mole frac ammonia of $y_{N+1} = 0.0058$ (the remainder is air) and flow rate of $V_{N+1} = 100$ kmol/h. The second gas stream is input as a feed in the column with a mole frac ammonia of $y_F = 0.003$ and flow rate of $V_F = 50$ kmol/h. We desire an outlet ammonia mole frac in the gas stream of $y_1 = 0.0004$. Find:

a. The outlet mole frac of ammonia in the liquid, x_N.

b. The optimum feed stage for the gas of mole frac y_F, and the total number of stages needed.

c. The minimum solvent flow rate, L_{min}.

D19.* Use the Kremser equation to check the McCabe-Thiele calculation done in Example 12-1.

D20. We need to remove H_2S and CO_2 from 1000 kmol/h of a water stream at 0°C and 15.5 atm. The inlet liquid contains 0.000024 mole frac H_2S and 0.000038 mole frac CO_2. We desire a 99% recovery of the H_2S in the gas stream. The gas used is pure nitrogen at 0°C and 15.5 atm. The nitrogen flow rate is 3.44 kmol/h. A staged countercurrent stripper will be used. Assume water flow rate and air flow rate are constant. Data are in Table 12-1. Note: Watch your decimals.

a. Determine the H_2S mole fractions in the outlet gas and liquid streams.

b. Determine the number of equilibrium stages required.

c. Determine the CO_2 mole frac in the outlet liquid stream.

D21. A gas processing plant has an absorber and stripper set up as shown in Figure 12-2 except both columns operate at 25°C but are at different pressures. The absorber is at 5.0 atm and the stripper is at 0.2 atm. The feed to the plant is V_{abs} = 100 kmol/h of air containing $y_{in,abs} = y_{N+1,abs}$ = 0.00098 mole fraction 1,2,3 trichloropropane. We want the outlet gas from the absorber $y_{out,abs} = y_{1,abs}$ = 0.000079 mole fraction 1,2,3 trichloropropane. The inlet liquid to the absorber is $x_{in,abs} = x_{0,abs}$ = 0.00001 mole fraction 1,2,3 trichloropropane. Note that because the absorber and stripper are connected, $x_{in,abs} = x_{out,stripper}$ and $x_{out,abs} = x_{in,stripper}$. The entering gas in the stripper is pure air. Determine the minimum liquid flow rate in the absorber and then operate with L_{abs} = 1.6($L_{abs,min}$). Note that $L_{abs} = L_{stripper}$. In the stripper determine the minimum gas flow rate and operate with $V_{stripper}$ = 1.5($V_{stripper,min}$). Equilibrium data are in Table 12-2.

a. Find $L_{abs,min}$, L_{abs}, $x_{out,abs} = x_{in,stripper}$, N_{abs}, $V_{stripper,min}$, $V_{stripper}$, $y_{out,stripper} = y_{1,stripper}$, $N_{stripper}$.

b. Do the mole fractions in the liquid ever exceed the solubility limits?

Suggestion: Easiest solution path is to roughly sketch McCabe-Thiele diagrams to help in calculation of $L_{abs,min}$ and $V_{stripper,min}$, use external balances to find $x_{out,abs} = x_{in,stripper}$ and $y_{out,stripper}$, and use Kremser equation to find N_{abs} and $N_{stripper}$. Watch your decimal points!

D22. We are absorbing n-butane from a light gas into a heavy oil at 1000 kPa and 15°C. The flow rate of the inlet gas is V_{N+1} = 150 kmol/h and the mole fraction n-butane in the inlet gas is y_{N+1} = 0.003. The inlet solvent flows at L_0 = 75 kmol/h and contains no n-butane, x_0 = 0. We want an exit vapor with y_1 = 0.0004 mole fraction n-butane. Use the DePriester chart for equilibrium data. Assume the light gas is insoluble and the heavy oil is nonvolatile.

a. Find the mole fraction of n-butane in the outlet liquid, x_N.

b. Find the number of equilibrium stages that is sufficient.

D23.* A complete gas treatment plant often consists of both an absorber to remove the solute and a stripper to regenerate the solvent. Some of the treated gas is heated and is used in the stripper. This is called stream B. In a particular application we wish to remove obnoxious impurity A from the inlet gas. The absorber operates at 1.5 atm and 24°C where equilibrium is given as y = 0.5x (units are mole fracs). The stripper operates at 1 atm and 92°C where equilibrium is y = 3.0x (units are mole fracs). The total gas flow rate is 1400 mol/day, and the gas is 15 mol% A. The nonsoluble carrier is air. We desire a treated gas concentration of 0.5 mol% A. The liquid flow rate into the absorber is 800 mol/day and the liquid is 0.5 mol% A.

a. Calculate the number of stages in the absorber and the liquid concentration leaving

b. If the stripper is an already existing column with four equilibrium stages, calculate the gas flow rate of stream B (concentration is 0.5 mol% A) and the outlet gas concentration from the stripper.

E. More Complex Problems

E1. A laboratory steam stripper with 11 real stages is used to remove 1000 ppm (wt) nitrobenzene from an aqueous feed stream which enters at 97°C. The flow rate of the liquid feed stream is $L_{in} = F = 1726$ g/h. The entering steam rate was measured as $S = 99$ g/hr. The leaving vapor rate was measured as $V_{out} = 61.8$ g/h. Column pressure was 1 atm. The treated water was measured at 28.1 ppm nitrobenzene. Data at 100°C are in Problem 12.D4. Molecular weights are 123.11 and 18.016 for nitrobenzene and water, respectively. What is the overall efficiency of this column?

Note: A significant amount of steam condenses in this system to heat the liquid feed to its boiling point and to replace significant heat losses. An approximate solution can be obtained by assuming all this condensation occurs on the top stage, ignoring condensation of nitrobenzene, and adjusting the liquid flow rate in the column.

F. Problems Requiring Other Resources

F1.* Laboratory tests are being made prior to the design of an absorption column to absorb bromine (Br_2) from air into water. Tests were made in a laboratory-packed column that is 0.1524 m in diameter, has 1.524 m of packing, and is packed with saddles. The column was operated at 20 °C and 5 atm total pressure, and the following data were obtained:
Inlet solvent is pure water.
Inlet gas is 0.02 mole frac bromine in air.
Exit gas is 0.002 mole frac bromine in air.
Exit liquid is 0.001 mole frac bromine in water.

What is the L/G ratio for this system? (Base your answer on flows of pure carrier gas and pure solvent.) What is the HETP obtained at these experimental conditions? Henry's law constant data are given in Perry's (4th ed) on pages 14-2 to 14-12. Note: Use mole ratio units. Assume that water is nonvolatile and air is insoluble.

F2.* (Difficult) An absorber with three equilibrium stages is operating at 1 atm. The feed is 10 mol/h of a 60 mol% ethane, 40 mol% n-pentane mixture and enters at 30°F. The solvent used is pure n-octane at 70°F, solvent flow rate is 20 mol/h. We desire to find all the outlet compositions and temperatures. The column is insulated. For a first guess assume that all stages are at 70°F. As a first guess on flow rates, assume:

$L_0 = 20.0$	$V_1 = 6.00$ mol/h
$L_1 = 20.2$	$V_2 = 6.18$
$L_2 = 20.8$	$V_3 = 6.66$
$L_3 = 24.0$	$V_4 = 10.0$

Then go through *one* iteration of the sum rates convergence procedure (Figure 12-13) using direct substitution to estimate new flow rates on each stage. You could use these new flow rates for a second iteration, but instead of doing a second iteration of the flow loop use a paired simultaneous convergence routine. To do this, use the new values for liquid and vapor flow rates to find compositions on each stage. Then calculate enthalpies and use the multivariable Newtonian method to calculate new temperatures on each stage. You will then be ready to recalculate K values and solve the mass balances for the second iteration. However, for purposes of this assignment stop after the new temperatures have been estimated. Use a DePriester chart for K values. Pure-component enthalpies are given in Maxwell (1950) and on pages 629 and 630 of Smith (1963). Assume ideal solution behavior to find the enthalpy of each stream.

F3. Global warming is very much in the news. Engineers will need to be heavily involved in methods to control global warming. One approach is to capture carbon dioxide and bury it. Read Socolow's (2005) article. Write a one page engineering analysis of the feasibility of using absorption to capture carbon dioxide. In your analysis explain why capturing carbon dioxide from the flue gases of large power plants is considered to be more feasible than capturing carbon dioxide from automobile exhaust or from air.

G. *Computer Problems*

G1. Do the following problem with a process simulator. A feed gas at 1 atm and 30°C is 90 mol% air and 10 mol% ammonia. Flow rate is 200 kmol/h. The ammonia is being absorbed in an absorber operating at 1 atm using water at 25°C as the solvent. We desire an outlet ammonia in the exiting air that is 0.0032 or less. The column is adiabatic.
  a. If N = 4, what L is required ($\pm$ 10 kmol/h)?
  b. If N = 8, what L is required ($\pm$ 10 kmol/h)?
  c. If N = 16, what L is required ($\pm$ 10 kmol/h)?
  d. Examine your answers. Why does N = 16 not decrease L more?

G2. Solve the following problem with a process simulator. We wish to absorb two gas streams in an absorber. The main gas stream (stream A) is at 15°C, 2.5 atm, and has a flow rate of 100 kmol/h. Stream A is 0.90 mole frac methane, 0.06 mole frac n-butane, and 0.04 mole frac n-pentane. The other gas stream (stream B) is at 10°C, 2.5 atm, and has a flow rate of 75 kmol/h. Stream B is 0.99 mole frac methane, 0.009 mole frac n-butane, and 0.001 mole frac n-pentane. The liquid solvent fed to the absorber is at 15°C, 2.5 atm and has a flow rate of 200 kmol/h. This stream is 0.999 mole frac n-decane and 0.001 mole frac n-pentane. Absorber pressure is 2.5 atm. We desire an outlet gas stream that is 0.999 mole frac methane or slightly higher.

Determine the total number of stages and the optimum feed stages for streams A and B required to just achieve the desired methane purity of the outlet gas stream. (Use "on stage" for all of the feeds.) Report the following information:
  a. Total number of stages required _____
  b. Feed stage location for the solvent _____
  c. Feed stage location for stream A _____
  d. Feed stage location for stream B _____
  e. Outlet mole fracs of gas stream leaving absorber_____
  f. Outlet mole fracs of liquid leaving absorber _____
  g. Outlet gas flow rate _____ kmol/h
  h. Outlet liquid flow rate _____ kmol/h
  i. Highest temperature in column _____°C and stage it occurs on_____

G3. Part a. We are stripping 200 kmol/h of a liquid feed that is 10 mol% isopropyl alcohol and 90 mol% water. The stripping gas is pure nitrogen. The stripper has five equilibrium stages. The column pressure can be varied between 1.0 and 5.0 atm (make the feed gas and feed liquid pressures equal to the column pressure). The feed gas can be between 25°C and 100°C, and the inlet liquid can be between 25°C and 75°C. Design the operating conditions (column pressure, inlet gas flow rate and temperature, and inlet liquid temperature) to recover 98% of the isopropyl alcohol (recover means in the gas phase), and we want as high an isopropyl alcohol concentration in the gas as possible. Report the column pressure, feed gas flow rate and temperature, liquid feed temperature, recovery of the isopropyl alcohol, mole fractions and flow rates of the leaving gas and liquid product streams, and temperature on every stage.

Part b. Determine the diameter of the column if it has sieve plates, uses a plate spacing of 0.60960 m, has an 85% flooding factor, and uses the Fair's flooding calculation method. Other values use the default values in Aspen Plus.

CHAPTER 12 APPENDIX.   COMPUTER SIMULATIONS FOR ABSORPTION AND STRIPPING

This appendix follows the instructions in the appendices to Chapters 2 and 6. Although the Aspen Plus simulator is referred to, other process simulators can be used. If difficulties are encountered while running the simulator, see Appendix: Aspen Plus Separations Troubleshooting Guide that follows Chapter 18.

Lab 11.   Aspen Plus uses RADFRAC for absorber calculations. If you want more information, once you are logged into Aspen Plus you can go to Help→Help Topics→index, and click on Absorbers-RADFRAC. Note: Help may be useful for other applications of Aspen Plus.

To draw an absorber, use the RADFRAC icon. Draw a system with a vapor feed at the bottom, a liquid bottoms product, a second feed (liquid) at the top, and a vapor distillate. In the configuration window for the RADFRAC block set:

Condenser	None
Reboiler	None
Convergence	Petroleum/wide boiling

Go to convergence for the block (probably easiest to do using the flow diagram in color on the left side of the Aspen Plus screen). Click on the name of the block for your absorber. Then click on the blue check next to Convergence. In the Basic window the algorithm should be listed as Sum Rates. Set the maximum iterations at 75.

Input your components (remember air is a component), pick the physical properties package, input conditions for the liquid and vapor feed streams (note that Aspen will determine if a stream is a liquid or a vapor), set the pressure and the number of stages for the absorber. Supply Aspen Plus the locations of the two feed streams:

In the streams window for the RADFRAC block set: Liquid feed *above* stage 1, and Vapor feed *on* the last stage.

For each simulation, determine the outlet gas and outlet liquid mole fracs, outlet flow rates, and temperatures on the top and bottom stages. Note if there is a temperature maximum.

The purpose of parts 1 and 2 is to explore how different variables affect the operation of absorbers and strippers. Desired specification is 0.003 or less mole frac acetone in the exiting air.

1. **Absorber: a.** You wish to absorb acetone from air into water. The column and feed streams are at 1 atm. The inlet gas stream is 3.2 mol% acetone. Flow of inlet gas is 100 kmol/h. Inlet gas is at 30°C. Water flow rate is 200 kmol/h. Inlet water is pure and temperature is 20°C. N = 6. Find the outlet concentrations and flow rates.

   Results: Look at the temperature profile and note the temperature maximum. Look at the concentration profiles. Does the outlet vapor meet the acetone requirement? (It should.) Where does all the water in the outlet gas come from? Note how dilute the outlet liquid is. What can we do to increase this mole frac?
   b. Decrease the water flow rate to 100 kmol/h and run again. Do you meet the required acetone concentration in the gas? (It shouldn't.) But note that the outlet water is more concentrated.
   c. Double the number of stages to 12 with L = 100. Does this help reduce the outlet vapor mole frac of acetone significantly? (Probably not.)
   d. Return to N = 6 with L = 100. Reduce the temperature of both the feed gas and the inlet water to 10°C. Did this allow you to meet the outlet specifications for air? (Answer depends on the vapor-liquid equilibrium package you used.) Did the amount of water in the air decrease? (Why?) Do you see much change in the outlet concentration of the liquid stream?
   e. Repeat part 1d but with feed, solvent, and absorber all at 2.0 atm.
2. **Stripper.** The same flowchart can be used as the stripper, or you can attach a heater and then a stripper to the liquid outlet from the absorber. The liquid feed to the stripper should be the liquid product from the absorber (part 1e), but heat it to 85°C (at 1 atm) first. The gas feed to the stripper should be pure steam, at 1 atm. Use 6 stages. Be sure to put your steam (the stripping gas) on stage 6 and the liquid from the absorber should be put in above stage 1. Stripper is at 1.0 atm.
   a. Set the steam flow rate at 20 kmol/h. The steam should be superheated to 101°C. If the specification is liquid leaving should have acetone mole fraction < 1.0 E −04, does this design satisfy the spec? (It should.) What is the acetone mole frac in the gas leaving the stripper? Note that this is high enough that pure acetone could be recovered by distillation.
   b. Repeat with saturated steam at 1 atm and 100°C (set V/F = 1.0).
   c. Using saturated steam (V/F = 1) at 1 atm, reduce the steam flow rate to just satisfy the specification on the outlet liquid composition.
   d. Try reducing the pressure of the stripper and the steam to 0.5 bar with steam at 83°C and 15 kmol/h of steam. What can you conclude about the effect of pressure?
3. **Complete gas plant.** Design a complete gas plant (similar to Figure 12-2 but without solvent recycle) to process the feed in Problem 1e. Gas temperature is 10°C. Absorber has 6 stages, pressure = 2.0 atm. The outlet mole frac of acetone leaving in the gas should be less than 0.003 mole frac. The solvent fed to the absorber is pure water at 10°C and 2.0 atm. Flow rate is 100 kmol/h. The liquid outlet from the absorber should be heated to 70°C at 0.5 atm and then be sent directly to a stripper (N = 6) operating at 0.5 atm with a gas feed of 0.5 atm saturated (V/F = 1) steam. The acetone mole frac leaving in the liquid from the stripper should be less than 0.0001. Adjust the steam flow rate until your outlet is slightly below this value. The flowsheet for the gas plant should be similar to Figure 12-A1. If the outlet gas specification is not met, adjust operating conditions (L, T, N, or p) until it is met.

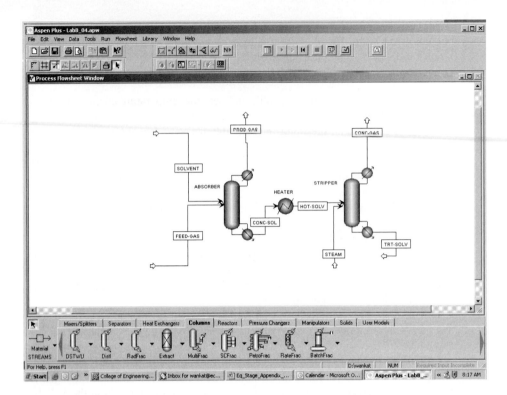

FIGURE 12-A1.   *Aspen Plus screenshot of gas plant without solvent recycle*

4. The treated solvent from part 3, which is close to pure water, should be recycled and connected to the absorber and fed to above stage 1; however, first cool this stream to 10°C in a heat exchanger (use Heater) and use a pump to increase the pressure to 2.0 atm. Now the *tricky* part. The fresh solvent has to be reduced to a flow rate at which the water mass balance can be valid, and the flow rate in the recycle loop must be controlled. In other words, at steady state the water in (makeup solvent + steam) must equal the water out (proc-gas + conc-gas + waste). But, you don't know what waste or makeup values to use. So, after the stripper on the TRT-SOLV line, put in a splitter (FSPLIT), and send TRT-SOLV as feed to the splitter. Hook up one of the product lines to the heat exchanger that cools the recycle liquid. Set the flow rate of this line to 100 kmol/h. The other product from the splitter is waste (XSH2O) and is dumped. After the recycle is hooked up (Figure 12-A2), use a fresh (or makeup) solvent rate of 100 and run the simulation. Decrease the makeup solvent flow in steps (try increments of −10 and then lower). Ultimately, the makeup solvent flow rate can be zero with an XSH2O flow rate > 0 (where does the water come from for the waste?). Note: Stream XSH2O also serves as a purge of liquid water. A small purge is needed to prevent nonvolatile impurities such as salts from building up in the solvent loop. Report the flow rates and mole fractions of the two outlet gas streams, the flow rate and mole fraction of the waste (XSH2O) stream, and the flow rate of the steam.

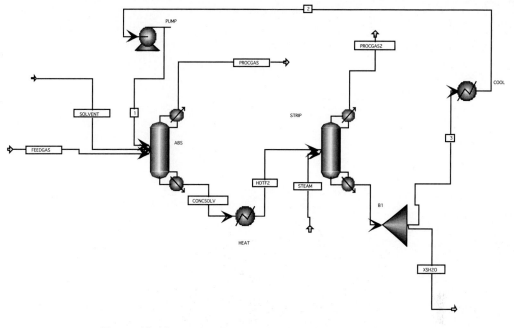

FIGURE 12-A2. *Aspen Plus screenshot of gas plant with solvent recycle*

CHAPTER 13

Liquid-Liquid Extraction

13.1 EXTRACTION PROCESSES AND EQUIPMENT

Extraction is a process where one or more solutes are removed from a liquid by transferring the solute(s) into a second liquid phase. The second liquid phase, the solvent, is a mass separating agent that must be recovered later. The two liquid phases must be immiscible (that is, insoluble in each other) or partially immiscible. In this chapter we discuss extraction equipment, immiscible extraction, partially miscible extraction, and equipment design. The separation is based on different solubilities of the solute in the two phases. Since vaporization is not required, extraction can be done at low temperature and is a gentle process suitable for unstable molecules such as proteins or DNA.

Extraction is a common laboratory and commercial unit operation. For example, in commercial penicillin manufacturing, after the fermentation broth is sent to a centrifuge to remove cell particles, the penicillin is extracted from the broth. Then the solvent and the penicillin are separated from each other by one of several techniques. In petroleum processing, aromatic hydrocarbons such as benzene, toluene, and xylenes are separated from the paraffins by extraction with a solvent such as sulfolane. The mixture of sulfolane and aromatics is sent to a distillation column, where the sulfolane is the bottoms product, and is recycled back to the extractor. Flowcharts for acetic acid recovery from water, for the separation of aromatics from aliphatics, and uranium recovery are given by Robbins (1997). Reviews of industrial applications of extraction are presented by Lo et al. (1983), Ritchy and Ashbrook (1979), and Frank et al. (2008); and extraction of biological compounds, particularly proteins, is discussed by Belter et al. (1988) and Harrison et al. (2003). Although extraction is fairly common, the ratio of distillation to extraction units in industry is about 20 to 1 (Alder et al., 1998).

As these commercial examples illustrate, the complete extraction process includes the extraction unit and the solvent recovery process. This is shown schematically in Figure 13-1. In many applications the downstream solvent recovery step (often distillation or a chemical stripping step) is more expensive than the actual extraction step. Woods (1995, Table 4-8) lists a variety of commercial processes and the solvent regeneration step employed. A variety of

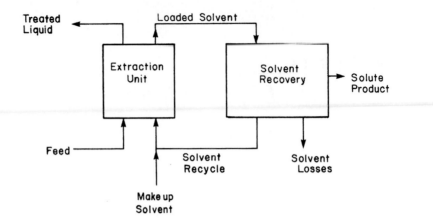

FIGURE 13-1. *Complete extraction process*

extraction cascades including single-stages, countercurrent cascades, and cross-flow cascades can be used; we discuss these later.

The variety of equipment used for extraction is much greater than for distillation, absorption, and stripping. Efficient contacting and separating of two liquid phases is considerably more difficult than contacting and separating a vapor and a liquid. In addition to plate and packed (random, structured and membrane) columns, many specialized pieces of equipment have been developed. Some of these are illustrated in Figure 13-2. Details of the different types of equipment are provided by Frank et al. (2008), Godfrey and Slater (1994), Humphrey and Keller (1997), Lo (1997), Lo et al. (1983), Reissinger and Schroeter (1978), Robbins and Cusack (1997), Schiebel (1978), and Woods (1995a). A summary of the features of the various types of extractors is presented in Table 13-1. Decision methods for choosing the type of extractor to use are presented by Frank et al. (2008), King (1980), Lo (1997), Reissinger and Schroeter (1978), and Robbins (1997). Detailed design of mixer-settlers is discussed in Section 13.14, and detailed design of Karr columns in Section 13.15.

The following heuristics are useful for making an initial decision on the type of extraction equipment to use:

1. If one or two equilibrium stages are required, use mixer-settlers.
2. If three equilibrium stages are required, use a mixer-settler, a sieve tray column, a packed column (random or structured), or a membrane contactor.
3. If four or five equilibrium stages are required, use a sieve tray column, a packed column (random or structured), or a membrane contactor.
4. If more than five equilibrium stages are required, use one of the systems that apply mechanical energy to a column (Figure 13-2).

Godfrey and Slater (1994) have detailed chapters on all the major types of extraction equipment with considerable detail on mass transfer rates. Humphrey and Keller (1997) discuss equipment selection in considerable detail and have height equivalent to a theoretical plate (HETP) and capacity data. Frank et al. (2008) discusses equipment selection, holdup, flooding, and mass transfer.

The number of equilibrium contacts required can be determined using the same stage-by-stage procedures for all the extractors. Determining stage efficiencies and hydrodynamic

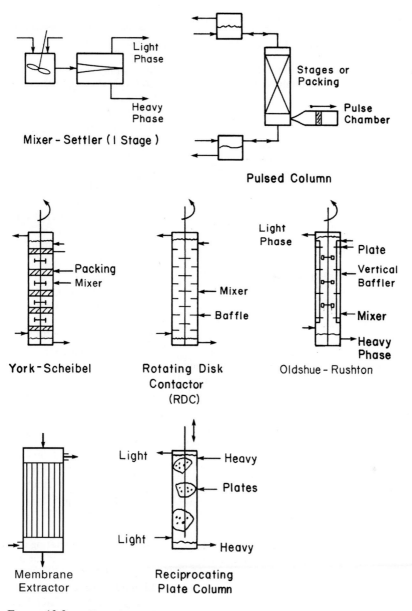

FIGURE 13-2. *Extraction equipment*

characteristics is more difficult. The more complicated systems are designed by specialists (e.g., see Lo, 1997, or Skelland and Tedder, 1987).

Solvent selection is critical for the development of an economical extraction system. The solvent should be highly selective for the desired solute and not very selective for contaminants. Both the selectivity, $\alpha = K_{desired} / K_{undesired}$, and $K_{desired}$ should be largelarge [K is defined later in Eq. (13-6)]. The solvent should be easy to separate from the diluent either as a totally

TABLE 13-1. *Extractor types*

Extractor	Examples	General Features
Unagitated columns	Plate columns; packed columns; spray columns	Low capital cost Low operating and maintenance cost Simplicity in construction Handles corrosive material Relatively low efficiency
Mixer-settlers	Mixer-settler; vertical mixer-settler; Morris contactor; static mixers	Higher-stage efficiency Handles wide solvent ratios High capacity Good flexibility Reliable scale-up Handles liquids with high viscosity Requires large area Expensive if large number of stages required
Pulsed columns	Perforated plate; packed	Low HETP No internal moving parts, but mechanically complex Many stages possible Karr column often less expensive
Rotary-agitation columns	Rotating disk contactor; Oldshue-Rushton; Scheibel (several types); Kuhni; asymmetric rotating disk	Reasonable capacity Reasonable HETP Many stages possible Reasonable construction cost Low operating & maintenance cost
Reciprocating-plate columns	Karr column; segmental passages; counter-moving plates	High throughput Low HETP Great versatility and flexibility Simplicity in construction Handles liquids containing suspended solids Handles mixtures with emulsifying tendencies
Centrifugal extractors and separators	Podbielniak; Quadronic; Alfa-Laval; Westfalia; Robatel	High capital & operating costs Short contacting time for unstable material Limited space required Handles easily emulsified material Handles systems with little liquid density difference Up to 7 equilibrium stages
Membrane extractors	Celgard membrane	High throughput—no flooding Emulsion not formed Chemical compatibility may be a problem Can operate very low liquid rate

Sources: Frank et al. (2008), Humphrey and Keller (1997), Lo (1997), Robbins and Cusack (1997)

immiscible system or a partially miscible system where separation by distillation is easy. In addition, the solvent should be nontoxic, noncorrosive, readily available, chemically stable, environmentally friendly, and inexpensive.

A useful, relatively simple approach to selecting a solvent with reasonably large selectivity is to use the solubility parameter (Giddings, 1991; King, 1980; Walas, 1985; Woods, 1995a). The solubility parameter δ is defined as

$$\delta_i = \left[\frac{(\Delta E_v)_i}{V_i} \right]^{1/2}$$

(13-1)

where $(\Delta E_v)_i = (\Delta H_v - p\Delta V)_i \sim (\Delta H_v - RT)_i$ is the latent energy of vaporization and V_i is the molal volume for component i. The solubility parameter has the advantage of being a property of only the pure components, it is easily calculated from parameters that are easy to measure and are often readily available, and tables of δ are available (Giddings, 1991; King, 1980; Walas, 1985; Woods, 1995b). The solubility parameter is useful for quick estimation of the miscibility of two liquids—the closer the values of δ_A and δ_B the more likely the liquids are miscible. For example, the δ values in (cal/ml)$^{1/2}$ are: water = 23.4, ethanol =12.7, and benzene = 9.2 (Giddings, 1991). Water and ethanol are miscible (although they are nonideal and the vapor-liquid equilibrium (VLE) shows an azeotrope), water and benzene are immiscible, and ethanol and benzene are miscible. Blumberg (1988), Frank et al. (2008), King (1981), Lo et al. (1983), and Treybal (1963) discuss solvent selection in detail.

The rules for selecting solvents are changing. Because of increased concern about the environment, global warming, and the desire to use "green" processes, solvents that used to be quite acceptable may be unacceptable now or in the future (Allen and Shonnard, 2002). There is a tendency to stop using chlorinated solvents and to use fewer hydrocarbon solvents. This has led to increased interest in ionic liquids, aqueous two-phase extraction, and the use of supercritical extraction (see Section 14-5). If a designer has a reasonable choice of solvents, consideration of life-cycle costs may lead to the use of a "greener" solvent even though it may be somewhat more expensive initially.

13.2 COUNTERCURRENT EXTRACTION

The most common type of extraction cascade is the countercurrent system shown schematically in Figure 13-3. In this cascade the two phases flow in opposite directions. Each stage is assumed to be an equilibrium stage so that the two phases leaving the stage are in equilibrium.

The solute, A, is initially dissolved in diluent, D, in the feed. Solute is extracted with solvent, S. The entering solvent stream is often presaturated with diluent. Streams with high

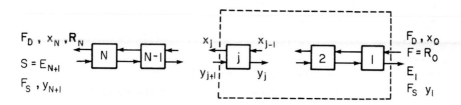

FIGURE 13-3. *Mass balance envelope for countercurrent cascade*

concentrations of diluent are called *raffinate,* while streams with high concentrations of solvent are the *extract.* The nomenclature in both weight fraction and weight ratio units is given in Table 13-2.

13.2.1 McCabe-Thiele Method for Dilute Systems

The McCabe-Thiele analysis for dilute immiscible extraction is very similar to the analysis for dilute absorption and stripping discussed in Chapter 12. It was first developed by Evans (1934) and is reviewed by Robbins (1997). In order to use a McCabe-Thiele type of analysis we must be able to plot a single equilibrium curve, have the energy balances automatically satisfied, and have one operating line for each section.

For equilibrium conditions, the Gibbs phase rule is: $F = C - P + 2$. There are three components (solute, solvent, and diluent) and two phases. Thus, there are three degrees of free-

TABLE 13-2. *Nomenclature for extraction*

Weight Fraction Units:

$x_{A,j}$	Raffinate weight fraction of solute leaving stage j
$x_{D,j}$	Raffinate weight fraction of diluent leaving stage j
$x_{S,j}$	Raffinate weight fraction of solvent leaving stage j
x_{A_F}, x_{A_M}	Weight fraction solute in feed and mixed streams
$y_{A,j}$	Extract weight fraction of solute leaving stage j
$y_{D,j}$	Extract weight fraction of diluent leaving stage j
$y_{S,j}$	Extract weight fraction of solvent leaving stage j
y_{A_S}	Weight fraction solute in solvent stream
z_A, z_B	Weight fraction in feed to fractional extractor
R_j	Raffinate flow rate leaving stage j, kg/h
E_j	Extract flow rate leaving stage j, kg/h
$S = E_{N+1}$	Solvent flow rate entering extractor (not necessarily pure), kg/h
$F = R_0$	Feed rate entering extractor, kg/h
$\hat{R}_j, \hat{E}_j, \hat{S}, \hat{F}$	Mass of raffinate, extract, solvent, feed, kg
M	Flow rate of mixed streams, kg/h

Weight Ratio Units (Solvent and Diluent Immiscible)

X_j	Weight ratio solute in diluent leaving stage j, kg A / kg D
Y_j	Weight ratio solute in solvent leaving stage j, kg A / kg S
F_D	Flow rate of diluent, kg diluent / h
F_S	Flow rate of solvent, kg solvent / h

Names:

A	Solute. Material being extracted
D	Diluent. Chemical that solute is dissolved in the feed
S	Solvent. Separating agent added for the separation

dom. In order to plot equilibrium data as a single curve, we must reduce this to one degree of freedom. The following two assumptions are usually made:

1. The system is isothermal.
2. The system is isobaric.

To have the energy balances automatically satisfied, we must also assume

3. The heat of mixing is negligible.

These three assumptions are usually true for dilute systems. The operating line will be straight, which makes it easy to work with, and the solvent and diluent mass balances will be automatically satisfied if the following assumption is valid:

4a. Diluent and solvent are totally immiscible.

When the fourth assumption is valid, then

$$F_D = \text{diluent flow rate, kg D/h} = \text{constant} \tag{13-2a}$$

$$F_S = \text{solvent flow rate, kg S/h} = \text{constant} \tag{13-2b}$$

These are flow rates of diluent only and solvent only and do *not* include the total raffinate and extract streams. Equations (13-2a and b) are the diluent and solvent mass balances, and they are automatically satisfied when the phases are immiscible.

Most solvent-diluent pairs that are essentially completely immiscible become partially miscible as more solute is added (see Section 13.7). This occurs because appreciable amounts of solute make the two phases chemically more similar. Since Eq. (13-2) is usually valid only for very dilute systems, we can usually analyze immiscible extraction systems using constant total flow rates. Thus, the fifth assumption we usually make is

5a. $$R = F_D = \text{const.} \tag{13-3a}$$
5b. $$E = F_s = \text{const.} \tag{13-3b}$$

For the mass balance envelope shown in Figure 13-3, the mass balance becomes,

$$Ey_{j+1} + Rx_0 = Ey_1 + Rx_j \tag{13-4}$$

Solving for y_{j+1} we obtain the operating equation

$$y_{j+1} = (R/E)x_j + [y_1 - (R/E)x_0] \tag{13-5}$$

Since R/E is assumed to be constant, this equation plots as a straight line on a y vs. x (McCabe-Thiele) graph.

Equilibrium data for dilute extraction are usually represented as a distribution ratio, K_d,

$$K_d = \frac{y_A}{x_A} \tag{13-6}$$

in weight fractions, mole fracs, or concentrations (e.g., kg/m³). For very dilute systems K_d will be constant, while at higher concentrations K_d often becomes a function of concentration.

Values of K_d are tabulated in Frank et al. (2008, pp. 15-29 to 15-31), Robbins and Cusack (1997, pp. 15-10 to 15-15), Hartland (1970, Chap. 6), Francis (1972), and Frank et al. (2008, pp. 15-29 to 15-31). A brief listing is given in Table 13-3. The value of K_d in the table in *Perry's Handbook* (Frank et al., 2008) ranges from 0.0012 to 181 for different solute-diluent-solvent combinations. Even for the same solute (phenol) in the same diluent (water), the value of K_d varies from 0.040 in isopropyl acetate (a nonselective solvent) to 39.8 in methyl isobutyl ketone.

It is common to think of K_d as dimensionless, but it has dimensions. For example, in weight fraction units the dimensions of $K_{d,wt}$ are (g A /g extract)/(g A /g raffinate), while in mole fraction units the dimensions of $K_{d,mol}$ are (mol A /mol extract)/(mol A /mol raffinate). If the molecular weights of extract and raffinate are not equal, then $K_{d,wt} \neq K_{d,mol}$.

The equilibrium "constant" is temperature- and pH-dependent. The temperature dependence is illustrated in Table 13-3 for the distribution of acetic acid between water (the diluent) and benzene (the solvent). Note that there is an optimum temperature at which K_d is a maximum. Benzene is not a good solvent for acetic acid because the K_d values are low, but water would be a good solvent if benzene were the diluent. As shown in Table 13-3, 1-butanol is a much better solvent than benzene for acetic acid. In addition, benzene would probably not be used as a solvent because it is carcinogenic. The use of extraction to fractionate components requires that the selectivity, $\alpha_{21} = K_{d2}/K_{d1}$, be large. An example where fractional

TABLE 13-3.   *Distribution coefficients for immiscible extraction*

Solute (A)	Solvent	Diluent	T, °C	$K_d = y_A/x_A$
Equilibrium in Weight Fraction Units (Perry and Green, 1984)				
Acetic acid	Benzene	Water	25	0.0328
Acetic acid	Benzene	Water	30	0.0984
Acetic acid	Benzene	Water	40	0.1022
Acetic acid	Benzene	Water	50	0.0588
Acetic acid	Benzene	Water	60	0.0637
Acetic acid	1-Butanol	Water	26.7	1.613
Furfural	Methylisobutyl ketone	Water	25	7.10
Ethyl benzene	β, β'-Thiodipropionitrile	n-Hexane	25	0.100
m-Xylene	β, β'-Thiodipropionitrile	n-Hexane	25	0.050
o-Xylene	β, β'-Thiodipropionitrile	n-Hexane	25	0.150
p-Xylene	β, β'-Thiodipropionitrile	n-Hexane	25	0.080
Equilibrium in Mass Ratio Units (Brian, 1972)				
Linoleic acid ($C_{17}H_{31}COOH$)	Heptane	Methylcellosolve + 10 vol % water		2.17
Abietic acid ($C_{19}H_{29}COOH$)	Heptane	Methylcellosolve + 10 vol % water		1.57
Oleic acid	Heptane	Methylcellosolve + 10 vol % water		4.14

extraction (see Section 13.3) is feasible is the separation of ethylbenzene and xylenes illustrated in Table 13-3. The ethylbenzene - p-xylene separation will be the most difficult of these, but is nonetheless feasible.

Extraction equilibria are dependent upon temperature, pH, and the presence of other chemicals. Although temperature dependence is small in the immiscible range (see Table 13-3) variations with temperature can be large when the solvents are partially miscible. Biological molecules, particularly proteins can have an order of magnitude change in K_d when the buffer salt is changed, and may have several orders of magnitude change in K_d when pH is varied (Harrison et al., 2003). Extraction processes often use shifts in temperature or pH to extract and then recover a solute from the solvent (Blumberg, 1988; Frank et al., 2008). For example, penicillin is extracted in a process with a pH swing.

The McCabe-Thiele diagram for extraction can be obtained by plotting the equilibrium data, which is a straight line if K_d is constant, and the operating line, which is also straight if assumption 5 (constant total flow rates) is valid. The operating line goes through the point (x_N, y_{N+1}), which are passing streams in the column. Since this procedure is very similar to stripping (e.g., Figure 12-5, but in wt frac units), we will consider a more challenging example problem.

EXAMPLE 13-1. Dilute countercurrent immiscible extraction
A feed of 100.0 kg/min of a 1.2 wt % mixture of acetic acid in water is to be extracted with 1-butanol at 1 atm pressure and 26.7°C. We desire an outlet concentration of 0.1 wt % acetic acid in the exiting water. We have available solvent stream 1 that is 44.0 kg/min of pure 1-butanol and solvent stream 2 that is 30.0 kg/min of 1-butanol that contains 0.4 wt % acetic acid. Devise a scheme to do this separation, find the outlet flow rate and concentration of the exiting 1-butanol phase, and find the number of equilibrium contacts needed.

Solution
A. Define. A feasible scheme is one that will produce exiting water with 0.001 wt frac acetic acid. We will assume that we want to use all of the solvent available; thus, the outlet butanol flow rate is 74 kg/min (the assumption that we want to use all of the butanol will be checked in Problem 13.D1). We need to find N and y_1.

B. Explore. The equilibrium data are given in Table 13-3, $y = 1.613x$ where y and x are the acetic acid wt fracs in the solvent and diluent phases, respectively. We learned in previous chapters that mixing is the opposite of separation; thus, we probably want to keep the two solvent streams separate. The extraction column will have an aqueous feed at the top. Since $x_N = 0.001 < y_{s2} = 0.004$, we probably want to input the pure solvent stream 1 at the bottom of the extractor and solvent 2 in the middle of the column (Figure 13-4A).

C. Plan. We can use the mass balance envelope in the bottom of the extractor to find the operating equation,

$$y = \frac{\overline{R}}{\overline{E}}x + \left(y_{N+1} - \frac{\overline{R}}{\overline{E}}x_N\right)$$

(13-7)

where the slope = $\overline{R}/\overline{E}$ = 100/44 = 2.273 and the operating line goes through the point (x_N, y_{N+1}) = (0.001, 0.0). If we use the mass balance envelope in the top of the extractor, we obtain Eq. (13-5) with

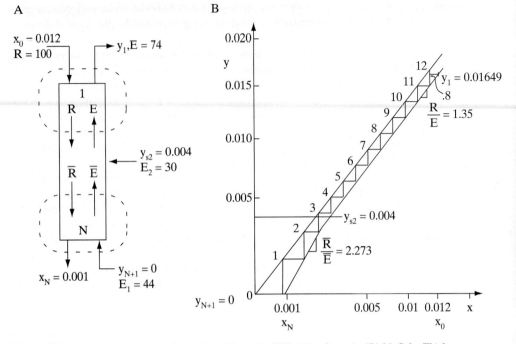

A

$$x_0 - 0.012$$
$$R = 100$$

$$y_1, E = 74$$

1

R   E

$$y_{s2} = 0.004$$
$$E_2 = 30$$

$$\overline{R} \quad \overline{E}$$

N

$$x_N = 0.001$$

$$y_{N+1} = 0$$
$$E_1 = 44$$

B

$$y_1 = 0.01649$$

$$\frac{R}{E} = 1.35$$

$$y_{s2} = 0.004$$

$$\frac{\overline{R}}{\overline{E}} = 2.273$$

$$y_{N+1} = 0$$

FIGURE 13-4. *Dilute extractor with two feeds (Example 13-1); (A) schematic, (B) McCabe-Thiele diagram*

$$\text{slope} = R/E = R/(E_1 + E_2) = 100/(44 + 30) = 1.35.$$

This operating line goes through point (x_0, y_1). Although y_1 is not immediately known, we can find it from an overall mass balance. We could also use the horizontal line $y = y_{\text{solvent2}} = 0.004$ as a feed line and use the point of intersection of the feed line and the bottom operating line to find a point on the top operating line.

Once we have plotted the operating lines, we step off stages until a stage crosses the feed line. At that point we switch operating lines.

D. Do it. The overall mass balance is

$$Rx_0 + E_1 y_1 + E_2 y_{s2} = Rx_N + Ey_1$$

Substituting in known values,

$$(100.0)(0.012) + (44.0)(0.0) + (30)(0.004) = (100.0)(0.001) + (74)y_1$$

and solving for y_1 we obtain $y_1 = 0.01649$.

The intersection point of the two operating lines occurs at the feed line, $y = y_{\text{solvent2}} = 0.004$. Substituting this value of y into Eq. (13-7) and solving for x we obtain $x_{\text{intersection}} = 0.00276$. Plotting the equilibrium line and the two operating lines and stepping off stages, we obtain Figure 13-4B. The optimum feed stage is the third from the bottom, and we need 11.8 equilibrium contacts.

E. Check. The calculation is probably more accurate than the assumptions involved. Since there is some miscibility between the organic and aqueous phases, the flow rates will not be entirely constant. The methods discussed in Sections 13.7 to 13.10 can be used for a more accurate calculation if sufficient data are available.

F. Generalization. The McCabe-Thiele procedure is quite general and can be applied to a number of different dilute extraction operations (Sections 13.3 and 13.4) and to other separation methods (Sections 14.1 to 14.3).

The McCabe-Thiele diagram shown in Figure 13-4 is very similar to the McCabe-Thiele diagrams for dilute stripping. This is true because the processes are analogous unit operations in that both contact two phases and solute is transferred from the x phase to the y phase. The analogy breaks down when we consider stage efficiencies and sizing the column diameter, since mass transfer characteristics and flow hydrodynamics are very different for extraction and stripping.

In stripping there was a maximum L/G. For extraction, a maximum value of R/E can be determined in the same way. This gives the minimum solvent flow rate, E_{min}, for which the desired separation can be obtained with an infinite number of stages.

For simulation problems the number of stages is specified but the outlet raffinate concentration is unknown. A trial-and-error procedure is required in this case. This procedure is essentially the same as the simulation procedure used for absorption or stripping.

Dilute multicomponent extraction can be analyzed on a McCabe-Thiele diagram if we add one more assumption.

6. Each solute is independent.

When these assumptions are valid, the entire problem can be solved in mole or weight fractions. Then for each solute the mass balance for the balance envelope shown in Figure 13-3 is

$$y_{i,j+1} = \frac{R}{E} x_{i,j} + \left(y_{i,1} - \frac{R}{E} x_{i,0} \right)$$

(13-8)

where i represents the solute and terms are defined in Table 13-2. This equation is identical to Eq. (13-5) when there is only one solute. The operating lines for each solute have the same slopes but different y intercepts. The equilibrium curves for each solute are independent because of assumption 6. Then we can solve for each solute independently. This solution procedure is similar to the one for dilute multicomponent absorption and stripping (Section 12.8). We first solve for the number of stages, N, using the solute that has a specified outlet raffinate concentration. Then with N known, a trial-and-error procedure is used to find x_{N+1} and y_1 for each of the other solutes.

Equations (13-5) or (13-8) can also be used, with care, if there is a constant small amount of solvent dissolved in the raffinate streams and a constant small amount of diluent dissolved in the extract streams. Flow rates R and E should include these small concentrations of the solvent or diluent. Unless entering streams are presaturated with solvent and diluent, $R = R_1 > R_0$ by the amount of solvent dissolved in the raffinate and $R_N < R_{N-1} = R$ by the amount of diluent that dissolves in entering solvent stream E_{N+1}. Similar adjustments need to be made to extract flow rates.

13.2.2 Kremser Method for Dilute Systems

If one additional assumption can be made, the Kremser equation can be used for dilute extraction of single or multicomponent systems.

7. Equilibrium is linear.

In this case, equilibrium has the form

$$y_i = m_i x_i + b_i \qquad (13\text{-}9)$$

The dilute extraction model now satisfies all the assumptions used to derive the Kremser equations in Chapter 12, so they can be used directly. Since we have used different symbols for flow rates, we replace L/V with R/E. Then Eq. (12-12) becomes

$$N = \frac{y_{N+1} - y_1}{y_1 - \dfrac{R}{E}x_o - b} \quad \text{for} \quad \frac{R}{mE} = 1 \qquad (13\text{-}10)$$

while if $R/(mE) \neq 1$, Eqs. (12-21) (inverted) and (12-22) become

$$\frac{y_1 - y_1^*}{y_{N+1} - y_1^*} = \frac{1 - \dfrac{R}{mE}}{1 - (\dfrac{R}{mE})^{N+1}} \qquad (13\text{-}11a)$$

with $y_1^* = mx_o + b$ and

$$N = \frac{\ln[(1 - \dfrac{mE}{R})(\dfrac{y_{N+1} - y_1^*}{y_1 - y_1^*}) + \dfrac{mE}{R}]}{\ln(\dfrac{R}{mE})} \qquad (13\text{-}11b)$$

Other forms of the Kremser equation can also be written with this substitution. When the Kremser equation is used, simulation problems are no longer trial-and-error.

The grouping mE/R is known as the extraction factor. Note that $mE/R = (yE)/(Rx)$. If $mE/R > 1$, then there is more solute in the extract phase and in Figure 13-3 net movement of solute is to the right. If $mE/R < 1$, then there is more solute in the raffinate phase and net movement of solute is to the left in Figure 13-3. If the goal is to remove solute from the diluent, then to have a reasonable solute recovery, we must have $mE/R > 1$, and the absolute theoretical minimum value of the extraction factor is $mE/R = 1$. A practical minimum value that is often used for the preliminary design of extractors is (Frank et al., 2008)

$$(mE/R)_{min} = 1.3 \Rightarrow (E/R)_{min} = (E/F)_{min} = 1.3/m \qquad (13\text{-}12)$$

As we will see when we discuss partially miscible extraction, there is also a maximum solvent to feed ratio when the solvent dissolves the entire feed and only one phase is formed.

Except for very dilute systems, the equilibrium will often be nonlinear and the various forms of the Kremser equation are not strictly valid. However, for modest curvature, a geometric average value of m (Frank et al., 2008)

$$m_{geometric\_average} = \sqrt{m_{low} m_{high}}$$

(13-13)

often gives a good fit. For the best fit, the value of b must also be fit and will not be zero.

Application of the Kremser equation to extraction is considered in detail by Hartland (1970), who also gives linear fits to equilibrium data over various concentration ranges. As the solution becomes more concentrated and more solute is transferred into the solvent, the total flow rate E increases while R decreases. The applicability of the Kremser equation can be extended by using solvent and diluent flow rates with ratio units (see Section 13.5).

13.3 DILUTE FRACTIONAL EXTRACTION

Very often, particularly in bioseparation, extraction is used to separate solutes from each other (Belter et al., 1988; Frank et al., 2008; Schiebel, 1978). In this situation we can use fractional extraction with two solvents as illustrated in Figure 13-5. In fractional extraction the two solvents are chosen so that solute A prefers solvent 1 and concentrates at the top of the column, while solute B prefers solvent 2 and concentrates at the bottom of the column. In Figure 13-5 solvent 2 is labeled as diluent so that we can use the nomenclature of Table 13-2. The column sections in Figure 13-5 are often separate so that each section can be at a different pH or temperature. This will make the equilibrium curve different for the two sections. It is also common to have reflux at both ends (Schiebel, 1978).

A common problem in fractional extraction is the center cut. In Figure 13-6 solute B is the desired solute while A represents a series of solutes that are more strongly extracted by solvent 1 and C represents a series of solutes that are less strongly extracted by solvent 1.

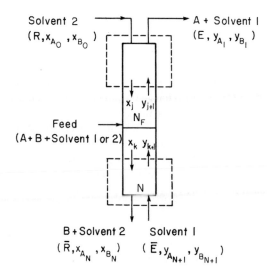

FIGURE 13-5.   *Fractional extraction*

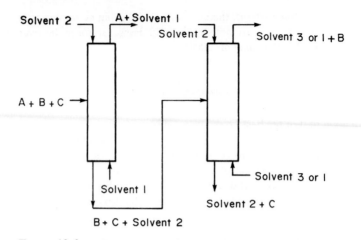

Solvent 2 ⟶ | A+Solvent I
| Solvent 2 ↓ | Solvent 3 or I + B

A + B + C ⟶

Solvent 3 or I

Solvent I

Solvent 2 + C

B + C + Solvent 2

FIGURE 13-6.   *Center-cut extraction*

Center cuts are common when pharmaceuticals are produced by fermentation, since a host of undesired chemicals are also produced.

Before looking at the analysis of fractional extraction it will be helpful to develop a simple criterion to predict whether a solute will go up or down in a given column. At equilibrium the solutes distribute between the two liquid phases. The ratio of *solute* flow rates in the column shown in Figure 13-5 is

$$\frac{\text{Solute A flow up column}}{\text{Solute A flow down column}} = \frac{y_{A,j} E_j}{x_{A,j} R_j} = \frac{K_{d,A,j} E_j}{R_j} \tag{13-14a}$$

where we have used the equilibrium expression, Eq. (13-6). If $(K_{d,A} E/R)_j > 1$, the net movement of solute A is up at stage j, while if $(K_{d,A} E/R)_j < 1$, the net movement of solute is down at stage j. In Figure 13-5 we want

$$(\frac{K_{d,A} E}{R})_j > 1 \text{ and } (\frac{K_{d,B} E}{R})_j < 1 \tag{13-14b}$$

in both sections of the column. By adjusting $K_{D,i}$ (say, by changing the solvents used or temperature) and/or the two solvent flow rates (E and R), we can change the direction of movement of a solute. This was done to solute B in Figure 13-6; B goes down in the first column and up in the second column. Ranges of $(K_{d,i} E/R)$ can be derived that will make the fractional extractor work (see Problem 13.A12). Since it is quite expensive to have a large number of equilibrium stages in a commercial extractor, the ratios in Eq. (13-14b) should be significantly different.

Analysis of fractional extraction is straightforward for dilute mixtures when the solutes are independent and total flow rates in each section are constant. The external mass balances for the fractional extraction cascade shown in Figure 13-5 are

$$R + \overline{E} + F = E + \overline{R} \tag{13-15a}$$

$$Rx_{A,0} + \overline{E}y_{A,N+1} + Fz_{A,F} = Ey_{A,1} + \overline{R}x_{A,N} \tag{13-15b}$$

$$Rx_{B,0} + \overline{E}y_{B,N+1} + Fz_{B,F} = Ey_{B,1} + \overline{R}x_{B,N} \qquad \textbf{(13-15c)}$$

If the feed is contained in solvent 1, the flow rates in the two sections are related by the expressions

$$E = \overline{E} + F \qquad \textbf{(13-16a)}$$

$$\overline{R} = R \qquad \textbf{(13-16b)}$$

while if the feed is in solvent 2,

$$E = \overline{E} \qquad \textbf{(13-17a)}$$

$$\overline{R} = R + F \qquad \textbf{(13-17b)}$$

The solute operating equations for the top section using the top mass balance envelope in Figure 13-5 is Eq. (13-8), which was derived earlier. For the bottom section the mass balances are represented by

$$y_{i,k+1} = \frac{\overline{R}}{\overline{E}} x_{i,k} - (\frac{\overline{R}}{\overline{E}} x_{i,N} - y_{i,N+1}) \qquad \textbf{(13-18)}$$

which is identical to the single component Eq. (13-7). Since the feed is usually dissolved in one of the solvents, the phase flow rates are usually slightly different in the two sections. A McCabe-Thiele diagram can be plotted for each solute. Each diagram will have two operating lines and one equilibrium line, as shown in Figure 13-7.

Figures 13-7A and B show the characteristics of both absorber and stripper diagrams. The solutes are being "absorbed" in the top section (that is, the solute concentration is increasing as we go down the column) while the solute is being "stripped" in the bottom section (solute concentration is increasing as we go up the column). Thus, the solute is most concentrated at the feed stage and diluted at both ends (because we add lots of extra solvent). The two operating lines will intersect at a feed line which is at the concentration of the solute in the feed. This feed concentration is usually much greater than the solute concentration on the feed stage. If there is no solvent in the feed (the feed is liquid A + B), then the effective feed concentrations (y_i or x_i) are very large and the operating lines are almost parallel.

The top section of the column in Figures 13-5 and 13-7 is removing ("absorbing") component B from A. The more stages in this section, the purer the A product will be (smaller $y_{B,1}$), and the higher the recovery of B in the B product will be. The bottom section of the column is removing ("stripping") component A from B. Extra stages in the bottom section increase the purity of the B product (reduce $x_{A,N}$) and increase the recovery of A in the A product.

Specifications would typically include temperature, pressure, feed composition (A, B, and solvents), feed flow rate, and both solvent compositions. Some of the ways of specifying the four remaining degrees of freedom are illustrated below.

Case 1 Specify $y_{A,1}$, R, $\overline{E}$, and N_F. Calculate $x_{A,N}$ from Eq. (13 15b), calculate $\overline{R}$ from Eq. (13-16b) or (13-17b), and calculate E from Eq. (13-16a) or (13-17a). Since pairs of passing streams and slopes are known, both operating lines can be plotted for solute A. The total number of stages can be determined from the solute A diagram. Solution for solute B is trial and error.

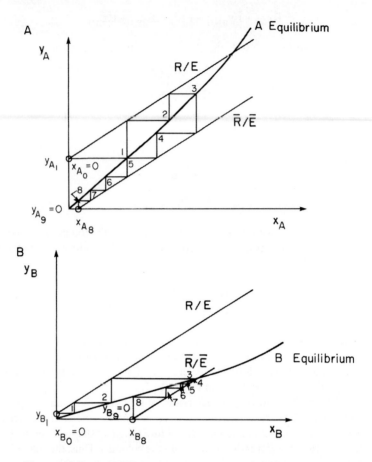

FIGURE 13-7. *McCabe-Thiele diagram for fractional extraction;
(A) solute A, (B) solute B*

Case 2. Specify $\overline{E}$, R, N_F, and N. This is trial and error for each solute, but the two solute problems are not coupled and can be solved separately.

Case 3. Specify R, $\overline{E}$, $y_{A,1}$, and $x_{B,N}$. After doing external mass balances, you can plot the operating lines as in Figure 13-7, but the problem is still trial-and-error. The feed stage must be varied until the total number of stages is the same for both solutes. Small changes in compositions or flow rates will probably be required to get an exact fit.

Because they are inherently trial-and-error, fractional extraction problems are naturals for computer solution (see Chapter 13 Appendix).

Brian (1972, Chapt. 3) explores fractional extraction calculations in detail. He illustrates the use of extract reflux and derives forms of the Kremser equation for multisection columns. An abbreviated treatment of the Kremser equation for fractional extraction is also presented by King (1980).

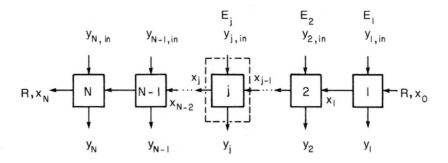

FIGURE 13-8.   *Cross-flow cascade*

13.4   IMMISCIBLE SINGLE-STAGE AND CROSS-FLOW EXTRACTION

Although countercurrent cascades are the most common, other cascades can be employed. One type occasionally used is the cross-flow cascade shown in Figure 13-8. Each stage is assumed to be an equilibrium stage. In this cascade, fresh extract streams are added to each stage and extract products are removed. For single-solute systems the same five assumptions made for countercurrent cascades are required for the McCabe-Thiele analysis. For dilute multicomponent systems, assumption 6 is again required.

To derive an operating equation, we use a mass balance envelope around a single stage as shown around stage j in Figure 13-8. For dilute systems the resulting steady-state mass balance is

$$Rx_{j-1} + E_j\, y_{j,in} = R\, x_j + E_j\, y_j \tag{13-19}$$

Solving for the outlet extract mass fraction, y_j, we get the operating equation,

$$y_j = -\frac{R}{E_j} x_j + \left(\frac{R}{E_j} x_{j-1} + y_{j,in}\right) \tag{13-20}$$

Each stage will have a different operating equation. On a McCabe-Thiele diagram plotted as y vs. x, this is a straight line of slope $-R/E_j$ and y intercept $(x_{j-1}\, R/E_j + y_{j,in})$. In these equations $y_{j,in}$ is the mass fraction of solute in the extract entering stage j and E_j is the flow rate of solvent entering stage j. The designer can specify all values of E_j and $y_{j,in}$ as well as x_0, R, and either x_N or N. If the calculation is started at the first stage (j = 1), $x_{j-1} = x_0$ is known and the operating equation can be plotted. Since the stage is an equilibrium stage, x_1 and y_1 are in equilibrium in addition to being related by operating equation (13-20). Thus, the intersection of the operating line and the equilibrium curve is at y_1 and x_1 (see Figure 13-9). This is the single-stage solution. For a cross-flow system, the raffinate input to stage 2 is x_1. Thus, the point $(y_{2,in}, x_1)$ on the operating line is known and the operating line for stage 2 can be plotted. The procedure is repeated until x_N is reached or N stages have been stepped off.

In general, each stage can have different solvent flow rates, E_j, and different inlet solvent mass fractions, $y_{j,in}$, as illustrated in Figure 13-9. This figure also shows that the point with the inlet concentrations $(y_{j,in}, x_{j-1})$ is on the operating line for each stage, which is easily proved.

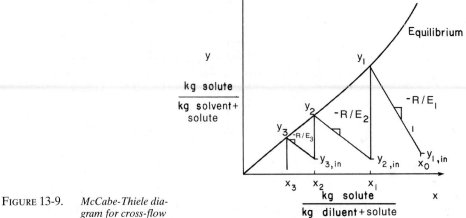

FIGURE 13-9. *McCabe-Thiele diagram for cross-flow*

If we let $y_j = y_{j,in}$ and $x_j = x_{j-1}$, Eq. (13-20) is satisfied. Thus, the point representing the inlet concentrations is on the operating line.

The operating lines in Figure 13-9 are similar to those we found for binary flash distillation. Both single-stage systems and cross-flow systems are arranged so that the two outlet streams are in equilibrium *and* on the operating line. This is not true of countercurrent systems.

The analysis for dilute multicomponent systems follows as a logical extension of the independent solution for each solute as was discussed for countercurrent systems. Total flow rates and mole or weight fractions are used in these calculations. Note that the simulation problems do not require trial-and-error solution for cross-flow systems.

EXAMPLE 13-2. Single-stage and cross-flow extraction of a protein

We wish to extract a dilute solution of the protein alcohol dehydrogenase from a aqueous solution of 5 wt % poly (ethylene glycol) (PEG) with an aqueous solution that is 10 wt % dextran. Aqueous two-phase extraction system is a very gentle method of recovering proteins that is unlikely to denature the protein since both phases are aqueous (Albertsson et al., 1990; Harrison et al., 2003). The two phases can be considered to be essentially immiscible. The dextran phase is denser and will be the cross-flow solvent. The entering dextran phases contain no protein. The entering PEG phase flow rate is 20 kg/h.

   a. If 10 kg/h of dextran phase is added to a single-stage extractor, find the total recovery fraction of alcohol dehydrogenase in the dextran solvent phase.
   b. If 10 kg/h of dextran phase is added to each stage of a cross-flow cascade with two stages, find the total recovery fraction of alcohol dehydrogenase in the dextran solvent phase.

The protein distribution coefficient is (Harrison et al., 2003)

$$K_d = (\text{wt frac protein in PEG, x})/(\text{wt frac protein in dextran, y}) = 0.12$$

Solution
A. Define. The two-stage cross-flow process is shown in Figure 13-10A. The single-stage system is the first stage of this cascade. For the single-stage extractor

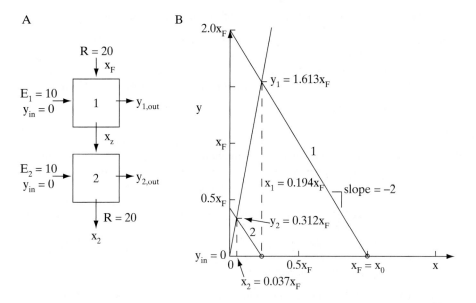

FIGURE 13-10.   *Single-stage and cross-flow extraction for Example 13-2; (A) schematic, (B) McCabe-Thiele diagram*

the fraction of the protein not extracted is $(Rx)_{out}/(Rx)_{in} = x_1/x_F$, and the fraction recovered $= 1 - x_1/x_F$. For the two stage system, the fraction of the protein not extracted is $(Rx)_{out}/(Rx)_{in} = x_2/x_F$, and the fraction recovered $= 1 - x_2/x_F$. If we find x_1/x_F and x_2/x_F, we have solved the problem.

B. Explore. Although the feed weight fraction is not specified, we can still plot the linear equilibrium with the abscissa x going from zero to x_F and the ordinate y from zero to an appropriate number of x_F units (see Figure 13-10B). This procedure works for linear systems since the slope of the equilibrium curve, $y/x = 1/0.12 = 8.33333$ is the same regardless of the value of x_F. The operating lines will also plot correctly regardless of the value of x_F.

C. Plan. Plot the equilibrium line, $y = 8.33333 \, x$, and plot the operating lines using Eq. (13-20) for each stage. Then calculate the recoveries.

D. Do it. Part a. Since $y_{in} = 0$, the operating line for stage 1 is

$$y = -(R/E_1)x + (R/E_1)x_0$$

Substituting in $R = 20$, $E_1 = 10$, and $x_0 = x_F$, the operating equation is

$$y = -2x + 2x_F$$

If we set $y = 0$ (for pure solvent addition to the stage), we find $x = x_F$. Figure 13-10B shows that the operating and equilibrium lines intersect at $y_1 = 1.613 \, x_F$, $x_1 = 0.194 \, x_F$. Thus, $x_1/x_F = 0.194$ and the fractional recovery for the single-stage system $= 1 - 0.194 = 0.806$.

Part b. The operating line for stage 2 of the cross-flow system is

$$y = -(R/E_2)x + (R/F_2)x_1 = -2x + 2x_1$$

The slope again $= -2$, and if we set $y = 0$ we obtain $x = x_1$. Since $x_1 = 0.194 \, x_F$ is known, we can plot this operating line (Figure 13-10B) and obtain $y_2 = 0.312 \, x_F$, $x_2 = 0.037 \, x_F$.

The fraction recovered $= 1 - x_2/x_F = 1 - 0.037 = 0.963$.

E. Check. This problem can also be solved analytically. We want to solve the linear equilibrium equation $y = mx$ simultaneously with the linear operating equation for each stage, $y = -(R/E_j)x + (R/E_j)x_{j-1} + y_{j,in}$. Do this sequentially starting with $j = 1$. This simultaneous solution is

$$x_j = [(R/E_j)x_{j-1} + y_{j,in}]/[m + (R/E_j)] \tag{13-21}$$

For example, for stage 1, $x_1 = [2x_F + 0]/[8.3333 + 2] = 0.19355 \, x_F$ which agrees with the graphical solution for part a. From the equilibrium expression, $y_1 = 8.33333x_1 = 1.6129 \, x_F$. For part b, stage 2 also agrees with the graphical solution.

F. Generalize. Although the problem statement involved a new type of system—aqueous two-phase extraction of proteins—we could apply the basic principles without difficulty. Since many of the details are often not necessary to solve the problem, it sometimes simplifies the problem to rewrite it as solute A being removed from diluent into solvent. Then see if you can solve the simplified version of the problem.

For linear equilibrium and operating lines an analytical solution is always relatively easy. The McCabe-Thiele diagram (even if only roughly sketched) is very useful to organize our thoughts and make sure we don't make any dumb algebraic errors. If the equilibrium line is curved, the analytical solution becomes considerably more complicated, but the McCabe-Thiele solution doesn't.

It is also interesting to compare the cross-flow system to a countercurrent system with 2 stages and a total flow rate of $E = 20 \, kg/h$. The result shows that the countercurrent process has a higher recovery.

Cross-flow systems have also been explored for stripping (Wnek and Snow, 1972). The analysis procedure for cross-flow stripping and absorption systems is very similar to the extraction calculations developed here (see Problem 12.D12).

In countercurrent systems the solvent is reused in each stage while in cross-flow systems it is not. Because solvent is reused, countercurrent systems can obtain more separation with the same total amount of solvent and the same number of stages. They can also obtain both high purity (x_{N+1} small) and high yield (high recovery of solute). Cross-flow systems can obtain either high purity or concentrated solvent streams but not both. For extraction they may have an advantage when flooding or slow settling of the two phases is a problem. For absorption and stripping the pressure drop may be significantly lower in a cross-flow system. Cascades that combine cross-flow and countercurrent operation have been studied by Thibodeaux et al. (1977) and by Bogacki and Szymanowski (1990).

13.5 CONCENTRATED IMMISCIBLE EXTRACTION

If the extraction system is relatively concentrated but still immiscible (this is almost a contradiction), we can extend the ratio unit mass balances developed in Chapter 12 for a single solute to extraction. Assumption 5, Eq. (13-2), no longer needs to be valid. This ratio-unit analysis is easily extended to cross-flow systems.

When the diluent and solvent are immiscible, weight ratio units are related to weight fractions as

$$X = \frac{x}{1-x} \text{ and } Y = \frac{y}{1-y} \tag{13-22}$$

where X is kg solute/kg diluent and Y is kg solute/kg solvent. Note that these equations require that the phases be immiscible.

The operating equation can be derived with reference to the mass balance envelope shown in Figure 13-3. In weight ratio units, the steady-state mass balance is

$$F_S Y_1 + F_D X_j = F_S Y_{j+1} + F_D X_0 \tag{13-23}$$

where weight fractions in Figure 13-3 have been replaced with weight ratios. Solving for Y_{j+1} we obtain the operating equation,

$$Y_{j+1} = (\frac{F_D}{F_S})X_j + (Y_1 - \frac{F_D}{F_S}X_0) \tag{13-24}$$

When plotted on a McCabe-Thiele diagram of Y vs. X, this is a straight line with slope F_D/F_S and Y intercept $(Y_1 - (F_D/F_S)X_0)$. Note that since F_D and F_S are constant, the operating line is straight. For the usual design problem, F_D/F_S will be known as will be X_0, X_N and Y_{N+1}. Since X_N and Y_{N+1} are the concentrations of passing streams, they represent the coordinates of a point on the operating line.

For the McCabe-Thiele diagram if flow rates are given as in Eq. (13-2), the equilibrium data must be expressed as weight or mole ratios. Equation (13-22) can be used to transform the equilibrium data, which is easy to do in tabular form. Usually equilibrium data are reported in fractions, not the ratio units given in the second part of Table 13-3.

With the equilibrium data and the operating equation known, the McCabe-Thiele diagram is plotted. First the point (Y_{N+1}, X_N) is plotted, and the operating line passes through this point with a slope of F_D/F_S. Y_1 can be found from the operating line at the inlet raffinate concentration X_0. Then the stages are easily stepped off.

When there is some partial miscibility of diluent and solvent, the McCabe-Thiele analysis can still be used if the following alternative assumption is valid.

4b. The concentration of solvent in the raffinate and the concentration of diluent in the extract are both constant.

The flow rates of the diluent and solvent streams are now defined as

$$F_D = \frac{\text{kg (D + S) in raffinate}}{h} = \text{constant} \tag{13-25a}$$

$$F_S = \frac{\text{kg (S + D) in extract}}{\text{h}} = \text{constant} \tag{13-25b}$$

The ratio units are defined as

$$X = \frac{\text{kg A in raffinate}}{\text{kg (D + S) in raffinate}} \tag{13-26a}$$

$$Y = \frac{\text{kg A in extract}}{\text{kg (S + D) in extract}} \tag{13-26b}$$

The calculation procedure now follows Eqs. (13-23) to (13-24).

For all of these situations, as long as F_D and F_S are constant and the equilibrium in ratio units can be approximated as a straight line, $Y = m_{ratio}X + b_{ratio}$, the Kremser equations [e.g., Eqs. (13-11a) or (13-11b)] are valid if written in terms of mass or mole ratios with F_D replacing R and F_S replacing E (Brian, 1972). Extension of the Kremser equation to ratio units extends the region of validity of this solution.

When the phases are partially miscible and assumption 4b is not valid, the methods developed in sections 13.8 to 13.10 should be used.

13.6   IMMISICIBLE BATCH EXTRACTION

In batch plants batch extraction will usually fit into the production scheme easier than continuous extraction. This is particularly true in the production of biochemicals (Belter et al., 1988). Batch extraction is very flexible, and there are a number of ways to do it. The simplest approach is to add solvent and diluent together in a tank, mix the two immiscible liquids, allow them to settle and then withdraw the solvent layer. If we define $\hat{F}, \hat{R}, \hat{E}$, and $\hat{S}$ as the mass (kg) of the streams, the resulting mass balance is

$$\hat{F}x_F + \hat{S}y_s = \hat{R}x + \hat{E}y \tag{13-27a}$$

For immiscible phases, $\hat{F} = \hat{R}$ and $\hat{S} = \hat{E}$. Solving for y, we obtain the operating equation

$$y = -(\hat{R}/\hat{E})x + (\hat{R}/\hat{E})x_F + y_s \tag{13-27b}$$

Except that this equation uses masses of raffinate and extract instead of flow rates, it is essentially the same as the continuous operating equation for a single-stage system (Eq. (13-19) with j = 1). Thus, the solution is identical to the continuous solution and can be obtained either graphically (Figures 13-9 and 13-10) or analytically Eq. (13-21) for linear isotherms with j = 1.

Additional purification can be attained by doing repeated batch extractions in the same tank. If the fresh feed is contacted with fresh solvent, the extract phase is removed, and then the raffinate is contacted again with fresh solvent, the operation is essentially identical to continuous cross-flow systems. The use of two-stage countercurrent contacting is more common (Frank et al., 2008). To start up this operation, the fresh feed is contacted first with fresh solvent. After removing the extract phase, the raffinate is contacted with fresh solvent again. The raffinate is removed as raffinate product, and the extract phase is saved in the tank. This completes the startup for the two-stage countercurrent batch contactor. For the next batch, the fresh feed is contacted with the saved extract. After settling and removal, the extract phase from this contacting is the extract product. The raffinate phase remaining in the tank is contacted with fresh solvent, and the mixture is allowed to settle. The raffinate phase is removed as the raffinate product, and the extract phase is left in the tank ready for the next batch. The operation can be repeated many times. Sketching the process will probably be helpful in understanding it (see Problem 13.A13).

If we want to totally remove the solute from the diluent and have it dissolved in a solvent, the continuous solvent addition batch extraction shown in Figure 13-11 will use less solvent than repeated single-stage batch extractions. A solvent that is pre-saturated with diluent is added continuously to a mixed tank that contains the feed. The raffinate and extract phases are separated in a settler with withdrawal of the extract product and recycle of the diluent.

This process is analogous to constant level batch distillation (section 9.3.) and the analysis for immiscible extraction is very similar to the analysis in that section. If we assume that the level in the mixed tank is constant, then the overall mass balance becomes in = out, and for $d\hat{S}$ kg of entering solvent,

$$d\hat{S} = d\hat{E} \qquad \text{(13-28a)}$$

For the component balance, since there is no entering solute, the equation is -out = accumulation,

$$-y d\hat{E} = \hat{R}_t dx_t + \hat{E}_t dy \qquad \text{(13-28b)}$$

where $\hat{R}_t$ is the mass of raffinate phase in the tank plus settler and x_t is the mass fraction solute in the raffinate phase. If we assume that the raffinate holdup is much greater than the solvent holdup $\hat{E}_t$, then $\hat{R}_t dx >> \hat{E}_t dy$, and Eq. (13-29b) simplifies to

$$-y d\hat{E} = \hat{R}_t dx_t \qquad \text{(13-28c)}$$

Assuming $\hat{R}_t$ is constant, substituting in Eq. (13-28a) and integrating, we obtain

$$-\hat{S}/\hat{R}_t = \int_{x_{t,feed}}^{x_{t,final}} \frac{dx_t}{y} \qquad \text{(13-29a)}$$

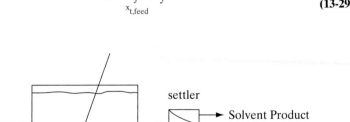

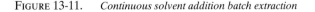

FIGURE 13-11.   *Continuous solvent addition batch extraction*

13.6 Immisicible Batch Extraction

where y and x_t are in equilibrium and $\hat{S}$ is the total mass of solvent added. In general, this equation can always be integrated numerically or graphically. If equilibrium is linear, $y = K_d x$, analytical integration is straightforward and the result is

$$-\hat{S}/\hat{R} = \frac{1}{K_d}\ell n\,(x_{t,final}/x_{t,feed})$$

(13-29b)

Numerical calculations for this process are in Problem 13.D20.

If the solvent and diluents are partially miscible, the methods in sections 13.7 to 13.13 need to be used. The development of Eqs. (13-28a) to (13-29b) is only valid after enough solvent has been added to the feed that two phases form. Since K_d is not usually constant, Eq. (13-29b) cannot be used. Integration of Eq. (13-29a) usually needs to be done numerically. The resulting amount of solvent added $\hat{S}_t$ is the amount added after addition of enough solvent to form two phases.

13.7 EXTRACTION EQUILIBRIUM FOR PARTIALLY MISCIBLE TERNARY SYSTEMS

All extraction systems are partially miscible to some extent. When partial miscibility is very low, as for toluene and water, we can treat the system as if it were completely immiscible and use McCabe-Thiele analysis or the Kremser equation. When partial miscibility becomes appreciable, it can no longer be ignored, and a calculation procedure that allows for variable flow rates must be used. In this case a different type of stage-by-stage analysis, which is very convenient for ternary systems, can be used. For multicomponent systems, computer calculations are required.

Extraction systems are noted for the wide variety of equilibrium behavior that can occur in them. In the partially miscible range utilized for extraction, two liquid phases will be formed. At equilibrium the temperatures and pressures of the two phases will be equal and the compositions of the two phases will be related. The number of independent variables that can be arbitrarily specified (i.e., the degrees of freedom) for a system at equilibrium can be determined from the Gibbs phase rule $F = C - P + 2$, which for a ternary extraction is $F = 3 - 2 + 2 = 3$ degrees of freedom. In an extraction, temperature and pressure are almost always constant so only one degree of freedom remains. Thus, if we specify the composition of one component in either phase, all other compositions will be set at equilibrium.

Extraction equilibrium data are easily shown graphically as either right triangular diagrams or equilateral triangular diagrams. Figure 13-12 shows the data listed in Table 13-4 for the system water-chloroform-acetone at 25°C on a right triangular diagram. We have chosen chloroform as solvent, water as diluent, and acetone as solute. We could also call water the solvent and chloroform the diluent if the feed was an acetone-chloroform mixture. Curved line AEBRD represents the *solubility envelope* for this system. Any point below this line represents a two-phase mixture that will separate at equilibrium into a saturated extract phase and a saturated raffinate phase. Line AEB is the saturated extract line, while line BRD is the saturated raffinate line. Point B is called the *plait point* where extract and raffinate phases are identical. Remember that the extract phase is the phase with the higher concentration of solvent. Tie line ER connects extract and raffinate phases that are in equilibrium.

Point N in Figure 13-12 is a single phase because the ternary system is miscible at these concentrations. Point M represents a mixture of two phases, since it is in the immiscible range

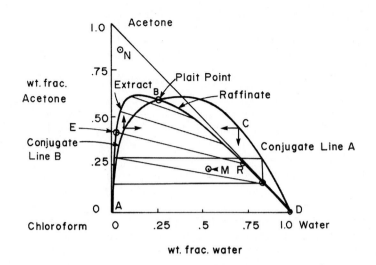

FIGURE 13-12.   *Equilibrium for water-chloroform-acetone at 25°C and 1 atm*

TABLE 13-4.   *Equilibrium data for the system water-chloroform-acetone at 1 atm and 25°C (Alders, 1959; Perry and Green, 1997, p. 2–33)*

Water Phase, wt %			Chloroform Phase, wt %		
x_D Water	x_S Chloroform	x_A Acetone	y_D Water	y_S Chloroform	y_A Acetone
99.19	0.81	0.00	0.5	99.5	0.00
82.97	1.23	15.80	1.3	70.0	28.7
73.11	1.29	25.60	2.2	55.7	42.1
62.29	1.71	36.00	4.4	42.9	52.7
45.6	5.1	49.3	10.3	28.4	61.3
34.5	9.8	55.7	18.6	20.4	61.0

Note: Water and chloroform phases on the same line are in equilibrium with each other.

for this ternary system. At equilibrium the mixture represented by M will separate into a saturated raffinate phase and a saturated extract phase in equilibrium with each other. Either of the conjugate lines shown in Figure 13-12 can be used to draw tie lines. Consider tie line ER, which was found by drawing a horizontal line from point E to the conjugate line (point C) and then a vertical line from point C to the saturated raffinate curve (point R). Points E and R are in equilibrium, so they are on the ends of a tie line. This construction is shown in Figure 13-13 for different equilibrium data. This procedure is analogous to the use of an auxiliary line on an enthalpy-composition diagram as illustrated in Figure 2-5.

To find the raffinate and extract phases that result when mixture M separates into two phases, we need a tie line through point M. This requires a simple eyeball trial-and-error calculation. Guess the location of the end point of the tie line on the saturated extract or raffinate curve, construct a tie line through this point, and check if the line passes through point

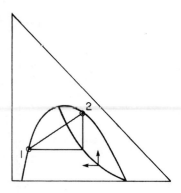

FIGURE 13-13.   *Construction of tie line using conjugate line*

M. If the first guess does not pass through M, repeat the process until you find a tie line that does. This is not too difficult because the tie lines that are close to each other are approximately parallel.

The solubility envelope, tie lines, and conjugate lines shown on the triangular diagrams are derived from experimental equilibrium data. To obtain these data a mixture can be made up and allowed to separate in a separatory funnel. Then the concentrations of extract and raffinate phases in equilibrium are measured. This measurement will give the location of one point on the saturated extract line, one point on the saturated raffinate line, and the tie line connecting these two points. One point on the conjugate line can be constructed from this tie line by reversing the procedure used to construct a tie line when the conjugate line was known (Figure 13-13).

The equilibrium data represented by Figure 13-12 are often called a type I system, since there is one pair of immiscible binary compounds. It is also possible to have systems with zero, two, and three immiscible binary pairs (Alders, 1959; Macedo and Rasmussen, 1987; Sorenson and Arlt, 1979, 1980; Walas, 1985). It is possible to go from a type I to a type II system as temperature decreases. This is shown in Figure 13-14 (Fenske et al., 1955) for the methylcyclohexane-toluene-ammonia system. At 10°F this is a type II system.

We will use right triangular diagrams exclusively in the remainder of this chapter because they are easy to read, they don't require special paper, the scales of the axes can be varied, and portions of the diagram can be enlarged. Although equilateral diagrams have none of these advantages, they are used extensively in the literature for reporting extraction data; therefore it is important to be able to read and use this type of extraction diagram.

Equilibrium data can be correlated and estimated with thermodynamic models that calculate activity coefficients. Although these calculations are similar to those for vapor-liquid equilibrium (VLE) they are more complicated and generally less accurate. An extensive compilation of data and UNIQUAC and NRTL parameters is given by Sorenson and Arlt (1979, 1980) and Macedo and Rasmussen (1987).

13.8   MIXING CALCULATIONS AND THE LEVER-ARM RULE

Triangular diagrams can be used for mixing calculations. In Figure 13-15A a simple mixing operation is shown, where streams F_1 and F_2 are mixed to form stream M. Streams F_1, F_2, and M can be either single-phase or two-phase. Operation of the mixer is assumed to be isothermal. For ternary systems there are three independent mass balances. With right triangular diagrams

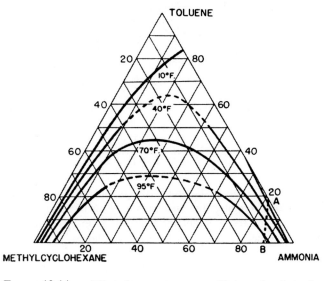

TOLUENE

METHYLCYCLOHEXANE

AMMONIA

FIGURE 13-14.   *Effect of temperature on equilibrium of methylcyclo-*
hexane-toluene-ammonia system from Fenske et al.,
AIChE *Journal, 1, 335 (1955), copyright 1955, AIChE*

it is convenient to use the diluent balance, the solute balance, and the overall mass balance. The solvent mass balance will be automatically satisfied if the three independent balances are satisfied. The nomenclature is in fraction units (see Table 13-2).

For the mixing operation in Figure 13-15A the flow rates F_1 and F_2 would be given as well as the concentration of the two feeds: x_{A,F_1}, x_{D,F_1}, x_{A,F_2}, x_{D,F_2}. The three independent mass balances used to solve for M, $x_{A,M}$ and $x_{D,M}$ are

$$F_1 + F_2 = M \qquad \qquad \textbf{(13-30a)}$$

A

B

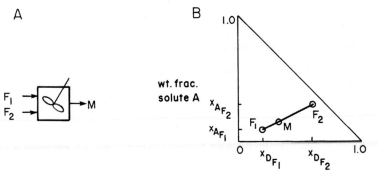

FIGURE 13-15.   *Mixing operation; (A) equipment, (B) triangular diagram*

$$F_1 x_{A,F_1} + F_2 x_{A,F_2} = M\, x_{A,M} \qquad \text{(13-30b)}$$

$$F_1 x_{D,F_1} + F_2 x_{D,F_2} = M\, x_{D,M} \qquad \text{(13-30c)}$$

The concentrations of the mixed stream M are

$$x_{A,M} = \frac{F_1 x_{A,F_1} + F_2 x_{A,F_2}}{F_1 + F_2} \qquad \text{(13-31a)}$$

$$x_{D,M} = \frac{F_1 x_{D,F_1} + F_2 x_{D,F_2}}{F_1 + F_2} \qquad \text{(13-31b)}$$

We will now show that points F_1, F_2, and M are collinear as shown in Figure 13-15B. We will first use Eq. (13-30a) to remove the mixed stream flow rate M from Eqs. (13-30b) and (13-30c). Next, we solve the resulting equations for the ratio F_1/F_2 and then set these two equations equal to each other. The manipulations are as follows:

$$F_1 x_{A,F_1} + F_2 x_{A,F_2} = (F_1 + F_2)\, x_{A,M}$$

$$F_1 x_{D,F_1} + F_2 x_{D,F_2} = (F_1 + F_2)\, x_{D,M}$$

Then

$$\frac{F_1}{F_2} = \frac{x_{A,M} - x_{A,F_2}}{x_{A,F_1} - x_{A,M}} \qquad \text{(13-32a)}$$

$$\frac{F_1}{F_2} = \frac{x_{D,M} - x_{D,F_2}}{x_{D,F_1} - x_{D,M}} \qquad \text{(13-32b)}$$

Finally, setting these equations equal to each other and rearranging, we have

$$\text{slope from point M to } F_2 = \frac{x_{A,M} - x_{A,F_2}}{x_{D,M} - x_{D,F_2}} \qquad \text{(13-33)}$$

$$= \frac{x_{A,F_1} - x_{A,M}}{x_{D,F_1} - x_{D,M}} = \text{slope from point M to } F_1$$

Equation (13-33), the three-point form of a straight line, states that the three points $(x_{A,M}, x_{D,M})$ (x_{A,F_2}, x_{D,F_2}) and (x_{A,F_1}, x_{D,F_1}) lie on a straight line. The manipulations used to derive Eq. (13-33) are very similar to those used to develop difference points for countercurrent calculations, and we will return to them shortly.

It will often prove convenient to be able to determine the location of the mixing point on the line between F_1 and F_2 without having to solve the mass balances analytically. This can be done using Eqs. (13-32a) or 13-32b), which relate the ratio of the feed rates to differences in the ordinate and abscissa, respectively. With F_1/F_2, x_{A,F_1}, x_{A,F_2} known, Eq. (13-31a) can be used to find $x_{A,M}$. Equation (13-31b) can be used in a similar way to find $x_{D,M}$.

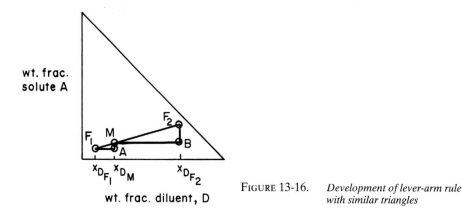

FIGURE 13-16.   *Development of lever-arm rule with similar triangles*

In Figure 13-16 similar triangles F_1AM and MBF_2 have been drawn. Since the triangles are similar,

$$\frac{\text{Distance from } F_1 \text{ to } M}{\text{Distance from } F_1 \text{ to } A} = \frac{\text{distance from } M \text{ to } F_2}{\text{distance from } M \text{ to } B} \qquad \textbf{(13-34a)}$$

Rearranging this formulation, we have

$$\frac{\overline{MF_2}}{\overline{F_1M}} = \frac{\overline{MB}}{\overline{F_1A}} = \frac{x_{D,M} - x_{D,F_2}}{x_{D,F_1} - x_{D,M}} \qquad \textbf{(13-34b)}$$

where the bar denotes distance. According to Eq. (13-32b), the right-hand side of this equation is equal to F_1/F_2. Thus, we have shown that

$$\frac{F_1}{F_2} = \frac{\overline{MF_2}}{\overline{F_1M}} \qquad \textbf{(13-35)}$$

Equation (13-35) is the lever-arm rule, which was first introduced in Figure 2-10. It may be helpful to review that material now. By measuring along the straight line between F_1 and F_2 we can find point M so that the lever-arm rule is satisfied. When you use the lever-arm rule, you don't need the individual values of the flow rates F_1 and F_2 to find the location of M.

Using Eq. (13-35), point M might be found by trial-and-error. Since this is a cumbersome procedure, it is worthwhile to develop the lever-arm rule in a different form. These alternative forms are

$$\frac{F_1}{M} = \frac{\overline{F_2M}}{\overline{F_1F_2}}, \quad \frac{F_2}{M} = \frac{\overline{F_1M}}{\overline{F_1F_2}} \qquad \textbf{(13-36)}$$

In this form the lever-arm rule is useful for finding the location of a stream M that is the sum of the two streams F_1 and F_2.

13.9   PARTIALLY MISCIBLE SINGLE-STAGE AND CROSS-FLOW SYSTEMS

Single-stage extraction systems can easily be solved with the tools we have developed. A batch extractor would consist of a single vessel equipped with a mixer. The two feeds would be charged to the vessel, mixed, and then allowed to settle into the two product phases. A continuous single-stage system requires a mixer and a settler as shown in Figure 13-2. Here the feed and solvent are fed continuously to the mixer, and the raffinate and extract products are continuously withdrawn from the settler. Figure 13-17 shows this schematically. The calculation procedures for batch and continuous operation are the same, the only real difference being that in batch operations S, F, M, E, and R are measured as total weight of material, whereas in continuous operation they are flow rates.

Usually the solvent and feed streams will be completely specified in addition to temperature and pressure. Thus, the known variables are S, F, $y_{A,S}$, $y_{D,S}$, $x_{A,F}$, $x_{D,F}$, T, and p. The values of E, R, $y_{A,E}$, $y_{D,E}$, $x_{A,R}$, and $x_{D,R}$ are usually desired. If we make the usual assumption that the mixer-settler combination acts as one equilibrium stage, then streams E and R are in equilibrium with each other.

The calculation method proceeds as follows. 1) Plot the locations of S and F on the triangular equilibrium diagram. 2) Draw a straight line between S and F, and use the lever-arm rule or Eqs. (13-31) to find the location of the mixed stream M. Now we know that stream M settles into two phases in equilibrium with each other. Therefore, 3) construct a tie line through point M to find the compositions of the extract and raffinate streams. 4) Find the ratio E/R using mass balances. We will follow this method to solve the following example.

> ### EXAMPLE 13-3. Single-stage extraction
> A solvent stream containing 10% by weight acetone and 90% by weight chloroform is used to extract acetone from a feed containing 55 wt % acetone and 5 wt % chloroform with the remainder being water. The feed rate is 250 kg/h, while the solvent rate is 400 kg/h. Operation is at 25 °C and atmospheric pressure. Find the extract and raffinate compositions and flow rates when one equilibrium stage is used for the separation.
>
> ### Solution
> A. Define. The equipment sketch is the same as Figure 13-17 with S = 400, $y_{A,S}$ = 0.1, $y_{S,S}$ = 0.9, $y_{D,S}$ = 0 and F = 250, $x_{A,F}$ = 0.55, $x_{S,F}$ = 0.05, $x_{D,F}$ = 0.40. Find $x_{A,R}$, $x_{D,R}$, $y_{A,E}$, $y_{D,E}$, R, and E.
> B. Explore. Equilibrium data are obviously required. They can be obtained from Table 13-4 and Figure 13-12.
> C. Plan. Plot streams F and S. Find mixing point M from the lever-arm rule or from Eqs. (13-31). Then a tie line through M gives locations of streams E and R. Flow rates can be found from mass balances.
> D. Do it. The graphical solution is shown in Figure 13-18. After locating streams F and S, M is on the line SF and can be found from the lever arm rule,

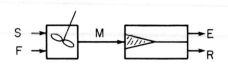

FIGURE 13-17.   *Continuous mixer-settler*

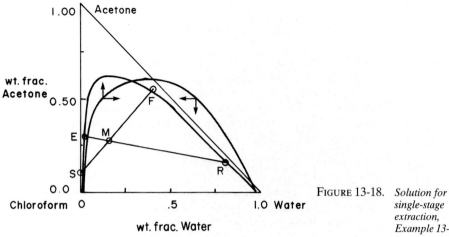

FIGURE 13-18. *Solution for single-stage extraction, Example 13-3*

$$\frac{F}{M} = \frac{\overline{SM}}{\overline{FS}} = \frac{250}{250 + 400} = 0.385$$

or from Eq. (13-31a),

$$x_{A,M} = \frac{Fx_{AF} + Sy_{A,S}}{F + S} = \frac{(250)(0.55) + 400(0.1)}{650} = 0.273$$

A tie line through M is then constructed by trial-and-error, and the extract and raffinate locations are obtained. Concentrations are

$$y_{A,E} = 0.30, \qquad y_{D,E} = 0.02, \qquad x_{A,R} = 0.16, \qquad x_{D,R} = 0.83$$

The flow rates can be determined from the mass balances
$$M = E + R \quad \text{and} \quad Mx_{A,M} = Ey_{A,E} + Rx_{A,R}$$

Solving for R, we obtain

$$R = M\,\frac{x_{A,M} - y_{A,E}}{x_{A,R} - y_{A,E}} \tag{13-37}$$

or $R = (650)\left(\dfrac{0.273 - 0.30}{0.16 - .30}\right) = 125.36$ kg/h

and $E = M - R = 650 - 125.36 = 524.64$.
   The lever arm rule can also be used but tends to be slightly less accurate.
E. Check. We can check the solute or diluent mass balances. For example, the solute mass balance is

$$Sy_{A,S} + Fx_{A,F} = Ey_{A,E} + Rx_{A,R}$$

which is

$$(400)(0.1) + (250)(0.55) = (524.64)(0.30) + (125.36)(0.16)$$

or 177.5 ~ 177.45, which is well within the accuracy of the calculation. The diluent mass balance also checks.

F. Generalize. This procedure is similar to the one we used for binary flash distillation in Figure 2-9. Thus, there is an analogy between distillation calculations on enthalpy-composition diagrams (Ponchon-Savarit diagrams) and extraction calculations on triangular diagrams.

From this example it is evident that a single extraction stage is sufficient to remove a considerable amount of acetone from water. However, quite a bit of solvent was needed for this operation, the resulting extract phase is not very concentrated, and the raffinate phase is not as dilute as it could be.

The separation achieved with one equilibrium stage can easily be enhanced with a cross-flow system as shown in Figure 13-19. Assume that a cross-flow stage is added to the problem given in Example 13-3 and another 400 kg/h of solvent (stream S_2 with 10% acetone, 90% chloroform) is used in stage 2. The concentrations of E_2 and R_2 are easily found by doing a second mixing calculation with streams S_2 and R_1. During this mixing calculation, R_{j-1} (the feed to stage j) is different for each stage. A tie line through the new mixing point M_2 (Figure 13-19B) gives the location of streams E_2 and R_2. Note that $x_{A,R_2} < x_{A,R_1}$ as desired.

In Section 13.4 we found that cross-flow systems are less efficient than countercurrent systems. In the next section the calculations for countercurrent cascades will be developed.

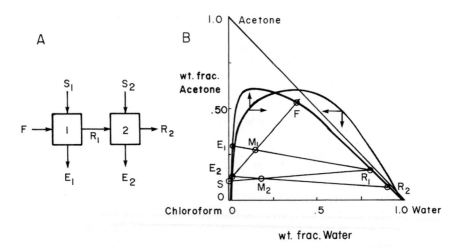

FIGURE 13-19.   *Cross-flow extraction; (A) cascade, (B) solution of triangular diagram*

13.10 COUNTERCURRENT EXTRACTION CASCADES FOR PARTIALLY MISCIBLE SYSTEMS

13.10.1 External Mass Balances

A countercurrent cascade allows for more complete removal of the solute, and the solvent is reused so less is needed. A schematic diagram of a countercurrent cascade is shown in Figure 13-20. All calculations will assume that the column is isothermal and isobaric and is operating at steady state. In the usual design problem, the column temperature and pressure, the flow rates and compositions of streams F and S, and the desired composition (or percent removal) of solute in the raffinate product are specified. The designer is required to determine the number of equilibrium stages needed for the specified separation and the flow rates and compositions of the outlet raffinate and extract streams. Thus, the known variables are T, p, R_{N+1}, E_0, $x_{A,N+1}$, $y_{A,0}$, $y_{D,0}$, and $x_{A,1}$, and the unknown quantities are E_N, R_1, $x_{D,1}$, $y_{A,N}$, $y_{D,N}$, and N.

For an isothermal ternary extraction problem, the outlet compositions and flow rates can be calculated from external mass balances used in conjunction with the equilibrium relationship. The mass balances around the entire cascade are

$$E_0 + R_{N+1} = R_1 + E_N \qquad (13\text{-}38a)$$

$$E_0 y_{A,0} + R_{N+1} x_{A,N+1} = R_1 x_{A,1} + E_N y_{A,N} \qquad (13\text{-}38b)$$

$$E_0 y_{D,0} + R_{N+1} x_{D,N+1} = R_1 x_{D,1} + E_N y_{D,N} \qquad (13\text{-}38c)$$

Since five variables are unknown (actually, there are seven, but $x_{S,1}$ and $y_{S,N}$ are easily found once $x_{A,1}$, $x_{D,1}$, $y_{A,N}$, and $y_{D,N}$ are known), a total of five independent equations are needed.

To find two additional relationships, note that streams R_1 and E_N are both leaving equilibrium stages. Thus, the compositions of stream R_1 must be related in such a way that R_1 is on the saturated raffinate curve. This gives a relationship between $x_{A,1}$ and $x_{D,1}$. Similarly, since stream E_N must be a saturated extract, $y_{A,N}$ and $y_{D,N}$ are related. If the saturated extract and saturated raffinate relationships are known in analytical form, these two equations can be added to the three mass balances, and the resulting five equations can be solved simultaneously for the five unknowns.

The procedure can also be carried out conveniently on a triangular diagram. Let us represent the cascade shown in Figure 13-20 as a mixing tank followed by a black box separation scheme that produces the desired extract and raffinate as shown in Figure 13-21A. In Figure 13-21A, streams E_N and R_1 are not in equilibrium as they were in Figure 13-17, but stream E_N

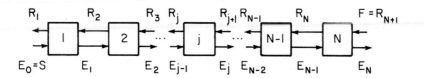

FIGURE 13-20. *Countercurrent extraction cascade*

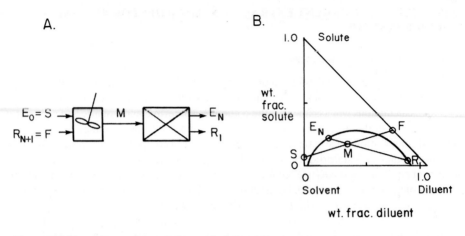

A.

$E_0 = S$
$R_{N+1} = F$
M
E_N
R_1

B.

1.0 Solute

wt. frac. solute

E_N
F
S,O
M
R
0

0 Solvent
1.0 Diluent

wt. frac. diluent

FIGURE 13-21.   *External mass-balance calculation; (A) mixer-separation representation, (B) solution on triangular diagram*

is a saturated extract and stream R_1 is a saturated raffinate. The external mass balances for Figure 13-21A are

$$E_0 + R_{N+1} = M = E_N + R_1 \tag{13-39a}$$

$$E_0 y_{A,0} + R_{N+1} x_{A,N+1} = M x_{A,M} = E_N y_{A,N} + R_1 x_{A,1} \tag{13-39b}$$

$$E_0 y_{D,0} + R_{N+1} x_{D,N+1} = M x_{D,M} = E_N y_{D,N} + R_1 x_{D,1} \tag{13-39c}$$

The coordinates of point M can be found from Eqs. (13-39):

$$x_{A,M} = \frac{E_0 y_{A,0} + R_{N+1} x_{A,N+1}}{E_0 + R_{N+1}} \tag{13-40a}$$

$$x_{D,M} = \frac{E_0 y_{D,0} + R_{N+1} x_{D,N+1}}{E_0 + R_{N+1}} \tag{13-40b}$$

Since Eq. (13-39) are the same type of mass balances as for a mixer, the points representing streams E_N, R_1, and M lie on a straight line given by

$$y_{D,N} = (y_{A,N} - x_{A,M})(\frac{x_{D,M} - x_{D,1}}{x_{A,M} - x_{A,1}}) + x_{D,M} \tag{13-41}$$

and the flow rates are related to the length of line segments by the lever-arm rule. We also know that E_N must lie on the saturated extract line and R_1 must lie on the saturated raffinate line. Since $x_{A,1}$ is known, the location of R_1 on the saturated raffinate line can be found. A straight line from R_1 extended through M (found from $x_{A,M}$ and $x_{D,M}$ or from the lever-arm rule) will intersect the saturated extract stream at the value of E_N. This construction is illustrated in Figure 13-21B. Note that this procedure is very similar to the one used for single equilibrium stages, but the line R_1ME_N is *not* a tie line. Mass balances can then be used to solve for the flow rates E_N and R_1.

The external mass balance and the equilibrium diagram in Figure 13-21B can be used to determine the effect of variation in the feed or solvent concentrations, the raffinate concentration, or the ratio F/S on the resulting separation. For example, if the amount of solvent is increased, the ratio of F/S will decrease. The mixing point M will move toward point S, and the resulting extract will contain less solute.

13.10.2   Difference Points and Stage-by-Stage Calculations

To determine the number of stages or flow rates and compositions inside the cascade, stage-by-stage calculations are needed after we use the external mass balances to find concentrations. But first we use the external mass balances to find concentrations $y_{A,N}$ and $y_{D,N}$ and flow rates E_N and R_1. Starting at stage 1 (Figure 13-20) we note that streams R_1 and E_1 both leave equilibrium stage 1. Therefore, these two streams are in equilibrium and the concentration of stream E_1 can be found from an equilibrium tie line.

Streams E_1 and R_2 pass each other in the diagram and are called passing streams. These streams can be related to each other by mass balances around stage 1 and the raffinate end of the extraction train. The unknown variables for these mass balances are concentrations $x_{A,2}$, $x_{D,2}$, and $x_{S,2}$ and flow rates E_1 and R_2. Concentration $x_{S,2}$ can be determined from the stoichiometric relation $x_{S,2} = 1.0 - x_{A,2} - x_{D,2}$. Taking this equation into account, there are four unknowns (E_1, R_2, $x_{A,2}$, and $x_{D,2}$) but only three independent mass balances. What is the fourth relation that must be used?

To develop a fourth relation we must realize that stream R_2 is a saturated raffinate stream. Thus, it will be located on the saturated raffinate line, and $x_{A,2}$ and $x_{D,2}$ are related by the relationship describing the saturated raffinate line. With four equations and four unknowns we can now solve for the variables E_1, R_2, $x_{A,2}$, and $x_{D,2}$.

To continue along the column, we repeat the procedure for stage 2. Since streams E_2 and R_2 are in equilibrium, a tie line will give the concentration of stream E_2. Streams E_2 and R_3 are passing streams; thus, they are related by mass balances. It will prove to be convenient if we write the mass balances around stages 1 and 2 instead of around stage 2 alone. The fourth required relationship is that stream R_3 must be a saturated raffinate stream. The stage-by-stage calculation procedure is then continued for stages 3, 4, etc. When the calculated solute concentration in the extract is greater than or equal to the specified concentration, that is, $y_{A,j_{calc}} \geq y_{A,N_{specified}}$, the problem is finished.

These stage-by-stage calculations can be done analytically and can be programmed for spreadsheet solution if equations are available for the tie lines and the saturated extract and saturated raffinate curves. If the equations are not readily available, either the equilibrium data must be fitted to an analytical form or a data matrix with a suitable interpolation routine must be developed. Graphical techniques can be employed and have the advantage of giving a visual interpretation of the process.

In a graphical procedure for countercurrent systems the equilibrium calculations can easily be handled by constructing tie lines. The relation between $x_{A,j}$ and $x_{D,j}$ is already shown as the saturated raffinate curve. All that remains is to develop a method for representing the mass balances graphically.

Referring to Figure 13-20, we can do a mass balance around the first stage. After rearrangement, this is

$$E_0 - R_1 = E_1 - R_2$$

If we now do mass balances around each stage and rearrange each balance as the difference between passing streams, we obtain

$$\Delta = E_0 - R_1 = \cdots = E_j - R_{j+1} = \cdots = E_N - R_{N+1} \tag{13-42a}$$

Thus, the difference in flow rates of passing streams is constant even though both the extract and raffinate flow rates are varying. The same difference calculation can be repeated for solute A,

$$(\Delta)x_{A,} = E_0 y_{A,0} - R_1 x_{A_1} = \cdots = E_j y_{A,j} - R_{j+1} x_{A,j+1} = \cdots = E_N y_{A,N} - R_{N+1} x_{A,N+1} \tag{13-42b}$$

and for diluent D,

$$(\Delta)x_{D,\Delta} = E_0 y_{D,0} - R_1 x_{D,1} = \cdots = E_j y_{D,j} - R_{j+1} x_{D,j+1} = \cdots = E_N y_{D,N} - R_{N+1} x_{D,N+1} \tag{13-42c}$$

The differences in flow rates (which is the net flow) of solute and diluent are constant.

Equations (13-42) define a difference or Δ (delta) point. The coordinates of this point are easily found from Eqs. (13-42b) and (13-42c).

$$x_{A,\Delta} = \frac{E_0 y_{A,0} - R_1 x_{A,1}}{\Delta} = \frac{E_N y_{A,N} - R_{N+1} x_{A,N+1}}{\Delta} \tag{13-43a}$$

$$x_{D,\Delta} = \frac{E_0 y_{D,0} - R_1 x_{D,1}}{\Delta} = \frac{E_N y_{D,N} - R_{N+1} x_{D,N+1}}{\Delta} \tag{13-43b}$$

where Δ is given by Eq. (13-42a). $x_{A,\Delta}$ and $x_{D,\Delta}$ are the coordinates of the difference point and are not compositions that occur in the column. Note that $x_{A,\Delta}$ and $x_{D,\Delta}$ can be negative.

The difference point can be treated as a stream for mixing calculations. Thus, Eqs. (13-42a), (13-42b), (13-42c) show that the following points are collinear.

$$\Delta(x_{A,\Delta}, x_{D,\Delta}),\ E_0(y_{A,0}, y_{D,0}),\ R_1(x_{A,1}, x_{D,1})$$
$$\Delta(x_{A,\Delta}, x_{D,\Delta}),\ E_1(y_{A,1}, y_{D,1}),\ R_2(x_{A,2}, x_{D,2})$$
$$\Delta(x_{A,\Delta}, x_{D,\Delta}),\ E_j(y_{A,j}, y_{D,j}),\ R_{j+1}(x_{A,j+1}, x_{D,j+1})$$
$$\Delta(x_{A,\Delta}, x_{D,\Delta}),\ E_N(y_{A,N}, y_{D,N}),\ R_{N+1}(x_{A,N+1}, x_{D,N+1})$$

The existence of these straight lines and the applicability of the lever-arm rule can be proved by deriving Eq. (13-44).

$$x_{D,j+1} = (x_{A,j+1} - x_{A,\Delta})(\frac{x_{D,\Delta} - y_{D,j}}{x_{A,\Delta} - y_{A,j}}) + x_{D,\Delta} \tag{13-44}$$

Since all pairs of passing streams lie on a straight line through the Δ point, the Δ point is used to determine operating lines for the mass balances. A difference point in each section replaces the single operating line used on a McCabe-Thiele diagram. The procedure for stepping off stages will be illustrated after we discuss finding the location of the Δ point.

There are three methods for finding the location of Δ:

Chapter 13   Liquid-Liquid Extraction

1. **Graphical construction.** Since the points Δ, E_0, and R_1; and Δ, E_N, and R_{N+1} are on straight lines, we can draw these two straight lines. The point of intersection must be the Δ point. For the typical design problem (see Figure 13-20), points R_{N+1}, E_0, and R_1 are easily plotted. E_N can be found from the external balances (Figure 13-21B). Then Δ is found as shown in Figure 13-22.

2. **Coordinates.** The coordinates of the difference point were found in Eq. (13-43). These coordinates can be used to find the location of Δ. It may be convenient to draw one of the straight lines in method 1 (such as line $\overline{E_0R_1}$) and use one of the coordinates (such as $x_{D,\Delta}$) to find Δ. This procedure is useful since accurate graphical determination of Δ can be difficult.

3. **Lever-arm rule.** The general form of the lever-arm rule for two passing streams is

$$\frac{R_{j+1}}{E_j} = \frac{\overline{E_j\Delta}}{\overline{R_{j+1}\Delta}} \tag{13-45}$$

The lever-arm rule can be used to find the Δ point. For instance, if flow rates R_1 and E_0 are known, then Δ can be found on the straight line through points R_1 and E_0 at a distance that satisfies Eq. (13-45) with $j = 0$.

Consider again the stage-by-stage calculation routine that was outlined previously. We start with the known concentration of raffinate product stream R_1 and use an equilibrium tie line (stage 1) to find the location of saturated extract stream E_1 (see Figure 13-23). The points representing Δ, E_1, and R_2 are collinear, since E_1 and R_2 are passing streams. If the location of Δ is known, the straight line from Δ to E_1 can be drawn and then be extended to the saturated raffinate curve. This has to be the location of stream R_2 (see Figure 13-23). Thus, the difference point allows us to solve the three simultaneous mass balances by drawing a single straight line—the *operating line*. The procedure may be continued by constructing a tie line (representing stage 2) to find the location of stream E_2 that is in equilibrium with stream R_2. Then the mass balances are again solved simultaneously by drawing a straight line from Δ through E_2 to the saturated raffinate line, which locates R_3. This process of alternating between equilibrium and mass balances is continued until the desired separation is achieved. The stages are counted along the tie lines, which represent extract and raffinate streams in equilibrium. To obtain an accurate solution, a large piece of graph paper is needed and care must be exercised in constructing the diagram.

Before continuing you should carefully reread the preceding paragraph; it contains the essence of the stage-by-stage calculation method.

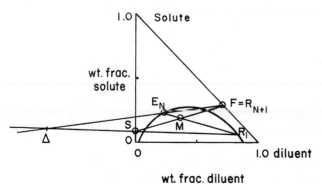

FIGURE 13-22.   *Location of difference point for typical design problem*

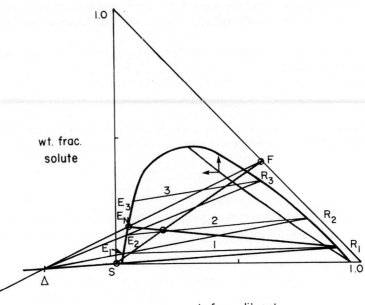

FIGURE 13-23.   *Stage-by-stage solution*

In Figure 13-23 we see that two equilibrium stages do not quite provide sufficient separation, and three equilibrium stages provide more separation than is needed. In a case like this, an approximate fractional number of stages can be reported.

$$\text{Fraction} = \frac{\overline{E_2 E_N}}{\overline{E_2 E_3}} \sim 0.2$$

This fraction can be measured along the curved saturated extract line. It should be stressed that the resulting number of stages, 2.2, is only approximate. The fractional number of stages is useful when the actual stages are not equilibrium stages. Thus, if a sieve-plate column with an overall plate efficiency of 25% were being used, the actual number of plates required would be

$$\frac{2.2}{0.25} = 8.8 \text{ or } 9 \text{ plates}$$

If a mixer-settler system were used, where each mixer-settler combination is approximately an equilibrium stage, then we would have three choices: 1) Use three stages, and obtain more separation than desired; 2) use two stages, and obtain less separation than desired; or 3) change the feed-to-solvent ratio to obtain the desired separation with exactly two or exactly three equilibrium stages.

One further important point should be stressed with respect to Figure 13-23. If two equilibrium stages were used with F/S = 2 as in the original problem statement, the saturated extract would *not* be located at the value E_2 shown on the graph and saturated raffinate would not be at the value R_1 shown. The streams R_1 and E_2 do *not* satisfy the external mass balance for this system. The values R_1 and E_N do, but R_1 and E_2 do not. The exact compositions of the product streams for a two-stage system require a trial-and-error solution.

What do we do if flow rate E_0 is less than R_1? The easiest solution is to define Δ so that it is now positive.

$$\Delta = R_1 - E_0 = R_{j+1} - E_j = R_{N+1} - E_N \qquad (13\text{-}46)$$

Δ is still equal to the difference between the flow rates of any pair of passing streams, but it is now raffinate minus extract. The corresponding lever-arm rule for any pair of passing streams is still Eq. (13-45), but the Δ point will be on the opposite side of the triangular diagram. This situation is shown in Figure 13-24. The stage-by-stage calculation procedure is unchanged when the location of Δ is on the right side of the diagram.

13.10.3 Complete Partially Miscible Extraction Problem

At this point you should be ready to solve a complete extraction problem.

EXAMPLE 13-4. Countercurrent extraction

A solution of acetic acid (A) in water (D) is to be extracted using isopropyl ether as the solvent (S). The feed is 1000 kg/h of a solution containing 35 wt % acid and 65 wt % water. The solvent used comes from a solvent recovery plant and is essentially pure isopropyl ether. Inlet solvent flow rate is 1475 kg/h. The exiting raffinate stream should contain 10 wt % acetic acid. Operation is at 20°C and 1 atm. Find the outlet concentrations and the number of equilibrium stages required for this separation. The equilibrium data are given by Treybal (1968) and are reproduced in Table 13-5.

Solution

A. Define. The extraction will be a countercurrent system as shown in Figure 13-20. $F = 1000$, $x_{A,F} = 0.35$, $x_{D,F} = 0.65$, $S = 1475$, $y_{A,S} = 0$, $y_{D,S} = 0$, $x_{A,1} = 0.1$. Find $x_{D,1}$, $y_{A,N}$, $y_{D,N}$, and N.

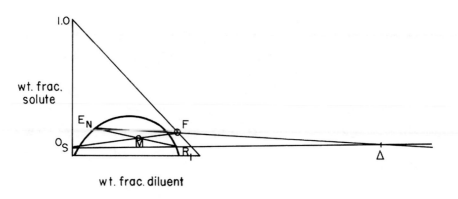

wt. frac. solute

wt. frac. diluent

FIGURE 13-24.   *Location of difference point when $R_1 > E_0$*

TABLE 13-5. *Equilibrium data for water-acetic acid-isopropyl ether at 20°C and 1 atm*

Water Layer, wt %			Isopropyl Ether Layer, wt %		
Acetic Acid	Water	Isopropyl Ether	Acetic Acid	Water	Isopropyl Ether
x_A	x_D	x_S	y_A	y_D	y_S
0.69	98.1	1.2	0.18	0.5	99.3
1.41	97.1	1.5	0.37	0.7	98.9
2.89	95.5	1.6	0.79	0.8	98.4
6.42	91.7	1.9	1.93	1.0	97.1
13.30	84.4	2.3	4.82	1.9	93.3
25.50	71.1	3.4	11.40	3.9	84.7
36.70	58.9	4.4	21.60	6.9	71.5
44.30	45.1	10.6	31.10	10.8	58.1
46.40	37.1	16.5	36.20	15.1	48.7

Points on the same horizontal line are in equilibrium. *Source:* Treybal (1968).

B, C. Explore and plan. This looks like a straightforward design problem. Use the method illustrated in Figures 13-22 and 13-24 to find Δ. Then step off stages as illustrated in Figure 13-23.

D. Do it. The solution is shown in Figure 13-25.

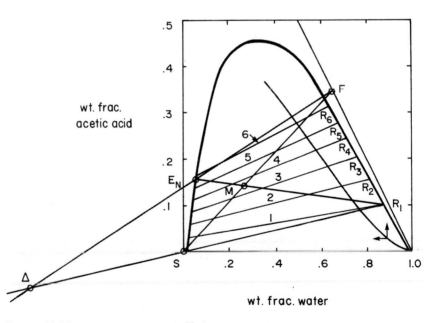

FIGURE 13-25. *Solution to Example 13-4*

1. Plot equilibrium data and construct conjugate line.
2. Plot locations of streams $E_0 = S$, $R_{N+1} = F$, and R_1.
3. Find mixing point M on line through points S and F at $x_{A,M}$ value calculated from Eq. (13-40a).

$$x_{A,M} = \frac{(1475)(0) + (1000)(0.35)}{1475 + 1000} = 0.1414$$

4. Line R_1M gives point E_N.
5. Find Δ point as intersection of straight lines E_0R_1 and E_NR_{N+1}
6. Step off stages, using the procedure shown in Figure 13-23. To keep the diagram less crowded, the operating lines, ΔE_j R_{j+1}, are not shown. You can use a straight edge on Figure 13-25 to check the operating lines. A total of 5.8 stages are required.

E. Check. Small errors in plotting the data or in drawing the operating and tie lines can cause fairly large errors in the number of stages required. If greater accuracy is required, a much larger scale and more finely divided graph paper can be used, or a McCabe-Thiele diagram (see the next section) or computer methods can be used.

13.11 RELATIONSHIP BETWEEN McCABE-THIELE AND TRIANGULAR DIAGRAMS FOR PARTIALLY MISCIBLE SYSTEMS

Stepping off a lot of stages on a triangular diagram can be difficult and inaccurate. More accurate calculations can be done with a McCabe-Thiele diagram. Since total flow rates are *not* constant, the triangular diagram and the Δ point are used to plot a curved operating line on the McCabe-Thiele diagram. This construction is illustrated in Figure 13-26 for a single point. For any arbitrary operating line (which must go through Δ), the values of the extract and raffinate concentrations of passing streams ($y_{A,op}$, $x_{A,op}$) are easily determined. These concentrations must represent a point on the operating line in the McCabe-Thiele diagram. Thus, the values of $y_{A,op}$ and $x_{A,op}$ are transferred to the diagram. Since the raffinate value is an x, the y = x line is used to find x. A very similar procedure was used in Figure 2-6B to relate McCabe-Thiele to enthalpy-composition diagrams.

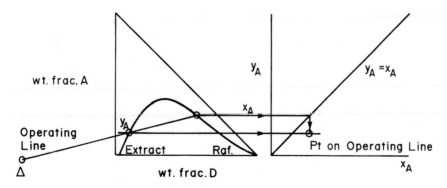

FIGURE 13-26. *Use of triangular diagram to plot operating line on McCabe-Thiele diagram*

When this construction is repeated for a number of arbitrary operating lines, a curved operating line is generated on the McCabe-Thiele diagram. The equilibrium data, y_A vs x_A, can also be plotted. Then stages can be stepped off on the diagram. This is shown in Figure 13-27 for Example 13-4. The equilibrium data were obtained from Table 13-5. The Δ point on the triangular diagram was used to find the operating line. The answer is 6 1/6 stages, which is reasonably close to the 5.8 found in Example 13-4.

Note that the operating line is close to straight. Thus, R_{j+1}/E_j is approximately constant. This occurs when the change in solubility of the solvent in the raffinate streams is approximately the same as the change in solubility of the diluent in the extract streams. When the changes in partial miscibility of the extract and raffinate phases are very unequal, R_{j+1}/E_j will vary significantly and the operating line will show more curvature. When the feed is not presaturated with solvent and the entering solvent is not presaturated with diluent, there can be a large change in flow rates on stages 1 and N. This is not evident in Figure 13-27, since the extract and raffinate phases are close to immiscible for the ranges of concentration shown.

The McCabe-Thiele diagrams are useful for more complicated extraction columns such as those with two feeds or extract reflux (Wankat, 1982, 1988).

13.12   MINIMUM SOLVENT RATE FOR PARTIALLY MISCIBLE SYSTEMS

As the solvent rate is increased, the separation should become easier, and the outlet extract stream, E_N, should become more diluted. The effect of increasing S/F is shown in Figure 13-28. As S/F increases, the mixing point moves toward the solvent and the solute concentration in stream E_N decreases. Starting with a low value of S/F, the difference point starts on the right-

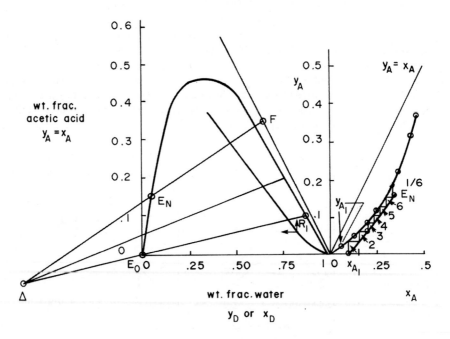

FIGURE 13-27.   *Use of triangular and McCabe-Thiele diagrams to solve Example 13-4*

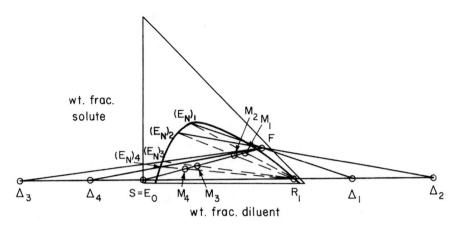

wt. frac.
solute

wt. frac. diluent

FIGURE 13-28.   *Effect of increasing S/F; $(S/F)_4 > (S/F)_3 > (S/F)_2 > (S/F)_1$*

hand side of the diagram $(R_{j+1} > E_j)$ and moves away from the diagram as S/F increases. When $R_{j+1} = E_j$, Δ is at infinity. A further increase in S/F puts Δ on the left-hand side of the diagram $(R_{j+1} < E_j)$. It now moves toward the diagram as S/F continues to increase. Some of these S/F ratios will be too low, and even a column with an infinite number of stages will not be able to do the desired separation.

It is often of interest to calculate the minimum amount of solvent that can be used and still obtain the desired separation. The minimum solvent rate (or minimum S/F) is the rate at which the desired separation can be achieved with an infinite number of stages. If less solvent is used, the desired separation is impossible; while if more solvent is used, the separation can be achieved with a finite number of stages. The corresponding Δ value, Δ_{min}, thus, represents the dividing point between impossible cases and possible solutions. This situation is analogous to minimum reflux in distillation.

To determine Δ_{min} and hence $(S/F)_{min}$, we note that to have an infinite number of stages, tie lines and operating lines must coincide (be parallel) somewhere on the diagram. This will require an infinite number of stages. The construction to determine Δ_{min} is shown in Figure 13-29 for Example 13-4, and is outlined here.

1. Draw and extend line R_1S.
2. Draw a series of arbitrary tie lines in the range between points F and R_1. Extend these tie lines until they intersect line R_1S.
3. Δ_{min} is located at the point of intersection of a tie line that is closest to the diagram on the left-hand side or furthest from the diagram if on the right-hand side. Thus, Δ_{min} is at the largest S/F that requires an infinite number of stages. Often the tie line that when extended goes through the feed point is the desired tie line. This is not the case in Figure 13-29.
4. Draw the line $\Delta_{min}F$. The intersection of this line with the saturated extract curve is $E_{N,min}$.
5. Draw the line from $E_{N,min}$ to R_1. Intersection of this line with the line from S to F gives M_{min}.
6. From the lever-arm rule, $(S/F)_{min} = \overline{FM}/\overline{SM}$. In Figure 13-29, $(S/F)_{min} = 1.296$ and $S_{min} = 1296$ kg/h. The actual solvent rate for Example 13-4 is 1475 kg/h, so the ratio $S/S_{min} =$

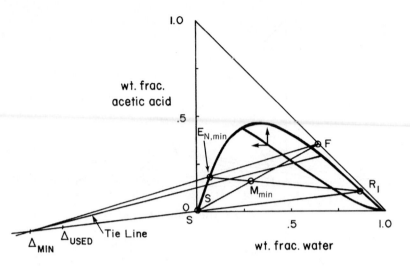

FIGURE 13-29. *Determination of minimum solvent rate*

1.138. Use of more solvent in Figures 13-25 or 13-27 would decrease the required number of stages.

On a McCabe-Thiele diagram the behavior will appear simpler. At low S/F the operating line will intersect the equilibrium curve. As S/F increases, the operating line will eventually just touch the equilibrium curve (minimum solvent rate). Then as S/F increases further, the operating line will move away from the equilibrium curve. Unfortunately, since the operating line is curved, it may be difficult to find an accurate value of S/F from the McCabe-Thiele diagram. An approximate value can easily be estimated.

13.13 EXTRACTION COMPUTER SIMULATIONS

Partially miscible ternary and multi-solute extraction systems can be set up for computer calculations. If we redraw and renumber the countercurrent extraction column (Figure 13-20) as shown in Figure 13-30 and relate R to L and E to V, the extraction column is analogous to a stripping or absorption column. The mass balances will be identical to those for absorption and stripping [Eqs. (12-45) to (12-48)] and they can be arranged into a tridiagonal matrix, Eq. (6-13). The new flow rates L_j (R_j) and V_j (E_j) can be determined from the summation Eqs. (12-49) and (12-50) using convergence check Eq. (12-51). The energy balances can again be written as Eq. (12-52) and the multivariate Newtonian solution method shown in Eqs. (12-53) to (12-58) is again applicable.

The flowsheet for the calculation shown in Figure 12-13 is applicable *except* that because of the nonideal behavior of extraction equilibrium a convergence loop for $x_{i,j}$ and $y_{i,j}$ *must* be added. This loop may cause convergence problems.

Computer calculations for extraction can also be applied to more complex extraction systems such as two-feed extractors or extraction with extract reflux. The triangular diagrams for these systems are considered in detail by Wankat (1982, 1988).

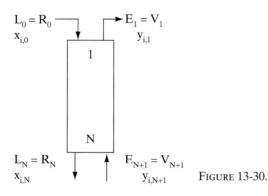

$L_0 = R_0$ $E_1 = V_1$
$x_{i,0}$ $y_{i,1}$

1

N

$L_N = R_N$ $F_{N+1} = V_{N+1}$
$x_{i,N}$ $y_{i,N+1}$

FIGURE 13-30. *Extraction column numbered for matrix calculations*

The analogy between stripping and extraction breaks down when one considers equilibrium (also hydraulics and efficiencies, but they do not affect the computer calculation). Our understanding of VLE is more advanced than our understanding of LLE and the data banks available in process simulators are much more complete and accurate for VLE than for LLE. Many of the correlations that work well for VLE will not work for LLE.

The correlations that are suggested for LLE are UNIQUAC and NRTL (Sorenson and Arlt, 1979, 1980; Macedo and Rasmussen, 1987; Walas, 1985). To obtain useful fits with experimental data specific parameters for the liquid-liquid system, not general parameters used for VLE, should be used (see Appendix B at back of book). If an extraction system will be used for which equilibrium data are unavailable, simulations can be used to determine if the system is worth investigating experimentally (Walas, 1985); however, the LLE must be measured experimentally before the system can be designed with confidence.

Application of Aspen Plus for extraction is delineated in the appendix to this chapter.

13.14 DESIGN OF MIXER-SETTLERS

Extraction systems have design conditions that often compete with each other. First, we would like a high surface area between the continuous and discontinuous phases (large number of small drops) to give a high mass transfer rate, and we want vigorous mixing to increase the continuous phase mass transfer coefficient. On the other hand, we would like a rapid settling (or rising if discontinuous phase is less dense) velocity for drops so that we can operate at higher flow rates. Since settling velocity is proportional to the drop diameter squared [shown later in Eq. (13-57)], large drops settle faster. Settling velocity is also directly proportional to the density difference between the two phases and is inversely proportional to viscosity. Finally, we do not want stable emulsions to form, which implies larger drops, lower viscosity, and less vigorous mixing are desired.

The mixer-settler shown in Figure 13-2 is one of the more common types of extraction equipment. The advantages of mixer-settlers are that they can be designed for any capacity, they can be used for batch or continuous operation, they have high stage efficiency (typically in the range from 0.8 to 1.0), they are flexible and have a high turndown ratio (can be operated with very different feed rates), they can be designed based on simple laboratory experiments, they can be staged for countercurrent (illustrated for washing in Figure 14-1A—extraction will be similar) or cross-flow operation, they can be designed as more compact multicompartment units that incorporate the mixer and settler in the same shell, and they work for a large variety of systems. Their disadvantages are that they have a relatively long residence time that is a problem for unstable compounds, the mixer may cause emulsion formation and the settler may be unable to separate the emulsion, they are not as compact as many other designs and require a large footprint (large floor area), they are relatively expensive for high feed rates, and they are expensive if many stages are required. In this section detailed design of mixers and settlers is discussed, and design of another type of extraction unit—Karr columns—is discussed in the next section. Mass transfer characteristics are discussed in Chapter 16.

The purpose of the mixer (Figure 13-31) is to disperse one of the two phases (raffinate = feed and extract = solvent) in the other phase to provide a large area for mass transfer. Most of the mass transfer occurs in the mixer. The settler's job is to undo the dispersion done in the mixer and produce two clear phases. To some extent, these two operations have opposite requirements. Mass transfer is more rapid if there are a large number of small droplets, while settling is much faster with large droplets. Thus, the design is a compromise that balances the needs of the mixer and the settler. The best way to do the design of both mixer and settler is to use results from simple miniplant experiments.

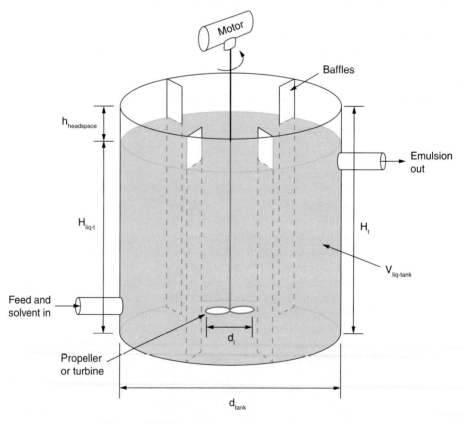

FIGURE 13-31.   *Schematic of mixer for mixer-settler*

Chapter 13   Liquid-Liquid Extraction

13.14.1 Mixer Design

Our goal is to achieve high efficiency with high rates of mass transfer while using drops that are large enough to settle quickly in the settler. According to Eqs. (1-4) and (1-5), the rate of mass transfer is increased if the area between the two phases is increased. This area can be increased with reasonably large drops (~ 300 μm or larger) by increasing the volumetric fraction of the dispersed phase, although volumetric fractions above 0.6 to 0.7 are usually unstable (Treybal, 1980). The designer has significant control over which phase is dispersed and over the volumetric fraction of the dispersed phase. Volumetric fraction φ_{phase_A} is defined as

$$\varphi_{phase\_A} = \text{Volume phase A / Total volume liquid in mixer} \qquad \textbf{(13-47a)}$$

Because of buoyancy or sedimentation forces, the dispersed phase tends to move faster than the continuous phase. Thus, the volumetric fraction of the dispersed phase φ_d is less than or equal to $\varphi_{d,feed}$, the volume fraction of dispersed phase in the feed.

$$\varphi_d \le \varphi_{d,feed} = Q_d/(Q_c + Q_d) \qquad \textbf{(13-47b)}$$

Q_d and Q_c are the volumetric flow rates of the dispersed and continuous phases in the feed. The value of φ_d is easy to determine experimentally. The mixer is run continuously until steady state is obtained. The feed and exit lines are closed, the mixer is turned off, and the phases are allowed to settle. The value of φ_d can then be determined by measuring the volume of the dispersed phase and the total volume and substitution of the values into Eq. (13-47a).

In general, the phase that is dispersed will conform to the following rules (Frank et al., 2008),

If $\varphi_A < 0.33$, then phase A is the dispersed phase.

If $0.33 < \varphi_A < 0.67$, then operation is in the *ambivalent* region.   **(13-48)**

If $0.67 < \varphi_A$, then phase A is the continuous phase.

These rules use the volume fractions in the mixer, not the volume fraction in the feed. In the ambivalent region, the dispersed phase depends on startup procedures, the rate of mixing and geometry of the mixer, and the materials of construction. In the ambivalent region, the dispersed phase is usually the phase that fills the mixer first during startup. In general, since metals are wet by aqueous phases, use of metals in the mixer promotes the aqueous phase becoming the continuous phase. Polymers tend to be wetted by organics and promote the organic phase becoming the continuous phase. In the ambivalent region, the dispersed phase may not be stable and phase inversion can occur. This is more likely if there are changes in the operation such as changes in the stirrer speed or flow rates. As phase inversion is approached, entrainment increases (Godfrey, 1994); thus, best practice is to not operate too close to phase inversion.

Since both mass transfer rates (see Chapter 16) and settling depend on which phase is dispersed, it is important to know which phase is dispersed. Mass transfer tends to be higher if the phase with the controlling resistance is the continuous phase (this may not be possible). A predictive test that is somewhat more conclusive than Eq. (13-48) can be developed by defining χ (Frank et al., 2008; Jacobs and Penney, 1987),

$$\chi = \frac{\phi_L}{\phi_{II}} \left(\frac{\rho_L \mu_H}{\rho_{II} \mu_L} \right)^{0.3} = \frac{\phi_L}{(1 - \phi_L)} \left(\frac{\rho_L \mu_H}{\rho_{II} \mu_L} \right)^{0.3}$$

(13-49)

φ_L and φ_H are the volume fractions of light and heavy phases, respectively, in the mixer. Then the test is:

If $\chi < 0.3$, light phase is always dispersed

If $0.3 < \chi < 0.5$, light phase probably dispersed

If $0.5 < \chi < 2.0$, either phase dispersed, phase inversion possible; design for worst case

(13-50)

If $2.0 < \chi < 3.3$, heavy phase probably dispersed

If $\chi > 3.3$, heavy phase always dispersed

If we have experimental data for φ_L and φ_H, Eqs. (13-49) and (13-50) will give us an idea of how stable the current arrangement of dispersed and continuous phases is.

To predict the behavior of the system without experimental data, we must first estimate the values of φ_L and φ_H. An approximate method is to assume φ_L and φ_H are both equal to their volumetric fractions in the feed. This assumption will assign a value that is too large for φ_d for the dispersed phase and too low for φ_c for the continuous phase. Thus, if the light phase is dispersed, the χ value calculated from Eq. (13-49) will be larger than the actual value. If the heavy phase is dispersed, the χ value calculated from Eq. (13-49) will be smaller than the actual value. As a result, the test in Eq. (13-50) is conservative if feed values are used. More accurate methods of predicting φ_d are considered later.

Mixer tanks (Figure 13-31) usually have baffles to improve the mixing. If the tank is enclosed and is operated full of liquid, baffles are not required. If there is a liquid-vapor interface, baffles are required to prevent vortex formation. Baffles typically have a width in the range 1/12 to 1/10 times the tank diameter (Treybal, 1980). The usual flow direction in a continuous mixer is to have both liquids enter together near the bottom of the mixer and exit near the top. The typical residence time in the mixer, $V_{liq\text{-}tank}/(Q_d + Q_c)$, where $V_{liq\text{-}tank}$ is the volume of liquid in the tank, is in the range of 1 to 3 minutes, although longer times may be required for reactive systems (Frank et al., 2008). Static mixers are also used, but the residence time may be too short for adequate mass transfer.

Many different types of agitators are used in mixers. For flat-blade impellers, the typical ratio of impeller diameter, d_i to tank diameter d_{tank} ranges from $d_i/d_{tank} = 0.25$ to 0.35, and the typical ratio of impeller width w_i to diameter is $w_i/d_i = 1/5$ to $1/8$ (Treybal, 1980). The use of a pump-mix combination that does both mixing and pumping will reduce costs and is highly desirable (Godfrey, 1994). Unfortunately, almost all of the data in the literature are on propellers or flat-blade impellers. This lack of literature data is easily overcome by studying the mixer in a miniplant equipped with a variable-speed mixer. The power P (Watts) required to completely disperse the system and operating conditions that provide reasonable stage efficiencies are determined by experiment. Murphree efficiencies are defined as

$$E_M = (x_{in} - x_{out})/(x_{in} - x^*)$$

(13-51)

where x* is the raffinate mass or mole fraction in equilibrium with the exiting extract mass or mole fraction. Efficiencies of at least 0.8 and preferably in the range from 0.9 to 0.95 are desirable.

Once appropriate experimental conditions have been found, the power number N_{Po} is calculated,

$$N_{Po} = (Pg_c)/(\rho_m \, \omega^2 \, d_i^5) \qquad (13\text{-}52)$$

In this equation, P is the power typically in Watts, ω is the revolutions per time typically in s^{-1}, and ρ_m is the mixture density calculated as

$$\rho_m = \phi_d \rho_d + (1 - \phi_d)\rho_c \qquad (13\text{-}53)$$

where subscript d is for the dispersed phase and c for the continuous phase. The usual scale-up procedure keeps N_{Po} constant. Since keeping N_{Po} constant often results in a slight increase in efficiency, this scale-up is a conservative design method (Frank et al., 2008).

Scale-up also assumes that the same entering feed and solvent concentrations and the same solvent-to-feed ratio, S/F, are used in the miniplant and the large unit. In addition, we need geometric similarity of the two mixers by keeping the ratios d_i/d_{tank}, H_t/d_{tank}, and (baffle width)/d_{tank} constant. Finally, the residence time $V_t/(Q_d + Q_c)$ must be equal in the two units. Based on equal power numbers and geometric similarity for the two units,

$$\frac{\omega(2)}{\omega(1)} = \left[\frac{d_i(1)}{d_i(2)} \right]^{2/3} = \left[\frac{d_t(1)}{d_t(2)} \right]^{2/3} \qquad (13\text{-}54)$$

When a new system is scaled up, the mixer on the large-scale unit should be equipped with a variable-speed drive so that the rpm can be tuned for optimum operation of both the mixer and the settler. Frank et al. (2008) briefly outline other scale-up approaches.

For preliminary estimates, the value of the power number for mixing can be estimated from correlations for propellers and flat-blade turbine impellers (Coulson et al., 1978; Frank et al., 2008; Treybal, 1980). A commonly used correlation is shown in Figure 13-32 (Treybal, 1980). This correlation was originally developed for mixing of homogeneous liquids. To use Figure 13-32 for extraction with two immiscible or partially miscible liquids, replace ρ_L with ρ_m from Eq. (13-53) and replace μ_L with μ_m from the following equation.

$$\mu_M = \frac{\mu_C}{\phi_C} \left(1 + \frac{1.5 \, \mu_D \phi_D}{\mu_C + \mu_D} \right) \qquad (13\text{-}55)$$

The viscosity of the two-phase mixture can be greater than the viscosity of both pure phases.

Propellers (curve a in Figure 13-32), which are typically used for liquid blending, not extraction, have lower power numbers at the same Reynolds numbers than the flat-blade turbines (curves b–g). The power number values for propellers depend on the geometry of the agitator and the tank. Typical propellers have $d_i/d_{tank} = 0.2$ or less (Treybal, 1980). At a Reynolds number ($Re_{mixer} = d_i^2 \, \rho_c \, \omega/\mu_c$) of 5000, N_{Po} is close to constant in the range of 0.3 to 0.7 for propellers (Coulson et al., 1978).

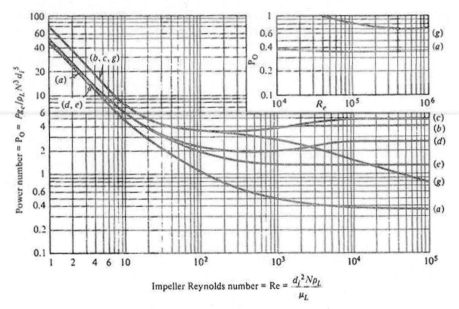

$$\text{Impeller Reynolds number} = Re = \frac{d_i^2 N \rho_L}{\mu_L}$$

FIGURE 13-32.   *Power for impellers immersed axially in single-phase liquids. Curves a, b, d, and e are for vessels with four baffles and with gas-liquid surface. (a) Marine impellers, d_i/d_{tank} = 1/3; (b) flat-blade impeller, $w = 0.2d_p$; (d) curved-blade turbines; (e) pitched-blade turbines. Curves c and g have no baffles. (c) Disk flat-blade turbines with or without a gas-liquid surface, (g) flat-blade turbines in unbaffled covered vessel with no interface and no vortex. Reprinted with permission from Treybal, R. E., Mass-Transfer Operations, McGraw-Hill, New York, 1980, p. 152. Copyright 1980 McGraw-Hill.*

If an experimental measurement is not available, an estimate of φ_d is needed (if φ_d is estimated, then $\varphi_c = 1 - \varphi_d$) to use Eqs. (13-53), (13-55) and Figure 13-32. As a first approximation, one can assume $\varphi_d = \varphi_{d,feed}$ for the calculation. A more accurate value of φ_d for mixers with a vapor-liquid interface and baffles can be estimated from the following empirical equation (Frank et al., 2008; Treybal, 1980).

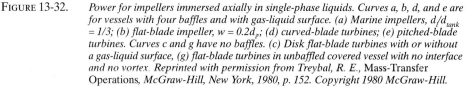

$$\frac{\phi_d}{\phi_{d,feed}} = 0.764 \left(\frac{PQ_D \, \mu_C^2}{V_{liq,tank} \, \sigma^3 \, g_C} \right)^{0.300} \left(\frac{\mu_C^3}{Q_D \, \rho_C^2 \, \sigma g_C} \right)^{0.178} \left(\frac{\rho_C}{\Delta \rho} \right)^{0.0741} \left(\frac{\sigma^3 \rho_C \, g_C^3}{\mu_C^4 \, g} \right)^{0.276} \left(\frac{\mu_D}{\mu_C} \right)^{0.136}$$

(13-56a)

This equation is valid for power/volume > 105 W/m³, where σ is the interfacial tension. The term $\Delta \rho = |\rho_c - \rho_D|$. Units are selected so that all terms in parenthesis are dimensionless. For unbaffled vessels run full of liquid with no vapor-liquid interface and no vortex, the following correlation is recommended (Frank et al., 2008; Treybal, 1980).

$$\frac{\phi_d}{\phi_{d,feed}} = 3.39 \left(\frac{PQ_D \, \mu_C^2}{V_{liq-tank} \, \sigma^3 \, g_C} \right)^{0.247} \left(\frac{\mu_C^3}{Q_D \, \rho_C^2 \, \sigma g_C} \right)^{0.427} \left(\frac{\rho_C}{\Delta \rho} \right)^{0.430} \left(\frac{\sigma^3 \rho_C \, g_C^3}{\mu_C^4 \, g} \right)^{0.401} \left(\frac{\mu_D}{\mu_C} \right)^{0.0987}$$ (13-56b)

For both correlations, if calculated value of $\varphi_d/\varphi_{d,feed} > 1.0$, use $\varphi_d = \varphi_{d,feed}$. Treybal (1980) notes that water dispersed in hydrocarbons may not follow these correlations.

Unfortunately, an estimate of the power P is needed to determine φ_d from Eq. (13-56a) or (13-56b), and an estimate of φ_d is needed to determine ρ_m and μ_m from Eqs. (13-53) and (13-55) to determine P from Figure 13-32. Thus, prediction of φ_d and power in the mixer is a classical trial-and-error problem. This solution is illustrated later in Example 13-5.

13.14.2 Settler (Decanter) Design

Settler design is just as important and just as empirical as mixer design. Although vertical settlers are used, horizontal settlers are much more common because the settling area is easily increased by making the settler longer. A schematic of a settler illustrating a number of design features is shown in Figure 13-33 (Frank, 2008; Jacobs and Penney, 1987). The feed pipe is increased in diameter to reduce the fluid velocity before entering the settler. Immediately after entering, the feed may be sent sideways with a tee. The baffles or diffuser (a sieve plate with ~ ½-inch holes) help calm the fluid motion. At the outlet end, the baffles for the light and heavy liquid also help calm the fluid and prevent entrainment. The overflow loop for the heavy phase is one method for controlling the location of the interface. In a well-designed settler, the height of the dispersion band ΔH is approximately constant. If a *rag layer*

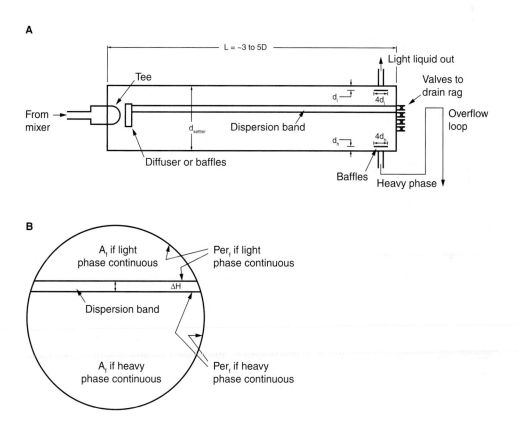

FIGURE 13-33. *Settler. (A) Schematic. (B) View showing terms to calculate hydraulic diameter*

(a stable emulsion containing finely divided solids and/or surface active agents) forms, a series of valves is useful to find and remove the rag (Jacobs and Penney, 1987). If not removed periodically, rag layers tend to grow and eventually contaminate one of the products. If the liquid feed contains gases, a vent for the gases (not shown in Figure 13-33) is also required (Frank et al., 2008). In this case, an overflow weir is usually used for collecting the light liquid.

Best design practice is to use experimental data for design of the settler. First, settling of the actual feed/solvent mixture, not a mixture of pure compounds, is characterized with a simple shaker test, which can be done with a 1-inch-diameter graduated cylinder (Frank et al., 2008). After shaking the cylinder, it is placed on a bench and the settling behavior is timed. In type I systems the dispersion band disappears in less than 5 minutes, resulting in two clear liquids with a clean interface. Systems showing this type of behavior typically have moderate to high interfacial tensions (> ~ 10 dyne/cm), density difference > 0.1 g/cm³, viscosities of each phase < 5 cP, and negligible fine solids and surfactants. Type I systems are controlled by the settling rate of drops and can be designed using Stokes law.

In type II systems the dispersion band disappears in less than 20 minutes, resulting in two clear liquids with a clean interface. Systems showing this type of behavior typically have moderate interfacial tensions (~ 10 dyne/cm), density difference > 0.1 g/cm³, viscosities of each phase < 20 cP, and negligible fine solids and surfactants. Type II systems are typically controlled by slow coalescence and are designed using experimental coalescence data.

Frank et al. (2008) also define systems with solids or surfactants (type III), or high viscosity (type IVa), or low interfacial tension (type IVb), or low density difference (type IVc), or formation of a stable emulsion stabilized by solids/surfactants (IVd). Successful settling of type III systems that have solids or surfactants present may be possible if solids are removed first by filtration and surfactants by adsorption (see Chapter 18). After the feed and solvent are pretreated, the shaker test should be repeated. Type IV systems probably require a different phase separation system, such as coalescers, centrifuges, hydrocyclones, ultrafiltration or electrotreatment; and pretreatment may also be necessary.

The design of rapid-settling type I systems is usually done with Stokes law or with a modified Stokes law. For dilute, low Reynolds number ($Re_{drop} = d_d \rho_c u_t / \mu_c < 0.3$) systems in a quiescent fluid with no interaction between drops, the terminal velocity u_t of dispersed rigid spherical drops of diameter d_d with gravity being the only body force (e.g., no electrostatic forces) is given by

$$u_{t,Stokes} = \frac{g d_d^2 \left(\rho_c - \rho_d \right)}{18 \mu_c} \tag{13-57}$$

where g is the acceleration of gravity. When Stokes law is applied outside its range of validity, care must be taken in using the results. The "settling" will be downwards if $\rho_d > \rho_c$ and upwards if $\rho_d < \rho_c$. Use of Eq. (13-57) requires knowledge of the smallest drops, which, because they have the smallest velocity, are most difficult to remove. If reliable experimental data on the smallest drop size are available, they can be used in Eq. (13-57). Although correlations for the largest drop size and the average drop size are available, reliable correlations for the smallest drop size are not available (Godfrey, 1994; Hartland and Jeelani, 1994). Thus, the usual empirical design procedure is to use a drop diameter $d_d = 150$ μm, which is well less than the typical minimum drop diameter of ~300 μm observed in industrial practice (Frank et al., 2008; Jacobs and Penney, 1987).

However, the assumption that drops do not interact is seldom valid. Frank et al. (2008) recommend the following empirical expression ignoring coalescence during settling (a conservative assumption) to calculate the reduction in the terminal settling velocity u_t.

$$u_{t,hindered} = u_{t,Stokes} (1 - \varphi_{d,initial}) \mu_c/\mu_d \qquad \textbf{(13-58)}$$

If $u_{t,hindered} > u_{t,Stokes}$, set $u_{t,hindered} = u_{t,Stokes}$ for remaining calculations.

In a settler the continuous phase is assumed to move vertically upwards or downwards from the inlet to the outlet in a uniform plug flow. At the interface the continuous phase velocity is Q_c/A_i, where $A_i = L \times$ (interface width) is the interface area. The width of the interface depends on its location in the settler. If droplets settle (or rise) faster than the continuous phase velocity, the droplets will be collected and the phases will be separated. Thus, design so that

$$Q_c/A_i \leq u_{t,hindered} /(1 + \text{safety factor}) \qquad \textbf{(13-59)}$$

A safety factor of about 0.2 (20%) is often used (Jacobs and Penney, 1987).

However, the fluid in a settler is not really quiescent. If we define $Re_{settler}$ as

$$Re_{settler} = v_c \, d_{hydraulic} \, \rho_c/\mu_c \qquad \textbf{(13-60a)}$$

the effect of turbulence is negligible if $Re_{settler} = \sim 5000$, and there is limited interference up to $Re_{settler} = \sim 20,000$ (Jacobs and Penney, 1987). To estimate $Re_{settler}$ we first estimate the velocity of the continuous phase in the settler. In a horizontal settler the continuous phase is flowing horizontally, which is perpendicular to the upward or downward movement of the drops. Thus, the velocity of the continuous phase is equal to the volumetric flow rate divided by the flow area

$$v_c = Q_c/A_f \qquad \textbf{(13-61)}$$

where the flow area A_f for the continuous phase depends on the location of the dispersion band (Figure 13-33B). The hydraulic diameter $d_{hydraulic}$ for continuous phase flow is

$$d_{hydraulic} = 4 \, A_f/Per_f \qquad \textbf{(13-62)}$$

The flow perimeter Per_f includes the interface (Figure 13-33B). Combining Eqs. (13-57a) to (13-59), if

$$Re_{settler} = 4Q_c \, \rho_c/(Per_f \, \mu_c) < 5000 - 20,000 \qquad \textbf{(13-60b)}$$

there will be minimal interference from turbulence. The location of the dispersion band needs to be known at least approximately to do this design check.

Type II slowly coalescing systems may be underdesigned if Stokes law, Eq. (13-57), or the modified Stokes law, Eq. (13-58), are used. Instead, data obtained with a miniplant coalescer on the steady-state height ΔH of the dispersion band are preferable. With constant mixture composition and phase ratio, ΔH can be related to changes in Q/A [A = L × (settler width) is the cross-sectional area of the settler] using either of the following equations to correlate the data (Frank et al., 2008; Hartland and Jeelani, 1994; Jacobs and Penney, 1987):

$$\frac{1}{Q/A} = \frac{1}{k_1 \Delta H} + \frac{1}{k_2} \tag{13-63a}$$

or

$$\Delta H \propto \left(\frac{Q}{A}\right)^b \propto \left(\frac{Q_c}{A}\right)^b \propto \left(\frac{Q_d}{A}\right)^b \tag{13-63b}$$

In these equations k_1, k_2, and b are empirical constants. The typical range of b is from 2.5 to 7. Jacobs and Penney (1987) recommend that $\Delta H < 0.1\, d_{settler}$, while Frank et al. (2008) recommend $\Delta H < 0.15\, d_{settler}$. These heuristics will prevent flooding of the settler if the flow rate increases, which can cause ΔH to increase drastically. As a rule of thumb, the residence time of the dispersed phase in the dispersion band should be greater than about 2 to 5 minutes (Jacobs and Penney, 1987). Because the dispersed phase occupies about half of the dispersion band,

$$t_{residence,dispersed} = 0.5\, A_i\, \Delta H / Q_d > 2 \text{ to 5 minutes} \tag{13-64}$$

EXAMPLE 13-5. Mixer-settler design

Design a baffled mixing vessel and a horizontal settler for extraction of benzoic acid from water (diluent) into toluene (solvent). The tank (Figure 13-31) should have $H_t/d_{tank} = 1.0$ and there should be a 1-inch (0.0254 m) head space (vapor space at top). The flow rate of the aqueous feed is 0.006 m³/s, and the volumetric solvent-to-feed ratio = 0.2. A 6-blade flat turbine impeller with impeller diameter $d_i = 0.25 d_{tank}$ and impeller width $w_i = d_i/5$ is operated at 1000 rpm. Simple settling experiments show that this is a type I (rapid settling) system.

A. For a mixture residence time of 1.0 minutes, find the mixer dimensions and estimate ϕ_d and the power P required.
B. If $L/d_{settler} = 4$, use Stokes law to size the horizontal settler.

Data (Treybal, 1980):

	Water Solution	Toluene Solution
Density ρ	998 kg/m³	865 kg/m³
Viscosity μ	0.95×10^{-3} kg/mgs	0.59×10^{-3} kg/mgs
Diffusivity D	2.2×10^{-9} m²/s	1.5×10^{-9} m²/s
Molecular wt	18.02	92.14

Interfacial tension, $\sigma = 0.0222$ N/m and distribution coefficient = $C_{extract}/C_{raffinate}$ = 20.8 with $C_{extract}$ in kmol benzoic acid/m³ extract and $C_{raffinate}$ in kmol benzoic acid/m³ raffinate.

Solution Part a: Mixer design

$$Q_{solvent} = (0.2)(0.006 m^3/s) = 0.0012 \text{ m}^3/s$$

and assuming toluene (the light phase) is dispersed,

$$\phi_{d,feed} = Q_d/(Q_d + Q_c) = 0.0012/(0.006 + 0.0012) = 0.167$$

Using the feed volume fractions, we obtain $\phi_L = \phi_{d,feed} = 0.167$ and $\phi_H = 1 - \phi_L = 0.833$. Then, from Eq. (13-49) at the feed conditions,

$$\chi_{feed} = \frac{\phi_L}{\phi_H}\left(\frac{\rho_L \mu_H}{\rho_H \mu_L}\right)^{0.3} = \frac{0.167}{0.833}\left[\frac{(865)(0.95 \times 10^{-3})}{(998)(0.59 \times 10^{-3})}\right]^{0.3} = 0.221$$

The test in Eq. (13-50) confirms our assumption that toluene is dispersed.

For a mixer residence time, $t_{res} = V_{liq-tank}/(Q_d + Q_c) = 1$ minute $= 60$s, we find $V_{liq-tank} = (60s)(0.006 + 0.0012 m^3/s) = 0.432$ m^3.

The liquid volume in the tank is $V_{liq-tank} = (H_t - 0.0254)\pi d^2_{tank}/4 = 0.432$m^3. Since we specified that $H_t = d_{tank}$, this is

$$(\pi/4)\left[d^3_{tank} - 0.0254\, d^2_{tank}\right] = 0.432 m^3$$

From Goal Seek, the solution of this equation is $d_{tank} = 0.8279$m, $H_t = 0.8279$m, and $h_{liq-t} = 0.8025$m.

To estimate ϕ_d and P, we must simultaneously solve Eqs. (13-53), (13-55), Figure 13-32, and Eq. (13-56). For use of Figure 13-32, we need the impeller Reynolds number $= d^2_i N\rho_L/\mu_L$. We know N = 1000 RPM = 16.67 RPS, $d_i = 0.25$ $d_{tank} = (0.25)(0.8279) = 0.2070$ m. Exact calculation requires substituting values for ρ_m and μ_m. As a first estimate, we use the properties of the liquid in excess (water). $\rho_w = 998$ kg/m^3 and $\mu_w = 0.95 \times 10^{-3}$ kg/m^3. Then,

$$Re_{L,estimate} = \frac{(0.2070)^2(16.67)(998)}{0.95 \times 10^{-3}} = 750,383$$

This is very high and clearly is in the range where curve b in Figure 13-32 predicts a constant power number, $N_{Po} = 4.0$. Since variations in ρ_m and μ_m will not affect N_{Po}, we do not need to calculate μ_m [it does not appear in Eq. (13-56)]. Since $\omega = N$ and in metric units $g_c = 1.0$, the power number definition, Eq. (13-52), is

$$P = N_{Po}\, \rho_m \omega^2 d^5_i / g_c = (4.0)\rho_m (16.67)^2 (0.2070)^5 / 1.0 = 0.42246\, \rho_m \qquad \textbf{(A)}$$

where Eq. (13-53) shows ρ_m is between $\rho_c = 998$ and $\rho_d = 865$ and is clearly closer to the density of the continuous phase.

We can now calculate the terms in Eq. (13-56a). We have $\varphi_{d,\text{feed}} = 0.167$, $Q_d = Q_{\text{solvent}} = 0.0012 \text{ m}^3/\text{s}$, $V_{\text{liq-tank}} = 0.432 \text{ m}^3$, $d_i = 0.2070 \text{ m}$, $\rho_c = \rho_w = 998 \text{ kg/m}^3$, $\rho_d = \rho_{\text{Tol}} = 865 \text{ kg/m}^3$, $\mu_c = \mu_w = 0.95 \times 10^{-3} \text{ kg/(m·s)}$, $g_c = 1.0$, $\mu_d = \mu_{\text{Tol}} = 0.59 \times 10^{-3}$ kg/(m s), s = 0.022 N/m, and g = 9.807 m/s^2. Then the dimensionless groups are

$$\left(\frac{P Q_d \mu_c^2}{V_{\text{liq-tank}} \sigma^3 g_c} \right) = \frac{0.0012 \left(0.95 \times 10^{-3} \right)^2 P}{(0.432)(0.022)^3 (1.0)} = (2.354 \times 10^{-4}) P$$

$$\frac{\mu_c^3}{Q_d \rho_c^2 \sigma g_c} = \frac{\left(0.95 \times 10^{-3} \right)^3}{(0.0012)(998)^2 (0.022)(1.0)} = 3.26 \times 10^{-11}$$

$$\frac{\rho_c}{|\rho_c - \rho_d|} = \frac{998}{|998 - 865|} = 7.5038$$

$$\frac{\sigma^3 \rho_c g_c^3}{\mu_c^4 g} = \frac{(0.022)^3 (998)(1.0)^3}{\left(0.95 \times 10^{-3} \right)^4 (9.807)} = 1.330 \times 10^9$$

$$\frac{\mu_d}{\mu_c} = \frac{0.59 \times 10^{-3}}{0.95 \times 10^{-3}} = 0.62105$$

Then Eq. (13-56a) is

$$\frac{\phi_d}{0.167} = (0.764) P^{0.3} \left(2.354 \times 10^{-4} \right)^{0.3} \left(3.26 \times 10^{-11} \right)^{0.178} (7.504)^{0.0741} \left(1.330 \times 10^9 \right)^{0.276} (.621)^{0.136}$$

or

$$\phi_d = 0.5076 \, P^{0.3} \qquad\qquad\qquad \textbf{(B)}$$

We must solve this expression simultaneously with previous Eq. (A), P = 0.42246 ρ_m and with $\rho_m = \phi_d \rho_d + (1 - \phi_d)\rho_c = 865\,\phi_d + 998(1 - \varphi_d)$.

Solving these three equations simultaneously in a spreadsheet using Goal Seek, we obtain $\phi_d = 0.302$ and since the ratio > 1, use $\phi_d = \phi_{d,\text{feed}} = 0.167$.

Then, from Eqs. (13-53) and (13-52),

$$\rho_m = \phi_d \rho_d + (1 - \phi_d)\rho_c = (0.167)(865) + (0.833)998 = 975.8$$

$$P = \frac{N_{Po} \rho_m \omega^2 d_i^5}{g_c} = \frac{4(975.8)(16.67)^2 (0.2070)^5}{1.0} = 412.2 \, \text{W}.$$

Note that we need to use Eq. (13-52) to calculate power, *not* Eq. (13-56), because Eq. (13-56a) gave a value of $\phi_d/\phi_{d,feed} > 1.0$.

Solution Part b: Settler design
Stokes law is given by Eq. (13-57).

$$u_t = \frac{g d_d^2 (\rho_c - \rho_d)}{18 \mu_c} = \frac{(9.807)(1.5 \times 10^{-4})^2 (998 - 865)}{18(0.95 \times 10^{-3})} = 0.00172 \text{ m/s}.$$

where here we use an assumed drop size $d_d = 150 \text{mm} = 1.5 \times 10^{-4}$ m, which is an arbitrary size often chosen because it includes a safety factor (Frank et al., 2008). A more accurate estimate of drop size can be made using the equations in Section 16.7.3 (see Example 16-6). As a check on the applicability of Stokes law, we can calculate the particle Reynolds number.

$$Re_p = \frac{d_d \rho_c u_t}{\mu_c} = \frac{(1.5 \times 10^{-4})(998)(0.00172)}{(0.95 \times 10^{-3})} = 0.27.$$

Since $Re_p < 0.3$ Stokes law is valid.
    The hindered settling velocity, Eq. (13-58), is

$$u_{t,hindered} = u_{t,Stokes}(1 - \varphi_{d,initial})\frac{\mu_c}{\mu_d} = (0.00172)(1 - 0.167)\frac{0.95 \times 10^{-3}}{0.59 \times 10^{-3}} = 0.0023 \text{ m/s}$$

Since $u_{t,hindered} > u_t$, we use the Stokes velocity = 0.00172 m/s.
    To calculate the settler size, first assume that the interface is held at the center of the settler. Then the interface area A_i (with $L/D_s = 4$) is

$$A_i = L(\text{width}) = LD_s = 4D_s^2$$

Then, from Eq. (13-59) (with safety factor = 0.2 and $u_{t,hindered} = u_{t,Stokes}$),

$$\frac{Q_c}{A_i} = \frac{0.006 \text{ m}^3/\text{s}}{4D_s^2} \le \frac{u_t}{(1 + \text{Safety factor})} = \frac{0.00172}{1.2} = 0.00143$$

Solving for the diameter of the settler D_s,

$$D_s \ge \left(\frac{0.006}{4(0.00143)}\right)^{1/2} = 1.023 \text{ m}.$$

We would probably use the smallest standard size that satisfies this restriction. We should also check that Eq. (13-60b) is satisfied. With the interface at the center, the flow area A_f for the horizontal flow of the continuous phase is half the area of a circle.

$$A_f = \left(\frac{1}{2}\right)\pi D_s^2 / 4 = 0.411 \text{ m}^2$$

And the flow perimeter Per_f for the horizontal flow of the continuous phase is half the perimeter of a circle plus the diameter (for the interface itself).

$$Per_f = \left(\frac{1}{2}\right)\pi D_s + D_s = 2.630 \text{ m}$$

Then, Eq. (13-60b) is

$$Re_{settler} = 4Q_c\rho_c / \left(Per_f\mu_c\right) = 4(0.006)(998) / \left[(2.63)(0.95\times10^{-3})\right] = 9586.6$$

There is likely to be some limited interference due to flow. One might want to increase the safety factor. A better approach would be to use experimental data for settling instead of Stokes law.

In this system, the denser water phase is continuous. Figure 13-33B shows that if the dispersion band is above the center point, the flow area and flow perimeter for the continuous phase will be larger. With a larger Per_f, $Re_{settler}$ will be decreased and interference from flow is unlikely. On the other hand, if the dispersion band is below the center point, the flow area and flow perimeter for the continuous phase will be smaller, $Re_{settler}$ will be increased, and interference from flow is more likely.

Although mixer-settlers are often designed as equilibrium staged systems with an assumed efficiency in the range from 80% to 95%, if mass transfer data are available, it should be used to predict the stage efficiency. This calculation is shown in Section 16.7.

Several methods have been developed to speed up coalescence (Frank et al., 2008; Hartland and Jeelani, 1994; Jacobs and Penney, 1987). Increasing the operating temperature usually increases coalescence mainly because the viscosity decreases. Coalescence-enhancing structures such as mesh or baffles are often employed. Coalescence can be further encouraged by using construction materials that are wetted by the dispersed phase (i.e., metals for aqueous and plastics for organics). Coalescence is also faster if mass transfer (assuming the mixer does not have an efficiency of 100%) is from the drops to the continuous phase; thus, one may want the raffinate phase to be dispersed. Settler performance can be improved by using two diffuser plates with increasing hole area at the settler inlet and by using overflow and underflow weirs to remove the light and heavy phases, respectively. If φ_d is high, settling will be hindered. Recycling continuous phase to the settler inlet will decrease φ_d and may increase coalescence. Recycling needs to be done with care, since Eq. (13-60b) may be violated by the increased velocity of the continuous phase.

The selection criteria listed in Section 13.1 indicate that mixer-settlers are a good choice if only a few equilibrium stages are needed; however, if, as is often the case, a few more stages are required, then a nonagitated (static) column such as a spray, sieve tray, or packed column should be used. These columns are commonly used for relatively large-scale systems in the chemical and petrochemical industries. Design methods for static columns are discussed in detail in *Perry's Chemical Engineering Handbook* (Frank et al., 2008). However if, as is often the case, more than five stages are required, then an agitated column system is often used. The design of one of the more popular types, the Karr column, is described in the next section.

13.15   INTRODUCTION TO DESIGN OF RECIPROCATING-PLATE (KARR) COLUMNS

Because the desired design conditions conflict (see Section 13-14), a number of different extractor designs (Figure 13-2 and Table 13-1) have been commercialized that attempt to find a good compromise between the competing design requirements. A number of column systems with external power applied to promote mixing have been developed. One generic design that is often a favorable compromise between the conflicting demands is the reciprocating-plate column (RPC) (Baird et al., 1994; Frank et al., 2008; Smith et al., 2008). In an RPC the plates move up and down (Figure 13-2) to promote mixing without forming a stable emulsion. Although first patented by Van Dijck in 1935, the RPC was not commercialized until 1959 when Karr developed a successful laboratory-scale RPC extractor without downcomers (Baird et al., 1994; Smith et al., 2008). Flow through the plates is upwards when the plates move down, and flow is downwards when the plates move up. Karr columns are currently the most popular design of RPC, and commercial columns are available in diameters greater than 2 meters. They are commonly used in pharmaceutical manufacturing. This short section focuses on the design of Karr columns.

The degree of mixing can be adjusted by changing the amplitude and frequency of the up-and-down reciprocal motion. Typical amplitudes are A_p up to 25 to 50 mm, with 25 mm being most common. The plates can be made of polymer or metal and typically have large holes from 10 to 16 mm in diameter, with 12.7 mm (0.5 inches) being common, with a plate-free area fraction e in the range from 0.5 to 0.6 for commercial systems (Baird et al., 1994; Smith et al., 2008). Plates with lower values of e have low capacities. Because of the large free area, Karr columns have high throughputs, up to 40 $m^3/(h\ m^2)$ [1000 gal/(h ft^2)] (Frank et al., 2008). Plates are typically spaced between 50 and 150 mm apart. The HETP values for the "difficult" (high interfacial tension system of o-xylene-acetic acid-water) ranged from 0.2 m to 10 m and depended on the frequency of reciprocation (Baird et al., 1994). Karr columns typically operate with a fairly low frequency f ($0.2 < f < 7$ strokes/s). There is normally an optimum frequency f that minimizes the HETP value for a given system. At the optimum frequencies, the HETP values for this difficult system varied from 0.2 to 0.6 m for different sized Karr columns. Thus, operating far from the optimum frequency results in a large increase in the column HETP.

Agitated column extractors are currently designed based on pilot plant tests (Frank et al., 2008). Karr and Lo (1971; 1976) developed a semi-empirical design procedure for Karr columns. Based on pilot plant runs, typically with a 25 to 76 mm id column, the system can be scaled up to a different diameter and frequency. This procedure (Baird et al., 1994; Frank et al., 2008; Smith et al., 2008) is to first optimize the variable plate spacing in the pilot plant column so that the approach to flooding is the same in all sections of the column. Additional pilot plant runs are used to determine $HETP_{pilot}$ at different frequencies of reciprocation f_{pilot} to determine the optimum frequency of operation. The appropriate values for the large column are given by

$$HETP_{large-scale} = HETP_{pilot} \, (D_{large}/D_{pilot})^{n1} \tag{13-65}$$

$$f_{large-scale} = f_{pilot} \, (D_{pilot}/D_{large})^{n2} \tag{13-66}$$

The exponents n_1 and n_2 depend on the chemical system. The original values for the "difficult" (high interfacial tension) system were $n_1 = 0.38$ and $n_2 = 0.14$. An average value of $n_1 = 0.25$ is often used. More recent data (Smith et al., 2008) with less difficult systems indicate that the exponent n_1 is too large and overpredicts the column height required. Their experiments showed that HETP and the optimum frequency are independent of column diameter; thus, they recommend using $n_1 = n_2 = 0$, which means the large-scale system has the same HETP and same frequency as the pilot plant. However, Frank et al. (2008) recommend use of the original exponents, $n_1 = 0.38$ and $n_2 = 0.14$. With difficult systems, the original exponents should probably be used, while with lower interfacial tension systems, smaller exponents are satisfactory. A donut baffle plate is often inserted every 5 regular plates to limit axial dispersion—this procedure also tends to reduce the values of n_1 and n_2. The extractor is usually supplied with a variable-speed motor so that the frequency of reciprocation can be tuned to obtain the best operation.

Flooding is usually correlated in terms of the sum of the continuous v_c and dispersed v_d phase superficial velocities (velocities based on the empty column cross-sectional area). Flooding data for a variety of systems was correlated in the range of plate fractional free area $e = 0.085$ to 0.58 by the following equation (Baird et al., 1994):

$$v_{c,flood} + v_{d,flood} = 0.8 \, e \, v_{characteristic} \tag{13-67}$$

where the characteristic velocity is

$$v_{characteristic} = 0.00507 \, \{[g^2 \, (\Delta\rho)^2 \, \gamma^{1.8}]/[\rho_c^{2.8} \, \mu_c]\}^{1/3} \, \{[H \, e^2]/[(1 - e^2)(A_p \, f)^3]\}^{0.4} \tag{13-68}$$

The fit to the data was good in the range from $e = 0.2$ to 0.58. In Eq. (13-68), g is the acceleration due to gravity in m/s^2; γ is the interfacial tension in $N/m = kg/s^2$; μ_c is the continuous phase viscosity $Pa \cdot s = kg/(m \cdot s)$; ρ_c and $\Delta\rho$ are the continuous phase density and the absolute difference in the two phase densities, both in kg/m^3; H = stage height in m; A_p is amplitude of the reciprocation in m; f is the frequency in 1/s; and e is the fraction plate-free area, which is dimensionless. The resulting value of $v_{characteristic}$ is in m/s. Note that the sum of the continuous and dispersed phase flooding velocities depends on $e^{1.8}/(1 - e^2)^{0.4}$. Because columns with low plate-free areas ($< \sim 0.2$) have very low capacities, they are not used commercially.

13.16   SUMMARY—OBJECTIVES

In this chapter, we considered what extraction is used for, developed McCabe-Thiele and Kremser methods for immiscible extraction, and explored methods for ternary partially miscible extraction systems. At the end of this chapter, you should be able to satisfy the following objectives:

1. Explain what extraction is and outline the types of equipment used
2. Apply the McCabe-Thiele and Kremser methods to immiscible extraction
3. Plot extraction equilibria on a triangular diagram. Find the saturated extract, saturated raffinate, and conjugate lines
4. Find the mixing point and solve single-stage and cross-flow extraction problems
5. For countercurrent systems, do the external mass balances, find the difference points, and step off the equilibrium stages
6. Use the difference points to plot the operating line(s) on a McCabe-Thiele diagram
7. Use a computer simulation program to solve extraction problems
8. Do detailed designs of mixer-settlers or Karr columns

REFERENCES

Adler, S., E. Beaver, P. Bryan, J. E. L. Rogers, S. Robinson, and C. Russomanno, *Vision 2020: 1998 Separations Roadmap*, Center for Waste Reduction Technologies of AIChE, New York, 1998.

Alders, L. *Liquid-Liquid Extraction,* 2nd ed., Elsevier, Amsterdam, 1959.

Albertsson, P.-A., G. Johansson, and F. Tjerneld, "Aqueous Two-Phase Separations," in J. A. Asenjo, Ed., *Separation Processes in Biotechnology,* Marcel Dekker, New York, 1990, Chap. 10.

Allen, D. T., and D. R. Shonnard, *Green Engineering: Environmentally Conscious Design of Chemical Processes*, Prentice Hall, Upper Saddle River, NJ, 2002.

Baird, M. H. I., N. V. Rama Rao, J. Prochazka, and H. Sovova, "Reciprocating-Plate Columns," in Godfrey, J. C. and M. J. Slater (Eds.), *Liquid-Liquid Extraction Equipment*, Wiley, Chichester, England, 1994, Chap. 11.

Belter, P. A., E. L. Cussler, and W.-S. Hu, *Bioseparations. Downstream Processing for Biotechnology,* Wiley-Interscience, New York, 1988.

Benedict, M., and T. Pigford, *Nuclear Chemical Engineering,* McGraw-Hill, New York, 1957.

Bogacki, M. B., and J. Szymanowski, "Modeling of Extraction Equilibrium and Computer Simulation of Extraction-Stripping Systems for Copper Extraction by 2-Hydroxy-5-Nonyl-benzaldehyde Oxime," *Ind. Engr. Chem. Research, 29,* 601 (1990).

Blumberg, R., *Liquid-Liquid Extraction,* Academic Press, London, 1988.

Brian, P. L. T., *Staged Cascades in Chemical Processes,* Prentice Hall, Upper Saddle River, NJ, 1972.

Coulson, J. M., J. F. Richardson, J. R. Backhurst, and J. H. Harker, *Chemical Engineering*, Vol. 2, 3rd ed., Pergamon Press, Oxford, 1978, Chap. 13.

Evans, T. W., "Countercurrent and Multiple Extraction," *Ind. Eng. Chem., 23,* 860 (1934).

Fenske, M. R., R. H. McCormick, H. Lawroski, and R. G. Geier, "Extraction of Petroleum Fractions by Ammonia Solvents," *AIChE J., 1,* 335 (1955).

Francis, A. W., *Handbook for Components in Solvent Extraction,* Gordon and Breach, New York, 1972.

Frank, T. C., L. Dahuron, B. S. Holden, W. D. Prince, A. F. Siebert, and L. C. Wilson, "Liquid-Liquid Extraction and Other Liquid-Liquid Operations and Equipment," in D. W. Green and R. H. Perry (Eds.), *Perry's Chemical Engineers' Handbook*, 8th ed, McGraw-Hill, New York, Section 15, 2008.

Giddings, J. C., *Unified Separation Science*, Wiley-Interscience, New York, 1991.

Godfrey, J. C., "Mixers," in Godfrey, J. C. and M. J. Slater (Eds.), *Liquid-Liquid Extraction Equipment*, Wiley, Chichester, England, 1994, Chap. 12.

Godfrey, J. C., and M. J. Slater, eds., *Liquid-Liquid Extraction Equipment*, Wiley, New York, 1994.

Harrison, R. G., P. Todd, S. R. Rudge, and D. P. Petrides, *Bioseparations Science and Engineering*, Oxford University Press, New York, 2003.

Hartland, S., *Countercurrent Extraction*, Pergamon, London, 1970.

Hartland, S. and S. A. K. Jeelani, "Gravity Settlers," in Godfrey, J. C. and M. J. Slater (Eds.), *Liquid-Liquid Extraction Equipment*, Wiley, Chichester, England, 1994, Chap. 13.

Humphrey, J. L., and G. E. Keller II, *Separation Process Technology*, McGraw-Hill, New York, 1997.

Jacobs, L. J., Jr., and W. R. Penney, "Phase Segregation," in R. W. Rousseau (Ed.), *Handbook of Separation Process Technology*, Wiley, New York, 1987, Chap. 3.

Karr, A. E., and T. C. Lo, "Performance and Scale-up of a Reciprocating Plate Extraction Column," *Proceedings of International Solvent Extraction Conference*, Vol. 1, 299–320 (1971).

Karr, A. E., and T. C. Lo, "Scaleup of Large Diameter Reciprocating-Plate Extraction Columns," *Chem. Engr. Progress, 72* (11), 68 (1976).

King, C. J., *Separation Processes*, 2nd ed., McGraw-Hill, New York, 1981.

Laddha, G.S., and T.E. Degaleesan, *Transport Phenomena in Liquid Extraction*, McGraw-Hill, New York, 1978.

Lo, T. C., "Commercial Liquid-Liquid Extraction Equipment," in P. A. Schweitzer (Ed.), *Handbook of Separation Techniques for Chemical Engineers*, 3rd ed., McGraw-Hill, New York, 1997, Section 1.10.

Lo, T. C., M. H. I. Baird, and C. Hanson (Eds.), *Handbook of Solvent Extraction*, Wiley, New York, 1983; reprinted by Krieger, Malabar, FL, 1991.

Macedo, M. E. A., and P. Rasmussen, *Liquid-Liquid Equilibrium Data Collection, Supplement* 1, vol. V, part 4, 1987, DECHEMA, Frankfurt (Main), Germany.

Perry, R. H., and D. W. Green, Eds., *Perry's Chemical Engineers' Handbook,* 7th ed, McGraw-Hill, New York, 1997.

Reissinger, K. H., and J. Schroeter, "Modern Liquid-Liquid Extractors: Review and Selection Criteria," in *Alternatives to Distillation*, Inst. Chem. Eng., No. 54, 1978, pp. 33–48.

Ritchy, G. M., and A. Ashbrook, "Hydrometallurgical Extraction," in P. A. Schweitzer (Ed.), *Handbook of Separation Techniques for Chemical Engineers*, McGraw-Hill, New York, 1979, pp. 2-105 to 2-130.

Robbins, L. A., "Liquid-Liquid Extraction," in P. A. Schweitzer (Ed.), *Handbook of Separation Techniques for Chemical Engineers*, 3rd ed., McGraw-Hill, New York, 1997, Section 1.9.

Robbins, L. A., and R. W. Cusack, "Liquid-Liquid Extraction Operations and Equipment," in R. H. Perry and D. W. Green, Eds., *Perry's Chemical Engineers' Handbook,* 7th ed, McGraw-Hill, New York, 1997, Section 15.

Scheibel, E. G., "Liquid-Liquid Extraction," in E. S. Perry and A. Weissberger (Eds.), *Separation and Purification,* 3rd ed., *Techniques of Chemistry*, Vol. XII, Wiley-Interscience, New York, 1978, Chap. 3.

Skelland, A. H. P., and D. W. Tedder, "Extraction-Organic Chemicals Processing," in R. W. Rousseau (Ed.), *Handbook of Separation Process Technology*, Wiley, New York, 1987, Chapter 7.

Smith, B. D., *Design of Equilibrium Stage Processes,* McGraw-Hill, New York, 1963.

Smith, K. H., T. Bowser, and G. W. Stevens, "Performance and Scale-up of Karr Reciprocating Plate Extraction Columns," *Ind. Eng. Chem. Research, 47,* 8368 (2008).

Sorenson, J. M., and W. Arlt, *Liquid-Liquid Equilibrium Data Collection, Binary Systems*, vol. V, part 1, 1979, *Ternary Systems,* vol. V, part 2, 1980, *Ternary and Quaternary Systems*, vol. 5, part 3, 1980, DECHEMA, Frankfurt (Main), Germany.

Thibodeaux, L. J., D. R. Daner, A. Kimura, J. D. Millican, and R. I. Parikh, "Mass Transfer Units in Single and Multiple Stage Packed Bed, Cross-Flow Devices," *Ind. Eng. Chem. Process Des. Develop., 16,* 325 (1977).

Treybal, R. E., *Liquid Extraction*, 2nd ed., McGraw-Hill, New York, 1963.

Treybal, R. E., *Mass Transfer Operations,* 3rd ed., McGraw-Hill, New York, 1980, Chapt. 13.

Varteressian, K.A., and M.R. Fenske, "The System Methylcyclohexane - Aniline - N-Heptane," *Ind. Eng. Chem., 29,* 270 (1937).

Wankat, P. C., "Advanced Graphical Extraction Calculations," in J. M. Calo and E. J. Henley (Eds.), *Stagewise and Mass Transfer Operations,* Vol. 3, *Extraction and Leaching,* AIChEMI, Series B, AICHE, New York, 1982, p. 17.

Wankat, P. C., *Equilibrium-Staged Separations,* Prentice Hall, Upper Saddle River, NJ, 1988, pp. 564–567.

Wnek, W. J., and R. H. Snow, "Design of Cross-Flow Cooling Towers and Ammonia Stripping Towers," *Ind. Eng. Chem. Process Des. Develop., 11,* 343 (1972).

Woods, D. R., *Process Design and Engineering Practice,* Prentice Hall, Upper Saddle River, NJ, 1995.

Woods, D. R., *Data for Process Design and Engineering Practice*, Prentice Hall, Englewood Cliffs, NJ, 1995.

HOMEWORK

A. *Discussion Problems*

A1. What is the designer trying to do in the extraction equipment shown in Figure 13-2 and listed in Table 13-1? Why are there so many types of extraction equipment and only two major types of equipment for vapor-liquid contact?

A2. Write your key relations chart for this chapter.

A3. For Figure 13-23 suppose we desired to obtain the desired raffinate concentration with exactly two equilibrium stages. This can be accomplished by changing the amount of solvent used. Would we want to increase or decrease the amount of solvent? Explain the effect this change will have on M, E_N, Δ, and the number of stages required.

A4. All the partially miscible calculations in this chapter use right triangle diagrams. What differences would result if equilateral triangle diagrams were used to solve the extraction problem?

A5. If the extract and raffinate phases are totally immiscible, extraction problems can still be solved using triangular diagrams. Explain how and describe what the equilibrium diagram will look like.

A6. What can be done to increase the concentration of solute in the extract (i.e., increase $y_{A,N}$)?

A7. What situation in countercurrent extraction is superficially analogous to total reflux in distillation? How does it differ?

A8. Study Figure 13-28. Explain what happens as S/F increases. What happens to M? What happens to E_N? What happens to Δ? How do you find Δ_{min} if it lies on the left-hand side? How do you find Δ_{min} if it lies on the right-hand side?

A9. Three analysis procedures were developed for extraction in Chapter 13: McCabe-Thiele, Kremser, and triangular diagrams. If you have an extraction problem, how do you decide which method to use? (In other words, explain when each is applicable.)

A10. Compare K_dE/R to KV/L used in distillation, absorption, and stripping. Do these quantities have the same significance?

A11. Which methods will give accurate results for concentrated, partially miscible extraction (circle all correct answers)?

    a. McCabe-Thiele diagram

    b. Kremser equation

    c. Aspen Plus

    d. None of the above.

A12. In fractional extraction what happens to solute C if:

    a. $(\dfrac{K_{d,c}E}{R})_{top} > 1$ and $(\dfrac{K_{d,c}E}{R})_{bottom} < 1$?

    b. $(\dfrac{K_{d,c}E}{R})_{top} < 1$ and $(\dfrac{K_{d,c}E}{R})_{bottom} > 1$?

    c. How would you adjust the extractor so that the conditions in parts a or b would occur?

A13. Sketch the two-stage countercurrent batch extraction process discussed in Section 13.6.

B. *Generation of Alternatives*

B1. For fractional extraction, list possible problems other than the three in the text. Outline the solution to these problems.

B2. How would you couple together cross-flow and countercurrent cascades? What might be the advantages of this arrangement?

C. *Derivations*

C1. Derive Eq. (13-10) starting with a McCabe-Thiele diagram (follow the procedure used to develop the Kremser equation in Chapter 12).

C2. Derive Eq. (13-18).

C3. For fractional extraction outline in detail a solution procedure for (a) case 1, (b) case 2, and (c) case 3.

C4. Develop the solution method for a dilute multicomponent extraction in a cross-flow cascade. Sketch the McCabe-Thiele diagrams.

C5. Single-stage systems ($N = 1$) can be designed as countercurrent systems, Figure 13-4, or as cross-flow systems, Figure 13-9. Develop the methods for both these designs. Which is easier? If the system is dilute, how can the Kremser equation be used?

C6. Prove that the locations of streams M, E_N, and R_1 in Figure 13-21A lie on a straight line as shown in Figure 13-21B. In other words, derive Eq. (13-41).

C7. Define Δ and the coordinates of Δ from Eqs. (13-42) and (13-43). Prove that points Δ, E_j, and R_{j+1} (passing streams) lie on a straight line by developing Eq. (13-44). While doing this, prove that the lever-arm rule is valid.

D. *Problems*

Answers to problems with an asterisk are at the back of the book.

D1. In Example 13-1 we assumed that we were going to use all of the solvent available. There are other alternatives. Determine if the following alternatives are capable of producing outlet water of the desired acetic acid concentration.

    a. Use only the pure solvent at the bottom of the extractor.

    b. Mix all of the pure and all of the impure solvent together and use them at the bottom of the column.

    c. Mix all of the pure and part of the impure solvent together and use them at the bottom of the column.

D2. Repeat Example 13-2 but use a countercurrent cascade with all of the solvent (E = 20 kg/h) flowing countercurrent to the feed through the two stages.

D3. Acetic acid (the solute) is being extracted from water (the diluent) into butanol (the solvent). A single mixer-settler that acts as one equilibrium stage is used. The flow rate of the entering feed stream is 100 kg/h, and this feed is 0.013 wt frac acetic acid with the remainder being water saturated with butanol. The entering solvent stream is 0.001 wt frac acetic acid with the remainder being butanol saturated with water. We want the outlet wt frac acetic acid in the exiting water stream (the raffinate) to be x = 0.007. Assume that that total flow rates of extract (butanol phase) and raffinate (water phase) are constant. Note that complete immiscibility is not required as long as total flow rates are constant, which will occur if there is little change in the amount of water in the extract phase and little change in the amount of butanol in the raffinate. Since these streams are presaturated, the assumption of constant flow rates is reasonable. Equilibrium is given in Table 13-3. Find the required solvent flow rate S in kg/h.

D4.* We have a mixture of acetic acid in water and wish to extract this with 3-heptanol at 25°C. Equilibrium is

$$\frac{\text{Wt frac acetic acid in solvent}}{\text{Wt frac acetic acid in water}} = 0.828$$

The inlet water solution flows at 550 lb/h and is 0.0097 wt frac acetic acid. We desire an outlet water concentration of 0.00046 wt frac acetic acid. The solvent flow rate is 700 lb/h. The entering solvent contains 0.0003 wt frac acetic acid. Find the outlet solvent concentration and the number of equilibrium stages required (use the Kremser equation). Is this an economical way to extract acetic acid?

D5. The equilibrium for extraction of acetic acid from water into 3-heptanol at 25°C is $y = 0.828 x$, where y is wt frac acetic acid in 3-heptanol and x = wt frac acetic acid in water. We have a feed with F = R = 400 kg/h that is $x_0 = 0.005$ wt frac acetic acid and 0.995 wt frac water. This feed is contacted in a counter-current extractor with E = 560 kg/h of solvent that is $y_{N+1} = 0.0001$ wt frac acetic acid and 0.9999 wt frac 3-heptanol. We desire an outlet raffinate concentration of $x_N = 0.0003$ wt frac acetic acid and 0.9997 wt frac water. Assume water and 3-heptanol are immiscible and that R and E are constant.
a. Determine the number of equilibrium stages N required.
b. What is the minimum solvent flow rate, E_{min}?

D6. The equilibrium for extraction of acetic acid from 3-heptanol into water at 25°C is $y = 1.208 x$, where y = wt frac acetic acid in water and x = wt frac acetic acid in 3-heptanol. We have a feed with F = R = 100 kg/h that is $x_0 = 0.005$ wt frac acetic acid and 0.995 wt frac 3-heptanol. This feed is contacted in a counter-current extractor with solvent that is $y_{N+1} = 0.0002$ wt frac acetic acid and 0.9998 wt frac water. We desire an outlet raffinate concentration of $x_N = 0.0005$ wt frac acetic acid and 0.9995 wt frac 3-heptanol. Assume water and 3-heptanol are immiscible and that R and E are constant. Note that parts a, b, and c can be solved independently of each other, or the value of y_1 from part a can be used in part b.
a. If solvent flow rate E = 140 kg/h, calculate the exiting extract wt frac y_1.
b. If solvent flow rate E = 140 kg/h, determine the number of equilibrium stages N required.
c. What are the minimum solvent flow rate, and the maximum exiting extract wt frac (N → infinity)?

d. Problems 13.D5. and 13.D6. are for the same system, but y = 1.208 x in one problem and y = 0.828 x in the other problem. Explain why.

D7.* We have an extraction column with 30 equilibrium stages. We are extracting acetic acid from water into 3-heptanol at 25°C. Equilibrium is given in Problem 13.D4. The aqueous feed flows at a rate of 500 kg/h. The feed is 0.011 wt frac acetic acid, and the exit water should be 0.00037 wt frac acetic acid. The inlet 3-heptanol contains 0.0002 wt frac acetic acid. What solvent flow rate is required? Assume that total flow rates are constant.

D8. We are extracting acetic acid from benzene (diluent) into water (solvent) at 25°C and 1.0 atm. 100.0 kg/h of a feed that is 0.00092 wt frac acetic acid and 0.99908 wt frac benzene is fed to a column. The inlet water (solvent) is pure and flows at 25.0 kg/h. We have an extractor that operates with 2 equilibrium stages.

$$K_d = \frac{y}{x} = \frac{\text{Wt frac. in extract (water)}}{\text{Wt frac. in raffinate (benzene)}} = 30.488. \text{ Find:}$$

a. The outlet wt frac of acetic acid in the benzene, x_{out}.
b. The outlet wt frac of acetic acid in the water, y_{out}.

D9.* We have a mixture of linoleic and oleic acids dissolved in methylcellosolve and 10% water. Feed is 0.003 wt frac linoleic acid and 0.0025 wt frac oleic acid. Feed flow rate is 1500 kg/h. A simple countercurrent extractor will be used with 750 kg/h of pure heptane as solvent. We desire a 99% recovery of the oleic acid in the extract product. Equilibrium data are given in Table 13-3. Find N and the recovery of linoleic acid in the extract product.

D10. Two feed solutions of dioxane in water are being extracted with benzene in the 4-stage cross-flow system. The entering benzene solvent to each stage is pure ($y_{dioxane} = 0$). The solvent flow rate to each stage is 50 kg/h. Entering feed to stage 1 has a total flow rate of 100 kg/h and is $x_{dioxane,F1} = 0.015$ wt frac dioxane. Entering feed 2 has a total flow rate of 70 kg/h and is $x_{dioxane,F2} = 0.005$ wt frac dioxane. Feed 2 is mixed with the raffinate leaving the second stage, and this mixture is fed to the third stage. Assume that benzene and water are completely immiscible and that the total flow rates of raffinate and extract entering and leaving each stage are constant.
The equilibrium for dioxane distributing between benzene and water is

$$y_{dioxane} = 1.02 \, x_{dioxane}$$

where $y_{dioxane}$ is the wt frac of dioxane in the extract (benzene phase) and $x_{dioxane}$ is the wt frac of dioxane in the raffinate (water phase).
Find the following dioxane wt frac: raffinate wt frac leaving stage 2, the raffinate wt frac entering stage 3, the raffinate wt frac leaving stage 4, and the extract wt frac leaving stage 4.

D11.* The fractional extraction system shown in Figure 13-5 is separating abietic acid from other acids. Solvent 1, heptane, enters at $\overline{F} = 1000$ kg/h and is pure. Solvent 2, methylcellosolve + 10% water, is pure and has a flow rate of R = 2500 kg/h. Feed is 5 wt % abietic acid in solvent 2 and flows at 1 kg/h. There are only traces of other acids in the feed. We desire to recover 95% of the abietic acid in the bottom raffinate stream. Feed is on stage 6. Assume that the solvents are completely immiscible and that the system can be considered to be very dilute. Equilibrium data are given in Table 13-3. Find N.

D12. We plan to recover acetic acid from water using 1-butanol as the solvent. Operation is at 26.7°C. The feed flow rate is 10.0 kmol/h of an aqueous solution that contains 0.0046 mole frac acetic acid. The entering solvent is pure and flows at 5.0 kmol/h. This operation will be done with three mixer-settlers arranged as a countercurrent cascade. Each mixer-settler can be assumed to be an equilibrium stage. Equilibrium data are available in Table 13-3. Find the exiting raffinate and extract mole fractions.

D13. We wish to extract p-xylene and o-xylene from n-hexane diluent using β, β' – Thiodipropionitrile as the solvent. The solvent and diluent can be assumed to be immiscible. The feed is 1000.0 kg/h. The feed contains 0.003 wt frac p-xylene and 0.005 wt frac o-xylene in n-hexane. We desire at least a 90% recovery of p-xylene and at least 95% recovery of o-xylene. The entering solvent is pure. Operation is at 25°C and equilibrium data are in Table 13-3. Use a simple countercurrent cascade.

    a. Calculate the value of $(R/E)_{max}$ for both p-xylene and for o-xylene to just meet the recovery requirements.

    b. The smaller $(R/E)_{max}$ value represents the controlling or key solute. Operate at $E = 1.5(E)_{min,controlling}$. Find the number of stages and solvent inlet flow rate.

    c. Determine the outlet raffinate concentration and the percent recovery of the non-controlling solute.

D14.* The equilibrium data for the system water-acetic acid-isopropyl ether are given in Table 13-5. We have a feed that is 30 wt % acetic acid and 70 wt % water. The feed is to be treated with pure isopropyl ether. A batch extraction done in a mixed tank will process 15 kg of feed.

    a. If 10 kg of solvent is used, find the outlet extract and raffinate compositions.

    b. If a raffinate composition of $x_A = 0.1$ is desired, how much solvent is needed?

D15.* The system shown in the figure is extracting acetic acid from water using benzene as the solvent. The temperature shift is used to regenerate the solvent and return the acid to the water phase.

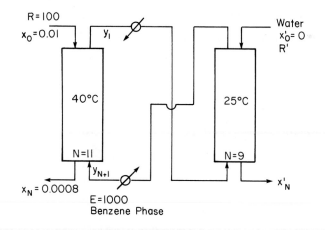

    a. Determine y_1 and y_{N+1} (units are wt fracs) for the column at 40°C.

    b. Determine R' and x_N' for the column at 25°C.

    c. Is this a practical way to concentrate the acid?

Data are in Table 13-3. Note: A similar scheme is used commercially for citric acid concentration using a more selective solvent.

D16. Acetic acid is being extracted from water with butanol as the solvent. Operation is at 26.7 °C and equilibrium data are in Table 13-3. The feed is 10.0 kg/min of an aqueous feed that contains 0.01 wt frac acetic acid. The entering solvent stream is butanol with 0.0002 wt frac acetic acid. The flow rate of the solvent stream is 8.0 kg/min. The column has 6 equilibrium stages. Find the outlet weight fractions. Assume butanol and water are immiscible.

a. Solve with the Kremser equation.

b. Check your answer with a McCabe-Thiele diagram (done in weight fractions). Note that if used as a check, the McCabe-Thiele diagram is *not* trial-and-error even when the number of stages is known.

D17. The feed in Problem 13.D16 is to be processed in a 6 stage cross-flow extractor. The same solvent as in Problem 13.D16 is increased to a total flow rate of 12.0 kg/min that is equally divided among the 6 stages, so that flow rate of solvent to each stage is 2.0 kg/min. Find the outlet diluent weight fraction.

D18. We contact 100 kg/h of a feed mixture that is 50 wt % methylcyclohexane and 50 wt % n-heptane with a solvent stream (15 wt % methylcyclohexane and 85 wt % aniline) in a mixer-settler that acts as a single equilibrium stage. We obtain an extract phase which is 10 wt % methylcyclohexane and a raffinate phase which is 61 wt % methylcyclohexane. Equilibrium data are in Table 13-6. What flow rate of the solvent stream was used?

D19. Many extraction systems are partially miscible at high concentrations of solute, but close to immiscible at low solute concentrations. At relatively low solute concentrations both the McCabe-Thiele and trianglar diagram analyses are applicable. This problem explores this. We wish to use chloroform to extract acetone from water. Equilibrium data are given in Table 13-4. Find the number of equilibrium stages required for a counter-

TABLE 13-6.   *Equilibrium data for methylcyclohexane, n-heptane-aniline*

Hydrocarbon Phase, wt %		Aniline-Rich Phase, wt %	
Methylcyclohexane	n-Heptane	Methylcyclohexane	n-Heptane
x_A	x_D	y_A	y_D
0.0	92.6	0.0	6.2
9.2	83.1	0.8	6.0
18.6	73.4	2.7	5.3
22.0	69.8	3.0	5.1
33.8	57.6	4.6	4.5
40.9	50.4	6.0	4.0
46.0	45.0	7.4	3.6
59.0	30.7	9.2	2.8
67.2	22.8	11.3	2.1
71.6	18.2	12.7	1.6
73.6	16.0	13.1	1.4
83.3	5.4	15.6	0.6
88.1	0.0	16.9	0.0

Source: Varteressian and Fenske (1937).

current cascade if we have a feed of 1000.0 kg/h of a 10.0 wt % acetone, 90.0 wt % water mixture. The solvent used is chloroform saturated with water (no acetone). Flow rate of stream $E_0 = 1371$ kg/h. We desire an outlet raffinate concentration of 0.50 wt % acetone. Assume immiscibility and use a weight ratio units graphical analysis. Compare results with Problem 13.D43.

Note: Use the lowest acetone wt % in Table 13-4 to estimate the distribution coefficient for acetone. Then convert this equilibrium to weight ratios.

D20. The aqueous two-phase system in Example 13-2 will be used in a batch extraction. We have 5.0 kg of PEG solution containing protein at mass fraction x_F. We will use 4.0 kg of pure dextran solution to extract the protein from the solution. Equilibrium data are in Example 13-2.

   a. Find the fractional recovery of the protein in the dextran phase if the two solutions are mixed together and then allowed to settle.

   b. Find the fractional recovery of the protein in the dextran phase if the continuous solvent addition batch extraction system shown in Figure 13-11 is used.

D21.* The equilibrium data for extraction of methylcyclohexane (A) from n-heptane (D) into aniline (S) are given in Table 13-6. We have 100 kg/h of a feed that is 60% methyl-cyclohexane and 40% n-heptane and 50 kg/h of a feed that is 20% methylcyclohexane and 80% n-heptane. These two feeds are mixed with 200 kg/h of pure aniline in a single equilibrium stage.

   a. What are the extract and raffinate compositions leaving the stage?

   b. What is the flow rate of the extract product?

D22. For the extraction in Example 13-5, the horizontal settler calculation was done for a settler diameter of $D_s = 1.023$ m and with the dispersion band assumed to be at the center of the circle. The conclusion was that there was likely to be a limited amount of interference due to flow. Repeat the calculation of $Re_{settler}$ and determine if interference is likely if the interface of the dispersion band is 0.1 m below the center of the circle.

D23. We have 500 kg/h of a feed that is 30 wt % pyridine and 70 wt % water. This will be extracted with pure chlorobenzene. Operation is at 1 atm and 25°C. Equilibrium data are in Table 13-7.

TABLE 13-7.   *Equilibrium data for pyridine, water, chlorobenzene at 25°C and 1 atm. (Treybal, 1980)*

Extract Phase		Raffinate Phase	
Wt frac pyridine	Wt frac water	Wt frac pyridine	Wt frac water
0	0.0005	0	0.9992
0.1105	0.0067	0.0502	0.9482
0.1895	0.0115	0.1105	0.8871
0.2410	0.0162	0.1890	0.8072
0.2860	0.0225	0.2550	0.7392
0.3155	0.0287	0.3610	0.6205
0.3505	0.0395	0.4495	0.5087
0.4060	0.0640	0.5320	0.3790
0.490	0.132	0.490	0.132

a. A single mixer-settler is used. Chlorobenzene flow rate is 300 kg/h. Find the outlet extract and raffinate flow rates and weight fractions.

b. We now want to convert the single-stage system in part a into a two-stage cross flow system. The raffinate from part a is fed to the second mixer-settler and is contacted with an additional 300 kg/h of pure chlorobenzene. Find the raffinate and extract flow rates and weight fractions leaving the second stage.

D24. We plan to remove methylcyclohexane from n-heptane using aniline as the solvent (see Table 13-6 for equilibrium data). The feed is 200 kg/h total. It is 40 wt % methylcyclo-hexane and 55 wt % n-heptane. The solvent is 95 wt % aniline and 5 wt % n-heptane.

a. With a single mixer-settler, set the total entering flow rate of solvent stream as 600 kg/h. Find the outlet wt fracs and the outlet flow rates.

b. Suppose we decide to use a two stage cross flow system with 300 kg/h of the solvent stream entering each stage. Find the outlet flow rates and wt fracs.

Note that the answers are very sensitive to how one draws the tie lines. This is typical of type II systems.

D25.* We wish to remove acetic acid from water using isopropyl ether as solvent. The opera-tion is at 20°C and 1 atm (see Table 13-5). The feed is 0.45 wt frac acetic acid and 0.55 wt frac water. Feed flow rate is 2000 kg/h. A countercurrent system is used. Pure sol-vent is used. We desire an extract stream that is 0.20 wt frac acetic acid and a raffinate that is 0.20 wt frac acetic acid.

a. How much solvent is required?

b. How many equilibrium stages are needed?

D26.* A countercurrent system with three equilibrium stages is to be used for water-acetic acid-isopropyl ether extraction (see Table 13-5). Feed is 40 wt % acetic acid and 60 wt % water. Feed flow rate is 2000 kg/h. Solvent added contains 1 wt % acetic acid but no water. We desire a raffinate that is 5 wt % acetic acid. What solvent flow rate is required? What are the flow rates of E_N and R_1?

D27. We have a total feed of 100 kg/h to a countercurrent extraction column. The feed is 40 wt % acetic acid and 60 wt % water. We plan to extract the acetic acid with isopropyl ether. Equilibrium data are in Table 13-5. The entering isopropyl ether is pure and has a flow rate of 111.2 kg/h. We desire a raffinate that is 20 wt % acetic acid.

a. Find the acetic acid wt. fraction in the outlet extract stream.

b. Determine the flow rates of the outlet raffinate and extract streams.

c. Find the number of equilibrium contacts.

D28.* We are extracting acetic acid from water with isopropyl ether at 20°C and 1 atm pres-sure. Equilibrium data are in Table 13-5. The column has three equilibrium stages. The entering feed rate is 1000 kg/h. The feed is 40 wt % acetic acid and 60 wt % water. The exiting extract stream has a flow rate of 2500 kg/h and is 20 wt % acetic acid. The entering extract stream (which is not pure isopropyl ether) contains no water. Find:

a. The exit raffinate concentration.

b. The required entering extract stream concentration.

c. Flow rates of exiting raffinate and entering extract streams.

Trial and error is *not* needed.

D29. We are recovering pyridine from water using chlorobenzene as the solvent in a coun-tercurrent extractor. The feed is 35 wt % pyridine and 65 wt % water. Feed flow rate is 1000 kg/h. The solvent used is pure. The desired outlet extract is 20 wt % pyridine, and the desired outlet raffinate is 4 wt % pyridine. Operation is at 25°C and 1 atm. Equilibrium data are in Table 13-7.

a. Find the number of equilibrium stages needed.

b. Determine the solvent flow rate required in kg/h.

D30. Suppose in Example 13-5 that we decide to build the settler with a diameter of 1.0 m and a length of 4.0 m. What safety factor are we employing?

D31. We have a mixer settler that operates as one equilibrium stage for an extraction separation. The feed is 1.0 wt. % acetic acid and 99.0 wt % water. Total feed flow rate is 50 kg/h. This feed is mixed with 100 kg/h of pure solvent (1-butanol). Equilibrium is given in Table 13-3.

Assume that 1-butanol and water are immiscible and that the system is dilute (constant total flow rates). Find the acetic acid wt. fractions in the extract, y_1, and the raffinate, x_1, products. Note: There are multiple solution paths for this problem.

D32. For the toluene-water system in Example 13-5 we found that toluene is the dispersed phase if $Q_{solvent}/Q_{feed} = 0.2$. Which phase is dispersed if

a. $Q_{solvent}/Q_{feed} = 0.6$

b. $Q_{solvent}/Q_{feed} = 1.0$

c. $Q_{solvent}/Q_{feed} = 2.0$

d. $Q_{solvent}/Q_{feed} = 5.0$

D33. For the extraction in Example 13-5, suppose we decide to have $H_{tank} = 2d_{tank}$ and want a 1.5 minute residence time. Find the tank dimensions.

D34. For the extraction in Example 13-5 with $H_{tank} = d_{tank}$ and a 1.0 minute residence time, a 6-blade flat turbine impeller with impeller diameter and impeller width $d_i = 0.20d_{tank}$ and impeller width $w_i = d_i/5$ is operated at 500 rpm. Estimate the value of φ_d in the tank and the power P required.

D35. For Example 13-4 set up the tridiagonal matrix for the mass balances assuming there are 6 stages in the column and the exit raffinate stream concentration is unknown. Develop the values for A, B, C, and D in only the acetic acid matrix using K values determined from Table 13-5. Do not invert the matrix.

D36. A 60 vol% tributyl phosphate (TBP) in kerosene solvent is used to extract $Zr(NO_3)_4$ from an aqueous solution of nitric and solution nitrate. The entering solvent is recycled from a solvent recovery system and contains 0.008 mol $Zr(NO_3)_4$/L. The aqueous feed contains 0.10 mol $Zr(NO_3)_4$/L and the outlet aqueous solution should contain 0.008 mol $Zr(NO_3)_4$/L. Feed rate of the aqueous solution is 200 L/h.

a. Find the minimum entering solvent rate $F_{solvent,min}$ in L/h.

b. If $F_{solvent,min} = 1.4 F_{solvent,min}$, find the number of equilibrium stages required.

c. Calculate the mole fraction and the mass fraction of $Zr(NO_3)_4$ in the feed and determine if the system is dilute. Assume the density of the feed is 1.0 g/ml.

Assume that the aqueous and organic phases are completely immiscible and that the densities of the two phases are both constant (volumetric flow rates are constant). For zirconium nitrate, $Zr(NO_3)_4$, the equilibrium data are listed below for 60 volume % TBP in kerosene (Benedict and Pigford, 1957):

mol $Zr(NO_3)_4$/L

Aqueous phase, C_{Aq}	Organic phase, C_{Org}
0.0	0.0
0.012	0.042
0.039	0.083
0.074	0.114
0.104	0.135
0.135	0.147

D37. Pilot plant data are obtained for a Karr column with 50 mm diameter plates with a plate free area of e = 0.52. The optimum frequency of operation that produces the lowest HETP value was 1.4 s^{-1} and the corresponding value of HETP = 0.24 m. We plan to scale up the Karr column to a column that is 1.1 m in diameter.

   a. Calculate the corresponding HETP and frequency if the exponents in Eqs. (13-65) and (13-66) are those recommended in *Perry's Chemical Engineers' Handbook* (Frank et al., 2008).

   b. Calculate the corresponding HETP and frequency if the exponents in Eqs. (13-65) and (13-66) are those recommended by Smith et al., (2008).

   c. Comment on the differences.

D38*. We are extracting benzoic acid from water into toluene in a single equilibrium stage system. The entering toluene is pure and the entering water contains 0.00023 mole fraction benzoic acid. The feed flow rate is 1.0 kmol/h and the solvent flow rate is 0.06 kmole/h. Find the outlet mole fractions of benzoic acid in the exiting raffinate and extract phases. Data are in Example 13-5. Assume water and toluene are completely immiscible.

D39. Plot the equilibrium data from Table 13A-1 for tri-ethylamine (solvent), carbon tetra-chloride (solute), acetic acid (diluent) on a right triangle diagram with ordinate = mole frac CCl_4 and abscissa = mole frac acetic acid. Solve the single-stage extraction problem for Lab 10, Problem 1 graphically. Find extract and raffinate mole fractions and flow rates.

D40. Plot the equilibrium data from Table 13A-1 for tri-ethylamine (solvent), carbon tetra-chloride (solute), acetic acid (diluent) on a right triangle diagram with ordinate = mole frac CCl_4 and abscissa = mole frac acetic acid. Solve the following two stage cross-flow problem graphically. F = 10 kmol/h, and is 10 mol% CCl_4 and 90 mol% acetic acid. Entering solvent is pure and 10 kmol/h are added to each stage. Find mole fractions and flow rates of both extract streams and the raffinate stream.

D41. Using the equilibrium data in Table 13A-1, find the number of stages needed for a countercurrent extractor if 10 kmol/h feed that is 10 mol% CCl_4 and 90 mol% acetic acid is processed at 293K and 1 atm. The solvent is pure tri-ethylamine with a flowrate of 14.5 kmol/h. We desire an exiting extract that contains 0.091 mole frac CCl_4. Also find flow rates and mole frac of exiting extract and raffinate streams.

D42. We are recovering pyridine from water using chlorobenzene as the solvent in a countercurrent extractor. The feed is 41 wt % pyridine and 59 wt % water. The solvent used is pure chlorobenzene and the solvent flow rate is 1000.0 kg/h. The desired outlet extract is 32 wt % pyridine, and the desired outlet raffinate is 5 wt % pyridine. Operation is at 25°C and 1.0 atm. Equilibrium data are in Table 13-7.

   a. Find the number of equilibrium stages needed.

   b. Determine the feed flow rate required in kg/h.

   NEATNESS COUNTS!

D43. We wish to use chloroform to extract acetone from water. Equilibrium data are given in Table 13-4. Find the number of equilibrium stages required for a countercurrent cascade if we have a feed of 1000.0 kg/h of a 10.0 wt % acetone, 90.0 wt % water mixture. The solvent used is chloroform saturated with water (no acetone). Flow rate of stream E_0 = 1371 kg/h. We desire an outlet raffinate concentration of 0.50 wt % acetone. Use a triangle diagram for solution and compare results with Problem 13.D19.

E. More Complex Problems

E1.* We have a liquid feed that is 48 wt % m-xylene and 52 wt % o-xylene, which are to be separated in a fractional extractor (Figure 13-5) at 25°C and 101.3 kPa. Solvent 1 is β,β′-thiodipropionitrile, and solvent 2 is n-hexane. Equilibrium data are in Table 13-3. For each kilogram of feed, 200 kg of solvent 1 and 20 kg of solvent 2 are used. Both solvents are pure when they enter the cascade. We desire a 92% recovery of o-xylene in solvent 1 and a 94% recovery of m-xylene in n-hexane. Find outlet composition, N, and N_f. Adjust the recovery of m-xylene if necessary to solve this problem.

E2. We are extracting meta, ortho and para xylenes from n-hexane using β, β′Thiodipropionitrile as solvent. Solvent and diluent (n-hexane) are immiscible. Feed flow rate is 1000.0 kg/h. Feed is $x_{m-xy} = 0.005$ wt frac m-xylene, $x_{o-xy} = 0.006$ wt frac o-xylene, and $x_{p-xy} = 0.004$ wt frac p-xylene in n-hexane. Solvent flow rate is 20,000 kg/h. The entering solvent is pure. We desire 96% recovery of p-xylene in the solvent. Operation is at 25°C and 1.0 atm. Equilibrium data are in Table 13-3. Use a simple countercurrent cascade.

    a. Find outlet p-xylene weight fraction, $x_{N,p-xy}$

    b. Find N

    c. Find $x_{N,o-xy}$

    d. Find $x_{N,m-xy}$

    e. What is minimum solvent flowrate (N approaches infinity)?

E3. The equilibrium data for extraction of methylcyclohexane (A) from n-heptane (D) into aniline (S) are given in Table 13-6. We wish to do a batch extraction of 20 kg of a feed that is 40 wt % methylcyclohexane and 60 wt % n-heptane. The solvent added is first presaturated with the diluent n-heptane so that according to the data in Table 13-6 it is 93.8 % aniline and 6.2 % n-heptane.

    a. We do a normal batch extraction adding 20 kg of the presaturated solvent. Find flow rates and compositions of the extract and raffinate products.

    b. We decide to process the same feed with pre-saturated solvent in a continuous solvent addition batch extraction. First, presaturated solvent is added with no removal of extract until a two phase mixture is obtained. How much pre-solvent has to be added to obtain two phases? Second, the continuous solvent addition batch extraction is conducted until the raffinate has a methylcyclohexane wt frac of 0.292. How much solvent is added during this step? What is the final raffinate composition? How much extract is collected, and what is its average composition? Assume $\hat{R}_t$ is constant and Eq. (13-27) is valid.

G. Computer Problems

G1. Do lab 12 in Appendix to chapter 13, but for a two-stage cross-flow system. Operation is at 293K, 1.0 atm, F = 10 kmol/h and is 10 mol% carbon tetrachloride and 90 mol% acetic acid. Entering solvent is pure triethylamine. Total amount of pure solvent is 20 kmol/h with 10 kmol/h fed to each stage. Find the flow rates and mole fractions in the three outlet streams. Then calculate the fraction of entering carbon tetrachloride that is extracted.

G2. Do lab 12 in Appendix to chapter 13, but for three stage systems. Operation is at 293K, 1.0 atm, F = 10 kmol/h and is 10 mol% carbon tetrachloride and 90 mol% acetic acid. Entering solvent is pure triethylamine.

    a. Do for a 3-stage cross-flow system with total amount of pure solvent is 30 kmol/h with 10.0 kmol/h fed to each stage. Find the total and component flow rates (kmol/h) in the four outlet streams. Then calculate the fraction of entering carbon tetrachloride that is extracted.

b. Do for a 3-stage countercurrent system using 3 decanters. 10 kmol/h of pure solvent is used. Find the total and component flow rates (kmol/h) in the two outlet streams. Then calculate the fraction of entering carbon tetrachloride that is extracted.

c. Repeat the 3-stage countercurrent system with different flow rates of pure solvent until the fraction of entering carbon tetrachloride extracted is the same as in part a. Which process uses less solvent?

CHAPTER 13 APPENDIX.   COMPUTER SIMULATION OF EXTRACTION

Lab 12.   The major purpose of this lab assignment is to have you learn how to simulate LLE. It will be helpful if you become familiar with the information resources available in Aspen Plus. The sources useful for this lab are: help, the Aspen Plus User Guide, and the Unit Operations document. The latter two are available as PDF files when you go to Documents instead of the Aspen User Interface. Find the documents, and read the few pages you need.

We will separate a feed that is 10 mol% carbon tetrachloride (CCl_4) and 90 mol% acetic acid $(C_2H_4O_2)$ using a solvent that is pure triethylamine $(C_6H_{15}N)$. Extractor, feed, and solvent are at 293 K and 1 atm. Data and NRTL parameters are available for this system in the DECHEMA data bank (Sorensen and Arlt, 1980). The DECHEMA parameters for NRTL are in Table 13-A1. Note: diluent = acetic acid, solute = carbon tct.

TABLE 13-A1.   *NRTL parameters and experimental tie line data for triethylamine (1), carbon tetrachloride (2), and acetic acid (3) at 293K and 1 atm (Sorensen and Arlt, 1980); alpha = 0.2, units are K*

i	j	Aij	Aji
1	2	−303.24	90.396
1	3	1162.9	−143.69
2	3	350.19	75.515

| *Experimental tie lines in mole frac:* | | | | | |
| *Triethylamine phase 2* | | | *Acetic acid phase 1* | | |
1	*2*	*3*	*1*	*2*	*3*
0.914	0.020	0.066	0.412	0.0090	0.579
0.861	0.0710	0.068	0.414	0.034	0.552
0.770	0.144	0.086	0.418	0.071	0.511
0.698	0.192	0.110	0.423	0.104	0.473
0.642	0.216	0.142	0.425	0.135	0.440
0.62	0.222	0.158	0.425	0.147	0.428

Hints on running Aspen Plus for extraction:

1. In Setup, choose liquid-liquid-vapor as the allowable phases.

2. For a mixer-settler, all you need in the flowsheet is a decanter with the feed and solvent streams input into the decanter feed port and two product streams. Select triethylamine as the key component for the second liquid phase, and use the default value for second liquid threshold.

3. For inputting equilibrium data, select NRTL as the property method. Aspen Plus has its own database for parameters, but the Aspen Plus method of inputting the parameter values is different than the DECHEMA method. On the input page for NRTL, select one of the columns for a binary pair and click on the DECHEMA button. Then input the Aij and Aji and alpha values from DECHEMA in Aspen Plus. Be sure that the button for LLE is clicked, not the button for VLE. This column in the table should now list "user" for source and units should be K. Repeat this procedure for the other two binary pairs.

4. To look at the predicted equilibrium data, go to Tools→Analysis→Property→Ternary, select Liquid-Liquid, input the temperature, Enter, and then click on Go. Compare the predicted results (the Table is probably easier to read than the graph) with Table 13-A1 experimental tie lines. The fit should be reasonable. If the fit is poor, go back to your NRTL parameters and input the values again, making sure that the LLE button is pressed for each binary pair.

5. For the countercurrent extractor, select "column" then "extract" in the bottom menu bar. The acetic acid feed should be input at the top of the column and the pure solvent at the bottom of the column. For thermal option select "specify temperature profile" and use 293 on the stages. Select acetic acid as the key component for the first liquid phase and triethylamine as the key component for the second liquid phase.

Do the following extraction problems using the NRTL parameters in Table 13-A1 and report your answer.

A. Feed flow rate is 10 kmol/h. The feed is mixed with 10 kmol/h of pure solvent in a decanter. All streams and the column are at 293 K_d and 1 atm.
    a. Determine the K value for carbon tetrachloride and compare to the data.
    b. Find the outlet flow rates and mole fracs in the two outlet streams.
    c. Try changing the ratio of S/F and the feed mole frac to determine what happens in this single-stage system. Note that it is possible for one of the phases to disappear.
B. Repeat Problem A, but in a countercurrent extraction column with two equilibrium stages.
    a. Determine the K_d value for carbon tetrachloride and compare to the data.
    b. Find the outlet flow rates and mole fracs in the two outlet streams.
    c. Try changing the ratio of S/F and the feed mole frac to determine what happens in this two-stage system. Note that it is possible for one of the phases to disappear.
    d. Try increasing the number of stages to determine what happens.
C. The countercurrent system can also be simulated with two decanters. Start with a 2-stage cross-flow system with F = 10, $S_1 = S_2 = 10$ and run the simulation. We do this first to give

Aspen Plus initial values for the streams that will enter decanter 1 when extract 2 is fed to decanter 1. Click in the flowsheet on extract stream from stage 2, then right click and choose Reconnect the Destination and put the destination at the feed to decanter 1. Run the simulation. Reduce the amount of S_1 to 1 and run the simulation. Reduce the amount of S_1 to zero and run the simulation. Because Aspen Plus is programmed to flag zero values of flow rates, you will get a warning that S_1 flow rate is not specified. Ignore this warning. The result is a 2 stage countercurrent system that should match the previous results. Do this comparison.

D. The LLE correlations in AspenPlus are not always as accurate as possible. If data are available, AspenPlus will find values of the parameters for any of the LLE correlations by doing a regression against the data you input. The purpose of this part is to obtain an improved fit for the NRTL correlation for the ternary system water, chloroform, acetone. Open a new AspenPlus file, draw an extractor with feed, solvent and two products, in Setup list valid phases as Liquid-Liquid-Vapor, and input these 3 components. Then use NRTL as LLE correlation and give any values for the feed and solvent and extractor conditions. Go to Analysis, list ternary and valid phases as liquid-liquid to obtain the predicted equilibrium results – save these for later comparison.

The experimental data for this system are in Table 13-4. To have Aspen Plus regress the NRTL parameters to fit the data, follow the instructions in Appendix B (at the end of the book) as modified for LLE in the last paragraph. Look at your results (View_Report) and at the ternary graph from Analysis (list valid phases as liquid-liquid). Compare to the experimental data and to the predicted equilibrium for the non-regressed parameters.

CHAPTER 14

Washing, Leaching, and Supercritical Extraction

A number of other separation processes can be analyzed, at least under limiting conditions, using McCabe-Thiele diagrams or the Kremser equation with an approach very similar to that used for absorption and stripping. These processes all require addition of a mass-separating agent. If flows are not constant, the trianglar diagram method developed for extraction can often be extended to these separations. This method is developed for leaching (or solid-liquid extraction) in Section 14.4.

14.1 GENERALIZED McCABE-THIELE AND KREMSER PROCEDURES

The McCabe-Thiele procedure has been applied to flash distillation, continuous countercurrent distillation, batch distillation, absorption, stripping, and extraction. What are the common factors for the McCabe-Thiele analysis in all these cases?

All the McCabe-Thiele graphs are plots of concentration in one phase vs. concentration in the other phase. In all cases there is a single equilibrium curve, and there is one operating line for each column section. It is desirable for this operating line to be straight. In addition, although it isn't evident on the graph, we want to satisfy the energy balance and mass balances for all other species.

In order to obtain a single equilibrium curve, we have to specify enough variables that only one degree of freedom remains. For binary distillation this can be done by specifying constant pressure. For absorption, stripping, and extraction we specified that pressure and temperature were constant, and if there were several solutes we assumed that they were independent. In general, we will specify that pressure and/or temperature are constant, and for multisolute systems we will assume that the solutes are independent.

To have a straight operating line for the more volatile component in distillation we assumed that constant molal overflow (CMO) was valid, which meant that in each section total flows were constant. For absorption, stripping, and extraction we could make the assumption that total flows were constant if the systems were very dilute. For more concentrated systems we assumed that there was one chemical species in each phase that did not transfer into the other phase; then the flow of this species (carrier gas, solvent, or diluent) was constant. In general, we have to assume either that total flows are constant or that flows of nontransferred species are constant.

These assumptions control the concentration units used to plot the McCabe-Thiele diagram. If total flows are constant, the solute mass balance is written in terms of *fractions*, and fractions are plotted on the McCabe-Thiele diagram. If flows of nontransferred species are constant, *ratio* units must be used, and ratios are plotted on the McCabe-Thiele diagram.

The McCabe-Thiele operating line satisfies the mass balance for only the more volatile component or the solute. In binary distillation the CMO assumption forces total vapor and liquid flow rates to be constant and therefore the overall mass balance will be satisfied. In absorption, when constant carrier and solvent flows are assumed, the mass balances for these two chemicals are automatically satisfied. In general, if overall flow rates are assumed constant, we are satisfying the overall mass balance. If the flow rates of nontransferred species are constant, we are satisfying the balances for these species.

The energy balance is automatically satisfied in distillation when the CMO assumption is valid. In absorption, stripping, and extraction the energy balances were satisfied by assuming constant temperature and a negligible heat of absorption, stripping or mixing. In general, we will assume constant temperature and a negligible heat involved in contacting the two phases.

The Kremser equation was used for absorption, stripping, and extraction. When total flows, pressure, and temperature are constant and the heat of contacting the phases is negligible, we can use the Kremser equation if the equilibrium expression is linear. When these assumptions are valid, the Kremser equation can be used for other separations.

Of course, the assumptions required to use a McCabe-Thiele analysis or the Kremser equation may not be valid for a given separation. If the assumptions are not valid, the results of the analysis could be garbage. To determine the validity of the assumptions, the engineer has to examine each specific case in detail. The more dilute the solute, the more likely it is that the assumptions will be valid.

In the remainder of this chapter these principles will be applied to generalize the McCabe-Thiele approach and the Kremser equation for a variety of unit operations. A listing of various applications is given in Table 14-1.

14.2 WASHING

When solid particles are being processed in liquid slurries, the solids entrain liquid with them. The removal of any solute contained in this entrained liquid is called *washing*. To be specific, consider an operation that mines sand from the ocean. The wet sand contains salt, and this salt can be removed by washing with pure water. The entrained liquid is called underflow liquid, because the solids are normally removed from the bottom of a settler as shown in Figure 14-1A. Washing is done by mixing solid (sand) and wash liquor (water) together in a mixer and sending the mixture to a settler or a thickener (Perry and Green, 1997, p. 18-64). The solids and entrained underflow liquid exit from the bottom of the settler, and clear overflow liquid without solids is removed from the top. In washing, the solute (salt) is not held up or attached to the inert solid (sand). The salt is assumed to be at the same concentration in the underflow liquid as it is in the overflow liquid. Thus, it can be removed by displacing it with clear water. The separation can be done in single-stage, cross-flow, and countercurrent cascades. A variety of different equipment have been developed for washing and leaching, which use essentially the same equipment (Coulson et al., 1978).

TABLE 14-1. *Applications of McCabe-Thiele and Kremser procedures*

Unit Operation	Feed	Const. Flow	Conc. Units	Comments
Absorption (Chapt. 12)	Gas phase	L = kg solvent/h G = kg carrier gas/h	Y,X, mass ratios	Op line above equil line
Dilute abs. (Chapt. 12)	Gas phase	Total flow rates	y,x, fractions	McCabe-Thiele or Kremser $(y = mx + b)$
Stripping (Chapt. 12)	Liquid	L = kg solvent/h G = kg carrier gas/h	Y,X, mass ratios	Op line below equil line Heat effects may be important
Dilute stripping (Chapt. 12)	Liquid	Total flow rates	y,x, fractions	McCabe-Thiele or Kremser
Dilute extraction (Chapt. 13)	Raffinate	F_D = kg diluent/h F_S = kg solvent/h	$Y = \dfrac{\text{kg solute}}{\text{kg solvent}}$ $X = \dfrac{\text{kg solute}}{\text{kg diluent}}$	McCabe-Thiele Op line below equil line
Very dilute extract (Chapt. 13)	Raffinate	E = total extract kg/h R = total raffinate	y = wt frac in extract x = wt frac in raffinate	McCabe-Thiele or Kremser $(y = mx + b)$
Washing (Chapt. 14)	Solids + underflow liquid	U = underflow liquid O = overflow liquid, both kg/h	y = wt frac in overflow x = wt frac in underflow	Equilibrium is y = x Op line under equil line or use Kremser
Leaching (Chapt. 14)	Solids & solutes	F_{solv} = kg solvent/h F_{solid} = kg insoluble solids/h	$Y = \dfrac{\text{kg solute}}{\text{kg solvent}}$ $X = \dfrac{\text{kg solute}}{\text{kg solid}}$	Op line under equil line If very dilute, can use Kremser $(y = mx + b)$
Three-phase (Wankat, 1980)	One or two phases	L, W, V	y = fraction x' (see reference)	Pseudo-equilibrium Kremser for linear equil

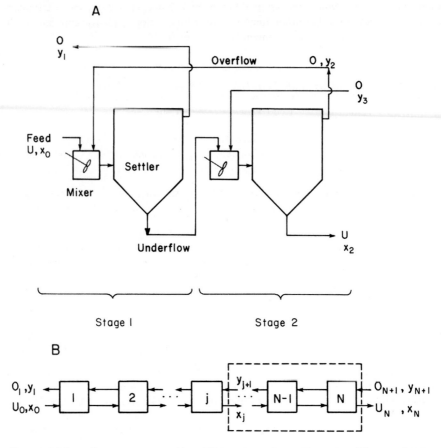

FIGURE 14-1.   *Countercurrent washing; (A) two-stage mixer-settler system, (B) general system*

The equilibrium condition for a washer is that solute concentration is the same in both the underflow and overflow liquid streams. This statement does not say anything about the solid, which changes the relative underflow and overflow flow rates but does not affect concentrations. Thus, the equilibrium equation is

$$y = x \tag{14-1}$$

where y = mass fraction solute in the overflow liquid and x = mass fraction solute in the underflow liquid.

For the mass balance envelope on the countercurrent cascade shown in Figure 14-1B, it is easy to write a steady-state mass balance.

$$O_{j+1} \, y_{j+1} + U_N \, x_N = O_{N+1} \, y_{N+1} + U_j \, x_j \tag{14-2}$$

Chapter 14   Washing, Leaching, and Supercritical Extraction

where O_j and U_j are the total overflow and underflow liquid flow rates in kg/h leaving stage j. The units for Eq. (13-30) are kg solute/h. To develop the operating equation, we solve Eq. (14-2) for y_{j+1}:

$$y_{j+1} = \frac{U_j}{O_{j+1}} x_j + \frac{O_{N+1}}{O_{j+1}} y_{N+1} - \frac{U_N}{O_{j+1}} x_N \qquad (14\text{-}3)$$

In order for this to plot as a straight line, the underflow liquid and overflow liquid flow rates must be constant.

Often the specifications will give the flow rate of dry solids or the flow rate of wet solids. The underflow liquid flow rate can be calculated from the volume of liquid entrained with the solids. Let ε be the porosity (void fraction) of the solids in the underflow. That is,

$$\varepsilon = \frac{\text{volume voids}}{\text{total volume}} = \frac{\text{volume liquid}}{\text{total volume}} \qquad (14\text{-}4a)$$

and then

$$1 - \varepsilon = \frac{\text{volume solids}}{\text{total volume}} \qquad (14\text{-}4b)$$

We can now calculate the underflow liquid flow rate, U_j. Suppose we are given the flow rate of dry solids. Then the volume of solids per hour is

$$\left(\frac{\text{kg dry solids}}{\text{h}}\right)\left(\frac{1}{\rho_s kg/m^3}\right) = m^3 \text{ dry solids/h} \qquad (14\text{-}5a)$$

and the total volume of underflow is

$$\left(\frac{\text{kg dry solids}}{\text{h}}\right)\left(\frac{1}{\rho_s \dfrac{kg}{m^3}}\right)\frac{1}{(1-\varepsilon)\left(\dfrac{m^3 \text{ solids}}{m^3 \text{ underflow}}\right)} = \frac{m^3 \text{ total underflow}}{h} \qquad (14\text{-}5b)$$

Then the volume of underflow liquid in m^3 liquid per hour is

$$\left(\frac{\text{kg dry solids}}{\text{h}}\right)\left(\frac{1}{\rho_s \dfrac{kg}{m^3}}\right)\left(\frac{1}{(1-\varepsilon)\left(\dfrac{m^3 \text{ solids}}{m^3 \text{ underflow}}\right)}\right)\left(\varepsilon \frac{m^3 \text{ liquid}}{m^3 \text{ total underflow}}\right) \qquad (14\text{-}5c)$$

and finally the kg/h of underflow liquid

$$U_j = (\text{rate dry solids, } \frac{kg}{h})\left(\frac{\varepsilon}{1-\varepsilon}\right)\frac{\rho_f}{\rho_s} \qquad (14\text{-}5d)$$

In these equations ρ_f is the fluid density in kg/m^3 and ρ_s is the density of dry solids in kg/m^3.

If the solids rate, ε, ρ_f, and ρ_s are all constant, then from Eq. (14-5d) $U_j = U = $ constant. If U is constant, then an overall mass balance shows that the overflow rate, O_j, must also be constant. Thus, to have constant flow rates we assume:

1. No solids in the overflow and solids do not dissolve. This ensures that the solids flow rate will be constant.
2. ρ_f and ρ_s are constant. Constant ρ_f implies that the solute has little effect on fluid density or that the solution is dilute.
3. Porosity ε is constant. Thus, the volume of liquid entrained from stage to stage is constant.

When these assumptions are valid, O and U are constant, and the operating equation simplifies to

$$y_{j+1} = \frac{U}{O} \, x_j + \left(y_{N+1} - \frac{U}{O} \, x_N\right) \tag{14-6}$$

which obviously represents a straight line on a McCabe-Thiele plot. Note that this equation is similar to all the other McCabe-Thiele operating equations we have developed. Only the nomenclature has changed.

An alternative way of stating the problem would be to specify the volume of *wet* solids processed per hour. Then the underflow volume is

$$\text{Underflow liquid volume, m}^3/\text{h} = (\text{volume wet solids}/\text{h}) \, \varepsilon \tag{14-7a}$$

and

$$U = (\text{Volume wet solids}/\text{h}) \, \varepsilon \, \rho_f \tag{14-7b}$$

If densities and ε are constant, volumetric flow rates are constant, and the washing problem can be solved using volumetric flow rates and concentrations in kg solute/m^3.

Note that we could have just assumed that overflow and underflow rates are constant and derived Eq. (14-6). However, it is much more informative to show the three assumptions required to make overflow and underflow rates constant. These assumptions show that this analysis for washing is likely to be invalid if the settlers are not removing all the solid, if for some reason the amount of liquid entrained changes, or if the fluid density changes markedly. The first two problems will not occur in well-designed systems. The third is easy to check with density data.

The McCabe-Thiele diagram can now be plotted as shown in Figure 14-2. This McCabe-Thiele diagram is unique, since temperature and pressure do not affect the equilibrium; however, temperature will affect the rate of attaining equilibrium and hence the efficiency, because at low temperatures more viscous solutions will be difficult to wash off the solid.

The analysis for washing can be extended to a variety of modifications. These include simulation problems, use of efficiencies, calculation of maximum U/O ratios, and calculations for cross-flow systems. The Kremser equation can also be applied to countercurrent washing with no additional assumptions. This adaptation is a straightforward translation of nomenclature and is illustrated in Example 14-1. Brian (1972) discusses application of the Kremser equation to washing in considerable detail.

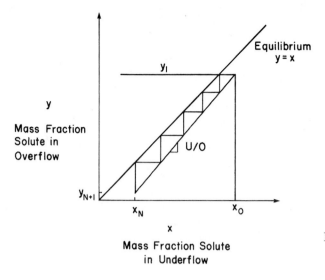

y_I

y

Mass Fraction
Solute in
Overflow

Equilibrium
y = x

U/O

y_{N+1}

x_N x_O

x

Mass Fraction Solute
in Underflow

FIGURE 14-2.   *McCabe-Thiele diagram for washing*

Washing is also commonly done by collecting the solids on a filter and then washing the filter cake. This approach, which is often used for crystals and precipitates that may be too small to settle quickly and is often a batch operation, is discussed by Mullin (2001) and Harrison et al. (2003).

EXAMPLE 14-1. Washing

In the production of sodium hydroxide by the lime soda process, a slurry of calcium carbonate particles in a dilute sodium hydroxide solution results. A four-stage countercurrent washing system is used. The underflow entrains approximately 3 kg liquid/kg dry calcium carbonate solids. The inlet water is pure water. If 8 kg wash water/kg dry calcium carbonate solids is used, predict the recovery of NaOH in the wash liquor.

Solution

A. Define. Recovery is defined as $1 - x_{out}/x_{in}$. Thus, recovery can be determined even though x_{in} is unknown.

B and C.   Explore and plan. If we pick a basis of 1 kg dry calcium carbonate/h, then O = 8 kg wash water/h and U = 3 kg/h. This problem can be solved with the Kremser equation if we translate variables. To translate: Since y = overflow liquid weight fraction, we set O = V. Then U = L. this translation keeps y = mx as the equilibrium expression. It is convenient to use the Kremser equation in terms of x. For instance, Eq. (12-31) becomes

$$\frac{x_N - x_N^*}{x_0 - x_N^*} = \frac{1 - mO/U}{1 - (mO/U)^{N+1}} \qquad \textbf{(14-8)}$$

D. Do it. Equilibrium is y = x; thus, m = 1. Since inlet wash water is pure, y_{N+1} = 0. Then $x_N^* = y_{N+1}/m = 0$, m O/U = (1)(8)/3, and N = 4. Then Eq. (14-8) is

$$\frac{x_N}{x_0} = \frac{1 - 8/3}{1 - (8/3)^5} = 0.01245$$

and

$$\text{Recovery} = 1 - x_N/x_0 = 0.98755$$

E. Check. This solution can be checked with a McCabe-Thiele diagram. Since the x_N value desired is known, the check can be done without trial and error.

F. Generalize. Recoveries for linear equilibrium can be determined without knowing the inlet concentrations. This can be useful for the leaching of natural products because the inlet concentration fluctuates. The translation of variables shown here can be applied to other forms of the Kremser equation.

The washing analysis presented here is for a steady state, completely mixed system where the wash water and the water entrained by the solid matrix are in equilibrium. When filter cakes are washed, the operation is batch, the system is not well mixed because flow is close to plug flow, and the operation is not at equilibrium since the entrained fluid has to diffuse into the wash liquid. This case is analyzed by Harrison et al. (2003).

14.3   LEACHING WITH CONSTANT FLOW RATES

Leaching, or solid-liquid extraction, is a process in which a soluble solute is removed from a solid matrix using a solvent to dissolve the solute. The most familiar examples are making coffee from ground coffee beans and tea from tea leaves. The complex mixture of chemicals that give coffee and tea their odor, taste, and physiological effects are leached from the solids by the hot water. An espresso machine just does the leaching faster into a smaller volume of water. Instant coffee and tea can be made by leaching ground coffee beans or tea leaves with hot water and then drying the liquid to produce a solid. There are many other commercial applications of leaching such as leaching soybeans to recover soybean oil (a source of biodiesel), leaching ores to recover a variety of minerals, and leaching plant leaves to extract a variety of pharmaceuticals (Rickles, 1965; Schwartzberg, 1980, 1987).

The equipment and operation of washing and leaching systems are often very similar. In both cases a solid and a liquid must be contacted, allowed to equilibrate, and then separated from each other. Thus, the mixer-settler type of equipment shown in Figure 14-1 is also commonly used for leaching easy-to-handle solids. A variety of other specialized equipment has been developed to move the solid and liquid countercurrently during leaching. Coulson et al. (1978), Lydersen (1983), McNulty (2008), Miller (1997), Prabhudesai (1997), and Schwartzberg (1980, 1987) present good introductions to this leaching equipment.

In leaching, the solute is initially part of the solid and dissolves into the liquid. In washing, which can be considered as a special case of leaching, the solute is initially retained in the pores of the solid and the solid itself does not dissolve. In leaching, the equilibrium equation is usually not y = x, and the total solids flow rate is usually not constant. Since diffusion rates in a solid are low, mass transfer rates are low. Thus, equilibrium may take days for large pieces such as pickles, where it is desirable to leach out excess salt, or even years for in-situ leaching of copper ores (Lydersen, 1983). A rigorous analysis of leaching requires that the changing solid and liquid flow rates be included. This situation is very similar to partially miscible extraction and is included in Section 14.4. In this section we will look at simple cases where a modified McCabe-Thiele or Kremser equation can be used.

A countercurrent cascade for leaching is shown in Figure 14-3A. We will consider the (idealized) case where entrainment of liquid with the solid underflow can be ignored. The assumptions are:

1. The system is isothermal.
2. The system is isobaric.
3. No solvent dissolves into solid.
4. No solvent entrained with the solid.
5. There is an insoluble solid backbone or matrix.
6. The heat of mixing of solute in solvent is negligible.
7. The stages are equilibrium stages.
8. No solid is carried with the overflow liquid.

With these assumptions the energy balance is automatically satisfied. A straight operating line is easily derived using the mass balance envelope shown in Figure 14-3A. Defining ratios and flow rates

$$Y = \frac{\text{kg solute}}{\text{kg solvent}}, X = \frac{\text{kg solute}}{\text{kg insoluble solid}}$$

$$F_{solv} = \text{kg solvent/h}, F_{solid} = \text{kg insoluble solid/h} \qquad \textbf{(14-9)}$$

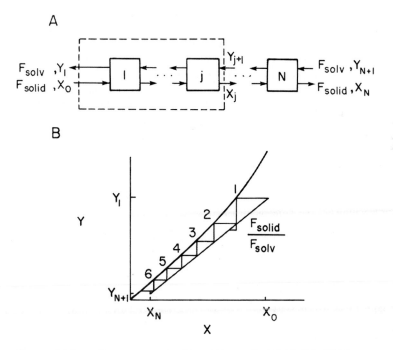

FIGURE 14-3. *Countercurrent leaching; (A) cascade, (B) McCabe-Thiele diagram*

the operating equation is

$$Y_{j+1} = \frac{F_{solid}}{F_{solv}} X_j + Y_1 - \frac{F_{solid}}{F_{solv}} X_o \qquad (14\text{-}10)$$

This represents a straight line as plotted in Figure 14-3B. The equilibrium curve is now the equilibrium of the solute between the solvent and solid phases. The equilibrium data must be measured experimentally. If the equilibrium line is straight, the Kremser equation can be applied.

In the previous analysis, assumptions 4 and 7 are often faulty. There is always entrainment of liquid in the underflow (for the same reason that there is an underflow liquid in washing). Since diffusion in solids is very slow, equilibrium is seldom attained in real processes. The combined effects of entrainment and nonequilibrium stages are often included by determining an "effective equilibrium constant." This effective equilibrium depends on flow conditions and residence times and is valid only for the conditions at which it was measured. Thus, the effective equilibrium constant is not a fundamental quantity. However, it is easy to measure and use. The McCabe-Thiele diagram will look the same as Figure 14-3B. Further simplification is obtained by assuming that the effective equilibrium is linear, $y = m_E x$.

14.4   LEACHING WITH VARIABLE FLOW RATES

Leaching, also known as solid-liquid extraction (SLE), and liquid-liquid extraction (LLE) have very different hydrodynamic and mass transfer characteristics. However, equilibrium staged analysis is almost identical for the two processes because it is not affected by these differences. In this section we will briefly consider the analysis of leaching systems using triangular diagrams (Coulson et al., 1978). In leaching the flow rates will generally not be constant. The variation in flow rates can be included by doing the analysis on a triangular diagram. A very similar technique uses ratio units in a Ponchon-Savarit diagram (Lydersen, 1983; McCabe et al., 2005; McNulty, 2008; Miller, 1997; Prabhudesai, 1997; Treybal, 1980).

A schematic of a countercurrent leaching system is shown in Figure 14-4 with the appropriate nomenclature. Since leaching is quite similar to LLE, the same nomenclature is used (see Table 13-2). Even if flow rates E and R vary, it is easy to show that the differences in total and component flow rates for passing streams are constant. Thus, we can define the difference point from these differences. This was done in Eqs. (13-42) and (13-43) for the LLE cascade of Figure 13-20. Since the cascades are the same (compare Figures 13-20 and 14-4), the results for leaching are the same as for LLE.

The calculation procedure for countercurrent leaching operations is exactly the same as for LLE:

1. Plot the equilibrium data.
2. Plot the locations of known points.
3. Find mixing point M.
4. Locate E_N.
5. Find the Δ point.
6. Step off stages.

This procedure is illustrated in Example 14-2.

Chapter 14   Washing, Leaching, and Supercritical Extraction

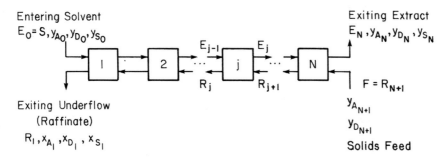

Entering Solvent
$E_0 = S, y_{A_0}, y_{D_0}, y_{S_0}$

Exiting Extract
$E_N, y_{A_N}, y_{D_N}, y_{S_N}$

Exiting Underflow
(Raffinate)
$R_1, x_{A_1}, x_{D_1}, x_{S_1}$

$F = R_{N+1}$
$y_{A_{N+1}}$
$y_{D_{N+1}}$
Solids Feed

FIGURE 14-4.   *Countercurrent leaching cascade and nomenclature*

The equilibrium data for leaching must be obtained experimentally since it will depend on the exact nature of the solids, which may change from source to source. If there is no entrainment, the overflow (extract) stream will often contain no inert solids (diluent). However, the raffinate stream will contain solvent. Test data for the extraction of oil from meal with benzene are given in Table 14-2 (Prabhudesai, 1997). In these data, inert solids are not extracted into the benzene. The data are plotted on a triangular diagram in Figure 14-5 (see Example 14-2). The conjugate line is constructed in the same way as for extraction.

EXAMPLE 14-2. Leaching calculations

We wish to treat 1000 kg/h (wet basis) of meal (D) that contains 0.20 wt frac oil (A) and no benzene(S). The inlet solvent is pure benzene and flows at 662 kg/h. We desire an underflow product that is 0.04 wt frac oil. Temperature and pressure are constant, and the equilibrium data are given in Table 14-2. Find the outlet extract concentration and the number of equilibrium stages needed in a countercurrent leaching system.

TABLE 14-2.   *Test data for extraction of oil from meal with benzene*

Mass Fraction Oil (Solute) in Solution	Mass Fraction Underflow (Raffinate)		
y_A	x_A	x_D	x_S
0	0	0.670	0.333
0.1	0.0336	0.664	0.302
0.2	0.0682	0.660	0.272
0.3	0.1039	0.6541	0.242
0.4	0.1419	0.6451	0.213
0.5	0.1817	0.6366	0.1817
0.6	0.224	0.6268	0.1492
0.7	0.268	0.6172	0.1148

Source: Prabhudesai (1997)

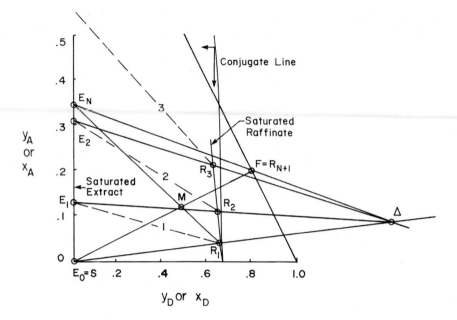

FIGURE 14-5. *Solution to leaching problem, Example 14-2*

Solution

A. Define. The system is similar to that of Figure 14-4 with streams E_0 and R_{N+1} specified. In addition, weight fraction $x_{A,1} = 0.04$ and is a saturated raffinate. We wish to find the composition of stream E_N and the number of equilibrium stages required.

B. Explore. This looks like a straightforward leaching problem, which can be solved like the corresponding extraction problem.

C. Plan. We will plot the equilibrium diagram on a scale that allows the Δ point to fit on the graph. Then we will plot points E_0, R_{N+1}, and R_1. We will use the lever-arm rule to find point M and then find point E_N. Next we find Δ and finally we step off the stages.

D. Do it. The diagram in Figure 14-5 shows the equilibrium data and the points that have been plotted. Point M was found along the line $E_0 R_{N+1}$ from Eq. (13-40a).

$$x_{A,M} = \frac{E_0 y_{A,0} + F x_{A,N+1}}{E_0 + F} = \frac{0 + (1000)(0.2)}{662 + 1000} = 0.120$$

Then points E_N and Δ were found as shown in Figure 14-5. Finally, stages were stepped off in exactly the same way as for a triangular diagram for extraction. The exit extract concentration is 0.305 wt frac oil, and 3 stages are more than enough. Two stages are not quite enough. Approximately 2.1 stages are needed.

E. Check. The outlet extract concentration can be checked with an overall mass balance. The number of stages could be checked by solving the problem using another method.

F. Generalize. Since the leaching example was quite similar to LLE, we might guess that the other calculation procedures developed for extraction would also be valid for leaching. Since this is true, there is little reason to reinvent the wheel and rederive all the methods.

McCabe-Thiele diagrams can also be used for leaching if flow rates are close to constant. If flow rates are not constant, curved operating lines can be constructed on the McCabe-Thiele diagram with the triangular diagram.

14.5 SUPERCRITICAL FLUID EXTRACTION

There has been an increasing amount of interest in the use of supercritical fluids (SCF) for leaching or extracting compounds from solids or liquids in the food and pharmaceutical industries because of the nontoxic nature of the primary SCF—carbon dioxide. In this section we will briefly consider the properties of SCFs that make them interesting for extraction. Then a typical process for SCF extraction will be explored and several applications will be discussed.

First, what is an SCF? Figure 14-6A shows a typical pressure-temperature diagram for a single component. Above the critical temperature T_c, it is impossible to liquefy the compound. The critical pressure p_c is the pressure required to liquefy the compound at the critical temperature. The critical temperatures and pressures for a large number of compounds have been determined (Paulaitis et al., 1983; Poling et al., 2001). SCFs of interest include primarily carbon dioxide ($p_c = 72.8$ atm, $T_c = 31°C$, $\rho = 0.47$ g/mL), but also propane ($p_c = 41.9$ atm, $T_c = 97°C$, $\rho = 0.22$ g/mL), and water ($p_c = 217.7$ atm, $T_c = 374°C$, $\rho = 0.32$ g/mL). An SCF behaves like a gas in that it will expand to fill the confines of the container.

As can be seen from this very short list, SCFs have densities much greater than those of typical gases and less than those of liquids by roughly a factor of 2 to 3. The viscosities of SCFs are about one-tenth those of liquids. This leads to low-pressure drops. The high diffusivities, roughly ten times those of liquids, plus the lack of a phase boundary leads to very high mass transfer rates and low HETP values in packed beds. The SCFs can often dissolve almost the same amount of solute as a good liquid solvent. Extraction can often be carried out at low temperatures, particularly when CO_2 is the SCF. This is particularly advantageous for extraction of foods and pharmaceuticals. Many SCFs are also completely natural or "green" and thus, are totally acceptable as additives in foods and pharmaceuticals (Allen and Shonnard, 2002). This is a major advantage to the use of supercritical CO_2 without additives.

The solubility of a solute in an SCF is a complex function of temperature and pressure. This is commonly illustrated with the solubility of naphthalene in CO_2, which is illustrated in Figure 14-6B (Hoyer, 1985; Paulaitis et al., 1983). As pressure is increased the solubility first decreases and then increases. At both high and low pressures the naphthalene is more soluble at high temperatures than at low temperatures. This is the expected behavior, because the vapor pressure of naphthalene increases with increasing temperature. Immediately above the critical pressure the solute is more soluble at the lower temperature. This is a *retrograde* phenomenon. If naphthalene solubility is plotted vs. CO_2 density, the retrograde behavior does not appear. (Gupta and Johnston, 2008). In addition to having high solubilities, the SCF

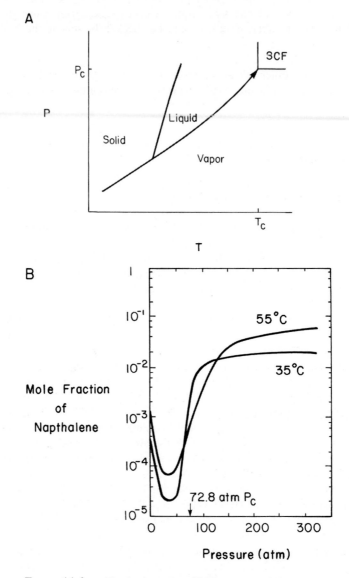

FIGURE 14-6. *Thermodynamics of SCF extraction; (A) pressure-temperature diagram for pure component, (B) solubility of naphthalene in CO_2*

should be selective for the desired solutes. Solute-solvent interactions can affect the solubility and the selectivity of the SCF, and therefore entrainers are often added to the SCF to increase solubility and selectivity.

Figure 14-6B also shows how the solute can be recovered from the CO_2. If the pressure is dropped, the naphthalene solubility plummets, and naphthalene will drop out as a finely divided solid. A typical process using pressure reduction is shown in Figure 14-7. Note that

Chapter 14   Washing, Leaching, and Supercritical Extraction

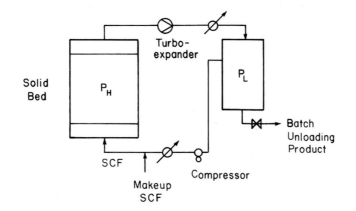

FIGURE 14-7.   *Batch SCF extraction; regeneration is by pressure swing*

this will probably be a batch process if solids are being processed because of the difficulty in feeding and withdrawing solids at the high pressures of supercritical extraction. Regeneration can also be achieved by changing the temperature, distilling the CO_2-solute mixture at high pressure or absorbing the solute in water (McHugh and Krukonis, 1994). In many cases one of these processes may be preferable because it will decrease the cost of compression.

Several current applications have been widely publicized. Kerr-McGee developed its ROSE (Residuum Oil Supercritical Extraction) process in the 1950s (Gupta and Johnston, 2008; Humphrey and Keller, 1997; Johnston and Lemert, 1997; McHugh and Krukonis, 1994). When oil prices went up, the process attracted considerable attention, since it has lower operating costs than competing processes. The ROSE process uses an SCF such as propane to extract useful hydrocarbons from the residue left after distillation. This process utilizes the high temperatures and pressures expected for residuum treatment to lead naturally to SCF extraction.

Much of the commercial interest has been in the food and pharmaceutical industries. Here, the major driving force is the desire to have completely "natural" processes, which cannot contain any residual hydrocarbon or chlorinated solvents (Humphrey and Keller, 1997). Supercritical carbon dioxide has been the SCF of choice because it is natural, nontoxic, and cheap, is completely acceptable as a food or pharmaceutical ingredient, and often has good selectivity and capacity. Currently, supercritical CO_2 is used to extract caffeine from green coffee beans to make decaffeinated coffee. Supercritical CO_2 is also used to extract flavor compounds from hops to make a hop extract that is used in beer production. The leaching processes that were replaced were adequate in all ways except that they used solvents that were undesirable in the final product.

A variety of other SCF extraction processes have been explored (Gupta and Johnston, 2008; Hoyer, 1985; McHugh and Krukonis, 1994; Paulaitis et al., 1983). These include extraction of oils from seeds such as soybeans, removal of excess oil from potato chips, fruit juice extraction, extraction of oxygenated organics such as ethanol from water, dry cleaning, removal of lignite from wood, desorption of solutes from activated carbon, and treatment of hazardous wastes. Not all of these applications were successful, and many that were technically successful are not economical.

The main problems in applying SCF extraction on a large scale have been scaling up for the high pressure required. The high-pressure equipment becomes quite heavy and expensive. In addition, methods for charging and discharging solids continuously have not been well developed for these high-pressure applications. Another problem has been the lack of design data for supercritical extraction. Supercritical extraction is not expected to be a cheap process. Thus, the most likely applications are extractions for which existing separation methods have at least one serious drawback and for which SCF extraction does not have major processing disadvantages.

14.6 APPLICATION TO OTHER SEPARATIONS

The McCabe-Thiele and Kremser methods can be applied to analyze other separation processes. Adsorption, chromatography, and ion exchange are occasionally operated in counter current columns. In some situations crystallization can be analyzed as an equilibrium stage separation. The application of the McCabe-Thiele procedure in these cases is explored by Wankat (1990).

The McCabe-Thiele and Kremser methods have also been applied to analyze less common separation methods. A modification of the McCabe-Thiele method has been applied to parametric pumping, which is a cyclic adsorption or ion exchange process (Grevillot and Tondeur, 1977). A similar modification can be used to analyze cycling zone adsorption (Wankat, 1986). McCabe-Thiele and Kremser methods can also be used to analyze three-phase separations (Wankat, 1980).

14.7 SUMMARY—OBJECTIVES

In this chapter we looked at the general applicability of the McCabe-Thiele and Kremser analysis procedures and applied them to washing and leaching. The methods are reviewed in Table 14-1. At the end of this chapter you should be able to achieve the following objectives:

1. Explain in general terms how the McCabe-Thiele and Kremser analyses can be applied to other separation schemes and delineate when these procedures are applicable
2. Explain what washing is and apply the McCabe-Thiele and Kremser procedures to washing problems
3. Explain what leaching is and apply both McCabe-Thiele and Kremser methods to leaching problems
4. Apply the triangular diagram analysis to leaching problems with variable flow rates.
5. Explain how supercritical extraction works, and discuss its advantages and disadvantages

REFERENCES

Allen, D. T., and D. R. Shonnard, *Green Engineering: Environmentally Conscious Design of Chemical Processes,* Prentice Hall, Upper Saddle River, NJ, 2002.

Brown, G. G., and Associates, *Unit Operations,* Wiley, New York, 1950.

Coulson, J. M., J. F. Richardson, J. R. Backhurst, and J. H. Harker, *Chemical Engineering*, Vol. 2, 3rd ed., Pergamon Press, Oxford, 1978, Chap. 10.

Grevillot, G., and D. Tondeur, "Equilibrium Staged Parametric Pumping. II. Multiple Transfer Steps per Half Cycle and Reservoir Staging," *AIChE J., 23*, 840 (1977).

Gupta, R. B., and K. P. Johnston, "Supercritical Fluid Separation Processes," in D. W. Green and R. H. Perry (Eds.), *Perry's Chemical Engineers' Handbook*, 8th ed., McGraw-Hill, New York, 2008, pp. 20–14 to 20–19.

Harrison, R. G., P. Todd, S. R. Rudge, and D. P. Petrides, *Bioseparations Science and Engineering*, Oxford University Press, New York, 2003.

Hoyer, G. G., "Extraction with Supercritical Fluids: Why, How, and So What," *Chemtech*, 440 (July 1985).

Humphrey, J. L., and G. E. Keller II, *Separation Process Technology*, McGraw-Hill, New York, 1997.

Johnston, K. P., and R. M. Lemert, "Supercritical Fluid Separation Processes," in R. H. Perry and D. W. Green, Eds., *Perry's Chemical Engineers' Handbook*, 7th ed., McGraw-Hill, New York, 1997, pp. 22–14 to 22–19.

Lyderson, A. L., *Mass Transfer in Engineering Practice*, Chichester, UK, Wiley, 1983.

McCabe, W. L., J. C. Smith, and P. Harriott, *Unit Operations of Chemical Engineering*, 7th ed., McGraw-Hill, New York, 2005.

McHugh, M. A., and V. J. Krukonis, *Supercritical Fluid Extraction: Principles and Practice*, 2nd ed., Butterworth-Heinemann, Boston, 1994.

McNulty, T. P., "Leaching," in D. W. Green and R. H. Perry (Eds.), *Perry's Chemical Engineers' Handbook*, 8th ed., McGraw-Hill, New York, 2008, pp. 18–59 to 18–66.

Miller, S. A., "Leaching," in R. H. Perry and D. W. Green, Eds., *Perry's Chemical Engineers' Handbook*, 7th ed., McGraw-Hill, New York, 1997, pp. 18–55 to 18–59.

Mullin, J. W., *Crystallization*, 4th ed., Oxford & Butterworth-Heinemann, Boston, 2001.

Paulaitis, M. E., J. M. L. Penniger, R. D. Gray, Jr, and P. Davidson, *Chemical Engineering at Supercritical Fluid Conditions*, Butterworths, Boston, 1983.

Perry, R. H., and D. Green (Eds.), *Perry's Chemical Engineer's Handbook*, 6th ed., McGraw-Hill, New York, 1984.

Poling, B. E., J. M. Prausnitz, and J. P. O'Connell, *The Properties of Gases and Liquids*, 5th ed., McGraw-Hill, New York, 2001.

Prabhudesai, R. K., "Leaching," in P.A. Schweitzer (Ed.), *Handbook of Separation Techniques for Chemical Engineers*, McGraw-Hill, New York, 1997, Section 5.1.

Rickles, R. H., "Liquid-Solid Extraction," *Chem. Eng., 72*, 157 (March 15, 1965).

Schwartzberg, H. G., "Continuous Countercurrent Extraction in the Food Industry," *Chem. Eng. Progress, 76* (4), 67 (April 1980).

Schwartzberg, H. G., "Leaching-Organic Materials," in R. W. Rousseau (Ed.), *Handbook of Separation Process Technology*, Wiley, New York, 1987, Chap. 10.

Treybal, R. E., *Mass Transfer Operations*, 3rd ed., McGraw-Hill, New York, 1980.

Wankat, P. C., "Calculations for Separations with Three Phases: 1. Staged Systems," *IEC Fundam. 19*, 358–363 (1980).

Wankat, P. C., *Large-Scale Adsorption and Chromatography*, CRC Press, Boca Raton, FL, 1986.

Wankat, P. C., *Mass-Transfer Limited Separations*, Kluwer, Amsterdam, 1990.

HOMEWORK

A. *Discussion Problems*

A1. Develop your key relations chart for this chapter. Remember that a key relations chart is *not* a core dump but is selective.

A2. How do the ideas of a general McCabe-Thiele procedure and the concept of unit operation relate to each other?

A3. In the unit operation called Washing (select the best answer):

    a. the solute is physically held to the solid particles

    b. at equilibrium the underflow liquid has the same solute concentration as the overflow liquid

    c. the underflow liquid is the liquid entrained with the particles

    d. all of the above

    e. a and b

    f. a and c

    g. b and c

    h. None of the above.

A4. In leaching the final saturated raffinate often contains a significant amount of solvent. This is the case for Example 14-3 and for Problems 14.E1 to 14.E3. How do you recover this solvent?

A5. How does the solid enter into washing calculations? Where does solids flow rate implicitly appear in Figure 14-1?

A6. Referring to Table 14-1, list similarities and differences between absorption, stripping, extraction, washing, and leaching.

A7. Show how Figure 14-7 could be modified to use a temperature swing instead of a pressure swing. What might be the advantage and disadvantage of doing this?

A8. What are some of the properties you would look for in a good solvent for extraction, leaching and supercritical extraction?

A9. Would you expect stage efficiencies to be higher or lower in leaching than in LLE? Explain.

C. *Derivations*

C1. Chapters 12 to 14 cover separations that use a mass-separating agent. Derive general operating and equilibrium equations for the separations in these chapters.

C2. Adapt the Kremser equation to leaching.

C3. Derive Eq. (14-10).

C4. Develop the procedures for single-stage and cross-flow systems for leaching using a triangular diagram.

C5. Batch washing will be similar to batch extraction except in continuous addition batch washing (Figure 13-11) there is no need to pre-saturate the wash liquid with the solid and a filter is used instead of a settler. If the terms are translated the same equations can be used for batch washing.

    a. Convert Eqs. (13-24b) and (13-21) to the equations for batch washing.

    b. Convert Eq. (13-28) to the equation for batch washing with continuous addition of wash liquid.

D. *Problems*

Answers to problems with an asterisk are at the back of the book.

D1. We plan to wash dilute sulfuric and hydrochloric acids from crushed rock in a counter current system. Operation is at 25°C and one atmosphere. 100.0 m³/day of wet rock are to be washed. After settling, the porosity is constant at 0.40. Thus, the underflow rate is 40.0 m³/day. The initial concentration of the underflow liquid is 1.0 kg sulfuric acid/m³ and 0.75 kg hydrochloric acid/m³. We desire an outlet underflow sulfuric acid concentration of 0.09 kg/m³. The wash liquid (overflow) rate is 50.0 m³/day, and the inlet wash water is pure. For these dilute solutions assume that the solution densities are constant, and are the same as pure water, 1000.0 kg/m³. Find:

a. The number of equilibrium stages required

b. The outlet concentration (kg/m³) of hydrochloric acid in the underflow liquid

D2.* You are working on a new glass factory near the ocean. The sand is to be mined wet from the beach. However, the wet sand carries with it seawater entrained between the sand grains. Several studies have shown that 40% by volume seawater is consistently carried with the sand. The seawater is 0.035 wt frac salt, which must be removed by a washing process.

Densities: Water, 1.0 g/cm³ (assume constant); Dry sand, 1.8 g/cm³ (including air in voids); Dry sand without air, 1.8/0.6 = 3.0 g/cm³.

a. We desire a final wet sand product in which the entrained water has 0.002 wt frac salt. For each 1000 cm³ of wet sand fed we will use 0.5 kg of pure wash water. In a countercurrent washing process, how many stages are required? What is the outlet concentration of the wash water?

b. In a cross-flow process we wish to use seven stages with 0.2 kg of pure wash water added to each stage for each 1000 cm³ of wet sand fed. What is the outlet concentration of the water entrained with the sand?

D3. Your boss has looked at the results of Example 14-1. He now asks you *what if* a four-stage cross-flow system is used instead of the countercurrent system. The pure wash water will be divided equally between the four stages. Thus, overhead flow is 2 kg wash water/kg calcium carbonate on each stage. Determine the recovery of NaOH and compare to the result in Example 14-1.

D4.* We wish to wash an alumina solids to remove NaOH from the entrained liquid. The underflow from the settler tank is 20 vol % solid and 80 vol % liquid. The two solid feeds to the system are also 20 vol % solids. In one of these feeds, NaOH concentration in the liquid is 5 wt %. This feed's solid flow rate (on a dry basis) is 1000 kg/hr. The second feed has a NaOH concentration in the liquid of 2 wt %, and its solids flow rate (on a dry basis) is 2000 kg/h. We desire the final NaOH concentration in the underflow liquid to be 0.6 wt % (0.006 wt frac) NaOH. A countercurrent operation is used. The inlet washing water is pure and flows at 4000 kg/h. Find the optimum feed location for the intermediate feed and the number of equilibrium stages required.

Data: $\rho_w = 1.0$ kg/liter (constant), $\rho_{alumina} = 2.5$ kg/liter (dry crushed)

D5. After reading the solution to Problem 14.D4, your boss thinks it will be just as good to take the two feeds and combine them rather than keeping the feeds separate. Calculate the number of equilibrium stages required to achieve the same outlet concentrations with the same flow rates if the two feeds are combined before being fed to the washing cascade. Compare with the answer to 14.D4.

D6. We are removing barium sulfide BaS from insoluble solids (see Problem 14.D14 for data) in a batch washing process. We initially have 10.0 kg of wet solids (each kg dry insoluble solids carries with it 1.5 kg underflow liquid). The initial underflow liquid is an aqueous solution with 0.05 mass fraction BaS. The wet solids are mixed with 10 kg of pure water, the solids are allowed to settle (each kg dry insoluble solids carries with it 1.5 kg underflow liquid), and the overflow liquid is removed. This process is then repeated twice using 10 kg of pure wash water each time. Assume that liquid density is constant and that the underflow liquid held by the solids has the same concentration of BaS as the overflow liquid.

a. Find mass fractions of BaS in the overflow and the underflow liquids after the first, second, and third washing steps.

b. If the operation is done continuously with 10.0 kg/h of wet solids feed in a 3 stage cross-flow system using 10.0 kg/h of pure wash water in each stage, find the mass fractions of BaS in the overflow and underflow liquids from each of the three stages.

c. If a continuous countercurrent 3 stage process is used with a total of 30 kg/h of wash water and 10.0 kg of wet solids/h, find the outlet underflow and overflow mass fractions of BaS.

D7.* You are working on a new glass factory near the ocean. The sand is to be mined wet from the beach. However, the wet sand carries with it seawater entrained between the sand grains. Data are given in Problem 14.D2. The salt must be removed by a washing process. A cross-flow process will be employed, with 0.2 kg of wash water added to each stage for each 1000 cm³ of wet sand fed. The wash water outlet from the last stage will be used as the wash water inlet for stage 3. Wash water outlet from stage N–1 will be used as wash inlet for stage 2, and wash water outlet from stage N–2 as wash water inlet to stage 1. All other stages have pure wash water inlet (see figure). We desire an outlet concentration of less than 0.002 wt frac salt in the entrained liquid. What is the minimum number of stages required to obtain this concentration? (This is *not* a minimum wash water flow rate problem. This specification means you don't have to obtain exactly 0.002 with an integer number of stages.) Note: This problem is *not* trial and error.

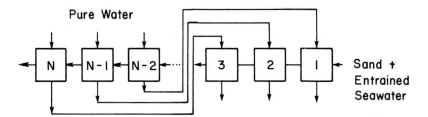

D8.* In the leaching of sugar from sugar cane, water is used as the solvent. Typically about 11 stages are used in a countercurrent Rotocel or other leaching system. On a volumetric basis liquid flow rate/solid flow rate = 0.95. The effective equilibrium constant is $m_E = 1.18$, where m_E = (concentration, g/liter, in liquid)/(concentration, g/liter, in solid) (Schwartzberg, 1980). If pure water is used as the inlet solvent, predict the recovery of sugar in the solvent.

D9. Experimental data for the leaching of sugar from sugar cane with water shows that a reasonable value for the effective equilibrium constant $y/x = m_E$ is 1.18 where y and x are the solute concentration in g/(L liquid) and g/(L solid), respectively (Schwartzberg, 1980). Batch leaching will be similar to batch extraction, and if the terms are translated the same equations can be used. We are doing a single stage batch leaching of sugar cane with water to recover sugar. The feed to the washing process is 1.0 kg of solids (on a dry basis). In this feed the cane contains 5.5 wt % sugar. The inlet solvent is pure water. Assume overflow and underflow amounts are constant.

Chapter 14 Washing, Leaching, and Supercritical Extraction

a. If 3.0 kg of wash liquid are used in a normal batch leaching operation, what are the outlet overflow liquid weight fraction sugar (value y) and the weight fraction sugar in the solid (value x)?

b. If we want the outlet overflow liquid to be 0.4 wt % sugar, how many kg of wash water are required in a normal batch leaching operation?

D10.* The use of slurry adsorbents has received some industrial attention because it allows for countercurrent movement of the solid phase. Your manager wants you to design a slurry adsorbent system for removing methane from a hydrogen gas stream. The actual separation process is a complex combination of adsorption and absorption, but the total equilibrium can be represented by a simple equation. At 5°C, equilibrium can be represented as

$$\text{Weight fraction CH}_4 \text{ in gas} = 1.2 \times (\text{weight fraction CH}_4 \text{ in slurry})$$

At 5°C, no hydrogen could be detected in the slurry and the heat of sorption was negligible. We wish to separate a gas feed at 5°C that contains 100 lb/h of hydrogen and 30 lb/h of methane. An outlet gas concentration of 0.05 wt frac methane is desired. The entering slurry will contain no methane and flows at a rate of 120 lb/h. Find the number of equilibrium stages required for this separation and the mass fraction methane leaving with the slurry.

D11. A washing operation is processing 10,000 kg/h of wet solids. The liquid is essentially water with a density of 1000.0 kg/m^3, the dry solids have a density of 1500.0 kg/m^3, and the porosity (also known as the volume fraction of void space) is measured as $\varepsilon = 0.40$. What is the flow rate of the underflow liquid in kg/h?

D12.* A countercurrent leaching system is recovering oil from soybeans. The system has five stages. On a volumetric basis, liquid flow rate/solids flow rate = 1.36. 97.5% of the oil entering with the nonsoluble solids is recovered with the solvent. Solvent used is pure. Determine the effective equlibrium constant, m_E, where m_E is (kg/m^3 of solute in solvent)/(kg/m^3 of solute in solid) and is given by the equation $y = m_E x$.

D13. Batch leaching will be similar to a batch extraction, and the equations developed in Section 13.6 can be adapted when the solution is dilute or there is an insoluble solid matrix. We have 12.5 liters of pure water that we will use to leach 10.0 liter of wet sugar cane solids. Equilibrium data are in Problem 14.D8.

a. Find the fractional recovery of the sugar in the water if the water and wet sugar cane solids are mixed together and after settling, the water layer is removed.

b. Find the fractional recovery of the sugar in the water if a continuous solvent addition batch leaching system analogous to Figure 13-11 is used.

D14. Barium sulfide is produced by reacting barium sulfate ore with coal. The result is barium black ash, which is BaS plus insoluble solids. Since BaS is soluble in water, it can be leached out with water. In thickeners the insoluble solids in the underflow typically carry with them 1.5 kg liquid per kg insoluble solids. At equilibrium the overflow and underflow liquids have the same BaS concentrations (Treybal, 1980). We want to process 350 kg/h of insoluble solids plus its associated underflow liquid containing 0.20 mass fraction BaS. Use a countercurrent system with 2075 kg/h of water as solvent. The entering water is pure. We desire the outlet underflow liquid to be 0.00001 mass fraction BaS. Find:

a. The BaS mass fraction in the exiting overflow liquid.

b. The number of equilibrium stages required.

D15. We are removing barium sulfide BaS from insoluble solids (see Problem 14.D14 for data). We are processing 1000 kg/h of dry insoluble solids plus the underflow liquid carried with the solids. The entering underflow liquid is 15 wt % BaS. A 99% recovery of the BaS is desired. The entering solvent (water) is pure. Use a countercurrent process.

   a. Find the minimum overflow rate (minimum entering water rate) $(Ov)_{min}$.

   b. If $Ov = 1.2 \, (Ov)_{min}$, find the number of steps required for the separation.

   c. If the separation actually requires 15 stages at $Ov = 2000$ kg/h, find the overall stage efficiency and alternatively, the effective equilibrium constant m_E. (m_E includes both equilibrium and efficiency. It represents the value of the equilibrium constant that will result in stages having 100% efficiency.)

D16. Batch washing will be similar to batch extraction, and if the terms are translated the same equations can be used. We are doing a single stage batch washing of alumina solids to remove NaOH from the entrained liquid. The feed to the washing process is 1.0 kg of solids (on a dry basis). Each kg of dry solids entrains 2.0 kg of underflow liquid. In this feed the entrained liquid contains 6 wt % NaOH. The inlet washing liquid is pure water. Assume overflow and underflow amounts are constant.

   a. If 2.0 kg of wash liquid are used in a normal batch washing operation, what is the outlet underflow liquid weight fraction NaOH?

   b. If we want the outlet underflow liquid to be 0.5 wt % NaOH, how many kg of wash water are required in a normal batch washing operation?

D17. Continuous water addition batch washing will be similar to continuous solvent addition batch extraction, and if the terms are translated the same equations can be used (see Problem 14.C5b). We are doing a single stage, continuous water addition batch washing of alumina solids to remove NaOH from the entrained liquid. The feed to the washing process is 1.0 kg of solids (on a dry basis). Each kg of dry solids entrains 2.0 kg of underflow liquid. In this feed the entrained liquid contains 6 wt % NaOH. The inlet washing liquid is pure water. Assume overflow and underflow amounts are constant.

   a. If 2.0 kg of wash liquid are used in a continuous water addition batch washing operation, what is the outlet underflow liquid weight fraction NaOH?

   b. If we want the outlet underflow liquid to be 0.5 wt % NaOH, how many kg of wash water are required in a continuous water addition batch washing operation?

   c. Compare results to results of Problem 14.D16.

D18.* Repeat Example 14-2 except for a single-stage system and unknown underflow product concentration.

D19.* Repeat Example 14-2 except for a three-stage countercurrrent system and unknown underflow product concentration.

D20.* Repeat Example 14-2 except for a three-stage cross-flow system, with pure solvent at the rate of 421 kg/h added to each stage and unknown underflow product concentration.

D21. A slurry of pure NaCl crystals, NaCl in solution, NaOH in solution and water is sent to a system of thickener(s) at a rate of 100 kg/min. The feed slurry is 45 wt % crystals. The mass fractions of the entire feed (including crystals and solution) are: $x_{NaCl} = 0.5193$, $x_{NaOH} = 0.099$, and $x_{water} = 0.3187$. We desire to remove the NaOH by washing with a saturated NaCl solution. The thickener(s) are operated so that the underflow is 80% solids (crystals) and 20% liquid. There are no solids in the overflow. Solubility of NaCl in caustic solutions is listed in the table (the y values) (Brown et al., 1950). The first two rows of x values (wt frac in underflow including both crystals and solution) have been calculated so that you can check your calculation procedure. The x values are specific for this operation where the underflow is 80% solids and the crystals are pure NaCl. [$x_{NaCl} = 0.8(1.0) + 0.2\, y_{NaCl}$, $x_{NaOH} = 0.8(0) + 0.2\, y_{NaOH}$]

x_{NaOH}	y_{NaOH}	y_{NaCl}	x_{NaCl}
0.0	0.0	0.270	0.854
0.004	0.02	0.253	0.8506
	0.04	0.236	
	0.06	0.219	
	0.10	0.187	
	0.14	0.156	
	0.18	0.126	

a. Complete the table and plot the saturated extract (y) and saturated raffinate (x) curves and construct a conjugate line.

b. If the feed is mixed with 20.0 kg/min of a saturated NaCl solution ($y_{NaCl} = 0.27$, $y_{NaOH} = 0.0$) in a single thickener, find the flow rates of the underflow and overflow, the compositions of these streams, and the weight of crystals in the underflow.

c. If a countercurrent cascade of thickeners is used with $S = 20$ kg/min of a saturated NaCl solution ($y_{NaCl} = 0.27$, $y_{NaOH} = 0.0$) and we desire a raffinate product that has $x_{NaOH} = 0.01$, determine the flow rates of R_1 and E_N, and the number of stages required.

Note: Although this is a washing problem, there is not a constant flow rate of solids. Thus, it needs to be solved like the leaching problems in this chapter.

E. *More Complex Problems*

E1. You are processing halibut livers that contain approximately 25.2 wt % fish oil and 74.8 wt % insoluble solids. The following data for leaching fish oil from halibut livers using diethyl ether solvent is given by Brown and Associates (1950):

$y_{oil} =$	0.0	0.1	0.2	0.3	0.4	0.5	0.6	0.65	0.70	0.72
$Z =$	0.205	0.242	0.283	0.339	0.405	0.489	0.600	0.672	0.765	0.810

where y_{oil} is the mass fraction oil in the overflow, which was found to contain no insoluble solids, and Z is the pounds of solution per pound of oil free solids in the underflow.

a. Convert the data to mass fraction oil and solvent in the overflow and to mass fraction oil, solvent and solids in the underflow. Plot the data as a saturated extract curve and a saturated raffinate curve with appropriate tie lines on a graph with y_{oil} or x_{oil} as the ordinate and y_{solids} or x_{solids} as the abscissa. Develop the conjugate line.

b. We have a total of 1000 lb of fish (oil + solids). We mix the fish with 500 lb of pure diethyl ether, let the mixture settle and draw off the solvent layer. Calculate the weight fraction of oil in the extract and raffinate layers, and the amounts of the extract and the raffinate.

c. The raffinate from step b is now mixed with 500 lb of pure diethyl ether. After settling, calculate the weight fraction of oil in the extract and raffinate layers, and the amounts of the extract and the raffinate.

d. We have a total of 1000 lb of fish (oil + solids)/hr. We continuously mix the fish with 500 lb of pure diethyl ether/hr in a thickener and draw off the solvent layer. Calculate the weight fraction of oil in the extract and raffinate layers, and the flow rates of the extract and the raffinate.

e. The raffinate from step d is now mixed with 500 lb/h of pure diethyl ether in a second thickener. Calculate the weight fraction of oil in the extract and raffinate layers, and the flow rates of the extract and the raffinate.

f. We have a total of 1,000.0 pounds of fish (oil + solids)/h, which we are going to continuously process in a countercurrent cascade of thickeners. Use 500 lb/h of pure diethyl ether. We want the outlet raffinate to contain 0.02 weight fraction fish oil. Find the number of stages required.

E2. Your company continues to have the stinky job of processing the halibut livers detailed in Problem 14.E1. Since the halibut arrive in batches on the fishing boats, you decide to evaluate a continuous solvent addition batch operation. The basic system will be similar to the extraction system shown in Figure 13-11 except a filter is used instead of a settler. The derivation in Eqs. (13-28a) through (13-29a) is valid, but since the equilibrium is not linear, Eq. (13-29b) is not valid. The feed is the same 25.2 wt % fish oil, 74.8 wt % insoluble solids feed as in Problem 14.E1. 1000 kg of this feed are to be processed using pure diethyl ether as the solvent. Operation proceeds by first adding enough solvent so that the mixture is in the two phase region. Then pure oil is skimmed off at the same rate as the solvent is added.

a. How much solvent must be added so that the mixture is in the two phase region? The two-phase region starts when the mixing line crosses the saturated curve. Note that since there can be no removal of extract until this amount of solvent is added, Eq. (13-29a) is not valid until this amount of solvent is added.

b. The continuous solvent addition batch operation continues adding solvent until the weight fraction of oil in the raffinate remaining in the tank is 0.02. Determine the total kg of solvent required, the weight of raffinate remaining in the tank, and the concentration of this raffinate. Also find the weight of extract collected and the average weight fraction of the oil and solvent in the extract. Note: Since Eq. (13-29a) assumes that $\hat{R}_t$ is constant, approximate constant $\hat{R}_t$ by using the average value of $\hat{R}_t$ in Eq. (13-29a) over the period when extract is removed.

E3. Solve Problem 14.E2, part b, but do not assume that $\hat{R}_t$ is constant. This requires deriving an alternative for Eq. (13-29a) for nonconstant $\hat{R}_t$, but with a constant amount of insoluble solids.

CHAPTER 15

Introduction to Diffusion and Mass Transfer

Except for the short introductory Section 1.3, to this point the entire analysis of separation processes has been equilibrium based. Effects of nonequilibrium operation have been lumped into either a stage efficiency (Sections 4.11, 10.2, 12.5, and 13.5) or to the height equivalent of a theoretical plate (HETP; Sections 10.9 and 10.11). We must move beyond an equilibrium analysis if we want to be able to predict values of the stage efficiency and the HETP (Chapter 16), to study membrane separators (Chapter 17), or to study sorption separations (Chapter 18). For all of these situations, we must look at the mass transfer occurring in the separator. This chapter presents the fundamentals of diffusion and mass transfer in sufficient detail so that the analysis in the remaining chapters is understandable. Additional information on mass transfer is presented as needed in Chapters 16 to 18. If you have already studied mass transfer and diffusion, most, but probably not all, of this chapter will be a review, and you will not have to spend much time studying the material.

Mass transfer is the movement of mass caused by species concentration differences in a mixture. Diffusion is the mass transfer caused by molecular movement, while convection is the mass transfer caused by bulk movement of mass. Large diffusion rates often cause convection. Because mass transfer can become intricate, at least five different analysis techniques have been developed to analyze it. Since they all look at the same phenomena, their ultimate predictions of the mass-transfer rates and the concentration profiles should be similar. However, each of the five has its place: they are useful in different situations and for different purposes. We start in Section 15.1 with a nonmathematical molecular picture of mass transfer (the first model) that is useful to understand the basic concepts, and a more detailed model based on the kinetic theory of gases is presented in Section 15.7.1. For robust correlation of mass-transfer rates with different materials, we need a parameter, the *diffusivity* that is a fundamental measure of the ability of solutes to transfer in different fluids or solids. To define and measure this parameter, we need a model for mass transfer. In Section 15.2, we discuss the second model, the *Fickian* model, which is the most common diffusion model. This is the diffusivity model usually discussed in chemical engineering courses. Typical values and correlations for the Fickian diffusivity are discussed in Section 15.3. Fickian diffusivity is convenient for binary mass transfer but has limitations for nonideal systems and for multicomponent mass transfer.

In Section 15.4, the engineering approach to mass transfer, the *linear driving-force model* introduced in Eq. (1-4), is explored in more detail, particularly for mass transfer between two phases. This third approach is applicable to any situation because correlations for the mass-transfer coefficients can be developed on the basis of dimensional analysis, and the constants in the correlations can be fit to experimental mass transfer data. In Section 15.5, a few correlations for the mass-transfer coefficient based on Fickian diffusivity are presented. Additional correlations are presented when needed in Chapters 16 to 18. If you have had a chemical engineering mass-transfer course, Sections 15.1 to 15.5 will contain familiar material. If you have not had a mass-transfer course, Sections 15.1 to 15.5 are the minimum material required to proceed to Chapter 16.

Section 15.6 describes the deficiencies in the Fickian model and points out why an alternative model (the fourth) is needed for some situations. The alternative *Maxwell-Stefan* model of mass transfer and diffusivity is explored in Section 15.7. The Maxwell-Stefan model has advantages for nonideal systems and multicomponent mass transfer but is more difficult to couple to the mass balances when designing separators. The fifth model of mass transfer, the irreversible thermodynamics model (de Groot and Mazur, 1984; Ghorayeb and Firoozabadi, 2000; Haase, 1990), is useful in regions where phases are unstable and can split into two phases, but it is beyond the scope of this introductory treatment. The advantages, disadvantages, and relationships among the models are delineated in Section 15.8. Applications of the mass-transfer theories to separations are covered in Chapters 16 to 18.

15.1 MOLECULAR MOVEMENT LEADS TO MASS TRANSFER

On a molecular level, all molecules move and collide because of thermal energy. These molecular collisions result in mass transfer by diffusion. At every temperature above absolute zero, molecules are always moving. When they bump into another molecule, the kinetic energy of the two molecules is redistributed and the molecules move away at different angles. With a large number of molecules, the motion of each molecule is random and the molecules tend to distribute throughout the volume available. At equilibrium there is an equal number density of molecules throughout the container.

The number of molecules present in a volume (e.g., 1.0 ml) can easily be estimated by remembering that a gram mole consists of Avogadro's number (6.023×10^{23}) of molecules. If we have liquid water at 20°C (molecular mass = 18.016 and density = 0.998 g/ml), there are 3.35×10^{22} molecules/ml—truly a large number! For an ideal gas (say, nitrogen with molecular weight = 28.0), a mole occupies 22.4 L at STP (0°C and 1.0 atm). In this case there are 2.69×10^{19} molecules/ml—fewer, but still a huge number. Molecules also bump into the walls of the container, but under normal pressures, so many molecules are present that the collisions with other molecules are much more likely than with the wall (an exception is Knudsen diffusion, discussed in Section 18.6.1). With an enormous number of molecules, the change in number density of molecules caused by the huge number of collisions appears to be continuous instead of discontinuous. This apparently continuous behavior allows us to use our normal continuous (differential) mathematics.

If we introduce a different type of molecule at one place in the container (e.g., a bit of helium in the nitrogen), both the helium and nitrogen molecules move randomly because of thermal energy. As a result, a huge number of collisions of molecules occurs, and both helium and nitrogen spread randomly throughout the container. At equilibrium there is not only an equal density of total molecules everywhere in the container but also an equal density of helium molecules everywhere. The net result of this random movement is that the helium molecules on average move from the location of high concentration to a location of low concentration. This process of movement of molecules from a region of high concentration to a region of low concentration is called *molecular diffusion*.

This picture is oversimplified, but it gives a reasonable starting point for studying binary diffusion. Since the velocity of molecules increases with higher thermal energy, we would expect that diffusion rates (however defined and measured) will increase as the temperature increases. Since gases have fewer molecules per volume than liquids, the random movement of the molecules will be less impeded in a gas. Thus, we would expect higher diffusion rates in gases than in liquids.

If we consider a slightly more complex arrangement, we could continually flow a stream relatively concentrated in ethanol on one side of the space and flow a stream relatively less concentrated in ethanol on the other side of the space, which is essentially what we do in many separation processes. If we flow fairly rapidly, we will prevent the entire space from ever reaching equilibrium (same ethanol concentration everywhere). However, there will be a net movement of ethanol from the concentrated region to the dilute region. You have experienced a somewhat analogous situation when you want to move through a crowd of people and have to bob and weave and sometimes step backwards to do so. The difference between the situations is that people have a purpose to their movement, while *molecules do not have a purpose and the motions are random*. Thus, in describing our models we must try to avoid assigning a purpose to the molecular diffusion.

Even though this picture is quite simple, it is relatively easy to consider complicating factors and qualitatively predict their effect on molecular diffusion. For example, if the system is quite concentrated, diffusion of a large number of molecules occurs. This diffusion leads to convection (movement by flow). This coupling of diffusion and convection complicates the analysis and is considered in Section 15.2.3. Although diffusion always causes convection, in dilute systems the convection is small enough that it can be ignored (Sections 15.2.1 and 15.2.2).

Consider another complication by analogy. Suppose you want to move through a crowd, but you have to take your little sister with you. You take her firmly by the hand, and the two of you zigzag through the crowd. Since the two of you together require a larger space to squeeze between people, your motion will be slower than if you were alone. A similar effect occurs if molecules agglomerate or stick together (but remember that the molecules do not have a purpose for their movement). The larger group of molecules moves more slowly, so the measured diffusivity is lower. This situation is discussed in Section 15.3.

Another situation we can explore by analogy occurs when you want to go in one direction, but a number of people are headed in another direction. The contact or "friction" with these other bodies will tend to carry you in the direction they are going, and if there are sufficient numbers of them, you may be swept along with them. The molecular equivalent can occur in a flow situation. Assume we have a water stream with a modest amount of methanol and a small amount of ethanol on the left, and another water stream with a small amount of methanol and a fairly large amount of ethanol on the right. If we allowed the system to come to equilibrium, we would have equal methanol concentrations everywhere. In the nonequilibrium flow situation, we can continually transfer ethanol from the right to the left. We (the people—not the molecules, which just move randomly) would expect random fluctuations to move methanol from the left to the right, but if there is sufficient ethanol movement, we may observe the reverse transfer direction of methanol. This case is discussed in Section 15.7.

The word pictures painted in this section have been quantified for gases in the kinetic theory of gases. An introduction to this theory is presented in Section 15.7.1.

In Chapter 1 of this book we presented Eqs. (1-5a and b) that relate the rate of mass transfer/volume to a mass-transfer coefficient, the area/volume, and a driving force. This mathematical model is an attempt to quantify a complicated situation. Although useful, this model and the other models presented later in this chapter can also be misleading. The term *driving force* implies purpose or desire to transfer, and as noted, there is no purpose—the molecules are just moving randomly. With this caveat, let's look at the various models used to analyze mass transfer, starting with the Fickian diffusion model.

15.2 FICKIAN MODEL OF DIFFUSIVITY

15.2.1 Fick's Law and the Definition of Diffusivity

In 1855 and 1856, physician Adolph Fick built on the previous work of Thomas Graham to develop a theory for the transfer of dilute solutes in physiological fluids. Since Fick was familiar with Fourier's analysis of thermal conduction and since his experimental apparatus was analogous to Fourier's apparatus, Fick modeled his theory on the analogous thermal conduction theory (Cussler, 2009). With constant density and heat capacity, Fourier had shown that for one-dimensional heat conduction with no convection and no thermal radiation,

$$(Q_z \text{ heat flux in z direction}) \propto (-\text{temperature difference})/(\text{distance in z direction}) \qquad \textbf{(15-1a)}$$

Defining the proportionality constant as the thermal conductivity $k_{conduction}$, the definition becomes

$$\text{Thermal conductivity } k_{conduction} = -(Q_z / A)/(\Delta T / \Delta z) = -(Q_z / A)/(dT / dz) \qquad \textbf{(15-1b)}$$

If Q_z is the heat-transfer rate by conduction, J/s, in the z direction for a material with an area for heat transfer of A m^2, over a distance measured in m, and a thermal gradient with units °C/m, then $k_{conduction}$ has units J/(s m °C). It was later realized that a slightly more general form of this equation is

$$\text{Thermal diffusivity } \alpha_{thermal} = -(Q_z / A)/[\Delta(\rho C_p T)/\Delta z] = -(Q_z / A)/[d(\rho C_p T)/dz] \qquad \textbf{(15-1c)}$$

where thermal diffusivity $\alpha_{thermal} = k_{conduction}/(\rho C_p)$ and $(\rho C_p T)$ is the thermal energy of the system. For the same units and with density ρ in kg/m^3 and heat capacity C_p in J/(kg °C), the thermal diffusivity $\alpha_{thermal}$ has units of m^2/s. For additional information on heat transfer, see Hottel et al. (2008) and Incropera et al. (2011).

Fick showed that for molecular diffusion of a dilute solution of solute A in solvent B (a binary mixture) with no convection in the z direction,

$$(\text{Molar flux of A in z direction}) \propto (-\text{concentration difference of A})/$$
$$(\text{distance in z direction}) \qquad \textbf{(15-2a)}$$

Defining the proportionality constant as the molecular diffusivity D_{AB}, the definition for D_{AB} becomes

Molecular diffusivity $D_{AB} = -J_{A,z} / (\Delta C / \Delta z) = -J_{A,z} / (dC / dz)$ **(15-2b)**

If $J_{A,z}$ is the molecular flux by diffusion, (mole A)/(s m^2), in the z direction over a distance measured in m and a concentration gradient with units (mole A)/m^3, then D_{AB} has units of m^2/s. Then the total amount transferred is

Total transfer of moles of A by diffusion = $J_{A,z}$ (Cross-sectional area for mass transfer) **(15-2c)**

To calculate the total transferred, we also need to know the area for mass transfer measured in m^2. For a given dilute binary system, the Fickian diffusivity D_{AB} depends, as expected, on temperature but is approximately independent of concentration. Typical values of D_{AB} for gas systems at atmospheric pressure are 10^{-5} m^2/s and for liquids are 10^{-9} m^2/s. Typical experimental values and correlations for predicting D_{AB} are presented in Section 15.3.

Note that the thermal diffusivity and the molecular diffusivity have identical units. In addition, if we consider that $1/\alpha_{thermal}$ is the resistance to transferring energy and $1/D_{AB}$ is the resistance to transferring mass, then Eqs. (15-1c) and (15-2b) can both be written in the form

Flux = (driving force)/(resistance) **(15-3)**

For mass transfer, the flux of moles of solute A is $J_{A,z}$ and the "driving force" is (dC_A / dz).

The usual form for writing Fick's law in one dimension, and with no convection, in the z direction is

$$J_{A,z} = -D_{AB} \frac{dC_A}{dz}$$ **(15-4a)**

This form or the equivalent Eq. (15-2b) is normally used to analyze experimental data to determine values of the Fickian diffusivity D_{AB}. However, Sherwood et al. (1975) and Bird et al. (2002) point out that the following form is fundamentally more correct.

$$J_{A,z} = -C_m D_{AB} \frac{dy_A}{dz}$$ **(15-4b)**

In this equation y_A is the mole fraction of A, and C_m is the mixture concentration (mol total mixture)/m^3. The equations are identical for dilute, isobaric, isothermal gases because $C_A = y_A C_m$ and C_m is constant. However, in nonisothermal situations, Eq. (15-4b) is the correct form (see Problem 15.A1).

Since Fick cast his equation in a familiar form and since Eqs. (15-2b) and (15-4a) fit data for isothermal dilute binary systems very well, this equation rapidly became enshrined as Fick's law (sometimes known as Fick's first law). However, problems arose when other researchers extended Fick's work to more concentrated systems. In Section 15.2.3 we will see that when there is significant convection in the diffusion direction, the diffusion flux J needs to be related to the flux N with respect to a fixed coordinate system (N is the flux needed to design equipment). This complicates the picture but does not invalidate Fick's law. As we shall see later, when extended to concentrated, nonideal systems or to multicomponent systems, Fick's law often requires very large adjustments of the molecular diffusivity—sometimes with negative values—as a function of concentration to predict behavior. Said in clearer terms, Fick's law no longer applies. We should not blame Fick for this lack of agreement. His "law" works fine for the conditions that he developed it for.

15.2.2 Steady-State Binary Fickian Diffusion and Mass Balances without Convection

For it to be useful, we need to couple Fick's law with mass balances. The first case considered is steady-state diffusion with no convection in the direction of diffusion. This is an important practical case for measuring diffusion coefficients, studying steady-state evaporation and steady-state permeation of gases and liquids in membranes, and in design of distillation and some other separation processes. The second case we consider is unsteady diffusion with no convection in the direction of diffusion, which is of practical significance in controlled-release drug delivery and in some batch reactors and separation processes.

The classic steady-state diffusion problem is diffusion across a thin film at constant pressure and temperature and with no convection in the direction of diffusion (z direction), as shown in Figure 15-1. At steady state, there is no accumulation in the film and the concentration profile does not change with time. Over a segment of the film of thickness Δz, the mass balance is input = output, which can be written in this case as $J_{A,z} = J_{A,z+\Delta z}$, and leads to

$$-D_{AB}(z)(dC_A / dz)\big|_z = -D_{AB}(z+\Delta z)(dC_A / dz)\big|_{z+\Delta z} \qquad \textbf{(15-5a)}$$

Rearranging Eq. (15-5a) and dividing by Δz, we obtain

$$[D_{AB}(z+\Delta z)(dC_A / dz)\big|_{z+\Delta z} - D_{AB}(z)(dC_A / dz)\big|_z]/\Delta z = 0 \qquad \textbf{(15-5b)}$$

Assuming that the limit as $\Delta z \to 0$ exists (and it does if there are no discontinuities) and taking this limit, we obtain

$$d(D_{AB}dC_A / dz)/ dz = 0 \qquad \textbf{(15-6a)}$$

If the diffusion coefficient does not depend on concentration, this simplifies to

$$D_{AB}d^2C_A / dz^2 = 0 \qquad \textbf{(15-6b)}$$

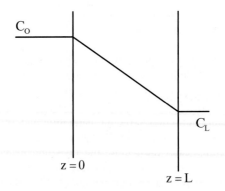

FIGURE 15-1.   *Steady-state diffusion across a thin layer of film*

Referring to Figure 15-1, the boundary conditions are

$$C_A(z=0)=C_0 \text{ and } C_A(z=L)=C_L \tag{15-7}$$

Integrating Eq. (15-6b) twice, we obtain

$$dC_A/dz = \text{Constant 1, and } C_A = (\text{Constant 1})z + \text{Constant 2}$$

From the first boundary condition in Eq. (15-7), Constant 2 = C_0. Then, from the second boundary condition, Constant 1 = $(C_L - C_0)/L$, and the equation for the concentration profile is

$$C_A = [(C_L - C_0)/L]z + C_0 \tag{15-8}$$

Why did the diffusivity disappear? With the problem defined in this way (C_0, C_L, and L defined), the concentration profile does not depend on the diffusivity. However, the flux does depend on the diffusivity. The flux is given by Eq. (15-4a) with dC_A/dz = Constant 1 = $(C_L - C_0)/L$; thus,

$$J_{A,z} = -D_{AB}\frac{dC_A}{dz} = -(D_{AB}/L)(C_{A,L} - C_{A,0}) \tag{15-9}$$

Obviously, the problem could be defined in a way that forces the concentration profile to depend on the diffusivity (see Problem 15.D1). Equations very similar to Eqs. (15-8) and (15-9) occur in a number of problems, such as steady-state evaporation and steady-state permeation of gases and liquids in membranes.

Example 15-1. Steady-state diffusion without convection: Low-temperature evaporation

Pure ethanol is contained at the bottom of a long vertical tube (cross-sectional area of 0.9 cm²) shown in Figure 15-2. Above the liquid is a quiescent layer of air. The liquid at the bottom of the tube is carefully adjusted so that the distance from the air-liquid interface to the open top of the tube is constant at 15.0 cm. No liquid is withdrawn from the tube. At the open end of the tube, air is blown perpendicular to the vertical tube so that the concentration of ethanol at the top of

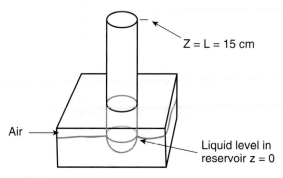

FIGURE 15-2.   *Tube and reservoir for Examples 15-1 and 15-2*

the tube is essentially zero. The entire apparatus is kept at 0°C, and $p_{tot} = 0.98$ atm. The tube is carefully arranged so that there is no convection in the tube. Over the course of several days, we find that the average evaporation rate is 0.9190×10^{-3} cm³/h. What is the value of the diffusion coefficient of ethanol in air at 0°C?

Solution

Liquid ethanol is evaporating at the bottom of the tube. If we assume equilibrium for the gas immediately above the liquid (z = 0), then

$$y_E(z=0) = \text{Vapor pressure ethanol } (0°C) / p_{tot} \tag{15-10a}$$

For an ideal gas $C = yC_m$, where the total molar concentration of the gas is

$$C_m = n / V = p_{tot} / (RT) \tag{15-10b}$$

Combining this result with Eq. (15-10a) for equilibrium evaporation, we obtain

$$C_0 = (\text{Vapor Pressure}) / RT \tag{15-10c}$$

The vapor pressure of ethanol can be accurately estimated by Antoine's equation,

$$\log_{10}(VP) = A - \frac{B}{T + C_{Antoine}} \tag{15-10d}$$

With VP in mm Hg, T in °C, $\Lambda = 8.32109$, B − 1718.10, and $C_{Antoine} = 237.52$ over the range from −2 to 100°C (Dean, 1985). Note: Although mm Hg is an obsolete unit for pressure, because a large amount of data is still in these units, engineers must be comfortable converting or using this unit.

The molecular diffusion coefficient can be determined from Eq. (15-9),

$$D_{AB} = -L J_{A,z} / (C_L - C_0) \tag{15-11a}$$

$C_L \sim 0$, C_0 is given by Eq. (15-10c), and $J_{A,Z}$ can be determined from the liquid feed rate,

$$J_{A,Z} = (\text{evaporation rate})/(\text{cross sectional area}) \tag{15-11b}$$

Because the evaporation rate is given in volumetric terms, we need a density. The molar density of liquid ethanol is 17040 mol/m³. Once the ethanol flux is known, we can determine D_{AB}.

Do it. Putting in the numbers: At 0°C, VP = 12.098 mm Hg. Since the gas constant $R = 8.20567 \times 10^{-5}$ (atm m³)/(mol K),

$$C_E(z=0) = \frac{(VP)_E}{RT} = \frac{\left(12.098 \text{mm Hg} \dfrac{1.0 \text{atm}}{760 \text{mm Hg}}\right)}{\left(8.20567 \times 10^{-5} \dfrac{m^3 \text{atm}}{\text{mol K}}\right)(273 \text{K})} = 0.7106 \frac{\text{mol}}{m^3}$$

For the liquid evaporation rate,

$$\text{evap.} = (0.9190 \times 10^{-3} \text{cm}^3/\text{h}) \left(\frac{1.0\text{m}^3}{10^6 \text{cm}^3} \right) \left(\frac{1.0\text{h}}{3600\text{s}} \right) (17040 \text{mol/m}^3) = 4.35 \times 10^{-9} \text{mol/s}$$

The flux of ethanol in the gas is

$$J_{\text{ethanol}} = \left(4.35 \times 10^{-9} \text{mol/s} \right) / \left(0.9\text{cm}^2 \left(\frac{1\text{m}^2}{10^4 \text{cm}^2} \right) \right) = 4.83 \times 10^{-5} \frac{\text{mol}}{\text{m}^2\text{s}}$$

Then, from Eq. (15-11) we obtain,

$$D_{\text{Ethanol-air}} = \frac{(0.15\text{m})[4.83 \times 10^{-5} \text{mol/(m}^2\text{s})]}{(0.7106 - 0)\text{mol/m}^3} = 0.102 \times 10^{-4} \text{m}^2/\text{s}$$

Comments
1. This result agrees with Table 15-1.
2. There is significant horizontal convection in this problem, or we would not be able to keep $C_L \approx 0$. However, the convection is not in the direction of the diffusion, and since the system is dilute, the convection induced in the z direction is quite small; thus, Eqs. (15-6b) and (15-7) are valid. We are also assuming that air does not enter the tube or cause turbulent flow.
3. The most accurate method of measuring the amount of liquid that evaporates is to weigh the tube and reservoir. Note that operation with a reservoir and no liquid addition is not at a true steady state. However, if a large reservoir is attached to the bottom of the tube so that $\Delta L/L$ is small, operation is almost at steady state (called *pseudo-steady state*) and the steady-state diffusion equations can be used. See Problem 15.B1 to brainstorm alternative operating procedures.
4. The assumption that convection can be ignored will be checked in Example 15-2.

15.2.3 Unsteady Binary Fickian Diffusion with No Convection (Optional)

We can also study unsteady diffusion in dilute systems. The classic unsteady-state diffusion problem is diffusion with no convection in the direction of diffusion in an infinitely thick slab. The entire slab is initially at concentration C_{initial}, and at t = 0, the z = 0 face of the slab is set to C = C_0 At the far end of the slab, z → ∞, the concentration is $C_{A,\infty} = C_{\text{initial}}$ for all times; thus, the slab is so thick that the far end is unaffected by the diffusion into the slab. We will see shortly that with ordinary liquid diffusion coefficients, the slab really does not need to be too thick to keep the far end at the initial concentration. For a segment of thickness Δz, the mass balance per unit area is Accumulation = Input – Output. The amount of material in the segment of thickness Δz at any time t is $C_A(t)\Delta z$. Since accumulation is the change in this amount of material, the mass balance becomes

$$d(C_A \Delta z)/dt = J_{A,z} - J_{A,z+\Delta z} \qquad \text{(15-12a)}$$

Dividing by Δz and taking the limit as $\Delta z \to 0$ (again, we need to assume this limit exists), we obtain

$$\partial C_A / \partial t = \partial J_A / \partial z \qquad \text{(15-12b)}$$

Since terms are now functions of both t and z, partial derivatives are required. Assuming constant diffusivity and substituting in Fick's law, Eq. (15-4a), we obtain

$$\partial C_A / \partial t = D_{AB} \partial^2 C_A / \partial z^2 \qquad \text{(15-12c)}$$

This equation is often known as Fick's second law. As expected, a very similar equation can be derived for unsteady-state heat conduction in an infinite slab (Incropera et al., 2011).

The boundary conditions for an infinitely thick slab are

$$C_A = C_{initial} \text{ for } t = 0, \text{ all } z; \ C_{A,} = C_{A0} \text{ for } t > 0 \text{ at } z = 0; \ C_A = C_{initial} \ z \to \infty \qquad \text{(15-13)}$$

Defining the dimensionless distance ς

$$\varsigma = \frac{z}{\sqrt{4D_{AB}t}} , \qquad \text{(15-14a)}$$

the solution for Eq. (15-12c) is

$$\frac{C_A - C_{A,0}}{C_{A,\infty} - C_{A,0}} = \text{erf}\varsigma \qquad \text{(15-14b)}$$

where erf is the error function. The definition and properties of the error function are discussed in Section 18.7.1 in conjunction with the solution for the unsteady dispersion equation for an adsorption column. Since the error function is supported by Excel, calculations with a spreadsheet are straightforward. If $C_{A,0} > C_{A,L} > C_{A,initial}$, a convenient form of Eq. (15-14b) for plotting the results is

$$\frac{C_A}{C_{A,0}} = 1 + \left(\frac{C_{A,\infty} - C_{A,0}}{C_{A,0}} \right) \text{erf}\varsigma \qquad \text{(15-14c)}$$

This equation was used to plot Figure 15-3 for diffusion of sucrose in liquid water. If $C_{A,0} > C_{A,L}$, a convenient form of Eq. (15-14b) for plotting the results is

$$\frac{C_A}{C_{A,L}} = 1 + \left(\frac{C_{A,\infty} - C_{A,L}}{C_{A,L}} \right) \text{erf}\varsigma \qquad \text{(15-14d)}$$

For different geometries, it can be useful to write Fick's second law in cylindrical coordinates for transfer in only the radial direction,

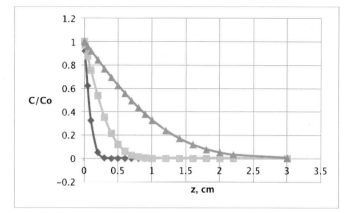

FIGURE 15-3.   *Unsteady diffusion of sucrose in an infinitely thick*
slab of water. Conditions are delineated in Problem
15.D19. Curves are plotted for t = 1000s, 10,000s,
and 100,000s.

$$\partial C_A / \partial t = \frac{D_{AB}}{r} \frac{\partial}{\partial r} \left(r \frac{\partial C_A}{\partial r} \right)$$

(15-12d)

and for spherical coordinates for transfer in the radial direction only,

$$\partial C_A / \partial t = \frac{D_{AB}}{r^2} \frac{\partial}{\partial r} \left(r^2 \frac{\partial C_A}{\partial r} \right)$$

(15-12e)

Except for this section and Section 18.7, the solutions of the unsteady diffusion equation in one to three dimensions are beyond the scope of this book. Solutions to Eqs. (15-12c, d, e), the corresponding two- and three-dimension equations, and the equivalent heat conduction equations have been extensively studied for a variety of boundary conditions (e.g., Crank, 1975; Cussler, 2009; Incropera et al., 2011). Readers interested in unsteady-state diffusion problems should refer to these or other sources on diffusion.

15.2.4 Steady-State Binary Fickian Diffusion and Mass Balances with Convection

To simplify the previous analyses, we assumed that there was no convection in the z direction. But in practical separation and other mass-transfer systems, this is obviously a special case. How do we analyze diffusion when there is also convection in the z direction?

A useful analogy is to consider a molecule (for example, sugar) moving in a person's body at the same time she walks across the room. The overall movement of the sugar will be dominated by her movement in the room (the room serves as a fixed frame of reference, and the flux measured in this frame is N). To study the movement of the sugar, we first look at the flux in her body, using the body as the reference frame (this means the sugar movement is measured with respect to the body, not the room). Once we know the flux J with respect to the moving reference frame, then we must find a way to add the flux J to the movement of her body to find N.

Since both convection and diffusion occur, we need to separate the terms in some fashion. The usual assumption in the Fickian model for binary systems is that the effects are additive,

$$\text{(Total Flux } N_A) = \text{(Fick's law diffusive flux A, } J_A) + \text{(convective flux A)} \quad \textbf{(15-15a)}$$

$$\text{(Total Flux } N_B) = \text{(Fick's law diffusive flux B, } J_A) + \text{(convective flux B)} \quad \textbf{(15-15b)}$$

This step certainly makes sense in the analogy with a person's body. Fick's law is applied in a reference frame (her body) with no convection, which allows use of the procedures of Section 15.2.3. Since we are doing a steady-state analysis the total fluxes, N_A and N_B [mol/(m²s)] are constant (e.g., not functions of z); however, the diffusive and convective fluxes do depend on z. To use this separation of terms, choose a reference velocity $v_{ref}(z)$ so that there is no convection in the reference frame. This is always possible in a steady-state system. The convective flux is defined in terms of the reference or basis velocity v_{ref},

$$\text{Convective flux of A} = C_A v_{ref} \quad \textbf{(15-15c)}$$

$$\text{Convective flux of B} = C_B v_{ref} \quad \textbf{(15-15d)}$$

where C_A and C_B are in mol/m³ and v_{ref} is in m/s. Since v_{ref}, C_A, and C_B can depend on z, the convective fluxes can depend on z.

The total flux (due to both diffusion and convection) can be defined in terms of currently unknown component transfer velocities $v_A(z)$ and $v_B(z)$,

$$N_A = C_A v_A \text{ and } N_B = C_B v_B \quad \textbf{(15-15e,f)}$$

The diffusive flux of A is given by Eq. (15-4a), since it is calculated in a reference frame with no convection, and there is a similar equation for B. The diffusive flux of A can also be determined from Eqs. (15-15a), (15-15c), and (15-15e) because J_A is the difference between the total flux and the convective flux (in the analogy, this step allows us to look at only what is happening within her body). Comparing the results from Eqs. (15-15) with Eq. (15-4a), we obtain

$$\text{Fick's law diffusive flux of A} = J_A = C_A(v_A - v_{ref}) = -D_{AB} dC_A / dz \quad \textbf{(15-16a)}$$

In a similar fashion, we obtain

$$\text{Fick's law diffusive flux of B} = J_B = C_B(v_B - v_{ref}) = -D_{BA} dC_B / dz \quad \textbf{(15-16b)}$$

Solving for the component velocities,

$$v_A = J_A / C_A + v_{ref} = -(D_{AB} / C_A) dC_A / dz + v_{ref} \quad \textbf{(15-16c)}$$

$$v_B = J_B / C_B + v_{ref} = -(D_{BA} / C_B) dC_B / dz + v_{ref} \quad \textbf{(15-16d)}$$

Combining these results with Eqs. (15-15e, f), we obtain the total fluxes,

$$N_A = C_A v_A = C_A(v_A - v_{ref}) + C_A v_{ref} = -D_{AB}dC_A / dz + C_A v_{ref} \qquad \textbf{(15-16e)}$$

$$N_B = C_B v_B = C_B(v_B - v_{ref}) + C_B v_{ref} = -D_{BA}dC_B / dz + C_B v_{ref} \qquad \textbf{(15-16f)}$$

To find the total fluxes, we have to decide on an appropriate reference or basis velocity. Because the reference velocity is defined as a velocity for which there is no convection, in the reference coordinate system the net flux of A plus B must be zero; otherwise there would be convection. Thus, in the reference coordinate system moving at reference velocity v_{ref}, by definition,

$$J_A = -J_B \text{ and } J_A + J_B = 0 \qquad \textbf{(15-17a)}$$

The reference velocity is calculated on the basis of the component velocities,

$$v_{ref} = x_{A,ref} v_A + x_{B,ref} v_B \qquad \textbf{(15-17b)}$$

which can be written in terms of volume, mole, or mass fractions as

$$v_{ref,vol} = x_{A,vol} v_A + x_{B,vol} v_B \qquad \textbf{(15-17b1)}$$

$$v_{ref,mol} = x_{A,mol} v_A + x_{B,mol} v_B \qquad \textbf{(15-17b2)}$$

or

$$v_{ref,mass} = x_{A,mass} v_A + x_{B,mass} v_B \qquad \textbf{(15-17b3)}$$

Here $x_{A,ref}$ and $x_{B,ref}$ are shown to be respectively the volume, molar, or mass fractions of the components. Remember that fractions are defined so that

$$x_{A,ref} + x_{B,ref} = 1 \qquad \textbf{(15-17c)}$$

Note that the reference velocity $v_{ref,vol}$, $v_{ref,mol}$, and $v_{ref,mass}$ can be identical or different depending on whether volume, mole, or mass fractions are the same. For example, volume and mole fractions are identical in an ideal gas, and $v_{ref,vol} = v_{ref,mol}$, and if the molecular weights are different, $\neq v_{ref,mass}$. However, we have not yet answered the question of which fractions and hence which velocity v_{ref} should be used for a reference velocity. Use the reference velocity that makes the problem as simple as possible! This is often a reference velocity that will be zero; however, in the analogy with a person the reference velocity was not zero, but the solution was not difficult. Cussler (2009) states that the volumetric average velocity $v_{ref,vol}$ is the reference velocity that will often result in the simplest diffusion problem. In separations it is common to use $v_{ref,mol}$ for distillation and absorption, $v_{ref,mass}$ for extraction, and $v_{ref,vol}$ for gas permeation through membranes. In fluid dynamics the mass average velocity is usually chosen as the reference velocity. The choice of reference velocity and the solution of Fickian diffusion problems are best illustrated with examples.

In the analogy with a person's body, we would probably use mass fractions, with A as the sugar and B as everything else, to calculate $v_{ref,mass}$. Since there is a lot more mass of her body than mass of the sugar, $x_{B,ref} = x_{Body,wt\ frac} \gg x_{A,ref}$ and $x_{A,ref} = x_{sugar,wt\ frac}$, $x_{B,ref} = 1$, and $v_{ref,mass} = v_{Body,mass}$. So to find the total flux of sugar, we add the velocity that she walks across the room to the diffusive flux of sugar within her body. We could use $v_{ref,vol}$ in this analogy with her body, but mass is more familiar.

For a second example, assume that we have two fixed, equal-volume chambers (Figure 15-4) that contain different ideal gases but are at the same pressure and temperature. With ideal gases at constant pressure and temperature, there is no volume change on mixing. Since $J_A = -J_B$, there will be equal and opposite volumetric flows of gases A and B, and the volumetric reference velocity $v_{ref,vol} = 0$. This case is an example of *equimolar counterdiffusion* and obviously simplifies Eq. (15-17b1). In this case use of a molar average reference velocity will also result in $v_{ref,vol} = 0$, but if the molecular weights are different, a mass average reference velocity will not be zero. Even if the gas is not ideal, use of a volumetric reference velocity will often make an approximate solution easier.

Most liquids have no or very little volume change on mixing—one of the ways to estimate the density of a liquid mixture is to assume volumes are additive. Even liquid systems with large volume changes on mixing rarely have more than a 10% change. Molar and mass densities are usually significantly less constant. Again, this points to use of the volumetric reference velocity $v_{ref,vol}$ as the simplest reference velocity to use.

On the other hand, calculation of diffusion in distillation columns tends to be easier if the molar average reference velocity $v_{ref,mol}$ is used. In distillation, constant molal overflow is often valid or close to valid (Section 4.2). The resulting equimolar counterdiffusion results in $N_A = -N_B$, and there is no convection in the reference frame with $v_{ref,mol} = 0$. If we choose the reference velocity as the molar average velocity, then Eq. (15-16e) becomes

$$N_A = -D_{AB} dC_A / dz = J_A \qquad (15\text{-}18)$$

A similar simplification occurs for the mass-transfer analysis in Section 15.4 and for the analysis of distillation in Section 16.1. If the molecular weights are not equal, then $v_{ref,mass} \neq v_{ref,mol}$ and solution of the diffusion equations for distillation will be more difficult if we use mass fractions (see Problem 15.C7).

Although this additive approach appears logical and works for calculating the total fluxes of A and B, this is certainly not the only way one could tackle the problem (see Problem 15.B2). For binary systems this approach has the advantages of forcing $D_{AB} = D_{BA}$ (see Problem 15.C1), and D_{AB} is the same for any choice of v_{ref} (see Problem 15.C2).

Diffusion problems are almost always solved as special cases. These solutions are tabulated in significant detail in Crank (1975), in Cussler (2009), and in Incropera et al. (2011). Example 15-2 presents the special case of steady-state diffusion of component A through a stagnant layer of B. This case is important for absorption and stripping.

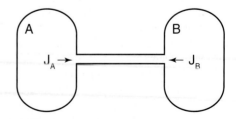

FIGURE 15-4.   *Counterdiffusion of ideal gases. Chambers have equal volumes, pressures, and temperatures.*

Example 15-2. Steady-state diffusion with convection: High-temperature evaporation

Repeat Example 15-1 (Figure 15-2), but operate at a temperature of 39.84°C = 313 K, use the diffusivity value from Table 15-1, and calculate both the diffusion flux $J_{Ethanol}$ at z = L and at z = 0 and the total flux $N_{Ethanol}$.

Solution

With a higher temperature, the vapor pressure will be significantly higher and the concentration C_0 given by Eq. (15-10b) will be higher. Thus, convection effects are much more likely to be significant.

Since operation is at steady state, N_A is constant and is not a function of either time or distance in the tube. Because the air (component B) is stagnant, $N_B = 0$ and $v_B = 0$. The volume average reference velocity, Eq. (15-17b1), is

$$v_{ref,vol} = y_{A,vol}v_A + y_{B,vol}v_B \qquad (15\text{-}19a)$$

Since $v_B = 0$, this becomes $v_{ref,vol} = y_{A,vol}v_A$. We will see shortly that $v_{ref,vol}$ is constant but not zero. In a reference frame moving at this velocity, the convective flux is zero.

For an ideal gas, volumetric and molar fractions are equal. Thus,

$$y_{A,vol} = C_A / C_m = C_A / (p_{tot} / RT) \qquad (15\text{-}19b)$$

The gas constant R = 8.20567 × 10⁻⁵ (atm m³)/(mol K). Note that $N_A = C_A v_A = (p_{tot}/RT)y_{A,vol}v_A$ = constant. Since pressure and temperature are constant, $y_{A,vol}$ is inversely proportional to v_A, which makes $v_{ref,vol}$ constant.

The boundary condition at z = L is $C_A = C_{A,L} \approx 0$, which is

$$y_{A,vol}(z=L) = C_{A,L} / C_m = y_{A,L} \approx 0 \qquad (15\text{-}19c)$$

and at z = 0, $C_A = C_{A,0}$ or

$$y_{A,vol}(z=0) = C_{A,0} / (p_{tot} / RT) = [VP(T)/(RT)]/[p_{tot}/(RT)] = VP(T)/p_{tot} \quad (15\text{-}19d)$$

Equation (15-16e) becomes

$$N_A = -D_{AB}dC_A / dz + C_A v_{ref,vol} = -D_{AB}dC_A / dz + C_A y_{A,vol}v_A = -D_{AB}dC_A / dz + N_A y_{A,vol}$$

We used $N_A = C_A v_A$ for the last equals sign. Solving for N_A, and substituting in $C_A = C_m y_A$, where the overall molar concentration is constant for an ideal gas at constant temperature and pressure, we obtain

$$N_A(1 - y_A) = -C_m D_{AB}\frac{dy_A}{dz} = J_A \qquad (15\text{-}20a)$$

The solution to Eq. (15-20a) for the constant total flux N_A is (Cussler, 2009)

$$N_A = \frac{D_{AB}C_m}{z} \ln\left(\frac{1-y_{A,z}}{1-y_{A,0}}\right) = \frac{D_{AB}C_m}{L} \ln\left(\frac{1-y_{A,L}}{1-y_{A,0}}\right) \qquad \text{(15-20b)}$$

For the boundary conditions of our example, the constant flux N_A is

$$N_A = \frac{D_{AB}C_m}{L} \ln\left(\frac{1-0}{1-y_{A,0}}\right) = \frac{D_{AB}C_m}{L} \ln\left(\frac{1}{1-VP(T)/p_{tot}}\right) \qquad \text{(15-20c)}$$

The concentration decreases from $z = 0$ to $z = L$.

$$\frac{1-y_A}{1-y_{A,0}} = \left(\frac{1-y_{A,L}}{1-y_{A,0}}\right)^{z/L} \qquad \text{(15-20d)}$$

Note that this equation satisfies the boundary conditions. The total flux is constant, but since the concentration profile is not linear, the diffusion flux depends on distance z.

$$J_A = -D_{AB}C_m \frac{dy_A}{dz} = D_{AB}C_m \left(\frac{1-y_{A,0}}{L}\right)\left(\frac{1-y_{A,L}}{1-y_{A,0}}\right)^{z/L} \ln\left(\frac{1-y_{A,L}}{1-y_{A,0}}\right) \qquad \text{(15-20e)}$$

Since by definition $J_A = -J_B$ in the moving reference frame, we have

$$J_B = -J_A = -D_{AB}C_m \left(\frac{1-y_{A,0}}{L}\right)\left(\frac{1-y_{A,L}}{1-y_{A,0}}\right)^{z/L} \ln\left(\frac{1-y_{A,L}}{1-y_{A,0}}\right) \qquad \text{(15-20f)}$$

but in the stationary reference frame, $N_B = 0$.

Do it. Putting in the numbers: From Antoine's equation (Example 15-1), $VP_{Ethanol} = 133.85$ mm Hg. Concentration of the ethanol in the gas next to the liquid surface is $C_{surface} = 6.857$ mol/m³. We can also find $y_{Ethanol,0} = VP/p_{total} = 0.1797$, and $y_{Ethanol,L} = 0$. $D_{Ethanol-air} - 0.145 \times 10^{-4}$ m²/s. The total concentration C_m is

$$C_m = \frac{P_{total}}{RT} = \frac{0.98\text{atm}}{\left(8.20567\times10^{-5}\,\dfrac{\text{atm m}^3}{\text{mol K}}\right)(313\text{K})} = 38.16\text{mol/m}^3$$

Note that, as expected, at $z = 0$, $C_{surface} = y_{ethanol,0}C_m$. From Eq. (15-20e) at $z = L = 0.15$m and at $z = 0$,

$$J_E(z=L) = (0.145\times10^{-4})(38.16)\left(\frac{1-.1797}{.15}\right)\left(\frac{1-0}{1-.1797}\right)^{15/.15} \ln\left[\frac{1-0}{1-.1797}\right] = 7.31\times10^{-4}\,\frac{\text{mol}}{\text{m}^2\text{s}}$$

$$J_E(z=0) = (0.145\times10^{-4})(38.16)\left(\frac{1-.1797}{.15}\right)\left(\frac{1-0}{1-.1797}\right)^{0/.15} \ln\left[\frac{1-0}{1-.1797}\right] = 5.99\times10^{-4}\,\frac{\text{mol}}{\text{m}^2\text{s}}$$

The total flux rate

$$N_E = \frac{D_{E-Air}C_m}{L}\ln\left(\frac{1-y_{E,L}}{1-y_{E,0}}\right) = \frac{(0.145\times10^{-4})(38.16)}{0.15}\ln\left(\frac{1-0}{1-0.1797}\right) = 7.31\times10^{-4}\frac{mol}{m^2s}$$

At $z = L$, where $y_A = 0$, Eq. (15-20a) requires the diffusion flux and the total flux to become equal. The numerical values satisfy this constraint.

Check on Example 15-1. Since the convective flux in Example 15-2 was largest at $z = 0$, we check if the convective flux is significant in Example 15-1 at $z = 0$. The convective flux is

$$\text{Convective flux } (z = 0) = N_E - J_E(z = 0)$$

Using the values from Example 15-1, we have

$$C_{m,Ex\ 15-1} = P_{tot}/(RT) = 0.98/[(8.20567\times10^{-5})(273)] = 43.747 mol/m^3$$

$$y_{E,Ex\ 15-1}(z=0) = C_E(z=0)/C_m = 0.7106/43.747 = 0.01624$$

$$N_{E,Ex\ 15-1} = \frac{D_{E-Air}C_m}{L}\ln\left(\frac{1-y_{E,L}}{1-y_{E,0}}\right) = \frac{(0.102\times10^{-4})(43.747)}{0.15}\ln\left(\frac{1-0}{1-0.01624}\right)$$

$$= 4.87\times10^{-5}mol/(m^2s)$$

At $z = 0$, Eq. (15-20e) becomes,

$$J_{E,Ex15-1}(z=0) = (0.102\times10^{-4})(43.747)\left(\frac{1-0.01624}{0.15}\right)(1)\ln\left(\frac{1-0}{1-0.01624}\right)$$

$$= 4.79\times10^{-5}\ mol/(m^2s)$$

Finally, the convective flux in Example 15-1 is

$$\text{Convective flux } (z = 0) = N_E - J_E(z = 0) = 0.08 \times 10^{-5}\ mol/(m^2s).$$

Thus, even at $z = 0$ where the convective flux is highest, it is only 1.6% of the total flux. When the convective flux is neglected, $N_E = J_E$, which was calculated as 4.83×10^{-5}. This is an error of 0.8% in the total flux. Thus, neglecting the convective flux in Example 15-1 is valid.

Neglecting flux in Example 15-2. Suppose we had neglected the flux in the second example. Then the solution for $J_{E,z}$ would be obtained from Eq. (15-9). With the ethanol concentration at $z = 0$ and ethanol diffusivity of Example 15-2, this is

$$J_{E,z} = -(D_{EW}/L)(C_{F,L} - C_{E,0}) = -\left(\frac{1.45\times10^{-5}m^2/s}{0.15m}\right)\left(0 - 6.857\frac{mol}{m^3}\right) = 6.57\times10^{-4}\frac{mol}{m^2s}$$

Since we are neglecting convection, we would set total flux $N_E = J_E$. Since the actual total flux was 7.31×10^{-4}, the error was 10%, and neglecting the convective flux in Example 15-2 is obviously not valid. If the temperature had been higher, the error would have been larger.

No choice of reference velocity simplifies concentrated absorbers and strippers; thus, we will use molar units in Section 16.4, since equilibrium data are often available in these units.

Students can find the development in this section separating the diffusive and convective fluxes confusing and even distressing. The choice of the reference velocity based on volume, mole, or mass is arbitrary and may result in different values for the convective flux. This in turn will result in different values for the diffusive flux. Even though the total fluxes of A and B will be the same, how can this be correct? The method works because the diffusive flux is defined with respect to a reference frame in which the convective flux is zero. If we change v_{ref}, we change the convective and diffusive fluxes by the exact amounts required to keep the total fluxes unchanged. If we report the diffusive flux and the convective flux, we need to be clear whether a volume, mole, or mass basis was used. Another way to think about this calculation is the total fluxes of A and B are state functions—they do not depend on the calculation path chosen. On the other hand, the diffusive and convective fluxes are path functions and depend on how the calculations are done.

15.3 VALUES AND CORRELATIONS FOR FICKIAN BINARY DIFFUSIVITIES

As noted earlier, determination of the diffusivity requires that a model be defined so that the concentration data collected in the experiment can be analyzed. Almost all of the diffusivity data tabulated in the literature (e.g., Cussler, 2009; Demirel, 2007; Marrero and Mason, 1972; Poling et al., 2008; Reid et al., 1987; Sherwood et al., 1975) were analyzed with the Fickian model.

15.3.1 Fickian Binary Gas Diffusivities

Based on the molecular argument in Section 15.1, we expect that diffusivities will increase with increasing temperature. This is indeed the case. Fortunately, in most cases the temperature dependence can be quite accurately predicted. Ideally, the diffusivity would not depend on concentration. For gases this is approximately true. Typical diffusivity data for gases are given in Table 15-1. At pressures below about 70 atm and at constant temperature, the product $[p_{tot} D_{AB}]$ is constant for each gas pair, and D_{AB} is independent of concentration.

Please do not consider the tables as an opportunity for a reading break. Instead of skipping Table 15-1, study the numbers and try to find patterns. Obviously, D_{AB} increases as temperature increases. By comparing the data for different alcohols in air at 298 K, we see that for the homologous series of alcohols, D_{AB} values are larger for molecules with lower molecular weights. This is generally true even for very different molecules if there is a big difference in molecular weights.

Diffusivities for gas pairs can be fairly accurately predicted from kinetic theories. A simple kinetic theory for hard spheres predicts (Cussler, 2009)

TABLE 15-1.	*Binary Fickian diffusivities for gases at 1.0 atm. For example, the diffusivity of air and ammonia at 273 K is 0.198×10^{-4} $m^2/s = 0.198$ cm^2/s. For each gas pair at low pressures and a given temperature, $p_{tot} D_{AB} = constant$. (Cussler, 2009; Demirel, 2007; Geankoplis, 2003; Poling et al., 2008; Reid et al., 1987; Sherwood et al., 1975; Treybal, 1980).*		

Gas A	Gas B	T Kelvin	$D_{AB} \times 10^4$, cm^2/s
Air	Ammonia	273	0.198
Air	Benzene	298.2	0.096
Air	2-butanol	299.1	0.089
Air	n-butanol	299.1	0.087
Air	Carbon dioxide	273	0.136
Air	Chlorine	273	0.124
Air	Ethanol	273	0.102
Air	Ethanol	298	0.132
Air	Ethanol	313	0.145
Air	n-hexane	294	0.080
Air	n-hexane	328	0.093
Air	Hydrogen	273	0.611
Air	Naphthalene	298	0.0611
Air	n-octane	298	0.0602
Air	n-pentane	294	0.071
Air	2-pentanol	299.1	0.071
Air	2-propanol	299.1	0.099
Air	Water	273	0.220
Air	Water	298.2	0.260
Air	Water	315	0.288
Air	Water	333.2	0.305
Carbon dioxide	Water	298	0.164
Carbon dioxide	Water	307.2	0.198
Carbon dioxide	Water	328.6	0.257
Helium	Water	298.2	0.908
Hydrogen	Ammonia	273	0.748
Hydrogen	Ammonia	358	1.093
Hydrogen	Ammonia	473	1.86
Hydrogen	Ammonia	533	2.149
Hydrogen	Methane	273	0.625
Hydrogen	Water	307.1	0.915
Propane	Iso-butane	378.2	0.0823
Propane	n-butane	378.2	0.107

$$D_{AB} \propto \frac{T^{3/2}(1/\overline{MW})^{1/2}}{p_{tot}\sigma^2} \tag{15-21}$$

T is the absolute temperature in kelvin, $\overline{MW}$ is an average molecular weight, p_{tot} is the total absolute pressure in atmospheres, and σ is average diameter of the spherical molecules in Å. The more detailed and accurate Chapman-Enskog kinetic theory is valid for nonpolar molecules to about 70 atm. This equation with D_{AB} in m^2/s (Cussler, 2009, Geankoplis, 2003; Wankat and Knaebel, 2008) is

$$D_{AB} = \frac{(1.858\times10^{-7})T^{3/2}(1/MW_A +1/MW_B)^{1/2}}{p_{tot}\sigma_{AB}^2\Omega_D} \tag{15-22a}$$

The collision diameter σ_{AB} is determined from the Lennard-Jones potential parameters for the two molecules

$$\sigma_{AB} =(\sigma_A +\sigma_B)/2 \tag{15-22b}$$

and the dimensionless Ω_D is a collision integral that is a function of $k_B T/\varepsilon_{AB}$, where k_B is Boltzmann's constant (1.38066×10^{-23} J/K) and ε_{AB} is the Lennard-Jones energy of interaction, which can be calculated from the values for the two molecules.

$$\varepsilon_{AB} / k_B = \sqrt{\frac{\varepsilon_A \varepsilon_B}{k_B^2}} \tag{15-22c}$$

Table 15-2 is a brief list of Lennard-Jones parameters for a few molecules and includes a brief list of values of Ω_D. Detailed tables are available for a variety of compounds to calculate these parameters (e.g., Cussler, 2009; Hirshfelder et al., 1954). An empirical fit for Ω_D is given by

TABLE 15-2. *Lennard-Jones potential parameters and values of the collision integral for ideal Fickian gas diffusivity calculation with Chapman-Enskog equation (15-22) (Cussler, 2009; Hirshfelder et al., 1954).*

Compound	σ (Å)	ε/k_B (K)	$k_B T/\varepsilon$	Ω_D
Air	3.711	78.6	0.30	2.662
Ammonia	2.900	558.3	0.50	2.066
Benzene	5.349	412.3	0.75	1.667
2-butane	5.278	330.1	1.00	1.439
1-butane	4.687	531.4	1.50	1.198
Carbon dioxide	3.941	195.2	2.00	1.075
Ethane	4.443	215.7	2.5	0.9996
Ethanol	4.530	362.6	5.0	0.8422
Methanol	3.626	481.8	20	0.6640
Propane	5.118	237.1	50	0.5756
Water	2.641	809.1	300	0.4360

Wankat and Knaebel (2008). Because of the temperature dependence of the collision integral, the gas diffusivities are proportional to T^2 at low temperatures and to $T^{1.66}$ at high temperatures. Wankat and Knaebel (2008) summarize other methods to predict gas diffusivities.

Although prediction methods are available, it is always better to have an experimental value for D_{AB} and then adjust for pressure or temperature differences. Both Eqs. (15-21) and (15-22) predict the same inverse dependence of diffusivity on pressure and a slightly different dependence on temperature. If a single experimental value is available, the value of D_{AB} can be accurately predicted at any temperature or pressure. This is illustrated in Example 15-3.

Example 15-3. Estimation of temperature effect on Fickian gas diffusivity
Estimate the diffusivity of carbon dioxide in air at 317.3 K and 1.0 atm.

Solution
From Table 15-1, $D_{air-CO2} = 0.136 \times 10^{-4}$ m²/s at 273 K. From Eq. (15-21), we have

$$\frac{D_{Air-CO2}(317.3K)}{D_{Air-CO2}(273K)} = \left(\frac{317.3}{273}\right)^{3/2} = 1.253 \rightarrow D_{Air-CO2}(317.3K) = 0.170 \times 10^{-4} \, m^2/s$$

From Eq. (15-22), at low temperatures the exponent on temperature is approximately 2.0, and we obtain

$$\frac{D_{Air-CO2}(317.3K)}{D_{Air-CO2}(273K)} = \left(\frac{317.3}{273}\right)^{2} = 1.351 \rightarrow D_{Air-CO2}(317.3K) = 0.184 \times 10^{-4} \, m^2/s$$

From Eq. (15-22), at high temperatures the exponent on temperature is approximately 1.66, and we obtain

$$\frac{D_{Air-CO2}(317.3K)}{D_{Air-CO2}(273K)} = \left(\frac{317.3}{273}\right)^{1.66} = 1.2835 \rightarrow D_{Air-CO2}(317.3K) = 0.175 \times 10^{-4} \, m^2/s$$

This compares to the experimental value of 0.177×10^{-4} m²/s. The predicted value with an exponent of 3/2 is 4.0% low, the predicted value with an exponent of 2.0 is 4.0% high, and the predicted value with an exponent of 1.66 is 1.1% low.

15.3.2 Fickian Binary Liquid Diffusivities

Experimental binary diffusivity data for liquid systems are presented in Table 15-3. The data for proteins in Table 15-3 show that, as expected, the diffusivities are significantly lower than for low molecular weight compounds, and the diffusivity decreases as the molecular weight of the protein increases. Diffusivities (shown for sucrose) decrease in aqueous gels as more solid is added and the gel becomes more viscous. As temperature increases (compare the chlorobenzene-bromobenzene data at the same concentrations or the infinite dilution ethanol-water data), the diffusivity increases. Pressure is not expected to affect liquid diffusivities. Most of the tabulations of liquid diffusivity data in the literature give values at the infinite dilution limits or for just a few concentrations. Compare the infinite dilution value for ethanol in water to the value for water in ethanol. In general, $D^o_{AB} \neq D^o_{BA}$ (see Problem 15.A3).

TABLE 15-3. *Binary Fickian diffusivities for liquids. For example, the diffusivity of air in water at 298.16 K is 2.00×10^{-9} m²/s = 2.00×10^{-5} cm²/s (Cussler, 2009; Demirel, 2007; Geankoplis, 2003; Poling et al., 2008; Sherwood et al., 1975; Treybal, 1980). x is mole fraction of solute A.*

Solute, A	Solvent, B	T Kelvin	Conc. Solute	$D_{AB} \times 10^9$, m²/s or $\times 10^5$, cm²/s
Air	Water	298.16	∞ dilution	2.00
Ammonia	Water	298.16	∞ dilution	1.64
Carbon dioxide	Water	298.16	∞ dilution	1.18
Chlorine	Water	298.16	∞ dilution	2.03
Helium	Water	298.16	∞ dilution	6.28
Hydrogen	Water	298.16	∞ dilution	4.50
Methanol	Water	288	∞ dilution	1.26
Ethanol	Water	283	∞ dilution	0.84
Ethanol	Water	298.16	∞ dilution	1.25
Ethanol	Water	298.15	x = 0.266	0.368
Ethanol	Water	298.15	x = 0.408	0.405
Ethanol	Water	298.15	x = 0.680	0.743
Ethanol	Water	298.15	x = 0.880	1.047
1-propanol	Water	298.16	∞ dilution	0.87
1-butanol	Water	298.16	∞ dilution	0.77
Fibrinogen human	Water	293	∞ dilution	0.0198 (MW 339700)
Γ-globulin human	Water	293	∞ dilution	0.040 (MW 153100)
Hemoglobin	Water	293	∞ dilution	0.0593 (MW 72300)
Sucrose	Water	298.16	C_A mol/L	$0.5228 - 0.65 C_A$
Sucrose	Water	278	∞ dilution	0.285
Sucrose 3	.8 wt % gelatin in water	278	∞ dilution	0.209
Sucrose	10.35 wt % gelatin in water	278	∞ dilution	0.107
Chlorobenzene	Bromobenzene	283.3	x = 0.0332	1.007
Chlorobenzene	Bromobenzene	283.3	x = 0.2642	1.069
Chlorobenzene	Bromobenzene	283.3	x = 0.5122	1.146
Chlorobenzene	Bromobenzene	283.3	x = 0.7617	1.226
Chlorobenzene	Bromobenzene	313.1	x = 0.0332	1.584
Chlorobenzene	Bromobenzene	313.1	x = 0.2642	1.691
KCl	Water	298	∞ dilution	1.870
NaCl	Water	291	0.05 mol/L	1.26
NaCl	Water	291	0.2 mol/L	1.21
NaCl	Water	291	1.0 mol/L	1.24
Water	Ethanol	298.16	∞ dilution	1.132
Water	1-propanol	288	∞ dilution	0.87
Water	Glycerol	293	∞ dilution	0.0083
Carbon dioxide	Ethanol	298	∞ dilution	3.42
Ethanol	Benzene	288	∞ dilution	2.25

There are a number of theories for predicting the diffusivity at infinite dilution (Cussler, 2009; Kirwan, 1987; Sherwood et al., 1975; Wankat and Knaebel, 2008). Many of these theories use the Stokes-Einstein equation as a starting point (Bird, et al., 1960; Cussler, 2009; Wankat and Knaebel, 2008).

$$D_{AB}^{o} = k_B T / (6\pi\mu_B R_A) \tag{15-23a}$$

In this equation μ_B is the viscosity of the solvent and $R_A = V_A^{1/3}$ is the radius of the solute, which is assumed to be a rigid sphere with gravity as the only body force. The Stokes-Einstein equation works best for unhydrated molecules with a molecular weight > 1000, and even there it is not very accurate. The most popular theory for the diffusivity of liquids is the Wilke-Chang theory (Cussler, 2009; Geankoplis, 2003; Sherwood et al., 1975; Wankat and Knaebel, 2008), which uses the Stokes-Einstein equation as a starting point and predicts the infinite dilution diffusivity of solute A in solvent B, D_{AB}^{o}.

$$D_{AB}^{o} = \frac{1.173 \times 10^{-16} [\varphi_B (MW_B)]^{1/2} T}{\mu_B V_A^{0.6}} \tag{15-23b}$$

In this equation V_A is the molar volume of solute in m^3/kmol at its normal boiling point, T is in kelvin, the solvent viscosity μ_B is in Pa·s [kg/(m s)], φ_B is a solvent interaction parameter, and D_{AB}^{o} is in m^2/s. There is some disagreement in the literature on the appropriate values of φ_B, particularly for water. The following values are recommended for different solvents (Wankat and Knaebel, 2008): water, $\varphi_B = 2.26$ [other authors such as Geankoplis (2003) recommend $\varphi_B = 2.6$]; methanol, $\varphi_B = 1.9$; ethanol, $\varphi_B = 1.5$; propanol, $\varphi_B = 1.2$; other solvents, $\varphi_B = 1.0$.

Although not obvious from the form of the equation, because of the variation of viscosity with temperature, the Wilke-Chang equation predicts an Arrhenius dependence on temperature (Kirwan, 1987).

$$D_{AB}^{o}(T) = D_o \exp\left(\frac{-E_o}{RT}\right) \tag{15-23c}$$

Typical values for the activation energy E_o are approximately 10,000 J/mol and R = 8.314 J/(mol K). This equation can also be used at constant mole fractions instead of infinite dilution if data are available (see Problem 15.D5). Alternatively, if the effect of temperature on solvent viscosity is known, Eq. (15-23b) can be used directly to determine the temperature dependence of D_{AB}^{o}.

Unfortunately, except in quite dilute mixtures, liquid diffusivity is rarely constant, and very large changes in D_{AB} are often observed going from a trace of A in almost pure B (the infinite dilution limit) to a trace of B in almost pure A (the other infinite dilution limit). For example, compare the change in values of D_{AB} in Table 15-3 for the relatively ideal chlorobenzene–bromobenzene system at 283.3 K to the change in values of D_{AB} for the nonideal ethanol–water system at 298.15 K. The former values increase modestly, while the latter values go through a minimum as the water mole fraction increases. If the diffusivity is not constant, we would at least hope that there is a relatively simple relationship that allows calculation of D_{AB} at any concentration based on the two infinite dilution limits, D_{AB}^{o} and D_{BA}^{o}. The Vignes correlation based on the activity coefficient γ_A is reasonably accurate (Kirwan, 1987; Sherwood et al., 1975; Treybal, 1980) for moderately nonideal systems.

$$D_{AB}(x_A) = (D_{AB}^o)^{1-x_A}(D_{BA}^o)^{x_A}\left(1 + x_A \frac{d \ln \gamma_A}{d \ln x_A}\right)$$ **(15-23d)**

The last term is a thermodynamic correction factor for nonideal solutions. Equation (15-23d) predicts that D_{AB} is not constant even for ideal systems [term in brackets in Eq. (15-23d) has a value of 1.0]. Equation (15-23d) can also predict a negative diffusion coefficient, which is an indication of the formation of two liquid phases and is very important for liquid-liquid extraction.

Again, the most accurate results are obtained not by predicting the diffusivity but by obtaining experimental values at two, preferably more, temperatures and then adjusting for temperature differences. Activity coefficient data is required to adjust for concentration changes in nonideal systems.

15.4 LINEAR DRIVING-FORCE MODEL OF MASS TRANSFER FOR BINARY SYSTEMS

Unfortunately, because of the complexity of the flow fields in most separators, reactors, and other devices where interfacial mass transfer occurs, solution of the Fickian diffusion equations for binary systems is not usually feasible. In these cases we typically use the empirical linear driving-force model for mass transfer that was briefly introduced in Section 1.3. This equation [a modification of Eq. (1-4)] is

Flux = Mass-transfer rate/area = (mass-transfer coefficient) × (driving force)   **(15-24a)**

As noted in Section 15.1, the use of *driving force* can be misleading, since the molecules have no brain or intention to transfer, and in molecular diffusion processes, transfer occurs because of random collisions. However, since *driving force* is embedded in the jargon of chemical engineering, we use the term here. The strength of Eq. (15-24a) is that it is very broad and flexible and can be applied to a wide variety of situations. The weakness of this equation is that it is empirical, and the theoretical background is weak. In some practical situations a few experiments will suffice to provide the mass-transfer coefficient in the range of interest. If temperature, pressure, and concentrations vary significantly in the separator, successful application of the equation requires a correlation for the mass-transfer coefficient based on extensive experimental data.

Equation (15-24a) can be written in terms of concentrations as

$$J_A = (\text{Rate of mass-transfer species A})/\text{Area} = k_c(C_{A,2} - C_{A,1})$$ **(15-24b)**

If the flux is desired in (kmol A)/(s m^2), the area across which the mass transfer occurs is measured in m^2, and concentration is in (kmol A)/m^3, then the mass-transfer coefficient k_c has units of velocity, m/s. In single-phase systems the concentrations $C_{A,2}$ and $C_{A,1}$ are the (kmol A)/m^3 at locations 1 and 2, which are usually the system boundaries. For mass transfer from one phase to another, one of the concentrations would be the concentration at the interface. Typical values for the mass-transfer coefficient are 0.1 m/s for gases and 10^{-4} m/s for liquids (Wesselingh and Krishna, 2000).

We can also write the mass-transfer equation in terms of liquid mole fractions:

$$J_A = \text{Rate transfer species A /Area} = k_x(x_{A,2} - x_{A,1}) \tag{15-24c}$$

In terms of vapor mole fractions, the equation is

$$J_A = k_y(y_{A,2} - y_{A,1}) \tag{15-24d}$$

In these cases the mass-transfer coefficients k_x and k_y have units of (kmol A)/[m^2 s (mole fraction A)] or equivalently (kmol fluid mixture)/(m^2 s). Occasionally, the flux is written in terms of a partial pressure driving force, $p_A = y_A p_{tot}$,

$$J_A = k_p(p_{A,2} - p_{A,1}) \tag{15-24e}$$

In this case the mass-transfer coefficient $k_p = k_y / p_{tot}$ has units (kmol fluid mixture)/(m^2 s bar).

In Section 15.2.1 the analogy between heat and mass transfer was discussed. This analogy also extends to linear driving-force models as long as heat transfer by radiation can be neglected. The heat-transfer equation analogous to Eq. (15-24a) is

Flux of heat = Heat-transfer rate/area = (heat-transfer coefficient) × (driving force) **(15-25a)**

The heat-transfer equation analogous to Eqs. (15-24b) to (15-24e) is

$$Q_z = h_{\text{heat transfer}}(T_2 - T_1) \tag{15-25b}$$

Here Q_z is the heat flux in J/(m^2s), $h_{\text{heat transfer}}$ is the heat-transfer coefficient in J/(m^2s·K) and T_2 and T_1 are the temperatures at the two system boundaries, K. The analogy between heat and mass transfer will prove useful in determining correlations for the mass- and heat-transfer coefficients in Section 15.5.4.

15.4.1 Film Theory for Dilute and Equimolar Transfer Systems

Figure 15-1 showed diffusion across a thin layer or film. Later in this chapter Figure 15-5 will show mass transfer to an interface through thin films of gas and liquid. After that, Figure 15-6 will show mass transfer in absorption when the liquid mass-transfer rate is very rapid, and Figure 15-7 will show mass transfer to a falling liquid film. The reason for this interest in mass transfer through films is that the most commonly used model for mass transfer in separators is a film model. The film model assumes that mass transfer to a surface or interface from the bulk fluid occurs across a stagnant film of unknown thickness δ. The mass-transfer rate in this film occurs by molecular diffusion alone and can be modeled with Fick's law and with one of the forms of Eq. (15-24). In Section 15.2 we saw that the flux J_A is the flux in a reference coordinate system moving at a reference velocity v_{ref} that made $J_A = -J_B$, so that the net flux in the reference coordinate system was zero. However, we really want the flux N_A calculated with respect to the fixed coordinates of the equipment. In this section we develop the film theory for dilute systems and systems with equimolar countertranster both of which have $N_A = J_A$. These systems are important for dilute absorbers and extractors and for distillation, which is usually very close to equimolar countertransfer.

Since the film theory postulates that mass transfer occurs across a stagnant film, we can use the diffusion solution from Eq. (15-9) for diffusion across a thin layer of fluid. If this thin layer is a gas film of unknown thickness δ and we convert from concentrations to mole fractions, we obtain

$$J_{A,z,gas} = (C_{m,gas}D_{AB,gas}/\delta_{gas})(y_{A,0}-y_{A,\delta}) \tag{15-26a}$$

For systems that are dilute or have equimolar countertransfer $N_{A,z} = J_{A,z}$. Since the film thickness δ_{gas} is unknown, we can define $(C_{m,gas}D_{AB,gas}/\delta_{gas})$ as the mass-transfer coefficient k_y and $(D_{AB,gas}/\delta_{gas})$ as the mass-transfer coefficient $k_{c,gas}$. Then the equation for mass transfer for dilute or equimolar countertransfer is

$$N_{A,z,gas} = J_{A,z,gas} = k_y(y_{A,0}-y_{A,\delta}) \tag{15-26b}$$

or in terms of concentration,

$$N_{A,z,gas} = J_{A,z,gas} = k_{c,gas}(C_{A,gas,0}-C_{A,gas,\delta}) \tag{15-26c}$$

For equilibrium staged and sorption separations, we are interested in mass transfer from one phase to another. This is illustrated schematically in Figure 15-5 for the transfer of component A from the liquid to a vapor phase. x_I and y_I are the interfacial mole fractions. For dilute absorbers and strippers and for distillation where there is equimolar countertransfer of the more volatile and less volatile components, the mass-transfer Eq. (15-26b) can be written for each stage in the following different forms:

$$\text{Rate of mass transfer in gas phase} = A_I N_{A,z,gas} = A_I k_y(y_{A,I}-y_A) \tag{15-27a}$$

$$\text{Rate of mass transfer in liquid phase} = A_I N_{A,z,liq} = A_I k_x(x_A-x_{A,I}) \tag{15-27b}$$

or

$$\text{Rate of mass transfer in gas phase} = A_I N_{A,z,gas} = A_I k_{c,gas}(C_{A,gas,I}-C_{A,gas}) \tag{15-27c}$$

$$\text{Rate of mass transfer in liquid phase} = A_I N_{A,z,liq} = A_I k_{c,liq}(C_{A,liq}-C_{A,liq,I}) \tag{15-27d}$$

where k_y and k_x are the individual mass-transfer coefficients with mole-fraction driving forces for the vapor and liquid phases, respectively, and $k_{c,gas}$ and $k_{c,liq}$ are the individual mass-trans-

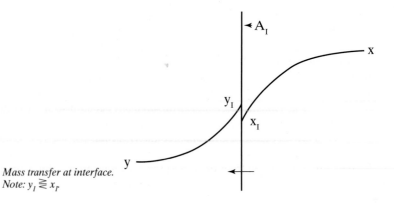

FIGURE 15-5. *Mass transfer at interface.*
Note: $y_I \gtreqless x_I$

fer coefficients with concentration driving forces for the vapor and liquid phases, respectively. At steady state, the rates of mass transfer in the gas and liquid phases have to be equal, and $N_{A,z,gas} = N_{A,z,liq}$. Additionally, the assumption is usually made that the gas and liquid at the interface are in equilibrium. If there is a resistance to obtaining equilibrium at the interface (for example, if a surface-active agent is present), then an equation for this resistance must be included.

Unfortunately, there are two major problems with these equations when they are applied to vapor-liquid and liquid-liquid contactors. First, the interfacial area A_I between the two phases is very difficult to measure. This problem is usually avoided by writing Eqs. (15-27) as

$$\text{Rate of mass transfer/Volume} = k_y a(y_{A,I} - y_A) \tag{15-28a}$$

$$\text{Rate of mass transfer/Volume} = k_x a(x_A - x_{A,I}) \tag{15-28b}$$

or

$$\text{Rate of mass transfer/Volume} = k_{c,vap} a(C_{A,vap,I} - C_{A,vap}) \tag{15-28c}$$

$$\text{Rate of mass transfer/Volume} = k_{c,liq} a(C_{A,liq} - C_{A,liq,I}) \tag{15-28d}$$

where a is the interfacial area per unit volume of the column (m^2/m^3). Since a is no easier to measure than A_I, we often measure and correlate the products $k_y a$, $k_x a$, $k_{c,vap} a$, and $k_{c,liq} a$. Typical units for $k_y a$ and $k_x a$ are kmol/(s-m^3) or lbmol/(h-ft^3), and for $k_c a$, typical units are $[(m/s)(m^2/m^3)] = s^{-1}$.

The second problem is that the interfacial mole fractions are also very difficult to measure. To avoid this problem, mass-transfer calculations often use a driving force defined in terms of hypothetical equilibrium mole fractions.

$$\text{Rate of mass transfer/Volume} = K_y a(y_A^* - y_A) \tag{15-29a}$$

$$\text{Rate of mass transfer/Volume} = K_x a(x_A - x_A^*) \tag{15-29b}$$

These equations [which are a repeat of Eqs. (1-5)] define the overall mass-transfer coefficients K_y and K_x. Typical units for $K_y a$ and $K_x a$ are kmol/(s-m^3) or lbmol/(h-ft^3). y_A^* is the vapor mole fraction, which would be in equilibrium with the bulk liquid of mole fraction x_A, and x_A^* is the liquid mole fraction that would be in equilibrium with the bulk vapor of mole fraction y_A.

To obtain the relationship between the overall and individual coefficients, we begin by assuming there is no resistance to mass transfer at the interface. This assumption implies that x_I and y_I must be in equilibrium. The mole fraction difference in Eq. (15-29a) can be written as

$$(y_A^* - y_A) = (y_A^* - y_{AI}) + (y_{AI} - y_A) = m(x_A - x_{AI}) + (y_{AI} - y_A) \tag{15-30a}$$

where m is the average slope of the equilibrium curve (y_A versus x_A) at x_A and x_{AI}.

$$m = (\partial y_A^* / \partial x_A)_{avg} \tag{15-30b}$$

Combining Eq. (15-30a) with Eqs. (15-28) and (15-29a), we obtain

$$\frac{\text{Rate/Volume}}{K_y a} = \left(\frac{\text{rate}}{\text{volume}}\right)\left[\frac{m}{k_x a} + \frac{1}{k_y a}\right] \tag{15-31a}$$

which leads to the result

$$\frac{1}{K_y a} = \frac{m}{k_x a} + \frac{1}{k_y a} \tag{15-31b}$$

Similar manipulations starting with Eq. (15-29b) lead to

$$\frac{1}{K_x a} = \frac{1}{k_x a} + \frac{1}{mk_y a} \tag{15-31c}$$

If there is a resistance at the interface, then this resistance must be added to Eqs. (15-31). This *sum-of-resistances* model shows that the overall coefficients will not be constant even if k_x and k_y are constant if the equilibrium is curved and m varies. In binary distillation m is the tangent to the equilibrium curve at $x_{A,avg} = (x_{A,I} + x_A)/2$, and since the equilibrium curve is never straight, m has to vary. For example, if we have a constant relative volatility system with equilibrium given by Eq. (2-22b), m from Eq. (15-30b) is

$$m = \frac{\alpha}{[1+(\alpha-1)x]^2}\bigg|_{x_{A,avg}} \tag{15-30c}$$

As an example of the variation in m, if the relative volatility $\alpha = 2.5$, at $x_{A,avg} = 0.01$, m = 2.43; at $x_{A,avg} = 0.5$, m = 0.82; and at $x_{A,avg} = 0.99$, m = 0.40. At the very least, average values of m need to be determined separately for the stripping and enriching sections.

Equations (15-31) also show the effect of equilibrium on the controlling resistance. If m is small, then from Eq. (15-31b) $K_y \sim k_y$ and the gas-phase resistance controls. If m is large, then Eq. (15-31c) gives $K_x \sim k_x$ and the liquid-phase resistance controls. Absorption of sparingly soluble gases that can have very large Henry's law constants (Section 12.1) are an example of a liquid-phase resistance-controlled separation, while absorption of very soluble gases (e.g., HCl in water) often have gas-phase resistance controlling. In distillation, m is often close to 1.0 and both resistances are important.

15.4.2 Transfer through Stagnant Films: Absorbers and Strippers

For absorbers and strippers (Section 16.4), the film model is often used. We again postulate a film of thickness δ, which has mass transfer by diffusion only. This is shown in Figure 15-6 for absorption with mass transfer of solute A in the gas-phase controlling. In absorption the carrier gas B often does not absorb; thus, there is a stagnant layer of B, $N_B = 0$. Then, from Eq. (15-19),

$$N_A = J_A/(1-y_A) \text{ when } N_B = 0 \tag{15-32a}$$

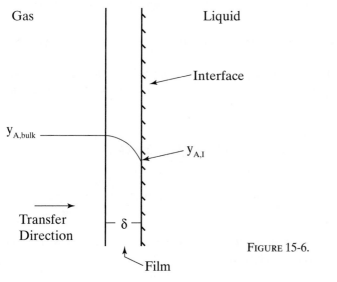

Gas

Liquid

Interface

$y_{A,bulk}$

$y_{A,I}$

Transfer
Direction

δ

Film

FIGURE 15-6. *Film model for absorption with gas-phase mass-transfer controlling*

The solution obtained in Example 15.2, Eq. (15-20a) can be written for transfer between phases as

$$N_A = \frac{D_{AB,gas} C_m}{\delta} \ln\left(\frac{1-y_{A,I}}{1-y_{A,bulk}}\right) \tag{15-32b}$$

In this equation $y_{A,I}$ is the mole fraction of A at the interface (see Figure 15-6), and $y_{A,bulk}$ is the mole fraction of A in the bulk of the gas. Equation (15-32b) can also be written as

$$N_A = \frac{D_{AB} C_m [(1-y_{A,I})-(1-y_{A,bulk})]}{\delta(1-y_A)_{lm}} = \frac{D_{AB} C_m [(y_{B,I})-(y_{B,bulk})]}{\delta(y_B)_{lm}} \tag{15-32c}$$

The log mean differences $(1-y_A)_{lm}$ and $(y_B)_{lm}$ are defined as

$$(1-y_A)_{lm} \equiv \frac{(1-y_{A,I})-(1-y_{A,bulk})}{\ln\left(\dfrac{1-y_{A,I}}{1-y_{A,bulk}}\right)} \text{ and } (y_B)_{lm} \equiv \frac{(y_{B,I})-(y_{B,bulk})}{\ln\left(\dfrac{y_{B,I}}{y_{B,bulk}}\right)} \tag{15-32d}$$

Note that as $y_A \to 0$ (very dilute systems), $(1-y_A)_{lm} \to 1$, and Eq. (15-32c) simplifies to (15-26a).

The reason for doing this rather convoluted algebra is that for concentrated systems with $N_B = 0$, if we define the concentrated mass-transfer coefficient as

$$k'_y = k_y / (1-y_A)_{lm} = (C D_{AB,gas} / \delta) / (1-y_A)_{lm} = (C D_{AB,gas} / \delta) / (y_B)_{lm} \tag{15-32e}$$

then the flux becomes

$$N_A = k'_y (y_{A,bulk} - y_{A,I})$$

(15-32f)

which mimics the flux for dilute systems, Eq. (15-26b). This analysis can also be done based on partial pressure differences and a log mean partial pressure difference (see Problem 15.C3). The concentrated system analysis is used in Section 16.4.

15.5 CORRELATIONS FOR MASS-TRANSFER COEFFICIENTS

Intuitively, we would expect that as temperature increases, molecules will move faster (higher diffusivity), and k will increase. In addition, if fluid flow is involved, we would expect that a higher velocity or more mixing will increase the mass-transfer rate. Both of these intuitive statements turn out to be valid.

In Section 15.5.2 we will see that for simple physical situations such as steady-state mass transfer across a stagnant film, the mass-transfer coefficient can be determined analytically by solving the appropriate diffusion equations. Because the actual process often does not match the assumptions of the theory, the resulting theoretical correlation is often not as accurate as we desire. However, with a modest amount of data, the coefficient can be adjusted to provide a significantly more accurate semi-empirical correlation. For the very common and practical complex situations that do not allow an analytical solution of the diffusion equation, empirical correlations for k or for ka are often developed on the basis of dimensional analysis and experimental data. Dimensional analysis can be used to predict probable forms for correlations, and the experimental data allow one to determine the exponents on the dimensionless groups in addition to the value of the coefficient. A few mass-transfer correlations are presented in this section, and a large number are summarized by Incropera et al. (2011) and in Section 5B of *Perry's Handbook of Chemical Engineering* (Wankat and Knaebel, 2008).

In simple geometries, determination of the area A in Eq. (15-1b), (15-1c), (15-2a), or (15-2b) is straightforward and is illustrated in this section. Membrane separators (Chapter 17) have the advantage that the area A and the area per volume a can be determined from straightforward geometric calculations. For more complex systems with interfacial mass transfer, the value of the interfacial area per volume a in Eqs. (15-3) and (15-4) and in the HTU terms in Table 16-1 is more difficult to determine. Although a can be determined separately, it is more common to determine the product (ka). Correlations for (ka) in packed beds for distillation and absorption are discussed in Section 16.3. Correlations to calculate k and a separately in extraction mixers are discussed in Section 16.7. Correlations to estimate k in membrane separators are given in Section 17.4.3. Correlations for (ka) in adsorption, chromatography, and ion exchange are discussed in Section 18.6.3.

15.5.1 Dimensionless Groups

Correlations for mass transfer are often arranged into equations where all terms occur as dimensionless groups. One advantage of doing this is that any set of units can be used as long as the group remains dimensionless. A second advantage is that if the form of the correlation is not suggested by theory, dimensional analysis can be used to determine appropriate forms for the dimensionless groups.

Mass-transfer correlations are often reported as the Sherwood number Sh, which is defined as

$$Sh_c \equiv (k_c \times length) / D_{AB}, \quad Sh_x = (k_x \times length) / (C_m D_{AB}),$$
$$Sh_y = (k_y \times length) / (C_m D_{AB}) \tag{15-33a}$$

The Sherwood number is the ratio of the actual rate of mass transfer to the mass transfer due to diffusion alone. The appropriate length to use is often a film thickness, a diameter, or a plate length. The mass-transfer coefficient can be either a *local* coefficient (e.g., at a given distance) or an *average* coefficient over the entire distance. This is discussed further in the next section. The dimensionless group defined in the first equation of Eq. (15-33a) is defined in terms of k_c, which uses rate Eq. (15-24b). For the second and third equations of Eq. (15-33a) that use k_x or k_y from rate Eqs. (15-24c) and (15-24d), the total molar concentration term C_m is needed in the definition of the Sherwood number to make it dimensionless.

The mass-transfer coefficient often depends upon the velocity. This is conveniently represented as a Reynolds number Re.

$$Re \equiv \frac{(length)(velocity)(density, \rho)}{(viscosity, \mu)} \tag{15-33b}$$

The Reynolds number is the ratio of inertial forces to viscous forces. In pipes the Reynolds number has the familiar form $Re = Dv_b\rho/\mu$, but in other configurations, such as the flow of a liquid film down a flat plate, the definition can look superficially very different [see Eq. (15-43a)].

In correlations of gas or liquid data in separators, we often use the Peclet number Pe, the ratio of flow velocity to the diffusion velocity.

$$Pe \equiv \frac{(velocity)}{(diffusivity)/(length)}, \quad Pe_{liq} = \left(\frac{vL}{D_{AB}}\right)_{liq}, \text{ or } Pe_{liq} = \left(\frac{vd_p}{D_{AB}}\right)_{liq}$$
$$Pe_{gas} = \left(\frac{vL}{D_{AB}}\right)_{gas} \text{ or } Pe_{gas} = \left(\frac{vd_p}{D_{AB}}\right)_{gas} \tag{15-33c}$$

Forms of the Peclet number are shown with both column length L and particle diameter d_p because both forms are used. It is obviously important to know whether L or d_p is being employed in the definition of Pe. Sometimes a dimensionless group called the Stanton number St, the ratio of mass-transfer velocity to flow velocity, is used.

$$St_c \equiv \frac{k_c}{v} = \frac{Sh_c}{Pe}, \quad St_x = \frac{k_x}{vC_{liq}} = \frac{Sh_x}{Pe_{liq}}, \quad St_y = \frac{k_y}{vC_{gas}} = \frac{Sh_y}{Pe_{gas}} \tag{15-33d}$$

The effect of the properties of the gas or liquid on mass transfer are taken into account in correlations by the Schmidt number Sc, which can be considered as the ratio of momentum diffusivity to mass diffusivity.

$$Sc \equiv \frac{(kinematic\ viscosity)}{(diffusivity)}, \quad Sc_{liq} = \left(\frac{\mu/\rho}{D_{AB}}\right)_{liq}, \quad Sc_{gas} = \left(\frac{\mu/\rho}{D_{AB}}\right)_{gas} \tag{15-33e}$$

In Section 15.5.4 we consider the analogy between heat and mass transfer. The dimensionless group for heat transfer analogous to Sc is the Prandtl number Pr, which is the ratio of momentum diffusivity to thermal diffusivity.

$$Pr \equiv \frac{(\text{kinematic viscosity})}{(\text{thermal diffusivity})} = \frac{\nu_{\text{kinematic}}}{\alpha_{\text{thermal}}} = \frac{(\mu/\rho)}{(k_{\text{conduct}}/(\rho\hat{C}_p))} = \frac{\mu\hat{C}_p}{k_{\text{conduct}}} \tag{15-33f}$$

The dimensionless group for heat transfer analogous to Sh is the Nusselt number Nu, which is the ratio of actual rate of heat transfer to the heat transfer due to thermal diffusion alone.

$$Nu \equiv \frac{h_{\text{heat transfer}}(\text{length})}{k_{\text{conduction}}} \tag{15-33g}$$

Since all of these dimensionless groups contain terms that depend on temperature and concentration, the dimensionless groups depend on both temperature and concentration. In dilute systems the concentration dependence is often negligible and the (temperature dependent) properties of the solvent can be used.

15.5.2 Theoretically Derived Mass-Transfer Correlations

If the physical situation closely matches the assumptions made in the development of theoretically derived mass-transfer correlations, these correlations can be quite accurate. If the physical situation is somewhat different than the assumptions, the correlation may still be useful by using a small amount of experimental data to tune the coefficient. Practicing engineers would seldom do the derivations, but since they commonly use the results, it is important to know both the assumptions and the limits of validity of the correlation. The best way to understand these is to closely follow the derivation.

As a simple first example, consider the case of diffusion in a stagnant film ($N_B = 0$) that was solved in Example 15-2. The general solution for the constant total flux through the film was given in modified form by Eq. (15-32c). This total flux has to be equal to the flux calculated from the linear driving-force model, Eqs. (15-32f) and (15-32g). Setting these results equal for a film of known thickness $\delta = L$, we obtain

$$N_A = \frac{D_{AB,\text{gas}}C_m[(1-y_{A,1})-(1-y_{A,2})]}{L(1-y_A)_{\text{lm}}} = k_y'(y_{A,2}-y_{A,1}) = [k_y/(1-y_A)_{\text{lm}}](y_{A,2}-y_{A,1}) \tag{15-34a}$$

Solving for k_y, we obtain the simple result that

$$k_y = D_{AB}C_m/L \quad \text{and} \quad k_y' = k_y/(1-y_A)_{\text{lm}} = (D_{AB}C_m/L)/(1-y_A)_{\text{lm}} \tag{15-34b}$$

For the stagnant film, the appropriate length for the Sherwood number Eq. (15-33a) is the film thickness L, and the correlation is

$$Sh = (k_y L/C_m D_{AB}) = 1 \tag{15-34d}$$

Use of the linear driving-force model for this single-phase system would be unusual, but it could certainly be done. For a concentrated system Eq. (15-32f) would be employed. For a dilute system this simplifies to Eq. (15-26b). Note that this development has been for a gas film, but we could easily repeat it for a liquid film with liquid mole fractions and the liquid-phase mass-transfer coefficient k_x.

Generally, the linear driving-force model is used when there is transfer between phases. We consider the case of mass transfer from a gas to a falling laminar liquid film as shown in Figure 15-7. This is a classic problem that practically every mass-transfer book includes. First, the problem has practical significance, since falling films can occur in absorption, distillation, and stripping. Second, the basic problem can be solved exactly with a series solution, and simple solutions are available for short residence times. Third, the mass transfer becomes more complicated, requiring a semi-empirical correlation, as the fluid velocity increases and ripples or waves form on the liquid surface. Finally, at even higher velocities, when the film becomes turbulent, empirical correlations are required. The theoretical model is considered in this section, and the semi-empirical and empirical models in Section 15.5.3.

The diffusion equation for the system in Figure 15-7 can be solved theoretically for the following rather restrictive set of conditions that can apply to absorption or stripping:

1. Flow of the liquid is fully developed and laminar (will be relaxed in Section 15.5.3).
2. The surface of the film is flat (will be relaxed in Section 15.5.3).
3. The entering liquid has a constant, uniform concentration $C_{A,init}$.
4. The gas-phase mass-transfer rate is much greater than the liquid-phase mass-transfer rate. Thus, the surface concentration (at $z = 0$) is constant at $C_{A,surf}$ (will be relaxed in Section 15.5.3).

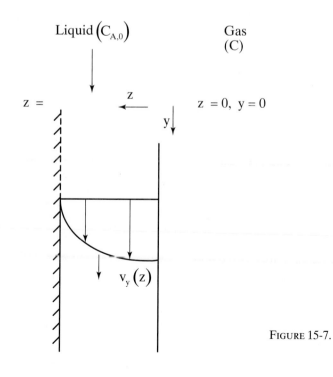

FIGURE 15-7. *Mass transfer into a falling liquid film*

5. The rate of diffusion in the liquid in the horizontal z direction is small, and convection in the z direction can be neglected.
6. In the vertical y direction, the rate of convection due to the liquid flow is much greater than diffusion, and diffusion in the y direction can be neglected.
7. The plate is wide enough that side-end effects can be ignored, and neither velocity nor concentration depend on the horizontal direction.
8. The diffusion coefficient D_{AB} is constant.
9. The system is very dilute (this assumption is not necessary but simplifies the problem).

The solution for the fluid velocity is a well-known result (Geankoplis, 2003; Treybal, 1980).

$$v_y = v_{y,max}\left[1-\left(\frac{z}{\delta}\right)^2\right] \text{ with } v_{y,max} = \frac{\rho g \delta^2}{2\mu} \qquad \text{(15-35a, b)}$$

The average velocity and film thickness δ are

$$v_{y,avg} = \frac{\rho g \delta^2}{3\mu} \text{ and } \delta = \left(\frac{3\mu q}{\rho g}\right)^{1/3} \qquad \text{(15-35c, d)}$$

where q is the volumetric flow rate divided by the width of the plate [$(m^3/s)/m = m^2/s$]. The average residence time of fluid on the plate, in seconds is

$$t_{res,avg} = L / v_{y,avg} \qquad \text{(15-35e)}$$

The steady-state mass balance states that (In – Out) of solute due to convection (in the y direction) equals (In – Out) due to diffusion (in the z direction). With constant diffusivity, this is

$$v_y(z)(C_A\big|_y - C_A\big|_{y+\Delta y}) = D_{AB}\left[\left(\frac{\partial C_A}{\partial z}\right)\bigg|_z - \left(\frac{\partial C_A}{\partial z}\right)\bigg|_{z+\Delta z}\right] \qquad \text{(15-36a)}$$

Taking the limits as $\Delta y \to 0$ and $\Delta x \to 0$ (assuming they exist), this equation becomes

$$v_y \frac{\partial C_A}{\partial y} = D_{AB}\frac{\partial^2 C_A}{\partial z^2} \qquad \text{(15-36b)}$$

This problem has a number of interesting characteristics. First, it has been solved for two different sets of boundary conditions. For relatively slow diffusion rates or short residence times (short length), the solute from the liquid surface never penetrates to the wall. The boundary conditions are

$$\text{At } z = 0, C_A = C_{A,perf}; \text{ at } y = 0, C_A = C_{A,init}; \text{ at } z = \infty, C_A = C_{a,init} \qquad \text{(15-37)}$$

These boundary conditions are the same as in Eq. (15-13) for unsteady diffusion. The answer is similar to Eq. (15-14) and to Eq. (18-69) for adsorption. This result (modified from Geankoplis, 2003; and Cussler, 2009) is

$$\frac{C_A - C_{A,init}}{C_{A,Surf} - C_{A,init}} = 1 - erf\left(\frac{z}{\sqrt{4D_{AB}y/v_{y,max}}}\right) \qquad \textbf{(15-38a)}$$

The definition of the error function is given in Eq. (18-70), and values are tabulated in Table 18-7. For our purposes here, note that

$$erf(0) = 0,\ erf(\infty) = 1,\ \frac{d[erf(a)]}{dz} = \frac{2\exp(-a^2)}{\sqrt{\pi}}\frac{da}{dz},\ \frac{d[erf(a)]}{dz}\bigg|_{a=0} = \frac{2}{\sqrt{\pi}}\frac{da}{dz}\bigg|_{a=0} \qquad \textbf{(15-38b)}$$

With these values, Eq. (15-38a) satisfies the boundary conditions in Eq. (15-37). For a dilute system, the flux can be calculated from Fick's law, Eq. (15-4a). From Fick's law and Eqs. (15-38a) and (15-38b), one obtains at z = 0,

$$J_{A,z}(y)\big|_{z=0} = -D_{AB}\frac{\partial C_A}{\partial z}\bigg|_{z=0} = (C_{A,surf} - C_{A,init})\sqrt{\frac{D_{AB}v_{y,max}}{\pi y}} \qquad \textbf{(15-39a)}$$

In general, for a system with $N_{B,z} = 0$, the relationship between N_A and J_A is given by Eq. (15-32a). In dilute systems this simplifies to $N_A = J_A$, and Eq. (15-29a) becomes

$$N_{A,z}(y)\big|_{z=0} = -D_{AB}\frac{\partial C_A}{\partial z}\bigg|_{z=0} = (C_{A,surf} - C_{A,init})\sqrt{\frac{D_{AB}v_{y,max}}{\pi y}} \qquad \textbf{(15-39b)}$$

For dilute systems the linear driving-force equation simplifies to

$$N_A(y)\big|_{z=0} = k_{c,liq,loc}(C_{A,surf} - C_{A,init}) \qquad \textbf{(15-39c)}$$

Setting Eqs. (15-39b) and (15-39c) equal, we obtain for dilute systems at z = 0,

$$k_{c,liq,local} = \sqrt{\frac{D_{AB}v_{y,max}}{\pi y}} \qquad \textbf{(15-39d)}$$

or in terms of the Sherwood number with the length chosen as the film thickness δ, at z = 0 this is

$$Sh_{local} = k_{c,liq,local}\delta/D_{AB} = \sqrt{\frac{\delta^2 v_{y,max}}{D_{AB}\pi y}} \qquad \textbf{(15-39e)}$$

The *local* mass-transfer coefficient depends on the value of y and is undefined at y = 0. For dilute systems $J_A = N_A$. Making this substitution in Eq. (15-39a) and integrating over the range from y = 0 to y = L (the plate length), we obtain the total amount of solute transferred into the liquid at the surface of the liquid (z = 0).

$$N_A \times (Lw) = w\int_0^L (C_{A,surf} - C_{A,init})\sqrt{\frac{D_{AB}v_{y,max}}{\pi y}}\,dy = Lw(C_{A,surf} - C_{A,init})\sqrt{\frac{4D_{AB}v_{y,max}}{\pi L}} \qquad \textbf{(15-40a)}$$

For dilute systems the linear driving-force model for the average mass-transfer rate is analogous to Eq. (15-39c),

$$N_{A,avg} = k_{c,liq,avg}(C_{A,surf} - C_{A,init})$$ (15-40b)

We can calculate the total transfer rate as $N_{A,avg} \times (Lw)$. Setting this value equal to Eq. (15-40a) and solving for the average mass-transfer coefficient and the average Sherwood number, we obtain

$$k_{c,liq,avg} = \sqrt{\frac{4D_{AB}v_{y,max}}{\pi L}}, \; Sh_{avg} = k_{c,liq,avg}\delta/D_{AB} = \sqrt{\frac{4\delta^2 v_{y,max}}{D_{AB}\pi L}}$$ (15-40c)

The alternative solution (Treybal, 1980; Sherwood et al., 1975) to Eq. (15-36) does not assume that contact time is short and uses the boundary conditions

$$\text{at } z = 0, \, C_A = C_{A,surf}; \text{ at } y = 0, \, C_A = C_{A,init}; \text{ at } z = \delta, \, \frac{\partial C_A}{\partial z} = 0$$ (15-41)

The last boundary condition at the solid wall implies there is no diffusion through the solid wall. Equation (15-36) is more difficult to solve with this set of boundary conditions. The solution is an infinite series for the average concentration leaving the film at $y = L$, $C_{A,avg,L}$.

$$\frac{C_{A,surf} - C_{A,avg,L}}{C_{A,surf} - C_{A,init}} = 0.7857 \exp(-5.1213\eta) + 0.1001 \exp(-39.318\eta) + \cdots$$ (15-42a)

$$\eta = \frac{D_{AB}L}{\delta^2 v_{y,max}}$$ (15-42b)

The total amount absorbed for a plate of width w is the amount of solute exiting in the liquid minus the amount entering,

$$\text{Total solute absorbed} = v_{y,avg} w\delta(C_{A,avg,L} - C_{A,init})$$ (15-42c)

This result can be related to $N_{A,avg} \times (Lw)$ where the average flux for dilute systems is determined from Eq. (15-40b). Since $C_{A,avg,L}$ is given by an infinite series, this calculation is complicated. However, we can find limiting solutions for short contact time (large Reynolds number) and long contact times (small Reynolds number), where the Reynolds number for the falling liquid film is

$$Re_{liquid\,film} = \frac{4\delta v_{y,avg}\rho}{\mu} = \frac{4\rho q}{\mu}$$ (15-43a)

For large values of Re, the result is Eq. (15-40c). Thus, both analyses agree for short contact times. For long contact times (Re < 100 and large L) the average mass-transfer coefficient and Sherwood number for a system with no surface ripples are

$$k_{c,liq,avg} \approx 3.41 \frac{D_{AB}}{\delta} \text{ and } Sh_{avg} = \frac{k_{c,liq,avg} \delta}{D_{AB}} \approx 3.41 \qquad \textbf{(15-43b)}$$

Remembering that both δ and the Reynolds number increase as q increases, this result predicts that $k_{c,liq,avg}$ decreases as the Reynolds number increases. This is the opposite of what normally occurs, which is $k_{c,liq,avg}$ increases as Reynolds number increases as shown, for example, by Eq. (15-45). Thus, in laminar flow, increasing velocity may not increase mass-transfer coefficients or rates.

If the ripples are suppressed with a surfactant, these results are valid at least to Re = 250. Treybal (1980) reports that Eq. (15-43b) is valid to Re = 1200 if ripples are suppressed, while Sherwood et al. (1975) report that transition to turbulence occurs before a Reynolds number of 1200. This type of disagreement between sources is not unusual in mass-transfer studies. Note that Eq. (15-40c) shows that at short contact times, the average mass-transfer coefficient is proportional to $\sqrt{D_{AB}}$, while Eq. (15-43b) shows that at long contact times $k_{c,liq,avg}$ is proportional to D_{AB}.

15.5.3. Semi-Empirical and Empirical Mass-Transfer Coefficient Correlations

The results in Eqs. (15-40c) and (15-43b) assume that the flow is laminar and the gas-liquid surface is flat. At Reynolds numbers less than 250, the flow is laminar (Sherwood et al., 1975). However, even if the fluid flow remains laminar, ripples can appear on the surface. Ripples cause local mixing, which increases the liquid-side mass-transfer coefficient markedly. On the other hand, the area for mass transfer increases only slightly. If even small quantities of surfactant (e.g., soap or proteins) are added, the ripples are eliminated and the previously derived correlations for $k_{c,liq,avg}$ remain valid.

For long contact times (large L), Eq. (15-43b) will be valid if a surfactant is added or if Re < 20 (Wankat and Knaebel, 2008). For 20 < Re < 100 with no surfactant, the effect of ripples can be approximated by increasing $k_{c,liq,avg}$ by 40% to 100%. Then the coefficient in Eq. (15-43b) ranges from 4.77 to 6.82. With no other information, the average value can be used.

$$k_{c,liq,avg} \approx (5.8) \frac{D_{AB}}{\delta} \text{ long contact time, laminar with ripples, } 20 < Re < 100 \qquad \textbf{(15-44a)}$$

For short contact times (small L), Eq. (15-40c) can be used with ripples by multiplying $k_{c,liq,avg}$ by an average value of 1.7.

$$k_{c,liq,avg} \approx 1.7 \sqrt{\frac{4 D_{AB} v_{y,max}}{\pi L}} \text{ short contact time, laminar with ripples, } 20 < Re < 100 \qquad \textbf{(15-44b)}$$

Obviously, there is a significant error possible (±18%) for both Eq. (15-44a) and (15-44b) from just the multiplier for ripple formation. This type of error is not unusual for semi-empirical and empirical mass-transfer correlations. These equations are semi-empirical because the basic form is from theoretical analysis, but the coefficient is from experimental data.

Sherwood et al. (1975) report that transition from laminar to turbulent occurs in the Reynolds number range from 250 to 500. Although they do not report any correlations for the liquid mass-transfer coefficient in turbulent flow, Treybal (1980) reports the following correlation for a liquid film with constant surface concentration at a somewhat higher range of Reynolds numbers,

$$Sh_{liq,avg} = \frac{k_{c,liq,avg}\delta}{D_{AB}} = 1.76 \times 10^{-5} Re_{liquid\ film}^{1.505} Sc_{liq}^{0.5} \quad (1300 < Re < 8300) \tag{15-45}$$

The Reynolds number $Re_{liquid\ film}$ for a liquid film is defined by Eq. (15-43a), and the Schmidt number that adjusts the mass-transfer coefficient for changes in the liquid properties is defined in Eq. (15-33e). In dilute systems, the properties of the solvent can be used to calculate the Schmidt number.

If the concentration of the liquid surface is not constant, there will be mass transfer and a mass-transfer resistance on the *gas* side also. *In separation processes, the gas-phase resistance often controls* (the gas-phase mass-transfer coefficient is often significantly smaller than the liquid-phase mass-transfer coefficient). For turbulent flow of the gas in a wetted wall tube, the following correlation was originally reported by Gilliland (see Sherwood et al., 1975; or Wankat and Knaebel, 2008).

$$Sh_{gas,avg} = \frac{k'_{y,avg}d_{tube}}{D_{AB,gas}} = 0.023\,Re_{gas}^{0.83} Sc_{gas}^{0.44}; \quad 2000 < Re_{gas} \leq 35000; 0.6 \leq Sc_{gas} \leq 2.5 \tag{15-46a}$$

Reevaluation of the data showed a better fit with (Wankat and Knaebel, 2008),

$$Sh_{gas,avg} = \frac{k'_{y,avg}d_{tube}}{D_{AB,gas}} = 0.0318\,Re_{gas}^{0.79} Sc_{gas}^{0.50}; \quad 2000 < Re_{gas} \leq 35000; 0.6 \leq Sc_{gas} \leq 2.5 \tag{15-46b}$$

Both of these equations use a log mean gas mole-fraction driving force, Eq. (15-32d). For mass transfer with a rippled surface, the gas-phase mass-transfer correlation (Sherwood et al., 1975; Wankat and Knaebel, 2008) is,

$$Sh_{gas,avg} = \frac{k'_{y,avg}d_{tube}}{D_{AB,gas}} = 0.00814\,Re_{gas}^{0.83} Sc_{gas}^{0.44} Re_{liq}^{0.15} \tag{15-46c}$$

This equation agrees with Eq. (15-46a) when $Re_{liq} = 1000$.

One additional common use of wetted wall columns has been to study the distillation of binary mixtures (Sherwood et al., 1975; Wankat and Knaebel, 2008). Johnstone and Pigford (1975) studied systems in which the gas mass-transfer coefficient controlled. If partial pressure is used as the driving force, their correlation is

$$\frac{k_p d_{tube}(p_B)_{lm}}{D_{AB}p_{tot}} = 0.0328(Re')^{0.77} Sc_{gas}^{0.33}; \quad (3000 < Re' < 40000; 0.5 < Sc_{gas} < 3) \tag{15-47a}$$

The modified Reynolds number is defined on the basis of the gas velocity relative to the surface of the liquid film. For countercurrent flow of gas and liquid, this is

$$Re' = \frac{d_{tube}\rho(|v_{gas}| - |v_{liq,y,max}|)}{\mu} \tag{15-47b}$$

The term $v_{liq.,ymax}$ is defined in Eq. (15-35b). This correlation was used as the basis for mass transfer in each flow element of a structured packing to develop the Bravo-Rocha-Fair correlation for mass transfer in structured packings (Wankat and Knaebel, 2008).

The purpose of showing these correlations is to show the types of forms that result, the need to know the range of validity, the need to know what difference or driving force is required, and the need to be very careful about the definition of terms. Obviously, if liquid terms are used where gas terms are supposed to be, the results are garbage. A very large number of correlations are available in the sources referenced in this and later chapters.

Example 15-4. Estimation of mass-transfer coefficients

Water at 25°C is flowing down a 2.0 m long vertical plate at a volumetric flow rate per meter of plate width of q = 0.000045 m^2/s (Figure 15-7). The entering (y = 0) water contains 0.9 g chlorine/kg water. The water is in contact with a gas phase of chlorine that is saturated with water vapor at 25°C (no water is evaporating). Chlorine (Cl_2) is transferring into the water. At the water-gas surface (z = 0), the water and chlorine are in equilibrium at the solubility limit of chlorine at $C_{C12,surf}$. The Fickian infinite dilution diffusivity is given in Table 15-3. At 25°C the solubility of chlorine in water is 6.5 g/kg water, the density of water is 997 kg/m^3, and the viscosity of water is 0.9 × 10^{-3} kg/(m s).

a. Determine the film thickness δ, the average vertical velocity of the film, and the Reynolds number. Determine the average mass-transfer coefficient when surfactant is present. Per meter of plate width, at what rate (kg/s) is chlorine transferred into the water over the length of the plate?

b. Repeat for q = 0.0018 m^2/s.

Solution part a

From Eqs. (15-35c and d), we obtain

$$\delta = \left(\frac{3\mu q}{\rho g}\right)^{1/3} = \left(\frac{3(0.0009 \text{kg/(m} \cdot \text{s)})(0.000045 \text{m}^2/\text{s})}{(997 \text{kg/m}^3)(9.81 \text{m/s}^2)}\right)^{1/3} = 0.0002316 \text{m}$$

$$v_{y,avg} = \frac{\rho g \delta^2}{3\mu} = \frac{(997 \text{kg/m}^3)(9.81 \text{m/s}^2)(0.0002316 \text{m})^2}{3(0.0009 \text{kg/(m} \cdot \text{s)})} = 0.1943 \text{m/s}$$

The average residence time of the liquid on the plate is

$$t_{res,avg} = L/v_{y,avg} = 2.0 \text{m/(0.1943m/s)} = 10.29 \text{s}$$

From Eq. (15-43a), the Reynolds number is

$$Re = \frac{4\rho q}{\mu} = \frac{4(997 \text{kg/m}^3)(0.000045 \text{m}^2/\text{s})}{0.0009 \text{kg/(m} \cdot \text{s)}} = 199.4$$

Since surfactant is added ripples are suppressed, residence time is fairly long and Re < 250; (Eq.15-43b) is applicable.

$$k_{c,liq,avg} \approx 3.41 \frac{D_{AB}}{\delta} = \frac{3.41(2.03 \times 10^{-9} \, m^2/s)}{0.0002316m} = 2.99 \times 10^{-5} \, m/s$$

The values of C_{Cl2} in kg/m^3 can be found from the specified mass ratios of chlorine to water.

$$C_{Cl2,init} = \left(\frac{0.9g \text{ chlorine}}{kg \text{ water}} \right) \left(\frac{997kg \text{ water}}{m^3} \right) \left(\frac{1.0kg}{1000g} \right) = 0.8973 kg/m^3$$

A similar calculation gives $C_{Cl2,surface} = 6.4805$ kg/m^3.

$$N_{A,avg} = k_{c,liq,avg} (C_{Cl2,surf} - C_{Cl2,init})$$

Chlorine absorbed/m width $= N_{A,avg} \times L$

$$= (2.99 \times 10^{-5} \, m/s)(6.4805 - 0.8973 kg/m^3)(2.0m) = 0.0003337 kg/(m \cdot s)$$

Solution part b

Repeating the calculations but with q = 0.0018 m^2/s, we obtain

$$\delta = 0.000792m, \, v_{y,avg} = 2.2729 m/s, \, t_{res,avg} = 0.8799s, \, Re = 7976.$$

This Reynolds number indicates turbulent flow. Equation (15-45) is appropriate to determine the mass-transfer coefficient. For this equation we need the value of the Schmidt number.

$$Sc_{liq} = \frac{\mu_w}{\rho_w D_{Cl2-w}} = \frac{0.0009 kg/(m \cdot s)}{(997 kg/m^3)(2.03 \times 10^{-9} \, m^2/s)} = 444.68$$

Then the Sherwood number is

$$Sh_{liq,avg} = \frac{k_{c,liq,avg} \delta}{D_{AB}} = 1.76 \times 10^{-5} \, Re_{liq}^{1.505} \, Sc_{liq}^{0.5} = 1.76 \times 10^{-5} (7976)^{1.505} (444.68)^{0.5} = 276.5$$

And $k_{c,liq,avg} = 0.0007087$ m/s. Then the Cl$_2$ absorbed per m of width = 0.007913 kg/m·s.

Comments

1. With turbulent flow, the mass-transfer coefficient is significantly higher and more chlorine is absorbed despite the thicker layer of liquid and lower residence time.
2. Obviously, one needs physical property values. The viscosity, density, gas solubility, and the acceleration due to gravity g were all obtained from http://www.engineeringtoolbox.com.
3. Carrying the dimensions through the calculations is good practice and helps to catch errors.

15.5.4. Correlations Based on Analogies

In Section 15.2.1 we noted that Fick derived his model for mass transfer partly by analogy to Fourier's law of heat transfer and that one reason Fick's model was rapidly accepted was this close analogy to Fourier's law. Shortly after Fick's developments, Osborne Reynolds (yes, the Reynolds number is named after him) stated that heat or mass transfer in a moving fluid should be the result of both normal diffusion processes and eddies caused by the fluid motion. At the time, he had not yet discovered the difference between laminar motion (only normal diffusion operates) and turbulent motion (both molecular and eddy diffusion occur). We now know that Reynolds was correct only for turbulent flow. Since eddies depend on fluid velocity, the easiest functional form is to assume that eddy diffusion is linearly dependent on velocity. Then the equation for mass transfer becomes

$$N_A = k_c(C_2 - C_1) = (a_{mass} + b_{mass}v)(C_2 - C_1) \tag{15-48a}$$

And in a similar fashion, the equation for heat transfer is

$$Q_z = h_{heat\ transfer}(T_2 - T_1) = (a_{heat} + b_{heat}v)(\rho\hat{C}_p)[T_2 - T_1] \tag{15-48b}$$

For fluid flow, we balance momentum to find shear stress τ,

$$\tau = f(\frac{\rho v^2}{2}) = \left(\frac{fv}{2}\right)\rho v = (a_{flow} + b_{flow}v)\rho v \tag{15-48c}$$

In highly turbulent flow, the a terms (molecular diffusion) are much less than the b terms (eddy diffusion) and can be neglected. Cussler (2009) notes that Reynolds now took an amazing intuitive leap and concluded that

$$b_{mass} = b_{heat} = b_{flow} \tag{15-49a}$$

This step says that the transfer of mass, heat, and momentum by eddies is the same. By inspection of Eqs. (15-48), we see that Eq. (15-49a) requires

$$\frac{k_c}{v} = \frac{h_{heat\ transfer}}{\rho\hat{C}_p v} = \frac{f}{2} \tag{15-49b}$$

Quite logically, this is known as the Reynolds analogy. The advantage of the analogy is if we know either the heat-transfer coefficient h or the friction factor f, we can estimate the mass-transfer coefficient k. For gases, the Reynolds analogy is reasonably accurate, but for liquids it often fails. This development of the Reynolds analogy follows Cussler's (2009) development.

Chilton and Colburn were able to empirically modify the Reynolds analogy into a much more robust analogy that works for gases and most liquids. Realizing that heat transfer would depend on the properties of the fluid through the Prandtl number, Eq. (15-33f), and that mass transfer would depend on the properties of the fluid through the Schmidt number, Eq. (15-33e), they included powers of the Prandtl and Schmidt numbers in the first two terms of Eq. (15-49b). Fitting the resulting equation to experimental data showed that a power of 2/3 was appropriate and easy to use in calculations (very important in slide rule days). The result is

$$\frac{k_c}{v}(Sc)^{2/3} = \frac{h_{heat\ transfer}}{\rho \hat{C}_p v}(Pr)^{2/3} = \frac{f}{2} \qquad \text{(15-50a)}$$

This equation is also written as

$$j_D = j_H = f/2 \qquad \text{(15-50b)}$$

The definition of j_D (the mass-transfer term) is $j_D = (k_c/v)(Sc)^{2/3}$, and the heat-transfer term j_H is defined as $j_H = [h_{heat\ transfer}/(\rho \hat{C}_p v)](Pr)^{2/3}$. Even when the analogy is not being used, literature sources often give mass-transfer correlations in terms of j_D. If the definitions of the dimensionless groups from Eqs. (15-33a, e, f, and g) are substituted in, the analogy can also be written as

$$Sh(Sc)^{-1/3} = Nu(Pr)^{-1/3} = vf/2 \qquad \text{(15-50c)}$$

Equations (15-50) apply to fully turbulent flow (although the equations also fit data for some laminar systems such as flow past a flat plate) with a range of Schmidt number from $0.6 \le Sc \le 3000$. Although not derived theoretically, there is some theoretical justification for these equations (Sherwood et al., 1975).

Because it is often easier to measure heat transfer than mass transfer, there are larger compilations of heat-transfer correlations (usually in the form of Nusselt numbers) than of mass-transfer correlations (e.g., Incropera et al., 2011). Thus, a major application of the analogy between heat and mass transfer in separations is to determine mass-transfer coefficients or Sherwood numbers from existing heat-transfer correlations. In distillation the situation is reversed (mass transfer is more studied than heat transfer) and the Chilton-Colburn analogy is used to estimate heat-transfer coefficients for the rate-based model of distillation (Section 16.8).

Although very useful, the Chilton-Colburn analogy, like all analogies, breaks down when the phenomena of heat transfer, mass transfer, and momentum transfer become different. For example, if radiation becomes important and is not separated from the heat-transfer term, then $j_D \ne j_H$. If both skin friction and form drag (due to flow past blunt objects) occurs, then $j_D \ne f/2$ and $j_H \ne f/2$.

15.6 DIFFICULTIES WITH FICKIAN DIFFUSION MODEL

One difficulty with the Fickian diffusion model should be obvious from the discussion in Section 15.2.3. One has to select a somewhat arbitrary basis velocity or plane of reference to calculate the convective and diffusive fluxes. The values and indeed the meaning of the convective and diffusive fluxes may change when the basis velocity is changed. Although irritating, this difficulty is not considered to be major because the total fluxes of A and B, which is the purpose of the calculation, do not change.

There is an additional irritating problem. First, for liquid binary systems, D_{AB} usually depends on concentration, and although Eq. (15-23) is often reasonably accurate, the activity coefficient data may not be readily available. Even if activity coefficients are known, accurate prediction of the infinite dilution diffusivities is difficult with current models (e.g., Sherwood et al., 1975; Wankat and Knaebel, 2008). Thus, a significant amount of data may be required to accurately determine the concentration dependence of binary diffusion coefficients in non-ideal systems. Although this difficulty is significant, it can be overcome by dedicated laboratory analysis.

Then, there is the very serious difficulty that empirical extension of Fick's law to systems with more than two components leads to logical inconsistencies and major calculation difficulties (Taylor and Krishna, 1993; Wesselingh and Krishna, 2000). For example, in ternary systems, the Fickian diffusion coefficient is not symmetrical, $D_{i,j} \neq D_{j,i}$, which means that additional constants are required. In addition, the values of $D_{i,j}$ depend on the value chosen for v^*. Even worse, in some ternary systems, such as acetone-benzene-methanol, a component can diffuse into a more concentrated instead of a less concentrated region (Taylor and Krishna, 1993). Accurately fitting this experimental data requires that some of the $D_{i,j}$ values be negative.

The conclusion is that use of the Fickian model for binary diffusivities is reasonable, although it may be awkward for nonideal liquid systems. For systems with more than two components, the Fickian model is not the best choice.

15.7 MAXWELL-STEFAN MODEL OF DIFFUSION AND MASS TRANSFER

In 1868, 12 years after Fick's definitive publication of his theory, James Clerk Maxwell published a paper on a different approach to studying the diffusivity of gases. In 1871 Josef Stefan extended Maxwell's theory and anticipated multicomponent effects (Cussler, 2009). Although the Maxwell-Stefan theory has had many strong adherents in the more than 140 years since its development, it always seems to be playing catch-up to the earlier Fickian theory. Three perceived difficulties have prevented wider acceptance of the Maxwell-Stefan theory. First, the Fickian model is well-entrenched in textbooks and diffusivity data collections, and it works well for many binary systems. Second, the Maxwell-Stefan theory gives one fewer flux N_i than is needed to completely solve the problem. However, this is really no different than choosing a reference velocity for Fick's law, and, as will be shown later, for most situations obtaining the additional flux equation is straightforward. Third, before digital computers made numerical solutions relatively simple (Taylor and Krishna, 1993), the differential form of the Maxwell-Stefan equations were considerably more difficult to solve than the Fickian equations. The development of difference forms of the Maxwell-Stefan equations (Wesslingh and Krishna, 1990, 2000) has also helped to minimize this difficulty. Thus, in reality, only the well-entrenched nature of the Fickian model has prevented acceptance of the Maxwell-Stefan model.

The proponents of Maxwell-Stefan theory claim it is a better model than Fickian theory. What makes one model better than another model? First, it is nice to have a model tied to the basic physics or chemistry. A simple explanation based on the physics or chemistry should explain the basic behavior. This is illustrated in Section 15.7.1. Second, the model should not conflict with well-accepted laws such as the first or second law of thermodynamics. The Maxwell-Stefan model incorporates thermodynamics for the analysis of nonideal systems. Third, we want a model that can explain and/or predict data and that can be extrapolated. Except for ideal gas behavior, diffusion models invariably have to use measured constants (diffusivity values). A good model will minimize the number of constants required and minimize the variation of these "constants." If the constants vary (say with concentration), the variation should be monotonic and preferably be close to linear. The constants for multicomponent systems should be predictable from binary pairs, and no impossible or improbable values (e.g., negative values) of the constants should be required to predict the data. Based on these criteria, the Maxwell-Stefan theory is a better theory than the Fickian model.

My advice is to learn to use the Maxwell-Stefan theory because its extra power allows relatively simple solutions of problems that are very difficult to solve with the Fickian model (e.g., very nonideal binary systems and multicomponent systems).

15.7.1 Introductory Development of the Maxwell-Stefan Theory of Diffusion

The Maxwell-Stefan theory can be derived from continuum mechanics (Datta and Vilekar, 2010), irreversible thermodynamics (Bird et al., 2002), or the kinetic theory of gases (Hirschfelder et al., 1954). Although the continuum mechanics approach is probably most powerful, for an introductory development, a simplified kinetic theory is easier to follow. The presentations of Taylor and Krishna (1993) and Wesselingh and Krishna (2000) are paraphrased in a somewhat loose manner here.

Consider an ideal gas at constant pressure and temperature consisting of gases A and B. The entire system can be moving at some average molar velocity $v_{z,ref,mol}$. The coordinate system and control volume shown in Figure 15-8 are being translated at $v_{z,ref,mol}$ so that there is no net flow. As noted in Section 15.1, at normal pressures there are a huge number of molecules, and collisions between molecules occur repeatedly. If the two gases are moving through each other in the z direction, we can do a force balance for the control volume of thickness Δz shown in Figure 15-8. There will be forces on molecule A due to the partial pressure of A ($p_A \times$ Area) at z, and at z + Δz

$$(\text{pressure force}) = [p_A(z) - p_A(z + \Delta z)][\text{Area}] \qquad \textbf{(15-51a)}$$

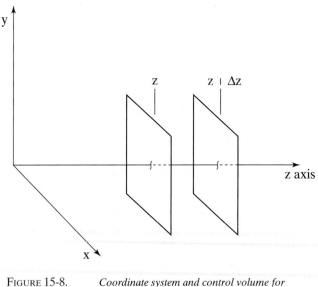

FIGURE 15-8.   *Coordinate system and control volume for Maxwell-Stefan derivation. The entire coordinate system is translating at $v_{z,ref,mol}$*

and there will be a frictional force caused by molecules of B flowing past molecules of A. Forces other than pressure and friction will be assumed to be negligible but are easily included in the continuum mechanics treatment (Datta and Vilekar, 2010). We know that the friction force is proportional to the difference in velocities of the two gases and the amounts of each gas per volume of the segment. For an ideal gas, the molar densities (amounts per volume) are

$$n_A / V = p_A / (RT), \, n_B / V = p_B / (RT) \tag{15-51b}$$

Then the friction force is

$$(\text{friction force}) \propto \frac{p_A p_B (v_B - v_A)}{(RT)^2} \tag{15-51c}$$

Setting the pressure force equal to the friction force and taking the limit as $\Delta z \to 0$, we obtain

$$-\frac{dp_A}{dz} \propto \frac{p_A p_B (v_A - v_B)}{(\text{Area})(RT)^2} \tag{15-51d}$$

Noting that $p_A = y_A p_{tot}$ and $p_B = y_B p_{tot}$, and for constant total pressure and constant temperature incorporating the term $p_{tot} / (RT)^2$ in a constant of proportionality $f_{A,B}$ this becomes

$$-\frac{1}{p_{tot}} \frac{dp_A}{dz} = f_{A,B} y_A y_B (v_A - v_B) = \frac{y_A y_B (v_A - v_B)}{\text{Ð}_{AB}} \tag{15-51e}$$

The term $f_{A,B}$ is a friction coefficient between molecules A and B. The inverse of the friction coefficient, $\text{Ð}_{A,B} = 1/f_{A,B}$, is the Maxwell-Stefan diffusivity. Since the system pressure is constant, the left-hand side becomes $-dy_A / dz$, and we obtain

$$\frac{dy_A}{dz} = -\frac{y_A y_B (v_A - v_B)}{\text{Ð}_{AB}} \quad (p_{total} = \text{constant}) \tag{15-52a}$$

The left-hand side of Eq. (15-52a) is the "driving force" for diffusion. The equation for component B is

$$\frac{dy_B}{dz} = -\frac{y_A y_B (v_B - v_A)}{\text{Ð}_{BA}} \quad (p_{total} = \text{constant}) \tag{15-52b}$$

If we substitute in the molar fluxes,

$$N_{A,z} = v_A y_A C_m \text{ and } N_{B,z} = v_B y_B C_m \tag{15-52c}$$

we obtain

$$\frac{dy_A}{dz} = -\frac{y_B N_{A,z} - y_A N_{B,z}}{C_m \text{Ð}_{AB}} \tag{15-52d}$$

$$\frac{dy_B}{dz} = -\frac{y_A N_{B,z} - y_B N_{A,z}}{C_m \mathcal{D}_{BA}} \qquad (15\text{-}52e)$$

It can be shown that these equations are equivalent to

$$J_{A,z} = -C_m D_{AB} \frac{dy_A}{dz} \text{ and } J_{B,z} = -C_m \mathcal{D}_{BA} \frac{dy_B}{dz} \qquad (15\text{-}52f)$$

Since the control volume in Figure 15-8 was set to move at the average molar velocity $v_{z,ref,mol}$, there is no net flow, $J_{B,z} = -J_{A,z}$ and

$$dy_A/dz + dy_B/dz = 0 \qquad (15\text{-}53a)$$

This equation requires

$$\mathcal{D}_{AB} = \mathcal{D}_{BA} \qquad (15\text{-}53b)$$

Thus, the binary ideal gas Maxwell-Stefan diffusivity is symmetric.

Equations (15-52) are different forms of the one-dimensional Maxwell-Stefan equations for a binary ideal gas at constant pressure and temperature. For ideal binary systems, Section 15.7.5 shows that $\mathcal{D}_{A,B} = D_{B,A}$ and the use of the Maxwell-Stefan equations is identical to the use of the Fickian diffusion equations. The equation can be written in three dimensions by using a gradient for the derivatives and vectors for the velocities.

15.7.2 Maxwell-Stefan Equations for Binary Nonideal Systems

For nonideal systems the generalized form of the driving forces in Eqs. (15-52) are based on the derivative of the chemical potential μ_i (Datta and Vilekar, 2010; Krishna and Wesselingh, 1997; Taylor and Krishna, 1993; Wesselingh and Krishna, 2000). Since nonidealities are most common in liquids, we will write the equations in terms of liquid mole fractions.

$$\frac{x_A}{RT} \frac{\partial \mu_A}{\partial z}\bigg|_{T,p} = -\frac{x_A x_B (v_A - v_B)}{\mathcal{D}_{AB}} \qquad (15\text{-}54a)$$

$$\frac{x_B}{RT} \frac{\partial \mu_B}{\partial z}\bigg|_{T,p} = -\frac{x_A x_B (v_B - v_A)}{\mathcal{D}_{BA}} \qquad (15\text{-}54b)$$

For ideal gases, the driving force on the left side of these equations reduces to $(1/P)dp_i/dz$, which for constant pressure systems is equivalent to dy_i/dz, and Eqs. (15-54) simplify to Eqs. (15-52a,b). For nonideal systems, the chemical potential can be written in terms of the activity coefficients γ_A and γ_B for liquids or the fugacity coefficients for gases. The results for component A for liquids are

$$\left(1 + \frac{x_A}{\gamma_A} \frac{\partial \ln \gamma_A}{\partial x_A}\right) \frac{dx_A}{dz} = -\frac{x_A x_B (v_A - v_B)}{\mathcal{D}_{AB}} \qquad (15\text{-}55a)$$

or the alternate form,

$$(1+\frac{x_A}{\gamma_A}\frac{\partial\gamma_A}{\partial x_A})\frac{dx_A}{dz} = -\frac{x_A x_B(v_A - v_B)}{Ð_{AB}} \tag{15-55b}$$

If we note that $v_A = N_{A,z}/(x_A\rho_m)$ and $v_B = N_{B,z}/(x_B\rho_m)$ where ρ_m is the molar density, Eq. (15-55b) becomes

$$(1+\frac{x_A}{\gamma_A}\frac{\partial\gamma_A}{\partial x_A})\frac{dx_A}{dz} = -\frac{x_B N_{A,z} - x_A N_{B,z}}{\rho_m Ð_{AB}} \tag{15-56a}$$

The corresponding equation for component B is

$$(1+\frac{x_B}{\gamma_B}\frac{\partial\gamma_B}{\partial x_B})\frac{dx_B}{dz} = -\frac{x_A N_{B,z} - x_B N_{A,z}}{\rho_m Ð_{BA}} \tag{15-56b}$$

These forms are often useful in solving mass-transfer problems. Obviously, detailed activity coefficient data or accurate correlations are required (Jakobsen, 2008).

15.7.3 Determining the Independent Fluxes $N_{i,z}$

Since the control volume in Figure 15-8 was set to move at the average molar velocity $v_{z,ref,mol}$, the sum of $J_{i,z}$ is equal to zero for any system. Thus, for a binary system, only one $J_{i,z}$ is independent. However, there are two independent fluxes $N_{i,z}$. In order to use the Stefan-Maxwell formulation in practical problems, we need another relationship (called a bootstrap equation) that allows us to determine the additional independent flux $N_{i,z}$. In other words, we need to tie the moving $J_{i,z}$ to the stationary $N_{i,z}$. The form of this additional equation depends on the situation. In general, $N_{i,z} = C_i v_i = C_m y_i v_i$, and if one of the v_i is known, we can calculate the required unknown flux $N_{i,z}$.

Situation 1. Equimolar counterdiffusion. In this case, the total flux $N_{tot,z} = 0$, $v_{z,ref,mol} = 0$, and $N_{i,z} = J_{i,z}$. For a binary system $N_{B,z} = -N_{A,z}$ and Eq. (15-56a) becomes

$$(1+\frac{x_A}{\gamma_A}\frac{\partial\gamma_A}{\partial x_A})\frac{dx_A}{dz} = -\frac{N_{A,z}}{\rho_m Ð_{AB}} \tag{15-57a}$$

Situation 2. Distillation. If constant molar flow (CMO) is valid, then $N_{tot,z} = 0$, $v_{z,ref,mol} = 0$, and $N_{i,z} = J_{i,z}$. For a binary system with CMO, Eq. (15-57a) is valid. If CMO is not valid, a reasonable approximation is $\sum N_{i,z}\lambda_i = 0$, which for a binary is

$$N_{A,z}\lambda_A + N_{B,z}\lambda_B = 0 \tag{15-57b}$$

Multicomponent distillation is treated in detail by Taylor and Krishna (1993).

Situation 3. Flow in a stagnant fluid. This situation can occur for condensation or evaporation when there is a noncondensing gas and for absorption or stripping. For a binary system, $N_{B,z} = 0$ is the additional relationship needed. This relationship leads to $N_A = J_A/x_B$, which is the same result as for Fickian diffusion. For a ternary system, see Example 15-6.

Situation 4. Trace component is stagnant. This happens for dilute membrane permeation with concentration polarization. The result is $N_{A,z} = 0$ and $v_{z,ref,mol} = 0$.

Situation 5. Flux ratios are specified. If $N_{A,z} = (frac_A)N_{tot,z}$, then for a binary system, Eq. (15-56a) becomes

$$(1 + \frac{x_A}{\gamma_A}\frac{\partial\gamma_A}{\partial x_A})\frac{dx_A}{dz} = -\frac{(1 - \frac{x_A}{frac_A})N_{A,z}}{\rho_m \mathcal{D}_{AB}} \tag{15-57c}$$

Situation 6. Chemical reaction. Reaction stoichiometry determines the relationship. This is outside the scope of this text. See Taylor and Krishna (1993) or Wesselingh and Krishna (1990, 2000).

15.7.4 Difference Equation Formulations

Wesselingh and Krishna (1990, 2000) note that difference equations instead of differential equations are much easier to solve and, in most cases, result in answers that are of acceptable accuracy. For example, the difference equations equivalent of Eqs. (15-55a) and (15-55b) are

$$(1 + \frac{\bar{x}_A}{\bar{\gamma}_A}\frac{\overline{\partial\gamma_A}}{\partial x_A})\frac{\Delta x_A}{\Delta z} = -\frac{\bar{x}_A\bar{x}_B(\bar{v}_A - \bar{v}_B)}{\bar{\mathcal{D}}_{AB}} \tag{15-58a}$$

$$(1 + \frac{\bar{x}_B}{\bar{\gamma}_B}\frac{\overline{\partial\gamma_B}}{\partial x_B})\frac{\Delta x_B}{\Delta z} = -\frac{\bar{x}_A\bar{x}_B(\bar{v}_B - \bar{v}_A)}{\bar{\mathcal{D}}_{BA}} \tag{15-58b}$$

The bars over terms mean they should be evaluated at the average conditions. If this is applied to mass transfer across a film, $\Delta z = \delta$ where δ is the film thickness, we can define the average mass-transfer coefficient in Maxwell-Stefan terms as $\bar{k} = \bar{\mathcal{D}}_{AB}/\Delta z$ Two additional useful forms can be obtained by substituting $v_A = N_A/(x_A\rho_m)$ and $v_B = N_B/(x_B\rho_m)$ (where ρ_m is the molar density) into Eq. (15-58a),

$$(1 + \frac{\bar{x}_A}{\bar{\gamma}_A}\frac{\overline{\partial\gamma_A}}{\partial x_A})\frac{\Delta x_A}{\Delta z} = -\frac{\bar{x}_B\bar{N}_{A,z} - \bar{x}_A\bar{N}_{B,z}}{\bar{\rho}_m\bar{\mathcal{D}}_{AB}} \quad or \quad (1 + \frac{\bar{x}_A}{\bar{\gamma}_A}\frac{\overline{\partial\gamma_A}}{\partial x_A})\Delta x_A = -\frac{\bar{x}_B\bar{N}_{A,z} - \bar{x}_A\bar{N}_{B,z}}{\bar{\rho}_m\bar{k}} \tag{15-58c,d}$$

and

$$(1 + \frac{\bar{x}_B}{\bar{\gamma}_B}\frac{\overline{\partial\gamma_B}}{\partial x_B})\frac{\Delta x_B}{\Delta z} = -\frac{\bar{x}_A\bar{N}_{B,z} - \bar{x}_B\bar{N}_{A,z}}{\bar{\rho}_m\bar{\mathcal{D}}_{BA}} \quad or \quad (1 + \frac{\bar{x}_B}{\bar{\gamma}_B}\frac{\overline{\partial\gamma_B}}{\partial x_B})\Delta x_B = -\frac{\bar{x}_A\bar{N}_{B,z} - \bar{x}_B\bar{N}_{A,z}}{\bar{\rho}_m\bar{k}} \tag{15-58e,f}$$

If the system is ideal, $\gamma_A = 1$ and the equations become

$$\frac{\Delta x_A}{\Delta z} = -\frac{\bar{x}_B\bar{N}_{A,z} - \bar{x}_A\bar{N}_{B,z}}{\bar{\rho}_m\bar{\mathcal{D}}_{AB}} \quad or \quad \Delta x_A = -\frac{\bar{x}_B\bar{N}_{A,z} - \bar{x}_A\bar{N}_{B,z}}{\bar{\rho}_m\bar{k}} \tag{15-58g,h}$$

These forms in terms of the fluxes can be convenient in solving mass-transfer problems.

The difference formula approximation can be made more accurate by doing several steps; however, since we need to know the concentrations for the ends and center of each step, the usual method of dividing Δz into equal segments is very awkward. The calculation is easier if the segments are selected to give known mole fractions of A, and the size of each step is an unknown. This is illustrated in Problem 15.D18.

15.7.5 Relationship between Maxwell-Stefan and Fickian Diffusivities

Since most of the diffusivity data collected in the literature are based on the Fickian diffusivity D_{AB} and the Fickian mass-transfer coefficient $k = D_{AB}/\delta$, if we want to solve problems using the Maxwell-Stefan model, we must relate the Maxwell-Stefan diffusivity $Ð_{AB}$ or mass-transfer coefficient k to the Fickian diffusivity D_{AB} or mass-transfer coefficient k. The relationship is relatively simple, although it requires activity coefficient information to calculate $Ð_{AB}$ and k in nonideal systems.

For binary systems, the relationships are

$$Ð_{AB} = \frac{D_{AB}}{\left(1 + x_A \dfrac{\partial \ln \gamma_A}{\partial x_A}\right)}, \text{ and } k_{AB} = \frac{k Ð_{AB}}{D_{AB}} = \frac{k_{AB}}{\left(1 + x_A \dfrac{\partial \ln \gamma_A}{\partial x_A}\right)} \text{ binary systems} \qquad \textbf{(15-59a,b)}$$

Since in the limit of infinite dilution $\gamma_A = 1$, Eqs. (15-59) require that the infinite dilution limits of the Maxwell-Stefan and the Fickian diffusivities and mass-transfer coefficients are equal.

$$Ð_{AB}^{\circ} = D_{AB}^{\circ}, \ Ð_{BA}^{\circ} = D_{BA}^{\circ}, k = k \text{ infinite dilution} \qquad \textbf{(15-59c)}$$

In nonideal systems over the range from $x_A = 0$ to $x_A = 1.0$, the binary Maxwell-Stefan diffusivity is much closer to linear between $Ð_{AB}^{\circ}$ (pure B) and $Ð_{BA}^{\circ}$ (pure A) than the binary Fickian diffusivity is between D_{AB}° (pure B) and D_{BA}° (pure A). Fewer data points are required to accurately fit an almost linear function than a more nonlinear one. The Maxwell-Stefan diffusivities for many binary nonlinear systems follow or approximately follow the empirical Vignes relationship (Krishna and Wesselingh, 1997; Reid et al., 1987; Taylor and Krishna, 1993; Wesselingh and Krishna, 1990).

$$Ð_{AB}(x_A) = (Ð_{AB}^{\circ})^{1-x_A} (Ð_{BA}^{\circ})^{x_A} \qquad \textbf{(15-60a)}$$

Wesselingh and Krishna (1990) state that Eq. (15-60a) is valid for approximately three-fourths of binary systems. Comparison of this equation with the corresponding equation for Fickian diffusivity, Eq. (15-23d), shows that the equation for Maxwell-Stefan diffusivity is simpler and, for nonlinear systems where $\gamma \neq 1$, Eq. (15-60a) will be closer to linear, which makes interpolation easier.

In ideal systems $\gamma_A = 1$ and

$$Ð_{AB} = D_{AB}, k = k \text{ ideal binary systems} \qquad \textbf{(15-60b,c)}$$

Ideality is most likely in low pressure gas systems.

Example 15-5. Maxwell-Stefan nonideal binary diffusion

A system of ethanol (E) and water (W) is undergoing equimolar counterdiffusion across a thin liquid layer ($\delta = 0.00068$ m) at 40°C and 1.0 atm. The boundary conditions are at $z = 0$, x_E (mole fraction) $= 0.2$ and at $z = \delta$, $x_E - 0.4$.

a. Estimate the flux of ethanol, N_E.
b. Calculate the Maxwell-Stefan diffusiveness in the range from $x_E = 0$ to $x_E = 1.0$.

Data: Tyn and Calus (1975) measured the Fickian diffusivity of the system ethanol-water at 40°C (all diffusivities $\times 10^{-9}$ m^2/s). The mole fraction of ethanol is x_E.

x_E	0.000	0.024	0.100	0.144	0.200	0.300	0.400	0.500	0.600	0.700	0.800	0.900	1.000
D_{EW}	1.700	1.510	1.000	0.780	0.680	0.610	0.640	0.730	0.865	1.060	1.275	1.475	1.640

The activity coefficients for this system can be fitted to the Van Laar equation:

$$\ln \gamma_E = \frac{A}{\left(1 + \dfrac{Ax_E}{Bx_W}\right)^2} \quad \text{and} \quad \ln \gamma_W = \frac{B}{\left(1 + \dfrac{Bx_W}{Ax_E}\right)^2} \tag{15-61a,b}$$

with $A = 1.4599$ and $B = 0.9609$ (Wesselingh and Krishna, 1990).

At 40°C the mass densities are approximately 922 kg/m^3 ($x_1 = 0.2$), 892 kg/m^3 ($x_1 = 0.3$) and 867 kg/m^3 ($x_1 = 0.4$).

Solution Part a

For equimolar counter diffusion, $N_{E,z} = -N_{M,z}$, Eq. (15-58d) becomes,

$$\bar{N}_{E,z} = -\frac{\bar{\rho}_m \Delta x_E}{\Delta z}\left[\bar{D}_{EW}\left(1 + \frac{\bar{x}_A}{\bar{\gamma}_A}\frac{\overline{\partial \gamma_A}}{\partial x_A}\right)\right] \tag{15-61c}$$

Substituting in the Maxwell-Stefan diffusivities from Eq. (15-59a), we find the term in brackets is the Fickian diffusivity, and the result is

$$\bar{N}_{E,z} = -\frac{\bar{\rho}_m \bar{D}_{EW}\Delta x_E}{\Delta z} \tag{15-61d}$$

The average MW is calculated from $MW_{avg} = x_E MW_E + x_W MW_W$. The values are 23.6, 26.4, and 29.2 for ethanol mole fractions of 0.2, 0.3, and 0.4, respectively. The molar densities can be determined from $\rho_m = \rho/MW_{avg}$. The values are $922/23.6 = 39.07$ kmol/m^3 for $x_E = 0.2$, $\rho_m = 33.79$ kmol/m^3 for $x_E = 0.3$, and 29.69 kmol/m^3 for $x_E = 0.4$. The average value of the Fickian diffusivity at $x_E = 0.3$ is $D_{EW} = 0.61 \times 10^{-9}$ m^2/s. Then,

$$\bar{N}_{E,z} = -\frac{\bar{\rho}_m \bar{D}_{EW} \Delta x_E}{\Delta z} = -(33.79)(0.61 \times 10^{-9})(0.4 - 0.2)/(0.00068) = -6.06 \times 10^{-3} \text{ mol/s}.$$

Part b

For the van Laar equation the derivative is (see Problem 15.C6)

$$x_E \frac{\partial \ln \gamma_E}{\partial x_E} = -2x_E x_W \frac{(AB)^2}{(Bx_W + Ax_E)^3} \tag{15-61e}$$

An example calculation gives,

$$\mathcal{D}_{EW}(x_E = 0.2) = \frac{D_{EW}(x_E = 0.2)}{1 - 2(0.2)(0.8)\dfrac{[(1.460)(0.961)]^2}{[(0.961)(0.8) + (1.460)(0.2)]^3}} = 1.44 \times 10^{-9} \text{ m}^2/\text{s}$$

Similar calculations give $\bar{\mathcal{D}}_{EW} = \mathcal{D}_{EW}(x_E = 0.3) = 1.54 \times 10^{-9}$ and $\mathcal{D}_{EW}(x_E = 0.4)$ = 1.62×10^{-9}. In addition, at the limits of $x_E = 0$ ($D_{EW} = 1.70 \times 10^{-9}$) and $x_E = 1.0$ ($D_{EW} = 1.64 \times 10^{-9}$), $\mathcal{D}_{EW} = D_{EW}$. Comparing the values of $\mathcal{D}_{EW}$ and D_{EW}, it is obvious that $\mathcal{D}_{EW}$ varies less than D_{EW}.

Comments

1. The minus sign indicates that the direction of ethanol transfer is opposite to the direction of the z axis (from z = 0 towards z = δ).
2. This is obviously an approximate solution, since we have estimated the derivative with a single finite step. If we divided the interval from z = 0 to z = δ into more parts, a more accurate answer would be generated. With two parts (see Problem 15.D18), the result is $\bar{N}_{E,z} = -6.375 \times 10^{-3}$ mol/s and with three parts the result is $\bar{N}_{E,z} = -6.35 \times 10^{-3}$ mol/s As the number of parts increases the answers are oscillating slightly, but the difference between the answers decreased significantly, which means that the answer is converging fairly rapidly.

15.7.6 Ideal Ternary Systems

The Maxwell-Stefan equations can be extended in a straightforward fashion to ternary and multicomponent systems. For an ideal ternary system, the basic equations are

$$\frac{dy_A}{dz} = -\frac{y_A y_B (v_A - v_B)}{\mathcal{D}_{AB}} - \frac{y_A y_C (v_A - v_C)}{\mathcal{D}_{AC}} \tag{15-62a}$$

$$\frac{dy_B}{dz} = -\frac{y_A y_B (v_B - v_A)}{\mathcal{D}_{BA}} - \frac{y_B y_C (v_B - v_C)}{\mathcal{D}_{BC}} \tag{15-62b}$$

$$\frac{dy_C}{dz} = -\frac{y_A y_C (v_C - v_A)}{\mathcal{D}_{CA}} - \frac{y_B y_C (v_C - v_B)}{\mathcal{D}_{CB}} \tag{15-62c}$$

Substituting in

$$v_i = N_{i,z} / (y_i \rho_m)$$ (15-63)

we obtain,

$$\frac{dy_A}{dz} = -\frac{y_B N_{A,z} - y_A N_{B,z}}{\rho_m Ð_{AB}} \quad \frac{y_C N_{A,z} - y_A N_{C,z}}{\rho_m Ð_{AC}}$$ (15-64a)

$$\frac{dy_B}{dz} = -\frac{y_A N_{B,z} - y_B N_{A,z}}{\rho_m Ð_{BA}} - \frac{y_C N_{B,z} - y_B N_{C,z}}{\rho_m Ð_{BC}}$$ (15-64b)

$$\frac{dy_C}{dz} = -\frac{y_A N_{C,z} - y_C N_{A,z}}{\rho_m Ð_{CA}} - \frac{y_B N_{C,z} - y_C N_{B,z}}{\rho_m Ð_{BC}}$$ (15-64c)

In difference notation, Eqs. (15-64a) and (15-64b) become

$$\frac{\rho_m \Delta y_A}{\Delta z} = -\frac{\bar{y}_B \bar{N}_{A,z} - \bar{y}_A \bar{N}_{B,z}}{\bar{Ð}_{AB}} - \frac{\bar{y}_C \bar{N}_{A,z} - \bar{y}_A \bar{N}_{C,z}}{\bar{Ð}_{AC}}$$ (15-65a)

$$\frac{\rho_m \Delta y_B}{\Delta z} = -\frac{\bar{y}_A \bar{N}_{B,z} - \bar{y}_B \bar{N}_{A,z}}{\bar{Ð}_{BA}} - \frac{\bar{y}_C \bar{N}_{B,z} - \bar{y}_B \bar{N}_{C,z}}{\bar{Ð}_{BC}}$$ (15-65b)

The bars over the terms means that the average values should be used. In the film model for mass transfer, the mass-transfer coefficient is $k_{AB} = Ð_{AB} / \delta$. Dividing the denominators of Eq. (15-65a) by Δz, we obtain

$$\rho_m \Delta y_A = -\frac{\bar{y}_B \bar{N}_{A,z} - \bar{y}_A \bar{N}_{B,z}}{\bar{Ð}_{AB} / \Delta z} - \frac{\bar{y}_C \bar{N}_{A,z} - \bar{y}_A \bar{N}_{C,z}}{\bar{Ð}_{AC} / \Delta z}$$

$$= -\frac{\bar{y}_B \bar{N}_{A,z} - \bar{y}_A \bar{N}_{B,z}}{k_{AB}} - \frac{\bar{y}_C \bar{N}_{A,z} - \bar{y}_A \bar{N}_{C,z}}{k_{AC}}$$ (15-65c)

A similar equation can be obtained for component B.

Unlike the Fickian diffusivities for ternary systems, the Maxwell-Stefan diffusivity is symmetric and is greater than or equal to zero for both ideal and nonideal systems. In ideal ternary systems, the Stefan-Maxwell diffusivities have the same values as the binary pairs, and as we already saw, the binary pairs have the same values as the Fickian binary pairs.

$$Ð_{ij(i \neq j), ternary} = Ð_{ij(i \neq j), binary} = D_{ij(i \neq j), binary}, \text{ ideal ternary}$$ (15-66)

Unfortunately, the same condition does not hold for the Fickian diffusivities. In an ideal ternary system, the values for the Fickian diffusivities are not equal to the values of the Fickian binary pairs.

$$D_{ij(i \neq j), ternary} \neq D_{ij(i \neq j), binary}$$ (15-67a)

For the limiting case of an ideal ternary that is quite dilute in one component (arbitrarily called component A), an effective Fickian diffusion coefficient for the dilute component can be calculated.

$$D_{effective,A} = \frac{(y_B + y_C)D_{AB}D_{AC}}{D_{AC}y_B + D_{AB}y_C} \qquad \text{(15-67b)}$$

Substitution of this equation to replace D_{AB} in Eq. (15-4a) allows calculation of the diffusion of the dilute component A. Note that the effective diffusion coefficient depends on the concentrations of the other species. Although the diffusion of the dilute component can be calculated accurately, the diffusion of the concentrated components B and C probably will not be accurate (see Problem 15.H3b). Even for this special limiting case (ideal ternary with dilute component A), the use of the Maxwell-Stefan equations is certainly preferable and not any more difficult than the Fickian method.

Example 15-6. Maxwell-Stefan ideal ternary system

A ternary mixture of water (A), ammonia (B), and air (C) is being condensed in a heat exchanger. The bulk gas is 0.6 mole fraction water, 0.1 mole fraction ammonia, and 0.3 mole fraction air. The surface of the heat exchanger is at 42°C, and for simplicity in evaluating properties, the entire operation is assumed to be at 42°C and 1.0 atm. The condensed and dissolved water and ammonia in the liquid are assumed to have a high mass-transfer coefficient so that gas-phase mass-transfer controls. The diffusion is across a stagnant gas film (Figure 15-1) that is $\delta = 0.01$m thick. Determine the fluxes (N values) of the three components.

Solution preliminaries

First, the air will not condense, and its solubility in the liquid is so low that it can be ignored. Thus, $N_{air,z} = N_{C,z} = v_c = 0$

At 42°C the vapor pressure of water is 0.1098 atm and $y_{w,surface} = VP_w/p_{tot} = 0.1098$.

This temperature is too hot for condensation of ammonia, but ammonia is very soluble in water. At 42°C the solubility of ammonia is 300 g NH_3/1000 g water. Thus, $x_{NH3,saturation} = 300/(1000 + 300) = 0.231$. This value would be at $y_{NH3,surface} = 1$. We can use this to estimate the Henry's law coefficient at 42°C. $H_{NH3} = y_{NH3}p_{tot}$ $/x_{NH3} = (1.0)(1.0)/(0.231) = 4.329$ atm. Then $y_{w,surface} = H_{NH3}x_{NH3}/p_{tot}$, but neither $y_{NH3,surface}$ nor x_{NH3} are known. However, since all of the liquid on the heat exchanger comes from the transfer of water and ammonia to the liquid,

$$x_{NH3} = N_{NH3,z}/(N_{NH3,z} + N_{water,z}) \qquad \text{(15-68)}$$

The problem is trial and error. We guess a value for x_{NH3}, calculate $y_{NH3,surface}$, complete the problem (conveniently formulated on a spread sheet shown in the appendix to Chapter 15), and use Eq. (15-68) to determine x_{NH3}. If this value matches our guess, we cheer. Otherwise, continue the loop, but this is easily done automatically with either Goal Seek or Solver in Excel.

Solution of Maxwell-Stefan equations

We will assume the gas is ideal and solve the difference Eqs. (15-65a and b). The parameters we need are ρ_m, D_{W-NH3}, D_{W-Air}, and $D_{NH3-Air}$. The molar density for an ideal gas is $\rho_m = p_{tot}/(RT) = 1/(0.820575)(315) = 0.03869 \text{ kmol/m}^3$.

The diffusivity D_{W-Air} is available at 315 K and $= 0.288 \times 10^{-4} \text{ m}^2/\text{s}$.

At 273 K, $D_{NH3-Air} = 0.198 \times 10^{-4} \text{ m}^2/\text{s}$. This was adjusted to 315 K by multiplying by $(315/273)^{1.66}$ with the result $D_{NH3-Air} = 0.251 \times 10^{-4} \text{ m}^2/\text{s}$.

D_{W-NH3} was estimated as $0.212 \times 10^{-4} \text{ m}^2/\text{s}$ from the Chapman-Enskog equation (see spreadsheet in appendix to Chapter 15).

For an ideal system, the Maxwell-Stefan binary-pair diffusivities are equal to the Fickian binary-pair diffusivities.

For the left-hand side of Eq. (15-65a),

$$\Delta y_w / \Delta z = (y_{w,surface} - y_{w,bulk})/(0.01\text{m}) = (0.1098 - 0.6)/.01 = -49.02$$

For NH_3 in Eq. (15-65b),

$$\Delta y_{NH3} / \Delta z = (y_{NH3,surface} - y_{NH3,bulk})/(0.01\text{m}) = ((H_{NH3}x_{NH3}/P_{tot}) - 0.1)/.01$$

will depend on the current value used for x_{NH3}.

On the right-hand sides of these equations,

$$\bar{y}_A = \bar{y}_w = (y_{w,bulk} + y_{w,surface})/2 = (0.6 + 0.1098)/2 = 0.3549$$

$\bar{y}_B = \bar{y}_{NH3} = (y_{NH3,bulk} + y_{NH3,surface})/2$ and will change for each trial as the surface mole fraction varies.

At the surface the air mole fraction is $y_{air,surface} = 1 - (y_{w,surface} + y_{NH3,surface})$, which varies for each trial. The average $\bar{y}_C = \bar{y}_{air} = (y_{air,bulk} + y_{air,surface})/2$ also varies for each trial.

For the final trial with $x_{NH3} = 0.04988$ (calculated value was 0.04987), $y_{NH3,surface} = 0.21593$ and $y_{air,surface} = 0.67409$. The average value for $NH_3 = 0.15797$ and for air $= 0.48704$.

$\Delta y_{water} = 0.10998 - 0.6 = -0.49002$ and $\Delta y_{NH3} = 0.21593 - 0.10 = 0.11593$.

The two equations to be solved with $N_C = 0$ are

$$\frac{\rho_m \Delta y_A}{\Delta z} = -\frac{\overline{y}_B \overline{N}_{A,z} - \overline{y}_A \overline{N}_{B,z}}{Ð_{AB}} - \frac{\overline{y}_C \overline{N}_{A,z}}{Ð_{AC}}, \text{ (15-65a), which becomes}$$

$$(0.03869)\frac{(-.49002)}{0.01} = -\frac{[(0.15797)\overline{N}_{A,z} - (0.35499)\overline{N}_{B,z}]}{0.212 \times 10^{-4}} - \frac{(0.48704)\overline{N}_{A,z}}{0.288 \times 10^{-4}}$$

and $\dfrac{\rho_m \Delta y_B}{\Delta z} = -\dfrac{\overline{y}_A \overline{N}_{B,z} - \overline{y}_B \overline{N}_{A,z}}{Ð_{BA}} - \dfrac{\overline{y}_C \overline{N}_{B,z}}{Ð_{BC}}$, (15-65b), which becomes

$$(0.03869)\frac{(0.11593)}{0.01} = -\frac{[(0.35499)\overline{N}_{B,z} - (0.15797)\overline{N}_{A,z}]}{0.212 \times 10^{-4}} - \frac{(0.48704)\overline{N}_{B,z}}{0.251 \times 10^{-4}}$$

We now have two equations with the two unknowns $\overline{N}_{A,z}$ and $\overline{N}_{B,z}$. Solution of these two equations with the spreadsheet shown in the appendix to Chapter 15 gives $\overline{N}_{A,z} = 8.072 \times 10^{-5}$ and $\overline{N}_{B,z} = 0.424 \times 10^{-5}$ kmol/(m²s). From Eq. (15-68) these values predict $x_{NH3} = 0.04988$, which matches the value used for the trial.

Comments

1. The actual solution of the difference form of the Maxwell-Stefan equations required less than a quarter of the total effort to solve this problem. Most of the difficulty arose because the need to know the surface concentration of ammonia in the vapor resulted in a trial-and-error problem.
2. A spreadsheet was used. Without it, the trial-and-error problem would have been laborious.
3. Some care must be made in choosing the trial values of x_{NH3}. If this value is too large, N_{NH3} will be negative, which does not fit the problem analysis or Eq. (15-68).
4. The water behaves as expected in that its flux is in the downhill direction to a lower water mole fraction. Air has a concentration gradient (caused by the changes in water and ammonia) even though there is no flux of air at steady state. The ammonia behaves counterintuitively (in a Fickian sense), since its flux direction is toward the higher mole fraction.
5. Since it is not dilute, this problem is quite difficult to solve using the Fickian model.
6. Because the difference form of the equations was used, the answer is approximate. A more accurate answer can be obtained by a more accurate numerical solution of the differential equations (Taylor and Krishna, 1993).
7. Changing the thickness of the gas layer is equivalent to changing the mass-transfer coefficient $(k = Ð / \delta)$ and will obviously affect the solution.

15.7.7 Nonideal Ternary Systems

In nonideal systems an effective Fickian diffusivity cannot be calculated, and some values of the Fickian ternary diffusivities can be negative. The extension of the Maxwell-Stefan equations to nonideal ternary systems is straightforward. Equations (15-54) were the most general form of the Maxwell-Stefan equations for binary systems. For ternary systems these equations become

$$\frac{x_A}{RT}\frac{\partial \mu_A}{\partial z}\bigg|_{T,p} = -\frac{x_A x_B (v_A - v_B)}{\text{Đ}_{AB}} - \frac{x_A x_C (v_A - v_C)}{\text{Đ}_{AC}} \tag{15-69a}$$

$$\frac{x_B}{RT}\frac{\partial \mu_B}{\partial z}\bigg|_{T,p} = -\frac{x_A x_B (v_B - v_A)}{\text{Đ}_{BA}} - \frac{x_B x_C (v_B - v_C)}{\text{Đ}_{BC}} \tag{15-69b}$$

Expanding the chemical potential μ in terms of activity coefficients, these equations become

$$(1+\frac{x_A}{\gamma_A}\frac{\partial \gamma_A}{\partial x_A})\frac{dx_A}{dz} = -\frac{x_A x_B (v_A - v_B)}{\text{Đ}_{AB}} - \frac{x_A x_C (v_A - v_C)}{\text{Đ}_{AC}} \tag{15-70a}$$

$$(1+\frac{x_B}{\gamma_B}\frac{\partial \gamma_B}{\partial x_B})\frac{dx_B}{dz} = -\frac{x_A x_B (v_B - v_A)}{\text{Đ}_{BA}} - \frac{x_B x_C (v_B - v_C)}{\text{Đ}_{BC}} \tag{15-70b}$$

If we note that $v_A = N_A/(x_A \rho_m)$ and $v_B = N_B/(x_B \rho_m)$ where ρ_m is the molar density, we obtain

$$(1+\frac{x_A}{\gamma_A}\frac{\partial \gamma_A}{\partial x_A})\frac{dx_A}{dz} = -\frac{x_B N_{A,z} - x_A N_{B,z}}{\rho_m \text{Đ}_{AB}} - \frac{x_C N_{A,z} - x_A N_{C,z}}{\rho_m \text{Đ}_{AC}} \tag{15-71a}$$

The corresponding equation for component B is

$$(1+\frac{x_B}{\gamma_B}\frac{\partial \gamma_B}{\partial x_B})\frac{dx_B}{dz} = -\frac{x_A N_{B,z} - x_B N_{A,z}}{\rho_m \text{Đ}_{BA}} - \frac{x_C N_{B,z} - x_B N_{C,z}}{\rho_m \text{Đ}_{BC}} \tag{15-71b}$$

Expanding the Maxwell-Stefan equations to systems with more components is straightforward. Matrix solution methods for these multicomponent systems were originally developed by Krishna and Standart (1976). The generalized form for diffusion in a film of thickness δ is,

$$(1+\frac{x_i}{\gamma_i}\frac{\partial \gamma_i}{\partial x_i})\frac{dx_i}{d(z/\delta)} = -\sum_{k=1,k\neq i}^{n} \frac{x_k N_{i,z} - x_i N_{k,z}}{\rho_m k_{ik}}, \quad i=1,2,...n \tag{15-72a}$$

where the Maxwell-Stefan mass-transfer coefficient for the binary pairs is

$$k_{ik} = \text{Đ}_{ik}/\delta \tag{15-72b}$$

Substituting in Eq. (15-59b), we obtain

$$k_{ik} = \frac{k_{ik}\text{Đ}_{ik}}{D_{ik}} = \frac{k_{ik}}{\left(1+x_i\frac{\partial \ln \gamma_i}{\partial x_i}\right)} \tag{15-72c}$$

Solutions to Eq. (15-72) are considered in detail by Krishna and Wesselingh (1997), Taylor and Krishna (1993), and Wesselingh and Krishna (1990, 2000). The Maxwell-Stefan matrix approach to multicomponent mass transfer is used in the Aspen Plus simulator to solve distillation problems. Use of the simulator to obtain rate-based solutions of distillation problems is considered in Section 16.8 and in the appendix of Chapter 16.

15.8 ADVANTAGES AND DISADVANTAGES OF DIFFERENT DIFFUSION AND MASS-TRANSFER MODELS

In this section we briefly summarize the advantages and disadvantages of each of the mass-transfer models.

Molecular model. A molecular model has the advantage of explaining physically what is happening during diffusion, but the detailed theory is complex. The results [e.g., Eqs. (15-22)] are useful for predicting the diffusivity of gases. In its current state of development, this model needs to be used in conjunction with another model for complicated mass-transfer calculations. Use this model for physical understanding.

Fickian model. The Fickian model is a widely accepted model for diffusion, which means no one will laugh if you use it. It works well for ideal and close to ideal binary systems and can be used for nonideal binary systems if data are available. Most diffusivity data and correlations for mass-transfer coefficients are based on the Fickian model. This model is very difficult to use for nonideal ternary systems and can require negative diffusion coefficients to predict data. This model works well for dilute binary systems.

Linear driving-force mass-transfer model. This model is a widely accepted (by chemical engineers) empirical formulation that can be extended to very difficult mass-transfer problems if empirical correlations for the mass-transfer coefficient are available. Since it is usually based on the Fickian formulation, can fail where the Fickian model fails. This is the model most commonly used for separation problems. Use of the Maxwell-Stefan formulation for this model is advised for ternary systems.

Maxwell-Stefan model. The Maxwell-Stefan model is generally agreed to be a better model than the Fickian model for nonideal binary and all ternary systems. However, it is not as widely understood by chemical engineers, data collected in terms of Fickian diffusivities need to be converted to Maxwell-Stefan values, and the model can be more difficult to use. Use this model, coupled with a mass-transfer model, when the Fickian model fails or requires an excessive amount of data.

Irreversible thermodynamics model. This model is useful in regions where phases are unstable and can split into two phases (de Groot and Mazur, 1984; Ghorayeb and Firoozabadi, 2000; Haase, 1990). However, this model is beyond the scope of this introductory treatment.

15.9 SUMMARY–OBJECTIVES

After completing this chapter, you should be able to satisfy the following objectives:

1. Explain qualitatively how molecular motion leads to diffusion
2. Describe Fick's model of diffusion in words and equations, and use the model to solve steady-state binary diffusion problems without convection
3. Choose an appropriate reference velocity v_{ref} and solve Fick's model for steady-state binary diffusion with convection

4. Estimate the diffusivity of gases and liquids in binary systems
5. Explain and use the linear driving-force model for mass transfer
6. Estimate mass transfer coefficients
7. Explain the deficiencies in the Fickian model of diffusion
8. Describe how the Maxwell-Stefan model differs from the Fickian model and use the Maxwell-Stefan model for ideal and nonideal binary and ideal ternary diffusion problems

REFERENCES

Bird, R. B., W. E. Stewart, and E. N. Lightfoot, *Transport Phenomena*, 2nd ed., Wiley, New York, 2002.

Crank, J., *The Mathematics of Diffusion*, 2nd ed., Clarendon Press, Oxford, 1975.

Cussler, E. L., *Diffusion. Mass Transfer in Fluid Systems*, 3rd ed, Cambridge University Press, Cambridge, 2009.

Datta, R., and S. A. Vilekar, "The Continuum Mechanical Theory of Multicomponent Diffusion in Fluid Mixtures," *Chem. Engr. Sci.*, 65, 5976–5989 (2010).

deGroot, S. R., and P. Mazur, *Non-Equilibrium Thermodynamics*, Dover, New York, 1984.

Dean, J. A. (Ed.), *Lange's Handbook of Chemistry*, 13th ed., McGraw-Hill, New York, 1985.

Demirel, Y., *Nonequilibrium Thermodynamics, Transport and Rate Processes in Physical, Chemical and Biological Systems*, 2nd ed., Elsevier, Amsterdam, 2007.

Ghorayeb, K., and A. Firoozabadi, "Molecular, Pressure, and Thermal Diffusion in Nonideal Multicomponent Mixtures, *AIChE J.*, 46 (5), 892–900 (2000).

Haase, R. *Thermodynamics of Irreversible Processes*, Dover, New York, 1990.

Hirschfelder, J., C. F. Curtiss, and R. Bird, *Molecular Theory of Gases and Liquids*, Wiley, New York, 1954.

Hottel, H. C., J. J. Noble, A. F. Sarofim, and G. D. Silcox, "Heat Transfer," in D. W. Green and R. H. Perry (Eds.), *Perry's Handbook of Chemical Engineering*, 8th ed., McGraw-Hill, New York, 2008, Section 5a.

Incropera, F. P., D. P. DeWitt, T. L. Bergman, and A. S. Lavine, *Fundamentals of Heat and Mass Transfer*, 7th ed., Wiley, New York, 2011.

Jakobsen, H. A., *Chemical Reactor Modeling. Multiphase Reactive Flows*, Springer-Verlag, Berlin, 2008.

Kirwan, D. J., "Mass Transfer Principles," in R. W. Rousseau (Ed.), *Handbook of Separation Process Technology*, Wiley-Interscience, New York, 1987, Chap. 2.

Krishna, R., and J. A. Wesselingh, "The Maxwell-Stefan Approach to Mass Transfer," *Chemical Engineering Science*, 52, 861–911 (1997).

Marrero, T. R., and E. A. Mason, "Gaseous Diffusion Coefficients," *J. Phys. Chem. Ref. Data*, 1, 3–118 (1972).

Poling, B. E., G. H. Thomson, and D. G. Friend, "Physical and Chemical Data," in D. W. Green and R. H. Perry (Eds.), *Perry's Handbook of Chemical Engineering*, 8th ed., McGraw-Hill, New York, 2008, Section 2, Tables 2-324 and 2-325.

Reid, R. C., J. M. Prausnitz, and B. Poling, *The Properties of Gases and Liquids*, 4th ed., McGraw-Hill, New York, 1987.

Sherwood, T. K., R. L. Pigford, and C. R. Wilke, *Mass Transfer*, McGraw-Hill, New York, 1975.

Taylor, R., and R. Krishna, *Multicomponent Mass Transfer*, Wiley-Interscience, New York, 1993.

Tyn, M. T., and W. F. Calus, "Temperature and Concentration Dependence of Mutual Diffusion Coefficients of Some Binary Liquid Systems," J. Chem. Eng. Data, 20, 310–316 (1975).

Wankat, P.C., and K. S. Knaebel, "Mass Transfer," in D. W. Green and R. H. Perry (Eds.), *Perry's Handbook of Chemical Engineering*, 8th ed., McGraw-Hill, New York, 2008, Section 5B, pp. 5-1, 5-2, 5-43 to 5-83.

Wesselingh, J. A., and R. Krishna, Mass Transfer, Ellis Horwood, Chichester, England, 1990.

Wesselingh, J. A., and R. Krishna, *Mass Transfer in Multicomponent Mixtures,* Delft University Press, Delft, Netherlands, 2000.

HOMEWORK

A. *Discussion Problems*

A1. Suppose we have a volume of nitrogen plus a small amount of water vapor at 1.0 atm. The walls of the container are at 25°C, and there is a hot pipe at 105°C running through the volume. Explain the behavior predicted by Eq. (15-4a), the behavior predicted by Eq. (15-10b), and the reasons that Eq. (15-4b) more closely predicts reality.

A2. When is $J_A = N_A$, and when are they not equal?

A3. Explain why the infinite dilution Fickian diffusivities for a binary liquid system are not equal, $D_{AB}^\circ \neq D_{BA}^\circ$.

A4. The constant in Eq. (5-23b) is 1.173×10^{-16}, which agrees with Geankoplis (2003). However, Cussler (2009) and Wankat and Knaebel (2008) use a constant of 7.4×10^{-8}. Both are correct. Explain.

A5. What is a controlling resistance? How do you determine which resistance, if either, is controlling?

B. *Generation of Alternatives*

B1. In Example 15-1 operation is at a pseudo-steady state. Brainstorm alternative designs for this diffusion measurement.

B2. Instead of treating diffusion and convection terms as additive Eq. (15-16a), what other approaches could be used to analyze simultaneous convection and diffusion?

C. *Derivations*

C1. For binary diffusion with convection, use Eqs. (15-16a, b, c) and the equivalent equations for component B to show that $D_{AB} = D_{BA}$.

C2. For binary diffusion, show that D_{AB} does not depend on the choice of v_{ref}.

C3. Derive the equation that is equivalent to Eqs. (15-32d) to (15-32e) in terms of a partial pressure driving force and a log mean partial pressure difference,

$$(p_B)_{lm} = \frac{(p_{B,2} - p_{B,1})}{\ln(p_{B,2}/p_{B,1})}.$$

C4. For a laminar falling film of liquid, relate the film thickness δ to the Reynolds number Re.

C5. For binary Maxwell-Stefan diffusion through a stagnant layer of B, show that $J_A = N_A/y_B$.

C6. Starting with Eqs. (15-61a) and (15-61b), derive Eq. (15-61e). Note: Because $x_W = 1 - x_E$, x_W is not a constant.

C7. For binary distillation with constant molar overflow (CMO), $v_{ref,mol} = 0$. If CMO is valid, show that $v_{ref,mass} \neq 0$ if $MW_A \neq MWB_A$, and calculate the functional form for $v_{ref,mass}$ that will make convection zero in this reference frame. Do this for diffusion in the vapor assuming an ideal gas.

D. *Problems*

D1. We have steady-state diffusion of 1-propanol across a liquid water film that is 0.10 mm thick. On one side of the film the 1-propanol concentration is $1.2 \, kg/m^3$. We desire a 1-propanol flux rate of $0.2 \times 10^{-5} \, kg/(m^2 s)$. The apparatus is at 25°C.

    a. * What is the concentration at the other side of the film? Depending on the direction of transfer, there are two answers.

    b. What are the equations for the concentration profiles?

Diffusivity value is given in Table 15-3.

D2. For the same system as in Problem 15.D1, we want the high concentration $C_{A,0} = 1.2$ kg/m^3 and the same value of $C_{A,L}$ ($= 0.9701$) as in Problem 15.D1, but we want a flux rate $= 0.35 \times 10^{-5} \, kg/(m^2 s)$. The apparatus temperature can be adjusted. The Fickian infinite dilution diffusivity at any temperature T can be estimated by adjusting the value from Table 15-3 for the temperature difference by assuming that $E_0 = 3000 \, cal/mol$ in Eq. (15-23c). What temperature is required?

D3. **a.** Estimate the Fickian diffusivity of a binary mixture of ammonia and air at 273 K and 1.0 atm pressure using the Chapman-Enskog theory and Table 15-2.

    b. Compare your result with the experimental value in Table 15-3.

    c. Estimate the Fickian diffusivity at 273K and 2.5 atm.

    d. Estimate the Maxwell-Stefan diffusivity of a binary mixture of ammonia and air at 273 K and 1.0 atm.

D4. **a.** Estimate the Fickian diffusivity of a binary mixture of benzene and air at 298.2 K and 1.0 atm pressure using the Chapman-Enskog theory and Table 15-2.

    b. Compare your result with the experimental value in Table 15-3.

    c. Estimate the Fickian diffusivity at 273 K and 0.5 atm.

    d. Estimate the Maxwell-Stefan diffusivity of a binary mixture of benzene and air at 298.2 K and 1.0 atm.

D5. What is the Fickian diffusivity of chlorobenzene in liquid bromobenzene at 300 K when the mole fraction of chlorobenzene is 0.0332? Assume that the diffusivity follows an Arrhenius form and use the data in Table 15-3 to determine E_0. Also report the value of E_0 in J/mol.

D6. Estimate the Fickian diffusivity of sucrose in liquid water at infinite dilution at 320 K. Assume that the diffusivity follows an Arrhenius form and use the data in Table 15-3. Also report the value of E_0 in J/mol.

D7. Assuming that the mixture is ideal, estimate the infinite dilution Fickian diffusivities at 283.3 K for chlorobenzene in liquid bromobenzene and for bromobenzene in liquid chlorobenzene from the data in Table 15-3.

D8. Use the Wilke-Chang theory to estimate the infinite dilution Fickian diffusivity of methanol in liquid water at 293.16 K. Data are available at http://www.engineeringtoolbox.com/ and http://www.cheric.org/research/kdb/ (click on Korean Physical Properties Data Bank).

    a. Use a value of $\varphi_B = 2.26$.

    b. Use a value of $\varphi_B = 2.6$.

D9. Determine the modified Sherwood number $Sh_{gas,partial\_pressure} = \dfrac{k_p d_{tube} (p_B)_{lm}}{D_{AB} p_{tot}}$ for the gas-side-controlled mass transfer for distillation in a wetted wall column. The tube diameter is 10.0 cm, and the distillation is ethanol and water. The measurement is made at very low ethanol concentrations where the flowing liquid can be assumed to be pure water. The total pressure is 1.0 bar. Liquid water at its boiling temperature is flowing down a vertical tube at a volumetric flow rate per meter of circumference of $q = 0.0000075 \ m^2/s$. Ethanol is diffusing through the vapor phase (almost pure water) at 1.0 bar and its boiling temperature. The upward flowing vapor velocity is 0.81 m/s. The densities and viscosities of pure water liquid and pure water vapor are available in *Perry's Handbook of Chemical Engineering*. Use the parameters in Table 15-2 to estimate the diffusivity of ethanol and water in the vapor.

D10. We have steady-state diffusion of ammonia in air across a film that is 0.15 mm thick. On one side of the film, the ammonia concentration is $0.00023 \ kmol/m^3$. We desire an ammonia flux rate of $0.25 \times 10^{-5} \ kmol/(m^2s)$. The apparatus is at 45°C and 1.2 atm.
  a. Use the Chapman-Enskog theory to estimate the diffusivity.
  b. What is the concentration at the other side of the film? Depending on the direction of transfer, there are two answers.

D11. We have steady-state diffusion of ammonia in air across a film that is 0.033 mm thick. On one side of the film, the ammonia concentration is 0.000180 and on the other side it is $0.000257 \ kmol/m^3$. We desire an ammonia flux rate of $9.60 \times 10^{-5} \ kmol/(m^2s)$. The apparatus is at 0.90 atm. Use the Chapman-Enskog theory to estimate the diffusivity. Find the required operating temperature.

D12. * Water at 20°C is flowing down a 3.0 m long vertical plate at a volumetric flow rate per meter of plate width of $q = 0.000005 \ m^2/s$. The entering $(y = 0)$ water is pure. The water is in contact with a gas phase of carbon dioxide that is saturated with water vapor at 20°C (no water is evaporating). Carbon dioxide is transferring into the water. At the water-gas surface $(z = 0)$, the water and carbon dioxide are in equilibrium at the solubility limit of carbon dioxide at $C_{CO2,surf}$. The Fickian infinite dilution diffusivity can be estimated by adjusting the value from Table 15-3 for the temperature difference by assuming that $E_o = 2000 \ cal/mol$ in Eq. (15-23c). At 20°C the solubility of carbon dioxide in water is 1.7 g/kg water.
  a. Determine the film thickness δ, the average vertical velocity of the film, and the Reynolds number.
  b. Is the flow laminar? If no surfactant is added, do you expect ripples on the surface? If surfactant is added, do you expect ripples on the surface?
  c. With no surfactant, determine the average mass-transfer coefficient and the average Sherwood number.
  d. Per meter of plate width, at what rate (kg/s) is carbon dioxide transferred into the water over the length of the plate?

D13. Repeat all parts of Problem 15.D12, but with a water rate of $q = 0.000015 \ m^2/s$.

D14. Repeat Problem 15.D12, but for $q = 0.0015 \ m^2/s$.
  a. Determine the film thickness δ, the average vertical velocity of the film, and the Reynolds number.
  b. Determine the average mass-transfer coefficient and the average Sherwood number.
  c. Per meter of plate width, at what rate (kg/s) is carbon dioxide transferred into the water over the length of the plate?

D15. Repeat Problem 15.D12 with $q = 0.000005 \ m^2/s$, but with a 0.01 m long vertical plate. Note that the theory assumes fully developed laminar flow that is unlikely with a very short plate. Thus, the results will be suspect.

D16. Repeat Problem 15.D13 ($q = .000015 \ m^2/s$, length = 3.0 m), but add a surfactant. Do the following:

a. Determine the film thickness δ, the average vertical velocity of the film, and the Reynolds number.

b. Determine the average mass-transfer coefficient and the average Sherwood number.

c. Per meter of plate width, at what rate (kg/s) is carbon dioxide transferred into the water over the length of the plate?

d. Compare your result to the answers for Problem 15.D13.

D17. Repeat Example 15-6, but with a mass-transfer coefficient that is 10 times larger (use δ = 0.001 m). Report x_{NH3}, $y_{NH3,surface}$, N_{water}, and N_{NH3}.

D18*. a. Repeat Example 15-5, but divide the film into two parts (from 0 to Δz and from Δz to δ = 0.00068m). Δz is unknown but is selected so that x_E = 0.30 (the previous average point) at the end of the interval. Now the average points for the first interval is x_E = 0.25 and for the second interval x_E = 0.35. Since operation is at steady state, N_E has to be the same in both intervals. Write Eq. (15-61c) for each interval (with thickness Δz for the first and δ – Δz for the second) and solve the resulting two equations for the two unknowns Δz and N_E.

b. Divide the film into three parts (0 to $Δz_1$, $Δz_1$ to $Δz_2$, and $Δz_2$ to δ). $Δz_1$ is unknown but is selected so that the value at the end of the interval is x_E = 0.25 (the previous average point). $Δz_2$ is also unknown but is selected so that the value at the end of the interval is x_E = 0.30 (the average point in example 15-5). The last interval is the same as in part a. Since operation is at steady state, N_E has to be the same in all three intervals. Write Eq. (15-61c) for each interval (with thickness $Δz_1$ for the first, $Δz_2$ for the second, and δ – $Δz_1$ – $Δz_2$ for the third) and solve the resulting three equations for the unknowns $Δz_1$, $Δz_2$, and N_E.

Data: At 40°C the density for 22.5 mol% ethanol is 914kg/m³, 25 mol% ethanol is 907kg/m³, 27.5 mol% ethanol is 899kg/m³, and for 35 mol% the density is 879 kg/m³.

D19. (Optional). Plot Figure 15-3 for the following unsteady diffusion problem. A thick layer of pure water ($C_{initial}$ = $C_∞$ = 0) at 25°C has a constant concentration of sucrose of C_0 = 0.002 mol/L placed on one surface at time t = 0. Assume the diffusivity (Table 15-3) is constant at the average concentration of 0.001 mol/L.

a. * Check the plot C/C_0 versus z (in cm) for t = 10,000 s. Selected values are in the answers at the back of the book.

b. If the threshold of a zero concentration is 1.0×10^{-6} mol/L, what is the minimum thickness the layer must have to appear infinite at t = 100,000s?

c. If the threshold of a zero concentration is 1.0×10^{-6} mol/L, what is the maximum exposure time the layer can have to appear infinite if the layer thickness is 0.1 cm?

H. *Spreadsheet Problems*

H1. Two identical large glass bulbs are filled with gases and connected by a capillary tube that is δ = 0.0100 m long. Bulb 1 at z = 0 contains the following mole fractions: y_{air} = 0.520, y_{H2} = 0.480, and y_{NH3} = 0.000. Bulb 2 at z = δ contains y_{air} = 0.540, y_{H2} = 0.000, and y_{NH3} = 0.460. Since the bulbs are quite large, operation is at pseudo- (or quasi-) steady state. [In other words, assume the mole fractions at the boundaries are constant (e.g., y_{air} = 0.520 at z = 0 and y_{air} = 0.540 at z = δ so that $Δy_{air}$ = 0.02) and the total flux of air + hydrogen + ammonia is zero.] The pressure is uniform at 2.00 atm, and the temperature is uniform at 273 K. Diffusivity values can be determined from Table 15-1. Assume the gases are ideal. Estimate the fluxes of the three components using the difference equation formulation of the Maxwell-Stefan method.

H2. Repeat Example 15-6 but for a bulk gas that is 40% air, 15% NH_3, and 45% water. Report x_{NH3}, $y_{NH3,surface}$, N_{water}, and N_{NH3}.

H3. **a.** Repeat Problem 15.H1 (use the Maxwell-Stefan equations), but bulb 1 at $z = 0$ contains the following mole fractions: $y_{air} = 0.500$, $y_{H2} = 0.500$, and $y_{NH3} = 0.000$. Bulb 2 at $z = \delta$ contains $y_{air} = 0.499$, $y_{H2} = 0.499$, and $y_{NH3} = 0.002$.

   b. Solve this problem using the effective Fickian diffusion coefficient for NH_3, Eq. (15-67b), and the usual Fickian equations for a very dilute system with no convection. Although there is not a simple approach for the two concentrated components, one is tempted to treat these as a binary with the Fickian binary diffusion coefficient for air and hydrogen. Try this, and then compare to the solution of part a.

H4. Repeat Problem 15.H1, but bulb 1 at $z = 0$ contains the following mole fractions: $y_{air} = 0.520$, $y_{H2} = 0.480$, and $y_{NH3} = 0.000$. Bulb 2 at $z = \delta$ contains $y_{air} = 0.520$, $y_{H2} = 0.000$, and $y_{NH3} = 0.480$.

CHAPTER 15 APPENDIX. SPREADSHEET FOR EXAMPLE 15-6

The spreadsheet below solves Example 15-6. The spreadsheet is shown first with the formulas. To estimate D_{W-NH3} from the Chapman-Enskog equation, $k_B T/\varepsilon_{AB}$ was calculated on the spreadsheet, and then the value of the collision integral Ω_D was determined by a hand calculation doing a linear interpolation on Table 15-2. After the value of the collision integral was inserted into the spreadsheet, the spreadsheet was run again. The spreadsheet could be totally automated by inputting the values for collision integrals from Table 15-2 and doing the linear interpolation on the spreadsheet.

The resulting values in the spreadsheet are shown below.

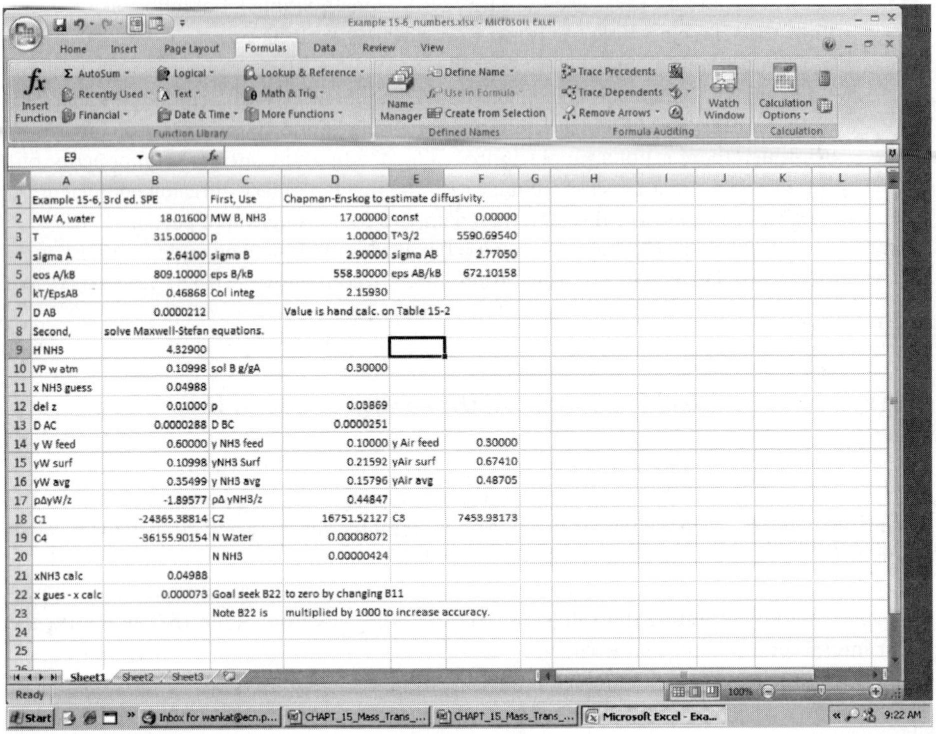

	A	B	C	D	E	F
1	Example 15-6, 3rd ed. SPE		First, Use	Chapman-Enskog to estimate diffusivity.		
2	MW A, water	18.01600	MW B, NH3	17.00000	const	0.00000
3	T	315.00000	p	1.00000	T^3/2	5590.69540
4	sigma A	2.64100	sigma B	2.90000	sigma AB	2.77050
5	eos A/kB	809.10000	eps B/kB	558.30000	eps AB/kB	672.10158
6	kT/EpsAB	0.46868	Col integ	2.15930		
7	D AB	0.0000212		Value is hand calc. on Table 15-2		
8	Second,	solve Maxwell-Stefan equations.				
9	H NH3	4.32900				
10	VP w atm	0.10998	sol B g/gA	0.30000		
11	x NH3 guess	0.04988				
12	del z	0.01000	p	0.03869		
13	D AC	0.0000288	D BC	0.0000251		
14	y W feed	0.60000	y NH3 feed	0.10000	y Air feed	0.30000
15	yW surf	0.10998	yNH3 Surf	0.21592	yAir surf	0.67410
16	yW avg	0.35499	y NH3 avg	0.15796	yAir avg	0.48705
17	pΔyW/z	-1.89577	pΔ yNH3/z	0.44847		
18	C1	-24365.38814	C2	16751.52127	C3	7453.93173
19	C4	-36155.90154	N Water	0.00008072		
20			N NH3	0.00000424		
21	xNH3 calc	0.04988				
22	x gues - x calc	0.000073	Goal seek B22 to zero by changing B11			
23			Note B22 is	multiplied by 1000 to increase accuracy.		
24						
25						

CHAPTER 16

Mass Transfer Analysis for Distillation, Absorption, Stripping, and Extraction

Up to now we have used an equilibrium stage analysis procedure even in packed columns where there are no stages. A major advantage of this procedure is that it does not require determination of the mass transfer rate.

In packed columns, it is conceptually incorrect to use the staged model even though it works if the correct height equivalent to a theoretical plate (HETP) is used. In this chapter we will develop a physically more realistic model for packed columns that is based on mass transfer between the phases. After developing the model for distillation, we will discuss mass transfer correlations that allow us to predict the required coefficients for common packings. Next, we will repeat the analysis for both dilute and concentrated absorbers and strippers and analyze cocurrent absorbers. A simple model for mass transfer on a stage will be developed for distillation, and the estimation of stage efficiency will be considered. After a mass transfer analysis of mixer-settler extractors, Section 16.8 and the appendix to Chapter 16 will develop the rate model for distillation.

It is assumed that readers have some knowledge of basic mass transfer concepts either from Chapter 15 or from other sources (e.g., Cussler, 1997; Geankoplis, 2003; McCabe et al., 2005; Taylor and Krishna, 1993).

16.1 HTU-NTU ANALYSIS OF PACKED DISTILLATION COLUMNS

Consider the packed distillation tower shown in Figure 16-1. Only binary distillation with constant molal overflow (CMO) will be considered. Let A be the more volatile component and B the less volatile component. In addition to making L/V constant and satisfying the energy balances, CMO automatically requires equimolal counterdiffusion, $N_A = -N_B$. Thus, CMO simplifies the mass balances, eliminates the need to solve the energy balances, and simplifies the mass transfer equations. We will also assume perfect plug flow of the liquid and vapor. This means that there is no eddy mixing to reduce the separation.

The mass transfer can be written in terms of the individual coefficients Eqs. (15-27) or overall coefficients Eqs. (15-28). For the differential height dz in the rectifying section, the mass transfer rate in terms of the individual coefficients is

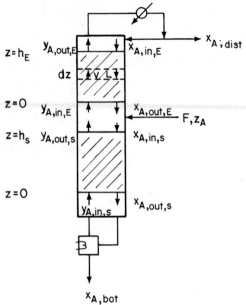

FIGURE 16-1.   *Packed distillation column*

$$N_A \, a \, A_c \, dz = k_y \, a \, (y_{AI} - y_A) \, A_c \, dz \qquad \text{(16-1)}$$

where N_A is the flux of A in kmol/m²-h or lbmol/ft²-h and A_c is the column cross-sectional area in m² or ft². This equation has units of kmol/h or lbmol/h. The mass transfer rate must also be equal to the changes in the amount of the more volatile component in the liquid and vapor phases.

$$N_A \, a \, A_c \, dz = V \, dy_A = L dx_A \qquad \text{(16-2)}$$

where L and V are constant molal flow rates. Combining Eqs. (16-1) and (16-2), we obtain

$$dz = \frac{V}{k_y a A_c \, (y_{AI} - y_A)} \, dy_A \qquad \text{(16-3)}$$

Integrating this from z = 0 to z = h, where h is the total height of packing in a section, we obtain

$$h = \frac{V}{k_y a A_c} \int_{y_{A,in}}^{y_{A,out}} \frac{dy_A}{y_{AI} - y_A} \qquad \text{(16-4)}$$

We have assumed that the term $V/(k_y a A_c)$ is constant. The limits of integration for y_A in each section are shown in Figure 16-1. Equation (16-4) is often written as

$$h = H_G \, n_G \qquad \text{(16-5)}$$

where the height of a gas-phase transfer unit H_G is

$$H_G = \frac{V}{k_y a A_c}$$

(16-6)

and the number of gas-phase transfer units n_G is

$$n_G = \int_{y_{A,in}}^{y_{A,out}} \frac{dy_A}{y_{AI} - y_A}$$

(16-7)

The height of transfer unit terms are commonly known as HTUs and the number of transfer units as NTUs. Thus, the model is often called the HTU-NTU model.

If we substitute Eq. (16-6) into (16-5) and solve for n_G, we obtain

$$n_G = k_y a (h A_c) / V = k_y a (h A_c) / (Q \rho_V / MW_V)$$

(16-8)

Since $h A_c$ is the volume of this section of the column, $(h A_c)/Q$ and $(h A_c)/V$ are measures of the residence time of the vapor. Thus, the number of transfer units is proportional to $(ka) \times$ (residence time). This is true of all definitions of number of transfer units in Table 16-1 with appropriate changes in ka and residence times.

An exactly similar analysis can be done in the liquid phase by starting with Eq. (15-28). The result for each section is

$$h = \frac{L}{k_x a A_c} \int_{x_{A,out}}^{x_{A,in}} \frac{dx_A}{x_A - x_{AI}}$$

(16-9)

which is usually written as

$$h = H_L n_L$$

(16-10)

where

$$H_L = \frac{L}{k_x a A_c}$$

(16-11)

$$n_L = \int_{x_{A,out}}^{x_{A,in}} \frac{dx_A}{x_A - x_{AI}}$$

(16-12)

In Section 15.4 we noted that although the basic Eq. (15-25a) is the same, several different combinations of mass transfer coefficient and driving force can be employed to analyze complicated mass transfer systems. If the driving force and mass transfer coefficient are changed, then the definition of HTU will also change. Table 16-1 lists the most commonly used definitions for driving force, mass transfer coefficient, and HTU.

Table 16-1.   Definitions of mass transfer coefficients and HTUs.

Basic Equation	Units on k	HTU Equation	References
$N_A = k_y \Delta y_A$	lbmol/ft^2–h or mol/cm^2–s or kmol/m^2–h	$H = \dfrac{V}{k_y a A_c}$	This book Bennett & Myers (1982) Bolles & Fair (1982) Hines & Maddox (1985) Treybal (1980) Greenkorn & Kessler (1972) Geankoplis (2003)
$\overline{N}_A = \overline{k}_y \Delta \overline{y}_A$	lb/h–ft^2 or kg/h–m^2	$\overline{H} = \dfrac{G}{\overline{k}_y a A_c}$	Transfer in mass units
$N_A = k_c \Delta C$	cm/s or ft/h or m/s or m/h	$H = \dfrac{V}{k_c a C_m A_c}$	This book Cussler (2009) Geankoplis (2003) Taylor & Krishna (1993)
$N_p = k_p \Delta p$	$k_p = k_y / p_{tot}$ lbmol/ft^2–h–atm or mol/cm^2–s–atm or kmol/m$_2$–h–bar	$H = \dfrac{V}{k_p a p_{tot} A_c}$	Sherwood et al. (1975) Geankoplis (2003) McCabe et al. (2005) Greenkorn & Kessler (1972)

In order to do the integrations to calculate n_G and n_L we must relate the interfacial mole fracs y_{AI} and x_{AI} to the bulk mole fracs y_A and x_A. To do this we start by setting Eqs. (15-27a) and (15-27b) equal to each other. After simple rearrangement, this is

$$\frac{y_{AI} - y_A}{x_{AI} - x_A} = -\frac{k_x a}{k_y a} = -\frac{L}{V}\frac{H_G}{H_L} \tag{16-13}$$

where the last equality on the right comes from the definitions of H_G and H_L. The left-hand side of this equation can be identified as the slope of a line from the point representing the interfacial mole fracs (y_{AI}, x_{AI}) to the point representing the bulk mole fracs (y_A, x_A). Since there is no interfacial resistance, the interfacial mole fracs are in equilibrium and must be on the equilibrium curve (Figure 16-2A). The bulk mole fracs are easily related by a mass balance through segment dz around either the top or the bottom of the column. This *operating line* in the rectifying section is

$$y_A = \frac{L}{V} x + \left(1 - \frac{L}{V}\right) x_{A,dist} \tag{16-14}$$

In the stripping section the operating line that relates y_A to x_A is

$$y_A = \frac{\overline{L}}{\overline{V}} x - (\frac{\overline{L}}{\overline{V}} - 1)x_{A,bot} \tag{16-15}$$

Since these operating equations are exactly the same as the operating equations for staged systems (Chapter 4), they intersect at the feed line.

We can now use a modified McCabe-Thiele diagram to determine x_{AI} and y_{AI}. From any point (y_A, x_A) on the operating line, draw a line slope $-k_x a/k_y a$. The intersection of this line with the equilibrium curve gives the interfacial mole fracs that correspond to y_A and x_A (see Figure 16-2A). After this calculation is done for a series of points, we can plot $1/(y_{AI} - y_A)$ vs y_A as shown in Figure 16-2B. The area under the curve is n_G. n_L is determined by plotting $1/(x_A - x_{AI})$ versus x_A. The areas can be determined from graphical integration or numerical integration such as Simpson's rule [see Eq. (9-12) and Example 16-1].

It will be most accurate to do the calculations for the stripping and enriching sections separately. For example, in the stripping section,

$$H_{G,S} = \frac{\overline{V}}{k_y a A_c} \tag{16-16}$$

$$n_{G,S} = \int_{y_{A,in,S}}^{y_{A,out,S}} \frac{dy_A}{y_{AI} - y_A} \tag{16-17}$$

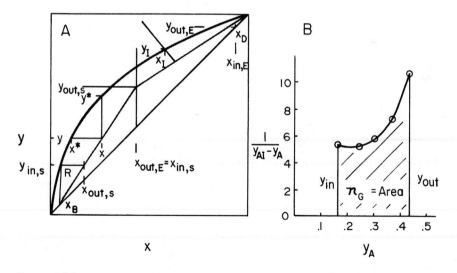

FIGURE 16-2.   *Analysis of number of transfer units; (A) determination of equilibrium or interfacial values, (B) graphical integration of Eq. (16-7) shown for stripping section of Example 16-1*

In the determination of n_G for the stripping section, $y_{A,in,S}$ is the vapor mole frac leaving the reboiler. This is illustrated in Figure 16-1 for a partial reboiler. Mole frac $y_{A,out,S}$ is the mole frac leaving the stripping section. This mole frac can be estimated at the intersection of the operating lines. This is shown in Figure 16-2A. Note that this estimate makes $y_{A,out,S} = y_{A,in,E}$.

Calculating the interfacial mole fracs adds an extra step to the calculation. Since it is often desirable to avoid this step, the overall mass transfer coefficients in Eq. (15-29) are often used. In terms of the overall driving force the mass transfer rate corresponding to Eq. (16-1) is

$$N_A a A_c dz = K_y a (y_A^* - y_A) A_c dz \tag{16-18}$$

In this equation y_A^* is the mole fraction A at value x_A (see Figure 16-2A). Setting this equation equal to Eq. (16-2), we obtain

$$dz = \frac{V}{K_y a A_c (y_A^* - y_A)} dy_A \tag{16-19}$$

Integration of this equation over a section of the column gives

$$h = \frac{V}{K_y a A_c} \int_{y_{A,in}}^{y_{A,out}} \frac{dy_A}{y_A^* - y_A} \tag{16-20}$$

This equation is usually written as

$$h = H_{OG} \, n_{OG} \tag{16-21}$$

where the height of an overall gas-phase transfer unit is

$$H_{OG} = \frac{V}{K_y a A_c} \tag{16-22}$$

and the number of overall gas-phase transfer units is

$$n_{OG} = \int_{y_{A,in}}^{y_{A,out}} \frac{dy_A}{y_A^* - y_A} \tag{16-23}$$

Exactly the same steps can be done in terms of the liquid mole fracs. The result is

$$h = H_{OL} \, n_{OL} \tag{16-24}$$

where

$$H_{OL} = \frac{L}{K_x a A_c} \tag{16-25}$$

$$n_{OL} = \int_{x_{A,in}}^{x_{A,out}} \frac{dx_A}{x_A - x_A^*}$$

(16-26)

The advantage of this formulation is that $y_A^* - y_A$ is easily found from vertical lines shown in Figure 16-2A. The value $x_A - x_A^*$ can be found from horizontal lines as shown in the figure. The number of transfer units, n_{OG} or n_{OL}, is then easily determined. Calculation of n_{OG} is similar to the calculation of n_G illustrated in Figure 16-2B. The disadvantage of using the overall coefficients is that the height of an overall transfer unit, H_{OG} or H_{OL}, is much less likely to be constant than H_G or H_L. This is easy to illustrate, since we can calculate the overall HTU from the individual HTUs. For example, substituting Eq. (15-31b) into Eq. (16-22) and rearranging, we obtain

$$H_{OG} = \frac{mV}{L} H_L + H_G$$

(16-27a)

H_{OL} can be found by substituting Eq. (15-31c) into Eq. (16-25)

$$H_{OL} = H_L + \frac{L}{mV} H_G$$

(16-27b)

Obviously, H_{OG} and H_{OL} are related:

$$H_{OL} = \frac{L}{mV} H_{OG}$$

(16-27c)

If H_G and H_L are constant, H_{OG} and H_{OL} cannot be exactly constant, since m, the slope of the equilibrium curve, varies in the column. The various NTU values must be related, since h in Eqs. (16-5), (16-10), (16-21), and (16-24) is obviously the same, but the HTU values vary. These relationships are derived in Problem 16.C1.

This approach can easily be extended to the more complex continuous columns discussed in Chapters 4 and 8 and to the batch columns discussed in Chapter 9. Any of these situations can be analyzed by plotting the appropriate operating lines and then proceeding with the HTU-NTU analysis. An alternative procedure is described in Problem 16.G1.

EXAMPLE 16-1. Distillation in a packed column

We wish to repeat Example 4-3 (distillation of ethanol and water) except that a column packed with 2-inch metal Pall rings will be used. F = 1000 kgmol/h, z = 0.2, T_F = 80°F, x_D = 0.8, x_B = 0.02, L/D = 5/3, and p = 1 atm. Use a vapor flow rate that is nominally 75% of flooding. In the enriching section H_G = 0.4054 m and H_L = 0.253 m, and in the stripping section H_G = 0.2835 m and H_L = 0.1067 m (see Example 16-2).

Solution

A. Define. Determine the height of packing in the stripping and enriching sections.

B and C. Explore and plan. The solution obtained in Example 4-3 can be used to plot the operating lines and the feed line. These are exactly the same as in Figure 4-13. Since the ethanol-water equilibrium is very nonlinear, the design will be more accurate if an individual mass transfer coefficient is used. Thus, use Eqs. (16-5) and (16-7) for the enriching and stripping sections separately. The term $(y_{AI} - y_A)$ can be determined as illustrated schematically in Figure 16-2A and for this example in Figure 16-3A. n_G can be found for each section as shown in Figures 16-2B and 16-3B.

D. Do it. The equilibrium and operating lines from Example 4-3 are plotted in Figure 16-3A. In the *stripping* section, Eq. (16-13) gives a slope of

$$\text{Slope} = -\frac{\overline{L}}{\overline{V}} \frac{H_G}{H_L} = -2.04 \frac{0.2835}{0.1067} = -5.37$$

where $\overline{L}/\overline{V} = 2.04$ from Example 4-3 or from mass balances. Lines with a slope = −5.37 are drawn in Figure 16-3A from arbitrary points on the stripping section operating line to the equilibrium curve. Values of y_A are on the operating line, while y_{AI} values are on the equilibrium line. The following table was generated.

y_{AI}	y_A	$y_{AI} - y_A$	$1/(y_{AI} - y_A)$
0.354	0.17	0.184	5.44
0.44	0.25	0.19	5.26
0.477	0.306	0.171	5.85
0.512	0.375	0.137	7.30
0.535	0.442	0.093	10.75

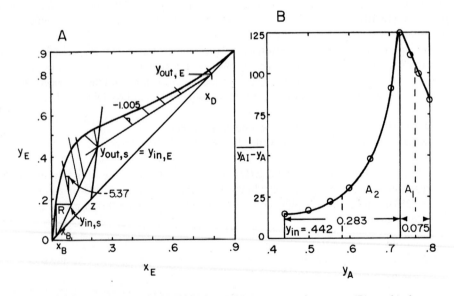

FIGURE 16-3. Solution to Example 16-1; (A) determination of $y_{AI} = y_{EI}$, (B) graphical integration for enriching section

Chapter 16 Mass Transfer Analysis for Distillation, Absorption, Stripping, and Extraction

From this table $1/(y_{AI} - y_A)$ vs. y_A is easily plotted, as shown in Figure 16-2B. n_G is the area under this curve from $y_{A,in,S} = 0.17$ to $y_{A,out,S} = 0.442$. $y_{A,in,S}$ is the vapor mole frac leaving the partial reboiler. Determination of $y_{A,in,S}$ is shown in Figure 16-3A. $y_{A,out,S}$ is the vapor mole frac at the intersection of the operating lines. The area in Figure 16-2B can be estimated from Simpson's rule (although the area will be overestimated since the minimum in the curve is not included) or other numerical integration schemes.

$$n_{G,S} = \frac{y_{A,out} - y_{A,in}}{6} [f(y_{A,in}) + 4f(y_{A,avg}) + f(y_{A,out})] \tag{16-28a}$$

where

$$f = \frac{1}{y_{AI} - y_A}, \; y_{A,avg} = \frac{y_{A,in} + y_{A,out}}{2} \tag{16-28b}$$

Note that Simpson's rule uses the end and middle points. For the stripping section, this is

$$n_{G,S} = \frac{0.272}{6} [5.44 + 4(5.85) + 10.75] = 1.79$$

And the height of packing in the stripping section is

$$h_S = H_{G,S} n_{G,S} = (0.2835)(1.79) = 0.507 \text{ m}$$

In the *enriching* section the slope is

$$\text{Slope} = -\frac{L}{V} \frac{H_G}{H_L} = -\left(\frac{5}{8}\right)\left(\frac{0.4054}{0.253}\right) = -1.005$$

Arbitrary lines of this slope are shown on Figure 16-3A. The following table was generated.

y_{AI}	y_A	$y_{AI} - y_A$	$1/(y_{AI} - y_A)$
0.505	0.442	0.063	15.9
0.557	0.5	0.057	17.5
0.594	0.55	0.044	22.7
0.632	0.60	0.032	31.25
0.671	0.65	0.021	47.6
0.711	0.7	0.011	90.9
0.733	0.725	0.008	125
0.759	0.75	0.009	111.1
0.785	0.775	0.01	100
0.812	0.8	0.012	83.3

The plot of $1/(y_{AI} - y_A)$ vs. y_A is shown in Figure 16-3B. An approximate area can be found using Simpson's rule, Eqs. (16-28a) and (16-28b), by splitting the total area into areas A_1 and A_2 in Figure 16-3B. For area A_1 the initial point is selected as the maximum point, and the middle point $y_A = 0.7625$ with f = 107 was calculated.

$$A_1 = \frac{0.075}{6}[125 + 4(107) + 83.3] = 7.95$$

$$A_2 = \frac{0.283}{6}[15.9 + 4(26.5) + 125] = 11.65$$

$n_{G,E}$ is the total area = 19.6. Then the height in the enriching section is

$$h_E = H_{G,E} n_{G,E} = (0.4054)(19.6) = 7.95 \text{ m (26.1 ft)}$$

E. Check. The operating and equilibrium curves were checked in Example 4-3. The areas can be checked by counting squares in Figures 16-2B and 16-3B. More accuracy could be obtained by dividing Figure 16-2B into two parts. The HTU values will be estimated and checked in Example 16-2.

The largest error in the calculation is usually caused by errors in the mass transfer coefficients. The values of $k_y a$ have an average error of 24.4% (Wankat and Knaebel, 2008). The error in $k_x a$ is probably similar. It is also not uncommon to have errors in the equilibrium data, although in this example the equilibrium data are known quite accurately. Finally, since the calculation of NTU in the enriching section involves determining the inverse of a small difference, rather large calculation errors can creep into the value of the integral (see Problem 16.D16). Usually, the most significant uncertainty results from the error in the mass transfer coefficients. A safety factor can be estimated by calculating the packing height needed with lower mass transfer coefficients. Bolles and Fair (1982) recommend multiplying the height by 1.70 in extreme cases. See also the discussion in part E of Example 16-2 and Problem 16.D17.

A check of the heights of the columns using Aspen Plus and a different integration of Eq. (16-19) agreed with the stripping section height, but not with the enriching section height (see Problem 16.G1). In addition, a staged analysis using HETP to determine packing heights (see Chapter 10) also showed a significantly smaller enriching section height. These differences may well be due to differences between the VLE correlation used in the simulation and the experimental data used in this example. Other calculation methods should always be used to check calculations whenever possible.

F. Generalize. The method illustrated here can obviously be used in other distillation systems. Since the curve for n_G can be very nonlinear, it is a good idea to plot the curve as shown in Figures 16-2B and 16-3B before doing the numerical integration.

16.2 RELATIONSHIP OF HETP AND HTU

In simple cases the HTU-NTU approach and the HETP approach discussed in Chapter 10 can be related with a derivation similar to that used for the Kremser equation (section 12.4). If the operating and equilibrium curves are straight and parallel, $mV/L = 1$, we have the situation shown in Figure 16-4A. The equilibrium equation is

$$y = mx + b \qquad (16\text{-}29a)$$

while a general equation for the straight operating line is

$$y = \frac{L}{V} x + y_{out} - \frac{L}{V} x_{in} \qquad (16\text{-}29b)$$

Now the integral in the definition of n_{OG} can easily be evaluated analytically. The difference between the equilibrium and operating lines, $y^* - y$, is

$$y^* - y = mx + b - \left[\frac{L}{V} x + y_{out} - \frac{L}{V} x_{in} \right] \qquad (16\text{-}30)$$

when $m = L/V$, this becomes

$$y^* - y = \frac{L}{V} x_{in} + b - y_{out}$$

Then Eq. (16-23b) becomes

$$n_{OG} = \int_{y_{in}}^{y_{out}} \frac{dy}{y^* - y_A} = \int_{y_{in}}^{y_{out}} \frac{dy}{\dfrac{L}{V} x_{in} + b - y_{out}}$$

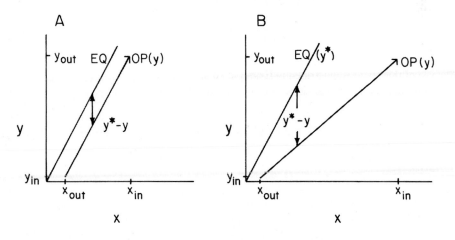

FIGURE 16-4.   *Calculation of $y^* - y$ with linear equilibrium and operating lines; (A) $mV/L = 1$, (B) $mV/L \neq 1$*

which is easily integrated.

$$n_{OG} = \frac{y_{in} - y_{out}}{y_{out} - \frac{L}{V} x_{in} - b}, \quad \text{for } \frac{mV}{L} = 1 \tag{16-31}$$

Since $h = H_{OG} n_{OG} = N \times (\text{HETP})$, we can solve for HETP.

$$\text{HETP} = \frac{n_{OG}}{N} H_{OG} \tag{16-32}$$

N can be obtained from Eq. (12-12). Comparison of Eqs. (16-31) and (12-12) shows that N = n_{OG} when mV/L = 1. Thus,

$$\text{HETP} = H_{OG}, \quad \text{for} \quad \frac{mV}{L} = 1 \tag{16-33}$$

If the operating and equilibrium lines are straight but not parallel, then we have the situation shown in Figure 16-4B. The difference between equilibrium and operating lines is still given by Eq. (16-30), but the terms with x do not cancel out. By substituting in x from the operating equation, y* – y in Eq. (16-23) can be determined as a linear function of y. After integration and considerable algebraic manipulation, n_{OG} is found to be

$$n_{OG} = \left(\frac{1}{1 - mV/L} \right) \ln \left[(1 - \frac{mV}{L})(\frac{y_{in} - y^*_{out}}{y_{out} - y^*_{out}}) + \frac{mV}{L} \right] \tag{16-34a}$$

where

$$y^*_{out} = mx_{in} + b \tag{16-35a}$$

The value of HETP can be determined from Eq. (16-32), where N is found from the Kremser Eq. (16-22) and n_{OG} from Eq. (16-34a). This result is

$$\text{HETP} = \frac{H_{OG} \ln(mV/L)}{mV/L - 1} \tag{16-36a}$$

The use of this result is illustrated in Example 16-2.

This analysis can also be done in terms of liquid mole fracs. The result is

$$n_{OL} = \left(\frac{1}{1 - L/mV} \right) \ln \left[(1 - L/mV) \left(\frac{x_{A,in} - x^*_{A,out}}{x_{A,out} - x^*_{A,out}} \right) + \frac{L}{mV} \right] \tag{16-34b}$$

where

$$x^*_{A,out} = (y_{in} - b)/m \tag{16-35b}$$

$$\text{HETP} = \frac{H_{OL} \ln(L/mV)}{L/mV - 1} \tag{16-36b}$$

Equations (16-34a) and (16-34b) are known as the Colburn equations.

Although it was derived for a straight operating line and straight equilibrium lines, Eqs. (16-36) will be approximately valid for curved equilibrium or operating lines. HETP should be determined separately for each section of the column, since mV/L is not usually the same in the enriching and stripping sections. If H_{OG} is approximately constant, then HETP must vary since mV/L varies. For maximum accuracy the HETP can be calculated for each stage in the column (Sherwood et al., 1975).

16.3 MASS TRANSFER CORRELATIONS FOR PACKED TOWERS

In order to use the HTU-NTU analysis procedure we must be able to predict the mass transfer coefficients or the HTU values. There has been considerable effort expended in correlating these terms (see Wang et al., 2005, for an extensive review). Care must be exercised in using these correlations since HTU values in the literature may be defined differently. The definitions given here are based on using mole fracs in the basic transfer equations (see Table 16-1). If concentrations or partial pressures are used, the mass transfer coefficients will have different units, which will lead to different definitions for HTU although the HTU will still have units of height. In working with these correlations, terms must be expressed in appropriate units.

16.3.1 Detailed Correlations for Random Packings

We will use the correlation of Bolles and Fair (1982), for which HTUs are defined in the same way as here. The Bolles-Fair correlation is based on the previous correlation of Cornell et al., (1960a, b) and a data bank of 545 observations and includes distillation, absorption, and stripping. This model and variations on it remain in common use (Wang et al., 2005).

The correlation for H_G is

$$H_G = \frac{\psi(D_{col}')^{b_1}(h_p/10)^{1/3}(Sc_v)^{1/2}}{[(3600)W_L(\mu_L/\mu_w)^{0.16}(\rho_L/\rho_w)^{-1.25}(\sigma_L/\sigma_w)^{-0.8}]^{b_2}} \qquad (16\text{-}37)$$

where ψ is a packing parameter that is given in Figure 16-5 (Bolles and Fair, 1982) for common packings, and other special terms are defined in Table 16-2. Viscosity, density, surface tension, and diffusivities should be defined in consistent units so that the Schmidt number and the ratios of liquid to water properties are dimensionless. The packing height h_p is the height of each packed bed; thus, the stripping and enriching sections should be considered separately.

The correlation for H_L is

$$H_L = \phi C_{fL}(h_p/10)^{0.15}(Sc_L)^{1/2} \qquad (16\text{-}38)$$

In this equation ϕ is a packing parameter shown in Figure 16-6, and C_{fL} is a vapor load coefficient shown in Figure 16-7 (Bolles and Fair, 1982). The value of u_{flood} in Figure 16-7 is from the packed bed flooding correlation in Figure 10-27.

The calculated H_G and H_L values can vary from location to location in each section. When this occurs, an integrated mean value should be used. The overall HTU values can be obtained from Eqs. (16-27). Even if H_G and H_L are constant, H_{OG} and H_{OL} will vary owing to the curvature of the equilibrium curve.

Bolles and Fair (1982) show that there is considerable scatter in modeled HETP data vs. experimental HETP data. HETP was calculated from Eq. (16-36). For 95% confidence in the results, Bolles and Fair suggest a safety factor of 1.70 in the determination of HETP. They note that this large a safety factor is usually not used, since there are often a number of hidden safety factors such as not including end effects and using nonoptimum operating conditions. However, if a tight design is used, then the 1.70 safety factor is required. This large a safety factor emphasizes that design of distillation systems is an art not a science.

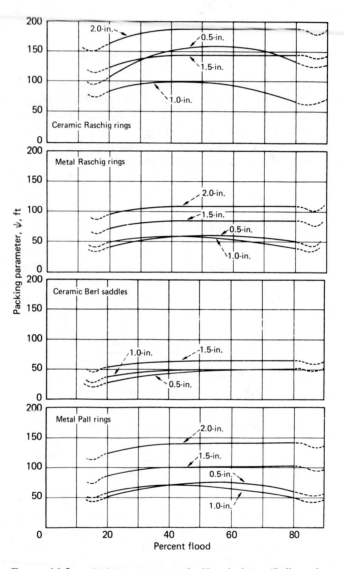

FIGURE 16-5.   *Packing parameter ψ for H_G calculation (Bolles and Fair, 1982) excerpted by special permission from Chemical Engineering, 89 (14), 109 (July 12, 1982), copyright 1982, McGraw-Hill, Inc., New York, NY 10020*

TABLE 16-2.   *Terms for Eqs. (16-37) and (16-38)*

D'_{col}	= lesser of column diameter in feet or 2
b_1	= 1.24 (rings) or 1.11 (saddles)
h_p	= height of each packed bed in feet
H_G, H_L	= HTU in feet
Sc_v	= Schmidt no. for vapor = $\mu_v / \rho_v D_v$
W_L	= Weight mass flux of liquid, lb/s-ft^2
b_2	= 0.6 (rings) or 0.5 (saddles)
Sc_L	= Schmidt no. for liquid = $\mu_L / \rho_L D_L$

EXAMPLE 16-2. *Estimation of H_G and H_L*

Estimate the values of H_G and H_L for the distillation in Examples 4-3 and 16-1 using 2-inch metal pall rings.

Solution

A. Define. We want to find H_G and H_L in both the stripping and enriching sections. This will be done as if we had completed Example 4-3 but not Example 16-1. Thus, we know the number of equilibrium stages required but we have not estimated packing heights.

B and C. Explore and Plan. We will use the Bolles and Fair (1982) correlation shown in Eqs. (16-37) and (16-38) and Figures 16-5 to 16-7. Obviously, we need to estimate the physical properties required in this correlation. We will do this for the striping and enriching sections separately. This estimation is easiest if a computer physical properties package is available. We will illustrate the estimation using values in the literature. The packing height, h_p, must be estimated for each section. These heights will be estimated from the number of stages in each section multiplied by an estimated HETP. Flow rates will be found from mass balances and then will be converted to weight units. A diameter calculation will be done to determine the actual percent flooding.

D. Do it. We will do calculations at the top of the column and assume that these values are reasonably accurate throughout the enriching section. Estimation of properties at the bottom of the column will be used for the stripping section. External balances give D = 230.8 kmol/h, B = 769.2 kmol/h. We will do the calculation in English units because the figures to determine parameters are in these units. The final answers will be converted to metric units.

Flooding at top.

$$x_D = y_1 = 0.8$$

$$\overline{MW}_v - y_1 MW_E + (1 - y_1) MW_w = 40.4$$

From the ideal gas law,

$$\rho_v = \frac{p(MW_v)}{RT} = 1.393 \times 10^{-3} \text{ g/cm}^3$$

where T = 78.4°C = 351.6 K from Figure 4-14.

Liquid density 80 mol% ethanol is 91.1 wt %. From Perry and Green (1984), ρ_L = 0.7976 g/mL at 40°C and ρ_L = 0.82386 at 0°C. By linear interpolation, ρ_L = 0.772 g/mL at 78.4°C. At 78.4°C, ρ_w = 0.973 g/mL.

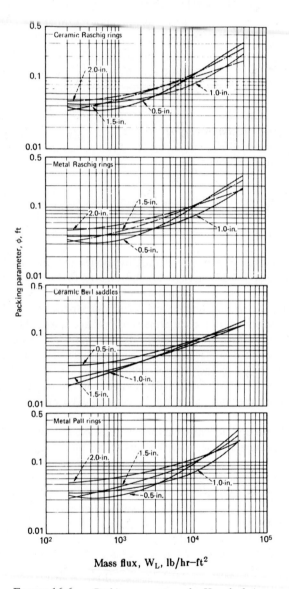

Mass flux, W_L, lb/hr–ft²

FIGURE 16-6.   *Packing parameter φ for H_L calculation (Bolles and Fair, 1982) excerpted by special permission from* Chemical Engineering, *89 (14), 109 (July 12, 1982), copyright 1982, McGraw-Hill, Inc., New York, NY 10020*

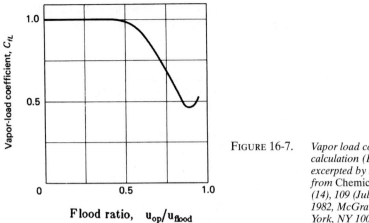

Left axis: Vapor-load coefficient, C_{FL}

Bottom axis: Flood ratio, u_{op}/u_{flood}

For the flooding curve in Figure 10-27 the abscissa is

$$\frac{L'}{G'}\left(\frac{\rho_v}{\rho_L}\right)^{1/2} = \frac{L}{V}\left(\frac{\rho_v}{\rho_L}\right)^{1/2} = \left(\frac{5}{8}\right)\left(\frac{1.393 \times 10^{-3}}{0.77}\right)^{1/2} = 0.27$$

Ordinate (flooding) = 0.197. Then

$$G'_{\text{flood}} = \left[\frac{(\text{ordinate})(\rho_G)(\rho_L)(g_c)}{F\psi\mu^{0.2}}\right]^{1/2}$$

$$G'_{\text{flood}} = \left[\frac{(0.197)(1.393 \times 10^{-3})(62.4)^2(0.77)(32.2)}{20\left(\frac{0.973}{0.772}\right)(0.52)^{0.2}}\right]^{1/2} = 1.16 \ \frac{\text{lb}}{\text{s} - \text{ft}^2}$$

The $(62.4)^2$ converts ρ_L and ρ_G to lb/ft^3. μ_L is estimated as 0.52. Then,

$$G'_{\text{actual}} = 0.75 \ G'_{\text{flood}} = 0.87 \ \frac{\text{lb}}{\text{s} - \text{ft}^2}$$

The molar vapor flow rate is

$$V = (L/D + 1)D = 615.4 \ \text{kmol/h}$$

which allows us to find the column cross-sectional area.

$$\text{Area} = \frac{V}{G'_{\text{actual}}}\left(\frac{1 \ \text{lbmol}}{0.46 \ \text{kmol}}\right)\left(\frac{40.4 \ \text{lb}}{\text{lbmol}}\right)\left(\frac{1 \ \text{h}}{3600 \ \text{s}}\right) = 17.2 \ \text{ft}^2$$

$$\text{Dia} = \left(\frac{4 \ \text{Area}}{\pi}\right)^{1/2} = 4.68 \ \text{ft}$$

Round this off to 5 feet (1.525 m), Area = 19.6 ft^2. This roundoff reduces the % flooding.
Actual fraction flooding = 0.75 (17.2/19.6) = 0.66

A repeat of the calculation at the bottom of the column shows that the column will flood first at the top since the molecular weight is much higher.

Estimation at top.

Liquid diffusivities. From Table 15-3, for very dilute systems $D^{\circ}_{we} = 1.25 \times 10^{-5}$ cm²/s and $D^{\circ}_{we} = 1.132 \times 10^{-5}$ at 25°C. The effect of temperature can be estimated since the ratio $D_L \mu_L / T \sim$ constant. At the top we want D_L at 78.4°C = 351.6 K.

$$D_L(78.4) = \frac{351.6}{\mu_L(78.4)} \left(\frac{D_L(25)\mu(25)}{298} \right)$$

Estimating viscosities from Perry and Green (1984) using 95% ethanol:

$$D_L(78.4) = \frac{351.6}{0.47cP} \left(\frac{(1.132 \times 10^{-5})(1.28cP)}{298} \right)$$

$$D_L = 3.64 \times 10^{-5} \text{ cm}^2/\text{s}$$

The liquid surface tension can be estimated from data in the *Handbook of Chemistry and Physics.*

$\sigma_L(78.4) = 18.2$ dynes/cm
$\sigma_W(78.4) = 62.9$ dynes/cm

For vapors the Schmidt number can be estimated from kinetic theory (Sherwood et al., 1975, pp. 17-24). The equation is

$$Sc_v = \left(\frac{\mu_v}{\rho_v D} \right) = 1.18 \frac{\Omega_D}{\Omega_v} \left(\frac{MW_A}{MW_A + MW_B} \right)^{1/2} \left(\frac{\sigma_{AB}}{\sigma_B} \right)^2 \tag{16-39}$$

where the collision integrals Ω_D, Ω_v and the Lennard-Jones force constants σ_{AB} and σ_B are discussed in Section 15.3.1. At the top of the column the result is $Sc_v = 0.355$.

The liquid flow rate at the top is

$$L = (L/D)D = 384.6 \text{ kmol/h.}$$

The liquid flux W_L is

$$W_L = \frac{(L \frac{\text{kmol}}{\text{h}})(\frac{1 \text{ lbmol}}{0.46 \text{ kmol}})(\frac{40.4 \text{ lb}}{\text{lbmol}})(\frac{1 \text{ h}}{3600 \text{ s}})}{\text{Area}} = 0.479 \frac{\text{lb}}{\text{s} - \text{ft}^2}$$

In Eq. (16-37) $D'_{col} = 2$, $\psi = 141$ from Figure 16-5 at 66% flood, $b_1 = 1.24$ and $b_2 = 0.6$. We can estimate h_p as (No. stages) × (HETP), where an average HETP is about 2 feet. Then

$$h_p = (11)(2) = 22 \text{ feet (6.7056 m)}$$

Equation (16-37) is then

$$H_{G,E} = \frac{(141)(2)^{1.24}(2.2)^{1/3}(0.355)^{1/2}}{[(3600)(.479)(.47/.36)^{0.16}(.772/.973)^{-1.25}(18.2/62.9)^{-0.8}]^{0.6}} = 1.33 \text{ ft } (0.4054 \text{ m})$$

For H_L we calculate W_L as 1724 lb/h-ft^2 and $\phi = 0.07$ from Figure 16-6. $C_{fl} = 0.81$ from Figure 16-7. Then from Eq. (16-38),

$$H_{L,E} = (0.07)(0.81)(2.2)^{.15}\left(\frac{0.0047}{(0.772)(3.64 \times 10^{-5})}\right)^{1/2} = 0.83 \text{ ft } (0.253 \text{ m})$$

Note that μ_L in Sc_L is in poise (0.01 P = 1 cP).

These calculations can be repeated for the bottom of the column. The results are: $H_{G,S} = 0.93$ feet (0.2835 m) and $H_{L,S} = 0.35$ feet (0.1067 m).

E. Check. One check can be made by estimating HETP using Eq. (16-36a). At the top of the column the slope of the equilibrium curve is m ~ 0.63. This will vary throughout the column. Then from Eq. (16-27a), at the top

$$H_{OG} = \frac{m}{L/V}H_L + H_G = \frac{0.63}{5/8}(0.827) + 1.33 = 2.16 \text{ ft } (0.6584 \text{ m})$$

Note that H_{OG} will vary in the enriching section since m varies. From Eq. (16-36a),

$$\text{HETP} = \frac{2.16}{0.63(8/5) - 1}\ln[0.63(8/5)] = 2.15 \text{ ft}$$

This is close to our estimated HETP, so our results are reasonable. The packing heights calculated in Example 16-1, $h_S = 0.507$ m and $h_E = 7.95$ m, differ from our initial estimates. A second iteration can be done to correct H_G and H_L. For example, from Eq. (16-37),

$$H_{G,E,cor} = \left(\frac{h_{p,calc}/10}{h_{p,initial}/10}\right)^{1/3} H_{G,E,initial} = \left(\frac{2.61\,\text{ft}}{2.2\,\text{ft}}\right)^{1/3}(1.33\,\text{ft}) = 1.41\,\text{ft}\,(0.43\,\text{m})$$

which is a 6% correction. Changing H_G and H_L will change the slopes of the lines used to calculate y_{AI}; thus, n_G will also change.

The largest likely error in the estimation of H_G, H_L, H_{OG}, and HETP is in the lack of accuracy of the mass transfer coefficients. Repeating the trial-and-error procedure to make more accurate predictions of these values does not help if the mass transfer coefficients are inaccurate. Since even careful predictions of $k_y a$ in randomly packed columns show average errors of $\pm 24.4\%$, a safety factor needs to be applied. Problem 16.D17 explores determination of this safety factor.

F. Generalize. This calculation is long and involved because of the need to estimate physical properties. This part of the problem is greatly simplified if a physical properties package is available on the computer. In this example m is close to one. Thus, both terms in Eqs. (15-31b, c) are significant and neither resistance controls. Thus, H_G and H_L are the same order of magnitude.

Models for structured packings are reviewed by Wang et al. (2005).

16.3.2 Simple Correlations for the Random Packings

The detailed correlation is fairly complex to use if a physical properties package is not available. Simplified correlations are available but will not be as accurate (Bennett and Myers, 1982; Greenkorn and Kessler, 1972; Perry and Green, 1984; Sherwood et al., 1975; Treybal, 1955): For H_G (in ft) the following empirical form has been used (Bennett and Myers, 1982; Greenkorn and Kessler, 1972; Treybal, 1955):

$$H_G = \frac{V}{k_y a A_c} = a_G W_G^b Sc_v^{0.5}/W_L^c \qquad (16\text{-}40a)$$

where W_G and W_L are the fluxes in lb/h-ft², and Sc_v is the Schmidt number for the gas phase. The constants are given in Table 16-3. The expression for H_L (in ft) developed by Sherwood and Holloway (1940) is

$$H_L = \frac{L}{k_x a A_c} = a_L \left(\frac{W_L}{\mu}\right)^d Sc_L^{0.5} \qquad (16\text{-}40b)$$

where Sc_L is the Schmidt number for the liquid. The constants are given in Table 16-3.

The correlations are obviously easier to use than Eqs. (16-37) and (16-38) since only the Schmidt number and the viscosity need to be estimated. However, Eqs. (16-40a) and (16-40b) will not be as accurate; thus, they should only be used for preliminary designs. These correlations were developed from absorption data and will be less accurate for distillation.

TABLE 16-3.   *Constants for determining H_G and H_L from Eqs. (16-40a) and (16-40b); range of W_L in Eq. (16-40b) is 400 to 15,000*

Packing	a_G	b	c	Range for Eq. (16-40a) W_G	W_L	a_L	D
Raschig rings							
3/8 inch	2.32	0.45	0.47	200-500	500-1500	0.0018	0.46
1	7.00	0.39	0.58	200-800	400-500	0.010	0.22
1	6.41	0.32	0.51	200-600	500-4500	—	—
2	3.82	0.41	0.45	200-800	500-4500	0.012	0.22
Berl saddles							
½ inch	32.4	0.30	0.74	200-700	500-1500	0.0067	0.28
½	0.811	0.30	0.24	200-700	1500-4500	—	—
1	1.97	0.36	0.40	200-800	400-4500	0.0059	0.28
3/2	5.05	0.32	0.45	200-1000	400-4500	0.0062	0.28

TABLE 16-4. *Approximate H_{OG} values for absorption in water (Reynolds, et al., 2002); the H_{OG} for ceramic packing is approximately twice the H_{OG} for plastic packing.*

Nominal Packing diameter, inch	1.0	1.5	2.0	3.0	3.5
Plastic packing H_{OG}, ft.	1.0	1.25	1.5	2.25	2.75

Water is frequently the solvent in absorption systems. The approximate values for H_{OG} for water as solvent are listed in Table 16-4 for random packings.

16.4 HTU-NTU ANALYSIS OF ABSORBERS AND STRIPPERS

The HTU-NTU analysis for concentrated absorbers and strippers with one solute is somewhat more complex than for distillation because total flow rates are not constant and solute A is diffusing through a stagnant film with no counterdiffusion, $N_B = 0$. We will assume that the system is isothermal. For stagnant films with $N_B = 0$, Eqs. (15-32a–f) are the appropriate mass transfer equations. The flux equation is (repeat of Eqs. 15-32),

$$N_A = J_A/(1 - y_A) \tag{16-41}$$

where J_A is the flux with respect to an axis moving at the molar average velocity of the fluid. As shown in Section 15.4.2 this leads to a transfer rate equation that is superficially similar to the previous equations (see Eq. (15-32g).

$$N_A a = k'_y\, a(y_A - y_{AI}) \tag{16-42}$$

Now the mass transfer coefficient times area/volume is defined as

$$k'_y a = k_y\, a/(1 - y_A)_{lm} \tag{16-43}$$

where the logarithmic mean mole frac is defined in the same manner as Eq. (15-32d).

$$(1 - y_A)_{lm} = \frac{(1 - y_A) - (1 - y_{AI})}{\ln\left(\dfrac{1 - y_A}{1 - y_{AI}}\right)} \tag{16-44}$$

For very dilute systems, $(1 - y_A)_{lm} = 1$ and $k'_y a = k_y a$.

We will now repeat the analysis of a packed section using Eq. (16 41) and including the nonconstant total flow rates. Figure 16-8A is a schematic diagram of an absorber. The absorber is assumed to be isothermal, and plug flow is assumed. The rate of mass transfer in a segment of the column dz is given by

$$N_A\, a\, A_c\, dz = k'_y\, a\, (y_A - y_{AI})\, A_c\, dz \tag{16-45}$$

Comparison of this equation with Eq. (16-1) shows that the sign on the mole fraction difference has been switched, since the direction of solute transfer in absorbers is opposite to that of transfer of the more volatile component in distillation. In addition, the modified mass transfer coefficient k'_y is used. The solute mass transfer can also be related to the change in solute flow rates in the gas or liquid streams.

$$N_A \, a \, A_c \, dz = -A_c \, d \, (Vy_A) = -A_c \, d(Lx_A) \qquad \text{(16-46)}$$

This equation differs from Eq. (16-2) derived for distillation since neither V nor L is constant.

The variations in V can be related to the constant flow rate of carrier gas, G.

$$V = \frac{G}{1 - y_A} \qquad \text{(16-47)}$$

which is the same as Eq. (12-41). Combining Eqs. (16-45) to (16-47), we obtain

$$-dz = \frac{d(\frac{Gy_A}{1 - y_A})}{k'_y aA_c \, (y_A - y_{AI})} \qquad \text{(16-48a)}$$

After taking the derivative, substituting in Eq. (16-47), and cleaning up the algebra, we obtain

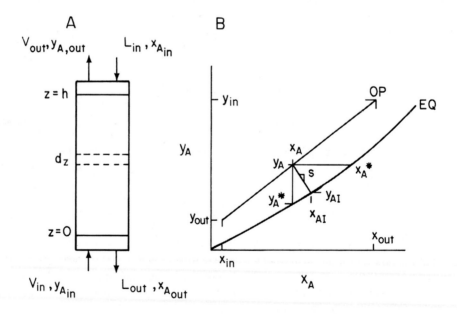

FIGURE 16-8.   Absorber calculation: (A) schematic of column, (B) calculation of interfacial mole fracs; slope, $s = -k'_x/k'_y$

$$-dz = \frac{Vdy_A}{k'_y aA_c (1 - y_A)(y_A - y_{AI})}$$

(16-48b)

Integrating this equation we obtain

$$h = \int_0^h dz = - \int_{y_{A,in}}^{y_{A,out}} \left(\frac{V}{k'_y aA_c}\right) \frac{dy_A}{(1 - y_A)(y_A - y_{AI})}$$

(16-49)

Substituting in Eq. (16-43), we obtain

$$h = - \int_{y_{A,in}}^{y_{A,out}} \left(\frac{V}{k_y aA_c}\right) \frac{(1 - y_A)_{lm} \, dy_A}{(1 - y_A)(y_A - y_{AI})}$$

(16-50)

The term $V/(k_y aA_c)$ is the height of a gas-phase transfer unit H_G defined in Eq. (16-6).

The variation in H_G can be determined from Eq. (16-37) and Figure 16-5, which are valid for both absorbers and distillation. The term that varies the most in Eq. (16-37) is the weight mass flux of liquid, W_L. H_G depends on W_L to the –0.5 to –0.6 power. In a single section of an absorber, a 20% change in liquid flow rate would be quite large. This will cause at most a 10% change in H_G. $k_y a$ is independent of concentration, since the concentration effect was included in k'_y in Eq. (16-42). Since the variation in H_G over the column section is relatively small, we will treat H_G as a constant. Then Eq. (16-50) becomes

$$h = H_G \int_{y_{A,out}}^{y_{A,in}} \frac{(1 - y_A)_{lm} \, dy_A}{(1 - y_A)(y_A - y_{AI})}$$

(16-51)

which is usually written as

$$h = H_G \, n_G,$$

$$n_G = \int_{y_{A,out}}^{y_{A,in}} \frac{(1 - y_A)_{lm} \, dy_A}{(1 - y_A)(y_A - y_{AI})}$$

(16-52)

Note that n_G for concentrated absorption is defined differently from n_G for distillation, Eq. (16-7b). The difference in the limits of integration in the two definitions for n_G occurs because the direction of transfer of component A in distillation is the opposite of the direction in absorption. There are additional terms inside the integral sign in absorption because the mass transfer takes place through a stagnant film and is not equimolar countertransfer as in distillation.

The method for finding the interfacial compositions is similar to that used to develop Eq. (16-13) and Figure 16-2 except that Eq. (16-42) and the corresponding equation in terms of liquid mole fracs are used as the starting point. The procedure is illustrated in Figure 16-8B. Use of this procedure lets us calculate the integrand in Eq. (15-52) at a series of points. The integral in Eq. (16-52) can be found either numerically or graphically.

Often, the integral in Eq. (16-52) can be simplified. The first simplification often employed is to replace the logarithmic mean with an arithmetic average.

$$(1-y_A)_{lm} \sim \frac{(1-y_A)+(1-y_{AI})}{2}$$

(16-53)

When Eq. (16-53) is substituted into Eq. (16-52), n_G can be simplified.

$$n_G \sim \int_{y_{A,out}}^{y_{A,in}} \frac{dy_A}{y_A - y_{AI}} + \frac{1}{2} \ln \left[\frac{1-y_{A,out}}{1-y_{A,in}} \right]$$

(16-54)

This equation shows that n_G for absorption is essentially the n_G for distillation plus a correction factor. The interfacial mole frac y_{AI} can be determined as shown in Figure 16-8B. The integral in Eq. (16-54) can then be determined graphically or numerically. For very dilute systems $1 - y_A$ is approximately 1 everywhere in the column. Then the correction factor in Eq. (16-54) will be approximately zero. Thus, n_G for *dilute* absorbers reduces to the same formula as for distillation.

For dilute absorbers and strippers, $(1 - y_A)_{lm} = 1$. Then $k'_y a = k_y a$ in Eq. (16-41). In this case we can use the overall gas-phase mass transfer coefficient. Following a development that parallels the analysis presented earlier for distillation, Eqs. (15-29a) and (16-18) to (16-23), we obtain for dilute absorbers

$$h = H_{OG} \, n_{OG}$$

(16-55)

where H_{OG} was defined in Eq. (16-22) and

$$n_{OG} = \int_{y_{A,out}}^{y_{A,in}} \frac{dy_A}{y_A - y_A^*}$$

(16-56)

This n_{OG} is essentially the same as for distillation in Eq. (16-23).

If the operating and equilibrium lines are straight, n_{OG} can be integrated analytically. The result is the Colburn equation given in Eqs. (16-31) and (16-34a). An alternative integration gives an equivalent equation.

$$n_{OG} = \frac{y_{A,in} - y_{A,out}}{(y_A - y_A^*)_{in} - (y_A - y_A^*)_{out}} \ln \left[\frac{(y_A - y_A^*)_{in}}{(y_A - y_A^*)_{out}} \right]$$

(16-57)

The development done here in terms of gas mole fracs can obviously be done in terms of liquid mole fracs. The development is exactly analogous to that presented here. The result for liquids is

$$h = H_L \, n_L$$

(16-58)

where H_L was defined in Eq. (16-11) and

$$n_L = \int_{x_{A,in}}^{x_{A,out}} \frac{(1-x_A)_{lm} dx_A}{(1-x_A)(x_{AI} - x_A)}$$

(16-59)

Equation (16-59) can often be simplified to

$$n_L \sim \int_{x_{A,in}}^{x_{A,out}} \frac{dx_A}{x_{AI} - x_A} + \frac{1}{2} \ln \left[\frac{1 - x_{A,out}}{1 - x_{A,in}} \right] \tag{16-60}$$

For dilute systems the correction factor in Eq. (16-60) becomes negligible. For dilute systems the analysis can also be done in terms of the overall transfer coefficient.

$$h = H_{OL} \, n_{OL} \tag{16-61}$$

where H_{OL} is defined in Eq. (16-25) and

$$n_{OL} = \int_{x_{A,in}}^{x_{A,out}} \frac{dx_A}{x_A^* - x_A} \tag{16-62}$$

If the operating and equilibrium lines are both straight, n_{OL} can be integrated analytically. The result is the Colburn Eq. (16-34b), or the equivalent expression,

$$n_{OL} = \frac{x_{A,out} - x_{A,in}}{\left(x_A^* - x_A\right)_{out} - \left(x_A^* - x_A\right)_{in}} \ln \left[\frac{\left(x_A^* - x_A\right)_{out}}{\left(x_A^* - x_A\right)_{in}} \right] \tag{16-63}$$

The development of the equations for concentrated systems presented here is not the same as those in Cussler (1997) and Sherwood et al. (1975). Since the assumptions have been different, the results are slightly different. However, the differences in these equations will usually not be important, since the inaccuracies caused by assuming an isothermal system with plug flow are greater than those induced by changes in the mass transfer equations. For dilute systems all the developments reduce to the same equations.

EXAMPLE 16-3. Absorption of SO_2

We are absorbing SO_2 from air with water at 20°C in a pilot-plant column packed with 0.5-in. metal Raschig rings. The packed section is 10-feet tall. The total pressure is 741 mm Hg. The inlet water is pure. The outlet water contains 0.001 mole frac SO_2, and the inlet gas concentration is $y_{in} = 0.03082$ mole frac. $L/V = 15$. The water flux $W_L = 1000$ lb/h-ft². The Henry's law constant is $H = 22,500$ mm Hg/mole frac SO_2 in liquid. Estimate H_{OL} for a 3.048 m high large-scale column operating at the same W_L and same fraction flooding if 2-inch metal Pall rings are used.

Solution

A. Define. Calculate H_{OL} for a large-scale absorber with 2-inch metal Pall rings.

B. Explore. We can easily determine n_{OL} for the pilot plant. Then $H_{OL} = h/n_{OL}$ for the pilot plant. Since the Henry's law constant H is large, m is probably large. This will make the liquid resistance control, and $H_L \sim H_{OL}$. Then Eq. (16-38) can be used to estimate $H_L = H_{OL}$ for the large-scale column. Only ϕ varies, and it can be estimated from Figure 16-6.

C. Plan. First calculate $m = H/P_{tot} = 22,500/741 = 30.36$. This is fairly large, and from Eq. (15-31c) the liquid resistance controls. For the pilot plant we can calculate n_{OL} from the Colburn Eq. (16-34b) since m is constant and L/V is approximately constant. Then $H_L = H_{OL} = h/n_{OL}$. The variation in ϕ with the change in packing can be determined from Figure 16-6, and $H_{OL} \sim H_L$ in the large column can be estimated from Eq. (16-38).

D. Do it. From Eq. (16-35b), $x^*_{out} = (y_{in} - b)/m$, so

$$x^*_{out} = \frac{0.03082 - 0.0}{30.36} = 0.001015$$

$(L/V)m = 15/30.36 = 0.4941$. From Eq. (16-34b) with $x_{in} = 0$ and $x_{out} = 0.001$,

$$n_{OL} = \left[\frac{1}{1 - 0.4941}\right] \ln\left[(1 - 0.4941)\ \frac{0 - 0.001015}{0.001 - 0.001015} + 0.4941\right] = 7.012$$

Then $H_L \sim H_{OL} = 3.048\ m/7.012 = 0.485\ m$. From Figure 16-6 at $W_L = 1000$, ϕ (0.5-in Raschig rings) = 0.32, while ϕ (2-inch Pall rings) = 0.62. Then taking the ratio of Eq. (16-38) for 2-inch Pall rings divided by Eq. (16-38) for .5-inch rings, $H_L(\text{2-inch}) = H_L(\text{0.5-inch})\ \phi(\text{2-inch})/\phi(\text{0.5-inch})$

$$H_{OL} \sim H_L\ (\text{2-in Pall rings}) = \frac{0.62}{0.32}(0.435) = 0.843\ m$$

since all other terms in Eq. (16-38) are constant.

E. Check. These results are the correct order of magnitude. A check of n_{OL} can be made by graphically integrating n_{OL}.

F. Generalization. This method of correlating H_L or H_G when packing size or type is changed can be used for scale-up. The large value of m in this problem allowed the assumption of liquid-phase control. This assumption simplifies the problem since $H_{OL} \sim H_L$. If liquid-phase control is not valid, this problem becomes significantly harder.

If there are multiple solutes transferring the analysis is significantly more complicated than the analysis shown here (Taylor and Krishna, 1993). These complications are beyond the scope of this chapter but can be solved with the Maxwell-Stefan approach.

16.5 HTU-NTU ANALYSIS OF CO-CURRENT ABSORBERS

In Section 12.10 we noted that co-current operation of absorbers was often employed when a single equilibrium stage was sufficient. Co-current operation has the advantage that flooding cannot occur. This means that high vapor and liquid flow rates can be used, which automatically leads to small-diameter columns.

A schematic of a co-current absorber is shown in Figure 16-9A. The analysis will be done for dilute systems using overall mass transfer coefficients. The system is assumed to be isothermal. The liquid and vapor are assumed to be in plug flow, and total flow rates are constant. The rate of mass transfer in segment dz is

$$N_A\ a\ A_c\ dz = K_y\ a\ (y_A - y^*_A)\ A_c\ dz \tag{16-64}$$

which can be related to the changes in solute flow rates

$$N_A \, a \, A_c \, dz = -d(Vy_A) = d(Lx_A) \qquad \textbf{(16-65)}$$

Combining these equations we obtain

$$dz = \frac{-d(Vy_A)}{k_y a A_c (y_A - y_A^*)} \qquad \textbf{(16-66)}$$

If $V/(k_y a A_c)$ is constant, Eq. (16-66) can be integrated to give

$$h = H_{OG} \, n_{OG} \qquad \textbf{(16-67)}$$

where H_{OG} is given in Eq. (16-22) and

$$n_{OG} = \int_{y_{A,out}}^{y_{A,in}} \frac{dy_A}{y_A - y_A^*} \qquad \textbf{(16-68)}$$

This development follows the development for countercurrent systems. The analyses differ when we look at the method for calculating $y_A - y_A^*$. The operating equation is [see Eq. (12-63)]

$$y = -\frac{L}{V} x + y_{in} - \frac{L}{V} x_{in} \qquad \textbf{(16-69)}$$

This operating line and the calculation of $y_A - y_A^*$ are shown in Figure 16-9B. When the operating and equilibrium lines are both straight, n_{OG} can be obtained analytically. The result corresponding to the Colburn equation is (King, 1980)

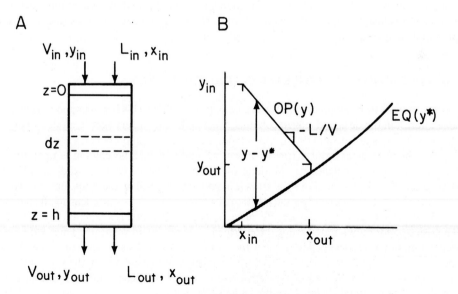

FIGURE 16-9. *Co-current absorber; (A) schematic of column, (B) calculation of y – y**

$$n_{OG} = \left(\frac{1}{1 + \frac{mV}{L}} \right) \ln \left[\left(1 + \frac{mV}{L} \right) \left(\frac{y_{A,in} - y^*_{A,out}}{y_{A,out} - y^*_{A,out}} \right) - \frac{mV}{L} \right] \tag{16-70}$$

where

$$y^*_{A,out} = mx_{A,out} + b \tag{16-71}$$

If a completely irreversible reaction occurs in the liquid phase, $y^*_A = 0$ everywhere in the column. Thus, the equilibrium line is the x axis, and the integration of Eq. (16-68) is straightforward.

$$n_{OG} = \int_{y_{A,out}}^{y_{A,in}} \frac{dy_A}{y_A} = \ln \left(\frac{y_{A,in}}{y_{A,out}} \right) \tag{16-72}$$

Exactly the same result is obtained for co-current and countercurrent columns with irreversible reactions, but co-current columns can have higher liquid and vapor flow rates.

H_{OG} is related to the individual coefficients by Eq. (16-27a). Unfortunately, it is dangerous to use Eqs. (16-37) and (16-38) to determine the values for H_L and H_G for co-current columns because the correlations are based on data in countercurrent columns at lower gas rates than those used in co-current columns. Reiss (1967) reviews co-current contactor data and notes that the mass transfer coefficients can be considerably higher than in countercurrent systems. Gianetto et al. (1973) operated with a 15-fold velocity increase and observed a 40-fold increase in k_L when liquid-phase resistance controlled. They recommended co-current operation for absorption with chemical reaction. Harmen and Perona (1972) did an economic comparison of co-current and countercurrent columns. For the absorption of CO_2 in carbonate solutions where the reaction is slow they concluded that countercurrent operation is more economical. For CO_2 absorption in monoethanolamine (MEA), where the reaction is fast, they concluded that countercurrent is better at low liquid fluxes whereas co-current was preferable at high liquid fluxes.

16.6 PREDICTION OF DISTILLATION TRAY EFFICIENCY

How does mass transfer affect the efficiency of a tray column? This is a question of considerable interest in the design of staged columns. We will develop a very simple model following the presentations of Cussler (1997), King (1980), Lewis (1936), and Lockett (1986).

A schematic diagram of a tray is shown in Figure 16-10. The column is operating at steady state. A mass balance will be done for the mass balance envelope indicated by the dashed outline. The vapor above the trays is assumed to be well mixed; thus, the inlet vapor mole frac $\bar{y}_{j+1}$ does not depend on the position along the tray, ℓ. The vapor leaving the balance envelope has not yet had a chance to be mixed and its composition is a function of position ℓ. The rising vapor bubbles are assumed to perfectly mix the liquid vertically. Thus, x does not depend upon the vertical position z, but the vapor fraction y does depend on z. The liquid mole frac can be a function of the distance ℓ along the tray measured from the start of the active region, $\ell = 0$, to the end of the active region, $\ell = \ell_a$. At steady state a solute or more volatile component mass balance for the *vapor* phase is

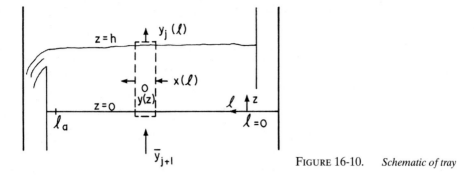

FIGURE 16-10. *Schematic of tray*

$$(\text{In} - \text{Out})_{\text{convection}} + (\text{Solute transferred from liquid}) = 0 \qquad \textbf{(16-73a)}$$

If we use the overall gas-phase mass transfer coefficient K_y, this equation is

$$(V/A_{\text{active}})\,[y(z) - y(z + \Delta z)] + K_y a\,\Delta z\,(y_\ell^* - y) = 0 \qquad \textbf{(16-73b)}$$

where y_ℓ^* is the vapor mole frac in equilibrium with the liquid of mole frac x_ℓ. A_{active} is the active area for vapor-liquid contact on the tray. Both A_{active} and V are assumed to be constant. Dividing Eq. (16-73b) by Δz and taking the limit as Δz goes to zero, we obtain

$$-\frac{dy}{dz} + \frac{K_y a A_{\text{active}}}{V}\,(y_\ell^* - y) = 0 \qquad \textbf{(16-74)}$$

This equation can now be integrated from $z = 0$ to $z = h$. The boundary conditions are

$$y = \bar{y}_{j+1}, \qquad z = 0 \qquad \textbf{(16-75a)}$$

$$y = y_\ell, \qquad z = h \qquad \textbf{(16-75b)}$$

After algebraic manipulation, the solution to Eqs. (16-74) and (16-75) is

$$E_{pt} = \frac{y_\ell - \bar{y}_{j+1}}{y_\ell^* - \bar{y}_{j+1}} = 1 - \exp(-K_y a A_{\text{active}} h/V) \qquad \textbf{(16-76a)}$$

The point efficiency E_{pt} was defined in Eq. (10-5). Comparing this equation to Eqs. (16-21) and (16-22), we obtain two alternative representations.

$$E_{pt} = 1 - \exp(-h/H_{OG}) \qquad \textbf{(16-76b)}$$

$$E_{pt} = 1 - \exp(-n_{OG}) \qquad \textbf{(16-76c)}$$

We would like to relate the point efficiency to the Murphree vapor efficiency given by Eq. (10-2). This relationship depends upon the liquid flow conditions on the tray. There are two limiting flow conditions that allow us to simply relate E_{pt} to E_{MV}. The first of these is a tray where the liquid is completely mixed. This means that x_ℓ is a constant and is equal to x_{out}, so that $y_\ell^* = y^*_{\text{out}}$ and $y_\ell = y_{\text{out}}$. Therefore $E_{MV} = E_{pt}$, and

$$E_{MV} = 1 - \exp\left(-K_y a A_{active} h / V\right) \qquad \textbf{(16-77)}$$

for a completely mixed stage.

The second limiting flow condition is plug flow of liquid with no mixing along the tray. By assuming that each packet of liquid has the same residence time, one can derive the relationship between E_{MV} and E_{pt} (Lewis, 1936; King, 1980; Lockett, 1986):

$$E_{MV} = \frac{L}{mV}\left[\exp\left(\frac{mV}{L}E_{pt}\right) - 1\right] \qquad \textbf{(16-78)}$$

where m is the local slope of the equilibrium curve, Eq. (15-30b). Since plug flow is often closer to reality than a completely mixed tray, Eq. (16-78) is more commonly used than Eq. (16-77).

Real plates often have mixing somewhere in between these two limiting cases. These situations are discussed elsewhere (AIChE, 1958; King, 1980; Lockett, 1986).

EXAMPLE 16-4. Estimation of stage efficiency

A small distillation column separating benzene and toluene gives a Murphree vapor efficiency of 0.65 in the rectifying section where $L/V = 0.8$ and $x_{benz} = 0.7$. The tray is perfectly mixed and has a liquid head of 2 inches. The vapor flux is 25 lbmol/h-ft^2. (a) Calculate $K_y a$. (b) Estimate E_{MV} for a large-scale column where the trays are plug flow and the liquid head h becomes 2.5 inches. Other parameters are constant.

Solution

A. From Eq. (16-77) assuming that the active area of the tray equals the area available for flow, $A_{active} = A_{flow}$ (this may be off by a few percent), we obtain,

$$K_y a = -\frac{V}{A_{active} h} \ln(1 - E_{MV}) \qquad \textbf{(16-79)}$$

since $V/A_c = 25$, $h = 2/12$ ft, and from Eq. (10-20c) $A_{active} \sim A_c (2\eta - 1) = 0.8 A_c$ with $\eta = 0.9$ this is

$$K_y a = \frac{-25}{(.8)(2/12)} \ln(0.35) = 196.9 \text{ lbmol/h} - \text{ft}^3$$

B. In the large-diameter system E_{pt} is given by Eq. (16-76a). Since $h = 2.5/12$ and $V/A_{active} = 25$,

$$E_{pt} = 1 - \exp\left[\frac{(196.9)(0.8)(2.5/12)}{25}\right] = 0.73$$

Increasing the liquid pool height increases the efficiency since the residence time is increased.

The Murphree vapor efficiency for plug flow is found from Eq. (16-78). The slope of the equilibrium curve, m, can be estimated. Since the equilibrium is

$$y_{benz} = \frac{\alpha\, x_{benz}}{1 + (\alpha - 1)\, x_{benz}}$$

the slope is

$$m = \frac{dy_{benz}}{dx_{benz}} = \frac{\alpha}{[1 + (\alpha - 1)x]^2}$$

With $\alpha = 2.5$ and $x = 0.7$, we obtain $m = 0.595$. Then Eq. (16-78) is

$$E_{MV} = \frac{0.8}{0.595} \left\{ \exp\left[\frac{(0.595)(0.73)}{(0.8)} \right] - 1 \right\} = 0.97$$

The plug flow system has a significantly higher Murphree plate efficiency than a well-mixed plate where $E_{MV} = E_{pt} = 0.73$. Note that $K_y a$ is likely to vary throughout the column since m varies (see Problem 16.D15). E_{MV} is also dependent upon m and will change from stage to stage. The effect of concentration changes can be determined by calculating K_y from Eq. (15-31b).

16.7 MASS-TRANSFER ANALYSIS OF EXTRACTION

One would expect that mass-transfer analysis of extraction would be very similar to the analysis of absorption and stripping (Section 16.4). However, there are significant differences, such as the frequent use of ratio units (Section 13.5), because they extend the region of validity of the analysis. Although an analysis similar to absorption with individual mass-transfer coefficients and concentrated analysis can be used, in practice the overall mass-transfer coefficient K_{O-ED} and the simplest form for NTU are used (Frank et al., 2008; Treybal, 1980). The analysis is kept simple because of large uncertainties in the mass-transfer data (discussed later). After developing the HTU-NTU analysis in Section 16.7.1, methods for determining the stage efficiency in mixers are developed in Section 16.7.2. Then, Section 16.7.3 discusses prediction of the area per volume and the drop diameter in mixers. Finally, methods to estimate the mass-transfer coefficients in mixers are discussed.

16.7.1 Mass-Transfer Equations and HTU-NTU Analysis

The transfer rate/volume in (kg A)/(s m³) [or (kmol A)/(s m³)] to or from the dispersed phase is

$$\text{Transfer rate/Volume} = K_{O-ED}\, a (x_D - x_D^*) \qquad \textbf{(16-80a)}$$

In these equations x_D is the weight or mole fraction of the solute A in the dispersed phase, and x_D^* is the weight or mole fraction that would be in equilibrium with the continuous-phase fraction of A, x_C. Note that x_D may refer to either raffinate or extract, which is different than the notation in Table 13-2. Determination of "a," the area/volume (m² dispersed phase/m³ total volume of mixer) for mass transfer, is discussed in Section 16.7.3. The units of K_{O-ED} are (kg solute A in dispersed phase)/[s (m² dispersed phase)(mass fraction solute in dispersed phase)] or the equivalent in molar units. Because mass fraction A in dispersed phase = (kg solute A in dispersed phase)/(total kg of dispersed phase), the units of K_{O-ED} can also be written as (total kg dispersed phase)/[s (m² dispersed phase)].

The overall mass-transfer coefficient is related to the individual coefficients for the continuous and dispersed phases with a sum of resistances model similar to that developed in Eq. (16-31c).

$$\frac{1}{K_{O-ED}} = \frac{1}{k_{xD}} + \frac{1}{k_{xC}m_{CD}} \tag{16-80b}$$

where the term m_{CD} is the average slope of the equilibrium curve

$$m_{CD} = \left(\partial x_C^* / \partial x_D\right)_{avg} \tag{16-80c}$$

If the equilibrium is linear, then m_{CD} is constant and is equal to the distribution coefficient, y/x, if the extract phase is continuous. If the raffinate is continuous, then m_{CD} is the inverse of the distribution coefficient. Remembering to use the proper value of m_{CD} can be a challenge.

The height of the extractor will be determined from h = (NTU)(HTU). Simplified equations for the number of extraction transfer units for the dispersed phase, n_{OE-D}, and continuous phase, n_{OE-C}, defined in the same way as n_{OL} in Eq. (16-26), are

$$n_{O-ED} = \int_{x_{D,in}}^{x_{D,out}} \frac{d\,x_D}{x_D - x_D^*}, \, n_{O-EC} = \int_{x_{C,in}}^{x_{C,out}} \frac{d\,x_C}{x_C - x_C^*} \tag{16-81a}$$

For packed columns and the countercurrent column systems shown in Figure 13-2, Eq. (16-81a) can be integrated by the methods developed in sections 16.1 and 16.4. Very convenient analytical solutions are obtained when the equilibrium and operating equations are both linear. Since the dispersed phase can be either the extract or the raffinate, we write these equations as n_{O-Ex} for the raffinate phase and n_{O-Ey} for the extract phase. If the solvent is the dispersed phase, then $n_{O-ED} = n_{O-Ey}$. These results, known as the Colburn equation, are

$$n_{O-Ex} = \frac{\ln\{[\frac{x_{in} - x_{out}^*}{x_{out} - x_{out}^*}][1 - F/(mS)] + F/(mS)\}}{1 - F/(mS)}, \, x_{out}^* = (y_{in} - b)/m \tag{16-81b}$$

$$\frac{x_{in} - x_{out}^*}{x_{out} - x_{out}^*} = \frac{\exp\{n_{O-Ex}[1 - F/(mS)]\} - F/(mS)}{1 - F/(mS)}, \, x_{out}^* = (y_{in} - b)/m \tag{16-81c}$$

$$n_{O-Ey} = \frac{1}{1 - mS/F}\ln[(1 - mS/F)(\frac{y_{in} - y_{out}^*}{y_{out} - y_{out}^*}) + mS/F], \, y_{out}^* = mx_{in} + b \tag{16-81d}$$

$$\frac{y_{in} - y_{out}^*}{y_{out} - y_{out}^*} = \frac{\{\exp[(1 - mS/F)n_{O-Ey}] - mS/F\}}{(1 - mS/F)}, \, y_{out}^* = mx_{in} + b \tag{16-81e}$$

These four equations are valid if $F/(mS) \neq 1$. The grouping $(mS)/F$ is the extraction factor that was introduced in Chapter 13. To understand some of the differences between a staged analysis and a mass-transfer analysis, these results should be compared with the Kremser Eqs. (13-11a,b) (see Problem 16.A1.)

The height of the transfer unit is defined in a similar way as in Eq. (16-22), which is

$$H \propto (\text{phase-velocity}) / (K_{\text{Overall}}a) \qquad \text{(16-82a)}$$

Although the exact form used depends on the equipment and whether $K_{\text{Overall}}a$ is in mass or molar units, a typical definition in columns for $H_{O-\text{raf}}$ in molar units is

$$H_{O-\text{raf}} = (Q_{\text{raf}} / A_c)(\rho_{\text{raf}} / MW_{\text{raf}}) / (K_{O-\text{raf}}a) \qquad \text{(16-82b)}$$

The term (Q_{raf}/A_c) is the superficial velocity of the raffinate phase in the column, and $(\rho_{\text{raf}} / MW_{\text{raf}})$ is a conversion factor because $(K_{O-\text{raf}}a)$ is in molar units.

16.7.2 Calculation of Stage Efficiency in Extraction Mixers

For mixers it is customary to work in terms of the dispersed phase values for NTU and HTU. For linear systems the continuous-phase equations will give identical final results. Following Eq. (16-82), the definition of H_{O-ED} in an extraction mixer is

$$H_{O\text{-}ED} = (Q_D / A_{\text{mixer}})(\rho_D / MW_D) / (K_{O\text{-}ED}a) \qquad \text{(16-83a)}$$

It is also often very useful to note that because $n_{O-ED} = h/H_{O-ED}$, we can write n_{O-ED} as

$$n_{O\text{-}ED} = (V_{\text{mixer}} / Q_D)(K_{O\text{-}ED}a) / (\rho_D / MW_D) \qquad \text{(16-83b)}$$

Q_D is the volumetric flow rate (m³/s) of the dispersed phase, and V_{mixer} is the total volume of the mixer (m³). The term V_{mixer}/Q_D is the residence time of the dispersed phase based on the superficial velocity of the dispersed phase. Since the value of n_{O-ED} can often be determined from integration of Eq. (16-81a), Eq. (16-83a) can provide a method for determining $(K_{O-ED}a)$ from experimental data.

It would probably make more sense to define

$$H_{O\text{-}ED} = (Q_D / \varphi_d A_{\text{mixer}})(\rho_D / MW_D) / (K_{O\text{-}ED}a)$$

and

$$n_{O\text{-}ED} = V_{\text{mixer}} \varphi_d (K_{O\text{-}ED}a) / [Q_D(\rho_D / MW_D)]$$

$V_{\text{mixer}}\varphi_d/Q_D$ is now the residence time of the dispersed phase and φ_d is the volumetric fraction of the dispersed phase in the mixer (see Section 13.4.1). However, we will follow the standard approach so that the values of $(K_{O-ED}a)$ from the literature agree with our formulation.

Because extraction mixer-settlers typically operate at stage efficiencies above 80% and often in the range from 95% to 100%, the equilibrium stage analysis in Section 13.14 is often used with an assumed value for the stage efficiency. However, a more accurate design will result if a mass-transfer analysis is used to estimate the stage efficiency. The purpose of the analysis will be to estimate the value of the dispersed-phase Murphree stage efficiency, E_{MD}.

$$E_{MD} = \frac{x_{D,in} - x_{D,out}}{x_{D,in} - x_{D,out}^*} \qquad (16\text{-}84)$$

As usual, $x_{D,out}^*$ is the mole fraction of solute in the dispersed phase that is in equilibrium with the actual mole fraction of solute in the continuous phase, $x_{C,out}$.

There are two different analyses for mass transfer in a mixer-settler extractor available in the literature. The simpler analysis (Seader and Henley, 2006; Treybal, 1980); assumes that the mixer is perfectly mixed. This means that the continuous phase is well mixed and the dispersed phase is well mixed. Thus, x_D, x_C, and x_D^* are all constant and equal to the values at the mixer outlet. Then Eq. (16-81a) simplifies to

$$n_{O\text{-}ED} = \frac{1}{x_{D,out} - x_{D,out}^*} \int_{x_{D,in}}^{x_{D,out}} dx_D = \frac{\left(x_{D,out} - x_{D,in} \right)}{\left(x_{D,out} - x_{D,out}^* \right)} \qquad (16\text{-}85)$$

We can manipulate Eq. (16-84) for the Murphree dispersed-phase stage efficiency so that it can be written in terms of $n_{OE\text{-}D}$. First,

$$E_{MD} = \frac{x_{D,in} - x_{D,out}}{x_{D,in} - x_{D,out}^*} = \frac{x_{D,in} - x_{D,out}}{\left(x_{D,in} - x_{D,out} \right) + \left(x_{D,out} - x_{D,out}^* \right)} = \frac{x_{D,out} - x_{D,in}}{\left(x_{D,out} - x_{D,in} \right) - \left(x_{D,out} - x_{D,out}^* \right)} \qquad (16\text{-}86a)$$

which can be written as

$$E_{MD} = \frac{\dfrac{\left(x_{D,out} - x_{D,in} \right)}{\left(x_{D,out} - x_{D,out}^* \right)}}{\dfrac{x_{D,out} - x_{D,in}}{x_{D,out} - x_{D,out}^*} + 1} = \frac{n_{O\text{-}ED}}{1 + n_{O\text{-}ED}} \qquad (16\text{-}86b)$$

Once we determine the mass-transfer coefficient from appropriate correlations, $n_{O\text{-}ED}$ can be determined from Eq. (16-83b) and E_{MD} from Eq. (16-86b). Alternatively, if the concentrations are measured, then E_{MD} and/or $n_{OE\text{-}D}$ can be determined from Eq. (16-86b) and $K_{O\text{-}ED}a$ from Eq. (16-83b). Note that the key assumption in this development is that all of the concentrations are constant. This requires not only that the droplets of dispersed phase are well mixed but also that the dispersed phase has the same average residence time everywhere in the mixer.

The second analysis procedure in the literature uses a differential equation approach (Frank et al, 2008; Laddha and Degaleesan, 1978) that parallels the analysis in Section 16.6. The continuous phase takes the role of the liquid phase in Figure 16-10, and the dispersed phase takes the part of the vapor in this figure. The continuous and dispersed phases exit together instead of separately as in Figure 16-10, but this does not change the analysis. For the dispersed phase, the steady-state mass balance is

$$(\text{In} - \text{Out})_{flow} + (\text{Solute transferred from continuous phase}) = 0 \qquad (16\text{-}87)$$

In the notation used for extraction with the same geometry as in Figure 16-10, this is

$$[Q_D(\rho_D/MW_D)/A_{mixer}][x_D(z)-x_D(z+\Delta z)]+K_{O-ED}a\Delta z(x_D-x^*_{D,l})=0 \quad \textbf{(16-88)}$$

Q_D/A_{mixer} is the superficial linear velocity of the dispersed phase in the mixer. Dividing this equation by Δz and taking the limit as Δz goes to zero,

$$-\frac{dx_D}{dz}+\frac{(K_{O-ED}a)A_{mixer}}{Q_D(\rho_D/MW_D)}\left(x_D-x^*_{D,l}\right)=0 \quad \textbf{(16-99)}$$

We can integrate from z = 0 to z = h with the boundary conditions,

$$x_D=x_{D,in} \text{ at } z=0 \quad \textbf{(16-90a)}$$

$$x_D=x_{D,l} \text{ at } z=h \quad \textbf{(16-90b)}$$

The result is the point efficiency,

$$E_{pt}=\frac{x_{D,in}-x_{D,l}}{x_{D,in}-x^*_{D,l}}=1-\exp\left(\frac{(-K_{O-ED}a)hA_{mixer}}{Q_D(\rho_D/MW_D)}\right) \quad \textbf{(16-91)}$$

For a mixer with a well-mixed continuous phase (but the dispersed phase is not well mixed), $x^*_{D,\ell}$ = constant = $x^*_{D,out}$, since it is in equilibrium with the continuous phase, which is the same everywhere. In addition, if the dispersed phase has the same average residence time everywhere in the mixer, the same amount of solute is transferred in or out of the dispersed phase and $x_{D,\ell}$ = constant = $x_{D,out}$. Thus, the Murphree efficiency equals the point efficiency, $E_{MD}=E_{pt}$. Substituting in $V_{mixer}=hA_{mixer}$, the Murphree efficiency is

$$E_{MD}=1-\exp\left(\frac{(-K_{O-ED}a)V_{mixer}}{Q_D(\rho_D/MW_D)}\right)=1-\exp(-n_{O-ED}) \quad \textbf{(16-92)}$$

Equation (16-92) is equivalent to Eq. (16-77) derived for a distillation tray. This derivation follows the change in concentration of a packet of the dispersed phase from z = 0 to z = h, but then requires mixing at z = h. This requires the assumption that the continuous phase, but not the dispersed phase, is well mixed. Equation (16-92) can be used either to estimate the value of E_{MD} if $K_{O-ED}a$ is known or to estimate the value of $K_{O-ED}a$ if concentrations are measured and E_{MD} is determined from Eq. (16-84). Frank et al. (2008) replace Q_D in Eq. (16-92) with $(Q_D+Q_C)\phi_d$. If $\phi_d=\phi_{d,feed}$, then $(Q_D+Q_C)\phi_d=Q_D$, and the results are identical.

Equations (16-86b) and (16-92) are not identical because they are based on different models. Both models assume that the continuous phase is well mixed, but their assumptions about the dispersed phase are different. These differences emphasize the comments in Chapter 15 that analysis of mass transfer remains a topic for research and discussion. With identical values of $K_{O-ED}a$, Eq. (16-92) always predicts a higher value for E_{MD} than does Eq. (16-86b). This is illustrated in Example 16.5. Another way of looking at this is that with the same E_{MD}, a larger $K_{O-ED}a$ value will be back-calculated from Eq. (16-86b) than from (16-92). Both equations predict that n_{O-ED} will be large and E_{MD} approach 1.0 if ($K_{O-ED}a$) (residence time) is large. Residence time is large if the mixer volume is large compared to the total liquid flow rate.

Chemical engineers always want to know which equation is correct. The answer is that the model that most closely models the physical situation is more correct, but it may not be the best model to use. If the continuous phase is not well mixed, then neither model is appropriate. If the mixer has good mixing of the continuous phase, then the appropriate model to use depends on mixing within the dispersed phase. In very clean systems there is often considerable internal circulation, and the drops tend to be well mixed. The well-mixed model is appropriate, and most of the data in the literature were analyzed with this model. In dirty systems with dust, crud, or surfactants (e.g., soap or proteins) present, the drop surface can be rigid and internal circulation is suppressed—then the differential model is more appropriate. Of course, we wish and hope that the choice of model should not depend on how the value of $K_{O-ED}a$ was determined. Unfortunately, this is not true. The same raw data (concentrations) will result in different $K_{O-ED}a$ values for the two models. The accuracy of these values to a large extent depends on the internal circulation of the dispersed phase. Mass-transfer coefficients obtained with clean systems do not apply to dirty systems, and vice versa. Using one model to determine $K_{O-ED}a$, then using a different model with these $K_{O-ED}a$ values, will not give correct answers. This is illustrated in Example 16.5. Thus, use the same model that was used to determine $K_{O-ED}a$.

Example 16.5. Conversion of mass-transfer coefficients and estimation of stage efficiency in mixer

Treybal (1980) estimated the overall mass-transfer coefficient and the stage efficiency for a mixer (0.5 m high and 0.5 m diameter, $A_{mixer} = 0.1963$ m^2, $V_{mixer} = 0.09817$ m^3) extracting benzoic acid from water into solvent pure toluene. The water plus benzoic acid flow rate was $Q_F = 0.003$ m^3/s, and the toluene flow rate was $Q_D = 0.0003$ m^3/s. The tank was well mixed. Toluene was the dispersed phase, φ_D was estimated as 0.0824. The estimated dispersed-phase surface to dispersed-phase volume ratio $a_D = 1940$ m^2 dispersed phase/m^3 dispersed phase. The estimated overall dispersed-phase mass-transfer coefficient $K_{LD} = 2.01 \times 10^{-5}$ kmol benzoic/[m^{2}s(kmol benzoic/m^3)]. Additional data: $\rho_{toluene} = 865$ kg/m^3, $MW_{toluene} = 92.14$ kg/kmol. Equilibrium is $C_{extract} = 20.8\, C_{raffinate}$ ($C_{extract}$ is kmol benzoic/m^3 extract); since the solvent is the dispersed phase, $m_{CD} = 1/20.8 = 0.0481$ [see Eq. (16-70c analog)].

a. Convert the mass-transfer coefficient to the units used in this section.
b. Calculate the stage efficiency using the completely mixed model.
c. Calculate the stage efficiency using the differential equation approach.
d. Calculate the exiting raffinate and extract mole fractions for a completely stirred mixer if the entering solvent contains no benzoic acid and the entering feed is 0.0003 mole fraction benzoic acid in water.

Solution

A. Since there is no standardized set of mass-transfer equations and units, converting terms in different units is a common task. The first thing to do is to look for the defining equation used for mass transfer or at the form of the mass-transfer coefficient correlation. Treybal (1980) developed the following equations:

$$\text{Transfer rate/Volume} = K_{LD}a(C_D - C_D^*) \qquad \textbf{(16-80a analog)}$$

$$N_{tOD} = \int_{C_{D,in}}^{C_{D,out}} \frac{dC_D}{C_D - C_D^*} \qquad \text{(16-81a analog)}$$

$$H_{tOD} = v_D / (K_{LD}a), \; v_D = Q_D / A_{mixer} \qquad \text{(16-83a analog)}$$

$$\frac{1}{K_{LD}} = \frac{1}{k_{LD}} + \frac{1}{k_{LC}m_{CD}} \qquad \text{(16-80b analog)}$$

$$m_{CD,conc\_units} = \left(\partial C_C^* / \partial C_D \right)_{avg} \qquad \text{(16-80c analog)}$$

In these equations C_D is the concentration of benzoic acid in the dispersed phase, (kmol benzoic acid)/(m³ dispersed phase), C_D^* is the dispersed-phase concentration in equilibrium with the continuous phase, v_D is the superficial linear velocity of the dispersed phase in the mixer m/s, k_{LD} and k_{LC} are the individual mass-transfer coefficients, and K_{LD} is the overall dispersed-phase mass-transfer coefficient, all in kmol benzoic/[m²s(kmol benzoic/m³)]. This set is an equally valid set of mass-transfer equations, but the units are different than used in this chapter.

One approach to convert units is to set the HTU values from the two approaches equal, $H_{O\text{-}ED} = H_{tOD}$. Solving for ($K_{O\text{-}ED}a$), we obtain

$$K_{O-ED}a = \rho_D K_{LD}a_D / MW_D \qquad \text{(16-93a)}$$

The numerical value is

$K_{O\text{-}ED}a = (865 \text{kg/m}^3)(2.01 \times 10^{-5} \text{kmol/[m}^2\text{s(kmol/m}^3)])(1940 \text{m}^2/\text{m}^3)/[(92.14 \text{ kg/kmol})]$
$= 0.3661 \text{ (kmol/[s m}^2(\text{mol frac dispersed})]) \text{ (m}^2/\text{m}^3\text{total volume)}$

Solving for $K_{O\text{-}ED}$,

$$K_{O-ED} = (K_{O-ED}a) / a \qquad \text{(16-93b)}$$

$K_{O\text{-}ED} = 0.3661/1940 = 0.0001887 \text{ (kmol/[s m}^2(\text{mol frac dispersed})])$

B. Note: Since Treybal used a completely mixed model to analyze his data, that is the appropriate model to use. Stage efficiency for the completely mixed model is given by Eq. (16-86) and $n_{O\text{-}ED}$ from Eq. (16-83b),

$$n_{O-ED} = V_{mixer}(K_{O-ED}a) / [Q_D(\rho_D / MW_D)]$$
$$= (0.09817)(0.3661)/[(0.0003)(865/92.14)] = 12.76$$

Then, $E_{MD} = \dfrac{n_{O-ED}}{1 + n_{O-ED}} = 12.76/(1 + 12.76) = 0.927.$

The values of $n_{O\text{-}ED}$ and E_{MD} agree with Treybal (1980).

C. If we inappropriately apply Treybal's value of $K_{O-ED}a$ and the resulting value for n_{O-ED} to the differential equation model, from Eq. (16-92) we obtain

$$E_{MD} = 1 - \exp(-n_{O-ED}) = 1 - \exp(-12.76) = 0.999997$$

Clearly, the differential model predicts significantly higher stage efficiency with the same value of $K_{O-ED}a$ than the completely mixed model because additional separation occurs along the path the fluid takes from inlet to outlet.

D. The mixed model is appropriate for a completely stirred mixer, and our mass-transfer coefficients were obtained with this model.

$$E_{MD} = \frac{x_{D,in} - x_{D,out}}{x_{D,in} - x^*_{D,out}} = 0.927$$

With $x_{D,in} = 0$ and $x^*_{D,out} = mx_{C,out}$, this becomes

$$E_{MD} = \frac{x_{D,in} - x_{D,out}}{x_{D,in} - x^*_{D,out}} = \frac{-x_{D,out}}{-x^*_{D,out}} = \frac{x_{D,out}}{mx_{C,out}} = 0.927$$

Equilibrium is $x^*_D = mx_{raffinate} = mx_{C,out}$. Value of m is unknown in mole fraction units, but in units of mol/m^3, $C_{extract} = m_{conc\ units}C_{raffinate}$, which gives

$$m_{conc\ units} = 20.8 \frac{kmol\ Benzoic\ /\ m^3\ extract}{kmol Benzoic\ /\ m^3\ raffinate}. \text{We need m in}$$

$$m_{mole\ fraction\ units} = \frac{mol\ Benzoic\ /\ mol\ extract}{mol\ Benzoic\ /\ mol\ raffinate}. \text{The resulting conversion is,}$$

$$m_{mole\ fraction\ units} = m_{conc\ units}(\frac{MW_{extract}}{\rho_{extract}})(\frac{\rho_{raffinate}}{MW_{raffinate}}) \tag{16-94}$$

Since the system is dilute, extract properties are essentially the same as pure solvent (toluene) and raffinate properties are essentially the same as pure diluent (water).

$$m_{mole\ fraction\ units} = (20.8)(92.14/865)(1000/18) = 123.1$$

We need a second equation in addition to the efficiency equation. The mass balance around the mixer can be used.

$$Sx_{D,in} + Fx_{feed} = Sx_{D,out} + Fx_{C,out} \tag{16-95}$$

The values of S and F can be calculated assuming that the solution properties are the same as pure toluene and pure water.

$$S = F_D = (0.0003m^3/s)(865kg/m^3)/(92.14kg/kmol) = 0.002816kmol/s$$

$$F = F_C = (0.003m^3/s)(1000kg/m^3)/(18kg/kmol) = 0.16652kmol/s$$

Then, $S/F = 0.016911$, and since $x_{D,in} = 0$,

$$x_{C,out} = x_{feed} - (S/F)x_{D,out} = 0.0003 - 0.016911x_{D,out}$$

Substituting this result into the equation for E_{MD}, we obtain

$$\frac{x_{D,out}}{123.1[0.0003 - 0.016911x_{D,out}]} = 0.927$$

Solving this equation we obtain

$x_{D,out} = 0.01685$ which from the definition of E_{MD} gives

$$x_{C,out} = x_{D,out}/mE_{MD} = 0.01685/[(123.1)(0.927)] = 1.477 \times 10^{-4}.$$

It is useful to draw some conclusions from this example.
1. Units are important.
2. Value of the equilibrium parameter depends on whether extract phase is in the numerator or denominator.
3. Value of the equilibrium parameter depends on the units.
4. The model used to determine efficiency makes a difference.

16.7.3 Area per Volume a and Average Drop Diameter in Mixers

To use these equations, we need to know the value of $K_{O\text{-}ED}a$. Although the best approach is to determine $K_{O\text{-}ED}a$ from experiments, it is less expensive and more convenient to estimate this value. To do this, values of a and the mass-transfer coefficients are often estimated separately. For spherical drops in a liquid,

$$\text{Surface area for n drops} = n\pi(\text{average diameter of drops})^2 \qquad \textbf{(16-96a)}$$

and the surface area per volume of the mixer is

$$a = n\pi(\text{average diameter of drops})^2/\text{Volume of vessel} \qquad \textbf{(16-96b)}$$

Since the volume of n drops $= n\pi(\text{average drop diameter})^3/6$, the volume fraction dispersed phase is

$$\varphi_D = n\pi(\text{average drop diameter})^3/[6(\text{Volume of vessel})] \qquad \textbf{(16-96c)}$$

Solving for volume of vessel and substituting into Eq. (16-96b), the surface area per volume is

$$a = n\pi(\text{diameter})^2 6\varphi_D/[n\pi(\text{diameter})^3] = 6\varphi_D/(\text{average drop diameter}) \quad \textbf{(16-96d)}$$

Methods for estimating φ_D in mixers are discussed in Section 13.14.1. The surface area per volume a of mixers can be estimated once the average diameter of the drops is known.

Several correlations for the average drop diameter in mixers have been published. Treybal (1980) recommends the following equation for baffled vessels:

$$\frac{d_p}{d_p^O} = 1 + 1.18 j_D \left(\frac{\sigma^2 g_c^2}{d_p^O \mu_C^2 g} \right) \left(\frac{\mu_C^4 g}{\Delta \rho \sigma^3 g_c^3} \right)^{0.62} \left(\frac{\Delta \rho}{\rho_C} \right)^{0.05}$$

(16-97a)

The term $\Delta \rho$ is the absolute value of $(\rho_C - \rho_D)$, and d_p^O is obtained from

$$\frac{d_p^{O^3} \rho_C^2 g}{\mu_C^2} = 29 \left(\frac{V_{Liq-tank}^3 \rho_C^2 \mu_C g^4}{P^3 g_c^3} \right)^{0.32} \left(\frac{\sigma^3 \rho_C g_c^3}{\mu_C^4 g} \right)^{0.14}$$

(16-97b)

Treybal notes that the terms in parentheses are dimensionless. For vessels that are operated full with no vapor-liquid interface and no baffles, Treybal recommends

$$d_p = 10^{-2.066 + 0.732 \phi_D} \left(\frac{\mu_C}{\rho_C} \right)^{0.0473} \left(\frac{Pg_c}{V_{Liq-tank} \rho_M} \right)^{-0.204} \left(\frac{\sigma g_c}{\rho_C} \right)^{0.274}$$

(16-98)

Godfrey (1994) recommends alternative forms.

16.7.4 Mixer Mass-Transfer Coefficients

Determination of mass-transfer coefficients in liquid-liquid extraction is fraught with more than the usual amount of uncertainty. Equation (16-80b) shows that the individual coefficients for both dispersed and continuous phases are required. Any uncertainties in the estimation of diffusivities will propagate error in the mass-transfer coefficients. For the typical ±20% error in liquid diffusivity, the result is a ±10% to 15% error in the mass-transfer coefficient (Slater, 1994). Also, because effects of coalescence, drop breakage, and time dependence are not well understood, they are usually ignored, which increases the potential error. The concentration of surface active agents and of small solids at the interface will affect coalescence and drop breakage and the internal circulation of drops. A few results for individual drops are presented in optional Section 16.7.4.1. For practical applications of mixers, the drops are in swarms, and the correlations in Section 16.7.4.2 must be used. A summary of a conservative (safe) design procedure for mixers is outlined in Section 16.7.4.3.

All of the dimensionless equations that follow give mass-transfer coefficients in units of m/s if D is in m²/s (or cm/s if D is in cm²/s). For these units, the typical driving force is a concentration difference (e.g., kmol/m³). To use the mass-transfer coefficients with mole or mass fraction driving forces, the units of the mass-transfer coefficients eventually must be adjusted using Eq. (16-93a). The correlations in this section assume that there is no diluent in the continuous phase and mass transfer is binary. Because in most cases there is significant partial miscibility of diluents and solvent, and mass transfer is in a ternary system (Taylor and Krishna, 1993), these methods are often approximate when applied in practice.

16.7.4.1 Mixer Mass-Transfer Coefficients for Individual Drops (Optional)

Unfortunately, even for individual drops, there is a relatively small amount of data and the number of correlations for the dispersed phase are particularly limited. Additional correlations are presented by Slater (1994) and Wankat and Knaebel (2008).

For clean systems, the mass-transfer coefficient in the continuous phase can be determined at low velocities for single drops (Slater, 1994). For stagnant conditions (velocity → 0), the theoretical result from solving the diffusion equation is

$$\text{Sh}_C \equiv (k_C \, d/D_{AC}) = 2.0 \qquad \textbf{(16-99a)}$$

The dimensionless term Sh_C is the Sherwood number for the continuous phase, and D_{AC} is the diffusivity of solute A in the continuous phase. With drop diameter d in meters and D_{AC} in m²/s, the units on the mass-transfer coefficient k_c are m/s.

For creeping flow in clean systems, the Sherwood number can be approximated as

$$\text{Sh}_C \equiv (k_C \, d/D_{AC}) = 1 + (1 + \text{Re}_{drop} \, \text{Sc}_C)^{0.33} \text{ for } \text{Re}_{drop} < 1 \qquad \textbf{(16-99b)}$$

The drop Reynolds number $\text{Re}_{drop} = (\rho_C d u_t/\mu_C)$ assumes the drop is at its terminal velocity u_t, which can often be estimated from Stokes's law, Eq. (13-57). The continuous-phase Schmidt number $\text{Sc}_C = [\mu_C /(\rho_C D_{AC})]$ includes the effect of molecular diffusivity. Equation (16-99b) simplifies to Eq. (16-99a) as u_t and $\text{Re}_{drop} \to 0$. At higher Reynolds numbers ($10 < \text{Re}_{drop} < 1200$, $190 < \text{Sc}_C < 241000$, and $1000 < \text{Pe}_C < 10^6$) Steiner's empirical results (Slater, 1994) for circulating drops are

$$\text{Sh}_{C,circ} \equiv (k_C \, d/D_{AC}) = (2/\pi^{0.5}) \, \text{Pe}_C^{0.5} \qquad \textbf{(16-100a)}$$

where the continuous-phase Peclet number $\text{Pe}_C = (d \, u_t/D_{AC})$. For rigid drops, Steiner obtained

$$\text{Sh}_{C,rigid} = 2.43 + \{0.773 \, \text{Re}_{drop}^{0.5} + 0.0103 \, \text{Re}_{drop}\}\text{Sc}_C^{0.33} \text{ for } 10^4 < \text{Pe}_C < 10^6 \qquad \textbf{(16-100b)}$$

The $0.0103 \, \text{Re}_{drop}$ term is a correction for the effect of wakes and is usually quite small. For conditions in between fully circulating and rigid, Steiner recommended the following equation (Slater, 1994):

$$(\text{Sh}_C - \text{Sh}_{C,rigid})/(\text{Sh}_{C,circ} - \text{Sh}_{C,rigid}) = 1 - \exp[-0.00418 \, \text{Pe}_C^{0.42}] \qquad \textbf{(16-100c)}$$

For clean systems, Eq. (16-100c) is preferred, while for dirty systems where the drops are often rigid, Eq. (16-100b), which predicts lower values of the mass-transfer coefficient, is more accurate.

Dispersed-phase coefficients are complicated, since mass transfer for individual drops is time dependent. If the drop life is fairly long, then a pseudo-steady state is reached and there is an asymptotic value of k_D. However, the exact state of the interface and the drop size have significant effects on the flow or lack of flow inside the drops. If the interfacial tension is low and the drop fairly large, there will be significant internal circulation and the rate of mass transfer is much larger than that predicted by molecular diffusion alone. This results in a significant increase in the dispersed-phase mass transfer coefficient k_D. A rough criterion for the critical drop diameter at which internal circulation starts in clean systems (Slater, 1994) is

$$d_{critical \, circulation} \approx [\sigma/(g\Delta\rho)]^{0.5} \qquad \textbf{(16-101)}$$

A third complicating factor is that mass transfer of solute usually lowers the interfacial tension σ and reduces $d_{\text{critical circulation}}$; however, the presence of small amounts of surfactants or dirt that collects at the interface will reduce internal circulation markedly and may introduce a resistance to mass transfer at the interface that is not included in Eq. (16-80b). Because industrial plants are usually not scrupulously clean, dispersed-phase mass-transfer coefficients in plant operations will often be significantly lower than the values obtained in scrupulously clean laboratories.

For large drops with toroidal internal circulation, Handlos and Baron solved the flow and mass-transfer equations. A simplified form of their result is in reasonable agreement with experimental results obtained under clean conditions for large drops (Slater, 1994),

$$k_D = 0.00375\, u_t\, /(1 + \mu_D/\mu_C) \tag{16-102a}$$

Note that with internal circulation (typical of large drops in clean systems), there is no dependence on the molecular diffusivity. For rigid drops (typically small drops or dirty interfaces) with no circulation, a limiting solution at long times (Slater, 1994) is

$$Sh_D \equiv (k_D\, d/D_{AD}) \approx 6.6 \tag{16-102b}$$

As expected, the molecular diffusivity of solute in dispersed phase D_{AD} is important when there is no internal circulation. Equation (16-102b) is conservative (predicted k_D is low).

16.7.4.2 Mass-Transfer Coefficients for Drop Swarms in Mixers

In practical extractors, drops do not occur individually but are in swarms of interacting drops. The results for individual drops are useful for predicting parameter effects, but correlations that include the effects of interacting drops are required for extractor design. In mixers, drops are not moving vertically at uniform velocities, and coalescence and drop breakage are important particularly in the vicinity of the impeller. Coalescence and drop breakage appear to enhance mass transfer and rates ~50% larger than those for rigid drops (Slater, 1994).

In extractions where the feed is the continuous phase and the distribution coefficient for transfer into the raffinate is large (m_{CD} in Eq. (16-80c) is small), the overall mass-transfer resistance is dominated by the continuous phase. In this case, the correlation of Skelland and Moeti can be used for mixer design (Frank et al., 2008; Slater, 1994; Wankat and Knaebel, 2008).

$$Sh_C = \frac{k_C d_p}{D_{AC}} = 0.00001237 \left(\frac{\mu_C}{\rho_C D_{AC}} \right)^{1/3} \left(\frac{d_i \omega^2}{g} \right)^{5/12} \left(\frac{d_i}{d_p} \right)^2 \left(\frac{d_p}{d_{\tan k}} \right)^{1/2} \left(\frac{\rho_D d_p g}{\sigma} \right)^{5/4} \varphi_d^{-1/2} \tag{16-103}$$

where d_i is the impeller diameter in meters and ω is the impeller speed in 1/s. This equation is restricted to low dispersed-phase holdup, $\phi_d < 0.06$. For rigid drops, correlations developed for solid particles can be used to provide conservative values (Treybal, 1980) for the continuous phase coefficient.

$$Sh_C = \frac{k_C d_p}{D_{AC}} = 2 + 0.47 \left[d_p^{4/3} \left(\frac{Pg_c}{V_{Liq-\tan k}} \right)^{1/3} \frac{\rho_C}{\mu_C}^{2/3} \right]^{0.62} \left(\frac{d_i}{d_{\tan k}} \right)^{0.17} \left(\frac{\mu_C}{\rho_C D_{AC}} \right)^{0.36} \tag{16-104}$$

There are few studies of dispersed-phase mass-transfer coefficients in mixers. Frank et al. (2008) recommend the correlation of Skelland and Xien for transfer from the dispersed phase to the continuous phase. Skelland and Xien (1990) studied batch extraction in a baffled mixer with six-flat-blade turbines. Their correlation is

$$\frac{k_D}{\left[D_{AD}/\left(t_{F,95}-t_0\right)\right]^{1/2}} = 5.0 \times 10^{-6} \phi_d^{-0.0204} \left(\frac{d_i N \rho_m}{\mu_m}\right)^{1.140} \left(\frac{\rho_C}{\left|\rho_C-\rho_D\right|}\right)^{0.518} \qquad \textbf{(16-105)}$$

In this eqquation, ρ_m is defined in Eq. (13-53), μ_m is defined in Eq. (13-55), N is the impeller speed in rps, t_0 is the initial time (s) that the dispersed phase is injected, and t_{F95} is the time (s) at which 95% of mass transfer has occurred. In a continuous mixer $(t_{F95}-t_0)$ can be considered the residence time of the dispersed phase in the mixer, which will result in 95% extraction of the solute. Additional correlations for mass-transfer coefficients are available (Treybal, 1980; Wankat and Knaebel, 2008).

16.7.4.3 Conservative Estimation of Mass-Transfer Coefficients for Extraction

Prediction of mass-transfer coefficients for extraction seems to follow Murphy's law (If anything can go wrong, it will) and O'Toole's corollary to Murphy's law (Murphy was an optimist). The best approach is to not predict the mass-transfer coefficients, but to measure the stage efficiency *using the exact solvent and feed from the plant.* Note that the use of clean solutions made up with high purity reagents will probably result in mass-transfer coefficients and stage efficiencies that are higher than observed in the plant.

However, obtaining data on the exact solutions to be used in the plant is often not possible, and a prediction of stage efficiency must be made. The following approach will result in a conservative estimate (stage efficiency will be low) because every equation is considered to be conservative and the additional mass transfer that occurs in the settler is neglected.

1. Assume that in the plant the extractor will contain particulates and surface active agents; thus, the drops are rigid. Estimate fraction dispersed phase ϕ_D and power P as in Example 13-5.
2. Use Eq. (16-97a and b) or (16-98) to estimate the drop diameter d_p.
3. Use Eq. (16-104) to estimate k_C.
4. Use Eq. (16-105) to estimate k_D.
5. Use Eq. (16-96d) to estimate a.
6. Use Eq. (16-80b analog) to estimate the overall mass-transfer coefficient k_{LD} in m/s.
7. Use Eq. (16-93a) to estimate K_{O-ED}.
8. Use Eq. (16-93b) to estimate n_{O-ED}.
9. Use the stirred tank model Eq. (16-86b) to determine the stage efficiency E_{MD}.
10. Check the results for any assumptions made during the calculation and repeat the calculation with a better assumption if necessary.

This approach is illustrated in Example 16-6.

> ### Example 16-6. Conservative estimation of mixer mass-transfer coefficients
> Estimate the mass-transfer coefficients and the mixer stage efficiency for the extraction of benzoic acid from water into toluene for the problem detailed in Example 13-5.

Solution

1. Follow the listed steps, and first assume drops are rigid. From Example 13-5, $\varphi_d - 0.167$ and power $P = 412.2$ W.

2. For a baffled vessel, we can use Eqs. (16-97a) and (16-97b) to find the drop diameter d_p. The terms for Eq. (16-97b) can now be determined:

$$\frac{\sigma^3 \rho_c \, g_c^3}{\mu_c^4 g} = 1.330 \times 10^9$$

$$\frac{V_{Liq-tank}^3 \rho_c^2 \mu_c g^4}{P^3 g_c^3} = \frac{(0.432)^3 (998)^2 (0.95 \times 10^{-3})(9.807)^4}{(412.2)^3 (1.0)^3} = 0.0010075$$

$$\frac{\rho_c^2 g}{\mu_c^2} = \frac{(998)^2 (9.807)}{(0.95 \times 10^{-3})^2} = 1.0823 \times 10^{13}$$

Then Eq. (16-97b) is

$$\left(d_p^o\right)^3 = (29)(0.0010075)^{0.32} (1.330 \times 10^9)^{0.14} \Big/ (1.08231 \times 10^{13}) = 5.5773 \times 10^{-12}$$

$$d_p^o = 0.0001774 \text{m}$$

For Eq. (16-97a), $1.18 \, \phi_d = 1.18(0.167) = 0.19706$,

$$\frac{\sigma^2 g_c^2}{d_p^o \mu_c^2 g} = \frac{(0.0222)^2 (1.0)^2}{(0.0001774)(0.9 \times 10^{-3})^2 (9.807)} = 313,962$$

$$\frac{\mu_c^4 g}{\Delta \rho \sigma^3 g_c^3} = \frac{(0.95 \times 10^{-3})^4 (9.807)}{(998 - 865)(0.0222)^3 (1.0)^3} = 5.4893 \times 10^{-9}$$

$$\frac{\Delta \rho}{\rho_c} = \frac{998 - 865}{998} = 0.13327$$

Eq. (16-97a) becomes

$$d_p = (0.0001774)\left[1 + (0.19706)(313962)(5.4893 \times 10^{-9})^{0.62} (0.13327)^{0.05}\right] = 0.0002524 \text{m}$$

3. For Eq. (16-104), $D_{benzoic-water} = 2.2 \times 10^{-9} \text{m}^2/\text{s}$,

$$\frac{\mu_c}{\rho_c D_{Ac}} = \frac{0.95 \times 10^{-3}}{(998)(2.2 \times 10^{-9})} = 432.68$$

$$\frac{d_i}{d_{tank}} = 0.25, \quad \frac{\rho_c}{\mu_c} = \frac{998}{0.95 \times 10^{-3}} = 1,050,526$$

$$\frac{Pg_c}{V_{Liq-tank}} = \frac{(412.2)(1.0)}{(0.432)} = 954.167$$

From Eq. (16-104), we obtain

$$k_c = \frac{2.2 \times 10^9}{(0.0002524)} \left[2 + 0.47 \left\{ \left[(0.0002524)^{4/3} (954.167)^{1/3} (105026)^{2/3} \right]^{.62} (.25)^{.17} (432.68)^{0.63} \right\} \right]$$

$$= 5.628 \times 10^{-5} \text{ m/s}$$

4. For Eq (16-105), from Example 13-5,

$$\phi_d = 0.167, \quad N = 1000 \text{ rpm} = 16.667 \text{ rps},$$

$$\rho_m = 975.8, \ d_i = 0.25 \ d_{tank} = 0.25 \ (0.8279) = 0.2070,$$

$$D_{benzoic-toluene} = 1.5 \times 10^{-9} \text{ m}^2/\text{s}$$

From Eq. (13-55),

$$\mu_m = \frac{\mu_c}{\phi_c}\left(1 + \frac{1.5\mu_d \phi_d}{\mu_c \mu_d}\right) = \left(\frac{0.95 \times 10^{-3}}{1-0.167}\right)\left(1 + \frac{(1.5)(0.59 \times 10^{-3})(0.167)}{0.95 \times 10^{-3} + 0.59 \times 10^{-3}}\right)$$

$$\mu_M = 0.0010589 \text{ kg/(ms)}$$

$$\frac{d_i N \rho_M}{\mu_M} = \frac{(0.2070)(16.667)(975.8)}{(0.0010589)} = 3179252$$

$$\frac{\rho_c}{|\rho_c - \rho_d|} = \frac{998}{998 - 965} = 7.504$$

The time $t_{F95} - t_0$ is the residence time that results in $E_{MD} = 0.95$. Example 16-5 resulted in $E_{MD} = 0.927$ for a 0.5 m tall by 0.5 m diameter mixer ($V_{Mixer} = 0.09817 \text{ m}^3$) at total flow rate of 0.0033 m³/s. Thus, residence time was = $V_{mixer}/Q_{total} = 29.755$. For the current example, $t_{res} = 60$s. We will assume 60s gives $E_{MD} = 0.95$ and then check our answer. $(t_{F95} - t_0) = 69$s. Then Eq. (16-105) becomes

$$k_D = \left(\frac{1.5 \times 10^{-9}}{60}\right)^{1/2} (5.0 \times 10^{-6})(0.167)^{-0.0204} (3179252)^{1.140} (7.504)^{0.518} = 0.001905 \, \text{m/s}$$

5. $a = 6\varphi_d/d_p = 6(0.167)/(0.0002524) = 3970 \, \text{m}^2/\text{m}^3$

6. $\dfrac{1}{K_{LD}} = \dfrac{1}{k_D} + \dfrac{1}{k_C m_{CD}} = \dfrac{1}{0.001905} + \dfrac{1}{(5.628 \times 10^{-5})(1/20.8)}$

   where $m_{CD} = C_C/C_D = 1/20.8$. Thus, $1/K_{LD} = 524.9 + 369581 = 370106$.

   $K_{LD} = 2.702 \times 10^{-6} \, \text{m/s}$ and $K_{LD}a = 0.01073 \, (1/\text{s})$

7. Eq. (16-93a) $K_{OED}a = \rho_O(K_{LD}a)/MW_D = (865)(0.01073)/(92.14) = 0.1007$

8. $n_{OED} = \dfrac{(V_{Mixer}/Q_D)(0.1007)}{(865/92.14)} = \dfrac{(0.432/0.0012)(0.1007)}{865/92.14} = 3.862$

9. $E_{MD} = \dfrac{3.862}{4.862} = 0.794$

The contribution of dispersed-phase resistance to total resistance is (524.9/370,106) × 100 = 0.142%, which is very small. If we ignore dispersed-phase resistance, $K_{LD} = k_C M_{CD} = 2.706 \times 10^{-6}$, which is very close to calculated K_{LD}.

10. This value of E_{MD} is low compared to the assumed value of 0.95 in step 4. However, since m_{CD} is quite low, the continuous-phase mass transfer controls ($K_{LD} \approx k_C M_{CD}$). Thus, the estimate for k_D is relatively unimportant, and the incorrect value for residence time in the estimation of k_D does not affect the final result.

Obviously, estimation of mass-transfer coefficients is challenging, and care must be taken to use a consistent set of units. As is usual with this sort of calculation, determination of the physical properties is often the most challenging part.

16.8 RATE-BASED ANALYSIS OF DISTILLATION

For binary distillation, the equilibrium stage analysis of distillation can be made to agree quite well with experimental results by predicting a Murphree efficiency for each stage in the column (Section 16.6). Commercial simulators will include efficiency calculations. Unfortunately, for multicomponent distillation, the Murphree efficiencies are, in general, not equal. To fit experimental results, it is sometimes necessary to use negative values for the Murphree efficiency. This is not satisfactory and is a sign that the assumption of equilibrium stages is not appropriate for this particular multicomponent distillation. A more fundamental analysis based on mass- and heat-transfer rates on each stage is required.

We saw in Section 15.6 that a Fickian mass-transfer analysis can lead to logical inconsistencies when extended to three or more components. Thus, a fundamental rate analysis of multicomponent distillation must be based on the Maxwell-Stefan mass-transfer model extended to nonideal multicomponent systems (Section 15.7.7). Since the significant detail required for these calculations is beyond the scope of an introductory textbook, the methods are summarized in enough detail to explain what the commercial simulator does (Lab 13 in appendix to Chapter 16) but not in enough detail to write a program of your own. Readers interested in the complete details are referred to Taylor and Krishna (1993) and Aspen Plus (2010).

The detailed rate model of distillation starts with material and energy balances for the vapor and liquid on each stage. If there are no reactions, the bulk vapor and liquid-phase component material balances are (Taylor and Krishna, 1993; Aspen Plus, 2010)

$$V_j y_{i,j} - V_{j+1} y_{i,j+1} - F_j^V y_{i,j,F} + N_{i,j}^V = 0 \qquad \text{(16-106a)}$$

$$L_j x_{i,j} - L_{j-1} x_{i,j-1} - F_j^L x_{i,j,F} - N_{i,j}^L = 0 \qquad \text{(16-106b)}$$

The transfer to the vapor from the liquid $N_{i,j}^V$ is arbitrarily considered to be positive. The transfer terms across the film are determined from the generalized matrix form of the Maxwell-Stefan equations (Krishna and Standart, 1976), Eq. (15-72a), with the mass-transfer coefficient given by Eq. (15-72c). Energy balances are required for both the bulk vapor and bulk liquid phases. For the bulk vapor phase, this equation is

$$V_j H_j^V - V_{j+1} H_{j+1}^V - F_j^V H_{j+1}^F + Q_j^V + E_j^V = 0 \qquad \text{(16-106c)}$$

where Q_j^V is the external heat load to the vapor and E_j^V is the energy transfer rate from the bulk liquid. There is a similar equation for the bulk liquid phase. The rate of energy transfer to the vapor from the liquid across the film is given by the following rate equation:

$$E_j^V = a_j^I h_j^V (T_j^V - T_j^I) + \sum_{i=1}^{n} N_{i,j}^V \overline{H}_{i,j}^V \qquad \text{(16-106d)}$$

In this equation, a_j^I is the interfacial area for heat transfer, h_j^V is the vapor-phase heat-transfer coefficient, T_j^V is the temperature bulk vapor phase, T_j^I is the temperature of the interface, and $\overline{H}_{i,j}^V$ is the partial molar enthalpy of component i, all on stage j. There is a similar equation for energy transfer from the vapor to the liquid. The film is assumed to have no accumulation of mass or energy. The film interface is assumed to be at equilibrium,

$$y_{i,j}^I = K_{i,j}(T_j^I) x_{i,j}^I \text{ and } T_j^{I,V} = T_j^{I,L} = T_j^I \qquad \text{(16-106e)}$$

These equations are written in matrix form.

The mass-transfer coefficients and interfacial area per volume a are obtained from correlations based on experimental data. Heat transfer coefficients are obtained from the Chilton-Colburn analogy, Eq. (15-50a), using the experimentally determined mass-transfer correlations. The most common correlation for interfacial area per volume a is the Zuiderweg (1982) correlation. The correlation depends on the regime of operation of the sieve plates. In the spray regime, the correlation is

$$a = \frac{40 A_{active}}{\left(A_{hole}/A_{active}\right)^{0.3}} \left(\frac{Q_L Q_V \rho_L^{0.5} \rho_V^{0.5} h_L}{\sigma A_{active}^2}\right)^{0.37} \quad \text{for } FP < 3.0 l_w h_{cl} A_{active} \qquad \textbf{(16-107a)}$$

where A_{active} is the active area of the tray and A_{hole} is the total hole area of a tray in m², Q_L and Q_V are the volumetric flow rates of liquid and vapor in m³/s, σ is the surface tension in N/m, FP is the flow parameter defined in Eq. (10-9), l_w is the weir length in m, and h_L is the calculated clear-liquid height in m on the tray. In the froth regime, the correlation is

$$a = \frac{43 A_{active}}{\left(A_{hole}/A_{active}\right)^{0.3}} \left(\frac{Q_L Q_V \rho_L^{0.5} \rho_V^{0.5} h_L}{\sigma A_{active}^2}\right)^{0.53} \quad \text{for } FP > 3.0 l_w h_{cl} A_{active} \qquad \textbf{(16-107b)}$$

The clear liquid height h_L for these correlations is calculated from

$$h_L = 0.6 h_w^{1/2} \left[\frac{p(\rho_V/\rho_L)^{1/2}}{l_w/A_{active}}\right]^{0.25} \quad \text{for 2.5 cm} < h_w < 10 \text{cm} \qquad \textbf{(16-107c)}$$

where p is the pitch of the sieve plate holes in m. Zuiderweg notes that these should be considered as apparent interfacial areas because they are based on the mass-transfer coefficients back-calculated from the operation of relatively large diameter Fractionation Research Institute distillation columns based on the somewhat improbable assumption that the mass-transfer coefficients are independent of flow regime or velocity. Thus, the effects of flow regime and velocity have been lumped into the calculation of a. Despite this difficulty, Eqs. (16-107a and b) are widely used.

Zuiderweg (1982) also determined correlations for the liquid and vapor mass-transfer coefficients (both in m/s):

$$k_{L,i} = 0.024 D_{L,i}^{0.25} \qquad \textbf{(16-108a)}$$

In this equation, D_L is in m²/s.

$$k_V = \frac{0.13}{\rho_V} - \frac{0.0065}{\rho_V^2} \quad \text{for } 1.0 < \rho_V < 80 \text{ kg/m}^3 \qquad \textbf{(16-108b)}$$

The equation for the vapor-phase mass-transfer coefficient is unique in that it does not depend on the diffusivity of the vapor and is the same for all components. Zuiderweg's correlation essentially assumes that liquid-phase resistances control; thus, it should not be extrapolated to cases where the vapor-phase and liquid-phase resistances are the same order of magnitude or the vapor-phase resistance controls.

Chan and Fair (1984) used the AIChE (1958) correlation for vapor mass transfer to determine $k_{L,i}a$. This correlation with $k_{L,i}a$ in (m/s)(m²/m³) is

$$k_{L,i}a = \left(0.40 U_a \rho_V^{1/2} + 0.017\right)\left(197 D_{L,i}^{1/2}\right) \qquad \textbf{(16-109a)}$$

The liquid diffusivity $D_{L,i}$ is in cm²/s, U_a = superficial vapor velocity in the active area of the tray, m/s, and ρ_V is the vapor density in kg/m³. The correlation they developed for V_{Vi} a with V_{Vi} in m/s and a in m²/m³ is

$$k_{V,i}a = \frac{D_{V,i}^{1/2}\left(1030f - 867f^2\right)}{h_L^{1/2}} \tag{16-109b}$$

The vapor diffusivity D_{Vi} is in cm²/s, f is the fractional approach to flooding, and h_L is the liquid holdup on the plate in cm. The liquid holdup is calculated from the correlation of Bennett et al. (1983),

$$h_L = \phi_e h_w + 1533\phi_e\left[0.0327 + 0.0286\exp(-1.378h_w)\right]\left(\frac{L}{\hat{\rho}_L l_{weir}\phi_e}\right)^{2/3} \tag{16-109c}$$

In this equation, h_w is the weir height in cm, L is the molar liquid flow rate in kmol/s, $\hat{\rho}_L$ is the molar liquid density in kmol/m³, l_{wier} is the weir length in m, and ϕ_e is the effective relative froth density given by

$$\phi_e = \exp\left[-12.55\left(U_a\sqrt{\frac{\rho_V}{\rho_L - \rho_V}}\right)^{0.91}\right] \tag{16-109d}$$

Note that different correlations use parameters in different units. It is obviously important to have the units correct when a particular correlation is used.

A more recent mass-transfer correlation was developed by Chen and Chuang (1993). They recommended using the clear liquid height calculated from Eq. (16-107c). Their correlations for mass-transfer coefficients are

$$k_{L,i} = \frac{14A_{active}}{\mu_L^{0.1}\left(A_{hole}/A_{active}\right)^{0.14}}\left(\frac{\rho_L\rho_V U_a^2}{\sigma^2}\right)^{1/3}\sqrt{D_{L,i}t_L}\frac{\rho_V}{\rho_L}\frac{U_a}{a} \tag{16-110a}$$

$$k_{V,i} = \frac{11A_{active}}{\mu_L^{0.1}\left(A_{hole}/A_{active}\right)^{0.14}}\left(\frac{\rho_L\rho_V U_a^2}{\sigma^2}\right)^{1/3}\sqrt{D_{V,i}t_V}\frac{U_a}{a} \tag{16-110b}$$

In these equations, σ is the interfacial surface tension, N/m; t_V = (froth height)/U_A and $t_L \approx \rho_L t_V$ /ρ_V are the average residence time per pass for the vapor and liquid, s. They recommend a correlation of Stichlmair for the interfacial area per volume a, but Aspen Plus (2010) recommends using the Zuiderweg (1982) correlations in conjunction with the Chen and Chuang mass-transfer correlation.

Which mass-transfer correlation should be used? One can simulate the distillation with each of these three correlations plus the other mass-transfer correlations supported by the simulator. If any of the correlations predict results that are clearly outliers, they probably should not be used. If a conservative design is desired, then use the correlation that predicts the least separation. If the column is operating outside the range of validity of a correlation, the correlation should be used with great caution. If a company has had good results using a particular correlation, they will probably keep using it. Obviously, all of the correlations for mass-transfer coefficients and for interfacial area per volume depend on the geometry of the plates, downcomers, and columns. Thus, the design requires the specification of these variables. It is highly recommended that the column be designed with equilibrium stages first (Chapter 6) and then the internals should be designed either by hand calculation (Chapter 10) or preferably with a tray-rating simulation (Lab 10). With the equilibrium staged design as a starting point, the rate-based design is more likely to converge.

All of the correlations provide mass-transfer coefficients at a point on the plate. To determine the overall amount transferred, a flow model for the stage is required—this is similar to the analysis in Section 16.6. Chan and Fair (1984) recommend calculating the Peclet number to determine which flow model is appropriate,

$$Pe = \frac{z_l^2}{D_{eddy} \bar{t}_{L,residence}} \qquad (16\text{-}111a)$$

The distance traveled z_l in m is from the exit of the downcomer to the overflow weir. The eddy diffusivity is determined from the Barker and Self correlation,

$$D_{eddy} = 6.675 \times 10^{-9} U_{active}^{1.44} + 0.000922 h_L - 0.00562 \qquad (16\text{-}111b)$$

The liquid holdup h_L in cm can be estimated from Eq. (16-109c), although if the only use of the calculation is to estimate the Peclet number, the weir height can be substituted for h_L. U_{active} is the vapor velocity in the active area of the plate, m/s. The residence time $\bar{t}_{L,residence}$ can be estimated from

$$\bar{t}_{L,residence} = \frac{h_L A_{active}}{100 Q_L} \qquad (16\text{-}111c)$$

In this equation, a_{active} is the active area of the plate in m², Q_L is the volumetric flow rate of the liquid in m³/s, and the 100 is required because the liquid holdup h_L is in cm. If Pe is low, then the completely mixed model is appropriate. This is the simplest flow model, and the mass-transfer coefficients are the same everywhere on the plate. Small diameter columns usually have low values of Pe. Thus, the completely mixed model is appropriate. Large columns with a large value of z_l often have a large value of Pe, and a plug flow model is more appropriate. With a plug flow model, the rate-based model can predict better separation than the equilibrium staged model (see Section 10.2). If a conservative result is desired, use the mixed model. The mixed model should always have results that show less separation than the equilibrium model.

16.9 SUMMARY—OBJECTIVES

At the end of this chapter you should be able to satisfy the following objectives:

1. Derive and use the mass transfer analysis (HTU-NTU) approach for distillation columns
2. Use HTU-NTU analysis for dilute and concentrated absorbers and strippers
3. Use the mass transfer correlations to determine the HTU

4. Derive and use HTU-NTU analysis for co-current flow
5. Use mass transfer analysis to determine tray efficiency of binary systems
6. Use mass transfer analysis for extraction mixer-settler design
7. Use rate-based simulators to design distillation columns

REFERENCES

AIChE, *Bubble Tray Design Manual,* AIChE, New York, 1958.

AIChE, *Tray Efficiencies in Distillation Columns*, Final Report from Univ. Delaware by J. A. Gerster, A. B. Hill , N. N. Hochgraf, and D. E. Robinson, AIChE, New York, 1958.

Aspen Plus, Process Simulator Help, "Rate-based Distillation," Version 7.2, Aspen Tech, Burlington, MA (2010).

Bennett, C. O,. and J. E. Myers, *Momentum, Heat and Mass Transfer,* 3rd ed., McGraw-Hill, New York, 1982.

Bennett, D. L., R. Agrawal, and P. J. Cook, "New Pressure Drop Correlation for Sieve-Tray Distillation Column," *AIChE J., 29*, 434–442 (1983).

Bolles, W. L., and J. R. Fair, "Improved Mass Transfer Model Enhances Packed-Column Design," *Chem. Eng., 89* (14), 109 (July 12, 1982).

Chan, H., and J. R. Fair, "Prediction of Point Efficiencies on Sieve Trays. 1. Binary Systems," *Ind. Engr. Chem. Process Des. Develop., 23*, 814–819 (1984).

Chen, G. X., and K. T. Chuang, "Prediction of Point Efficiency for Sieve Trays in Distillation," *Ind. Engr. Chem. Research, 32*, 701-708 (1993).

Cornell, D., W. G. Knapp, H. J. Close, and J. R. Fair, "Mass Transfer Efficiency-Packed Columns. Part II," *Chem. Eng. Prog., 56(8)*, 48 (1960).

Cornell, D., W. G. Knapp, and J. R. Fair, "Mass Transfer Efficiency-Packed Columns. Part I," *Chem. Eng. Prog., 56*(7), 68 (1960).

Cussler, E. L., *Diffusion. Mass Transfer in Fluid Systems,* 2nd ed., Cambridge Univ. Press, Cambridge, UK, 1997.

Geankoplis, C. J., *Transport Processes and Separation Process Principles,* 4th ed., Prentice Hall, Upper Saddle River, NJ, 2003.

Gianetto, A., V. Specchia, and G. Baldi, "Absorption in Packed Towers with Concurrent Downward High Velocity Flows—II, Mass Transfer," *AIChE Journal, 19* (5), 916 (Sept. 1973).

Godfrey, J. C., "Mixers," in J. C. Godfrey and M. J. Slater (Eds.), *Liquid-Liquid Extraction Equipment,* Wiley, Chichester, UK, 1994, Chap. 12.

Greenkorn, R. A., and D. P. Kessler, *Transfer Operations,* McGraw-Hill, New York, 1972.

Harmen, P., and J. Perona, "The Case for Co-Current Operation," *Brit. Chem. Eng., 17,* 571 (1972).

Hines, A. L., and R. N. Maddox, *Mass Transfer Fundamentals and Applications,* Prentice Hall, Upper Saddle River, NJ, 1985.

King, C. J., *Separation Processes,* 2nd ed., McGraw-Hill, New York, 1980.

Krishna, R., and G. L. Standart, "A Multicomponent Film Model Incorporating a General Matrix Method of Solution to the Maxwell-Stefan Equations," *AIChE Journal, 22,* 383–389 (1976).

Lewis, W. K. Jr., "Rectification of Binary Mixtures. Plate Efficiency of Bubble-Cap Columns," *Ind. Eng. Chem., 28,* 399 (1936).

Lockett, M. J., *Distillation Tray Fundamentals,* Cambridge University Press, Cambridge, UK, 1986.

McCabe, W. L., J. C. Smith, and P. Harriott, *Unit Operations in Chemical Engineering,* 7th ed., McGraw-Hill, New York, 2005.

Mickley, H. S., T. K. Sherwood, and C. E. Reed, *Applied Mathematics in Chemical Engineering,* 2nd ed., McGraw-Hill, New York, 1957, pp. 187–191.

Perry, R. H., and D. Green, *Perry's Chemical Engineer's Handbook,* 6th ed., McGraw-Hill, New York, 1984.

Reid, R. C., J. M. Prausnitz, and T. K. Sherwood, *The Properties of Gases and Liquids,* 3rd ed., McGraw-Hill, New York, 1977.

Reiss, L. P., "Co-current Gas-Liquid Contacting in Packed Columns," *Ind. Eng. Chem. Process Design Develop., 6,* 486 (1967).

Reynolds, J., J. Jeris, and L. Theodore, *Handbook of Chemical and Environmental Engineering Calculations,* Wiley, New York, 2002.

Sherwood, T. K., and F. A. L. Holloway, "Performance of Packed Towers-Liquid Film Data for Several Packings," *Trans. AIChE, 36,* 39 (1940).

Sherwood, T. K., R. L. Pigford, and C. R. Wilke, *Mass Transfer,* McGraw-Hill, New York, 1975.

Skelland, A. H. P., and H. Xien, "Dispersed-Phase Mass Transfer in Agitated Liquid-Liquid Systems," *Ind Eng. Chem. Research, 29,* 415 (1990).

Slater, M. J., "Rate Coefficients in Liquid-Liquid Extraction Systems" in J. C. Godfrey and M. J. Slater (Eds.), *Liquid-Liquid Extraction Equipment,* Wiley, Chichester, UK, 1994, Chap. 14.

Taylor, R., and R. Krishna, *Multicomponent Mass Transfer*, Wiley, New York, 1993.

Treybal, R. E., *Mass Transfer Operations,* McGraw-Hill, New York, 1955, p. 239.

Treybal, R. E., *Mass Transfer Operations,* 3rd ed., McGraw-Hill, New York, 1980.

Wang, G. Q., X. G. Yuan, and K. T. Yu, "Review of Mass Transfer Correlations for Packed Columns," *Ind. Engr. Chem. Res., 44,* 8715 (2005).

Wankat, P. C., and K. S. Knaebel, "Mass Transfer," Section 5B, in *Perry's Handbook of Chemical Engineering*, 8th ed., Green, D. (Ed.), McGraw-Hill, New York, (2008), p. 5-1, 5-2, 5-43 to 5-83.

Zuiderweg, F. J., "Sieve Trays. A View on the State of the Art," *Chem. Engr. Sci., 37,* 1441-1464 (1982).

HOMEWORK

A. *Discussion Problems*

A1. Compare Colburn Eq. (16-81d) to the equivalent Kremser Eq. (13-11b) and compare Eq. (16-81e) to (13-11a). If we relate n_{O-Ey} to N and set S = E, F = R, $y_{N+1} = y_{in}$, $y_1 = y_{out}$, and $y_1^* = y_{out}^*$ what terms are similar and what terms are different? [Note that the LHS of Eq. (13-11a) is inverted compared to Eq. 16-81e).]

A2. The mass transfer models include transfer in only the packed region. Mass transfer also occurs in the ends of the column where liquid and vapor are separated. Discuss how these "end effects" will affect a design. How could one experimentally measure the end effects?

A3. Is a stage with a well-mixed liquid less or more efficient than a stage with plug flow of liquid across the stage (assume $K_G a$ is the same)? Explain your result with a physical argument.

A4. **a.** The Bolles and Fair (1982) correlation indicates that H_G is more dependent on liquid flux than on gas flux. Explain this on the basis of a simple physical model.

  b. Why do H_G and H_L depend on the packing depth?

  c. Does H_G increase or decrease as μ_G increases? Does H_G increase or decrease as μ_L increases?

A5. Why is the mass transfer analysis for concentrated absorbers considerably more complex than the analysis for binary distillation or for dilute absorbers?

A6. Construct your key relations chart for this chapter.

A7. Suppose you are designing a mixer-settler extraction system and you obtain a mass transfer correlation from a book. Unfortunately, the book does not explain which model was used to obtain the values of the mass transfer coefficient. Which model would you use to determine the stage efficiency? Why?

A8. You are designing a mixer-settler extraction system and you obtain a mass transfer coefficient from the Internet, but the site does not give the units of the mass transfer coefficient. What should you do?

A9. Why are mass transfer coefficients from clean drops higher than mass transfer coefficients in dirty systems? What is the practical significance of this?

B. *Generation of Alternatives*

B1. Develop contactor designs that combine the advantages of co-current, cross-flow, and countercurrent cascades.

C. *Deriviations*

C1. Derive the relationships among the different NTU terms for binary distillation.

C2. Derive Eq. (16-94).

C3. Derive the following equation to determine n_{OG} for distillation at total reflux for systems with constant relative volatility:

$$n_{OG} = \frac{1}{1-\alpha} \ln\left[\frac{(y_{out} - 1)(y_{in})}{(y_{in} - 1)(y_{out})}\right] + \ln\left(\frac{y_{in} - 1}{y_{out} - 1}\right)$$

C4. If a mixer-settler is used for an extraction, we can often estimate the stage efficiency of the mixer E_{MD} and of the settler E_{SD} separately (see Example 16-5 and Problem 16.D21) where,

$$E_{SD} = \frac{y_{S,IN} - y_{S,OUT}}{y_{S,IN} - y_{S,OUT}^*} = \frac{y_{M,OUT} - y_{S,OUT}}{y_{M,OUT} - y_{S,OUT}^*}$$

Of course, we are really most interested in the overall stage efficiency of the mixer-settler combination $E_{total,D}$. For a dilute extraction with linear equilibrium, y = mx, completely immiscible solvent and diluent, the extract as the dispersed phase, and an entering pure solvent $y_{in} = 0$, the definition of $E_{total,D}$ is

$$E_{total,D} = \frac{y_{IN} - y_{S,OUT}}{y_{IN} - y_{S,OUT}^*} = \frac{y_{S,OUT}}{y_{S,OUT}^*}$$

Derive the following equation relating $E_{total,D}$ to mS/F, E_{MD}, and E_{SD}.

$$E_{total,D} = \frac{E_{MD} + E_{SD}\left[1 + E_{MD}\left(\frac{mS}{F} - 1\right)\right]}{\left(1 + \left(\frac{mS}{F}\right)E_{MD}\right)\left(1 + \left(\frac{mS}{F}\right)E_{SD}\right) - \left(\frac{mS}{F}\right)\left\{E_{MD} + E_{SD}\left[1 + E_{MD}\left(\frac{mS}{F} - 1\right)\right]\right\}}$$

C5. If the mixer in Problem 16.C4 is at equilibrium, then $E_{total,D} = E_{MD} = 1.0$. Show that the result in Problem 16.C4 satisfies this condition and determine the value of E_{SD}.

D. *Problems*

Answers to problems with an asterisk are at the back of the book.

D1.* For Examples 16-1 and 16-2, estimate an average H_{OG} in the enriching section. Then calculate n_{OG} and $h_E = H_{OG,avg} \, n_{OG}$.

D2.* If 1-in metal Pall rings are used instead of 2-inch rings in Example 16-2:
   a. Recalculate the flooding velocity and the required diameter.
   b. Recalculate H_G and H_L in the enriching section.

D3. In part E of Example 16-2 a HETP value of 2.15 feet is calculated for the top of the enriching section. Since the average error in the individual mass transfer coefficients k_y and k_x can be ±24.4% (Wankat and Knaebel, 2008), calculate the range of HETP values at the top of the column (m = 0.63) and for a geometric average of m over the enriching section (m = 0.577). Determine the safety factor that should be used compared to the 2.15 feet originally calculated.

D4.* A distillation column is separating a feed that is 40 mol% methanol and 60 mol% water. The two-phase feed is 60% liquid. Distillate product should be 92 mol% methanol, and bottoms 4 mol% methanol. A total reboiler and a total condenser are used. Reflux is a saturated liquid. Operation is at 101.3 kPa. Assume CMO, and use L/D = 0.9. Under these conditions $H_G = 1.3$ feet and $H_L = 0.8$ feet in both the enriching and stripping sections. Determine the required heights of both the enriching and stripping sections. Equilibrium data are given in Table 2-7.

D5. We have a column that has a 6-foot section of packing. The column can be operated as a stripper with liquid feed, as an enricher with a vapor feed or at total reflux. We are separating methanol from isopropanol at 101.3 kPa. The equilibrium can be represented by a constant relative volatility, $\alpha = 2.26$.
   a. At total reflux we measure the vapor mole frac methanol entering the column, $y_{in} = 0.650$ and the vapor mole frac methanol leaving, $y_{out} = 0.956$. Determine n_{OG}, the average value of H_{OG}, and the value of HETP.
   b. We operate the system as an enricher with L/D = 2. The vapor mole frac methanol entering the column, $y_{in} = 0.783$ (this is the feed) and the vapor mole frac methanol leaving, $y_{out} = 0.940$. Determine n_{OG}, the average value of H_{OG}, and the value of HETP.
   c. This problem was generated with a constant HETP. Why do the estimates in parts a and b differ?

D6.* A distillation column operating at total reflux is separating methanol from ethanol. The average relative volatility is 1.69. Operation is at 101.3 kPa. We obtain methanol mole fracs of $y_{out} = 0.972$ and $y_{in} = 0.016$.
   a. If there is 7.47 m of packing, determine the average H_{OG} using the result of Problem 16.C3.
   b. Check your results for part a, using a McCabe-Thiele diagram.

D7. A distillation column operating at total reflux is separating acetone and ethanol at 1 atm. There is 2.0 m of packing in the column. The column has a partial reboiler and a total condenser. We measure the bottoms composition in the partial reboiler as x = 0.10 and the liquid composition in the total condenser as x = 0.9. Equilibrium data are in Problem 4.D7. Estimate the average value of H_{OG}.

D8.* We wish to strip SO_2 from water using air at 20°C. The inlet air is pure. The outlet water contains 0.0001 mole frac SO_2, while the inlet water contains 0.0011 mole frac SO_2. Operation is at 855 mm Hg, and L/V = 0.9 (L/V)$_{max}$. Assume $H_{OL} = 0.84$ m and that the Henry's law constant is 22,500 mm Hg/mole frac SO_2. Calculate the packing height required.

D9. Calculate the stage efficiency for the mixer in an extraction system for values of n_{O-ED} varying from 0.1 to 100 for both the completely mixed staged model and the differential equation model. Compare the results..

D10.* A packed tower is used to absorb ammonia from air using aqueous sulfuric acid. The gas enters the tower at 31 lbmol/(h-ft²) and is 1 mol% ammonia. Aqueous 10 mol% sulfuric acid is fed at a rate of 24 lbmol/(h-ft²). The equilibrium partial pressure of ammonia above a solution of sulfuric acid is zero. We desire an outlet ammonia composition of 0.01 mol% in the gas stream.
    a. Calculate n_{OG} for a countercurrent column.
    b. Calculate n_{OG} for a co-current column.
    c. What is the importance of the gas and liquid flow rates?

D11. An air stream containing 50 ppm (mole) of H_2S is to be absorbed with a dilute NaOH solution. The base reacts irreversibly with the acid gas H_2S so that at equilibrium there is no H_2S in the air. An outlet gas that contains 0.01 ppm (mole) of H_2S is desired. $L/V = 0.32$.
    a. Calculate n_{OG} for a co-current system.
    b. Calculate n_{OG} for a countercurrent system.

D12.* We wish to absorb ammonia into water at 20°C. At this temperature $H = 2.7$ atm/mole frac. Pressure is 1.1 atm. Inlet gas is 0.013 mole frac NH_3, and inlet water is pure water.
    a. In a countercurrent system we wish to operate at $L/G = 15$ $(L/G)_{min}$. A $y_{out} = 0.00004$ is desired. If $H_{OG} = 0.75$ ft at $V/A_c = 5.7$ lbmol air/(h-ft²), determine the height of packing required.
    b. For a co-current system a significantly higher V/A_c can be used. At $V/A_c = 22.8$, $H_{OG} = 0.36$ ft. If the same L/G is used as in part a, what is the lowest y_{out} that can be obtained? If $y_{out} = 0.00085$, determine the packing height required.

D13. Repeat Example 16-3 to determine n_{OG} except use a co-current absorber.
    a. Same conditions as Example 16-3. If specifications can be met, find y_{out} and n_{OG}. If specifications cannot be met, explain why not.
    b. Same conditions as Example 16-3 except $x_{out} = 0.002$. If specifications can be met, find y_{out} and n_{OG}. If specifications cannot be met, explain why not.
    c. Same conditions as Example 16-3 except $x_{out} = 0.0003$ and $L/V = 40$. If specifications can be met, find y_{out} and n_{OG}. If specifications cannot be met, explain why not.

D14.* We are operating a staged distillation column at total reflux to determine the Murphree efficiency. Pressure is 101.3 kPa. We are separating methanol and water. The column has a 2-inch head of liquid on each well-mixed stage. The molar vapor flux is 30 lbmol air/(h-ft²). Near the top of the column, when $x = 0.8$ we measure $E_{MV} = 0.77$. Near the bottom, when $x = 0.16$, $E_{MV} = 0.69$. Equilibrium data are given in Table 2-7.
    a. Calculate $k_x a$ and $k_y a$.
    b. Estimate E_{MV} when $x = 0.01$.

D15. The large-scale column in Example 16-4 has a feed that is a saturated liquid with a feed mole frac $z = 0.5$, and separation is essentially complete ($x_{dist} \sim 1$ and $x_{bot} \sim 0$). The Murphree vapor efficiency is often approximately constant in columns. Assume the value calculated in Example 16-4, $E_{MV} = 0.97$, is constant in the large-scale column (plug flow trays). Calculate E_{pt} and $K_y a$ in the stripping section at $x = 0.10$ and $x = 0.30$, and in the enriching section at $x = 0.9$. Repeat the enriching section calculation at $x = 0.7$ (shown in Example 16-4) as a check on your procedure.

D16. Although the largest errors in the calculation of the height of a packed column are 1) errors in the mass transfer coefficients and 2) errors in the VLE data, calculation errors can also be significant because the calculation of n_G and of n_{OG} both require subtracting y values that are close to each other and then taking the inverse of this difference. Suppose that all of the values of $y_{AI} - y_A$ in the table in Example 16-1 are too low by 0.001 (thus, for $y_A = 0.8$ the value of $y_{AI} - y_A$ should be 0.013). Recalculate the area for $n_{G,enriching}$ and the required height of packing in the enriching section.

D17. Errors in the mass transfer coefficients obviously affect the value of H_G and hence the height of the packed section. These errors also affect the calculation of y_{AI} and thus the calculation of n_G and the height of the packed section. Return to the calculation in Example 16-1. We want to calculate the change in $y_{AI} - y_A$ and in $1/(y_{AI} - y_A)$. We will do this calculation for a value of $y_A = 0.8$. Assume that the equilibrium values given in Table 2-1 follow a straight line between x = 0.7472 and x = 0.8943 and determine the equation for this straight line.

    a. Calculate the range of values of $1/(y_{AI} - y_A)$ for a value of $y_{AI} = 0.8$ if H_G varies by ±24.4% compared to the 0.4054 m used in Example 16-1 but H_L does not vary from the 0.253 m value. Then, assuming that the % error in $1/(y_{AI} - y_A)$ is constant throughout the enriching section, estimate the range in values of n_G and in the height of packing (remember that H_G varies).

    b. Calculate the range of values of $1/(y_{AI} - y_A)$ for a value of $y_A = 0.8$ if both H_G and H_L vary by ±24.2% compared to the 0.4054 m and 0.253 m used in Example 16-1 for H_G and H_L, respectively. Then, assuming that the % error in $1/(y_{AI} - y_A)$ is constant throughout the enriching section, estimate the range in values of n_G and in the height of packing (remember that H_G varies).

D18. For extraction of benzoic acid from water into toluene with toluene as the dispersed phase, we measure the following mole fractions of benzoic acid: $x_{D,in} = 0$, $x_{D,out} = 0.00023$, and $x_{C,out} = 1.99 \times 10^{-6}$. The mixer is 0.75 m tall and 0.75 m diameter. Flow rate of the dispersed phase is $Q_D = 0.0012$ m³/s and $Q_C = 0.097$ m3/s. Data for density, equilibrium, and molecular weight are in Example 16-5.

    a. Determine the stage efficiency $E_{MD,mole\_frac}$. Note: The unit conversion in Eq. (16-94) is required to calculate m for equilibrium, y = mx, in mole fractions.

    b. Calculate the value of $k_{O-ED}a$ using the completely mixed staged model.

    c. Calculate the value of $k_{O-ED}a$ using the differential equation model.

    d. If by accident the value of $k_{O-ED}a$ calculated using the completely mixed staged model is used to predict the stage efficiency using the differential equation model, what value (incorrect) of $E_{MD,mole\_frac}$ is obtained?

D19. For extraction of benzoic acid from water into toluene with toluene as the dispersed phase, we measure the following concentrations of benzoic acid: $C_{D,in} = 0$, $C_{D,out} = 0.00023$, and $C_{C,out} = 0.00536$ with concentrations in mol/m³. The mixer is 0.75 m tall and 0.75 m diameter. Flow rate of the dispersed phase is $Q_D = 0.0012$ m³/s and $Q_C = 0.097$ m³/s. $\varphi_d = 0.09$. Data for density, equilibrium, and molecular weight are in Example 16-5.

    a. Determine the stage efficiency $E_{MD,Conc}$.

    b. Calculate the value of $k_{O-ED}a$ in concentration units using the completely mixed staged model.

    c. Calculate the value of $k_{O-ED}a$ in concentration units using the differential equation model.

d. If by accident the value of $k_{O-ED}a$ calculated using the completely mixed staged model is used to predict the stage efficiency using the differential equation model, what value (incorrect) of $E_{MD,Conc}$ is obtained?
This problem is the mirror of 16.D18, but with a different value for continuous-phase concentration and in different units.

D20. **a.** Revisit Example 16-6 and predict E_{MD} if the system is assumed to have low dispersed phase holdup so that Eq. (16-103) can be used.
b. Determine the % contribution of k_D to the sum of resistances $(1/K_{LD})$ for this problem and compare to the % contribution to the sum of resistances in Example 16-6.
c. What is value of K_{LD} if k_D is ignored?
d. Would you recommend doing a design based on the answer for E_{MD} from Example 16-6 or from this problem? Why?

D21. Mass transfer continues in settlers. Estimate the overall dispersed phase mass transfer coefficient for the settler in Example 13-5. Use the drop diameter calculated for the mixer in Example 16-6 as the drop diameter in the settler instead of the 150 μm assumed in Example 13-5. Recalculate the Stokes velocity, estimate an average ϕ_d in the settler, estimate the value of a in the settler, estimate k_C and K_{LD}, then determine $K_{O-settler-D}a_{settler}$. Since the settler is not well mixed, use the equivalent of Eq. (16-82) to estimate a Murphree efficiency for the settler, E_{SD}. When the extract is the dispersed phase, E_{SD} is defined as,

$$E_{SD} = \frac{y_{S,IN} - y_{S,OUT}}{y_{S,IN} - y_{S,OUT}^*} = \frac{y_{M,OUT} - y_{S,OUT}}{y_{M,OUT} - y_{S,OUT}^*}$$

Note that the answer for E_{SD} will depend on the approach used to estimate ϕ_d in the settler. An answer for E_{SD} based on one estimate of ϕ_d is given in Problem 16.D22.

D22. For the problem in Examples 13-5 and 16-6 with $E_{MD} = 0.794$, use an estimate for $E_{SD} = 0.82$.

a. If the entering solvent is pure and the mole fraction of benzoic acid in the aqueous feed is 0.00026, calculate $y_{out,settler}$ from the mixer-settler mass balance, equilibrium, and these values of E_{MD} and E_{SD}. Then calculate the stage efficiency of the overall mixer-settler combination, $E_{total,D}$ which for the case where the extract is the dispersed phase and the entering solvent is pure is defined as

$$E_{total,D} = \frac{y_{IN} - y_{S,OUT}}{y_{IN} - y_{S,OUT}^*} = \frac{y_{S,OUT}}{y_{S,OUT}^*}$$

b. Compare your answer with the answer obtained from the equation derived in Problem 16.C4.
Note: This problem does not require solving Problem 16.D21.

G. *Computer-Simulation Problems*
G1. Aspen Plus is not programmed to use a mass transfer approach (HTU-NTU) for binary packed bed distillation. However, it is easy to obtain accurate values of (y*-y) and m for distillation of binary mixtures from Aspen Plus. Then, Eq. (16-19) can be integrated without assuming that $H_{OG} = V/(K_y aA_c)$ is constant. If we assume that H_L and H_G in Eq. (16-27a) are constant in each section of the column and we expand Eq. (16-19) before integrating, we obtain after integration the height of the section,

$$h=\frac{H_L}{(L/V)}\int_{y_{A,in}}^{y_{A,out}}\frac{mdy_A}{(y_A*-y_A)}+H_G\int_{y_{A,in}}^{y_{A,out}}\frac{dy_A}{(y_A*-y_A)}$$

(16-21a)

The integrals can be determined numerically external to Aspen Plus once the appropriate values have been determined. Note that vapor and liquid leaving a stage in Aspen Plus results are in equilibrium (thus, the y_3 value is y_3* in equilibrium with x_3). Vapor stream y_4 is a passing stream with liquid stream x_3 (they are on the operating line). For example, if we pick y_4 on Aspen's tray compositions in the results, x_3 is on the operating line with y_4, and y_3 is the y* value needed to calculate y*-y in Eq. (16-21a) above [in other words, (y*-y) for this example = the Aspen values for (y_3-y_4)]. The appropriate value of m is the slope of the equilibrium curve at y (= y_4 from Aspen). Probably the easiest method to find m is to run Analysis in Aspen with 101 points. Then numerically determine the

slope = $(y_{n+1} - y_n)/(x_{n+1} - x_n)$ if y is near the center of $(y_{n+1} - y_n)$ or as
slope = $(y_{n+1} - y_{n-1})/(x_{n+1} - x_{n-1})$ if y is near to the value of y_n.

Then m is the average of the slopes at x (corresponds to y_A*) and x_1 (corresponds to $y_{A,1}$); however, x_1 is unknown, although it has to be between y_A and y_A*. To simplify the calculations, we will calculate m at the average value, $y_{A,avg} = (y_A+y_A*)/2$

a. Why is the use of 101 points in Analysis convenient? Try It!
b. Use Aspen Plus with NRTL for VLE to simulate the distillation problem in Example 16-1. First, find the optimum feed plate and the total number of stages that give $x_D = 0.80 \pm 0.0012$ and $x_B = 0.02 \pm 0.0004$. Then, use Eq. (16-21a) and the values obtained from Aspen Plus to determine the height of the stripping section. Compare to Example 16-1.

G2. Continue the calculation in Problem 16.G1, but determine the height of the enriching section.

G3. [Fairly involved problem.] We wish to distill 80.0 mol/s of a saturated vapor feed at 15.0 atm. The feed is 0.100 mole fraction ethane, 0.300 mole fraction propane, 0.500 mole fraction n-butane, and 0.100 mole fraction n-pentane. The column operates at 15.0 atm, has a partial condenser, and produces a vapor distillate. A kettle type reboiler is used. Our goal is to design a column using mass transfer rate analysis that will have a maximum n-butane mole fraction in the vapor distillate of $y_{D,C4,max} = 0.00875$ and a maximum mole fraction of propane in the liquid bottoms of $x_{Bot, C3,max} = 0.005833$.

a. Since this feed is the same as in Lab 13 except that the feed is a saturated vapor, start with N (Aspen notation) = 35 and the feed on stage 16. Increase reflux ratio to L/D = 5.0, and Distillate flow rate = 32.0 mol/s. Do an equilibrium run with Tray Sizing with one section to see if one section works.
b. Then, find $(L/D)_{min}$ with equilibrium runs and operate at L/D = 1.2(L/D)$_{min}$. Find the optimum feed stage, the number of stages needed, and the column diameter at 75% flooding calculated with Jim Fair's method for an equilibrium-staged model.
c. Convert to a rate model at 75% flood. Use default values for tray and downcomer design parameters. With Aspen Plus use the Chen and Chuang mass transfer rate correlations, the Chilton-Colburn analogy for heat transfer, the Zuiderweg interfacial area correlation, and VPLUG flow model. Have the design mode in tray rating-design calculate the column diameter. Check that the DC backup/Tray spacing, and weir loading are okay. Find the optimum feed stage with the rate-based model and the number of stages that just gives the desired purities.

Chapter 16   Mass Transfer Analysis for Distillation, Absorption, Stripping, and Extraction

d. Do a rate model again at 75% flood. Use default values for tray and downcomer design parameters. With Aspen Plus use the Chen and Chuang mass transfer rate correlations, the Chilton-Colburn analogy for heat transfer, the Zuiderweg interfacial area correlation, and Mixed flow model. Have the design mode in tray rating-design calculate the column diameter. Check that the DC backup/Tray spacing, and weir loading are okay. Find the optimum feed stage with the rate-based model and the number of stages that just gives the desired purities.

e. Compare and explain results in parts b, c, and d.

H. *Spreadsheet Problems*

H1. Use a spreadsheet program with a sixth-order polynomial fit (see Appendix 2B [at the end of Chapter 2]) for the ethanol-water VLE to determine n_{OG}, H_{OG}, and the height of the enriching section for Problem 16.D1.

H2. Use a spreadsheet program with a sixth-order polynomial fit (see Appendix 2B [at the end of Chapter 2]) for the ethanol-water VLE to determine n_{OG}, H_{OG}, and the height of the stripping section for Examples 16-1 and 16-2.

CHAPTER 16 APPENDIX.   COMPUTER RATE-BASED SIMULATION OF DISTILLATION

This appendix shows how the Aspen Plus simulator can be used to do detailed rate-based analysis of distillation using the Maxwell-Stefan approach outlined in Sections 15.7 and 16.8. Lab 10 should be done before this lab. NOTE: If you have convergence problems, reinitialize and try running again.

Lab 13. Aspen Plus uses RADFRAC with the Tray Rating option to do detailed heat- and mass transfer design calculations and tray and downcomer design. Start by setting up the problem below with RADFRAC.

1. We wish to distill 80.0 mol/s of a feed at 25°C and 15.0 atm. The feed is 0.100 mole fraction ethane, 0.300 mole fraction propane, 0.500 mole fraction n-butane, and 0.100 mole fraction n-pentane. The column operates at 15.0 atm, has a partial condenser, and produces a vapor distillate. A kettle type reboiler is used. Our goal is to design a column using mass transfer rate analysis that will have a maximum n-butane mole fraction in the vapor distillate of $y_{D,C4,max} = 0.00875$ and a maximum mole fraction of propane in the liquid bottoms of $x_{Bot, C3,max} = 0.005833$.*

Preliminary analysis with an equilibrium model using the Peng-Robinson VLE was done in Lab 10. Because the flow rate in Lab 10 was significantly higher, the diameter was scaled to the reduced feed rate by using the ratio

$$Dia_{Col2} = Dia_{Col1} \sqrt{\frac{F_2}{F_1}}$$

* This problem was adapted from Example 14.3.1 in R. Taylor and R. Krishna, *Multicomponent Mass Transfer*, Wiley, New York (1993).

After doing this scaling, Lab 10 shows that the following variables are a reasonable starting point: N (Aspen notation) = 35 and the feed on stage 16. Reflux ratio is $L/D = 2.5$, and Distillate flow rate = 32.0 mol/s. The column has two sections, section 1 is stages 2 to 15 and section 2 is stages 16 to 34. Each section has one pass. For initial diameters use Dia_1 = 1.36 m and Dia_2 = 1.745 m. In both sections use sieve plates with the Fair flooding calculation method with flooding at 70%. Choose tray spacing = 0.60 m, weir height = 0.0508 m, hole diameter = 0.0127 m, fraction sieve hole area to active area = 0.12, and downcomer clearance = 0.0381 m. Leave other variables in Tray Rating blank, which uses the default numbers for the variables in both sections.

2. If we had not already done Lab 10, we would do a number of equilibrium runs to find reasonable operating conditions and diameters for the column sections. Since we have reasonable starting conditions, this step can be skipped.

3. Draw the column using RADFRAC. The Setup, Components, Properties, and Streams are input in exactly the same way as for other Aspen Plus simulations. In the Block→Setup→Configuration Tab→Calculation Type menu, select Rate-Based. Complete the other items in the Configuration and other tabs as normal.

4. Now click on Tray Rating. In the Object Manager, click on New and call this section 1. In the Specs tab, input the stages for this section (2 to 15), pick Sieve from the Tray Type menu, and input the other variables as listed in item 1. In the Design/Pdrop tab, pick the "Fair" flooding calculation method and do NOT check Update section pressure drop. In the Layout tab, the tray type is sieve, and use the defaults for the other items. At this point do not put any numbers into the Downcomers tab (this means default values will be used).

5. In the menu on the left side, you will see Tray Rating, Section 1, and below Setup there will be Rate-Based. Click on this. In the Rate-Based tab, click on the box labeled "Rate-based calculations" (Aspen Plus allows you to have sections with equilibrium calculations as long as at least one section is done rate-based). Clicking on this box activates the menus below. Use the default values for Calculation Parameters. The Mixed flow model (called "Mixed-Mixed" in the report) assumes that vapor and liquid are well mixed so that the bulk properties are the same as the exit properties. This model is appropriate for trays (not packing) and was used in Section 16.6 to derive Eq. (16-77) for binary distillation. The effect of flow model will be looked at in item 10. Select film for both Liquid and Vapor in the Film Resistance section, and select No for both nonideality corrections.

6. In the Correlations tab are three menus. Unfortunately, there is no single best choice for the mass transfer correlation to use. We will first use the Chen and Chuang (1993) correlation and try others later. For the Heat transfer coefficient method, the Chilton and Colburn correlation is standard, and for the Interfacial area method, the Zuiderweg (1982) correlation is most commonly used.

7. In the Design tab, click on the box labeled "Design mode to calculate column diameter." This means a different diameter than you specified will be used. The "Base Stage" is usually selected as the stage that floods first—from the equilibrium runs, this is stage 4. "Base Flood" is 0.7, since we are designing at 70% of flooding.

8. Repeat steps 4 to 7, but for section 2. Return to and click on Tray Rating→Object Manager→New, and call it section 2. In the tabs, section 2 is from stages 16 to 34 and has a different diameter than section 1. Other values are the same as in section 1. Items 5 and 6 are the same as for section 1. In item 7 the Base Stage for section 2 is stage 34.

9. Click Next, run the simulation, and look at the Report (View→Report→check block). Check the distillate (y values for stage 1) and bottoms (x values for stage 35) mole fractions to be sure the specifications are met. Note that the report now contains a Tray Rating section, and that under the heading "Rating Results" different diameters have been calculated for each section. Look at the Downcomer (DC) backup for both sections. Both sections should be OK (DC backup/Tray spacing < 1/2). Next check the weir loading. It should be < 70 m^2/h = 0.01944 m^2/s (Torzewski, 2009). Both of the sections should be OK. Also look at the velocity in the downcomers. With a minimal foaming system it should be less than 0.21 m/s in both sections.

10. We will now look at the effect of changing the flow model. Go to Tray Rating, click on the + by section 1, and click on Rate-Based. Then in the Rate-Based tab under Calculation parameters in the menu for Flow Model, select VPLUG (called "Mixed-Plug" in the report). Repeat these steps for section 2. In the VPLUG model the vapor is assumed to be in plug flow on the tray while the liquid is mixed. The vapor concentration is calculated as an average of the plug flow concentrations. VPLUG can be used with either trays or packing. Click Next and run the simulation. Compare results ($y_{D,C4}$, $x_{Bot,C3}$, DC backup/Tray spacing, and weir loading) with the results from item 9. You should see very little difference in DC backup/Tray spacing and weir loading, and a better separation with the VPLUG model. Data on the trays (either direct observation data of flow patterns or comparison of results with models) can be used to determine which model is more appropriate. Otherwise, the Peclet number can be estimated from Eq. (16-111) and used to determine if a mixed or plug flow model is more appropriate. The conservative approach is to select the Mixed model.

11. We will now look at other mass transfer correlations. Use the Mixed model for both sections.
   a. In both sections select the Chan and Fair (1984) mass transfer correlation. Click on Next and run the simulation. Compare $y_{D,C4}$, $x_{Bot,C3}$, DC backup/Tray spacing and weir loading with the run from item 9.
   b. In both sections select the Zuiderweg (1982) mass transfer correlation. Click on Next and run the simulation. Compare $y_{D,C4}$, $x_{Bot,C3}$, DC backup/Tray spacing and weir loading with the runs from items 9 and 11a.
   The DC backup/Tray spacing and weir loading should be acceptable for all three runs. For a conservative design, select the worst separation (highest values of $y_{D,C4}$, $x_{Bot,C3}$). Check if this separation meets the specifications.

12. The column is a workable design but is not optimized. Choose the Mixed model and the mass transfer correlation with the worst separation (this will be most conservative and hence a safe design) and partially optimize. Find the optimum feed stage based on lowest values of $y_{D,C4}$, $x_{Bot,C3}$, and then find the minimum number of stages that will just give the desired separation or slightly better. Report your partially optimized design (N, N_F, $y_{D,C4}$,

$x_{Bot,C3}$, Q_c, Q_R, and for both sections the diameters, values of DC backup/Tray spacing, and weir loadings). NOTE: As the feed stage and total number of stages are changed, the starting and ending stages for the sections in Tray Rating have to be adjusted accordingly. When changing the total number of stages, it is usually necessary to reinitialize to obtain convergence.

CHAPTER 17

Introduction to Membrane Separation Processes

Membrane separation processes such as gas permeation, pervaporation, reverse osmosis (RO), and ultrafiltration (UF) are not operated as equilibrium-staged processes. Instead, these separations are based on the rate at which solutes transfer though a semipermeable membrane. The key to understanding these membrane processes is the rate of mass transfer not equilibrium. Yet, despite this difference we will see many similarities in the solution methods for different flow patterns with the solution methods developed for equilibrium-staged separations. Because the analyses of these processes are often analogous to the methods used for equilibrium processes, we can use our understanding of equilibrium processes to help understand membrane separators. These membrane processes are usually either complementary or competitive with distillation, absorption, and extraction.

This chapter presents an introduction to the four membrane separation methods most commonly used in industry: gas permeation, RO, UF, and pervaporation. At the level of this introduction the mathematical sophistication needed to understand the membrane processes is approximately the same as that needed for the equilibrium-staged processes. A background in mass transfer (Chapter 15) will be helpful but is not essential. Detailed descriptions of these membrane separation processes are found in Baker et al. (1990), Eykamp (1997), Geankoplis (2003), Kucera (2010). Noble and Stern (1985), Mohr et al. (1988), Mulder (1996), Osada and Nakagawa (1992), Hagg (1998), Ho and Sirkar (1992), and Wankat (1990).

Some knowledge of the membrane separations will prove to be very helpful even if the engineer will usually design equilibrium-based processes. In gas permeation components selectively transfer through the membrane. Gas permeation competes with cryogenic distillation as a method to produce nitrogen gas. Absorption and gas permeation are competitive methods for removing carbon dioxide from natural gas streams (Baker, 2002). In RO a tight membrane that rejects essentially all dissolved components is used. The water dissolves in the membrane and passes through under a pressure difference up to 6000 kPa (800 psi) (Li and Kulkarni, 1997). RO has in many cases displaced distillation as a method for desalinating seawater and is extensively used to make waste water potable (Reisch, 2007). UF membranes are fabricated to pass

Note: A nomenclature list for this chapter is included in the front matter of this book.

TABLE 17-1. *Properties of membrane separation systems (Drioli and Romano, 2001; Noble and Terry, 2004; Wankat, 1990)*

	Membrane Processes								
	Gas Permeation	*Vapor Permeation*	*Reverse Osmosis*	*Ultra-filtration*	*Micro-filtration*	*Per-vaporation*	*Liquid Membranes*	*Dialysis*	*Electro-Dialysis*
Separation Mechanism	Solution-Diffusion	Solution-Diffusion	Solution-Diffusion	Sieving	Sieving	Solubility & Volatility	Solution-Diffusion	Solution-Diffusion	Ion transfer
Driving force Eq (17-1)	Partial Press.	Partial Press.	$\Delta p - \Delta \pi$	Δp	Δp	Partial press. conc.	Conc.	Conc.	Electric Potential
Flow driven by	Δp 0.1-10 Mpa	Δp	Δp 1-10 Mpa	Δp 100-800 kPa	Δp 100-500 kPa	Δp & ΔC	ΔC	ΔC	Elec. Pot.
Feed/product	Gas/Gas	Vapor/vapor	Liq/Liq	Liq/Liq	Liq/Liq	Liq/vapor	Liq/Liq	Liq/Liq	Liq/Liq
Material removed	Molecules	Molecules	Ions Molecules	Macro-Molecules	Small particles	Molecules	Molecules	Molecules ions	Ions
Approx. size	1-10 Å	3-20 Å	3-20 Å	30-1100 Å	400-20,000 Å	3-20 Å	3-20 Å		3-10 Å
Equipment	Hollow fiber, spiral wound	Plate & frame	Hollow fiber spiral wound	Hollow fiber, spiral wound, plate & frame	Plate & frame, spiral wound	Plate & frame hollow fiber	Extraction equipment Supported	Plate & frame hollow fiber	Plate & frame
Application	N_2, H_2 Dehumidify CO_2 & H_2S	Dehydration	Desalination Water purific.	Food processing Water purific.	Food processing Water purific.	Azeotropes Organics	Chemical processing	Hemodialysis	Desalination Water Treatment
Membrane materials	Polysulfone CA	Composite polysulfone	CA composite polyamides	Polysulfone CA acrylonitrile	Polysulfone	Composite ceramics silicone rubber	Immiscible liquids with carrier	CA Cellulose polysulfone	Ion-Exchange Membrane

Key: CA = cellulose acetate

low molecular weight molecules and to retain high molecular weight molecules and particulates. The pressure difference, 70 to 1400 kPa (10–100 psi) is more modest than in RO. UF is a useful method for separating proteins and other large molecules that essentially have no vapor pressure and thus, cannot be distilled. UF competes with extraction as a separation method for biochemicals. In pervaporation the feed is a liquid while the permeate product is removed as a vapor. Pervaporation is used as a method to break azeotropes and is often coupled with distillation columns. Since the membrane separations are based on different physicochemical properties than the equilibrium-staged separations, the membrane methods can often perform separations such as separation of azeotropic mixtures or separation of nonvolatile components, which cannot be done by distillation or other equilibrium-based separations.

There are several other membrane processes that are in commercial use but are not covered in this chapter. In *dialysis,* small molecules in a liquid diffuse through a membrane because of a concentration driving force (Wankat, 1990). The major application is hemodialysis or the artifical kidney developed by Kolff and Beck in 1944 for treatment of people whose kidneys do not function properly (Lonsdale, 1982). *Electrodialysis* (ED) uses an electrical field to force cations through cation exchange membranes and anions through anion exchange membranes (Wankat, 1990; Lonsdale, 1982). The membranes are alternated in a stack, and every alternate region becomes concentrated or diluted. ED is used for desalination of brackish water and in the food industry. *Vapor permeation* is similar to gas permeation except vapors that are easily condensed are processed (Huang, 1991). This process has not met its potential partly because of difficulties with condensation of liquid. *Liquid membranes* use a layer of liquid instead of a solid polymer to achieve the separation (Wankat, 1990). Liquid membrane systems can be operated as countercurrent processes, and, to some extent, compete with extraction. *Microfiltration* is similar to UF but is used for particles between the sizes processed by UF and normal filtration (Noble and Terry, 2004). *Nanofiltration* removes particles between those removed by RO and UF and is essentially a loose RO membrane (Wankat, 1990). The design procedures developed for RO and UF can be applied to microfiltration and nanofiltration. The entire spectrum of membrane separations is summarized in Table 17-1. Nanofiltration is not listed separately but is the same as RO except that Δp is from 0.3 to 3 MPa and the approximate size retained is 8-50Å.

17.1 MEMBRANE SEPARATION EQUIPMENT

A membrane is a physical barrier between two fluids (feed side and product side) that selectively allows certain components of the feed fluid to pass. The fluid that passes through the membrane is called *permeate* and the fluid retained on the feed side is called *retentate.* The equipment needed for the separation is deceptively simple. It consists of the membrane plus the container to hold the two fluids. The simplest arrangement is to use a stirred tank that is separated into two volumes via a membrane. Stirred-tank systems are used in laboratories but not commonly in large-scale separations.

The common commercial geometries are shown in Figure 17-1 (Leeper et al., 1984). More extensive construction details are shown by Baker et al. (1990) and Eykamp (1997). The plate-and-frame system (Figure 17-1A) is similar to a parallel plate heat exchanger or a plate-and-frame filter press, except that the filter cloth is replaced by flat sheets of membranes. This design is used for food processing applications where rigorous cleaning by disassembly may be required, for electrodialysis (which involves passing a current through the membrane), and for membrane materials that are difficult to form into more complicated shapes.

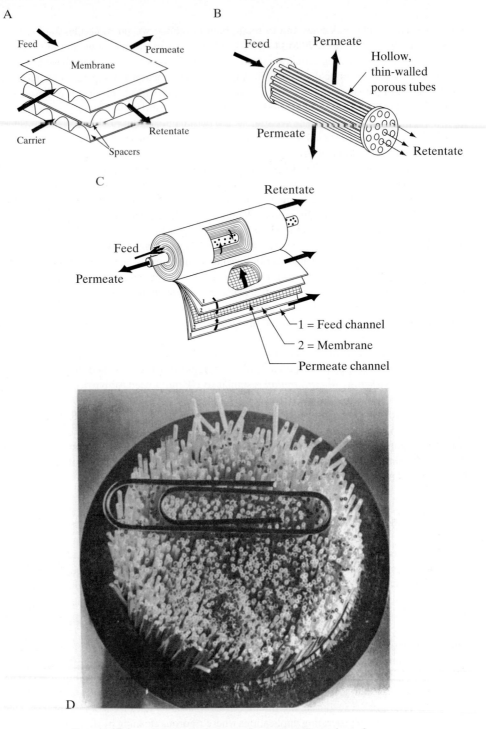

FIGURE 17-1. *Schematic diagrams of common industrial membrane
modules; (A) plate-and-frame, (B) tube-in-shell,
(C) spiral-wound, (D) details of hollow-fiber module.
Reprinted from Leeper et al. (1984), pp. 36–37*

The tube-in-shell system (Figure 17-1B) is occasionally used. This configuration is very similar to a shell-and-tube heat exchanger. The membrane would be coated on a porous support. The main advantage of these systems is that they can be cleaned by passing sponge balls through the separators. The surface area per unit volume is more than for plate-and-frame but less than for spiral-wound.

The spiral-wound configuration (Figure 17-1C) is more complicated but has a significantly higher surface area per unit volume. With proper design of the channels there will be significant turbulence at the membrane surface that promotes mass transfer. These systems have been used for carbon dioxide recovery, UF of relatively clean solutions and RO.

The hollow-fiber configuration (Figure 17-1D) looks schematically very similar to the tube-in-shell system, except the tubes are replaced by a very large number of hollow fibers made from the membrane. This configuration has the largest area-to-volume ratio. The hollow fibers can be optimized for a particular separation. For RO the inner diameter is about 42 microns while the outer diameter is about 85 microns. The separation is done by a 0.1 to 1.0 micron skin on the outer surface. The remainder of the membrane is a structural support. For gas permeation the requirement of a small pressure drop inside the tubes dictates a larger inner fiber diameter. For UF, where the feed can be dirty, the membranes are 500 to 1100 microns inside diameter. Typically the feed is inside the tubes, and the thin membrane skin is on the inside of the fibers. Care must be taken that particulates do not clog the fibers. Hollow-fiber membranes are technologically the most difficult to make, and the typical user buys complete modules from the manufacturer.

The most common systems are spiral-wound and hollow-fiber. Systems are usually purchased as off-the shelf modules in a limited number of sizes. Since a standard size is unlikely to provide the required separation and flow rate required for a given problem, a series of modules is cascaded to obtain the desired separation (see Figure 17-2). The parallel configuration (Figure 17-2A) allows one to increase the feed flow rate and is used with all of the membrane separations. Parallel operation is roughly analogous to increasing the diameter of a distillation or absorption column.

The retentate-in-series system (Figure 17-2B) will increase the purity of the more strongly retained components and simultaneously increase the recovery of the more permeable species. Unfortunately, the purity of permeate and the recovery of retained components both decrease. This configuration is used for production of less-permeable nitrogen from air. This process produces a product nitrogen gas at relatively high pressure since the nitrogen has not permeated through the membrane. The system is roughly analogous to a cross-flow extractor or stripper. Parallel and retentate-in-series systems are often combined (Figure 17-2C).

The permeate-in-series system (Figure 17-2D) is used when the permeate product is not of high enough purity. One or more additional stages of separation are required (Baker, 2002). Designers try to avoid this configuration if possible. The major cost of operating most membrane separators is the pressure difference required to force permeate through the membrane. This configuration will require an additional compressor or pump to repressurize permeate for the next membrane separator. One of the major disadvantages of the membrane separators is that additional stages for permeate do *not* reuse the energy (pressure) separating agent. The equilibrium-staged separations have the advantage that the energy (heat and cooling) separating agents can easily be reused. Membrane separators are often used because a membrane can be developed which produces permeate of desired purity in one stage.

A final common membrane cascade is the retentate-recycle or feed-and-bleed mode (Figure 17-2E). The recycle allows for a very high flow rate in the membrane module to increase

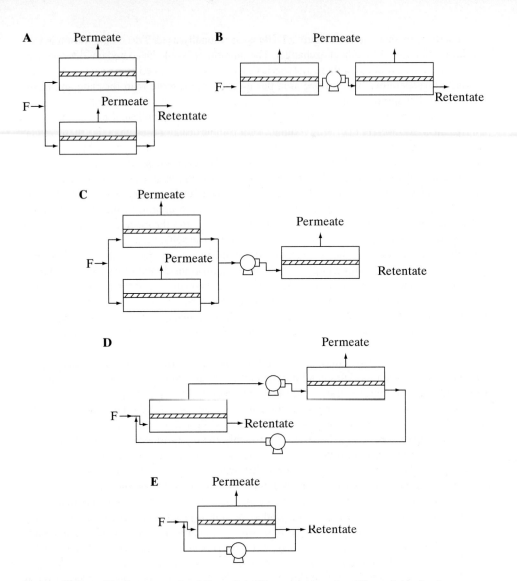

FIGURE 17-2. *Membrane cascades; (A) parallel, (B) retentate-in-series, (C) parallel and series, (D) permeate-in-series, (E) retentate recycle or feed-and-bleed*

mass transfer rates and minimize fouling. The same high flow rates can be obtained without the recycle but a very long membrane system would be required to have a sufficiently long residence time. Recycle is used extensively with continous and batch UF systems. Note that the recycle in Figure 17-2E is similar to the recycle in Figure 17-2D since the high-pressure retentate is recycled in both cases. Thus, the compressor or pump only needs to boost the pressure (e.g., because of friction losses for flow inside a hollow-fiber membrane), not over-come the large pressure difference that results when fluid permeates through the membrane.

The flow patterns on the retentate (feed) and permeate sides of the membrane have major effects on the mass balances and the separation. These flow patterns can be perfectly

Chapter 17   Introduction to Membrane Separation Processes

mixed on both sides, plug flow on one side and perfectly mixed on the other side, plug flow in the same direction on both sides (co-current), plug flow in opposite directions on the two sides (countercurrent), mixed on one side and cross-flow on the other, and somewhere in-between these ideal regimes. For example, if we consider the hollow-fiber module shown in Figure 17-1D, the flow inside the hollow fibers is very close to plug flow. Depending on the shell side design, the flow on the shell side could be approximately well-mixed, co-current plug flow or countercurrent plug flow. Countercurrent plug flow usually gives the most separation.

In most of this chapter we will make the assumption that both sides of the membrane are perfectly mixed. This assumption greatly simplifies the mathematics. The resulting design will be conservative in that the actual apparatus will result in the same or better separation than predicted. The effect of other flow patterns will be explored in section 17.7.

17.2 MEMBRANE CONCEPTS

Clearly the key to the membrane separators is the membrane. The membrane needs to have a high permeability for permeate and a low permeability for retentate. It helps if the membrane has high temperature and chemical resistances, is mechanically strong, resists fouling, can be formed into the desired module shapes, and is relatively inexpensive. Commercial membranes are made from polymers, ceramics, and metals (e.g., palladium for helium purification), but polymer membranes are by far the most common. Membrane development is done by a few large chemical companies and by several specialty firms. Most chemical engineers will use membrane separators, but they will never be involved in membrane development. However, some understanding of the polymer membrane will be useful when specifying and operating membrane separators.

A large number of polymers have been used to make membranes for membrane separators. Cellulose, ethyl cellulose, and cellulose acetate [actually a polymer blend of cellulose, cellulose acetate, and cellulose triacetate (Kesting and Fritzsche, 1993)] were the first commercially successful membranes and are still used. These materials have relatively poor chemical resistance, but their low cost makes them attractive when they can be used. The most common commercial membrane is polysulfone, which has excellent chemical and thermal resistance. Various polyamide polymers are also commonly used in RO. Silicone rubber, which has very high fluxes but low selectivity, is often used in pervaporation to preferentially permeate organics and as a coating on composite membranes for gas permeation. A large variety of grafted polymers, specialty polymers, and composite membranes has been developed to optimize flux and selectivity particularly for pervaporation (Huang, 1991) but also other membrane separations. Kesting and Fritzsche (1993) is an excellent source for information on the relationship between polymer structure and membrane properties for the polymers used for gas separation. The repeating units for several polymers used to form membranes are shown in Figure 17-3.

The basic equation for flux of permeate through the membrane is

$$\text{Flux} = \frac{\text{Transfer Rate}}{\text{Transfer Area}} = \frac{\text{Permeability}}{\text{Separation Thickness}} \text{(Driving Force)} \qquad \textbf{(17-1a)}$$

Although this equation is derived from experiments and is *not* a fundamental law, it is so basic and powerful that you need to memorize it and explore its implications in as many ways as

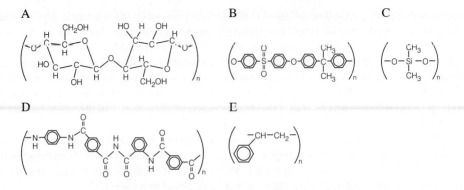

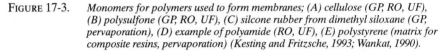

FIGURE 17-3. *Monomers for polymers used to form membranes; (A) cellulose (GP, RO, UF), (B) polysulfone (GP, RO, UF), (C) silcone rubber from dimethyl siloxane (GP, pervaporation), (D) example of polyamide (RO, UF), (E) polystyrene (matrix for composite resins, pervaporation) (Kesting and Fritzsche, 1993; Wankat, 1990).*

possible. The exact form of the equation depends upon the type of flux and the type of separation.

The flux can be written as a volumetric flux,

$$J = (Volume/time)/(Area) \tag{17-1b}$$

or as a mass flux,

$$J' = (Mass/time)/(Area) \tag{17-1c}$$

or as a molar flux,

$$\hat{J} = (Mol/time)/(Area) \tag{17-1d}$$

These fluxes are obviously related to each other,

$$J\rho = J', \hat{J} = \hat{J}\hat{\rho} \tag{17-1e, f}$$

where $\hat{\rho}$ and ρ are the molar and mass densities, respectively. Once the flux is known, it is used in conjunction with the feed rate and the feed concentration to determine the membrane area required.

Equation (17-1a) can be used to determine the flux once the appropriate terms on the right hand side of the equation are known. The *permeability* P is a transport coefficient that can be determined directly from experiment. In some cases the permeability can be estimated from more fundamental variables such as solubility and diffusivity. This is briefly discussed in the next section. The *membrane separation thickness* t_{ms} is the thickness of the portion of the membrane that is actually doing the separation. This is often a thin skin with a thickness less than 1 micron and, in polymer hollow-fiber gas permentation membranes, can be as low as 0.1 micron. Since t_{ms} can be difficult to measure, the ratio (P/t_{ms}), the *permeance*, is often reported from experimental data. The units of P and (P/t_{ms}) depend upon the flux and driving force employed.

The *driving force* depends upon the type of membrane separation. For gas permeation the driving force is the difference in partial pressure of the transferring species across the membrane. For RO the driving force is the pressure difference minus the osmotic pressure difference across the membrane. For UF the driving force is the same as for RO, but this usually simplifies to the pressure difference across the membrane since osmotic pressures are small. These driving force effects are specific to each membrane separation and are discussed in more detail in the later sections.

Since flux is flow rate per membrane area, increasing the flux (while maintaining the desired separation) will decrease the membrane area. The result will be a more compact, less expensive device. Equation (17-1a) shows that flux can be increased by increasing permeability or driving force, or by decreasing the separation thickness. Permeability of the membrane depends upon interactions between the molecular structure of the polymers and the solutes. These effects will be briefly discussed in the sections on each membrane separation. More details on molecular structure-permeability effects are in the books by Kesting and Fritzsche (1993), Ho and Sirkar (1992), Mulder (1996), Noble and Stern (1995), and Osada and Nakagawa (1992).

For many years attempts were made to make very thin membranes, but there was always difficulty in making the membranes mechanically strong enough. Sidney Loeb and Srinivasa Sourirajan (Loeb and Sourirajan, 1960, 1963) made the breakthrough discovery. They found that anisotropic (asymmetric) membranes with a very thin skin (0.1 to 1.0 μm) on a much thicker porous support had the necessary mechanical strength while having a very small separation thickness. Schematics of isotropic and anisotropic membranes are shown in Figure 17-4. Another method for producing an asymmetric membrane is to cast a thin layer of one polymer onto a porous supporting layer of another polymer to form a composite membrane. Although not all membrane separators use asymmetric membranes, they are the most common.

17.3 GAS PERMEATION

We will consider gas permeation first, since, in addition to being commercially important, in many ways these systems are closest to being ideal. Once the ideal operation is understood, we can look at deviations from ideality for other membrane separations. Gas permeation membrane systems are used commercially for separation of permanent gases. Some common applications are purification of helium, hydrogen, and carbon dioxide, production of high purity nitrogen from air, and production of air enriched in oxygen (low purity oxygen).

Gas permeation systems typically use hollow-fiber or spiral-wound membranes, although hollow-fiber systems are more common (Baker, 2004). Cellulose acetate membranes are used for carbon dioxide recovery, polysulfone coated with silicone rubber is used for hydrogen purification, and composite membranes are used for air separation. The feed gas is forced into the membrane module under pressure. Retentate, which does not go through the membrane, will become concentrated in the less permeable gas. Retentate exits at a pressure that will be close to the input pressure. The more permeable species will be concentrated in permeate. Permeate, which has passed through the membrane, exits at low pressure. The operating cost for a gas permeator is the cost of compression of the feed gas and the irreversible pressure difference that occurs for the gas that permeates the membrane. A typical hollow-fiber unit will contain 5000 m^2 membrane area per m^3 at a cost of approximately \$200/m^2.

For gas permeation the steady-state general flux Eq. (17-1a) for gas A becomes

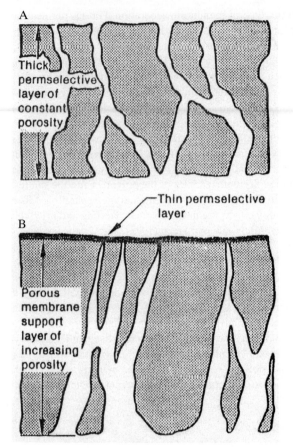

A

Thick
permselective
layer of
constant
porosity

Thin permselective
layer

B

Porous
membrane
support
layer of
increasing
porosity

FIGURE 17-4. *Schematics of (A) isotropic and (B) anisotropic membranes, reprinted from Leeper et al. (1984)*

$$J_A = \frac{\hat{F}_t y_{t,A}}{\hat{\rho}_A A} = P_A \frac{\Delta p_A}{t_{ms}} \tag{17-2a}$$

where J_A is the volumetric flux of gas A, $\hat{F}_t$ is the molar transfer rate of gas which permeates through the membrane, $y_{t,A}$ is the mole frac A in the gas that transfers (or permeates) through the membrane, $\hat{\rho}_A$ is the molar density of the solute in the gas transferring through the membrane, A is the membrane area available for mass transfer, P_A is the permeability of species A in the membrane, the driving force for the separation Δp_A is the difference in the partial pressure of A across the portion of the membrane which does the separation, and t_{ms} is the thickness of the membrane skin which actually does the separation. The ratio P_A/t_{ms} is known as the *permeance*, and is often the variable that is measured experimentally. Since partial pressure = (mole frac)(total pressure) or $p_A = y_A p$, Eq. (17-2a) can be expanded to

$$J_A = \frac{\hat{F}_t y_{t,A}}{\hat{\rho}_A A} = \frac{P_A(p_r y_{r,wA} - p_p y_{p,w,A})}{t_{ms}} \tag{17-2b, c}$$

where p_r is the total pressure on the retentate (high pressure) side, $y_{r,w,A}$ is the mole frac A on the retentate side at the membrane wall, and p_p and $y_{p,w,A}$ are the pressure and mole frac at the wall on the permeate (low pressure) side. The flux equation for other components will look exactly the same as for component A except the subscript A will be changed to B, C, ... as appropriate. The relationship between the mole frac transferring through the membrane $y_{t,A}$ and the permeate mole frac at the wall $y_{p,w,A}$ depends upon the flow patterns in the permeate.

17.3.1 Gas Permeation of Binary Mixtures

Commercial separation of binary mixtures by gas permeation is common. In binary mixtures we can use the substitution

$$y = y_A = 1 - y_B \tag{17-3a}$$

where y without the subscript A or B is understood to refer to the mole frac of the faster permeating species A. In a binary system, the flux of the slower moving component B becomes

$$J_B = \frac{P_B}{t_{ms}} [p_r(1 - y_{r,w}) - p_p(1 - y_p)] \tag{17-3b}$$

Since pressure and mole fracs can vary along the membrane, Eqs. (17-2a), (17-2c), and (17-3b) are applied point by point along the membrane. In almost all systems there is no concentration gradient on the permeate side perpendicular to the membrane; thus, we replaced $y_{p,w,A}$ with $y_{p,A}$, which can vary along the membrane. In addition, if mass transfer on the retentate side is not rapid, the mole fracs and partial pressures at the surface of the membrane will not be the same as in the bulk gas or in the feed gas (see concentration polarization in Section 17.4). Usually mass transfer rates in gas systems are high enough that the mole frac at the membrane surface is almost equal to the mole frac in the bulk gas.

Most gas permeation membranes work by a solution-diffusion mechanism. On the high-pressure side of the membrane the gas first dissolves into the membrane. It then diffuses through the thin skin of the membrane to the low-pressure side where permeate reenters the gas phase. For this type of membrane the permeability is the product of the gas solubility H_A in the membrane times the diffusivity D_m in the membrane

$$P_A = H_A D_{m,A} \tag{17-4a}$$

The solubility parameter H_A is very similar to a Henry's law coefficient. Representative permeabilities for several polymer membranes are shown in Figure 17-5 and in Table 17-2. Small molecules such as helium have high diffusivities but low solubilities while large gas molecules such as carbon dioxide have high solubilities but low diffusivities. The resulting product, the permeability P is relatively large for both small and large molecules, but has a minimum for molecules around the size of nitrogen.

If the driving forces are equal, gases with higher permeabilities transfer through the membrane at higher rates. It is useful to define a selectivity α_{AB}

$$\alpha_{AB} = P_A/P_B \tag{17-4b}$$

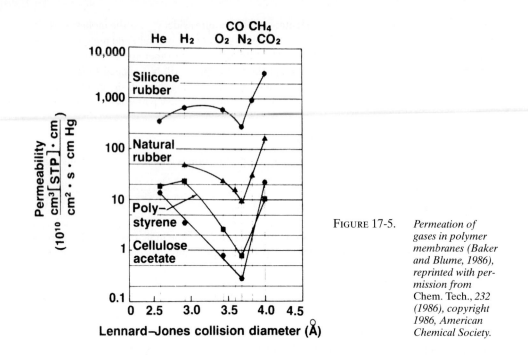

FIGURE 17-5. *Permeation of gases in polymer membranes (Baker and Blume, 1986), reprinted with permission from Chem. Tech., 232 (1986), copyright 1986, American Chemical Society.*

as a short-hand method of comparing the ease of separation of gases. The selectivity of two solutes can be estimated from Figure 17-5 and Table 17-2. For example, from Figure 17-5, $\alpha_{CO_2-N_2} = P_{CO_2}/P_{N_2} = 24/0.3 = 80$. The selectivity is a function of concentration, pressure and temperature since the individual permeabilities depend on concentration, pressure, and temperature. However, α_{AB} tends to be less dependent on these factors than the individual permeabilities. Note that α_{AB} is analogous to the relative volatility defined in Chapter 2. Since we prefer to operate gas permeation as a one-stage system, α_{AB} values of 20 and higher are preferred instead of the much more modest values commonly employed in distillation.

In the early 1990s Robeson (1993) found an upper limit to the performance of polymer membranes in the commercially important separation of oxygen and nitrogen from air. On a log-log plot of selectivity versus oxygen permeability (a Robeson plot), the upper bound plots as a straight line (see Problem 17.D17 for more details). Although theoretical reasons for this limit have not been found, very few new membranes have been developed that are able to perform better than Robeson's limit. Membrane research has focused on ways to do better than Robeson's upper limit.

17.3.2 Binary Permeation in Perfectly Mixed Systems

The mass balances for a gas permeation membrane module that is completely mixed on both sides of the membrane will have the simplest form since the mole fracs and pressures on each side of the membrane are constant. Equations (17-2) and (17-3b) can then be used in algebraic mass balances with $y_p = y_t$ and integration is not required to find the average retentate and permeate mole fracs.

We can obtain a *rate transfer* (RT) equation for a perfectly mixed permeator by writing the equation for transfer through the membrane, Eq. (17-2) for both the more permeable and

TABLE 17-2. *Permeabilities of gases in various membranes; $cm^3(STP)cm/[cm^2 \cdot s \cdot cm\ Hg] \times 10^{10}$; Reference Code: N = Nakagawa, 1992; DR = Drioli and Romano, 2001; G = Geankoplis, 2003*

Membrane	Temp. (°C)	He	H_2	CO_2	O_2	CH_4	N_2	Ref
Poly[1-(trimethylsilyl)-1-propyne]				28,000	7730		4970	DR
Poly(dimethylsiloxane)	25	230		3240	605		300	N
Poly(dimethylsiloxane)	35			4550	781		351	DR
Poly(4-methyl pentene)	25	100		93	32		8.0	N
Natural rubber	25	23.7	90.8	99.6	17.7		6.12	N
Natural rubber	25			153.0			9.43	DR
Natural rubber	25	31	49	131	24	30	8.1	G
Silicone rubber	25	300	550	2700	500	800	250	G
Ethyl cellulose	25	53.4		113	15		3.0	N
Ethyl cellulose	35			5.0	12.4		3.4	DR
Ethyl cellulose	30	35.7	49.2	47.5	11.2	7.47	3.29	G
Poly(2.6-dimethylphenylene oxide)	25			75	15		4.43	N
Polystyrene	20	16.7		10.0	2.01		0.32	N
Polystyrene	35			12.4	2.9		0.52	DR
Polystyrene	30	40.8	56.0	23.3	7.47	2.72	2.55	G
Polycarbonate	25	19		8.0	1.4		0.30	N
Butyl rubber	25	8.42		5.2	1.30		0.33	N
Butyl rubber	35			5.18			0.324	DR
Acetyl cellulose	22	13.6			0.43		0.14	N
Cellulose acetate	35			4.75	0.82		0.15	DR
Nylon 6	30	0.53		0.16	0.038			N
Nylon 66	25	1.0		0.17	0.034		0.008	G

the less permeable species. The perfectly mixed assumption makes $y_p = y_t$. We will also assume that the module has a high rate of mass transfer from the bulk fluid to the membrane surface. This assumption makes $y_r = y_{r,w}$ (the mole frac of A at the membrane wall). These two consequences of the perfectly mixed assumption greatly simplify the analysis. The second assumption is reasonable for many gas permeators since diffusivities and mass transfer rates are high. From Eq. (17-2c), the transfer equation for the more permeable component A is

$$\hat{F}_p y_p = \frac{P_A \hat{A} \hat{\rho}_A}{t_{ms}} (p_r y_r - p_p y_p) \qquad \text{(17-5a)}$$

while from Eq. (17-3b) for the less permeable component B in a binary separation,

$$\hat{F}_p(1-y_p) = \frac{P_B A \hat{\rho}_B}{t_{ms}} [p_r(1-y_r) - p_p(1-y_p)]$$

(17-5b)

Because permeate is typically at relatively low pressures where gas molecules are far apart, we will assume that transfer of A is independent of transfer of B and vice-versa. Thus, P_A and P_B are constant. We now solve these two equations for $\hat{F}_p$, and set them equal to each other.

$$\frac{P_A A \hat{\rho}_A}{y_p t_{ms}}(p_r y_r - p_p y_p) = \frac{P_B A \hat{\rho}_B}{(1-y_p) t_{ms}} [p_r(1-y_r) - p_p(1-y_p)]$$

(17-5c)

This equation can be solved for either y_r or for y_p. Since solving for the former is easier, we will solve for y_r. After some algebra, the result is

$$y_r = \frac{y_p\left[\left(\dfrac{\hat{\rho}_A \alpha_{AB}}{\hat{\rho}_B} - 1\right)\left(\dfrac{p_p}{p_r}\right)(1-y_p) + 1\right]}{\alpha_{AB}\dfrac{\hat{\rho}_A}{\hat{\rho}_B} - \left(\alpha_{AB}\dfrac{\hat{\rho}_A}{\hat{\rho}_B} - 1\right)y_p}$$

(17-6a)

Equation (17-6a), which we will call the rate transfer or *RT curve*, relates the mole fracs on both sides of the membrane based on the RT parameters and the driving force. In deriving Eq. (17-6a) we invoked the assumption that the membrane separator is perfectly mixed when we replaced y_t with y_p. However, this equation is applicable to other flow configurations if written in terms of y_t and applied point-by-point on the membrane. Equation (17-6a) is not an equilibrium expression since it was derived based on transfer rates, but as we will see, it can take the place of an equilibrium expression for binary systems.

The average outlet concentration of the permeate can be found by dividing the permeation rate of component A integrated over the entire membrane by the total permeation rate. For a perfectly mixed separator values are constant, and $y_{r,out} = y_{r,average} = y_r$; $y_{p,out} = y_{p,average} = y_p$ and $F_r = F_{out}$.

We can now do external balances for the completely mixed system shown in Figure 17-6. Note that this system is the rate equivalent of flash distillation (Hoffman, 2003). A single feed goes into a well-mixed chamber and two products are withdrawn. The more permeable

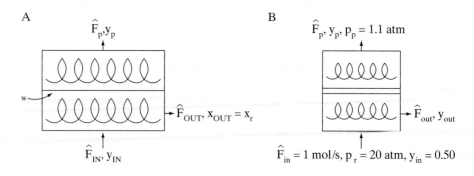

FIGURE 17-6.   *Completely mixed membrane module; (A) general case, (B) sketch for Example 17-1*

species concentrates in the permeate product which is analogous to the more volatile component concentrating in the vapor. However, the two products are not in equilibrium but are related by the RT expression. Because of the geometric similarity, we can use an analysis procedure analogous to that used for binary flash distillation.

The overall mole balance is

$$\hat{F}_{in} = \hat{F}_p + \hat{F}_r \tag{17-7a}$$

and the mole balance on the more permeable component is

$$\hat{F}_{in}y_{in} = \hat{F}_p y_p + \hat{F}_r y_r \tag{17-7b}$$

Equations (17-7) can be combined and solved for either y_{out} or for y_p. Since solving for y_p keeps the analogy with flash distillation intact, we will solve for y_p. In molar units,

$$y_p = -\frac{(1 - \hat{F}_p/\hat{F}_{in})}{(\hat{F}_p/\hat{F}_{in})}\, y_r + \frac{y_{in}}{(\hat{F}_p/\hat{F}_{in})} \tag{17-8}$$

Equation (17-8) is the operating equation for the well-mixed gas permeator. The ratio $\theta = (\hat{F}_p/\hat{F}_{in})$, known as the "cut," is an important operating parameter. The RT Eq. (17-6a) and the operating Eq. (17-8) are the two equations needed to solve problems with perfectly mixed, binary gas permeators. This is illustrated in Example 17-1 when y_p is known and in Example 17-2 when θ is specified.

Gas permeators often operate at temperatures and pressures where the gases are very close to ideal. Ideal gas behavior simplifies the calculations. At constant pressure and temperature the molar densities of all ideal gases are equal, $\hat{\rho}_A = \hat{\rho}_B$ and the densities cancel in Eq. (17-6a). In addition the volumetric flow rate units of the permeabilities in Table 17-2 are cm^3 (STP)/s. For ideal gases this is equivalent to a molar flow rate since 1.0 mol occupies 22400 cm^3 at Standard Temperature and Pressure (STP). These simplifications are illustrated in Example 17-1.

> ### EXAMPLE 17-1. Well-mixed gas permeation—sequential, analytical solution
> A perfectly mixed gas permeation module is separating carbon dioxide from nitrogen using a poly (2,6 – dimethylphenylene oxide) membrane. The feed is 20.0 mol% carbon dioxide and is at 25 °C. The module has 50.0 m^2 of membrane. The module is operated with a retentate pressure of 5.5 atm and a permeate pressure of 1.01 atm. We desire a permeate that is 40.0 mol% carbon dioxide. The membrane thickness is $t_{ms} = 1.0 \times 10^{-4}$ cm. Find $y_{r,CO2} = y_{out,CO2}$, J_{CO2}, $\hat{F}_P$ (in mol/min), cut θ and $\hat{F}_{in}$ (in mol/min).
>
> ### Solution
> A. Define. This is basically a simulation problem. We want to find how much gas can be processed by a well-mixed module with a known membrane area at specified operating conditions.
>
> B and C. Explore and plan. The permeabilities of carbon dioxide and nitrogen for this membrane are listed in Table 17-2. The selectivity, α, is the ratio of these permeabilities. The value of $y_{r,CO2}$ can be found from the RT Eq. (17-6a). The CO_2 flux and F_p can be calculated from Eq. (17-2). The cut and F_{in} can then be determined from mass balances, Eqs. (17-7).

D. Do it. From Table 17-2,

$P_{CO2} - 75$ and $P_{N2} - 4.43$ both cm^3 (STP)cm/[cm^2s cmHg] $\times 10^{-10}$.
Then $\alpha_{CO2-N2} = 75 \times 10^{-10}/4.43 \times 10^{-10} = 16.9$.
The pressure ratio $p_r/p_p = 5.5/1.01 = 5.446$. For ideal gases, which is a
reasonable assumption, $\hat{\rho}_{CO2}/\hat{\rho}_{N2} = 1.0$ and $\hat{\rho}_{CO2} = 1$ mol/22.4L(STP).

For an ideal gas the RT Eq. (17-6a) simplifies to

$$y_r = \frac{y_p[(\alpha - 1)(p_p/p_r)(1 - y_p) + 1]}{\alpha - (\alpha - 1)y_p}$$ (17-6b)

Substituting in values for the parameters with $y_p = 0.4$, we obtain

$$y_{r,CO2} = \frac{0.40[(15.9)(1/5.446)(1 - 0.4) + 1]}{16.9 - (15.9)(0.4)} = 0.1044$$

The CO_2 flux can be determined from Eq. (17-2) for a perfectly mixed module.

$$J_A = \frac{F_p y_p}{\hat{\rho}_A A} = \frac{P_A(p_r y_{r,A} - p_p y_{p,A})}{t_{ms}},$$

$$J_{CO2} = \frac{75 \times 10^{-10}}{(1.0 \times 10^{-4} cm)}\left(\frac{cm^3 (STP)cm}{cm^3 \text{ s cmHg}}\right)\left[(5.5atm)\left(\frac{76.0cmHg}{atm}\right)(0.1044) - (1.01atm)(\frac{76.0cmHg}{atm})0.4\right]$$

$$= 9.7014 \times 10^{-4} cm^3 (STP)/(cm^2 s)$$

Then $\hat{F}_p = J_{CO2}\hat{\rho}_{CO2} A/y_{p,CO2}$

$$= (9.7014 \times 10^{-4} \frac{cm^3 STP}{cm^2 s})(\frac{1 mol}{22.4L \text{ STP}})(\frac{1 L}{1000 cm^3})(\frac{60s}{min})(50m^2)(\frac{10,000cm^2}{m^2})/0.40 = 3.248 \text{ mol/min}$$

Note that $\hat{\rho}_{CO2}$ is the molar density of an ideal gas at STP.

The mass balances can now be solved several different ways. For example, by
solving Eqs. (17-7a) and (17-7b), simultaneously we obtain,

$$cut = \theta = \hat{F}_p/\hat{F}_{in} = \left(\frac{y_{in} - y_{out}}{y_p - y_{out}}\right)$$ (17-7c)

Then $\theta = (0.20 - 0.1044)/(0.40 - 0.1044) = 0.3234$, and
$\hat{F}_{in} = \hat{F}_p/\theta = 3.248/0.3234 = 10.044$ mol/min.

E. Check. Recalculating all the numbers gave the same result.
F. Generalize. Notes: 1. Because one of the outlet mole fracs was known, we
could calculate the other from the RT equation; thus, simultaneous solution
of Eqs. (17-6b) and (17-8) was not required. If neither outlet mole frac is
known, then simultaneous solution is required (see Example 17-2).

2. Units are obviously very important and should be carried throughout the solution.
3. Using the conversion 1 mol = 22.4 L (STP), is equivalent to using the ideal gas law.

A graphical solution to perfectly mixed gas permeators is also straightforward. On a graph of y_p vs. y_{out}, Eq. (17-8) plots as a straight line with a slope of $-(1-\theta)/\theta$. Note that the cut θ is analogous to $f = V/F$ in a flash distillation system. The intersection with the $y_p = y_{out}$ line occurs at $y_p = y_{out} = y_{in}$ which is analogous to $y = x = z$ in a flash system. The intersections with the axes can also be determined (see Example 17-2 and Figure 17-7). The simultaneous solution is obtained at the point of intersection of the straight line representing Eq. (17-8) and the curve representing Eq. (17-6a). This method is illustrated in Figure 17-7 for a number of values of θ. Figure 17-7 is plotted for the separation of carbon dioxide and methane in a cellulose acetate membrane (see Example 17-2).

EXAMPLE 17-2. Well-mixed gas permeation—simultaneous analytical and graphical solutions

Chern et al. (1985) measured the permeability of carbon dioxide and methane as $P_{CO2} = 15.0 \times 10^{-10}$ and $P_{CH4} = 0.48 \times 10^{-10}$ [cc(STP) cm]/[cm^2 s cm Hg] in a cellulose acetate membrane at 35° C and $p_r = 20$ atm. We wish to separate a feed gas that is 50 mol% carbon dioxide and 50.0 mol% methane at 35°C. The high pressure is 20.0 atm and the permeate pressure is $p_p = 1.1$ atm. The completely mixed membrane module shown in Figure 17-6 will be used.

a. Determine the values of y_p and y_{out} for cut values of 0, 0.25, 0.5, 0.75 and 1.0 using both analytical ($\theta = 0.25$ only) and graphical solutions.

b. If the effective membrane thickness is $t_{ms} = 1.0 \times 10^{-6}$ m, determine the fluxes for $\theta = 0.25$.

c. For $\theta = 0.25$ determine the membrane area needed for a feed gas flow rate of $\hat{F}_{in} = 1$ mol/s.

Solution

A. Define. The module is sketched in Figure 17-6B. We need to solve the equations analytically and plot a graph that allows us to find y_p and y_{out} for different values of θ, and when $\theta = 0.25$ find the fluxes of carbon dioxide and methane and the membrane area.

B and C. Explore and plan. The selectivity $\alpha_{AB} = P_A/P_B$. We will assume the gases are ideal.

Analytical solution: We need to solve Eqs. (17-6a) and (17-8a) simultaneously.

Graphical solution: If we pick values of y_p, Eq. (17-8b) can be used to calculate the corresponding values of y_r for the RT curve. For a well-mixed permeator this RT curve and the operating line, Eq. (17-8), can then be plotted on a y_p vs. y_r graph for different values of θ. Since the system is well mixed, the points of intersection give the solutions for y_p and $y_r = y_{out}$. The fluxes can then be calculated from Eq. (17-4).

D. **Do it.** Preliminary calculations: For an ideal gas the molar densities are equal, and their ratio equals 1.0. The selectivity is

$$\alpha_{CO2-CH4} = P_{CO2}/P_{CH4} = (15.0 \times 10^{-10})/(0.48 \times 10^{-10}) = 31.25$$

$p_p/p_r = 1.1/20 = 0.055$ (Note that any set of consistent units can be used for this ratio.)

Analytical Solution: Solving Eq. (17-7a) and (17-7b) for y_r,

$$y_r = -\frac{\theta}{1-\theta} y_p + \frac{y_{in}}{1-\theta} \tag{17-9}$$

After substituting Eq. (17-9) in (17-6b) (in Example 17-1) and doing considerable algebra, we obtain the quadratic equation,

$$ay_p^2 + by_p + c = 0 \tag{17-10a}$$

where

$$a = \left[\frac{\theta}{1-\theta} + \frac{P_p}{P_r}\right](\alpha - 1) \tag{17-10b}$$

$$b = (1 - \alpha)\left[\frac{P_p}{P_r} + \frac{\theta}{1-\theta} + \frac{y_{in}}{1-\theta}\right] - \frac{1}{1-\theta} \tag{17-10c}$$

$$c = \alpha y_{in}/(1 - \theta) \tag{17-10d}$$

This quadratic equation can be solved numerically or analytically. An analytical solution is obtained if we use the quadratic formula

$$y_p = \frac{-b \pm \sqrt{b^2 - 4ac}}{2a} \tag{17-10e}$$

Substituting in values $\theta = 0.25$, $\alpha = 31.25$, $y_{in} = 0.5$, and $p_p/p_r = 0.055$, we obtain $a = 11.747$, $b = -33.247$, $c = 20.833$, and $y_p = 0.9365$ or 1.894. The second value for y_p can be eliminated since the mole frac cannot be greater than 1.0. The value of y_r can be found from Eq. (17-9), $y_r = 0.3545$. Solutions obtained with a spreadsheet using Solver will be identical.

Graphical Solution: The RT curve, Eq. (17-6b), that relates y_r and y_p becomes

$$y_r = \{[(31.25 - 1)(0.055)(1 - y_p) + 1] y_p\}/[31.25 - 30.25 y_p]$$

The following table for the RT curve is generated by selecting arbitrary values of y_p.

y_p	y_r
0	0
0.05	0.004339
0.1	0.008848
0.2	0.0185
0.3	0.029285
0.4	0.04185
0.5	0.05680
0.6	0.07628
0.7	0.1042
0.8	0.1512
0.9	0.2608
0.95	0.4096
1.0	1.0000

These values were plotted in Figure 17-7 for the RT curve.

a. For operating Eq. (17-8) the intersection with $y_p = y_r$ line occurs at $y_r = y_{in}$ = 0.5 and the slope = $-[1 - \theta]/\theta$. For example, when $\theta = 0.25$, the slope = $-0.75/0.25 = -3$.

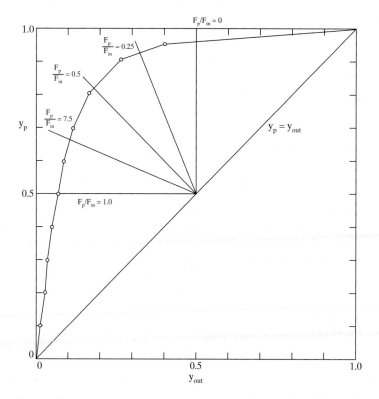

FIGURE 17-7.   *Solution for well-mixed gas permeation system for Example 17-2*

The solutions can be obtained at the intersections of the operating lines and the RT curve. For $\theta = 0.25$, Figure 17-7 shows that $y_p = 0.93$ and $y_r = y_{out} = 0.35$.

b. When $\theta = 0.25$, the flux can be determined from Eq. (17-4), using the y_p and y_r values from part a,

$$J_{CO2} = \frac{15 \times 10^{-10} \left[\dfrac{\text{cc STP cm}}{\text{cm}^2 \text{ s cm Hg}} \right]}{1\mu m (10^{-4} \text{ cm}/1\mu m)} \frac{76 \text{ cm Hg}}{1 \text{ atm}} [(20)(0.35) - (1.1)(0.93)]\text{atm}$$

$$J_{CO2} = 0.00681 \frac{\text{cc STP}}{\text{cm}^2 \text{ s}}$$

For $CH_4 : y_{p,CH4} = 1 - y_p = 0.07$, $y_{out,CH4} = 1. - 0.35 = 0.65$

$$J_{CH4} = \frac{0.48 \times 10^{-10}}{1(10^{-4})} \frac{76}{1} [20(0.65) - (1.1)(0.07)] = 0.000474 \frac{\text{cc STP}}{\text{cm}^2 \text{ s}}$$

Note that care must be taken with the units.

c. $\theta = 0.25$ and $\hat{F}_{in} = 1$ mol/s, $\hat{F}_p = 0.25$ mol/s. Since $\hat{F}_p = \hat{J}A$ or $A = \hat{F}_p / \hat{J}$, we can find the area once the molar flux $\hat{J}$ is known. The volumetric flux is, $J = J_{CO2} + J_{CH4} = 0.007284 \frac{\text{cc STP}}{\text{cm}^2 \text{ s}}$

Taking care to properly calculate units, the membrane area is,

$$A = \frac{(0.25 \text{ mol /s})}{0.007284 \left(\dfrac{\text{cc STP}}{\text{cm}^2 \text{ s}} \right)} \left(\frac{1000 \text{cc STP}}{\text{L STP}} \right) \left(\frac{22.4 \text{ L STP}}{1 \text{ mol}} \right) = 768{,}800 \text{ cm}^2$$

$$= 76.88 \text{ m}^2$$

E. Check. The analytical and graphical solutions for $\theta = 0.25$ gave identical results.

F. Generalize. 1. The RT curve contains the variables that affect mass transfer rates across the membrane (α_{AB}, p_p/p_r). If either of these variables vary (for example, if the temperature changes α_{AB} will change) a new RT curve must be generated. The operating equation depends on the cut and the feed mole frac. If either of these variables is changed, then a new operating line needs to be drawn. One advantage of this graphical approach is it separates the rate and operating terms; thus, making it easier to determine the effect of varying the conditions.

2. Units are important.

3. Although this is *not* an equilibrium process it looks very similar to a flash distillation with a large relative volatility (Hoffman, 2003). Membrane separators are useful because they are a practical way of generating favorable RT curves.

4. A commercial gas permeator would probably be close to cross-flow or countercurrent flow (see section 17.7). This would result in more separation than is predicted for the perfectly mixed system.

5. Although the permeate product is fairly pure (93 % carbon dioxide) the retentate product is impure (65% methane). Unlike distillation systems, simple membrane separators cannot produce two pure products simultaneously. More complicated membrane cascades can do this, but repressurization is required (e.g., see Wankat, 1990, Chapter 12).

6. The driving force for methane transfer is significantly higher than for carbon dioxide (check the flux calculations). Yet, because of the high selectivity the carbon dioxide transfer rate is considerably greater. The carbon dioxide is being transferred "uphill" to a higher carbon dioxide mole frac but "downhill" to a lower partial pressure of carbon dioxide. The increased pressure forces this to happen.

7. At low values of y_{in} the methane flux can be greater than the carbon dioxide flux even though the system is concentrating carbon dioxide. At these low feed concentrations and with a relatively high cut the retentate product can be almost pure methane, but permeate will be impure. (Try doing some of these calculations. Since the RT curve is not changed, only the operating line needs to be changed. Because the analytical solution is easily set up in a spreadsheet, a large number of examples can easily be run.)

8. Greater accuracy can be obtained by expanding the scales for the portion of the diagram needed for the calculation; however, this is seldom necessary since the graphs are probably at least as accurate as the experimental values of the permeabilities.

9. Once y_p and y_{out} have been calculated, determination of the fluxes and the membrane area are straightforward. Note that the RT curve and hence y_p and y_r depend upon the ratio p_r/p_p. The fluxes and the area depend upon the difference $(p_r y_r - p_p y_p)$. Higher pressures at the same pressure ratio will produce the same products but require less membrane area.

10. If the membrane area is specified instead of the cut, the problem is a simulation, not a design problem. Although the analytical solution shown in Example 17-1 is simpler, a graphical solution can be used. $\hat{F}_p$ is related to the membrane area and the flux through Eq. (17-4); however, calculation of J_A requires knowing y_p and $y_r = y_{out}$, and we need to know θ to calculate the mole fracs. This is a classical trial-and-error situation. It can be solved as follows: Generate the RT curve. Guess y_p and find y_{out} from the RT curve. Calculate J_A and $\hat{F}_p$ from Eq. (17-4). Calculate F_p from the mass balance Eq. (17-7c). If the values of $\hat{F}_p$ calculated from the flux and from the mass transfer expressions are not equal, continue the calculation.

Membrane separations are often most effective at low concentrations. This is exactly where distillation is most expensive. Thus, *hybrid* systems where a membrane separator is combined with distillation are often used commercially.

There is one other significant difference between membrane separators and the equilibrium-staged separations we have studied. Companies are much more likely to buy "off-the-shelf" or "turnkey" membrane units. Off-the-shelf systems are modules in standard sizes that are connected together to more-or-less perform the desired separation. The engineer needs to determine the performance of these units since they will not be exactly the same as the design desired. With a turnkey unit the company buys from a manufacturer who guarantees a given performance level. This situation arose because only a limited number of companies have the technical expertise necessary to make membrane separators. A large number of companies are capable of making distillation and absorption systems. If your company decides to buy a turnkey unit, knowledge of membrane separations will enable you to include all pertinent items in the contract and to negotiate a contract that is more favorable for your company. After delivery, you will be able to perform the appropriate tests to determine if the membrane system meets the specifications in the contract. If the specifications are not met, your company can demand that the vendor fix the problems.

17.3.3 Multicomponent Permeation in Perfectly Mixed Systems

In the previous section we saw that the RT curve took the place of the equilibrium curve in binary flash distillation. For multicomponent flash distillation the y-x equilibrium curve is replaced by equilibrium expressions of the form $y_i = K_i x_i$ and the combination of the operating equations and equilibrium are summed using $\Sigma x_i = 1$, $\Sigma y_i = 1$, or $\Sigma y_i - \Sigma x_i = 0$. Since the operating equations for well-mixed permeators are similar to the operating equations for flash distillation (Hoffman, 2003), if we can write the rate expression in the form $y_i = K_{m,i} x_i$, then the mathematics to solve the permeation problem will be very similar to that used for flash distillation. Of course, $K_{m,i}$ has a totally different meaning than K_i in flash distillation.

The operating equation is essentially Eq. (17-8a) written for each component,

$$y_{p,i} = - \frac{1 - \theta}{\theta} y_{out,i} + \frac{y_{in,i}}{\theta}$$

(17-8c)

For ideal gases since molar ratios are equal to the volume ratios, θ is the same in molar and volume units. In addition, for ideal gases mole frac equals volume fraction; thus, Eq. (17-8c) can be used with either molar or volumetric flow rates. The rate expression Eq. (17-4) can be written for molar flows as,

$$\hat{F}_p y_{p,i} = \frac{P_i A \hat{\rho}_i (p_r y_{r,i} - p_p y_{p,i})}{t_{ms}}$$

(17-11a)

or in volumetric flow rates,

$$F_p y_{p,i} = \frac{P_i A (p_r y_{r,i} - p_p y_{p,i})}{t_{ms}}$$

(17-11b)

Solving Eq. (17-11b) for $y_{p,i}$, we obtain

$$y_{p,i} = K_{m,i} y_{r,i}$$

(17-11c)

where, after some rearrangement,

$$K_{m,i} = \frac{(P_i / t_{ms}) p_r}{F_p / A + (P_i / t_{ms}) p_p}$$

(17-11d)

In a design problem permeate volumetric flow rate F_p will be known but the area A is unknown. If Eqs. (17-8c) and (17-11d) undergo exactly the same algebraic steps used for flash distillation [from Eq. (2-36) to (2-37)], the resulting equation for $y_{r,i}$ (equivalent to x_i in flash) is,

$$y_{r,i} = \frac{y_{in,i}}{1 + (K_{mi} - 1)\theta}$$

(17-11e)

and the summation equation $\sum y_{r,i} = 1$, (equivalent to $\sum x_i = 1$ in flash) is

$$\sum \frac{y_{in,i}}{1 + (K_{mi} - 1)\theta} = 1$$

(17-11f)

which is essentially the same as flash distillation Eq. (2-40) (see Problem 17.C5).

To find the area required for a multicomponent permeator, we estimate the value of F_p/A (or equivalently of $y_{p,i}$ for one of the components, which becomes the key component), calculate all the values of $K_{m,i}$, check to see if Eq. (17-11f) is satisfied, and if not repeat with a new value of F_p/A. Unlike the flash distillation counter part, this procedure converges rapidly using $\sum y_{r,i} = 1$ as the basic equation, and it is easy to solve on a spread sheet or in a mathematical program such as MATLAB, Mathematica or Mathcad. Once correct values for $K_{m,i}$ and F_p/A are known, we can calculate area A, $y_{r,i}$ from Eq. (17-11e), and $y_{p,i}$ from Eq. (17-11c). This procedure is illustrated in Example 17-3. Although shown only for gas permeation, this approach can be used with other membrane separators.

EXAMPLE 17-3. Multicomponent, perfectly mixed gas permeation

A perfectly mixed gas permeation unit is separating a mixture that is 20 mol% carbon dioxide, 5 mol% oxygen and 75 mol% nitrogen using a poly(dimethyl-siloxane) membrane at 25°C. Feed flow rate is 20,000 cm^3 (STP)/s. The membrane thickness is 1 mil (0.00254 cm). Pressure on the feed side is 3.0 atm and on the permeate side is 0.40 atm. We desire a cut fraction = 0.225. Find the membrane area needed, and the outlet mole fracs of permeate and retentate.

Solution

Permeabilities (Table 17-2) are $P_{CO2} = 3240 \times 10^{-10}$, $P_{O2} = 605 \times 10^{-10}$, $P_{N2} = 300 \times 10^{-10}$, all in [cm^3 (STP)cm/(cm^2 s cm Hg)]. So that we don't forget, change the units of the pressures: $p_r = 3.0$ atm (76 cm Hg/atm) = 228 cm Hg, $p_p = 0.40$ atm = 30.4 cm Hg.

The K values can be calculated from Eq. (17-11d). For example, if we choose carbon dioxide as the key component,

$$K_{m,CO2} = \frac{p_r(P_{CO2}/t_{ms})}{F_p/A + p_p(P_{CO2}/t_{ms})} = \frac{(228)(3240 \times 10^{-10})/(0.00254)}{F_p/A + (30.4)(3240 \times 10^{-10})/(0.00254)}$$

and similarly for oxygen and nitrogen. To use these terms in a trial and error procedure, we need to start with a value for F_p/A. How do we find a reasonable first guess for F_p/A?

If we guess a value for $y_{p,i}$ for the key component, we can then calculate $y_{r,key}$ from the rearranged component operating Eq. (17-8c),

$$y_{r,key} = \frac{\theta}{1-\theta} y_{p,key} + \frac{y_{in,key}}{1-\theta}$$

and determine our guess for F_p/A by rearranging Eq. (17-11b).

$$F_p/A_{guess} = [P_{key}/(t_{ms}y_{p,key})](p_r y_{r,key} - p_p y_{p,key})$$

For example, we can arbitrarily guess that $y_{p,CO2} = 0.50$. Then the first guess for $F_p/A = 0.002689$. With this guess we use Eq. (17-11d) to find $K_{m,CO2} = 4.42857$, $K_{m,O2} = 1.6827$, $K_{m,N2} = 0.8834$ and then from Eq. (16-11e) find $y_{r,O2}$ and $y_{r,N2}$. Then the check is Eq. (17-11f), which becomes,

$$\sum \frac{y_{in,i}}{1 + (K_{mi} - 1)\theta} = 0.638.$$ We need a higher value of F_p/A or a lower guess for $y_{p,key}$.

The following values were generated with a spreadsheet, by guessing the next value for $y_{p,CO2}$. (If desired, use Solver or Goal Seek.)

$y_{p,CO2}$	F_p/A	$\Sigma(y_{r,i})$
0.50	0.002689	0.638
0.45	0.004357	1.006
0.455	0.004174	0.999428
0.4549	0.004178	0.999563
0.4546	0.004189	0.999968

Then $A = F_p/(F_p/A) = [\theta(F_{in})]/(F_p/A) = [(0.225)(20,000)]/(0.004189) = 1,074,360 \text{ cm}^2$. From Eq. (17-11c) we find $K_{m,CO2} = 3.60554$, $K_{m,O2} = 1.1748$, and $K_{m,N2} = 0.5922$. Since F_{in} and F_p are cm³(STP)/s of ideal gases, $F_p/F_{in} = \hat{F}_p/\hat{F}_{in} = \theta$.

The converged values of the retentate mole fracs from Eq. (17-11e) are: $y_{r,CO2} = 0.1261$, $y_{r,O2} = 0.0481$, $y_{r,N2} = 0.8258$; and the permeate mole fracs from Eq. (17-11c) are: $y_{p,CO2} = 0.4546$, $y_{p,O2} = 0.0565$, $y_{p,N2} = 0.4890$.

To be sure you know how to do multicomponent permeation problems, set up your own spreadsheet and solve either this example or Problem 17.H1. Geankoplis (2003) solves the multicomponent permeator system by a different method, but the results are identical (see Problem 17.H1).

Note that permeate is not very pure. Although nitrogen has the lowest permeability, there is still more nitrogen in the permeate than carbon dioxide. This occurs because the large amount of nitrogen in the feed produces a large driving force to push nitrogen through the membrane. Retentate is significantly purer than the feed. To obtain a still purer nitrogen stream in the retentate with this feed, we can increase the cut and the retentate pressure; however, a better approach would be to use a membrane with higher selectivity.

17.4 REVERSE OSMOSIS (RO)

Reverse osmosis (RO) is a process commonly used to purify and desalinate water (Reisch, 2007). RO is also an integral part of industrial water management plans to reach zero-liquid discharge goals (Kucera, 2010). The liquid water is forced under pressure through a non-porous membrane in the opposite direction to osmosis (osmosis is defined shortly). Most salts and uncharged molecules are retained by the membrane. Thus, permeate is much purer water and retentate becomes significantly more concentrated. The commonly used membranes shown in Figure 17-3 are: 1) a blend of cellulose acetate and cellulose triacetate, 2) aromatic polyamides (aramids), and 3) cross-linked aromatic polyamides (Eykamp, 1997). Currently, the best membranes are thin-film interfacial composite membranes (Baker, 2004). Both hollow-fiber and spiral-wound modules are used for RO (Figure 17-1), but about 85% of applications use spiral-wound membranes (Baker, 2004). Spiral-wound membranes are used because techniques for making hollow-fiber composite membranes have not been developed. In addition, spiral-wound membranes do not clog as easily as hollow fibers; thus, less pretreatment is needed for spiral-wound membranes.

The feed to an RO system usually requires pretreatment to remove any particulates that would clog the membrane. If there are ions or solutes in solution that have limited solubility, the design must include a solubility calculation to determine if they will precipitate onto the membrane when retentate is concentrated. If precipitation is likely, these ions or solutes must either be removed or made more soluble to prevent them from precipitating. A simple schematic of a simple RO system including the most important auxiliary equipment is shown in Figure 17-8. In practice, large-scale systems may have hundreds of membrane modules arranged both in series and in parallel (Figure 17-2C) in what is often called a "Christmas-tree" pattern (Baker, 2004). More details on equipment are available in Baker et al. (1990), Eykamp (1997), Ho and Sirkar (1992), and Noble and Stern (1995). Kucera (2010) discusses details of operation and maintenance of RO systems.

17.4.1 Analysis of Osmosis and Reverse Osmosis

The driving force for solvent flux in RO is the difference between the pressure drop across the membrane and the osmotic pressure difference across the membrane ($\Delta p - \Delta \pi$). Then the mass flux of solvent is

$$J'_{solv} = \frac{F'_{solv}}{A} = \frac{K'_{solv}}{t_{ms}} \left(\Delta p - \Delta \pi \right), g/m^2 s \qquad \text{(17-12a,b)}$$

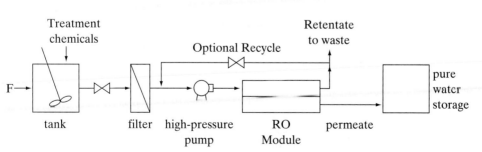

FIGURE 17-8.   *Schematic of RO system*

In these equations K'_{solv} is the permeability of the solvent (usually water) through the membrane with an effective membrane skin thickness of t_{ms}. The pressure drop across the membrane, $\Delta p = p_r - p_p$. $\Delta\pi$ is the difference in the osmotic pressure across the membrane (see description below). The mass solute flux across the membrane J'_A can be written as

$$J'_A = \frac{K'_A}{t_{ms}} \Delta x, g/m^2 s$$
(17-13)

where K'_A is the solute permeability and Δx is the difference in mass frac of solute across the membrane. For a membrane with perfect solute retention, $K'_A = 0$. Typical units are pressure p and osmotic pressure π in atm, the solvent permeance K'_{solv}/t_{ms} in g/(atm·s·m²), the solvent flux J'_{solv} in g/(m²s), the salt permeance K'_A/t_{ms} in g/(m²s·mass fraction), and the salt flux J' in g/(m²s).

To study RO, we must first study osmosis. In *osmosis* solvent flows through a *semipermeable* membrane (one which passes solvent but not solutes or ions) from the less concentrated region to the more concentrated region. For example, if we have pure water on one side of a semipermeable membrane and a sugar solution on the other side, the water will flow into the sugar solution to dilute it. Thus, in Figure 17-9 osmotic flow is from the right side (pure water) to the left side (sugar solution). This natural direction of solvent flow is the direction that will cause chemical potentials to equalize. We can stop or reverse the flow by increasing the pressure on the sugar solution using the piston shown on the left. Osmotic pressure π is the additional pressure required on the concentrated side to stop osmotic flow assuming the permeate side is pure water that contains no solute.

Osmotic pressure is a thermodynamic property of the solution. Thus, π is a state variable that depends upon temperature, pressure, and concentration but does *not* depend upon the membrane as long as the membrane is semipermeable. Osmotic equilibrium requires that the chemical potentials of the solvent on the two sides of the membrane be equal. Note that the solutes are not in equilibrium since they cannot pass through the membrane. Although osmotic pressure can be measured directly, it is usually estimated from other measurements (e.g., Reid, 1966). For an incompressible liquid osmotic pressure can be estimated from vapor pressure measurements,

$$\pi V_{solvent} - RT \ell n \left[\frac{VP_{solvent}}{VP_{solution}} \right]$$
(17-14)

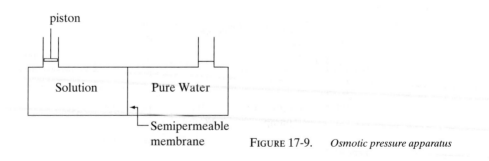

piston

Solution | Pure Water

Semipermeable membrane

FIGURE 17-9.   *Osmotic pressure apparatus*

Chapter 17   Introduction to Membrane Separation Processes

where $v_{solvent}$ is the partial molar volume of the solvent. Another common method is to relate osmotic pressure to freezing point depression (Reid, 1966). For dilute systems the osmotic pressure is often a linear function of concentration,

$$\pi = \hat{a}\, x \text{ (mole frac) or } \pi = a'x \text{ (wt frac)} \qquad \textbf{(17-15a,b)}$$

where $\hat{a}$ or a' are determined by plotting the data as a straight line. As the solution becomes more concentrated, the osmotic pressure increases more rapidly than predicted by the linear relationship. For some dilute systems the linear constant can be estimated from the van't Hoff equation,

$$\hat{a} = RT \qquad \textbf{(17-15c)}$$

Since the van't Hoff equation assumes that Raoult's law is valid, this result will be incorrect if the solution associates or dissociates even though empirical Eq. (17-15a) may still be accurate.

The osmotic pressure for natural waters containing salts can be determined from the following empirical relation with T in °C and salt concentration C_s in kg/m³ (Geraldes et al., 2005).

$$\pi = 206.4 * (320 + T) * C_s \text{ (for } C_s < 20 kg/m^3) \qquad \textbf{(17-15d)}$$

$$\pi = 206.4 * (320 + T) * (1.17 * C_s - 3.4) \text{ (for } C_s > 20 kg/m^3) \qquad \textbf{(17-15e)}$$

Unfortunately, natural osmotic flow is in the opposite direction to what we want to do (produce pure water). In RO we push the solvent out of the concentrated solution into the dilute solution. Thus, it takes energy to concentrate the retentate. Because RO is reversing the natural flow direction, RO is inherently a nonequilibrium process. The increase in osmotic pressure as retentate becomes more concentrated also puts a natural limit on the recovery of pure solvent by RO. If one tries to recover too much solvent, retentate becomes very concentrated, the osmotic pressure difference becomes extremely large, and the pressure drop required by Eq. (17-12) for a reasonable flux rate becomes too large for practical operation.

The analysis procedure developed in the previous section for gas permeation forms the basis for analyzing RO. However, the RO analysis is more complicated because of 1) osmotic pressure, which is included in Eq. (17-12), and 2) mass transfer rates are much lower in liquid systems. Since the mass transfer rates are relatively low, the wt frac of solute at the membrane wall x_w will be greater than the wt frac of solute in the bulk of the retentate x_r. This buildup of solute at the membrane surface occurs because the movement of solvent through the membrane carries solute with it to the membrane wall. Since the solute does not pass through the semipermeable membrane, its concentration will build up at the wall and it must back diffuse from the wall to the bulk solution. This phenomenon, *concentration polarization*, is illustrated in Figure 17-10. Concentration polarization has a major effect on the separations obtained in RO and UF (see next section). Since concentration polarization causes $x_w > x_r$, the osmotic pressure becomes higher on the retentate side and, following Eq. (17-12), the flux declines. Concentration polarization will also increase Δx in Eq. (17-13) and flux of solute may increase, which is also undesirable. In addition, since concentration polarization increases solute concentration, precipitation becomes more likely.

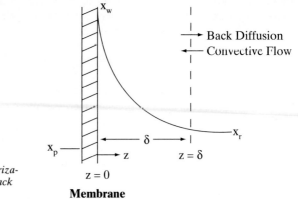

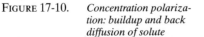

FIGURE 17-10. *Concentration polariza-*
tion: buildup and back
diffusion of solute

Since $J'_{solv} = F'_p(1 - x_p)$ [where x is mass frac in RO], we can expand Eq. (17-12) using Eq. (17-3a) as,

$$J'_{solv} = F'_p \frac{(1-x_p)}{A} = \frac{K'_{solv}}{t_{ms}} [(p_r - p_p) - (\pi_r(x_w) - \pi_p(x_p))] \tag{17-16a}$$

The x values are mass fractions, and the osmotic pressure on the retentate side depends on the concentration of solute at the membrane wall. To simplify the analysis, we will temporarily assume that the osmotic pressure data are linear and are satisfactorily fit by Eq. (17-15b). Then the flux equation becomes

$$J'_{solv} = \frac{F'_p(1-x_p)}{A} = \frac{K'_{solv}}{t_{ms}} [(p_r - p_p) - a'(x_w - x_p)] \tag{17-16b}$$

To relate the wall concentration to retentate concentration, we define the *concentration polarization modulus M* in terms of the wt frac,

$$M - x_w/x_r \tag{17-17}$$

Methods to measure or predict M will be developed shortly. Substituting the definition for M into Eq. (17-16b), we obtain

$$J'_{solv} = \frac{F'_p(1-x_p)}{A} = \frac{K'_{solv}}{t_{ms}} [(p_r - p_p) - a'(Mx_r - x_p)] \tag{17-16c}$$

The solute flux Eq. (17-13) can also be expanded and written in terms of the concentration polarization modulus.

$$J'_A = \frac{F'_p x_p}{A} = \frac{K'_A}{t_{ms}} (x_w - x_p) = \frac{K'_A}{t_{ms}} (Mx_r - x_p) \tag{17-18}$$

Essentially the same procedure used to solve for the concentrations in gas permeators will be used. That is, after assuming that K'_A is not zero and is independent of the solvent

Chapter 17 Introduction to Membrane Separation Processes

transfer rate, and that K'_{solv} is independent of the solute transfer rate, we will solve Eqs. (17-16c) and (17-18) for F'_p, set the two equations equal to each other, and solve for the desired concentration. Setting the equations equal, we obtain

$$\frac{A(K'_{solv}/K'_A)}{t_{ms}(1-x_p)}[(p_r-p_p)-(Mx_r-x_p)a'] = \frac{A}{x_p t_{ms}}(Mx_r-x_p) \tag{17-19}$$

It is convenient to define the selectivity of the membrane α' as,

$$\alpha' = K'_{solv}/K'_{solute} = K'_{solv}/K'_A \tag{17-20}$$

We can now solve for either x_r as a function of x_p, or vice versa. Since it is easier to solve for x_r, we will do that here.

$$x_r = \frac{x_p[x_p(\alpha'a'-1)+\alpha'(p_r-p_p)+1]}{M[1+(\alpha'a'-1)x_p]} \tag{17-21a}$$

This RT equation represents the transfer rate of solvent and solute through the membrane in mass frac units. If $K'_A = 0$, α' will be infinite, x_p will be zero, and x_r must be determined from a mass balance. If the selectivity is not constant, α' will depend upon $x_w = Mx_r$. It will then be convenient for calculation purposes to solve for x_p as a function of x_r.

Since RO often operates with high retention of solute and thus very low values of x_p, it is useful to simplify RT Eq. (17-21a) for small x_p. As $x_p \to 0$, Eq. (17-21a) becomes

$$x_r \approx x_p\left[\alpha'\rho_{solv}(p_r-p_p)+1\right]/M \tag{17-21b}$$

Usually, $\alpha'\rho_{solv}(p_r-p_p) \gg 1$ and

$$x_r \approx x_p\alpha'\rho_{solv}(p_r-p_p)/M \tag{17-21c}$$

The simplified linear equations are easier to use. Note that if the modules are well mixed, then $x_r = x_{out}$.

The RT Eq. (17-21) needs to be solved simultaneously with an expression for the mass balance. We will again assume the membrane module is well mixed. The module is identical to Figure 17-6A except the y terms are replaced by liquid wt frac. The external mass balances for this well-mixed module (in mass units) are

$$F'_{in} = F'_p + F'_{out}, \quad F'_{in}x_{in} = F'_p x_p + F'_{out}x_{out} \tag{17-22 a,b}$$

Solving for x_p, we obtain the operating equation

$$x_p = -\frac{1-\theta}{\theta}x_{out} + \frac{x_{in}}{\theta} \tag{17-23a}$$

or the alternative,

$$x_r = \frac{x_{in} - \theta' x_p}{1 - \theta'}$$

(17-23b)

where $\theta' = F_p'/F_{in}'$ is the cut in mass units. Equation (17-23) is analogous to Eq. (17-8) obtained for the well-mixed gas permeator.

In the well-mixed module x_r is constant and is equal to x_{out}. Thus, Eq. (17-21) is valid with x_r replaced by x_{out}; however, since there is usually concentration polarization, $x_w = Mx_{out}$ and M must be determined. Once M is known, simultaneous solution of Eqs. (17-21) and (17-23) can be obtained either analytically (see Example 17-3) or by plotting both equations on a graph of x_p vs. x_{out}. Equation (17-23) is a straight line that is identical to the operating line obtained for gas permeation. The curve representing Eq. (17-21) is plotted by calculating values in the same way as in Example 17-2. Note that this curve will be below the $x_p = x_{out}$ line on a graph of x_p vs. x_{out}. This equation serves as an alternate RT equation for RO.

RO data are often reported as the rejection coefficient R.

$$R = \frac{x_r - x_p}{x_r} = \frac{x_{out} - x_p}{x_{out}} = 1 - \frac{x_p}{x_{out}}$$

(17-24a)

The rejection coefficient when there is no concentration polarization (M = 1) is called the inherent rejection coefficient R^o. By manipulating Eqs. (17-21c) and (17-24a), we can find a relationship between retention R at one set of conditions and R at another set of conditions for well-mixed systems. Often the base case (Case A) condition is an experiment with no concentration polarization ($M_{Case\ A} = 1$ and $R_{Case\ A} = R^o$). Substitute Eq. (17-21c) into the definition of R, Eq. (17-24a), for both Cases A and B. Solve for $1 - R$ in both equations and then divide the Case B equation by the Case A equation. The ratio is

$$\left(1 - R_{Case\ B}\right)/\left(1 - R_{Case\ A}\right) = \left[\left(p_r - p_p\right)_{Case\ A}/\left(p_r - p_p\right)_{Case\ B}\right]\left[M_{Case\ B}/M_{Case\ A}\right]$$ **(17-24b)**

This equation is valid for the linearized RT equation (17-21c). Equation (17-24b) allows calculation of R for a different set of conditions, M and $(p_r - p_p)$ if R is known at the Case A conditions. Equation (17-24b) is only valid for systems where the RT equation can be linearized to Eq. (17-21c). If the pressure difference $(p_r - p_p)$ is the same in the two runs, then

$$\left(1 - R_{Case\ B}\right)/\left(1 - R_{Case\ A}\right) = \left[M_{Case\ B}/M_{Case\ A}\right], \quad \text{constant } \left(p_r - p_p\right)$$ **(17-24c)**

Remarkably, this equation can be obtained exactly without linearization (except for osmotic pressure) by starting with Eq. (17-21a) and requiring constant $(p_r - p_p)$ and constant permeate mass fraction y_p. Since Eq. (17-21a) required that osmotic pressure is a linear function of weight fraction (the a' term), the requirement of a linear relationship between π and x restricts these equations to relatively low concentrations.

When R is known, we can solve the RT Eq. (17-24a) simultaneously with operating Eq. (17-23) for a completely mixed system,

$$x_{out} = x_{in} / \left[\theta'(1-R) + (1-\theta') \right]$$ (17-25)

which is a very convenient and simple solution. Alternatively, we can solve for x_p

$$x_p = \frac{(1-R)x_{in}}{1-R\theta'}$$ (17-26)

In these equations $\theta' = F_p'/F_{in}'$ is the cut in mass units. Use of these equations is illustrated in Example 17-5.

17.4.2 Determination of Membrane Properties from Experiments

Experimental values of x_p and x_{out} can be used to determine α' and M. First, an experiment is done in a very well mixed cell under conditions where there is no concentration polarization. Solving Eq. (17-19) for the selectivity, we obtain

$$\alpha' = \frac{(1-x_p)(Mx_r - x_p)}{x_p[p_r - p_p - (Mx_r - x_p)a']}$$ (17-27)

where under the conditions of the experiment M = 1 and all other terms on the right hand side are known. This method is illustrated in Example 17-4.

EXAMPLE 17-4. Determination of RO membrane properties

We test a new composite membrane. Both experiments were done with a retentate pressure of 15.0 atm and a permeate pressure of 1.0 atm. The temperature is 25°C. The following data were obtained:

Experiment a. The pure water flux was 1029 L/(m² day).

Experiment b. In a perfectly mixed laboratory system with wt frac sodium chloride, $x_{in} = 0.0023$, the rejection coefficient R was measured as 0.983. Operation was with $\theta' = 0.30$. The system was highly stirred and you can assume there was no concentration polarization (M = 1.0).

Find the water permeance and the water-salt selectivity.

Solution

A. Define. Find K'_{solv}/t_{ms} and $\alpha'_{water-salt}$.

B and C. Explore and plan. Experiment a allows us to calculate K'_{solv}/t_{ms} with x_r = x_p = 0.0 from Eq. (17-16c). Experiment b allows us to calculate x_p and x_{out} from Eqs. (17-25) and (17-26). Then, Eq. (17-27) (with $x_r = x_{out}$ and M = 1.0) can be used to find the selectivity, $\alpha'_{water-salt}$.

D. Do it. *Experiment a.* With $x_r = x_p = 0.0$, we can solve Eq. (17-16c) for K'_{solv}/t_{ms}.

$$\frac{K'_{solv}}{t_{ms}} = \frac{J'_{pure\ water}}{(p_r - p_p)}$$ (17-28)

$$\frac{K'_{solv}}{t_{ms}} = \left(\frac{1029\,\text{L}/\text{m}^2\text{day}}{(15.0-1.0)\text{atm}}\right)\left(\frac{997\,\text{g water}}{1.0\,\text{L}}\right)\left(\frac{1\,\text{day}}{86400\,\text{s}}\right) = 0.8483\,\frac{\text{g}}{\text{atm}\cdot\text{s}\cdot\text{m}^2}$$

Experiment b.
Plugging in numbers to Eqs. (17-26) and (17-25) we obtain

$$x_p = \frac{(1-0.983)(0.0023)}{1-(0.983)(0.30)} = 0.00005545 \text{ and } x_{out} = 0.003262$$

Equation (17-27) with M = 1 becomes,

$$\alpha'_{w-salt} = \frac{(1-x_p)(x_{out}-x_p)}{x_p[(p_r-p_p)-(x_{out}-x_p)a']} \quad (for\ M=1) \tag{17-29}$$

The term a' is the linear coefficient for the effect of concentration on the osmotic pressure in mass frac units. Osmotic pressures are given for aqueous sodium chloride solutions in Perry and Green, 6th edition (1984, p. 16–23). At 25°C and 0.001 mole frac the osmotic pressure is 0.05 atm. Thus, in molar units $\hat{a} = 0.05$ atm/0.001 mole frac = 50 atm/mole frac.

We must convert $\hat{a}$ to a' in mass frac units.
$\pi = a'x$ or $a' = \hat{a}m/x$ where m is mole fraction. For very dilute solutions,

(m, mole frac solute) = (x, mass frac solute) (MW water)/(MW solute)

Then $a' = \hat{a}\,\dfrac{m}{x} = \hat{a}\,\dfrac{MW_{water}}{MW_{solute}} = 50\,\dfrac{18.016}{58.45} = 15.446$ atm/mass frac

The selectivity α' can be determined from Eq. (17-29),

$$\alpha'_{w-salt} = \frac{(1-0.00005545)(0.003262-0.00005545)}{(0.00005545)[(15-1)-(0.003262-0.00005545)15.446]} = 4.145\,\frac{\text{mass frac}}{\text{atm}}$$

E. Check. The numerical values are within reasonable ranges.
F. Generalize. This example illustrates how to determine parameter values from experiments. The selectivity determined from the experiments will be used in Example 17-5 to calculate expected behavior of a membrane module when there is no concentration polarization.
 1. The pure water flux is often measured and used to find the value of K'_{solv}/t_{ms}. This is a preferred method because it is very easy and is quite reproducible. Note that the flux declines significantly when there is salt present. Thus, the pure water flux values should *never* be used as a direct estimate of the feed capacity of the system.
 2. Rejection data are commonly used to find the value of selectivity, α'_{w-salt}. This selectivity does *not* have the same meaning as the selectivity in gas permeation. Here the selectivity is given by Eq. (17-20), but the permeabilities refer to equations with different driving forces, and the permeabilities have different units.

EXAMPLE 17-5. RO without concentration polarization

We continue testing the new composite membrane from Example 17-4 in a perfectly mixed membrane system but now with a retentate pressure of 10.0 atm and a permeate pressure of 1.0 atm. Find the outlet wt fracs of the permeate and the retentate streams. There is a very high mass-transfer coefficient and $M = 1.0$.

a. $\theta' = 0.30$ (in mass flow rate units) and the inlet stream is 0.009 wt frac sodium chloride.

b. $\theta' = 0.40$ (in mass flow rate units) and the inlet stream is 0.010 wt frac sodium chloride.

Solution

A. Define. Find x_p and x_{out} for both parts a and b.

B and C. Explore and plan. Equations (17-21a) and (17-23) can be solved simultaneously to find x_p and x_{out}. Since the change in operating conditions only affects the operating Eq. (17-23), the RT equation is the same for parts a and b.

D. Do it. The RT Eq. (17-21a) in weight units is,

$$x_r = \frac{x_p[x_p(\alpha'a'-1) + \alpha'(p_r - p_p) + 1]}{M[1 + (\alpha'a'-1)x_p]}$$

(17-21a repeated)

If we use Eq. (17-23b) to substitute for x_r in Eq. (17-21a), we obtain

$$\frac{x_{in} - \theta'x_p}{1-\theta'} = \frac{x_p[x_p(\alpha'a'-1) + \alpha'(p_r - p_p) + 1]}{M[1 + x_p(\alpha'a'-1)]}$$

With θ', x_{in}, α', a', p_p, p_r, and M known, this is a quadratic equation in x_p. The equation can be solved by the quadratic formula, although this requires a fair amount of algebra. The equation can also be solved with an Excel spreadsheet using Goal Seek or Solver.

Part a. $\theta' = 0.30$ and $x_{in} = 0.009$. The spreadsheet result is $x_p = 0.000339$ and $x_r = 0.0127$.

Part b. $\theta' = 0.40$ and $x_{in} = 0.010$. The spreadsheet result is $x_p = 0.000439$ and $x_r = 0.0164$.

E. Check. Graphical solution and linearization give the same results. For example, the linearized equations using the retention [Eqs. (17-24b) and (17-24c)] can be used with the retention reported in Example 17-4. Calling Example 17-4 Case A, we have the following values:

$$R^o_{CaseA} = 0.983, \ p_r = 15 \ atm, \ (p_r - p_p)_{CaseA} = 14, \text{ and } M_A = 1$$

The problem in Example 17-5 part a is now Case B with $p_r = 10$ atm, $p_p = 1$ atm, $(p_r - p_p) = 9$, and $M = 1.0$. Then from Eq. (17-24b),

$$R_B = 1 - \left[(p_r - p_p)_{Case A} / (p_r - p_p)_{Case B}\right][M_B / M_A](1 - R_{Case A})$$
$$= 1 - (14/9)(1.0/1.0)(1 - 0.983) = 0.974$$

For part a of Example 17-5, $x_{in} = 0.009$, $p_r = 10$ atm, $p_p = 1$ atm, $\theta = 0.30$, and $R_B = 0.974$. Then from Eq. (17-25),

$$x_{out} = x_{in} / \left[\theta'(1-R) + (1-\theta') \right] = 0.009 / \left[0.3 (1-0.974) + (1-0.3) \right] = 0.0127$$

and from Eq. (17-26),

$$x_p = \frac{(1-R)x_{in}}{1-R\theta'} = \frac{(1-0.974)(.009)}{1-(0.974)(0.3)} = 0.000330$$

These results agree closely with the part a results obtained previously from the solution of the complete quadratic equation. Thus, at these low mass fractions, linearization is valid.

F. Generalize. This example showed that at low concentrations the linearization procedure used for Eqs. (17-24b) to (17-26) is valid. This will prove to be particularly useful when there is concentration polarization (see Example 17-7).

17.4.3 Mass-Transfer Analysis to Determine Concentration Polarization

Once values for α' are known, experiments can be done under conditions where concentration polarization is expected. M can be determined by solving Eq. (17-19) for M.

$$M = \frac{x_p \left[x_p (\alpha'a' - 1) + \alpha'(p_r - p_p) + 1 \right]}{x_r \left[1 + (\alpha'a' - 1)x_p \right]} \tag{17-30}$$

All terms on the right-hand side are known, and M can be calculated. However, estimation of M will allow us to avoid doing expensive and time consuming experiments, and is illustrated in Example 17-6.

Concentration polarization was shown schematically in Figure 17-9. This figure applies to the simplest situation that is steady-state, one-dimensional back diffusion of one solute into the bulk retentate stream with a perfectly rejecting membrane ($R = 1$). [If rejection is not almost complete, more detailed theories are required (e.g., Ho and Sirkar, 1992; Noble and Stern, 1995; Wankat, 1990).] The differential mass balance for this simple situation using a Fickian analysis is (Problem 17.C3)

$$(J'_{solv} / \rho_{solv})x + D\frac{dx}{dz} = 0 \tag{17-31}$$

where D is the Fickian diffusivity in the liquid solution in m²/s and ρ_{solv} is in g/m³. This equation is subject to the boundary conditions that concentration equals the wall concentration at $z = 0$,

$$x = x_w \text{ at } z = 0 \tag{17-32a}$$

and the concentration becomes the bulk concentration x_r when z is greater than the boundary layer thickness δ. The value of δ depends on the operating conditions (geometry, velocity, T).

$$x = x_r \text{ at } z \geq \delta \tag{17-32b}$$

Defining the mass-transfer coefficient as

$$k = D/\delta \tag{17-33}$$

the solution is

$$M = \frac{x_w}{x_r} = \exp[(J'_{solv}/\rho_{solv})/k] \tag{17-34}$$

The mass transfer coefficient depends on the operating conditions and has units m/s. If there are multiple solutes, a Maxwell-Stefan analysis (Section 15.7) is recommended.

This short development is useful to determine what affects concentration polarization. If the solvent flux J'_{solv} increases, M increases. If the mass transfer coefficient k increases, concentration polarization decreases. Increasing the diffusivity will increase k. Thus, operating at a higher temperature will decrease M although there are obvious limits based on the membrane thermal stability and the thermal stability of the solutes (e.g., most proteins are not thermally stable). Decreasing the boundary layer thickness δ by promoting turbulence or operating at very high shear rates in thin channels or narrow tubes will also increase k.

The quantitative use of Eq. (17-34) requires either experimentally determined values of the mass transfer coefficient k or a correlation for k (which is ultimately based on experimental data). If experimental data are available which allow the calculation of M from Eq. (17-30), then Eq. (17-34) can be used to find k. A large number of mass transfer correlations are available for a variety of geometries and flow conditions (Wankat and Knaebel, 2008). Four that are useful for membrane separators are the correlations for turbulent flow in tubes, for laminar flow in tubes and between parallel plates, and for well-mixed tanks (Blatt et al., 1970; Wankat, 1990; Wankat and Knaebel, 2008). For *turbulent* flow in tubes the mass transfer coefficient can be estimated from

$$Sh = \frac{d_t k}{D} = 0.023 Re^{0.83} Sc^{1/3}, \text{turbulent flow} \tag{17-35a}$$

where Sh is the Sherwood number, and the Reynolds number Re, and Schmidt number Sc, are defined as

$$Re = (d_t u_b \rho)/\mu, \ Sc = \mu/(\rho D) \tag{17-35b,c}$$

and u_b is the bulk velocity in the tube. Fully developed turbulent flow will certainly occur for Re > 20,000, and usually appears in UF devices for Re > 2000.

Spiral-wound membranes (Figure 17-1C) are most commonly used for water treatment. The spacers in the feed channel are designed to promote turbulence and increase mass transfer rates. Schock and Miquel (1987) experimentally determined the following mass transfer correlation for typical spiral-wound modules.

$$\text{Sh} \equiv \frac{k_c d}{D_s} = 0.065 * \text{Re}^{0.875} * \text{Sc}^{0.25}$$

(17-35d)

Here D_s is the diffusivity of the salt, and the Reynolds number is defined by Eq. (17-35b) with d = height of feed channel. This equation is similar to Eq. (17-35a), but predicts a higher mass transfer coefficient.

For *laminar* flow in a tube of length L and radius R with a bulk velocity u_b, the average mass transfer coefficient is,

$$k = 1.295 \left(\frac{u_b D^2}{RL} \right)^{1/3}$$

(17-36a)

For *laminar* flow between parallel plates with a spacing of 2h, the average mass transfer coefficient is,

$$k = 1.177 \left(\frac{u_b D^2}{hL} \right)^{1/3}$$

(17-36b)

For flat membranes in a well-stirred *turbulent* tank, the mass transfer coefficient can be estimated from

$$\frac{k d_{tank}}{D} = 0.04433 \left[\frac{\omega d_{tank}^2}{v} \right]^{0.75} (\text{Sc})^{0.33}$$

(17-37a)

In Eqs. (17-37) k is in cm/s, ω is the stirrer speed in radians/s, d_{tank} is the tank diameter in cm, D is the diffusivity in cm²/s, and the kinematic viscosity $v = \mu/\rho$ is in cm²/s. Stirred tanks are a convenient laboratory configuration.

EXAMPLE 17-6. RO with concentration polarization

We continue testing the new composite membrane explored in Examples 17-4 and 17-5. An additional experiment was done at 25°C with a retentate pressure of 15.0 atm and a permeate pressure of 1.0 atm. In addition to the data reported in Example 17-4, the following new data are obtained:

Experiment c. Experiments were done in a baffled stirred-tank system with a 35 cm diameter tank. We expect that Eq. (17-37a) will be valid, except the coefficient 0.04433 has to be adjusted. The stirrer was operated at 900 rpm. The inlet solution was 0.005 wt frac sodium chloride. The measured wt frac were $x_p = 0.0001287$, and $x_{out} = 0.006218$.

Based on these experiments determine the value for the coefficient in the correlation for mass transfer in turbulent stirred tanks (Eq. 17-37a).

Solution

A, B, and C: Define, Explore, and Plan. We can calculate the polarization modulus M from Eq. (17-30), J'_{solv} from Eq. (17-16c), the value of the mass transfer coefficient from Eq. (17-34), and a new coefficient for the correlation in Eq. (17-37a).

D. Do it. *Experiment c.* Since x_p and x_{out} were measured, and α', a', p_r, and p_p are all known, a straightforward plug-and-chug in Eq.(17-30),

$$M = \frac{x_p\left[x_p\left(\alpha'a'-1\right)+\alpha'\left(p_r-p_p\right)+1\right]}{x_r\left[1+\left(\alpha'a'-1\right)x_p\right]} \quad \text{gives,}$$

$$M = \frac{0.0001287\{0.0001287[(4.145)(15.446)-1]+(4.145)(14)+1\}}{0.006218\{1+[(4.145)(15.446)-1]0.0001287\}} = 1.21$$

We can also calculate M from Eq. (17-24c) since $(p_r - p_p)$ is constant. Calling experiment C case B, we have

$$R_B = 1 - \frac{x_p}{x_{out}} = 1 - \frac{0.0001287}{0.006218} = 0.979$$

Eq. (17-24c) is

$$M_B = M_A\left(1-R_A\right)/\left(1-R_B\right) = (1.0)\frac{(1-0.9793)}{(1-0.983)} = 1.22$$

Very close to result from Eq. (17-30).

The solvent flux can then be determined from Eq. (17-16c) using $K'_{solv}/t_{ms} = 0.8483$ determined in Example 17-4, and because $x_r = x_{out}$ for a well-mixed tank,

$$J'_{solv} = \frac{K_{solv}}{t_{ms}}\left[(p_r-p_r)-a\left(Mx_r-x_p\right)\right] \text{ becomes}$$

$$J'_{solv} = 0.8483\{(15-1)-15.446[(1.212)(0.006218)-0.0001287]\} = 11.78 \text{ g/(m}^2\cdot\text{s)}$$

Note that this is smaller than the pure water flux (285.0) because the driving force is reduced by the osmotic pressure difference.

Solving Eq. (17-34) for k,

$$k = (J'_{solv}/\rho_{solv})/\ln M = \{[11.78\text{g/(m}^2\text{s)}]/(9.97\times10^5\text{g/m}^3)\}/\ln(1.21) = 6.21\times10^{-5}\text{m/s} = 0.00621\text{cm/s}$$

Eq. (17-37a) can be written with an unknown constant instead of 0.0443. Solving for the constant in this equation, we obtain

$$\text{Constant} = \left(\frac{kd_{tank}}{D}\right)\left(\frac{\omega d_{tank}^2}{v}\right)^{-0.75}\left(\frac{\mu}{\rho D}\right)^{-0.33} \tag{17-37b}$$

Since the solution is quite dilute, the properties of water can be used.

$$\rho = 0.997 \text{ g/cm}^3, \mu = 1.0 \text{cp} = 0.01 \text{poise} = 0.01 \text{ g/(cm·s)}$$

$$\omega = (900 \text{rpm})\left(\frac{2\pi \text{ radians}}{\text{revolution}}\right)\left(\frac{1\text{min}}{60\text{s}}\right) = 94.25 \text{radians/s}$$

The diffusivity of NaCl in water at 25°C is 1.53×10^{-5} cm²/s.

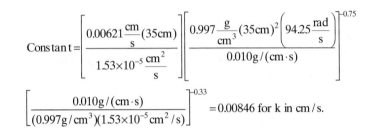

$$\left[\frac{0.010\text{g/(cm·s)}}{(0.997\text{g/cm}^3)(1.53\times10^{-5}\text{cm}^2/\text{s})}\right]^{-0.33} = 0.00846 \text{ for k in cm/s.}$$

E. Check. It is not unusual to have a general mass transfer correlation differ from experiment by this much. We will use the measured value of the constant in Example 17-7.

F. Generalization. This example illustrates how we can take experimental data, and fine tune mass transfer correlations by adjusting the constants.

Experiment c required a number of steps to eventually find the mass transfer coefficient k. This is invariably the case since k is not a directly measured variable but depends upon interpretation of the data using a model. Once k was obtained, we have a single data point to compare to the correlation Eq. (17-37a). They disagreed. You may be tempted to use the correlation and ignore the data point. However, mass transfer correlations are not very accurate. They usually predict the trends well (such as the effect of Reynolds and Schmidt numbers), but the absolute value predicted can be significantly off. A single data point can be used to adjust the constant in the correlation for application to this particular system. If more data were available, we could check the entire correlation.

EXAMPLE 17-7. Prediction of RO performance with concentration polarization

Predict the values of x_p, x_{out}, and J_{solv} if a 0.01 wt frac sodium chloride in water solution is separated by the membrane studied in Examples 17-4 to 17-6 in a stirred tank that is geometrically similar to the one in Example 17-6 except the tank is 40 cm in diameter and the stirrer speed is 500 rpm. The retentate pressure is 10 atm and the permeate pressure is 1 atm. Operate at a cut $\theta' = 0.40$ and at 25°C.

Solution

A, B, and C. Define, Explore, and Plan. We can now calculate mass transfer coefficient k using D, ρ, μ, and the coefficient = 0.00846 from experiment c and Example 17-6.

$$\omega = 500 \text{ rpm}\left[\frac{2\pi \text{ radians}}{\text{revolution}}\right]\frac{1\text{ min}}{60\text{ s}} = 52.36 \text{ radians/s}$$

From Eq. (17-37a) with the modified constant,

$$k = \left[\frac{(0.00846)(1.53 \times 10^{-5})}{(40)} \right] \left[\frac{(0.997)(40)^2 (52.36)}{0.010} \right]^{0.75} \left[\frac{0.010}{(0.997)(1.53 \times 10^{-5})} \right]^{0.33}$$

$$= 0.00428 \text{ cm/s} = 4.28 \times 10^{-5} \text{ m/s}$$

At this point we can proceed to use either a permeation analysis or the linearized retention analysis. Both will be trial and error, but the easier retention analysis will be demonstrated.

Use Example 17-4 as Case A. $R_A = 0.983$, $M_A = 1.0$, $(p_r - p_p)_A = 14$.

Case B is the current problem: $(p_r - p_p)_B = 9$, $\theta' = 0.04$, $x_{in} = 0.01$, $a' = 15.446$,

$(K'_{solv}/t_{ms}) = 0.8483 \dfrac{g}{\text{atm} \cdot \text{s} \cdot \text{m}^2}$. With a guessed value, $M_{B,guess}$, we can use

Eq. (17-24b) to find retention:

$$R_B = 1 - (1 - R_A) \frac{(p_r - p_p)_A}{(p_r - p_p)_B} \frac{M_B}{M_A}$$

With R_B known Eq. (17-26) gives x_p

$$x_p = \frac{(1 - R_B) x_{in}}{1 - R_B \theta'}$$

and then from Eq.(17-24a),

$$x_{out} = x_r = \frac{x_p}{(1 - R_B)}$$

We now have enough information to find J'_{solv} from Eq. (17-16c).

$$J'_{solv} = (K'_{solv}/t_{ms}) \left[(p_r - p_p)_B - a'(M_B x_r - x_p) \right]$$

Then Eq. (17-34) can be used to calculate M_B,

$$M_{Bcalc} = \exp\left((J'_{solv}/\rho_{solv})/k \right)$$

To have consistent units, k has to be in m/s. If $M_{B,calc} - M_{B,guess} = 0$, then the calculation is correct. If *not*, use a new value of $M_{B,guess}$ and repeat. This is relatively straightforward on a spreadsheet. Goal Seek or Solver can be used to converge M_B.

D. Calculations. A spreadsheet was set up to calculate the five equations in order. With a first guess of $M_B = 1.0$, we found $R_B = 0.974$, $x_p = 0.000433$, $x_r = 0.01638$, $J'_{solv} = 7.45$, and $M_{B,calc} = 1.19$. Using Goal Seek the converged solution was $M_B = 1.19$, $R_B = 0.969$, $x_p = 0.000513$, $x_r = 0.01632$, and $J'_{solv} = 7.39$.

E. Check. The results obtained are consistent with what we expect. This is a helpful and quick check but does not guarantee there are no errors. We can also check the results for the limiting case when $M = 1.0$ with the results obtained in Example 17-5 for the same feed concentration, pressures and cut as this problem ($x_{out} = 0.0164$ and $x_p = 0.000439$). The spreadsheet gives $x_{out} = 0.0164$ and $x_p = 0.000433$. The agreement is quite good.

F. Generalization.
1. The prediction of performance was trial-and-error because the unknown variables were needed to calculate other unknowns that were needed to calculate the first unknown. This circle is broken by guessing a variable, doing the calculation, and then checking the guess. Convergence was rapid. In more concentrated systems or with less vigorous stirring with much larger M and larger π values (which may be nonlinear functions of wt frac), convergence can be slower.

2. RO is commonly used to produce ultrapure water in the electronics and pharmaceutical industries. In these applications $R°$ is much closer to 1.0 and x_{in} is smaller. These systems have little concentration polarization (M ~ 1.0) and produce very pure permeate.

3. All of these calculations assume an undamaged membrane with no holes. Even a tiny pinprick can cause a large increase in x_p. The liquid will pass through a hole as convective flow at a salt concentration of x_w. This flux will be quite large because of the large pressure drop. In addition, undesired large molecules can also pass through holes in the membrane. Performance of RO systems needs to be monitored continuously, and damaged membranes need to be plugged or replaced.

4. Membrane life will depend upon the membrane material and the operating conditions. Membrane replacement costs should be included in the operating expenses.

5. If a limiting case does not agree with an independent calculation, then there must be an error in either the original calculation, the method to produce the limiting case or the independent calculation. If the limiting case agrees with an independent calculation (as it does for this problem for M = 1), we have not proved that the calculation for M ≠ 1 is correct. As the number of limiting cases that agree with independent calculations increase, our confidence in the general solution increases.

6. The problem statement stated the tanks are geometrically similar. If they aren't, the constant in mass transfer correlation, Eq. (17-37a) is probably different in the two tanks. Geometric similarity allows one to scale-up.

17.4.4 RO with Concentrated Solutions

A more complicated situation occurs when x_r, and hence x_w, are concentrated and osmotic pressure depends upon x_p in a nonlinear fashion. An *Advanced RT* equation can be developed if the osmotic pressure of permeate is linear in x_p, $\pi_p = a'x_p$. Since permeate is quite dilute, this equation is often valid even if π is not a linear function of x at x_{in} and x_{out}. Start with Eq. (17-16a) and substitute in $x_w = Mx_r$ and $\pi_p(x_p) = a'x_p$. Solve resulting equation and Eq. (17-18) for F'_p. Then set equal and rearrange.

$$a'x_p[(p_r - p_p) - (\pi(Mx_r) - a'x_p)] = (1 - x_p)(Mx_r - x_p) \qquad \textbf{(17-38)}$$

We assume the osmotic pressure $\pi(Mx_r)$ is available in tabular or equation form. Expand in terms of x_p and collect terms.

$$(a'a' - 1)x_p^2 + [a'(p_r - p_p) - a'\pi(Mx_r) + (1 + Mx_r)]x_p - Mx_r = 0 \qquad \textbf{(17-39)}$$

If we pick a value of Mx_r we can determine $\pi(Mx_r)$ and calculate x_p. Then we can generate the RT curve including nonlinear osmotic pressure and concentration polarization. If the membrane module is perfectly mixed, the operating equation is

$$x_p = -\frac{1-\theta'}{M\theta'}(Mx_r) + \frac{x_{in}}{\theta'} \qquad \textbf{(17-40)}$$

The solution can then be done trial and error in a spreadsheet in a fashion similar to Example 17-7.

17.5 ULTRAFILTRATION (UF)

Ultrafiltration (UF) is another membrane separation method used to purify liquids. UF is commonly used for recovery of proteins and in food and pharmaceutical applications. It is useful for separating "permanent" emulsions since the oil droplets will not pass through the membrane. UF is used for the removal of fine colloidal particles, and for recovery of dyes from wastewater. In many applications such as whey processing UF and RO are used in series. The valuable proteins are recovered by UF, and permeate from the UF system is sent to the RO system. The remaining sugars and salts are concentrated in the RO system by removing water. The concentrated permeate can then be fermented to produce ethanol, lactic acid or other products.

The equipment for UF systems often looks very similar to RO systems although they operate at lower pressures. However, this similarity does not extend to the molecular level. Remember that RO membranes are nonporous and separate based on a solution-diffusion mechanism. UF membranes are porous and separate based on size exclusion. Large molecules are excluded from pores in the thin membrane skin and thus, the large molecules are retained in the retentate. Small molecules fit into the pores and pass through to the permeate. Since there is usually a distribution of pore sizes, molecules within the range of pore sizes partially permeate and are partially retained. In a somewhat oversimplified picture, UF is cross-flow filtration at the molecular level.

Because of the different separation mechanism, UF membranes have significantly higher fluxes than RO membranes. Thus, concentration polarization is usually worse in UF than in RO because there is a much greater solvent flow from the bulk fluid through the wall. This concentration polarization can cause membrane *fouling* which not only decreases the flux but also can drastically decrease the membrane life. Hydrophilic membranes tend to foul less rapidly but have shorter lives than the more stable hydrophobic membranes. The choice of the best membrane thus depends upon the operating conditions. Cellulose acetate (Figure 17-3) membranes were the first commercial membranes and are still used where their low level of interaction with proteins is more important than their relatively short life. Polymeric membranes are used where more basic conditions are encountered. The most common polymeric membrane is polysulfone (Figure 17-3). Membranes are tailor-made to sieve molecules in different size ranges depending on the purpose of the separation (Figure 17-11). The nominal molecular weight exclusion (shown by x in Figure 17-11) is often reported by manufacturers, but it is not nearly as useful as the complete retention curve.

UF can be initially analyzed by the same procedures used for RO; thus, Eqs. (17-12) and (17-16) are valid. However, in UF the molecules being retained are often very large and the resulting osmotic pressure is very low. For most UF applications the osmotic pressure difference can be ignored and the solvent flux equation can be simplified.

In mass units the solvent flux is

$$J'_{solv} = \frac{F'_p\left(1-x_p\right)}{A} = \frac{K'_{solv}}{t_{ms}}\left(p_r - p_p\right)$$

(17-41)

where F'_p is the mass transfer rate of permeate (e.g., kg/s) and x_p is the wt frac of solute in the permeate. As we will see, ignoring the osmotic pressure difference is a more important simplification than it looks at first.

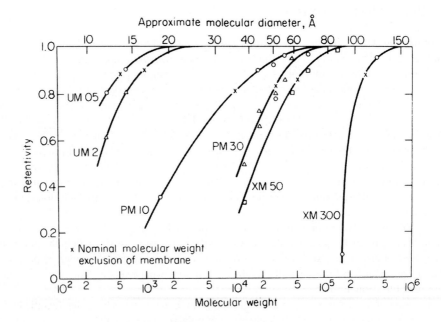

FIGURE 17-11. *Solute retention on Amicon Diaflo membranes (Porter, 1997), reprinted with permission from P.A. Schweitzer (Ed.),* Handbook of Separation Techniques for Chemical Engineers, *3rd ed., (1997), copyright 1997, McGraw-Hill.*

For sieve type membranes if a pore does not exclude solute, it will carry solute at the wall wt frac, $x_w = Mx_r$, through the pore. In sieve type membranes the inherent rejection R^o ($M = 1$) can be interpreted as the fraction of flux carried by pores which exclude solute. Then $1 - R^o$ is the fraction of flux carried by pores that do not exclude solute. Ideally, R^o is independent of $(p_r - p_p)$ and by definition R^o is independent of M.

The permeate wt frac, x_p, is then

$$x_p = (1 - R^o)\, x_w = (1 - R^o)\, M\, x_r \tag{17-42}$$

This equation can also be solved for the retentate wt frac,

$$x_r = x_p/[(1 - R^o)\, M] \tag{17-43}$$

Equation (17-42) or (17-43) are the RT equations for UF. They are particularly simple because the inherent rejection R^o is based on experimental data. Note that the retention definition for UF is microscopic, which is different than the macroscopic definition used for RO (Eq. (17-24a)). The RT equations for UF depend only on the inherent solute rejection and M. For a perfectly mixed membrane module we assume that $x_r = x_{out}$. Then either Eq. (17-42) or (17-43) written in terms of x_{out} can be solved simultaneously with the mass balances, Eqs. (17-22) or operating Eq. (17-23). Equations (17-23) and (17-43) or (17-44) can also be solved analytically or numerically to obtain,

$$x_{out} = \frac{x_{in}}{\left(1 - R^o\right) M\theta + \left(1 - \theta\right)} \tag{17-44}$$

Note differences between UF Equation (17-44) and RO equation (17-25). The value of R^o in UF is approximately constant, whereas in RO, the value of R depends on both M and $(p_r - p_p)$.

Experimental results with low concentration feeds or under conditions where M is close to 1.0 are in good agreement with the theoretical predictions. However, when the wall concentration becomes high, the solvent flux J_{solv} often cannot be controlled by adjusting the pressure difference. Thus, Eq. (17-40) no longer holds! Some other phenomenon must be controlling the solvent flux. Careful examination of the membrane surface after these experiments shows a gel-like layer covering the membrane surface. This gel layer alters the flux-pressure drop relationship and controls the solvent flow rate.

The effect of gel formation at the wall can be studied using the diffusion equation. Return to Figure 17-10. We implicitly assumed that the wall concentration was a variable that could increase without bound. In gelling systems once the wall concentration equals the gel concentration, x_g, the concentration at the wall becomes constant at $x_w = x_g$. As additional solute builds up at the wall, the gel concentration is unchanged, but the thickness of the gel layer increases; thus, x_w becomes a constant set by the solute gelling behavior. This is illustrated in Figure 17-12. The value of x_g can vary from less than 1 wt % for polysaccharides to 50 vol % for polymer latex suspensions (Blatt et al., 1970).

To analyze gelling systems we can again use Eq. (17-31) when R = 1. (Once a gel forms it is usually quite immobile and 100% retention is reasonable.) With the coordinate system redefined as in Figure 17-12, the boundary conditions are the same as Eqs. (17-32). The solution has the same form as in Eq.(17-34); however, since x_w and x_r are fixed, the solvent flux is the variable. If we solve for the solvent flux, the result is,

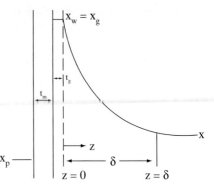

FIGURE 17-12. *Concentration polarization with gel layer at wall*

$$J'_{solv}/\rho_{solv} = \frac{D}{\delta} \, \ell n \, \frac{x_g}{x_r} = k \ell n \, \frac{x_g}{x_r} \qquad \textbf{(17-45)}$$

The consequence of this equation is that once a gel has formed the solvent flux is set by the rate of back diffusion of the solute. We no longer control the solvent flux!

What happens if we increase the pressure drop across the membrane for a system with a gel present? The solvent flux will temporarily increase, but more solute is carried to the membrane by the flow through the membrane than can be removed by back diffusion. The extra solute is deposited on the gel layer, which increases the thickness of the gel layer. This increases the resistance to flow, which reduces the flux through the membrane until the steady state solvent flux given in Eq. (17-45) is again obtained. Note that this effect is usually not reversible. Reduction of the pressure will result in a flow rate less than that predicted by Eq. (17-45) because the thicker gel layer remains on the membrane. The gel layer can often be removed by shutting down the system and backflushing (running pure solvent at a higher pressure on the permeate side) or by mechanical scrubbing of the membrane surface. Unfortunately, the membrane may become fouled after a gel forms and it may be very difficult to return the membrane to its original flux behavior. The use of fouling-resistant membranes is highly recommended under gelling conditions.

The engineer does have some control since decreasing the boundary layer thickness, δ, with thin channels or turbulence will increase the mass transfer coefficient k and hence the flux. To a limited extent D can be increased by raising the temperature. The effect of these variables on the mass transfer coefficient can be explored using correlation Eqs. (17-35) to (17-37). If fouling is severe after gel formation, it may be necessary to operate so that a gel will never form: $x_w = M x_r < x_g$ at all times. This can be done by operating under conditions that reduce M (high values of k) and keeping x_r low (low feed concentrations and low cuts). The feed-and-bleed system in Figure 17-2 is frequently used because very high velocities and hence larger k can be achieved. When a gel does not form the solvent flux in UF is controlled by Eqs. (17-41), the concentration polarization modulus can be calculated from Eq. (17-34), and solution is straightforward.

Most feeds processed by UF contain a number of different solutes. How should we analyze these separations? The almost overwhelming temptation is to study each solute individually and then assume that *superposition* is valid. That is, we assume that each solute in the mixture will behave in the same way as it does alone. After all, this is what we did for gas permeation and RO. Thus, we would predict that large molecules will be retained and small molecules will pass through the membrane. If no gel forms, this behavior is often observed; however, if a gel forms, the gel is usually much tighter (less porous) than the membrane, and the gel layer will usually capture the small molecules. Thus, the separation behaviors with and without the gel are quite different. If the purpose of the UF operation is to separate the large and small molecules, then we must operate under conditions where a gel will not form. Gel formation and fouling have prevented UF from achieving its full potential because they often limit both the separation and the flux.

RO membranes will also cause gelling and will foul if they are operated with feeds that can gel. To prevent this it is common to use an UF system before an RO system. Then the UF system will remove particulates and large molecules that could foul the RO membrane. This procedure is followed in the processing of whey where the UF system retains proteins and the RO system retains sugars and salts.

EXAMPLE 17-8. UF with gel formation

We are ultrafiltering latex particles which are known to form a gel when $x_w = x_g = 0.5$.

The well-mixed system is operated with permeate pressure of 1.0 bar and retentate pressure of 4.5 bar. We do a series of experiments with different values of the inlet concentration with a cut θ' of 1/3. The permeate wt frac is zero for all of the experiments. We find that gelling occurs when $x_{in} = 0.1466$, and the measured flux is $J' = 51.93$ g/(m²·s). Predict the solvent flux for inlet wt frac $x_{in} = 0.20$.

Solution

A. Define. Find J'_{solv} when $x_{in} = 0.20$, $\theta' = 1/3$, $p_p = 1.0$ bar, and $p_r = 4.5$ bar.

B and C. Explore and plan. Use a mass balance to determine $x_r = x_{out}$ for $x_{in} = 0.1466$, and Eq. (17-45) to determine k using the measured flux at this inlet wt frac. Then for $x_{in} = 0.20$ find $x_r = x_{out}$ (since x_p is reliably zero, we can assume it remains zero for $x_{in} = 0.20$). Use Eq. (17-45) to find J'_{solv} at this higher feed wt frac. Since $x_p = 0$, $\rho_{solv} = 997,000$ g/m³.

D. Do it. Experimental conditions:

$$\theta' = 1/3, \ F'_{out} = F'_{in} - F'_p, \ F'_{out} = (2/3)F'_{in}$$

With $x_p = 0$, $F'_{out} x_{out} = F'_{in} x_{in}$ or $x_{out} = \dfrac{F'_{in}}{F'_{out}} x_{in} = \dfrac{3}{2}x_{in}$

When $x_{in} = 0.1466$, $x_{out} = 1.5 \ (0.1466) = 0.220$.

Rearranging Eq. (17-45), $\quad k = \dfrac{J'_{solv}/\rho_{solv}}{\ell n(x_g/x_r)} = \dfrac{51.93/997000}{\ell n(0.5/0.220)} = 6.471 \times 10^{-5} \dfrac{g}{s \cdot m^2}$

Now solve Eq. (17-45) with $x_{in} = 0.20$ and $x_{out} = 1.5 \ x_{in} = 0.30$.

Final result is $J'_{solv} = 6.471 \times 10^{-5} \ \ell n(0.5/0.30)(997000 \text{ g/m}^3) = 33.0 \ \dfrac{g}{m^2 \cdot s}$

E. Check: Qualitatively we would expect a lower flux since there are more latex particles carried toward the wall per liter of fluid which permeates through the membrane. Other than checking the equations and calculations, a check is difficult.

F. Generalize. 1. We have assumed k is not concentration dependent. This is reasonable for particles.

2. If a correlation for k was available, we could use Eq. (17-45) and J'_{solv} to estimate x_g.

3. The general procedure was to:

a. Use rearranged design equation to find a design parameter (k) using experimental conditions. Then

b. Use design equation to predict flux under design conditions.

This general procedure is very common in all types of separation problems.

4. It is critically important to determine if a gel layer forms.

Warning! Experimental results and equations for both UF and RO are reported in many different units. It is easy to make a unit mistake if you do not carefully carry units in the equations. It is especially easy to make a subtle mistake in the appropriate solution density to use when converting from concentration c in g/L or mole/L to wt frac x. The correct conversion from g/L to wt frac is,

$$x = \left(c, \frac{\text{g solute}}{\text{L solution}} \right) \Big/ \left(\rho_{solution}, \frac{\text{g solution}}{\text{L solution}} \right) \qquad \textbf{(17-46)}$$

Unfortunately, the solution density $\rho_{solution}$ is a function of concentration. For a relatively pure permeate (R close to 1.0) the permeate solution density is approximately the solvent density. For more concentrated streams such as the feed or the retentate the density needs to be known or estimated as a function of concentration.

When concentrations are given in units of g/L, approximations are often made to calculate fluxes. For example, the correct equation for the solute mass flux is

$$J'_A \left(\frac{\text{g solute}}{\text{m}^2\text{s}} \right) = J_{solv} \left(\frac{\text{L solvent}}{\text{m}^2\text{s}} \right) + J_A \left(\frac{\text{L solute}}{\text{m}^2\text{s}} \right) \times c_p \left(\frac{\text{g solute}}{\text{L solution}} \right) \qquad \textbf{(17-47a)}$$

This is often approximated as

$$J'_A \left(\frac{\text{g solute}}{\text{m}^2\text{s}} \right) \approx J_{solv} \left(\frac{\text{L solvent}}{\text{m}^2\text{s}} \right) \times c_p \left(\frac{\text{g solute in permeate}}{\text{L solution}} \right) \qquad \textbf{(17-47b)}$$

For relatively high rejection coefficients permeate is quite pure, $J_{solv} \gg J_A$ and the approximation is quite good. In other cases the approximation may not be as accurate.

The inherent rejection and concentration polarization modulus are often defined in concentration units,

$$R_c^o = 1 - (c_p/c_{out}), \text{ for } M_c = 1 \qquad \textbf{(17-48)}$$

With these definitions the solute mass flux is

$$J_A' = M_c c_{out}(1 - R_c^o)J_{solution} \approx M_c c_{out}(1 - R_c^o)J_{solv} \qquad \textbf{(17-49)}$$

Equations (17-48) and (17-49) are valid ways to formulate the problem. Unfortunately, it may be assumed that $R^o = R_c^o$ and $M = M_c$ when the exact equations are

$$R^o = 1 - \frac{x_p}{x_{out}} = 1 - \frac{(c_p/\rho_{solution,p})}{(c_{out}/\rho_{solution,out})} \approx 1 - \frac{c_p}{c_{out}} = R_c^o \qquad \textbf{(17-50a)}$$

$$M = \frac{x_w}{x_{out}} = \frac{(c_w/\rho_{solution,p})}{(c_{out}/\rho_{solution,out})} \approx \frac{c_w}{c_{out}} = M_c \qquad \textbf{(17-50b)}$$

If $\rho_{solution,out}$ and $\rho_{solution,p}$ are very different, confusing R with R_c and M with M_c can result in significant error. To be exact, check how all terms are defined, use the appropriate solution densities of permeate and retentate to convert to wt frac units, and then solve in wt frac units.

17.6 PERVAPORATION (PERVAP)

Pervaporation can be traced back to Graham's work in the 1860s, but definite work was done by Binning in the 1950s and 1960s, with first commercialization for ethanol purification in the 1980s (Baker et al., 1990). Pervaporation is currently a rapidly expanding membrane separation technique because very high selectivities with reasonable fluxes are often obtained. In pervaporation a high-pressure liquid is fed to one side of the membrane, one component preferentially permeates the membrane, then evaporates on the downstream side, and a vapor product, the permeate, is withdrawn (Figure 17-13). The word "pervaporation" is a contraction of permeation and evaporation. The retentate, which does not permeate through the membrane, is a high-pressure liquid product. Either permeate or retentate may be the desired product.

In pervaporation the driving force for separation in Eq. (17-1) is the difference in activities across the membrane. The flux equation is (Eykamp, 1997),

$$J_i = -D_i c_i \frac{d \ln a_i}{dz} \qquad \textbf{(17-51)}$$

where a_i is the activity. Unfortunately, neither this equation nor Eq. (17-1) is of great practical use. Since a detailed analysis is usually not solvable in practical situations (e.g., see Neel, 1992) we will use a greatly simplified theory, which relies heavily on experimental selectivity data. The modeling procedures used earlier for gas permeation do *not* work for pervaporation because the fluxes of the two components interact significantly, the membrane swells, and permeabilities are very concentration dependent. Detailed models are available in the books by Dutta et al. (1996-97), Ho and Sirkar (1992), Huang (1991), Mulder (1996), and Noble and Stern (1995).

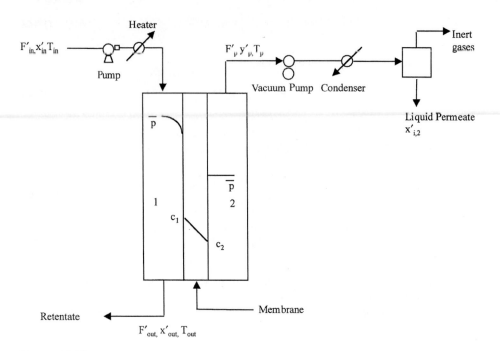

Heater

F'_{in}, x'_{in}, T_{in}

Pump

F'_p, y'_p, T_p

Vacuum Pump   Condenser

Inert
gases

Liquid Permeate
$x'_{i,2}$

$\overline{p}$

1

c_1

$\overline{p}$

2

c_2

Retentate

$F'_{out}, x'_{out}, T_{out}$

Membrane

FIGURE 17-13.   *Simplified schematic of single-pass pervaporation*

Since both selective membrane permeation and evaporation occur, pervaporation both separates and concentrates. Evaporating the liquid at the downstream side of the membrane also increases the driving force. Assume that the driving force is adequately represented by the partial pressure difference across the membrane. The local partial pressure on the upstream side, labeled 1 in Figure 17-13, is

$$p_{i,1} = x_{i,1} \, (VP)_{i,1} \qquad \textbf{(17-52a)}$$

while the local partial pressure on the downstream side, labeled 2, is

$$p_{i,2} = y_{i,2} \, p_{tot,2} \qquad \textbf{(17-52b)}$$

and the local driving force is

$$(driving\ force)_i = \Delta p_i = x_{i,1} \, (VP)_{i,1} - y_{i,2} \, p_{tot,2} \qquad \textbf{(17-52c)}$$

Since $x_{i,1}$, $y_{i,2}$ and $p_{tot,2}$ vary, the local values of the partial pressures and the driving force depend upon the flow patterns and pressure drop in the membrane module. The driving force can be increased by lowering $p_{tot,2}$ either by drawing a vacuum as shown in Figure 17-13, or, less frequently, reducing $y_{i,2}$ by using a sweep gas on the permeate side. The driving force will also be increased if the upstream partial pressure of component i is increased. This occurs for more concentrated feeds (higher $x_{i,1}$) and at higher upstream temperatures (larger $VP_{i,1}$). Use of higher upstream temperatures may require a higher upstream pressure to prevent vapor-

Chapter 17   Introduction to Membrane Separation Processes

ization of liquid on the upstream side. The higher upstream pressure does not increase the permeation rate significantly (Neel, 1991). Although the feed concentration is not usually a variable under the control of the designer in a single pass system (see Figure 17-13), it is a design variable in hybrid systems (see Figure 17-14).

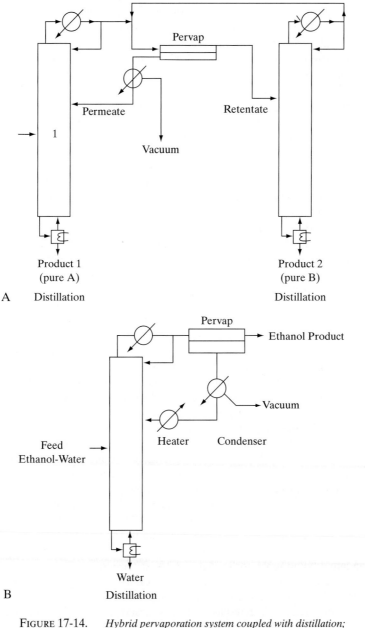

FIGURE 17-14. FIGURE 17-14.   *Hybrid pervaporation system coupled with distillation; (A) general two-column system, (B) simplified one-column system shown for dehydration of ethanol*

As usual with membrane separations, the membrane is critical for success. Currently, two different classes of membranes are used commercially for pervaporation. To remove traces of organics from water a *hydrophobic* membrane, most commonly silicone rubber is used. To remove traces of water from organic solvents a *hydrophilic* membrane such as cellulose acetate, ion exchange membrane, polyacrylic acid, polysulfone, polyvinyl alcohol, composite membrane, and ceramic zeolite is used. Both types of membranes are nonporous and operate by a solution-diffusion mechanism. Selecting a membrane that will preferentially permeate the more dilute component will usually reduce the membrane area required. Membrane life is typically about four years (Baker, 2004).

The use of evaporation to increase the driving force allows one to use a highly selective membrane and still retain a reasonable flux. However, evaporation complicates both the equipment and the analysis. Typical permeate pressures are quite low (0.1-100 Pa) (Leeper, 1992). Because of this low pressure, permeate needs a large cross-sectional area for flow or the pressure drop caused by the flow of the permeate vapor will be large. Thus, plate-and-frame, spiral-wound, and hollow fibers with feed inside the fibers are used commercially. Figure 17-13 shows that a vacuum pump and a condenser are required to recover the dilute low-pressure vapor. Unless the cut is small, an additional energy source is required to supply the heat of evaporation. If the cut is small, the energy in the hot liquid can supply this energy. For larger cuts a portion of the retentate can be heated and then recycled. More details on equipment and equipment design are given in Huang's (1991) book.

Although selectivity in pervaporation units can be quite high, (e.g., Leeper (1992) reports values from 1.0 to 28,000) the values are not infinite. Thus, there will always be some of the less permeable species in the permeate and some of the more permeable species in the retentate. If the feed concentration and the selectivity are high enough, the single pass system shown in Figure 17-13 can produce either a permeate or a retentate that meets product specifications. However, the other stream will contain a significant amount of valuable product. The single pass system can produce high purity or high recovery but not both simultaneously.

If both high purity and high recovery are desirable or the single pass system cannot meet the product specifications, a *hybrid* system with recycle is often used (Huang, 1991; Suk and Matsura, 2006; Wankat, 1990). Hybrid systems use two different types of separation to achieve the desired total separation. The most common pervaporation hybrid is to combine it with distillation with either two columns (Figure 17-14A) or a single column (Figure 17-14B). In Figure 17-14A the feed to distillation column 1 forms a minimum-boiling azeotrope. This distillation produces essentially pure component A as the bottoms product. The distillate product from column 1, which approaches the azeotrope concentration, is sent to the pervaporation unit. If component A preferentially permeates through the membrane, permeate will be more concentrated in A than the distillate. Permeate is then recycled to column 1 to recover the A product. Component B is retained in the pervaporation unit and is concentrated in retentate, which is the feed to distillation column 2. The distillate from column 2 also approaches the azeotrope concentration, and is part of the feed to the pervaporation unit. The bottoms product from column 2 is essentially pure B. The two distillation columns in Figure 17-14A will be similar to the columns in Figure 8-4A. Since the pervaporation unit replaces the liquid-liquid separator in Figure 8-4A, the azeotropic system does not have to have a heterogeneous azeotrope; thus, Figure 17-14A is more generally applicable.

If the selectivity is high enough, retentate may meet the purity specifications. Then distillation column 2 is not needed which results in obvious savings in capital and operating costs. This is illustrated with a simplified one-column flowchart used for breaking the ethanol-water azeotrope (Figure 17-14B). Open steam heating may be used instead of a reboiler (Leeper,

1992). A hydrophilic membrane that selectively permeates water is used. This figure is essentially the same as Figure 8-1B with the pervaporation system replacing the unidentified separator.

Hybrid systems are usually designed with a low cut per pass to prevent large temperature drops in the system. The total cut for the entire unit can be any desired value. Typically the vacuum pump is the highest operating expense. The load on the vacuum pump can be decreased by refrigerating the final stage of the permeate condenser. Appropriate design of the heat exchanger system and thermally integrating the pervaporation and distillation systems can significantly reduce the energy costs.

If selectivity and flux data are available, pervaporation units can be designed without knowing the details of the concentration polarization, diffusion, and evaporation steps. For a binary separation the selectivity, $\alpha_{A,B}$ is defined as,

$$\alpha_{A,B} = \frac{y_A/x_A}{y_B/x_B} = \frac{y_A/(1-y_A)}{x_A/(1-x_A)} \qquad \text{(17-53a,b)}$$

Where x and y are the wt frac in the liquid and vapor, respectively. Molar units can also be used, and care is necessary to keep units correct. This definition is superficially similar to the definition of relative volatility in distillation. Of course, here the selectivity is for a rate process and represents the RT curve, not equilibrium. Selectivity data are usually obtained under conditions where concentration polarization is negligible. Experimentally determined selectivities are complex functions of temperature and liquid mole frac (e.g., see Figure 17-15).

If data are available as selectivities, we can convert it to a y vs. x format by solving for y_A in Eq. (17-53a)

$$y_A = \frac{\alpha_{AB} x_A}{1 + (\alpha_{AB} - 1)x_A} \qquad \text{(17-53c)}$$

Since the selectivity depends on the liquid concentration, Eqs. (17-53a,b,c) are valid locally.

We will again consider the simplest case—membrane modules that are well-mixed on both the retentate and permeate sides. In this situation Eq. (17-53c) is valid with y replaced by y_p and x replaced by x_{out},

$$y_p = \frac{\alpha_{AB} x_{out}}{1 + (\alpha_{AB} - 1)x_{out}} \qquad \text{(17-54)}$$

where y_p and x_{out} refer to the wt frac of the more permeable component in the permeate and retentate, respectively. The selectivity in Eq. (17-54) must be determined at the operating temperature T and the liquid wt frac x_{out}. If data are available in the form of Figure 17-16, it must be at the operating temperature T of the pervaporation system.

The overall mass balance for the single pass system shown in Figure 17-13 is

$$F'_{in} = F'_p + F'_{out} \qquad \text{(17-55a)}$$

and the mass balance for the more permeable species is

$$F'_{in} x_{in} = F'_p y_p + F'_{out} x_{out} \qquad \text{(17-55b)}$$

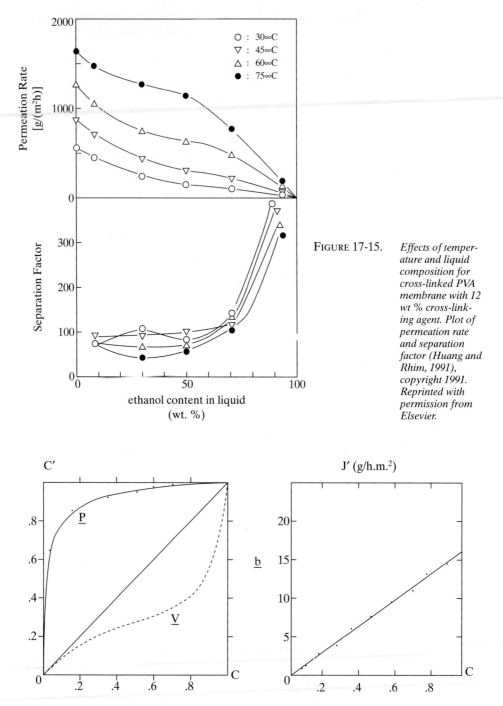

FIGURE 17-15. *Effects of temperature and liquid composition for cross-linked PVA membrane with 12 wt % cross-linking agent. Plot of permeation rate and separation factor (Huang and Rhim, 1991), copyright 1991. Reprinted with permission from Elsevier.*

Figure 17-16. *Vacuum-pervaporation of water-ethanol mixtures through homogeneous films made from hydrophilic polymer. c, c': Water concentration (by weight in liquid (c) and in permeate (c')); J': Permeation flux. Polyacrylonitrile film (20 μm thick), T = 25°C. (Neel, 1991), copyright 1991. Reprinted with permission from Elsevier.*

These equations are identical to the balances for gas permeation, Eqs. (17-7a) and (17-7b), except that x has replaced y for the feed and retentate. These two equations can be solved simultaneously for y_p. The resulting operating equation is

$$y_p = -\frac{(1 - \theta')x_{out}}{\theta'} + \frac{x_{in}}{\theta}$$

(17-56)

where the cut $\theta' = F'_p/F'_{in}$. This result is essentially the same as Eq. (17-8). If we want to use a graphical procedure, the operating equation will plot as a straight line on a graph of y_p vs. x_{out}.

If x_{out} and T are specified (which means α_{AB} is known), Eqs. (17-54) and (17-55a,b) can be solved simultaneously. The resulting equation is quadratic in y_p and linear in θ' (Wankat, 1990, pp. 707–709). We can easily solve for y_p. Unfortunately, this procedure is more complicated and less useful in real situations than it appears at first glance. Since neither x_{out} nor T is usually known, the selectivity is not known and the calculation becomes complicated. We will use a simultaneous graphical solution to develop a somewhat simpler procedure that will be illustrated in Examples 17-9 and 17-10.

An energy balance is needed to estimate the temperature of the pervaporation system. We will assume that the pervaporation system is operating at steady state and that it is adiabatic. Then the energy balance for the system shown in Figure 17-13 is

$$F'_{in}h_{in} = F'_p H_p + F'_{out}h_{out}$$

(17-57a)

We choose the reference point as pure liquid component A at temperature T_{ref}. Assuming that heat of mixing is negligible, the enthalpies of the liquid streams are

$$h_{in} = C_{PL,in}(T_{in} - T_{ref}), \qquad h_{out} = C_{PL,out}(T_{out} - T_{ref})$$

(17-57b)

and the vapor enthalpy is

$$H_p = C_{PV,p}(T_p - T_{ref}) + \lambda_p(y_p, T_{ref})$$

(17-57c)

where λ_p is the mass latent heat of vaporization of the permeate determined at T_{ref} and the heat capacities are also in mass units. Combining Eqs. (17-57) we obtain

$$F'_{in}C_{PL,in}(T_{in} - T_{ref}) = F'_p C_{PV,p}(T_p - T_{ref}) + F'_p\lambda_p + F'_{out}C_{PL,out}(T_{out} - T_{ref})$$

(17-58)

Since the membrane module is well mixed, it is reasonable to assume that the system is in thermal equilibrium, $T_{out} = T_p$. If we arbitrarily set the reference temperature equal to the outlet temperatures, $T_{ref} = T_{out} = T_p$, we obtain the simplified forms of the energy balance.

$$F'_{in}C_{PL,in}(T_{in} - T_{out}) = F'_p\lambda_p$$

(17-59a)

$$\theta' = \frac{C_{PL,in}}{\lambda_p}(T_{in} - T_{out})$$

(17-59b)

$$T_{out} = T_{in} - \frac{\theta' \lambda_p}{C_{PL,in}}$$

(17-59c)

From Eq. (17-59b) we can determine the cut necessary to obtain a specified outlet temperature, while Eq. (17-59c) allows us to determine the outlet temperature for any specified cut. Note that the relationship between cut and the temperature drop of the liquid stream is linear. Equation (17-59c) is instructive. Since the latent heat is significantly greater than the heat capacity, the outlet temperature drops rapidly as the cut is increased. To prevent this drop in temperature a recycle system with a low cut per pass is often used.

We are now ready to develop the solution procedure for completely mixed pervaporation systems. This procedure is straightforward if either the outlet temperature or the cut are specified. If the cut is specified, calculate T_{out} from Eq. (17-59c). To plot the RT curve, pick arbitrary values for x_{out}, calculate α_{AB} from appropriate data such as Figures 17-16 or 17-17, calculate y_p from the RT Eq. (17-54), and plot the point on the curve. If experimental y vs. x data at operating temperature T is available, plot it directly as y_p vs. x_{out}. The simultaneous solution to the selectivity equation and the mass balances is at the point of intersection of the RT curve and the straight operating line, Eq. (17-56) (see Figure 17-17 in Example 17-9).

EXAMPLE 17-9. Pervaporation: feasibility calculation

We wish to separate water from ethanol using a single-pass pervaporation system that uses a 20-micron thick polyvinyl alcohol film. The perfectly mixed pervaporation system will operate at 60°C with 1,000 kg/h of hot feed. The feed is 60 wt % water. If the retentate product is 20 wt % water, is this process feasible?

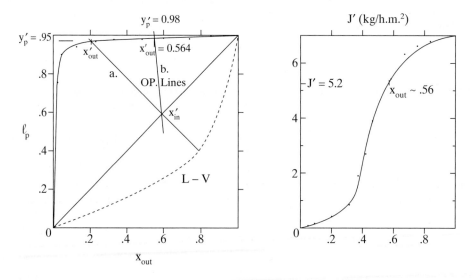

FIGURE 17-17. *Solution for Example 17-9; RT curve for water-ethanol through 20 μm polyvinyl alcohol at 60°C. (Neel, 1991). Flux curve from Neel (1991), copyright 1991. Reprinted with permission from Elsevier.*

Chapter 17   Introduction to Membrane Separation Processes

Solution

A. Define. In this case, feasibility means: can the process be conducted with the feed at a reasonable temperature? Thus, to determine feasibility we need to determine the required inlet temperature.

B and C. Explore and Plan. The rate transfer information and the flux data for this film are given by Neel (1991), and are shown in Figure 17-17. The data required for the energy balances is available in Perry and Green (1997) on pages 2-235 and 2-306:

Ethanol: $C_{PL,E} = 2.78$ kJ/(kg K) and $\lambda_E = 985$ kJ/kg. Estimated at 60°C.
Water: $C_{PL,W} = 4.185$ kJ/(kg K) and $\lambda_W = 2359$ kJ/kg. Estimated at 60°C.

We will solve for the cut and the feed temperature required and then discuss feasibility.

D. Do it. The data from Neel (1991) is plotted in Figure 17-17. The operating line (not shown by Neel), Eq. (17-56) goes through the point $y_p = x_{out} = x_{in}$ which is on the y = x line. The operating line must intersect the RT curve at $x_{out} = 0.20$. The permeate wt frac can be read at this point as $y_p = 0.95$ wt frac water.

Since all the wt frac are now known, it is easiest to simultaneously solve Eqs. (17-55a) and (17-55b) for the cut, θ'.

$$\theta' = \frac{x_{in} - x_{out}}{y_p - x_{out}} \qquad (17\text{-}60)$$

For this problem, $\theta' = (0.6 - 0.2)/(0.95 - 0.2) = 0.533$.

This cut appears to be too high. To check for feasibility, we can use the energy balance to estimate the required inlet temperature. Rearranging Eq. (17-59c), we obtain

$$T_{in} = T_{out} + \frac{\theta' \lambda_p}{C_{PL,in}} \qquad (17\text{-}61)$$

The latent heat of the permeate can be estimated as

$$\lambda_p = y_{p,W} \lambda_W + y_{p,E} \lambda_E \qquad (17\text{-}62)$$

$$\lambda_p = 0.95 \lambda_W + 0.05 \lambda_E = 2290 \text{ kJ/kg}$$

The heat capacity of the feed at 60 °C can be estimated as

$$C_{PL,in} = x_{in,W} C_{PL,W}(T = 60) + x_{in,E} C_{PL,E}(T = 60) \qquad (17\text{-}63)$$

$$C_{PL,in} = 0.60 \, C_{PL,W} + 0.40 \, C_{PL,E} = 3.62 \text{ kJ/(kg °C)}$$

Then from Eq. (17-61), $T_{in} = 60 + (0.533)(2290)/(3.62) = 396.9$°C.

This is obviously too hot. The process is not feasible because the membrane will be thermally degraded and the feed will not be a liquid. Too large a cut is used.

E. Check. Since typical values for the cut are approximately 0.1 or less, our conclusion that the process is not feasible is reasonable.
F. Generalization. Example 17-9 illustrates that operation of single-pass pervaporation units is controlled by the simplified energy balance, Eq. (17-59c). Essentially, there must be sufficient energy in the feed liquid to vaporize the desired fraction of the feed. High cuts require very hot feeds to supply the energy needed for vaporization.

EXAMPLE 17-10. Pervaporation: development of feasible design

Since the operation in Example 17-9 was not feasible, choose an appropriate feed temperature, and develop a feasible pervaporation process for the feed in Example 17-9. A permeate that is greater than 95 wt % water is desired. Determine the cut, permeate and retentate wt frac and the membrane area.

Solution

A. Define. An appropriate feed temperature is within the temperature limits of the membrane, and the feed must be liquid at a reasonable pressure. There is no single correct answer.

B. and C. Explore and plan. A pressurized feed will still be liquid at 120 °C. With a smaller cut, permeate will be greater than 95% water. If the cut were zero, the operating line will be vertical. From Figure 17-17, $y_p = 0.98$ in this case. We can use $y_p = 0.98$ to estimate λ_p, which is needed to calculate the cut. The accuracy of this estimate can be checked when we are finished.

D. Do it. We can estimate the latent heat of vaporization of the permeate product $(y_p - 0.98)$ as, $\lambda_p = 0.98 \lambda_w + 0.02 \lambda_E = 2331$ kJ/kg. The value of $C_{PL,in} = 3.62$ kJ/(kg °C) was determined in Example 17-9. Then from Eq. (17-59b),

$$\theta' = \frac{C_{PL,in}}{\lambda_p} (T_{in} - T_{out}) = (3.62)(120 - 60)/(2331) = 0.093, \text{ and}$$

$$F_p' = \theta' F_{in}' = (0.093)(1{,}000 \text{ kg/h}) = 93 \text{ kg/h}$$

From Eq. (16-59) the operating line has a slope = $-(1 - 0.093)/0.093 = -9.75$. The operating line is drawn in Figure 17-17. The permeate wt frac is about 0.98 (thus, the approximation was accurate) and the retentate is slightly above 0.56.

The membrane flux can be determined from the flux part of Figure 17-17. J′ is about 5.2 kg/(h m²). Then, membrane area = $F_p'/J' = 93/5.2 = 16.9$ m².

F. Generalization. At this retentate liquid concentration the membrane has a high flux. At the original retentate value, $x_{out} = 0.2$ used in Example 17-9, the flux is much lower and significantly more membrane area would be required.

Of course, there are many different feasible designs. If a higher permeate concentration is desired, the ethanol-water feed mixture in Example 17-10 would probably be concentrated in an ordinary distillation column to a concentration much closer to the concentration of the azeotrope (see Figure 17-14B). The retentate would typically be recycled to the distillation column. Pervaporation, azeotropic distillation (Chapter 8) and adsorption (Chapter 18) are all used commercially to break the ethanol-water azeotrope.

17.7 BULK FLOW PATTERN EFFECTS

Completely mixed membrane systems are relatively common in laboratory units since the effects of concentration polarization and gel formation can be minimized or eliminated. Commercial units usually use one of the modules shown in Figure 17-1 since a large membrane area can be packaged in a small volume. The flow patterns of the bulk fluid in these modules are unlikely to be perfectly mixed but are more likely to approximate cross-flow (Figure 17-18A), co-current (Figure 17-18B) or countercurrent (Figure 17-18C) flow. Since these bulk flow patterns all result in better separation than a perfectly mixed module, our previous results are conservative estimates. In this section we will analyze ideal examples (that is with no axial mixing and no concentration polarization) for gas permeation for these other flow patterns to determine the improvement in separation. Because real systems usually have some axial mixing, the separation is often between that predicted for the ideal case and the perfectly mixed case.

EXAMPLE 17-11. Flow pattern effects in gas permeation
The increase in separation for the different flow patterns can be compared for separation of oxygen and nitrogen. Geankoplis (2003) presents a problem with a feed

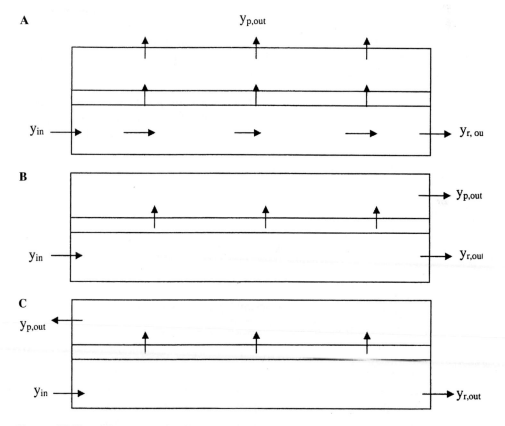

FIGURE 17-18. *Flow patterns in gas permeation systems; (A) cross-flow with unmixed permeate, (B) co-current flow, (C) countercurrent flow*

that is 20.9% oxygen, $p_r = 190$ and $p_p = 19$ cm Hg, $P_{O2}/t_{ms} = 1.9685 \times 10^{-5}$, $\alpha_{O2-N2} = 10.0$, $F_{in} = 1,000,000$ cm^3 (STP)/s, and cut = 0.2. The models discussed later in this section were implemented using the spreadsheets in the chapter appendix. The results shown in Table 17-3 illustrate that a countercurrent flow pattern provides better separation with less membrane area.

Binous (2006) solved the system with Mathematica and obtained essentially identical values for y_p and y_r, and areas that agree to the first three significant figures.

Geankoplis (2003) presents a complicated analytical solution for the completely mixed and cross-flow cases. He obtained identical results for the completely mixed case and almost identical results for cross-flow.

It is convenient to use staged models (Figure 17-19) to analyze the different flow patterns in permeation systems. Essentially, this represents a numerical integration of the differential equations representing the three configurations shown in Figure 17-18 (Coker et al., 1998).

17.7.1 Binary Cross-Flow Permeation

Spiral-wound membrane units are very close to cross-flow operation with an unmixed permeate (Figure 17-18A). We can approximate this cross-flow system as a series of N well-mixed stages as shown in Figure 17-19A. Since we know how to solve each of the well-mixed stages in Figure 17-19A, we can add all the results to find the behavior of the cross-flow system. Since a complicated analytical solution exists for binary cross-flow gas permeation (Geankoplis, 2003; Hwang and Kammermeyer, 1984), we can check the accuracy of our numerical solution.

For a design problem (cut is specified and membrane area is unknown) solution is easiest if we make the permeate flow rate $F_{p,j}$ for each small well-mixed stage the same $F_{p,j} = F_p$; thus, each stage has different areas. Then the flow rates from the cross-flow system are

$$F_{p,total} = \sum_{j=1}^{N} F_{p,j} = NF_p, \quad F_{r,N} = F_{r,out}$$

(17-64a, b)

and the area is

$$A_{total} = \sum_{j=1}^{N} A_j$$

(17-65)

TABLE 17-3. *Results for oxygen mole fracs for different flow patterns in gas permeation systems.*

Flow Pattern	y_p	y_r	Area, m^2
completely mixed*	0.507	0.135	32310
co-current	0.558	0.122	29550
cross-flow	0.568	0.119	29000
countercurrent	0.576	0.117	28600

* Result for N = 1 from co-current or cross-flow spreadsheets.

A

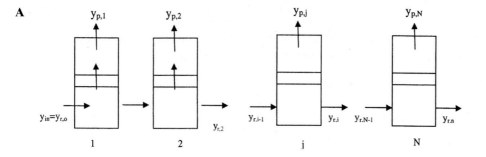

B

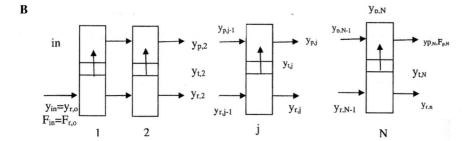

C

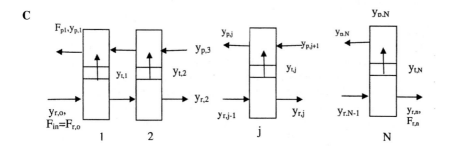

FIGURE 17-19. *Staged models for different flow patterns in gas permeation; (A) cross-flow system, (B) co-current system, (C) countercurrent system*

Since the stages are well-mixed and rapid mass transfer is assumed, $y_p = y_t$ and $y_r = y_{r,w}$. Thus, we can write Eqs. (17-6a) and (17-8a) for each small, well-mixed stage j using variables $y_{p,j}$, $y_{r,j}$ (both for the more permeable component), $\hat{F}_{p,j}$, and $\hat{F}_{r,j}$. For constant selectivity α, these equations can be solved simultaneously by substituting Eq. (17-8a) into (17-6a), rearranging, and solving with the quadratic equation. The resulting solutions for each stage are identical to Eqs. (17-10) (in Example 17-2) except y_{in} is replaced with $y_{r,j-1}$ and θ is replaced by $\theta_j = F_p / F_{r,j-1}$ because volumetric flow rates are in cm³(STP)/s. Retentate flow rates and retentate mole fracs for each stage can be found from the external mass balances Eq. (17-7), which can be written in terms of volumetric flow rates in cm³(STP)/s. (Note: For ideal gases, mole fraction = volume fraction.)

$$F_{r,j} = F_{r,j-1} - F_{p,j} = F_{r,j-1} - F_p \qquad \textbf{(17-66a)}$$

$$y_{r,j} = (F_{r,j-1}y_{r,j-1} - F_{p,j}y_{p,j})/F_{r,j} \qquad \textbf{(17-66b)}$$

Equations (17-10) and (17-66) are solved for each stage starting with j = 1. The area A_j for each stage can be determined from a rearrangement of Eq. (17-4b),

$$A_j = F_{p,j}y_{p,j}/[(P_A/t_{ms})(p_r y_{r,j} - p_p y_{p,j})] \qquad \textbf{(17-67)}$$

Once every stage has been calculated, the mole fracs for the cross-flow system are

$$y_{p,total} = (F_{p,j})\left(\sum_{j=1}^{N} y_{p,j}\right)\Big/ F_{p,total}, \; y_{r,out} = y_{r,N} \qquad \textbf{(17-68a, b)}$$

and the total area is given by Eq. (17-65). This set of equations is easy to program within a mathematical package. If N is large enough (in most cases N = 100 is sufficient) very good agreement will be obtained with the analytical solution, with less effort and less time. The numerical solution has the advantage that variable selectivity can easily be included. A spreadsheet program and example are in the chapter appendix. Shindo et al. (1985) show an alternate numerical solution method based on solution of the differential equations and apply it to multicomponent separations.

The cross-flow system with unmixed permeate increases the separation compared to a completely mixed permeation system. The cross-flow system works better than the well-mixed system because the average driving force, $[\Sigma(p_r y_{r,j} - p_p y_{p,j})]/N$, is considerably larger in the cross-flow system than the single driving force, $(p_r y_{r,out} - p_p y_p)$, in the well-mixed system. A real system probably has some but not complete mixing on both permeate and retentate sides; thus, the permeate mole frac is probably between the values calculated for the ideal cases.

The analysis for cross-flow can easily be extended to multicomponent systems. Now each stage is a perfectly mixed multicomponent system with a feed from the previous stage. The results for each stage can be calculated using the procedure in Section 17.3.3. Thus, a trial-and-error is needed on each stage, but the entire cascade is only calculated once. Coker et al. (1998) illustrate a different procedure for multicomponent cross-flow.

17.7.2 Binary Co-current Permeation

Tubular and hollow-fiber membrane systems can be operated in co-current flow (Figure 17-18B). Binary gas permeation with co-current flow can be analyzed with a staged model similar to that used for cross-flow except now $y_p \neq y_t$, and the flow pattern must be used to relate the permeate mole frac to the mole frac transferring through the membrane. The co-current model is shown in Figure 17-19B. This model again represents a numerical integration of the differential equations. Coker et al. (1998) present a solution method for multicomponent permeation when the membrane area is known but the cut is unknown, and show that N = 100 is usually sufficient.

The solution for a design problem (cut is specified, and membrane area is unknown) is again easiest if we make the total permeate mass transfer rate through the membrane for each small well-mixed stage, F_t, constant, which is the same as using the same fraction of the total cut on each stage.

$$F_t = F_{p,N}/N \text{ or } F_t/F_{in} = (F_{p,N}/F_{in})/N = \theta/N \qquad \textbf{(17-69a, b)}$$

thus, each stage has a different area. The flow rates leaving each stage are

$$F_{p,j} = F_{p,j-1} + F_t,$$
$$F_{r,j} = F_{in} - F_{p,j} \qquad \text{(17-69c,d)}$$

Because volumetric flow rates are in $cm^3(STP)/s$ and gases are ideal, these equations are equivalent to mole balances. The transfer rates for gases A and B are similar to Eqs. (17-5a) and (17-5b), respectively, except $y_p \neq y_t$,

$$F_t y_{t,j} = (P_A/t_{ms})(A_j)(p_r y_{r,j} - p_p y_{p,j}) \qquad \text{(17-70a)}$$
$$F_t(1 - y_{t,j}) = (P_B/t_{ms})(A_j)(p_r(1 - y_{r,j}) - p_p(1 - y_{p,j})) \qquad \text{(17-70b)}$$

The mass balance for the permeate is

$$F_{p,j} y_{p,j} = F_{p,j-1} y_{p,j-1} + y_{t,j} F_t \qquad \text{(17-71a)}$$

Solving for the mole frac of the transferred gas,

$$y_{t,j} = (F_p y_{p,j} - F_{p,j-1} y_{p,j-1})/F_t \qquad \text{(17-71b)}$$

Dividing Eq. (17-70a) by (17-70b) and simplifying we obtain

$$\frac{y_{t,j}}{1 - y_{t,j}} = \frac{\alpha[y_{r,j} - (p_p/p_r)y_{p,j}]}{((1 - y_{r,j}) - (p_p/p_r)(1 - y_{p,j}))} \qquad \text{(17-72)}$$

After substituting in Eq. (17-71b) and simplifying, this becomes

$$a_j y_{p,j}^2 + b_j y_{p,j} + c_j = 0 \qquad \text{(17-73)}$$

with

$$a_j = (1 - \alpha)[F^2_{p,j}/F_{r,j} + F_{p,j} p_p/p_r] \qquad \text{(17-74a)}$$

$$b_j = F_{p,j}[(\alpha - 1)F_{in} y_{in}/F_{r,j} + 1 - p_p/p_r] + [(\alpha - 1)F_{p,j-1} y_{p,j-1} + \alpha F_t]$$
$$\times [F_{p,j}/F_{r,j} + p_p/p_r] \qquad \text{(17-74b)}$$

$$c_j = (F_{in} y_{in}/F_{r,j})[(1 - \alpha)F_{p,j-1} y_{p,j-1} - \alpha F_t] + F_{p,j-1} y_{p,j-1}(p_p/p_r - 1) \qquad \text{(17-74c)}$$

In the limiting case of N = 1 Eqs. (17-74a) to (17-74d) simplify to Eqs. (17-10a) to (17-10d). The solution, as expected, is

$$y_{pj} = \frac{-b_j \pm \sqrt{b_j^2 - 4a_j c_j}}{2a_j} \qquad \text{(17-75)}$$

To use Eqs. (17-74) and (17-75), we must know $y_{p,j-1}$. Then after $y_{p,j}$ has been calculated, we can obtain $y_{r,j}$ from a mass balance around the permeator.

$$y_{r,j} = -F_{p,j}y_{p,j}/F_{r,j} + F_{in}y_{in}/F_{r,j} \qquad (17\text{-}76)$$

Now we can find $y_{t,j}$ from Eq. (17-71b), and A_j from either Eq. (17-70a) or (17-70b). The value of j is increased by one and the calculation is repeated for the next stage. When stage $j = N$ has been calculated, the outlet mole fracs are $y_{p,N}$ and $y_{r,N}$. The total membrane area can then be calculated by summing all the A_j, Eq. (17-65). Since no trial-and-error is involved, the calculation is very quick. A spreadsheet program and example are in the chapter appendix.

The analysis for co-current flow can easily be extended to multicomponent systems. This extension is very similar to the extension for cross-flow systems.

17.7.3 Binary Countercurrent Flow

Tubular and hollow-fiber modules can be arranged to have approximately countercurrent flow patterns (Figure 17-18c). Since these flow patterns are an advantage in equilibrium-staged systems, we would expect that they might be an advantage in membrane separations also. Example 17-11 showed that this is true. The staged model for countercurrent flow is shown in Figure 17-19c. Coker et al. (1998) present a solution method for a staged model for countercurrent, multicomponent permeation when the membrane area is known but the cut is unknown, and show that $N = 100$ is usually sufficient.

Developing a design method for a binary, countercurrent permeator is challenging. Typically the values of F_{in}, $y_{r,0}$, p_p, p_r, P_A/t_{ms}, P_B/t_{ms}, temperature, and one additional variable, typically the cut θ, will be specified. The difficulty with countercurrent flow is deciding how to get started, since generally not everything needed to start is known at either end of the system. We will again assume that F_t is the same for every stage, which requires the area of each stage to be different. Since $y_{r,N}$ is unknown, we will *assume* a value for $y_{r,N}$ so that we can step off stages from the retentate product end of the column. This makes the calculation trial-and-error, but convergence to the correct value of $y_{r,N}$ appears to be rapid if an appropriate damping factor is selected.

To start stepping off stages in Figure 17-19c, we know or have assumed F_{in}, $y_{r,N}$, $y_{r,0}$, N (~ 100), and cut θ. Then because volumetric flow rates are in cm³(STP)/s, $F_{p,1} = \theta F_{in}$, $F_{r,N} = F_{in} - F_{p,1}$, and the transfer rate F_t is,

$$F_t = F_{p,N}/N = \frac{F_{r,N}\theta/N}{(1-\theta)} \qquad (17\text{-}77)$$

The procedure to step off stages is similar to the procedure to step off stages for the co-current system, Eqs.(17-69) to (17-76), but the mass balance equations change since the permeate flow direction has changed and we are stepping off stages backwards. The steps in the derivation starting with the equation equivalent to Eq. (17-69c) and ending with Eq. (17-76) are given in Table 17-4.

For the stage by stage procedure, start with stage N, since $y_{r,N}$ is known. Counter i goes from 1 to N. Then for any i, stage $j = N - i + 1$.

1. Calculate flow rates $F_{r,j-1}$ and $F_{p,j}$ from Eq. ($\overline{\text{17-69c, d}}$) and $y_{r,j-1}$ from Eq. ($\overline{\text{17-69 e}}$) (Table 17-4).
2. Calculate $y_{p,j}$ from Eq. ($\overline{\text{17-75}}$) using a, b, and c values from Eqs. ($\overline{\text{17-74 a, b, c}}$) (Table 17-4).
3. Calculate $y_{t,j}$ from Eq. ($\overline{\text{17-71, b}}$) (Table 17-4), and A_j from either Eq. (17-70a) or (17-70b).
4. If $i < N$, let $i = i + 1$ and repeat starting with step 1.
5. Check if correct value of $y_{r,N}$ was used.

Eqs. $(\overline{17\text{-}69a,b})$ are identical to Eqs. (17-69a, b). The flow rate equations differ,

$$F_{r,j} = F_{r,j+1} + F_t \qquad (\overline{17\text{-}69c})$$

$$F_{p,j} = F_{p,j+1} - F_t \qquad (\overline{17\text{-}69d})$$

The mass balance around stages j+1 to N (Figure 17-20c) is,

$$y_{r,j} = -y_{p,j+1}F_{p,j+1}/F_{r,j} + y_{r,N}F_{r,N}/F_{r,j} \qquad (\overline{17\text{-}69e})$$

Eqs. $(\overline{17\text{-}70a, b})$ are identical to Eqs. (17-70a, b), but permeate mass balance becomes

$$F_{p,j}y_{p,j} = F_{p,j+1}y_{p,j+1} + F_t y_{t,j} \qquad (\overline{17\text{-}71a})$$

$$y_{t,j} = (F_{p,j}y_{p,j} - F_{p,j+1}, y_{p,j+1})/F_t \qquad (\overline{17\text{-}71b})$$

Eqs. $(\overline{17\text{-}72})$ and $(\overline{17\text{-}73})$ are identical to Eqs. (17-72) and (17-73), but the a, b, and c terms in Eq. (17-73) are different.

$$a_j = (F_{p,j}p_p/p_r)(1 - \alpha) \qquad (\overline{17\text{-}74a})$$

$$b_j = (\alpha - 1)(p_p/p_r)F_{p,j+1}y_{p,j+1} + \alpha(p_p/p_r)F_t + F_{p,j}(\alpha - 1)y_{r,j} + F_{p,j}(1 - p_p/p_r) \qquad (\overline{17\text{-}74b})$$

$$c_j = (F_{p,j+1}y_{p,j+1})[-1 - (\alpha - 1)y_{r,j} + p_p/p_r] - \alpha y_{r,j}F_t \qquad (\overline{17\text{-}74c})$$

Eq. $(\overline{17\text{-}75})$ is identical to Eq. (17-75). Once $y_{p,j}$ is known, we can calculate $y_{r,j-1}$ from the mass balance around stages j to N (Figure 17-19C) by setting $j = j-1$ in Eq. $(\overline{17\text{-}69e})$.

a. The calculated value of $y_{in,calc} = y_{r,0}$. If $y_{in,calc} \neq y_{in}$, adjust the guess for $y_{r,N}$. The following empirical approach appears to work well

$$y_{r,N,new} = y_{r,N,old} + (\text{damping factor})(y_{in} - y_{in,calc}) \qquad (17\text{-}78)$$

where the damping factor is to prevent excessive oscillation. A value of 0.9 is reasonable for small cuts such as 0.2 in Example 17-11. For larger cuts lower values may be required, and some experimentation may be required to obtain convergence.

b. When $y_{in,calc} = y_{in}$ within some tolerance ε, go to step 6 and finish the calculation.

6. Calculate the area from Eq. (17-65).

This calculation is fast on a spreadsheet. A spreadsheet program and example for the binary design procedure are in the chapter appendix.

Coker et al. (1998) present a staged model for multicomponent simulation problems (membrane area known) that develops a tri-diagonal matrix similar to those developed previously for distillation and for absorption. All internal flows need to be assumed initially, and a sum-rates method was used to find new flow rates. A good initial guess of flow rates is required for convergence. This procedure is recommended for multicomponent countercurrent modules instead of using a stage-by-stage calculation. The difficulty with the stage-by-stage calculation is C – 1 mole fracs (C = number of components) need to be estimated to specify the exiting retentate, and then trial and error is needed for these C 1 mole fracs.

Countercurrent membrane systems can also use a sweep gas or retentate reflux on the permeate side to increase separation (Baker, 2002). The spreadsheet in Appendix 17.A.3 is easily modified for these systems.

17.8 SUMMARY—OBJECTIVES

In this chapter we have studied membrane separations including gas permeation, RO, UF and pervaporation. At the end of this chapter you should be able to satisfy the following objectives:

1. Explain the differences and similarities between the different membrane separation systems
2. For a perfectly mixed system with no concentration polarization, predict the performance of an existing membrane separation system and design a new membrane separation system using both analytical and graphical analysis procedures
3. Explain and analyze the effect of concentration polarization and include it in the design of perfectly mixed membrane separators
4. Explain and analyze the effect of gel formation and include it in the design of UF systems
5. Use energy balances to determine appropriate operating conditions for pervaporation systems
6. Explain and analyze the effects of flow patterns on the separation achieved in membrane systems

REFERENCES

Baker, R. W., "Future Directions of Membrane Gas Separation Technology," *Ind. Engr. Chem. Research, 41,* 1393 (2002).

Baker, R. W., "Membrane Technology," in A. Seidel (Ed.), *Kirk-Othmer Encyclopedia of Chemical Technology,* 5th ed., Vol. 15, Wiley-Interscience, New York, 2004, pp. 796–852

Baker, R. W., and I. Blume, "Permselective Membranes Separate Gases," *Chem. Tech., 16,* 232 (1986).

Baker, R. W., E. L. Cussler, W. Eykamp, W. J. Koros, R. L. Riley, and H. Strathmann, *Membrane Separation Systems,* U.S. Department of Energy, DOE/ER30133-H1, April 1990. Reprinted by Noyes Data Corp., Park Ridge, NJ, 1991.

Binous, H., "Gas Permeation Computations with Mathematica," *Chem. Eng. Educ.,* 140 (Spring 2006).

Blatt, W. F., A. Dravid, A. S. Michaels, and L. Nelson, "Solute Polarization and Cake Formation in Membrane Ultrafiltration: Causes, Consequences and Control Techniques," in J. E. Flinn, (Ed.), *Membrane Science and Technology,* Plenum Press, New York, 1970, pp. 47–97.

Chern, R. T., W. J. Koros, H. B. Hopfenberg, and V. T. Stannett, "Material Selection for Membrane-Based Gas Separations," in D. R. Lloyd (Ed.), *Materials Science of Synthetic Membranes,* Am. Chem. Soc., Washington, D. C., Chap. 2, 1985.

Coker, D. T., B. D. Freeman, and G. K. Fleming, "Modeling Multicomponent Gas Separation Using Hollow-Fiber Membrane Contactors," *AChE Journal, 44,* 1289 (1998).

Conlee, T. D., H. C. Hollein, C. H. Gooding, and C. S. Slater, "Ultrafiltration of Dairy Products in a ChE Laboratory Experiment," *Chem. Eng. Educ., 32* (4), 318 (Fall 1998).

Drioli, E., and M. Romano, "Progress and New Perspectives on Integrated Membrane Operations for Sustainable Industrial Growth," *Ind. Engr. Chem. Research, 40,* 1277 (2001).

Dutta, B. K., W. Ji, and S. K. Sikdar, "Pervaporation: Principles and Applications," *Sep. Purif. Methods, 25,* 131–224 (1996–97).

Eykamp, W., "Membrane Separation Processes," R. H. Perry and D. H. Green (Eds.), *Perry's Chemical Engineers' Handbook*, 7th ed., McGraw-Hill, New York, pp. 22-37 to 22-69, 1997.

Geankoplis, C. J., *Transport Processes and Separation Process Principles (Includes Unit Operations)*, 4th ed., Prentice Hall, Upper Saddle River, NJ, 2003, Chap. 13.

Geraldes, V., N. E. Pereira, and M. Norberta de Pinho, "Simulation and Optimization of Medium-Sized Seawater Reverse Osmosis Processes with Spiral-Wound Modules," *Ind. Engr. Chem. Research, 44*, 1897 (2005).

Hagg, M.-B., "Membranes in Chemical Processing. A Review of Applications and Novel Developments," *Sep. Purif. Methods, 27,* 51–168 (1998).

Ho, W. S. W., and K. K. Sirkar (Eds.), *Membrane Handbook*, Van Nostrand Reinhold, New York, 1992.

Hoffman, E. J., *Membrane Separations Technology: Single-Stage, Multistage, and Differential Permeation,* Gulf Professional Publishing (Elsevier Science), Amsterdam, 2003.

Huang, R. Y. M. (Ed.), *Pervaporation Membrane Separation Processes*, Elsevier, Amsterdam, 1991.

Huang, R. Y. M., and J. W. Rhim, "Separation Characteristics of Pervaporation Membrane Separation Processes," in R. Y. M. Huang (Ed.), *Pervaporation Membrane Separation Processes*, Elsevier, Amsterdam, 1991, pp. 111–180.

Hwang, S.-T., and K. Kammermeyer, *Membranes in Separations,* Reprint edition, Robert E. Kreiger Publishing, Malabar, FL, 1984. Original edition, Wiley, 1975.

Kesting, R. E. and A. K. Fritzsche, *Polymeric Gas Separation Membranes*, Wiley, New York, 1993.

Kucera, J., *Reverse Osmosis: Industrial Applications and Processes*, Scrivener Publishing, Salem, MA, 2010.

Leeper, S. A., "Membrane Separations in the Recovery of Biofuels and Biochemicals: An Update Review," in N. N. Li and J. M. Calo (Eds.), *Separation and Purification Technology*, Marcel Dekker, New York, 1992, pp. 99–194.

Leeper, S. A., D. H. Stevenson, P. Y.-C. Chiu, S. J. Priebe, H. F. Sanchez, and P. M. Wikoff, *Membrane Technology and Applications: An Assessment,* U. S. Department of Energy, Feb. 1984.

Li, N. N., and S. S. Kulkarni, "Membrane Separations," in *McGraw-Hill Encyclopedia of Science and Technology,* 8th ed., McGraw-Hill, New York, 1997, Vol. 10, pp. 670–671.

Loeb, S., and S. Sourirajan. "Seawater Demineralization by Means of a Semipermeable Membrane," University of California at Los Angeles Engineering Report No. 60-60, 1960.

Loeb, S., and S. Sourirajan, "Seawater Demineralization by Means of an Osmotic Membrane," *Advan. Chem. Ser., 38,* 117 (1963).

Mohr, C. M., S. A. Leeper, D. E. Englegau, and B. L. Charboneau, *Membrane Applications and Research in Food Processing: An Assessment*, U.S. Department of Energy, August 1988.

Mulder, M., *Basic Principles of Membrane Technology*, 2nd ed., Kluwer Academic Publishers, Dordrecht, 1996.

Nakagawa, T., "Gas Separation and Pervaporation," in Y. Osada and T. Nakagawa (Eds.), *Membrane Science and Technology*, Marcel Dekker, New York, 1992, Chap. 7.

Neel, J., "Introduction to Pervaporation," in R. Y. M. Huang (Ed.), *Pervaporation Membrane Separation Processes*, Elsevier, Amsterdam, 1991, pp. 1–109.

Noble, R. D., and A. Stern (Eds.), *Membrane Separations Technology,* Elsevier, Amsterdam, 1995.

Noble, R. D., and P. A. Terry, *Principles of Chemical Separations with Environmental Applications,* Cambridge University Press, Cambridge, UK, 2004.

Osada, Y., and T. Nakagawa (Eds.), *Membrane Science and Technology*, Marcel Dekker, New York, 1992.

Perry, R. H., and D. W. Green (Eds.), *Perry's Chemical Engineers' Handbook,* 7th ed., McGraw-Hill, New York, 1997.

Porter, M. C., "Membrane Filtration," in P. A. Schweitzer (Ed.), *Handbook of Separation Techniques for Chemical Engineers,* 3rd ed., McGraw-Hill, New York, 1997, Section 2.1.

Reid, C. E., "Principles of Reverse Osmosis," in U. Merten (Ed.), *Desalination by Reverse Osmosis,* MIT Press, Cambridge, MA, Chapt. 1, 1996.

Reisch, M. C., "Filtering Out the Bad Stuff," *Chem. & Eng. News, 21* (April 23, 2007).

Robeson, L. M., "Correlation of Separation Factor Vesus Permeability of Polymeric Membranes," *J. Membrane Sci., 62,* 165 (1993).

Schock, G., and A. Miquel, "Mass-Transfer and Pressure Loss in Spiral-Wound Modules," *Desalination, 64,* 339 (1987).

Shindo, Y., T. Hakuta, H. Yoshitome, and H. Inoue, "Calculation Methods for Multicomponent Gas Separations by Permeation," *Separ. Sci Technol., 20,* 445 (1985).

Suk, D. E., and T. Matsura, "Membrane-based Hybrid Processes: A Review," *Separ. Sci Technol., 41,* 595 (2006).

Wankat, P. C., *Rate-Controlled Separations,* Kluwer, Amsterdam, 1990, Chaps. 12 & 13.

Wankat, P. C., and K. S. Knaebel, "Mass Transfer," in D. W. Green and R. H. Perry (Eds.), *Perry's Chemical Engineers' Handbook,* 8th ed., McGraw-Hill, New York, 2008, pp. 5-43 to 5-83.

Warashina, T., Y. Hashino, and T. Kobayashi, "Hollow-fiber Ultrafiltration," *Chem. Tech., 15,* 558 (1985).

HOMEWORK

A. *Discussion Problems*

A1. Membrane systems are rate processes and flash distillation is an equilibrium process. Explain why the solution methods are so similar for well-mixed membrane separators and flash distillation.

A2. How would you use the countercurrent spreadsheet program in Appendix A.17.3 as a simulator instead of as a designer?

A3. What are the advantages of asymmetric membranes compared to symmetric membranes?

A4. What are the advantages of hollow-fiber membranes compared to other geometries? When would one use other geometries?

A5. In large membrane systems it is common to have membrane cascades with membranes arranged both in parallel and in series. What are the advantages of this? The arrangements after the first set of membranes in series usually will have fewer membranes in parallel. This has been called a Christmas tree pattern. Draw this.

A6. Two membranes may be made of the same polymer but have very different behaviors. Explain why.

A7. We are operating a stirred-tank UF system to concentrate an intermediate molecular weight polypeptide. Unfortunately, for the membrane available, the retention is significantly lower than desired. Two options for improving the retention have been proposed. Explain the conditions for which each option will or will not improve retention.
   a. Increase the stirrer speed.
   b. Decrease the stirrer speed.

A8. We try decreasing the stirrer speed to increase the retention of the intermediate molecular weight polypeptide and find it works. Unfortunately, the separation between the intermediate molecular weight polypeptide and a low molecular weight compound becomes much worse. Explain what happened.

A9. The Fly-by-Night Membrane Separation Company claims to have the solution for the osmotic pressure limitation that occurs in RO of aqueous solution of concentrated salts. Their proposal is basically to use UF instead of RO. Your boss wants to know if your company should invest in the Fly-by-Night Membrane Separation Company. What do you tell your boss, and what is the reason?

A10. Some azeotropic mixtures can be separated by sending the vapor mixture to a gas permeation system—designated as vapor permeation if the mixture is easily condensed (Huang, 1991; Neel, 1991)—and some (probably different) azeotropic mixtures can be separated by sending a liquid mixture to an RO system. Why is pervaporation a much more popular method of separating azeotropic mixtures? Note: In all cases a hybrid membrane-distillation system will probably be used.

B. *Generation of Alternatives*

B1. There are a number of ways distillation and membrane separators can be combined as hybrid systems. Brainstorm as many methods as you can.

B2. Devise schemes that will increase separation of the intermediate molecular weight polypeptide from a low molecular weight compound if the retention of the intermediate molecular weight polypeptide is initially too low.

C. *Derivations*

C1. Derive Eqs. (17-10) for a gas permeator.

C2. Solve Eq. (17-19) for x_p as a function of x_r.

C3. Derive Eq. (17-31) from shell balances. You should obtain a second-order equation. Do the first integration and apply the boundary condition $R = 1.0$ at $z = 0$.

C4. For an RO system where osmotic pressure is a linear function of weight fraction of solute, $\pi = a'x$, show that at the same values of x_p, the relationship between R, R^o, and M is

$$R = 1 - M(1 - R^o)$$

What assumptions have to be made to use this relationship for a different RO module?

C5. Derive Eq. (17-11e).

D. *Problems*

Answers to problems with an asterisk are at the back of the book.

D1.* A gas permeation system with a cellulose acetate membrane will be used to purify a carbon dioxide-methane stream. The permeabilities are given in Example 17-2. The effective membrane thickness $t_{ms} = 1.0$ μm (a micron $= 1.0 \times 10^{-6}$ meter). Operation is at 35°C. $p_H = 12$ atm and $p_L = 0.2$ atm (a vacuum). The feed is 15 mol% carbon dioxide. Assume ideal gases.

   a.* A single-stage, well-mixed membrane separator will be used. Operate with a cut, θ $= F_p/F_{in} = 0.32$. Find y_p, y_{out}, and flux of carbon dioxide. If we feed 1 kmol/h of feed, calculate the membrane area needed, F_p and F_{out}.

   b. Design the two-step gas permeation system shown in the figure on the following page. Each stage is well mixed. Use the same feed to the first stage as in part a. Pressures and feed mole frac are the same as in part a. Set $F_{p1} = F_{p2} = 0.5\,F_{p,parta}$ (then $F_{out,2} = F_{out,parta}$). Find y_{p1}, y_{p2}, $y_{out1} = y_{in2}$, y_{out2} and F_{out1}. Find the carbon dioxide fluxes in both stages and the membrane areas required in both stages.

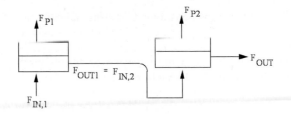

D2. We are separating a gas stream by gas permeation in a perfectly mixed system. The feed is 20 mol% carbon dioxide and 80 mol% methane. The membrane permeabilities are $P_{CO2} = 15.0 \times 10^{-10}$ and $P_{CH4} = 0.48 \times 10^{-10}$ [ccSTP cm]/[cm² s cm Hg]. The selectivity = 31.25. The effective thickness of the membrane skin is $t_{ms} = 1.0$ microns. We operate with $p_p = 3.3$ atm, $p_r = 60$ atm, $F_{in} = 2$ mol/s, and $F_p/F_{in} = 0.3$. Find:

a. y_p, y_{out}.

b. Membrane area needed in m².

D3. We are doing RO experiments with a completely mixed laboratory unit.

Experiment A. With a very high stirrer speed so that there is no concentration polarization, we do an experiment with a feed that is 0.0077 wt frac sodium chloride in water. The cut θ' (in mass units) is 0.22 and the measured inherent retention is $R^\circ = 0.9804$. The permeate pressure was 1.1 atm and the solvent flux was 415.4 g/(m²·s). $K_{solv}/t_{ms} = 33.29$ g/(atm m²·s). Data: Assume solutions have same density as water = 997 kg/m³. For dilute systems the osmotic pressure of sodium chloride is $\pi = a' x$ where x is in wt frac, π in atm, and $a' = 15\,446$ atm/(wt frac).

For experiment A, the retentate pressure was not recorded. Calculate the value of retentate pressure that was used for this experiment.

D4*. A cellulose acetate membrane is being used for RO of aqueous sucrose solutions at 25°C.

Data: Density of solvent (water) is $\rho = 0.997$ kg/L.

Density (kg/L) of dilute aqueous sucrose solutions is $\rho = 0.997 + 0.4$ x where x is the wt frac sucrose.

At low sucrose wt frac the osmotic pressure (in atms) can be estimated as $\pi = 59.895$ x where x is the wt frac sucrose.

Molecular weight of water is 18.016. Molecular weight of sucrose is 342.3.
(Note: Some of this data may not be needed to solve this problem.)

Experiment A. This experiment is done in a well-mixed stirred tank. At 1000 rpm with a 3 wt % solution of sucrose in water using $p_r = 75$ atm and $p_p = 2$ atm, we obtain $J'_{solv} = 4.625$ g/(m²s). The mass-transfer coefficient $k = 6.94 \times 10^{-5}$ m/s. We measure $x_r = x_{out} = 0.054$ and $x_p = 3.6 \times 10^{-4}$.

a. Calculate the concentration polarization modulus M.

b. Calculate selectivity α', K'_{solv}/t_{ms}, and K'_A/t_{ms}.

c. Estimate the mass transfer coefficient k if the rpm is increased to 2000.

D5. We are using a cellulose acetate RO membrane to concentrate a dilute sucrose mixture. Operation is at 25°C.

a. We measure the pure water flux (no sucrose present) and find at a pressure drop of 102 atm across the membrane, $J'_{solv} = 1.5 \times 10^{-3}$ [g/(cm² s)]. An experiment in a highly stirred system ($M = 1$) with a dilute sucrose solution gives a rejection $R^o = 0.997$ for an inlet wt frac of 0.050 and cut, $\theta' = 0.45$ (in weight units). Find (K'_{solv}/t_{ms}) and the selectivity.

b. We will design a perfectly mixed RO system to operate at conditions where $M = 3.0$. The feed is 2.0 wt % sucrose. We want a cut, $\theta' = 2/3$. Feed rate will be 5 kg/s. Pressure drop across the membrane is 78 atm. Find x_p, $x_r = x_{out}$, J'_{solv}, and membrane area.

Data: At low concentrations the osmotic pressure (in atm.) of an aqueous sucrose solution can be estimated as $\pi = 59.895\ x$, where x is wt frac sucrose, and the density (in g/ml) can be estimated as $\rho = 0.997 + 0.4\ x$.

D6. UF of a dextran solution in a well-mixed system with gel formation gives the following data:

J', g/(cm² min)	0.052	0.037	0.026	0.0134
x_{out}, wt frac dextran	0.012	0.03	0.06	0.135

Determine k and the wt frac at which dextran gels.

D7. An UF membrane is first tested in a stirred cell where there is no concentration polarization. The experimental values obtained are listed in the table. Then the same membrane is used in a spiral-wound module where there is concentration polarization. Values listed are obtained. Assume membrane thickness and permeabilities are the same. Estimate R^o_c from the stirred cell data and assume the inherent membrane rejection is unchanged. Calculate the expected solvent flux and the concentration polarization modulus M_c in the spiral-wound membrane system. No gel layer forms. Use concentration units, g/L. Assume density = 0.997 kg/L.

	c_{out}(conc. solute)	c_p(conc. solute)	J'_{solv}, g/m²s	Δp, bars
Stirred Cell	10 g/L	0.30 g/L	69.25	3.0
Spiral-Wound	8 g/L	1.0 g/L		3.5

D8. A perfectly mixed membrane module is used to concentrate helium in the retentate. The feed is 5 vol % helium and 95 vol % hydrogen at 5 atm. The membrane is natural rubber of the type used by Nakagawa (Table 17-2) and operation is at 25°C. Active portion of membrane is 1.0 micron thick. Permeate pressure is 1.0 atm. Assume gases are ideal.

a. If $y_{pHe} = 0.025$, what are cut θ and y_{rHe}?
b. If $\theta = 0.75$, what are y_{pHe} and y_{rHe}?
c. If $y_{rHe} = 0.06$, what are cut θ and y_{pHe}?
d. For part a, what membrane area is required for a feed of 100m³(STP)/h?

D9. The following data were obtained for UF of skim milk in a spiral-wound system (Con-lee, et al., 1998).

x_r (wt % solids)	J'_{solv} (kg/(m² h))
21	21.9
6.6	54.8
2.7	67.8
1.9	73.8
1.4	74.8
1.2	75.8
1.05	79.8

a. Estimate the gelling wt frac, x_g, and the mass transfer coefficient, k.

b. If you were the technician's supervisor and one more run could be done, what data would you want the technician to obtain to improve the estimate of x_g?

D10. We are separating CO_2 from CH_4 in a perfectly mixed membrane module using a silicone rubber membrane at 25°C. The permeabilities are $P_{CO_2} = 2700 \times 10^{-10}$ and $P_{CH_4} = 800 \times 10^{-10}$ cm^3 (STP) cm/[cm^2·s·cm Hg]. The thickness of the active layer of the membrane is $t_{ms} = 1.2$ microns. The permeate pressure $p_p - 0.5$ atm, and the retentate pressure is $p_r = 2.0$ atm. Both gases can be assumed to be ideal gases. The feed is 25 mol% CO_2 and 75 mol% CH_4.

a. Find y_p and y_r if the cut $\theta = 0.30$.

b. Find the area for part a if $\hat{F}_{in} = 1.0$ mol/s of feed gas.

c. For $\theta = 0.4$, $y_p = 0.325$, and $y_r = 0.175$, find the area if $\hat{F}_{in} = 1.0$ mol/s of feed gas.

D11. We are testing a UF membrane in a perfectly stirred system. With pure distilled water we obtain a pure water flux of $J = 5000.0$ L/(m^2 day) for a permeate pressure of 1.0 bar and a retentate pressure of 4.0 bar. The density of water is 997 kg/m^3.

Next we do experiments in the completely mixed system with a constant stirrer speed to separate an aqueous feed that is 0.001 wt frac polysaccharide. Previous experiments showed that with this stirred speed, the mass transfer coefficient $k = 2.89 \times 10^{-5}$ m/s. Since the system is quite dilute, the solvent density is the same as pure water. The inherent rejection coefficient is $R^o = 1.0$. The permeate pressure is 1.0 bar, and a cut θ' (in weight units) of 0.4 is used. The retentate pressure was increased in steps, and the following data were collected:

Retentate pressure, bar	Solvent flux, L/(m^2 day)
2.0	1670
3.0	3330
4.0	5000
5.0	5200
6.0	5200
7.0	5200

Find the wt frac of the polysaccharide at which gelling occurs.

D12. A perfectly mixed membrane module is concentrating helium in the retentate. The feed is 20 vol % helium and 80 vol % hydrogen at 4.5 atm. Permeate pressure is 1.0 atm and gases are ideal. If $y_{p,He} = 0.1$, find cut θ' and $y_{r,He}$. The membrane is the natural rubber studied by Nakagawa at 25°C with the permeabilities listed in Table 17-2.

D13. We are running an RO system with NaCl in the aqueous feed at 0.035 wt frac. This approximates the concentration of sea water, but sea water is much more complex. Typically, the highest cut for RO of sea water is approximately 0.50 to 0.55. In the United States, the EPA lists the preferred salt weight frac. in drinking water as 0.00050. In other countries, a higher drinking water maximum salt concentration is used.

a. If $x_{feed} = 0.035$, $x_p = 0.00050$, and $\theta' = 0.55$, what value of the retention R (including concentration polarization effects) is required?

b. If $x_{feed} = 0.035$, $x_p = 0.00100$ and $\theta' = 0.55$, what value of the retention R (including concentration polarization effects) is required?

c. For part b if the inherent rejection coefficient $R^o = 0.992$ (R^o is with M = 1, no concentration polarization), what was the value of M?

Data: For dilute systems the osmotic pressure of sodium chloride is $\pi = a' x$ where x is in wt frac, π in atm, and $a' = 15.446$ atm/(wt frac). Although this is not a dilute system, assume that it is.

D14. A pervaporation system will be used to remove water from n-butanol using a cellulose 2.5 acetate membrane in a perfectly mixed module. Feed is 90 mol% n-butanol. Feed rate is 100 lb/h. Flux is 0.2 lb/ft²h. The pervaporation unit operates at $T_p = T_{out} = 30°C$ where the selectivity of water compared to butanol is 43 (in mole fraction units. WATCH UNITS!).

Data: Latent heats: butanol = 141.6 cal/g; water = 9.72 kcal/gmole.
$C_{p,B}$ (45°C) = 0.635 cal/g °C, $C_{p,w}$ (45°C) = 1.0 cal/g °C, $MW_B = 74.12$, $MW_w = 18.016$.
Assume that the heat capacities and latent heats are independent of temperature.

a. If the feed is at 60°C, find the cut, permeate mole frac, outlet liquid mole frac and the membrane area.

b. If a cut of $\theta = 0.08$ is used, find the permeate mole frac, the outlet liquid mole frac, and the feed inlet temperature.

c. If the outlet water liquid mole frac is 0.05, find the permeate mole frac, cut, and inlet feed temperature.

D15. An UF system is being used to concentrate latex particles in an aqueous suspension. The membrane system is perfectly mixed on the retentate side and operates with a permeate pressure of 1.0 bar and a retentate pressure of 2.2 bar. Because of the extensive stirring, the concentration polarization modulus is M = 1.2 and no gel forms ($x_{wall} < x_{gel}$ = 0.5). The osmotic pressure of latex particles is negligible. The density of the solutions can be assumed to equal the density of pure water = 0.997 kg/L. All of the latex particles are retained by the membrane ($R° = 1.0$). The feed to the UF module is $x_{r,in} = 0.10$ wt frac latex (for all parts). We operate with a cut $\theta' = 0.2$. The feed rate of the suspension, $F'_{r,in}$, is 100 kg/h. The flux rate of the membrane with pure water is 2500 L/(m² day).

**a.*Find $F'_{r,out}$, x_p and $x_{r,out}$.

b. What membrane area is required?

c. If we decrease stirring and M increases, at what value of M will gel formation occur ($\theta' = 0.2$)?

d. At M = 1.2, what value of the cut θ' will cause gel formation to occur?

e. If M = 1.2 and $\theta' = 0.2$, at what feed wt frac will gel formation occur? What is $x_{r,out}$?

f. We are doing an experiment of slowly decreasing the amount of stirring. If gel formation first occurs with $x_F = 0.20$ and $\theta' = 0.25$, what are the values of M and the mass transfer coefficient k with this amount of stirring? $J_{solv} = 2500$, but the area has changed.

g. With the same amount of stirring as in part f, but otherwise new conditions, we do an experiment with $x_F = 0.20$, $p_r = 3.4$ bar, $p_p = 1.0$ bar, and $\theta' = 0.2$. Does a gel form? What is the solvent flux in L/(m²day)?

h. With the same amount of stirring as in part f, but otherwise new conditions, we do an experiment with $x_F = 0.20$, $p_r = 3.4$ bar, $p_p = 1.0$ bar, and $\theta' = 0.26$. Does a gel form? What is the solvent flux in L/(m²day)?

D16. We are doing RO experiments with a completely mixed laboratory unit. $K'_{solv}/t_{ms} = 1.387$ g/(atm m²s). Data: Assume solutions have same density as water = 997 kg/m³. The osmotic pressure of sodium chloride is $\pi = a'x$ where x is in wt frac, π in atm, and $a' = 15.446$ atm/(wt frac).

Experiment B. With a very high stirrer speed so that there is no concentration polarization, we do an experiment with a feed that is 0.0077 wt frac sodium chloride in water. The cut θ' (in wt units) is 0.22, and the measured inherent retention is $R^\circ = 0.976$. The permeate pressure was 1.1 atm, and the retentate pressure was 12.06 atm.

Experiment C. In this experiment a feed with 0.0093 wt frac sodium chloride was separated with a permeate pressure of 1.1 atm and a retentate pressure of 15.2 atm. A cut (in wt units) of $\theta' = 0.26$ was used. Since operation was at a lower stirring speed, concentration polarization is expected. A retention of $R = 0.939$ was measured.

    a. What is the value of the concentration polarization modulus M in experiment C? If you are unable to calculate a value of M, use M = 3.0 for part b.

    b. What are the values of x_{out}, x_p, and the solvent flux J'_{solv} in Experiment C?

D17. On a Robeson plot (a log-log plot of selectivity versus oxygen permeability in Barrers) the upper bound for separation of oxygen from nitrogen plots as a straight line. Approximate values of the end points are for $P_{O2} = 0.0001$ Barrers, $\alpha_{O2\text{-}N2} = 42$; and for $P_{O2} = 10,000$ Barrers, $\alpha_{O2\text{-}N2} = 1.9$. A Barrer is 10^{-10} [ccSTP cm]/[cm^2 s cm Hg].

    a. Find the equation for the straight line.

    b. If we need an $\alpha_{O2\text{-}N2} = 8.0$, what is the maximum value of P_{O2} that can be obtained?

D18. A silicone rubber membrane at 25°C is used to separate a feed that is 25 vol% N_2, 40 vol% CO_2, 5 vol% He, and 30 vol% H_2. Feed rate is 1.0 m^3(STP)/s. The thickness of the active layer of the membrane is 0.8 mil. The membrane module is perfectly mixed and a cut fraction $\theta = 0.40$ is desired. Data are in Table 17-2. Assume ideal gases. Note: 1.0 mil = 0.00254 cm.

    a. If $p_r = 2.5$ atm and $p_p = 1.0$ atm, find membrane area and outlet mole fractions of permeate and retentate.

    b. If $p_r = 1.0$ atm and $p_p = 0.4$ atm, find membrane area and outlet mole for fractions of permeate and retentate.

Suggestion: Solve with Spreadsheet.

E. *More Complex Problems*

E1. We are doing RO of dilute aqueous sucrose solutions at 25°C. A feed that is 2.2 wt % sucrose is separated in a very well-stirred system (M = 1.0) with $p_r = 60.0$ atm and $p_p = 1.1$ atm. We measure the flux as $J'_{water} = 3.923$ g/(m^{2}s) when $x_p = 0.00032$ and $x_{r,out} = 0.056$.

    a. Calculate the cut θ', the inherent rejection R°, and the sucrose flux $J'_{sucrose}$.

    b. Calculate the selectivity $\alpha'_{water\text{-}sucrose}$, K'_{water}/t_{ms}, $K'_{sucrose}/t_{ms}$ and J'_{water}.

    c. If we now repeat the experiment with less mixing so that M = 2.1 with $x_{feed} = 0.022$, $p_r = 60.0$ atm, $p_p = 1.1$ atm, and $\theta' = 0.61$, calculate x_r, x_p, J'_{water}, and $J'_{sucrose}$.

Data: For dilute systems the osmotic pressure of sucrose is $\pi = a'x$ where x is in wt frac, π in atm and a' and other data for sucrose are in Problem 17.D4.

E2. We wish to use pervaporation to increase the ethanol concentration in an ethanol water mixture to 98.5 wt %. The feed from the distillation column to the pervaporation unit is 90 wt % ethanol. This feed is pressurized and heated to 85°C. It is then fed to a system similar to Figure 17-14, except the retentate can be reheated to 85°C and be sent to a second pervaporation stage. Additional stages can be added if needed. A 20-micron thick polyacrylonitrile film membrane is used (see Figure 17-16). Assume that each stage of pervaporation is perfectly mixed and operates at 25°C. The final retentate (the product) needs to be at 98.5 wt % ethanol or higher. The permeate streams are pressurized and recycled to the distillation column. If we desire to process 100 kg/h of the 90 wt % ethanol distillate, find:

a. Number of stages needed in the pervaporation system.

b. For each stage: the cut, the permeate flow rate and the product flow rate. Also the wt frac of the combined permeate streams and the retentate wt frac.

c. The membrane area required for each stage.

F. *Problems Requiring Other Resources*

F1. We are using pervaporation to separate a benzene–isopropyl alcohol mixture. The pervaporation unit is perfectly mixed. Figure 12-23 in Wankat (1990) shows the separation factor of benzene with respect to isopropyl alcohol vs. the wt. fraction of benzene in the liquid for pervaporation. If operation is at $50°C = T_p = T_{out}$, $x_{in} = 0.30$ wt frac benzene and the $\theta = 0.10$, find y_p, x_{out}, and T_{in}. (Assume heat capacities and latent heats are independent of temperature.)

Note: Use the lines on the selectivity diagram not the data points.

H. *Spreadsheet Problems*

H1.* In Section 17.3.3 the following statement appears, "Geankoplis (2003) solves the multicomponent permeator system by a different method but the results are identical." Solve Example 13.5-1 in Geankoplis (2003) using the method in Section 17.3.3 with a spreadsheet and show that the results agree with Geankoplis' solution.

H2. Repeat Problem 17.D2, except do it as a cross-flow system on a spreadsheet.
a. Find solution for $N = 1$, which is the perfectly mixed answer.
b. Find solution for $N = 1000$, which is the cross-flow answer.

H3. A countercurrent membrane module with a poly(dimethylsiloxane) membrane operates at $35°C$ to separate air (assume air is 20.9% oxygen and 79.1% nitrogen). We want a permeate product that is 23.5% oxygen (above this limit for safety reasons, stainless steel has to be used in all later equipment). $F_{in} = 100,000 \text{ cm}^3(\text{STP})/s$, $p_r = 1.5$ atm, $p_p = 1.0$ atm, and $t_{ms} = 0.00002$ cm,

Find θ, $y_{r,out}$, F_p and A.

WATCH YOUR UNITS!

The spreadsheet in Appendix 17A.3 is not set up for this type of problem. The easiest approach is to use the spreadsheet as is and guess thetatot. Change thetatot until you get the desired y_p.

H4. Pervaporation problems are trial and error but can be conveniently solved on a spreadsheet if the selectivity is constant.
a. Solve Problem 17.D14a with a spreadsheet and check your answer with the graphical solution. The graphical and spreadsheet solutions should agree.
b. Repeat Problem 17.D14a but with the feed at $80°C$.

H5. RO experiments are done in a laboratory stirred-tank membrane system. With pure water we measure the pure water flux $= 17.89 \text{ g}/(\text{m}^2 \text{ s})$ when $p_r = 15.2$ atm and $p_p = 1.1$ atm. We then do an experiment with a highly stirred system ($M = 1$) with $p_r = 13.4$ atm and $p_p = 1.1$ atm and find $R° = 0.971$. We now plan on doing an experiment with $p_r = 11.3$ atm and $p_p = 1.1$ atm, $\theta' = 0.55$, $x_{in} = 0.006$, and a mass transfer coefficient of $k = 4.63 \times 10^{-5}$ m/s. Density of water is 0.997 kg/L. Predict M, R, J'_{solv}, x_p, and x_r for the new experiment. Use of a spreadsheet or other computer solver is highly recommended.

H6. Repeat Problem 17.H5, but find the value of k that gives $M = 1.1$. Report M, R, J'_{solv}, x_p, and x_r. Use of a spreadsheet or other computer solver is highly recommended.

H7. Repeat Problem 17.H5 ($x_{in} = 0.006$) but doing an experiment with $p_r = 21.4$ atm and $p_p = 1.1$ atm, $\theta' = 0.45$, and a mass transfer coefficient of $k = 4.63 \times 10^{-5}$m/s. Predict M, R, J'_{solv}, x_p, and x_r for the new experiment.

CHAPTER 17 APPENDIX.   SPREADSHEETS FOR FLOW PATTERN CALCULATIONS FOR GAS PERMEATION

The flow pattern calculations can be done with an Excel spreadsheet using Visual Basic (VBA). In the spreadsheets shown here, data are input into the spreadsheet and all calculations are done in the VBA program. All programs are done for constant values of the selectivity α, although it would not be difficult to use an α that depended on gas mole frac. These programs are meant to provide examples of VBA spreadsheet programs to analyze the three flow configurations shown in Figure 17-19. The best way to learn this material is to write your own spreadsheet, not copy these spreadsheets. If you are unfamiliar with VBA, read Appendix 4B, Part 1.

17.A.1 Cross-Flow Spreadsheet and VBA Program

The cross-flow calculation requires no trial-and-error and is very fast even with n = 20,000. The spreadsheet inputs the required values and records results. All calculations of the equations in Section 17.7.2 are done in the Visual Basic program. The spreadsheet is shown with input values for the separation of oxygen and nitrogen in Example 17-11 and the final results.

	A	B	C	D	E	F	G	H	I	J
1	Table 17.A.1	Cross-flow	Input	Flowrates are cm3 STP/s		Table 17A2 is VBA program/				
2	Fin	1000000	yin,0	0.209	thetatot	0.2	permB	1.9685E-06		
3	permA	0.000019685	pr	190	pp	19	Units are cm Hg			
4	Fr,out	800000	N	20000						
5	Fptot	200000	Fp	10		These values returned by VBA calc.				
6										
7										
8	ypavg	0.568853177								
9	Areatot	289965875.4								
10	yrout	0.119036706								

```vba
Option Explicit
Sub Main()
'     Table 17A1. VBA program for Table 17A1 cross-flow gas permeation.
      Dim N, i As Integer
      Dim sumyp, Area, Areatot, yr, yp, ypavg, alpha, thetatot, theta As Double
      Dim pr, pp, a, b, c, Fp, Fin, Fptotal, permA, permB, Fr, Frold As Double
'     Clean cells where results will be given, then read values from spreadsheet 2.
      Sheets("Sheet2").Select
      Range("B5", "B11").Clear
      Fin = Cells(2, 2).Value
      yr = Cells(2, 4).Value
      pr = Cells(3, 4).Value
      pp = Cells(3, 6).Value
      permA = Cells(3, 2).Value
      permB = Cells(2, 8).Value
      thetatot = Cells(2, 6).Value
      N = Cells(4, 4).Value
'     Calculate values & initialize loop. Loop starts with For
      alpha = permA / permB
      Fptotal = Fin * thetatot
      Fp = Fptotal / N
      Fr = Fin
      sumyp = 0
      Areatot = 0
For i = 1 To N
      theta = Fp / Fr
'     Quadratic solution Eq (17-10e),calculate yp & Calculate new Fr and yr
      a = (theta / (1 - theta) + pp / pr) * (alpha - 1)
      b = (1 - alpha) * (pp / pr + theta / (1 - theta) + yr / (1 - theta)) - 1 / (1 - theta)
      c = alpha * yr / (1 - theta)
      yp = (-b - (b * b - 4 * a * c) ^ 0.5) / (2 * a)
      sumyp = sumyp + yp
      Frold = Fr
      Fr = Frold - Fp
      yr = (Frold * yr - Fp * yp) / Fr
'     Calculate area of this stage, total area,Increment stage number & return to line below For
      Area = Fp * yp / (permA * (pr * yr - pp * yp))
      Areatot = Areatot + Area
Next i
'     When loop done, calculate average permeate mole fraction & print values
      ypavg = sumyp / N
      Cells(5, 4).Value = Fp
      Cells(5, 2).Value = Fptotal
      Cells(8, 2).Value = ypavg
      Cells(9, 2).Value = Areatot
      Cells(10, 2).Value = yr
End Sub
```

17.A.2 Co-current Flow Spreadsheet and VBA Program

The co-current calculation requires no trial-and-error and is very fast. The spreadsheet inputs the required values. All calculations of the equations in Section 17.7.3 are done in the Visual Basic program. The spreadsheet is shown with values for the separation of oxygen and nitrogen in Example 17-11.

	A	B	C	D	E	F	G	H	I	J
1	Table 17A.3	Co-Current		Table 17.A4 is the VBA program for this spreadsheet						
2	Fin ml STP/s	1000000	yin	0.209	thetatot	0.2	permB/tms	1.97E-06		
3	permA/tms	0.000019685	pr	190	pp	19				
4			N	20000						
5										
6	Values below from VBA calculation - must enable Macros.									
7	Massbal	1.45519E-11								
8	ypout	0.558452716								
9	Areatot	295527395.9								
10	yrout	0.121636821								
11	Frout	800000								
12	Fptotal	200000								
13										
14										
15										
16										
17										
18										
19										
20										
21										
22										
23										
24										
25										
26										
27										
28										
29										
30										
31										
32										
33										
34										

Sheet1 / Sheet2 / Sheet3

```
Option Explicit
Sub Main()
'    Table 17A4. VBA program for spreadsheet in Table 17A3.
Dim N, i As Integer
Dim permA, permB, Fr, Fin, Area, Areatot, yt, yin, yr, yp, ypold, thetatot, alpha As Double
Dim pr, pp, a, b, c, ConA, ConB, Cj, Cjold, Ft, Fp, Fptotal, Fpold, Frout, Massbal As Double
'    Use sheet 1 from spreadsheet & clean cells. Read values from spreadsheet for input values
     Sheets("Sheet1").Select
     Range("B7", "B12").Clear
     Fin = Cells(2, 2).Value
     yin = Cells(2, 4).Value
     pr = Cells(3, 4).Value
     pp = Cells(3, 6).Value
     thetatot = Cells(2, 6).Value
     permA = Cells(3, 2).Value
     permB = Cells(2, 8).Value
     N = Cells(4, 4).Value
'    Calculate values, Initialize variables, For loop for calculation of each stage
     alpha = permA / permB
     Ft = Fin * thetatot / N
     Fptotal = Ft * N
     Frout = Fin - Fptotal
     Fp = 0
     Fpold = 0
     yp = 0
     Areatot = 0
For i = 1 To N
'    Initialize for this stage, do mass balance, calculate constants & solve quadratic
     ypold = yp
     Fpold = Fp
     Fp = Fpold + Ft
     Fr = Fin - Fp
     ConA = Fin * yin / Fr
     ConB = Fpold * ypold
     a = (1 - alpha) * (Fp * Fp / Fr + Fp * pp / pr)
     b = Fp * ((alpha - 1) * ConA + (1 - pp / pr)) + (alpha - 1) * ConB * (Fp / Fr + pp / pr)
     + alpha * Ft * (Fp / Fr + pp / pr)
     c = (1 - alpha) * ConA * ConB + ConB * (pp / pr - 1) - alpha * Ft * ConA
     yp = (-b + (b * b - 4 * a * c) ^ 0.5) / (2 * a)
'    Calculate mass balance, stage area, total area, Increment stage number & return to line below For
     yr = -yp * Fp / Fr + ConA
     yt = (Fp * yp - Fpold * ypold) / Ft
     Area = Ft * yt / (permA * (pr * yr - pp * yp))
     Areatot = Areatot + Area
Next i
'    Do external mass balance & Print final values on spreadsheet
     Massbal = yin * Fin - Fptotal * yp - yr * (Fin - Fptotal)
     Cells(7, 2).Value = Massbal
     Cells(8, 2).Value = yp
     Cells(9, 2).Value = Areatot
     Cells(10, 2).Value = yr
     Cells(11, 2).Value = Frout
     Cells(12, 2).Value = Fptotal
End Sub
```

17.A.3 Countercurrent Flow Spreadsheet and VBA Program

This calculation is trial-and-error since $y_{l,N}$ must be guessed. A loop using Eq. (17-78) to converge on the correct value for $y_{r,N}$ is used with an arbitrary number of trials M (in this example M = 5). The spreadsheet inputs the required values. All calculations of the equations in Section 17.7.4 are done in the Visual Basic program. The spreadsheet is shown with values for the separation of oxygen and nitrogen in Example 17-11. Note that a very poor first guess for $y_{r,N} = 0.001$ (not shown here) converged with M = 10 to the correct y_{in}.

	A	B	C	D	E	F	G	H	I	J
1	Table 17A5	Counter-current		Table 17A6 is VBA program for this spreadsheet						
2	Fin	1000000	yin	0.209	thetatot	0.2	permB	1.9685E-06		
3	permA	0.000019685	pr	190	pp	19				
4	M	5	N	20000	yroutguess	0.116				
5	df	0.95								
6										
7	yp	0.576313456								
8	Areatot	285963883.9								
9	yincalc	0.208997727								
10	Fincalc	1000000								
11	Massbal	2.63555E-06								
12	yrout	0.117168342								
13										
14										
15										
16										
17										
18										
19										
20										
21										
22										
23										
24										
25										
26										
27										
28										
29										
30										
31										
32										
33										
34										

Sheet1 Sheet2 Sheet3

```
(General)                                            ▼   (Declarations)                          ▼
Option Explicit
Sub Main()
'   Table 17A6. VBA program for spreadsheet in Table 17A5.
Dim N, M, k, j, i As Integer
Dim pr, pp, a, b, c, ConA, ConB, Cj, Cjold, df, erroracc, Ft, Fp, Fpold, Fpout, permA, permB As Double
Dim Fr, Frout, Fin, Area, Areatot, Massbal, yt, yin, yincalc, yr, yrout, yroutold, yrold, yp As Double
Dim ypold, alpha, thetatot, thetaj As Double
'   Use sheet 1 of spreadsheet, clean answer space, and input values.
    Sheets("Sheet1").Select
    Range("B7", "B12").Clear
    Fin = Cells(2, 2).Value
    thetatot = Cells(2, 6).Value
    yin = Cells(2, 4).Value
    yrout = Cells(4, 6).Value
    pr = Cells(3, 4).Value
    pp = Cells(3, 6).Value
    permA = Cells(3, 2).Value
    permB = Cells(2, 8).Value
    M = Cells(4, 2).Value
    df = Cells(5, 2).Value
    N = Cells(4, 4).Value
'   Calculate external values, Initialize variables for trial and error loop, For loop
    Fpout = thetatot * Fin
    Frout = Fin - Fpout
    Ft = Fpout / N
    alpha = permA / permB
    yincalc = yin
    yroutold = yrout
For k = 1 To M
'   Initialize values for stage calculations
    yr = yrout
    Fr = Frout
    Fp = 0
    Fpold = 0
    yp = 0
    Areatot = 0
'   Loop for calculations for each stage. j is backward counting index, Initialize for stage
For i = 1 To N
    j = N - i + 1
    ypold = yp
    Fpold = Fp
'   Mass balance, quadratic equation, calculate area
    Fp = Fpold + Ft
    a = (1 - alpha) * Fp * pp / pr
    b = Fp * (1 - pp / pr) + (alpha - 1) * Fpold * ypold * pp / pr + alpha * Ft * pp / pr + Fp * (alpha - 1) * yr
    c = Fpold * ypold * (-1 - (alpha - 1) * yr + pp / pr) - alpha * Ft * yr
    yp = (-b + (b * b - 4 * a * c) ^ 0.5) / (2 * a)
    yr = Fp * yp / Fr + yrout * Frout / Fr
    yt = (Fp * yp - Fpold * ypold) / Ft
    Area = Ft * yt / (permA * (pr * yr - pp * yp))
    Areatot = Areatot + Area
'   Mass balance calculations and value of theta for next stage, increment i & do For loop
    Fr = Fp + Frout
    thetaj = Fp / Fr
Next i
'   Initialize & do damped direct substitution for yrout. Return to line after For k=1 To M
    yincalc = yr
    yroutold = yrout
    yrout = yroutold + df * (yin - yincalc)
Next k
'   Run M iterations.  External mass balance. Print final values on spreadsheet
    Massbal = (yin * Fin - Fpout * yp - yrout * Frout) / Fin
    Cells(11, 2).Value = Massbal
    Cells(7, 2).Value = yp
    Cells(8, 2).Value = Areatot
    Cells(9, 2).Value = yincalc
    Cells(10, 2).Value = Fr
    Cells(12, 2).Value = yrout
End Sub
```

Appendix.   Spreadsheets for Flow Pattern Calculations for Gas Permeation   **803**

CHAPTER 18

Introduction to Adsorption, Chromatography, and Ion Exchange

In the first 14 chapters we looked at separation techniques such as distillation and extraction that are often operated as equilibrium-staged separations, and even when they are not operated this way can be analyzed as equilibrium-staged separations. Except for Chapter 9, "Batch Distillaton," and Section 13.6, which discussed batch extraction, all separations were steady state. In Chapter 17 we studied membrane separations that are not operated as equilibrium processes; however, well-mixed membrane separators are analogous to flash distillation, and staged models are useful for integrating the mass balances and rate expressions for more complex flow patterns. Since the membrane processes are normally operated at steady state, the analysis is usually straightforward. In this chapter we will study three closely related processes that are rarely operated or analyzed as steady state, equilibrium-staged systems. These sorption processes are usually operated in packed columns in a cycle that includes feed and regeneration steps; thus, as normally operated, these processes are inherently unsteady state.

Adsorption (note the "d" not a "b") involves contacting a fluid (liquid or gas) with a solid (the adsorbent). One or more of the components of the fluid are attracted to the surface of the adsorbent. These components can be separated from components that are less attracted to the surface. Adsorption is commonly used to clean fluids by removing components from the fluid or to recover the components. Many homes and apartments use a carbon "filter" (actually an adsorber) for water purification. *Chromatography* is a similar process that uses a solid packing material (an adsorbent or other solid that preferentially attracts some of the components in the mixture), but the operation is devised to separate components from each other. You may have analyzed the composition of samples with analytical chromatography (gas or liquid) in chemistry or engineering labs. In *ion exchange* the solid contains charged groups that interact with charged ions in the liquid. The best-known application of ion exchange is water softening to remove calcium and magnesium ions and replace them with sodium ions. These separation methods are complementary to the equilibrium-staged processes. They are often used for chemical analysis, separation of dilute mixtures, and separation of difficult mixtures where the equilibrium-staged separations either do not work or are too expensive.

Note: A nomenclature list for this chapter is included in the front matter of this book.

The three separation techniques studied in this chapter are similar since a solid phase causes the separation. When we want to lump them together we will call them *sorption* processes. The general term for an adsorbent, ion exchange resin or chromatographic packing is *sorbent*. The most common equipment for sorption processes is a stationary packed bed of the solid. The solid in sorption systems directly causes the separation, which is different than packed beds for the equilibrium separations studied earlier where the solid was inert but increased the interfacial area and mass transfer coefficients between the gas and liquid. If feed is introduced continuously into the sorption packed bed, the bed will eventually saturate (e.g., approach the feed concentration) and separation will cease. Much of the art and expense of designing these systems is in the *regeneration* step that removes the component from the packing material. Regeneration is so important that different processes are often named based on the regeneration method used.

This chapter is a simplified introduction to the fascinating and valuable sorption separation methods. The first three-fourths of the chapter relies on equilibrium analysis—mass transfer is introduced in Section 18.6. The development is similar, but at a more introductory level, to that in Wankat (1986; 1990). Once you understand this chapter, you will be able to discuss these techniques with experts and will be prepared to begin more detailed explorations of these methods in more advanced books (e.g., Do, 1998; Ruthven, 1984; Ruthven et al., 1994; Yang, 1987; 2003).

18.1 SORBENTS AND SORPTION EQUILIBRIUM

Since sorption and sorbents are quite different from the equilibrium-staged processes and the membrane separations studied earlier, we need to first carefully define the terms needed to study and design these systems. After a short description of the different sorbents, the equilibrium behavior of sorbent systems will be introduced. Then the last fundamental piece required is the mass transfer characteristics of sorbents and sorption processes.

18.1.1 Definitions

The most common contacting device for adsorption, chromatography, and ion exchange is the packed bed shown schematically in Figure 18-1. The particles are packed in the cylindrical column of cross-sectional area A_c and the length of the packed section is L. Some type of support netting or frit is used at the bottom of the packed section and a hold down device such as a net or frit is used at the top of the packed section. Figure 18-1 illustrates a number of important variables. The external porosity ε_e is the fraction of the column volume that is outside the particles. To some extent the value of ε_e depends on the shape of the particles (e.g., ε_e is smaller for spheres than for irregular shaped particles) and the packing procedure used. It is important to have a uniformly packed column with a constant value of ε_e. The internal porosity ε_p is the fraction of the volume of the pellets that consists of pores and thus, is available to the fluid. The total volume available to the fluid is

$$V_{fluid} = [\varepsilon_e + (1 - \varepsilon_e)\varepsilon_p]V_{column} \qquad \textbf{(18-1a)}$$

From this equation we can define the total porosity ε_T as the sum of all the voids.

$$\varepsilon_T = \varepsilon_e + (1 - \varepsilon_e)\varepsilon_p \qquad \textbf{(18-1b)}$$

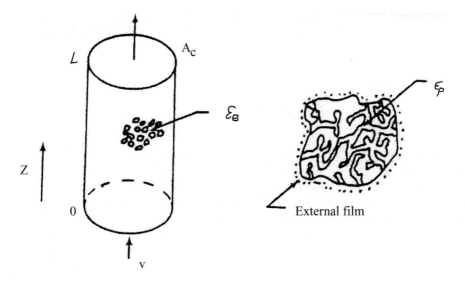

FIGURE 18-1.   *Schematic of adsorption column and particle (Wankat, 1986), reprinted with permission, copyright 1986, Phillip C. Wankat*

Porosities are dimensionless quantities.

Although the fluid can fit into all the pores, molecules such as proteins may be too big to fit into some or all of the pores. This *size exclusion* can be quantified in terms of a dimensionless parameter K_d where $K_d = 1.0$ if the molecule can penetrate all of the pores and $K_d = 0$ if the molecule can penetrate none of the pores. The value of $K_{d,i}$ for a given molecule i can also be between 0 and 1 since the pores are not of uniform size. The volume available to a molecule is,

$$V_{available} = [\varepsilon_e + (1 - \varepsilon_e)\varepsilon_p K_{d,i}]V_{column} \qquad \textbf{(18-1c)}$$

This picture is useful but does not match all adsorbents. Gel-type ion-exchange resins have no permanent pores. Instead they consist of a tangled network of interconnected polymer chains into which the solvent dissolves. In effect, $\varepsilon_p = 0$. Macroporous ion-exchange resins have permanent pores and $\varepsilon_p > 0$, but often $K_d < 1.0$ for large molecules. Many activated carbons have both macropores and micropores; thus, there are two internal porosities. Molecular sieve zeolite adsorbents are used as pellets that are agglomerates of zeolite crystals and a binder such as clay. In this case, there is an interpellet porosity (typically, $\varepsilon_e \sim 0.32$), an intercrystal porosity ($\varepsilon_{p1} \sim 0.23$) and an intracrystal porosity ($\varepsilon_{p2} \sim 0.19$), which has $K_{d,i} \leq 1.0$ (Lee, 1972).

Two different velocities are typically defined for the column shown in Figure 18-1. The *superficial velocity*, which is easy to measure, is the average velocity the volumetric flow of fluid would have in an *empty* column. Thus,

$$v_{super} = Q/A_c \qquad \textbf{(18-2a)}$$

where Q is the volumetric flow rate (e.g., in m^3/s), the cross sectional area $A_c = \pi D_{col}^2/4$ is in m^2 and v_{super} is in m/s. The *interstitial velocity* v_{inter} (also in m/s) is the average velocity the fluid has flowing in between the particles. Since the cross-sectional area actually available to the fluid is $\varepsilon_e A_c$, a mass balance on the flowing fluid is

$$v_{inter} \, \varepsilon_e \, A_c = v_{super} \, A_c$$

which gives

$$v_{inter} = v_{super} / \varepsilon_e \qquad (18\text{-}2b)$$

Since ε_e is less than 1.0, $v_{inter} > v_{super}$. Very large molecules that are totally excluded from the pores and are not adsorbed move at an average velocity of v_{inter}.

There are also different densities of interest. The first, the fluid density ρ_f (e.g., in kg/m³) is familiar. The second density is the *structural density* ρ_s (e.g., also in kg/m³) of the solid. This is the density of the solid if it is crushed and compressed so that there are no pores and all of the air is removed. The *particle or pellet density* ρ_p is the average density of the particles consisting of solid plus the fluid in the pores.

$$\rho_p = (1 - \varepsilon_p) \, \rho_s + \varepsilon_p \, \rho_f \qquad (18\text{-}3a)$$

Manufacturers often report a *bulk density* ρ_b of the adsorbent. This density is the weight of the adsorbent as delivered, which includes fluid in the pores and between the particles, divided by the volume of the container. The bulk density can be calculated from the other densities.

$$\rho_b = (1 - \varepsilon_e) \, \rho_p + \varepsilon_e \, \rho_f \qquad (18\text{-}3b)$$

The bulk and particle densities will also be in kg/m³. If the fluid is a gas with $\rho_f \ll \rho_p$,

$$\rho_b = (1 - \varepsilon_e) \, \rho_p = (1 - \varepsilon_e)(1 - \varepsilon_p)\rho_s \qquad \text{(when fluid is a gas)} \qquad (18\text{-}3c)$$

Unfortunately, it is often unclear as to which density is being referred to. By comparing the values given by the manufacturer to the approximate values listed in Table 18-1, one may be able to determine which density is being referred to.

One last useful definition is the *tortuosity* τ. The tortuosity relates the effective diffusivity in the pores $D_{effective}$ to the molecular diffusivity in free solution, $D_{molecular}$

$$\tau = \varepsilon_p \, D_{molecular} / D_{effective} \qquad (18\text{-}4)$$

Note that τ is dimensionless. Equation (18-4) was originally derived from geometric considerations using a simple geometric model. However, measured values of τ are often much larger than expected from purely geometric arguments because the diffusion is hindered by the walls or at the pore mouth. Thus, we will treat τ and $D_{effective}$ as empirical (experimentally measured) quantities. The molecular diffusivity can be determined experimentally, but there are also a number of well-known methods to predict $D_{molecular}$ (See Section 15.3).

18.1.2 Sorbent Types

A variety of sorbents are used commercially for separations. We will first present a very short introduction to commercially used adsorbents. If more detail is required, refer to Yang (2003) who presents very detailed analyses. The commercially important adsorbents are highly porous and have high surface areas per gram. This high surface area greatly increases the capacity for adsorption at the surface. Typically, 98% of the adsorption occurs in the pores inside the particles and only 2% on the external surface. Molecules that adsorb are called the *adsorbate.* Adsorbed molecules typically have a density close to that of a liquid. Thus, there are major density and volume changes when gases are adsorbed, but very little change when liquids are adsorbed.

Activated carbon is a very porous adsorbent with a carbon backbone but a number of other species such as oxides of carbon on the surface. Since activated carbon is inexpensive, strongly adsorbs organic compounds, and has a large number of applications, it is the most commonly used adsorbent (Bonsal et al., 1988; Faust and Aly, 1987). It is produced by carbonizing a material such as wood, coke, or coconut shells. Activation is typically done with carbon dioxide or steam to create the porous structure and to oxidize the surface. Additional chemical treatments such as with iodine can be used to produce specialty carbons (Yin et al., 2007). Carbons are produced for both liquid and gas separations. Because the starting materials and the chemical treatments vary, different activated carbons can have very different properties. Thus, the average values reported in Table 18-1 should be used only for very preliminary calculations and approximate designs. Experimental data on the particular brand, grade, and size of activated carbon must be used for more detailed designs.

Because activated carbon has essentially a nonpolar surface, water is adsorbed weakly, often by capillary condensation in gas systems (Yang, 1987). Thus, many organic compounds are much more strongly adsorbed than water. This makes activated carbon the usual adsorbent of choice for processing aqueous solutions and humid gases. Activated carbon is commonly used for pollution control to remove organic compounds from water. Equilibrium isotherms have been measured for a large number of compounds (Dobbs and Cohen, 1980). Activated carbon is also frequently used to adsorb small amounts of organics from gases, process sugar, purify alcohol, provide personal protection as part of the complex mixture of adsorbents included in gas masks, and many other applications.

Carbon molecular sieves (CMS) have very tightly controlled pore structures. They are prepared in a manner that is similar to activated carbon except there is often an additional step where a hydrocarbon is cracked or polymerized on the surface to create the desired uniform pore size (Ruthven et al., 1994). Because of the extra care required in processing and because of patent protection, CMS are significantly more expensive than activated carbon. Currently, CMS are commonly used for producing pure nitrogen from air. Unlike the vast majority of commercial adsorbents, CMS can separate based on different diffusion rates instead of different equilibrium behavior. The design of these *kinetic* adsorption processes is highly specialized and is beyond the scope of this chapter (e.g., see Ruthven et al., 1994).

Zeolite molecular sieves are crystalline aluminosilicates with the general formula

$$M_{x/n}[(AlO_2)_x (SiO_2)_y]z \; H_2O$$

where M represents a metal cation such as lithium, sodium, potassium or calcium of valence n; x and y are integers ($y \geq x$); and z is the number of water molecules per unit cell. Since zeolites are crystals, the pores have exact dimensions (Ruthven, 1984; Ruthven et al., 1994;

TABLE 18-1. Properties of common adsorbents (Humphrey and Keller, 1997; Reynolds et al., 2002; Ruthven et al., 1994; Wankat, 1990; Yang, 1987)

Properties	Activated Carbon		Zeolite Molecular Sieves				Silica Gel	Activated Alumina
	Gas Phase	Liquid Phase (coal base)	Major cation					
			3A K^+	4A Na^+	5A Ca^+	13X Na^+		
	—	—					—	—
Particle Size	−6 to +14 Tyler mesh	−8 to +30 Tyler mesh					0.1 to 3.0 mm	14/28 mesh to 0.5 inch spheres
Bulk density ρ_b, (kg/m³)	515 (avg)	500 (avg)	670–740	660–720	670–720	610–710	700–820 720 typical	~800
ε_p	0.6 to 0.85	0.6 ~ 0.85	0.3	0.32	0.34	0.38	0.38 to 0.55	0.5 to 0.57
Surface Area m²/g	200–2000	—	←——— 500 to 800 ———→				750–800	200–390 250 to 320 normal
Reactivation T, °C	100–140 (steaming)	—	←——— 200–350 ———→				120–280	150–315
Heat Cap. Btu/(lb °F)	0.27–0.36	—	—	0.19	0.19	—	0.22–0.26	0.21–0.25
Tortuosity	—	5–65	←——— 1.7–4.5 ———→				2–6	2–6
Equil. Water Capacity kg H₂O/kg ads. @4.6 mm Hg, 25°C	0.01		0.25	0.30	0.27	0.33	0.12	0.07 to 0.25
@25°C	0.05–0.07 @250 mm Hg						0.54 @ 17.5 mm Hg	0.19 @ 17.5 mm Hg

Sherman, 1999; Yang, 1987, 2003). Thus, zeolites can be used to separate based partially on steric exclusion although separations where steric exclusion is not employed are more common. A large number of synthetic and naturally occurring zeolites are known (Vaughan, 1988) although not all are used commercially as adsorbents. Commercial applications of zeolites as adsorbents include drying air and natural gas, drying organic liquids, removal of carbon dioxide, separation of ethanol and water to break the azeotrope, and separation of oxygen from the nitrogen in air. The major steric exclusion application is the separation of straight-chain hydrocarbons (used for biodegradable detergents) from branched-chain hydrocarbons. Some of the properties of zeolites are reported in Table 18-1.

Silica gel is an amorphous solid made up of colloidal silica SiO_2 that is normally used in a dry granular form. [The name "gel" arises from the jellylike form of the material during one stage of its production. (Reynolds et al., 2002).] Silica gel is commonly used for drying gases and liquids since it has a high affinity for water. Silica gel is complimentary to zeolites since it is cheaper and has a higher capacity at water vapor pressures greater than about 10 mm Hg (Humphrey and Keller, 1997) but cannot dry the fluid to as low a water content. Columns with a layer of silica gel at the feed end and a layer of zeolite at the product end are commonly used for drying since they can combine the best properties of both adsorbents. Note that silica gel can be damaged by liquid water. Some of the properties of silica gel are reported in Table 18-1.

Activated alumina Al_2O_3 is also commonly used for drying gases and liquids and is not damaged by immersion in liquid water. It is produced by dehydrating aluminum trihydrate, $Al(OH)_3$ by heating. Activated alumina has properties that are similar to silica gel although it is physically more robust. It competes with silica gel in drying applications although its capacity is a bit lower at water vapor pressures greater than about 1 mm Hg (Humphrey and Keller, 1997). Activated alumina is also used in water treatment to selectively remove excess fluoride. Some of the properties of activated alumina are listed in Table 18-1.

A large number of other materials have been used commercially as adsorbents. These include an uncharged form of the organic polymer resins commonly used for ion exchange (see Section 18.5). Although considerably more expensive, these resins compete with activated carbon for the recovery of organics. Activated bauxite is an impure form of activated alumina. A number of clays are used for purification of vegetable oils. Bone char, which has both adsorptive and ion exchange capacities, is used in purification of sugar.

When the common adsorbents are used in chromatography applications (Cazes, 2005), they are used as much smaller particles with a much tighter particle size distribution. A number of specialized packing materials have been developed for chromatographic applications. In gas-liquid chromatography a high-boiling, nonvolatile liquid (the stationary phase) is coated onto an inert solid such as diatomaceous earth. A similar method called liquid-liquid chromatography coats an immiscible liquid on an inert solid. This packing is now often replaced with *bonded* packing where the stationary phase, often a C_8 or a C_{18} compound, is chemically bonded to the inert solid, which is usually silica gel. The equilibrium behavior of these specialized packings is usually similar to gas-liquid absorption or liquid-liquid extraction, but because of the presence of the inert solid, the equipment and operating principles are similar to adsorption-chromatography.

18.1.3 Adsorption Equilibrium Behavior

Equilibrium behavior of adsorbents is usually determined as constant temperature *isotherms*. Valenzuela and Myers (1984, 1989) present the most extensive compilations of isotherm data, and are the best entry point into the very large literature on adsorption equilibrium measurements and theories. Basmadjian (1986) has extensive data on water isotherms from gases and liquids. Dobbs and Cohen (1980) and Faust and Aly (1987) present adsorption data for common pollutants.

Relatively typical isotherm data for gas adsorption of single components are shown in Figures 18-2A and 18-2B. Figure 18-2A shows the effect of changing the gas. Hydrogen is least strongly adsorbed, followed by methane. Adsorption peaks for ethane and ethylene and then the maximum is less for propane although its adsorption is stronger at low partial pressures. Each adsorbent has an optimum size molecule for peak adsorption. Figure 18-2B compares the adsorption of ethylene on different adsorbents.

At low concentrations or partial pressures isotherms are often linear. As the partial pressure of the gas increases the isotherm becomes *nonlinear*, that is, it curves. Equilibrium data for adsorption of single gases are often fit with the *Langmuir* isotherm,

$$q_A = q_{A,max} K_{A,p} p_A/(1 + K_{A,p} p_A) \qquad \text{(18-5a)}$$

where q_A is the amount of species A adsorbed and $q_{A,max}$ is the maximum amount of species A that can adsorb (kg/kg adsorbent or mol/kg adsorbent), p_A is the partial pressure of species A (mm Hg, kPa, or other pressure units), and K_A is the adsorption equilibrium constant in suitable units. Note that for very small partial pressures, $K_{A,p} p_A \ll 1.0$ and Eq. (18-5a) simplifies to the linear form

$$q_A = q_{A,max} K_{A,p} p_A = K'_{A,p} p_A \qquad \text{(18-5b)}$$

while at very high partial pressures, $K_{A,p} p_A \gg 1.0$, Eq. (18-5a) simplifies to

$$q_A = q_{A,max} \qquad \text{(18-5c)}$$

This is a horizontal line that represents *saturation* of the adsorbent.

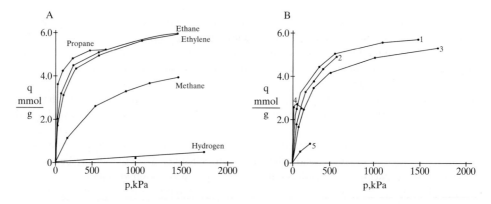

FIGURE 18-2. *Adsorption isotherms for pure gases: (A) Gases on Columbia grade L activated carbon at 310.92 K (Ray and Box, 1950; Valenzuela and Myers, 1989), (B) Ethylene on different adsorbents at varying temperatures (Valenzuela and Myers, 1989). Key: 1 – Columbia grade L activated carbon at 310.92 K. 2 = Taiyo Kaken Co., Japan, attrition resistant activated carbon at 310.95 K. 3 = Pittsburgh Chemical Co. BPL activated carbon at 301.4 K. 4. Union Carbide 13X Linde zeolite at 298.15 K. 5 = Davison Chemical Co. silica gel at 298.15 K.*

For liquid systems the isotherm is usually written in terms of the liquid concentration c_A (mol/m^3 or kg/m^3),

$$q_A = q_{A,max} K_{A,c} c_A/(1 + K_{A,c} c_A) \tag{18-6a}$$

In the linear limit when $K_A c_A << 1.0$, this simplifies to

$$q_A = q_{A,max} K_{A,c} c_A = K'_{A,c} c_A \tag{18-6b}$$

The adsorption equilibrium constants $K_{A,c}$ and $K'_{A,c}$ will be in different units for liquid systems than for gas systems. Equilibrium constants for several systems are listed in Table 18-2.

Isotherm data at different temperatures invariably shows that there is less material adsorbed as the temperature increases. The adsorption equilibrium constant *often* follows an Arrhenius form

$$K_A = K_{Ao} \exp(-\Delta H/RT) \tag{18-7a}$$

where K_{Ao} is a pre-exponential factor, ΔH is the heat of adsorption (e.g., in J/kg), R is the gas constant and T is the absolute temperature (e.g., in K). Since adsorption is invariably exothermic, ΔH is negative. *If* the Arrhenius form is followed, a plot of $\ln K_A$ vs. 1/T will be a straight line with a slope of $-\Delta H/R$. Don't automatically assume that data follow an Arrhenius form. Plot the data and check if they are on a straight line. Typically, $q_{A,max}$ slowly decreases as temperature increases, perhaps in a linear fashion.

Langmuir used a simple kinetic argument (e.g., Wankat, 1990) to derive Eqs. (18-5a) and (18-6a). When this argument is used, $q_{A,max}$ is the coverage obtained with a monolayer. Langmuir's isotherm can also be derived with a statistical mechanics argument (e.g., Ruthven, 1984). The Langmuir isotherm is used to correlate data even when there is reason to believe that the mechanism postulated by Langmuir is incorrect. For liquid systems it is common to write the Langmuir isotherm as

$$q_A = a\, c_A/(1 + b\, c_A) \tag{18-6c}$$

and there is no implication that a or b have physical interpretations. Correlation of data is best done by multivariable regression techniques, but is often done by plotting c/q vs. c (or p/q vs. p). In these plots the Langmuir isotherm plots as a straight line (see Example 18-1 and homework Problem 18.D1).

The Langmuir isotherm is thermodynamically correct for single component systems. It has also been used as the basis for a variety of other isotherms such as the BET isotherm (multiple layers of adsorbate), the Langmuir-Freundlich isotherm, and the linear-Langmuir isotherm (add a linear isotherm and a Langmuir isotherm Eq. (18-6c) with different values of a). The Langmuir isotherm is also commonly extended to the adsorption of multicomponent mixtures. For example, for the simultaneous adsorption of components A and B Eq. (18-6c) becomes

$$q_A = a_A c_A/(1 + b_A c_A + b_B c_B), \quad q_B = a_B c_B/(1 + b_A c_A + b_B c_B) \tag{18-8a,b}$$

This equation correctly predicts that the two adsorbates *compete* for adsorption sites. That is, if the concentration of B increases the amount of A adsorbed will decrease. Unfortunately, very few systems follow Eqs. (18-8) exactly, and if a_A does not equal a_B Eq. (18-8) is

TABLE 18-2. *Equilibrium Constants*

Gas Systems

Adsorbate	carrier gas	adsorbent	form	equation	conditions	ref.
Methane	none	act. carbon	~ Langmuir	See data Table 18-3	296, 373, 480 K	Ritter & Yang (1987)
Methane	helium	5A zeolite	linear	$q = 23.90c$ (q & c gmol/ml)		Cen & Yang (1986)
Methane	H_2, N_2, CO_2	BPL act. carbon	Langmuir	$q = \dfrac{q_{max} K_A y}{1 + K_A y}$, $q_{max} = 3.65 \times 10^{-3}$	$p = 1 atm$ $q \dfrac{mol}{g}$, c mole frac.	Sircar & Kumar (1983)
cont.			$K_A = 0.4179(25°C)$	$K_A = 0.2550(45°C)$	$K_A = 0.1343(75°C)$	$K_A = 0.0928(95°C)$
acetylene	H_2		linear	$q = 35.6c \left(q\ \dfrac{qmol}{g} , c\ \dfrac{qmol}{m} \right)$	150°C	
nitrogen	none	5A zeolite	linear	$q = 5.40c \left(q\ \dfrac{gmol}{m^3} , c\ \dfrac{gmol}{m^3} \right)$	30°C	Kayser & Knaebel (1986)

TABLE 18-2. *Equilibrium Constants (continued)*

Liquid Systems

Adsorbate	Solvent	adsorbent	form	equation	conditions	ref.
acetic acid	water	activated carbon	Freundlich	$q = 3.646\ c^{1/3.277}$	4°C	Baker & Pigford (1971)
"	"	"	Freundlich	$q = 3.019\ c^{1/2.428}$	60°C	"
bovine serum albumin	buffered water	DEAE Sephadex A-50	Langmuir	$q\ \dfrac{mg}{g\ dry\ resin} = \dfrac{254{,}259\ c}{1 + 30.79\ c},\ c\ \dfrac{mg}{g\ soln}$	20°C pH = 6.9	Tsou & Graham (1985)
fructose	water	IEX resin Ca^{+2} form	linear	$q = 0.88\ c$ (q & c in g/100 ml)	30°C, <5.0g/100 ml	Ching & Ruthven (1985)
glucose	water	IEX resin Ca^{+2} form	linear	$q = 0.51\ c$ (q & c in g/100 ml)	30°C, <5.0g/100 ml	
anthracene	cyclohexane	activated alumina	Langmuir	$q\left(\dfrac{gmol}{kg}\right)\dfrac{22\ c}{1 + 375\ c}\ c\ (gmol/L)$	~25°C	Thomas (1948)
acetonaphthalene	23% methylene chloride	silica gel	linear	$k' = K\rho_p\ (1 - \varepsilon_e)/\varepsilon_e = 5.5$		
dinitronaphtahlene	77% pentane	"	linear	$k' = 5.8$		
human serum albumin	buffered water	Ca-alginate magnetite bead	linear	$K = \dfrac{c}{q} = 0.08\ \dfrac{g}{ml}$	$\dfrac{0.2\ mg\ HSA}{ml\ soln.}$	Wankat (1990)

not thermodynamically consistent (LeVan and Vermeulen, 1981). Despite these problems Eq. (18-8) and its extensions to more components are commonly used for theories for multi-component adsorption because this is the simplest form that shows competition.

A thermodynamically correct approach, the Ideal Absorbed Solution (IAS) theory, uses the Langmuir isotherm as the basic single component isotherm for the adsorption of mixtures. Since the details for these additional isotherms are beyond the scope of this book, readers who need to use more complex isotherms are referred to Do (1998), Ruthven (1984), Valenzuela and Myers (1989), and Yang (1987; 2003).

EXAMPLE 18-1. Adsorption equilibrium

Experimental equilibrium data for the adsorption of methane on Calgon Carbon Corp. PCB activated carbon are listed in Table 18-3. Determine if the Langmuir isotherm, Eq. (18-5a), is a good fit to the data at $T = 373$ K and if the adsorption equilibrium constant $K_{A,p}$ follows the Arrhenius form, Eq. (18-7). The other values for $K_{A,p}$ are $K_{A,p}$ (296 K) = 2.045×10^{-3} (kPa)$^{-1}$ and $K_{A,p}$ (480 K) = 1.888×10^{-4} (kPa)$^{-1}$ (see Problem 18.D1).

Solution

A. Define. Find the best-fit *parameter* values for $K_{A,p}$ and $q_{A,max}$ in Eq. (18-5a) for the 373 K data. Then plot the isotherm data and the Langmuir isotherm to determine if this is a good fit. Finally, determine if the $K_{A,p}$ data satisfies Eq. (18-7) and find ΔH.

B and C. Explore and plan. Equation (18-5a) can be rearranged so that it will be a straight line. Multiply both sides by $(1 + K_{A,p}p_A)$ and divide by q_A

$$\frac{p_A}{q_A} = \left(\frac{1}{q_{max}}\right)p_A + \left(\frac{1}{q_{max}K_A}\right) \tag{18-5c}$$

If the Langmuir isotherm is valid, a plot of p_A/q_A vs. p_A will be linear. Direct nonlinear fitting of the raw data can also be done instead of linearization.

Take the natural log of both sides of the Arrhenius relationship, Eq. (18-7a)

$$\ln K_{Ap} = -\left(\frac{\Delta H}{R}\right)\frac{1}{T} + \ln K_{A,o} \tag{18-7b}$$

A plot of $\ln K_{A,p}$ vs. $(1/T)$ will be a straight line if the Arrhenius equation is followed. We can also check to see if q_{max} follows an Arrhenius form.

D. Do it. The values of p/q at 373 K are listed in Table 18-3. The plot of p/q vs. p is shown in Figure 18-3A.

Intercept = $360 = 1/(q_{max}K_{Ap})$

Slope = $\left(\frac{1000 - 360}{3000 - 0}\right) = 0.213 = 1/q_{max}$, and $q_{max} = 4.69$ m mol/g.

Thus, $K_{Ap} = \dfrac{\left(1/q_{max}\right)}{\left[1/(q_{max}K_{Ap})\right]} = \dfrac{0.213}{360} = 5.916 \times 10^{-4}$ (kPa)$^{-1}$

TABLE 18-3. *Equilibrium data for methane on Calgon PCB activated carbon (Ritter and Yang, 1987; Valenzuela and Myers, 1989)*

T = 296 K		T = 373 K			T = 480 K	
p (kPa)	q (m mol/g)	p (kPa)	q (m mol/g)	p/q	p(kPa)	q (m mol/g)
275.7880	2.0299	482.6290	1.1153	432.735	637.7598	0.3569
1137.6255	4.0822	1123.8361	1.8515	606.987	1296.2036	0.7584
2413.1450	5.0414	1620.2545	2.2307	726.344	2378.6716	1.2046
3757.6116	5.3983	1999.4630	2.5430	786.262	3709.3486	1.6061
5239.9722	5.5767	3447.3501	3.1676	1088.316	5329.6030	1.9184
6274.1772	5.6213	4929.7104	3.4799	1416.624	6246.5981	2.0745
6687.8589	5.6213	6156.9673	3.6806	1672.816	6687.8589	2.1415
		6584.4385	3.7699	1746.582		

and the Langmuir isotherm at 373 K is

$$q = \frac{(4.69)(5.916 \times 10^{-4})p_A}{1 + (5.916 \times 10^{-4})p_A}$$

The plot of this isotherm is shown in Figure 18-3B. Agreement between the Langmuir curve and the data are quite good.
The values for the Arrhenius plot are,

T	$1/T$	$K_{A,p}$	$\ln K_{A,p}$	q_{max}	$\ln q_{max}$
296	3.378 E-3	2.045 E-3	−6.192	6.11	1.810
373	2.681 E-3	5.916 E-4	−7.433	4.69	1.545
480	2.083 E-3	1.888 E-4	−8.575	3.84	1.345

The plots of $\ln K_{A,p}$ vs. $1/T$ and $\ln q_{max}$ vs. $1/T$ are both shown in Figure 18-3C. Clearly the data for both plots are well fit by straight lines and the Arrhenius relation is satisfied for both $K_{A,p}$ and q_{max}. Then from Eq. (18-7b) for the $K_{A,p}$ plot, slope = $-\Delta H/R$

Slope = 1840

$$\Delta H = -(1840 \text{ K}) 1.987 \frac{\text{cal}}{\text{mol K}} = -3656 \text{ cal/mol}$$

E. Check. Figure 18-3B is a check on the fit of the Langmuir constants. Since the agreement between the curve and the data points is quite good, the analysis is confirmed. The close agreement of the Arrhenius plot is another check on the analysis procedure.

F. Generalize. The fit to the straight line in Figure 18-3C is closer than for many other adsorption systems. Remember that although the amount adsorbed generally decreases as temperature increases, adsorption does not always

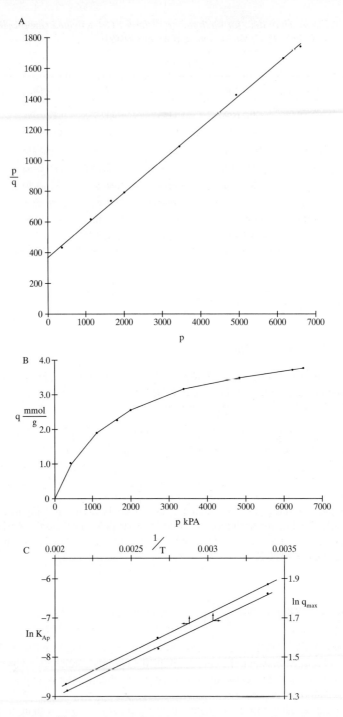

FIGURE 18-3. *Plots of equilibrium data for Example 18-1; (A) plot to give straight line for Langmuir isotherm, (B) Langmuir isotherm, (C) check on Arrhenius relationship*

follow an Arrhenius relationship. Note that the values of q_{max} are even less likely to follow the Arrhenius relationship although this system does.

18.2 SOLUTE MOVEMENT ANALYSIS FOR LINEAR SYSTEMS: BASICS AND APPLICATIONS TO CHROMATOGRAPHY

Packed columns similar to Figure 18-1 are the most common contacting devices used for adsorption and chromatography. Although there are exceptions, they are usually operated vertically with the flow parallel to the axis of column. Adsorption, chromatography, and ion exchange in packed columns are inherently unsteady state or batch type processes. Since the sorbent is stationary, it will saturate at the feed concentration if feed enters the column continuously. Thus, there must also be a *regeneration* step that removes most of the sorbate from the packing. The commonly used regeneration methods are to use an inert purge stream, change the temperature, change the pressure, and use a desorbent. After the regeneration step, there may be an optional cooling or drying step. Then the next cycle starts with the feed step. These processes will be analyzed in Sections 18.2 to 18.8 using increasingly complex analysis procedures. In this section we will start with the simplest theory, solute movement theory for linear isotherms, applied to elution chromatography, the simplest process to analyze.

Complete analysis of sorption processes requires computer simulation with a rather complex simulation program to solve the coupled algebraic and partial differential equations. Unfortunately, simulators often do not provide a physical picture of why the separation occurs and once a result is obtained simulators don't tell what to do to improve the separation. A relatively simple tool that is based on physical arguments and can be solved with pencil and paper (or a spreadsheet) will prove to be very useful even if it is not completely accurate. *Solute movement analysis* is a tool that allows engineers to use physical reasoning and understanding so that they can understand the results from experiments or simulations. The role of solute movement analysis in sorption processes is thus, analogous to the role of McCabe-Thiele diagrams in distillation, absorption, and extraction. Solute movement theory is used to understand the separations and for troubleshooting, not for final design.

18.2.1 Movement of Solute in a Column

Solute or sorbate within the packed section of a column can be in one of three locations. The solute can be in the interstitial void space ε_e and be moving at the interstitial velocity v_{inter}. If the solute is in the intraparticle voids $(1 - \varepsilon_e)\varepsilon_p$ or sorbed to the stationary solid it will have a net axial velocity of zero. In other words, the solute molecules are either scooting forward axially at a high velocity (remember that v_{inter} is greater than v_{super}), or they aren't moving at all. The average solute velocity u_s is

$$u_s = \text{(fraction of solute in the mobile phase, e.g., intraparticle voids) } v_{inter} \qquad \textbf{(18-9)}$$

$$\text{Fraction solute in } \varepsilon_e = (\text{Amt in } \varepsilon_e)/[\text{Amts in: } (\varepsilon_e + (1 - \varepsilon_e)\varepsilon_p + \text{sorbed})] \qquad \textbf{(18-10)}$$

This fraction can be calculated by considering the distribution of an incremental change in solute concentration Δc (e.g., in kmol/m^3) and its corresponding change in the amount sorbed Δq (e.g., in kmol/kg adsorbent). The amount in each location can be determined by calculating the inventory of mass.

$$\text{Amt in } \varepsilon_e = \varepsilon_e \text{ (Vol. Column segment)(conc. change)} = \varepsilon_e (\Delta z \, A_c)(\Delta c) \quad \textbf{(18-11a)}$$

$$\text{Amt. in pores} = (1 - \varepsilon_e)\varepsilon_p K_d \text{ (Vol. col. segment)(conc. change)}$$
$$= (1 - \varepsilon_e)\varepsilon_p K_d (\Delta z A_c)(\Delta c) \quad \textbf{(18-11b)}$$

where K_d is the fraction of pores that molecules can squeeze into. This term becomes important in *size exclusion chromatography* which separates molecules based on size and ideally has no adsorption, $\Delta q = 0$ (Wu, 2004).

Both Eqs. (18-11a) and (18-11b) are in kmoles adsorbate if c is $kmol/m^3$.

$$\text{Amt. sorbed} = (1 - \varepsilon_e)(1 - \varepsilon_p)\text{(Vol. col. segment)(change amt. sorbed)}$$

$$= (1 - \varepsilon_e)(1 - \varepsilon_p)(\Delta z A_c) \, \rho_s \frac{\text{kg solid}}{m^3} \; (\Delta q, \, kmol/kg \, ads.) \quad \textbf{(18-11c)}$$

The K_d term is not included in Eq. (18-11c) since we assume Δq is based on a measurement that *automatically* includes any steric hindrance. Equation (18-11c) will also be in kmoles adsorbate. If q and c are in different units than in this derivation, the mass balances will be slightly different (compare Eqs. 18-14a, b and c to 18-13).

Inserting Eq. (18-11) into Eq. (18-10) one obtains,

$$\text{Fraction} = \varepsilon_e (\Delta z \, A_c)(\Delta c)/\{(\Delta z \, A_c)([\varepsilon_e \, \Delta c + (1 - \varepsilon_e)\varepsilon_p K_d \, \Delta c + (1 - \varepsilon_e)(1 - \varepsilon_p) \, \rho_s \, \Delta q)]\}$$
$$\textbf{(18-12)}$$

After Eq. (18-12) is substituted into Eq. (18-9), the result can be simplified to

$$u_s = \frac{v_{inter}}{1 + \dfrac{(1 - \varepsilon_e)}{\varepsilon_e}\varepsilon_p K_d + \dfrac{(1 - \varepsilon_e)(1 - \varepsilon_p)}{\varepsilon_e} \, \rho_s \dfrac{\Delta q}{\Delta c}} \quad \textbf{(18-13)}$$

The exact form of Eq. (18-13) depends upon the units of the equilibrium data. For example, if q is in kg solute/kg solid and x is in kg solute/kg fluid so that the isotherm expression is $q = f(x)$, then there must be a ρ_F (kg fluid/m^3) term in Eqs. (18-11a) and (18-11b) and Δx replaces Δc in these equations. Then Eq. (18-13) becomes

$$u_s = \frac{v_{inter}}{1 + \dfrac{(1 - \varepsilon_e)}{\varepsilon_e}\varepsilon_p K_d + \dfrac{(1 - \varepsilon_e)(1 - \varepsilon_p)}{\varepsilon_e} \, \dfrac{\rho_s}{\rho_f} \dfrac{\Delta q}{\Delta x}} \quad \textbf{(18-14a)}$$

If q and c are both in mol/m^3, the equation for u_s is obtained by eliminating ρ_s from Eq. (18-13).

For gas systems equilibrium is usually expressed in terms of the partial pressure, p_A (e.g., Eqs., (18-5a) and (18-5b)). The solute velocity for gases is then

$$u_s = \frac{v_{inter}}{1 + \dfrac{(1 - \varepsilon_e)}{\varepsilon_e}\varepsilon_p K_d + \dfrac{(1 - \varepsilon_e)(1 - \varepsilon_p)}{\varepsilon_e} \, \rho_s \dfrac{\Delta q}{\Delta p_A} \dfrac{p_{ToT}}{\bar{\rho}_f}} \quad \textbf{(18-14b)}$$

where $\bar{\rho}_f$ is the molar density. For an ideal gas $\bar{\rho}_f = p_{tot}/RT$ and $p_{tot}/\bar{\rho}_f = RT$. Thus, for ideal gases,

$$u_s = \frac{v_{inter}}{1 + \dfrac{(1 - \varepsilon_e)}{\varepsilon_e}\varepsilon_p K_d + \dfrac{(1 - \varepsilon_e)(1 - \varepsilon_p)}{\varepsilon_e}\rho_s \dfrac{\Delta q}{\Delta p_A} RT} \qquad \textbf{(18-14c)}$$

In some ion exchange and biochemical systems c and q are both in g/L or mol/L. Then Δq is also in g/L or mol/L and the ρ_s term in Eqs. (18-11c), (18-12) and (18-13) does not appear. The result is,

$$u_s = \frac{v_{inter}}{1 + \dfrac{1 - \varepsilon_e}{\varepsilon_e}\varepsilon_p K_d + \dfrac{(1 - \varepsilon_e)(1 - \varepsilon_p)}{\varepsilon_e}\dfrac{\Delta q}{\Delta c}} \qquad \textbf{(18-14d)}$$

Equations (18-13) and (18-14) allow us to calculate the average velocity of the solute if we can calculate $(\Delta q / \Delta c)$, $(\Delta q / \Delta x)$, or $(\Delta q / \Delta p_A)$. If we assume that mass transfer is very rapid so that the solid and fluid are locally (at each z value) in equilibrium, the solute velocity depends only upon the equilibrium behavior $(\Delta q / \Delta c)$, $(\Delta q / \Delta x)$, or $(\Delta q / \Delta p_A)$ and the properties of the packed column, not upon mass transfer rates. Mass transfer will be critically important to determine how the solute spreads from the average (see Sections 18.6 to 18.8). In Sections 18.2 to 18.5 we will insert the appropriate isotherm into Eqs. (18-13) and (18-14).

18.2.2 Solute Movement Theory for Linear Isotherms

The theory becomes simplest when the linear isotherm, Eq. (18-5b) or (18-6b), is used. Since almost all equilibrium data becomes linear at low enough concentrations or partial pressures, there are a number of real applications of linear isotherms.

When Eq. (18-6b) is valid, $\Delta q / \Delta c = K_i'$, and Eq. (18-13) becomes

$$u_{s,i} = \frac{v_{inter}}{1 + \dfrac{(1 - \varepsilon_e)}{\varepsilon_e}\varepsilon_p K_{d,i} + \dfrac{(1 - \varepsilon_e)(1 - \varepsilon_p)}{\varepsilon_e}\rho_s K_i'} \qquad \textbf{(18-15a)}$$

for any adsorbate i. Note that none of the terms depend upon the concentration of adsorbate i. Thus, at low concentrations where the linear isotherm is valid the solute velocity becomes constant. The solute velocity $u_{s,i}$ depends upon temperature since K_i' depends upon temperature (e.g., following Eq. (18-7)). If we have a number of different solutes with different values of the equilibrium constant, the weakest sorbed solute (lowest value of K_i') moves the fastest, and the strongest sorbed solute (highest value of K_i') moves the slowest. Since they move at different speeds, they can be separated. A single-porosity form of this equation is also commonly used (see Problem 17.C5, which includes Eq. (18-15b).

To visualize what this looks like, we will start with a packed column that is initially clean ($c_A = 0$). At time t_0, we start adding a feed with a concentration $c_{A,F}$ at a known interstitial velocity. Equation (18-15a) can be used to calculate the solute velocity $u_{s,A}$ (the numerical calculation procedure is illustrated in Example 18-2). In Figure 18-4A the *solute movement* or *characteristic* diagram for this process is plotted. Solute starts at z = 0 at time t_0, and moves upward at velocity $u_{s,A}$, which is the slope of the characteristic line shown in the figure. The procedure will probably be easiest to understand after you study Example 18-2. The concentrations in the column are shown at four times in Figures 18-4 B, C, D, and E. The solute moves upward in a wave at a constant velocity $u_{s,A}$. If adsorption is strong the wave moves slowly while if adsorption is weak it moves quickly. Waves for nonlinear systems are shown later in Figure 18-15.

For systems in wt frac, $q_A = K'_{A,x} x_A$, or $q_A = K'_{A,y} y_A$ (q in kg solute/total kg solid and x and y in kg solute/kg fluid), Eq. (18-14a) becomes

$$u_{s,i} = \frac{v_{inter}}{1 + \dfrac{(1-\varepsilon_e)}{\varepsilon_e} \varepsilon_p K_{d,i} + \dfrac{(1-\varepsilon_e)}{\varepsilon_e}(1-\varepsilon_p)\dfrac{\rho_s}{\rho_f} K'_{A,x}}$$

(18-15c)

For ideal gases with $\Delta q / \Delta p_A = K'_{A,p}$ Eq. (18-14c) becomes

$$u_{s,i} = \frac{\varepsilon_e v_{inter}}{\varepsilon_e + (1-\varepsilon_e)\varepsilon_p K_{d,i} + (1-\varepsilon_e)(1-\varepsilon_p)\,\rho_s RT K'_{A,p}}$$

(18-15d)

With linear isotherms the denominator in Eq. (18-15a) through Eq. (18-15d) is independent of concentration. Thus, for purposes of algebraic manipulation it is convenient to write these equations as

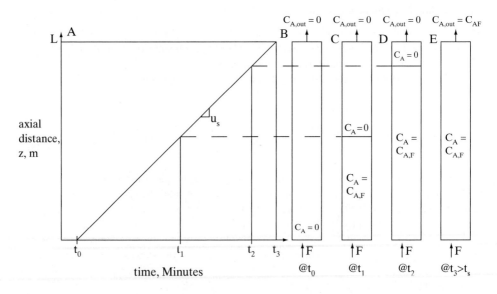

FIGURE 18-4. *Wave movement for step change in feed concentration. (A) Solute movement diagram for linear isotherm. B, C, D, E are concentrations in column at t = t_0, t_1, t_2, and t_3, respectively. Since t_3 = breakthrough time t_{br} = $L/u_{s,A}$, entire column and outlet are at $c_{A,F}$*

Chapter 18   Introduction to Adsorption, Chromatography, and Ion Exchange

$$u_{si} = C_i v_{inter} \qquad \text{where } C_i \text{ is a constant} \qquad \textbf{(18-15e)}$$

Although very convenient, linear isotherms find limited applications in industrial processes. To decrease the diameter of the columns and thus, reduce costs, most industrial separations are done at high concentrations or very close to the solubility limit. Since the isotherms are probably nonlinear and use of linear isotherms can lead to large errors, one needs to use the nonlinear theories in Section 18.6 for concentrated systems.

18.2.3 Application of Linear Solute Movement Theory to Purge Cycles and Elution Chromatography

The simplest regeneration cycle is a *purge* cycle using an inert carrier gas for gas systems or an inert solvent for liquid systems. This cycle can be operated co-flow (Figure 18-5A) or *counterflow* (Figure 18-5B). When the purge gas or liquid enters the column, the partial pressure or concentration of the adsorbate will drop since it is being diluted. This causes adsorbate to desorb (see Figure 18-2), thus, allowing it to be flushed from the column. The ideal carrier gas or solvent has the following properties: easy to separate from the adsorbate, easy to remove from the bed, nontoxic, nonflammable, available, and inexpensive. For purge gas systems nitrogen and hydrogen are close to ideal as carrier gases. Since the adsorbate is diluted, purge cycles are not commonly used for large-scale commercial adsorption processes.

Purge cycles are commonly used in *elution chromatography*, particularly in analytical chemistry. Elution chromatography involves input of a feed pulse into the packed column followed by co-flow (Figure 18-5A) of an inert solvent or carrier gas. (If the solvent or carrier gas also adsorbs, the process can become gradient or displacement chromatography, which are discussed later.) The column can be packed with any of the adsorbents or chromatography packings mentioned previously. If the solutes have different equilibrium isotherms, the solutes will move in the column at different velocities and will be separated (Figure 18-5C). Both gas and liquid chromatography are commonly operated in this elution mode to determine the compositions of unknown samples. Large-scale elution chromatography systems are also becoming more common, particularly in the pharmaceutical and fine chemical industries. Detailed design considerations for large-scale biochromatography systems are discussed in detail by Bonnerjea and Terras (1994), Ladisch (2001), and Rathore and Velayudhan (2004). The dilution that is inherent in purge operations tends to make these processes expensive for large-scale separation, but with complicated feeds there may be no better alternative separation method.

> ### EXAMPLE 18-2. Linear solute movement analysis of elution chromatography
>
> A 1-meter column is packed with activated alumina. The column initially contains pure liquid cyclohexane solvent. At time t = 0 a feed pulse that contains 0.0001 mol/L anthracene and 0.0002 mol/L naphthalene in cyclohexane is input for 10.0 minutes. The superficial velocity is 20 cm/min for both feed and solvent steps. Use the solute movement theory to predict the outlet concentrations.
>
> Data: Bulk density (fluid is air) = 642.6 kg/m³, $\varepsilon_p = 0.51$, $\varepsilon_e = 0.39$, ρ_f (cyclohexane) = 0.78 kg/L. $K_A' = 22.0$ L/kg adsorbent and $K_N' = 16.0$ L/kg adsorbent where A = anthracene and N = naphthalene. $K_d = 1.0$ for both anthracene and naphthalene.

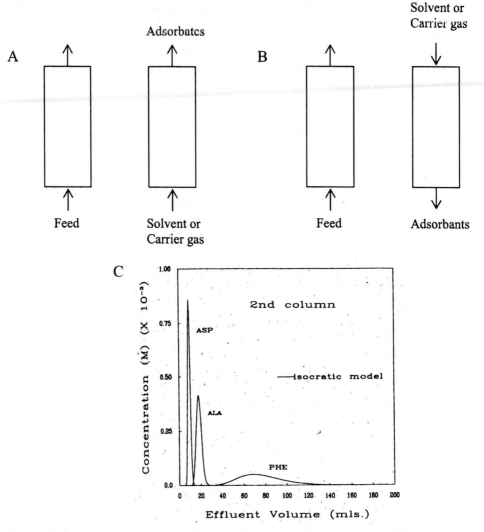

FIGURE 18-5.   *Purge systems. (A) co-flow (e.g., elution chromatography), (B) counterflow, (C) outlet concentrations for elution chromatography of aspartic acid (asp), alanine (ala), and phenylalanine (phe) on cation exchange resin (Agosto et al., 1989). Reprinted with permission from* Ind. Eng. Chem. Research, 28, *1358 (1989), copyright 1989 American Chemical Society.*

Solution

A. Define. The apparatus is sketched in Figure 18-5a. We want to find when the anthracene and naphthalene pulses exit the column.

B. Explore. Equation (18-15) can be used to determine numerical values $u_{s,A}$ and $u_{s,N}$. Lines drawn on a solute movement diagram with these slopes from the start (t = 0 min) and end (t = 10 min) of the feed pulse outline the movement of average molecules of anthracene and naphthalene. Since $K'_N < K'_A$, $u_{s,N} > u_{s,A}$, and the naphthalene should exit the column first.

C. Plan. Calculate $u_{s,N}$ and $u_{s,A}$. Plot the solute movement diagram for a 10 minute feed pulse.

D. Do it. The interstitial velocity can be determined from Eq. (18-2b),

$$v_{inter} = v_{super} / \varepsilon_e = (0.2 \text{ m/min})/0.39 = 0.513 \text{ m/min}.$$

The structural density can be determined by rearranging Eq. (18-3c),

$$\rho_s = \rho_b /[(1 - \varepsilon_e)(1 - \varepsilon_p)] = (642.6 \text{ kg/ m}^3)/(0.61)(0.49) = 2150 \text{ kg/m}^3$$

The solute velocities can now be calculated from Eq. (18-15)

$$u_{s,A} = \cfrac{0.513 \text{ m/min}}{1 + \cfrac{0.61}{0.39}(0.51)(1.0) + \cfrac{(0.61)(0.49)}{0.39}(2150 \text{ kg/m}^3)(22 \text{ L/kg})\left(\cfrac{m^3}{1000 \text{ L}}\right)}$$

Note that we had to convert K'_A from L/kg to m³/kg. Then,

$$u_{s,A} = \frac{0.513}{1 + 0.798 + 36.251} = 0.0135 \text{ m/min}$$

$$\text{Similarly, } u_{s,N} = \frac{0.513}{1.798 + 26.368} = 0.0182 \text{ m/min}$$

These solute velocities are the slopes for each solute on the solute movement diagram (Figure 18-6A). The lines for each solute can be drawn from the start and finish of the feed pulse. The resulting solute pulses or waves exit the column at z = L = 1.00 m as shown in Figure 18-6B. The naphthalene exits from 54.95 to 64.95 minutes and the anthracene exits from 74.07 to 84.07 minutes. The naphthalene and anthracene peaks are both at their feed concentrations, 0.0002 and 0.0001 mol/L, respectively.

E. Check. Naphthalene input at t = 0 will exit at t = $L/u_{s,N}$ = 1.0 m/(0.0182 m/min) = 54.95 minutes. Similarly the anthracene input at t = 0 exits at t = 74.07 minutes. Since the peaks both last 10 minutes the naphthalene exits at 64.95 minutes and the anthracene at 84.07 minutes. The peak centers start at $t_{feed}/2$ = 5 minutes, and are at 59.95 and 79.07 minutes. These results agree with the graph in Figure 18-6A.

F. Generalize. Since no spreading phenomena are included in the simple solute movement theory, the outlet peaks are predicted to be square waves (Figure 18-6B). When mass transfer and axial dispersion are included, the curves are spread out more as was illustrated in Figure 18-5C. The solute movement theory correctly predicts the behavior of an *average* molecule. Thus, the time for the center of the peaks is correctly predicted. Note that the dominant term in the denominator for both solutes is the adsorption term. This will be the case when there is relatively strong adsorption.

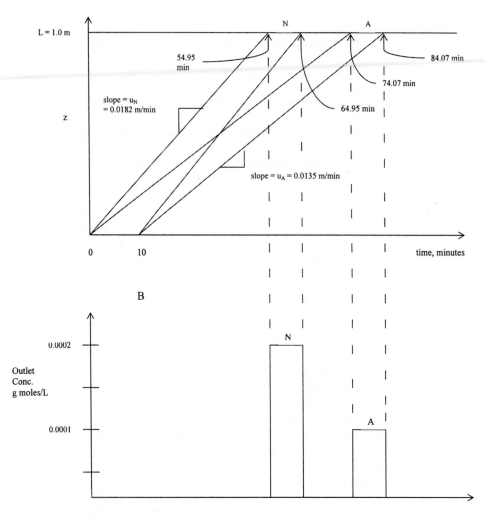

FIGURE 18-6.   *Results for Example 18-2 for elution chromatography; (A) solute movement diagram,
(B) outlet concentration profiles*

Basmadjian (1997) reports that a typical linear velocity for the feed step for
adsorption in liquid systems is 0.001 m/s, which is 6 cm/min. The purge step may
be roughly ten times faster. Thus, the flow rates in this problem are reasonable.

There is an analogy that may be useful in understanding this solute movement analysis.
The problems are similar to algebra problems where two trains start to leave a station at the
same time, but with different velocities (u_A and u_B in chromatography). You want to calculate
when each train arrives at a second station (a distance L away) and when the tail end of each
train (analogous to the feed time, t_F) leaves the second station.

In actual practice it is much easier to calculate the exit times as shown in step E of Example 18-2 rather than drawing the solute movement diagram exactly to scale. This can easily be done with a spreadsheet even for much more complex processes. However, a sketch of the solute movement diagram should always be made since it will guide the calculations and provide a visual check on the calculations.

In real systems the results predicted by solute movement theory are spread considerably by axial dispersion, mass transfer resistances and mixing in column dead volumes, valves and pipes. Thus, the predictions for the simple elution chromatography system shown in Figure 18-6B would spread and the two solute peaks would overlap as shown in Figure 18-5C. This calculation will be illustrated later in Example 18-10. If high product purities are required, this zone spreading requires either a very short feed step and/or a long column to move the peaks apart. In addition, the next feed pulse must be delayed for a considerable time or zone spreading will result in too much overlap of the slow peak with the fast peak from the next feed pulse. The net result is that elution chromatography can be very expensive for large-scale applications. Simulated moving bed (SMB) systems are often less expensive for binary separations, and are analyzed in Section 18.3.3.

A number of variations of elution chromatography have been developed. In *flow programming* the flow rate is increased to more rapidly elute the late components in Figure 18-5C. This technique does not change the volume of elutant required, but reduces the time of operation. *Temperature programming* increases the temperature of the entire column during the elution. Increased temperature decreases the adsorption of the solutes so that they exit both sooner and more concentrated. The related *temperature gradient* method increases the temperature of the fluid entering an adiabatic column. Temperature changes usually have more effect in gas systems than in liquid systems. In liquid systems it is common to use a *solvent gradient* to change the solvent to decrease the sorption of the solutes. Common changes in the solvent are to change ionic strength, polarity, pH, the fraction of organic solvent, and the addition of a strongly sorbed material. Gradients, which can be done as continuous changes or as step functions, are commonly used in bioseparations (Ladisch, 2001). If a step change is made in the concentration of a chemical that is more strongly sorbed than all the components of the feed, the process is called *displacement chromatography* [see Cramer and Subramanian (1990) for an extensive review].

A major commercial problem is the purification of strongly adsorbed species such as moderate molecular weight (C_{10} to C_{20}) straight-chain hydrocarbons (Ruthven, 1984). Purge systems require excessive purge gas or solvent for these strongly adsorbed species. Pressure swing cycles (Section 18.3.2) also require too much purge gas for strongly adsorbed materials. Thermal cycles (Section 18.3.1) can cause excessive thermal decomposition since very high temperatures are required. Displacement cycles using a desorbent (e.g., n-pentane, n-hexane, and ammonia) that is adsorbed have proven to be effective for this otherwise intractable problem (Wankat, 1986). (Unfortunately, the nomenclature is confusing since displacement cycles for adsorption have a desorbent that can adsorb less or more than the adsorbate while in displacement chromatography the desorbent is the most strongly adsorbed compound.) The cycles will be basically the same as the counterflow cycle shown in Figure 18-5B. Since displacement adsorption requires that the desorbent be recovered from the product—often by distillation—they are relatively expensive processes that are only used when other adsorption processes fail. Both purge and displacement cycles are commonly used in SMB systems (Section 18.3.3).

18.3 SOLUTE MOVEMENT ANALYSIS FOR LINEAR SYSTEMS: THERMAL AND PRESSURE SWING ADSORPTION AND SIMULATED MOVING BEDS

Since purge cycles use large amounts of solvent, other regeneration methods have been developed. These methods and their analysis with the solute movement theory is the topic of this section.

18.3.1 Temperature Swing Adsorption (TSA)

Temperature swing adsorption (TSA) is commonly used for gas systems particularly for the recovery or removal of trace components that are strongly adsorbed (Ruthven, 1984; Yang, 1987). Typical applications are removal of pollutants such as volatile organic compounds (VOC) (Fulker, 1972; Reynolds et al., 2002) and drying gases (Basmadjian, 1984, 1997). The basic cycle using counterflow of the hot regenerant gas is shown in Figure 18-7A. The cycle can be thought of as a purge operation with a hot purge gas. The feed step can be done either with upward or downward flow. If the feed gas flows continuously, two or more units are operated in parallel. Although the pure gas is usually the desired product, in some cases the concentrated adsorbate is the valuable product. After the regeneration step, an optional cooling step may be inserted. Cooling is required if the feed gas can react with the hot adsorbent. Insertion of the cooling step tends to produce a purer product, but lowers the productivity (kg feed processed)/(hour × kg adsorbent).

The basis for this regeneration method is the large reduction in the equilibrium constant observed in most systems when the temperature is increased, Eq. (18-7), and the simultaneous reduction in the partial pressure or concentration with the addition of the purge gas. Both of these effects lead to removal of the adsorbate from the column. Since large increases in the adsorbate concentration can occur, TSA systems are also used to concentrate dilute gas streams. When cooling is not required to prevent chemical reactions, a number of modifications of the basic cycle have been developed (Natarajan and Wankat, 2003). These modifications are explored as homework problems.

One major disadvantage of the TSA system shown in Figure 18-7A is that large amounts of pure regeneration gas may be required to heat the adsorption column and adsorbent. This occurs because at normal pressures the volumetric heat capacity of the gas is quite low compared to the volumetric heat capacity of the adsorbent and the metal shell of the column. Thus, regeneration may be relatively slow, expensive, and not produce the desired concentrated adsorbate product. This disadvantage tends to be minor when the feed gas is quite dilute and the adsorption is strong. Since the feed step will be quite long, the relative amount of hot regenerant gas used is reasonable. However, if the feed gas is concentrated (above a few percent) the adsorbent will saturate fairly quickly and the feed step will be relatively short. Since the same amount of hot regenerant gas is required to heat the column, the ratio (hot regenerant gas/feed gas) becomes excessive.

To study TSA systems with the solute movement analysis we must determine the effect of temperature changes on the solute waves, the rate at which a temperature wave moves in the column, and the effect of temperature changes on concentration. The first of these is easy. As temperature increases the equilibrium constants, K_A and K'_A, both decrease, often following an Arrhenius type relationship as shown in Eq. (18-7). If the effect of temperature on the equilibrium constants is known, new values of the equilibrium constants can be calculated and new solute velocities can be determined.

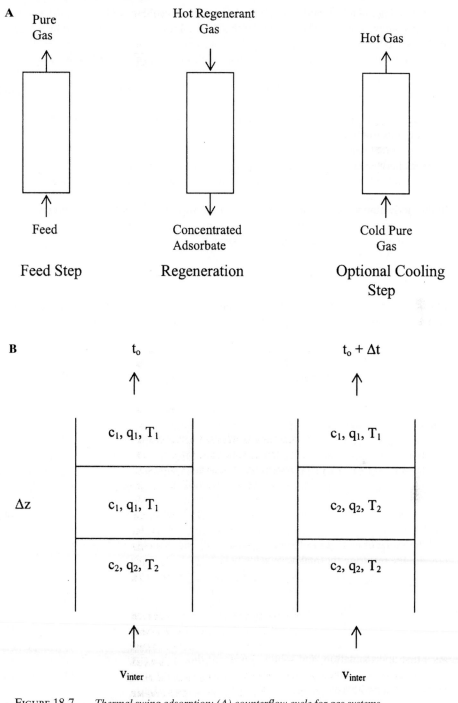

A   Pure
      Gas

Hot Regenerant
      Gas

Hot Gas

Feed

Concentrated
Adsorbate

Cold Pure
    Gas

Feed Step

Regeneration

Optional Cooling
      Step

B t_o $t_o + \Delta t$

Δz

c_1, q_1, T_1 c_1, q_1, T_1

c_1, q_1, T_1 c_2, q_2, T_2

c_2, q_2, T_2 c_2, q_2, T_2

v_{inter} v_{inter}

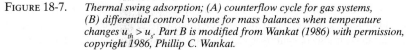

FIGURE 18-7.   *Thermal swing adsorption; (A) counterflow cycle for gas systems,
(B) differential control volume for mass balances when temperature
changes $u_{th} > u_s$. Part B is modified from Wankat (1986) with permission,
copyright 1986, Phillip C. Wankat.*

Changing the temperature of the feed to a sorption column will cause a *thermal wave* to pass through the column. The velocity of this thermal wave can be calculated by a procedure analogous to that used for solute waves. The thermal wave velocity will be the fraction of the change in thermal energy in the mobile phase multiplied by the interstitial velocity,

$$u_{th} = \text{(fraction of energy change in the mobile phase) } v_{inter} \qquad \textbf{(18-16)}$$

Since changes in the energy contained in the fluid in the pores, in the solid, and the walls are stagnant, this fraction is,

$$\frac{\text{frac.energy change}}{\text{in mobile phase}} = \frac{\text{energy change in mobile phase}}{\text{Energy change in:(mobile + stagnant fluid + solid + wall)}} \qquad \textbf{(18-17)}$$

In this derivation we are assuming a pure thermal wave with no adsorption, no reactions and no phase changes. Thus, energy changes are totally due to specific heat. For example, the amount of energy change in the mobile phase is

$$\text{Amt. Energy Change in Mobile Phase} = \varepsilon_e(\Delta z A_c)\rho_f C_{p,f}\Delta T_f \qquad \textbf{(18-18)}$$

where C_{pf} is the heat capacity of the fluid and ΔT_f is the change in fluid temperature. Substituting in the appropriate terms, the fraction of energy change in the mobile phase is

$$\text{fraction} = \frac{\varepsilon_e(\Delta z A_c)\rho_f C_{p_f}\Delta T_f}{\{\Delta z A_c[\varepsilon_e\Delta T_f + (1-\varepsilon_c)\varepsilon_p\Delta T_{pf}\rho_f C_{p_f} + (1-\varepsilon_e)(1-\varepsilon_p)\rho_s C_{p_s}\Delta T_s] + (\Delta z W)C_{p_w}\Delta T_w\}} \qquad$$

$$\textbf{(18-19)}$$

where W is the weight of the column per length (kg/m),and ΔT_{pf}, ΔT_s and ΔT_w are the changes in pore fluid, solid and wall temperatures induced by the change in fluid temperature.

If we divide the numerator and denominator of Eq. (18-19) by ΔT_f, we will have the ratios of ΔT_{pf}, ΔT_s and ΔT_w to ΔT_f. If heat transfer is very rapid, the system will be in thermal equilibrium, $T_f = T_{pf} = T_s = T_w$. This equality requires that the changes in temperature all be equal

$$\Delta T_f = \Delta T_{pf} = \Delta T_s = \Delta T_w \qquad \textbf{(18-20)}$$

and the ratios of changes in temperatures are all one. Combining Eqs. (18-17), (18-19) and (18-20), the resulting thermal wave velocity is

$$u_{th} = \frac{v_{inter}\rho_f C_{p_f}}{\left[\left[1 + \left(\frac{(1-\varepsilon_e)}{\varepsilon_e}\right)\varepsilon_p\right]\rho_f C_{p_f} + \frac{(1-\varepsilon_e)(1-\varepsilon_p)}{\varepsilon_e} C_{p_s}\rho_s + \frac{W}{\varepsilon_e A_c} C_{p_w}\right]} \qquad \textbf{(18-21)}$$

As a first approximation, and solute movement theory is a first approximation, the thermal wave velocity is independent of concentration and temperature. Temperature is constant along the lines with a slope equal to the numerical value of u_{th}.

In TSA processes the purpose of increasing the temperature is to remove the adsorbate from the adsorbent. This happens when the thermal wave intersects the solute wave. Assume that the column is initially at a uniform temperature T_1, concentration c_1, and adsorbent loading

q_1. Fluid at temperature T_2 is fed into the column. This temperature change causes the concentration and adsorbent loading to change to c_2 and q_2 (currently both unknown). Since the solute movement theory assumes local equilibrium, c_2 and q_2 are in equilibrium at T_2. The control volume shown in Figure 18-7B (Wankat, 1986) will be used to develop the mass balance for this temperature change. Initially, the thermal wave is at the bottom of the control volume. The thermal wave is assumed to move faster than the solute wave, which is true for most dilute liquid and some dilute gas systems. For a differential slice of column of arbitrary height Δz, the temperature of the slice will change from the initial temperature T_1 to the final temperature T_2 if $\Delta t = \Delta z / u_{th}$.

The mass balance for the differential slice over the time interval Δt is,

$$\varepsilon_e v_{inter} \Delta t (c_2 - c_1) - [\varepsilon_e + K_d \varepsilon_p (1 - \varepsilon_e)](c_2 - c_1) \Delta z$$
$$-(1 - \varepsilon_e)(1 - \varepsilon_p)\rho_s (q_2 - q_1)\Delta z = 0 \qquad \textbf{(18-22)}$$

Since $\Delta t = \Delta z / u_{th}$, this simplifies to,

$$\left(\varepsilon_e + \varepsilon_p (1 - \varepsilon_e) K_d - \frac{\varepsilon_e v_{inter}}{u_{th}} \right)(c_2 - c_1)$$

$$+ (1 - \varepsilon_e)(1 - \varepsilon_p)\rho_s [q_2(c_2,T_2) - q_1(c_1,T_1)] = 0 \qquad \textbf{(18-23)}$$

Equation (18-23) and the appropriate isotherm equation can be solved simultaneously for the unknowns c_2 and q_2. For a linear isotherm, Eq. (18-6b), the simultaneous solution is

$$\frac{c(T_2)}{c(T_1)} = \left(\frac{1}{u_s(T_1)} - \frac{1}{u_{th}} \right) \Big/ \left(\frac{1}{u_s(T_2)} - \frac{1}{u_{th}} \right) \qquad \textbf{(18-24)}$$

For a liquid system where $u_{th} > u_s$, the use of a hot feed liquid ($T_2 > T_1$) will increase the outlet concentration. This is illustrated in Example 18-3.

EXAMPLE 18-3. Thermal regeneration with linear isotherm

Use a thermal swing adsorption process to remove traces of xylene from liquid n-heptane using silica gel as adsorbent. The adsorber operates at 1.0 atm. The feed is 0.0009 wt frac xylene and 0.9991 wt frac n-heptane at 0°C. Superficial velocity of the feed is 8.0 cm/min. The adsorber is 1.2 meters long and during the feed step is at 0°C. The feed step is continued until xylene breakthrough occurs. To regenerate use counterflow of pure n-heptane at 80°C, and continue until all xylene is removed. Superficial velocity during purge is 11.0 cm/min.

Data: At low concentrations isotherms for xylene : q = 22.36x @ 0°C, q = 2.01x @ 80°C, q and x are in g solute/g adsorbent and g solute/g fluid, respectively (Matz and Knaebel, 1991). $\rho_s = 2100$ kg/m³, $\rho_f = 684$ kg/m³, $C_{ps} = 2000$ J/kg °C, $C_{pf} = 1841$ J/kg °C, $\varepsilon_e = 0.43$, $\varepsilon_p = 0.48$, $K_d = 1.0$.

Assume: Wall heat capacities can be ignored, heat of adsorption is negligible, no adsorption of n-heptane, and system is at cyclic steady state. Using the solute movement theory

a) Determine the breakthrough time for xylene during the feed step.
b) Determine the time for the thermal wave to breakthrough during both the feed and purge steps.
c) Determine the xylene outlet concentration profile during the purge step.

Solution

A. **Define.** The process is similar to the sketch in Figure 18-7A but without the optional cooling step. The breakthrough time for solute is the time that xylene first appears at the column outlet, $z = L$. The thermal wave breakthrough times occur when the temperature starts to decrease (during the feed step) or increase (during the purge step). The desired outlet concentration profile is xylene concentration vs. time.

B. **Explore.** Since operation is at cyclic steady state (each cycle is an exact repeat of the previous cycle), the column will be hot when the cold feed is started. The cold feed causes a cold thermal wave during the feed step. We expect that this wave will move faster than the xylene wave (this expectation will be checked while doing the calculations); thus, the waves are independent. When the flow direction is reversed, the xylene wave concentration is unchanged until the xylene wave intersects the thermal wave. This causes a period when the regenerated liquid exiting the bottom of the column is at the feed concentration (study Figure 18-8 to understand this.) When the two waves intersect the temperature changes, the isotherm parameter changes, and the xylene wave velocity changes. At the same time, xylene is desorbed and the xylene concentration in the fluid increases.

C. **Plan.** Since mass fractions are used in the equilibrium expression, we use Eq. (18-15c) to calculate the velocity of the solute at both 0 and 80°C. The thermal wave velocity is determined from Eq. (18-21) with $W = 0$. The effect of the temperature change on the fluid concentration can be determined either from a mass balance over one cycle or from Eq. (18-24).

D. **Do it.**

$$V_{inter,F} = \frac{v_{super,F}}{\varepsilon_e} = \frac{8.0}{0.43} = 18.60 \text{ cm/min}$$

$$V_{inter,purge} = \frac{-11.0}{0.43} = -25.58 \text{ cm/min}$$

To calculate solute velocity,
Use Eq. (18-15c),

$$u_s = \frac{v_{inter}}{1 + \frac{(1-\varepsilon_e)}{\varepsilon_e}\varepsilon_p K_d + \frac{(1-\varepsilon_e)(1-\varepsilon_p)}{\varepsilon_e}\frac{\rho_s}{\rho_f}K'(0°C)}$$

$$u_{s,F}(0°C) = \frac{18.60}{1 + \frac{(0.57)(0.48)(1.0)}{(0.43)} + \frac{(0.57)(0.52)}{0.43}\frac{2100}{684}(22.36)} = 0.3799 \text{ cm/min}$$

Where $K'(0°C) = 22.36$.

And xylene breakthrough time is,

$$t_{br,xy} = \frac{120cm}{0.3799 \text{ cm/min}} = 315.85 \text{ min}$$

The thermal wave velocity from Eq.(18-24) with $W = 0$ is

$$u_{th,F} = \frac{v_{inter}}{(1 + \dfrac{(1-\varepsilon)}{\varepsilon_e}\varepsilon_p) + \dfrac{(1-\varepsilon_e)(1-\varepsilon_p)}{\varepsilon_e}\dfrac{\rho_s C_{ps}}{\rho_f C_{pf}}}$$

$$= \frac{18.60}{1.636 + \dfrac{(.57)(.52)}{.43}\dfrac{2100}{684}\dfrac{2000}{1841}} = 4.701 \text{ cm/min}$$

And thermal breakthrough time is,

$$t_{br,th,F} = \frac{120cm}{4.701 \text{ cm/min}} = 25.53 \text{ min}$$

For the purge step $v_{inter,purge} = -25.58 \text{ cm/min}$ and $u_{th,purge} = -6.466 \text{ cm/min}$. They are negative because the flow direction is reversed. This breakthrough time is

$$t_{th,br,purge} = \frac{-120}{-6.466} = 18.56 \text{ min}$$

Note that the thermal wave moves considerably faster than the solute wave and breakthrough is quicker. After the temperature change $K'(80°C) = 2.01$, and the solute velocity and breakthrough time during purge are

$$u_s(80°) = \frac{-25.58}{1.636 + \dfrac{(0.57)(0.52)}{.43}\dfrac{(2100)}{684}(2.01)} = \frac{-25.58}{1.636 + 4.253} = -4.343 \text{ cm/min}$$

$$t_{br,xy,purge} = \frac{-120}{-4.343} = 27.629 \text{ min}$$

The xylene mass balance on one cycle at cyclic steady-state is, In = Out.

$$\text{In} = (8.0 \text{cm}/\text{min}) (A_c, \text{cm}^2)(315.85 \text{min})\rho_f (\text{g solv}/\text{cm}^3) (0.0009 \text{ g xyl/g soln})$$

The outlet stream shown in Figure 18-8, consists of one part at x_F and one part that is concentrated.

$$\text{Out} = (11.0 \text{cm}/\text{min})(A_c, \text{cm}^2)(18.56 \text{ min})(\rho_f \text{ g solvent/cm}^3)(0.0009 \text{ g xyl/g soln})$$
$$+ (11.0 \text{cm}/\text{min})(A_c \text{cm}^2)(27.629 - 18.56 \text{ min})(\rho_f \text{ g solvent/cm}^3)(x_{conc})$$

Set In = Out, divide out A_c and ρ_f (assumed to be constant), and solve for x_{conc},

$$X_{conc} = \frac{(8.0)(315.85)(0.0009) - (11.0)(18.56)(0.0009)}{(11.0)(27.629 - 18.56)} = 0.02095 \text{ g xyl}\Big/\text{g soln}$$

The xylene exiting the bottom of the column is at $x_{out} = x_F = 0.0009$ for the first 18.56 minutes of the regeneration step. Then from 18.56 to 27.629 minutes of the regeneration $x_{out} = x_{conc} = 0.02095$. If regeneration continues for times longer than 27.629 minutes, $x_{out} = 0$. The average wt frac during regeneration is $x_{out,avg} = 0.00748$.

E. Check. Equation (18-24) can be used to check the outlet xylene wt frac. This equation is applied at the point where the adsorbent changes temperature. From Figure 18-8, this occurs during the purge step; thus, all velocities should be calculated at the purge velocity. Since solute velocity is directly proportional to interstitial velocity,

$$u_s(v_{purge}, T=0) = u_s(v_F, T=0)\,(v_{purge}/v_F) = (0.3799)(-11.0/8.0) = -0.5224\,\text{cm/min.}$$

$$\frac{x_{xy}(T_2 = 80)}{x_{xy}(T_1 = 0)} = \left(\frac{1}{u_s(T=0)} - \frac{1}{u_{th}}\right)\Big/\left(\frac{1}{u_s(T=80)} - \frac{1}{u_{th}}\right)$$

$$= \left(\frac{1}{-0.5224} - \frac{1}{-6.466}\right)\Big/\left(\frac{1}{-4.343} - \frac{1}{-6.466}\right) = 23.277$$

and $x_{xy}(T = 80°C) = (23.277)(0.0009) = 0.02095$, which agrees with the mass balance result.

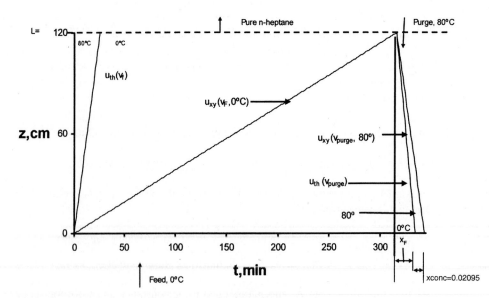

FIGURE 18-8. *Solute movement solution for counterflow TSA in Example 18-3*

Chapter 18 Introduction to Adsorption, Chromatography, and Ion Exchange

F. Generalize. This large increase in the solute concentration during thermal regeneration is a general phenomenon for strongly adsorbed solutes if the feed is dilute. The energy required to concentrate the dilute xylene in the n-heptane by adsorption is significantly less than the energy required to do the same concentration by distillation. (Going from 0.0009 to 0.02095 wt frac xylene may not seem like much change, but removal of a very large amount of pure n-heptane is necessary to obtain this amount of concentration.)

If the solute waves move faster than the thermal wave, which may occur in dilute gas systems, a mass balance equation and solution similar but subtly different than Eqs. (18-22) to (18-24) can be derived. In this situation the concentrated solute exits ahead of the thermal wave instead of behind it as predicted by Eqs. (18-22) to (18-24). One other case that can occur but is rare in dilute systems is when $u_s(T_{hot}) > u_{th} > u_s(T_{cold})$. In this case, which is beyond the scope of this introductory treatment, the solute concentrates, or *focuses,* at the temperature boundary (Wankat, 1990).

A number of different thermal cycles are used commercially. Figure 18-9 shows an alternate TSA cycle commonly used for recovery of solvents (typically volatile organic compounds (VOC) of intermediate molecular weight ~45 to 200) from drying and curing operations (Basmadjian, 1997; Fulker, 1972; Wankat, 1986). Activated carbon is used as the adsorbent and steam is used as the regeneration gas. Horizontal beds with a depth of one to two meters are often employed since strong adsorption of the solvent on the activated carbon allows for quite short beds and large gas flows require a large cross-sectional area to avoid excessive pressure drop. If feed gas needs to be treated continuously, two or more adsorbers are used in parallel, with a typical feed time of approximately two hours. Because the latent heat of steam is high, a large amount of energy can be rapidly transferred into the adsorber heating it quickly. A "heel" of leftover solvent is usually left in the bed since complete regeneration of the bed would require excessive amounts of steam. Because of incomplete regeneration and competition with water vapor for adsorption sites, the typical design capacity used for the activated carbon is about 25% to 30% of the maximum capacity of the carbon. Bed capacity can be increased by reducing the relative humidity of the feed gas to less than 50%.

In ideal applications of this process (e.g., removing small amounts of toluene from air) the peak mole fractions of toluene are close to 1.0 (Basmadjian, 1997) and the toluene is almost completely immiscible with water. Thus, the adsorbate can be recovered from the steam by condensing the concentrated adsorbate stream and allowing the liquid to separate into an organic layer and a water layer. If the adsorbate is miscible with water (e.g., ethanol), the condenser/settler shown in Figure 18-9 must be replaced with a distillation column, which greatly increases capital and operating costs.

Note that there can be safety hazards in the operation of the activated carbon solvent recovery equipment shown in Figure 18-9. If the solvent being recovered is flammable, care must be taken to prevent a fire. If the feed gas is air, then the concentration of the solvent in the feed gas must be kept below the lower explosion limit, and is often kept below ¼ of the lower explosion limit to provide a safety margin. This requirement invariably means that the feed gas must be quite dilute and the flow rates are large. If the feed gas is hot, it is often cooled before the adsorber to increase safety and to increase the adsorber's capacity. An alternative is to operate at much higher concentrations using nitrogen or carbon dioxide as the gas for the drying or curing operation, but then the carrier gas must be recovered and recycled. If the hot activated carbon can catalyze a reaction with the feed, a cooling step is added to the process. Sometimes a drying step is added before the cooling step since water

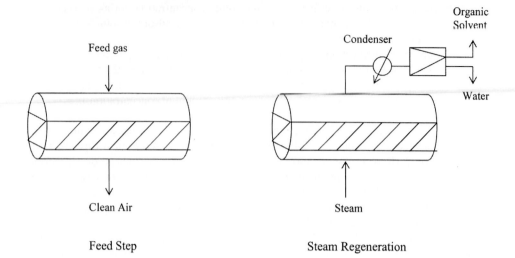

Feed Step   **Steam Regeneration**

FIGURE 18-9. *Solvent recovery with activated carbon and steam regeneration*

may interfere with the adsorption or react with the feed. If the feed gas is concentrated, the adsorbent can become quite hot because of the large heat of adsorption. Unfortunately, carbon beds occasionally catch on fire when this happens. This can be prevented by significant cooling of the feed gas, incorporating a cooling step in the cycle, or replacing air in the process with an inert gas.

Since several companies provide package units for activated carbon solvent recovery, new engineers are more likely to be involved in the purchase and installation of a unit than in designing a new unit. The more you know about solvent recovery with activated carbon, the better choice of unit and better bargain you will be able to make for your company.

Various thermal cycles are also employed for liquid systems although they tend to be somewhat different than those used for gases. The largest application of liquid adsorption is the use of activated carbon to treat drinking water and wastewater (Faust and Aly, 1987). Since contaminant levels are very low and adsorption tends to be very strong, the feed portion of the cycle may last for several months. Regeneration of the activated carbon is difficult and is usually done by removing the carbon from the column and sending it to a kiln to burn off the adsorbates. In small units (e.g., those used to purify tap water in homes) the carbon is discarded after use. Activated carbon is commonly used in bottling plants to remove chlorine from water by reacting with the carbon to produce HCl (Wankat, 1990). The slightly acidic water should be used immediately after use since it no longer contains chlorine to stop microbial growth.

A major industrial application of liquid adsorption is the drying of organic solvents (Basmadjian, 1984). A typical process is shown in Figure 18-10. Upward flow is used during the refilling and feed steps to avoid trapping gas in the bed. Since the water content in the organic is usually low, the feed step may be relatively long. Once breakthrough occurs (substantial amounts of water appear in the exiting solvent), the feed is turned off and the column is drained. Regeneration is done with downward flow of hot gas and is usually followed by a cooling step. Adsorptive drying competes with drying by distillation (Chapter 8). Operating expenses for adsorptive drying are dominated by the cost of energy to evaporate residual liq-

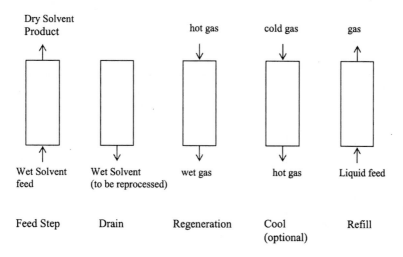

| Feed Step | Drain | Regeneration | Cool (optional) | Refill |

FIGURE 18-10.   *Drying liquid solvents by adsorption*

uid and desorb water. Adsorptive drying usually has an economic advantage compared to distillation when the water concentrations in the solvent are low.

In *concentrated* systems the energy generated by adsorption can be as large or significantly larger than the sensible heat from the temperature change. This causes a coupling of the concentration and temperature waves, and they often travel together. Basmadjian (1997) presents a simple way to estimate the maximum temperature rise in the system.

$$(\Delta T)_{max} = y_{adsorbate,feed} \, |\Delta H_{ads}| / C_{P,f} \tag{18-25}$$

The typical range for the heat of adsorption $|\Delta H_{ads}|$ is in the range from 1000 to 4000 kJ/kg (average ~ 2500) and typically the gas heat capacity $C_{P,f}$ is approximately 1.0 kJ/kg. This estimate gives a maximum temperature rise of approximately 25°C for a feed containing 1.0 wt % adsorbate and a maximum temperature rise of 1.25°C for a feed containing 0.05 wt % adsorbate. Basmadjian (1997) recommends that an isothermal analysis can be used if the predicted maximum temperature increase is less than 1°C or 2°C. Situations with large temperature increases obviously need to be designed for and are economically important, but the detailed theoretical treatment is beyond the scope of this introductory chapter. Interested readers should consult Basmadjian (1997), LeVan et al. (1997), Ruthven (1984) or Yang (1987). These more concentrated systems can also be simulated with commercial simulators.

18.3.2 Pressure Swing Adsorption

Pressure swing adsorption (PSA) and *vacuum swing adsorption (VSA)* cycles are alternatives to thermal cycles for gas systems. They are particularly useful for more concentrated feeds and/or adsorbates that are not strongly adsorbed. Figure 18-11A shows the steps in the basic Skarstrom cycle (named after Charles Skarstrom, the inventor of the process) for PSA (Ruthven et al., 1994; Wankat, 1986). Usually, the desired product is the pure product gas after adsorbate removal. Typical applications include drying gases, purifying hydrogen, and producing oxygen or nitrogen from air.

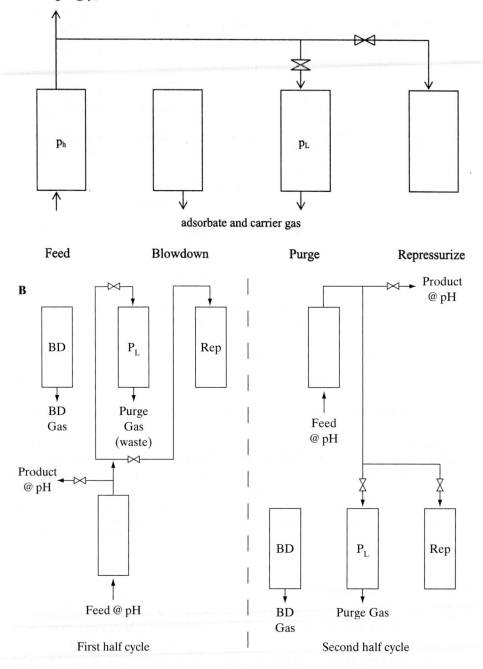

A
Product gas @ p_h

adsorbate and carrier gas

Feed   Blowdown   Purge   Repressurize

B

First half cycle

Second half cycle

FIGURE 18-11.   *Pressure swing adsorption; (A) steps for single column in Skarstrom cycle,*
(B) use of two columns in parallel for continuous feed and product. Period of
feed step = period of blowdown + purge + repressurization steps.

Following the feed step at the higher pressure, the pressure in the column is reduced to the lower pressure by counterflow blowdown. At this reduced pressure the column is purged (counterflow to the feed) using part of the pure product gas. Since pure product gas is used as a purge, the purge product (or waste) gas contains both adsorbate and carrier gas. The volume of purge gas required for the Skarstrom cycle is,

$$V_{purge} = \gamma V_{feed} \qquad \qquad \textbf{(18-26)}$$

where γ typically is between 1.15 and 1.5. Because of the volumetric expansion of the product gas from p_h to purge gas at p_L, significantly less moles of purge gas are needed than feed gas. Dropping the pressure also reduces the partial pressure, which helps desorb the adsorbate. The final step in the Skarstrom cycle is repressurization of the column. This step was originally done with fresh feed gas although with concentrated systems it is now much more common to use high-pressure product gas as shown in Figure 18-11. Usually two or more columns are operated in parallel, but with cycles out of phase so that one column is producing product when the other needs to be purged or repressurized. One method of doing this is shown in Figure 18-11B. PSA systems with from one to twelve columns are used commercially. PSA has the advantage that cycles can be very fast—a minute or two is common in industry and some cycles are as short as a few seconds. These short cycles lead to high productivity and hence relatively small adsorbers.

Figure 18-12 illustrates a simple vacuum swing cycle. The feed enters at the high pressure, which may be essentially atmospheric pressure. If p_h is significantly above atmospheric pressure, a short optional blowdown step is included. Then a vacuum pump is used to reduce

Pure product gas

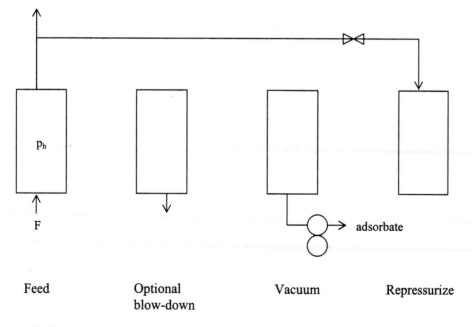

| Feed | Optional blow-down | Vacuum | Repressurize |

FIGURE 18-12.   *Basic vacuum swing adsorption cycle*

the pressure to very low pressures. At very low pressures the partial pressure is very low and very little adsorbate can be adsorbed (see Figure 18-2). Unfortunately, this step is slow and productivities of VSA systems are low. However, VSA has the advantage that a relatively pure product gas and a relatively pure adsorbate product can be produced. For example, VSA units can separate air into an oxygen product and a nitrogen product. The final step is repressurization of the column. VSA units are usually operated with several columns in parallel. A large number of variations of PSA, VSA, and combinations of PSA and VSA cycles have been invented (Kumar, 1996; Ruthven et al., 1994; Tondeur and Wankat, 1985). For example, it is common for the purge step in a PSA cycle to be operated at a pressure less than atmospheric pressure.

The simple Skarstrom cycle for PSA shown in Figure 18-11A has constant pressure (isobaric) periods and periods when pressure is changing. We will assume that a very dilute gas stream containing trace amounts of adsorbate A in an weakly adsorbed carrier gas is being processed and that over the concentration range of interest the linear isotherm, Eq. (18-5b), is accurate. If mass transfer is very rapid, then the solute movement theory can be applied. Since the system is very dilute, the gas velocity is constant and the system is assumed to be isothermal. In more concentrated PSA systems neither of these assumptions are true, and a more complicated theory must be used (Ruthven et al., 1994).

During the isobaric periods (feed at p_h and purge at p_L) the solute moves at a velocity u_s. For a gas that can be assumed to be ideal and an isotherm in partial pressure units, the solute velocity is given by Eq. (18-15d). Normally, $K_{d,i} = 1.0$ and all of the adsorption sites are accessible to the small gas molecules.

During blowdown the mole fraction of adsorbate in the gas increases because of desorption as the pressure drops. During repressurization the opposite occurs and the adsorbate mole fraction in the gas decreases. When the pressure changes, the solute waves also shift location. Determining the mole fraction changes and shifts in location for these steps requires solution of the partial differential equations for the system (Chan et al., 1981). The results for linear isotherms are relatively simple. Define parameter β_{strong} for the strongly adsorbed component as

$$\beta_{strong} = \frac{\varepsilon_e + (1 - \varepsilon_e)\varepsilon_p K_{d,i} + (1 - \varepsilon_e)(1 - \varepsilon_p)\rho_s K'_{weak} RT}{\varepsilon_e + (1 - \varepsilon_e)\varepsilon_p K_{d,i} + (1 - \varepsilon_e)(1 - \varepsilon_p)\rho_s K'_{strong} RT} = u_{s,strong}/u_{s,weak} \qquad \textbf{(18-27)}$$

This parameter measures the ratio of the amount of weakly adsorbed to the amount of strongly adsorbed adsorbate in a segment of the column. If the weakly adsorbed component does not adsorb, then $K'_{weak} = 0$ in Eq. (18-27). Since $K'_{weak} < K'_{strong}$, $\beta_{strong} < 1.0$. Then the shift in mole fraction of the strongly adsorbed species A is

$$\left(\frac{y_{A,after}}{y_{A,before}}\right) = \left[\frac{p_{after}}{p_{before}}\right]^{(\beta_{strong} - 1)} \qquad \textbf{(18-28a)}$$

Equation (18-28a) predicts an increase in mole fraction y_A for a decrease in pressure (try it to convince yourself). The shift in location of solute waves can be found from

$$\left(\frac{y_{after}}{y_{before}}\right) = \left[\frac{p_{after}}{p_{before}}\right]^{-\beta_{strong}} \qquad \textbf{(18-28b)}$$

where the axial distance z *must* be measured from the end of the column that is closed (this can vary during the PSA cycles). Note that if $z_{before} = 0$, then $z_{after} = 0$ also. Solute waves at the closed end of the column cannot shift.

Application of the solute movement theory will be illustrated in Example 18-4. Before doing this, we note that axial dispersion is normally significant in gas systems. Thus, we expect that the solute movement theory will over-predict the separation that occurs. Alternatively, the value of γ required in Eq. (18-26) for a given product purity will be larger in a real system than predicted by the solute movement theory ($\gamma = 1$ for linear systems). For separations based on differences in equilibrium isotherms, if the solute movement theory predicts that a separation is not feasible or will not be economical, more detailed calculations will rarely improve the results.

EXAMPLE 18-4. PSA system

A 0.50 m. long column is used to remove methane (M) from hydrogen using Calgon Carbon PCB activated carbon. The feed gas is 0.002 mole fraction methane. Superficial velocity is 0.0465 m/s during the feed step. The high pressure is 3.0 atm while the low pressure is 0.5 atm. A standard 2-column Skarstrom cycle is used. The symmetric cycle is:

Repressurize with feed	0 to 1 sec.
Feed step at p_H	1 to 30 sec.
Blowdown	30 to 31 sec.
Purge at p_L	31 to 60 sec.

The operation is at 480 K. Use a pure purge gas with a purge to feed ratio of $\gamma = 1.1$. Carbon properties: $\rho_s = 2.1$ g/cc, $K_d = 1.0$, $\varepsilon_p = 0.336$, $\varepsilon_e = 0.43$. Equilibrium data are available in Table 18-3 and has been analyzed in Example 18-1. Draw the characteristic diagram for the first cycle assuming the bed is initially clean at 0.5 atm and predict the outlet concentration profile.

Solution

A. Define. Plot the movement of solute during the four steps shown in Figure 18-11A and use this diagram plus appropriate equations to predict outlet mole fractions.

B. Explore. The Table in Example 18-1 gives $K_M = 1.888 \times 10^{-4}$ (kPa)$^{-1}$ and $q_{max} = 3.84$ m mol/g at 480 K. Since the feed mole fraction is very low, the isotherm will be in the linear range ($q_A = K'_{M,p}p_m$) where $K'_{M,p} = q_{max}K_M = (3.84)$ $(1.888 \times 10^{-4}) = 0.000725$ mmol/(K g Pa). Since hydrogen does not adsorb, $K'_{weak} = 0$. We will assume operation is isothermal.

C. Plan. Start by repressurizing with the feed. We can calculate β_M from Eq. (18-27) and the distance the wave moves in the column can be determined from Eq. (18-28b). During the feed step at 3.0 atm the methane travels at a constant solute velocity given by Eq. (18-15d). There will be two waves as shown in Figure 18 13A. During blowdown Eq. (18-28a) is used to determine the new mole fraction. Waves during the purge step again follow Eq. (18-15d) but with $v_{purge} = \gamma v_{feed}$.

D. Do it. <u>Repressurization Step</u>: From Eq. (18-27)

$$\beta_M = \frac{0.43 + (0.57)(.336) + 0}{0.43 + (0.57)(0.336) + (0.57)(0.664)\left[2.1 \times 10^6 \, \frac{g}{m^3}\right]\left[7.25 \times 10^{-10} \, \frac{mol}{Pa\,g}\right](480K)\left(8.314 \, \frac{m^3 Pa}{mol\,k}\right)}$$

$$\beta_M = \frac{0.6215}{2.921} = 0.2128$$

The units in the last term in the denominator are a little tricky. The gas constant used is $R = 8.314 \, \frac{m^3 Pa}{mol\,K}$. If a different gas constant is used, the units on the other terms have to be adjusted.

Equation (18-28b) is used for the shift in location of solute waves. Because the top of the column in Figure 18-11A is the closed end, z is measured for this step from that end. Then the feed end is z = 0.50 m. From Eq. (18-28b)

$$z_{after} = (0.50 \, m)\left(\frac{(3.0 \, atm)}{0.5 \, atm}\right)^{-0.2128} = 0.3415 \, m$$

This is $0.50 - 0.3415 = 0.1585$ m from feed end of column (point 1 in Fig. 18-13A).

The mole fraction methane at this location can be determined from Eq. (18-28a)

$$y_{M,after} = (0.002)\left(\frac{(3 \, atm)}{0.5 \, atm}\right)^{0.2128-1} = 0.000488$$

where $y_{M,\,before} = 0.002$ is the feed mole fraction from the bottom of the column which shifts during repressurization to point 1. The mole fraction $y_{M,\,after}$ is lower since methane is adsorbed as the column is pressurized.

Feed Step: Equation (18-15d) is used to determine the methane solute velocity u_M,

$$u_{s,i} = \frac{\varepsilon_e v_{inter}}{\varepsilon_e + (1 - \varepsilon_e)\varepsilon_p K_{d,i} + (1 - \varepsilon_e)(1 - \varepsilon_p) \, \rho_s RTK'_{A,p}} \qquad \textbf{(18-15d)}$$

(Note the denominator is the same as the denominator for β_M).

$$u_M = \varepsilon_e \frac{v_{inter}}{2.921}$$

Since $\varepsilon_e v_{inter} = v_{super} = 0.0465$ m/s,

$$u_m = \frac{0.0465}{2.921} = 0.01592 \, m/s$$

In the 29 seconds of the feed step the methane waves can move 0.462 m. Thus, one of the waves breaks through while the other does not (see points 2 and 3 on Figure 18-13A).

Chapter 18 Introduction to Adsorption, Chromatography, and Ion Exchange

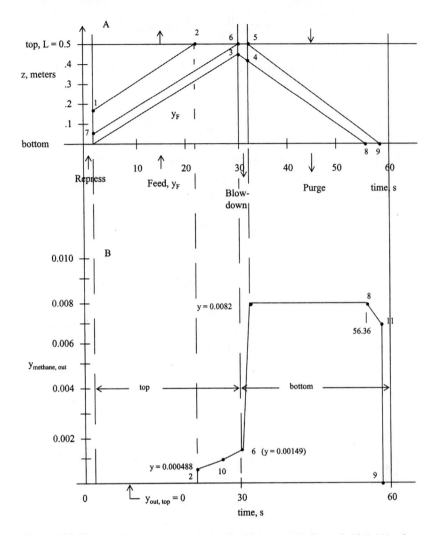

FIGURE 18-13. *Solute movement solution for PSA system in Example 18-4; (A) solute movement diagram, (B) outlet concentration profiles*

The wave that breaks through travels 0.3415 m, which requires (0.3415 m)/(0.01592 m/s) = 21.45 s. Including the 1.0 seconds for repressurization this is 22.45 s after the start of the cycle. Point 3 is at z = 0.462 m.

Blowdown: Point 3 will shift according to Eq. (18-28b) with the closed end again at the top. Thus, measuring from the top we have

$$z_{before} = 0.5 - 0.462 = 0.038 \text{ m}$$

and from Eq. (18-28b)

$$z_{after} = 0.038 \left(\frac{0.5}{3.0} \right)^{-0.2128} = 0.056 \text{ m}$$

or $0.50 - 0.056 = 0.444$ m from the bottom. With $y_{before} = y_F = 0.002$, Eq. (18-28a) gives

$$y_{after} = 0.002 \left(\frac{0.5 \text{ atm}}{3.0 \text{ atm}} \right)^{0.2128-1} = 0.0082$$

As expected, methane mole fraction increases as methane desorbs.

Purge Step. The methane velocity during the purge step is again given by Eq. (18-15d); however, the interstitial velocity is increased, since $\gamma > 1.0$.

$$v_{super, purge} = \gamma v_{super, feed} = (1.1)(0.0465) = 0.05115 \text{ m/s}$$

Then

$$u_{M, purge} = \gamma u_{M, feed} = (1.1)(0.01592) = 0.01751 \text{ m/s}$$

There are two waves, (from top of column, point 5, and from point 4). They can both travel a distance 0.508 m in 29 s; thus, they both exit the column.

Wave from Point 4 exit time: 0.444 m/0.01751 m/s + 31 s = 56.36 s (Point 8)

Wave from Point 5 exit time: 0.5 m/0.01751 m/s + 31 s = 59.57 s (Point 9)

Outlet Mole Fraction Profile: At the top of the column
   0 to 1 s, No product
   1 to 22.45 s (point 2), y = 0
   At point 2, y = 0.000488

At point 6, to estimate mole fraction, follow solute back to point 7 at t = 1 second (end of repressurization).

$$z_7 = t_{feed}/u_{M, feed} = (29 \text{ s})(0.01592 \text{ m/s}) = 0.462 \text{ m from top}$$

The final mole fraction at point 7 follows solute that enters during repressurization at a specific, but unknown, pressure between $p_L = 0.5$ and $p_H = 3.0$ atm. Equation (18-28) can be employed with $z_{after} = 0.462$, $z_{before} = 0.50$, $p_{after} = 3.0$, $\beta_M = 0.2128$ to calculate this unknown pressure, p_{before}. Then

$$p_{before} = p_{after} \left(\frac{z_{after}}{z_{before}} \right)^{1/\beta_M} \tag{18-28c}$$

$$p_{before} = (3.0 \text{ atm}) \left(\frac{0.462}{0.50} \right)^{\frac{1}{0.2128}} = 2.069 \text{ atm}$$

Since the feed entered at $y_{M, before} = 0.002$, Eq. (18-28a) can be used to estimate $y_{M, after}$.

$$y_{M, \text{after}} = (0.002)\left(\frac{3.0 \text{ atm}}{2.069 \text{ atm}}\right)^{0.2128-1} = 0.00149, \text{ Point 7}$$

Since concentrations are constant along the trace of the solute movement, this is also the mole fraction at point 6.

A similar procedure can be used (see Problem 18.D10) to find intermediate point 10 (shown at 26.126 s and $y_M = 0.000876$). The outlet profile is not linear.

During blowdown, gas exits (at bottom of column) initially at $y_F = 0.002$ and increases to $y_{\text{after, BD}} = 0.0082$. The exact shape can be estimated by the procedure used above. The mole fraction is constant at 0.0082 until gas from point 4 exits at Point 8 at 56.36 s. Gas mole fraction drops to $y_{\text{out}} = 0$ and the column is completely regenerated at point 9 (59.57 s). The intermediate point 11 shown in Figure 18-13B is estimated (see Problem 18.D10).

E. Check. Because the flow rates vary in unknown ways during repressurization and blowdown steps, a complete mass balance check is not possible. However, an approximate check balancing methane flows in the feed and purge steps can be done.

$$\text{Methane during feed} = v_{\text{super,feed}} A_c y_{M,F} \overline{\rho}_{\text{feed}} t_f = v_{\text{super}} A_c y_{M,F} t_F p_F / RT$$

where the molar density is $\overline{\rho}_{\text{feed}} = \rho_F / RT$ for an ideal gas and $A_c = $ cross-sectional area.

$$\text{Methane in} = (0.0465 \text{ }^{\text{m}}/_{\text{s}})(A_c \text{m}^2)(0.002)(29 \text{ s})(3.0 \text{ atm}) \bigg/ \left(8.206 + 10^{-5} \frac{\text{m}^3 \text{atm}}{\text{mol K}}\right)(480 \text{ K})$$

$$= 0.2054 A_c$$

$$\text{Methane out} = \frac{v_{\text{super,feed}} A_c p_F}{RT} \int_{22.45}^{30} y_{M,\text{out,top}} dt + \frac{v_{\text{super,purge}} A_c p_{\text{purge}}}{RT} \int_{31}^{60} y_{M,\text{out,bot}} dt$$

The integrals can be estimated by assuming the variation in y_M is linear.

Then, $$\int_{22.45}^{30} y_{M,\text{out,top}} dt = y_{M,\text{out,top,Avg}}(30-22.45) = (0.000989)(7.55) = 0.007467$$

$$\int_{31}^{60} y_{M,\text{out,bot}} dt = (0.082)(56.36 - 31) + \left(\frac{0.0082 - 0}{2}\right)(59.57 - 56.36) = 0.22103$$

$$\text{Methane out} = \frac{[(0.0465)(3.0)(0.007467) + (0.05115)(0.5)(0.22103)]}{(8.206 \times 10^{-5})(480)} A_c = 0.1700 A_c$$

The inlet and outlet amounts are reasonably close.

F. Generalization Notes: 1. The ratio of moles gas fed to the purge gas used

$$\frac{(v_{\text{super,feed}} A_c p_F / RT)}{(v_{\text{super,purge}} A_c p_{\text{purge}} / RT)} = \frac{p_F}{p_{\text{purge}}} \frac{V_{\text{feed}}}{V_{\text{purge}}} = \frac{p_F}{p_{\text{purge}}} \frac{1}{\gamma}$$

is $\dfrac{3.0}{(0.5)(1.1)} = 5.455$ and (ignoring repressurization and blowdown) the ratio of product gas (hydrogen) to feed gas is

$$\frac{\text{Moles Prod. Gas}}{\text{Moles Feed}} \approx \frac{V_{\text{feed}}p_F - V_{\text{purge}}p_{\text{purge}}}{V_{\text{feed}}p_F} = 1 - \gamma\,\frac{p_{\text{purge}}}{p_F} = 1 - \frac{(1.1)(0.5)}{(3.0)} = 0.8167$$

PSA produces a significant amount of high-pressure pure product because the gas is expanded before it is used for purging.

2. This design is inappropriate if pure hydrogen is desired during the entire feed step. Breakthrough can be prevented by changing the design (see Problem 18.B2).
3. Repressurization with feed causes the methane to penetrate the bed a significant distance during this step. Repressurization with product works better (see Problem 18.D11).
4. This example uses complete regeneration (the column is clean at the end of the cycle). Incomplete regeneration (leaving a heel) allows for more production of pure product and is employed in industrial systems.

This section illustrates PSA calculations for the simplest possible case—the local equilibrium theory for trace components for an isothermal system. If the mole fraction of the strongly adsorbed component is higher in the feed, the isotherm is likely to be nonlinear and the velocity will vary along the length of the column. In addition, operation is much more likely to be adiabatic instead of isothermal. It is also common to have both components adsorb or to have more than two components. If dispersion and mass transfer resistances are important, detailed simulations will be required. In addition, PSA has spawned a large number of inventive cycles to accomplish different purposes. If you need to understand any of these situations, Ruthven et al. (1994) provide an advanced treatment. White and Barkley (1989) and White (2008) discuss practical aspects of PSA design such as pressure drops, the velocity limit to prevent fluidization of the bed, retaining the heat of adsorption in the bed and start-up.

18.3.3 Simulated Moving Beds (SMB)

Most adsorption processes remove all adsorbed solutes from a nonadsorbed or weakly adsorbed carrier gas or solvent. Elution chromatography, on the other hand, was developed to separate a number of solutes from each other with all of the products containing the carrier gas or solvent. *Simulated moving bed* (SMB) technology is a melding of purge or displacement adsorption and chromatographic methods that was developed by UOP in the late 1950s and early 1960s (Broughton and Gerhold, 1961; Broughton et al., 1970). SMBs are currently extensively used for binary separations. Although gas-phase systems have been studied, all current commercial applications are liquid-phase (see Problem 18.A6). Common industrial applications of SMBs are separation of aqueous solutions of glucose from fructose to make sweeteners (heavily used in soft drinks), separation of p-xylene (used to make polyesters) from m-xylene, and the separation of optical isomers in the pharmaceutical industry. The SMB process is relatively expensive since the products must be separated from the solvent or desorbent usually by evaporation or distillation (Ruthven, 1984; Wankat, 1986). Chin and

Wang (2004) discuss practical aspects of SMB systems such as pump placement, pressure drops, and valve requirements.

The most efficient approach to separating a binary mixture by adsorption appears to be to have a counter-current process similar to extraction. Since it is convenient to regenerate the adsorbent within the device, the result is the *true moving bed (TMB)* system (Figure 18-14A). Zones 2 and 3 do the actual separation of the two solutes. (*Zones* are packed regions between inlet and outlet ports.) Zone 1, which is optional but almost always included, adsorbs the weakly adsorbed solute A onto the solid so that desorbent D can be recycled. Zone 4 regenerates the solid by removing strongly adsorbed solute B using a purge or

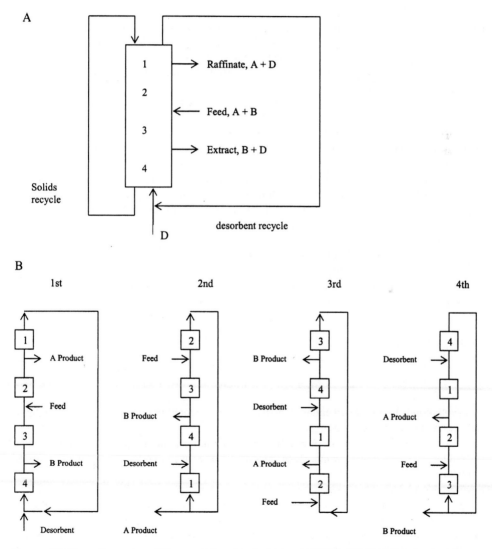

FIGURE 18-14. *Separation of binary mixtures by adsorption; (A) TMB, (B) SMB showing complete cycle with four time steps*

displacement with desorbent D. This figure is loosely analogous to a distillation unit: A is the light key, B is the heavy key, zones 2 and 3 are the enriching and stripping sections, respectively; and zones 1 and 4 are analogous to a total condenser and a total reboiler, respectively. The TMB system would work *if* chemical engineers had the technology to build large-scale systems that could move a solid countercurrent to a fluid with *no* axial mixing. Since this goal has proved to be elusive, UOP developed the SMB.

An SMB system is shown in Figure 18-14B for four different time steps. The system is arranged in a continuous loop with inlet and outlet ports. After a set time period, the switching time t_{sw}, the port locations are all advanced by one column. From the viewpoint of an observer at the extract port the solid moves downward when the switch is done while the fluid continues to move continuously upwards. Thus, an intermittent, counter-current movement of the solid and liquid has been simulated. The port switching is continued indefinitely. In Figure 18-14B with four columns the cycle repeats after every four switches. If there are N columns, the cycle repeats after N switches. A large number of modifications to SMB systems have been developed. The most common is to have two or more columns per zone instead of the one column per zone shown in Figure 18-14B. This makes the operation closer to the TMB and improves the purity of products.

The SMB system shown in Figure 18-14B is quite a complicated system, particularly if compared to the simple elution chromatographic system shown in Figure 18-5B. The SMB is used in industry for high purity separations of binary feeds since much less desorbent and adsorbent are required. The solute movement analysis helps to explain how this complicated process works.

Each column in Figure 18-14B acts as a chromatographic column that undergoes a series of steps. The solute velocity in each column of the SMB can be determined in the same way as for an elution chromatography system. Thus, the solute velocity in any column of the SMB is given by Eqs. (18-15). The chromatography and SMB processes differ only in the way columns are coupled, which are their boundary conditions!

In order for the SMB in Figure 18-14B to separate solutes A and B completely the following conditions must be met:

Zone 1. To produce pure desorbent for recycle, solute A must not breakthrough (that is, appear in the outlet of zone 1) during the switching period, t_{sw}. Thus,

$$u_{A,1} \, t_{sw} \leq L$$

Since the average port velocity is defined as $u_{port} = L/t_{sw}$, this condition becomes

$$u_{A,1} = M_1 \, u_{port} \qquad \text{where} \qquad M_1 \leq 1.0 \qquad \textbf{(18-29a)}$$

The equation is written as an equality instead of as an inequality since it is easier to manipulate equations than inequalities.

Zones 2 and 3. To separate A and B we want net movement of A up the column and net movement of B down the column. Since velocity is higher in zone 2, the condition on solute B controls.

$$u_{B,2} = M_2 \, u_{port} \quad (M_2 \leq 1.0) \qquad \textbf{(18-29b)}$$

These conditions require that solute A will break through and solute B will not break through from zones 2 and 3 during the switching period.

In zone 3 upward movement of solute A controls.

$$u_{A,3} = M_3 \, u_{port} \qquad (M_3 \geq 1.0) \qquad \textbf{(18-29c)}$$

Zone 4. To regenerate the column, all solute B must be removed in zone 4. This requires that solute B have a net upward movement.

$$u_{B,4} = M_4 \, u_{port} \qquad (M_4 \geq 1.0) \qquad \textbf{(18-29d)}$$

Equations (18-29a) to (18-29d) can all be simultaneously satisfied by changing the flow rates of feed, desorbent and the two product streams in Figure 18-14B to change the velocities in each zone. For example, since the A product stream is withdrawn between zones 1 and 2, $v_1 < v_2$. By proper selection of the velocities and the port velocity, we can satisfy these four equations.

Usually the desorbent must be removed from the A and B product streams. Increasing the amount of desorbent will increase the cost for this removal and will also increase the diameters of the columns requiring more adsorbent. Thus, the ratio of desorbent to feed, D/F, often controls the cost of SMB systems. For an ideal system with no zone spreading (no axial dispersion and very fast mass transfer rates) the solute movement theory can be used to calculate D/F by solving Eqs. (18-29a) to (18-29d) simultaneously with Eq. (18-15) and the mixing mass balances with constant density.

$$v_2 = v_1 + v_{A,product}, \ \ v_2 = v_3 + v_{Feed}, \ \ v_4 = v_3 + v_{B,product}, \ \ v_4 = v_1 + v_D \ \ \textbf{(18-30a,b,c,d)}$$

where $v_{Feed} = F/(\varepsilon_e \, A_c)$ is the interstitial velocity the feed would have in the column. F is the volumetric flow rate of feed and A_c is the cross-sectional area of the columns with similar definitions for the desorbent and product velocities. If F and A_c are known, then we can first solve for u_{port}, v_1, v_2, v_3, and v_4, then for $v_{A,product}$, $v_{B,product}$, and v_D. Then $D/F = v_D / v_{Feed}$. These calculations are developed in Problem 18.C10. The minimum D/F ratio, $(D/F)_{min}$ can be calculated by setting all $M_i = 1.0$. This minimum has a significance similar to that of $(L/D)_{min}$ in distillation. For linear systems $(D/F)_{min} = 1.0$, which is also the thermodynamic minimum.

In actual practice there is considerable spreading due to mass transfer resistances, axial dispersion, and mixing in the transfer lines and valves. If the SMB is operated at $(D/F)_{min}$, the raffinate and extract products will not be pure. To obtain higher purities D/F is usually increased; however, the SMB is more complicated than binary distillation. The additional desorbent must be distributed throughout the four zones to give the optimum velocities in each zone. One approach to this optimization is to pick values of the multipliers $M_1 < 1, M_2 < 1, M_3 > 1$, and $M_4 > 1$. Then the value of velocities and D/F can be determined by solving the solute movement equations (see Problem 18.C10 for the equation for D/F). The experiment or simulation is then run again with these new flow rates. The procedure is repeated until the desired purities are achieved.

EXAMPLE 18-5. SMB system

Ching and Ruthven (1985) found that the equilibrium of fructose and glucose on ion exchange resin in the calcium form was linear for concentrations below 5 g/100 ml. Their equilibrium expressions are: $q_{gluc} = 0.51 \, c_{gluc}$, $q_{fruc} = 0.88 \, c_{fruc}$, at 30°C where both q and c are in g/L. For this resin, $\varepsilon_p = 0$ and $\varepsilon_e = 0.4$.

We want to design an SMB system to separate fructose and glucose. If the switching time t_{sw} = 5 min and D_{col} = 0.4743 m, design an SMB system with one column per zone for this separation at the minimum D/F = 1.0.

Solution

A. Define. To design the system we need to determine L, and all feed and product flow rates. Since feed flow rate F is not specified, we can find flow rates as functions of F.

B. Explore. Since q and c are in mass/volume, $u_{s,i}$ is obtained from Eq. (18-15a) by removing ρ_s. Thus, in terms of Eq. (18-15e) the solute velocity constant is

$$C_i = \frac{1}{1 + \dfrac{1-\varepsilon_e}{\varepsilon_e}\, \varepsilon_p K_d + \dfrac{(1-\varepsilon_e)}{\varepsilon_e}\,(1-\varepsilon_p)K_i'} \quad \text{and } u_i = C_i\, v_{inter}$$

where K'_{Gluc} = 0.51 and K'_{Fruc} = 0.88. The resulting velocities can be used in Eq. (18-29), (18-30) and $u_{port} = L/t_{sw}$. The value of u_{port} (derived in Problem 18.C10) is

$$u_{port} = \frac{v_F}{\dfrac{M_2}{C_B} - \dfrac{M_3}{C_A}} \tag{18-31a}$$

where C_B is for the solute with stronger adsorption (fructose).

C. Plan. We can calculate the solute velocity constants C_{gluc} and C_{fruc} (fructose is more strongly adsorbed). Then u_{port} can be found as a function of v_F or F. Equations (18-29) and (18-30) can be used to calculate all other flow rates and $L = u_{port} t_{sw}$.

D. Do it. From the expression for C_i

$$C_{gluc} = \frac{1}{1 + 0 + \dfrac{(0.6)(1.0)}{0.4}\,(0.51)} = 0.5666$$

$$C_{fruc} = \frac{1}{1 + 0 + \dfrac{(0.6)(1.0)}{0.4}\,(0.88)} = 0.4310$$

The feed velocity $v_F = \dfrac{F}{\varepsilon_e \pi D_{col}^2/4} = 14.147$ F m/min if F is in m³/min.
With M_i = 1.0, Eq. (18-31a) is,

$$u_{port} = \frac{v_F}{\dfrac{1}{C_{fruc}} - \dfrac{1}{C_{gluc}}} = \frac{14.147\ F}{\dfrac{1}{0.4310} - \dfrac{1}{0.5666}} = 25.478\ F \text{ m /min}$$

$L = u_{port} t_{sw}$ = (25.478 F m/min) (5 min.) = 127.39 F (in m).
From Eqs. (18-29b) and (18-15e) we obtain

$$v_2 = M_2\, u_{port}/C_{fruc} = (1.0)\,(25.478\ F)/0.4310 = 59.113\ F$$

From Eq. (18-30b)

$$v_3 = v_2 - v_F = 59.113 \, F - 14.147 \, F = 44.966 \, F$$

Similarly,

$$v_1 = M_1 \, u_{port}/C_{gluc} = (1.0)(25.478 \, F)/0.5666 = 44.966 \, F = v_{recycle}$$
$$v_4 = M_4 \, u_{port}/C_{fruc} = (1.0)(25.478 \, F)/0.4310 = 59.113 \, F$$

Then, from Eq. (18-30a), Eq. (18-30c), and Eq. (18-30d)

$$v_{gluc, prod} = v_{A, prod} = v_2 - v_1 = 14.147 \, F$$
$$v_{fruc, prod} = v_{B, prod} = v_4 - v_3 = 14.147 \, F$$
$$v_D = v_4 - v_1 = 14.147 \, F \qquad \text{and} \qquad D/F = v_D/v_F = 1.0.$$

E. Check. As expected, $D/F = 1.0$ since all the $M_i = 1.0$. Also $A_{prod}/F = v_{A, prod}/v_F = 1.0$ and $B_{prod}/F = 1.0$.

F. Generalization. The equal values for feed, desorbent and products; and $v_1 = v_3$, $v_2 = v_4$ occurs only for all $M_i = 1$. Usually, v_4 is the highest velocity. Further development of equations is done in Problem 18.C10. Note: From Eq. (18-31a) with $u_{port} > 0$, there are limits to how small M_2 and how large M_3 can be.

To complete the design we need a value for F in m^3/min. Then the velocities and pressure drops can be calculated (e.g., with the Ergun equation). If pressure drop is too large, F or D_{col} need to be adjusted. Because of dispersion and mass transfer resistances the two products will not be 100% pure. The actual purities can be determined by experiment or detailed simulation. If more separation is needed, one can reduce M_1 and M_2 while increasing M_3 and M_4.

The TMB shown in Figure 18-14A is also of interest and can be analyzed using solute movement theory.

18.4 NONLINEAR SOLUTE MOVEMENT ANALYSIS

Since most adsorption and ion exchange separations of commercial significance operate in the nonlinear region of the isotherm, the previous analysis needs to be expanded to nonlinear systems. Nonlinear behavior is distinctly different than linear behavior since one usually observes *shock* or *constant pattern* waves during the feed step and *diffuse* or *proportional pattern* waves during regeneration. Experimental evidence for constant pattern waves is shown in Figure 18-15A. The isotherm for carbon dioxide on activated carbon is a Langmuir-type shape. During loading we expect shock or constant pattern waves. This is clearly shown in the top figure since the waves can be moved along the time axis and easily be superimposed on each other. During desorption (elution) a diffuse or *proportional pattern* wave is expected (Figure 18-15B). These waves cannot be superimposed on each other. The width of proportional pattern waves is directly proportional to the distance the wave travels in the column.

Equation (18-14), which was derived for any type of isotherm, is the starting point for the analysis of movement of solutes with nonlinear isotherms. We now need to substitute the

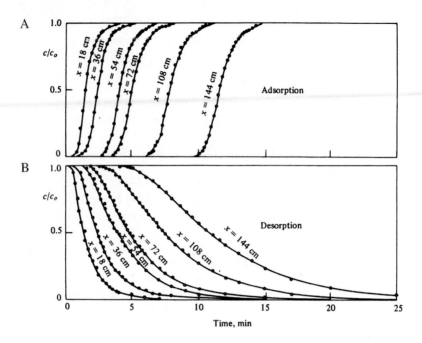

FIGURE 18-15. *Adsorption and desorption of CO$_2$ on activated carbon; (A) adsorption breakthrough curves illustrating constant pattern behavior, (B) desorption (elution) curves illustrating proportional pattern behavior, v_{super} = 4.26 cm/s. Reprinted from Weyde and Wicke (1940).*

desired nonlinear equilibrium expression into the last term in the denominator for $\Delta q/\Delta c$. This insertion differs from inserting $\Delta q/\Delta c = K'$ in the linear case because two separate forms of the equation result depending on the operation and because the result will be concentration dependent. There are a huge number of expressions for nonlinear isotherms. To be specific, we will focus on the Langmuir isotherm, Eq. (18-5a), Eq. (18-6a), and Eq. (18-6c), which are probably the most popular forms.

18.4.1 Diffuse Waves

For an isotherm of Langmuir shape, if the column is initially loaded at some high concentration, c_{high}, and is fed with a fluid of low concentration, c_{low}, the result is a diffuse or proportional pattern wave. Since the derivative of q with respect to c exists,

$$\lim_{\Delta c \to 0} \frac{\Delta q}{\Delta c} = \left. \frac{\partial q}{\partial c} \right|_T \tag{18-32a}$$

We can now calculate this derivative for any desired nonlinear isotherm [for linear isotherms $dq/dc = K'$ and the result is Eq. (18-15)]. Specifically for the Langmuir isotherm in Eq. (18-6c)

$$\frac{\partial q}{\partial c} = \frac{a}{(1 + bc)^2}$$

(18-32b)

and the solute wave velocity is

$$u_s = \frac{v_{inter}}{1 + \dfrac{(1 - \varepsilon_e)}{\varepsilon_e} K_d \varepsilon_p + \dfrac{(1 - \varepsilon_e)(1 - \varepsilon_p)}{\varepsilon_e} \rho_s \dfrac{a}{(1 + bc)^2}}$$

(18-32c)

Note that for nonlinear isotherms the solute wave velocity depends upon the concentration. For Langmuir isotherms as the solute concentration increases the denominator in Eq. (18-32b) will decrease and the solute wave velocity increases. Another way to look at this is that dq/dc is the local slope or tangent of the isotherm. Figure 18-2 shows that for a Langmuir isotherm dq/dc is largest and thus, the velocity u_s is smallest as c approaches zero. Values for u_s can be calculated from Eq. (18-32c) for a number of specific concentrations and the diffuse wave can be plotted as shown in Figure 18-16A. The times at which these solute waves, which are at known concentrations, exit the column (at z = L) can be determined (see Figure 18-16A) and the outlet concentration profile can be plotted as in Figure 18-16B. Note that the outlet wave varies continuously and is diffuse. This result agrees with experiments that show zone spreading is proportional to the column length and have a smooth even spread in concentration. The analysis procedure is illustrated in Example 18-6.

EXAMPLE 18-6. Diffuse wave

A 100.0 cm long column is packed with activated alumina. The column is initially totally saturated at c = 0.011 mol/L anthracene in cyclohexane solvent. It is then eluted with pure cyclohexane solvent (c = 0) at a superficial velocity of 30.0 cm/min. Predict and plot the outlet concentration profile using solute movement theory.

Data: $\varepsilon_e = 0.42$, $\varepsilon_p = 0$, $K_d = 1.0$, ρ_f (cyclohexane) = 0.78 kg/L, $\rho_p = 1.465$ kg/L, Equilibrium $q = \dfrac{22\,c}{1 + 375\,c}$ where q = mol/kg and c = mol/L.

Assume operation is isothermal.

Solution

A. Define. Find the values of the outlet concentration at different times using the solute movement theory.

B. Explore. The most important decision is whether a diffuse or a shock wave (Section 18.4.2) will result. With a Langmuir isotherm a diffuse wave results when a more concentrated solution (c = 0.011) is eluted with a dilute solution (c = 0.0). If this is incorrect, the analysis will show us that there is an error.

C. Plan. Since $\varepsilon_p = 0$, the solute velocity (Eq. 18-32c) for a diffuse wave becomes,

$$u_s = \frac{v_{inter}}{1 + \dfrac{1 - \varepsilon_e}{\varepsilon_e} \rho_p \dfrac{(22)}{(1 + 375\,c)^2}} \quad \text{and} \quad v_{inter} = \frac{v_{super}}{\varepsilon_e} = \frac{30}{0.42} = 71.4$$

Substituting in the parameter values this becomes

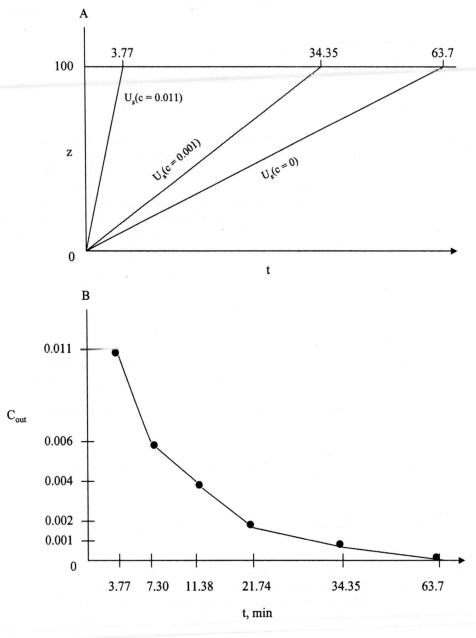

FIGURE 18-16.   *Diffuse wave analysis; (A) solute movement graph for Example 18-6, (B) predicted outlet concentration profile*

Chapter 18   Introduction to Adsorption, Chromatography, and Ion Exchange

$$u_s = \frac{71.43}{1 + \dfrac{2.023\,(22)}{(1 + 375\,c)^2}} = \frac{71.43}{1 + \dfrac{44.5}{(1 + 375\,c)^2}}$$

Since u_s depends upon the concentration, we can select arbitrary values of the concentration ranging from 0.011 to zero, calculate u_s, and determine $t_{out} = L/u_s$.

D. Do it. The values are tabulated below for selected values of concentration.

c, mol/L	u_s, cm/min	t_{out}, min
0.011	26.51	3.77
0.008	17.89	5.29
0.006	13.70	7.30
0.004	8.79	11.38
0.002	4.60	21.74
0.001	2.91	34.35
0.000	1.57	63.70

The solute movement solution using these values is shown in Figure 18-16A and the outlet concentration profile plotted from the tabulated values is shown in Figure 18-16B.

E. Check. A check can be made with a mass balance over the entire elution time.

$$-\text{Outlet} - \text{Accumulation} = 0$$

where the outlet concentration $= \int A_c v_{super} c_{out}(t)$, and accumulation $= A_c \rho_s [q(c{=}0) - q(c{=}0.011)]$ with q determined from the isotherm.

F. Generalize. The shape shown in Figure 18-16B, particularly the strong tailing at very low concentrations is typical of elution behavior with highly nonlinear isotherms. Complete removal of anthracene with an isothermal purge step will take a large amount of solute.

The width of the wave (in time units) at $z = L$ is easily determined as

$$\text{Width of wave} = L/u_s\,(c = c_{low}) - L/u_s\,(c = c_{high}) \tag{18-33}$$

Since the width is proportional to the column length L, this result agrees with experimental observations.

18.4.2 Shock Waves

For an isotherm with a Langmuir shape, if the column is initially loaded at some low concentration, c_{low}, ($c_{low} = 0$ if the column is clean) and is fed with a fluid of a higher concentration, c_{high} (see Figure 18-17A), the result will be a *shock* wave. The feed step in adsorption

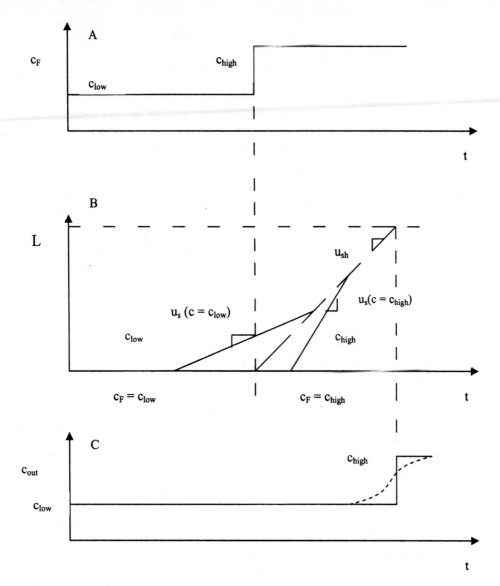

FIGURE 18-17.   *Shock wave analysis: (A) inlet concentration; (B) shock wave following Eq. (18-34); (C) outlet concentrations with solid line predicted by solute movement theory, and dashed line representing experimental result (modified from Wankat, 1986). Reprinted with permission, copyright, 1986 Phillip C. Wankat.*

processes usually results in shock waves. Experiments show that when a shock wave is predicted the zone spreading is constant regardless of the column length (a *constant pattern* wave). With the assumptions of the solute movement theory (infinitely fast rates of mass transfer and no axial dispersion), the wave becomes infinitely sharp (a shock) and the derivative dq/dc does not exist. Thus, the $\Delta q/\Delta c$ term in the denominator of Eq. (18-14) must be retained as discrete jumps in q and c, and the shock wave velocity is,

$$u_{sh} = \cfrac{v_{inter}}{1 + \cfrac{(1-\varepsilon_e)}{\varepsilon_e} K_d \varepsilon_p + \cfrac{(1-\varepsilon_e)(1-\varepsilon_p)}{\varepsilon_e} \rho_s \cfrac{(q_{after} - q_{before})}{(c_{after} - c_{before})}} \qquad \textbf{(18-34)}$$

where the subscripts "before" and "after" refer to the conditions immediately before and immediately after the shock wave. The fluid and solid before the shock wave (c_{before} and q_{before}) are assumed to be in equilibrium as are the values of c_{after} and q_{after} after the shock wave. For a general Langmuir isotherm, Eq. (18-6c), the shock wave velocity is

$$u_{sh} = \cfrac{v_{inter}}{1 + \cfrac{(1-\varepsilon_e)}{\varepsilon_e} K_d \varepsilon_p + \cfrac{(1-\varepsilon_e)(1-\varepsilon_p)}{\varepsilon_e} \rho_s \cfrac{\left[\cfrac{a\, c_{after}}{1 + b\, c_{after}} - \cfrac{a\, c_{before}}{1 + b\, c_{before}}\right]}{(c_{after} - c_{before})}} \qquad \textbf{(18-35)}$$

Now the shock wave velocity depends upon the concentrations on both sides of the shock wave. The resulting shock wave is shown in Figure 18-17B and the outlet concentration is shown in Figure 18-17C.

Superficially, the outlet concentration profile in Figure 18-17C looks like the result from a linear isotherm (concentration jumps to c_{feed}). However, for the linear isotherm this outlet step occurs at $t = L/u_s$, which is constant regardless of the feed and initial concentrations. The shock wave outlet step occurs at $t = L/u_{sh}$, which depends upon both the feed and initial concentrations. In addition, shock waves are *self-sharpening*, a concept explored in Example 18-7. Waves in systems with linear isotherms are not self-sharpening.

EXAMPLE 18-7. Self-sharpening shock wave

A 100.0 cm long column is packed with activated alumina. The column is initially filled with pure cyclohexane solvent (c = 0.0 mol/L anthracene). At t = 0 a feed containing 0.0090 mol/L anthracene in cyclohexane solvent is input. At t = 10 minutes a feed containing c = 0.011 mol/L anthracene in cyclohexane solvent is input. Superficial velocity is 20.0 cm/min. Predict and plot the outlet concentration profile using solute movement theory.

Data: $\varepsilon_e = 0.42$, $\varepsilon_p = 0$, $K_d = 1.0$, ρ_f (cyclohexane) = 0.78 kg/L, $\rho_p = 1.465$ kg/L, Equilibrium q = 22c/(1 + 375c) where q = mol/kg and c = mol/L.

Solution

A. Define. Find the values of the outlet concentration at different times using the solute movement theory.

B. Explore. The most important decision is whether a diffuse or a shock wave (Section 18.4.2) will result for each feed step. With a Langmuir isotherm a shock wave results when a concentrated feed (first c = 0.009 then c = 0.011) is fed to a column that is initially more dilute (first c = 0.0 then c = 0.009). Thus, we expect a first shock wave followed after 10 minutes by a second, which is faster. If they intersect, there will be a third shock wave (feed $c_F = 0.011$ and initial c = 0.0). (Realizing that there could be a third shock wave is probably the hardest part of this problem.)

C. Plan. The shock velocity can be calculated from (Eq. 18-35) with $\varepsilon_p = 0$. With $\varepsilon_p = 0$, the solid density ρ_s is equal to the particle density ρ_p. The interstitial velocity is

$$v_{inter} = v_{super}/\varepsilon_e = 20.0/0.42 = 47.62 \text{ cm/min}$$

Substituting in the values of parameters, except for $\Delta q/\Delta c$, the shock wave velocity is

$$u_{sh} = \cfrac{v}{1 + \cfrac{1 - \varepsilon_e}{\varepsilon_e} \rho_p \cfrac{\Delta q}{\Delta c}} = \cfrac{47.62}{1 + 2.023\left(\cfrac{\Delta q}{\Delta c}\right)}$$

where

$$\frac{\Delta q}{\Delta c} = \frac{q(c_F) - q(c_{intial})}{c_F - c_{intial}}$$

For the first shock wave, since $c_{initial} = 0$, $q\,(c_{initial}) = 0$. Since $c_F = 0.009$, we can calculate $q_{sh1}\,(c_F)$ from the Langmuir isotherm, which allows us to calculate $\Delta q/\Delta c$ and u_{sh1}. A similar calculation can be done to calculate u_{sh2}. Then we can calculate at what distance the two shock waves intersect. If this occurs before the end of the column, we can determine the velocity of the third shock wave.

D. Do it.

Shock wave 1 ($c_{initial} = 0$, $c_F = 0.009$)

$$q_{sh1}(c_f) = \frac{22(0.009)}{1 + 375(0.009)} = 0.04526$$

$$u_{sh,1} = \cfrac{47.62}{1 + 2.023\left(\cfrac{0.04526 - 0}{0.009 - 0}\right)} = 4.262 \text{ cm/min}$$

Shock wave 2 ($c_{initial} = 0.009$, $c_F = 0.011$). $q_{2initial} = 0.04526$

$$q_{2feed} = \frac{22(0.011)}{1 + 375(0.011)} = 0.0472$$

$$u_{sh,2} = \cfrac{47.62}{1 + 2.023\cfrac{(0.472 - 0.04526)}{(0.011 - 0.009)}} = 16.075 \text{ cm/min}$$

The solute movement diagram is shown in Figure 18-18A. The intersection of the two shock waves occurs at $z_{intersect}$, $t_{intersect}$. The first shock wave travels

$$z_{intersect} = u_{sh,1}\, t_{intersect}$$

while the second shock wave travels

$$z_{intersect} = u_{sh,2}\,(t_{intersect} - 10)$$

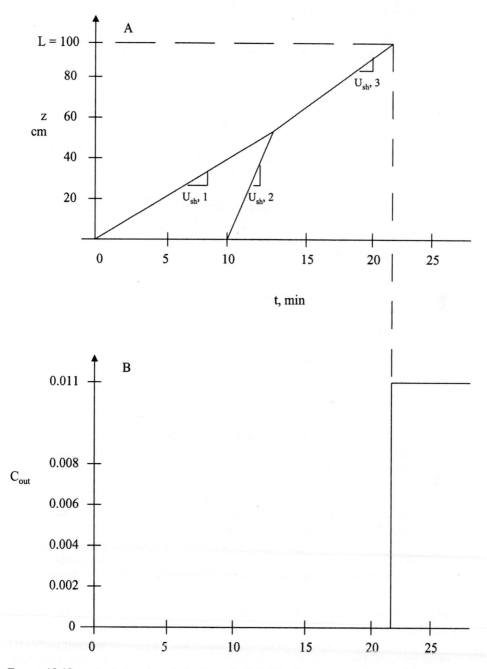

FIGURE 18-18. *Analysis and results for Example 18-7: (A) solute movement diagram showing inter-section of two shock waves, (B) outlet concentration profile*

Setting these equal and solving for $t_{intersect}$

$$t_{intersect} = \frac{10\, u_{sh,2}}{u_{sh,2} - u_{sh,1}} = \frac{(10)(16.075)}{16.075 - 4.262} = 13.608 \text{ min}$$

Then

$$z_{intersect} = (4.262)(13.608 \text{ min}) = 58.0 \text{ cm}$$

After this distance there is a third shock wave with $c_{initial} = 0.0$ (the initial column concentration at $t = 0$) and $c_F = 0.011$ (feed concentration after 10 minutes).

$$u_{sh,3} = \frac{47.62}{1 + 2.023\left(\dfrac{0.0472 - 0}{0.011 - 0}\right)} = 4.919 \text{ cm/min}$$

This shock exits at

$$t_{out} = \frac{L - z_{intersect}}{u_{sh,3}} + t_{intersect} = \frac{(100 - 58.0)}{4.919} + 13.608 = 22.146 \text{ min}$$

The complete solute movement diagram is shown in Figure 18-18A and the outlet concentration profile is in Figure 18-18B.

E. Check. As expected $u_{sh,2} > u_{sh,3} > u_{sh,1}$. We can also check the mass balance until breakthrough occurs

$$\text{Inlet} - \text{Accumulation} = 0$$

The inlet consists of feed 1 ($c_F = 0.009$) for ten minutes and feed 2 ($c_F = 0.011$) until breakthrough at $t_{out} = 22.146$ minutes. The accumulation $= \Delta q\, \rho_s\, (\pi D^2/4)$ L where the change in the amount adsorbed, $\Delta q = q(c = 0.011) - q(c=0) = 0.0472$. This mass balance is satisfied.

F. Generalize. The single shock wave that results when two steps are input occurs because of the *self-sharpening* behavior of shock waves. Diffuse waves can also be sharpened if they are followed with a shock wave. This phenomenon is used in commercial cycles. The regeneration cycle will require much less purge if an adsorbate tail (called a *heel*) is left in the column (see Figure 18-16B). This diffuse wave is then sharpened when the next feed step forms a shock wave.

The wave interactions shown in Figure 18-18A are a very important part of the study of nonlinear systems. Shock waves and diffuse waves can also interact (Wankat, 1990, Chapter 7). If there are two or more adsorbates that compete for sites (e.g., with the multicomponent Langmuir isotherm, Eq. (18-8)), interactions often occur between shock waves and diffuse waves from the different components (e.g., Ruthven, 1984; Yang, 1987). The theories for two or more interacting solutes are beyond the scope of this introductory chapter.

In experiments (Figure 18-15A) the outlet concentration profiles are not sharp as shown in Figures 18-17B and 18-18B. Instead the finite mass transfer rates and finite amounts of axial dispersion spread the wave while the isotherm effect (illustrated in Example 18-7) counteracts this spreading. The final result is a dynamic equilibrium where the wave spreads a certain amount and then stops spreading. Once formed, this *constant pattern wave* has a constant width regardless of the column length.

18.5 ION EXCHANGE

Ion exchange is a unit operation in which ions held on a solid resin are exchanged for ions in the feed solution. For most people the most familiar ion exchange system is water softening, which replaces calcium and magnesium ions ("hard" water ions) in the feed water with sodium ions ("soft" water ions). If the feed water containing calcium and magnesium is continued, eventually the resin will become saturated with these ions and no additional exchange occurs. To produce soft water the resin needs to be regenerated, which can be done with a concentrated solution of sodium chloride salt. A number of other ion exchange separations are done commercially.

The most common materials used for ion exchange are polymer resins with charged groups attached (Alexandratos, 2009; Anderson, 1979; Dechow, 1989; Dorfner, 1991; LeVan et al., 1997; Wankat, 1990). *Cation-exchange resins* have fixed negative charges while *anion-exchange resins* have fixed positive charges. The resin is called *strong* if the resin is fully ionized and *weak* if the resin is not fully ionized. The most commonly used strong resins are based on polystyrene (see Figure 17-3E) cross-linked with divinyl benzene. The most common strong cation exchange resin uses benzene-sulfonic acid groups while the most common strong anion-exchange resins have a quaternary ammonium structure. These resins are commonly used for water treatment and have very good chemical resistance although they are attacked by chlorine. Copolymers of divinylbenzene and acrylic or methacrylic acid are used for weak acid resins. No single type of weak base resin is dominant although the use of a tertiary amine group on a polystyrene-DVB resin is common. Although the capacity of the weak exchangers is lower than for strong resins, they also require less regenerant. The weak resins tend to be less robust than the strong resins and need to be protected from chemical attack. Typical properties of ion-exchange resins are listed in Table 18-4.

Ion exchange involves a reversible reaction between ions in solution and ions held on the resin. An example of monovalent cation exchange is the removal of sodium ions from the resin using hydrochloric acid.

$$H^+ + R^-Na^+ + Cl^- = R^-H^+ + Na^+ + Cl^- \tag{18-36}$$

In this equation R^- represents the fixed negative charges such as SO_3^- on the resin. The hydrogen and sodium ions that are exchanging are called *counter ions.* The chloride ion, which has the same charge as the fixed SO_3^- groups, is called the *co-ion.* Although the chloride anion does not directly affect the reaction, at high concentrations it does affect the equilibrium characteristics. Exchange of a divalent cation with a monovalent cation is also common, and is exemplified by removal of calcium ions from water and replacement with sodium ions (water softening).

$$Ca^{+2} + 2R^-Na^+ + 2\,X^- = R_2^-Ca^{+2} + 2\,Na^+ + 2\,X^- \tag{18-37}$$

where X^- is any anion. Of course, there are a variety of other possibilities.

TABLE 18-4. Properties of common ion-exchange resins (LeVan et al., 1997; Wankat, 1990)

Resin	ρ_b, wet (drained) kg/L	Moisture Content (drained) % by wt.	% Swelling upon exchange	Max. Oper. T, °C	pH range	Wet Exchange cap. eq/L	Max. flow m/h	Regenerant
Polystyrene – Sulfonic Acid								
4% DVB	0.75–0.85	64–70	10–12	120	0–14	1.2–1.6	30	HCl, H_2SO_4 or Na_2Cl_4
8–10% DVB	0.77–0.87	48–60	6–8	to 150	0–14	1.5–1.9	30	
Polyacrylic acid gel	0.70–0.75	45–50	20–80	120	4–14	3.3–4.0	20	HCl, H_2SO_4
Polystyrene Quaternary Ammonium	0.70	46–50	~20	60–80	0–14	1.3–1.5	17	NaOH
Polystyrene Tert-amine gel	0.67	~45	8–12	100	0–7	1.8	17	NaOH

Standard practice is to define the equivalent fractions of ions in solution x_i and on the resin y_i

$$x_i = c_i/c_T, \quad y_i = c_{Ri}/c_{RT} \qquad \textbf{(18-38)}$$

where c_i is the concentration of ion i in solution (e.g., in equivalents/m³), c_T is the total concentration of cations or anions in solution, c_{Ri} is the concentration of ion i on the resin (in volume units such as equivalents/m³), and c_{RT} is the total concentration of ions on the resin. The total concentration of ions on the resin c_{RT}, the resin capacity, is a constant equal to the concentration of fixed negative sites set when the resin is made. One advantage of using equivalent fractions is they must sum to one.

$$\sum_{i=1}^{N} x_i = 1, \quad \sum_{i=1}^{N} y_i = 1.0 \qquad \textbf{(18-39)}$$

18.5.1 Ion Exchange Equilibrium

For a simplified view of ion exchange equilibrium for binary ion exchange assume the equilibrium constant can be determined by the law of mass action. Let A represent the hydrogen ion and B the sodium ion. For monovalent ion exchange

$$A^+ + RB + X^- = AR + B^+ + X^-$$

illustrated in Eq. (18-36), the mass action equilibrium expression simplifies to

$$K_{AB} = \frac{c_{RA}\, c_B}{c_A\, c_{RB}} = \frac{y_A x_B}{y_B x_A} = \frac{y_A(1 - x_A)}{(1 - y_A)x_A} \qquad \textbf{(18-40a)}$$

where we have assumed the local concentrations of co-ion X^- in the interior of the resin are both identical (and very low) in the numerator and denominator and hence cancel. (These concentrations are very low in the resin because of a phenomenon known as Donnan exclusion—the very high concentrations of fixed charges on the resin exclude the free anion X^- from the resin.) The equilibrium constant K_{AB} isn't really constant (e.g., see Wankat, 1990), but in dilute solutions it will be very close to constant. Solving Eq. (18-40a) for y_A,

$$y_A = \frac{K_{AB} x_A}{1 + (K_{AB} - 1)x_A} \qquad \textbf{(18-40b)}$$

Note that the order of the subscripts is important and $K_{BA} = 1/K_{AB}$. These equations can be applied to any monovalent exchange by substituting in the appropriate symbols for the exchanging ions. Experimental values for the equilibrium constant are required. A few representative values are given in Table 18-5. If we know the equilibrium constants K_{AB} and K_{CB}, we can calculate the value of K_{CA} from,

$$K_{CA} = K_{CB}/K_{AB} \qquad \textbf{(18-40c)}$$

TABLE 18-5. *Approximate equilibrium constants for ion exchange (Anderson, 1997)*

Strong-Acid Resin with 8% cross linking. Ion B = Li$^+$				Strong-Base Resin with Medium Moisture. Ion B = Cl$^-$			
Ion A	K_{AB}	Ion D	K_{DB}	Ion A	K_{AB}	Ion D	K_{DB}
Li$^+$	1.0	Mg^{++}	3.3	F$^-$	0.1	SO$_4^{--}$	0.15
H$^+$	1.3	Zn^{++}	3.5	CH$_3$COO$^-$	0.2	CO$_3^{--}$	0.03
Na$^+$	2.0	Co^{++}	3.7	HCO$_3^-$	0.4	HPO$_4^{--}$	0.01
NH$_4^+$	2.6	Cu^{++}	3.8	BrO$_3^-$	1.0		
K$^+$	2.9	Cd^{++}	3.9	Cl$^-$	1.0		
Cs$^+$	3.3	Ni^{++}	3.9	CN$^-$	1.3		
Ag$^+$	8.5	Mn^{++}	4.1	NO$_2^-$	1.3		
		Ca^{++}	5.2	HSO$_4^-$	1.6		
		Sr^{++}	6.5	Br$^-$	3		
		Pb^{++}	9.9	NO$_3^-$	4		
		Ba^{++}	11.5	I$^-$	8		

This result expands the usefulness of Table 18-5. However, since the values in Table 18-5 are approximate [e.g., Dechow (1989) lists the following values for strong acid resins: H$^+$ = 1.26, Na$^+$ = 1.88, NH$_4^+$ = 2.22, K$^+$ = 2.63, Cs$^+$ = 2.91, and Ag$^+$ = 7.36], they should not be used for detailed design calculations.

The equilibrium expression for a divalent ion exchanging with a monovalent ion [e.g., the reaction in Eq. (18-37)] is not as simple. If we let D represent the divalent calcium ion and B the monovalent sodium ion, the reaction is

$$D^{+2} + RB_2 + 2X^- = RD + 2B^+ + 2\,X^-$$

and the equilibrium constant from the mass action expression is

$$K_{DB} = \frac{c_{RD}c_B^2}{c_D c_{RB}^2} = \left(\frac{c_T}{c_{RT}}\right)\frac{y_D x_B^2}{x_D y_B^2} \tag{18-41}$$

Selected values of divalent-monovalent equilibrium constants are given in Table 18-5. If K_{DA} and K_{BA} are known, then the desired constant K_{DB} is

$$K_{DB} = K_{DA}/(K_{BA})^2 \tag{18-42}$$

Substituting the summation Eqs. (18-39) into Eq. (18-41), we obtain

$$\frac{y_{DB}}{(1 - y_{DB})^2} = \frac{K_{DB}c_{RT}}{c_T}\frac{x_D}{(1 - x_D)^2} \tag{18-43}$$

This equation can conveniently be solved for y_D for any specified value of x_D using the formula for solution of quadratic equations, or by using Goal Seek or Solver in a spreadsheet.

Note that the effective equilibrium parameter in Eq. (18-43) is $(K_{DB} \, c_{RT} / c_T)$. Since the total concentration in the fluid can easily be changed, this effective equilibrium parameter can be changed. This behavior is illustrated in Example 18-8 and Figure 18-19.

The equilibrium parameters are temperature dependent. However, ion exchange is usually operated at a constant temperature near the ambient temperature.

18.5.2 Movement of Ions

The solute movement theory developed in Sections 18.3. and 18.4. is easily extended to ion movement. For gel-type ion-exchange resins, which are most popular, there are no permanent pores and $\varepsilon_p = 0$. The development of the solute movement theory from Eqs. (18-10) to (18-14) is modified by setting $\varepsilon_p = 0$, expressing concentrations in terms of the equivalent fractions x and y, and including a Donnan exclusion factor K_{DE}. The result is

$$u_{ion,i} = \frac{v_{inter}}{1 + \dfrac{1}{\varepsilon_e} \dfrac{c_{RT}}{c_T} \dfrac{\Delta y_i}{\Delta x_i} K_{DE,i}} \tag{18-44}$$

Co-ions (ions with the same charge as the ions fixed to the resin) are excluded and they have $K_{DE} = 0$. Exchanging ions are not excluded and $K_{DE} = 1$. Note that the $(1 - \varepsilon_e) \, \rho_s$ term does not appear in Eq. (18-44) because volumetric units are commonly used for ion concentrations in solution and on the resin.

Equation (18-44) can be applied to either diffuse or shock waves. Diffuse waves occur if a column that is concentrated in ion A (or D) is fed a solution of low concentration A (or D) and $K_{AB} > 1.0$ [or $(K_{DB} \, c_{RT} / c_T) > 1.0$]. If $K_{AB} < 1.0$ [or $(K_{DB} \, c_{RT} / c_T) < 1.0$] diffuse waves occur when a column that has a dilute amount of ion A (or D) is fed a solution that is concentrated in A (or D). For diffuse waves the ion velocity is

$$u_{s,ion,i} = \frac{v_{inter}}{1 + \dfrac{1}{\varepsilon_e} \dfrac{c_{RT}}{c_T} K_{DE,i} \left(\dfrac{dy_i}{dx_i} \right)} \tag{18-45a}$$

The derivative (dy_i/dx_i) can be determined from the appropriate equilibrium expression such as Eq. (18-40b) or Eq. (18-42b). These derivatives are

$$\frac{dy_A}{dx_A} = \frac{K_{AB}}{(1 + (K_{AB} - 1)x_A)^2} \tag{18-45b}$$

for monovalent-monovalent exchange and

$$\frac{dy_D}{dx_D} = \left(\frac{K_{DB} C_{RT}}{C_T} \right) \frac{(1 - y_D)^3 (1 + x_D)}{(1 - x_D)^3 (1 + y_D)} \tag{18-45c}$$

for divalent-monovalent exchange.

Shock waves occur when the conditions are the opposite of those for diffuse waves. For example, if $K_{AB} > 1.0$ [or $(K_{DB} c_{RT}/c_T) > 1.0$] and a solution concentrated in A (or D) is fed to a column containing a dilute solution of the ion, a shock wave would be expected. The shock wave equation is

$$u_{sh,ion,i} = \frac{v_{inter}}{1 + \dfrac{1}{\varepsilon_e} \dfrac{c_{RT}}{c_T} K_{DE,i} \dfrac{(y_{i,after} - y_{i,before})}{(x_{i,after} - x_{i,before})}} \qquad \text{(18-46)}$$

The equivalent concentrations of ion A (or D) in solution x_i and on the resin y_i are assumed to be in equilibrium both before and after the shock wave.

The ion velocities depend upon the ratio (c_{RT}/c_T) regardless of the form of the isotherm. This agrees with our physical intuition. If the resin capacity c_{RT} is high while the concentration of ion in solution c_T is low, we would expect that waves would move slowly.

Because changes in the total ion concentration can affect both equilibrium and ion velocities, we need to balance the total ion concentration. When the total ion concentration in the feed is changed, an *ion wave* passes through the column. For relatively dilute solutions the resin is already saturated with counter-ions, and more ions cannot be retained. For a balance on all ions, the total ions are excluded, $K_{DE} = 0$, and Eq. (18-44) becomes

$$u_{total\ ion} = v_{inter} \qquad \text{(18-47a)}$$

This is the same result that is obtained for co-ions. The total ion and co-ion waves move rapidly through the column. The total ion wave affects counter-ion (u_A or u_D) velocities for all ion exchange systems. For exchange of ions of equal charges (e.g., monovalent-monovalent), the equilibrium is not affected and $x_{i,after} = x_{i,before}$ and $y_{i,after} = y_{i,before}$. The equilibrium for the exchange of ions with different charges (e.g., divalent-monovalent or trivalent-monovalent) is changed. A mass balance on a segment of column (see Problem 18.C9) shows that

$$y_{after\_total\_ion\_wave} = y_{before\_total\_ion\_wave} \qquad \text{(18-47b)}$$

For monovalent-monovalent exchange,

$$x_{after\_total\_ion\_wave} = x_{before\_total\_ion\_wave} \qquad \text{(18-47c)}$$

while for divalent-monovalent exchange,

$$x_{after\_total\_ion\_wave} \neq x_{before\_total\_ion\_wave} \qquad \text{(18-47d)}$$

but $x_{after\_total\_ion\_wave}$ is in equilibrium with $y_{after\_total\_ion\_wave}$.
The calculation of these effects is illustrated in Example 18-8.

EXAMPLE 18-8. Ion movement for divalent-monovalent exchange

An ion exchange column is filled with a strong acid resin ($c_{RT} = 2.0$ equivalents/L, $\varepsilon_e = 0.40$). The column initially is at a total cation concentration of $c_T = 0.01$N with $x_{Na} = 0.90$, $x_{Cu} = 0.10$. Chloride is the co-ion. At t = 0 we feed a 2.5 N aqueous solution of NaCl ($x_{Na} = 1.0$). The selectivity constants can be calculated from Table 18-5. The column is 50 cm long. The counter ions are not excluded ($K_E = 1.0$). The superficial velocity throughout the experiment is 20.0 cm/min.

Predict:

a. Equilibrium behavior at $c_T = 0.01$ N and at $c_T = 2.5$ N
b. The time the total ion wave exits
c. The values of x_{Cu} and y_{Cu} after the total ion wave exits
d. The time and shape of the exiting sodium wave

Solution

A. Define. We first want to find the equilibrium parameters at $c_T = 0.01$ N and at $c_T = 2.5$ N, and then plot the equilibrium results. Then, find the breakthrough time for the total ion wave, the equivalent fractions of copper at $c_T = 2.5$ N, and the outlet concentration profile for sodium.

B. Explore. The determination of the equilibrium behavior, the ion wave, and the values of x_{Cu} and y_{Cu} follows the equations, but the order can be a bit confusing. One reason for showing this example is to clarify how the calculations proceed. Sometimes, the biggest challenge is determining if the sodium wave is a shock or diffuse wave.

C. Plan. The selectivity can be found from Eq. (18-42) and the equilibrium parameter is $(K_{DB} c_{RT}/c_T)$. The equilibrium curves can be found at arbitrary x values from Eq. (18-43). The velocity of the total ion wave is equal to the interstitial velocity. The equivalent fractions of copper can be found by solving Eq. (18-43) with $x_{Cu,before} = 0.10$ and $c_T = 0.01$ N to find $y_{Cu,before}$. When the total ion wave passes, set $y_{Cu,after} = y_{Cu,before}$, and solve Eq. (18-43) with this value of $y_{Cu,after}$ and $c_T = 2.50$ N for $x_{Cu,after}$. Finally, the sodium breakthrough time can be calculated from either Eq. (18-45) or Eq. (18-46) after we decide if it is a diffuse or shock wave, respectively. To do this we will look at the shape of the isotherm for this decrease in copper concentration in the feed.

D. Do it.

First, find equilibrium curve at $c_T = 0.01$ N. Since copper is divalent and sodium is monovalent, use Eqs. (18-42) and (18-43). From Eq. (18-42)

$$K_{CuNi} = \frac{K_{CuLi}}{(K_{NaLi})^2} = \frac{3.8}{(2.0)^2} = 0.95$$

At $c_T = 0.01$ N, $\dfrac{K_{CuNi}c_{RT}}{c_T} = \dfrac{0.95(2.0)}{(0.01)} = 190$

Then Eq. (18-43) becomes

$$\frac{y_{Cu}}{(1 - y_{Cu})^2} = 190\frac{x_{Cu}}{(1 - x_{Cu})^2}$$

At $x_{Cu} = 0.1$ (the initial concentration)

$$\frac{y_{Cu}}{(1 - y_{Cu})^2} = 190\left[\frac{0.1}{(0.9)^2}\right] = 23.46$$

Solving for y_{Cu} with a spreadsheet, we obtain $y_{Cu} = 0.814$. The equilibrium table below at $c_T = 0.01$ N was generated using this spreadsheet.

xCu	0.002	0.005	0.01	0.02	0.05	0.10	0.20	0.30	0.4	0.6	0.8
yCu	0.228	0.379	0.495	0.608	0.736	0.814	0.878	0.911	0.936	0.963	0.984

When $c_T = 2.5$, $\dfrac{K_{CuNi}c_{RT}}{c_T} = \dfrac{0.95(2.0)}{2.5} = 0.760$ and Eq. (18-43) is

$$\frac{y_{cu}}{(1 - y_{cu})^2} = 0.760 \frac{x_{cu}}{(1 - x_{cu})^2} \text{ and the equilibrium results are}$$

xCu	0.1	0.2	0.3	0.4	0.5	0.6	0.7	0.8	0.9	0.95
yCu	0.0795	0.165	0.257	0.353	0.454	0.558	0.665	0.774	0.886	0.943

These two tables are plotted in Figure 18-19. Note that with $c_T = 0.01$ N a very favorable isotherm results while with $c_T = 2.5$ N the isotherm is unfavorable.

The velocity of the total ion wave is $u_{total\_ion} = v_{super}/\varepsilon_e = 20.0/0.4 = 50$ cm/min, and the breakthrough time of this wave is $t_{br} = L/u_{total\_ion} = 50.0/50.0 = 1.0$ min.

From the table of equilibrium values at $c_T = 0.01$ N, when $x_{Cu,before} = 0.10$, $y_{Cu,before} = 0.814$. Then, when the total ion wave passes (now $c_T = 2.5$ N), $y_{Cu,after} = y_{Cu,before} = 0.814$. At $c_T = 2.5$ N and $y_{Cu,after} = 0.814$, we can solve Eq. (18-41c) for $x_{Cu,after}$.

$$\frac{x_{Cu}}{(1 - x_{Cu})^2} = \frac{1}{\dfrac{K_{DB}c_{RT}}{c_T}} \frac{y_{Cu}}{(1 - y_{Cu})^2} = \frac{1}{0.76} \frac{0.814}{(0.186)^2}$$

The result is $x_{Cu, after} = 0.836$.

Note the large increase in copper equivalent fraction in the liquid is caused by the change in equilibrium at the higher total ion concentration. The increase in copper is supplied by removing copper from the resin when the sodium cloride wave passes.

When the 2.5 N sodium chloride solution is fed to the column, the copper equivalent fraction, and concentration in the feed drop to zero. Since the column is at 2.5 N when the sodium wave reaches any part of the column, we use that equilibrium curve in Figure 18-19. This is an unfavorable isotherm for copper; thus, with a drop in copper concentration in the feed a *shock wave results*. Then use Eq.(18-46) to calculate the shock wave velocity.

$$u_{sh} = \frac{v}{1 + \dfrac{c_{RT}}{\varepsilon_e c_T} \dfrac{\Delta y_{Cu} K_{DE}}{\Delta x_{Cu}}} \text{ where } \Delta y_{Cu} = y_{Cu, \text{ after NaCl}} - y_{Cu, \text{ before NaCl}}.$$

$\Delta y_{cu} = 0.0 - 0.814 = -0.814$, $\Delta x_{cu} = 0 - 0.836 = -0.836$, and

$$u_{sh} = \frac{50}{1 + \dfrac{2.0}{(0.4)(2.5)}\left(\dfrac{-0.814}{-0.836}\right)} = 16.96 \ \frac{cm}{min}$$

This shock wave exits at $t_{NaCl} = L/u_{sh} = 50/16.96 = 2.95$ minutes

The solute movement diagram is plotted in Figure 18-20.

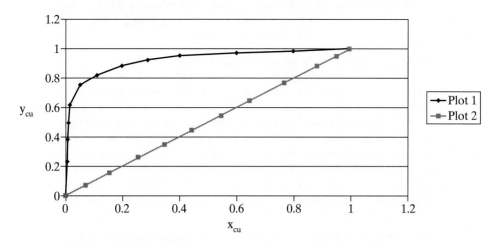

FIGURE 18-19.   *Equilibrium for copper-sodium exchange for Example 18-8;* ◆ $c_T = 0.01$ N, ▩ $c_T = 2.5$ N

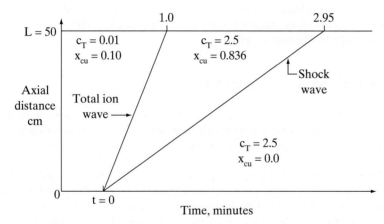

FIGURE 18-20.   *Solute movement diagram for Example 18-8*

E. Check. A copper mass balance on the entire cycle can be used as a check,
Cu in – Cu out = Cu accumulation

From t = 0 to t = 2.95 minutes there is no copper flowing into the system. The outlet copper amount consists of one minute at ($c_T = 0.01$ N, $x_{Cu} = 0.10$) and 1.95 minutes at ($c_T = 2.5$ N, $x_{Cu} = 0.836$) as shown in Figure 18-20. Since we know the conditions at the beginning of the cycle ($c_T = 0.01$ N, $x_{Cu,before} = 0.10$, $y_{Cu,before} = 0.814$) and at the end of the cycle ($c_T = 2.5$ N, $x_{Cu,after\_NaCl} = 0.0$, $y_{Cu,after\_NaCl} = 0.0$), we can calculate the accumulation term for the cycle. The mass balance then becomes

$$0 - (v_{super} = 20)A_c[(1.0 \text{ min})(c_T = 0.01)(x_{cu} = 0.10) + 1.95 \text{ min } (c_T = 2.5)(x_{cu} = 0.836)]$$
$$= -A_c \text{ } (L = 50)(c_{RT} = 2.0)[(y_{cu} = 0.814) - (y_{cu} = 0)]$$
$$\text{or } -81.5 \text{ } A_c - -81.4 \text{ } A_c$$

This is certainly within the accuracy of the calculations.

F. Generalization. The change in equilibrium behavior from a favorable isotherm to an unfavorable isotherm when c_T is increased only occurs for exchange of ions with unequal charges. This change can result in shock waves during both the feed and regeneration steps. Since shock waves show significantly less spreading than diffuse waves, significantly less regenerant is required. This phenomenon is used to advantage in water softeners that exchange "hard" Ca^{++} and Mg^{++} ions (they precipitate when heated and foul cooking utensils or heat exchangers and interfere with soap) at very low c_T values with Na^+ ions. Regeneration is done with a concentrated salt (NaCl) solution at high c_T.

This introductory presentation on ion exchange has been restricted to dilute solutions with the exchange of two ions. A variety of more complex situations including complex equilibria and mass transfer, partial ion exclusion, swelling of the resin (see Table 18-4), and the need for backwashing to remove dirt often occur in practice, but are beyond the scope of this introductory section. Information about the variety of phenomena, different equipment, equilibrium theory and mass transfer is discussed by Helferrich (1962), Dechow (1989), and Tondeur and Bailly (1986) respectively.

18.6 MASS AND ENERGY TRANSFER IN PACKED BEDS

For detailed predictions and understanding of sorption separations we need to do a detailed analysis of diffusion rates, mass and energy transfer, and mass and energy balances in the column. In order to make the results somewhat tractable, we will make all the usual assumptions listed in Table 18-6.

The fluid flowing in the external void volume in Figure 18-1 is usually assumed to have a constant concentration (or partial pressure) at each axial distance z; thus, this bulk concentration is not a function of the radial distance from the center of the bed. During the course of separation the sorbates are first transferred down the column by bulk transfer. The sorbates then transfer across the external film and diffuse in the pores until they reach the sorbent sites. At these sites they are sorbed in a usually rapid step. For an equilibrium system, some sorbates will always be desorbing and diffusing back into the bulk fluid. Desorption is favored during the regeneration step by increasing temperature or dropping the concentration of the sorbate.

Since the first step in the separation process, bulk transport, is assumed to be rapid enough to keep the bulk concentration (at any given value of z, t) constant, we will start with an analysis of film mass transfer.

18.6.1 Mass Transfer and Diffusion

Mass transfer across the film (see Figure 18-1) occurs by a combination of diffusion and convection. As is usual in film mass transfer, the driving force is assumed to be $(c-c_{pore})$ where the concentration in the pores c_{pore} is calculated at the surface of the particles. The film transfer

TABLE 18-6. *Assumptions for mass and energy transfer analysis*

Assumption	Comments
1. Homogeneous packing (no channeling)	Valid if carefully packed and $D_{col}/d_p > \sim30$.
2. Negligible radial gradients	May not be valid if high heat losses.
3. Neglect thermal and pressure diffusion	Usually OK.
4. No chemical reactions except for sorption	Sorbents can act as catalysts. Need to check.
5. Neglect kinetic and potential energy	Usually OK.
6. No radiant heat transfer	Absolutely true only for isothermal. Usually lumped with convective heat transfer.
7. No electrical or magnetic fields	Usually OK.
8. No phase changes except sorption	Usually OK (watch for solute precipitation).
9. Velocity constant across cross-section	Reasonable approximation if: no channeling, no radial gradients, no viscous fingering.
10. No irreversible adsorption	Usually OK once sorbent has been used and regenerated.
11. Rigid packing	OK except for soft packings (e.g., polymers).
12. Constant sorption properties	OK if no slow degradation or poisoning.
13. No breakage or dissolution of packing	Can be problem with moving beds or when there is chemical attack.
14. Negligible conduction in column walls	OK for large diameter columns.

term becomes $-k_f a_p (c-c_{pore})$ where k_f is the film mass transfer coefficient (m/s) and a_p is the surface area of the particles per volume (m^{-1}). For spherical particles

$$a_p = 6/d_p \tag{18-48a}$$

With porous particles the film equation is a bit more complicated than in Chapter 15 since the accumulation of mass in the particle is distributed between the pores and the solid. The resulting equation is

$$\rho_s (1 - \varepsilon_p)(1 - \varepsilon_e) \frac{\partial \overline{q}_i}{\partial t} + K_{d_i} \varepsilon_p (1 - \varepsilon_e) \frac{\partial \overline{c}_i}{\partial t} = -k_f a_p (c_{i_{pore}} - c_i) \tag{18-48b}$$

The left-hand side of this equation is the accumulation of solute on the solid and in the fluid within the pores. The right-hand side is the mass transfer rate across the film. Since the amount adsorbed in the particles and the concentration of the pore fluid are functions of r, $\overline{q}_i$ and $\overline{c}_i$ are the average amount adsorbed in the particle and the average pore fluid concentration, respectively.

A more familiar looking version of the film transport equation is obtained for a single-porosity model. This result can be formally obtained by setting $\varepsilon_p = 0$ and $\varepsilon = \varepsilon_e$ in Eq. (18-48b).

$$\rho_p(1-\epsilon)\frac{\partial \overline{q_i}}{\partial t} = -k_f a_p (c_{i,surface} - c_i)$$

(18 49)

Since mass transfer rates in the film are usually quite high compared to diffusion in the particles, film resistance is rarely important in commercial processes (Basmadjian, 1997).

After passing through the film, solute then diffuses in the pores by normal diffusion (large pores), Knudsen diffusion (small pores), or surface diffusion. In polymer resins where there are no permanent pores, the solute diffuses in the polymer phase. For spherical particles with a radial coordinate r, the diffusion equation in pores is

$$(1-\epsilon_p)\frac{\partial \overline{q}}{\partial t} = \epsilon_p D_{effective}\left(\frac{2}{r}\frac{\partial c}{\partial r} + \frac{\partial^2 c}{\partial r^2}\right)$$

(18-50)

For ordinary or Fickian diffusion $D_{effective}$ is related to the diffusivity in free solution through the tortuosity factor, Eq. (18-4). Equation (18-50) assumes that the mean free paths of the molecules are significantly less than the radius of the pores. This is the usual case with liquids and gases at high pressures.

Knudsen diffusion occurs when the mean free path of the molecules is significantly greater than the radius of the pores. In this case instead of colliding with other molecules, a molecule collides with the pore walls. The same diffusion equation can be used, but now $D_{effective}$ is determined from Eq. (18-4) with D_K, the Knudsen diffusivity replacing $D_{molecular}$. The Knudsen diffusivity can be estimated from (Yang, 1987)

$$D_K = 9.7 \times 10^3 \, r_p \left(\frac{T}{M}\right)^{1/2}$$

(18-51)

where r_p the pore radius is in cm, the absolute temperature T is in Kelvin, M is the molecular weight of the solute, and D_K is in cm²/s. If the mean free path of the molecules is the same order of magnitude as the pore diameter, both ordinary and Knudsen diffusion mechanisms are important. The diffusivity can be estimated as

$$D \approx \frac{1}{\left(1/D_{molecular}\right) + \left(1/D_K\right)}$$

(18-52)

This value of D is then used instead of $D_{molecular}$ in Eq. (18-4) to estimate $D_{effective}$. The use of Eqs.(18-51) and (18-52) is often required for the adsorption of gases in adsorbents with small pores.

In *surface diffusion* the adsorbate does not desorb from the surface but instead diffuses along the surface. This mechanism can be important in gas systems and when gel-type (non-porous) resins are used. The surface diffusion flux is

$$\text{Flux by surface diffusion} = -D_s \frac{dq}{dr}$$

(18-53)

where the surface diffusion coefficient D_s depends strongly on the surface coverage, q. Currently, values of D_s must be back-calculated from diffusion or adsorption experiments. Ordinary, Knudsen, and surface diffusion may occur simultaneously. More detailed descriptions of the diffusion terms are available in the books by Do (1998), Ruthven (1984) and Yang (2003).

18.6.2 Column Mass Balances

The equations for film diffusion and diffusion inside the particle tell us what is happening at a given location inside the column. To determine what is happening for the entire column we need a mass balance on the solid and fluid phases. In order to write reasonably simple balance equations, we usually make a number of "common-sense" assumptions such as those listed in Table 18-6. The resulting equation for the two-porosity model is

$$\varepsilon_e \frac{\partial c_i}{\partial t} + K_{d_i}(1 - \varepsilon_e)\,\varepsilon_p \frac{\partial \overline{c}_i}{\partial t} + \rho_s(1 - \varepsilon_e)(1 - \varepsilon_p)\frac{\partial \overline{q}_i}{\partial t}$$

$$+ \varepsilon_e \frac{\partial (v_{inter}c_i)}{\partial z} - \varepsilon_e E_D \frac{\partial^2 c_i}{\partial z^2} = 0 \qquad \textbf{(18-54)}$$

The first three terms are the accumulation in the fluid between particles, within the pore fluid with an average concentration $\overline{c}_i$, and sorbed on the solid, respectively. The fourth term represents convection while the fifth term is axial dispersion. Coefficient E_D is the axial dispersion coefficient due to both eddy and molecular effects. The value of $\overline{q}_i$ in Eq. (18-54) can be related to c_i and $\overline{c}_i$ through Eqs. (18-48b) and (18-50). A slightly simpler equation is obtained using the single-porosity model,

$$\varepsilon \frac{\partial c_i}{\partial t} + \rho_p(1 - \varepsilon)\frac{\partial \overline{q}_i}{\partial t} + \varepsilon \frac{\partial (c_i v_{inter})}{\partial z} - \varepsilon E_D \frac{\partial^2 c_i}{\partial z^2} = 0 \qquad \textbf{(18-55)}$$

Solution of the appropriate set of equations: equilibrium, 18-48b or 18-49, 18-50 (with diffusivities calculated from $D_{molecular}$, 18-51, or 18-52 inserted into 18-4 and/or use of 18-53), and either 18-54 or 18-55 plus a suitable set of boundary conditions is very difficult. This calculation is so difficult that even with detailed simulators a simplified procedure is usually employed.

18.6.3 Lumped Parameter Mass Transfer

One very common simplification is to assume that the film diffusion and diffusion in the particles can be lumped together in a *lumped parameter* mass transfer expression. In this form the total of all the mass transfer is assumed to be proportional to the driving force caused by the concentration difference $(c_i^* - c_i)$ *or* by the driving force caused by the difference in amount adsorbed $(q_i^* - q_i)$. The value c_i^* is the concentration that would be in equilibrium with $\overline{q}_i$ and q_i^* is the amount adsorbed that would be in equilibrium with fluid of concentration c_i. Note that neither c_i^* nor q_i^* actually exist in the column—they are hypothetical constructs. The two resulting equations are

$$\rho_s(1 - \varepsilon_p)(1 - \varepsilon_e)\frac{\partial \overline{q}_i}{\partial t} + K_{d_i}\varepsilon_p(1 - \varepsilon_e)\frac{\partial c_i^*}{\partial t} = -k_{m,c}a_p(c_i^* - c_i) \qquad \textbf{(18-56a)}$$

$$\rho_s(1 - \varepsilon_p)(1 - \varepsilon_e)\frac{\partial \overline{q}_i}{\partial t} + K_{d_i}\varepsilon_p(1 - \varepsilon_e)\frac{\partial c_i}{\partial t} = -k_{m,q}a_p(q_i^* - q_i) \qquad \textbf{(18-56b)}$$

Note that Eq. (18-56a) is very similar to Eq. (18-48b) except that c* replaces the concentration of the pore fluid c_{pore} and the lumped parameter mass transfer coefficient $k_{m,c}$ replaces the film coefficient k_f. As expected, the lumped parameter expressions using the single-porosity model are simpler.

$$\rho_p(1-\varepsilon)\frac{\partial \overline{q}_i}{\partial t} = -k_{m,c}a_p(c_i^* - c_i) \tag{18-57a}$$

$$\rho_p(1-\varepsilon)\frac{\partial \overline{q}_i}{\partial t} = -k_{m,q}a_p(q_i^* - q_i) \tag{18-57b}$$

Equations (18-56a) and (18-56b) or (18-57a) and (18-57b) can be converted into each other if the equilibrium is linear ($k_{m,q} = k_{m,c}/K'$). Since both Eqs. (18-57a) and (18-57b) are commonly used, it is necessary to be clear which lumped parameter coefficient is reported—unfortunately, authors usually don't put a subscript denoting the driving force on this term. Look for their driving force equation to determine which form they used.

The lumped parameter models are useful because they simplify the theory and the coefficients can often be estimated with reasonable accuracy. The most common way to determine $k_{m,q}$ is with a sum of resistances approach (Ruthven et al., 1994) that is similar to the approach used in Section 15.1.

$$\frac{1}{K'k_{m,q}\alpha_p} = \frac{1}{k_f a_p} + \frac{d_p^2}{60 D_{effective}(1-\varepsilon_e)\rho_p} \tag{18-58a}$$

The value of a_p, the mass transfer area per volume, is usually estimated as $6/d_p$. A number of correlations have been developed to estimate the film coefficient. The Wakoa and Funazkri (1978) correlation appears to be quite accurate

$$Sh = 2.0 + 1.1\, Sc^{1/3}\, Re^{0.6}, 3 < Re < 10^4 \tag{18-59}$$

where the Sherwood, Schmidt and Reynolds numbers are defined as,

$$Sh = \frac{k_f d_p}{D_m}, Sc = \frac{\mu}{\rho_f D_m}, Re = \frac{\rho_f v_{inter}\varepsilon_e d_p}{\mu} \tag{18-60}$$

A final value needed to solve the complete set of equations is the eddy dispersion coefficient, E_D. The Chung and Wen (1968) correlation is commonly used to determine E_D.

$$\varepsilon_e N_{Pe} = 0.2 + 0.011\, Re^{0.48} \tag{18-61}$$

where the Reynolds number was defined in Eq. (18-60) and the Peclet number is defined as

$$N_{Pe} = d_p v_{inter}/E_D \tag{18-62}$$

In many systems, particularly liquid systems, the resistance due to diffusion in the pores is much more important than the resistance due to mass transfer across the film ($D_{effective}$ is small and hence the second term on the right hand side of Eq. (18-58a) is much larger than the first term). Then,

$$K'k_{m,q}a_p = \frac{60 D_{effective}(1-\varepsilon_e)\rho_s}{d_p^2} \tag{18-58b}$$

Thus, when pore diffusion controls the mass transfer coefficient is independent of fluid velocity and proportional to $(1/d_p)^2$. Basmadjian (1997) suggests that initial estimates can be made with $k_{m,c}a_p$ values of 10^{-1} 1/s for gases and 10^{-2} to 10^{-3} 1/s for liquids. More detailed discussions of the kinetics and mass transfer in adsorbents can be found in the books by Do (1998), Ruthven (1984), and Yang (2003).

18.6.4 Energy Balances and Heat Transfer

Since there are usually significant heat effects in gas systems, energy balances will be required. For the single-porosity model the energy balance (based on the assumptions in Table 18-6) for the fluid, particles, and column wall is

$$\rho_f C_{P,f} \varepsilon \frac{\partial T}{\partial t} + \rho_p C_{P,p} (1 - \varepsilon) \frac{\partial T_s}{\partial t} + \rho_f C_{p,f} v_{inter} \, \varepsilon \frac{\partial T}{\partial z}$$

$$- E_{DT} \rho_f C_{P,f} \varepsilon \frac{\partial^2 T}{\partial z^2} = h_w A_w (T_{amb} - T_w) - \frac{C_{P,w} W}{A_c} \frac{\partial T_w}{\partial t} \qquad \textbf{(18-63)}$$

The first two terms and the last term represent accumulation of energy in the fluid, the particle, and the column wall. The third term is convection of energy while the fourth term is the axial dispersion of energy. The fifth term (first term on right hand side) represents the heat transfer from the column walls. Because industrial scale systems have a small ratio of wall area to column volume, the fifth term is often negligible (the column is adiabatic), and the sixth term is often negligible because the mass of the column wall is small compared to the mass of adsorbent.

The transfer of energy from the fluid to the solid can often be represented as a lumped parameter expression of the following form:

$$\rho_p C_{P,p} (1 - \varepsilon) \frac{\partial T_s}{\partial t} = - h_p a_p (T^* - T) + (1 - \varepsilon) \rho_p \Delta H_{ads} \frac{\partial q}{\partial t} \qquad \textbf{(18-64)}$$

The first term represents accumulation of energy in the solid, the second term is the heat transfer rate from the fluid to the solid, and the last term is the heat generated by adsorption. This last term can be quite large. Increases in gas temperature of over 100°C can occur in gas adsorption systems, and if oxygen is present activated carbon beds can catch on fire.

18.6.5 Derivation of Solute Movement Theory

The solute movement equations can be derived rigorously by solving the mass and heat transfer equations with a set of limiting assumptions. Start with the column balance on fluid and solid, Eq. (18-54). Assuming that mass transfer is very rapid, the bulk fluid and solid will be in equilibrium. Thus, $c = c^*$ which is in equilibrium with $q = q^*$, and the lumped parameter expression, Eq. (18-56a) or (18-56b) is not required. In addition, assuming that axial dispersion is negligible, Eq. (18-54) becomes

$$[\varepsilon_e + K_d (1 - \varepsilon_e) \varepsilon_p] \frac{\partial c}{\partial t} + \rho_s (1 - \varepsilon_e)(1 - \varepsilon_p) \frac{\partial q}{\partial t} + \varepsilon_e v_{inter} \frac{\partial c}{\partial z} = 0 \qquad \textbf{(18-65a)}$$

Since the solid and fluid are in equilibrium, q is related to c and T through the isotherm. After assuming that solid (ε_e, ε_p, K_d, and ρ_s) properties are constant, applying the chain rule and simplifying, Eq. (18-65a) becomes

$$\frac{\partial c}{\partial t} + u_s \frac{\partial c}{\partial z} = -\frac{u_s}{v_{inter}}\left(\frac{1-\varepsilon_e}{\varepsilon_e}\right)(1-\varepsilon_p)\rho_s \frac{\partial q}{\partial t}\frac{\partial T}{\partial t} \tag{18-65b}$$

The total derivative dT/dt = 0 for isothermal systems, systems with instantaneous temperature changes, and systems with square wave changes in the temperature. With these simplifications Eq. (18-65b) can be solved by the method of characteristics (e.g., Ruthven, 1984; Sherwood et al., 1975). The result for constant interstitial velocity (valid for liquids, exchange adsorption and dilute gases) is that concentration is constant along lines of constant solute velocity where the solute velocity is given by Eq. (18-14).

A similar analysis can be applied to the energy balance equation by assuming very rapid energy transfer, negligible axial thermal diffusion, constant solid and fluid properties (e.g., densities and heat capacities), constant interstitial velocity and the heat of adsorption can be neglected. The result is that temperature is constant along lines of constant thermal velocity where the thermal velocity is given by Eq. (18-21).

The solute movement analysis is thus a physically based analysis that can be derived rigorously with appropriate limiting assumptions. If mass transfer is slow and the velocity is high or the column is short, the solute may not have sufficient residence time in the column to diffuse into the solid. The solute then skips the separation mechanism (equilibrium between solid and fluid) and exits with the void volume of the fluid. In this situation the predictions of solute movement are not useful. Basmadjian (1997) states that one of the following conditions must be satisfied to avoid this "instantaneous breakthrough,"

$$L > 4.5\, v_{super}/(k_{m,c}a_p) \text{ or } v_{super} < 0.22/(L\, k_{m,c}a_p) \tag{18-66}$$

Local equilibrium analysis can be extended to systems with variable interstitial velocity (concentrated gases), interacting solute isotherms such as Eq. (18-8), or finite heats of adsorption (most important for concentrated gases), but this extension is beyond the scope of this chapter (e.g., see Ruthven, 1984; Yang, 1987).

18.6.6 Detailed Simulators

Simultaneous solution of the combined mass and energy balances, the pressure drop equation [see Eq. (18-90)], and the equilibrium expressions is a formidable task, particularly for multi-component, nonlinear systems. Until the 1990s solution of this set of differential, algebraic equations (DAE) was typically a task done by Ph.D. students for their thesis or by a few industrial experts who devoted their careers to simulation of sorption processes. This situation changed with the development of fairly general solvers for DAEs such as SPEEDUP, gPROMS, and Aspen Custom Modeler, and later the development of simulators such as ADSIM and Aspen Chromatography designed specifically for adsorption, chromatography and ion exchange. Current versions of the DAE solvers and simulators are reasonably user friendly. Currently, almost all designs of distillation and absorption processes are designed using simulators. In the future the design of sorption systems will follow down the simulation path, and most designs will be based on simulation programs. Teaching the use of these simulation packages is discussed by Wankat (2006), and the introductory computer laboratories for Aspen Chromatography are included in the appendix to this chapter.

In the next two sections the mass and energy transfer equations will be used to obtain realistic solutions for a variety of simplified adsorption problems.

18.7 MASS TRANSFER SOLUTIONS FOR LINEAR SYSTEMS

The solution of differential equations is much simpler when the equations are linear. The various sets of differential equations for mass transfer discussed in Section 18.6 are all linear *if* the equilibrium isotherm is linear and the system is isothermal. (Note that nonisothermal operation introduces the Arrhenius relationship, Eq. (18-7), which is decidedly nonlinear.) This section is limited to isothermal operation of systems with linear isotherms Eqs. (18-5b) or (18-6b).

One characteristic of the solutions for Eqs. (18-54) and (18-55) for linear isotherms is mass transfer resistances and axial dispersion both cause zone spreading that *looks identical if* the mass transfer parameters or axial dispersion parameters are adjusted. Thus, from an experimental result it is impossible to determine if the spreading was caused solely by mass transfer resistances, solely by axial dispersion, or by a combination of both. This property of linear systems allows us to use simple models to predict the behavior of more complex systems.

18.7.1 Lapidus and Amundson Solution for Local Equilibrium with Dispersion

In a classic paper Lapidus and Amundson (1952) studied liquid chromatography for isothermal operation with linear, independent isotherms when mass transfer is very rapid, but axial dispersion is important. Although the two-porosity model can be used (Wankat, 1990), the solution was originally obtained for the single-porosity model. Starting with Eq. (18-55), we substitute in the equilibrium expression Eq. (18-6a) to remove the variable q (solid and fluid are assumed to be in local equilibrium). Since the fluid density is essentially constant in liquid systems, the interstitial fluid velocity v_{inter} can be assumed to be constant. The resulting equation for each solute is

$$\left(1 + \frac{(1-\varepsilon)}{\varepsilon}\rho_p K\right)\frac{\partial c}{\partial t} + v_{inter}\frac{\partial c}{\partial z} - E_{eff}\frac{\partial^2 c}{\partial z^2} = 0 \qquad \textbf{(18-67)}$$

An effective axial dispersion constant E_{eff} has been employed in Eq. (18-67) since it includes the effects of both mass transfer and axial dispersion. Under most experimental conditions mass transfer resistances are important and axial dispersion effects are rather small. If we use the value of E_D predicted from the Chung and Wen correlation Eq. (18-61), the spreading of the wave will be significantly under-predicted. How do we determine the effective axial dispersion coefficient? Dunnebier et al. (1998) compared solutions that included mass transfer and axial dispersion to the results of the Lapidus and Amundson solution. The effective axial dispersion coefficient can be estimated from

$$E_{eff} = E_D + \frac{u_s^2 d_p (K')^2}{6\,k_{m,c}}\frac{(1-\varepsilon)}{\varepsilon} \qquad \textbf{(18-68)}$$

where we have assumed that q and c are in the same units and K' is dimensionless. If q and c are in different units, then appropriate density(s) need to be included with K' [the procedure is similar to that used to derive Eqs. (18-14a) and (18-14b)]. The value of $k_{m,c}$ can be estimated from Eqs. (18-58) to (18-60) since $k_{m,c} = K'k_{m,q}$. For linear systems the Lapidus and Amundson solution with E_{eff} gives *identical* results as solving Eqs. (18-55) and (18-57a).

For a step input from c=0 to c=c_F, the boundary conditions used by Lapidus and Amundson were

$$c = c_F \text{ for } z - 0 \text{ and } t > 0$$
$$c = 0 \text{ for } t = 0 \text{ and } z > 0$$
$$c = 0 \text{ as } z \text{ approaches infinity and } t > 0.$$

The last boundary condition, the infinite column boundary condition, greatly simplifies the solution. It is approximately valid for long columns.

For sufficiently long columns the solution for each solute is

$$c = \frac{c_F}{2} \left\{ 1 - \text{erf} \left[\frac{z - u_s t}{\left[\frac{4E_{eff} u_s t}{v_{inter}} \right]^{1/2}} \right] \right\}$$

$$\text{(18-69)}$$

where u_s is the single-porosity form of the solute velocity, Eq. (18-15b) (included in Problem 18.C5). The term erf is the error function, which is the definite integral

$$\text{erf}[a] = - \text{erf}[-a] = \frac{2}{\pi^{1/2}} \int_0^a \exp(-\zeta^2) d\zeta$$

$$\text{(18-70)}$$

Since the error function is a definite integral, for any value of the argument (the value within the brackets) the error function is a *number*. The values of the error function can be calculated from the normal curve of error available in most handbooks, are tabulated in Wankat (1990), and are available in many computer and calculator packages including Excel. A brief tabulation of values is presented in Table 18-7. Spreadsheets are a convenient method for solving linear problems with the Lapidus and Amunelson solution. These are explored in Problem 18.H1.

Note that when $z = u_s t$, which is the solute movement solution, the argument of the error function is zero, erf is zero, and $c/c_F = \frac{1}{2}$. Thus, the solute movement theory predicts the center of the spreading wave.

Equation (18-69) is the solution for the breakthrough curve for linear isotherms. Example 18-9 will illustrate that the result is an S-shaped curve, which matches experimental

TABLE 18-7.   *Values for error function*

a	erf(a)	a	erf(a)	a	erf(a)	a	erf(a)
0	0.0000	0.6	0.6039	1.10	0.8802	1.60	0.9764
0.1	0.1124	0.7	0.6778	1.1632	0.9000	1.70	0.9838
0.2	0.2227	0.8	0.7421	1.20	0.9103	1.80	0.9891
0.3	0.3286	0.9	0.797	1.30	0.934	2.00	0.9953
0.4	0.4284	0.9063	0.800	1.40	0.9523	2.50	0.9996
0.5	0.5206	1.000	0.8427	1.50	0.9661	3.24	0.99999

results. S-shaped breakthrough curves are also predicted by other solutions for linear sorption systems (e.g., Carta, 1988; Rosen, 1954).

18.7.2 Superposition in Linear Systems

Another characteristic of linear systems is that *superposition* is valid. In other words, solutions can be added and subtracted to give the solution for a combined process. Note that superposition is *not* valid for nonlinear isotherms. This was illustrated in Example 18-7 where two shock waves combined to form a single shock wave. For linear systems the two waves remain separate and form a staircase type arrangement.

We have already employed superposition in some examples. In Example 18-2 we solved a linear chromatography system for the separation of anthracene and naphthalene. This solution was derived by obtaining the solution for anthracene and the solution for naphthalene and then superimposing these two solutions (Figure 18-6). We inherently assumed that the two adsorbates are independent. This type of procedure cannot usually be applied to nonlinear systems since the solutes interact [e.g., as shown by the multicomponent Langmuir isotherm, Eq. (18-8)].

Figure 18-6 also illustrates another aspect of superposition. The first step increase for naphthalene in Figure 18-6 is the breakthrough solution (a feed of concentration c_F is introduced to an initially clean column) using the solute movement theory for a system with a linear isotherm. The solution to a step-down in feed concentration from c_F to zero is the feed concentration minus the breakthrough solution. Although the solutions including mass transfer and/or dispersion are more complicated than the simple solute movement solutions shown in Figure 18-6, the superposition principle remains valid for any linear system.

Consider an adsorption column that initially contains no solute. At $t = t_{feed}$ a feed with a concentration c_F is introduced. The breakthrough solution, which is the behavior of c_{out}, is

$$c_{out,breakthrough} = X_{breakthrough} (z = L, t\text{-}t_{feed}) \qquad (18\text{-}72)$$

where $X_{breakthrough}$ is any solution for breakthrough with linear isotherms including the solution in Figure 18-6 or in Eq. (18-69). Now suppose that we want the solution for elution of a column fully loaded at a fluid concentration of c_F using pure solvent to remove the adsorbate. Based on superposition this solution is the solution for the loaded column $[c (z,t) = c_F]$ minus the breakthrough solution started at time $t = t_{elution}$ (we now have a step down in concentration instead of a step up). Thus, at the outlet $(z = L)$

$$c_{out,elution} = c_F - X_{breakthrough} (z = L, t\text{-}t_{elution}) \qquad (18\text{-}73)$$

If we use the Lapidus and Amundson breakthrough solution, Eq. (18-69), then the elution solution is

$$c_{out,elution} = \frac{c_F}{2} \left\{ 1 + erf \left[\frac{L - u_s (t\text{-}t_{elution})}{\left(\frac{4E_{eff} u_s (t\text{-}t_{elution})}{v_{inter}} \right)^{\frac{1}{2}}} \right] \right\} \qquad (18\text{-}74)$$

The advantage of using superposition is this result is obtained with little effort.

As another example, suppose a pulse of feed is input at $t = t_{start}$ and stopped at $t = t_{end}$ and pure solvent is fed to the column after the pulse is stopped. The solution for this pulse is the breakthrough solution (step-up from $t = t_{start}$) minus a breakthrough solution (step-down from $t = t_{end}$).

$$c_{out,pulse} = X_{breakthrough} (L, t\text{-}t_{start}) - X_{breakthrough} (L, t\text{-}t_{end}) \qquad (18\text{-}75)$$

The period of this pulse is $t_F = t_{end} - t_{start}$. Any solution for breakthrough for linear isotherms, such as Eq. (18-69) can be substituted into Eq. (18-75).

EXAMPLE 18-9. Lapidus and Amundson solution for elution

A column is packed with ion exchange resin in Ca^{+2} form. The column is initially saturated with glucose at a concentration of 10.0 g/L. It is then eluted with pure water starting at $t = 0$ at a velocity, $v_{inter} = 20$ cm/min. Column: $L = 75.0$ cm, $D_{col} = 4.0$ cm. Properties: $\varepsilon_e = 0.39$, $\varepsilon_p = 0$, $E_{eff} = 5.0$ cm^2/min, and $K_{d,i} = 1.0$. Determine the elution curve.

A. Define. Plot c_{out} vs. time.

B. and C. Explore and plan. Note that this problem is not ion exchange, but is a chemical complexing of the glucose with the Ca^{+2} on the resin. Equilibrium data are given in Table 18-2, $q = 0.51c$. For step down (elution) the Lapidus and Amundson solution is given by Eq. (18-74) where u_s is given by Eq. (18-15b) but without the ρ_p term (because q and c are in same units).

D. Do it. The solute velocity is,

$$u_s = \frac{v_{inter}}{1 + \dfrac{(1-\varepsilon_e)}{\varepsilon_e} K_i}$$

$$u_s = \frac{20}{1 + \dfrac{(0.61)(0.51)}{0.39}} = 11.125 \; ^{cm}/_{min}$$

From Eq. (18-74) the argument $a = \dfrac{L - u_s(t - t_{elution})}{\left(\dfrac{4E_{eff} u_s (t - t_{elution})}{v_{inter}} \right)^{1/2}}$ where $t_{elution} = 0$.

Thus, $a = \dfrac{75 - 11.125t}{\left[\dfrac{4(5.0)(11.125)t}{20} \right]^{1/2}} = \dfrac{22.486}{t^{1/2}} - 3.3354 t^{1/2}$ \qquad (A)

Given a Table of Values, the easiest solution method is to pick a value of the argument "a" listed in the table and then calculate both c_{out} and t for this value.

For example, if we select $a = 0$, which is in Table 18-7,

$$c_{out} = \frac{c_F}{2}[1 + erf(0)] = \frac{c_F}{2} = 5.0 \text{ g/L}$$

The equation for time becomes

$$0 = \frac{22.486}{t^{\frac{1}{2}}} - 3.3354t^{\frac{1}{2}} \implies t = 6.74 \text{ min.}$$

For the general case (a ≠ 0) we can solve for t either directly or by multiplying both sides of Eq. (A) by $(t^{1/2})$ and rearranging

$$3.3354(t^{1/2})^2 - a(t^{1/2}) - 22.486 = 0$$

For example, if a = 0.4 [erf (0.4) = 0.4284 is listed in Table 18-7], and we find t = 6.437 min. Then at this time

$$c_{out} = \frac{c_F}{2} [1 + \text{erf } (0.4)] = 5.0[1 + 0.4284] = 7.142 \text{g/L}$$

If a = –0.4, t = 7.060 min, c_{out} = 2.858 g/L.

If this is done for other values of "a", we can generate the table below and the curve shown in Figure 18-21.

a	2	1.4	1.2	0.8	0.4	0	–0.4	–0.8	–1.2	–2
t	5.354	5.736	5.87	6.147	6.437	6.742	7.06	7.39	7.74	8.49
c	9.977	9.762	9.552	8.711	7.142	5	2.858	1.29	0.45	0.02

E. Check. The center of the pattern (c=c_F/2 for this symmetric curve) should occur at the time calculated by the solute movement solution.

$$t_{center} = \frac{L}{u_s} = \frac{75}{11.125} = 6.74 \text{ min.}$$

Thus, the overall mass balance checks.

F. Generalization. The symmetric S-shaped curve shown in Figure 18-21 is characteristic of linear systems. In linear systems, the elution curve (Fig. 18-21)

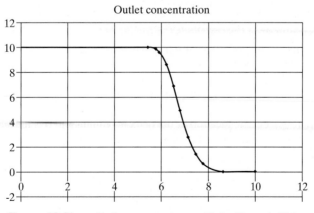

Outlet concentration

FIGURE 18-21.   *Outlet concentration profile for Example 18-9*

and the breakthrough curve (feed to a clean column) are also symmetric. The shape of the breakthrough curve can be seen if the page is flipped over and you look through the backside of the paper. This transformation comes from comparison of the elution solution, Eq. (18-74), with the breakthrough solution, Eq. (18-69). Spreadsheets' solution is also convenient.

18.7.3 Linear Chromatography

In analytical applications of elution chromatography very small feed pulses are used. Thus, concentrations are invariably very low and the isotherms are almost always linear. If the pulse is differential (t_F is very small compared to the time required to elute the components), Eq. (18-75) for the Lapidus and Amundson solution applied to a differential pulse can be simplified to (Wankat, 1990)

$$c_{differ,pulse} = \frac{c_F}{2\pi^{\frac{1}{2}}} u_s t_F \left(\frac{v_{inter}}{z\,E_{eff}}\right)^{\frac{1}{2}} \exp\left[-\left(\frac{v_{inter}}{z\,E_{eff}}\right)\frac{(t - z/u_s)^2\,u_s^2}{4}\right] \tag{18-76}$$

where t is the time after the pulse is fed to the column. Equation (18-76) is the *Gaussian* solution for linear chromatography, which in various forms is extensively used to predict outcomes for analytical chromatography. Outlet concentrations are determined by setting z = L. The maximum outlet concentration of the peak occurs at z = L and t = L/u_s

$$c_{out,differ,pulse,max} = \frac{c_F}{2\pi^{\frac{1}{2}}} u_s t_F \left(\frac{v_{inter}}{L\,E_{eff}}\right)^{\frac{1}{2}} \tag{18-77}$$

The classic paper by Martin and Synge (1941) on liquid-liquid chromatography used an equilibrium-staged model with linear isotherms for the chromatographic column. Comparison of staged solutions with Eq. (18-76) shows that the number of stages N is

$$N = (1/2)(Lv_{inter}/E_{eff}) = (1/2)\,Pe_L \tag{18-78a}$$

where N is also related to L through the height of an equilibrium plate (HETP).

$$L = N(HETP) \tag{18-78b}$$

and Pe_L is the Peclet number based on the column length. These results allow one to calculate the value of N or HETP if E_{eff} is known, or vice-versa calculate E_{eff} if N or HETP is known. This conversion is useful since easy methods to estimate N from chromatographic results are available (see Example 18-10 and homework problems).

The Gaussian solution can be written in shorthand notation as

$$c_{out} = c_{max} \exp(-x^2/2\sigma^2) \tag{18-79}$$

where x is the deviation from the location of the peak maximum and σ is the standard deviation. The terms for x and σ must be in the same units—time, length, or volume. For example in time units

$$x_t = t - z/u_s, \quad \sigma_t = \frac{1}{u_s}\left(\frac{2LE_{eff}}{v_{inter}}\right)^{\frac{1}{2}} = \frac{L}{u_s}\left(\frac{1}{N}\right)^{\frac{1}{2}} = \frac{1}{u_s}\,[L(HETP)]^{\frac{1}{2}} \tag{18-80a}$$

and in length units

$$x_1 = z - u_s t_R, \quad \sigma_1 = \left(\frac{2t_R E_{eff} u_s}{v_{inter}}\right)^{\frac{1}{2}} = \frac{L}{\sqrt{N}} = \sqrt{L(HETP)}$$

(18-80b)

where t_R is the molecule's retention time, $t_r = L/u_s$.

In linear systems the variances (σ^2) from different sources add. This is equivalent to stating that the amount of zone spreading from different sources is additive. Mathematically, this ability to add variances is the reason we can use an effective diffusion coefficient to model a system where mass transfer resistances are important.

Equations (18-79) and (18-80a) can be used to analyze experimental peaks, which are essentially plots of concentration vs. time, to determine the value of u_s and N. From Eq. (18-79) the peak maximum must occur when $x_t = 0$. Since the outlet concentration profile is being measured at x = L, the peak maximum occurs when $t_{max} = t_r = L/u_s$, which is identical to the solute movement result. Thus, u_s and hence an experimental value for the equilibrium constant K′ can be determined from the time the peak maximum exits the column. This procedure is illustrated in Example 18-10.

The value of N can be determined from (Bidlingmeyer and Warren, 1984; Giddings, 1965; Jönsson, 1987; Wankat, 1990),

$$N = (constant)\left(\frac{peak\ maximum}{width}\right)^2$$

(18-81a)

where the peak maximum and width are measured in time units. The value of the constant depends upon what width is used. The easiest derivation uses the width as the distance between intersections of the two tangent lines with the base line, which is 13.4% of the total height. The simplest to use experimentally is the width of the pulse at the half height (c = $c_{max}/2$).

constant = 5.54 for width at half height, constant = 16.0 for tangents   (18-81b)

The standard deviation σ_t of an experimental Gaussian peak can be estimated from,

$$\sigma_t \approx 0.425\ (width\ of\ peak\ at\ half\ height\ in\ time\ units)$$

(18-81c)

There are a number of other methods for determining N and σ_t of Gaussian peaks (Bidlingmeyer and Warren, 1984; Giddings, 1965; Jönsson, 1987; Wankat, 1990) that all give essentially the same results, although the method presented here is less sensitive to peak asymmetry than some of the other methods (Bidlingmeyer and Warren, 1984). The use of these equations is illustrated in Example 18-10.

The purpose of chromatography is to separate different compounds. Separation occurs because compounds travel at different solute velocities. At the same time axial dispersion and mass transfer resistances spread the peaks. If two peak maxima are separated by more than the spreading of the two peaks then they are said to be resolved. As a measure of how well the peaks are separated, chromatographers use *resolution*, defined as (Giddings, 1965; Jönsson, 1987; Wankat, 1990),

$$R = \frac{(t_{R,B} - t_{R,A})}{2(\sigma_{t,A} + \sigma_{t,B})}$$

(18-82)

where $t_{R,A}$ and $t_{R,B}$ are the retention times when the peak maxima exit. When $R = 1.0$, the maxima of the two peaks are separated by $2(\sigma_{t,A} + \sigma_{t,B}) \approx 4\sigma_t$, and there is about a 2 % overlap in the two peaks. An $R = 1.5$ is considered to be complete baseline resolution of the two peaks.

The resolution can also be predicted by substituting in the expressions for retention times and the standard deviations. Assuming that the N values are the same for the two components (a reasonable assumption since resolution is usually calculated for similar compounds) the resulting *fundamental equation of chromatography* (Giddings, 1965; Wankat, 1990) is

$$R = \frac{1}{4} N^{\frac{1}{2}} (u_{s,A} - u_{s,B}) / \bar{u}_s, \qquad \bar{u}_s = (u_{s,A} + u_{s,B}) / 2 \qquad \text{(18-83)}$$

This equation indicates how we can increase resolution if the separation is inadequate. For example, increasing column length, which increases N according to Eq. (18-78b), increases resolution, but only as $L^{1/2}$. More effect can be obtained by changing the equilibrium isotherms (Schoenmakers, 1986) to increase the distance between the two peaks (see Homework Problem 18.B3) or increasing N by decreasing the particle diameter, which decreases H in Eq. (18-78b).

EXAMPLE 18-10. Determination of linear isotherm parameters, N, and resolution for linear chromatography

A chromatogram is run in a preparative chromatographic system to separate acetonaphthalene (AN) from dinitronaphthalene (DN). The results are shown in Figure 18-22. Find K'_{AN}, K'_{DN}, N and the resolution. The K' values should be in units m^3/kg adsorbent = L/g adsorbent.

Data: $L = 50.0$ cm, Solvent flow rate = 100.0 cm^3/min, Pulse time = 0.02 minutes, feed concentration AN = 2.0 g/L, feed concentration DN = 1.0 g/L, Internal diameter of column = 2.0 cm, $\varepsilon_e = 0.4$, $\varepsilon_p = 0.46$, $K_{d,i} = 1.0$, $\rho_s = 2222$ kg/m^3.

Solution

A. Define: Find K'_{AN}, K'_{DN}, N and the Resolution.

B and C. Explore and plan. We can use the peak maxima in Figure 18-22 to find the times for the two peak maxima, $t_{max,i}$. The solute velocities can then be determined as $u_{s,i} = L / t_{max,i}$. This allows us to determine K'_{AN} and K'_{DN} by solving Eq. (18-15a) for the K' values. This equation is

$$K_i' = \frac{\left[\dfrac{v_{inter}}{u_{s,i}} - 1 - \dfrac{(1 - \varepsilon_e)\varepsilon_p K_{d,i}}{\varepsilon_e} \right]}{(1 - \varepsilon_e)(1 - \varepsilon_p)\rho_s / \varepsilon_e} \qquad \text{(18-84)}$$

The value of N for each solute can be calculated from Eqs. (18-81a) and (18-81b). Since the width at the half height is easier to measure, we will use this approach.

The resolution can be determined from Eq. (18-82) using Eq. (18-81c) to estimate the values of σ_{AN} and σ_{DN}.

D. Do it. The values of the peak maximum concentrations and the half height values are given in Figure 18-22.

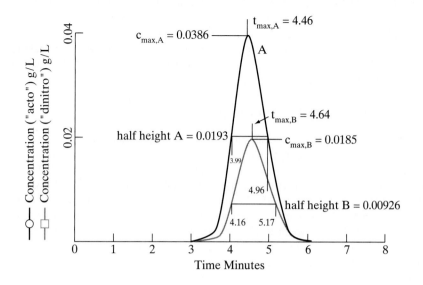

FIGURE 18-22.   *Analysis of Gaussian chromatography peaks for Example 18-10*

The solute velocity is

$u_{s,AN} = L/t_{max,AN}$ = 50.0 cm/ 4.46 min = 11.21 cm/min. The interstitial velocity is,
$v_{inter} = v_{super}/\varepsilon_e$ = volumetric flow rate/$(\pi D^2/4)$ /ε_e
= $(100 \text{ cm}^3/\text{min})/(\pi(2.0 \text{ cm})^2/4)/0.4$ = 79.58 cm/min

Then K'_{AN} can be determined from Eq. (18-84)

$$K'_{AN} = \frac{\dfrac{79.58}{11.21} - 1 - \dfrac{(1-0.4)(0.46)(1.0)}{0.4}}{\dfrac{(1-0.4)(1-0.46)2222}{0.4}} = 0.00301 \text{ m}^3/\text{kg}$$

Combining Eqs. (18-81a,b), we have

$$N_{AN} = (5.54)\left(\frac{\text{peak maximum}}{\text{width at half height}}\right)^2 = 5.54\,[4.46/(4.96-3.99)]^2 = 117$$

From Eq. (18-81c)
$\sigma_{AN} = 0.425\,(4.96 - 3.99) = 0.412$

Similar calculations give $K'_{DN} = 0.00316$, N = 117, and $\sigma_{AN} = 0.429$

The resolution can be calculated from Eq. (18-82)

$$R = \frac{(t_{R,DN} - t_{R,AN})}{2(\sigma_{AN} + \sigma_{DN})} = (4.64 - 4.46)/[2(0.412 + 0.429)] = 0.107$$

E. Check: The values K' of determined are within 2% of the values in the literature, $K'_{AN} = 0.00306$ and $K'_{DN} = 0.00322$.

F. Generalization: This is a very low resolution, which agrees with Figure 18-22 since the peaks are clearly not separated. To obtain better resolution in an analytical system, much smaller particles would probably be used to drastically increase N. An alternative is to use different chromatographic packing that has a higher selectivity. In a preparative system these methods could be employed and the column length would probably be increased.

18.8 LUB APPROACH FOR NONLINEAR SYSTEMS

In general, we cannot obtain analytical solutions of the complete mass and energy balances for nonlinear systems. One exception to this is for isothermal systems when a *constant pattern* wave occurs. Constant pattern waves are concentration waves that do not change shape as they move down the column. They occur when the solute movement analysis predicts a shock wave.

Experimental results (Figure 18-15) and the shock wave analysis showed that the wave shape for constant pattern waves is independent of the distance traveled. This allows us to decouple the analysis into two parts. First, the center of the wave can be determined by analyzing the shock wave with solute movement theory. Second, the partial differential equations for the column mass balance can be simplified to an ordinary differential equation by using a variable $= t - z/u_{sh}$ that defines the deviation from the center of the wave. This approach is detailed in more advanced sources (e.g., Ruthven, 1984; Sherwood et al., 1975; Wankat, 1990).

A simplified analysis procedure called the *length of unused bed* (*LUB*), or mass transfer zone (MTZ), method that uses experimental data to design columns during constant pattern operation is used in industry. This method is based on the work of Michaels (1952). The constant pattern wave inside the bed is shown schematically in Figure 18-23A. After being fully developed, the pattern does not change as it moves through the bed. The width of this pattern (called the length of the MTZ, L_{MTZ}) is usually arbitrarily measured from 0.05 c_F to 0.95 c_F. The reason for not using zero or the feed concentration is it is very difficult to determine exactly when these values are left or attained. During operation, the feed step is stopped at t_{br} when the outlet concentration reaches a predetermined level, usually 5% of the feed concentration. Note that a fraction of the bed is not fully used for adsorption since the feed step was stopped before the bed was fully saturated.

Of course, it is difficult to measure what is happening inside the bed; however, if we run a column to saturation and measure the outlet concentrations we can infer what happened inside the bed. The outlet concentration profile is shown schematically in Figure 18-23B. The width of the MTZ t_{MTZ} (which is again arbitrarily measured from 0.05 c_F to 0.95 c_F) is now easy to measure. The length of the MTZ inside the bed L_{MTZ} can be calculated as,

$$L_{MTZ} = u_{sh}t_{MTZ} \tag{18-85a}$$

The shock wave velocity can be calculated from Eq. (18-34) or from experimental data.

$$u_{sh} = L/t_{center} \tag{18-85b}$$

where t_{center} is the time the center of the pattern, 0.5 ($c_F - c_{initial}$), exits the column.

All of the bed up to the MTZ is fully utilized for adsorption. Within the MTZ the fraction of bed not used is (Area unused bed in MTZ)/(Total area of MTZ). This ratio can be determined from Figures 18-23A, or from the measurements shown in Figure 18-23B. Thus, the frac bed use is

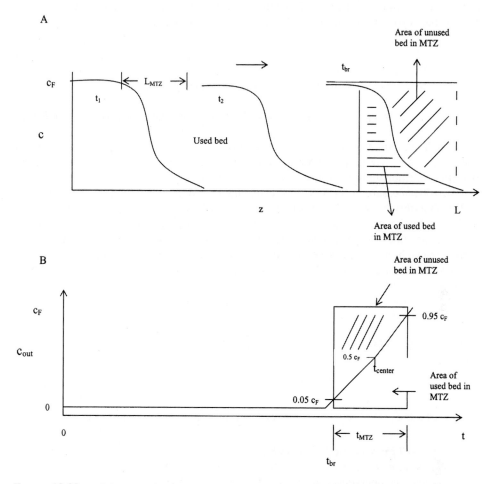

FIGURE 18-23. *Schematic of constant pattern profiles and unused portion of bed; (A) inside column, (B) outlet concentration profile*

$$\text{Frac. bed use} = 1 - \left(\frac{\text{Area unused in MTZ}}{\text{Total Area in MTZ}}\right)\frac{L_{MTZ}}{L} \qquad \textbf{(18-86a)}$$

Measuring the ratio of the areas isn't necessary if the adsorption system produces a symmetric breakthrough curve (e.g., as shown in Figure 18-23B). For symmetric breakthrough curves the ratio of areas is always one half. Thus, for symmetric breakthrough curves the frac bed use is

$$\text{Frac. bed use} = 1 - 0.5\, L_{MTZ}/L \qquad \textbf{(18-86b)}$$

For symmetric breakthrough curves if $L/L_{MTZ} = 1.0$, frac bed use = 0.5; if $L/L_{MTZ} = 2.0$, frac bed use = 0.75; if $L/L_{MTZ} = 3.0$, frac bed use = 0.833; if $L/L_{MTZ} = 4.0$, frac bed use = 0.875, and so forth. The optimal bed length for adsorption is often between two to three times the L_{MTZ}. If a frac bed use is chosen, the column length can be determined.

$$L = 0.5 \, L_{MTZ}/(1 - \text{frac bed use}) \qquad \textbf{(18-86c)}$$

Once the frac bed use is known, we can find the bed capacity.

$$\text{Bed capacity} = (\text{Frac. bed use})(\text{Mass or volume adsorbent})(q_F) \qquad \textbf{(18-86d)}$$

where q_F is the amount adsorbed at the feed concentration in the appropriate units. Of course, q_F can be determined from the equilibrium isotherm for an isothermal adsorber or from the experimentally determined value of u_{sh}.

We often measure the pattern velocity u_{sh} and the width of the MTZ t_{MTZ} experimentally with a laboratory column. This measurement needs to be done with the design values for the initial and final concentrations. It is most convenient if the measurement is done with the same velocity and same particle sizes as in the large-scale unit; however, if pore diffusion controls, we can adjust the results for changes in velocity and particle diameter. For constant pattern systems the width of the MTZ t_{MTZ} is inversely proportional to $k_{m,q}a_p$ (Wankat, 1990),

$$t_{MTZ} \propto \frac{1}{k_{m,q}a_p} \qquad \textbf{(18-87a)}$$

If pore diffusion controls $k_{m,q}a_p$ can be estimated from Eq. (18-58b). The result is,

$$t_{MTZ} \propto \frac{d_p^2}{D_{eff}} \text{ , independent of } v_{inter} \qquad \textbf{(18-87b)}$$

The use of these equations in the LUB analysis is illustrated in Example 18-11.

EXAMPLE 18-11. LUB approach

A laboratory column that is 25.0 cm long is packed with 0.10 cm BPL activated carbon. The operation is at 1.0 atm and 25°C. The column initially contains pure hydrogen. At t = 0 we introduce a feed gas that is 5.0 vol % methane and 95.0% hydrogen. The inlet superficial velocity is 25.0 cm/sec. We measure the outlet wave and find that the center exits at 18.1 s and the width (from 0.05 c_F to 0.95 c_F) is 9.6 s. The breakthrough curve appears to be symmetric. Assume the hydrogen does not adsorb and that pore diffusion controls methane mass transfer.
 a. Determine the shock velocity u_{sh}, L_{MTZ}, and frac bed use in the lab unit.
 b. We now want to design a larger unit with the same particle size. The superficial velocity will be increased to 50.0 cm/s and the frac bed use will be increased to 0.90. Determine the new column length and new breakthrough time (when c = 0.05 c_F).

Solution
Part a. This part is straightforward.
From Eq. (18-85b) the pattern velocity is

$$u_{sh} = L/t_{center} = 25.0 \text{ cm}/18.1 \text{ s} = 1.38 \text{ cm/s}$$

Then from Eq. (18-85a)

$$L_{MTZ} = u_{sh} t_{MTZ} = (1.38 \text{ cm/s})(9.6 \text{ s}) = 13.3 \text{ cm}$$

And for a symmetrical breakthrough curve the frac bed use can be obtained from Eq. (18-86b)

$$\text{Frac. bed use} = 1 - 0.5 \, L_{MTZ}/L = 1 - 0.5 \, (13.3 \text{ cm})/25.0 \text{ cm} = 0.73$$

Part b.
B. and C. Explore and plan. We need to relate L_{MTZ} to velocity. Starting with Eq. (18-85a), we can substitute in Eqs. (18-85c) and (18-87b) to obtain

$$L_{MTZ} \propto \frac{d_p^2 v_{super}}{D_{eff}} \tag{18-88}$$

The effective pore diffusion controls D_{eff} is not a function of velocity. Taking the ratio of the flow rates, we can find both u_{sh} and L_{MTZ} for the large-scale unit. Since the desired frac bed use is known, Eq. (18-86c) can be solved for length L.

The breakthrough time can be calculated from the time the center exits and t_{MTZ}. Referring to Figure 18-23B, for a symmetric breakthrough curve,

$$t_{br} = t_{center} - 0.5 \, t_{MTZ} \tag{18-89}$$

The required values for the large-scale system can now all be calculated.
D. Do it. The values of u_{sh} and L_{MTZ} for the large-scale unit are obtained from the values of the laboratory unit by multiplying the lab scale values by the ratio of velocities.

$$u_{sh,large\ scale} = u_{sh,lab} \, (v_{super,large\ scale}/v_{super,lab}) = (1.38 \text{ cm/s})(50/25) = 2.76 \text{ cm/s}$$

$$L_{MTZ,large\ scale} = L_{MTZ,lab} \, (v_{super,large\ scale}/v_{super,lab}) = (13.3 \text{ cm})(50/25) = 26.6 \text{ cm}$$

From Eq. (18-85c),

$$L = 0.5 \, L_{MTZ}/(1 - \text{frac bed use}) = (0.5) \, (26.6 \text{ cm})/(1 - 0.9) = 133.0 \text{ cm}$$

Equation (18-89) can be used to find the breakthrough time by using Eq. (18-85a) to solve for t_{MTZ} and (18-84b) to solve for t_{center}.

$$t_{MTZ} = L_{MTZ} / u_{sh} = (26.6 \text{ cm})/(2.76 \text{ cm/s}) = 9.6 \text{ s (unchanged)}$$

$$t_{center} = L/ u_{sh} = (133.0 \text{ cm})/(2.76 \text{ cm/s}) = 48.2 \text{ s}$$

$$t_{br} = t_{center} - 0.5 \, t_{MTZ} = 48.2 - (0.5)(9.6) = 43.4 \text{ s}$$

E. Check. As expected, both L_{MTZ} and the value of t_{center} scale proportionally to the velocity ratio. The values for L and t_{br} are difficult to check independently, but the values are reasonable.

F. Generalization. If the frac bed use had been kept constant and only the velocity was changed, the column length would double in the large-scale system (this comes from Eq. (18-86c) since L_{MTZ} doubles). The large increase in the required bed length is mainly caused by increasing the frac bed use in the large-scale column. Since t_{MTZ} is independent of velocity when pore diffusion controls, there was no change in the value of t_{MTZ}; however, the breakthrough time did change, but not proportionally to the velocity change.

The LUB approach is used for the adsorption step. During desorption a proportional pattern (diffuse) wave usually results as shown in Figure 18-15B (monovalent-divalent ion exchange can be an exception to this—see Example 18-8). Since the shape of the pattern changes with length, the LUB approach cannot be used for desorption. However, the results of the solute movement theory for diffuse waves are often quite accurate. Thus, the diffuse wave predictions can be used for preliminary design. The desorption step should be checked with experimental data.

18.9 CHECKLIST FOR PRACTICAL DESIGN AND OPERATION

There are always practical considerations in the design of separation systems that may not be obvious based on the theories. Since the practical considerations for adsorption, chromatography, and ion exchange are very different than for the other separations considered in this book, they have been collected here.

1. Broadly speaking, the sorption (feed) step makes money, and the regeneration step costs money. The optimum sorbent is often a trade-off between these two steps (relatively strong sorption to process the feed, but not so strong that regeneration is not feasible).
2. Regeneration (broadly defined) is always a key cost and is often the controlling cost. These costs are: TSA—energy for heating, PSA—energy for compression, Chromatography and SMB—solvent or desorbent recovery (ultimately energy if distillation or evaporation are used), IEX—regeneration chemicals.
3. Sorption methods of separation always compete with other separation methods such as distillation or gas permeation. Thus, always need to consider alternatives.
4. Compression costs are significant in large-scale adsorption systems for gases; thus, pressure drop is critically important since it controls compression costs. A resonable gas velocity is 100 ft/min (Seider et al., 2009). Pressure drop can be calculated from well-known correlations for flow in packed beds. For example, the pressure drop in a packed bed of rigid particle in laminar flow is (Bird et al., 1960)

$$\Delta p = \frac{150 \mu L (1-\varepsilon_e)^2 v_{superficial}}{\varepsilon_e^3 d_p^2}$$

(18-90)

5. Pressure drop is also important in liquid systems but for different reasons than in gas systems. Because liquids are almost incompressible, obtaining high pressures, and thus from Eq. (18-90) high velocities and high throughputs ($A_c \rho_f v_{superficial}$) is easy and inexpensive. High throughput reduces the cross-sectional area and hence the cost of the

adsorber. A common industrial design procedure is to operate at the highest maximum pressure that the equipment is designed for. This is the highest allowable Δp and hence the highest possible throughput. Of course, the increase in length of the MTZ has to be accounted for in the design.

6. Theories all assume that the column is well packed. If it isn't, it won't work well. Want $D_{col}/d_p > \sim 30$. This limit is often important in lab columns.

   Special equipment is needed to pack small particles. Packing large diameter columns with small particles needs to be done by experts.

   If a wet column is allowed to dry out, the packing will often crack, which will cause channeling.

   Packed beds are efficient depth filters. To prevent clogging, feeds containing particulates must be filtered. It is common, particularly in IEX, to use upward flow wash steps to remove particulates from the column.

   Movement of the bed will cause attrition of brittle packings such as activated carbon and zeolites that will result in fines that can clog the bed or frits. Use a hold down plate, frit, or net to help prevent attrition.

   Soft packing materials (e.g., Sephadex and agarose) require different packing procedures than rigid packings. The swelling and contracting of polymer packings, particularly ion-exchange resins, must be designed for.

7. Simulations and other solutions to the theories can only include phenomena that were built into the model. Experiments are usually needed to find the unexpected.

8. If the fluid velocity is high and the mass transfer rate is low, there may not be enough residence time for much of the solute to diffuse into the sorbent. This solute, which *bypasses* the packing, does not undergo separation and exits at the feed concentration. If separation problems are observed, try reducing the fluid velocity by one or more orders of magnitude. A reasonable liquid velocity is 1.0 ft/min (Seider et al., 2009).

9. Many adsorbents, particularly activated carbon, show a very high initial adsorption capacity. After regeneration, this capacity is not fully regained. When testing adsorbents do extensive cleaning and/or washing first, and then do several complete cycles. Do not use initial results for design of cyclic processes.

10. Slow decay of adsorbents due to irreversible adsorption of trace components or thermal deactivation of active sites is also common. When this occurs, operating conditions must be adjusted accordingly. Because of this poisoning, adsorption processes, which use surface phenomena, are often much more sensitive to trace chemicals than distillation and other separation techniques that rely on bulk properties. An occasional wash step or extreme regeneration step may be needed. A short life for the sorbent, which can be a problem in biological operations, often makes the process uneconomical. Long-term pilot plant tests with the actual feed from the plant are useful to determine the seriousness of these problems.

11. The surface properties and surface morphology of sorbents is critically important. Thus, different activated carbons are different adsorbents and are not interchangeable. Different batches of what is supposed to be the same sorbent may differ significantly. Thus, batches must be sampled and tested before being used on a large scale.

12. Pressure and flow spikes can be very detrimental if they cause the bed to shift since this can result in channeling or attrition. Unfortunately, these spikes naturally occur when concentrated feeds are adsorbed. They are greatly reduced if the concentrated feed is introduced in steps (e.g., go from 0% to 45% and then to 90% in two steps instead of a single step from 0% to 90%).

13. Temperature increases must be controlled. Adsorbates may be thermally sensitive and some adsorbents such as activated carbon readily burn. Hot adsorbents are also more likely to catalyze unwanted reactions.

14. Beds in series are often treated as if they were equivalent to a single long bed. However, their transient behavior is different and depends on the connecting pipes and valves.

15. For safely reasons, personnel must always wear respirators when entering chromatography or adsorption columns. Many adsorbents adsorb oxygen, and others may desorb toxic gases. Strong acid and strong base ion exchange resins are essentially solid forms of acid and base. Thus, they can cause chemical burns, particularly in the eyes. The use of eye protection and normal protective clothing is recommended (Shuey, 1990). In addition, spills of any adsorbent or ion exchange resin should be cleaned up immediately because they "can act as miniature ball bearings" (Shuey, 1990, p. 278) and cause falls. The resin expansion can also fracture vessels if sufficient room for expansion is not available.

16. Ion exchange columns need to be backwashed periodically to remove solids and fines from the resin and to relieve pressure built up by periodic contraction and expansion. Backwashing will reduce the pressure drop and extend the life of the resin (Shuey, 1990).

18.10 SUMMARY—OBJECTIVES

In this chapter the basic concepts for adsorption, chromatography, and ion exchange separations were developed. At the end of this chapter, you should be able to satisfy the following objectives:

1. Determine equilibrium constants from data and use the equilibrium equations in calculations
2. Explain in your own language how the different sorption processes (e.g., elution chromatography, adsorption with thermal regeneration, PSA, SMB, monovalent-monovalent ion exchange, and water softening) work
3. Explain the meaning of each term in the development of the solute movement equations and use this theory for both linear and nonlinear isotherms to predict the outlet concentration and temperature profiles for a variety of different operations including elution chromatography, adsorption with thermal regeneration, PSA, SMB, and ion exchange
4. Explain the meaning of each term in the column mass and energy balances, and in the mass and heat transfer equations
5. Use the Lapidus and Amundson solution plus superposition to determine the outlet concentration profiles for linear adsorption and chromatography problems
6. Use the theory of linear chromatography with very small pulses to analyze chromatography systems
7. Use the LUB theory in combination with experimental data to design columns

REFERENCES

Agosto, M., N.-H. L. Wang and P. C. Wankat, "Moving-Withdrawal Liquid Chromatography of Amino Acids," *Ind. Eng. Chem. Research*, *28*, 1358 (1989).

Alexandratos, S. D., "Ion-Exchange Resins: A Retrospective from *Industrial and Engineering Chemistry Research*," *Ind. Eng. Chem. Research*, *48*, 388 (2009).

Anderson, R. E., "Ion-exchange separations," in P. A. Schweitzer (Ed.), *Handbook of Separation Techniques for Chemical Engineers*, 3rd ed., McGraw-Hill, NY, 1997, Section 1.12.

Basmadjian, D., "The Adsorption Drying of Gases and Liquids," in A. S. Mujumdar (Ed.), *Advances in Drying,* Vol. 3, Hemisphere Pub. Co., Washington, DC, 1984, Chap. 8.

Basmadjian, D., *The Little Adsorption Book: A Practical Guide for Engineers and Scientists,* CRC Press, Boca Raton, FL, 1997.

Bidlingmeyer, B. A., and F. V. Warren, Jr., "Column Efficiency Measurement," *Analytical Chemistry, 56,* 1583A (1984).

Bird, R. B., W. J, Stewart, and E. N. Lightfoot, *Transport Phenomena,* Wiley, New York, 1960.

Bonnerjea, J., and P. Terras, "Chromatography Systems," in Lydersen, B. K., N. A. D'Elia, and K. L. Nelson (Eds.), *Bioprocess Engineering: Systems, Equipment and Facilities,* Wiley, New York, 1994, pp. 159–186.

Bonsal, R. C., J.-B. Donnet, and F. Stoeckli, *Active Carbon,* Marcel Dekker, New York, 1988.

Broughton, D. B., and C. G. Gerhold, U.S. Patent 2,985,589 (May 23, 1961).

Broughton, D. B., R. W. Neuzil, J. M. Pharis, and C. S. Brearley, "The Parex Process for Recovering Paraxylene," *Chem. Eng. Prog., 66,* (9), 70 (1970).

Carta, G., "Exact Analytic Solution of a Mathematical Model for Chromatographic Operations" *Chem. Engr. Sci., 43,* 2877 (1988).

Cazes, J., *Encyclopedia of Chromatography,* 2nd ed., CRC Press, Boca Raton, FL, 2005.

Chin, C. Y., and N.-H. L. Wang, "Simulated Moving Bed Equipment Designs," *Sep. Purif. Rev. 33,* 77 (2004).

Ching, C. B., and D. M. Ruthven, *AIChE Symp. Ser., 81,* (242), 1 (1985).

Chung, S. F., and C. Y. Wen, "Longitudinal Dispersion of Liquid Flowing through Fixed and Fluidized Beds," *AIChE Journal, 14,* 857 (1968).

Cramer, S. M., and G. Subramanian, "Recent Advances in the Theory and Practice of Displacement Chromatography," *Separation and Purification Methods, 19,* 31–91 (1990).

Dechow, F. J., *Separation and Purification Techniques in Biotechnology,* Noyes, Park Ridge, NJ, 1989.

Do, D. D., *Adsorption Analysis: Equilibria and Kinetics,* Imperial College Press, London, 1998.

Dobbs, R. A., and J. A. Cohen, *Carbon Adsorption Isotherms for Toxic Organics,* EPA Cincinnati, OH, 45268, EPA-600/8-80-023, April 1980.

Dorfner, K., *Ion Exchangers,* de Gruyter, New York, NY, 1991.

Dunnebier, G., I. Weirich, and K. U. Klatt, "Computationally Efficient Dynamic Modeling and Simulation of Simulated Moving Bed Chromatographic Processes with Linear Isotherms," *Chem. Engr. Sci., 53,* 2537 (1998).

Faust, S. D., and D. M. Aly, *Adsorption Processes for Water Treatment,* Butterworth, Boston, 1987.

Fulker, R. D., "Adsorption," in G. Nonhebel (Ed.), *Processes for Air Pollution Control,* CRC Press, Boca Raton, FL, 1972, Chap. 9.

Giddings, J. C., *Dynamics of Chromatography, Part I, Principles and Theory,* Marcel Dekker, New York, 1965.

Helfferich, F., *Ion Exchange,* McGraw-Hill, New York, 1962.

Humphrey, J. L., and G. E. Keller II, *Separation Process Technology,* McGraw-Hill, New York, 1997.

Jönsson, J. Å. (Ed.), *Chromatographic Theory and Basic Principles,* Marcel Dekker, New York, 1987.

Kim, J.-K., N. Abunasser, P. C. Wankat, A. Stawarz, and Y.-M. Koo, "Thermally Assisted Simulated Moving Bed Systems," *Adsorption, 11,* 579–584 (2005).

Kumar, R., "Vacuum Swing Adsorption Process for Oxygen Production—A Historical Perspective," *Separ. Sci. Technol., 31,* 877–893 (1996).

Ladisch, M. R., *Bioseparations Engineering—Principles, Practice, and Economics,* Wiley-Interscience, New York, 2001.

Lee, M. N. Y., "Novel Separation with Molecular Sieve Adsorption," in N. N. Li (Ed.), *Recent Developments in Separation Science,* Vol I, CRC Press, Boca Raton, FL, 1972, p. 75.

LeVan, M. D., G. Carta, and C. M. Yon, "Adsorption and Ion Exchange," in D. W. Green (Ed.), *Perry's Chemical Engineers' Handbook*, 7th ed., McGraw-Hill, New York, 1997, Section 16.

LeVan, M. D., and T. Vermeulen, "Binary Langmuir and Freundlich Isotherms for Ideal Adsorbed Solutions," *J. Phys. Chem., 85,* 3247 (1981).

Martin, A. J. P., and R. L. M. Synge, "A New Form of Chromatogram Employing Two Liquid Phases," *Biochem. J., 35,* 1358 (1941).

Matz, M. J., and K. S. Knaebel, "Recycled Thermal Swing Adsorption: Applied to Separation of Binary and Ternary Mixtures," *Ind. Engr. Chem. Research, 30,* 1046 (1991).

Michaels, A. S., "Simplified Method of Interpreting Kinetic Data in Fixed Bed Ion Exchange," *Ind. Engr. Chem., 44,* 1922 (1952).

Natarajan, G., and P. C. Wankat, "Thermal-Adsorptive Concentration," *Adsorption, 9,* 67 (2003).

Perry, R. H., and D. W. Green (Eds.), *Perry's Chemical Engineers' Handbook,* 7th ed., McGraw-Hill, New York, 1997.

Rathore, A. S., and A. Velayudhan, *Scale-Up and Optimization in Preparative Chromatography: Principles and Biopharmaceutical Applications,* CRC Press, Boca Raton, FL, 2004.

Reynolds, J. P., J. S. Jeris, and L. Theodore, *Handbook of Chemical and Environmental Engineering Calculations,* Wiley-Interscience, New York, 2002.

Rosen, J. B., "General Numerical Solution for Solid Diffusion in Fixed Beds," *Ind. Engr. Chem., 46,* 1590 (1954).

Ruthven, D. M., *Principles of Adsorption and Adsorption Processes,* Wiley-Interscience, New York, 1984.

Ruthven, D. M., S. Farooq, and K. S. Knaebel, *Pressure Swing Adsorption,* VCH Publishers, Inc., New York, 1994.

Schiesser, W. E., *The Numerical Method of Lines: Integration of Partial Differential Equations,* Academic Press, San Diego, 1991.

Schoenmakers, P. J., *Optimization of Chromatographic Selectivity: A Guide to Method Development,* Elsevier, Amsterdam, 1986.

Seader, J. D., and E. J. Henley, *Separation Process Principles,* 2nd ed., Wiley, New York, 1998.

Seider, W. D., J. D. Seader, D. R. Lewis, and S. Widago, *Product & Process Design Principles,* 3rd ed, Wiley, New York, 2009.

Sherman, J. D., "Synthetic Zeolites and Other Microporous Oxide Molecular Sieves," *Proc. Natl. Acad. Sci. USA, 96,* 3471–3478 (1999).

Sherwood, T. K., R. L. Pigford, and C. R. Wilke, *Mass Transfer,* McGraw-Hill, New York, 1975, Chapter 10.

Shuey, C. O., "Ion-Exchange Processes," in *Separation Processes in Biotechnology,* Asenjo, J. A. (Ed.), Dekker, New York, 1990.

Thomas, H. C., "Chromatography: A Problem in Kinetics," *Ann. N.Y. Acad. Sci., 49,* 161 (1948).

Tondeur, D., and M. Bailly, "Design Methods for Ion Exchange Processes Based on the Equilibrium Theory," in A. Rodrigues (Ed.), *Ion Exchange: Science and Technology,* Martinus Nijhoff Publishers BV, Dordrecht, The Netherlands, 1986, 147–198.

Tondeur, D. and P. C. Wankat, "Gas Purification by Pressure Swing Adsorption," *Separation and Purification Methods, 14,* 157–212 (1985).

Valenzuela, D., and A. L. Myers, "Gas Adsorption Equilibria," *Separ. Purific. Methods, 13,* 153–183 (1984).

Valenzuela, D., and A. L. Myers, *Adsorption Equilibrium Data Handbook,* Prentice Hall, Englewood Cliffs, NJ, 1989.

Wakao N., and T. Funazkri, "Effect of Fluid Dispersion Coefficients on Particle-to-fluid Mass Transfer Coefficients in Packed Beds," *Chem. Engr. Sci., 33,* 1375 (1978).

Wankat, P. C., *Large-Scale Adsorption and Chromatography,* CRC Press, Boca Raton, FL, 1986.

Wankat, P. C., *Rate-Controlled Separations,* Kluwer, Amsterdam, 1990.

Wankat, P. C., "Using a Commercial Simulator to Teach Sorption Separations," *Chem. Engr Educ., 40* (3), 165–172 (2006).

Weyde and Wicke, *Kolloid Z., 90,* 156 (1940).

White, D. H., Jr., "Compressed Air and Gas Purification and Fractionation for High Purity Applications by Improved PSA Processes," *Separ. Sci. Technol., 43,* 2298–2306 (July 2008).

White, D. H., Jr., and P. G. Barkley, "The Design of Pressure Swing Adsorption Systems," *Chem. Engr. Progress, 85,* (1), 25–33 (January 1989).

Wu, C.-S., *Handbook of Size Exclusion Chromatography and Related Techniques,* 2nd ed., CRC Press, Boca Raton, FL, 2004.

Yang, R. T., *Gas Separation by Adsorption Processes,* Butterworth, Boston, 1987.

Yang, R. T., *Adsorbents: Fundamentals and Applications,* Wiley-Interscience, New York, 2003.

Yin, C. Y., M. K. Aroua, and W. M. A. W. Daud, "Review of Modifications of Activated Carbon for Enhancing Contaminant Uptakes from Aqueous Solution," *Separation Purification Techniques, 52,* 403–415 (2007).

HOMEWORK

A. *Discussion Problems*

A1. Feed Step. Column is initially clean ($c_{init}=0$). Feed at concentration $c_F>0$.
   1. If solute follows a linear isotherm, the resulting solute wave will be
      a) diffuse wave, b) shock wave, c) simple wave.
   2. If solute follows a favorable isotherm, the resulting solute wave will be
      a) diffuse wave, b) shock wave, c) simple wave.
   3. If solute follows an unfavorable isotherm, the resulting solute wave will be
      a) diffuse wave, b) shock wave, c) simple wave.

A2. Elution. Column initially is saturated with solute at concentration $c_{init} > 0$. The feed to the column is $c_F = 0$.
   1. If solute follows a linear isotherm, the resulting solute wave will be
      a) diffuse wave, b) shock wave, c) simple wave.
   2. If solute follows a favorable isotherm, the resulting solute wave will be
      a) diffuse wave, b) shock wave, c) simple wave.
   3. If solute follows an unfavorable isotherm, the resulting solute wave will be
      a) diffuse wave, b) shock wave, c) simple wave.

A3. Sorption separations are not always used in industry in cases where they would be most economical. What are some of the non-economic barriers to more use of sorption separations?

A4. Explain why an adsorption isotherm that is too steep may not work well in a PSA process.

A5. Briefly explain why the SMB system is much more efficient (i.e., uses less solvent and less adsorbent) than an elution chromatograph doing the same binary separation. Assume that both systems are operating in the migration mode using isocratic elution. Both systems are optimized. The elution chromatograph uses repeated pulses of feed.

A6. There are a number of important industrial separations such as separation of meta and para xylene that can be operated as either gas or liquid. All of the current commercial adsorption applications are operated as liquids. What are the advantages of operation as a liquid as compared to operation as a gas?

A7. In an SMB suppose the A product purity is OK, but the B product purity is too low. We can increase the B product purity while maintaining the A product purity by,
a) Increasing all M_i, b) Decreasing all M_i, c) Increasing M_4 and increasing M_3, d) Decreasing M_1 and increasing M_3, e) Increasing M_4 and decreasing M_2, f) Decreasing M_4 and decreasing M_2.

A8. Why doesn't the LUB approach work for linear systems?

A9. We desire to separate a dilute ternary mixture consisting of A, B, and C dissolved in D (A is least strongly adsorbed component and C is most strongly adsorbed component) in a normal SMB (modify Figure 18-14b for the ternary separation). One product is A (plus D) while the other product is B + C (plus D). The four SMB design Eqs. (18-29) are:
a) unchanged, b) the same except Eq. (18-29c) becomes $u_{c,2} = M_{2c} u_{port}$ ($M_{2c} \le 1.0$), c) the same except Eq. (18-29c) becomes $u_{B,3} = M_{3B} u_{port}$ ($M_{3B} \ge 1.0$), d) the same except Eq. (18-29d) becomes $u_{C,4} = M_{4C} u_{port}$ ($M_{4C} > 1.0$), e) the same except Eq. (18-29b) is changed as in answer b *and* Eq. (18-29d) is changed as in answer d.

A10. For a differential pulse with a linear system you have calculated the standard deviation of the Gaussian peak σ_t in time units. You now want to calculate σ_l inside the column. The formula you would use is:
a) $\sigma_l = \sigma_t$, b) $\sigma_l = \sigma_t/v_{inter}$, c) $\sigma_l = \sigma_t(v_{inter})$, d) $\sigma_l = \sigma_t/u_s$, e) $\sigma_l = \sigma_t(u_s)$.

A11. Many variations of the basic Skarstrom cycle shown in Figure 18-11 have been developed. One variation is to include a partial co-flow blowdown step before the counter-flow blowdown shown in Figure 18-11. What are the likely advantages and disadvantages of doing this?

B. *Generation of Alternatives*

B1. Brainstorm some possible alternative adsorbents made from common agricultural and/or forest products or wastes.

B2. The PSA cycle used in Example 18-4 does not produce pure hydrogen throughout the entire feed step. Brainstorm what can be done to change the cycle so that it will produce pure hydrogen.

B3. Separation in chromatography is often limited because the selectivity is too close to 1.0. For both liquid and gas systems, list as many ways as possible that the equilibrium constants can be changed.

B4. What can be done to develop more economical sorption separations?

C. *Derivations*

C1. A molecular sieve zeolite adsorbent consists of pellets that are agglomerates of zeolite crystals with density $\rho_{crystal}$ scattered in a continuous phase of clay binder with a density ρ_{clay}. In this case, there is an interpellet porosity ε_e (between pellets—this is normal ε_e) an intercrystal porosity ε_{p1}, (which is the porosity in the binder), and an intracrystal porosity ε_{p2}, (inside the crystals). If the fraction of the particle volume that is crystals (including porosity within the crystals) is f_{cry}, derive formulas for the total porosity, $V_{available}$, and the particle and bulk densities.

C2. Derive the value for the limiting conditions in the Langmuir isotherm, Eq. (18-6a), when c_A becomes very large ($K_{A,c} c_A \gg 1.0$).

C3. Use the mass action expression to prove Eqs. (18-40a) and (18-41).

C4. Derive Eq. (18-44).

C5. Derive the equation for the solute velocity for linear isotherms for systems using a single-porosity model. The result obtained should be,

$$ u_{s,i} = \frac{v_{inter}}{1 + \dfrac{(1 - \varepsilon_e)}{\varepsilon_e} \rho_p K'_i} \qquad \textbf{(18-15b)} $$

C6. Derive the appropriate mass balance equation and solutions equivalent to Eqs. (18-22) to (18-24) but for the case where $u_s(T_{hot}) > u_s(T_{cold}) > u_{th}$. Hint: Start by redrawing the differential control volume in Figure 18-7B noting that the concentration, amount adsorbed and temperature above c_2, q_2, and T_2 are c_{change}, q_{change} and T_1, where c_{change} is the concentration caused by the temperature change, which moves *ahead* of the thermal wave.

C7. Derive an equation for t_{MTZ} for a step input for a system with a linear isotherm using the Lapidus and Amundson solution. The time for the MTZ is defined as the time required to go from $0.05 (c_{feed} - c_{initial})$ to $0.95 (c_{feed} - c_{initial})$.

Show that t_{MTZ} is approximately proportional to $L^{1/2}$.

Note: As far as I know the solution for t_{MTZ} for a breakthrough curve is not available in any books. The ability to derive this sort of solution (starting with a known solution and obtaining the equation for a specific case) will be very valuable in industry and will help you stand out from most engineers.

C8. Show that if the entire column is heated (or cooled) simultaneously (known as direct heating) Eq. (18-24) simplifies to,

$$ c(T_2)/c(T_1) = u(T_2)/u(T_1) \qquad \textbf{(18-24 direct heating)} $$

Explain why direct heating is practical for laboratory-sized columns (small diameter) but not for commercial units with large diameters. (Hint: consider the heat transfer characteristics.)

C9. For divalent-monovalent exchange, show that when a total ion wave passes through a segment of the column, $y_{i,after} = y_{i,before}$. Hint: Use the balance envelope in Figure 18-7B and Eq. (18-22), but with $\Delta z / \Delta t = u_{total\ ion} = v_{inter}$.

C10. We want to operate an SMB (Figure 18-14B) doing a binary separation at $(D/F)_{min}$. Assume that the volumetric feed rate F and the cross sectional area of the columns A_c are specified. Derive the equations for u_{port}, v_1, v_2, v_3, v_4, $v_{A,product}$, $v_{B,product}$, v_D, and $D/F = v_D / v_{Feed}$ that will satisfy Eqs. (18-29) and (18-30) using Eq. (18-15e). Show that Eq. (18-31a) is correct. After simplification you should obtain the following result for D/F,

$$D/F = \left[\frac{M_4}{C_B} - \frac{M_1}{C_A}\right] \Big/ \left[\frac{M_2}{C_B} - \frac{M_3}{C_A}\right] \qquad \textbf{(18-31b)}$$

Show that this result becomes D/F = 1.0 when all the M_i = 1.0.

Hint: Start with the equations for zones 2 and 3 and $v_2 = v_3 + v_{Feed}$, and solve for u_{port}, v_2 and v_3.

C11. Derive the mass action equilibrium expression for trivalent-trivalent ion exchange and derive the derivative of y_{tri} with respect to x_{tri}.

D. *Problems*

Answers to problems with an asterisk are at the back of the book.

D1.* Find the values of K_A and q_{max} for methane adsorption on PCB activated carbon at 296 and 480 K (Table 18-3).

D2.* The adsorption of anthracene from cyclohexane on activated alumina follows a Langmuir isotherm,

q = 22c /(1 + 375 c) where c is in mol/L and q is in mol/kg. (Thomas, 1948)

Convert this to the form of Eq. (18-6a) in terms of q_{max} and K_{Ac} in units of g/L for c and g/kg for q. The range of validity should be from c = 0.0 to 0.012 mol/L.

Data: density of cyclohexane = 0.78 kg/L, molecular weight cyclohexane = 84, molecular weight anthracene = 178.22, ρ_p = 1.47 kg/L, ε_e = 0.4, ε_p = 0.0.

D3. We are separating acetonaphthalene (AN) from dinitronaphthalene (DN). The linear equilibrium parameters and column parameters are:

$K'_{AN} = 0.00301, K'_{DN} = 0.00316, \varepsilon_e = 0.40, \varepsilon_p = 0.46, K_d = 1.0, v_{super} = 100 \text{cm}/\text{min}, \rho_s = 2200 \text{kg}/\text{m}^3, L = 25 \text{cm}.$

K' are in units L/g adsorbent. For a very small pulse of feed, we desire a resolution between the two peaks of R = 0.97.

a. At what time does the AN peak exit the column?

b. At what time does the DN peak exit the column?

c. What values of N and HETP are required to achieve the desired resolution?

d. What is the width of the AN peak at half height and σ_t in minutes?

D4. We wish to use a thermal swing adsorption process to remove traces of toluene from n-heptane using silica gel as adsorbent. The adsorber operates at 1.0 atm. The feed is 0.0011 wt frac toluene and 0.9989 wt frac n-heptane at 0°C. Superficial velocity of the feed is 10.0 cm/min. The absorber is 2.0 meters long and during the feed step is at 0°C. The feed step is continued until breakthrough occurs. To regenerate, use counterflow of pure n-heptane at 80°C. Superficial velocity during purge is 10.0 cm/min. Column is cooled to 0°C before the next feed step.

Data: At low concentrations isotherms for toluene: q = 17.46x @ 0°C, q = 1.23x @ 80°C, q and x are in g solute/g adsorbent and g solute/g fluid, respectively (Matz and Knaebel, 1991). ρ_s = 2100 kg/m³, ρ_f = 684 kg/m³, C_{ps} = 2000 J/(kg °C), C_{pf} = 1841 J/(kg °C), ε_e = 0.43, ε_p = 0.48, K_d = 1.0. Note: Use Eq. (18-15c) for solute velocities. Assume that wall heat capacities can be ignored, heat of adsorption is negligible, no adsorption of n-heptane.

Using the solute movement theory,

a) determine the breakthrough time for toluene during the feed step.

b) determine time for thermal wave to break through.

c) determine time to remove all toluene from column.

d) determine the outlet concentration profile of the regeneration fluid.

D5. We have a column packed with a resin that immobilizes a liquid stationary phase. The column is initially clean, $c_A = 0$. At time $t = 0$, we input a feed that is $c_{A,feed} = 1.5$ g/L. The superficial velocity is 20 cm/min. The column is 50 cm long. The packing has $\varepsilon_e = 0.4$, $\varepsilon_p = 0.54$, $K_d = 1.0$, $\rho_s = 1.124$ kg/L, and the equilibrium for component A is an unfavorable isotherm,

$$q = 1.2\, c_A / (1 - 0.46\, c_A) \text{ where q is in g/kg and } c_A \text{ is in g/L.}$$

Use solute movement theory to predict the outlet concentration profile of A (c_{out} vs. time). (You can report this as a graph or as a table or as both.)

D6. We are adsorbing p-xylene from n-heptane on silica gel. The n-heptane does not adsorb, and the p-xylene follows a linear isotherm with Arrhenius temperature dependence. The calculated values of the equilibrium are,

$$q = 12.1090\, c \text{ at } 300 \text{ K and } q = 4.423\, c \text{ at } 350 \text{ K where q and c are both in g xylene/L.}$$

The column is initially clean ($c = q = 0$) and is at 300 K. At $t = 0$ we input a feed that contains 0.010 g/L p-xylene at 300K. At $t = 20$ min, we input a clean solvent ($c = 0$) at 350K.

The interstitial velocity is constant throughout at $v_{inter} = 30$ cm/min. $L = 50$ cm, $\varepsilon_e = 0.43$, $\varepsilon_p = 0.50$, $K_d = 1.0$, $\rho_s = 1.80$ kg/L, $\rho_f = 0.684$ kg/L, $C_{p,s} = 920$ J/kg/K, $C_{p,f} = 2240$ J/kg/K, and heat storage in the wall can be neglected. The column is adiabatic.

a. Find u_s (T = 300 K), u_s (T = 350 K), and u_{th}.

b. Find the breakthrough time for the thermal wave.

c. Predict the outlet concentration profile (c_{out} vs. time).

D7.* A 60.0 cm long column contains activated carbon. The column is initially clean. At $t = 0$ we start feeding the column in an upwards direction a dilute aqueous solution of acetic acid at 4°C. The feed concentration is 0.01 kmol/m³. The superficial velocity of the feed is 15.0 cm/min. After a very long time (1200 min) and when the column is certainly totally saturated ($c = 0.01$ everywhere), the feed is stopped, the flow direction is reversed, and the column is eluted with pure water at 60°C at a superficial velocity of 15.0 cm/min. This elution continues for another 1200 minutes.

Data: Equilibrium at 4°C, $q = 0.08943\, c$

Equilibrium at 60°C, $q = 0.045305\, c$, c is in kmol/m³ and q is kmol/kg carbon, $\rho_f = 1000$ and $\rho_s = 1820.0$ kg/m³, $K_d = 1.0$, $\varepsilon_e = 0.434$, $\varepsilon_p = 0.57$, $C_{ps} = 0.25$ and $C_{pf} = 1.00$ cal/(g°C). Ignore wall heat capacity effects.

a. From $t = 0$ to 1200 minutes predict the outlet concentration profile (top of column).

b. Predict the outlet concentration profile (bottom of column) for the 1200 minutes of elution.

D8. Redo Example 18-3 but with co-current thermal regeneration. Input the hot thermal wave (co-current purge step at superficial velocity of 11.0 cm/min) so that it exits at the same time that the xylene breaks through at the product end. The hot thermal wave is stopped as soon as thermal breakthrough occurs. After the hot thermal wave, cold feed (at superficial velocity of 8.0 cm/min) is input immediately. Find the feed time, the time when the last solute leaves the column, the average mass fraction of the peak from initial breakthrough to the last solute leaves, and the time that the exit from the column becomes cold again. Compare the average outlet mass fraction of the peak to the average mass fraction of the regeneration in Example 18-3.

D9. A very simple PSA cycle consists of the following three steps: 1) Pressurize the column with feed. 2) Feed step withdrawing product gas at p_H. 3) Counterflow blowdown to pressure p_L. This cycle is used with a single column to process a feed that contains 0.003 mol frac methane in hydrogen using Calgon Carbon PCB activated carbon. Operation is at 480 K. The feed step has a superficial velocity of 0.05 m/s, the column is 0.75 m long, $p_H = 4.0$ atm, $p_L = 1.0$ atm, and data are available in Example 18-4. Assume the column is initially clean (filled with hydrogen that does not adsorb).

    a. The cycle is: Repressurize with feed: 0 to 1 sec.
        Feed step at p_H: 1 sec to 8 sec
        Blowdown: 8 sec to 10 sec
    Is there breakthrough of methane in the first cycle? Is all of the methane removed in the blowdown step? What are the highest and the lowest mole fractions of methane in the blowdown step?

    b. With the same repressurization and blowdown steps, how long should the feed step be to just remove all of the methane in the blowdown step? Interpret your result.

D10.* Intermediate concentrations in the outlet concentration profile for trace PSA systems can be estimated using Eqs. (18-28a) to (18-28c). For Example 18-4:

    a. Show that point 10 in Figure 18-13B was determined by starting with $t = 1$ s (end of blowdown) at $z_{after} = 0.40$ m (0.10 m from feed end). This can be done by using Eq. (18-28c) to find p_{before} and Eq. (18-28a) to find $y_{M,after}$. Since solute of this mole fraction moves as a wave at the known velocity u_M during the feed step, the time it exits the top of the column can be determined.

    b. Starting at the point $t = 1$ s, $z_{after} = 0.48$m, calculate point 11 in Figure 18-13B.

D11. Repeat Example 18-4 but using repressurization with pure product.

D12. We have a column packed with activated carbon that is used for adsorbing acetone from water at 25°C. The isotherm can be approximated as a Langmuir isotherm (Seader and Henley, 1998), $q = (0.190\ c)/(1 + 0.146c)$, q is mol/kg carbon and c is mol/m^3. The bed properties are: $K_d = 1.0$, $\rho_s = 1820$ kg/m^3, $\rho_f = 1000$ kg/m^3, $\varepsilon_e = 0.434$, $\varepsilon_p = 0.57$. The bed is 0.50 m long and has an internal diameter of 0.08 m. The flow rate is 0.32 L/minute for both parts a and b.

    a. If the bed is initially clean ($c = q = 0$), determine the outlet concentration curve (c_{out} vs. time) if a feed containing 50 mol/m^3 of acetone is fed to it starting at $t = 0$.

    b. If the bed is initially saturated with a fluid containing 50 mol/m^3 of acetone, predict the outlet concentration profile when the bed is eluted with pure water ($c = 0$) starting at $t = 0$. Find outlet times for concentrations of c = 50, 40, 30, 15, 5, and 0 mol/m^3. Plot the curve of concentration out vs. time.

Note: The solute movement theory works just as well for mole balances as for mass balances.

D13. We wish to use the local equilibrium model to estimate reasonable flow rates for the separation of dextran and fructose using an SMB. The isotherms are linear and both q and c are in g/L. The linear equilibrium constants are: dextran, 0.23 and fructose, 0.69. The interparticle void fraction = 0.4 and the intraparticle void fraction = 0.0. The columns are 40.0 cm in diameter. We want a feed flow rate of 1.0 L/min. The feed has 50.0 g/L of each component. The desorbent is water and the adsorbent is silica gel. The columns are each 60.0 cm long. The lumped parameter mass transfer coefficients using fluid concentration differences as the driving force are 2.84 1/min for both dextran and fructose. Operation is isothermal. Use multiplier values (see notation in Figure 18-14) of $M_1 = 0.97$, $M_2 = 0.99$, $M_3 = 1.01$, and $M_4 = 1.03$. Determine the flow rates of desorbent, dextran product, fructose product, and recycle rate; and find the ratio D/F.

D14. We have a 50 cm long column packed with a strong acid cation exchange resin ($c_{RT} = 2.2$ eq/L, $\varepsilon_e = 0.42$). Fluid superficial velocity is 25.0 cm/min. Cations are not excluded. Anions are excluded. Anderson (1997) lists $K_{H-Li} = 1.3$ and $K_{K-Li} = 2.9$. DeChow (1989) lists $K_{H-Li} = 1.26$ and $K_{K-Li} = 2.63$.

   a. If the resin is initially in the H^+ form in equilibrium with a 0.1 eq/L solution of HCl and an aqueous feed that is 0.1 eq/L KCl is fed to the column, calculate the predicted times that the K^+ shock wave exits the column for both sets of equilibrium parameters.

   b. If the resin is initially in equilibrium with a 1.0 eq/L solution of HCl and an aqueous feed that is 1.0 eq/L KCl is fed to the column, calculate the predicted times that the K^+ shock wave exits the column for both sets of equilibrium parameters.

   c. If the resin is initially in equilibrium with a 1.0 eq/L solution of HCl and KCl ($x_H = 0.8$, $x_K = 0.2$) and an aqueous feed that is 1.0 eq/L solution of HCl and KCl ($x_H = 0.15$, $x_K = 0.85$) is fed to the column, calculate the predicted times that the K^+ shock wave exits the column for both sets of equilibrium parameters.

   d. Discuss your results.

D15. We have a 50 cm long column packed with a strong acid cation exchange resin ($c_{RT} = 2.2$ eq/L, $\varepsilon_e = 0.42$). Fluid superficial velocity is 25.0 cm/min. Cations are not excluded. Anions are excluded. Anderson (1997) lists $K_{H-Li} = 1.3$ and $K_{K-Li} = 2.9$ and $K_{Na-Li} = 2.0$.

   a. The resin is initially in the H^+ form in equilibrium with a 0.5 eq/L solution of HCl. Your technician does the following two experiments: (1) an aqueous feed that is 0.5 eq/L KCl is fed to the column, and (2) after re-equilibrating the column with 0.5 eq/L HCL, an aqueous feed that is 0.5 eq/L NaCl is fed to the column. The results show that the K^+ shockwave and the Na^+ shockwave in the two separate experiments exit the columns at the same times. The technician believes this result cannot be correct. Determine from calculations if this result is correct.

   b. The technician then takes a column in equilibrium with $x_{Na} = 0.4$ ($C_T = 0.5$) and feeds with a solution with $x_{Na} = 0.9$ ($C_T = 0.5$). He repeats the experiment with $x_K = 0.4$ ($C_T = 0.5$) and feeds with a solution with $x_K = 0.9$ ($C_T = 0.5$). At what time do these two waves exit the column?

   c. The technician then decides to take a column saturated with K^+ and elute it with 0.5 N HCl and to separately take a column saturated with Na^+ and elute it with 0.5 N HCl. Much to the surprise of the technician, the center ($x_K = 0.5$) of the K^+ and the center ($x_{Na} = 0.5$) of the Na^+ waves do not occur at the same time, and the K^+ center exits earlier than the Na^+ center. Calculate when the centers of these diffuse waves exit the column.

   d. The technician then looks at when the very tails of the two elution curves exit ($x_K \rightarrow$ 0.0) and ($x_{Na} \rightarrow$ 0.0). The Na^+ tail exits before the K^+ tail. Calculate the times that these two tails exit.

   e. Next, the technician looks at the times of the two elution curves when ($x_K = 0.90$) and ($x_{Na} = 0.90$). The K^+ at this concentration exits before the Na^+ exits. Calculate the times that these two concentrations exit.

f. Finally, the technician notices that for the two diffuse waves there is a concentration of the solute (Na^+ or K^+) that exits from the column at the same time. What is this concentration ($x_{Na} = x_K$)?

g. The technician is thoroughly confused and does not understand why Na^+ and K^+ come out at the same time in one experiment, the K^+ comes out ahead in some experiments, and the Na^+ comes out ahead in different experiments. Explain the results to the technician.

D16. A column packed with gas-phase activated carbon is initially filled with clean air. At t = 0 a feed gas containing y = 0.0005 wt frac toluene in air is started. This feed continues until t = 10.0 hours at which time a feed that is y = 0.0015 wt frac toluene is introduced and continued throughout the remainder of the operation. The superficial velocity is always 15.0 cm/s. Find the minimum column length required to have a single shock wave exit the column.

Data: Equilibrium $q = \dfrac{2000\ y}{1 + 2200\ y}$ (Basmadjian, 1997)

q = kg toluene/kg carbon and y = wt frac, which is essentially kg toluene/kg air. T = 298 K, P_{tot} = 50 kPa. Assume gas has density of pure air, which acts as an ideal gas.

$MW_{air} = 28.9\ \dfrac{g}{mol}$, $R = 0.008314\ \dfrac{m^3 kPa}{mol\ K}$, $K_d = 1.0$, $\varepsilon_e = 0.40$, $\varepsilon_p = 0.65$, $\rho_s = 1500 kg/m^3$.

Watch your units.

D17. A 25 cm long column packed with gas-phase activated carbon is initially filled with air containing y = 0.0010 wt frac toluene. At t = 0 a feed gas containing y = 0.0005 wt frac toluene in air is started. This feed continues until t = 20.0 hours at which time a feed that is pure air (y = 0.0000) is introduced and continued throughout the remainder of the operation. The superficial velocity is always 21.0 cm/s.

Data: Equilibrium $q = \dfrac{2000\ y}{1 + 2200\ y}$ (Basmadjian, 1997)

q = kg toluene/kg carbon and y = wt frac, which is essentially kg toluene/kg air. T = 298 K, P_{tot} = 50 kPa. Assume gas has density of pure air, which acts as an ideal gas.

$MW_{air} = 28.9\ \dfrac{g}{mol}$, $R = 0.008314\ \dfrac{m^3 kPa}{mol\ K}$, $K_d = 1.0$, $\varepsilon_e = 0.40$, $\varepsilon_p = 0.65$, $\rho_s = 1500 kg/m^3$.

Predict the outlet concentration profile. Specifically, find when the following concentrations exit: y = 0.0010, 0.00075, 0.0005, 0.00025 and 0.00000. Sketch the outlet concentration profile (y vs. t) and label the times when these concentrations exit the column. *Watch your units.*

D18. We are separating glucose (G) from fructose (F) on an ion exchange resin in the Ca^{+2} form. (Note: This is NOT an ion exchange problem—the sugars complex with the Ca^{+2} and there is no exchange of ions.) The linear equilibrium parameter for glucose and values of ε_e, ε_p, E_{eff}, K_d are given in Example 18-9. The equilibrium for fructose (Table 18-2) is q = 0.88 c with q and c in g sugar/100 ml. The interstitial velocity is 20.0 cm/min. For a very small pulse of feed, we desire a resolution between the two peaks of R = 1.05.

a. How long should the column be?

b. At what times do the glucose and fructose peaks exit the column?

c. What are the values of σ_t for glucose and fructose?

D19. We are exchanging Ag^+ and K^+ on a strong acid resin with 8% DVB. The total resin capacity is $c_{RT} = 2.0$ eq/L and the ionic concentration of the feed solution is $c_T = 1.2$ eq/L, all of which is Ag^+. The column is initially at a solution concentration of $c_T = 0.2$ eq/L all of which is K^+.

a. How long does it take for the total ion wave to breakthrough?

b. At what time does the Ag^+ shock wave breakthrough?

c. After the Ag^+ shock wave breaks through, the column is regenerated with pure K^+ solution with $c_T = 1.2$ eq/L. Predict the shape of the ensuing diffuse wave. (Reset t = 0 when start the K^+ regeneration solution.)

Data: $\varepsilon_e = 0.4$, $\varepsilon_p = 0.0$, $K_E = 1.0$, $v_{super} = 3.0$ cm/min, L = 50 cm.

D20. A column is packed with a strong cation exchanger. The column is initially in the K^+ form and $c_T = 0.02$ eq/L. The column is 75.0 cm long, the superficial velocity is 20.0 cm/min and the flow is in the same direction for all steps, $\varepsilon_p = 0$, $\varepsilon_e = 0.4$, $K_E = 1.0$ (for the cations), and $c_{RT} = 2.0$ eq/L. Equilibrium data are listed in Table 18-5.

a. At t = 0, feed the column with a solution with $c_T = 0.02$ eq/L, $x_{Ca} = 0.80$, $x_K = 0.20$. Plot the outlet value of x_{Ca} vs. time.

b. At t = 500 minutes, the column is regenerated with a pure aqueous solution of K^+ with $x_K = 1.0$, $x_{Ca} = 0.0$ and $c_T = 1.0$ eq/L. Plot the outlet value of c_T vs. time and the outlet value of x_{Ca} vs. time.

Note: If either of these steps require you to calculate a diffuse wave, calculate velocities and breakthrough times at three values of x_{Ca}: at the highest mole fraction, at the lowest mole fraction and at $x_{Ca} = 0.5$.

D21. The isotherms for dilute amounts of toluene and xylene adsorbed on silica gel from n-heptane are linear at low concentrations: $q_{Tol} = K'_{Tol}(T)c_{Tol}$ and $q_{xy} = K'_{xy}(T)c_{xy}$ where q and c are in g/L (Kim et al., 2005; Matz and Knaebel, 1991). The linear "constants" are functions of temperature and can be fit to Arrhenius relationships: $K'_{Tol} = 0.0061\exp(2175.2695/T)$ and $K'_{xy} = 0.0105\exp(2115.1052/T)$ with T in degrees Kelvin. Find the appropriate zone, product, and desorbent interstitial velocities (based on the column diameter) for an SMB separating toluene and xylene if the interstitial feed velocity = 1.0 cm/min and the temperature is constant at 300 K. Switching time is 100 min. Choose $M_1 = M_2 = 0.95$, $M_3 = M_4 = 1.05$.
Physical Properties: $\varepsilon_e = 0.43$, $\varepsilon_p = 0.48$. $\rho_s = 2100$ kg/m³, $\rho_f = 684$ kg/m³, $K_d = 1.0$.

D22. Use the Lapidus and Amundson solution to predict the behavior of fructose in a column packed with silica gel. The column is initially clean (contains no fructose). The feed is 50 g/L, the feed pulse lasts for 8 minutes, and then it is eluted with water. The flow rate is 20 ml/min. The other values are:

Value	Units	Description
Hb	200.0 cm	Height of adsorbent layer
Db	2.0 cm	Internal diameter of adsorbent layer
$\varepsilon_e = 0.4$	m³ void/m³ bed	Interparticle voidage
$\varepsilon_p = 0.0$	m³ void/m³ bed	Intraparticle voidage
$d_p = 0.01$	cm	particle diameter (needed to find $E_{effective}$)
$E_D = 0.15$	cm²/min	

Lumped parameter with concentration driving force, $k_{m,c}a_p = 5.52$ 1/min.
Isotherm is linear, $K'_{fructose} = 0.69$. Isotherm parameter, q and c in g/L.
Calculate $E_{effective}$, and then calculate enough points on the curve to plot it. Note that this is a step up followed 8 minutes later by a step down.

D23. For a linear system the breakthrough solution is $c_{out}/c_{feed} = X(z,t)$. At $t = 0$ we have an adsorption column that is initially at a uniform concentration $c_{initial}$, and the feed concentration is also $c_{initial}$. At $t = 17.5$ minutes, the feed concentration is reduced to $c_{F1}(c_{F1} < c_{initial})$. At $t = 28$ minutes the feed concentration is reduced to 0.0. Write the solution for c_{out} in terms of $c_{initial}$, c_{F1}, and the breakthrough solution X.

D24. The isotherms for dilute amounts of toluene and xylene adsorbed on silica gel from n-heptane are linear at low concentrations: $q_{Tol} = K'_{Tol}(T)c_{Tol}$ and $q_{xy} = K'_{xy}(T)c_{xy}$ where q and c are in g/L (Kim et al., 2005; Matz and Knaebel, 1991). The linear "constants" are functions of temperature and can be fit to Arrhenius relationships: $K'_{Tol} = 0.0061\exp(2175.2695/T)$ and $K'_{xy} = 0.0105\exp(2115.1052/T)$ with T in degrees Kelvin. Find the appropriate zone, product, and desorbent interstitial velocities (based on the column diameter) for an SMB separating toluene and xylene if the interstitial feed velocity = 1.0 cm/min and the temperature is constant at 300 K. Switching time is 100 min. Choose $M_1 = M_2 = 0.9$, $M_3 = M_4 = 1.10$. Compare results with Problem 18.D21. Physical Properties: $\varepsilon_e = 0.43$, $\varepsilon_p = 0.48$. $\rho_s = 2100$ kg/m³, $\rho_f = 684$ kg/m³. $K_d = 1.0$

D25. TSA regeneration can be combined with an SMB to reduce desorbent usage (Kim et al., 2005; Wankat, 1986). For the separation of toluene and xylene, repeat Problem 18.D24 except operate zones 2 and 3 at 300 K, operate zone 4 at 350 K with $M_4 = 2.0$, and operate zone 1 at 273 K with $M_1 = 0.5$. This will give a better separation than the M values in Problem 18.D24.

D26. A chromatograph is separating acetonaphthalene (A) from dinitronaphthalene (Dinitro) on 20-micron silica gel. For a single-porosity model, $[(1 - \varepsilon)\rho_s K'_A/\varepsilon] = 5.5$, and $[(1 - \varepsilon)\rho_s K'_{Dinitro}/\varepsilon] = 5.8$ in a solvent with 23% methylene chloride, and 77% n-pentane. When the interstitial velocity $v = 1.0$ cm/s, HETP is 0.05 cm.
a. If we desire a resolution of $R = 1.5$ for an infinitesimal pulse of feed, what column length is required?
b. Plot the outlet A curve as c/c_{max}, vs. time using the Gaussian solution for $R = 1.5$.

Note: In the chromatography literature the parameter $k'_i = [(1 - \varepsilon)\rho_s/K'_A/\varepsilon]$ is known as the *relative retention*. The relative retention is easily determined from experiments.

D27. For a linear system the breakthrough solution is $c_{out} = X(z,t)$. We have an adsorption column that is initially at a uniform concentration $c_{initial} = 0$, and the feed concentration is also $c_{initial} = 0$. At $t = 0$ minutes, the feed concentration is increased to $0.33c_F (c_F > 0)$. At $t = 0.4\ t_F$ minutes the feed concentration is increased to c_F. At $t = 0.8\ t_F$ the feed concentration is decreased to $0.55\ c_F$. Finally, at $t = t_F$ the feed concentration is reduced to 0.0. Write the solution for c_{out} in terms of c_F, t_F, and the breakthrough solution X.

D28.* A 25.0 cm long laboratory column is packed with particles that have an average diameter of 0.12 mm. At a superficial velocity of 9.0 cm/min we measure the breakthrough curve for a step input of solute. The column was initially clean. The center of the symmetrical breakthrough curve exits at 35.4 minutes while the width (measured from 0.05 c_F to 0.95 c_F) is 2.8 minutes. Pore diffusion controls and the isotherm has a Langmuir type shape.
a. What is L_{MTZ} in the lab unit?
b. We want to design a large-scale unit with 0.80 frac bed use. The average particle diameter will be 1.0 mm. The superficial velocity will be 12 cm/min. How long should this unit be? What is $t_{br} (c_{out} = 0.05\ c_F)$? Assume ε_e is same in both units.

D29. A 90 cm long laboratory column packed with a strong acid cation exchange resin ($c_{RT} = 2.5$ eq/L, $\varepsilon_e = 0.39$). Cations are not excluded. Anions are excluded. Anderson (1997) lists $K_{H-Li} = 1.3$ and $K_{K-Li} = 2.9$ and $K_{Ca-Li} = 5.2$. The resin is initially in the K^+ form in equilibrium with a solution of KCl that is 0.03 eq/L.

a. First, a 0.03 eq/L solution with $x_{Ca+2} = 0.70$ is fed to the column at a fluid superficial velocity of 25.0 cm/min at time t = 0. Does a shock or diffuse wave occur? What is the velocity and breakthrough time of this wave?

b. After the column is totally saturated with the 0.03 eq/L solution with $x_{Ca+2} = 0.70$, reset the clock to t = 0 and feed in counterflow the regenerant fluid that is 1.1 eq/L of KCl (no calcium) with a superficial velocity of 35.0 cm/min. How long does it take for the total ion wave to breakthrough?

A diffuse calcium wave occurs during elution. Calculate the velocity and breakthrough time of the fastest C_a^{+2} wave, the velocity and breakthrough time of the slowest C_a^{+2} wave, and the velocity and breakthrough time of the C_a^{+2} wave with x_{Ca+2} arbitrarily set equal to 0.5.

c. What concentration of regenerant fluid is required to *not* have a diffuse wave during the regeneration step?

D30. We have a 500 cm long column packed with a strong acid resin ($c_{RT} = 2.2$ eq/L, $\varepsilon_e = 0.42$). Fluid superficial velocity is 25.0 cm/min. Counter-ions are not excluded. Co-ions are excluded. DeChow (1989) lists $K_{H-Li} = 1.26$ and $K_{K-Li} = 2.63$. Note: The questions ask for three exit times—if there is a shockwave, these times will be identical.

a. If the resin is initially in equilibrium with a 1.0 eq/L solution of HCl and KCl ($x_H = 0.8$, $x_K = 0.2$) and an aqueous feed that is 1.0 eq/L solution of HCl and KCl ($x_H = 0.15$, $x_K = 0.85$) is fed to the column, calculate the predicted times that the K^+ wave exits the column (give the exit times for K^+ concentrations of 0.01, 0.5, and 0.85 eq/L).

b. If the resin is initially in equilibrium with a 1.0 eq/L solution of HCl and KCl ($x_H = 0.2$, $x_K = 0.8$) and an aqueous feed that is 1.0 eq/L solution of HCl and KCl ($x_H = 0.85$, $x_K = 0.15$) is fed to the column, calculate the predicted times that the K^+ wave exits the column (give the exit times for K^+ concentrations of 0.15, 0.5, and 0.8 eq/L).

D31. We have a 30.0 cm long laboratory column packed with a strong acid cation exchange resin ($c_{RT} = 2.4$ eq/L, $\varepsilon_e = 0.40$). Cations are not excluded. Anions are excluded. Anderson (1997) lists $K_{H-Li} = 2.9$ and $K_{Na-Li} = 2.0$. The resin is initially in the Na^+ form in equilibrium with a 1.10 eq/L solution of NaCl. Then a 1.10 eq/L solution of KCl is fed to the column at a fluid superficial velocity is 10.0 cm/min. The breakthrough starts at ($x_K = 0.005$) at 7.06 minutes, the center of the breakthrough wave exits at 7.31 minutes, and the end of the breakthrough wave ($x_K = 0.95$) at 7.57 minutes.

a. Calculate the time that the center of the wave is expected to leave the column (this is when the shockwave exits) and the error between the experimental time and the calculated time.

b. The error could be caused by a number of small errors. Use the experimental data to determine an experimental value of the shock velocity.

c. From the experimental values, calculate the L_{MTZ} and the fractional bed use of the lab unit.

d. We desire to use a large-scale column (L = 200 cm) with the same resin ($c_{RT} = 2.4$ eq/L, $\varepsilon_e = 0.40$), but with beads that are 4 times the diameter of the beads in lab column. The superficial velocity is increased to 20 cm/min. Initial and feed concentrations are the same as in the laboratory column. Calculate L_{MTZ}, the fractional bed use, and the time that breakthrough starts ($x_K = 0.05$) for the large-scale column. Use the experimental values.

F. *Problems Requiring Other Resources*

F1. An adsorbent gas drier has been sized to contain 2580 cubic feet of adsorbent. The drier will be an horizontal cylindrical vessel that has a circular cross-section and flow from top to bottom of the vessel (across the circular cross-section). The adsorbent will be supported above the bottom of the vessel. The area (width × length) of the adsorbent layer at the adsorbent support should be 860 ft². Determine the vessel dimensions that will minimize the weight of the vessel (minimum weight will be minimum cost) subject to the constraints that the vessel is designed for an operating pressure of 6 atm, a maximum temperature of 500°C, a maximum diameter of 16 feet, and a maximum length of 60 feet.

  a. Report diameter, length, and shell thickness of the vessel and the width of adsorbent layer. (Note: A spreadsheet is suggested for determining the vessel dimensions. Sizing pressure vessels is covered in Chapter 16 of Seider et al., 2004.)

  b. What was the purchase cost of the vessel FOB in mid-2000? (Use the weight method in Seider et al., 2004.)

  c. What was the purchase cost of molecular sieve adsorbent for this dryer in mid-2000? Ignore curvature when finding the amount of adsorbent.

  d. What are the purchase cost FOB and the installed cost of the vessel at current prices? (Update cost with Chemical Engineering Plant Cost index.)

  e. What is the current purchase cost of molecular sieve adsorbent for this dryer?

G. *Simulator Problems*

G1. **a.** If you have access to a simulator such as Aspen Chromatography, repeat Problem 18.D22 on the simulator.

  b. Rerun Problem 18.D22 on the simulator, but with $k_{m,c}a_p = 100,000$ 1/min and $E_D = E_{effective}$.

  c. Compare your two simulator runs.

  d. Compare the Lapidus and Amundson solution to the Aspen solutions. Note: you do need to consider convergence and accuracy of the Aspen runs.

G2. Set up a chromatographic column with one feed, a column and a product. The components that adsorb are acetonaphthalene (A) and dinitronaphthalene (DN). Use a model with convection with constant axial dispersion coefficients (0.25 cm²/min for both A and DN), constant pressure, and velocity. Use a linear lumped parameter model with driving force of ($c - c^*$) and constant mass transfer coefficients ($k_{m,c}a_p = 50.0$ 1/min for A and 45.0 1/min for DN). The isotherms are both linear, $q = 0.003056\ c$ (for A) and $q = 0.003222\ c$ (for DN) where q is in g adsorbed / kg adsorbent and c is in g solute/m³ of solution. Operation is isothermal.

The column will be 50.0 cm long with a 2.0 cm diameter. The adsorbent has the following properties: $\varepsilon_e = 0.40$, $\varepsilon_p = 0.46$, $K_D = 1.0$, $\rho_s = 2222$ kg/m³. The feed flow rate is 0.10 L/min, feed pressure = 3.0 bar, and the feed concentration for A is 2.0 g/L, while the feed concentration of D is 1.0 g/L.

  a. Run a breakthrough curve for 10 minutes. Print, label, and turn in your plot. Accurately determine the t_{MTZ} for component A where t_{MTZ} is measured from 0.05 times the A feed concentration to 0.95 times the A feed concentration. Show this calculation.

b. Input a 0.010-min feed pulse, and develop with pure solvent for a total time of 10 minutes. Print your plot.

G3. We want to separate component 1 from component 2 using an SMB. In the feed both compounds are dissolved in water, component 1 is 40 g/L and component 2 is 60 g/L. Feed rate is 141.55 ml/min. The components both have linear isotherms where both q and c are in g/L. Component 1: $q_1 = 1.5 c_1$; Component 2: $q_2 = 3.5 c_2$. The SMB has six columns: two columns between the feed and the extract (B) product (zone 3 in Figure 18-14B) and two columns between the feed and the raffinate (A) product (zone 2). There is one column between the desorbent addition and the extract (B) product (zone 4). The SMB has closed recycle, and there is one column between the raffinate (A) product and the desorbent addition (zone 1). Use constant pressure and volume (pressure = 3.0 bars), linear lumped parameter with (c – c*) driving force, and isothermal operation. Each column is 50.0 cm long and has a diameter of 10.0 cm. Data: $\varepsilon_e = 0.40$, $\varepsilon_p = 0.45$, $K_D = 1.0$, $\rho_{bulk} = 800$ kg/m^3. The density of the fluid can be assumed to be 1000 g/L. The values of both axial dispersion coefficients are 0.35 cm^2/min, the mass transfer coefficient ($k_{m,c}a_p$) for component 1 is 150 1/min and for component 2 is 100 1/min.

a. First, design your SMB to operate at the optimum point for D/F = 1.0 (all $M_i = 1.0$). Calculate D, E, and R, and the switching time and the recycle rate, and report these values. Run the system for at least 10 complete cycles (a cycle is 6 switching times). Turn in a plot of raffinate concentrations (components 1 and 2) and a plot of extract concentrations (components 1 and 2). Also, report the average mass fraction over the last cycle for component 1 in the raffinate, and the average mass fraction over the last cycle for component 2 in the extract.

b. Next, increase D/F to 2.0 (there are, of course, many ways to do this). Calculate D, E, and R, and the switching time and the recycle rate, and report these values. Briefly, explain your rationale for how you did the design to increase D/F to 2. Run the system for at least 10 complete cycles. Turn in a plot of raffinate concentrations (components 1 and 2) and a plot of extract concentrations (components 1 and 2). Also, report the average mass fraction over the last cycle for component 1 in the raffinate and the average mass fraction over the last cycle for component 2 in the extract.

G4. Ion Exchange. Use a bed that is 50 cm long and 20 cm in diameter. External porosity = 0.4, internal porosity = 0.0. Use axial dispersion coefficients of 0.2 cm^2/min for all components. Mass transfer coefficients ($k_{m,c}a_p$) are 10.0 1/min for all components. For a strong acid resin, use $c_{RT} = 2.0$ eq/L. $K_{H-H} = 1.0$, $K_{Na-H} = 1.54$. Operate with a feed rate of 5.0 L/min. Pressure is 2.0 bar.

a. Set the feed component concentration of H$^+$ equal to 0.0. Use a feed concentration of 1.0 eq/L for Na$^+$. The column is initially in the H$^+$ form with $c_T = 1.0$ eq/L. Do a breakthrough run.

b. After the column is converted to the Na$^+$ form, feed it with pure acid (H$^+$) at 1.0 eq/L (Na$^+$ is 0.0 and 1.0 eq/L H$^+$). Do a breakthrough curve.

c. Now do a wash step with pure water (feed concentrations of all components are zero). Continue the run until concentrations all become zero. Explain why this is so quick.

G5. Water Softening. Use the same column, same resin, same mass transfer and axial dispersion coefficients as in Problem 18.G4, except exchange Mg^{++} with H$^+$. $K_{Mg-H} = 1.9527$. The column is initially in the H$^+$ form with $c_T = 0.10$ eq/L. We want a feed that is 0.0 eq/L of H$^+$ and 0.10 eq/L of Mg^{++} ($c_T = 0.10$ eq/L). The value of $c_{RT} = 2.0$ eq/L is unchanged. (There is no Na^+ involved in this operation.) To reduce run time, operate with a column length of 10.0 cm. Do a breakthrough run and report results.

Unfortunately, most simulators will find this problem to be difficult since the equations are very stiff. Change the convergence settings and the discretization procedure to obtain convergence.

G6. Separation of ternary mixture with a feed consisting of dextran, fructose and a heavy impurity. A 25.0 cm long column with a diameter of 2.0 cm is used. External porosity = 0.4, internal porosity = 0.0. Pressure is constant at 2.0 bar. Use axial dispersion coefficients of 0.35 cm^2/min for all components. Mass transfer coefficients $(k_{m,c}a_p)$ are 2.84 1/min for dextran and fructose, and 1.5 1/min for the heavy; and the driving force for mass transfer is concentration differences. The isotherms are linear with q and c both measured in g/L. The isotherm parameters are 0.23 for dextran, 0.69 for fructose and 10.0 for the heavy impurity. The feed flow rate is 20.0 ml/min, the overall concentration of the feed is 100.0 g/L, and the feed is 48 wt % dextran, 48 wt % fructose, and 4.0 wt % heavy.

  a. Input a 1.5-minute feed pulse followed by pure solvent (water). Set up a plot with the outlet concentrations of dextran, fructose and heavy and run for 100 minutes.
       Where is the heavy? If your plot uses the same scale for all three components, it will be difficult to see the heavy. To find it set up the plot with a separate scale for heavy concentration. Note that the heavy has an initial peak and then a main, very broad peak (all at low concentrations). The initial peak occurs because with the low mass transfer coefficient some of the heavy exits the column without ever entering the packing material. This is called "bypassing" or "instantaneous breakthrough" and is obviously undesirable. Check to see if Eq. (18-66) is satisfied, and interpret your results.

  b. Change the heavy mass transfer coefficient $(k_{m,c}a_p)$ to 10.0 1/min and run again. Now it should look more normal, except the concentrations of the heavy are very low. This type of problem (at least two components that are difficult to separate plus very slow component(s)) is known as the "general elution problem." The column length needs to be set to separate the difficult separation, but then slow components take a long time to come out.

  c. (All mass transfer coefficients are $k_{m,c}a_p$ = 10.0 1/min for this step.) Input a 1.0-minute pulse of feed followed by 4.5 minutes of pure solvent. Then reverse the flow of solvent and elute for 15 minutes. Develop plots for both the forward flow and for the reverse flow.

G7. Thermal effects. We wish to adsorb toluene from n-heptane and then use co-current desorption with a hot liquid. Set the feed rate = 16.0 cm^3/min. Use a feed concentration of toluene of 0.008 g/L. Pressure = 3.0 atm. Feed temperature is 25.0°C. Column is initially pure n-heptane and is at 25.0°C.

  a. Do a breakthrough curve with a run time of 125 minutes. Create plots of toluene concentration and temperature.

  b. We now want to develop this saturated column co-flow using hot pure solvent. Increase the feed temperature to 80.0°C, set the toluene concentration = 0.0 and do the regeneration run. Print the plots of outlet toluene concentration and temperature. Explain the results.

  Column conditions and data: L= 100.0 cm, D_{col} = 2.0 cm, ε_e = 0.43, ε_p = 0.5, d_p = 0.0335 cm, ρ_s = 1800.0 kg/m^3, E_D = 0.2 cm^2/s for both toluene and n-heptane, $k_{m,c}a_s$ = 5.5656 1/min. The isotherm is linear with Arrhenius temperature dependence: pre-exponential factor $K_{Toluene,o}$ = 0.0061, and exponential factor $(-\Delta H/R)$ = 2175.27. C_{P_s} = 920.0 J/kg/K, and the effective liquid-solid heat transfer coefficient, = 0.005 J/(sm²K).

H. *Spreadsheet Problems*

H1. Set up a spreadsheet and solve Problem 18.D12.

H2. Solve Problem 18.D22 with a spreadsheet. Note: For unexplained reasons, the argument for the error function in Excel must be positive. An error is returned if the argument is negative. In this case make the argument positive and use the equality erf (−a) = −erf (a).

CHAPTER 18 APPENDIX.   INTRODUCTION TO THE ASPEN CHROMATOGRAPHY SIMULATOR

The Aspen Chromatography V7.2 simulator is fairly complicated. This appendix contains material from two of the nine labs that were developed for an elective course (Wankat, 2006). The complete set of labs is available from the author by sending a request to wankat@ecn.purdue.edu. The numerical method used by Aspen Chromatography is the method of lines, (e.g., see Schiesser, 1991).

Lab 1. The goal of this lab is to get you started in Aspen Chromatography V7.2. It consists of a cookbook on running Aspen Chromatography, some helpful hints, and the simulation of a real separation. The assistance of Dr. Nadia Abunasser in developing the original version of this lab was critically helpful.

1. Log in to the computer. Use your local operating system method of getting into Aspen Chromatography. Most likely, Aspen Tech → Aspen Process Development → Aspen Chromatography.

2. We will first develop a simple chromatography (or adsorption) column system. To do this, go to the menu bar and on the left hand side (LHS) select File. Go to Templates. In that window click on "Blank trace liquid batch flowsheet," and then click on Copy. It will ask for a directory name. Use something like, "column1." This will be saved in your working file. *NOTE: In all file names and names for components, columns, streams, and so forth there must be **no** spaces.*

3. On the "Exploring simulation" box (LHS), click on "component list." Then in the Contents box below double-click on Default. A and B are listed. Change these names to the names of the components to be separated (fructose and dextran T6). First, click the "Remove all" button. In the window below, type in the first component name (e.g., fructose), and click on the "add" button. Do the same for all the other components. Click "OK."

4. Now to draw the column. Click on the + to the left of "Libraries" and click on the plus left of "Chromatography" in the Exploring Simulation box. This opens other possibilities. Click on the word "chromatography." This should give "Contents of Chromatography" in a box below. Double-click on the model you want to use (Chrom_Reversible—since it is most up-to-date). Click and drag the specific model you want—in this case "chrom_r_column"—and move to the center of the Process Flowsheet Window. This gives a column

labeled B1. Left-click on B1, then right-click to open a menu. Click on "rename." Call the block something like "column."

5. Now for some confusion. Go to the Nonreversible model by clicking on the word "Chrom_Nonreversible" in the upper box of "Exploring." This will get you new contents. Click and drag a feed stream ("chrom_feed"), and put it near the top of the column (since the nonreversible column has flow downwards by arbitrary convention). Now click and drag a product stream ("chrom_product"), and put it near the bottom of the column. Rename as feed and prod. (We are using the code in the reversible model since it is up to date but are using nonreversible for feed, product, and connecting lines as it is simpler and less likely to cause problems.)

6. Now to connect everything together. In Exploring Simulation under "Chromatography" click on the word "Stream Types." This gives contents below. Click and drag "chrom_Material_Connection" (avoid the one with "r"). Click on the blue arrow of a source (e.g., the feed), move over to the inlet blue arrow for the column, and left-click. Repeat for column outlet and product arrow. (Note for Aspen Plus users: In Aspen Plus we would click on the Next button, it would tell us that the connections were complete, and would then lead us through the necessary steps. Unfortunately, Aspen Chromatography does not have this feature.)

7. We will now set up the column operating conditions. Double-click on the column. This gives a box labeled, "Configure Block/Stream Column." For now keep the UDS1 PDE discretization scheme, but change the number of nodes to 50. Note that Aspen Chromatography has a number of integration schemes that can be used for more complex systems or longer columns, but they are not required for this lab. The important issue of accuracy of numerical integration will be explored in detail in lab 2.

   a. Go to the Material balance tab. Select Convection with constant dispersion. The box for trace liquid assumption should be checked!

   b. Under the Kinetic tab, the fructose-dextran T6 data we have use $(c - c^*)$ as the driving force; thus, choose "fluid" for film model assumption. For Kinetic model assumption use linear lumped resistance. For mass transfer coefficient use "constant."

   c. In the Isotherm tab, the fructose-dextran T6 system has a *linear* isotherm that is based on fluid concentration (select "Volume base, g/L") for loading base.

   d. Energy balance tab—isothermal should be checked.

8. Now click on the Specify button. This opens an imposing table where you specify the column dimensions, isotherms, and so forth. Note that it lists fructose and dextran T6 because in step 3 you told it to do this. Use the following values to start with: L = Hb = 25 cm, column diameter Db = 2.0 cm, external porosity Ei = 0.4, particle porosity Ep = 0.0, dispersion coefficients Ez = 0.15 for both components. Mass transfer coefficient MTC = 5.52 min^{-1} (fructose) and = 2.84 min^{-1} (dextran T6). Isotherm coefficients: Aspen uses the formula for linear isotherm: q = (IP1) c + IP2.

   Set IP2 = 0.0 for both components. The adsorbent used was silica gel, and the solvent was water. (This information is never entered into Aspen when a template is used.) For fructose IP1 = 0.69, and for dextran T6, IP1 = 0.23. (Units on q and c are in g/volume solid and g/volume fluid.) Hit Enter, and then close this window (click on X).

Chapter 18   Introduction to Adsorption, Chromatography, and Ion Exchange

9. Click on the button for "Presets/Initials." For an initially clean column (which is what we want) all values should be zero. If they aren't, insert 0.0 values, hit enter, and close window.

10. Click on the "Initialize "button. This makes all concentrations the proper values. Close the Configure Block/Streams window.

11. To set the feed concentrations and flow rates, double-click on the arrow for the feed. This opens a window labeled Configure Block/Stream Feed. Click on the Specify button. Use a feed concentration of 50 g/L for both components. Set pressure at 2 bar (this has little effect). Flow rate of 20.0 cm^3/min is reasonable (Note: You have to change the units in the menu to the right of the number. Change the units first, and then enter the desired value.) Hit enter. Check that your values are correct, and close the window. Close the Configure block/stream feed window.

12. Do *not* touch the product stream—specifying something here will over specify system.

13. Set integration step size: Go to Run in the toolbar. In the Run menu click on "solver options, which opens the Solver Properties window. In the Integrator tab, pick the "Implicit Euler" method, and click on the Fixed radio button. Typically use a step size that is approximately (run time)/200. Might try 0.025 as a first try. For now, ignore the other tabs. Click OK.

14. We now need to set up the plots we want. Left-click and then right-click on the name of your product stream. In the menu that opens go to Forms, and click on "All Variables." This opens a table for this product stream. We will use this for dragging variables. Click on the icon of a star on top of a square donut (in toolbar). Name the plot (e.g., "Chromatography1"). Click on plot. Click OK. Drag the variable "Process_in.C("fructose")" to the y axis. Do this for dextran T6 as well. Double-click on the plot, which opens a window, "Pfs Plot 24.0 Control Properties." Click on the second tab, Axis Map, and click on "all in one." Then click OK. This puts the two concentrations on the same scale. (The problem with two scales is Aspen has a tendency to use different ranges, which makes comparisons rather difficult.) Close the product line table.

15. We will now set up a breakthrough run: Go to Run on the toolbar, and click on Run Options in the menu. Click on "Pause at." Try 10 minutes. Click OK.

16. To initialize the separation, in the center of the toolbar in the window choose Initialize. (This is not strictly necessary, but it is good practice because it allows you to use the rewind button later.) Click on the Play button (to the right of the menu). Click on OK after it runs. The run is initialized. Now, go back to the window, and select Dynamic. Cross your fingers, and click on the Play button. After it runs, click OK. Right-click on the figure. Select "Zoom Full." You can print the figure if you want. Right-click on the figure again. In the menu select, "Show as history." This gives a table of both concentrations. This table can be cut and pasted into Excel if further manipulation of the numbers is desired. It is better to *not* close the plot—we will reuse the plot. If you minimize it, you can find it under Window. Note: If you close the plot (or close and later reopen the entire file), go to the All Items box in the "Exploring Simulation," and click on the word "flowsheet." The

plot will be next to an icon that looks like a square donut and will be listed by name in the contents box. Double-click on the name to recover the plot.

17. Now do a pulse input of 1.0 minute. Start with a clean column. If you initialized, then you can simply restart to the initial state (this is the rewind button on a VCR—fourth button over from the menu in toolbar). Now you will have a clean column. To check this, double-click on the column, then click on the results button. All concentrations should be zero. If it is not zero, input zero values, and click on the Initialize button. Close the windows. There are a number of ways to add a pulse. One *easy* way is as follows: First, go to Run in toolbar, and click on Run Options. Select "Pause at," and insert your pulse length (1.0 minute). Initialize the simulation again (using the menu in the center of toolbar). Select "dynamic," and click on the Start button. (Since a plot was already drawn, you will get one automatically.) The run for this pulse is very short (lasts one minute). We have now input the pulse, and we must now develop the pulse with pure solvent. Double-click on the feed, click on specify, set the concentrations to zero (the template, since it is for a dilute system, "knows" there is solvent present), hit enter, close the windows. Go back to Run (toolbar), click on Run Options, and change the "Pause at" time to the desired run time (10 minutes will work for this system). Click OK. Do *not* initialize this time. We want the pulse to remain in the column. The menu on the toolbar should read "Dynamic." Click on the start button. When run is finished, click on OK, and look at the plot and the history.

18. When you look at the figure, it will probably appear to be a series of connected straight lines instead of a smooth curve. We can fix this by using more points in the plot. Click on rewind to get a clean column. Go to Run (toolbar) and then Run Options. Change "Communication" to 0.1 minute. Click OK. Follow the procedure for running the pulse [input the feed for one minute (use "Pause at") and Run. Then set the feed concentration to zero, set "Pause at" to 10 minutes, and Run.] Look at your plot. It should be smooth curves. Print this plot, and label it. Note: "Communication" just sets the number of points that are plotted but does not affect the integration. Thus, you can keep communication at 0.1 minutes for the remaining runs. You can also obtain a useful report by right-clicking on the column and selecting Forms, then Chromatography Report.

19. The two peaks are not completely separated. There are a number of ways they can be separated more completely. For example, double the value of L to L = 50 cm. Hit rewind, change L in the column dimensions table, and then rerun the one minute pulse input. When you run pure solvent, a pause time greater than 10 minutes is needed since doubling column length will double time for material to exit. Do this run and look at the result. Separation is better, but still not complete. Save your file (remember the file name), and exit Aspen.

Lab 2. The goal of this lab is to explore numerical convergence and accuracy of the simulator. Open your file from Lab 1.

1. You need to increase accuracy of the discretization procedure. We will first do this by increasing the number of nodes with UDS1.
   a. Rerun the breakthrough curve from Lab 1 (step 15), but use UDS1 with 200 nodes. To do this, double-click on the column, and in the Configure Block/ Stream Column win-

dow use the General tab. Change the number of nodes and then hit Enter. The integration method is Implicit Euler with time step at 0.025 min. Check that preset/initials are all zero, and initialize in this block. Feed is 50g/L both components. Then initialize the run. Do the run. Compare to your previous result with 50 nodes. If there is significant change, 50 nodes was not enough.

b. Make the comparison quantitative by calculating the width of the MTZ for each solute. The time of the MTZ, t_{MTZ}, is measured from $c = c_{initial} + 0.05(c_{feed} - c_{initial})$ to $c = c_{initial} + 0.95(c_{feed} - c_{initial})$. If the initial concentration is zero, measure from $0.05c_{feed}$ to $0.95c_{feed}$. (The reason for using 5% and 95% of the change is that with experiments it is very difficult to tell when one is exactly at $c_{initial}$ or at c_{feed}.) Note: The graphs are not accurate enough—use the *history* for the calculation of t_{MTZ}.

c. Try UDS1 with 800 nodes. Calculate t_{MTZ} values and compare to previous runs.

d. Change to either the BUDS or the QDS integration schemes with 50 nodes. Integration is still Implicit Euler with a time step of 0.025. [Change the number of nodes (no need to hit Enter), and then change to BUDS or QDS using the menu. Hit Enter (probably not necessary), and reinitialize.] Compare. Both BUDS and QDS work pretty well on this linear problem. Calculate the t_{MTZ} for both components and compare to previous.

e. Repeat the breakthrough runs for L = 50 cm and 100 cm, but using BUDS or QDS with 100 nodes and integration time step of 0.025. Calculate t_{MTZ} for both components for each run. Theory says that the widths should be proportional to $L^{1/2}$. (Runs with UDS1 with 50 nodes will not satisfy this. If you have time, try this.) Plot your widths (for each component separately) vs. $L^{1/2}$.

Note: BUDs and QDS are great for linear isotherms but often have convergence and/or oscillation problems with nonlinear isotherms.

APPENDIX A

Aspen Plus Troubleshooting Guide for Separations

Although specific for Aspen Plus, many of these suggestions will be helpful for other simulators. If Aspen Plus is *too slow,* close other programs and close extra windows within Aspen Plus. If none of the correlations fit the equilibrium data, see Appendix B. If things are *not working,* check the following:

1. Probably *the most common error* occurs when specifying streams. You must specify total flow and the composition of each component. The default for composition is molar flows. If the sum of the component molar flows is not equal to the total flow, Aspen Plus still lets you run, but results are often fouled up. Best practice: Always use mole or mass *fractions.* Another common error is to input the wrong numbers or with the wrong units (e.g., using a feed temperature of 30K when you mean 30°C. Separation modules won't work with a feed this cold.)

2. If you expect two liquid phases and one vapor phase, you must enable this condition. In Setup, for "valid phases" choose liquid-liquid-vapor from the menu. Do the same for Analysis. For Flash, use Flash3 not Flash2.

3. A common error in distillation is to connect a product to the vapor stream from the condenser, and then specify "total condenser" in the configuration page of the setup for the column. Aspen Plus gives an error message for this one. You need to disconnect the product and connect it to the liquid distillate product of the distillation column.

4. Aspen may ask you for conditions in an intermediate or product stream, and you do not know or want to specify these. If this persists and Aspen won't let you run when you push the Next button, delete the block and then add a new block and reconnect it to the streams. An alternative is to delete the stream and then add a new stream. One of these procedures will usually solve this problem.

5. Convergence problems:

   a. Open up the block in the data browser. (You can always get to the data browser from the menu—go to data and look at the bottom of the menu.) After opening up blocks and then the desired block, click on "convergence." Increase the number of iterations from 25 to 75. This often helps, but does not always solve the problem. Try making a change in column specifications in Aspen Plus, change back to the desired value (this tricks Aspen Plus into letting you run again) and then run again. If close to convergence, this may work. An alternative is to reinitialize and run again.

   b. Make sure that your numbers are compatible. For example, if the feed to a distillation column is 100, D cannot = 110. If D = 100, then B = 0, which is also probably not what you want to do. Selection of reflux *rate* and boilup *rate* (not ratios) will be compatible and should result in a successful run. Conditions can then be adjusted to obtain desired purities.

    c. For absorbers and strippers go to Block, Setup, Configuration, Convergence. Select "petroleum/wide boiling" from the menu. This is also necessary for distillation if the boiling range is very long (e.g., a feed containing both methane and n-octane).

    d. For very nonlinear systems such as those forming two liquid phases (e.g., butanol and water) select Convergence as "strongly non-ideal liquid."

    e. For recycle problems, convergence often requires a trick:

        Recycle Example 1, Extractive distillation. First, solve without solvent recycle by using all fresh solvent. Then connect the recycle and reduce solvent in steps down to steady state value. There must be a place for the "excess" solvent to leave the system.

        Recycle Example 2, Two-pressure distillation system to break azeotrope: Initially set B in one column and a low D value in the other. Then, without reinitializing, increase D until it approaches the desired recycle value.

        Recycle Example 3, Two-column binary separation with heterogeneous azeotrope. First, see Note 2 above. Then set up the system with total condensers and modest reflux ratios (say 0.5 or 1.0). Run the system and reduce L/D in both columns in steps to very low values that approximate all of the reflux coming from the decanter. Specify the value of B for one column, but not the other (try boilup ratio in the stripping column).

6. In absorbers and strippers you *need* to draw RADFRAC with a vapor distillate product. In the configuration page for the block pick condenser = none and reboiler = none in the menus. The operating specifications section (where you would normally pick L/D and D) should turn gray—no menu items will be available. If you try to change to no condenser and no reboiler *after* picking operating specifications, Aspen Plus becomes confused and does not adjust (the menu items for L/D and D remain in place). To correct, delete the block for the absorber/stripper, install a new block, and reconnect all lines. Now pick condenser = none and reboiler = none, and everything should work.

7. Can't get desired purity.

    a. In distillation, this is often due to inappropriate choice of D (or B). Don't select both! For high purities redo your external mass balances (accurately!) assuming complete separation and use these values of D and B.

    b. In absorption or stripping you may need to make the equilibrium more favorable by changing the column pressure or by changing the temperature of input streams. An alternative is to use a purer solvent stream.

    c. If mole fractions of impurity you are trying to obtain are too low, Aspen Plus may never reach these values if the convergence parameters are not set tight enough. A more stringent value for mole fraction (say E−06) will probably work.

8. Try Help in Aspen Plus.

Although often not too helpful, Help sometimes explains what the difficulty is.

9. Miscellaneous.

There are a number of other things that can happen. Probably 98% to 99% of the Aspen Plus problems are operator error. Although hard to decipher, the Aspen Plus error messages often give a hint of where to look for items to change. These errors may occur when you try to do something you have not done before (e.g., set a pressure drop on every stage). If you are trying to do something different and get an error message, stop doing what is different (e.g., return to constant pressure in the column) and see if Aspen Plus will run. If it will, figure out a way to approach what you want to do in small steps (e.g., use very small pressure drop on each stage). This approach may show you what the problem is.

10. ***As a last resort,*** log out of Aspen Plus, and then log back in using a blank simulation. Rebuild your entire system. Do not use a saved file as it is probably corrupted. Don't do this step on "automatic pilot." If you do, you are likely to repeat the error without thinking about it.

Instructions for Fitting VLE and LLE Data with Aspen Plus

The vapor-liquid equilibrium (VLE) and liquid-liquid extraction (LLE) correlations in Aspen Plus are not always as accurate as possible. This can cause significant errors, particularly near pinch points in distillation columns. If data is available, Aspen Plus will find values of the parameters for any of the VLE or LLE correlations by doing a regression against the data you input. This is illustrated to obtain an improved fit for the non-random two-liquid (NRTL) VLE correlation for the binary system water and isopropanol (IPA). VLE data for water and isopropanol is listed in Table B-1. This system has a minimum boiling azeotrope at ~80.46°C. The Aspen Plus fit to the data with NRTL is not terrible but can be improved.

Open Aspen Plus with either a template or a blank flowsheet. When you get to the process flowsheet window, draw either a distillation column (RADFRAC) or a flash distillation (Flash2) with the appropriate feed and product streams. Then click Next and click OK to display the next input.

In the Setup window, change run type to Data Regression, and phases should be Vapor-Liquid. After you click the Next button again, the components selection window appears. Enter as components "water" and "isopropanol." Click Next and select NRTL as the property method. Click Next, and then OK for message "Enter the data sets to be regressed." In the Properties Data-Data Browser, click New. Then select a name (or use the default name), and in the dropdown menu, select Mixture as the type. Click OK. In the setup window, the components listed should be water and IPA. Click the >> button to select both. Then click the Data Type dropdown menu and select Txy. In the Category dropdown menu, select Phase equilibrium. At the bottom of the form, input the desired pressure. Click Next.

On the Aspen Plus data page on the line with an *, input the temperature (you can select units) and the x and y values for water or IPA (input the one listed first in your Aspen Plus table). When all data is input, click Next. Click Specify the data regressions cases, and click OK. Click Next. Click on new and select a name or use the default name, and click OK. In the Properties Regression window, the method should be NRTL, and if you click the data set window, you will get the name of your data set. Calculation type should be regression. Click Next. On the table that asks for parameters, input Binary, NRTL (the "Element" to the left of name is 1), components in order of your table. Click Next, click Go to Next Required Input Step, then click OK to run the regression now. Click OK to run the regression name. If you already have NRTL parameters, you will be asked if you want to replace NRTL parameters: click Yes to All. Look at your results (click the blue box with the checkmark in it), particularly the Profiles, to compare with the data. To look at the VLE in graphical form, use Analysis. Now you can go to Setup-Specifications and change the Run type to Flowsheet and run any simulation you desire.

Fitting LLE data is similar except start by drawing a decanter or extraction column. Then in the dropdown menu for data type, select TPXX (temperature, pressure, and two liquid fractions). Input the temperature (you can select units), the pressure, and the x_1 and x_2 values for two of the components. Aspen automatically enters the third component. Ignore any error messages that three-phase flash results do not agree and that constraints cannot be met.

Table B-1.   *VLE Data for isopropanol-water at 1.0 atm.* Perry's Chemical Engineers Handbook *(Perry and Green, 1984, p. 13–13)*

Temperature, °C	x, water (mole fraction)	y, water (mole fraction)
100.00	1.0	1.0
97.57	.9955	.9185
96.2	.9931	.8595
93.66	.9873	.7815
87.84	.9643	.6308
84.28	.9322	.5353
82.84	.8670	.4964
82.52	.8349	.4847
81.52	.6796	.4544
81.45	.6664	.4511
81.19	.6248	.4385
80.77	.5280	.4140
80.73	.5244	.4144
80.58	.4803	.3967
80.52	.4055	.3670
80.46	.2120	.2454
80.55	.1980	.2320
81.32	.0697	.0990
81.85	.0340	.0475
82.39	.0000	.0000

APPENDIX C

Unit Conversions and Physical Constants

UNIT CONVERSIONS[1]

Density and Weight
1000 g = 1.0 kg
1.0 g/cm^3 = 1000 kg/m^3
1.0 lbm = 0.45359 kg
1.0 lbm/ft^3 = 16.01846 kg/m^3

Transport Properties: Diffusivity, Viscosity, and Mass-Transfer Coefficient
1.0 ft^2/s = 0.009290 m^2/s
1.0 cm^2/s = 0.0001 m^2/s
1.0 Pa s = 1.0 (N s)/m^2 = 1.0 kg/(m·s)
1.0 cP = 0.001 Pa s = 0.01 g/(cm·s)
1.0 lbm/(ft s) = 1.48816 Pa s
1.0 mol/[s m^2(mol/L)] = 10.0 m/s
1.0 lbmol/[h ft^2(lbmol/ft^3)] = 0.00008467 m/s

Energy and Specific Heat
1.0 Btu = 1.055056 kJ
1.0 Btu = 0.0002931 kWh
1.0 cal = 4.184 J
1.0 kcal = 4.184 kJ
1.0 Btu/(lbm °F) = 4.1868 kJ/(kg K)
1.0 kcal/(kg °C) = 4.184 kJ/(kg K)

Length and Volume
1.0 μm (1.0 micron) = 10^{-6} m
100 cm = 1.0 m
1.0 feet = 0.3048 m
1000 liters = 1.0 m^3
1.0 ft^3 = 0.02832 m^3
1.0 ft^3 = 7.481 US gal

[1] Source: Green & Perry, *Perry's Chemical Engineers' Handbook*, 8th ed., Tables 1-4 and 1-7, 2008.

1.0 US gal = 0.003785 m³
1.0 US gal = 3.785 L

Pressure
100,000 Pa = 100 kPa = 1.0 bar
101.3 kPa = 1.013 bar = 1.0 atm
760 mm Hg at 0°C (density 13.5951 g/cm³) = 760 torr = 1.0 atm
33.899 ft water at 3.94°C = 1.0 atm

Temperature
T in °C = (5/9) (T in °F − 32)
T in °C = T in K − 273.15
T in K = (5/9) (T in °R)
T in °F = T in °R + 459.7

Time
24 h = 1 day
60 min = 1 h
3600 s = 1 h
60 s = 1 min

IDEAL GAS VALUES

R   Gas constant, 8.314 J/(mol·K) = 1.9872 cal/(mol·K) = 1.9872 Btu/(lbmol °R) or 8.314 m³ Pa/(mol·K) or 0.008314 m³ kPa/(mol·K) or 82.057 cm³ atm/(mol·K) or 1.314 atm·ft³/(lbmol·K) or 0.7302 atm ft³/(lbmol °R)

22.4 liters of gas = 1.0 mol at STP (1.0 atm and 0°C)
379 ft³ of gas = 1.0 lbmol at STP

USEFUL PHYSICAL CONSTANTS

g   Acceleration due to gravity at sea level, 9.8066 m²/s = 32.174 ft/s²
g_c   Conversion factor in English units, 32.2 ft·lbm/(lbf·s²)
k_B   Boltzmann's constant, 1.380 J/(K·molecule)
N   Avogadro's number, 6.024×10^{23} molecules/mol

APPENDIX D

Data Locations

This Appendix gives the location of the large amount of data scattered throughout the textbook.

abietic acid-heptane-methylcellosolve+10% water: Distribution coefficient, Table 13-3

acetic acid: Adsorption isotherm on activated carbon, Table 18-2 and Problem 18.D7

acetic acid-benzene-water: Distribution coefficients, Table 13-3

acetic acid-1-butanol-water: Distribution coefficient, Table 13-3

acetic acid-3-heptanol-water: Distribution coefficient, Problems 13.D5 and 13.D6

acetic acid-isopropyl ether-water: Equilibrium data, Table 13-5

acetonaphthalene: Adsorption isotherm on silica gel, Table 18-2

acetone: Henry's law constant in water, Problem 12.D1; adsorption isotherm on activated carbon, Problem 18.D12

acetone-benzene-chloroform: Distillation curves, Figure 8-8

acetone-chloroform-water: LLE, Table 13-4 and Figure 13-12

acetone-ethanol: VLE, Problem 4.D7

activated alumina: Properties as adsorbent, Table 18-1 and Examples 18-2 and 18-6 and 18-7

activated carbon: Properties as adsorbent, Table 18-1 and Example 18-4 and Problem 18.D7; isotherms for adsorption of hydrogen, methane, ethane, ethylene, and propane, Figure 18-2; hydrogen and methane adsorption isotherms, Example 18-4

alcohol dehydrogenase-aqueous 5% PEG-aqueous 10% dextran: Distribution coefficient, Example 13-2

ammonia-water: Solubility, Table 12-3 and Problem 12.D10 and Example 15-6 and Problem 16.D12

anions: Approximate equilibrium constants, Table 18-5

anthracene: MW = 178.22; adsorption isotherm on activated alumina, Table 18-2 and Examples 18-2, 18-6, and 18-7 argon: Solubility in liquid ammonia, Problem 12.D14

benzene: Vapor pressure, Example 8-1; solubility parameter, paragraph following Eq. (13-1)

benzene-acetone-chloroform: Distillation curves, Figure 8-8

benzene-ethanol-water: Residue curves, Figure 8-12

benzene-ethylene dichloride relative volatility, $\alpha_{B\text{-}ED} = 1.11$

benzene-toluene relative volatility: Problem 4.G2

benzoic acid-toluene-water: Diffusivity benzoic acid in water and in toluene and Distribution coefficient between water and toluene, Examples 13-5 and 16-5; K_{LD} and $K_{O\text{-}ED}$, Example 16-5

bovine serum albumin: Adsorption isotherm on DEAE Sephadex A-50, Table 18-2

butane, iso-: K values, Figures 2-11 and 2-12, Table 2-3

butane, n-: K values, Figures 2-11 and 2-12, Table 2-3; vapor pressure, Problems 2.D14 and 5.D2

butanol, n-: MW = 74.12; λ, $C_{P,L}$ Problem 17.D14

butanol, n- n-propanol relative volatility, $\alpha_{nB\text{-}nP} = 0.412$

butanol, n- and water: VLE, Figure 8-2 and Table 8-2 (in Problem 8.D2); Selectivity in pervap, Problem 17.D14

carbon dioxide: Henry's Law constants in water, Table 12-1; normal CO_2 % in air, Problem 12.D16; Permeability, Figure 17-5 and Table 17-2

carbon monoxide: Henry's Law constants in water, Table 12-1; Permeability: Figure 17-5

carbon tetrachloride-acetic acid-triethylamine: Equilibrium data, Table 13-A1 (appendix to Chapter 13)

cations: Approximate equilibrium constants, Table 18-5

chlorine: Solubility in water, Example 15-4

chlorinated compounds: Henry's Law constants and solubilities in water, Table 12-2

chloroethane: 1,2-dichloroethane and 1,1,2-trichloroethane relative volatility, $\alpha_{di\text{-}tri} = 2.24$

chloroform-acetone-benzene: Distillation curves, Figure 8-8

copper-sodium ion exchange: Equilibrium, Example 18-4

cost indices: Table 11-1

cost of
    condensers: Eq. (11-3) and Figure 11-3
    materials of construction: Table 11-2
    packing: Figure 11-2
    reboilers: Figure 11-3
    towers: Figure 11-1
    trays: Figure 11-2

cresol-phenol relative volatility, $\alpha_{PC} = 1.76$

cumene-toluene relative volatility, $\alpha_{TC} = 0.21$

cyclohexane: MW = 84; liquid density, Example 18-2 and Problem 18.D2

cyclohexane-n-heptane: Distillation tray efficiencies, Figure 10-15

decane,-n: MW = 142.28; λ and vapor pressure, Example 8-2

diffusivities
    gases, Table 15-1
    liquids, Table 15-3

dinitronaphthalene: Adsorption isotherm on silica gel, Table 18-2

dioxane-benzene-water: Distribution coefficient, Problem 13.D10

efficiencies, trays: O'Connell correlation for distillation, Figure 10-14 and Eq. (10-6)
    O'Connell correlation for absorbers, Figure 12-8 and Eq. (12-36)

ethane: MW = 30.07; K values: Figures 2-11 and 2-12, Table 2-3; liquid density, Problem 2.D21; adsorption isotherm on activated carbon, Figure 18-2a

ethanol: MW = 46; λ, $C_{P,L}$, $C_{P,V}$, liquid density and MW, Problem 2.D9; λ, Example 4-1; $C_{P,L}$ Problem 4.D20; solubility parameter, paragraph below Eq. (13-1); molar liquid density and vapor pressure, Example 15-1; ϕ_B (solvent interaction parameter) = 1.5; surface tension, Example 16-2; λ, $C_{P,L}$ Example 17-9

ethanol-acetone: VLE, Problem 4.D7

ethanol-benzene-water: Residue curves, Figure 8-12

ethanol-ethylene glycol-water: Residue curves, Figure 8-14

ethanol and n-propanol relative volatility, $\alpha_{E\text{-}nP} = 2.17$

ethanol-water: Activity coefficients, Example 15-5

ethanol-water: VLE, Table 2-1, Figures 2-2 to 2-4; appendix B of Chapter 2, Eqs. (2.B-1) and (2.B-2)

ethanol-water: Liquid densities, Example 15-5

ethanol-water: Diffusivity, Example 15-5

ethanol-water: Liquid viscosity, Example 16-2

ethanol-water: Vapor Schmidt number, Example 16-2

ethanol-water: Pervaporation data, Figures 17-15 and 17-16

ethyl benzene-β,β'-Thiodipropionitrile-n-hexane: Distribution coefficient, Table 13-3

ethylene: K values, Figures 2-11 and 2-12, Table 2-3; adsorption isotherms on activated carbon, Figures 18-2a,b; adsorption isotherm on zeolite, Figure 18-2b

ethylene dibromide and propylene dibromide relative volatility, $\alpha_{EP} = 1.30$

ethylene dichloride-benzene relative volatility, $\alpha_{B\text{-}ED} = 1.11$

ethylene glycol-ethanol-water: Residue curves, Figure 8-14

fish oil leached from halibut livers with diethyl ether: Equilibrium data, Problem 14.E1 fructose: Adsorption isotherm on IEX resin in Ca^{2+} form, Table 18-2 and Example 18-5

furfural-MIBK-water: Distribution coefficient, Table 13-3

glucose: Adsorption isotherm on IEX resin in Ca^{2+} form, Table 18-2 and Example 18-5

heat transfer coefficients (approximate): Table 11-3

helium: Permeability, Figure 17-5 and Table 17-2

heptane: , n-: MW = 100.2; K values: Figures 2-11 and 2-12, Table 2-3; viscosity, Example 10-1 and Problem 10.D2;: normal bp = 98.4°C; specific gravity and surface tension, Problem 10.D2; λ, $C_{P,L}$, Problem 11.D2 and Example 18-3; liquid density, Example 18-3 and Problem 18.D21

HETP: see packings

hexane, n-: MW = 86.17; K values: Figures 2-11 and 2-12, Table 2-3; λ, $C_{P,L}$ $C_{P,V}$ and boiling point, Table 2-5; liquid density, Example 2-4; vapor pressure, Problems 2.D14 and 2.D.27 and 5.D2; λ, $C_{P,L}$ $C_{P,V}$, Problem 3.D6; viscosity, Example 10-1; liquid specific gravity and surface tension, Example 10-2

human serum albumin: Adsorption isotherm on Ca-alginate magnetite bead, Table 18-2

H^+ ion: Equilibrium values on cation exchange resins, Table 18-4 and Problems 18.D14 and 18.D15

hydrogen: Permeability, Figure 17-5 and Table 17-2; adsorption isotherm on activated carbon, Figure 18-2A

hydrogen chloride: MW = 36.46; Solubility in water, Problem 12.D8

hydrogen sulfide: Henry's Law constants in water, Table 12-1

ion exchange resin: Properties, Table 18-4 and Examples 18-5 and 18-8; approximate equilibrium constants, Table 18-5

K^+ ion: Equilibrium values on cation exchange resins, Table 18-4 and Problems 18.D14 and 18.D15

Lennard-Jones potential parameters: Table 15-2

linoleic acid-heptane-methylcellosolve + 10% water: Distribution coefficient, Table 13-3

mercury, Hg: liquid density = 845.3 lbm/ft³. Example 10-3

methane: K values, Figures 2-11 and 2-12, Table 2-3; λ, $C_{P,L}$ $C_{P,V}$ and boiling point Table 2-5; Solubility in liquid ammonia, Problem 12.D14; Permeability: Figure 17-5 and Table 17-2; adsorption isotherms on activated carbon, Figures 18-2a and 18-3, and Tables 18-2 and 18-3, and Example 18-2; adsorption isotherm on zeolite, Table 18-2

methanol: MW = 32.04; liquid density, Problem 2.D1; λ, $C_{P,L}$ $C_{P,V}$, Problem 3.E1; λ, Example 4-2; surface tension, Problem 10.D9; ϕ_B (solvent interaction parameter) = 1.9

methanol-ethanol relative volatility, α_{M-E} = 1.69

methanol-isopropanol relative volatility, α_{M-iP} = 2.26

methanol-methyl butyrate-toluene: Residue curves, Figure 11-11

methanol-n-propanol relative volatility, α_{M-nP} = 3.58

methanol-water: VLE, Table 2-7 in Problem 2.D1

methyl butyrate-methanol-toluene: Residue curves, Figure 11-11

methylcyclohexane-n-heptane-water: Equilibrium data, Table 13-6 (in Problem 13.D18)

methylcyclohexane-toluene-ammonia: Equilibrium data, Figure 13-14

Na^+ ion: Equilibrium values on cation exchange resins, Table 18-4 and Problem 18.D15

naphthalene: Solubility in carbon dioxide, Figure 14-6; adsorption isotherm on activated alumina, Example 18-2

nitrobenzene: MW = 123.11; solubility in water, Problem 12-D4

nitrogen: Permeability, Figure 17-5 and Table 17-2; adsorption isotherm on zeolite, Table 18-2

nitromethane-water: VLE, Table 8-3 (in Problem 8.E1)

nonane, n-: K values, Figures 2-11 and 2-12, Table 2-3

octane, n-: MW = 114.22; K values, Figures 2-11 and 2-12, Table 2-3; liquid density, Example 2-4

octanol: MW = 130.23; vapor pressure, Problems 8.D11 and 9.D19

oil leaching from meal with benzene: Equilibrium data, Table 14-2 and Figure 14-5

oleic acid-heptane-methylcellosolve + 10% water: Distribution coefficient, Table 13-3

oxygen: Permeability, Figure 17-5 and Table 17-2

packings, random
    absorption in water, approx. H_{OG}: Table 16-3
    capacity factor Nutter rings: Figure 10-29
    F factor: Table 10-3
    flooding: Figure 10-27 and Eq. (10-39a)
    HETP (approximate): Paragraph above Eq. (10-37c)
    HETP Nutter rings: Figure 10-28
    HETP Pall rings: Figure 10-26
    HTU estimation: Eqs. (16-37) and (16-38) and Figures 16-5 to 16-7
        Alternate HTU estimation: Eqs. (16-40) and Table 16-2
    pressure drop: Figure 10-27 and Eq. (10-39b) and Table 10-3
    sizing factor: Eq. (10-38) and Table 10-5

packings, structured
    capacity factor Intalox 2T: Figure 10-29
    F factor: Table 10-4
    HETP (approximate): Eq. (10-37c)
    HETP Norton Intalox 2T: Figure 10-28

pentane, iso-: K values, Figures 2-11 and 2-12, Table 2-3

pentane, n-: MW = 72.15; K values: Figures 2-11 and 2-12, Table 2-3; λ, $C_{P,L}$ $C_{P,V}$ and boiling point Table 2-5; vapor pressure, Problem 2.D27; liquid density, Problem 2.D21; Vapor pressure, Problems 2.D27 and 5.D2; λ, $C_{P,L}$ $C_{P,V}$, Problem 3.D6

permeability of gases: Figure 17-5 and Table 17-2

phenol-cresol relative volatility, $\alpha_{PC} = 1.76$

polarities of compounds: Table 8-1

polymer latex suspensions: Gel formation in UF, $x_g \sim 50$ vol %

polysaccharides: Gel formation in UF, $x_g < 1$ wt %

propane: K values, Figures 2-11 and 2-12, Table 2-3; adsorption isotherm on activated carbon, Figure 18-2a

propanol, iso-: Vapor pressure, Problem 2.D22

propanol, iso and n-propanol relative volatility, $\alpha_{iso-n} = 1.86$

propanol, iso and water VLE, Table B-1 (in Appendix B)

propanol, n-: Vapor pressure, Problems 2.D22 and 6.D4; ϕ_B (solvent interaction parameter) = 1.2

propylene: K values, Figures 2-11 and 2-12, Table 2-3

propylene dibromide and ethylene dibromide relative volatility, $\alpha_{EP} = 1.30$

pyridine-chlorobenzene-water: Equilibrium data, Table 13-7

sand, dry: Density, Problem 14.D2

sea water: Approximate wt fraction total dissolved salts = 0.035

silica gel: Properties as adsorbent, Table 18-1 and Example 18-3 and Problems 18.D4 and 18.D6

sodium chloride: MW = 58.45; osmotic pressure coefficient, Example 17-4

sulfur dioxide: Henry's law constant in water, Example 16-3

sucrose: MW = 342.3

sucrose, dilute aqueous solution: Density and osmotic pressure, Problem 17.D4

sugar: Effective equilibrium constant for leaching from sugar cane into water, Problem 14.D8 and 14.D9

toluene: MW = 92.14; liquid density and viscosity, Example 13-5; interfacial tension with water, Example 13-5; liquid density, C_{PL}, and adsorption isotherm on silica gel, Problem 18.D4; Adsorption isotherm on silica gel, Problem 18.D21; adsorption isotherm on gas-phase activated carbon, Problem 18.D16

toluene-benzene relative volatility, Problem 4.G2

toluene-cumene relative volatility, $\alpha_{TC} = 1/0.21$

toluene-methanol-methyl butyrate: Residue curves, Figure 11-11

toluene-xylene relative volatility, $\alpha_{T-X} = 3.03$

trays
    Capacity factor: Figure 10-29
    Crest height over weir: Eq. (10-26)
    Downcomer pressure drop: Eq. (10-27)
    Dry pressure drop: Eq. (10-24) and Figure 10-23

Efficiencies: O'Connell correlation for distillation, Figure 10-14 and Eq. (10-6)
Equivalent HETP: Figure 10-28
K_v valve trays: Eq. (10-34b)
O'Connell correlation for absorbers, Figure 12-8 and Eq. (12-36)
Orifice coefficient: Eq. (10-25)

ultrafiltration solute retention: Figure 17-11

washing equilibrium: Eq. (14-1)

water: MW = 18.016; liquid density, Problem 2.D.2; λ, $C_{P,L}$ $C_{P,V}$, liquid density, Problem 2.D9; λ, $C_{P,L}$ $C_{P,V}$, Problem 3.E1; λ, Example 4-1; λ, $C_{P,L}$ $C_{P,V}$, Problem 4.D23; vapor pressure, Problems 8.D10, 8.D15 and 9.D19; viscosity, Problem 10.D14a and Example 15-4; surface tension, Problem 10.D15 and Example 16-2; solubility parameter, paragraph below Eq. (13-1); ϕ_B (solvent interaction parameter) = 2.26 (2.6 also recommended); λ, $C_{P,L}$ Example 17-9

water-benzene-ethanol: Residue curves, Figure 8-12

water and n-butanol: VLE, Figure 8-2 and Table 8-2 (in Problem 8.D2)

water-ethanol: VLE, Table 2-1, Figures 2-2 to 2-4; Eqs. (2.B-1) and (2.B-2) in appendix B of Chapter 2

water-ethanol-ethylene glycol: Residue curves, Figure 8-14

water and iso propanol VLE, Table B-1 (in Appendix B)

water-methanol: VLE, Table 2-7 (in Problem 2.D1)

water-nitromethane: VLE, Table 8-3 (in Problem 8.E1)

xylene: Adsorption isotherms on silica gel, Example 18-3 and Problem 18.D21

xylene, p-: Adsorption isotherms on silica gel, liquid density and $C_{P,L}$, Problem 18.D6

xylene (meta, ortho, para)-β,β'-Thiodipropionitrile-n-hexane: Distribution coefficients, Table 13-3

xylene-toluene relative volatility, $\alpha_{T-X} = 3.03$

zeolite molecular sieve: Adsorbent properties, Table 18-1; ethylene adsorption isotherm, Figure 18-2B

$Zr(NO_3)_4$: Extraction equilibrium data from aqueous solution into TBP-kerosene, Problem 13.D36

Answers to Selected Problems

4.D3. a. slope – –2.54

    b. $q = 0.58$

4.D4. a. slope = 0.6 and goes through $y = x = z = 0.6$

    b. $q = 0.6$, slope = –1.5

    c. $q = -1/5$, slope = 1/6

4.D5. $L_1/V_2 = 0.55$

4.D7. a. $x_5 = 0.515$

    b. $y_2 = 0.515$. Note: It is an accident these values are the same.

    c. Optimum feed is seventh or eighth from the top; need 8 stages + partial reboiler. $q = 0.692$.

4.D8. a. $N_{min} \sim 5.67$

    b. $(L/D)_{min} = 1.941$

    c. Multiplier = 2.06

    d. 11 real stages + partial reboiler

4.D10. a. $y_3 = 0.9078$

    b. $x_6 = 0.66$

4.D16. $q = 1.13$

4.D19. $x_D = z = 0.75$, $x_B \sim 0.01$ to 0.02

4.D20. Optimum feed stage = first above partial reboiler

    Need ~8 equilibrium stages + partial reboiler.

4.D22. $y = (L/V)x + (D/V)y_D - (W/V)x_w = 0.75x + 0.17$

    Intersects $y = x = 0.68$ (not at y_D).

4.D23. $L/V = 0.77$

4.D27. Optimum feed is 3rd above partial reboiler. Need ~ 5.5 equilibrium stages + partial reboiler.

4.D30. a. N ~ 5 equilibrium stages

    b. $\overline{V}/B = 20$

4.D31. $L/D = 0.636$

4.D32. $x_B = z = 0.4$

4.D33. a. $(L/D)_{min} = 0.659$

    b. Optimum feed is third real stage from bottom; need 9 real stages + partial condenser.

    c. S = 760 lb mol/h = 13,680 lb steam/h

4.D34. Trial-and-error. $x_B \sim 0.058$

4.E2. $x_{side} = 0.0975$, Trial-and-error to find $x_D \sim 0.85$

4.E3. Optimum feed is tenth below condenser, vapor from intermediate reboiler is returned at stage 11; need 12 ½ equilibrium stages.

4.F3. a. See Example 4-4.

    b. CMO is OK.

    c. CMO not valid. Latent heat of acetic acid is 5.83 kcal/gmole compared to 9.72 for water.

    d. n-butane $\lambda = 5.331$ kcal/gmole and n-pentane $\lambda = 6.16$. This 15% error is marginal. Constant *mass* overflow works better.

    e. benzene $\lambda = 7.353$ kcal/gmole and toluene $\lambda = 8.00$. CMO is within ~ 6%

4G1. a.*See Example 4-4, part E.

4.G3. a. Optimum feed is eleventh below condenser; need 19.43 equilibrium contacts including the partial reboiler.

b. Optimum feed is eight from the top; need 21.9 equilibrium contacts, including the partial reboiler.

Chapter 5

5.D3. a. D = 2217.8 and B = 7782.2 kmol/day

    b. Bottoms mole fractions: Methanol = 0.0006, Ethanol = 0.6011, n-Propanol = 0.2313, n-Butanol = 0.1670

5.D4. a. D = 402 and B = 598 kmol/h.

    b. Distillate mole fractions: isopentane = 0.9851, n-hexane = 0.0149, n-C7 = 0. Bottoms mole fractions: isopentane = 0.0067, n-hexane = 0.4916, n-heptane = 0.5017

    c. $L = 1005$, $V = 1407$, $\overline{L} = 1605$, $\overline{V} = 1007$ kg moles/hr

5.D13. T = 28.8°C

Chapter 6

6.F1. $T_1 = 203.0$ °F, $T_2 = 212.5$ °F, $T_3 = 227.6$ °F, $T_4 = 248.7$ °F using data in Maxwell (1950) (see Table 2-2). The exact answer will vary depending on the data used.

Chapter 7

7.D1. a. $N_{min} = 5.97$,

    b. $(L/D)_{min} = 1.75$,

    c. N = 24.6 (including reboiler) and $N_{feed} = 14$ from top

7.D3. $\alpha = 1.287$

7.D5. $x_B = 0.229$

7.D6. $N_{min} = 10.8$, $(L/D)_{min} = 1.75$, N = 25.3 (including partial reboiler)

7.D8. a. $(L/D)_{min} = 22.83$,

    b. $N_{min} = 96.9$,

    c. N = 181.9. This separation would probably not be done by distillation.

7.D12. a. $N_{min} = 10.47$ and $FR_{cumene,bot} = 1.0$

    b. $(L/D)_{min} = 2.71$

    c. N = 20.24 (including partial reboiler) and optimum feed is stage 10 or 11

7.D13. $(L/D)_{min} = 0.2993$ if $\alpha = 2.5$ and $(L/D)_{min} = 0.3073$ if $\alpha = 2.25$. $(L/D)_{min}$ will be more sensitive to α for sharper separations.

7.D15. N = 9.45 (including reboiler) and use stage 3 as the feed

7.D16. a. $N_{min} = 12.7$

    b. $(L/D)_{min} = 2.13$

    c. $(L/D)_{actual} = 2.4$

7.D18 Part a. N = 13.2 if use original Gilliland curve or N = 14.1 if use Liddle's curve.

Chapter 8

8.D2. a. Butanol product = 3743.45 and water product = 1256.55 kmol/h

    b. Column 1: Optimum feed is stage 3 below condenser, and need 3 stages + partial reboiler.

    Column 2: Need ~ 2/3 stage + partial reboiler.

8.D8. Must convert wt frac to mole Frac, $\alpha_{water\text{-}ether\_in\_ether\_layer} = 7.026$

    $(L/D)_{min} = 7.467$, $L/D = 11.2$, Top stage is opt. feed and need ~4 3/5 equilibrium contacts.

8.D10. a. T = 97.5°C

    b. $n_{water}/n_{organic} = 10.81$

8.F1. a. $T = 95.0°C$
   b. $n_{water}/n_{organic} = 5.045$
   c. mole water condensed/mole nonane vaporized = 1.009
   d. $t = 99.9°C$ and $n_{water}/n_{organic} = 245.75$

Chapter 9

9.C2. a. $\ln(D_{final}/F) = - \int_{x_f}^{x_{D,final}} [dx_D/(x_D - x_w)]$

9.D2. $W_{final} = 35.2$ and $D = 64.8$ moles. $x_{D,avg} = 0.86$
9.D3. $W_{final} = 39.7$ and $D = 60.3$ kmoles. $x_{D,avg} = 0.85$
9.D9. a. $D = 3.45$
   b. $x_{w,final,min} = 0.21$
9.D14. a. $T_{final} = 99.7$ °C
   b. $W_{final} = 1.11$ and $D = 8.89$ moles
   c. $n_{water}/n_{organic} = 107.2$
9.D16. a. $(L_0/D)_{initial} = 0.47$
   b. $(L_0/D)_{final} = 5.46$
   c. $W_{final} = 5.79$ and $D = 4.21$ kmoles

Chapter 10

10.D1. $E_o = 0.73$
10.D2. $D = 12.35$ feet
10.D4. At balance point: $h_{\Delta p,valve} = 1.34$ inches of liquid
   Closed: $h_{\Delta p,valve} = (0.0287) v_o^2$ inches, for $v_o < 6.83$ ft/sec
   Open: $h_{\Delta p,valve} = (0.00478) v_o^2$ inches, for $v_o > 16.73$ ft/sec
10.D5. $HETP = 0.31$ m.
10.D6. a. For $\alpha = 2.315$, $HETP = 0.32$ m
   b. For $\alpha = 2.61$, $HETP = 0.37$ m
   c. For $\alpha_{avg} = 2.46$, $HETP = 0.34$ m
10.D9. $D = 10.27$ ft.
10.D11. a. $D = 6.05$ inches
   b. $D = 8.01$ inches
   c. $D = 19.14$ inches
10.D12. $HETP = 2.5$ feet; $x_B \sim 0.65$
10.D16. $D = 10$ feet
10.D17. $D = 15.4$ feet
10.F3. Maximum diameter is 2.1 feet in enriching section. Probably use 2.5 feet since there is almost no cost penalty. Need 22 real stages plus partial reboiler. Height is approximately 36 feet.
10.F4. Maximum diameter is 2.1 feet in enriching section. Need ~ 12 feet of packing.

Chapter 11

11.D1. a. $(L/D)_{min} = 2.44$
   b. $N_{min} = 23.1$
   c. $N_{equil} = 36.3$ + P.R.
   d. $N_{actual} = 50$ stages + P.R.
   e. \$1,054,000 as of Sept. 2001

11.D2. $Q_c = -3.39 \times 10^7$ BTU/h, $A_{cond} = 2850$ ft^2, $A_{Reb} = 32,800$ ft^2, total cost = \$811,000. Areas and costs are very sensitive to the values of U used.

Chapter 12

12.D5. HETP = 1.7 feet

12.D6. McCabe-Thiele gave 5 real stages, and Kremser gave 5.07 real stages. In practice, use 6 real stages.

12.D10. $y_{out} = 0.1267$, $x_{out} = 4.93 \times 10^{-4}$

12.D11. m = 1.414. m = 1.2 is incorrect (L/mV = 1)

12.D12. Need 4 equilibrium stages.

12.D15. N = 2.39, $y_{1,C4} = 7.2 \times 10^{-6}$, $y_{1,C3} = 2.98 \times 10^{-4}$

12.D17. L = 41.1 kmol/h

12.D23. a. Absorber: N = 8 equilibrium stages, $X_{out} = 0.2614$ mole ratio
   b. Stripper: G (stream B) = 723.7 mol carrier gas/day, $Y_{out} \approx 0.287$ mole ratio

12.F1. L/G = 18.4, HETP = 1.52 feet

12.F2. T = 99.1°F, T = 80.9°F, T = 73.9°F

Chapter 13

13.D4. $y_{out} = 0.0075$, N = 33.6

13.D7. E = 639.6 kg/h.

13.D9. N = 5.44, Recovery linoleic acid = 87.7%

13.D11. N ~ 8.5 equilibrium stages

13.D14. a. $y_A = 0.115$, $y_w = 0.04$, $x_A = 0.23$, $x_w = 0.73$
   b. S = 85.7 kg/h

13.D15. a. Column 1, $y_1 = 0.00092693$, $y_{N+1} = 6.929$ E-6
   b. Column 2: R = 50.35, $x_N = 0.0183$

13.D21. a. $y_{AE} = 0.06$, $y_{DE} = 0.05$, $x_{AR} = 0.42$, $x_{DR} = 0.48$
   b. E = 214 kg/h

13.D25. a. S = 2444 kg/h
   b. N = 2

13.D26.S = 5600, $E_N = 6830$, $R_1 = 770$ kg/h. Extract: 10.5% acetic acid and 3.5% water
   Raffinate: 5% acetic acid and 93% water

14.D14.   a.   Exit raffinate 27.5% acetic acid and 67.5% water
   b. Entering extract 13% acetic acid and 0.0% water
   c. $R_1 = 655$ kg/h and Entering extract flow rate = 2155 kg/h

13.D38. $x_{out} = 0.0000275$ and $y_{out} = 0.0037$. If your answer is $x_{out} = 0.000102$, which is incorrect, you have not been careful with your units.

13.E1. ~ 93.5 % m-xylene recovery with 8 stages with feed on stage 4

Chapter 14

14.D2. a. 6-2/3 equilibrium stages, $y_{out} = 0.0264$

   b. $x_{out} = 0.002$

14.D4. Optimum feed is fourth stage and need 5.4 stages.

14.D7. Need 8 equilibrium stages.

14.D8. Recovery = 95.9%

14.D10. Solve problem in mass ratio units. Need ~ 5 1/8 equilibrium stages, $x_{out} = 0.171$ (mass fraction).

14.D12. $m_E = 1.313$

14.D18. Extract: $y_{oil} = 0.238$, $y_{solid} = 0.0$; Raffinate: $x_{oil} = 0.078$, $x_{solid} = 0.656$

14.D19. Outlet Extract: $y_{oil} = 0.38$, $y_{solid} = 0.0$; Outlet Raffinate: $x_{oil} = 0.026$, $x_{solid} = 0.66$

14.D20. Stage 1 extract: $y_{oil} = 0.35$, $y_{solid} = 0.0$; Stage 2 extract: $y_{oil} = 0.18$, $y_{solid} = 0.0$; Stage 3 extract: $y_{oil} = 0.09$, $y_{solid} = 0.0$; Stage 3 raffinate: $x_{oil} = 0.03$, $x_{solid} = 0.66$

Chapter 15

15.D1. a. $C_{A,L}$ can be larger or smaller than $C_{A,0}$. For smaller, $C_{A,L} = 0.9701$ kg/m^3.

15.D12. At 298.16 K, $D_{AB} = 1.114$ m^2/s. Calculate $\delta = 0.000115282$ m, $v_{y,avg} = 0.04338$ m/s, Re = 19.966. This is a long residence time with Re< 20, so there are no ripples. $Sh_{avg} = 3.41$ and $k_{avg} = 3.295 \times 10^{-05}$ m/s, and 0.000168 kg/s carbon dioxide are absorbed.

15.D18. Answers are compared in Example 15-5. Obviously, additional intervals can be added for more accuracy.

15.D19. a. At t = 10,000 s, obtain following values of C/C_0:

z,cm	0	.01	.05	.10	.2	.3	.4	.5	.6	.8	1.0	1.2
C/C_0	1.0	.9753	.8771	.7571	.5662	.3535	.2161	.1220	.0635	.0134	.0020	.0002

Chapter 16

16.D1. Average $H_{OG} = 2.1$, $n_{OG} = 12.1$, height = 25.4

16.D2. a. $G'_{flood} = 0.75$ lb/ft^2, D = 5.8 feet

   b. $H_G = 0.40$, $H_L = 0.47$ feet

16.D4. Height of stripping section = 9.8 feet, height of enriching section = 14.1 feet

16.D6. a. $H_{OG} = 1.67$ feet

16.D8. height = 26.2 feet

16.D10. a. $n_{OG} = 4.6$ feet

   b. $n_{OG} = 4.6$ feet

16.D12. a. h = 4.59 feet

   b. h = 1.98 feet, lowest $y_{out} = 0.00081$

16.D14. a. $k_x a = 1408.19$, $k_y a = 366.32$

   b. $E_{MV} = 0.47$

Chapter 17

17.D1. a. A = 2.80×10^6 cm^2, $F_p = 0.32$, $F_{out} = 0.68$ kmol/h

17.D4. a. M = 1.069

   b. $\alpha'_{AB} = 2.29$ atm^{-1}, $K'_{solv}/t_{ms} = 0.0665$ g/(m^2·s·atm), $K'_A/t_{ms} = 0.0665$ g/(m^{2}s)

   c. k = 0.000117 m/s

17.D15. a. $x_p = 0$, $x_{r,out} = 0.125$, $F'_{out} = 80$ kg/h

17.H1. The solution is in Example 13.5-1 in Geankoplis (2003).

Chapter 17

17.D1. a. $A = 2.80 \times 10^6$ cm^2, $F_p = 0.32$, $F_{out} = 0.68$ kmol/h

17.D4. a. $M = 1.069$

    b. $\alpha'_{AB} = 2.29$ atm^{-1}, $K'_{solv}/t_{ms} = 0.0665$g/(m^2·s·atm), $K'_A/t_{ms} = 0.0665$g/(m^{2}s)

    c. $k = 0.000117$ m/s

17.D15. a. $x_p = 0$, $x_{r,out} = 0.125$, $F'_{out} = 80$ kg/h

17.H1. The solution is in Example 13.5-1 in Geankoplis (2003).

Chapter 18

18.D1. The answers are given in Example 18-1.

18.D2. $q_{max} = 0.10456$ g anthracene/g adsorbent, $K_{Ac} = 2.104$ L/g anthracene.

18.D7. a. From $t = 0$ to 161.49 minutes $c_{out} = 0$. Then $c_{out} = c_F = 0.01$ until 1200 minutes.

    b. For downflow ($t = 0$ is start of downflow), $c_{out} = c_F = 0.01$ from $t = 0$ to $t = 3.496$ minutes; from this time until $t = 83.3$ minutes, $c_{out} = 0.019796$ kmol/m^3.

18.D10. The answers are in Example 18-4.

18.D28. a. $L_{MTZ,lab} = 1.9774$ cm

    b. $L = 4.576$ m, $t_{br} = 389$ min

Index

reaching an answer, 5
rules of thumb, 6
steps involved, 5–6
Profiles. *See* Composition profiles; Flow profiles;
    Temperature profiles
Propane data, 33, 618, 812
Propanol data, 240, 440, 620
Proportional pattern waves, 851–852
Propylene data, 33
Publications
    annotated bibliography, 8–9
    batch distillation, 329
    Chemical Engineering magazine, 420
    Das New gross Distiller Buch, 329
    *How to Model It: Problem Solving for the
        Computer Age,* 6
    Index of Learning Styles, 7
    Liber de arte distillandi, 329
    VLE data sources, 16–17
Punched hole pattern, sieve trays, 378
Purge cycles, 823–827
Purity levels, 152
Pyridine data, 566

Q

q (feed quality), 118–120
    examples, 121–124
q-line. *See* feed lines

R

Rachford Rice equations, 37–38, 39–42
RADFRAC, 237–242, 721–724
Raffinate, 504, 522–523, 544
Random packings, 388–390, 393, 403
Raoult's law, 33
Rate-based analysis, of distillation, 708–712,
    721–724
Rate-transfer (RT) equation, *see* RT equation
Rayleigh equations, 331–332
Reboilers
    intermediate, 143–144, 377–378
    partial, 105
    total, 141, 219
Reciprocating-plate columns (RPC), 557–558
Rectifying section, 83–84
Recycling *versus* reflux and boilup, 83
Reference books. *See* Publications
Reflux
    class distillation, 81–84
    definition, 81
    minimum, 146–148
    versus recycling, 83
    subcooled, 153–155

total, 146–148
Reflux ratio
    batch distillation, 340–344
    constant, 340–344
    cost effects, 427, 428, 430, 433
    determining, 88
    finite, 233–237
    packed columns, 401
    Underwood equations, 228–233
    variable, 344
Regeneration steps, 819
Relative volatility, 27–28, 225
    stage-by-stage calculations for, 189–193
Residue curves, 285–290
    example, 445–447
Residuum Oil Supercritical Extraction (ROSE),
    589
Resins, 861–862
Resources. *See* Publications
Retentate, 727
Retentate-in-series system, 729–730
Retentate-recycle mode, 729–730
Reusing distillation columns, 151–153
Reverse osmosis (RO)
    with concentrated solutions, 764–765
    concentration polarization, 758–764
    examples, 755–756, 755–758, 760–762,
        762–764
    membrane properties, determining, 755–756
    versus osmosis, 749–755
    overview, 749
Reynolds analogy, 639
Reynolds number, 629
RO (reverse osmosis). *See* Reverse osmosis (RO)
ROSE. *See* Residuum Oil Supercritical Extraction
    (ROSE)
RPC. *See* Reciprocating-plate columns (RPC)
RT (rate-transfer) equation, 736–740, 754, 757,
    764–765, 767, 775, 778–779

S

Safety hazards
    absorption column failure, 403
    activated carbon solvent recovery, 835
    bed fires, 892
    fire prevention, packed columns, 403–404
    respirators in adsorbers, 892
    total reflux distillation, 146
Salt data. *See* sodium chloride data
Sandwich components, 197
Saturated extract, 524
Saturated raffinate, 522–523
Scaling up tray efficiencies, 369
SCFs (supercritical fluids), 587–589

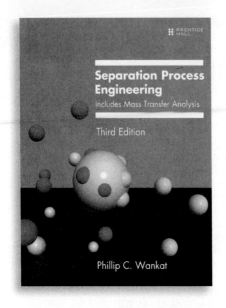